Circles

The standard form of an equation of a circle with center (h, k) and radius r:

$$(x - h)^2 + (y - k)^2 = r^2$$

The standard form of an equation of a circle with center $(0, 0)$ and radius r:

$$x^2 + y^2 = r^2$$

Quadratic Functions

A **quadratic function** is a second-degree polynomial function in one variable of the form

$$f(x) = ax^2 + bx + c \quad \text{or} \quad y = ax^2 + bx + c,$$

where a, b, and c are real numbers and $a \neq 0$.

The graph of a quadratic function is a **parabola** with vertex at

$$\left(-\frac{b}{2a}, c - \frac{b^2}{4a}\right) \quad \text{or} \quad \left[-\frac{b}{2a}, f\left(-\frac{b}{2a}\right)\right]$$

- If $a > 0$, the parabola **opens upward.**
- If $a < 0$, the parabola **opens downward.**

The **standard form of an equation of a quadratic function** is

$$y = f(x) = a(x - h)^2 + k \quad (a \neq 0)$$

The vertex is at (h, k).
- The parabola opens upward when $a > 0$ and downward when $a < 0$.
- The axis of symmetry of the parabola is the vertical line graph of the equation $x = h$.

Tests for Symmetry

Test for x-axis symmetry:
To test for x-axis symmetry, replace y with $-y$. If the resulting equation is equivalent to the original one, the graph is symmetric about the x-axis.

Test for y-axis symmetry:
To test for y-axis symmetry, replace x with $-x$. If the resulting equation is equivalent to the original one, the graph is symmetric about the y-axis.

Test for origin symmetry:
To test for symmetry about the origin, replace x with $-x$ and y with $-y$. If the resulting equation is equivalent to the original one, the graph is symmetric about the origin.

Graphs of Basic Functions

$f(x) = b$

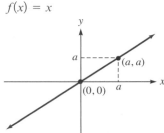

Domain: $(-\infty, \infty)$
Range: $\{b\}$

$f(x) = x$

Domain: $(-\infty, \infty)$
Range: $(-\infty, \infty)$

$f(x) = x^2$

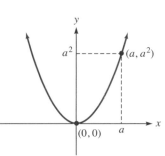

Domain: $(-\infty, \infty)$
Range: $[0, \infty)$

$f(x) = x^3$

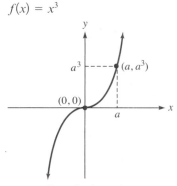

Domain: $(-\infty, \infty)$
Range: $(-\infty, \infty)$

$f(x) = \sqrt{x}$

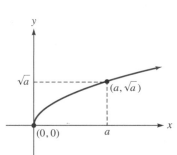

Domain: $[0, \infty)$
Range: $[0, \infty)$

$f(x) = \sqrt[3]{x}$

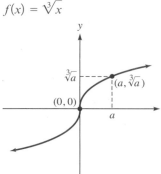

Domain: $(-\infty, \infty)$
Range: $(-\infty, \infty)$

$f(x) = |x|$

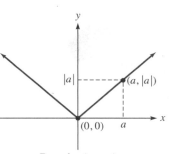

Domain: $(-\infty, \infty)$
Range: $[0, \infty)$

$f(x) = \dfrac{1}{x}$

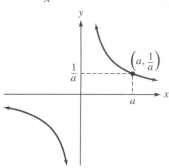

Domain: $(-\infty, 0) \cup (0, \infty)$
Range: $(-\infty, 0) \cup (0, \infty)$

Greatest-Integer Function

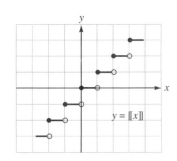

$y = [\![x]\!]$

Interest Formulas

Compound interest formula:
If P dollars are deposited in an account earning interest at an annual rate r, compounded n times each year, the amount A in the account after t years is given by

$$A = P\left(1 + \frac{r}{n}\right)^{nt}$$

Continuous compound interest formula:
If P dollars are deposited in an account earning interest at an annual rate r, compounded continuously, the amount A after t years is given by the formula

$$A = Pe^{rt}$$

Properties of Logarithms
If b is a positive number and $b \neq 1$,

1. $\log_b 1 = 0$
2. $\log_b b = 1$
3. $\log_b b^x = x$
4. $b^{\log_b x} = x$
5. **Product Rule:**
   $\log_b MN = \log_b M + \log_b N$
6. **Quotient Rule:**
   $\log_b \dfrac{M}{N} = \log_b M - \log_b N$
7. **Power Rule:**
   $\log_b M^p = p \log_b M$
8. **One-to-One Property:**
   If $\log_b x = \log_b y$, then $x = y$.

Graphs of $f(x) = e^x$ and $f(x) = \ln x$

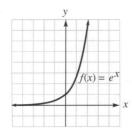
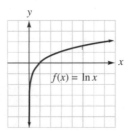

Natural Logarithm Properties

1. $\ln 1 = 0$
2. $\ln e = 1$
3. $\ln e^x = x$
4. $e^{\ln x} = x$

Theorems

Remainder Theorem:
If $P(x)$ is a polynomial, r is any number, and $P(x)$ is divided by $x - r$, the remainder is $P(r)$.

Factor Theorem:
If $P(x)$ is a polynomial and r is any number, then

If $P(r) = 0$, then $x - r$ is a factor of $P(x)$.

If $x - r$ is a factor of $P(x)$, then $P(r) = 0$.

Binomial Theorem:
If n is any positive integer, then

$$(a + b)^n = a^n + \frac{n!}{1!(n-1)!}a^{n-1}b$$

$$+ \frac{n!}{2!(n-2)!}a^{n-2}b^2 + \frac{n!}{3!(n-3)!}a^{n-3}b^3$$

$$+ \frac{n!}{r!(n-r)!}a^{n-r}b^r + \cdots + b^n$$

Parabolas
A **parabola** is the set of all points in a plane equidistant from a line l (called the **directrix**) and fixed point F (called the **focus**) that is not on line l.

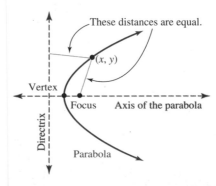

Parabola opening	Vertex at origin	
Right	$y^2 = 4px$	$(p > 0)$
Left	$y^2 = 4px$	$(p < 0)$
Upward	$x^2 = 4py$	$(p > 0)$
Downward	$x^2 = 4py$	$(p < 0)$

- For a parabola that opens right or left with vertex at the origin, the directrix is $x = -p$ and the focus is $(p, 0)$.
- For a parabola that opens upward or downward with vertex at the origin, the directrix is $y = -p$ and the focus is $(0, p)$.

Parabola opening	Vertex at $V(h, k)$	
Right	$(y - k)^2 = 4p(x - h)$	$(p > 0)$
Left	$(y - k)^2 = 4p(x - h)$	$(p < 0)$
Upward	$(x - h)^2 = 4p(y - k)$	$(p > 0)$
Downward	$(x - h)^2 = 4p(y - k)$	$(p < 0)$

- For a parabola that opens right or left with vertex at (h, k), the directrix is $x = -p + h$ and the focus is $(h + p, k)$.
- For a parabola that opens upward or downward with vertex at (h, k), the directrix is $y = -p + k$ and the focus is $(h, k + p)$.

Precalculus

Karla V. Neal
Louisiana State University

R. David Gustafson
Rock Valley College

Jeffrey D. Hughes
Hinds Community College

BROOKS/COLE
CENGAGE Learning·

Australia • Brazil • Japan • Korea • Mexico • Singapore • Spain • United Kingdom • United States

BROOKS/COLE
CENGAGE Learning·

Precalculus
Karla V. Neal, R. David Gustafson,
Jeffrey D. Hughes

Executive Editor: Liz Covello

Acquisitions Editor: Gary Whalen

Developmental Editor: Leslie Lahr

Assistant Editor: Cynthia Ashton

Editorial Assistant: Sabrina Black

Media Editor: Lynh Pham

Senior Marketing Manager: Danae April

Marketing Communications Manager: Mary Anne
Payumo

Content Project Manager: Jennifer Risden

Art Director: Vernon Boes

Manufacturing Planner: Karen Hunt

Rights Acquisitions Specialist: Tom McDonough

Production Service and Composition:
Graphic World Inc.

Photo Researcher: Bill Smith Group

Text Researcher: Isabel Saraiva

Copy Editor: Graphic World Inc.

Illustrator: Lori Heckelman

Text Designer: Kim Rokusek

Cover Designer: Denise Davidson

Cover Image: © Van Evan Fuller,
www.vanevanfuller.com

For product information and technology assistance, contact us at
Cengage Learning Customer & Sales Support, 1-800-354-9706.

For permission to use material from this text or product,
submit all requests online at **www.cengage.com/permissions.**
Further permissions questions can be emailed to
permissionrequest@cengage.com.

Library of Congress Control Number: 2011943459

ISBN-13: 978-0-495-82662-0

ISBN-10: 0-495-82662-6

Brooks/Cole
20 Davis Drive
Belmont, CA 94002-3098
USA

Cengage Learning is a leading provider of customized learning solutions with office locations around the globe, including Singapore, the United Kingdom, Australia, Mexico, Brazil, and Japan. Locate your local office at **www.cengage.com/global.**

Cengage Learning products are represented in Canada by Nelson Education, Ltd.

To learn more about Brooks/Cole, visit
www.cengage.com/brookscole.

Purchase any of our products at your local college store or at our preferred online store **www.CengageBrain.com.**

Printed in the United States of America
1 2 3 4 5 6 7 16 15 14 13 12

To the memory of my
parents, Carl and Mary Lou,
who taught me that a life
of faith is a life well lived
KVN

To my wife, Carol,
and my children, Kristy and Steven
RDG

To my mother, Joyce, and
to the memory of my father, Earl
JDH

About the Authors

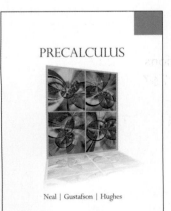

Karla Neal received two degrees in mathematics from Louisiana State University and has been an instructor in the Department of Mathematics for 31 years. During that time, she has served as the coordinator for College Trigonometry for more than 15 years. She has received three campus-wide teaching awards and has served as a faculty advisor for several Christian campus organizations. Karla is an avid fan of the LSU Tigers. When she is not on campus, she is very active in her church, Healing Place Church in Baton Rouge, and can often be found on the golf course. At home, there are two dogs, Winston and Josie, waiting patiently.

Jeff Hughes is a mathematics instructor at Hinds Community College in Mississippi and has degrees from Mississippi College and the University of Mississippi. He has worked at Hinds CC for 23 years, and his favorite courses to teach are College Algebra, the Calculus Sequence, and Differential Equations. He sees teaching as a privilege and enjoys the challenge of finding new ways to present topics while engaging students. He has been the recipient of Hinds CC's Outstanding Academic Instructor Award twice and is active in AMATYC. He currently serves as president of LaMsMATYC, the Mississippi-Louisiana affiliate of AMATYC. He has worked part-time as Minister of Students at his church for over 15 years and enjoys traveling. He taught English in China for seven summers with ELIC, the English Language Institute/China.

R. David Gustafson is professor emeritus of mathematics at Rock Valley College in Illinois and coauthor of several best-selling math texts, including Gustafson/Frisk's *Beginning Algebra, Intermediate Algebra, Beginning and Intermediate Algebra: A Combined Approach, College Algebra,* and the Tussy/Gustafson developmental mathematics series. His numerous professional honors include Rock Valley Teacher of the Year and Rockford's Outstanding Educator of the Year. He earned a Master of Arts from Rockford College in Illinois, as well as a Master of Science from Northern Illinois University.

About the Cover

PRECALCULUS

Neal | Gustafson | Hughes

The wonderful cover art is by Van Evan Fuller. Van is not only a brilliant artist but also a gifted writer. He lives in Baton Rouge, Louisiana, and is a graduate of LSU. The cover art was initially created as a traditional, though digital, abstract. The artist then processed it through deforming software into which he programmed a series of variables. After creating about a dozen such compositions, all based on the same abstract painting, the final design was created using four of these. In the art, numerous variations of mathematical curves such as circles, parabolas, ellipses, and hyperbolas can be seen intertwined in an amazing, colorful array. In the words of the artist, "Every beautiful thing is somehow rooted in mathematics, and the expression of mathematics is always beautiful." The authors were struck by this beauty and the way Van was able to capture it. You can see more of Van's creative designs at his online gallery at www.vanevanfuller.com.

Contents

© Istockphoto/MarcuzClackson

© Istockphoto.com/Ryan Lindsay

© Istockphoto.com/nullplus

Nomad_Soul/Shutterstock.com

Chapter 10

Sequences, Series, and Probability   889

© Istockphoto.com/Steve Cole

Chapter 11

Introduction to Calculus (Available online at www.cengagebrain.com)

Appendix A

A Review of Basic Algebra (Available online at www.cengagebrain.com)

Appendix B

Selected Proofs (Available online at www.cengagebrain.com)

Preface

To the Instructor

Using our combined experience of more than 80 years of teaching as a guide, we have written this textbook with both the instructor and student in mind. Instructors are always faced with the challenge of getting their students to actually read and study from a mathematics textbook. Therefore, we have written a text that has good, solid mathematics, along with features that will catch the eye of a student. Because this is a precalculus textbook, we have included not only the basics of algebra and trigonometry but also exercises that will prepare the student for calculus and courses beyond. Every instructor knows that a student cannot succeed in calculus without solid algebra and trigonometry skills. Likewise, every instructor has heard a student say, "Why do I have to learn this?" In this textbook, we have given you, the instructor, a feature to quickly answer this question. At the end of every chapter, the Looking Ahead to Calculus section ties the subject matter in that chapter to a topic in calculus. In addition, throughout the textbook a icon will point to concepts and topics that are addressed in calculus. It is important for students to see the connection between what they are learning now and how it will be applied in the future. This provides students the motivation to grasp the skills and concepts presented in the course.

In this textbook, you will find:

- solid mathematics written so that students with a broad range of abilities and backgrounds should find it easy to understand;
- functions that are defined and used in ways they are used in calculus;
- exercises that challenge students to use what they have learned in ways not often seen in algebra and trigonometry, emphasizing skills that are necessary to succeed in calculus; and
- a variety of applications that are both current and relevant.

All of this is written in a style that is pedagogically sound, yet interesting to both the instructor and the student. While the use of technology is blended into the textbook, it is presented in a manner that allows the instructor flexibility in determining how much it will be used.

Features

- **Looking Ahead to Calculus** *To demonstrate the relevance of topics in algebra and trigonometry,* Looking Ahead to Calculus offers students a brief but interesting glimpse into calculus. By connecting chapter content with topics in calculus, students will see how essential their precalculus course is to their future success. Each Looking Ahead to Calculus includes stated objectives that reference the related chapter material, a brief but detailed discussion of the topics, and Connect to Calculus exercises. This optional material is placed at the end of the chapter to allow the greatest flexibility for instructors.

LOOKING AHEAD TO **CALCULUS**

1. Using the Difference Quotient to Find a Tangent Line and Define the Derivative
   - Connect the difference quotient to tangents to a curve—Section 1.2
2. Derivative Rules—Using the Algebra and Composition of Functions
   - Connect the algebra and composition of functions to the derivative—Section 1.6

In addition, there are references to the Looking Ahead to Calculus feature in both the exposition and exercises throughout each chapter to help reinforce the connection between the material at hand and future courses. These references are clearly highlighted so that instructors may easily choose to teach or skip this material at their discretion.

4. Evaluate the Difference Quotient, Given a Function

We saw in Example 7 that it is possible to evaluate a function with an expression and not just real numbers. There is a quotient that is very important in calculus. It is called the **difference quotient**:

$$\frac{f(x + h) - f(x)}{h}, h \neq 0$$

 Evaluate the difference quotient for each function $f(x)$.

91. $f(x) = 3x + 1$ **92.** $f(x) = 5x - 1$

93. $f(x) = x^2 + 1$ **94.** $f(x) = x^2 - 3$

• **Accents on Technology and Calculators**   *To encourage students to become intelligent users of technology and grasp concepts graphically,* Accents on Technology appear throughout the textbook. These illustrate and guide the use of a TI-84 graphing calculator for specific problems. Although graphing calculators are incorporated into the book, their use is not required. All graphing topics are fully discussed in traditional ways.

ACCENT ON TECHNOLOGY   **Graphing Polar Equations**

A graphing calculator can be used to graph some polar equations. To graph a polar equation with a graphing calculator, we have to be able to write the equation in terms of r. Graphing a polar equation requires an additional step in setting up the window.

To graph in polar coordinates, we need to set the proper modes. While the graphs can be done in degree mode, we will use the radian mode. We will also need to set the calculator to polar mode, POL.

When we go to the graph menu, we will notice that instead of the functions being entered in terms of y, they are written as functions of r. The variable key will enter θ instead of x.

Setting the window in polar mode involves more than just setting the values for the axes. We must also set the boundaries and increments for the variable. In radian mode, we can set as shown and get a good graph.

Since polar graphs often have symmetry, using a square window will ensure that our graph is a true representation of the shape.

It is important to set the proper step for the variable. Otherwise, we may get a really strange-looking function.

- **Student-Friendly Writing Style** *To alleviate student anxiety about reading a mathematics textbook*, the exposition is clear, concise, and reader-friendly. The writing level is informal yet accurate. Students and instructors alike should find the reading both interesting and inviting.

- **Careers and Mathematics Chapter Openers** *To encourage students to explore careers that use mathematics and make a connection between math and real life,* each chapter opens with Careers and Mathematics. New, exciting careers are featured in this edition. These snapshots include information on how professionals use math in their work and on who employs them. Most information is taken from the *Occupational Outlook Handbook.* A web address is provided for students to learn more about the career.

CAREERS AND MATHEMATICS: **Forensic Analyst**

Forensic analysts investigate crimes by collecting and analyzing physical evidence. By performing tests on weapons or on substances such as fiber, hair tissue, and body fluids, they can determine the significance of the substances to the investigation. Some specialize in DNA analysis or firearm examination.

Education and Mathematics Required
- Forensic-analyst positions usually require a bachelor's degree, in either forensic science or another natural science.
- College Algebra, Trigonometry, Geometry, Calculus I and II, and Statistics are required.

- **Section Openers** *To pique interest and motivate students to read the material,* each section begins with a contemporary photo and a real-life application that will appeal to students of varied interests.

Guests aboard the Royal Caribbean's cruise ship *Freedom of the Seas* can now "hang 10" while out to sea. The FlowRider surf simulator allows riders to surf a wavelike water flow of 34,000 gallons per minute.

It is important that water in the FlowRider have the proper pH value. For example, if the pH is too low, the water will be acidic and will make riders' eyes and noses burn. It will also make their skin dry and itchy. The pH of the water is one of the most important factors in pool-water balance and should be tested frequently. To calculate pH, we need to understand *logarithms,* the topic of this section. As we continue through this chapter, we will see several real-life applications of logarithms.

- **Numbered Objectives** *To keep students focused,* numbered learning objectives are given at the beginning of each section and appear as subheadings in the section.

3.3 Logarithmic Functions and Their Graphs

In this section, we will learn to

1. Solve logarithmic equations with base 10, base e, and other bases.

2. Graph logarithmic functions.

3. Determine the domain of a logarithmic function.

4. Use transformations to graph logarithmic functions.

- **Example Structure** *To help students gain a deeper understanding of how to solve each problem,* solutions begin with a stated approach. The examples are engaging, and step-by-step solutions with annotations are provided.

EXAMPLE 1   Using Intercepts to Graph an Equation

Graph the equation $y = x^2 - 4$ by finding the intercepts and plotting points.

SOLUTION   To graph the equation $y = x^2 - 4$, we first find the x- and y-intercepts.

To find any x-intercepts, we let $y = 0$ and solve for x.

$$y = x^2 - 4$$
$$0 = x^2 - 4 \qquad \text{Substitute 0 for } y.$$
$$0 = (x - 2)(x + 2) \qquad \text{Factor.}$$
$$x = 2 \quad \text{or} \quad x = -2 \qquad \text{Solve for the } x\text{-intercepts.}$$

Since $y = 0$ when $x = -2$ and $x = 2$, these are the x-intercepts. The points where the graph intersects the x-axis are $(-2, 0)$ and $(2, 0)$. (See Figure 1-2.)

- **Application Examples** *To answer the student question, When will I ever use this math?* applications from a wide range of disciplines demonstrate how mathematics is used to solve real problems. These applications motivate the material and help students become better problem solvers. An eye-catching photo accompanies many of these modern examples.

EXAMPLE 5   An Application of the Position Function

A baseball is thrown vertically upward from the top of the Sears Tower in Chicago. The tower is 1,353 feet tall. If the initial velocity of the ball is 200 feet per second, for what interval of time will the ball have a height greater than the tower's?

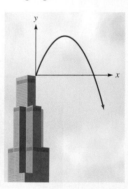

Dave Newman/Shutterstock.com

SOLUTION   Using the position function $s(t) = -gt^2 + v_0 t + h_0$, we substitute $g = 16$, $v_0 = 200$, and $h = 1{,}353$. Our height function is

$$s(t) = -16t^2 + 200t + 1{,}353.$$

- **Self Checks** *To actively reinforce student understanding of concepts and example solutions,* each example is followed immediately by a Self Check exercise. The answers for students are offered at the end of each section.

Self Check 7   Convert each equation written in rectangular form to an equation in polar form.
   a. $4x^2 y = 3$ **b.** $x = -6$ **c.** $y^2 = 2x$

Now Try Exercise 77.

- **Now Try Exercises** *To provide students an additional opportunity to assess their understanding of the concept related to each worked example,* a reference to an exercise follows all Examples and Self Check problems. These references also show students a correspondence between the examples in the book and the exercise sets.

- **Strategy Boxes** *To enable students to build on their mathematical reasoning and approach problems with confidence,* Strategy boxes offer problem-solving techniques and steps at appropriate points in the material.

> **Strategy for Solving Logarithmic Equations**
>
> 1. For logarithmic equations containing one logarithm, we can use the definition of the logarithm. We write the logarithm in exponential form and then solve for the variable.
>   **Note:** This strategy can be used to solve the following logarithmic equations:
> $$\log_3(x + 5) = 2, \ \log(x^2 - 4) = 5, \text{ and } 3 \ln 4x = 7.$$
>
> 2. For more complicated logarithmic equations, we can combine and isolate the logarithmic expressions. We can then use the definition of the logarithm and write the logarithm in exponential form to solve for the variable.

- **Caution Boxes** *To alert students to common errors and misunderstandings and reinforce correct mathematics,* Caution boxes appear throughout the text.

> **Caution**
>
> $\sin 2\theta = 2 \sin \theta$ Do NOT divide by 2 at this point.
>
> $\sin \theta = \sin \theta$ These are NOT equivalent equations.

- **Comments** *To provide additional insights into specific content,* Comment boxes appear throughout the textbook. These noteworthy statements provide clarification on a specific step or concept in an example. Some offer a tip for studying the material.

> **Comment**
>
> These identities are used extensively in many types of problems. They establish relationships between the functions, and, in other forms, can be modified for substitutions and used in other applications. For example, $\sin^2\theta + \cos^2\theta = 1$ can be written as $\cos^2\theta = 1 - \sin^2\theta = (1 - \sin\theta)(1 + \sin\theta)$. It is important that you learn these identities.

- **Comprehensive Exercise Sets** *To improve mathematical skills and cement understanding,* the exercise sets progress from routine to more challenging. The mathematics in each exercise set is sound, but not so rigorous that it will confuse students. All exercise sets include Getting Ready, Practice, Application, and Discovery and Writing problems. The book contains over 7,000 exercises.

- **Getting Ready Exercises** *To test student understanding of concepts and proper use of mathematical vocabulary,* each problem set begins with Getting Ready exercises. Students should be able to answer these vocabulary and concept statements before moving on to the Practice exercises.

> **Getting Ready**
> *You should be able to complete these vocabulary and concept statements before you proceed to the practice exercises.*
>
> *Fill in the blanks.*
>
> 1. The polar coordinate system is based on a ray, called the _____, and its endpoint, called the _____.
>
> 2. A point in the polar coordinate system is labeled _____.
>
> 3. True or False: The value of r is always positive.
>
> 4. True or False: A point in the polar plane has only one set of coordinates in rectangular form.
>
> 5. For a point with polar coordinates (r, θ) and any integer n, the polar coordinates $(r, \theta + \underline{\hspace{1cm}})$ and $(-r, \theta + \underline{\hspace{1cm}} + \underline{\hspace{1cm}})$ represent the same point in the polar coordinate system.

- **Interesting and Contemporary Application Exercises**   *To give students a sense of how math is applied in the real world,* emphasis is placed on solving problems through realistic applications. All application problems have descriptive titles.

---

49. Projectile motion   A quarterback passes a football to a receiver directly downfield at an initial velocity of 66 feet per second. The pass is released at a height of 7 feet from the surface of the field, at an angle of 65° to the horizontal, and the receiver catches the ball at a height of 5 feet.

Andy Lyons/Getty Images

  a. Write a set of parametric equations for the path of the football.
  b. How long did it take for the receiver to get to the point of the reception?
  c. How far down the field does the receiver catch the ball?

- **Discovery and Writing Exercises**   *To challenge students to think about what they have learned,* these interesting exercises require an understanding of mathematical concepts and a healthy dose of creativity.

Discovery and Writing

135. Show that the area of a sector of the unit circle is given by the formula $A = \frac{1}{2}\theta$, where θ is the radian measure of the central angle of the sector.

Use this figure to complete Exercises 136–138.

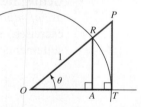

136. Show that the area of triangle OAR is given by the formula $A = \frac{1}{2}\cos\theta\sin\theta$.

• **Chapter Reviews and Chapter Tests** *To give students the best opportunity for study and exam preparation,* each chapter closes with a Chapter Review and Chapter Test. Chapter Reviews are comprehensive. Each Chapter Review is divided by section and includes all the important definitions, properties, and theorems from that section with a reference to a matching example. Review problems follow the summary of each section. Chapter Tests cover all the important topics yet are brief enough to emulate a real-time test, so students can practice not only the math but also their test-taking aptitude.

CHAPTER REVIEW

SECTION **5.1**   Inverse Trigonometric Functions

Definitions and Concepts	Examples
Inverse sine function: $y = \sin^{-1} x$ if and only if $x = \sin y$ and $-\dfrac{\pi}{2} \le y \le \dfrac{\pi}{2}$. Domain $= \{x \mid -1 \le x \le 1\}$; range $= \left\{ y \left\mid -\dfrac{\pi}{2} \le y \le \dfrac{\pi}{2} \right. \right\}$.	See Example 1, pages 512–513
Composition of sine and inverse sine functions: $\sin(\sin^{-1} x) = x$ when $-1 \le x \le 1$. $\sin^{-1}(\sin x) = x$ when $-\dfrac{\pi}{2} \le x \le \dfrac{\pi}{2}$.	See Example 3, pages 514–515

EXERCISES

Find the exact value of each expression, if any. Do not use a calculator. Note that each answer should be a real number.

1. $\sin^{-1} \dfrac{\sqrt{2}}{2}$

2. $\cos^{-1}\left(-\dfrac{\sqrt{3}}{2}\right)$

3. $\tan^{-1} 0$

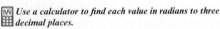 *Use a calculator to find each value in radians to three decimal places.*

4. $\sin^{-1} 0.56$

5. $\cos^{-1}(-0.34)$

6. $\tan^{-1}(-25)$

Find the exact value of each expression or state why the value does not exist. Do not use a calculator.

7. $\cos\left(\cos^{-1}\dfrac{4}{5}\right)$

8. $\sin^{-1}\left(\sin\dfrac{7\pi}{4}\right)$

9. $\tan^{-1}\left(\tan\dfrac{5\pi}{6}\right)$

10. $\sin(\sin^{-1} 1.5)$

Find the exact value of each expression. Do not use a calculator.

11. $\tan\left[\cos^{-1}\left(-\dfrac{5}{13}\right)\right]$

12. $\cos\left(\tan^{-1}\dfrac{4}{7}\right)$

13. $\sin\left(\cos^{-1}\dfrac{5}{13}\right)$

14. $\sec\left[\sin^{-1}\left(-\dfrac{2}{3}\right)\right]$

Applications

15. Dubai skyscraper The tallest building in the world is the Burj Khalifa, in Dubai. It is 2,716.5 feet high. If the angle of depression from the top of the building to a point on the ground 800 feet away from its base is represented by $\tan^{-1}\dfrac{2,716.5}{800}$ radians, find the angle of depression. Round to two decimal places.

CHAPTER TEST

Find the exact value of each expression, if any. Do not use a calculator. Note that each answer should be a real number.

1. $\cos^{-1}\left(-\dfrac{\sqrt{2}}{2}\right)$

2. $\tan^{-1}\dfrac{\sqrt{3}}{3}$

Find the exact value of each expression. Do not use a calculator.

3. $\csc\left(\cos^{-1}\dfrac{4}{5}\right)$

4. $\sin\left[\tan^{-1}\left(-\dfrac{2}{7}\right)\right]$

Verify each identity.

5. $\tan^2 x - \cot^2 x = \sec^2 x - \csc^2 x$

6. $\dfrac{(\sin x + \cos x)^2}{\sin x \cos x} = 2 + \sec x \csc x$

10. Given that $\sin\theta = \dfrac{24}{25}$ (θ is in QII), find the exact value of each of the following:

   (a) $\sin 2\theta$ **(b)** $\cos 2\theta$ **(c)** $\tan 2\theta$

   (d) $\sin\dfrac{\theta}{2}$ **(e)** $\cos\dfrac{\theta}{2}$ **(f)** $\tan\dfrac{\theta}{2}$

Find the exact solutions of each equation over the interval $[0, 2\pi)$.

11. $\sin\theta = \cos 2\theta$

12. $4\sin^2\theta + 2\cos^2\theta = 3$

13. $2\cos 3\theta = 1$

14. $2\sin\dfrac{\theta}{3} - \sqrt{3} = 0$

Solve each equation over the interval $[0, 2\pi)$. *Round to three decimal places.*

15. $5\sin^2\theta - 1 = 0$

16. $3\sin\theta = 7\cos\theta$

Organization and Coverage

This text can be used in a variety of ways. To maintain optimum flexibility, many chapters are sufficiently independent to allow you to pick and choose topics that are relevant to your students. Chapter R contains many topics that may be considered review material for a precalculus course and can be covered in its entirety or just a few sections. After teaching Chapters 1 and 2 in order, you can teach Chapters 3, 8, 9, or 10 in any order. The trigonometry content in Chapters 4–7 is designed to be taught in order, with the exception of Sections 6.1 and 6.2 (right triangle applications), which can be taught following Section 4.3.

Ancillaries for the Instructor

Enhanced WebAssign® (ISBN-10: 0-538-73810-3; ISBN-13: 978-0-538-73810-1)

Exclusively from Cengage Learning, Enhanced WebAssign offers an extensive online program for Precalculus to encourage the practice that's so critical for concept mastery. The meticulously crafted pedagogy and exercises in this text become even more effective in Enhanced WebAssign, supplemented by multimedia tutorial support and immediate feedback as students complete their assignments. Algorithmic problems allow you to assign unique versions to each student. The Practice Another Version feature (activated at your discretion) allows students to attempt the questions with new sets of values until they feel confident enough to work the original problem. Students benefit from a new YouBook with highlighting and search features; Personal Study Plans (based on diagnostic quizzing) that identify chapter topics they still need to master; and links to video solutions, interactive tutorials, and even live online help.

PowerLecture with ExamView (ISBN-10: 1-111-98986-9; ISBN-13: 978-1-111-98986-6)

This CD-ROM provides the instructor with dynamic media tools for teaching. Create, deliver, and customize tests (both print and online) in minutes with Exam-View® Computerized Testing Featuring Algorithmic Equations. Easily build solution sets for homework or exams using Solution Builder's online solutions manual. Microsoft® PowerPoint® lecture slides and figures from the book are also included on this CD-ROM.

Solution Builder (www.cengage.com/solutionbuilder)

This online instructor database offers complete worked solutions to all exercises in the text, allowing you to create customized, secure solution printouts (in PDF format) matched exactly to the problems you assign in class.

Complete Solutions Manual (ISBN-10: 0-495-82665-0; ISBN-13: 978-0-495-82665-1)

This manual contains solutions to all exercises from the text, including Chapter Review Exercises, Chapter Tests, and Cumulative Review Exercises.

Test Bank (ISBN-10: 1-111-98989-3; ISBN-13: 978-1-111-98989-7)

The test bank includes six tests per chapter as well as three final exams. The tests are made up of a combination of multiple-choice, free-response, true/false, and fill-in-the-blank questions.

Ancillaries for the Student

Enhanced WebAssign® (ISBN-10: 0-538-73810-3; ISBN-13: 978-0-538-73810-1)

Exclusively from Cengage Learning, Enhanced WebAssign offers an extensive online program for Precalculus to encourage the practice that's so critical for concept mastery. You'll receive multimedia tutorial support as you complete your assignments. You'll also benefit from a new Premium eBook with highlighting and search features; Personal Study Plans (based on diagnostic quizzing) that identify chapter topics you still need to master; and links to video solutions, interactive tutorials, and even live online help.

Student Solutions Manual (ISBN-10: 0-495-82666-9; ISBN-13: 978-0-495-82666-8)
Go beyond the answers—see what it takes to get there and improve your grade! This manual provides worked-out, step-by-step solutions to the odd-numbered problems in the text. This gives you the information you need to truly understand how these problems are solved.

Text-Specific DVDs (ISBN-10: 0-495-82664-2; ISBN-13: 978-0-495-82664-4)
These text-specific instructional videos provide students with visual reinforcement of concepts and explanations given in easy-to-understand terms with detailed examples and sample problems. A flexible format offers versatility for quickly accessing topics or catering lectures to self-paced, online, or hybrid courses. Closed captioning is provided for the hearing impaired.

CengageBrain.com
Visit www.cengagebrain.com to access additional course materials and companion resources. At the CengageBrain.com home page, search for the ISBN of your title (from the back cover of your book) using the search box at the top of the page. This will take you to the product page, where free companion resources can be found.

To the Student

Have you ever said to yourself, "Every time I try to read a math book I don't understand it, and besides that, it's boring"? If that's the case, we feel that if you take the time to read this textbook and take advantage of all the learning tools we've created for you, you will discover that it is both easy to follow and filled with problems that relate to *you* and *your* world. Every section opens with an applied problem taken from popular culture and real life. These applications continue throughout the section and into the exercises.

Have you ever thought about a math course and said, "I'm never going to use this"? Many times students feel that what they are learning has absolutely no connection with what they will do in the future. Every chapter opens with Careers and Mathematics, which gives a snapshot of a current career that requires the use and understanding of math. You may be surprised by the broad range of interesting (and well-paying) jobs that use math.

Have you ever asked yourself, "Why do I need to learn this?" This is a precalculus textbook and is written to help you develop the algebraic and trigonometric skills necessary to succeed in calculus and courses beyond. With that in mind, we have included references to highlight the concepts you are learning that will be applied in calculus. At the end of each chapter is a section titled Looking Ahead to Calculus. These sections correlate the practical use of what you have learned to specific topics in calculus.

We understand that doing math is the only way to really learn math. Each section is filled with interesting examples that show step-by-step solutions. Every example is followed by a corresponding Self Check exercise that allows you to determine whether you have fully grasped the example. Each Self Check is then followed by a Now Try reference that points you to a related problem in the exercise set that will give you more practice to master the concept.

We wish you great success in this course and as you move on to calculus. Best of luck!

Acknowledgments

We wish to thank all of those who were so helpful to us as we produced the manuscript for this text.

Your input over the course of writing this book was extremely valuable, and many of the suggestions have been incorporated into this text.

Paul Ache, Kutztown University of Pennsylvania

Arun K. Agarwal, Grambling State University

Camillia Smith Barnes, Sweet Briar College

Scott Barnett, Henry Ford Community College

Chris Bendixen, Lake Michigan College

Kevin Bolan, Everett Community College

Douglas Burkholder, Lenoir-Rhyne College

Gregory L. Cameron, Brigham Young University–Idaho

Ana-Maria Croicu, Kennesaw State University

Scott Dixon, University of Alabama at Birmingham

Marcia Drost, Texas A&M University

Douglas Dunbar, Northwest Florida State College

Hamidullah Farhat, Hampton University

Jayanthi Ganapathy, University of Wisconsin Oshkosh

Nikki Grantham, Florida Community College at Jacksonville–Downtown

Lisa Grilli, Northern Illinois University

Klara Grodzinsky, Georgia Institute of Technology

Donald Harden, Georgia State University

Miles Hubbard, St. Cloud State University

Vera Hu-Hyneman, Suffolk County Community College

Ionut Emil Iacob, Georgia Southern University

Jon M. James, Tarrant County College–Northwest Campus

Anedra Jones, University of Akron

Mike Kirby, Tidewater Community College

Katy Koe, Lincoln College

Dr. Yelena Kravchuk, University of Alabama at Birmingham

Mulatu Lemma, Savannah State University

Meagan McNamee, Central Piedmont Community College–North Campus

Jennifer McNeilly, University of Illinois at Urbana–Champaign

Mary Ann Moore, Florida Gulf Coast University

John Nadig, Mercer County Community College

Cao Nguyen, Central Piedmont Community College

Faith Peters, Miami-Dade College–Wolfson Campus

David Price, Tarrant County College–Southeast

Kevin Reeves, East Texas Baptist University

Martha L. Robertson, San Jacinto College

Rachel Schwell, Central Connecticut State University

Alicia Serfaty, Miami-Dade College–Kendall Campus

Sally Shao, Cleveland State University

Dr. Parashu R. Sharma, Grambling State University

Derek Smith, Nashville State Community College

Professor James D. Strange, Western Nevada Community College

Tirtha Timsina, Georgia State University

Laura Tucker, Central Piedmont Community College

Steve White, Jacksonville State University

Judith Wood, College of Central Florida

We wish to thank the staff at Cengage Learning, especially Gary Whalen, who inspired the creation of this text and supported us throughout the process, and Jennifer Risden for her tireless efforts during its production. Also special thanks go to Cynthia Ashton, Sabrina Black, and Lynh Pham, who worked so diligently to put the many pieces together. Leslie Lahr, our developmental editor, was there with us every step of the way, providing support, humor, encouragement, and valuable input. We appreciate you so much for staying by our side throughout this endeavor and giving us the help we needed to complete it. We also want to thank Lori Heckelman for the wonderful artwork in the text, and Van Fuller, who created the beautiful cover art. Another big thank you goes to Rhoda Bontrager at Graphic World for her tremendous copyediting skills, and for her patience and willingness to work with us at every turn and change. We also are most appreciative of the important work of accuracy reviewing done by John Samons.

Karla Von Neal

Jeffrey D. Hughes

R. David Gustafson

Equations, Inequalities, and the Rectangular Coordinate System

R

In this chapter, we learn to solve equations—one of the most important concepts in mathematics. Equations are used in almost every academic discipline, especially in chemistry, physics, medicine, and business. We will also learn about inequalities and the rectangular coordinate system.

CAREERS AND MATHEMATICS: Pharmacist

© Istockphoto.com/Lucas Rucchin

Pharmacists distribute prescription drugs to individuals. They also advise patients, physicians, and other health workers on the selection, dosages, interactions, and side effects of medications. They also monitor patients to ensure that they are using their medications safely and effectively. Some pharmacists specialize in oncology, nuclear pharmacy, geriatric pharmacy, or psychiatric pharmacy.

Education and Mathematics Required
• Pharmacists are required to possess a Pharm.D. degree from an accredited college or school of pharmacy. This degree generally takes four years to complete. For admission to a Pharm.D. program, at least two years of college must be completed, which includes courses in natural sciences, mathematics, humanities, and the social sciences. A series of examinations must also be passed to obtain a license to practice pharmacy.
• College Algebra, Trigonometry, Statistics, and Calculus I are courses required for admission to a Pharm.D. program.

How Pharmacists Use Math and Who Employs Them
• Pharmacists use math throughout their work to calculate dosages of various drugs. These dosages are based on weight and whether the medication is given in pill form, by infusion, or intravenously.
• Most pharmacists work in a community setting, such as a retail drugstore, or in a health-care facility, such as a hospital.

Career Outlook and Salary
• Employment of pharmacists is expected to grow by 17% between 2008 and 2018, which is faster than the average for all occupations.
• The median annual wage of wage and salary pharmacists is approximately $106,410.

For more information see http://www.bls.gov/oco.

R.1 Linear Equations and Applications

In this section, we will learn to

1. Use the properties of equality.
2. Solve linear equations.
3. Solve rational equations.
4. Solve formulas for a specific variable.
5. Solve application problems.

stocknadia/Shutterstock.com

It's been said that "weddings today are as beautiful as they are expensive." Suppose a couple budgets $3,000 for their wedding reception at a historic home. If $600 is charged for renting the home and there is a $24 per-person fee for food and beverages, how many guests can the couple accommodate at their reception?

If we let the **variable** x represent the number of guests, the expression $600 + 24x$ represents the cost for the reception. That is, $600 for the home rental plus $24 times the number of guests, x. We want to know the value of x that makes the expression equal $3,000. We can write the statement $600 + 24x = 3,000$ to indicate that the two quantities are equal. A statement indicating that two quantities are equal is called an **equation.** In this section, we will review how to solve equations of this type. If $x = 100$, the equation is true, because when we substitute 100 for x, we obtain a true statement.

$$600 + 24(\mathbf{100}) = 3,000$$

$$600 + 2,400 = 3,000$$

$$3,000 = 3,000$$

The couple can accommodate 100 guests at their wedding reception.

It is essential that we develop the skills necessary to solve many types of equations. Solving equations is one of the most important concepts in algebra. Equations are used in almost every academic discipline and vocational area, especially in chemistry, physics, medicine, economics, and business.

An equation can be either true or false, depending on the value(s) substituted for the variable(s). Whether an equation such as $3x - 2y = 12$ is true or false is dependent on the values that are substituted for the variables.

Any number that satisfies an equation is called a **solution** or **root** of the equation. The set of all solutions of an equation is called its **solution set.** To **solve** an equation means to find its solution set.

There can be restrictions on the values of a variable. For example, in the fraction

$$\frac{x^2 + 4}{x - 2}$$

we cannot replace x with 2, because that would make the denominator equal to 0.

EXAMPLE 1 **Finding the Restrictions on the Value of a Variable**

Find the restrictions on the values of b in the equation $\sqrt{b} = \dfrac{2}{b - 1}$.

SOLUTION For $\sqrt{b}$ to be a real number, b must be nonnegative, and for $\dfrac{2}{b - 1}$ to be a real number, b cannot be 1. Thus, the values of b are restricted to the set of all nonnegative real numbers except 1.

Self Check 1   Find the restrictions on a: $\sqrt{a} = \dfrac{3}{a-2}$.

Now Try Exercise 13.

There are three types of equations: identities, contradictions, and conditional equations. These are defined and illustrated in the following table:

Type of Equation	Definition	Example
Identity	Every acceptable real-number replacement for the variable is a solution.	$x^2 - 9 = (x + 3)(x - 3)$ Every real number x is a solution.
Contradiction	No real number is a solution.	$x = x + 1$ The equation has no solution. No real number can be 1 greater than itself.
Conditional equation	Solution set contains some but not all real numbers.	$3x - 2 = 10$ The equation has one solution, the number 4.

If two equations have the same solution set, they are called **equivalent equations.**

1. Use the Properties of Equality

There are certain properties of equality we can use to transform equations into equivalent but less complicated equations. If we use these properties, the resulting equations will be equivalent and will have the same solution set.

Properties of Equality

Addition and Subtraction Properties

If a, b, and c are real numbers and $a = b$, then

$$a + c = b + c \quad \text{and} \quad a - c = b - c.$$

Multiplication and Division Properties

If a, b, and c are real numbers and $a = b$, then

$$ac = bc \quad \text{and} \quad \frac{a}{c} = \frac{b}{c} \quad (c \neq 0).$$

Substitution Property

In an equation, a quantity may be substituted with an equivalent expression without changing the truth of the equation.

2. Solve Linear Equations

First-degree or **linear equations** are usually very easy to solve. They are also called **first-degree polynomial equations,** since these equations involve a first-degree polynomial.

Linear Equation	A **linear equation in one variable** (say, x) is any equation that can be written in the form $ax + b = 0$ (a and b are real numbers and $a \neq 0$).

EXAMPLE 2   **Solving a Linear Equation**

Find the solution set: $3(x + 2) = 5x + 2$.

SOLUTION   We proceed as follows:

$$3(x + 2) = 5x + 2$$

$$3x + 6 = 5x + 2 \qquad \text{Use the Distributive Property and remove parentheses.}$$

$$3x + 6 - \mathbf{3x} = 5x - \mathbf{3x} + 2 \qquad \text{Subtract } 3x \text{ from both sides.}$$

$$6 = 2x + 2 \qquad \text{Combine like terms.}$$

$$6 - \mathbf{2} = 2x + 2 - \mathbf{2} \qquad \text{Subtract 2 from both sides.}$$

$$4 = 2x \qquad \text{Simplify.}$$

$$\frac{4}{2} = \frac{2x}{2} \qquad \text{Divide both sides by 2.}$$

$$2 = x \qquad \text{Simplify.}$$

Because all of the above equations are equivalent, the solution set of the original equation is $\{2\}$. ∎

Self Check 2   Find the solution set: $4(x - 3) = 7x - 3$.

Now Try Exercise 37.

ACCENT ON TECHNOLOGY   **Checking Solutions to Linear Equations**

Using the table feature on a graphing calculator, we can easily check the solution of a linear equation. We enter each side of the equation into the graph editor, go to the table, and then enter the value of x that we found as the solution. The value of both entries in the table should be the same. These steps are shown in Figure R-1 for Example 2. We see that both Y_1 and Y_2 equal 12 when $x = 2$. This verifies the solution.

Enter each side of the equation.

(a)

Go to the table and enter the value $x = 2$.

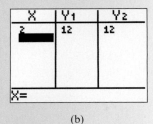

(b)

FIGURE R-1

EXAMPLE 3   **Solving a Linear Equation with Fractions**

Find the solution set: $\dfrac{3}{2}y - \dfrac{2}{3} = \dfrac{1}{5}y$.

SOLUTION To clear the equation of fractions, we multiply both sides by 30—the least common denominator (LCD) of the three fractions $\frac{3}{2}$, $\frac{2}{3}$, and $\frac{1}{5}$—and proceed as follows:

$$\frac{3}{2}y - \frac{2}{3} = \frac{1}{5}y$$

$$30\left(\frac{3}{2}y - \frac{2}{3}\right) = 30\left(\frac{1}{5}y\right) \qquad \text{Multiply both sides by the LCD 30.}$$

$$45y - 20 = 6y \qquad \text{Simplify.}$$

$$45y - 20 + 20 = 6y + 20 \qquad \text{Add 20 to both sides.}$$

$$45y = 6y + 20 \qquad \text{Simplify.}$$

$$45y - 6y = 6y - 6y + 20 \qquad \text{Subtract } 6y \text{ from both sides.}$$

$$39y = 20 \qquad \text{Combine like terms.}$$

$$\frac{39y}{39} = \frac{20}{39} \qquad \text{Divide both sides by 39.}$$

$$y = \frac{20}{39} \qquad \text{Simplify.}$$

The solution set is $\left\{\frac{20}{39}\right\}$. ∎

Self Check 3   Find the solution set: $\frac{2}{3}p - 3 = \frac{p}{6}$.

Now Try Exercise 35.

EXAMPLE 4   **Solving a Linear Equation**

Solve: **a.** $3(x + 5) = 3(1 + x)$ **b.** $5 + 5(x + 2) - 2x = 3x + 15$

SOLUTION   **a.**
$$3(x + 5) = 3(1 + x)$$

$$3x + 15 = 3 + 3x \qquad \text{Remove parentheses.}$$

$$3x - 3x + 15 = 3 + 3x - 3x \qquad \text{Subtract } 3x \text{ from both sides.}$$

$$15 = 3 \qquad \text{Combine like terms.}$$

Since $15 = 3$ is false, the equation has no roots. Its solution set is the empty set, which is denoted as $\varnothing$. This equation is a **contradiction.**

b. $5 + 5(x + 2) - 2x = 3x + 15$

$$5 + 5x + 10 - 2x = 3x + 15 \qquad \text{Remove parentheses.}$$

$$3x + 15 = 3x + 15 \qquad \text{Simplify.}$$

Because both sides of the final equation are identical, every value of x will make the equation true. The solution set is the set of all real numbers. This equation is an **identity.** ∎

Self Check 4   Solve: **a.** $3(x + 1) + 4 = 3(x + 3) - 2$ **b.** $-2(x - 4) + 6x = 4(x + 1)$

Now Try Exercise 17.

3. Solve Rational Equations

Rational equations are equations that contain rational expressions. Some examples of rational equations are

$$\frac{2}{x-3} = 7, \qquad \frac{x+1}{x-2} = \frac{3}{x-2}, \qquad \text{and} \qquad \frac{x+2}{x+3} + \frac{1}{x^2+2x-3} = 1.$$

In a rational equation, substituting certain values of x may result in the value of the denominator being 0. Since division by 0 is undefined, we must exercise care when solving rational equations and note any restrictions on the value of the variable.

Generally, the easiest method of solving a rational equation is to clear the equation of fractions by multiplying both sides of the equation by the LCD. When we do this, the new equation may have a solution or solutions that would result in a zero denominator when substituted back into the original equation. In this case, we have found an **extraneous solution.** These solutions do not satisfy the original equation and must be discarded.

EXAMPLE 5   **Solving a Rational Equation**

Solve the equation $\dfrac{x+1}{x-2} = \dfrac{3}{x-2}$.

SOLUTION   We begin by noting any restrictions on the variable. If we substitute 2 for x we obtain 0s in the denominators; therefore, 2 cannot be a solution of the equation. We proceed as follows:

$$\frac{x+1}{x-2} = \frac{3}{x-2}$$

$$(x-2)\left(\frac{x+1}{x-2}\right) = (x-2)\left(\frac{3}{x-2}\right) \qquad \text{Multiply both sides by } x-2.$$

$$x+1 = 3 \qquad\qquad\qquad \text{Simplify.}$$

$$x = 2 \qquad\qquad\qquad \text{Subtract 1 from both sides.}$$

Since 2 cannot be a solution of the equation, 2 is an extraneous solution and must be discarded. There is no solution to this equation. The solution set is $\varnothing$, the empty set.

Self Check 5   Solve: $\dfrac{2x}{x+1} + \dfrac{2}{x+1} = 1$.

Now Try Exercise 45.

EXAMPLE 6   **Solving a Rational Equation**

Solve: $\dfrac{x+2}{x+3} + \dfrac{1}{x^2+2x-3} = 1$.

SOLUTION   We will use several steps as outlined below and proceed as follows:

$$\frac{x+2}{x+3} + \frac{1}{x^2+2x-3} = 1$$

Step 1:   We begin by factoring $x^2 + 2x - 3$ and noting any restrictions on x.

$$\frac{x+2}{x+3} + \frac{1}{(x+3)(x-1)} = 1$$

We note that x cannot be -3 or 1.

Step 2: Clear the fractions by multiplying both sides by $(x + 3)(x - 1)$.

$$(x + 3)(x - 1)\left[\frac{x + 2}{x + 3} + \frac{1}{(x + 3)(x - 1)}\right] = (x + 3)(x - 1)1$$

Step 3: Use the Distributive Property, removing the brackets.

$$\cancel{(x + 3)}(x - 1)\frac{(x + 2)}{\cancel{(x + 3)}} + \cancel{(x + 3)(x - 1)}\frac{1}{\cancel{(x + 3)(x - 1)}} = (x + 3)(x - 1)$$

Step 4: Solve the resulting equation.

$$(x - 1)(x + 2) + 1 = (x + 3)(x - 1)$$

$$x^2 + x - 2 + 1 = x^2 + 2x - 3 \qquad \text{Simplify.}$$

$$x - 1 = 2x - 3 \qquad \text{Subtract } x^2 \text{ from both sides and combine like terms.}$$

$$2 = x$$

Because 2 is not a restricted value for x, it is a root. However, we will check it.

$$\frac{2 + 2}{2 + 3} + \frac{1}{2^2 + 2(2) - 3} = \frac{4}{5} + \frac{1}{5} = 1$$

Since 2 satisfies the equation, it is the solution (root) of the equation.   ■

Self Check 6   Solve: $\dfrac{3}{5} + \dfrac{7}{x + 2} = 2$.

Now Try Exercise 49.

4. Solve Formulas for a Specific Variable

Many equations, called **formulas,** contain several variables. For example, the formula that converts degrees Celsius to degrees Fahrenheit is $F = \frac{9}{5}C + 32$. If we want to change a large number of Fahrenheit readings to degrees Celsius, it would be tedious to substitute each value of F into the formula and then repeatedly solve it for C. It is better to solve the formula for C, substitute the values for F, and evaluate C directly.

EXAMPLE 7   **Solving a Formula for a Variable**

Solve $F = \dfrac{9}{5}C + 32$ for C.

SOLUTION   We use the same methods as for solving linear equations.

$$F = \frac{9}{5}C + 32$$

$$F - 32 = \frac{9}{5}C \qquad \text{Subtract 32 from both sides.}$$

$$\frac{5}{9}(F - 32) = \frac{5}{9}\left(\frac{9}{5}C\right) \qquad \text{Multiply both sides by } \frac{5}{9}.$$

$$\frac{5}{9}(F - 32) = C \qquad \text{Simplify.}$$

This result can be written in the alternate form $C = \dfrac{5F - 160}{9}$.

Self Check 7   Solve $C = \dfrac{5}{9}(F - 32)$ for F.

Now Try Exercise 63.

EXAMPLE 8   **Solving a Formula for a Variable**

The formula $A = P + Prt$ is used to find the amount of money in a savings account at the end of a specified time. A represents the amount, P represents the principal (the original deposit), r represents the rate of simple interest per unit of time, and t represents the number of units of time. Solve this formula for P.

SOLUTION   We factor P from both terms on the right-hand side of the equation and proceed as follows:

$$A = P + Prt$$

$$A = P(1 + rt) \qquad \text{Factor out } P.$$

$$\frac{A}{1 + rt} = P \qquad \text{Divide both sides by } 1 + rt.$$

$$P = \frac{A}{1 + rt}$$

Self Check 8   Solve $pq = fq + fp$ for f.

Now Try Exercise 69.

5. Solve Application Problems

Now we will apply the equation-solving techniques we covered to solve applied problems (often called word problems). To solve these problems, we must translate the verbal description of the problem into an equation. The process of finding the equation that describes the words of the problem is called **mathematical modeling.** The equation itself is often called a **mathematical model** of the situation described in the word problem.

Tim McGraw is one of the most successful country-music singers. He is married to country singer Faith Hill and is the son of former baseball player Tug McGraw. His album sales have exceeded 40 million copies and his "Soul to Soul" concert tour has been attended by thousands of fans.

© Everett Collection Inc/Alamy

Suppose we hear that Best Buy has Tim McGraw's CD *Southern Voice* on sale for 30% off the original price. Knowing this is a bargain, we immediately drive to Best Buy, quickly pay the $12.99 selling price for the CD, and listen to the great music as we drive. Later that day, this question comes to mind: What was the original price of the CD?

A linear equation can be used to model this problem. We can let x represent the original price of the CD and subtract 30% of x (the discount) to get the selling price of $12.99. The linear equation is $x - 0.3x = 12.99$. We can solve this equation to determine the original price of the CD.

$$x - 0.3x = 12.99$$

$$0.7x = 12.99 \qquad \text{Combine like terms.}$$

$$x = \frac{12.99}{0.7} \qquad \text{Divide both sides by 0.7.}$$

$$x = 18.56$$

The original price of the CD is $18.56.

The following list of steps provides a strategy to follow when we try to find an equation that models an applied problem:

Strategy for Modeling with Equations	**1.** Analyze the problem to see what you are to find. Often, drawing a diagram or making a table will help you visualize the facts.
	2. Pick a variable to represent the quantity that is to be found, and write a sentence telling what that variable represents. Express all other quantities mentioned in the problem as expressions involving this single variable.
	3. Find a way to express a quantity in two different ways. This might involve a formula from geometry, finance, or physics.
	4. Form an equation indicating that the two expressions found in Step 3 are equal.
	5. Solve the equation.
	6. Answer the questions asked in the problem.
	7. Check the answers in the words of the problem.

This list does not apply to all situations, but it can be used for a wide range of problems with only slight modifications.

EXAMPLE 9   **Solving a Number Problem**

A student scores 74%, 78%, and 70% on three college algebra exams. What score is needed on a fourth exam for the student to earn an average grade of 80%?

SOLUTION   To find an equation that models the problem, we can let x represent the required grade on the fourth exam. The average grade will be the sum of the four grades divided by 4. We want to find the value that yields an average of 80.

The average of the four grades	equals	the required average grade.
$\dfrac{74 + 78 + 70 + x}{4}$	=	80

We can solve this equation for x.

$$\frac{222 + x}{4} = 80 \qquad 74 + 78 + 70 = 222.$$

$$222 + x = 320 \qquad \text{Multiply both sides by 4.}$$

$$x = 98 \qquad \text{Subtract 222 from both sides.}$$

To earn an average of 80%, the student must score 98% on the fourth exam.   ∎

Self Check 9   A student scores 82%, 96%, 91%, and 92% on four college algebra exams. What score is needed on a fifth exam for the student to earn an average grade of 90%?

Now Try Exercise 77.

EXAMPLE 10   **Solving a Geometry Problem**

A city ordinance requires a man to install a fence around the swimming pool shown in Figure R-2. He wants the border around the pool to be of uniform width. If he has 154 feet of fencing, find the width of the border.

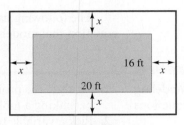

FIGURE R-2

SOLUTION   We can let x represent the width of the border. The distance around the large rectangle, called its **perimeter,** is given by the formula $P = 2l + 2w$. Writing the length and width in terms of x, we have length $l = 20 + 2x$ and width $w = 16 + 2x$. Since the man has 154 feet of fencing, the perimeter will be 154 feet. To find an equation that models the problem, we substitute these values into the formula for perimeter.

$P = 2l + 2w$	This is the perimeter formula of a rectangle.
$154 = 2(20 + 2x) + 2(16 + 2x)$	Substitute 154 for P, $20 + 2x$ for l, and $16 + 2x$ for w.
$154 = 40 + 4x + 32 + 4x$	Use the Distributive Property to remove parentheses.
$154 = 72 + 8x$	Combine like terms.
$82 = 8x$	Subtract 72 from both sides.
$\dfrac{82}{8} = \dfrac{41}{4} = x$	Divide both sides by 8.

This border will be $\frac{41}{4} = 10\frac{1}{4}$ feet wide. ∎

Self Check 10   In Example 10, if 168 feet of fencing is available, find the border's width.

Now Try Exercise 81.

EXAMPLE 11   Solving an Investment Problem

A woman invested \$10,000, part at 9% and the rest at 14%. If the annual income from these investments is \$1,275, how much was invested at each rate?

SOLUTION   We can let x represent the amount invested at 9%. Then $10,000 - x$ represents the amount invested at 14%. Since the annual income from any investment is the product of the interest rate and the amount invested, we have the following information:

Type of Investment	Rate	Amount Invested	Interest Earned
9% investment	0.09	x	$0.09x$
14% investment	0.14	$10,000 - x$	$0.14(10,000 - x)$

The total income from these two investments can be expressed in two ways: as \$1,275 and as the sum of the incomes of the two investments.

The income from the 9% investment	plus	the income from the 14% investment	equals	the total income.
$0.09x$	$+$	$0.14(10,000 - x)$	$=$	1,275

We can solve this equation for x.

$$0.09x + 0.14(10,000 - x) = 1,275$$

$$9x + 14(10,000 - x) = 127,500 \qquad \text{Multiply both sides by 100 to eliminate the decimal points.}$$

$$9x + 140,000 - 14x = 127,500 \qquad \text{Use the Distributive Property to remove parentheses.}$$

$$-5x + 140,000 = 127,500 \qquad \text{Combine like terms.}$$

$$-5x = -12,500 \qquad \text{Subtract 140,000 from both sides.}$$

$$x = 2,500 \qquad \text{Divide both sides by } -5.$$

The amount invested at 9% was $2,500, and the amount invested at 14% was $7,500 ($10,000 − $2,500). These amounts are correct, because 9% of $2,500 is $225, 14% of $7,500 is $1,050, and the sum of these amounts is $1,275.

Self Check 11 A man invests $12,000, part at 7% and the rest at 9%. If the annual income from these investments is $965, how much was invested at each rate?

Now Try Exercise 87.

EXAMPLE 12 **Solving a Shared-Work Problem**

The Tollway Authority needs to pave 100 miles of interstate highway before freezing temperatures come in about 60 days. Sanders and Sons has estimated that it can do the job in 110 days. Churchill and Sons has estimated that it can do the job in 140 days. If the authority hires both contractors, will the job get done in time?

SOLUTION We can let n represent the number of days it will take to pave the highway if both contractors are hired. In one day, the contractors working together can do $\frac{1}{n}$ of the job. In one day, Sanders can do $\frac{1}{110}$ of the job. In one day, Churchill can do $\frac{1}{140}$ of the job. The work that they can do together in one day is the sum of what each can do in one day.

The part Sanders can pave in one day	plus	the part Churchill can pave in one day	equals	the part they can pave together in one day.
$\dfrac{1}{110}$	$+$	$\dfrac{1}{140}$	$=$	$\dfrac{1}{n}$

We can solve this equation for n.

$$\frac{1}{110} + \frac{1}{140} = \frac{1}{n}$$

First, we multiply both sides by $(110)(140)n$.

$$(110)(140)n\left(\frac{1}{110} + \frac{1}{140}\right) = (110)(140)n\left(\frac{1}{n}\right)$$

$$\frac{(110)(140)n}{110} + \frac{(110)(140)n}{140} = \frac{(110)(140)n}{n} \qquad \text{Simplify.}$$

$$140n + 110n = 15,400$$

$$250n = 15,400 \qquad \text{Combine like terms.}$$

$$n = 61.6 \qquad \text{Divide both sides by 250.}$$

It will take the contractors about 62 days to pave the highway. If the Tollway Authority is lucky, the job will be done in time.

Self Check 12   John and Eric both work at Firestone Auto Care. John can install a new set of tires in 45 minutes. Eric is faster and can install a set in 30 minutes. If John and Eric work together, how long will it take them to install one set of tires?

Now Try Exercise 95.

EXAMPLE 13   **Solving a Mixture Problem**

A container is partially filled with 20 liters of whole milk containing 4% butterfat. How much 1% milk must be added to obtain a mixture that is 2% butterfat?

SOLUTION   Since the first container shown in Figure R-3(a) contains 20 liters of 4% milk, it contains 0.04(20) liters of butterfat. To this amount, we will add the contents of the second container, which holds 0.01l liters of butterfat. The sum of these two amounts will equal the amount of butterfat in the third container, which is 0.02(20 + l) liters of butterfat. This information is presented in table form in Figure R-3(b).

The butterfat in the 4% milk	plus	the butterfat in the 1% milk	equals	the butterfat in the 2% milk.
4% of 20 liters	+	1% of l liters	=	2% of (20 + l) liters

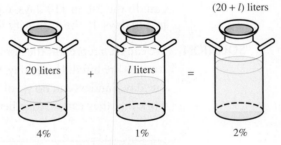

FIGURE R-3(a)

	Percentage of Butterfat	·	Amount of Milk	=	Amount of Butterfat
4% milk	0.04		20		0.04(20)
1% milk	0.01		l		0.01(l)
2% milk	0.02		20 + l		0.02(20 + l)

FIGURE R-3(b)

We can solve this equation for l.

$$0.04(20) + 0.01(l) = 0.02(20 + l)$$

$$4(20) + l = 2(20 + l)$$ Multiply both sides by 100.

$$80 + l = 40 + 2l$$ Remove parentheses.

$$40 = l$$ Subtract 40 and l from both sides.

To dilute the 20 liters of 4% milk to a 2% mixture, 40 liters of 1% milk must be added. To check, we note that the final mixture contains 0.02(60) = 1.2 liters of pure butterfat, and that this is equal to the amount of pure butterfat in the 4% milk and the 1% milk; 0.04(20) + 0.01(40) = 1.2 liters.

Self Check 13   Milk containing 4% butterfat is mixed with 8 gallons of milk containing 1% butterfat to make a low-fat cottage cheese mixture containing 2% butterfat. How many gallons of the richer milk are used?

Now Try Exercise 101.

EXAMPLE 14   **Solving a Uniform-Motion Problem**

A man leaves home driving his Ford F-150 truck at the rate of 50 mph. When his daughter discovers that he has forgotten his wallet, she drives after him in her Ford Mustang at the rate of 65 mph (Figure R-4). How long will it take her to catch her dad if he has a 15-minute head start?

SOLUTION   Uniform-motion problems are based on the formula $d = rt$, where d is the distance, r is the rate, and t is the time. In the chart, t represents the number of hours the daughter must drive to overtake her father. Because the father has a 15-minute, or $\frac{1}{4}$-hour, head start, he has been on the road for $(t + \frac{1}{4})$ hours.

Patrick Mezirka/Shutterstock.com

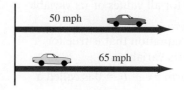

50 mph

65 mph

FIGURE R-4

We can set up the following equation and solve it for t, as shown in Figure R-5.

The distance the man drives	equals	the distance the daughter drives.
$50\left(t + \dfrac{1}{4}\right)$	$=$	$65t$

FIGURE R-5

$$50\left(t + \frac{1}{4}\right) = 65t \qquad \text{Set up equation.}$$

$$50t + \frac{25}{2} = 65t \qquad \text{Clear parentheses.}$$

$$\frac{25}{2} = 15t \qquad \text{Simplify.}$$

$$\frac{5}{6} = t \qquad \text{Solution.}$$

It will take the daughter $\frac{5}{6}$ hour, or 50 minutes, to overtake her father.

Self Check 14   On a reality television show, an officer traveling in a police cruiser at 90 miles per hour pursues Jennifer, who has a three-minute head start. If the officer overtakes Jennifer in 12 minutes, how fast is Jennifer traveling?

Now Try Exercise 113.

Exercises R.1

Getting Ready

You should be able to complete these vocabulary and concept statements before you proceed to the practice exercises.

Fill in the blanks.

1. If a number satisfies an equation, it is called a _____ or a _____ of the equation.

2. If an equation is true for all values of its variable, it is called an _____.

3. A contradiction is an equation that is true for _____ values of its variable.

4. An equation of the form $ax + b = 0$ is called a _____ equation.

5. If an equation contains rational expressions, it is called a _____ equation.

6. The _____ of a fraction can never be 0.

Practice

For each equation, determine any restrictions on x.

7. $x + 3 = 1$

8. $\dfrac{1}{2}x - 7 = 14$

9. $\dfrac{1}{x} = 12$

10. $\dfrac{3}{x - 2} = x$

11. $\sqrt{x} = 4$

12. $\sqrt[3]{x} = 64$

13. $\dfrac{1}{x - 3} = \dfrac{5}{x + 2}$

14. $\dfrac{24}{\sqrt{x - 3}} = \dfrac{3}{x + 4}$

Solve each equation, if possible. Classify each one as an identity, a conditional equation, or a contradiction.

15. $2x + 5 = 15$

16. $3x + 2 = x + 8$

17. $2(n + 2) - 5 = 2n$

18. $3(m + 2) = 2(m + 3) + m$

19. $\dfrac{x + 7}{2} = 7$

20. $\dfrac{x}{2} - 7 = 14$

21. $3(x - 3) = \dfrac{6x - 18}{2}$

22. $x(x + 2) = (x + 1)^2$

23. $\dfrac{3}{b - 3} = 1$

24. $x^2 - 8x + 15 = (x - 3)(x + 5)$

25. $2x^2 + 5x - 3 = (2x - 1)(x + 3)$

26. $2x^2 + 5x - 3 = 2x\left(x + \dfrac{19}{2}\right)$

Solve each equation. If an equation has no solution, so indicate.

27. $2x + 7 = 10 - x$

28. $9a - 3 = 15 + 3a$

29. $\dfrac{5}{3}z - 8 = 7$

30. $\dfrac{4}{3}y + 12 = -4$

31. $\dfrac{3x - 2}{3} = 2x + \dfrac{7}{3}$

32. $\dfrac{7}{2}x + 5 = x + \dfrac{15}{2}$

33. $5(x - 2) = 2x + 8$

34. $5(r - 4) = -5(r - 4)$

35. $2(2x + 1) - \dfrac{3x}{2} = \dfrac{-3(4 + x)}{2}$

36. $(x - 2)(x - 3) = (x + 3)(x + 4)$

37. $7(2x + 5) - 6(x + 8) = 7$

38. $(t + 1)(t - 1) = (t + 2)(t - 3) + 4$

39. $\dfrac{3}{2}(3x - 2) - 10x - 4 = 0$

40. $a(a - 3) + 5 = (a - 1)^2$

41. $\dfrac{(y + 2)^2}{3} = y + 2 + \dfrac{y^2}{3}$

42. $2x - \dfrac{7}{6} + \dfrac{x}{6} = \dfrac{4x + 3}{6}$

43. $\dfrac{3}{x} + \dfrac{1}{2} = \dfrac{4}{x}$

44. $\dfrac{2}{x + 1} + \dfrac{1}{3} = \dfrac{1}{x + 1}$

45. $\dfrac{3}{x - 2} + \dfrac{1}{x} = \dfrac{3}{x - 2}$

46. $\dfrac{9t + 6}{t(t + 3)} = \dfrac{7}{t + 3}$

47. $x + \dfrac{2(-2x + 1)}{3x + 5} = \dfrac{3x^2}{3x + 5}$

48. $\dfrac{2}{(a - 7)(a + 2)} = \dfrac{4}{(a + 3)(a + 2)}$

49. $\dfrac{2}{n - 2} + \dfrac{1}{n + 1} = \dfrac{1}{n^2 - n - 2}$

50. $\dfrac{2x + 3}{x^2 + 5x + 6} + \dfrac{3x - 2}{x^2 + x - 6} = \dfrac{5x - 2}{x^2 - 4}$

51. $\dfrac{3x}{x^2 + x} - \dfrac{2x}{x^2 + 5x} = \dfrac{x + 2}{x^2 + 6x + 5}$

52. $\dfrac{3x + 5}{x^3 + 8} + \dfrac{3}{x^2 - 4} = \dfrac{2(3x - 2)}{(x - 2)(x^2 - 2x + 4)}$

53. $\dfrac{1}{n + 8} - \dfrac{3n - 4}{5n^2 + 42n + 16} = \dfrac{1}{5n + 2}$

54. $\dfrac{5}{y + 4} + \dfrac{2}{y + 2} = \dfrac{6}{y + 2} - \dfrac{1}{y^2 + 6y + 8}$

55. $\dfrac{6}{2a - 6} - \dfrac{3}{3 - 3a} = \dfrac{1}{a^2 - 4a + 3}$

56. $\dfrac{3y}{6 - 3y} + \dfrac{2y}{2y + 4} = \dfrac{8}{4 - y^2}$

57. $\dfrac{3 + 2a}{a^2 + 6 + 5a} - \dfrac{2 - 3a}{a^2 - 6 + a} = \dfrac{5a - 2}{a^2 - 4}$

58. $\dfrac{x - 1}{x + 3} + \dfrac{x - 2}{x - 3} = \dfrac{1 - 2x}{3 - x}$

Solve each formula for the specified variable.

59. $P = 2l + 2w; \ w$

60. $V = \dfrac{1}{3}\pi r^2 h; \ h$

61. $V = \dfrac{1}{3}\pi r^2 h; \ r^2$

62. $z = \dfrac{x - \mu}{\sigma}; \ \mu$

63. $P_n = L + \dfrac{si}{f}; \ s$

64. $P_n = L + \dfrac{si}{f}; \ f$

65. $F = \dfrac{mMg}{r^2}; \ m$

66. $\dfrac{1}{f} = \dfrac{1}{p} + \dfrac{1}{q}; \ f$

67. $\dfrac{x}{a} + \dfrac{y}{b} = 1; \ y$

68. $\dfrac{x}{a} - \dfrac{y}{b} = 1; \ a$

69. $\dfrac{1}{r} = \dfrac{1}{r_1} + \dfrac{1}{r_2}; \ r$

70. $\dfrac{1}{r} = \dfrac{1}{r_1} + \dfrac{1}{r_2}; \ r_1$

71. $l = a + (n - 1)d; \ n$

72. $l = a + (n - 1)d; \ d$

73. $a = (n - 2)\dfrac{180}{n}; \ n$

74. $S = \dfrac{a - lr}{1 - r}; \ a$

75. $R = \dfrac{1}{\dfrac{1}{r_1} + \dfrac{1}{r_2} + \dfrac{1}{r_3}}; \ r_1$

76. $R = \dfrac{1}{\dfrac{1}{r_1} + \dfrac{1}{r_2} + \dfrac{1}{r_3}}; \ r_3$

Applications

77. **Test scores** Tate scored 5 points higher on his statistics midterm and 13 points higher on his statistics final than he did on his first exam in the course. If his mean (average) score was 90, what was his score on the first statistics exam?

78. **Golfing** Par on most golf courses is 72. If a professional golfer shot rounds of 76, 68, and 70 in a tournament, what will she need to shoot on the final round to average par?

79. **Replacing locks** A locksmith at Pop-A-Lock charges $40 plus $28 for each lock installed. How many locks can be replaced for $236?

80. **Delivering ads** A University of Florida student earns $20 per day delivering advertising brochures door-to-door, plus 75¢ for each person he interviews. How many people did he interview on a day when he earned $56?

81. **Width of a picture frame** The picture frame with the dimensions shown in the illustration was built with 14 feet of framing material. Find its width.

$(x + 2)$ ft

x ft

82. **Fencing a garden** If a gardener fences in the total rectangular area shown in the illustration instead of just the square area, he will need twice as much fencing to enclose the garden. How much fencing will he need?

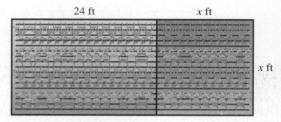

24 ft x ft

x ft

83. **Swimming-pool dimensions** The area of the triangular swimming pool shown in the illustration is doubled by adding a rectangular wading pool. Find the dimensions of the wading pool. (*Hint:* The area of a triangle is $\frac{1}{2}bh$, and the area of a rectangle is lw.)

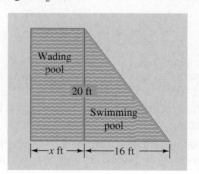

Wading pool

20 ft

Swimming pool

$\leftarrow x$ ft $\rightarrow|\leftarrow$ 16 ft $\rightarrow|$

84. House construction A builder wants to install a triangular window with the angles shown in the illustration. What angles will he have to cut to make the window fit? (*Hint:* The sum of the angles in a triangle equals 180°.)

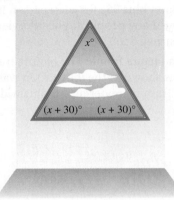

85. Length of a living room If a carpenter adds a porch with dimensions shown in the illustration to the living room, the living area will be increased by 50%. Find the length of the living room.

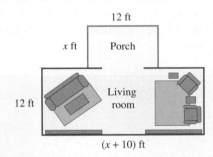

86. Depth of water in a trough The trough in the illustration has a cross-sectional area of 54 square inches. Find the depth d of the trough. (*Hint:* Area of a trapezoid is $\frac{1}{2}h(b_1 + b_2)$.)

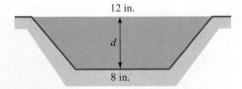

87. Investment An executive invests $22,000, some at 7% and some at 6% annual interest. If he receives an annual return of $1,420, how much is invested at each rate?

88. Financial planning After inheriting some money, a woman wants to invest enough to have an annual income of $5,000. If she can invest $20,000 at 9% annual interest, how much more will she have to invest at 7% to achieve her goal? (See the table.)

Type	Rate	Amount	Income
9% investment	0.09	20,000	.09(20,000)
7% investment	0.07	x	.07x

89. Ticket sales A full-price ticket for a college basketball game costs $15, and a student ticket costs $3.50. If 1,545 tickets were sold, and the total receipts were $17,712.50, how many tickets were student tickets?

90. Ticket sales Of the 800 tickets sold to a movie, 480 were full-price tickets costing $8 each. If the gate receipts were $5,440, what did a student ticket cost?

91. Discounts After being discounted 20%, a weather radio sells for $63.96. Find the original price.

92. Markups A merchant increases the wholesale cost of a Maytag washing machine by 30% to determine the selling price. If the washer sells for $588.90, find the wholesale cost.

93. Computer sales A computer store has fixed costs of $8,925 per month and a unit cost of $850 for every computer it sells. If the store can sell all the computers it can get for $1,275 each, how many must be sold for the store to break even? (*Hint:* The break-even point occurs when costs equal income.)

94. Restaurant management A restaurant has fixed costs of $137.50 per day and an average unit cost of $4.75 for each meal served. If a typical meal costs $6, how many customers must eat at the restaurant each day for the owner to break even?

95. Roofing houses Kyle estimates that it will take him 7 days to roof his house. A professional roofer estimates that it will take him 4 days to roof the same house. How long will it take if they work together?

96. Sealing asphalt One crew can seal a parking lot in 8 hours and another in 10 hours. How long will it take to seal the parking lot if the two crews work together?

97. Mowing lawns Julie can mow her lawn with a lawn tractor in 2 hours, and her husband can mow the same lawn with a push mower in 4 hours. How long will it take to mow the lawn if they work together?

98. Filling swimming pools A garden hose can fill a swimming pool in 3 days, and a larger hose can fill the pool in 2 days. How long will it take to fill the pool if both hoses are used?

99. Filling swimming pools An empty swimming pool can be filled in 10 hours. When full, the pool can be drained in 19 hours. How long will it take to fill the empty pool if the drain is left open?

100. Preparing seafood Kevin stuffs shrimp in his job as a seafood chef. He can stuff 1,000 shrimp in 6 hours. When his sister helps him, they can stuff 1,000 shrimp in 4 hours. If Kevin gets sick, how long will it take his sister to stuff 500 shrimp?

101. Diluting solutions How much water should be added to 20 ounces of a 15% solution of alcohol to dilute it to a 10% solution?

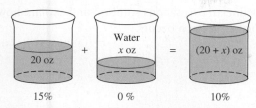

102. Increasing concentrations The beaker shown below contains a 2% saltwater solution.

a. How much water must be boiled away to increase the concentration of the salt solution from 2% to 3%?

b. Where on the beaker would the new water level be?

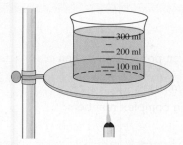

103. Winterizing cars A car radiator has a 6-liter capacity. If the liquid in the radiator is 40% antifreeze, how much liquid must be replaced with pure antifreeze to bring the mixture up to a 50% solution?

104. Mixing milk If a bottle holding 3 liters of milk contains $3\frac{1}{2}$% butterfat, how much skim milk must be added to dilute the milk to 2% butterfat?

105. Cleaning swimming pools A swimming pool contains 15,000 gallons of water. How many gallons of chlorine must be added to "shock the pool" and bring the water to a $\frac{3}{100}$% solution?

106. Mixing fuels A new automobile engine can run on a mixture of gasoline and a substitute fuel. If gas costs $3.50 per gallon and the substitute fuel costs $2 per gallon, what percentage of a mixture must be substitute fuel to bring the cost down to $2.75 per gallon?

107. Evaporation How many liters of water must evaporate to turn 12 liters of a 24% salt solution into a 36% solution?

108. Increasing concentrations A beaker contains 320 milliliters of a 5% saltwater solution. How much water should be boiled away to increase the concentration to 6%?

109. Mixing solutions How many gallons of a 5% alcohol solution must be mixed with 90 gallons of a 1% solution to obtain a 2% solution?

110. Preparing medicines A doctor prescribes an ointment that is 2% hydrocortisone. A pharmacist has 1% and 5% concentrations in stock. How much of each should the pharmacist use to make a 1-ounce tube?

111. Driving rates John drove to Daytona Beach, Florida, in 5 hours. When he returned, there was less traffic, and the trip took only 3 hours. If John averaged 26 mph faster on the return trip, how fast did he drive each way?

112. Distance Allison drove home at 60 mph, but her brother Austin, who left at the same time, could drive at only 48 mph. When Allison arrived, Austin still had 60 miles to go. How far did Allison drive?

113. Distance Two cars leave Baton Rouge Community College traveling in opposite directions. One car travels at 60 mph and the other at 64 mph. In how many hours will they be 310 miles apart?

114. Bank robberies Some bank robbers leave town, speeding at 70 mph. Ten minutes later, the police give chase, traveling at 78 mph. How long after the robbery takes place will it take the police to overtake the robbers?

115. Boating A Johnson motorboat goes 5 miles upstream in the same time it requires to go 7 miles downstream. If the river flows at 2 mph, find the speed of the boat in still water.

116. Wind velocity A Beechcraft airplane can fly 340 mph in still air. If it can fly 200 miles downwind in the same amount of time it can fly 140 miles upwind, find the velocity of the wind.

Use a calculator to help solve each problem.

117. Machine tool design 712.51 cubic millimeters of material was removed by drilling the blind hole as shown in the illustration. Find the depth of the hole. (*Hint:* The volume of a cylinder is given by $V = \pi r^2 h$.)

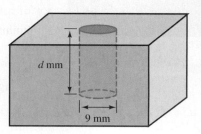

118. Architecture The Norman window with dimensions shown is a rectangle topped by a semicircle. If the area of the window is 68.2 square feet, find its height h.

6 ft

119. Explain why a conditional linear equation always has exactly one root.
120. Define an extraneous solution and explain how such a solution occurs.

R.2 Complex Numbers

In this section, we will learn to

1. Define and simplify imaginary numbers.
2. Define and perform operations on complex numbers.
3. Simplify powers of i.

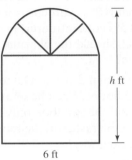

William Thomas Cain/Getty Images

The Blue Man Group is a highly successful group of creative performers who cover themselves in blue greasepaint, wear latex bald caps, and dress in black. They combine rock music, comedy, multimedia theatrics, and sophisticated lighting to entertain their audiences.

There are a variety of themes used in Blue Man Group performances, and the group uses fractals in their shows. Fractals are beautiful art designs that are computer generated from numbers called *complex numbers*. We will study these numbers in this section.

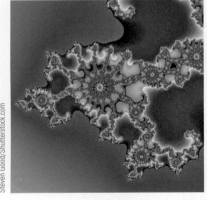

Steven Good/Shutterstock.com

1. Define and Simplify Imaginary Numbers

For many years, mathematicians considered such numbers as $\sqrt{-1}$, $\sqrt{-5}$, and $\sqrt{-9}$ to make no sense. Sir Isaac Newton (1642–1727) called them "impossible numbers." In the 17th century, these symbols were called **imaginary numbers** by René Descartes. Today, they have important uses, such as describing the behavior of alternating current in electronics. The imaginary numbers are based on the imaginary unit; the imaginary unit is denoted by the letter i.

Imaginary Unit i   The **imaginary unit** is denoted by the letter i and is defined as

$$i = \sqrt{-1}.$$

From the definition it follows that $i^2 = -1$.

Because imaginary numbers follow the rules for exponents, we have

$$(3i)^2 = 3^2 i^2 = 9(-1) = -9.$$

Since $(3i)^2 = -9$, $3i$ is the square root of -9, and we can write

$$\sqrt{-9} = \sqrt{9(-1)}$$

$$= \sqrt{9}\sqrt{-1} \qquad \sqrt{ab} = \sqrt{a}\sqrt{b}$$

$$= 3i \qquad \sqrt{-1} = i$$

The Multiplication Property of Radicals can be used to simplify imaginary numbers. Four examples are shown in the following table:

Comment

Since it is easy to confuse $\sqrt{b}i$ and $\sqrt{bi}$, we usually write i first so that it is clear that i is not under the radical symbol.

Imaginary Numbers Written in Terms of i
$\sqrt{-25} = \sqrt{25(-1)} = \sqrt{25}\sqrt{-1} = 5i$
$-\sqrt{-169} = -\sqrt{169(-1)} = -\sqrt{169}\sqrt{-1} = -13i$
$\sqrt{-24} = \sqrt{24(-1)} = \sqrt{24}\sqrt{-1} = 2\sqrt{6}i$ or $2i\sqrt{6}$
$\sqrt{\dfrac{-8}{49}} = \sqrt{\dfrac{8}{49}(-1)} = \sqrt{\dfrac{8}{49}}\sqrt{-1} = \dfrac{\sqrt{8}}{\sqrt{49}}i = \dfrac{2\sqrt{2}}{7}i$ or $\dfrac{2i\sqrt{2}}{7}$

Caution

If a and b are both negative, then $\sqrt{ab} \neq \sqrt{a}\sqrt{b}$. For example, the correct simplification of $\sqrt{-16}\sqrt{-4}$ is

$$\sqrt{-16}\sqrt{-4} = (4i)(2i) = 8i^2 = 8(-1) = -8.$$

The following simplification is incorrect. Since both numbers are negative, the Multiplication Property of Radicals does not apply.

$$\sqrt{-16}\sqrt{-4} = \sqrt{(-16)(-4)} = \sqrt{64} = 8$$

ACCENT ON TECHNOLOGY   **Simplifying Imaginary Numbers**

If we place our graphing calculator in $a + bi$ mode, it can be used to simplify imaginary numbers. Please note that on a graphing calculator i is located above the decimal-point key.

In Figure R-6, we see the graphing-calculator results for $\sqrt{-81}$, $\sqrt{-\dfrac{4}{25}}$, and $\sqrt{-108}$.

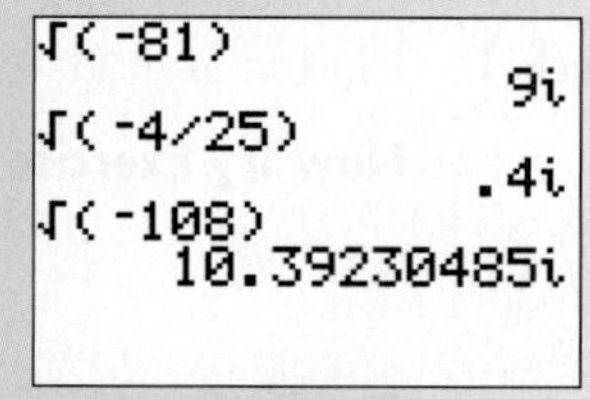

```
√(-81)
                    9i
√(-4/25)
                   .4i
√(-108)
        10.39230485i
```

FIGURE R-6

Numbers that are the sum or difference of a real number and an imaginary number, such as $3 + 4i$, $-5 + 7i$, and $-1 - 9i$, are called *complex numbers*.

Complex Numbers   A **complex number** is a number that can be written in the form $a + bi$, where a and b are real numbers and $i = \sqrt{-1}$. The number a is called the **real part**, and b is called the **imaginary part**.

A complex number written in the form $a + bi$ is said to be in **standard form**. If $b = 0$, the complex number $a + bi$ is the real number a. If $a = 0$ and $b \neq 0$, the complex number $a + bi$ is the imaginary number bi. It follows that the set of real numbers and the set of imaginary numbers are subsets of the set of complex numbers. Figure R-7 illustrates how the various sets of numbers are related.

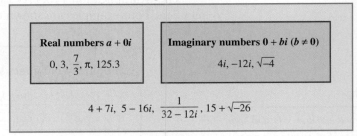

Complex numbers

Real numbers $a + 0i$	Imaginary numbers $0 + bi$ $(b \neq 0)$
$0, 3, \dfrac{7}{3}, \pi, 125.3$	$4i, -12i, \sqrt{-4}$

$4 + 7i, \ 5 - 16i, \ \dfrac{1}{32 - 12i}, \ 15 + \sqrt{-26}$

FIGURE R-7

To determine whether two complex numbers are equal, we can use the following property.

Equality of Complex Numbers   Two complex numbers are equal if their real parts are equal and their imaginary parts are equal. If $a + bi$ and $c + di$ are two complex numbers, then

$$a + bi = c + di \quad \text{if and only if} \quad a = c \quad \text{and} \quad b = d.$$

EXAMPLE 1   **Finding the Values of x and y**

For what numbers x and y is $3x + 4i = (2y + x) + xi$?

SOLUTION   Since the numbers are equal, their imaginary parts must be equal: $x = 4$. Since their real parts are equal, $3x = 2y + x$. We can solve the system

$$\begin{cases} x = 4 \\ 3x = 2y + x \end{cases}$$

by substituting 4 for x in the second equation and solving for y. We find that $y = 4$. The solution is $x = 4$ and $y = 4$.

Self Check 1   Find x: $a + (x + 3)i = a - (2x - 1)i$.

Now Try Exercise 15.

2. Define and Perform Operations on Complex Numbers

Complex numbers can be added and subtracted as if they were binomials.

Adding and Subtracting Complex Numbers   Two complex numbers such as $a + bi$ and $c + di$ are added and subtracted as if they were binomials:

$$(a + bi) + (c + di) = (a + c) + (b + d)i$$

$$(a + bi) - (c + di) = (a - c) + (b - d)i$$

Because of the preceding definition, the sum or difference of two complex numbers is another complex number.

EXAMPLE 2   **Adding and Subtracting Complex Numbers**

Simplify: **a.** $(3 + 4i) + (2 + 7i)$ **b.** $(-5 + 8i) - (2 - 12i)$

SOLUTION   **a.** $(3 + 4i) + (2 + 7i) = 3 + 4i + 2 + 7i$

$$= 3 + 2 + 4i + 7i$$

$$= 5 + 11i$$

b. $(-5 + 8i) - (2 - 12i) = -5 + 8i - 2 + 12i$

$$= -5 - 2 + 8i + 12i$$

$$= -7 + 20i$$

Self Check 2   Simplify: **a.** $(5 - 2i) + (-3 + 9i)$ **b.** $(2 + 5i) - (6 + 7i)$

Now Try Exercise 21.

Complex numbers can also be multiplied as if they were binomials.

Multiplying Complex Numbers   The numbers $a + bi$ and $c + di$ are multiplied as if they were binomials, with $i^2 = -1$:

$$(a + bi)(c + di) = (ac - bd) + (ad + bc)i$$

Because of this definition, the product of two complex numbers is another complex number.

EXAMPLE 3   **Multiplying Complex Numbers**

Multiply: **a.** $(3 + 4i)(2 + 7i)$ **b.** $(5 - 7i)(1 + 3i)$

SOLUTION   **a.** $(3 + 4i)(2 + 7i) = 6 + 21i + 8i + 28i^2$

$$= 6 + 21i + 8i + 28(-1) \qquad i^2 = -1$$

$$= 6 - 28 + 29i$$

$$= -22 + 29i$$

b. $(5 - 7i)(1 + 3i) = 5 + 15i - 7i - 21i^2$

$$= 5 + 15i - 7i - 21(-1) \qquad i^2 = -1$$

$$= 5 + 21 + 8i$$

$$= 26 + 8i$$

Self Check 3   Multiply: $(2 - 5i)(3 + 2i)$.

Now Try Exercise 31.

To avoid errors in determining the sign of the result, always express numbers in $a + bi$ form before attempting any algebraic manipulations.

EXAMPLE 4   **Multiplying Complex Numbers**

Multiply: $\left(-2 + \sqrt{-16}\right)\left(4 - \sqrt{-9}\right)$.

SOLUTION   We change each number to $a + bi$ form:

$$-2 + \sqrt{-16} = -2 + \sqrt{16}\sqrt{-1} = -2 + 4i$$
$$4 - \sqrt{-9} = 4 - \sqrt{9}\sqrt{-1} = 4 - 3i$$

and then find the product.

$$(-2 + 4i)(4 - 3i) = -8 + 6i + 16i - 12i^2$$
$$= -8 + 6i + 16i - 12(-1) \qquad i^2 = -1$$
$$= -8 + 12 + 22i$$
$$= 4 + 22i$$

Self Check 4   Multiply: $\left(3 + \sqrt{-25}\right)\left(2 - \sqrt{-9}\right)$.
Now Try Exercise 35.

Before we discuss the division of complex numbers, we introduce the concept of a complex conjugate.

Complex Conjugates   The complex numbers $a + bi$ and $a - bi$ are called **complex conjugates** of each other.

For example,

- $2 + 5i$ and $2 - 5i$ are complex conjugates.
- $-\frac{1}{2} + 4i$ and $-\frac{1}{2} - 4i$ are complex conjugates.
- $13i$ and $-13i$ are complex conjugates, because $13i$ is $0 + 13i$ and $-13i$ is $0 - 13i$.

What makes this concept important is the fact that the product of two complex conjugates is always a real number. For example,

$$(2 + 5i)(2 - 5i) = 4 - 10i + 10i - 25i^2$$
$$= 4 - 25(-1) \qquad i^2 = -1$$
$$= 4 + 25$$
$$= 29$$

In general, we have

$$(a + bi)(a - bi) = a^2 - abi + abi - b^2i^2$$
$$= a^2 - b^2(-1) \qquad i^2 = -1$$
$$= a^2 + b^2$$

To divide complex numbers, we use the concept of complex conjugates to rationalize the denominator.

EXAMPLE 5   **Dividing Complex Numbers**

Divide and write the result in $a + bi$ form: $\dfrac{3}{2 + i}$.

SOLUTION To divide, we rationalize the denominator and simplify.

$$\frac{3}{2+i} = \frac{3(2-i)}{(2+i)(2-i)}$$

To make the denominator a real number, multiply the numerator and denominator by the complex conjugate of $2+i$, which is $2-i$.

$$= \frac{6-3i}{4-2i+2i-i^2}$$ Multiply.

$$= \frac{6-3i}{4+1}$$ Simplify the denominator.

$$= \frac{6-3i}{5}$$

$$= \frac{6}{5} - \frac{3}{5}i$$

It is common to accept $\frac{6}{5} - \frac{3}{5}i$ as a substitute for $\frac{6}{5} + \left(-\frac{3}{5}\right)i$.

Self Check 5 Divide and write the result in $a+bi$ form: $\dfrac{3}{3-i}$.

Now Try Exercise 43.

EXAMPLE 6 **Dividing Complex Numbers**

Divide and write the result in $a+bi$ form: $\dfrac{2-\sqrt{-16}}{3+\sqrt{-1}}$.

SOLUTION $\dfrac{2-\sqrt{-16}}{3+\sqrt{-1}} = \dfrac{2-4i}{3+i}$ Change each number to $a+bi$ form.

$$= \frac{(2-4i)(3-i)}{(3+i)(3-i)}$$

To make the denominator a real number, multiply the numerator and denominator by $3-i$.

$$= \frac{6-2i-12i+4i^2}{9-3i+3i-i^2}$$ Remove parentheses.

$$= \frac{2-14i}{9+1}$$ Combine like terms; $i^2 = -1$.

$$= \frac{2}{10} - \frac{14i}{10}$$

$$= \frac{1}{5} - \frac{7}{5}i$$

Self Check 6 Divide and write the result in $a+bi$ form: $\dfrac{3+\sqrt{-25}}{2-\sqrt{-1}}$.

Now Try Exercise 51.

Examples 5 and 6 illustrate that the quotient of two complex numbers is another complex number.

3. Simplify Powers of i

The powers of i with natural-number exponents produce an interesting pattern, as shown in the table.

Powers of i	
$i^1 = \sqrt{-1} = i$	$i^5 = i^4 i = 1i = i$
$i^2 = \left(\sqrt{-1}\right)^2 = -1$	$i^6 = i^4 i^2 = 1(-1) = -1$
$i^3 = i^2 i = -1i = -i$	$i^7 = i^4 i^3 = 1(-i) = -i$
$i^4 = i^2 i^2 = (-1)(-1) = 1$	$i^8 = i^4 i^4 = 1(1) = 1$
The pattern continues: $i, -1, -i, 1, \ldots .$	

Comment

i^0 is 1 because any number except 0 raised to the 0th power is 1.

EXAMPLE 7   **Simplifying Powers of i**

Simplify: i^{365}.

SOLUTION   Since $i^4 = 1$, each occurrence of i^4 is a factor of 1. To determine how many factors of i^4 are in i^{365}, we divide 365 by 4. The quotient is 91, and the remainder is 1.

$$i^{365} = (i^4)^{91} \cdot i^1$$

$$= 1^{91} \cdot i^1 \qquad i^4 = 1$$

$$= i \qquad\qquad 1^{91} = 1 \text{ and } 1 \cdot i = i$$

■

Self Check 7   Simplify: $i^{1,999}$.

Now Try Exercise 57.

The result of Example 7 illustrates the following theorem:

Powers of i   If n is a natural number that has a remainder of r when divided by 4, then

$$i^n = i^r.$$

When n is divisible by 4, the remainder r is 0 and $i^0 = 1$.

We can also simplify powers of i that involve negative-integer exponents.

$$i^{-1} = \frac{1}{i} = \frac{1 \cdot i}{i \cdot i} = \frac{i}{-1} = -i \qquad\qquad i^{-2} = \frac{1}{i^2} = \frac{1}{-1} = -1$$

$$i^{-3} = \frac{1}{i^3} = \frac{1 \cdot i}{i^3 \cdot i} = \frac{i}{i^4} = \frac{i}{1} = i \qquad\qquad i^{-4} = \frac{1}{i^4} = \frac{1}{1} = 1$$

ACCENT ON TECHNOLOGY   **Performing Operations on Complex Numbers**

If we place our graphing calculator in $a + bi$ mode we can use it to perform operations on complex numbers. Addition, subtraction, and multiplication of $7 - 2i$ and $2 + i$ are shown in Figure R-8(a). Division of $7 - 2i$ and $2 + i$ is shown in Figure R-8(b). We can also find powers of i using our graphing calculator. i^{24} and

i^{25} are shown in Figure R-8(c). Since $4\text{E}{-}13$ (or 4×10^{-13}) can be approximated as 0, $i^{24} = 1$. Since $-5\text{E}{-}13$ (or -5×10^{-13}) can be approximated as 0, $i^{25} = i$.

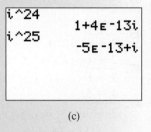

(a)   (b)   (c)

FIGURE R-8

Self Check Answers   1. $-\dfrac{2}{3}$ 2. a. $2 + 7i$ b. $-4 - 2i$ 3. $16 - 11i$ 4. $21 + i$

5. $\dfrac{9}{10} + \dfrac{3}{10}i$ 6. $\dfrac{1}{5} + \dfrac{13}{5}i$ 7. $-i$

Exercises **R.2**

Getting Ready
You should be able to complete these vocabulary and concept statements before you proceed to the practice exercises.

Fill in the blanks.

1. $\sqrt{-3}$, $\sqrt{-9}$, and $\sqrt{-12}$ are examples of _____ numbers.

2. In the complex number $a + bi$, a is the _____ part and b is the _____ part.

3. If $a = 0$ and $b \neq 0$ in the complex number $a + bi$, the number is an _____ number.

4. If $b = 0$ in the complex number $a + bi$, the number is a _____ number.

5. The complex conjugate of $2 + 5i$ is _____.

6. The product of two complex conjugates is a _____ number.

Practice
Simplify the imaginary numbers.

7. $\sqrt{-144}$

8. $-\sqrt{-225}$

9. $-2\sqrt{-24}$

10. $7\sqrt{-48}$

11. $\sqrt{-\dfrac{50}{9}}$

12. $-\sqrt{-\dfrac{72}{25}}$

13. $-7\sqrt{-\dfrac{3}{8}}$

14. $5\sqrt{-\dfrac{5}{27}}$

Find the values of x and y.

15. $x + (x + y)i = 3 + 8i$

16. $x + 5i = y - yi$

17. $3x - 2yi = 2 + (x + y)i$

18. $\begin{cases} 2 + (x + y)i = 2 - i \\ x + 3i = 2 + 3i \end{cases}$

Perform all operations. Give all answers in a + bi form.

19. $(2 - 7i) + (3 + i)$

20. $(-7 + 2i) + (2 - 8i)$

21. $(5 - 6i) - (7 + 4i)$

22. $(11 + 2i) - (13 - 5i)$

23. $(14i + 2) + \left(2 - \sqrt{-16}\right)$

24. $\left(5 + \sqrt{-64}\right) - (23i - 32)$

25. $\left(3 + \sqrt{-4}\right) - \left(2 + \sqrt{-9}\right)$

26. $\left(7 - \sqrt{-25}\right) + \left(-8 + \sqrt{-1}\right)$

27. $-5(3 + 5i)$

28. $5(2 - i)$

29. $7i(4 - 8i)$

30. $-2i(3 - 7i)$

31. $(2 + 3i)(3 + 5i)$

32. $(5 - 7i)(2 + i)$

33. $(2 + 3i)^2$

34. $(3 - 4i)^2$

35. $\left(11 + \sqrt{-25}\right)\left(2 - \sqrt{-36}\right)$

36. $\left(6 + \sqrt{-49}\right)\left(6 - \sqrt{-49}\right)$

37. $\left(\sqrt{-16} + 3\right)\left(2 + \sqrt{-9}\right)$

38. $\left(12 - \sqrt{-4}\right)\left(-7 + \sqrt{-25}\right)$

39. $\dfrac{1}{-i}$

40. $\dfrac{3}{i}$

41. $\dfrac{-4}{3i}$

42. $\dfrac{10}{7i}$

43. $\dfrac{1}{2+i}$

44. $\dfrac{-2}{3-i}$

45. $\dfrac{2i}{7+i}$

46. $\dfrac{-3i}{2+5i}$

47. $\dfrac{2+i}{3-i}$

48. $\dfrac{3-i}{1+i}$

49. $\dfrac{4-5i}{2+3i}$

50. $\dfrac{34+2i}{2-4i}$

51. $\dfrac{5-\sqrt{-16}}{-8+\sqrt{-4}}$

52. $\dfrac{3-\sqrt{-9}}{2-\sqrt{-1}}$

53. $\dfrac{2+i\sqrt{3}}{3+i}$

54. $\dfrac{3+i}{4-i\sqrt{2}}$

Simplify each expression.

55. i^9

56. i^{26}

57. i^{38}

58. i^{99}

59. i^{87}

60. i^{44}

61. i^{100}

62. i^{201}

63. i^{-6}

64. i^0

65. i^{-10}

66. i^{-31}

67. $\dfrac{1}{i^3}$

68. $\dfrac{3}{i^5}$

69. $\dfrac{-4}{i^{10}}$

70. $\dfrac{-10}{i^{24}}$

Applications

In electronics, the formula V = IR is called Ohm's Law. It gives the relationship in a circuit between the voltage V (in volts), the current I (in amperes), and the resistance R (in ohms).

71. Electronics Find V when $I = 3 - 2i$ amperes and $R = 3 + 6i$ ohms.

72. Electronics Find R when $I = 2 - 3i$ amperes and $V = 21 + i$ volts.

73. Electronics The impedance Z in an AC (alternating current) circuit is a measure of how much the circuit impedes (hinders) the flow of current through it. The impedance is related to the voltage V and the current I by the formula $V = IZ$. If a circuit has a current of $(0.5 + 2.0i)$ amperes and an impedance of $(0.4 - 3.0i)$ ohms, find the voltage.

74. Fractals Complex numbers are fundamental in the creation of the intricate geometric shape shown here, called a *fractal*. The process of creating this image is based on the following sequence of steps, beginning with picking any complex number, which we will call z.

1. Square z, and then add that result to z.
2. Square the result from Step 1, and then add it to z.
3. Square the result from Step 2, and then add it to z.

If we begin with the complex number i, what is the result after performing Steps 1, 2, and 3?

Kostyantyn Ivanyshen/Shutterstock.com

Discovery and Writing

75. Show that the addition of two complex numbers is commutative by adding the complex numbers $a + bi$ and $c + di$ in both orders and observing that the sums are equal.

76. Show that the multiplication of two complex numbers is commutative by multiplying the complex numbers $a + bi$ and $c + di$ in both orders and observing that the products are equal.

77. Show that the addition of complex numbers is associative.

78. Find three examples of complex numbers that are reciprocals of their own conjugates.

79. Explain how to determine whether two complex numbers are equal.

80. Define the complex conjugate of a complex number.

R.3 Quadratic Equations and Applications

In this section, we will learn to

1. Solve quadratic equations by factoring and by using the Square-Root Property.

2. Solve quadratic equations by completing the square and by using the Quadratic Formula.

3. Define and use the discriminant.

4. Solve application problems.

Eric Broder Van Dyke/Shutterstock.com

Fenway Park, "America's Most Beloved Ballpark," is home to the Boston Red Sox baseball club. The park opened in 1912 and is currently the oldest major-league baseball stadium.

In baseball, the distance between home plate and first base is 90 feet and the distance between first base and second base is 90 feet. To find the distance between home plate and second base, we can use the *Pythagorean Theorem*, which states:

The sum of the squares of the two legs of a right triangle is equal to the square of its hypotenuse.

Because home plate, first base, and second base form a right triangle, we can let x represent the distance between home plate and second base (the hypotenuse of the right triangle) and 90 feet represent the length of each leg. We can then apply the Pythagorean Theorem and write the equation $90^2 + 90^2 = x^2$. To find the distance between home plate and second base, we must solve this equation.

$$90^2 + 90^2 = x^2$$

$$8,100 + 8,100 = x^2 \qquad \text{Square 90 two times.}$$

$$16,200 = x^2 \qquad \text{Simplify.}$$

To find x, we must determine what number squared gives 16,200. We know that this number is the square root of 16,200.

$$\sqrt{16,200} \approx 127.3 \qquad \text{Use a calculator and round to the nearest tenth.}$$

To the nearest tenth, the distance between home plate and second base is 127.3 feet.

Since this equation contains the term x^2, it is an example of a new type of equation, called a *quadratic* equation. In this section, we will learn several strategies for solving these equations.

1. Solve Quadratic Equations by Factoring and by Using the Square-Root Property

Polynomial equations such as $2x^2 - 11x - 21 = 0$ and $3x^2 - x + 2 = 0$ are called *quadratic* or *second-degree* equations.

Quadratic Equation   A **quadratic equation** is an equation that can be written in the form $ax^2 + bx + c = 0$, where a, b, and c are real numbers and $a \neq 0$.

To solve quadratic equations by factoring, we can use the following theorem:

Zero-Factor Theorem   If a and b are real numbers, and if $ab = 0$, then $a = 0$ or $b = 0$.

A proof of the Zero-Factor Theorem is shown in online Appendix B.

EXAMPLE 1   **Solving a Quadratic Equation by Factoring**

Solve: $2x^2 - 9x - 35 = 0$.

SOLUTION   The left side can be factored and written as

$$(2x + 5)(x - 7) = 0.$$

This product can be 0 if and only if one of the factors is 0. So we can use the Zero-Factor Theorem and set each factor equal to 0. We can then solve each equation for x.

$$2x + 5 = 0 \quad \text{or} \quad x - 7 = 0$$
$$2x = -5 \qquad\qquad x = 7$$
$$x = -\frac{5}{2}$$

Because $(2x + 5)(x - 7) = 0$ only if one of its factors is zero, $x = -\frac{5}{2}$ and $x = 7$ are the only solutions of the equation.   ∎

Self Check 1   Solve: $6x^2 + 7x = 3$.

Now Try Exercise 15.

In many quadratic equations, the quadratic expression does not factor over the set of integers. For example, the left side of $x^2 - 5x + 3 = 0$ is a prime polynomial and cannot be factored over the set of integers. To develop a method to solve these equations, we consider the equation $x^2 = c$. A quadratic equation of this form can also be solved by factoring.

$$x^2 = c$$
$$x^2 - c = 0 \qquad\qquad \text{Subtract } c \text{ from both sides.}$$
$$x^2 - \left(\sqrt{c}\right)^2 = 0 \qquad\qquad \left(\sqrt{c}\right)^2 = c$$
$$\left(x - \sqrt{c}\right)\left(x + \sqrt{c}\right) = 0 \qquad\qquad \text{Factor the difference of two squares.}$$
$$x - \sqrt{c} = 0 \quad \text{or} \quad x + \sqrt{c} = 0 \qquad \text{Set each factor equal to 0.}$$
$$x = \sqrt{c} \qquad\qquad x = -\sqrt{c}$$

The roots of $x^2 = c$ are $x = \sqrt{c}$ and $x = -\sqrt{c}$. This fact is summarized in the **Square-Root Property.**

Square-Root Property   If $c > 0$, the equation $x^2 = c$ has two real roots:

$$x = \sqrt{c} \quad \text{or} \quad x = -\sqrt{c}.$$

The solutions are often written in a shorter form, $x = \pm\sqrt{c}$, and we say x *equals plus or minus the square root of c.* Both solutions are real if $c > 0$. If $c < 0$, the solutions are imaginary, and we write $x = i\sqrt{c}$ or $x = -i\sqrt{c}$. In the abbreviated form we write $x = \pm i\sqrt{c}$.

EXAMPLE 2   **Using the Square-Root Property to Solve a Quadratic Equation**

Solve: $x^2 - 8 = 0$.

SOLUTION   We solve for x^2 and use the Square-Root Property.

$$x^2 - 8 = 0$$
$$x^2 = 8$$

$$x = \sqrt{8} \qquad \text{or} \quad x = -\sqrt{8}$$
$$x = 2\sqrt{2} \qquad \quad x = -2\sqrt{2} \qquad \sqrt{8} = \sqrt{4}\sqrt{2} = 2\sqrt{2}$$

The solutions are $x = \pm 2\sqrt{2}$.   ∎

Self Check 2   Solve: $x^2 - 12 = 0$.

Now Try Exercise 23.

EXAMPLE 3   **Using the Square-Root Property to Solve a Quadratic Equation**

Solve: $(x + 4)^2 = -36$.

SOLUTION   Again we will use the Square-Root Property.

$$(x + 4)^2 = -36$$

$$x + 4 = \sqrt{-36} \qquad \text{or} \quad x + 4 = -\sqrt{-36}$$
$$x + 4 = 6i \qquad \qquad \quad x + 4 = -6i$$
$$x = -4 + 6i \qquad \qquad x = -4 - 6i$$

The solutions are $x = -4 \pm 6i$.   ∎

Self Check 3   Solve: $(x + 5)^2 = -4$.

Now Try Exercise 29.

2. Solve Quadratic Equations by Completing the Square and by Using the Quadratic Formula

Another way to solve quadratic equations is called **completing the square.** This method is based on the following products:

$$x^2 + 2ax + a^2 = (x + a)^2 \qquad \text{and} \qquad x^2 - 2ax + a^2 = (x - a)^2$$

The trinomials $x^2 + 2ax + a^2$ and $x^2 - 2ax + a^2$ are perfect-square trinomials, because each one factors as the square of a binomial. In each case, the coefficient of the first term is 1. If we take one-half of the coefficient of x in the middle term and square it, we obtain the third term.

$$\left[\frac{1}{2}(2a)\right]^2 = a^2 \qquad \text{and} \qquad \left[\frac{1}{2}(-2a)\right]^2 = (-a)^2 = a^2$$

This suggests that to make $x^2 + bx$ a perfect-square trinomial, we find one-half of b, square it, and add the result to the binomial. For example, to make $x^2 + 10x$ a perfect-square trinomial, we find one-half of 10 to get 5, square 5 to get 25, and add 25 to $x^2 + 10x$.

$$x^2 + \mathbf{10}x + \left[\frac{1}{2}(\mathbf{10})\right]^2 = x^2 + 10x + (5)^2$$

$$= x^2 + 10x + 25 \qquad \text{Note that } x^2 + 10x + 25 = (x + 5)^2.$$

To make $x^2 - 11x$ a perfect-square trinomial, we find one-half of -11 to get $-\frac{11}{2}$, square $-\frac{11}{2}$ to get $\frac{121}{4}$, and add $\frac{121}{4}$ to $x^2 - 11x$.

$$x^2 - \mathbf{11}x + \left[\frac{1}{2}(\mathbf{-11})\right]^2 = x^2 - 11x + \left(-\frac{11}{2}\right)^2$$

$$= x^2 - 11x + \frac{121}{4} \qquad \text{Note that } x^2 - 11x + \frac{121}{4} = \left(x - \frac{11}{2}\right)^2.$$

To solve a quadratic equation in x by completing the square, we follow these steps:

Strategy for Completing the Square

1. If the coefficient of x^2 is not 1, make it 1 by dividing both sides of the equation by the coefficient of x^2.
2. If necessary, add a number to both sides of the equation to get the constant on the right side of the equation.
3. Complete the square on x:
   a. Identify the coefficient of x, take one-half of it, and square the result.
   b. Add the number found in part (a) to both sides of the equation.
4. Factor the perfect-square trinomial and combine like terms.
5. Solve the resulting quadratic equation by using the Square-Root Property.

EXAMPLE 4 **Solving a Quadratic Equation by Completing the Square**

Solve: $x^2 - 4x - 6 = 0$.

SOLUTION We will use the steps outlined above to complete the square and solve the quadratic equation. Since the coefficient of x^2 is already 1, we move to Step 2 and add 6 to both sides to isolate the binomial $x^2 + 4x$.

$$x^2 + 4x = 6$$

We then find the number to add to both sides by completing the square. Since one-half of 4 (the coefficient of x) is 2 and $2^2 = 4$, we add 4 to both sides.

$$x^2 + 4x + \mathbf{4} = 6 + \mathbf{4} \qquad \text{Add 4 to both sides.}$$

$$x^2 + 4x + 4 = 10$$

$$(x + 2)^2 = 10 \qquad \text{Factor } x^2 + 4x + 4.$$

$$x + 2 = \sqrt{10} \quad \text{or} \quad x + 2 = -\sqrt{10} \qquad \text{Use the Square-Root Property.}$$

$$x = -2 + \sqrt{10} \quad \Big| \quad x = -2 - \sqrt{10}$$

The solutions are $x = -2 \pm \sqrt{10}$.

Self Check 4 Solve: $x^2 - 2x - 9 = 0$.

Now Try Exercise 45.

˙EXAMPLE 5   **Solving a Quadratic Equation by Completing the Square**

Solve: $6x^2 + 5x + 6 = 0$.

SOLUTION   We begin by dividing both sides of the equation by 6 to make the coefficient of x^2 equal to 1. Then we proceed as follows:

$$6x^2 + 5x + 6 = 0$$

$$x^2 + \frac{5}{6}x + 1 = 0 \qquad\qquad \text{Divide both sides by 6.}$$

$$x^2 + \frac{5}{6}x = -1 \qquad\qquad \text{Subtract 1 from both sides.}$$

$$x^2 + \frac{5}{6}x + \frac{25}{144} = -1 + \frac{25}{144} \qquad \text{Add } \left(\frac{1}{2}\cdot\frac{5}{6}\right)^2, \text{ or } \frac{25}{144}, \text{ to both sides.}$$

$$\left(x + \frac{5}{12}\right)^2 = -\frac{119}{144} \qquad\qquad \text{Factor } x^2 + \frac{5}{6}x + \frac{25}{144}.$$

We now apply the Square-Root Property.

$$x + \frac{5}{12} = \sqrt{-\frac{119}{144}} \qquad \text{or} \qquad x + \frac{5}{12} = -\sqrt{-\frac{119}{144}}$$

$$x + \frac{5}{12} = \frac{\sqrt{-119}}{\sqrt{144}} \qquad\qquad\qquad x + \frac{5}{12} = -\frac{\sqrt{-119}}{\sqrt{144}}$$

$$x + \frac{5}{12} = i\frac{\sqrt{119}}{12} \qquad\qquad\qquad x + \frac{5}{12} = -i\frac{\sqrt{119}}{12}$$

$$x = -\frac{5}{12} + i\frac{\sqrt{119}}{12} \qquad\qquad\qquad x = -\frac{5}{12} - i\frac{\sqrt{119}}{12}$$

The solutions are $x = -\dfrac{5}{12} \pm i\dfrac{\sqrt{119}}{12}$.

Self Check 5   Solve: $2x^2 - 5x + 7 = 0$.

Now Try Exercise 51.

We can solve the equation $ax^2 + bx + c = 0$ $(a \neq 0)$ by completing the square. The result will be a formula that we can use to solve quadratic equations.

$$ax^2 + bx + c = 0$$

$$\frac{ax^2}{a} + \frac{b}{a}x + \frac{c}{a} = \frac{0}{a} \qquad\qquad \text{Divide both sides by } a.$$

$$x^2 + \frac{b}{a}x = -\frac{c}{a} \qquad\qquad \text{Simplify and subtract } \frac{c}{a} \text{ from both sides.}$$

$$x^2 + \frac{b}{a}x + \frac{b^2}{4a^2} = \frac{b^2}{4a^2} - \frac{4ac}{4aa} \qquad \text{Add } \frac{b^2}{4a^2} \text{ to both sides and multiply the numerator and denominator of } \frac{c}{a} \text{ by } 4a.$$

$$\left(x + \frac{b}{2a}\right)^2 = \frac{b^2 - 4ac}{4a^2} \qquad \text{Factor the left side and add the fractions on the right side.}$$

We can now use the Square-Root Property.

$$x + \frac{b}{2a} = \sqrt{\frac{b^2 - 4ac}{4a^2}} \qquad \text{or} \qquad x + \frac{b}{2a} = -\sqrt{\frac{b^2 - 4ac}{4a^2}}$$

$$x = -\frac{b}{2a} + \frac{\sqrt{b^2 - 4ac}}{2a} \qquad\qquad x = -\frac{b}{2a} - \frac{\sqrt{b^2 - 4ac}}{2a}$$

$$x = \frac{-b + \sqrt{b^2 - 4ac}}{2a} \qquad\qquad x = \frac{-b - \sqrt{b^2 - 4ac}}{2a}$$

These values of x are the two roots of the equation $ax^2 + bx + c = 0$. They are usually combined into a single expression, called the **Quadratic Formula**.

Quadratic Formula   The solutions of the general quadratic equation, $ax^2 + bx + c = 0$, are

$$x = \frac{-b \pm \sqrt{b^2 - 4ac}}{2a} \quad (a \neq 0).$$

Caution

Be sure to write the Quadratic Formula correctly. Do **not** write it as

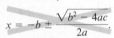

The Quadratic Formula should be read twice, once using the plus sign and once using the minus sign. The Quadratic Formula implies that

$$x = \frac{-b + \sqrt{b^2 - 4ac}}{2a} \qquad \text{or} \qquad x = \frac{-b - \sqrt{b^2 - 4ac}}{2a}.$$

EXAMPLE 6   **Solving a Quadratic Equation Using the Quadratic Formula**

Solve: $2x^2 + 8x - 7 = 0$.

SOLUTION   In this equation, $a = 2$, $b = 8$, and $c = -7$. We will substitute these values into the Quadratic Formula.

$$x = \frac{-b \pm \sqrt{b^2 - 4ac}}{2a} \qquad\qquad \text{This is the Quadratic Formula.}$$

$$x = \frac{-8 \pm \sqrt{8^2 - 4(2)(-7)}}{2(2)} \qquad\qquad \text{Substitute 2 for } a, \text{ 8 for } b, \text{ and } -7 \text{ for } c.$$

$$x = \frac{-8 \pm \sqrt{120}}{4} \qquad\qquad 8^2 - 4(2)(-7) = 64 + 56 = 120$$

$$x = \frac{-8 \pm 2\sqrt{30}}{4} \qquad\qquad \sqrt{120} = \sqrt{4 \cdot 30} = 2\sqrt{30}$$

$$x = \frac{2(-4 \pm \sqrt{30})}{4} \qquad\qquad \text{Factor out 2 in the numerator.}$$

$$x = \frac{-4 + \sqrt{30}}{2} \quad \text{or} \quad x = \frac{-4 - \sqrt{30}}{2} \qquad \text{Simplify.}$$

The solutions are $x = \dfrac{-4 \pm \sqrt{30}}{2}$.

Self Check 6   Solve: $4x^2 + 16x - 13 = 0$.

Now Try Exercise 61.

So far, we have solved quadratic equations by *factoring*, by the *square-root method*, by *completing the square*, and by the *Quadratic Formula*. With so many methods available, it is useful to think about which one will be the easiest way to solve a specific quadratic equation. Although we have used completing the square to develop the Quadratic Formula, it is usually the most complicated way to solve a quadratic equation. Therefore, unless specified, we will usually not use this method. However, we will complete the square again later in the book to write certain equations in specific forms.

3. Define and Use the Discriminant

We can predict what type of roots a quadratic equation will have before we solve it. Suppose that the coefficients a, b, and c in the equation $ax^2 + bx + c = 0$ $(a \neq 0)$ are real numbers. Then the two roots of the equation are given by the Quadratic Formula

$$x = \frac{-b \pm \sqrt{b^2 - 4ac}}{2a} \quad (a \neq 0).$$

The value of $b^2 - 4ac$, called the **discriminant**, determines the nature of the roots. The possibilities are summarized in the table as follows:

Discriminant	Number and Type of Roots
0	One repeated rational root
Positive and a perfect square	Two different rational roots
Positive and not a perfect square	Two different irrational roots
Negative	Two different imaginary roots that are complex conjugates

EXAMPLE 7   **Determining the Nature of the Roots of a Quadratic Equation**

Determine the number and type of roots of the equation $3x^2 + 4x + 1 = 0$.

SOLUTION   We calculate the discriminant $b^2 - 4ac$.

$$b^2 - 4ac = 4^2 - 4(3)(1) \qquad \text{Substitute 4 for } b, 3 \text{ for } a, \text{ and 1 for } c.$$

$$= 16 - 12$$

$$= 4$$

Since a, b, and c are rational numbers and the discriminant is a positive perfect square, the two roots will be different rational numbers. ■

Self Check 7   Determine the number and nature of the roots of $4x^2 - 3x - 2 = 0$.

Now Try Exercise 85.

© Bennie Thornton/Alamy

The Grand Canyon Skywalk is a tourist attraction located along the Colorado River in the state of Arizona. The glass walkway is shaped like a horseshoe and is 4,000 feet above the floor of the canyon. The Grand Canyon, known for its overwhelming size and beautiful landscape, is awe inspiring and one of our nation's most astounding natural wonders.

If a Clif Bar energy bar is accidentally dropped over the side of the skywalk, how long will it take to hit the canyon floor?

If t represents the time in seconds, the quadratic equation $-16t^2 + 4,000 = 0$ models the time it takes the energy bar to fall to the canyon floor. We can solve this equation by using the Square-Root Property.

$$-16t^2 + 4,000 = 0$$

$$-16t^2 = -4,000 \qquad \text{Subtract 4,000 from both sides.}$$

$$t^2 = 250 \qquad \text{Divide both sides by } -16.$$

$$t = \pm\sqrt{250} \qquad \text{Use the Square-Root Property.}$$

$$\approx \pm 15.8 \qquad \text{Round to the nearest tenth.}$$

Because time cannot be negative, we disregard the negative answer. The time it will take the energy bar to reach the canyon floor is about 15.8 seconds.

As this example illustrates, the solutions of many problems involve quadratic equations.

4. Solve Application Problems

EXAMPLE 8   **Solving a Geometry Application Problem**

On a college campus, a sidewalk 85 meters long (represented by the red line in Figure R-9) joins a dormitory building D with the student center C. However, the students prefer to walk directly from D to C. If segment DC is 65 meters long, how long is each piece of the existing sidewalk?

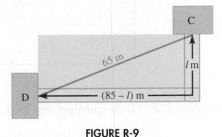

FIGURE R-9

SOLUTION   We note that the triangle shown in the figure is a right triangle, with a hypotenuse that is 65 meters long. If we let the shorter leg of the triangle be l meters long, the length of the longer leg will be $(85 - l)$ meters. By the Pythagorean Theorem, we know that the sum of the squares of the two legs of a right triangle is equal to the square of the hypotenuse. Thus, we can form the equation

$$l^2 + (85 - l)^2 = 65^2 \qquad \text{In a right triangle, } a^2 + b^2 = c^2.$$

which we can solve as follows:

$$l^2 + 7,225 - 170l + l^2 = 4,225 \qquad \text{Expand } (85 - l)^2.$$

$$2l^2 - 170l + 3,000 = 0 \qquad \text{Combine like terms and subtract 4,225 from both sides.}$$

$$l^2 - 85l + 1,500 = 0 \qquad \text{Divide both sides by 2.}$$

Since the left side is difficult to factor, we will solve this equation using the Quadratic Formula.

$$l = \frac{-b \pm \sqrt{b^2 - 4ac}}{2a}$$

$$l = \frac{-(-85) \pm \sqrt{(-85)^2 - 4(1)(1,500)}}{2(1)}$$

$$l = \frac{85 \pm \sqrt{1{,}225}}{2}$$

$$l = \frac{85 \pm 35}{2}$$

$$l = \frac{85 + 35}{2} \quad \text{or} \quad l = \frac{85 - 35}{2}$$

$$= 60 \qquad\qquad\quad = 25$$

The length of the shorter leg is 25 meters. The length of the longer leg is $(85 - 25)$ meters, or 60 meters. ∎

Self Check 8   The length of a video screen is 14 feet longer than its height. If the diagonal of the screen is 26 feet, find the height of the screen.

Now Try Exercise 109.

EXAMPLE 9   **Solving a Falling-Body Application Problem**

If a ball is hit straight up into the air with an initial velocity of 144 feet per second, its height is given by the formula $h = 144t - 16t^2$, where h represents its height (in feet) and t represents the time (in seconds) since it was thrown. How long will it take for the ball to return to the point from which it was thrown?

SOLUTION   When the ball returns to its starting point, its height is again 0. Thus, we can set h equal to 0 and solve for t.

$$\boldsymbol{h} = 144t - 16t^2$$

$$\boldsymbol{0} = 144t - 16t^2 \qquad \text{Let } h = 0.$$

$$0 = 16t(9 - t) \qquad \text{Factor.}$$

$$16t = 0 \quad \text{or} \quad 9 - t = 0 \qquad \text{Set each factor equal to 0.}$$

$$t = 0 \qquad\qquad t = 9$$

At $t = 0$, the ball's height is 0, because it was just released. When $t = 9$, the height is again 0, and the ball has returned to its starting point. ∎

Self Check 9   How long does it take the ball in Example 9 to reach a height of 324 feet?

Now Try Exercise 115.

Self Check Answers   1. $\dfrac{1}{3}, -\dfrac{3}{2}$ 2. $\pm 2\sqrt{3}$ 3. $-5 \pm 2i$ 4. $1 \pm \sqrt{10}$ 5. $\dfrac{5}{4} \pm i\dfrac{\sqrt{31}}{4}$

6. $\dfrac{-4 \pm \sqrt{29}}{2}$ 7. two different irrational roots   8. 10 feet

9. 4.5 seconds

Exercises R.3

Getting Ready

You should be able to complete these vocabulary and concept statements before you proceed to the practice exercises.

Fill in the blanks.

1. A quadratic equation is an equation that can be written in the form _____, where $a \neq 0$.

2. If a and b are real numbers and _____, then $a = 0$ or $b = 0$.

3. If $c > 0$, the equation $x^2 = c$ has two roots. They are $x =$ _____ and $x =$ _____.

4. The Quadratic Formula is _____ ($a \neq 0$).

5. The Discriminant Formula is _____.

6. If a, b, and c are real numbers and $b^2 - 4ac < 0$, the roots of the quadratic equation are _____.

7. The formula for the area of a rectangle is _____.

8. The formula that relates distance, rate, and time is _____.

Practice

Solve each equation by factoring.

9. $x^2 - x - 6 = 0$
10. $x^2 + 8x + 15 = 0$
11. $x^2 - 144 = 0$
12. $x^2 + 4x = 0$
13. $2x^2 + x - 10 = 0$
14. $3x^2 + 4x - 4 = 0$
15. $15x^2 + 16x = 15$
16. $6x^2 - 25x = -25$
17. $12x^2 + 9 = 24x$
18. $24x^2 + 6 = 24x$

Use the Square-Root Property to solve each equation.

19. $x^2 = 9$
20. $x^2 = -64$
21. $y^2 + 50 = 0$
22. $x^2 - 75 = 0$
23. $2y^2 = 36$
24. $4x^2 = -108$
25. $4y^2 = -63$
26. $49x^2 = 99$
27. $(x - 1)^2 = 4$
28. $(y + 2)^2 - 49 = 0$
29. $(x + 1)^2 + 8 = 0$
30. $(y + 2)^2 - 98 = 0$

Complete the square to make each binomial a perfect-square trinomial.

31. $x^2 + 6x$
32. $x^2 + 8x$
33. $x^2 - 4x$
34. $x^2 - 12x$
35. $a^2 + 5a$
36. $t^2 + 9t$
37. $r^2 - 11r$
38. $s^2 - 7s$
39. $y^2 + \dfrac{3}{4}y$
40. $m^2 - \dfrac{2}{3}m$

Solve each equation by completing the square.

41. $x^2 - 8x + 15 = 0$
42. $x^2 + 10x + 21 = 0$
43. $x^2 + 8x - 5 = 0$
44. $x^2 - 4x = 11$
45. $x^2 - 16x = -40$
46. $x^2 + 12x = 4$
47. $x^2 = -2x - 10$
48. $x^2 = 12x - 72$
49. $x^2 + 6x + 57 = 0$
50. $x^2 - 14x + 69 = 0$
51. $2x^2 - 20x = -55$
52. $4x^2 + 8x = -7$
53. $3x^2 = 1 - 4x$
54. $2x^2 = 3x + 1$

Use the Quadratic Formula to solve each equation.

55. $x^2 - 12 = 0$
56. $x^2 - 20 = 0$
57. $2x^2 - x - 15 = 0$
58. $6x^2 + x - 2 = 0$
59. $2x^2 + 2x - 4 = 0$
60. $3x^2 + 18x + 15 = 0$
61. $-3x^2 = 5x + 1$
62. $2x(x + 3) = -1$
63. $5x\left(x + \dfrac{1}{5}\right) = 3$
64. $7x^2 = 2x + 2$
65. $x^2 + 2x + 2 = 0$
66. $a^2 + 4a + 8 = 0$
67. $y^2 + 4y + 5 = 0$
68. $x^2 + 2x + 5 = 0$
69. $x^2 - \dfrac{2}{3}x = -\dfrac{2}{9}$
70. $x^2 + \dfrac{5}{4} = x$

Solve each formula for the indicated variable.

71. $h = \dfrac{1}{2}gt^2$; t
72. $x^2 + y^2 = r^2$; x
73. $h = 64t - 16t^2$; t
74. $y = 16x^2 - 4$; x
75. $\dfrac{x^2}{a^2} + \dfrac{y^2}{b^2} = 1$; y
76. $\dfrac{x^2}{a^2} - \dfrac{y^2}{b^2} = 1$; x
77. $\dfrac{x^2}{a^2} - \dfrac{y^2}{b^2} = 1$; a
78. $\dfrac{x^2}{a^2} - \dfrac{y^2}{b^2} = 1$; b
79. $x^2 + xy - y^2 = 0$; x
80. $x^2 - 3xy + y^2 = 0$; y

Use the discriminant to determine the number and nature of the roots of each equation. Do not solve the equation.

81. $x^2 + 6x + 9 = 0$
82. $9x^2 + 42x + 49 = 0$
83. $10x^2 + 29x = 21$
84. $10x^2 + x = 21$
85. $x^2 + 3x = 2$
86. $-8x^2 - 2x = 13$

87. Does $1{,}492x^2 + 1{,}984x - 1{,}776 = 0$ have any roots that are real numbers?

88. Does $2{,}004x^2 + 10x + 1{,}994 = 0$ have any roots that are real numbers?

89. Find two values of k such that $x^2 + kx + 3k - 5 = 0$ will have two roots that are equal.

90. For what value(s) of b will the solutions of $x^2 - 2bx + b^2 = 0$ be equal?

Change each equation to quadratic form and solve it by the most efficient method.

91. $x + 1 = \dfrac{12}{x}$

92. $x - 2 = \dfrac{15}{x}$

93. $\dfrac{5}{x} = \dfrac{4}{x^2} - 6$

94. $\dfrac{6}{x^2} + \dfrac{1}{x} = 12$

95. $x\left(30 - \dfrac{13}{x}\right) = \dfrac{10}{x}$

96. $x\left(20 - \dfrac{17}{x}\right) = \dfrac{10}{x}$

97. $\dfrac{4 + x}{2x} = \dfrac{x - 2}{3}$

98. $\dfrac{36}{x} - 17 = \dfrac{-24}{x + 1}$

99. $\dfrac{1}{x} + \dfrac{3}{x + 2} = 2$

100. $\dfrac{1}{x - 1} + \dfrac{1}{x - 4} = \dfrac{5}{4}$

101. $\dfrac{1}{x + 1} + \dfrac{5}{2x - 4} = 1$

102. $\dfrac{x(2x + 1)}{x - 2} = \dfrac{10}{x - 2}$

103. $x + 1 + \dfrac{x + 2}{x - 1} = \dfrac{3}{x - 1}$

104. $\dfrac{1}{4 - y} = \dfrac{1}{4} + \dfrac{1}{y + 2}$

Applications

105. Geometry The base of a triangle is one-third as long as its height. If the area of the triangle is 24 square meters, how long is its base?

106. Imax screens A large movie screen is in the Panasonic Imax theater at Darling Harbor, Sydney, Australia. The rectangular screen has an area of 11,349 square feet. Find the dimensions of the screen if it is 20 feet longer than it is wide.

107. Dallas Cowboys video screen Cowboys Stadium in Dallas has the world's largest video screen. The rectangular screen's length is 88 feet more than its height. If the video screen has an area of 11,520 square feet, find the dimensions of the screen.

Ken Durden/Shutterstock.com

108. Flags In 1912, an order by President Taft fixed the height and width of the U.S. flag in the ratio 1 to 1.9. If 100 square feet of cloth are to be used to make a U.S. flag, estimate its dimensions to the nearest $\tfrac{1}{4}$ foot.

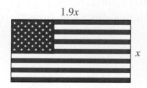

1.9x

x

109. Manufacturing A manufacturer of television sets received an order for sets with a 46-inch screen (measured along the diagonal, as shown in the illustration). If the screens are to be rectangular in shape and $17\tfrac{1}{2}$ inches wider than they are high, find the dimensions of a screen to the nearest tenth of an inch.

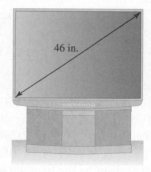

46 in.

110. Metal fabrication A piece of tin 12 inches on a side is to have four equal squares cut from its corners, as in the illustration. If the edges are then to be folded up to make a box with a floor area of 64 square inches, find the depth of the box.

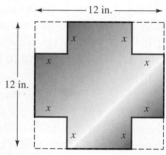

12 in.

12 in.

111. Cycling rates A cyclist rides from DeKalb to Rockford, a distance of 40 miles. His return trip takes 2 hours longer, because his speed decreases by 10 mph. How fast does he ride each way?

112. Travel times A farmer drives a tractor from one town to another, a distance of 120 kilometers. He drives 10 kilometers per hour faster on the return trip, cutting 1 hour off the time. How fast does he drive each way?

113. Uniform motion If the speed were increased by 10 mph, a 420-mile trip would take 1 hour less time. How long would the trip take at the slower speed?

114. Uniform motion By increasing her usual speed by 25 kilometers per hour, a bus driver decreases the time on a 25-kilometer trip by 10 minutes. Find the usual speed.

115. Ballistics The height of a projectile fired upward with an initial velocity of 400 feet per second is given by the formula $h = -16t^2 + 400t$, where h is the height in feet and t is the time in seconds. Find the time required for the projectile to return to earth.

116. Ballistics The height of an object tossed upward with an initial velocity of 104 feet per second is given by the formula $h = -16t^2 + 104t$, where h is the height in feet and t is the time in seconds. Find the time required for the object to return to its point of departure.

117. Falling coins An object will fall s feet in t seconds, where $s = 16t^2$. How long will it take for a penny to hit the ground if it is dropped from the top of the Sears Tower in Chicago? (*Hint:* The tower is 1,454 feet tall.)

118. Movie stunts According to the *Guinness Book of World Records, 1998,* stuntman Dan Koko fell a distance of 312 feet into an airbag after jumping from the Vegas World Hotel and Casino. The distance d in feet traveled by a free-falling object in t seconds is given by the formula $d = 16t^2$. To the nearest tenth of a second, how long did the fall last?

119. Accidents The height h (in feet) of an object that is dropped from a height of s feet is given by the formula $h = s - 16t^2$, where t is the time the object has been falling. A 5-foot-tall woman on a sidewalk looks directly overhead and sees a window washer drop a bottle from 4 stories up. How long does she have to get out of the way? Round to the nearest tenth. (A story is 12 feet.)

120. Ballistics The height of an object thrown upward with an initial velocity of 32 feet per second is given by the formula $h = -16t^2 + 32t$, where t is the time in seconds. How long will it take the object to reach a height of 16 feet?

121. Architecture A **golden rectangle** is one of the most visually appealing of all geometric forms. The front of the Parthenon, built in Athens in the fifth century BC and shown in the illustration, is a golden rectangle. In a golden rectangle, the length l and the height h of the rectangle must satisfy the equation $\dfrac{l}{h} = \dfrac{h}{l - h}$. If a rectangular billboard is to have a height of 15 feet, how long should it be if it is to form a golden rectangle? Round to the nearest tenth of a foot.

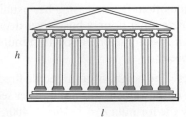

l

122. Golden ratio Rectangle $ABCD$, shown here, will be a **golden rectangle** if $\frac{AB}{AD} = \frac{BC}{BE}$, where $AE = AD$. Let $AE = 1$ and find the ratio of AB to AD.

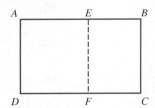

123. Filling storage tanks Two pipes are used to fill a water storage tank. The first pipe can fill the tank in 4 hours and the two pipes together can fill the tank in 2 hours less than the second pipe alone. How long would it take for the second pipe to fill the tank?

124. Filling swimming pools A hose can fill a swimming pool in 6 hours. Another hose needs 3 more hours to fill the pool than the two hoses combined. How long would it take the second hose to fill the pool?

125. Mowing lawns Kristy can mow a lawn in 1 hour less time than her brother Steven. Together they can finish the job in 5 hours. How long would it take Kristy if she worked alone?

126. Milking cows Working together, Sarah and Heidi can milk the cows in 2 hours. If they work alone, it takes Heidi 3 hours longer than it takes Sarah. How long would it take Heidi to milk the cows alone?

127. Geometry Is it possible for a rectangle to have a width that is 3 units shorter than its diagonal and a length that is 4 units longer than its diagonal?

128. Geometry If two opposite sides of a square are increased by 10 meters and the other sides are decreased by 8 meters, the area of the rectangle that is formed is 63 square meters. Find the area of the original square.

129. Investment Maude and Matilda each have a bank CD. Maude's is $1,000 larger than Matilda's, but the interest rate is 1% less. Last year Maude received interest of $280, and Matilda received $240. Find the rate of interest for each CD.

130. Investment Scott and Laura have both invested some money. Scott invested $3,000 more than Laura and at a 2% higher interest rate. If Scott received $800 annual interest and Laura received $400, how much did Scott invest?

131. Automobile engines As the piston shown moves upward, it pushes a cylinder of a gasoline/air mixture that is ignited by the spark plug. The formula that gives the volume of a cylinder is $V = \pi r^2 h$, where r is the radius and h the height. Find the radius of the piston (to the nearest hundredth of an inch) if it displaces 47.75 cubic inches of gasoline/air mixture as it moves from its lowest to its highest point.

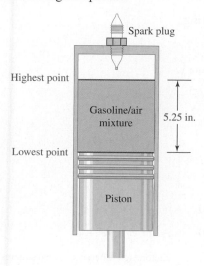

Spark plug

Highest point

Gasoline/air mixture   5.25 in.

Lowest point

Piston

132. History One of the important cities of the ancient world was Babylon. Greek historians wrote that the city was square-shaped. Its area numerically exceeded its perimeter by about 124. Find its dimensions in miles. (Round to the nearest tenth.)

Discovery and Writing

133. If r_1 and r_2 are the roots of $ax^2 + bx + c = 0$, show that $r_1 + r_2 = -\frac{b}{a}$.

134. If r_1 and r_2 are the roots of $ax^2 + bx + c = 0$, show that $r_1 r_2 = \frac{c}{a}$.

In Exercises 135 and 136, a stone is thrown straight upward, higher than the top of a tree. The stone is even with the top of the tree at time t_1 on the way up and at time t_2 on the way down. If the height of the tree is h feet, both t_1 and t_2 are solutions of $h = v_0 t - 16t^2$.

135. Show that the tree is $16t_1 t_2$ feet tall.

136. Show that v_0 is $16(t_1 + t_2)$ feet per second.

137. Explain why the Zero-Factor Theorem is true.

138. Explain how to complete the square for $x^2 - 17x$.

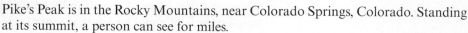

R.4   Other Types of Equations

In this section, we will learn to

1. Solve polynomial equations by factoring.
2. Solve other equations by factoring.
3. Solve radical equations.
4. Solve applications of radical equations.
5. Solve absolute-value equations.

John Hoffman/Shutterstock.com

Pike's Peak is in the Rocky Mountains, near Colorado Springs, Colorado. Standing at its summit, a person can see for miles.

The distance a person can see from the peak of the mountain is called the **horizon distance**. If this distance d is measured in miles and the height of the observer h is measured in feet, d and h are related by the formula $d = \sqrt{1.5h}$.

Since the height of Pike's Peak is approximately 14,000 feet, we can substitute 14,000 for h in the formula and simplify.

$$d = \sqrt{1.5\textbf{h}}$$

$$d = \sqrt{1.5(\textbf{14,000})} \qquad \text{Substitute 14,000 for } h.$$

$$= \sqrt{21,000}$$

$$\approx 144.9137675$$

From the top of Pike's Peak, a person can see about 145 miles.

Since this equation contains a radical, it is part of a group called *radical equations*—one of the topics of this section.

1. Solve Polynomial Equations by Factoring

The equation $ax^2 + bx + c = 0$ is a polynomial equation of second degree, because its left side contains a second-degree polynomial. Many polynomial equations of higher degree can be solved by factoring.

EXAMPLE 1 **Solving Polynomial Equations by Factoring**

Solve: **a.** $6x^3 - x^2 - 2x = 0$ **b.** $x^4 - 5x^2 + 4 = 0$

SOLUTION We will solve each equation by factoring.

a.
$$6x^3 - x^2 - 2x = 0$$
$$x(6x^2 - x - 2) = 0 \qquad \text{Factor out } x.$$
$$x(3x - 2)(2x + 1) = 0 \qquad \text{Factor } 6x^2 - x - 2.$$

We set each factor equal to 0.

$$x = 0 \quad \text{or} \quad 3x - 2 = 0 \quad \text{or} \quad 2x + 1 = 0$$
$$x = \frac{2}{3} \qquad\qquad x = -\frac{1}{2}$$

b.
$$x^4 - 5x^2 + 4 = 0$$
$$(x^2 - 4)(x^2 - 1) = 0 \qquad \text{Factor } x^4 - 5x^2 + 4.$$
$$(x + 2)(x - 2)(x + 1)(x - 1) = 0 \qquad \text{Factor each difference of two squares.}$$

We set each factor equal to 0.

$$x + 2 = 0 \quad \text{or} \quad x - 2 = 0 \quad \text{or} \quad x + 1 = 0 \quad \text{or} \quad x - 1 = 0$$
$$x = -2 \qquad\quad x = 2 \qquad\quad x = -1 \qquad\quad x = 1$$

The solutions to the equation are $x = -2$, $x = -1$, $x = 1$, and $x = 2$. ∎

Self Check 1 Solve: $2x^3 + 3x^2 - 2x = 0$.

Now Try Exercise 11.

2. Solve Other Equations by Factoring

To solve other types of equation by factoring, we use a property that states that equal powers of equal numbers are equal.

Power Property of Real Numbers   If a and b are real numbers, n is an integer, and $a = b$, then $a^n = b^n$.

When we raise both sides of an equation to the same power, the resulting equation might not be equivalent to the original one. For example, if we raise both sides of $x = 4$ to the second power, we obtain $x^2 = 16$. The two equations have different solution sets: the solution set of the first equation is $\{4\}$, but the solution set of the second equation is $\{4, -4\}$. Because raising both sides of an equation to the same power often introduces **extraneous solutions** (false solutions that don't satisfy the original equation), we must check all suspected roots to be certain that they satisfy the original equation.

The following equation has an extraneous solution:

$$x - x^{1/2} - 6 = 0$$

$$(x^{1/2} - 3)(x^{1/2} + 2) = 0 \qquad\qquad \text{Factor } x - x^{1/2} - 6.$$

$$x^{1/2} - 3 = 0 \quad \text{or} \quad x^{1/2} + 2 = 0 \qquad \text{Set each factor equal to 0.}$$

$$x^{1/2} = 3 \qquad\qquad\quad x^{1/2} = -2$$

Because equal powers of equal numbers are equal, we can square both sides of the previous equations to get

$$(x^{1/2})^2 = (3)^2 \quad \text{or} \quad (x^{1/2})^2 = (-2)^2$$

$$x = 9 \qquad\qquad\qquad x = 4$$

The number 9 satisfies the equation $x - x^{1/2} - 6 = 0$, but 4 does not, as the following check shows:

If $x = 9$	**If $x = 4$**
$x - x^{1/2} - 6 = 0$	$x - x^{1/2} - 6 = 0$
$9 - 9^{1/2} - 6 \overset{?}{=} 0$	$4 - 4^{1/2} - 6 \overset{?}{=} 0$
$9 - 3 - 6 \overset{?}{=} 0$	$4 - 2 - 6 \overset{?}{=} 0$
$0 = 0$	$-4 \neq 0$

Thus, $x = 9$ is the only solution.

EXAMPLE 2   Solving Equations by Factoring

Solve: **a.** $2x^{2/5} - 5x^{1/5} - 3 = 0$ **b.** $3(3x - 2x^{1/2}) = -1$

SOLUTION We will solve each equation by factoring.

a.
$$2x^{2/5} - 5x^{1/5} - 3 = 0$$

$$(2x^{1/5} + 1)(x^{1/5} - 3) = 0 \qquad\qquad \text{Factor } 2x^{2/5} - 5x^{1/5} - 3.$$

$$2x^{1/5} + 1 = 0 \quad \text{or} \quad x^{1/5} - 3 = 0 \qquad \text{Set each factor equal to 0.}$$

$$2x^{1/5} = -1 \qquad\qquad\quad x^{1/5} = 3$$

$$x^{1/5} = -\frac{1}{2}$$

We can raise both sides of each of the previous equations to the fifth power to obtain

$$(x^{1/5})^5 = \left(-\frac{1}{2}\right)^5 \quad \text{or} \quad (x^{1/5})^5 = (3)^5$$

$$x = -\frac{1}{32} \qquad\qquad\quad x = 243$$

Verify that the solutions $x = -\frac{1}{32}$ and $x = 243$ satisfy the original equation.

b.
$$3(3x - 2x^{1/2}) = -1$$

$$9x - 6x^{1/2} + 1 = 0 \qquad\qquad \text{Remove parentheses and add 1 to both sides.}$$

$$(3x^{1/2} - 1)(3x^{1/2} - 1) = 0 \qquad\qquad \text{Factor.}$$

$$3x^{1/2} - 1 = 0 \quad \text{or} \quad 3x^{1/2} - 1 = 0 \qquad \text{Set each factor equal to 0.}$$

$$x^{1/2} = \frac{1}{3} \qquad\qquad\quad x^{1/2} = \frac{1}{3} \qquad\qquad \text{Solve each equation.}$$

We can square both sides of each of the previous equations to obtain

$$(x^{1/2})^2 = \left(\frac{1}{3}\right)^2 \quad \text{or} \quad (x^{1/2})^2 = \left(\frac{1}{3}\right)^2$$
$$x = \frac{1}{9} \qquad\qquad\qquad x = \frac{1}{9}$$

Here, the solutions are the same. Verify that $x = \frac{1}{9}$ satisfies the equation. ∎

Self Check 2   Solve: $x^{2/5} - x^{1/5} - 2 = 0$.

Now Try Exercise 23.

3. Solve Radical Equations

Radical equations are equations containing radicals with variables in the radicand. To solve such equations, we use the Power Property of Real Numbers.

EXAMPLE 3   **Solving a Radical Equation**

Solve: $\sqrt{x + 3} - 4 = 7$.

SOLUTION   We will isolate the radical on the left side and then square both sides.

$$\sqrt{x + 3} - 4 = 7$$
$$\sqrt{x + 3} = 11 \qquad \text{Add 4 to both sides to isolate the radical.}$$
$$\left(\sqrt{x + 3}\right)^2 = (11)^2 \qquad \text{Square both sides.}$$
$$x + 3 = 121 \qquad \text{Simplify.}$$
$$x = 118 \qquad \text{Subtract 3 from both sides.}$$

Since squaring both sides might introduce extraneous roots, we must check the result of 118.

$$\sqrt{x + 3} - 4 = 7$$
$$\sqrt{118 + 3} - 4 \overset{?}{=} 7 \qquad \text{Substitute 118 for } x.$$
$$\sqrt{121} - 4 \overset{?}{=} 7$$
$$11 - 4 \overset{?}{=} 7$$
$$7 = 7$$

Comment

Remember to check all roots when solving radical equations, because raising both sides of an equation to a power can introduce extraneous roots.

Because it checks, $x = 118$ is the solution of the equation. ∎

Self Check 3   Solve: $\sqrt{x - 3} + 4 = 7$.

Now Try Exercise 29.

EXAMPLE 4   **Solving a Radical Equation**

Solve: $\sqrt{x + 3} = 3x - 1$.

SOLUTION We will square both sides of the equation to eliminate the radicals, and solve the resulting equation by factoring.

$$\sqrt{x+3} = 3x - 1$$

$$\left(\sqrt{x+3}\right)^2 = (3x-1)^2 \qquad \text{Square both sides.}$$

$$x + 3 = 9x^2 - 6x + 1 \qquad \text{Remove parentheses.}$$

$$0 = 9x^2 - 7x - 2 \qquad \text{Add } -x - 3 \text{ to both sides.}$$

$$0 = (9x+2)(x-1) \qquad \text{Factor } 9x^2 - 7x - 2.$$

$$9x + 2 = 0 \quad \text{or} \quad x - 1 = 0 \qquad \text{Set each factor equal to 0.}$$

$$x = -\frac{2}{9} \qquad\qquad x = 1$$

Since squaring both sides can introduce extraneous roots, we must check each result.

$$\sqrt{x+3} = 3x - 1 \qquad \text{or} \qquad \sqrt{x+3} = 3x - 1$$

$$\sqrt{-\frac{2}{9} + 3} \stackrel{?}{=} 3\left(-\frac{2}{9}\right) - 1 \qquad\bigg|\qquad \sqrt{1+3} \stackrel{?}{=} 3(1) - 1$$

$$\sqrt{\frac{25}{9}} \stackrel{?}{=} -\frac{2}{3} - 1 \qquad\qquad \sqrt{4} \stackrel{?}{=} 3 - 1$$

$$\frac{5}{3} \neq -\frac{5}{3} \qquad\qquad\qquad 2 = 2$$

Since $x = -\frac{2}{9}$ does not satisfy the equation, it is extraneous. Since $x = 1$ checks, it is the only solution.

Self Check 4　Solve: $\sqrt{x-2} = 2x - 10$.

Now Try Exercise 43.

EXAMPLE 5　**Solving a Radical Equation That Contains a Cube Root**

Solve: $\sqrt[3]{x^3 + 56} = x + 2$.

SOLUTION　To eliminate the radical, we cube both sides of the equation.

$$\sqrt[3]{x^3 + 56} = x + 2$$

$$\left(\sqrt[3]{x^3 + 56}\right)^3 = (x+2)^3 \qquad \text{Cube both sides.}$$

$$x^3 + 56 = x^3 + 6x^2 + 12x + 8 \qquad \text{Remove parentheses.}$$

$$0 = 6x^2 + 12x - 48 \qquad \text{Simplify.}$$

$$0 = x^2 + 2x - 8 \qquad \text{Divide both sides by 6.}$$

$$0 = (x+4)(x-2) \qquad \text{Factor } x^2 + 2x - 8.$$

$$x + 4 = 0 \quad \text{or} \quad x - 2 = 0 \qquad \text{Set each factor equal to 0.}$$

$$x = -4 \qquad\qquad x = 2$$

We check each possible solution to see whether either is extraneous.

For $x = -4$	For $x = 2$
$\sqrt[3]{x^3 + 56} = x + 2$	$\sqrt[3]{x^3 + 56} = x + 2$
$\sqrt[3]{(-4)^3 + 56} \stackrel{?}{=} -4 + 2$	$\sqrt[3]{2^3 + 56} \stackrel{?}{=} 2 + 2$
$\sqrt[3]{-64 + 56} \stackrel{?}{=} -2$	$\sqrt[3]{8 + 56} \stackrel{?}{=} 4$
$\sqrt[3]{-8} \stackrel{?}{=} -2$	$\sqrt[3]{64} \stackrel{?}{=} 4$
$-2 = -2$	$4 = 4$

Since both values satisfy the equation, $x = -4$ and $x = 2$ are roots.

Self Check 5 Solve: $\sqrt[3]{x^3 + 7} = x + 1$.

Now Try Exercise 51.

EXAMPLE 6 **Solving a Radical Equation with Two Radicals**

Solve: $\sqrt{2x + 3} + \sqrt{x - 2} = 4$.

SOLUTION We can write the equation in the form

$$\sqrt{2x + 3} = 4 - \sqrt{x - 2} \qquad \text{Subtract } \sqrt{x - 2} \text{ from both sides.}$$

so that the left side contains one radical. We then square both sides to get

$$\left(\sqrt{2x + 3}\right)^2 = \left(4 - \sqrt{x - 2}\right)^2$$

$$2x + 3 = 16 - 8\sqrt{x - 2} + x - 2$$

$$2x + 3 = 14 - 8\sqrt{x - 2} + x \qquad \text{Combine like terms.}$$

$$x - 11 = -8\sqrt{x - 2} \qquad \text{Subtract 14 and } x \text{ from both sides.}$$

We then square both sides again to eliminate the radical.

$$(x - 11)^2 = \left(-8\sqrt{x - 2}\right)^2$$

$$x^2 - 22x + 121 = 64(x - 2) \qquad \text{Don't forget to square } -8.$$

$$x^2 - 22x + 121 = 64x - 128$$

$$x^2 - 86x + 249 = 0$$

$$(x - 3)(x - 83) = 0$$

$$x - 3 = 0 \quad \text{or} \quad x - 83 = 0 \qquad \text{Set each factor equal to 0.}$$

$$x = 3 \qquad\qquad x = 83$$

Substituting these results into the equation will show that 83 doesn't check; it is extraneous. However, 3 does satisfy the equation and is a root.

Self Check 6 Solve: $\sqrt{2x + 1} + \sqrt{x + 5} = 6$.

Now Try Exercise 59.

Caution When expanding a term such as $\left(4 - \sqrt{x - 2}\right)^2$, note that $\left(4 - \sqrt{x - 2}\right)^2 \neq 16 - (x - 2)$.

4. Solve Applications of Radical Equations

EXAMPLE 7   Solving an Application Problem

A highway curve banked at 8° will accommodate traffic traveling s mph if the radius of the curve is r feet, according to the formula $s = 1.45\sqrt{r}$. Find what radius is necessary to accommodate 70 mph traffic (see Figure R-10).

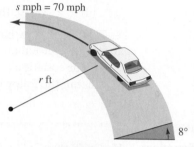

s mph = 70 mph

r ft

8°

FIGURE R-10

SOLUTION   We can substitute 70 for s in the formula and solve for r.

$$s = 1.45\sqrt{r}$$

$$70 = 1.45\sqrt{r} \qquad \text{Substitute 70 for } s.$$

$$\frac{70}{1.45} = \sqrt{r} \qquad \text{Divide both sides by 1.45.}$$

$$\left(\frac{70}{1.45}\right)^2 = r \qquad \text{Square both sides.}$$

We can use a calculator to find r. The radius of the curve is approximately 2,331 feet. ∎

Self Check 7   Find the radius necessary to accommodate 65 mph traffic.

Now Try Exercise 101.

5. Solve Absolute-Value Equations

We will now review absolute value and then learn about techniques used to solve absolute-value equations.

Absolute Value   The **absolute value** of the real number x, denoted by $|x|$, is defined as follows:

If $x \geq 0$, then $|x| = x$.

If $x < 0$, then $|x| = -x$.

This definition provides a way to associate a nonnegative real number with any real number.

*   If $x \geq 0$, then x (which is positive or 0) is its own absolute value.
*   If $x < 0$, then $-x$ (which is positive) is the absolute value.

Either way, $|x|$ is positive or 0: $|x| \geq 0$ for all real numbers x.

For example:

- Because 7 is positive, $|7| = 7$.
- Because -3 is negative, $|-3| = -(-3) = 3$.
- The expression $-|-7|$ means "the negative of the absolute value of -7." Thus, $-|\mathbf{-7}| = -(\mathbf{7}) = -7$.
- To denote the absolute value of a variable quantity, we must give a conditional answer:

$$\begin{cases} \text{If } x - 2 \geq 0, \text{ then } |x - 2| = x - 2. \\ \text{If } x - 2 < 0, \text{ then } |x - 2| = -(x - 2) = 2 - x. \end{cases}$$

Now that we have reviewed absolute value, we can turn our attention to absolute-value equations.

In the equation $|x| = 8$, x can be either 8 or -8, because $|8| = 8$ and $|-8| = 8$. In general, the following is true:

Absolute-Value Equations   If $k > 0$, then $|x| = k$ is equivalent to $x = k$ or $x = -k$.

If $k = 0$, then $|x| = k$ is equivalent to $x = 0$.

If $k < 0$, then $|x| = k$ has no solution.

The absolute value of a number represents the distance on the number line from that number to the origin. The solutions of $|k|$ are the coordinates of the two points that lie exactly k units from the origin (see Figure R-11).

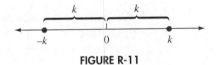

FIGURE R-11

The equation $|x - 3| = 7$ indicates that a point on the number line with a coordinate of $(x - 3)$ is 7 units from the origin. Thus, $(x - 3)$ can be 7 or -7.

$$x - 3 = 7 \quad \text{or} \quad x - 3 = -7$$
$$x = 10 \qquad\qquad x = -4$$

The solutions of the equation are $x = 10$ and $x = -4$. Both of these numbers satisfy the equation.

EXAMPLE 8   **Solving an Absolute-Value Equation with One Absolute Value**

Solve: $|3x - 5| + 3 = 10$.

SOLUTION   We begin by subtracting 3 from both sides and writing the equation in the form $|x| = k$.

$$|3x - 5| + 3 - \mathbf{3} = 10 - \mathbf{3}$$
$$|3x - 5| = 7$$

The equation $|3x - 5| = 7$ is equivalent to two equations,

$$3x - 5 = 7 \quad \text{or} \quad 3x - 5 = -7$$

which can be solved separately:

$$3x - 5 = 7 \quad \text{or} \quad 3x - 5 = -7$$
$$3x = 12 \qquad\qquad 3x = -2$$
$$x = 4 \qquad\qquad x = -\frac{2}{3}$$

The solutions are $x = 4$ and $x = -\frac{2}{3}$.

Self Check 8   Solve: $|2x + 3| + 5 = 12$.

Now Try Exercise 79.

Caution

> We must isolate the absolute value term to apply the property. $|3x - 5| + 3 = 10$ is not equivalent to $3x - 5 + 3 = 10$ or $3x - 5 + 3 = -10$.

The equation $|a| = |b|$ is true when $a = b$ or when $a = -b$. For example:

$$|3| = |3| \quad \text{or} \quad |3| = |-3|$$
$$3 = 3 \qquad\qquad 3 = 3$$

In general, the following is true:

Equations with Two Absolute Values

If a and b represent algebraic expressions, the equation $|a| = |b|$ is equivalent to

$$a = b \quad \text{or} \quad a = -b.$$

EXAMPLE 9   **Solving an Absolute-Value Equation with Two Absolute Values**

Solve: $|2x| = |x - 3|$.

SOLUTION   The equation $|2x| = |x - 3|$ will be true when $2x$ and $(x - 3)$ are equal or when they are negatives. This gives two equations, which can be solved separately:

$$2x = x - 3 \quad \text{or} \quad 2x = -(x - 3)$$
$$x = -3 \qquad\qquad 2x = -x + 3$$
$$3x = 3$$
$$x = 1$$

Self Check 9   Solve: $|3x + 1| = |5x - 3|$.

Now Try Exercise 97.

Self Check Answers   1. $0, \frac{1}{2}, -2$ 2. $-1, 32$ 3. 12 4. 6 5. $1, -2$ 6. 4

7. $2{,}010$ feet   8. $-5, 2$ 9. $2, \frac{1}{4}$

Exercises R.4

Getting Ready

You should be able to complete these vocabulary and concept statements before you proceed to the practice exercises.

Fill in the blanks.

1. Equal powers of equal real numbers are _____.

2. If a and b are real numbers and $a = b$ then $a^2 =$ _____.

3. False solutions that don't satisfy the equation are called _____ solutions.

4. _____ equations contain radicals with variables in their radicands.

5. If $x \geq 0$, then $|x| =$ _____.

6. If $x < 0$, then $|x| =$ _____.

7. $|x| = k$ is equivalent to _____.

8. $|a| = |b|$ is equivalent to $a = b$ or _____.

Practice

Use factoring to solve each equation for real values of the variable.

9. $x^3 + 9x^2 + 20x = 0$

10. $x^3 + 4x^2 - 21x = 0$

11. $6a^3 - 5a^2 - 4a = 0$

12. $8b^3 - 10b^2 + 3b = 0$

13. $y^4 - 26y^2 + 25 = 0$

14. $y^4 - 13y^2 + 36 = 0$

15. $x^4 - 37x^2 + 36 = 0$

16. $x^4 - 50x^2 + 49 = 0$

17. $2y^4 - 46y^2 = -180$

18. $2x^4 - 102x^2 = -196$

19. $z^{3/2} - z^{1/2} = 0$

20. $r^{5/2} - r^{3/2} = 0$

21. $2m^{2/3} + 3m^{1/3} - 2 = 0$

22. $6t^{2/5} + 11t^{1/5} + 3 = 0$

23. $x - 13x^{1/2} + 12 = 0$

24. $p + p^{1/2} - 20 = 0$

25. $2t^{1/3} + 3t^{1/6} - 2 = 0$

26. $z^3 - 7z^{3/2} - 8 = 0$

27. $6p + p^{1/2} - 1 = 0$

28. $3r - r^{1/2} - 2 = 0$

Find all real solutions of each radical equation.

29. $\sqrt{x - 2} = 5$

30. $\sqrt{a - 3} - 5 = 0$

31. $3\sqrt{x + 1} = \sqrt{6}$

32. $\sqrt{x + 3} = 2\sqrt{x}$

33. $\sqrt{5a - 2} = \sqrt{a + 6}$

34. $\sqrt{16x + 4} = \sqrt{x + 4}$

35. $2\sqrt{x^2 + 3} = \sqrt{-16x - 3}$

36. $\sqrt{x^2 + 1} = \dfrac{\sqrt{-7x + 11}}{\sqrt{6}}$

37. $\sqrt[3]{7x + 1} = 4$

38. $\sqrt[3]{11a - 40} = 5$

39. $\sqrt[4]{30t + 25} = 5$

40. $\sqrt[4]{3z + 1} = 2$

41. $\sqrt{x^2 + 21} = x + 3$

42. $\sqrt{5 - x^2} = -(x + 1)$

43. $\sqrt{y + 2} = 4 - y$

44. $\sqrt{3z + 1} = z - 1$

45. $x - \sqrt{7x - 12} = 0$

46. $x - \sqrt{4x - 4} = 0$

47. $x + 4 = \sqrt{\dfrac{6x + 6}{5}} + 3$

48. $\sqrt{\dfrac{8x + 43}{3}} - 1 = x$

49. $\sqrt{\dfrac{x^2 - 1}{x - 2}} = 2\sqrt{2}$

50. $\dfrac{\sqrt{x^2 - 1}}{\sqrt{3x - 5}} = \sqrt{2}$

51. $\sqrt[3]{x^3 + 7} = x + 1$

52. $\sqrt[3]{x^3 - 7} + 1 = x$

53. $\sqrt[3]{8x^3 + 61} = 2x + 1$

54. $\sqrt[3]{8x^3 - 37} = 2x - 1$

55. $\sqrt{2p + 1} - 1 = \sqrt{p}$

56. $\sqrt{r} + \sqrt{r + 2} = 2$

57. $\sqrt{x + 3} = \sqrt{2x + 8} - 1$

58. $\sqrt{x + 2} + 1 = \sqrt{2x + 5}$

59. $\sqrt{y + 8} - \sqrt{y - 4} = -2$

60. $\sqrt{z + 5} - 2 = \sqrt{z - 3}$

61. $\sqrt{2b + 3} - \sqrt{b + 1} = \sqrt{b - 2}$

62. $\sqrt{a + 1} + \sqrt{3a} = \sqrt{5a + 1}$

63. $\sqrt{\sqrt{b} + \sqrt{b + 8}} = 2$

64. $\sqrt{\sqrt{x + 19} - \sqrt{x - 2}} = \sqrt{3}$

Write each expression without absolute-value symbols.

65. $|7|$

66. $|-9|$

67. $|0|$

68. $|3 - 5|$

69. $|5| - |-3|$

70. $|-3| + |5|$

71. $|\pi - 2|$

72. $|\pi - 4|$

73. $|x - 5|$ and $x \geq 5$

74. $|x - 5|$ and $x \leq 5$

75. $|x^3|$

76. $|2x|$

Solve each absolute-value equation for x.

77. $|x + 2| = 2$

78. $|2x + 5| = 3$

79. $|3x - 1| + 4 = 9$

80. $|7x - 5| - 6 = -3$

81. $\left|\dfrac{3x - 4}{2}\right| = 5$

82. $\left|\dfrac{10x + 1}{2}\right| = \dfrac{9}{2}$

83. $\left|\dfrac{2x - 4}{5}\right| = 2$

84. $\left|\dfrac{3x + 11}{7}\right| = 1$

85. $\left|\dfrac{x - 3}{4}\right| = -2$

86. $\left|\dfrac{x - 5}{3}\right| = 0$

87. $\left|\dfrac{4x - 2}{x}\right| = 3$

88. $\left|\dfrac{2(x - 3)}{3x}\right| = 6$

89. $|x| = x$

90. $|x| + x = 2$

91. $|x + 3| = |x|$

92. $|x + 5| = |5 - x|$

93. $|x - 3| = |2x + 3|$

94. $|x - 2| = |3x + 8|$

95. $|x + 2| = |x - 2|$

96. $|2x - 3| = |3x - 5|$

97. $\left|\dfrac{x + 3}{2}\right| = |2x - 3|$

98. $\left|\dfrac{x - 2}{3}\right| = |6 - x|$

99. $\left|\dfrac{3x-1}{2}\right| = \left|\dfrac{2x+3}{3}\right|$ **100.** $\left|\dfrac{5x+2}{3}\right| = \left|\dfrac{x-1}{4}\right|$

Applications

101. Height of a bridge The distance d (in feet) that an object will fall in t seconds is given by the formula $t = \sqrt{\dfrac{d}{16}}$. To find the height of a bridge above a river, a man drops a stone into the water (see the illustration). If it takes the stone 5 seconds to hit the water, how high is the bridge?

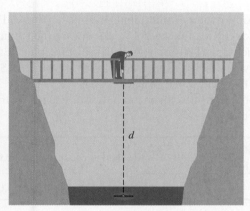

102. Horizon distance The higher a lookout tower, the farther an observer can see (see the illustration). The distance d (called the **horizon distance,** measured in miles) is related to the height h of the observer (measured in feet) by the formula $d = \sqrt{1.5h}$. How tall must a tower be for the observer to see 30 miles?

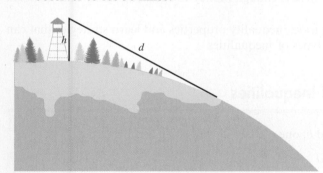

103. Carpentry During construction, carpenters often brace walls, as shown in the illustration. The appropriate length of the brace is given by the formula $l = \sqrt{f^2 + h^2}$. If a carpenter nails a 10-foot brace to the wall 6 feet above the floor, how far from the base of the wall should he nail the brace to the floor?

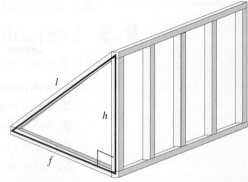

104. Windmills The power generated by a windmill is related to the velocity of the wind by the formula $v = \sqrt[3]{\dfrac{P}{0.02}}$, where P is the power (in watts) and v is the velocity of the wind (in mph). To the nearest 10 watts, find the power generated when the velocity of the wind is 31 mph.

105. Diamonds The effective rate of interest r earned by an investment is given by the formula $r = \sqrt[n]{\dfrac{A}{P}} - 1$, where P is the initial investment that grows to value A after n years. If a diamond buyer got \$4,000 for a 1.03-carat diamond that he had purchased 4 years earlier, and earned an annual rate of return of 6.5% on the investment, what did he originally pay for the diamond?

106. Theater productions The ropes, pulleys, and sandbags shown in the illustration are part of a mechanical system used to raise and lower scenery for a stage play. For the scenery to be in the proper position, the following formula must apply: $w_2 = \sqrt{w_1^2 + w_3^2}$. If $w_2 = 12.5$ pounds and $w_3 = 7.5$ pounds, find w_1.

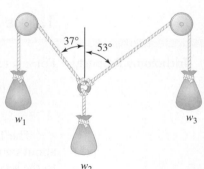

Discovery and Writing

107. Explain why squaring both sides of an equation might introduce extraneous roots.

108. Can cubing both sides of an equation introduce extraneous roots? Explain.

109. Explain how to find the absolute value of a number.

110. Explain why the equation $|x| + 9 = 0$ has no solution.

R.5 Inequalities

In this section, we will learn to

1. Use properties of inequalities.
2. Solve linear inequalities.
3. Solve applications of linear inequalities.
4. Solve compound inequalities.
5. Solve absolute-value inequalities.

Eduardo Rivero/Shutterstock.com

Studying abroad is a wonderful opportunity for students. Suppose you read about a program to study Spanish in Costa Rica for several weeks during the summer.

The cost of the program includes $900 for round-trip airfare plus $350 per week, which covers tuition, meals, and living accommodations. If you have $3,000, how many weeks can you afford to spend in Costa Rica?

If x represents the number of weeks you can spend in Costa Rica, we can write an inequality that represents the cost.

Airfare	plus	cost of $350 per week for x weeks	is less than or equal to	$3,000.
900	+	$350x$	$\leq$	3,000

To solve this inequality, we can proceed as follows:

$$900 + 350x \leq 3,000$$

$$350x \leq 2,100 \qquad \text{Subtract 900 from both sides.}$$

$$x \leq 6 \qquad \text{Divide both sides by 350.}$$

Since x can be equal to 6, there is enough money to spend up to 6 weeks in Costa Rica.

In this section, we will learn inequality properties and learn strategies that can be applied to solve several types of inequalities.

1. Use Properties of Inequalities

Trichotomy Property   For any real numbers a and b, one of the following statements is true:

$$a < b, \quad a = b, \quad \text{or} \quad a > b.$$

The Trichotomy Property indicates that one of the following statements is true about two real numbers: Either the first is less than the second, or the first is equal to the second, or the first is greater than the second.

Transitive Property   Let a, b, and c represent real numbers.

If $a < b$ and $b < c$, then $a < c$.

If $a > b$ and $b > c$, then $a > c$.

The first part of the Transitive Property indicates that if a first number is less than a second and the second number is less than a third, then the first number is less than the third.

The second part of the Transitive Property is similar, with the words "is greater than" substituted for "is less than."

Addition and Subtraction Properties of Inequality

Let a, b, and c represent real numbers.

If $a < b$, then $a + c < b + c$.

If $a < b$, then $a - c < b - c$.

Similar properties exist for $>$, $\leq$, and $\geq$.

These properties state that *any real number can be added to (or subtracted from) both sides of an inequality to obtain another inequality with the same order (direction)*. For example, if we add 4 to (or subtract 4 from) both sides of $8 < 12$, we get

$$8 < 12 \qquad\qquad 8 < 12$$

$$8 + 4 < 12 + 4 \qquad\qquad 8 - 4 < 12 - 4$$

$$12 < 16 \qquad\qquad 4 < 8$$

and the $<$ symbol is unchanged.

Multiplication and Division Properties of Inequality

Let a, b, and c represent real numbers.

Part 1:   If $a < b$ and $c > 0$, then $ca < cb$.

If $a < b$ and $c > 0$, then $\dfrac{a}{c} < \dfrac{b}{c}$.

Part 2:   If $a < b$ and $c < 0$, then $ca > cb$.

If $a < b$ and $c < 0$, then $\dfrac{a}{c} > \dfrac{b}{c}$.

Similar properties exist for $>$, $\leq$, and $\geq$.

This property has two parts. Part 1 states that *both sides of an inequality can be multiplied (or divided) by the same positive number to obtain another inequality with the same order*. For example, if we multiply (or divide) both sides of $8 < 12$ by 4, we get

$$8 < 12 \qquad\qquad 8 < 12$$

$$8(4) < 12(4) \qquad\qquad \frac{8}{4} < \frac{12}{4}$$

$$32 < 48 \qquad\qquad 2 < 3$$

and the $<$ symbol is unchanged.

Comment

Unless we are multiplying or dividing by a negative number, the properties of inequalities are the same as the properties of equality.

Part 2 states that *both sides of an inequality can be multiplied (or divided) by the same negative number to obtain another inequality with the opposite order.* For example, if we multiply (or divide) both sides of $8 < 12$ by -4, we get

$$8 < 12 \qquad\qquad 8 < 12$$

$$8(-4) > 12(-4) \qquad\qquad \frac{8}{-4} > \frac{12}{-4}$$

$$-32 > -48 \qquad\qquad -2 > -3$$

and the $<$ symbol is changed to a $>$ symbol.

2. Solve Linear Inequalities

Linear inequalities are inequalities such as $ax + b < 0$ or $ax + b \geq 0$, where $a \neq 0$. Numbers that make an inequality true when substituted for the variable are solutions of the inequality. Inequalities with the same solution set are called **equivalent inequalities.** Because of the previous properties, we can solve inequalities as we do equations. However, we must always remember to change the order of an inequality when multiplying (or dividing) both sides by a negative number.

EXAMPLE 1 **Solving a Linear Inequality**

Solve: $3(x + 2) < 8$.

SOLUTION We proceed as with equations.

$$3(x + 2) < 8$$

$$3x + 6 < 8 \qquad \text{Remove parentheses.}$$

$$3x < 2 \qquad \text{Subtract 6 from both sides.}$$

$$x < \frac{2}{3} \qquad \text{Divide both sides by 3.}$$

All numbers that are less than $\frac{2}{3}$ are solutions of the inequality. The solution set can be expressed in interval notation as $\left(-\infty, \frac{2}{3}\right)$ and can be graphed as in Figure R-12.

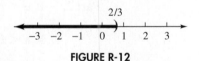

FIGURE R-12

Self Check 1 Solve: $5(p - 4) > 25$.

Now Try Exercise 15.

EXAMPLE 2 **Solving a Linear Inequality**

Solve: $-5(x - 2) \leq 20 + x$.

SOLUTION We proceed as with equations.

$$-5(x - 2) \leq 20 + x$$

$$-5x + 10 \leq 20 + x \qquad \text{Remove parentheses.}$$

$$-6x + 10 \leq 20 \qquad \text{Subtract } x \text{ from both sides.}$$

$$-6x \leq 10 \qquad \text{Subtract 10 from both sides.}$$

We now divide both sides of the inequality by –6, which changes the order of the inequality.

$$x \geq \frac{10}{-6}$$ Divide both sides by -6.

$$x \geq -\frac{5}{3}$$ Simplify the fraction.

The graph of the solution set is shown in Figure R-13. It is the interval $\left[-\frac{5}{3}, \infty\right)$.

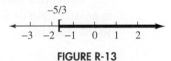

FIGURE R-13

Self Check 2   Solve: $-4(x + 3) \geq 16$.

Now Try Exercise 29.

3. Solve Applications of Linear Inequalities

EXAMPLE 3   Solving an Application Problem

An empty truck with driver weighs 4,350 pounds. It is loaded with feed corn weighing 31 pounds per bushel. Between the farm and the market is a bridge with a 10,000-pound load limit. How many bushels can the truck legally carry?

Robert Kyllo/Shutterstock.com

SOLUTION   The empty truck with driver weighs 4,350 pounds, and the corn weighs 31 pounds per bushel. If we let b represent the number of bushels in a legal load, the weight of the corn will be $31b$ pounds. Since the combined weight of the truck, driver, and cargo cannot exceed 10,000 pounds, we can form the following inequality:

The weight of the empty truck with driver	plus	the weight of the corn	must be less than or equal to	10,000 pounds.
4,350	+	31b	≤	10,000

We can solve the inequality as follows.

$$4,350 + 31b \leq 10,000$$

$$31b \leq 5,650$$ Subtract 4,350 from each side.

$$b \leq 182.2580645$$ Divide both sides by 31.

The truck can legally carry $\approx 182\frac{1}{4}$ bushels or less.

Self Check 3   If the empty truck and driver in Example 3 weigh 3,800 pounds, how many bushels can the truck legally carry?

Now Try Exercise 89.

4. Solve Compound Inequalities

The statement that x is between 2 and 5 implies two inequalities,

$$x > 2 \quad \text{and} \quad x < 5.$$

It is customary to write both inequalities as one **compound inequality:**

$$2 < x < 5. \qquad \text{Read as "2 is less than } x \text{ and } x \text{ is less than 5."}$$

To express that x is not between 2 and 5, we must convey the idea either that x is greater than or equal to 5 or that x is less than or equal to 2. This is equivalent to the statement

$$x \geq 5 \quad \text{or} \quad x \leq 2.$$

This inequality is satisfied by all numbers x that satisfy one or both of its parts.

Comment

Remember that $2 < x < 5$ means that $x > 2$ and $x < 5$. The word *and* indicates that both inequalities must be true at the same time.

Caution

> It is incorrect to write $x \geq 5$ or $x \leq 2$ as $2 \geq x \geq 5$, because this would mean that $2 \geq 5$, which is false.

EXAMPLE 4   Solving a Compound Inequality

Solve: $5 < 3x - 7 \leq 8$.

SOLUTION   We can isolate x between the inequality symbols by adding 7 to each part of the inequality to get

$$5 + 7 < 3x - 7 + 7 \leq 8 + 7 \qquad \text{Add 7 to each part.}$$
$$12 < 3x \leq 15 \qquad \text{Do the additions.}$$

and dividing all parts by 3 to get

$$4 < x \leq 5.$$

The solution set is the interval $(4, 5]$, whose graph appears in Figure R-14.

FIGURE R-14

Self Check 4   Solve: $-5 \leq 2x + 1 < 9$.

Now Try Exercise 33.

EXAMPLE 5   Solving a Compound Inequality

Solve: $3 + x \leq 3x + 1 < 7x - 2$.

SOLUTION   Because it is impossible to isolate x between the inequality symbols, we must solve each inequality separately.

$$3 + x \leq 3x + 1 \quad \text{and} \quad 3x + 1 < 7x - 2$$
$$3 \leq 2x + 1 \qquad\qquad\qquad 1 < 4x - 2$$
$$2 \leq 2x \qquad\qquad\qquad\quad 3 < 4x$$
$$1 \leq x \qquad\qquad\qquad\quad \frac{3}{4} < x$$
$$x \geq 1 \qquad\qquad\qquad\quad x > \frac{3}{4}$$

Since the connective in this inequality is *and*, the solution set is the intersection (or overlap) of the intervals $[1, \infty)$ and $\left(\frac{3}{4}, \infty\right)$, which is $[1, \infty)$. The graph is shown in Figure R-15.

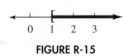

FIGURE R-15

Self Check 5   Solve: $x + 1 < 2x - 3 \leq 3x - 5$.

Now Try Exercise 47.

Comment

It is possible for an inequality to be true for all or no real numbers. For example:

• $x < x + 1$ is true for all numbers x.

• $x > x + 1$ is not true for any numbers x.

5. Solve Absolute-Value Inequalities

Cancún, Mexico, is a popular spring-break destination for college students. The beautiful beaches, exciting nightlife, and tropical weather make it a pleasurable experience.

The average annual temperature in Cancun is 78°F, with fluctuations of approximately 7°. We can represent this temperature range using absolute-value notation. If we let x represent the temperature at a given time, the absolute value of the difference between x and 78 is less than or equal to 7. We can write this as $|x - 78| \leq 7$.

In this section, we will continue our study of inequalities and solve inequalities containing absolute values. There are three types of absolute-value inequalities we will consider:

• Inequalities of the form $|x| < k$
• Inequalities of the form $|x| > k$
• Inequalities with two absolute values

The inequality $|x| < 5$ indicates that a point with coordinate x is less than 5 units from the origin. Thus, x is between -5 and 5, and $|x| < 5$ is equivalent to $-5 < x < 5$ (see Figure R-16).

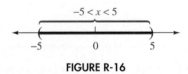

FIGURE R-16

In general, the inequality $|x| < k$ $(k > 0)$ indicates that a point with coordinate x is less than k units from the origin (see Figure R-17).

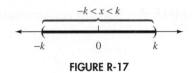

FIGURE R-17

Inequalities of the Forms | If $k > 0$, then $|x| < k$ is equivalent to $-k < x < k$.
$|x| < k$ and $|x| \leq k$ | If $k > 0$, then $|x| \leq k$ is equivalent to $-k \leq x \leq k$.

EXAMPLE 6 **Solving an Absolute-Value Inequality of the Form $|x| < k$**

Solve: $|x - 2| < 7$.

SOLUTION The inequality $|x - 2| < 7$ is equivalent to

$$-7 < x - 2 < 7.$$

We can add 2 to each part of this inequality to get

$$-5 < x < 9.$$

The solution set is the interval $(-5, 9)$, shown in Figure R-18.

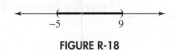

FIGURE R-18

Self Check 6 Solve: $|x + 3| - 9 < 0$.

Now Try Exercise 59.

The inequality $|x| > 5$ indicates that a point with coordinate x is more than 5 units from the origin. Thus, $x < -5$ or $x > 5$ (see Figure R-19).

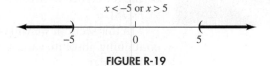

FIGURE R-19

In general, the inequality $|x| > k$ $(k > 0)$ indicates that a point with coordinate x is more than k units from the origin (see Figure R-20).

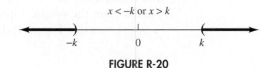

FIGURE R-20

Inequalities of the Forms
$|x| > k$ and $|x| \geq k$

If $k > 0$, then $|x| > k$ is equivalent to $x < -k$ or $x > k$.

If $k > 0$, then $|x| \geq k$ is equivalent to $x \leq -k$ or $x \geq k$.

EXAMPLE 7 **Solving an Absolute-Value Inequality of the Form $|x| > k$**

Solve: $\left| \dfrac{2x + 3}{2} \right| + 7 \geq 12$.

SOLUTION We begin by subtracting 7 from both sides of the inequality to isolate the absolute value on the left side.

$$\left| \frac{2x + 3}{2} \right| \geq 5$$

This result is equivalent to two inequalities that can be solved separately.

$$\frac{2x + 3}{2} \le -5 \quad \text{or} \quad \frac{2x + 3}{2} \ge 5$$

$$\begin{array}{rcl}
2x + 3 \le -10 & \bigg| & 2x + 3 \ge 10 \\
2x \le -13 & \bigg| & 2x \ge 7 \\
x \le -\dfrac{13}{2} & \bigg| & x \ge \dfrac{7}{2}
\end{array}$$

The solution set is the union of the intervals $\left(-\infty, -\frac{13}{2}\right]$ and $\left[\frac{7}{2}, \infty\right)$. Its graph appears in Figure R-21.

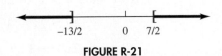

FIGURE R-21

Self Check 7   Solve: $\left|\dfrac{3x - 6}{3}\right| + 2 \ge 12$.

Now Try Exercise 65.

Caution

> We must isolate the absolute value expression before we apply the property. For example, $\left|\dfrac{2x + 5}{2}\right| + 7 \ge 12$ cannot be solved by using $\left|\dfrac{2x + 5}{2}\right| + 7 \ge 12$ or $\left|\dfrac{2x + 5}{2}\right| + 7 \le -12$.

EXAMPLE 8   **Solving a Compound Inequality That Contains an Absolute Value**

Solve: $0 < |x - 5| \le 3$.

SOLUTION   The inequality $0 < |x - 5| \le 3$ consists of two inequalities that can be solved separately. The solution will be the intersection of the inequalities

$$0 < |x - 5| \quad \text{and} \quad |x - 5| \le 3.$$

The inequality $0 < |x - 5|$ is true for all x except 5. The inequality $|x - 5| \le 3$ is equivalent to the inequality

$$-3 \le x - 5 \le 3$$

$$2 \le x \le 8 \qquad \text{Add 5 to each part.}$$

The solution set is the intersection of these two solutions, which is the interval $[2, 8]$, except 5. This is the union of the intervals $[2, 5)$ and $(5, 8]$, as shown in Figure R-22.

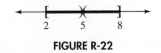

FIGURE R-22

Self Check 8   Solve: $0 < |x + 2| \le 5$.

Now Try Exercise 69.

We have seen that $|a|$ could be defined as $|a| = \sqrt{a^2}$. We will use this fact in the next example.

EXAMPLE 9   Solving an Inequality with Two Absolute Values

Solve $|x + 2| > |x + 1|$ and give the result in interval notation.

SOLUTION

$$|x + 2| > |x + 1|$$

$$\sqrt{(x + 2)^2} > \sqrt{(x + 1)^2} \qquad \text{Use } |a| = \sqrt{a^2}.$$

$$(x + 2)^2 > (x + 1)^2 \qquad \text{Square both sides.}$$

$$x^2 + 4x + 4 > x^2 + 2x + 1 \qquad \text{Expand each binomial.}$$

$$4x > 2x - 3 \qquad \text{Subtract } x^2 \text{ and 4 from both sides.}$$

$$2x > -3 \qquad \text{Subtract } 2x \text{ from both sides.}$$

$$x > -\frac{3}{2} \qquad \text{Divide both sides by 2.}$$

The solution set is the interval $\left(-\frac{3}{2}, \infty\right)$. Check several numbers in this interval to verify that this interval is the solution. ∎

Self Check 9   Solve $|x - 3| \le |x + 2|$ and give the result in interval notation.

Now Try Exercise 81.

Self Check Answers   1. $(9, \infty)$ 2. $(-\infty, -7]$ 3. 200 bushels or less   4. $[-3, 4)$
5. $(4, \infty)$ 6. $(-12, 6)$ 7. $(-\infty, -8] \cup [12, \infty)$ 8. $[-7, -2) \cup (-2, 3]$
9. $\left[\frac{1}{2}, \infty\right)$

Exercises R.5

Getting Ready
You should be able to complete these vocabulary and concept statements before you proceed to the practice exercises.

Fill in the blanks.

1. If $x > y$, then x lies to the _____ of y on a number line.

2. $a < b$, _____, or $a > b$.

3. If $a < b$ and $b < c$, then _____.

4. If $a < b$, then $a + c <$ _____.

5. If $a < b$, then $a - c <$ _____.

6. If $a < b$ and $c > 0$, then ac ____ bc.

7. If $a < b$ and $c < 0$, then ac ____ bc.

8. If $a < b$ and $c < 0$, then $\dfrac{a}{c}$ ____ $\dfrac{b}{c}$.

9. $3x - 5 < 12$ and $ax + c > 0$ $(a \ne 0)$ are examples of _____ inequalities.

10. If two inequalities have the same solution set, they are called _____ inequalities.

11. $|x| < k$ is equivalent to _____.

12. $|x| > k$ is equivalent to _____.

Practice
Solve each linear or compound inequality, graph the solution set, and write the answer in interval notation.

13. $3x + 2 < 5$

14. $-2x + 4 < 6$

15. $3x + 2 \ge 5$

16. $-2x + 4 \ge 6$

17. $-5x + 3 > -2$

18. $4x - 3 > -4$

19. $-5x + 3 \le -2$

20. $4x - 3 \le -4$

21. $2(x - 3) \le -2(x - 3)$

22. $3(x + 2) \le 2(x + 5)$

23. $\dfrac{3}{5}x + 4 > 2$

24. $\dfrac{1}{4}x - 3 > 5$

25. $\dfrac{x + 3}{4} < \dfrac{2x - 4}{3}$

26. $\dfrac{x + 2}{5} > \dfrac{x - 1}{2}$

27. $\dfrac{6(x-4)}{5} \ge \dfrac{3(x+2)}{4}$

28. $\dfrac{3(x+3)}{2} < \dfrac{2(x+7)}{3}$

29. $\dfrac{5}{9}(a+3) - a \ge \dfrac{4}{3}(a-3) - 1$

30. $\dfrac{2}{3}y - y \le -\dfrac{3}{2}(y-5)$

31. $\dfrac{2}{3}a - \dfrac{3}{4}a < \dfrac{3}{5}\left(a + \dfrac{2}{3}\right) + \dfrac{1}{3}$

32. $\dfrac{1}{4}b + \dfrac{2}{3}b - \dfrac{1}{2} > \dfrac{1}{2}(b+1) + b$

33. $4 < 2x - 8 \le 10$

34. $3 \le 2x + 2 < 6$

35. $9 \ge \dfrac{x-4}{2} > 2$

36. $5 < \dfrac{x-2}{6} < 6$

37. $0 \le \dfrac{4-x}{3} \le 5$

38. $0 \ge \dfrac{5-x}{2} \ge -10$

39. $-2 \ge \dfrac{1-x}{2} \ge -10$

40. $-2 \le \dfrac{1-x}{2} < 10$

41. $-3x > -2x > -x$

42. $-3x < -2x < -x$

43. $x < 2x < 3x$

44. $x > 2x > 3x$

45. $2x + 1 < 3x - 2 < 12$

46. $2 - x < 3x + 5 < 18$

47. $2 + x < 3x - 2 < 5x + 2$

48. $x > 2x + 3 > 4x - 7$

49. $3 + x > 7x - 2 > 5x - 10$

50. $2 - x < 3x + 1 < 10x$

51. $x \le x + 1 \le 2x + 3$

52. $-x \ge -2x + 1 \ge -3x + 1$

Solve each absolute-value inequality, graph the solution set, and write the answer in interval notation.

53. $|x - 3| < 6$

54. $|x - 2| \ge 4$

55. $|x + 3| > 6$

56. $|x + 2| \le 4$

57. $|2x + 4| \ge 10$

58. $|5x - 2| < 7$

59. $|3x + 5| + 1 \le 9$

60. $|2x - 7| - 3 > 2$

61. $|x + 3| > 0$

62. $|x - 3| \le 0$

63. $\left|\dfrac{5x + 2}{3}\right| < 1$

64. $\left|\dfrac{3x + 2}{4}\right| > 2$

65. $3\left|\dfrac{3x - 1}{2}\right| > 5$

66. $2\left|\dfrac{8x + 2}{5}\right| \le 1$

67. $\dfrac{|x - 1|}{-2} > -3$

68. $\dfrac{|2x - 3|}{-3} < -1$

69. $0 < |2x + 1| < 3$

70. $0 < |2x - 3| < 1$

71. $8 > |3x - 1| > 3$

72. $8 > |4x - 1| > 5$

73. $2 < \left|\dfrac{x - 5}{3}\right| < 4$

74. $3 < \left|\dfrac{x - 3}{2}\right| < 5$

75. $10 > \left|\dfrac{x - 2}{2}\right| > 4$

76. $5 \ge \left|\dfrac{x + 2}{3}\right| > 1$

77. $2 \le \left|\dfrac{x + 1}{3}\right| < 3$

78. $8 > \left|\dfrac{3x + 1}{2}\right| > 2$

Solve each inequality, write the answer in interval notation.

79. $|x + 1| \ge |x|$

80. $|x + 1| < |x + 2|$

81. $|2x + 1| < |2x - 1|$

82. $|3x - 2| \ge |3x + 1|$

83. $|x + 1| < |x|$

84. $|x + 2| \le |x + 1|$

85. $|2x + 1| \ge |2x - 1|$

86. $|3x - 2| < |3x + 1|$

Applications
Solve each problem.

87. **Long distance** A long-distance telephone call costs 36¢ for the first three minutes and 11¢ for each additional minute. How long can a person talk for less than $2?

88. **Buying a computer** A student who can afford to spend up to $2,000 sees the ad shown in the illustration. If she buys a computer, how many games can she buy?

Big Sale!!!!

$1,695.95

Games
$19.95

89. **Buying albums** Andy can spend up to $275 on an iPod and some albums. If he can buy an iPod for $150 and download albums for $9.75, what is the greatest number of albums that he can buy?

90. **Graduation party** Mary wants to spend less than $600 on her daughter's graduation party. If the graduation gift costs $425 and food costs $7.50 for each person, how many guests can Mary invite to the party?

91. **Buying a refrigerator** A woman who has $1,200 to spend wants to buy a refrigerator. Refer to the following table and write an inequality that shows how much she can pay for the refrigerator.

State sales tax	6.5%
City sales tax	0.25%

92. **Renting a rototiller** The cost of renting a rototiller is $17.50 for the first hour and $8.95 for each additional hour. How long can a person have the rototiller if the cost must be less than $75?

© Istockphoto.com/
Keith Webber Jr.

93. Real-estate taxes A city council has proposed the following two methods of taxing real estate:

Method 1	$2,200 + 4% of assessed value
Method 2	$1,200 + 6% of assessed value

For what range of assessments a would the first method benefit the taxpayer?

94. Medical plans A college provides its employees with a choice of the two medical plans shown in the following table. For what size hospital bills is Plan 2 better for the employee than Plan 1? (*Hint:* The cost to the employee includes both the deductible payment and the employee's coinsurance payment.)

Plan 1	Plan 2
Employee pays $100	Employee pays $200
Plan pays 70% of the rest	Plan pays 80% of the rest

95. Medical plans To save costs, the college in Exercise 94 raised the employee deductible, as shown in the following table. For what size hospital bills is Plan 2 better for the employee than Plan 1? (*Hint:* The cost to the employee includes both the deductible payment and the employee's coinsurance payment.)

Plan 1	Plan 2
Employee pays $200	Employee pays $400
Plan pays 70% of the rest	Plan pays 80% of the rest

96. Geometry The perimeter of a rectangle is to be between 180 inches and 200 inches. Find the range of values for its length l when its width is 40 inches.

97. Geometry The perimeter of an equilateral triangle is to be between 50 centimeters and 60 centimeters. Find the range of lengths of one side s.

98. Geometry The perimeter of a square is to be from 25 meters to 60 meters. Find the range of values for its area.

99. Temperature ranges The temperatures on a summer day in Maui satisfy the inequality $|t - 78°| \le 8°$, where t is the temperature in degrees Fahrenheit. Express this range without using absolute-value symbols.

100. Operating temperatures A car CD player has an operating temperature of $|t - 40°| < 80°$, where t is the temperature in degrees Fahrenheit. Express this range without using absolute-value symbols.

101. Range of camber angles The specifications for a certain car state that the camber angle c of its wheels should be $0.6° \pm 0.5°$. Express this range with an inequality containing an absolute value.

102. Tolerance of a sheet of steel A sheet of steel is to be 0.25 inch thick, with a tolerance of 0.015 inch. Express this specification with an inequality containing an absolute value.

103. Humidity level A Steinway piano should be placed in an environment where the relative humidity h is between 38% and 72%. Express this range with an inequality containing an absolute value.

© Istockphoto.com/ilexx

104. Light bulbs A light bulb is expected to last h hours, where $|h - 1,500| \le 200$. Express this range without using absolute-value symbols.

105. Error analysis In a lab, students measured the percentage of copper p in a sample of copper sulfate. The students know that copper sulfate is actually 25.46% copper by mass. They are to compare their results to the actual value and find the amount of *experimental error*.

a. Which measurements shown in the illustration satisfy the absolute-value inequality $|p - 25.46| \le 1.00$?

b. What can be said about the amount of error for each of the trials from part (a)?

Lab 4	Section A
Title: "Percent copper (CU) in copper sulfate (CuSO$_4$·5H$_2$O)"	
Results	
	% Copper
Trial #1:	22.91%
Trial #2:	26.45%
Trial #3	26.49%
Trial #4:	24.76%

106. Error analysis See Exercise 105.

a. Which measurements satisfy the absolute-value inequality $|p - 25.46| > 1.00$?

b. What can be said about the amount of error for each of the trials from part (a)?

Discovery and Writing

107. Express the relationship $20 < l < 30$ in terms of P, where $P = 2l + 2w$.

108. Express the relationship $10 < C < 20$ in terms of F, where $F = \frac{9}{5}C + 32$.

109. If we multiply or divide an inequality by a negative number, then we must change the direction of the inequality symbol. Explain why.

110. Explain why the relation ≥ is transitive.

111. Explain the use of parentheses and brackets when graphing inequalities.

112. If $k > 0$, explain the differences between the solution sets of $|x| < k$ and $|x| > k$.

R.6 The Rectangular Coordinate System

In this section, we will learn to

1. Plot points in the rectangular coordinate system.
2. Graph linear equations.
3. Graph horizontal and vertical lines.
4. Solve applications using linear equations.
5. Find the distance between two points.
6. Find the midpoint of a line segment.

We often say that a picture is worth a thousand words. In fact, pictures and graphs are an effective way to present information. For this reason, they appear frequently in newspapers and magazines. For example, the graph shown in Figure R-23 provides a visual representation of the manatee population in Florida after the year 2003.

From the graph we can note many facts about manatees. Among them are:

- The number of manatees in Florida declined from 2005 to 2007.
- In 2005, the population was about 3,100 animals.
- In this time period, the lowest population of manatees occurred in 2004.

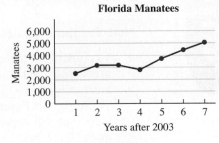

Florida Manatees

FIGURE R-23

In mathematics, graphs are also an effective way to present information. In this chapter, we will draw graphs of equations containing two variables and then discuss the information that we can derive from graphs.

The solutions of an equation with variables x and y, such as $y = -\frac{1}{2}x + 4$, are ordered pairs of real numbers (x, y) that satisfy the equation. To find some ordered pairs that satisfy the equation, we substitute **input values** of x into the equation and find the corresponding **output values** of y. For example, if we substitute 2 for x, we obtain

$$y = -\frac{1}{2}x + 4$$

$$y = -\frac{1}{2}(2) + 4 \qquad \text{Substitute 2 in for } x.$$

$$= -1 + 4$$

$$= 3$$

Since $y = 3$ when $x = 2$, the ordered pair $(2, 3)$ is a solution of the equation. The first coordinate, 2, of the ordered pair is usually called the **x-coordinate**. The second coordinate, 3, is usually called the **y-coordinate**. The solution $(2, 3)$ and several other solutions are listed in the table of values shown in Figure R-24.

Comment

To complete the table of solutions, we first pick values for x. Next we compute each y value. Then we write each solution as an ordered pair. Note that we choose x values that are multiples of the denominator, 2. This makes the computations easier when multiplying the x value by $-\frac{1}{2}$ to find the corresponding y value.

$y = -\dfrac{1}{2}x + 4$		
x	y	(x, y)
-4	6	$(-4, 6)$
-2	5	$(-2, 5)$
0	4	$(0, 4)$
2	3	$(2, 3)$
4	2	$(4, 2)$

FIGURE R-24

ACCENT ON TECHNOLOGY   Generating Table Values

© Texas Instruments images used with permission

If an equation in x and y is solved for y, we can use a graphing calculator to generate a table of solutions.

The instructions in this discussion are for a TI-84 Plus graphing calculator. For details about other brands, please consult the owner's manual.

To construct a table of solutions for $x + 2y = 8$, we first solve the equation for y.

$$x + 2y = 8$$

$$2y = -x + 8 \qquad \text{Subtract } x \text{ from both sides.}$$

$$y = -\frac{1}{2}x + 4 \qquad \text{Divide both sides by 2 and simplify.}$$

Note that this is the equation shown in Figure R-24.

- To construct a table of values for $y = -\frac{1}{2}x + 4$, we first enter the equation. We press $\boxed{\text{Y=}}$ and enter $-(1/2)x + 4$, as shown in Figure R-25(a).
- Next, we press $\boxed{\text{2nd}}$ $\boxed{\text{WINDOW}}$ and enter one value for x on the line labeled TblStart=. In Figure R-25(a), -4 has been entered on this line. Other values for x that will appear in the table are determined by setting an **increment value** on the line labeled $\triangle$Tbl =. In Figure R-25(b), an increment value of 2 has been entered. This means that each x value in the table will be 2 units larger than the previous one.
- Finally, we press $\boxed{\text{2nd}}$ $\boxed{\text{GRAPH}}$ to obtain the table of values shown in Figure R-25(c). This table contains all of the solutions listed in Figure R-24, plus the two additional solutions $(6, 1)$ and $(8, 0)$. To see other values, we simply scroll up and down the screen by pressing the up and down arrow keys.

(a)

(b)

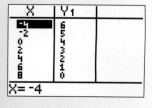

(c)

FIGURE R-25

Before we can present the table of solutions shown in Figure R-25 in graphical form, we need to discuss the rectangular coordinate system.

1. Plot Points in the Rectangular Coordinate System

The **rectangular coordinate system** consists of two perpendicular number lines that divide the plane into four **quadrants,** numbered as shown in Figure R-26. The horizontal number line is called the **x-axis,** and the vertical number line is called the **y-axis.** These axes intersect at a point called the **origin,** which is the 0 point on each axis. The positive direction on the x-axis is to the right, the positive direction on the y-axis is upward, and the same unit distance is used on both axes, unless otherwise indicated.

To plot (or graph) the point associated with the ordered pair $x = 2$ and $y = 3$, denoted as $(2, 3)$, we start at the origin, count 2 units to the right, and then count 3 units up (see Figure R-27). Point P (which lies in the first quadrant) is the graph of the ordered pair $(2, 3)$. The ordered pair $(2, 3)$ gives the **coordinates** of point P.

To plot point Q with coordinates $(-4, 6)$, we start at the origin, count 4 units to the left, and then count 6 units up. Point Q lies in the second quadrant. Point R with coordinates $(6, -4)$ lies in the fourth quadrant.

Caution

The ordered pairs $(-4, 6)$ and $(6, -4)$ represent **different** points. $(-4, 6)$ is in the second quadrant and $(6, -4)$ is in the fourth quadrant.

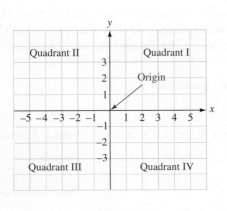

FIGURE R-26

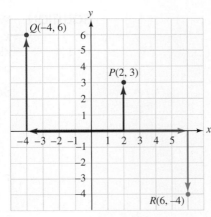

FIGURE R-27

2. Graph Linear Equations

The **graph of the equation** $y = -\frac{1}{2}x + 4$ is the graph of all points (x, y) on the rectangular coordinate system whose coordinates satisfy the equation. To graph $y = -\frac{1}{2}x + 4$, we plot the pairs listed in the table of solutions shown in Figure R-28. These points lie on the line shown in the figure. This line is the graph of the equation.

Comment

When we say that the graph of an equation is a line, we imply two things:

1. Every point with coordinates that satisfy the equation will lie on the line.
2. Every point on the line will have coordinates that satisfy the equation.

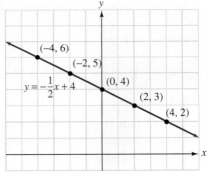

$y = -\dfrac{1}{2}x + 4$		
x	y	(x, y)
-4	6	$(-4, 6)$
-2	5	$(-2, 5)$
0	4	$(0, 4)$
2	3	$(2, 3)$
4	2	$(4, 2)$

FIGURE R-28

When the graph of an equation is a line, we call the equation a **linear equation.** These equations are often written in **standard form** as $Ax + By = C$, where A, B, and C are specific numbers (called **constants**) and x and y are variables. Either A or B can be 0, but A and B cannot both be 0. Here are four examples of linear equations written in standard form:

Linear Equation	Values of A, B, and C
$3x + 2y = 6$	$A = 3, B = 2, C = 6$
$5x - 2y = -10$	$A = 5, B = -2, C = -10$
$2y = 7$	$A = 0, B = 2, C = 7$
$x = -4$	$A = 1, B = 0, C = -4$

EXAMPLE 1   **Graphing a Linear Equation**

Graph: $x + 2y = 5$.

SOLUTION   We will solve the equation for y and form a table of solutions by picking values for x, substituting them into the equation, and solving for the other variable y. We will plot the points represented in the table of solutions and draw a line through the points.

Solve the equation for y.

$$x + 2y = 5$$

$$x - x + 2y = 5 - x \qquad \text{Subtract } x \text{ from both sides.}$$

$$2y = -x + 5 \qquad \text{Simplify.}$$

$$y = -\frac{1}{2}x + \frac{5}{2} \qquad \text{Divide both sides by 2.}$$

Pick values for x and solve for y.

If we pick $x = 0$, we can find y as follows:

$$y = -\frac{1}{2}x + \frac{5}{2}$$

$$y = -\frac{1}{2}(0) + \frac{5}{2} \qquad \text{Substitute 0 for } x.$$

$$y = \frac{5}{2} \qquad \text{Simplify.}$$

The ordered pair $\left(0, \frac{5}{2}\right)$ satisfies the equation.

To find another ordered pair, we pick $x = 1$ and find y.

$$y = -\frac{1}{2}x + \frac{5}{2}$$

$$y = -\frac{1}{2}(1) + \frac{5}{2} \qquad \text{Substitute 1 for } x.$$

$$y = 2 \qquad \text{Simplify.}$$

The ordered pair $(1, 2)$ satisfies the equation.

These pairs and others that satisfy the equation are shown in Figure R-29. We plot the points and join them with a line to get the graph of the equation.

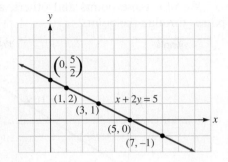

FIGURE R-29

$x + 2y = 5$		
x	y	(x, y)
0	$\dfrac{5}{2}$	$\left(0, \dfrac{5}{2}\right)$
1	2	$(1, 2)$
3	1	$(3, 1)$
5	0	$(5, 0)$
7	-1	$(7, -1)$

Comment

Even though there are infinitely many points that lie on a line, only two are required to graph a line. However, it is a good idea to find a third point as a check.

Self Check 1   Graph: $3x - 2y = 6$.

Now Try Exercise 35.

EXAMPLE 2   **Graphing a Linear Equation**

Graph: $3(y + 2) = 2x - 3$.

SOLUTION   We will solve the equation for y and find ordered pairs (x, y) that satisfy the equation. Then we will plot the points and graph the line.

Solve the equation for y.

$$3(y + 2) = 2x - 3$$

$$3y + 6 = 2x - 3 \qquad \text{Use the Distributive Property to remove parentheses.}$$

$$3y = 2x - 9 \qquad \text{Subtract 6 from both sides.}$$

$$y = \frac{2}{3}x - 3 \qquad \text{Divide both sides by 3.}$$

Pick values for x and solve for y.
We now substitute numbers for x to find the corresponding values of y. If we let $x = 0$ and find y, we get

$$y = \frac{2}{3}x - 3$$

$$y = \frac{2}{3}(0) - 3 \qquad \text{Substitute 0 for } x.$$

$$y = -3 \qquad \text{Simplify.}$$

The point $(0, -3)$ lies on the graph.
If we let $x = 3$, we get

$$y = \frac{2}{3}x - 3$$

$$y = \frac{2}{3}(3) - 3 \qquad \text{Substitute 3 for } x.$$

$$y = 2 - 3 \qquad \text{Simplify.}$$

$$y = -1$$

The point $(3, -1)$ lies on the graph.

We plot these points and others, as in Figure R-30, and draw the line that passes through the points.

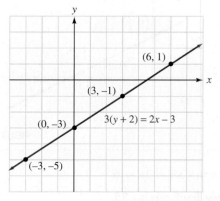

$3(y + 2) = 2x - 3$		
x	y	(x, y)
-3	-5	$(-3, -5)$
0	-3	$(0, -3)$
3	-1	$(3, -1)$
6	1	$(6, 1)$

FIGURE R-30

Self Check 2   Graph: $2(x - 1) = 6 - 8y$.

Now Try Exercise 37.

ACCENT ON TECHNOLOGY   Graphing Equations

We can graph equations with a graphing calculator. To see a graph, we must choose the minimum and maximum values of the x- and y-coordinates that will appear in the calculator's window. A window with standard settings of

$$\text{Xmin} = -10, \qquad \text{Xmax} = 10, \qquad \text{Ymin} = -10, \qquad \text{and} \qquad \text{Ymax} = 10$$

will produce a graph where the value of x is in the interval $[-10, 10]$ and the value of y is in the interval $[-10, 10]$.

- To use a graphing calculator to graph $3x + 2y = 12$, we first solve the equation for y.

$$2y = -3x + 12 \qquad \text{Subtract } 3x \text{ from both sides.}$$

$$y = -\frac{3}{2}x + 6 \qquad \text{Divide both sides by 2.}$$

- Next, we press ⬚ Y= and enter the right side of the equation. The screen is shown in Figure R-31(a).
- We then press ⬚ GRAPH to obtain the graph shown in Figure R-31(b).

(a)

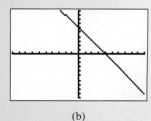

(b)

FIGURE R-31

In Figure R-29, the graph intersects the y-axis at the point $\left(0, \frac{5}{2}\right)$. $\frac{5}{2}$ is called the **y-intercept.** The graph intersects the x-axis at the point $(5, 0)$. 5 is called the **x-intercept.**

Intercepts of a Line

x-intercept: The value(s) of x where the graph of the line intersects or touches the x-axis. To find the x-intercept, substitute 0 in for y in the equation of the line and solve for x.

y-intercept: The value(s) of y where the graph of the line intersects or touches the y-axis. To find the y-intercept, substitute 0 in for x in the equation of the line and solve for y.

EXAMPLE 3   **Graphing a Line by Finding the Intercepts**

Use the x- and y-intercepts to graph the equation $3x + 2y = 12$.

SOLUTION   To find the y-intercept, we substitute 0 for x and solve for y. To find the x-intercept, we substitute 0 for y and solve for x. We will also find a third point as a check and then plot the points and draw the graph.

Find the y-intercept.
To find the y-intercept, we substitute 0 for x and solve for y.

$$3x + 2y = 12$$
$$3(0) + 2y = 12 \qquad \text{Substitute 0 for } x.$$
$$2y = 12 \qquad \text{Simplify.}$$
$$y = 6 \qquad \text{Divide both sides by 2.}$$

The y-intercept is 6. The point $(0, 6)$ lies on the graph of the line.

Find the x-intercept.
To find the x-intercept, we substitute 0 for y and solve for x.

$$3x + 2y = 12$$
$$3x + 2(0) = 12 \qquad \text{Substitute 0 for } y.$$
$$3x = 12 \qquad \text{Simplify.}$$
$$x = 4 \qquad \text{Divide both sides by 3.}$$

The x-intercept is 4. The point $(4, 0)$ lies on the graph of the line.

Find a third point as a check.
If we let $x = 2$, we will find that $y = 3$.

$$3x + 2y = 12$$
$$3(2) + 2y = 12 \qquad \text{Substitute 2 for } x.$$
$$6 + 2y = 12 \qquad \text{Simplify.}$$
$$2y = 6 \qquad \text{Subtract 6 from both sides.}$$
$$y = 3 \qquad \text{Divide both sides by 2.}$$

The point $(2, 3)$ satisfies the equation.

We plot each pair (as in Figure R-32) and join them with a line to get the graph of the equation.

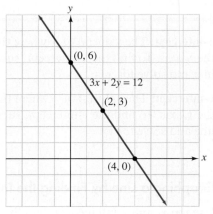

$3x + 2y = 12$		
x	y	(x, y)
0	6	$(0, 6)$
2	3	$(2, 3)$
4	0	$(4, 0)$

FIGURE R-32

Self Check 3   Graph: $2x - 3y = 12$.

Now Try Exercise 43.

ACCENT ON TECHNOLOGY   **Finding Intercepts**

We can use the trace feature on a graphing calculator to find the approximate coordinates of any point on a graph. When we press [TRACE], a flashing cursor will appear on the screen. The coordinates of the cursor will also appear at the bottom of the screen.

- To find the y-intercept of the graph of $2y = -5x - 7$ (or $y = -\frac{5}{2}x - \frac{7}{2}$), we graph the equation, using $[-10, 10]$ for x and $[-10, 10]$ for y, and press [TRACE] to get Figure R-33(a). We see from the figure that the y-intercept is -3.5.
- We can approximate the x-intercept by using the left arrow key and moving the cursor toward the x-intercept until we arrive at a point with the coordinates shown in Figure R-33(b). We see from the graph that the x-intercept is approximately -1.489362.
- To get better results, we can zoom in to get a magnified picture, trace again, and move the cursor to the point with coordinates shown in Figure R-33(c). Since the y-coordinate is almost 0, we now have a good approximation for the x-intercept. The x-intercept is approximately -1.382979. We can achieve better results with repeated zooms.

Comment

In Chapter 1, we will use another graphing calculator feature, the zero feature, to find x-intercepts.

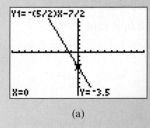

(a)

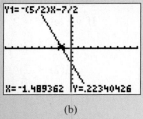

(b)

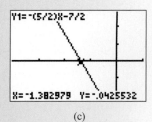

(c)

FIGURE R-33

3. Graph Horizontal and Vertical Lines

In the next example, we will graph a horizontal and a vertical line.

EXAMPLE 4   Graphing Horizontal and Vertical Lines

Graph: **a.** $y = 2$ **b.** $x = -3$

SOLUTION   In each case, we will plot a few ordered pairs that satisfy the equation and then draw the graph of the line.

a. In the equation $y = 2$, the value of y is always 2. Any value can be used for x. If we pick x values of $-3, 0, 2,$ and 4, we get the ordered pairs $(-3, 2), (0, 2), (2, 2),$ and $(4, 2)$. Plotting the pairs shown in Figure R-34, we see that the graph is a horizontal line, parallel to the x-axis and having a y-intercept of 2. The line has no x-intercept.

b. In the equation $x = -3$, the value of x is always -3. Any value can be used for y. If we pick y values of $-2, 0, 2,$ and 3, we get the ordered pairs $(-3, -2), (-3, 0), (-3, 2),$ and $(-3, 3)$. After plotting the pairs shown in Figure R-34, we see that the graph is a vertical line, parallel to the y-axis and having an x-intercept of -3. The line has no y-intercept.

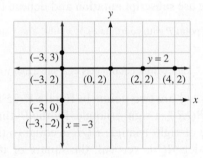

$y = 2$		
x	y	(x, y)
-3	2	$(-3, 2)$
0	2	$(0, 2)$
2	2	$(2, 2)$
4	2	$(4, 2)$

$x = -3$		
x	y	(x, y)
-3	-2	$(-3, -2)$
-3	0	$(-3, 0)$
-3	2	$(-3, 2)$
-3	3	$(-3, 3)$

FIGURE R-34

Self Check 4   Graph: **a.** $x = 2$ **b.** $y = -3$

Now Try Exercise 45.

Example 4 suggests the following facts:

Equations of Vertical and Horizontal Lines

If a and b are real numbers, then:

1. The graph of the equation $x = a$ is a vertical line with an x-intercept of a. If $a = 0$, the line $x = 0$ is the y-axis.
2. The graph of the equation $y = b$ is a horizontal line with a y-intercept of b. If $b = 0$, the line $y = 0$ is the x-axis.

4. Solve Applications Using Linear Equations

EXAMPLE 5   Solving an Application Problem

A computer purchased for \$2,750 is expected to depreciate according to the formula $y = -550x + \$2,750$, where y is the value of the computer after x years. When will the computer be worth nothing?

SOLUTION   The computer will be worth nothing when its value y is 0. To find x when $y = 0$, we substitute 0 for y and solve for x.

$$y = -550x + 2{,}750$$

$$0 = -550x + 2{,}750$$

$$-2{,}750 = -550x \qquad \text{Subtract 2,750 from both sides.}$$

$$5 = x \qquad \text{Divide both sides by } -550.$$

The computer will have no value in 5 years.

Self Check 5   When will the value of the computer be $1,650?

Now Try Exercise 91.

5. Find the Distance between Two Points

To derive the formula used to find the distance between two points in the rectangular coordinate system, we use **subscript notation** and denote the points as

$$P(x_1, y_1) \qquad \text{Read as "point } P \text{ with coordinates of } x \text{ sub 1 and } y \text{ sub 1."}$$

and

$$Q(x_2, y_2). \qquad \text{Read as "point } Q \text{ with coordinates of } x \text{ sub 2 and } y \text{ sub 2."}$$

If $P(x_1, y_1)$ and $Q(x_2, y_2)$ are two points in Figure R-35 and point R has coordinates (x_2, y_1), triangle PQR is a right triangle. By the Pythagorean Theorem, the square of the hypotenuse of right triangle PQR is equal to the sum of the squares of the two legs. Because leg RQ is vertical, the square of its length is $(y_2 - y_1)^2$. Since leg PR is horizontal, the square of its length is $(x_2 - x_1)^2$. Thus we have

$$(1) \quad d^2 = (x_2 - x_1)^2 + (y_2 - y_1)^2.$$

Because equal positive numbers have equal positive square roots, we can take the positive square root of both sides of Equation (1) to obtain the **distance formula.**

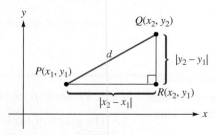

FIGURE R-35

Distance Formula   The distance d between points (x_1, y_1) and (x_2, y_2) is given by

$$d = \sqrt{(x_2 - x_1)^2 + (y_2 - y_1)^2}.$$

EXAMPLE 6   **Finding the Distance between Two Points**

Find the distance between $P(-1, -2)$ and $Q(-7, 8)$.

SOLUTION   We use the distance formula, $d = \sqrt{(x_2 - x_1)^2 + (y_2 - y_1)^2}$, to find the distance between $P(-1, -2)$ and $Q(-7, 8)$.

If we let $P(-1, -2) = P(x_1, y_1)$ and $Q(-7, 8) = Q(x_2, y_2)$, we can substitute -1 for x_1, -2 for y_1, -7 for x_2, and 8 for y_2 into the formula and simplify.

$$d(PQ) = \sqrt{(x_2 - x_1)^2 + (y_2 - y_1)^2}$$ Read $d(PQ)$ as "the length of segment PQ."

$$d(PQ) = \sqrt{[-7 - (-1)]^2 + [8 - (-2)]^2}$$

$$= \sqrt{(-6)^2 + (10)^2}$$

$$= \sqrt{36 + 100}$$

$$= \sqrt{136}$$

$$= \sqrt{4 \cdot 34}$$

$$= 2\sqrt{34}$$ $\sqrt{4 \cdot 34} = \sqrt{4}\sqrt{34} = 2\sqrt{34}$

■

Self Check 6   Find the distance between $P(-2, -5)$ and $Q(3, 7)$.

Now Try Exercise 67.

6. Find the Midpoint of a Line Segment

If point M in Figure R-36 lies midway between points $P(x_1, y_1)$ and $Q(x_2, y_2)$, point M is called the **midpoint** of segment PQ. To find the coordinates of M, we find the average of the x-coordinates and the average of the y-coordinates of P and Q.

Midpoint Formula   The midpoint of the line segment with endpoints at $P(x_1, y_1)$ and $Q(x_2, y_2)$ is the point M with coordinates

$$M = \left(\frac{x_1 + x_2}{2}, \frac{y_1 + y_2}{2} \right).$$

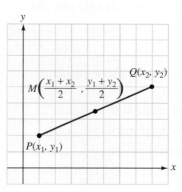

FIGURE R-36

We can prove this formula by using the distance formula to show that $d(PM) + d(MQ) = d(PQ)$.

EXAMPLE 7   **Finding the Midpoint of a Line Segment Joining Two Points**

Find the midpoint of the segment joining $P(-7, 2)$ and $Q(1, -4)$.

SOLUTION We use the midpoint formula, $M = \left(\dfrac{x_1 + x_2}{2}, \dfrac{y_1 + y_2}{2} \right)$, to find the midpoint of the line segment joining $P(-7, 2)$ and $Q(1, -4)$. To do so, we substitute $P(-7, 2)$ for $P(x_1, y_1)$ and $Q(1, -4)$ for $Q(x_2, y_2)$ into the midpoint formula to get

$$x_M = \frac{x_1 + x_2}{2} \quad \text{and} \quad y_M = \frac{y_1 + y_2}{2}$$

$$= \frac{-7 + 1}{2} \qquad\qquad = \frac{2 + (-4)}{2}$$

$$= \frac{-6}{2} \qquad\qquad\quad = \frac{-2}{2}$$

$$= -3 \qquad\qquad\quad = -1$$

The midpoint is $M(-3, -1)$.

Self Check 7 Find the midpoint of the segment joining $P(-7, -8)$ and $Q(-2, 10)$.

Now Try Exercise 75.

Self Check Answers

1.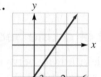

2.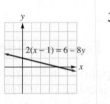

3.

4.

5. 2 years   6. 13   7. $M\left(-\dfrac{9}{2}, 1\right)$

Exercises **R.6**

Getting Ready

You should be able to complete these vocabulary and concept statements before you proceed to the practice exercises.

Fill in the blanks.

1. The coordinate axes divide the plane into four _____ .

2. The coordinate axes intersect at the _____ .

3. The positive direction on the x-axis is _____ .

4. The positive direction on the y-axis is _____ .

5. The x-coordinate is the _____ coordinate in an ordered pair.

6. The y-coordinate is the _____ coordinate in an ordered pair.

7. A _____ equation is an equation whose graph is a line.

8. The y value on the point where a line intersects the _____ is called the y-intercept.

9. The x value on the point where a line intersects the x-axis is called the _____ .

10. The graph of the equation $x = a$ will be a _____ line.

11. The graph of the equation $y = b$ will be a _____ line.

12. Complete the distance formula: $d = $ _____ .

13. If a point divides a segment into two equal segments, the point is called the _____ of the segment.

14. The midpoint of the segment joining $P(x_1, y_1)$ and $Q(x_2, y_2)$ is _____ .

Practice

Refer to the illustration and determine the coordinates of each point.

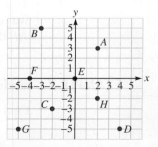

15. A **16.** B
17. C **18.** D
19. E **20.** F
21. G **22.** H

Indicate the quadrant in which the point lies or the axis on which it lies.

23. $(2, 5)$ **24.** $(-3, 4)$
25. $(-4, -5)$ **26.** $(6, 2)$
27. $(5, 2)$ **28.** $(3, -4)$
29. $(4, 0)$ **30.** $(0, 2)$

Solve each equation for y and graph the equation. Then check your graph with a graphing calculator.

31. $y - 2x = 7$ **32.** $y + 3 = -4x$
33. $y + 5x = 5$ **34.** $y - 3x = 6$
35. $6x - 3y = 10$ **36.** $4x + 8y - 1 = 0$
37. $2(x + y + 1) = x + 2$ **38.** $5(x + 2) = 3y - x$

Use the x- and y-intercepts to graph each equation.

39. $x + y = 5$ **40.** $x - y = 3$
41. $2x - y = 4$ **42.** $3x + y = 9$
43. $4x - 5y = 20$ **44.** $3x - 5y = 15$

Graph each equation.

45. $3x + 5 = -1$ **46.** $7y - 1 = 6$
47. $3(y + 2) = y$ **48.** $4 + 3y = 3(x + y)$
49. $3(y + 2x) = 6x + y$ **50.** $5(y - x) = x + 5y$

Use a graphing calculator to find the x-intercept to the nearest hundredth.

51. $y = 3.7x - 4.5$ **52.** $y = \dfrac{3}{5}x + \dfrac{5}{4}$
53. $1.5x - 3y = 7$ **54.** $0.3x + y = 7.5$

Find the distance between P and the origin.

55. $P(4, -3)$ **56.** $P(-5, 12)$
57. $P(-3, 2)$ **58.** $P(5, 0)$
59. $P(1, 1)$ **60.** $P(6, -8)$
61. $P(\sqrt{3}, 1)$ **62.** $P(\sqrt{7}, \sqrt{2})$

Find the distance between P and Q.

63. $P(3, 7); Q(6, 3)$ **64.** $P(4, 9); Q(9, 21)$
65. $P(4, -6); Q(-1, 6)$ **66.** $P(0, 5); Q(6, -3)$
67. $P(-2, -15); Q(-9, -39)$ **68.** $P(-7, 11); Q(3, -13)$
69. $P(3, -3); Q(-5, 5)$ **70.** $P(6, -3); Q(-3, 2)$
71. $P(\pi, -2); Q(\pi, 5)$ **72.** $P(\sqrt{5}, 0); Q(0, 2)$

Find the midpoint of the line segment PQ.

73. $P(2, 4); Q(6, 8)$ **74.** $P(3, -6); Q(-1, -6)$
75. $P(2, -5); Q(-2, 7)$ **76.** $P(0, 3); Q(-10, -13)$
77. $P(-8, 5); Q(8, -5)$ **78.** $P(3, -2); Q(2, -3)$
79. $P(0, 0); Q(\sqrt{5}, \sqrt{5})$ **80.** $P(\sqrt{3}, 0); Q(0, -\sqrt{5})$

One endpoint P and the midpoint M of line segment PQ are given. Find the coordinates of the other endpoint, Q.

81. $P(1, 4); M(3, 5)$ **82.** $P(2, -7); M(-5, 6)$
83. $P(5, -5); M(5, 5)$ **84.** $P(-7, 3); M(0, 0)$

Use the theory from this section and formulas from geometry to solve each problem.

85. Show that a triangle with vertices at $(13, -2)$, $(9, -8)$, and $(5, -2)$ is isosceles.

86. Show that a triangle with vertices at $(-1, 2)$, $(3, 1)$, and $(4, 5)$ is isosceles.

87. In the illustration, points M and N are the midpoints of AC and BC, respectively. Find the length of MN.

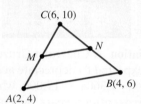

88. In the illustration, points M and N are the midpoints of AC and BC, respectively. Show that $d(MN) = \frac{1}{2}[d(AB)]$.

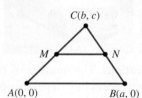

89. In the illustration, point M is the midpoint of the hypotenuse of right triangle AOB. Show that the area of rectangle $OLMN$ is one-half of the area of triangle AOB.

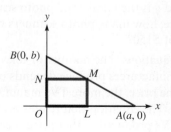

90. Rectangle $ABCD$ in the illustration is twice as long as it is wide, and its sides are parallel to the coordinate axes. If the perimeter is 42, find the coordinates of point C.

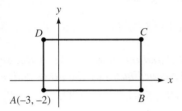

Applications

91. Condo appreciation A condo purchased for $225,000 is expected to appreciate according to the formula $y = 17,500x + 225,000$, where y is the value of the condo after x years. Find the value of the condo after 5 years.

92. Car depreciation A Chevrolet Cruze purchased for $17,000 is expected to depreciate according to the formula $y = -1,360x + 17,000$, where y is the value of the car after x years. When will the car be worthless?

93. Demand equations The number of photo scanners that consumers buy depends on the price. The higher the price, the fewer photo scanners people will buy. The equation that relates price to the number of photo scanners sold at that price is called a **demand equation.** If the demand equation for a photo scanner is $p = -\frac{1}{10}q + 170$, where p is the price and q is the number of photo scanners sold at that price, how many photo scanners will be sold at a price of $150?

94. Supply equations The number of television sets that manufacturers produce depends on price. The higher the price, the more TVs manufacturers will produce. The equation that relates price to the number of TVs produced at that price is called a **supply equation.** If the supply equation for a 25-inch TV is $p = \frac{1}{10}q + 130$, where p is the price and q is the number of TVs produced for sale at that price, how many TVs will be produced if the price is $150?

95. Meshing gears The rotational speed V of a large gear (with N teeth) is related to the speed v of a smaller gear (with n teeth) by the equation $V = \dfrac{nv}{N}$.

If the larger gear in the illustration is making 60 revolutions per minute, how fast is the smaller gear spinning?

96. Crime prevention The number n of incidents of family violence requiring police response appears to be related to d, the money spent on crisis intervention, by the equation $n = 430 - 0.005d$. What expenditure would reduce the number of incidents to 350?

97. Navigation See the illustration. An ocean liner is located 23 miles east and 72 miles north of Pigeon Cove Lighthouse, and its home port is 47 miles west and 84 miles south of the lighthouse. How far is the ship from port?

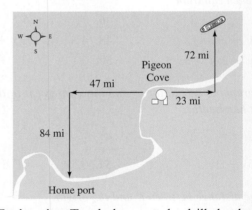

98. Engineering Two holes are to be drilled at locations specified by the engineering diagram shown in the illustration. Find the distance between the centers of the holes.

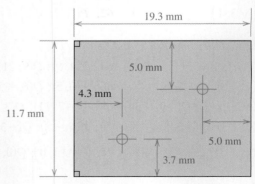

Discovery and Writing

99. Explain how to graph a line using the intercept method.

100. Explain how to determine the quadrant in which the point $P(a, b)$ lies.

101. In Figure R-36, show that $d(PM) + d(MQ) = d(PQ)$.

102. Use the result of Exercise 101 to explain why point M is the midpoint of segment PQ.

R.7 Equations of Lines

In this section, we will learn to

1. Find the slope of a line and use slope to solve applications.
2. Find slopes of horizontal and vertical lines.
3. Find slopes of parallel and perpendicular lines.
4. Write equations of lines.
5. Graph linear equations using the slope and y-intercept.
6. Write equations of parallel and perpendicular lines.
7. Write an equation of a line that models a real-life problem.
8. Use linear curve fitting to model a real-life problem.

© dk/Alamy

The world's steepest passenger railway is the Lookout Mountain Incline Railway in Chattanooga, Tennessee. Passengers experience breathtaking views of the city and surrounding mountains as the trolley-style railcars travel up Lookout Mountain. The grade, or steepness, of the track is 72.7% near the top, giving it the special distinction of being the steepest in the world.

Mathematicians use the term **slope** to represent the measure of the steepness of a line. We will begin this section by exploring the topic of slope. It has many real-life applications.

1. Find the Slope of a Line and Use Slope to Solve Applications

Suppose that a college student rents a room for $300 per month, plus a $200 nonrefundable deposit. The table shown in Figure R-37(a) gives the cost (y) for different numbers of months (x). If we construct a graph from these data, we get the line shown in Figure R-37(b).

Number of Months	Total Cost
x	y
0	200
1	500
2	800
3	1,100
4	1,400

(a)

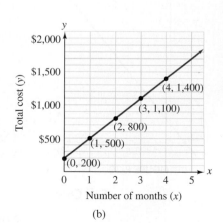

(b)

FIGURE R-37

From the graph, we can see that if x changes from 0 to 1, y changes from 200 to 500. As x changes from 1 to 2, y changes from 500 to 800, and so on. The ratio of the change in y to the change in x is the constant 300.

$$\frac{\text{Change in } y}{\text{Change in } x} = \frac{500 - 200}{1 - 0} = \frac{800 - 500}{2 - 1} = \frac{1{,}100 - 800}{3 - 2} = \frac{1{,}400 - 1{,}100}{4 - 3} = \frac{300}{1} = 300$$

The ratio of the change in y divided by the change in x between any two points on any line is always a constant. This constant rate of change is called the **slope** of the line.

Slope of a Nonvertical Line

The **slope of the nonvertical line** (see Figure R-38) passing through points $P(x_1, y_1)$ and $Q(x_2, y_2)$ is

$$m = \frac{\text{change in } y}{\text{change in } x} = \frac{y_2 - y_1}{x_2 - x_1} \qquad (x_2 \neq x_1).$$

Comment

Slope is often considered to be a measure of the steepness or tilt of a line. Note that we can use the coordinates of any two points on a line to compute the slope of the line.

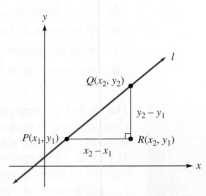

FIGURE R-38

EXAMPLE 1   **Finding the Slope of a Line, Given Two Points**

Find the slope of the line passing through $P(-1, -2)$ and $Q(7, 8)$. (See Figure R-39.)

SOLUTION   We will substitute the points $P(-1, -2)$ and $Q(7, 8)$ into the slope formula, $m = \dfrac{\text{change in } y}{\text{change in } x} = \dfrac{y_2 - y_1}{x_2 - x_1}$, to find the slope of the line.

Let $P(x_1, y_1) = P(-1, -2)$ and $Q(x_2, y_2) = Q(7, 8)$. Then we substitute -1 for x_1, -2 for y_1, 7 for x_2, and 8 for y_2 to get

$$m = \frac{\text{change in } y}{\text{change in } x}$$

$$m = \frac{y_2 - y_1}{x_2 - x_1} = \frac{8 - (-2)}{7 - (-1)} = \frac{10}{8} = \frac{5}{4}$$

The slope of the line is $\frac{5}{4}$. We would have obtained the same result if we had let $P(x_1, y_1) = P(7, 8)$ and $Q(x_2, y_2) = Q(-1, -2)$.

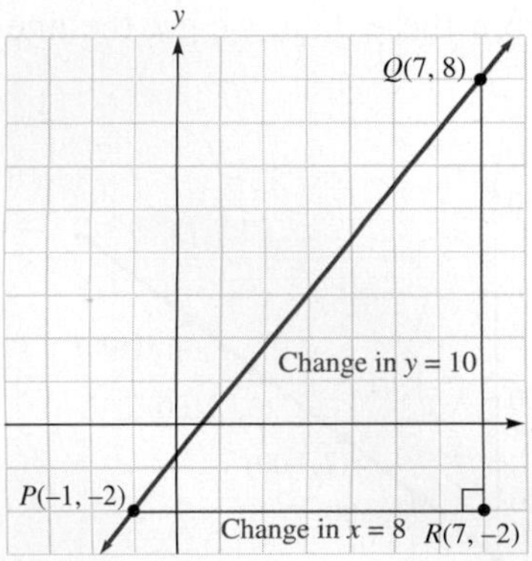

FIGURE R-39

Self Check 1   Find the slope of the line passing through $P(-3, -4)$ and $Q(5, 9)$.

Now Try Exercise 13.

Caution

When calculating slope, always subtract the y values and the x values in the same order:

$$m = \frac{y_2 - y_1}{x_2 - x_1} \quad \text{or} \quad m = \frac{y_1 - y_2}{x_1 - x_2}.$$

Otherwise, you will obtain an *incorrect* result.

A slope can be a positive real number, 0, or a negative real number. If the denominator of the slope formula is 0, slope is not defined.

The change in y (often denoted as Δy) is the **rise** of the line between points P and Q. The change in x (often denoted as Δx) is the **run.** Using this terminology, we can define slope to be the ratio of the rise to the run:

$$m = \frac{y_2 - y_1}{x_2 - x_1} = \frac{\Delta y}{\Delta x} = \frac{\text{rise}}{\text{run}} \qquad (\Delta x \neq 0).$$

EXAMPLE 2   **Finding the Slope of a Line, Given an Equation or Graph of the Line**

Find the slope of the line determined by $5x + 2y = 10$ (see Figure R-40).

SOLUTION   First we will find the x- and y-intercepts, and then we will use the slope formula,

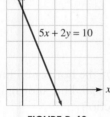

FIGURE R-40

$m = \dfrac{\text{change in } y}{\text{change in } x} = \dfrac{y_2 - y_1}{x_2 - x_1}$, to find the slope of the line.

• If $y = 0$, then $x = 2$ and the point $(2, 0)$ lies on the line.
• If $x = 0$, then $y = 5$ and the point $(0, 5)$ lies on the line.

We then find the slope of the line between $P(2, 0)$ and $Q(0, 5)$.

$$m = \frac{\text{change in } y}{\text{change in } x}$$

$$m = \frac{y_2 - y_1}{x_2 - x_1} = \frac{5 - 0}{0 - 2} = -\frac{5}{2}$$

The slope is $-\frac{5}{2}$.   ■

Self Check 2   Find the slope of the line determined by $3x - 2y = 9$.

Now Try Exercise 23.

EXAMPLE 3   **Solving an Application Problem**

It takes a skier 25 minutes to complete the course shown in Figure R-41. Find the skier's average rate of descent in feet per minute.

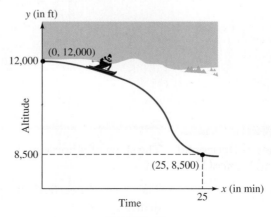

FIGURE R-41

SOLUTION To find the average rate of descent, we will find the ratio of the change in altitude a to the change in time t. To find this ratio, we will calculate the slope of the line passing through the points (0, 12,000) and (25, 8,500).

$$\begin{aligned} \text{Average rate} \atop \text{of descent} &= \frac{12,000 - 8,500}{0 - 25} \\ &= \frac{3,500}{-25} \\ &= -140 \end{aligned}$$

The average rate of descent is $\frac{\Delta a}{\Delta t} = -140$, so the rate of descent is 140 ft/min. ■

Self Check 3 If it takes the skier 20 minutes to complete the course, find the average rate of descent in feet per minute.

Now Try Exercise 105.

2. Find Slopes of Horizontal and Vertical Lines

If $P(x_1, y_1)$ and $Q(x_2, y_2)$ are points on the horizontal line shown in Figure R-42(a), then $y_1 = y_2$, and the numerator of the fraction

$$\frac{y_2 - y_1}{x_2 - x_1} \qquad \text{On a horizontal line, } x_2 \neq x_1.$$

is 0. Thus the value of the fraction is 0, and the slope of the horizontal line is 0.

If $P(x_1, y_1)$ and $Q(x_2, y_2)$ are points on the vertical line shown in Figure R-42(b), then $x_1 = x_2$, and the denominator of the fraction

$$\frac{y_2 - y_1}{x_2 - x_1} \qquad \text{On a vertical line, } y_2 \neq y_1.$$

is 0. Since the denominator of a fraction cannot be 0, the slope of a vertical line is not defined.

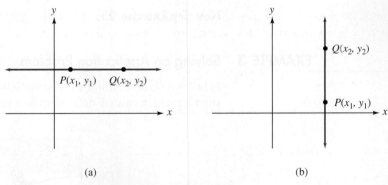

(a)   (b)

FIGURE R-42

Slopes of Horizontal and Vertical Lines

The slope of a horizontal line (a line with an equation of the form $y = b$) is 0.

The slope of a vertical line (a line with an equation of the form $x = a$) is not defined.

Here are a few facts about slope (see Figure R-43):

- If a line rises as we follow it from left to right, as in Figure R-43(a), its slope is positive.
- If a line drops as we follow it from left to right, as in Figure R-43(b), its slope is negative.
- If a line is horizontal, as in Figure R-43(c), its slope is 0.
- If a line is vertical, as in Figure R-43(d), it has no defined slope.

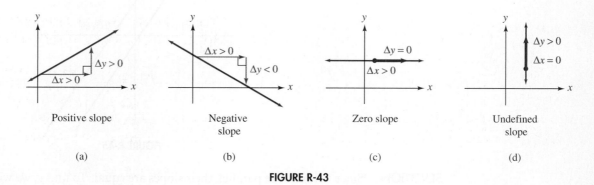

| Positive slope | Negative slope | Zero slope | Undefined slope |
| (a) | (b) | (c) | (d) |

FIGURE R-43

3. Find Slopes of Parallel and Perpendicular Lines

To see a relationship between parallel lines and their slopes, we refer to the parallel lines l_1 and l_2 shown in Figure R-44, with slopes of m_1 and m_2, respectively. Because right triangles ABC and DEF are similar, it follows that

$$m_1 = \frac{\Delta y \text{ of } l_1}{\Delta x \text{ of } l_1} = \frac{\Delta y \text{ of } l_2}{\Delta x \text{ of } l_2} = m_2.$$

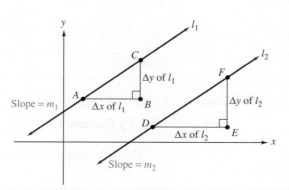

FIGURE R-44

This shows that if two nonvertical lines are parallel, they have the same slope. It is also true that when two lines have the same slope, they are parallel.

Slopes of Parallel Lines   Nonvertical parallel lines have the same slope, and lines having the same slope are parallel.

Since vertical lines are parallel, lines with undefined slopes are parallel.

EXAMPLE 4   **Solving a Slope Problem That Involves Parallel Lines**

The lines in Figure R-45 are parallel. Find y.

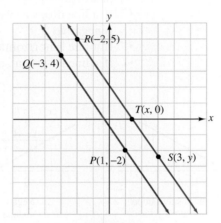

FIGURE R-45

SOLUTION   Since the lines are parallel, their slopes are equal. To find y, we will find the slope of each line, set them equal, and solve the resulting equation.

Slope of PQ = Slope of RS

$$\frac{-2-4}{1-(-3)} = \frac{y-5}{3-(-2)}$$

$$\frac{-6}{4} = \frac{y-5}{5} \qquad \text{Simplify.}$$

$$-30 = 4(y-5) \qquad \text{Multiply both sides by 20.}$$

$$-30 = 4y-20 \qquad \text{Remove parentheses and simplify.}$$

$$-10 = 4y \qquad \text{Add 20 to both sides.}$$

$$-\frac{5}{2} = y \qquad \text{Divide both sides by 4 and simplify.}$$

Thus, $y = -\frac{5}{2}$.

Self Check 4   Find x in Figure R-45.

Now Try Exercise 45.

The following theorem relates perpendicular lines and their slopes:

Slopes of Perpendicular Lines   If two nonvertical lines are perpendicular, the product of their slopes is -1.

If the product of the slopes of two lines is -1, the lines are perpendicular.

A proof of this property is shown in online Appendix B.

Comment   If the product of two numbers is -1, the numbers are called **negative reciprocals.**

Comment   It is also true that a horizontal line is perpendicular to a vertical line.

EXAMPLE 5 **Solving a Slope Problem That Involves Perpendicular Lines**

Are the lines shown in Figure R-46 perpendicular?

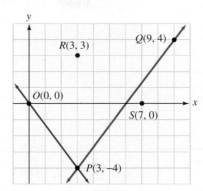

FIGURE R-46

SOLUTION We will determine the slopes of the lines and see whether their product is –1.

$$\text{Slope of } OP = \frac{\Delta y}{\Delta x} \qquad\qquad \text{Slope of } PQ = \frac{\Delta y}{\Delta x}$$

$$= \frac{y_2 - y_1}{x_2 - x_1} \qquad\qquad = \frac{y_2 - y_1}{x_2 - x_1}$$

$$= \frac{-4 - 0}{3 - 0} \qquad\qquad = \frac{4 - (-4)}{9 - 3}$$

$$= -\frac{4}{3} \qquad\qquad = \frac{8}{6}$$

$$\qquad\qquad\qquad\qquad\qquad = \frac{4}{3}$$

Since the product of the slopes is $-\frac{16}{9}$ and not -1, the lines are not perpendicular.

Self Check 5 Is either line in Figure R-46 perpendicular to the line passing through R and S?

Now Try Exercise 47.

In mathematics, it is important to learn to write equations of lines. Suppose we purchase a Kawasaki motorcycle for $10,000 and know that it depreciates $1,200 in value each year. We can use the facts given to write the linear equation that represents the value y of the motorcycle x years after it was purchased. Because the motorcycle's value decreases $1,200 each year, the slope of the line is $-1,200$. Since the purchase price is $10,000, we know that when we let $x = 0$, the value of y will equal 10,000. The linear equation that satisfies these two conditions is $y = -1,200x + 10,000$. This equation represents the straight-line depreciation of the motorcycle.

We will now learn to write equations of lines, given specific characteristics or features of the lines. We will begin with the point–slope form of an equation of a line.

4. Write Equations of Lines

Suppose that line l in Figure R-47 has a slope of m and passes through the point $P(x_1, y_1)$. If $Q(x, y)$ is any other point on line l, we have

$$m = \frac{y - y_1}{x - x_1}.$$

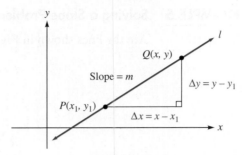

FIGURE R-47

Comment

$y - y_1 = m(x - x_1)$ is often referred to as the **point–slope formula**.

Comment

Equations of lines are not unique. For example, $y = \frac{1}{2}x + 3$ and $x - 2y + 3 = 0$ are both equations of the same line.

If we multiply both sides by $x - x_1$, we have

(2) $y - y_1 = m(x - x_1).$

Since Equation (2) displays the coordinates of the point (x_1, y_1) on the line and the slope m of the line, it is called the **point–slope form** of an equation of a line.

Point–Slope Form of an Equation of a Line

An equation of a line passing through $P(x_1, y_1)$ and with slope m is

$$y - y_1 = m(x - x_1).$$

EXAMPLE 6 **Finding an Equation of a Line Passing through Two Points**

Find an equation of the line passing through $P(3, 7)$ and $Q(-5, 3)$.

SOLUTION We will find the slope of the line and then choose either point P or point Q and substitute both the slope and coordinates of the point into the point–slope form. First we find the slope of the line.

$$m = \frac{y_2 - y_1}{x_2 - x_1} \qquad \text{This is the slope formula.}$$

$$= \frac{3 - 7}{-5 - 3} \qquad \text{Substitute 3 for } y_2, \text{ 7 for } y_1, \text{ } -5 \text{ for } x_2, \text{ and 3 for } x_1.$$

$$= \frac{-4}{-8}$$

$$= \frac{1}{2}$$

We can choose either point P or point Q and substitute its coordinates into the point–slope form. If we choose $P(3, 7)$, we substitute $\frac{1}{2}$ for m, 3 for x_1, and 7 for y_1.

$$y - y_1 = m(x - x_1) \qquad \text{This is the point–slope form.}$$

$$y - 7 = \frac{1}{2}(x - 3) \qquad \text{Substitute } \frac{1}{2} \text{ for } m, \text{ 3 for } x_1, \text{ and 7 for } y_1.$$

$$y = \frac{1}{2}x - \frac{3}{2} + 7 \qquad \text{Remove parentheses and add 7 to both sides.}$$

$$y = \frac{1}{2}x + \frac{11}{2} \qquad -\frac{3}{2} + 7 = -\frac{3}{2} + \frac{14}{2} = \frac{11}{2}$$

An equation of the line is $y = \frac{1}{2}x + \frac{11}{2}$. We will see soon that this line is written in a form known as *slope-intercept form*.

Self Check 6 Find an equation of the line passing through $P(-5, 4)$ and $Q(8, -6)$.

Now Try Exercise 61.

Since the y-intercept of the line shown in Figure R-48 is b, we can write an equation of the line by substituting 0 for x_1 and b for y_1 into the point–slope form and simplifying.

$$y - y_1 = m(x - x_1) \qquad \text{This is the point–slope form.}$$

$$y - b = m(x - 0) \qquad \text{Substitute 0 for } x_1 \text{ and } b \text{ for } y_1.$$

$$y - b = mx \qquad\qquad x - 0 = x$$

(3) $\qquad y = mx + b \qquad\qquad \text{Add } b \text{ to both sides.}$

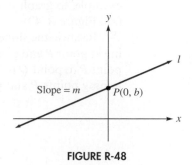

FIGURE R-48

Because Equation (3) displays the slope m and the y-intercept b, it is called the **slope-intercept form** of an equation of a line.

| **Slope-Intercept Form of an Equation of a Line** | An equation of the line with slope m and y-intercept b is $y = mx + b$. |

Three examples of linear equations written in slope-intercept form are shown here:

Comment

Note that when a line is written in slope-intercept form, the coefficient of x is the slope and the constant term is the y-intercept.

Example	Slope	y-intercept
$y = 2x + 7$	$m = 2$	$b = 7$
$y = \dfrac{2}{3}x - 5$	$m = \dfrac{2}{3}$	$b = -5$
$y = -4x + \dfrac{1}{5}$	$m = -4$	$b = \dfrac{1}{5}$

EXAMPLE 7   **Writing an Equation of a Line Using Slope-Intercept Form**

Use slope-intercept form to write an equation of the line with slope 4 that passes through $P(5, 9)$.

SOLUTION   Since we know that $m = 4$ and that the ordered pair $(5, 9)$ satisfies the equation, we substitute 4 for m, 5 for x, and 9 for y in the equation $y = mx + b$ and solve for b.

Comment

When we are given a point and slope, we can determine an equation of the line by substituting into point–slope form or slope-intercept form.

$$y = mx + b \qquad \text{This is the slope-intercept form.}$$

$$9 = 4(5) + b \qquad \text{Substitute 4 for } m, \text{ 5 for } x, \text{ and 9 for } y.$$

$$9 = 20 + b \qquad \text{Simplify.}$$

$$-11 = b \qquad \text{Subtract 20 from both sides.}$$

Because $m = 4$ and $b = -11$, an equation of the line is $y = 4x - 11$. ∎

Self Check 7   Use slope-intercept form to write an equation of the line with slope $\frac{7}{3}$ passing through (3, 1).

Now Try Exercise 67.

5. Graph Linear Equations Using the Slope and y-Intercept

It is easy to graph a linear equation when it is written in slope-intercept form. For example, to graph $y = \frac{4}{3}x - 2$, we note that $b = -2$ and that the y-intercept is -2 (see Figure R-49).

Because the slope is $\frac{\Delta y}{\Delta x} = \frac{4}{3}$, we can locate another point Q on the line by starting at point P and counting 3 units to the right and 4 units up. The change in x from point P to point Q is $\Delta x = 3$, and the corresponding change in y is $\Delta y = 4$. The line joining points P and Q is the graph of the equation.

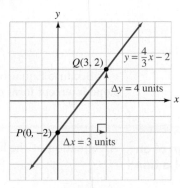

FIGURE R-49

EXAMPLE 8   **Finding the Slope and y-Intercept and Graphing the Line**

Find the slope and the y-intercept of the line with equation $3(y + 2) = 6x - 1$, and graph it.

SOLUTION   We will write the equation in the form $y = mx + b$ to find the slope m and the y-intercept b. Then we will use m and b to graph the line.

$$3(y + 2) = 6x - 1$$

$$3y + 6 = 6x - 1 \qquad \text{Remove parentheses.}$$

$$3y = 6x - 7 \qquad \text{Subtract 6 from both sides.}$$

$$y = 2x - \frac{7}{3} \qquad \text{Divide both sides by 3.}$$

The slope of the graph is 2, and the y-intercept is $-\frac{7}{3}$. We plot the y-intercept. Then we find a second point on the line by moving 1 unit to the right and 2 units up, to the point $\left(1, -\frac{1}{3}\right)$. To get the graph, we draw a line through the two points, as shown in Figure R-50.

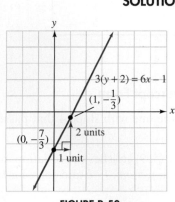

FIGURE R-50

Self Check 8   Find the slope and the y-intercept of the line with equation $2(x - 3) = -3(y + 5)$. Then graph it.

Now Try Exercise 79.

6. Write Equations of Parallel and Perpendicular Lines

EXAMPLE 9   **Writing an Equation of a Line Parallel to a Given Line**

Write an equation of the line passing through $P(-2, 5)$ and parallel to the line $y = 8x - 3$.

SOLUTION   We will substitute the coordinates of $P(-2, 5)$ and the slope of the line parallel to $y = 8x - 3$ into point–slope form and simplify the results to write an equation of the parallel line.

The slope of the line given by $y = 8x - 3$ is 8, the coefficient of x. Since the graph of the desired equation is to be parallel to the graph of $y = 8x - 3$, its slope must also be 8. We will substitute -2 for x_1, 5 for y_1, and 8 for m in the point–slope form and simplify.

$$y - y_1 = m(x - x_1)$$
$$y - 5 = 8[x - (-2)] \qquad \text{Substitute 5 for } y_1, \text{ 8 for } m, \text{ and } -2 \text{ for } x_1.$$
$$y - 5 = 8(x + 2) \qquad -(-2) = 2$$
$$y - 5 = 8x + 16 \qquad \text{Use the Distributive Property to remove parentheses.}$$
$$y = 8x + 21 \qquad \text{Add 5 to both sides.}$$

An equation of the desired line is $y = 8x + 21$.

Self Check 9   Write an equation of the line passing through $Q(1, 2)$ and parallel to the line $y = 8x - 3$.

Now Try Exercise 89.

EXAMPLE 10   **Writing an Equation of a Line Perpendicular to a Given Line**

Write an equation of the line passing through $P(-2, 5)$ and perpendicular to the line $y = 8x - 3$.

SOLUTION   We will substitute the coordinates of $P(-2, 5)$ and the slope of the line perpendicular to $y = 8x - 3$ into the point–slope form and simplify to write an equation of the parallel line.

Because the slope of the given line is 8, the slope of the desired perpendicular line must be $-\frac{1}{8}$. We substitute -2 for x_1, 5 for y_1, and $-\frac{1}{8}$ for m into the point–slope form and simplify.

$$y - y_1 = m(x - x_1)$$
$$y - 5 = -\frac{1}{8}[x - (-2)] \qquad \text{Substitute 5 for } y_1, -\frac{1}{8} \text{ for } m, \text{ and } -2 \text{ for } x_1.$$
$$y - 5 = -\frac{1}{8}(x + 2) \qquad -(-2) = 2$$
$$y = -\frac{1}{8}x - \frac{1}{4} + 5 \qquad \text{Remove parentheses and add 5 to both sides.}$$
$$y = -\frac{1}{8}x + \frac{19}{4} \qquad -\frac{1}{4} + 5 = -\frac{1}{4} + \frac{20}{4} = \frac{19}{4}$$

An equation of the line is $y = -\frac{1}{8}x + \frac{19}{4}$.

Self Check 10   Write an equation of the line passing through $Q(1, 2)$ and perpendicular to $y = 8x - 3$.

Now Try Exercise 95.

We have shown that the graph of any equation of the form $y = mx + b$ is a line with slope m and y-intercept b. In Section R.6 we saw that the graph of any equation of the form $Ax + By = C$ (where A and B are not *both* zero) is also a line. $Ax + By = C$ is called the **standard form** of an equation of a line.

Standard Form of an Equation of a Line	The standard form of an equation of a line is $Ax + By = C,$ where A, B, and C are real numbers, and A and B are not both zero.

Comment	When writing equations in $Ax + By = C$ form, we usually clear the equation of fractions and make A positive. For example, the equation $-x + \frac{5}{2}y = 2$ can be changed to $2x - 5y = -4$ by multiplying both sides by -2. We will also divide out any common integer factors of A, B, and C. For example, we would write $4x + 8y = 12$ as $x + 2y = 3$.

We summarize the various forms of an equation of a line as follows:

Summary of Forms of Equations of Lines	
Standard form	$Ax + By = C$ A and B cannot both be 0.
Slope-intercept form	$y = mx + b$ The slope is m, and the y-intercept is b.
Point–slope form	$y - y_1 = m(x - x_1)$ The slope is m, and the line passes through (x_1, y_1).
A horizontal line	$y = b$ The slope is 0, and the y-intercept is b.
A vertical line	$x = a$ There is no defined slope, and the x-intercept is a.

7. Write an Equation of a Line That Models a Real-Life Problem

For tax purposes, many businesses use *straight-line depreciation* to find the declining value of aging equipment.

EXAMPLE 11   **Solving a Straight-Line Depreciation Problem**

A business purchases a digital multimedia projector for $1,970 and expects it to last for 10 years. It can then be sold as scrap for a *salvage value* of $270. If y is the value of the projector after x years of use, and y and x are related by an equation of a line,

a. Find an equation of the line in slope-intercept form.

b. Find the value of the projector after $2\frac{1}{2}$ years.

c. Find the economic meaning of the y-intercept of the line.

d. Find the economic meaning of the slope of the line.

SOLUTION   **a.** We will find the slope and use the point–slope form to write an equation of the line.

When the projector is new, its age x is 0 and its value y is \$1,970. When the projector is 10 years old, $x = 10$ and $y = \$270$. Since the line passes through the points (0, 1,970) and (10, 270), the slope of the line is

$$m = \frac{y_2 - y_1}{x_2 - x_1} \qquad \text{This is the slope formula.}$$

$$= \frac{270 - 1,970}{10 - 0} \qquad \text{Substitute 270 for } y_2, \text{ 1,970 for } y_1, \text{ 10 for } x_2, \text{ and 0 for } x_1.$$

$$= \frac{-1,700}{10}$$

$$= -170$$

To find an equation of the line, we substitute -170 for m, 0 for x_1, and 1,970 for y_1 in the point–slope form and simplify.

$$y - y_1 = m(x - x_1)$$

$$y - 1,970 = -170(x - 0)$$

$$(4) \qquad\qquad y = -170x + 1,970$$

The value y of the projector is related to its age x by the equation $y = -170x + 1,970$.

b. To find the value after $2\frac{1}{2}$ years, we will substitute 2.5 for x in Equation (4) and solve for y.

$$y = -170x + 1,970$$

$$= -170(\mathbf{2.5}) + 1,970 \qquad \text{Substitute 2.5 for } x.$$

$$= -425 + 1,970$$

$$= 1,545$$

In $2\frac{1}{2}$ years, the projector will be worth \$1,545.

c. The y-intercept is $b = 1,970$ and represents the value of a 0-year-old projector. This is the projector's original cost, \$1,970. Note that when we substitute 0 in for x, we get 1,970.

$$y = -170x + 1,970$$

$$y = -170(\mathbf{0}) + 1,970 \qquad \text{Substitute 0 for } x.$$

$$y = 1,970$$

d. Each year, the value decreases by \$170, because the slope of the line is -170. The slope of the depreciation line is called the *annual depreciation rate*.

Problems that have an annual appreciation rate can be worked similarly. ∎

Self Check 11   A business purchases a Canon copier for \$2,700 and expects it to last for 10 years. It can then be sold for \$300. Write the straight-line depreciation equation for the copier.

Now Try Exercise 113.

8. Use Linear Curve Fitting

In statistics, the process of using one variable to predict another is called **regression.** For example, if we know a woman's height, we can make a good prediction about her weight, because taller women usually weigh more than shorter women. To write a **prediction equation** (sometimes called a **regression equation**), we must find an equation of the line that comes closer to all of the points in the scattergram than any other possible line. There are statistical methods to find such an equation, but we can only approximate it here.

EXAMPLE 12   Using Linear Curve Fitting

Figure R-51 shows the result of sampling 10 women and finding their heights and weights. The graph of the ordered pairs (h, w) is called a **scattergram** and is also shown.

a. Write an approximation of the regression equation.

b. Use the regression equation to approximate the weight of a woman 66 inches tall.

Woman	Height (h) in inches	Weight (w) in pounds
1	60	100
2	61	105
3	62	120
4	62	130
5	63	135
6	64	120
7	64	125
8	65	155
9	67	155
10	69	160

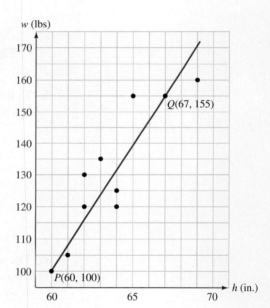

FIGURE R-51

SOLUTION   a. We will place a straightedge on the scattergram as shown in Figure R-51 and draw the line joining two points that seems to best fit all the points. In the figure, line PQ is drawn, where point P has coordinates of $(60, 100)$ and point Q has coordinates of $(67, 155)$.

Our approximation of the regression equation will be an equation of the line passing through points P and Q. To find an equation of this line, we first find its slope.

$$m = \frac{y_2 - y_1}{x_2 - x_1} \qquad \text{This is the slope formula.}$$

$$= \frac{155 - 100}{67 - 60} \qquad \text{Substitute 155 for } y_2, \text{ 100 for } y_1, \text{ 67 for } x_2, \text{ and 60 for } x_1.$$

$$= \frac{55}{7}$$

We can now use the point–slope form to find an equation of the line.

$$y - y_1 = m(x - x_1)$$ This is the point–slope form.

$$y - 100 = \frac{55}{7}(x - 60)$$ Choose (60, 100) for (x_1, y_1).

$$y = \frac{55}{7}x - \frac{3,300}{7} + 100$$ Remove parentheses and add 100 to both sides.

(5) $$y = \frac{55}{7}x - \frac{2,600}{7}$$ Simplify.

Our approximation of the regression equation is $y = \frac{55}{7}x - \frac{2,600}{7}$.

b. To predict the weight of a woman who is 66 inches tall, we substitute 66 for x in Equation (5) and simplify.

$$y = \frac{55}{7}x - \frac{2,600}{7}$$

$$y = \frac{55}{7}(66) - \frac{2,600}{7}$$

$$y \approx 147.1428571$$

We would predict that a 66-inch-tall woman chosen at random will weigh about 147 pounds.

Self Check 12   Use the regression equation found in Example 13 to predict the weight of a 67-inch-tall woman.

Now Try Exercise 125.

ACCENT ON TECHNOLOGY   **Linear Regression**

We can use the linear-regression feature on a graphing calculator to find an equation of the line that best fits a given set of data points. We will do that for the data given in Example 13.

- First, we press STAT to enter the statistics menu on the calculator. This screen is shown in Figure R-52(a). Next we press ENTER to input our data. We can input our heights into the L1 column and our weights into the L2 column. This is shown in Figure R-52(b).

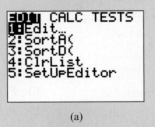

(a)

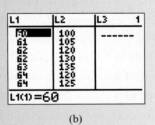

(b)

FIGURE R-52

- To obtain an equation of the regression line, we press STAT and then the right arrow key once to enter the calculate menu. This screen is shown in Figure R-53(a). To calculate a linear regression, we press 4 to select LinReg $(ax + b)$ and then ENTER to obtain an line. The screen is shown in Figure R-53(b).

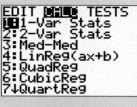

(a) (b)

FIGURE R-53

Note that the regression line is of the form $y = ax + b$, where a is the slope and b is the y-intercept. If we substitute the values shown in Figure R-53(b) for a and b and round to hundredths, we can write an equation of the line that best fits the data. The regression line is $y = 6.78x - 300.18$.

Self Check Answers 1. $\dfrac{13}{8}$ 2. $\dfrac{3}{2}$ 3. 175 feet per minute rate of descent 4. $\dfrac{4}{3}$

5. Yes, PQ is perpendicular to RS. 6. $y = -\dfrac{10}{13}x + \dfrac{2}{13}$ 7. $y = \dfrac{7}{3}x - 6$

8. $-\dfrac{2}{3}, -3$

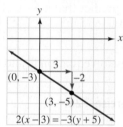

9. $y = 8x - 6$ 10. $y = -\dfrac{1}{8}x + \dfrac{17}{8}$

11. $y = -240x + 2{,}700$ 12. 155 pounds

Exercises R.7

Getting Ready
You should be able to complete these vocabulary and concept statements before you proceed to the practice exercises.

Fill in the blanks.

1. The slope of a nonvertical line is defined to be the change in y _____ by the change in x.

2. The change in _____ is often called the rise.

3. The change in x is often called the _____.

4. The slope of a _____ line is 0.

5. The slope of a _____ line is undefined.

6. If the slopes of two lines are equal, the lines are _____.

7. If two lines are perpendicular, the product of their slopes is _____.

8. The formula for the point–slope form of a line is _____.

9. In the equation $y = mx + b$, ____ is the slope of the graph of the line and b is the _____.

10. The standard form of an equation of a line is _____.

Practice
Find the slope of the line passing through each pair of points, if possible.

11.

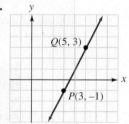

12.

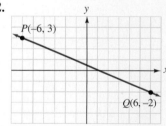

13. $P(3, -2)$; $Q(-1, 5)$ 14. $P(3, 7)$; $Q(6, 16)$

15. $P(8, -7)$; $Q(4, 1)$ 16. $P(5, 17)$; $Q(17, 17)$

17. $P(-4, 3)$; $Q(-4, -3)$ 18. $P(2, \sqrt{7})$; $Q(\sqrt{7}, 2)$

19. $P\left(\dfrac{3}{2}, \dfrac{2}{3}\right)$; $Q\left(\dfrac{5}{2}, \dfrac{7}{3}\right)$ 20. $P\left(-\dfrac{2}{5}, \dfrac{1}{3}\right)$; $Q\left(\dfrac{3}{5}, -\dfrac{5}{3}\right)$

Find the slope of each line. To do so, find any two points on the line and then use the slope formula.

21. $y = 3x + 2$ 22. $y = 5x - 8$

23. $5x - 10y = 3$ 24. $8y + 2x = 5$

25. $3(y + 2) = 2x - 3$ **26.** $4(x - 2) = 3y + 2$

27. $3(y + x) = 3(x - 1)$ **28.** $2x + 5 = 2(y + x)$

Find the slope of the line, if possible.

29. $y = 7$ **30.** $2y = 5$

31. $x = -\dfrac{1}{2}$ **32.** $x - 7 = 0$

Determine whether the slope of the line is positive, negative, 0, or undefined.

33.   **34.**

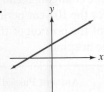   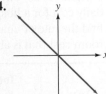

35.   **36.**

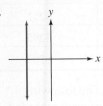   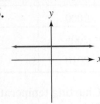

Determine whether the lines with the given slopes are parallel, perpendicular, or neither.

37. $m_1 = 3; m_2 = -\dfrac{1}{3}$ **38.** $m_1 = \dfrac{2}{3}; m_2 = \dfrac{3}{2}$

39. $m_1 = \sqrt{8}; m_2 = 2\sqrt{2}$ **40.** $m_1 = 1; m_2 = -1$

41. $m_1 = -\sqrt{2}; m_2 = \dfrac{\sqrt{2}}{2}$ **42.** $m_1 = 2\sqrt{7}; m_2 = \sqrt{28}$

43. $m_1 = -0.125; m_2 = 8$ **44.** $m_1 = 0.125; m_2 = \dfrac{1}{8}$

Determine whether the line through the given points and the line through $R(-3, 5)$ and $S(2, 7)$ are parallel, perpendicular, or neither.

45. $P(2, 4); Q(7, 6)$ **46.** $P(-3, 8); Q(-13, 4)$

47. $P(-4, 6); Q(-2, 1)$ **48.** $P(0, -9); Q(4, 1)$

Use the point–slope formula to find an equation of the line with the given properties. Write each equation in standard form.

49. $m = 2$ passing through $P(2, 4)$

50. $m = -3$ passing through $P(3, 5)$

51. $m = 2$ passing through $P\left(-\dfrac{3}{2}, \dfrac{1}{2}\right)$

52. $m = -6$ passing through $P\left(\dfrac{1}{4}, -2\right)$

53. $m = \dfrac{2}{5}$ passing through $P(-1, 1)$

54. $m = -\dfrac{1}{5}$ passing through $P(-2, -3)$

55. $m = 0$ passing through $P(-6, -3)$

56. m is undefined passing through $P(-6, -3)$

57. $m = \pi$ passing through $P(\pi, 0)$

58. $m = \pi$ passing through $P(0, \pi)$

Use the point–slope formula to find an equation of the line passing through the two given points. Write each equation in slope-intercept form.

59. $P(0, 0); Q(4, 4)$ **60.** $P(-5, -5); Q(0, 0)$

61. $P(3, 4); Q(0, -3)$ **62.** $P(4, 0); Q(6, -8)$

Use the point–slope formula to find an equation of each line. Write each equation in slope-intercept form.

63.   **64.**

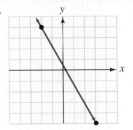   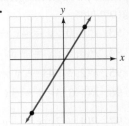

Write an equation of the line with the given properties in slope-intercept form.

65. $m = 3; b = -2$ **66.** $m = -\dfrac{1}{3}; b = \dfrac{2}{3}$

67. $m = 5; b = -\dfrac{1}{5}$ **68.** $m = \sqrt{2}; b = \sqrt{2}$

Find the slope and the y-intercept of the line determined by the given equation.

69. $3x - 2y = 8$ **70.** $-2x + 4y = 12$

71. $-2(x + 3y) = 5$ **72.** $5(2x - 3y) = 4$

73. $x = \dfrac{2y - 4}{7}$ **74.** $3x + 4 = -\dfrac{2(y - 3)}{5}$

Find the slope and the y-intercept of the given linear equation. Then use the slope and y-intercept to graph the line.

75. $x - y = 1$ **76.** $x + y = 2$

77. $x = \dfrac{3}{2}y - 3$ **78.** $x = -\dfrac{4}{5}y + 2$

79. $3(y - 4) = -2(x - 3)$

80. $-4(2x + 3) = 3(3y + 8)$

Determine whether the graphs of each pair of equations are parallel, perpendicular, or neither.

81. $y = 3x + 4; y = 3x - 7$

82. $y = 4x - 13; y = \dfrac{1}{4}x + 13$

83. $x = y + 2; y = x + 3$

84. $2x + 3y = 9; 3x - 2y = 5$

85. $y = 3; x = 4$

86. $x = \dfrac{y - 2}{3}; 3(y - 3) + x = 0$

Write an equation, in slope-intercept form, of the line that passes through the given point and is parallel to the given line.

87. $P(0, 0)$; $y = 4x - 7$

88. $P(0, 0)$; $x = -3y - 12$

89. $P(2, 5)$; $4x - y = 7$

90. $P(-6, 3)$; $y + 3x = -12$

91. $P(4, -2)$; $x = \dfrac{5}{4}y - 2$

92. $P(1, -5)$; $x = -\dfrac{3}{4}y + 5$

Write an equation, in slope-intercept form, of the line that passes through the given point and is perpendicular to the given line.

93. $P(0, 0)$; $y = 4x - 7$

94. $P(0, 0)$; $x = -3y - 12$

95. $P(2, 5)$; $4x - y = 7$

96. $P(-6, 3)$; $y + 3x = -12$

97. $P(4, -2)$; $x = \dfrac{5}{4}y - 2$

98. $P(1, -5)$; $x = -\dfrac{3}{4}y + 5$

99. Find an equation of the line perpendicular to the line $y = 3$ and passing through the midpoint of the segment joining $(2, 4)$ and $(-6, 10)$.

100. Find an equation of the line parallel to the line $y = -8$ and passing through the midpoint of the segment joining $(-4, 2)$ and $(-2, 8)$.

101. Find an equation of the line parallel to the line $x = 3$ and passing through the midpoint of the segment joining $(2, -4)$ and $(8, 12)$.

102. Find an equation of the line perpendicular to the line $x = 3$ and passing through the midpoint of the segment joining $(-2, 2)$ and $(4, -8)$.

Applications

103. Rate of growth When a college started an aviation program, the administration agreed to predict enrollments using a straight-line method. If the enrollment during the first year was 12, and the enrollment during the fifth year was 26, find the rate of growth per year (the slope of the line). See the illustration.

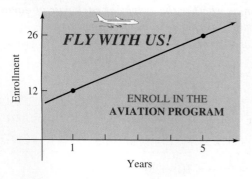

104. Rate of growth A small business predicts sales according to a straight-line method. If sales were $50,000 in the first year and $110,000 in the third year, find the rate of growth in dollars per year (the slope of the line).

105. Rate of decrease The price of computers has been dropping steadily for the past 10 years. If a desktop PC cost $6,700 10 years ago, and the same computing power cost $2,200 three years ago, find the rate of decrease per year. (Assume a straight-line model.)

106. Hospital costs The table shows the changing mean daily cost for a hospital room. For the 10-year period, find the rate of change per year of the portion of the room cost that is absorbed by the hospital.

Year	Total Cost to the Hospital	Amount Passed On to Patient
2000	$459	$212
2005	$670	$295
2010	$812	$307

107. Charting temperature changes The following Fahrenheit temperature readings were recorded over a four-hour period:

Time	12:00	1:00	2:00	3:00	4:00
Temperature	47°	53°	59°	65°	71°

Let t represent the time (in hours), with 12:00 corresponding to $t = 0$. Let T represent the temperature. Plot the points (t, T) and draw the line through those points. Explain the meaning of $\frac{\Delta T}{\Delta t}$.

108. Tracking the Dow The Dow Jones Industrial Average at the close of trade on three consecutive days was as follows:

Day	Monday	Tuesday	Wednesday
Close	12,981	12,964	12,947

Let d represent the day, with $d = 0$ corresponding to Monday, and let D represent the Dow average. Plot the points (d, D) and draw the graph. Explain the meaning of $\frac{\Delta D}{\Delta d}$.

109. Speed of an airplane A pilot files a flight plan indicating her intention to fly at a constant speed of 590 mph. Write an equation that expresses the distance traveled in terms of the flying time. Then graph the equation and interpret the slope of the line. (*Hint:* $d = rt$.)

Ilja Mašík/Shutterstock.com

110. **Growth of savings** A student deposits $25 each month in a Holiday Club account at her bank. The account pays no interest. Write an equation that expresses the amount A in her account in terms of the number of deposits n. Then graph the line and interpret the slope of the line.

In Exercises 111–117, assume straight-line depreciation or straight-line appreciation.

111. **Depreciation** A Toyota Tundra truck was purchased for $24,300. Its salvage value at the end of its 7-year useful life is expected to be $1,900. Find the depreciation equation.

112. **Depreciation** A business purchases the laptop computer shown. It will be depreciated over a 4-year period, after which its salvage value will be $300. Find the depreciation equation.

$2,700

113. **Depreciation** Find the depreciation equation for the TV in the following want ad:

> *For Sale*: 3-year-old 54-inch TV, $1,900 new. Asking $1,190. Call 875-5555. Ask for Mike.

114. **Depreciation** A Bose Wave radio costs $555 when new and is expected to be worth $80 after 5 years. What will it be worth after 3 years?

115. **Value of an antique** An antique table is expected to appreciate $40 each year. If the table will be worth $450 in 2 years, what will it be worth in 13 years?

116. **Value of an antique** An antique clock is expected to be worth $350 after 2 years and $530 after 5 years. What will the clock be worth after 7 years?

117. **Purchase price of real estate** A cottage that was purchased 3 years ago is now appraised at $47,700. If the property has been appreciating $3,500 per year, find its original purchase price.

118. **Computer repair** A computer-repair company charges a fixed amount, plus an hourly rate, for a service call. Use the information in the illustration to find the hourly rate.

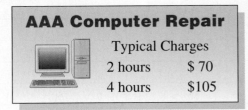

AAA Computer Repair

Typical Charges

| 2 hours | $ 70 |
| 4 hours | $105 |

119. **Predicting fires** A local fire department recognizes that city growth and the number of reported fires are related by a linear equation. City records show that 300 fires were reported in a year when the local population was 57,000 people, and 325 fires were reported in a year when the population was 59,000 people. How many fires can be expected in the year when the population reaches 100,000 people?

120. **Estimating the cost of rain gutter** A neighbor tells you that an installer of rain gutter charges $60, plus a dollar amount per foot. If the neighbor paid $435 for the installation of 250 feet of gutter, how much will it cost you to have 300 feet installed?

121. **Converting temperatures** Water freezes at 32° Fahrenheit, or 0° Celsius. It boils at 212°F, or 100°C. Find a formula for converting a temperature from degrees Fahrenheit to degrees Celsius.

122. **Converting units** A speed of 1 mile per hour is equal to 88 feet per minute, and of course 0 miles per hour is 0 feet per minute. Find an equation for converting a speed x in miles per hour to the corresponding speed y in feet per minute.

123. **Smoking** The percentage y of 18- to 25-year-old smokers in the United States has been declining at a constant rate since 1974. If about 47% of this group smoked in 1974 and about 29% smoked in 1994, find a linear equation that models this decline. If this trend continues, estimate what percentage will smoke in 2014.

124. **Forensic science** Scientists believe there is a linear relationship between the height h (in centimeters) of a male and the length f (in centimeters) of his femur bone. Use the data in the table to find a linear equation that expresses the height h in terms of f. Round all constants to the nearest thousandth. How tall would you expect a man to be if his femur measures 50 centimeters? Round to the nearest centimeter.

Person	Length of Femur (f)	Height (h)
A	62.5 cm	200 cm
B	40.2 cm	150 cm

125. Crickets The table shows the approximate chirping rate at various temperatures for one type of cricket.

Temperature (°F)	Chirps Per Minute
50	20
60	80
70	115
80	150
100	250

a. Construct a scattergram of the data.

b. Assume a linear relationship and write a regression equation.

c. Estimate the chirping rate at a temperature of 90°F.

126. Fishing The table shows the lengths and weights of seven muskies captured by the Department of Natural Resources in Catfish Lake in Eagle River, Wisconsin.

Muskie	Length (in)	Weight (lb)
1	26	5
2	27	8
3	29	9
4	33	12
5	35	14
6	36	14
7	38	19

a. Assume a linear relationship and write a regression equation that represents the data.

b. Estimate the weight of a muskie that is 32 inches long.

127. Use the linear-regression feature on a graphing calculator to determine the line that best fits the data given in Exercise 125. Round to the hundredths.

128. Use the linear-regression feature on a graphing calculator to determine the line that best fits the data given in Exercise 126. Round to the hundredths.

Discovery and Writing

129. Prove that an equation of a line with an x-intercept of a and a y-intercept of b can be written in the form

$$\frac{x}{a} + \frac{y}{b} = 1.$$

130. Find the x- and y-intercepts of the line $bx + ay = ab$.

Investigate the properties of slope and the y-intercept by experimenting with the following problems.

131. Graph $y = mx + 2$ for several positive values of m. What do you notice?

132. Graph $y = mx + 2$ for several negative values of m. What do you notice?

133. Graph $y = 2x + b$ for several increasing positive values of b. What do you notice?

134. Graph $y = 2x + b$ for several decreasing negative values of b. What do you notice?

135. How will the graph of $y = \frac{1}{2}x + 5$ compare to the graph of $y = \frac{1}{2}x - 5$?

136. How will the graph of $y = \frac{1}{2}x - 5$ compare to the graph of $y = \frac{1}{2}x$?

CHAPTER REVIEW

SECTION R.1   Linear Equations and Applications

Definitions and Concepts	Examples
An **equation** is a statement indicating that two quantities are equal.	See Example 1, page 2
There can be restrictions on the variable in an equation. We cannot take the square root of a negative number or divide by 0.	

Definitions and Concepts	Examples
Properties of equality: If $a = b$ and c is a number, then $\quad a + c = b + c$ and $a - c = b - c$; $\quad ac = bc \quad$ and $\quad \dfrac{a}{c} = \dfrac{b}{c} \quad (c \neq 0)$. A **linear equation** is an equation that can be written in the form $ax + b = 0$, $a \neq 0$. To solve a linear equation, use the properties of equality to isolate x on one side of the equation. • An **identity** is an equation that is true for all acceptable replacements for its variable. • A **contradiction** is an equation that is false for all acceptable replacements for its variable.	See Examples 3 and 4, pages 4–5
Rational equations are equations that contain rational expressions. To solve rational equations, multiply both sides of the equation by an expression that will remove the denominators and solve the resulting equation. Be sure to check the answers to identify any **extraneous solutions**.	See Example 5, page 6
Solving an application problem: Use the following steps to solve an application problem: 1. Analyze the problem. 2. Pick a variable to represent the quantity to be found. 3. Form an equation. 4. Solve the equation. 5. Check the solution in the words of the problem.	See Example 11, pages 10–11

EXERCISES

Find the restrictions on x, if any.

1. $3x + 7 = 4$

2. $x + \dfrac{1}{x} = 2$

3. $\sqrt{x} = 4$

4. $\dfrac{1}{x - 2} = \dfrac{2}{x - 3}$

Solve each equation and classify it as an identity, a conditional equation, or a contradiction.

5. $3(9x + 4) = 28$

6. $\dfrac{3}{2}a = 7(a + 11)$

7. $8(3x - 5) - 4(x + 3) = 12$

8. $\dfrac{x + 3}{x + 4} + \dfrac{x + 3}{x + 2} = 2$

9. $\dfrac{3}{x - 1} = \dfrac{1}{2}$

10. $\dfrac{8x^2 + 72x}{9 + x} = 8x$

11. $\dfrac{3x}{x - 1} - \dfrac{5}{x + 3} = 3$

12. $x + \dfrac{1}{2x - 3} = \dfrac{2x^2}{2x - 3}$

13. $\dfrac{4}{x^2 - 13x - 48} - \dfrac{1}{x^2 + x - 6} = \dfrac{2}{x^2 - 18x + 32}$

14. $\dfrac{a - 1}{a + 3} + \dfrac{2a - 1}{3 - a} = \dfrac{2 - a}{a - 3}$

Solve each formula for the indicated variable.

15. $C = \dfrac{5}{9}(F - 32); F$

16. $P_n = l + \dfrac{si}{f}; f$

17. $\dfrac{1}{f} = \dfrac{1}{f_1} + \dfrac{1}{f_2}; f_1$

18. $S = \dfrac{a - lr}{1 - r}; l$

Applications

19. Test scores Carlos took four tests in an English class. On each successive test, his score improved by 4 points. If his mean score was 66%, what did he score on the first test?

20. Fencing a garden A homeowner has 100 feet of fencing to enclose a rectangular garden. If the garden is to be 5 feet longer than it is wide, find its dimensions.

21. Driving rates Two women leave a shopping center by car traveling in opposite directions. If one car averages 45 mph and the other 50 mph, how long will it take for the cars to be 285 miles apart?

22. Driving rates Two taxis leave an airport and travel in the same direction. If the average speed of one taxi is 40 mph and the average speed of the other taxi is 46 mph, how long will it take before the taxis are 3 miles apart?

23. Preparing a solution A liter of fluid is 50% alcohol. How much water must be added to dilute it to a 20% solution?

24. Washing windows Scott can wash 37 windows in 3 hours, and Bill can wash 27 windows in 2 hours. How long will it take the two of them to wash 100 windows?

25. Filling a tank A tank can be filled in 9 hours by one pipe and in 12 hours by another. How long will it take both pipes to fill the empty tank?

26. Producing brass How many ounces of pure zinc must be alloyed with 20 ounces of brass that is 30% zinc and 70% copper to produce brass that is 40% zinc?

27. Lending money A bank lends $10,000, part of it at 11% annual interest and the rest at 14%. If the annual income is $1,265, how much was lent at each rate?

28. Producing Oriental rugs An Oriental-rug manufacturer can use one loom with a setup cost of $750 that can weave a rug for $115. Another loom, with a setup cost of $950, can produce a rug for $95. How many rugs are produced if the costs are the same on each loom?

SECTION R.2   Complex Numbers

Definitions and Concepts	Examples
A **complex number** is a number that can be written in the form $a + bi$, where a and b are real numbers and $i = \sqrt{-1}$. • $a + bi = c + di$ if and only if $a = c$ and $b = d$. • $(a + bi) + (c + di) = (a + c) + (b + d)i$. • $(a + bi) - (c + di) = (a - c) + (b - d)i$. • $(a + bi)(c + di) = (ac - bd) + (ad + bc)i$, or multiply them as if they were binomials.	See Examples 2 and 3, page 21
The **complex conjugate** of $a + bi$ is $a - bi$. To divide complex numbers, rationalize the denominator.	See Example 6, page 23
If n is a natural number that has a remainder of r when divided by 4, then $i^n = i^r$.	See Example 7, page 24

EXERCISES

Perform all operations and express all answers in $a + bi$ form.

29. $(2 - 3i) + (-4 + 2i)$ **30.** $(2 - 3i) - (4 + 2i)$

31. $\left(3 - \sqrt{-36}\right) + \left(\sqrt{-16} + 2\right)$

32. $\left(3 + \sqrt{-9}\right)\left(2 - \sqrt{-25}\right)$

33. $\dfrac{3}{i}$ **34.** $-\dfrac{2}{i^3}$

35. $\dfrac{3}{1 + i}$

36. $\dfrac{2i}{2 - i}$

37. $\dfrac{3 + i}{3 - i}$

38. $\dfrac{3 - 2i}{1 + i}$

39. Simplify: i^{53}.

40. Simplify: i^{103}.

SECTION **R.3**   Quadratic Equations and Applications

Definitions and Concepts	Examples
A **quadratic equation** is an equation that can be written in the form $ax^2 + bx + c = 0$, where a, b, and c are real numbers and $a \neq 0$.	See page 27
Zero-Factor Theorem: If $ab = 0$, then $a = 0$ or $b = 0$.	See Example 1, page 28
Square-Root Property: If $c > 0$, $x^2 = c$ has two real roots: $x = \sqrt{c}$ or $x = -\sqrt{c}$.	See Example 3, page 29
To complete the square: 1. Make the coefficient of x^2 equal to 1. 2. Get the constant on the right side of the equation. 3. Complete the square for x. Take one-half of the coefficient of x, square it, and add it to both sides of the equation. 4. Factor the resulting perfect-square trinomial and combine like terms. 5. Solve the resulting quadratic equation by using the Square-Root Property.	See Example 5, page 31
Quadratic Formula: $x = \dfrac{-b \pm \sqrt{b^2 - 4ac}}{2a}$ $(a \neq 0)$	See Example 6, page 32
The **discriminant** is $b^2 - 4ac$. • If the discriminant is positive and a perfect square, then the quadratic equation will have two different rational roots. • If the discriminant is positive and not a perfect square, then the quadratic equation will have two different irrational roots. • If the discriminant is 0, then the quadratic equation will have one repeated rational root. • If the discriminant is negative, then the quadratic equation will have two different imaginary roots that are complex conjugates.	See Example 7, page 33
Many real-life problems are modeled by quadratic equations.	See Example 9, page 35

EXERCISES

Solve each equation by factoring.

41. $2x^2 - x - 6 = 0$ **42.** $12x^2 + 13x = 4$

43. $5x^2 - 8x = 0$ **44.** $27x^2 = 30x - 8$

Solve each equation by using the Square-Root Property.

45. $4x^2 = 27$ **46.** $(5x - 7)^2 = -25$

Solve each equation by completing the square.

47. $x^2 - 8x + 15 = 0$ **48.** $3x^2 + 18x = -24$

49. $5x^2 - x - 1 = 0$ **50.** $5x^2 - x = 0$

Use the Quadratic Formula to solve each equation.

51. $x^2 + 5x - 14 = 0$ **52.** $3x^2 - 25x = 18$

53. $5x^2 = 1 - x$ **54.** $-5 = a^2 + 2a$

55. Solve: $3x^2 - 2x + 1 = 0$.

56. Solve: $3x^2 + 4 = 2x$.

57. Calculate the discriminant associated with the equation $6x^2 + 5x + 1 = 0$.

58. Determine the nature of the root(s) of the equation in Exercise 57.

59. Find the value of k that will make the roots of $kx^2 + 4x + 12 = 0$ equal.

60. Find the values of k that will make the roots of $4y^2 + (k + 2)y = 1 - k$ equal.

61. Solve: $\dfrac{1}{a} - \dfrac{1}{5} = \dfrac{3}{2a}$.

62. Solve: $\dfrac{4}{a - 4} + \dfrac{4}{a - 1} = 5$.

Applications

63. Fencing a field A farmer wishes to enclose a rectangular garden with 300 yards of fencing. A river runs along one side of the garden, so no fencing is needed there. Find the dimensions of the rectangle if the area is 10,450 square yards.

64. Flying rates A jet plane, flying 120 mph faster than a propeller-driven plane, travels 3,520 miles in 3 hours less than the propeller plane requires to fly the same distance. How fast does each plane fly?

65. Flight of a ball A ball thrown into the air reaches a height h (in feet) according to the formula $h = -16t^2 + 64t$, where t is the time elapsed (in seconds) since the ball was thrown. Find the shortest time it will take the ball to reach a height of 48 feet.

66. Width of a walk A man built a walk of uniform width around a rectangular pool. If the area of the walk is 117 square feet and the dimensions of the pool are 16 feet by 20 feet, how wide is the walk?

SECTION R.4   Other Types of Equations

Definitions and Concepts	Examples
Many polynomial equations of higher degree can be solved by factoring.	See Example 1, page 40
Power Property of Real Numbers:	See Example 6, page 44
$\quad$ If $a = b$, then $a^2 = b^2$	
To solve radical equations, use the Power Property of Real Numbers. Remember to check all roots.	
$\lvert x \rvert = \begin{cases} x \text{ when } x \geq 0 \\ -x \text{ when } x < 0 \end{cases}$	See page 45
If $k > 0$, then $\lvert x \rvert = k$ is equivalent to $x = k$ or $x = -k$.	See Example 8, pages 46–47
If a and b are algebraic expressions, $\lvert a \rvert = \lvert b \rvert$ is equivalent to $a = b$ or $a = -b$.	See Example 9, page 47

EXERCISES

Solve each equation.

67. $\dfrac{3x}{2} - \dfrac{2x}{x - 1} = x - 3$

68. $\dfrac{12}{x} - \dfrac{x}{2} = x - 3$

69. $x^4 - 2x^2 + 1 = 0$

70. $x^4 + 36 = 37x^2$

71. $a - a^{1/2} - 6 = 0$

72. $x^{2/3} + x^{1/3} - 6 = 0$

73. $\sqrt{x - 1} + x = 7$

74. $\sqrt{a + 9} - \sqrt{a} = 3$

75. $\sqrt{5 - x} + \sqrt{5 + x} = 4$

76. $\sqrt{y + 5} + \sqrt{y} = 1$

77. $\lvert x + 1 \rvert = 6$

78. $\lvert 2x - 1 \rvert = \lvert 2x + 1 \rvert$

79. $\left\lvert \dfrac{3x + 11}{7} \right\rvert - 1 = 0$

80. $\left\lvert \dfrac{2a - 6}{3a} \right\rvert - 6 = 0$

SECTION **R.5**   Inequalities

Definitions and Concepts	Examples
If a, b, and c are real numbers: • If $a < b$, $a + c < b + c$ and $a - c < b - c$. • If $a < b$ and $c > 0$, $ac < bc$ and $\dfrac{a}{c} < \dfrac{b}{c}$. • If $a < b$ and $c < 0$, $ac > bc$ and $\dfrac{a}{c} > \dfrac{b}{c}$. **Trichotomy Property:** $a < b$, $a = b$, or $a > b$	See page 51
Solving inequalities: Use the same steps to solve inequalities as you would use to solve equations. However, remember to reverse the order of the inequality when you multiply (or divide) both sides of an inequality by a negative number.	See Example 2, pages 52–53
If $k > 0$, then $\|x\| < k$ is equivalent to $-k < x < k$.	See Example 6, page 56
If $k > 0$, then $\|x\| > k$ is equivalent to $x > k$ or $x < -k$.	See Example 7, pages 56–57

EXERCISES

Graph each inequality, and write the solution using interval notation.

81. $2x - 9 < 5$

82. $5x + 3 \geq 2$

83. $\dfrac{5(x - 1)}{2} < x$

84. $\dfrac{1}{4}x + \dfrac{2}{3}x - x > \dfrac{1}{2} + \dfrac{1}{2}(x + 1)$

85. $0 \leq \dfrac{3 + x}{2} < 4$

86. $2 + a < 3a - 2 \leq 5a + 2$

87. $|x + 3| < 3$

88. $|3x - 7| \geq 1$

89. $\left|\dfrac{x + 2}{3}\right| < 1$

90. $\left|\dfrac{x - 3}{4}\right| > 8$

SECTION **R.6**   The Rectangular Coordinate System

Definitions and Concepts	Examples
The rectangular coordinate system divides the plane into four quadrants.	See page 63
The graph of an equation in x and y is the set of all points (x, y) that satisfy the equation.	See Example 1, pages 64–65
Intercepts: • The **y-intercept** of a line is the y value on the point where the line intersects the y-axis. To find the y-intercept, substitute 0 for x and solve the equation for y. • The **x-intercept** of a line is the x value on the point where the line intersects the x-axis. To find the x-intercept, substitute 0 for y and solve the equation for x.	See Example 3, pages 67–68
Vertical and horizontal lines: • Equation of a vertical line through (a, b): $x = a$ • Equation of a horizontal line through (a, b): $y = b$	See Example 4, page 69

Definitions and Concepts	Examples
Distance formula:	See Example 6, pages 70–71
The distance d between points (x_1, y_1) and (x_2, y_2) is given by $d = \sqrt{(x_2 - x_1)^2 + (y_2 - y_1)^2}$.	
Midpoint formula:	See Example 7, pages 71–72
The midpoint of the line segment joining (x_1, y_1) and (x_2, y_2) is the point M with coordinates $\left(\dfrac{x_1 + x_2}{2}, \dfrac{y_1 + y_2}{2}\right)$.	

EXERCISES

Refer to the illustration and find the coordinates of each point.

91. A **92.** B

93. C **94.** D

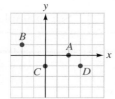

Graph each point. Indicate the quadrant in which the point lies or the axis on which it lies.

95. $(-3, 5)$ **96.** $(5, -3)$

97. $(0, -7)$ **98.** $\left(-\dfrac{1}{2}, 0\right)$

Solve each equation for y and graph the equation. Then check your graph with a graphing calculator.

99. $2x - y = 6$ **100.** $2x + 5y = -10$

Use the x- and y-intercepts to graph each equation.

101. $3x - 5y = 15$ **102.** $x + y = 7$

103. $x + y = -7$ **104.** $x - 5y = 5$

Graph each equation.

105. $y = 4$ **106.** $x = -2$

107. Depreciation A Ford Mustang purchased for $18,750 is expected to depreciate according to the formula $y = -2{,}200x + 18{,}750$. Find its value after 3 years.

108. House appreciation A house purchased for $250,000 is expected to appreciate according to the formula $y = 16{,}500x + 250{,}000$, where y is the value of the house after x years. Find the value of the house after 5 years.

Find the length of the segment PQ.

109. $P(-3, 7); Q(3, -1)$ **110.** $P(-8, 6); Q(-12, 10)$

111. $P\left(\sqrt{3}, 9\right); Q\left(\sqrt{3}, 7\right)$ **112.** $P(a, -a); Q(-a, a)$

Find the midpoint of the segment PQ.

113. $P(-3, 7); Q(3, -1)$ **114.** $P(0, 5); Q(-12, 10)$

115. $P\left(\sqrt{3}, 9\right); Q\left(\sqrt{3}, 7\right)$ **116.** $P(a, -a); Q(-a, a)$

SECTION R.7   Equations of Lines

Definitions and Concepts	Examples
The **slope of a nonvertical line** passing through points $P(x_1, y_1)$ and $Q(x_2, y_2)$ is $m = \dfrac{\text{change in } y}{\text{change in } x} = \dfrac{y_2 - y_1}{x_2 - x_1} \ (x_2 \neq x_1).$	See Example 1, page 76

Definitions and Concepts	Examples
Slopes of horizontal and vertical lines: • The slope of a horizontal line (a line with an equation of the form $y = b$) is 0. • The slope of a vertical line (a line with an equation of the form $x = a$) is not defined.	See page 78
Slopes of parallel lines and perpendicular lines: • Nonvertical parallel lines have the same slope. • The product of the slopes of two perpendicular lines is -1, provided neither line is vertical.	See Examples 4 and 5, pages 80–81
Point–slope form: An equation of the line passing through $P(x_1, y_1)$ with slope m is $y - y_1 = m(x - x_1)$.	See Example 6, page 82
Slope-intercept form: An equation of the line with slope m and y-intercept b is $y = mx + b$. Slope-intercept form can be used to find the slope and the y-intercept from an equation of a line.	See Example 8, page 84
Standard form of an equation of a line: $Ax + By = C$	See page 86

Equations of horizontal and vertical lines:
• **Horizontal line:** $y = b$
  The slope is 0, and the y-intercept is b.
 	
• **Vertical line:** $x = a$
  There is no defined slope, and the x-intercept is a.
	See page 78

EXERCISES

Find the slope of the line PQ, if possible.

117. $P(3, -5)$; $Q(1, 7)$ **118.** $P(2, 7)$; $Q(-5, -7)$

119. $P(b, a)$; $Q(a, b)$ **120.** $P(a + b, b)$; $Q(b, b - a)$

Find the slope of each line. To do so, find any two points on the line and then use the slope formula.

121. $y = 3x + 6$ **122.** $y = 5x - 6$

Determine whether the slope of each line is 0 or undefined.

123.   **124.**

Determine whether the slope of each line is positive or negative.

125.   **126.**

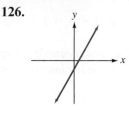

Determine whether the lines with the given slopes are parallel, perpendicular, or neither.

127. $m_1 = 5$; $m_2 = -\dfrac{1}{5}$ **128.** $m_1 = \dfrac{2}{7}$; $m_2 = \dfrac{7}{2}$

129. A line passes through $(-2, 5)$ and $(6, 10)$. A line parallel to it passes through $(2, 2)$ and $(10, y)$. Find y.

130. A line passes through $(-2, 5)$ and $(6, 10)$. A line perpendicular to it passes through $(-2, 5)$ and $(x, -3)$. Find x.

131. **Rate of descent** If an airplane descends 3,000 feet in 15 minutes, what is the average rate of descent in feet per minute?

132. **Rate of growth** A small business predicts sales according to a straight-line method. If sales were $50,000 in the first year and $147,500 in the third year, find the rate of growth in dollars per year (the slope of the line).

Use the point–slope formula to find an equation of each line. Write each equation in standard form.

133. The line passes through the origin and the point $(-5, 7)$.

134. The line passes through $(-2, 1)$ and has a slope of -4.

135. The line passes through $(2, -1)$ and has a slope of $-\frac{1}{5}$.

136. The line passes through $(7, -5)$ and $(4, 1)$.

Use the slope-intercept form to write an equation of each line.

137. The line has a slope of $\frac{2}{3}$ and a y-intercept of 3.

138. The slope is $-\frac{3}{2}$ and the line passes through $(0, -5)$.

Use the slope-intercept form to graph each equation.

139. $y = \frac{3}{5}x - 2$ **140.** $y = -\frac{4}{3}x + 3$

Find the slope and y-intercept of the graph of each line.

141. $y = 2x$ **142.** $x = -2y$

143. $-2y = -3x + 10$ **144.** $2x = -4y - 8$

145. $5x + 2y = 7$ **146.** $3x - 4y = 14$

Write an equation of each line.

147. The line has a slope of 0 and passes through $(-5, 17)$.

148. The line has no defined slope and passes through $(-5, 17)$.

Write an equation of each line in slope-intercept form.

149. The line is parallel to $3x - 4y = 7$ and passes through $(2, 0)$.

150. The line passes through $(7, -2)$ and is parallel to the line segment joining $(2, 4)$ and $(4, -10)$.

151. The line passes through $(0, 5)$ and is perpendicular to the line $x + 3y = 4$.

152. The line passes through $(7, -2)$ and is perpendicular to the line segment joining $(2, 4)$ and $(4, -10)$.

Determine whether the graphs of each pair of equations are parallel, perpendicular, or neither.

153. $y = 3x + 8$; $2y = 6x - 19$

154. $2x + 3y = 6$; $3x - 2y = 15$

CHAPTER TEST

Find all restrictions on x.

1. $\dfrac{x}{x(x-1)} = 2$ **2.** $\sqrt{x} = 1$

Solve each equation.

3. $7(2a + 5) - 7 = 6(a + 8)$

4. $\dfrac{3}{x^2 - 5x - 14} = \dfrac{4}{x^2 + 5x + 6}$

5. Solve for x: $z = \dfrac{x - \mu}{\sigma}$.

6. Solve for a: $\dfrac{1}{a} = \dfrac{1}{b} + \dfrac{1}{c}$.

7. A student's average on three tests is 75. If the final is to count as two one-hour tests, what grade must the student make to bring the average up to 80?

8. A woman invested part of $20,000 at 6% interest and the rest at 7%. If her annual interest is $1,260, how much did she invest at 6%?

Solve each equation.

9. $4x^2 - 8x + 3 = 0$ **10.** $2b^2 - 12 = -5b$

11. Write the Quadratic Formula.

12. Use the Quadratic Formula to solve $3x^2 - 5x - 9 = 0$.

13. Find k such that $x^2 + (k + 1)x + k + 4 = 0$ will have two equal roots.

14. The height of a projectile shot up into the air is given by the formula $h = -16t^2 + 128t$. Find the time t (in seconds) required for the projectile to return to its starting point.

Perform each operation and write all answers in a + bi form.

15. $(4 - 5i) - (-3 + 7i)$ **16.** $(4 - 5i)(3 - 7i)$

17. $\dfrac{2}{2 - i}$ **18.** $\dfrac{1 + i}{1 - i}$

Simplify each expression.

19. i^{13} **20.** i^0

Solve each equation.

21. $z^4 - 13z^2 + 36 = 0$ **22.** $2p^{2/5} - p^{1/5} - 1 = 0$

23. $\sqrt{x + 5} = 12$

24. $\sqrt{2z + 3} = 1 - \sqrt{z + 1}$

Solve each inequality, graph the solution set, and write the solution using interval notation.

25. $5x - 3 \le 7$ **26.** $\dfrac{x + 3}{4} > \dfrac{2x - 4}{3}$

Solve each equation.

27. $\left|\dfrac{3x+2}{2}\right| = 4$

28. $|x+3| = |x-3|$

Solve each inequality, graph the solution set, and write the solution using interval notation.

29. $|2x-5| > 2$

30. $\left|\dfrac{2x+3}{3}\right| \le 5$

Indicate the quadrant in which the point lies or the axis on which it lies.

31. $(-3, \pi)$

32. $(0, -8)$

Find the x- and y-intercepts and use them to graph the equation.

33. $x + 3y = 6$

34. $2x - 5y = 10$

Graph each equation.

35. $2(x+y) = 3x + 5$

36. $3x - 5y = 3(x-5)$

37. $\dfrac{1}{2}(x-2y) = y - 1$

38. $\dfrac{x+y-5}{7} = 3x$

Find the distance between points P and Q.

39. $P(1, -1);\ Q(-3, 4)$

40. $P(0, \pi);\ Q(-\pi, 0)$

Find the midpoint of the line segment PQ.

41. $P(3, -7);\ Q(-3, 7)$

42. $P\left(0, \sqrt{2}\right);\ Q\left(\sqrt{8}, \sqrt{18}\right)$

Find the slope of the line PQ.

43. $P(3, -9);\ Q(-5, 1)$

44. $P\left(\sqrt{3}, 3\right);\ Q\left(-\sqrt{12}, 0\right)$

Determine whether the two lines are parallel, perpendicular, or neither.

45. $y = 3x - 2;\ y = 2x - 3$

46. $2x - 3y = 5;\ 3x + 2y = 7$

Write an equation of the line with the given properties. Write in slope-intercept form, if possible.

47. Passing through $(3, -5);\ m = 2$

48. $m = 3;\ b = \dfrac{1}{2}$

49. Parallel to $2x - y = 3;\ b = 5$

50. Perpendicular to $2x - y = 3;\ b = 5$

51. Passing through $\left(2, -\dfrac{3}{2}\right)$ and $\left(3, \dfrac{1}{2}\right)$

52. Parallel to the *y*-axis and passing through $(3, -4)$

 LOOKING AHEAD TO **CALCULUS**

1. **Using Tables of Values to Help Understand the Definition of a Limit at a Point**
   - Connect the completion of tables of values used to graph equations to understanding the definition of a limit—Section R.6

2. **Using Algebraic Techniques to Evaluate a Limit at a Point**
   - Connect algebraic techniques such as factoring to the evaluation of limits—Sections R.3 and R.4

1. Using Tables of Values to Help Understand the Definition of a Limit at a Point

In calculus, a limit is very important. It is used to define essential elements of calculus, as we will see in later chapters. For this discussion, we will examine just the basics. We all have an intuitive idea of what the word *limit* means. When we drive, there are posted speed limits; elevators have weight limits; a rubber band has a limit as to how much it can be stretched.

In mathematics, it is often important to know where the value of an algebraic expression is headed as the value of the independent variable gets closer and closer to a specific number. Let's start with the simple linear equation $y = 3x + 4$. What happens to the value of y as x gets closer and closer to the value 2? Look at the table below for values of x as the values get closer and closer to 2.

x	1.000	1.900	1.990	1.999	$\cdots$	$\rightarrow$	2
y	7.000	9.700	9.970	9.997	$\cdots$	$\rightarrow$	10

The values of y seems to be approaching (getting closer and closer to) 10 as x gets closer and closer to 2 from the left (from the "less than 2" side), and we write this symbolically as

$$\lim_{x \to 2^-} (3x + 4) = 10.$$

This is read "the limit of $3x + 4$ as x approaches 2 from the left is 10." Looking at our table, we could say "for any value of x just less than 2 (no matter how close to 2), y is just less than 10." That is, as $x \to 2^-$, $y \to 10$. This is called the left-hand limit.

Now let's examine the behavior from the right of 2, meaning close to 2, but just a little greater than 2.

x	2	$\leftarrow$	$\cdots$	2.0001	2.001	2.01	2.1
y	10	$\leftarrow$	$\cdots$	10.0003	10.003	10.03	10.3

We write this symbolically as $\lim_{x \to 2^+} (3x + 4) = 10$, which indicates that as x gets closer and closer to 2 from the right side (the "greater than 2" side), the value of y gets closer and closer to 10. This is read "the limit of $3x + 4$ as x approaches 2 from the right is 10." Looking at our table, we could say "for any value of x just greater than 2 (no matter how close to 2), y is just a little greater than 10." That is, as $x \to 2^+$, $y \to 10$. This is called the right-hand limit.

Since the left-hand limit and the right-hand limit are the same, we can say that $\lim_{x \to 2} (3x + 4) = 10$. A value of x that is just less than 2 will result in a value of y that is just less than 10. Likewise, a value of x that is just greater than 2 will result in a value of y that is just greater than 10. This is demonstrated in the figure.

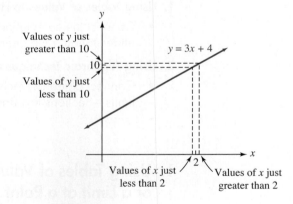

Of course there are many other types of equations to consider with limits that are beyond the scope of this section. Some will not have limits at certain values of x, and there are ways to determine that. For our discussion, we are going to only concern ourselves with limits that exist and can be evaluated with algebraic techniques. You will see how the algebraic skills you have learned are used in calculus to evaluate limits. These skills include factoring, reducing rational expressions, and simplifying complex fractions.

Connect to Calculus:
Use a Table to Evaluate a Function and to Estimate a Limit at a Point

1. For the equation $y = 3x - 5$, set up a table of function values to estimate $\lim_{x \to 3}(3x - 5)$.

2. For the equation $y = 2x^2 - 1$, set up a table of function values to estimate $\lim_{x \to 2}(2x^2 - 1)$.

2. Using Algebraic Techniques to Evaluate a Limit at a Point

There can be a left-hand limit and a right-hand limit, and they do not have to be the same. There can be a left-hand limit and not a right-hand limit, or vice versa. There are several possibilities.

Many limits can be evaluated by simply substituting the value of the variable. For example, $\lim_{x \to 3}(x^2 - 2x + 4) = (3^2 - 2(3) + 4) = 9 - 6 + 4 = 7$. Values of x just less than 3 will result in a value of y that is just less than 7. Values of x just greater than 3 will result in a value of y that is just greater than 7. This is illustrated in the following graph.

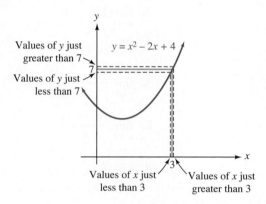

Sometimes just substituting the value of the variable will not immediately produce an answer. For example, suppose you want to evaluate $\lim_{x \to 2}\dfrac{x^2 - 4}{x - 2}$. Just substituting $x = 2$ into the expression $\dfrac{x^2 - 4}{x - 2}$ results in $\dfrac{2^2 - 4}{2 - 2} = \dfrac{0}{0}$, which is indeterminate. However, if we factor the numerator and reduce the rational expression, we can evaluate the limit.

$$\lim_{x \to 2}\frac{x^2 - 4}{x - 2} = \lim_{x \to 2}\frac{(x - 2)(x + 2)}{(x - 2)} = \lim_{x \to 2}(x + 2) = 2 + 2 = 4.$$

The graph for this function, shown in the following figure, shows that there is a "hole" in the graph at the point $(2, 4)$, but there is still a limit. The value of a limit is not determined by what the value of the expression is at a specific value of the independent variable. Rather, the value of the limit is determined by the value of the expression as the independent variable gets closer and closer to a particular value.

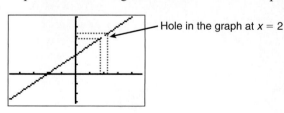

Hole in the graph at $x = 2$

Now you can see why it is so important to establish your basic algebraic skills in order to succeed in calculus.

Sometimes limits just do not exist. For example $\lim\limits_{x \to 3} \dfrac{x + 3}{x - 3}$ does not exist. The expression cannot be reduced, and when $x = 3$ is substituted, the denominator is 0 and the expression is undefined. The following graph demonstrates this.

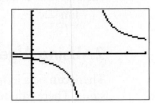

Connect to Calculus:
Using Algebraic Techniques to Evaluate a Limit

Evaluate each limit if possible. When necessary, reduce the expression to evaluate the limit. If the limit does not exist, state why.

3. $\lim\limits_{x \to 3} (2x^2 - 4)$

4. $\lim\limits_{x \to 0} \dfrac{7x}{2x + 3}$

5. $\lim\limits_{x \to 0} \dfrac{9x^2 - x + 1}{x - 2}$

6. $\lim\limits_{x \to 0} \dfrac{4x^2 - 3x}{x}$

7. $\lim\limits_{x \to 0} \dfrac{\frac{2}{x} + 1}{\frac{3}{x}}$

Graphs and Functions

1

CAREERS AND MATHEMATICS: Computer Scientist

Benis Arapovic/Shutterstock.com

Computer scientists are highly trained, innovative workers who design and invent new technology. They solve complex business, scientific, and general computing problems. Computer scientists conduct research on a variety of topics, including computer hardware, virtual reality, and robotics.

Education and Mathematics Required
- Most computer scientists are required to possess a PhD in computer science, computer engineering, or a closely related discipline. An aptitude for math is important.
- College Algebra, Trigonometry, Calculus, Linear Algebra, Ordinary Differential Equations, Theory of Analysis, Abstract Algebra, Graph Theory, Numerical Methods, and Combinatorics are math courses required.

How Computer Scientists Use Math and Who Employs Them
- Computer scientists use mathematics as they span a range of topics from theoretical studies of algorithms to the computation of implementing computing systems in hardware and software.
- Many computer scientists are employed by Internet service providers; web search portals; and firms in data processing, hosting, and related services. Others work for government, manufacturers of computer and electronic products, insurance companies, financial institutions, and universities.

Career Outlook and Earnings
- Employment of computer scientists is expected to grow by 24% through 2018, which is much faster than the average for all occupations.
- The median annual wage of computer and information scientists is approximately $98,000. Some earn more than $150,000 a year.

For more information see http://www.bls.gov/oco.

In this chapter, we will discuss one of the most important concepts in mathematics—the concept of a function.

1.1 Graphs of Equations

In this section, we will learn to

1. Find the x- and y-intercepts of a graph.
2. Determine whether a graph has x-axis, y-axis, or origin symmetry.
3. Graph an equation using intercepts and symmetry.
4. Write an equation of a circle.
5. Graph an equation of a circle.
6. Solve an equation graphically.

The London Eye opened in 2000 and is one of the world's tallest and most beautiful observation wheels. It stands 443 feet high and was the vision of architects David Marks and Julia Barfield. Its circular design was used as a metaphor for the turning of the century. The Eye has been described as a breathtaking feat of design and engineering.

The graphs of many equations are curves and circles. If we plot several points (x, y) that satisfy such an equation, the shape of the graph will usually become evident. We can then sketch the graph by joining these points with a smooth curve.

1. Find the x- and y-Intercepts of a Graph

In Figure 1-1(a), the **x-intercepts** of the graph are **a** and **b**, and the points where the graph intersects the x-axis are $(a, 0)$ and $(b, 0)$. In Figure 1-1(b), the **y-intercept** is **c** and the point where the graph intersects the y-axis is $(0, c)$.

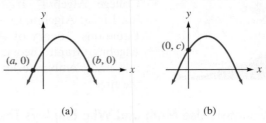

(a)   (b)

FIGURE 1-1

EXAMPLE 1   **Using Intercepts to Graph an Equation**

Graph the equation $y = x^2 - 4$ by finding the intercepts and plotting points.

SOLUTION   To graph the equation $y = x^2 - 4$, we first find the x- and y-intercepts.

To find any x-intercepts, we let $y = 0$ and solve for x.

$$y = x^2 - 4$$

$$0 = x^2 - 4 \qquad \text{Substitute 0 for } y.$$

$$0 = (x - 2)(x + 2) \qquad \text{Factor.}$$

$$x = 2 \quad \text{or} \quad x = -2 \qquad \text{Solve for the } x\text{-intercepts.}$$

Since $y = 0$ when $x = -2$ and $x = 2$, these are the x-intercepts. The points where the graph intersects the x-axis are $(-2, 0)$ and $(2, 0)$. (See Figure 1-2.)

To find the y-intercept, we let $x = 0$ and solve for y.

$$y = x^2 - 4$$

$$y = 0 - 4 \qquad \text{Substitute 0 for } x.$$

$$y = -4 \qquad \text{Solve for the } y\text{-intercept.}$$

Since $x = 0$ when $y = -4$, the y-intercept is -4, and the point where the graph intersects the y-axis is $(0, -4)$.

We can find other coordinate pairs (x, y) that satisfy the equation by substituting values for x and finding the corresponding values of y. For example, if $x = -3$ then $y = (-3)^2 - 4 = 5$, and the point $(-3, 5)$ lies on the graph.

The coordinates of the intercepts and other points appear in Figure 1-2. If we plot the points and draw a curve through them, we obtain the graph of the equation $y = x^2 - 4$.

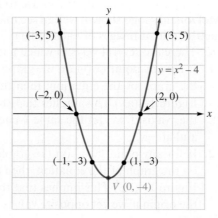

$y = x^2 - 4$		
x	y	(x, y)
-3	5	$(-3, 5)$
-2	0	$(-2, 0)$
-1	-3	$(-1, -3)$
0	-4	$(0, -4)$
1	-3	$(1, -3)$
2	0	$(2, 0)$
3	5	$(3, 5)$

FIGURE 1-2

Self Check 1   Find the intercepts of the graph of the equation $y = 6x^2 + x - 1$.

Now Try Exercise 15.

The graph of the equation in Example 1 is called a **parabola.** Its lowest point, $V(0, -4)$, is called the **vertex.** Because the y-axis divides the parabola into two congruent halves, it is called an **axis of symmetry.** We say that the parabola is **symmetric about the y-axis.**

ACCENT ON TECHNOLOGY   Graphing an Equation

A graphing calculator can be used to draw the graph of the equation $y = x^2 - 4$. When using a graphing calculator, we must enter the equation and then set the window.

- To enter the equation, press ⬚ Y= , which opens the window shown in Figure 1-3(a), and enter the right side of the equation.
- To set the window, press the ⬚WINDOW key as shown in Figure 1-3(b). This will determine the axes for the graph. We can experiment with different windows, or we can use the ⬚ZOOM menu to select some default windows.
- Press ⬚GRAPH , and the equation is graphed as shown in Figure 1-3(c).

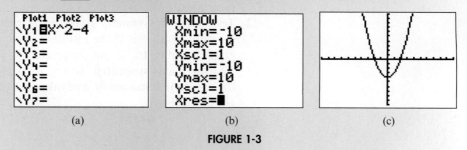

(a)   (b)   (c)

FIGURE 1-3

An equation does not necessarily have any intercepts, but if it does, they should always be found, if possible. Knowing the intercepts will greatly facilitate our ability to graph an equation.

2. Determine Whether a Graph Has *x*-Axis, *y*-Axis, or Origin Symmetry

Symmetry is a tool that we can use to help graph equations. When we look at the McDonald's logo shown, we can see that the left and right sides of the logo are symmetric (mirror images of each other).

There are several ways in which a graph can have symmetry.

1. If for any point (x, y) on the graph of an equation, the point $(-x, y)$ is also on the graph of the equation, then the graph is **symmetric about the *y*-axis.** (See Figure 1-4(a).) This implies that the graph of an equation is symmetric about the *y*-axis if we get the same *y*-coordinate when we evaluate its equation at x and at $-x$.

2. If for any point (x, y) on the graph of an equation, the point $(-x, -y)$ is also on the graph of the equation, then the graph is **symmetric about the origin.** (See Figure 1-4(b).) This implies that the graph of an equation is symmetric about the origin if we get opposite values of y when we evaluate its equation at x and at $-x$.

3. If for any point (x, y) on the graph of an equation, the point $(x, -y)$ is also on the graph of the equation, then the graph is **symmetric about the *x*-axis.** (See Figure 1-4(c).) This implies that the graph of an equation is symmetric about the *x*-axis if we get the same *x*-coordinate when we evaluate its equation at y and at $-y$.

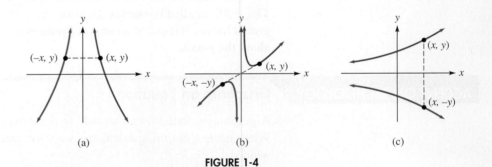

(a)   (b)   (c)

FIGURE 1-4

Tests for Symmetry for Graphs of Equations in *x* and *y*

1. **Test for *y*-axis symmetry** To test for *y*-axis symmetry, replace x with $-x$. If the resulting equation is equivalent to the original, the graph is symmetric about the *y*-axis.

2. **Test for origin symmetry** To test for symmetry about the origin, replace x with $-x$ and y with $-y$. If the resulting equation is equivalent to the original, the graph is symmetric about the origin.

3. **Test for *x*-axis symmetry** To test for *x*-axis symmetry, replace y with $-y$. If the resulting equation is equivalent to the original, the graph is symmetric about the *x*-axis.

EXAMPLE 2   **Testing for Symmetries**

Determine whether the graphs of the equations have any symmetry by using the tests for symmetry.

a. $y = 2x^2 - 3$ **b.** $x^2 + 2y^2 = 10$

SOLUTION   **a.** For the equation $y = 2x^2 - 3$, to test for y-axis symmetry, we replace x with $-x$.

(1) $y = 2x^2 - 3$ The original equation

$y = 2(-x)^2 - 3$ Replace x with $-x$.

(2) $y = 2x^2 - 3$ $(-x)^2 = x^2$

Since Equations (1) and (2) are the same, the graph is symmetric about the y-axis. To test for symmetry about the origin, we replace x with $-x$ and y with $-y$.

(1) $y = 2x^2 - 3$ The original equation

$-y = 2(-x)^2 - 3$ Replace x with $-x$ and y with $-y$.

(3) $-y = 2x^2 - 3$ $(-x)^2 = x^2$

Since Equations (1) and (3) are different, the graph is not symmetric about the origin. To test for x-axis symmetry, we replace y with $-y$.

(1) $y = 2x^2 - 3$ The original equation

(4) $-y = 2x^2 - 3$ Replace y with $-y$.

Since Equations (1) and (4) are different, the graph is not symmetric about the x-axis.

b. For the equation $x^2 + 2y^2 = 10$, to test for y-axis symmetry, we replace x with $-x$.

(5) $x^2 + 2y^2 = 10$ The original equation

$(-x)^2 + 2y^2 = 10$ Replace x with $-x$.

(6) $x^2 + 2y^2 = 10$ $(-x)^2 = x^2$

Since Equations (5) and (6) are the same, the graph is symmetric about the y-axis. To test for symmetry about the origin, we replace x with $-x$ and y with $-y$.

(5) $x^2 + 2y^2 = 10$ The original equation

$(-x)^2 + 2(-y)^2 = 10$ Replace x with $-x$ and y with $-y$.

(7) $x^2 + 2y^2 = 10$ $(-x)^2 = x^2, (-y)^2 = y^2$

Since Equations (5) and (7) are the same, the graph is symmetric about the origin. To test for x-axis symmetry, we replace y with $-y$.

(5) $x^2 + 2y^2 = 10$ The original equation

$x^2 + 2(-y)^2 = 10$ Replace y with $-y$.

(8) $x^2 + 2y^2 = 10$ $(-y)^2 = y^2$

Since Equations (5) and (8) are the same, the graph is symmetric about the x-axis.

Self Check 2   Determine whether the equation $y = 3x^3 - 2x$ has any symmetry.

Now Try Exercise 33.

3. Graph an Equation Using Intercepts and Symmetry

EXAMPLE 3   **Graphing an Equation Using Intercepts and Symmetry**

Graph $y = |x|$ by finding all intercepts, testing for symmetry, plotting a few points, and joining the points with a smooth curve.

SOLUTION   **Find the x-intercept(s).** To find the x-intercept(s), we let $y = 0$ and solve for x.

$$y = |x|$$

$$0 = |x| \qquad \text{Substitute 0 for } y.$$

$$x = 0 \qquad x\text{-intercept}$$

The x-intercept is 0, and the graph intercepts the x-axis at $(0, 0)$. (See Figure 1-5.)

Find the y-intercept(s). To find the y-intercept(s), we let $x = 0$ and solve for y.

$$y = |x|$$

$$y = |0| \qquad \text{Substitute 0 for } x.$$

$$y = 0 \qquad y\text{-intercept}$$

The y-intercept is 0, and the graph intercepts the y-axis at $(0, 0)$.

Test for symmetry. To test for y-axis symmetry, we replace x with $-x$.

(1) $y = |x|$ \qquad The original equation

$y = |-x|$ \qquad Replace x with $-x$.

(2) $y = |x|$ \qquad $|-x| = |x|$

Since Equations 1 and 2 are the same, the graph is symmetric about the y-axis.
To test for symmetry about the origin, we replace x with $-x$ and y with $-y$.

(1) $y = |x|$ \qquad The original equation

$-y = |-x|$ \qquad Replace x with $-x$ and y with $-y$.

(3) $-y = |x|$ \qquad $|-x| = |x|$

Since Equations 1 and 3 are different, the graph is not symmetric about the origin.
To test for x-axis symmetry, we replace y with $-y$.

(1) $y = |x|$ \qquad The original equation

(4) $-y = |x|$ \qquad Replace y with $-y$.

Since Equations 1 and 4 are different, the graph is not symmetric about the x-axis.
To graph the equation, we plot the x- and y-intercepts and several other pairs (x, y) with positive values of x. We can use the property of y-axis symmetry to draw the graph for negative values of x. (See Figure 1-5.)

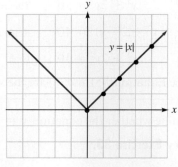

\multicolumn{3}{c}{$y = |x|$}		
x	y	(x, y)
0	0	$(0, 0)$
1	1	$(1, 1)$
2	2	$(2, 2)$
3	3	$(3, 3)$
4	4	$(4, 4)$

FIGURE 1-5

Self Check 3   Graph $y = -|x|$.

Now Try Exercise 51.

ACCENT ON TECHNOLOGY   **Graphing an Absolute Value Equation**

A graphing calculator can be used to draw the graph of the equation $y = |0.5x|$.

- To enter the equation, press [Y=], which opens the window shown in Figure 1-6(a). To enter the absolute value function into equation Y1, you need to access the absolute value function on your calculator. On the TI-84, press [MATH] and move cursor to NUM as shown in Figure 1-6(b). Press [ENTER] and input the function in the graph window [Figure 1-6(c)].
- To set the window, press [ZOOM] [6] [Figure 1-6(d)].
- The equation is graphed as shown in Figure 1-6(e).

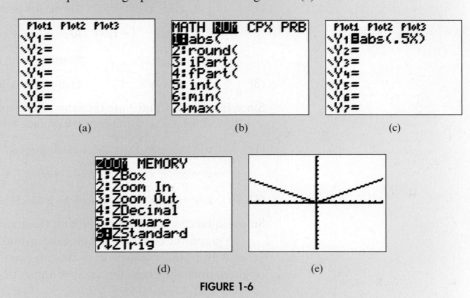

(a)   (b)   (c)

(d)   (e)

FIGURE 1-6

EXAMPLE 4   **Graphing an Equation Using Intercepts and Symmetry**

Graph $y = x^3 - x$ by finding all intercepts, testing for symmetry, plotting a few points, and joining the points with a smooth curve.

SOLUTION   **Find the x-intercept(s).** To find the x-intercept(s), we let $y = 0$ and solve for x.

$$y = x^3 - x$$

$$0 = x^3 - x \qquad \text{Substitute 0 for } y.$$

$$0 = x(x^2 - 1) \qquad \text{Factor out } x.$$

$$0 = x(x + 1)(x - 1) \qquad \text{Factor } x^2 - 1.$$

$$x = 0 \quad \text{or} \quad x + 1 = 0 \quad \text{or} \quad x - 1 = 0 \qquad \text{Set each factor equal to 0.}$$

$$x = -1 \qquad x = 1$$

The x-intercepts are 0, 1, and -1, and the graph intersects the x-axis at the points $(0, 0)$, $(-1, 0)$, and $(1, 0)$.

Find the y-intercept(s). To find the y-intercept(s), we let $x = 0$ and solve for y.

$$y = x^3 - x$$

$$y = 0^3 - 0 \qquad \text{Substitute 0 for } x.$$

$$y = 0$$

The y-intercept is 0, and the graph intersects the y-axis at the point $(0, 0)$.

Test for symmetry. To test for y-axis symmetry, we replace x with $-x$.

(1) $y = x^3 - x$ The original equation

       $y = (-x)^3 - (-x)$ Replace x with $-x$.

(2) $y = -x^3 + x$ Simplify.

Since Equations 1 and 2 are different, the graph is not symmetric about the y-axis. To test for symmetry about the origin, we replace x with $-x$ and y with $-y$.

(1) $y = x^3 - x$ The original equation

       $-y = (-x)^3 - (-x)$ Replace x with $-x$ and y with $-y$.

       $-y = -x^3 + x$ Simplify.

(3) $y = x^3 - x$ Multiply both sides by -1.

Since Equations 1 and 3 are the same, the graph is symmetric about the origin. To test for x-axis symmetry, we replace y with $-y$.

(1) $y = x^3 - x$ The original equation

       $-y = x^3 - x$ Replace y with $-y$.

(4) $y = -x^3 + x$ Multiply both sides by -1.

Since Equations 1 and 4 are different, the graph is not symmetric about the x-axis.

To graph the equation, we plot the x- and y-intercepts and several other pairs (x, y) with positive values of x. We can use the property of symmetry about the origin to draw the graph for negative values of x. (See Figure 1-7.)

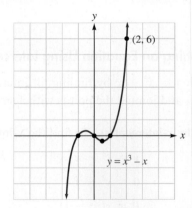

$y = x^3 - x$		
x	y	(x, y)
-1	0	$(-1, 0)$
0	0	$(0, 0)$
$\dfrac{1}{2}$	$-\dfrac{3}{8}$	$\left(\dfrac{1}{2}, -\dfrac{3}{8}\right)$
1	0	$(1, 0)$
2	6	$(2, 6)$

FIGURE 1-7

Self Check 4   Graph: $y = x^3 - 9x$.

Now Try Exercise 49.

ACCENT ON TECHNOLOGY   **Finding the *x*-Intercepts of a Graph**

A graphing calculator can find the intercepts of an equation, such as
$y = 2x^3 - 3x$, as shown in Section R.6. The easiest way to find the *x*-intercepts is
to use the ZERO feature. An *x*-intercept is also called a **zero**.

- Enter the function into Y_1 [Figure 1-8(a)].
- Set the window. Use [ZOOM] [4] for this graph [Figures 1-8(b) and (c)].
- To find the zeros, you need to access the CALC menu, which is found by press-
  ing [2nd] [TRACE] [2] [Figure 1-8(d)].
- Move the cursor until it is to the left of the zero you are trying to find and press
  [ENTER] [Figure 1-8(e)].
- Repeat the previous step by moving the cursor to the right of the zero. Press
  [ENTER] **two times**. Do not stop when the screen shows "Guess?" [Figures 1-8
  (f)−(h)].
- The zero is $x \approx -1.224745$.
- Repeat the steps to find the zero $x \approx 1.224745$.
- We can see that another zero is $x = 0$.

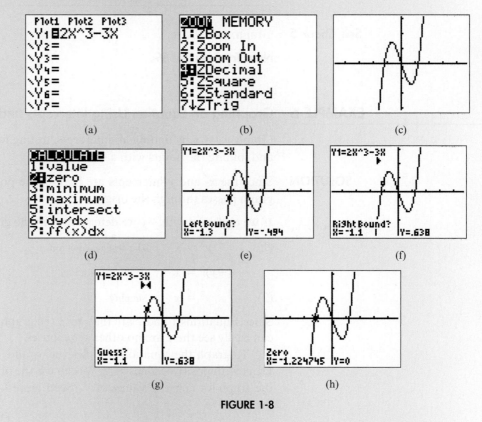

(a)   (b)   (c)

(d)   (e)   (f)

(g)   (h)

FIGURE 1-8

EXAMPLE 5   **Graphing an Equation Using Intercepts and Symmetry**

Graph $y = \sqrt{x}$ by finding all intercepts, testing for symmetry, plotting a few points,
and joining the points with a smooth curve.

SOLUTION   We can see that the x- and y-intercepts are both 0. We now test for symmetries. Since $x \geq 0$ in the radical $\sqrt{x}$, there is no need to replace x with $-x$. Since $y = \sqrt{x}$ and $\sqrt{x} \geq 0$, there is also no need to replace y with $-y$. The graph of the equation has no symmetries. We plot several points to obtain the graph in Figure 1-9.

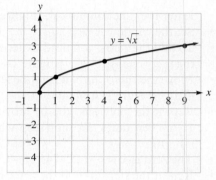

$y = \sqrt{x}$		
x	y	(x, y)
0	0	$(0, 0)$
1	1	$(1, 1)$
4	2	$(4, 2)$
9	3	$(9, 3)$

FIGURE 1-9

Self Check 5   Graph: $y = -\sqrt{x}$.

Now Try Exercise 59.

EXAMPLE 6   **Graphing an Equation Using Intercepts and Symmetry**

Graph $y^2 = x$ by finding all intercepts, testing for symmetry, plotting a few points, and joining the points with a smooth curve.

SOLUTION   Since the x- and y-intercepts are both 0, the point $(0, 0)$ is on the graph, and the graph passes through the origin.

If we replace y with $-y$ we determine that the graph has x-axis symmetry.

(1) $y^2 = x$ The original equation

        $(-y)^2 = x$ Replace y with $-y$.

(2) $y^2 = x$ Simplify.

Since Equations 1 and 2 are the same, the graph is symmetric about the x-axis. We can easily see there are no other symmetries.

   To graph the equation, we plot the x- and y-intercept and several other pairs (x, y) with positive values of y. We can use the property of x-axis symmetry to draw the graph for negative values of y. (See Figure 1-10.)

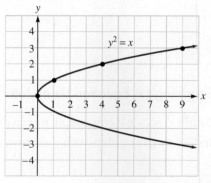

$y^2 = x$		
x	y	(x, y)
0	0	$(0, 0)$
1	1	$(1, 1)$
4	2	$(4, 2)$
9	3	$(9, 3)$

FIGURE 1-10

Self Check 6   Graph: $y^2 = -x$.

Now Try Exercise 57.

ACCENT ON TECHNOLOGY   **Graphing an Equation with y^2**

A graphing calculator is programmed to graph equations with y defined in terms of x. However, $y^2 = 2x$ has x defined in terms of y. Therefore, it will be necessary to solve for y and then enter the resulting equations to see the graph of $y^2 = 2x$.

$$y^2 = 2x$$

$$y = \pm\sqrt{2x}$$

$$y = \sqrt{2x} \quad \text{or} \quad y = -\sqrt{2x}$$

- Enter the two equations in the graph menu [Figure 1-11(a)].
- Then set the WINDOW to show the graph [Figure 1-11(b)].
- Graph the equation(s) [Figure 1-11(c)].

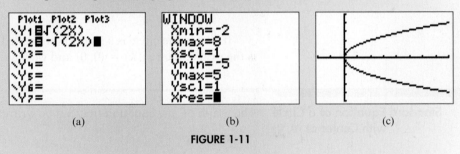

(a)   (b)   (c)

FIGURE 1-11

The graphs in Examples 5 and 6 are related. We have solved the equation $y^2 = x$ for y, and two equations resulted:

$$y = \sqrt{x} \quad \text{and} \quad y = -\sqrt{x}$$

The first equation, $y = \sqrt{x}$, was graphed in Example 5. It is the top half of the parabola shown in Example 6. The second equation, $y = -\sqrt{x}$, is the bottom half.

4. Write an Equation of a Circle

Circle   A **circle** is the set of all points in a plane that are a fixed distance from a point called its **center**. The fixed distance is the **radius of the circle**.

To find an equation of a circle with radius r and center at $C(h, k)$, we must find all points $P(x, y)$ in the xy plane such that the length of the line segment PC is r. (See Figure 1-12.) We can use the distance formula to find the length of CP, which is r:

$$r = \sqrt{(x - h)^2 + (y - k)^2}.$$

After squaring both sides, we get

$$r^2 = (x - h)^2 + (y - k)^2.$$

This equation is called the **standard equation of a circle.**

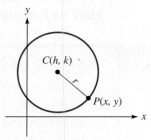

FIGURE 1-12

Standard Equation of a Circle with Center at (h, k)

The graph of any equation that can be written in the form

$$(x - h)^2 + (y - k)^2 = r^2$$

is a circle with radius r and center at point (h, k).

If $r = 0$, the circle is a single point called a **point circle.** If the center of a circle is the origin, then $(h, k) = (0, 0)$ and we have the following result:

Standard Equation of a Circle with Center at $(0, 0)$

The graph of any equation that can be written in the form

$$x^2 + y^2 = r^2$$

is a circle with radius r and center at the origin.

Equations of four circles written in standard form, and their graphs, are shown in Figure 1-13.

Examples of Circles in Standard Form

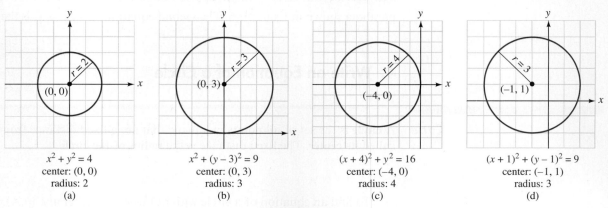

$x^2 + y^2 = 4$
center: $(0, 0)$
radius: 2
(a)

$x^2 + (y - 3)^2 = 9$
center: $(0, 3)$
radius: 3
(b)

$(x + 4)^2 + y^2 = 16$
center: $(-4, 0)$
radius: 4
(c)

$(x + 1)^2 + (y - 1)^2 = 9$
center: $(-1, 1)$
radius: 3
(d)

FIGURE 1-13

If we square the binomials in $(x - h)^2 + (y - k)^2 = r^2$, we obtain an equation of the form

$$x^2 + y^2 + cx + dy + e = 0,$$

where c, d, and e are real numbers. This form is called the **general form of an equation of a circle.**

EXAMPLE 7   **Finding the General Form of an Equation of a Circle**

Find an equation in general form of the circle with radius 5 and center at $(3, 2)$.

SOLUTION   We substitute 5 for r, 3 for h, and 2 for k in the standard equation of a circle and simplify:

$$(x - \mathbf{h})^2 + (y - k)^2 = r^2$$

$$(x - \mathbf{3})^2 + (y - 2)^2 = \mathbf{5}^2$$

$$x^2 - 6x + 9 + y^2 - 4y + 4 = 25 \qquad \text{Remove parentheses.}$$

$$x^2 + y^2 - 6x - 4y - 12 = 0 \qquad \text{Subtract 25 from both sides and simplify.}$$

The general form is $x^2 + y^2 - 6x - 4y - 12 = 0$. ∎

Self Check 7   Find an equation in general form of a circle with radius 6 and center at $(-2, 5)$.

Now Try Exercise 83.

EXAMPLE 8   **Finding an Equation of a Circle in General Form**

Find an equation in general form of the circle with endpoints of its diameter at $(8, -3)$ and $(-4, 13)$.

SOLUTION   We can find the center $C(h, k)$ of the circle by finding the midpoint of its diameter. Using the midpoint formula and $(x_1, y_1) = (8, -3)$ and $(x_2, y_2) = (-4, 13)$, we have

$$h = \frac{\mathbf{x_1} + x_2}{2} \qquad\qquad k = \frac{\mathbf{y_1} + y_2}{2}$$

$$h = \frac{\mathbf{8} + (-4)}{2} \qquad\qquad k = \frac{\mathbf{-3} + 13}{2}$$

$$= \frac{4}{2} = 2 \qquad\qquad = \frac{10}{2} = 5$$

The center is $C(h, k) = C(2, 5)$.

To find the radius, we find the distance between the center and one endpoint of the diameter. The center is $C(2, 5)$ and one endpoint is $(8, -3)$.

$$r = \sqrt{(\mathbf{x_2} - x_1)^2 + (\mathbf{y_2} - \mathbf{y_1})^2} \qquad \text{This is the distance formula.}$$

$$r = \sqrt{(\mathbf{2} - 8)^2 + [5 - (-\mathbf{3})]^2} \qquad \text{Substitute 8 for } x_1, -3 \text{ for } y_1, 2 \text{ for } x_2, \text{ and 5 for } y_2.$$

$$= \sqrt{(-6)^2 + (8)^2}$$

$$= \sqrt{36 + 64}$$

$$= 10$$

To find the general form of an equation of a circle with center at $(2, 5)$ and radius 10, we substitute 2 for h, 5 for k, and 10 for r in the standard equation of the circle and simplify:

$$(x - \mathbf{h})^2 + (y - k)^2 = r^2$$

$$(x - \mathbf{2})^2 + (y - 5)^2 = \mathbf{10}^2$$

$$x^2 - 4x + 4 + y^2 - 10y + 25 = 100 \qquad \text{Remove parentheses.}$$

$$x^2 + y^2 - 4x - 10y - 71 = 0 \qquad \text{Subtract 100 from both sides and simplify.}$$

∎

Self Check 8   Find the general form of an equation of a circle with endpoints of its diameter at $(-2, 2)$ and $(6, 8)$.

Now Try Exercise 87.

5. Graph an Equation of a Circle

If we look at the standard form of an equation of a circle, $(x - h)^2 + (y - k)^2 = r^2$, we can immediately identify the coordinates of the center (h, k) and the radius r. Consequently, it is easy to sketch the graph. However, what do we do when graphing a circle if we are given an equation in general form, $x^2 + y^2 + cx + dy + e = 0$? Our strategy is to transform the equation from general form to standard form by completing the square on both x and y.

EXAMPLE 9   **Graphing a Circle with an Equation in General Form**

Graph the circle whose equation is $2x^2 + 2y^2 - 8x + 4y - 40 = 0$.

SOLUTION   We first add 40 to both sides and then divide both sides of the equation by 2 to make the coefficients of x^2 and y^2 equal to 1.

$$2x^2 + 2y^2 - 8x + 4y = 40 \qquad \text{Add 40 to both sides.}$$
$$x^2 + y^2 - 4x + 2y = 20$$

To find the coordinates of the center and the radius, we write the equation in standard form by completing the square on both x and y:

$$x^2 + y^2 - 4x + 2y = 20$$
$$x^2 - 4x + y^2 + 2y = 20$$
$$x^2 - 4x + \mathbf{4} + y^2 + 2y + \mathbf{1} = 20 + \mathbf{4} + \mathbf{1} \qquad \text{Add 4 and 1 to both sides to complete the square.}$$
$$(x - 2)^2 + (y + 1)^2 = 25 \qquad \text{Factor } x^2 - 4x + 4 \text{ and } y^2 + 2y + 1.$$
$$(x - 2)^2 + [y - (-1)]^2 = 5^2$$

From this equation of the circle, we see that its radius is 5 and the coordinates of its center are $h = 2$ and $k = -1$. The graph is shown in Figure 1-14.

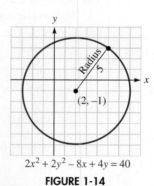

$$2x^2 + 2y^2 - 8x + 4y = 40$$

FIGURE 1-14

Self Check 9   Graph: $2x^2 + 2y^2 + 4x - 8y = -2$.

Now Try Exercise 99.

ACCENT ON TECHNOLOGY   **Graphing a Circle**

When we were looking at how to graph $y^2 = 2x$ using a graphing calculator, we saw that it is necessary to solve the equation for y in terms of x. The same is true when we graph a circle on a graphing calculator. To graph $(x - 2)^2 + (y + 1)^2 = 25$ we must solve the equation for y:

$$(x - 2)^2 + (y + 1)^2 = 25$$

$$(y + 1)^2 = 25 - (x - 2)^2$$

$$y + 1 = \pm\sqrt{25 - (x - 2)^2}$$

$$y = -1 \pm \sqrt{25 - (x - 2)^2}$$

This last expression represents two equations: $y = -1 + \sqrt{25 - (x - 2)^2}$ and $y = -1 - \sqrt{25 - (x - 2)^2}$. We graph both of these equations separately on the same coordinate axes by entering the first equation as Y₁ and the second as Y₂, as shown in Figure 1-15(a). Depending on the settings of the maximum and minimum values of x and y, the graph may not appear to be a circle. However, if you use a "square" window [Figure 1-15(b)], the graph will be circular. If it appears that there are gaps in the graph [Figure 1-15(c)], that is due to the way the calculator draws graphs by darkening pixels.

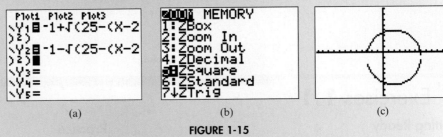

| (a) | (b) | (c) |

FIGURE 1-15

6. Solve an Equation Graphically

We can solve many equations using the graphing concepts discussed in this section and a graphing calculator. For example, the solutions of $x^2 - x - 3 = 0$ will be the numbers x that will make $y = 0$ in the equation $y = x^2 - x - 3$. These numbers will be the x-intercepts of the graph of the equation. Some equations can be solved algebraically using factoring and other techniques that we will study. For many equations, there are no algebraic techniques that can be used to solve them, and a graphing calculator is recommended. A graphing calculator can be used to solve most equations that we will encounter in this course. To use a graphing calculator to solve an equation, we graph the equation and find the x-intercepts. The x-intercepts will be the solutions of the equation. To find these intercepts, refer to the Accent on Technology earlier in this section (Finding the Intercepts of a Graph).

Self Check Answers   **1.** x-intercepts $\dfrac{1}{3}$ and $-\dfrac{1}{2}$, y-intercept -1 **2.** origin symmetry

3.

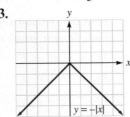

4.

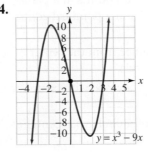

5.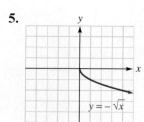

$y = -\sqrt{x}$

6.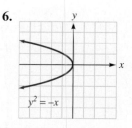

$y^2 = -x$

7. $x^2 + y^2 + 4x - 10y - 7 = 0$ **8.** $x^2 + y^2 - 4x - 10y + 4 = 0$

9.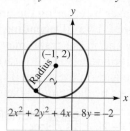

$(-1, 2)$

Radius 2

$2x^2 + 2y^2 + 4x - 8y = -2$

Exercises **1.1**

Getting Ready

You should be able to complete these vocabulary and concept statements before you proceed to the practice exercises.

Fill in the blanks.

1. The x-coordinate of the point where a graph intersects the x-axis is called the _____.

2. The y-intercept is the y-coordinate of the point where a graph intersects the _____.

3. If a line divides a graph into two congruent halves, we call the line an _____.

4. If the point $(-x, y)$ lies on a graph whenever (x, y) does, the graph is symmetric about the _____.

5. If the point $(x, -y)$ lies on a graph whenever (x, y) does, the graph is symmetric about the _____.

6. If the point $(-x, -y)$ lies on a graph whenever (x, y) does, the graph is symmetric about the _____.

7. A _____ is the set of all points in a plane that are a fixed distance from a point called its _____.

8. A _____ is the distance from the center of a circle to a point on the circle.

9. The standard equation of a circle with center at the origin and radius r is _____.

10. The standard equation of a circle with center at (h, k) and radius r is _____.

Practice

Find the x- and y-intercepts of each graph. Do not graph the equation.

11. $y = x^2 - 4$

12. $y = x^2 - 9$

13. $y = 4x^2 - 2x$

14. $y = 2x - 4x^2$

15. $y = x^2 - 4x - 5$

16. $y = x^2 - 10x + 21$

17. $y = x^2 + x - 2$

18. $y = x^2 + 2x - 3$

19. $y = x^3 - 9x$

20. $y = x^3 + x$

21. $y = x^4 - 1$

22. $y = x^4 - 25x^2$

Graph each equation. Check your graph with a graphing calculator.

23. $y = x^2$

24. $y = -x^2$

25. $y = -x^2 + 2$

26. $y = x^2 - 1$

27. $y = x^2 - 4x$

28. $y = x^2 + 2x$

29. $y = \dfrac{1}{2}x^2 - 2x$

30. $y = \dfrac{1}{2}x^2 + 3$

Find the symmetries, if any, of the graph of each equation. Do not graph the equation.

31. $y = x^2 + 2$

32. $y = 3x + 2$

33. $y^2 + 1 = x$

34. $y^2 + y = x$

35. $y^2 = x^2$

36. $y = 3x + 7$

37. $y = 3x^2 + 7$

38. $x^2 + y^2 = 1$

39. $y = 3x^3 + 7$

40. $y = 3x^3 + 7x$

41. $y^2 = 3x$

42. $y = 3x^4 + 7$

43. $y = |x|$

44. $y = |x + 1|$

45. $|y| = x$

46. $|y| = |x|$

Graph each equation. Be sure to find any intercepts and symmetries. Check your graph with a graphing calculator.

47. $y = x^2 + 4x$ **48.** $y = x^2 - 6x$

49. $y = x^3$ **50.** $y = x^3 + x$

51. $y = |x - 2|$ **52.** $y = |x| - 2$

53. $y = 3 - |x|$ **54.** $y = 3|x|$

55. $y^2 = -x$ **56.** $y^2 = 4x$

57. $y^2 = 9x$ **58.** $y^2 = -4x$

59. $y = \sqrt{x} - 1$ **60.** $y = 1 - \sqrt{x}$

61. $xy = 4$ **62.** $xy = -9$

Identify the center and radius of each circle written in standard form.

63. $x^2 + y^2 = 100$ **64.** $x^2 + y^2 = 81$

65. $x^2 + (y - 5)^2 = 49$ **66.** $x^2 + (y + 3)^2 = 8$

67. $(x + 6)^2 + y^2 = \dfrac{1}{4}$ **68.** $(x - 5)^2 + y^2 = \dfrac{16}{25}$

69. $(x - 4)^2 + (y - 1)^2 = 9$

70. $(x + 11)^2 + (x + 7)^2 = 121$

71. $\left(x - \dfrac{1}{4}\right)^2 + (y + 2)^2 = 45$

72. $\left(x + \sqrt{5}\right)^2 + (y - 3)^2 = 1$

Write an equation in standard form of each circle with the given properties.

73. Center at the origin; $r = 5$

74. Center at the origin; $r = \sqrt{3}$

75. Center at $(0, -6)$; $r = 6$

76. Center at $(0, 7)$; $r = 9$

77. Center at $(8, 0)$; $r = \dfrac{1}{5}$

78. Center at $(-10, 0)$; $r = \sqrt{11}$

79. Center at $(2, 12)$; $r = 13$

80. Center at $\left(\dfrac{2}{7}, -5\right)$; $r = 7$

Find an equation in general form of each circle with the given properties.

81. Center at the origin; $r = 1$

82. Center at the origin; $r = 4$

83. Center at $(6, 8)$; $r = 4$

84. Center at $(5, 3)$; $r = 2$

85. Center at $(3, -4)$; $r = \sqrt{2}$

86. Center at $(-9, 8)$; $r = 2\sqrt{3}$

87. Ends of diameter at $(3, -2)$ and $(3, 8)$

88. Ends of diameter at $(5, 9)$ and $(-5, -9)$

89. Center at $(-3, 4)$ and passing through the origin

90. Center at $(-2, 6)$ and passing through the origin

Graph each equation.

91. $x^2 + y^2 - 25 = 0$ **92.** $x^2 + y^2 - 8 = 0$

93. $(x - 1)^2 + (y + 2)^2 = 4$ **94.** $(x + 1)^2 + (y - 2)^2 = 9$

95. $x^2 + y^2 + 2x - 24 = 0$ **96.** $x^2 + y^2 - 4y = 12$

97. $9x^2 + 9y^2 - 12y = 5$ **98.** $4x^2 + 4y^2 + 4y = 15$

99. $4x^2 + 4y^2 - 4x + 8y + 1 = 0$

100. $9x^2 + 9y^2 - 6x + 18y + 1 = 0$

Use a graphing calculator to find the coordinates of the vertex to the nearest hundredth.

101. $y = 2x^2 - x + 1$ **102.** $y = x^2 + 5x - 6$

103. $y = 7 + x - x^2$ **104.** $y = 2x^2 - 3x + 2$

Use a graphing calculator to solve each equation. Round to the nearest hundredth.

105. $x^2 - 7 = 0$ **106.** $x^2 - 3x + 2 = 0$

107. $x^3 - 3 = 0$ **108.** $3x^3 - x^2 - x = 0$

Applications

109. Golfing Tiger Woods's tee shot follows a path given by $y = 64t - 16t^2$, where y is the height of the ball (in feet) after t seconds of flight. How long will it take for the ball to strike the ground?

110. Golfing Halfway through its flight, the golf ball of Exercise 109 reaches the highest point of its trajectory. How high is that?

111. Stopping distances The stopping distance D (in feet) for a Ford Fusion car moving V miles per hour is given by $D = 0.08V^2 + 0.9V$. Graph the equation for velocities between 0 and 60 mph.

112. Stopping distances See Exercise 111. How much farther does it take to stop at 60 mph than at 30 mph?

113. Basketball court The center circle of the Kansas Jayhawks' basketball court has a 12-foot diameter. If the center of the circle is located at the origin, find an equation in standard form that models the circle.

114. Oil spill Oil spills from a tanker in the Gulf of Mexico and surfaces continuously at coordinates $(0, 0)$. If oil spreads in a circular pattern for 10 hours and the circle's radius increases at a rate of 2 inches per hour, write an equation of the circle that models the range of the spill's effect.

115. Roller coaster The Fire Ball Super Loop is a roller coaster that is shaped like a circle. Find a general equation of the loop if it is positioned 5 feet off of the ground and has a diameter of 60 feet and a center at coordinates (0, 35).

Kenneth William Caleno/Shutterstock.com

116. Hurricane As a hurricane strengthens, an eye begins to form at the center of the storm. At a wind speed of 80 mph, the eye of a hurricane is circular when viewed from above and is 30 miles in diameter. If the eye is located at map coordinates (5, 10), write an equation of the circle that models the eye of the hurricane.

117. CB radios The CB radio of a trucker covers the circular area shown in the illustration. Find an equation of that circle, in general form.

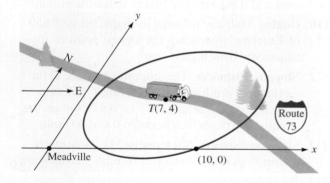

118. Firestone tires Two 24-inch-diameter Firestone tires stand against a wall, as shown in the illustration. Find an equation in general form of the circular boundaries of the tires.

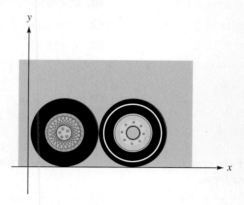

Discovery and Writing

The solution of the inequality $y < 0$ consists of those numbers x for which the graph of y lies below the x-axis. To solve $y < 0$ we graph y and trace to find numbers x that produce negative values of y. Solve each inequality.

119. $x^2 + x - 6 < 0$ **120.** $x^2 - 3x - 10 > 0$

When writing an equation of a circle in standard form, it is possible to obtain a constant term that is zero or negative. If the constant term is zero, the graph is a single point. If the constant term is negative, the graph is nonexistent. Determine whether the graph of the equation is a single point or nonexistent.

121. $x^2 - 4x + y^2 - 8y + 20 = 0$

122. $x^2 - 12x + y^2 + 4y + 43 = 0$

1.2 Functions and Function Notation

In this section, we will learn to

1. Identify ordered pairs, domain, and range of a relation.
2. Define a function and determine the domain and range of a function.
3. Use functional notation and evaluate a function, given a specific input.
4. Evaluate the difference quotient, given a function.
5. Solve applications involving functions.

This table shows the four highest-grossing movies of all time and the year each movie was released:

Movie	Year
Avatar	2009
The Lord of the Rings: The Return of the King	2003
Pirates of the Caribbean: Dead Man's Chest	2006
The Dark Knight	2008

The information shown in the table sets up a correspondence between a movie and the year it was released. Note that for each movie listed, there corresponds exactly one year in which it was released. The phase *there corresponds exactly one* is extremely important in our understanding of the concept of a function, the topic of this section.

1. Identify Ordered Pairs, Domain, and Range of a Relation

Functional relationships are studied throughout mathematics because they exist in the world around us. There is a relationship between supply and demand, between IQ and accomplishment, and between the amount of interest earned and both the amount invested and the rate of return. When we fill up our car with gas, the amount we pay is determined by the price per gallon and how many gallons we buy.

A **relation** is a correspondence between a set of input values (called the *domain*) and a set of output values (called the *range*). For example, a golf course has 18 holes, and each hole has a length in yards. The domain of the relation would be the numbers 1 through 18, and the range would be the list of the yardages.

Relation   A **relation** is a set of ordered pairs. The set created by the first components in the ordered pairs is called the **domain** (the inputs). The set created by the second components in the ordered pairs is the **range** (the outputs).

EXAMPLE 1   **Writing the Ordered Pairs in a Relation and Identifying the Domain and Range**

The scorecard for a golf course is shown in Figure 1-16. List the set of ordered pairs {Hole number, Yards}, and then state the domain and range created by this set of ordered pairs.

Hole	1	2	3	4	5	6	7	8	9	10	11	12	13	14	15	16	17	18
Yards	269	315	140	390	365	120	451	358	375	465	298	165	315	495	135	330	358	420

FIGURE 1-16

SOLUTION   The set of ordered pairs is {(1, 269), (2, 315), (3, 140), (4, 390), (5, 365), (6, 120), (7, 451), (8, 358), (9, 375), (10, 465), (11, 298), (12, 165), (13, 315), (14, 495), (15, 135), (16, 330), (17, 358), (18, 420)}.

The domain is the set created by the first components in the ordered pairs:

$$D = \{1, 2, 3, 4, 5, 6, 7, 8, 9, 10, 11, 12, 13, 14, 15, 16, 17, 18\}$$

The range is the set created by the second components in the ordered pairs:

$$R = \{269, 315, 140, 390, 365, 120, 451, 358, 375, 465, 298, 165, 315, 495, 135, 330, 358, 420\}$$

Self Check 1   Five Ford Fusion sedan styles and prices are shown.

I-4 S	$19,696
I-4 SE	$21,225
I-4 SEL	$24,655
Sport	$26,505
V-6 SEL	$28,115

Write the set of ordered pairs {Style, Price} represented by the data. State the domain and range created by this set of ordered pairs.

Now Try Exercise 9a.

EXAMPLE 2   **Determining the Ordered Pairs in a Relation and Identifying the Domain and Range**

Use Figure 1-17 to list the ordered pairs in each relation illustrated. State the domain and range of each relation.

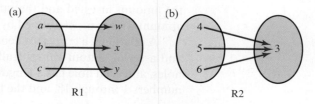

(a)   R1   (b)   R2

FIGURE 1-17

SOLUTION   (a) $R1 = \{(a, w), (b, x), (c, y)\}$; Domain $= \{a, b, c\}$; Range $= \{w, x, y\}$ (b) $R2 = \{(4, 3), (5, 3), (6, 3)\}$; Domain $= \{4, 5, 6\}$, Range $= \{3\}$

Self Check 2   Use Figure 1-18 to list the ordered pairs in the relation illustrated. State the domain and range of the relation.

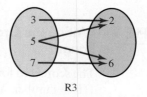

R3

FIGURE 1-18

Now Try Exercise 11a.

Now let's look at a special and very important type of relation called a **function**.

2. Define a Function and Determine the Domain and Range of a Function

Any relation that assigns exactly one value of y to each value of x is called a **function**.

Function   A *function* f is a correspondence between a **set of input values** x (called the **domain**) and a **set of output values** y (called the **range**), where to each x value in the domain, there corresponds exactly one y value in the range.

There are common occurrences of a relation of this type in everyday life.

*   To every person, there is only one birthday.
*   To every house, there is only one address.
*   To every car, there is only one license plate.
*   The solar-powered aircraft *Pathfinder* recorded the following data relating altitude to temperature. This chart is an example of a function. For each altitude, there is exactly one temperature. The set of ordered pairs created can be plotted on a graph. (The graph in Figure 1-19 was created on a graphing calculator using LIST and STAT Plot.)

Altitude (1,000 ft)	Temperature (°C)
15	4.5
20	−5.9
25	−16.1
30	−27.9
35	−39.8
40	−50.2
45	−62.9

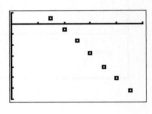

FIGURE 1-19

Caution

Every function is a relation, but **not** every relation is a function.

In Example 2, the relations established in parts (a) and (b) are also functions, because to every value in the domain, there is assigned only one value in the range. However, the relation in Self Check 2 is not a function, because the value 5 in the domain is assigned two values in the range, 2 and 6. In a relation that is also a function, a number in the range (the second component in the ordered pair) can appear in more than one ordered pair, as in part (b) in Example 2. However, a number in the domain can be in only one ordered pair.

A function can be demonstrated by *Input* ⇒ *Rule* ⇒ *Output*. Remember, **the domain is the set of all possible inputs**, and **the range is the set of all possible outputs**.

If an equation is used to define a relation, how can we determine whether the equation also defines a function? One way is to write the equation so that y is defined in terms of x. By examining the equation, we can determine whether each value of x is mapped to exactly one value of y.

For example, the equation $y = x^2 - 1$ sets up a correspondence where each value of x determines only one value of y. The variable x is the **input,** $x^2 - 1$ is the **rule,** and y is the **output**. For the equation $y = x^2 - 1$, if the input x is 2, the output y is $y = 2^2 - 1 = 3$. This is illustrated in Figure 1-20.

Input

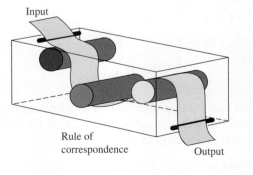

Rule of
correspondence

Output

(a)

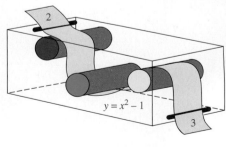

$y = x^2 - 1$

(b)

FIGURE 1-20

By examining the equation, we easily see that it represents a function.

To find the domain of a function, it is often helpful to determine any restrictions on the input values for x. Three equations representing y as a function of x are shown in the table. Let's consider this information to gain a better understanding of domain.

Understanding Domain

Equation	Restriction	Domain
$y = x^3 - x^2 + x - 5$	There are no restrictions on x, because any real number can be cubed, squared, and combined with constants.	All real numbers, $(-\infty, \infty)$
$y = \sqrt{x}$	The restriction that x cannot be a negative real number is placed on x because the square root of a negative number is an imaginary number.	$x \geq 0, [0, \infty)$
$y = \dfrac{1}{x}$	The restriction that x cannot be 0 is placed on x because division by 0 is not defined.	$x \neq 0, (-\infty, 0) \cup (0, \infty)$

To determine the range of a function, we must carefully analyze the function and identify the possible outputs. Consider $y = |x|$. Because the absolute value of any real number is always greater than or equal to 0, the range, or set of outputs of the given function, would be $y \geq 0$, or $[0, \infty)$. If we consider $y = |x| + 5$, we must also take into account the addition of 5. The range would be $y \geq 5$, or $[5, \infty)$.

One strategy we can sometimes use to determine the range is to solve the equation for x in terms of y and note any restrictions on y. Consider $y = x^2$. Solving for x, we note that $x = \pm\sqrt{y}$. The restriction that y cannot be a negative real number implies that the range would be $y \geq 0$, or $[0, \infty)$. It is not always possible to use this strategy; sometimes other techniques, such as graphing, will have to be used.

EXAMPLE 3 **Determining Whether Equations Are Functions and Determining the Domain and Range of Functions**

Decide whether the following equations define y to be a function of x. If it is a function, determine the domain and range.

a. $7x + y = 5$ **b.** $y = \dfrac{3}{x - 2}$ **c.** $y^2 = x - 4$

SOLUTION **a.** First, we solve $7x + y = 5$ for y to get $y = 5 - 7x$. Since there is exactly one value of y for each number we substitute for x, the equation determines y to be a function of x. Because x can be any real number, the domain is the set of all real numbers, which we can express in interval notation as $D = (-\infty, \infty)$. To determine the range, solve for x in terms of y to see whether there are restrictions.

$$7x + y = 5$$
$$7x = 5 - y \qquad \text{Subtract } y \text{ from both sides.}$$
$$x = \frac{5 - y}{7} \qquad \text{Divide by 7.}$$

We can see that there are no restrictions on the value of y. Because y can be any real number, the range is the set of all real numbers, which we can express in interval notation as $R = (-\infty, \infty)$.

b. Since each input x (except for 2) gives exactly one output y, the equation $y = \frac{3}{x-2}$ defines y to be a function of x. The domain is $D = (-\infty, 2) \cup (2, \infty)$, which is the set of all real numbers except 2. We cannot substitute 2 for x, because division by 0 is undefined. To determine the range of the function, we will solve for x in terms of y to see if there are any restrictions.

$$y = \frac{3}{x - 2}$$

$$y(x - 2) = 3 \qquad \text{Multiply by } (x - 2) \text{ to clear the denominator.}$$

$$x - 2 = \frac{3}{y} \qquad \text{Divide by } y.$$

$$x = \frac{3}{y} + 2 \qquad \text{Add 2 to both sides.}$$

$$y \neq 0 \qquad \text{The denominator cannot be 0.}$$

We note that the value of y cannot be 0. Therefore, we determine that the range is $R = (-\infty, 0) \cup (0, \infty)$, which is the set of all real numbers except 0.

ACCENT ON TECHNOLOGY   ## Determining Domain and Range

If we have a graphing calculator, we can graph $y = \frac{3}{x - 2}$ to help us determine the domain and range. The window we select for the graph will determine how much of the graph we see. For the graph in Figure 1-21, a standard window was used. Notice that the graph does not cross the x-axis, because y is never zero. Likewise, if we were to draw a line at $x = 2$, we would see that the graph does not touch that line. Remember, the domain is the set of values that can be substituted for x, and the range is the set of values that result, which are the values of y.

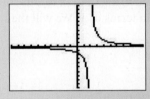 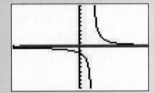 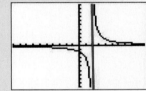

FIGURE 1-21

c. If we solve the equation $y^2 = x - 4$ for y, we get $y = \pm\sqrt{x - 4}$. Since we do not get a unique value of y for each x, the equation is not a function. For example, if $x = 5$, then $y = \pm\sqrt{5 - 4} = \pm 1$ and two ordered pairs result, $(5, 1)$ and $(5, -1)$. Therefore, the equation is a relation but is not a function. ■

Self Check 3   Decide whether each equation defines y to be a function of x. If so, find the domain and range. **a.** $y = |x|$ **b.** $x^2 + y^2 - 4 = 0$ **c.** $y = \sqrt{x}$

Now Try Exercise 27.

EXAMPLE 4   ## Determining the Domain and Range of a Function

Find the domain and range of the function defined by the equation $y = \sqrt{3x - 2}$.

SOLUTION   Since the radicand must be nonnegative, we have

$$3x - 2 \geq 0$$

$$3x \geq 2 \qquad \text{Add 2 to both sides.}$$

$$x \geq \frac{2}{3} \qquad \text{Divide both sides by 3.}$$

The domain is $D = \left[\frac{2}{3}, \infty\right)$. Because the radical sign calls for the nonnegative square root, the range is the interval $R = [0, \infty)$. The graph of the equation is shown in Figure 1-22 to demonstrate what we have just found.

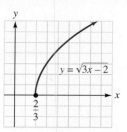

$y = \sqrt{3x - 2}$

$\frac{2}{3}$

FIGURE 1-22

Self Check 4   Find the domain and range of the function defined by $y = \sqrt{x + 2}$.

Now Try Exercise 61.

EXAMPLE 5   **Determining the Domain and Range of a Function**

Find the domain and range of the function defined by the equation $y = \dfrac{x + 1}{x - 2}$.

SOLUTION   In this equation, x cannot be 2, because the denominator cannot be 0. So the domain is $D = (-\infty, 2) \cup (2, \infty)$. For this function we can find the range by solving the equation for x in terms of y. We will then find the restrictions on y, to determine the range.

$$y = \frac{x + 1}{x - 2}$$

$y(x - 2) = x + 1$ If $x \neq 2$, we can multiply both sides by $x - 2$.

$xy - 2y = x + 1$ Use the Distributive Property to remove parentheses.

$xy - x = 2y + 1$ Add $2y$ and subtract x from both sides.

$x(y - 1) = 2y + 1$ Factor out x.

$x = \dfrac{2y + 1}{y - 1}$ Divide both sides by $y - 1$.

From the last equation, we determine that the value of y cannot be 1. So the range is the set of numbers $R = (-\infty, 1) \cup (1, \infty)$.

Self Check 5   Find the domain and range of the function defined by $y = \dfrac{x - 2}{x + 1}$.

Now Try Exercise 55.

3. Use Functional Notation and Evaluate a Function, Given a Specific Input

Since functions are used extensively in mathematics, it is necessary to use a notation to facilitate evaluating and manipulating them. We can define a function using an equation such as $y = 5 - 7x$. However, if we want to evaluate the function at different values of x, it is inconvenient to say, "Evaluate the function at $x = 2$, $x = 4$, $x = a + b$, etc." While the use of function notation is not required, it is often much more efficient.

To indicate that y is a function of x, we use **function notation** and write

$y = f(x)$, which is read as "y equals f of x."

The notation $y = f(x)$ provides a way of denoting the value of y (the **dependent variable**) that corresponds to some number x (the **independent variable**). For example, if $y = f(x)$, the value of y that is determined when $x = 2$ is denoted by $f(2)$, read as "f of 2."

EXAMPLE 6   Evaluating a Function

For the function $f(x) = 5 - 7x$, evaluate: **a.** $f(2)$ **b.** $f(-5)$

SOLUTION   **a.** Evaluate $f(2)$ by substituting 2 for x.

$$f(x) = 5 - 7x$$

$$f(2) = 5 - 7(2) \qquad \text{Substitute 2 for } x.$$

$$f(2) = -9$$

If $x = 2$, then $y = f(2) = -9$.

b. To evaluate $f(-5)$, we substitute -5 for x.

$$f(x) = 5 - 7x$$

$$f(-5) = 5 - 7(-5) \qquad \text{Substitute } -5 \text{ for } x.$$

$$f(-5) = 40$$

If $x = -5$, then $y = f(-5) = 40$.   ∎

Self Check 6   For the function $f(x) = 3 - 2x^3$, evaluate: **a.** $f(2)$ **b.** $f(-5)$

Now Try Exercise 79.

It is important that we understand that the value of the function is the value of y, as shown in Figure 1-23. Remember, in function notation we can write $y = f(x)$. The notations y and $f(x)$ both represent the output of a function, and can be used interchangeably.

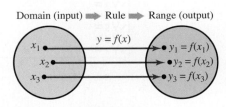

FIGURE 1-23

Sometimes functions are denoted by letters other than f. The notations $y = g(x)$ and $y = h(x)$ also denote functions involving the independent variable x. Also, we can evaluate functions using expressions and not just real numbers. We shall see later that we can even evaluate another function in a function. It is very important that we become comfortable with evaluating functions.

EXAMPLE 7   **Evaluating a Function**

Let $g(x) = 3x^2 + x - 4$. Find **a.** $g(-3)$ **b.** $g(k)$ **c.** $g(-t^3)$ **d.** $g(k + 1)$

SOLUTION   **a.** $g(x) = 3x^2 + x - 4$

$g(-3) = 3(-3)^2 + (-3) - 4$

$= 3(9) - 3 - 4$

$= 20$

b. $g(x) = 3x^2 + x - 4$

$g(k) = 3k^2 + k - 4$

c. $g(x) = 3x^2 + x - 4$

$g(-t^3) = 3(-t^3)^2 + (-t^3) - 4$

$= 3t^6 - t^3 - 4$

d. $g(x) = 3x^2 + x - 4$

$g(k + 1) = 3(k + 1)^2 + (k + 1) - 4$

$= 3(k^2 + 2k + 1) + k + 1 - 4$

$= 3k^2 + 6k + 3 + k + 1 - 4$

$= 3k^2 + 7k$

∎

Self Check 7   For $g(x) = 3x^2 + x - 4$, evaluate: **a.** $g(0)$ **b.** $g(2)$ **c.** $g(k - 1)$

Now Try Exercise 69.

ACCENT ON TECHNOLOGY   **Evaluating a Function**

Functions can be evaluated on a graphing calculator. There are several ways to do this, but one of the easiest is to use the graph editor and the table. Enter the function in the graph editor. Go to TABLE. Make sure that the table is set up to allow the input of the values of x. Enter values for x.

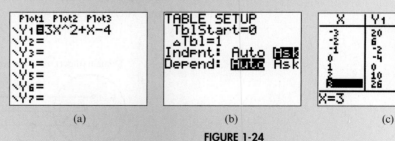

(a)   (b)   (c)

FIGURE 1-24

The function $y = 3x^2 + x - 4$, shown in Figure 1-24, demonstrates that a function is a mapping from one set to another. This function maps -3 to 20, -2 to 6, -1 to -2, etc. A set of ordered pairs is created by the mapping. The set of ordered pairs is very important in the process of graphing functions.

4. Evaluate the Difference Quotient, Given a Function

We saw in Example 7 that it is possible to evaluate a function with an expression and not just real numbers. There is a quotient that is very important in calculus. It is called the **difference quotient:**

$$\frac{f(x + h) - f(x)}{h}, h \neq 0$$

EXAMPLE 8   **Evaluating the Difference Quotient**

If $f(x) = x^2 - 2x - 5$, evaluate the difference quotient $\dfrac{f(x + h) - f(x)}{h}, h \neq 0$.

SOLUTION   We will evaluate the difference quotient in three steps: Find $f(x + h)$; then subtract $f(x)$; then divide by h.

Step 1: Find $f(x + h)$.

$$f(x) = x^2 - 2x - 5$$

$$f(x + h) = (x + h)^2 - 2(x + h) - 5 \qquad \text{Substitute } x + h \text{ for } x.$$

$$= x^2 + 2xh + h^2 - 2x - 2h - 5$$

Step 2: Find $f(x + h) - f(x)$. We can use the result from step 1.

$$f(x + h) - f(x) = x^2 + 2xh + h^2 - 2x - 2h - 5 - (x^2 - 2x - 5) \qquad \text{Subtract } f(x).$$

$$= x^2 + 2xh + h^2 - 2x - 2h - 5 - x^2 + 2x + 5 \qquad \text{Remove parentheses.}$$

$$= 2xh + h^2 - 2h \qquad \text{Combine like terms.}$$

Step 3: Find the difference quotient $\dfrac{f(x + h) - f(x)}{h}, h \neq 0$. We can use the result from step 2.

$$\frac{f(x + h) - f(x)}{h} = \frac{2xh + h^2 - 2h}{h} \qquad \text{Divide both sides by } h.$$

$$= \frac{h(2x + h - 2)}{h} \qquad \text{In the numerator, factor out } h.$$

$$= 2x + h - 2 \qquad \text{Divide out } h\text{: } \frac{h}{h} = 1.$$

The difference quotient is $2x + h - 2$.

Comment

After the completion of Step 2 in Example 8, when $f(x + h) - f(x)$ is simplified, each term will always include an h. We can use that fact to check our work as we progress through the problem.

Self Check 8   If $f(x) = x^2 + 2$, evaluate the difference quotient.

Now Try Exercise 97.

5. Solve Applications Involving Functions

Functions are used in many real-life situations.

EXAMPLE 9   **Solving an Application Problem**

The target heart rate $f(x)$, in beats per minute, at which a person should train to get an effective workout is a function of age x in years. If $f(x) = -0.6x + 132$, find the target heart rate for a 20-year-old college student.

SOLUTION   We can evaluate the function at $x = 20$ to determine the target heart rate.

$$f(x) = -0.6x + 132$$

$$f(20) = -0.6(20) + 132 \qquad \text{Substitute 20 in for } x.$$

$$= -12 + 132$$

$$= 120$$

The target heart rate would be 120 beats per minute.

Self Check 9   Find the target heart rate for a 30-year-old person.

Now Try Exercise 109.

Self Check Answers
1. The set of ordered pairs is {(I-4 S, $19,696), (I-4 SE, $21,225), (I-4 SEL, $24,655), (Sport, $26,505), (V-6 SEL, $28,115)}; D = {I-4 S, I-4 SE, I-4 SEL, Sport, V-6 SEL}; R = {$19,696, $21,225, $24,655, $26,505, $28,115}.
2. R3 = {(3, 2), (5, 2), (5, 6), (7, 6)}; D = {3, 5, 7}; R = {2, 6}
3. **a.** yes, $D = (-\infty, \infty)$; $R = [0, \infty)$ **b.** no
   c. yes, $D = [0, \infty)$; $R = [0, \infty)$ **4.** $D = [-2, \infty)$; $R = [0, \infty)$
5. $D = (-\infty, -1) \cup (-1, \infty)$; $R = (-\infty, 1) \cup (1, \infty)$ **6. a.** -13 **b.** 253
7. **a.** -4 **b.** 10   **c.** $3k^2 - 5k - 2$ **8.** $2x + h$ **9.** 114 beats per minute

Exercises **1.2**

Getting Ready
You should be able to complete these vocabulary and concept statements before you proceed to the practice exercises.

Fill in the blanks.

1. A correspondence that assigns exactly one value of y to any number x is called a _____.

2. A correspondence that assigns one or more values of y to any number x is called a _____.

3. The set of input numbers x in a function is called the _____ of the function.

4. The set of all output values y in a function is called the _____ of the function.

5. The statement "y is a function of x" can be written as the equation _____.

6. In the function of Exercise 5, _____ is called the independent variable.

7. In the function of Exercise 5, y is called the _____ variable.

8. The expression $\dfrac{f(x + h) - f(x)}{h}$, $h \neq 0$, is called the _____.

Practice
In Exercises 9–12, a relation is given. (a) Write the ordered pairs of the relation and state the domain and the range. (b) Determine whether the relation is a function.

9.
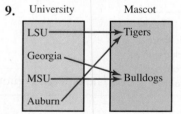

10.
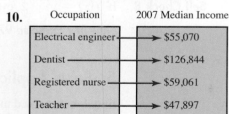

11.
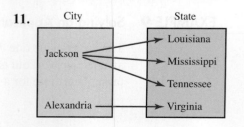

12.

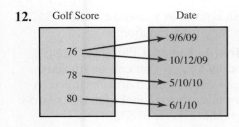

Determine whether the relation shown is a function.

13. $\{(2, 3), (3, 4), (4, 5), (5, 6)\}$

14. $\{(5, 4), (6, 4), (7, 4), (8, 4)\}$

15. $\{(1, 3), (1, 4), (2, 5), (5, 2)\}$

16. $\{(-1, 2), (2, -1), (0, 1), (0, 3)\}$

Determine whether each equation defines y to be a function of x. Assume that all variables represent real numbers.

17. $y = x$

18. $y - 2x = 0$

19. $y^2 = x$

20. $|y| = x$

21. $y = x^2$

22. $3y - 7 = 4x$

23. $y^2 - 4x = 1$

24. $x^2 + y^2 - 3y = 0$

25. $|x| = |y|$

26. $x = 7$

27. $|x - 2| = y$

28. $y = 4$

29. $y = \dfrac{4}{x}$

30. $y = \dfrac{2x + 1}{x - 4}$

Find the domain of each function $f(x)$.

31. $f(x) = 4x^2 - 5x + 2$

32. $f(x) = -5x^4 + 2$

33. $f(x) = \dfrac{5}{x + 6}$

34. $f(x) = \dfrac{x}{x - 9}$

35. $f(x) = \dfrac{x}{x^2 - 4x - 5}$

36. $f(x) = -\dfrac{1}{x^2 - 9}$

37. $f(x) = \sqrt{2x - 5}$

38. $f(x) = \sqrt{4 - x}$

39. $f(x) = \sqrt[3]{x - 1}$

40. $f(x) = -2\sqrt[3]{x}$

41. $f(x) = |x| + 9$

42. $f(x) = |x - 8| + 2$

Find the range of each function $f(x)$.

43. $f(x) = 4x^2$

44. $f(x) = x^2 + 2$

45. $f(x) = |x| - 3$

46. $f(x) = |x| + 2$

47. $f(x) = \sqrt{x + 1}$

48. $f(x) = \sqrt[3]{x}$

49. $f(x) = \dfrac{2}{x}$

50. $f(x) = \dfrac{1}{x^2}$

Find the domain and range of each function $f(x)$.

51. $f(x) = 3x + 5$

52. $f(x) = -5x + 2$

53. $f(x) = x^2$

54. $f(x) = x^3$

55. $f(x) = \dfrac{3}{x + 1}$

56. $f(x) = \dfrac{-7}{x + 3}$

57. $f(x) = \sqrt{x}$

58. $f(x) = \sqrt{x^2 - 1}$

59. $f(x) = \dfrac{x}{x + 3}$

60. $f(x) = \dfrac{x}{x - 3}$

61. $f(x) = \sqrt{2x - 1}$

62. $f(x) = \sqrt{3x + 1}$

63. $f(x) = \dfrac{x - 2}{x}$

64. $f(x) = \dfrac{x + 3}{x}$

Let the function f be defined by $y = f(x)$. Find $f(2)$, $f(-3)$, $f(k)$, and $f(k^2 - 1)$.

65. $f(x) = 3x - 2$

66. $f(x) = 5x + 7$

67. $f(x) = \dfrac{1}{2}x + 3$

68. $f(x) = \dfrac{2}{3}x + 5$

69. $f(x) = x^2$

70. $f(x) = 3 - x^2$

71. $f(x) = |x^2 + 1|$

72. $f(x) = |x^2 + x + 4|$

73. $f(x) = \dfrac{2}{x + 4}$

74. $f(x) = \dfrac{3}{x - 5}$

75. $f(x) = \dfrac{1}{x^2 - 1}$

76. $f(x) = \dfrac{3}{x^2 + 3}$

77. $f(x) = \sqrt{x^2 + 1}$

78. $f(x) = \sqrt{x^2 - 1}$

Given $f(x) = 3x^2 - 5x + 1$, evaluate each of the following.

79. $f(4)$

80. $f(-3)$

81. $f\left(-\dfrac{1}{2}\right)$

82. $f\left(\dfrac{1}{3}\right)$

83. $f(x^3)$

84. $f(-2x^2)$

Use a calculator to evaluate the function $f(x) = 2x^2 - 3x + 1$. Round to two decimals.

85. $f(1.2)$

86. $f(3.9)$

87. $f(-2.1)$

88. $f(-1.5)$

89. $f\left(\dfrac{4}{5}\right)$

90. $f\left(\dfrac{3}{7}\right)$

Evaluate the difference quotient for each function $f(x)$.

91. $f(x) = 3x + 1$

92. $f(x) = 5x - 1$

93. $f(x) = x^2 + 1$

94. $f(x) = x^2 - 3$

95. $f(x) = 4x^2 - 6$

96. $f(x) = -5x^2 + 3$

97. $f(x) = x^2 + 3x - 7$

98. $f(x) = x^2 - 5x + 1$

99. $f(x) = 2x^2 - 4x + 2$

100. $f(x) = 3x^2 + 2x - 3$

101. $f(x) = x^3$

102. $f(x) = \dfrac{1}{x}$

Use a calculator to determine the domain and range of the function.

103. $f(x) = \sqrt{2x - 5}$

104. $f(x) = \sqrt{4 - x}$

105. $f(x) = \dfrac{2x}{x - 1}$

106. $f(x) = \dfrac{3x}{x + 4}$

107. $f(x) = \sqrt[3]{5x - 1}$

108. $f(x) = -\sqrt[3]{3x + 2}$

Applications

109. Target heart rate The target heart rate $f(x)$, in beats per minute, at which a person should train to get an effective workout is a function of age x in years. If $f(x) = -0.6x + 132$, find the target heart rate for a 25-year-old person.

110. Temperature conversion The Fahrenheit temperature reading F is a function of the Celsius reading C. The function can be written as $F(C) = \frac{9}{5}C + 32$.

    a. Identify the independent variable and the dependent variable.

    b. Find the Fahrenheit temperature for the Celsius temperatures: $C = 0$; $C = -40$; $C = 10$.

111. Go green The typical American family uses about 300 gallons of water a day. If the number of gallons g used expressed in terms of days d is $g(d) = 300d$, find the number of gallons used in one year.

112. Free-falling objects The velocity v of a falling object is a function of the time t it has been falling. If v as a function of t can be expressed as $v(t) = -32t + 15$, where v is in feet per second and t is in seconds, when will the velocity be 0?

113. Cliff divers The velocity v, in feet per second, of a cliff diver is a function of the time t, in seconds, the diver has been falling. If v as a function of t can be expressed as $v(t) = -32t + 2$, what is the initial velocity of the diver?

114. Volume of a basketball The volume V of a sphere can be expressed in terms of its radius r according to the function $V(r) = \frac{4}{3}\pi r^3$. Find the volume of a men's NCAA basketball if the diameter of the ball is 29.5 centimeters.

ssuaphotos/Shutterstock.com

115. Formulas The area A of a rectangle is determined by the length and width. If the length of a rectangle is x and the width is 5 more than the length, express the area as a function of the length.

116. Formulas The volume V of a rectangular box is determined by the length, the width, and the height. For a particular set of boxes, the height

is 4, the length is given as x, and the width is $3x$. Express the volume as a function of x.

117. Cost of T-shirts A chapter of Phi Theta Kappa, an honors society for two-year college students, is purchasing T-shirts for each of its members. A local company has agreed to make the shirts for $8 each plus a graphic-arts fee of $75.

    a. Write a function that describes the cost C for the shirts in terms of x, the number of T-shirts ordered.

    b. Find the total cost of 85 T-shirts.

118. Service projects The Circle K club is planning a service project at a local children's home. They plan to rent a *Dora the Explorer* moonwalk for the event. The cost of the moonwalk will include a $60 delivery fee and $45 for each hour it is used. Express the total bill b in terms of the hours used h.

119. Cell-phone plans A grandmother agrees to purchase a cell phone for emergency use only. AT&T now offers such a plan for $9.99 per month and $0.07 for each minute t the phone is used.

    a. Write a function that describes the monthly cost (C) in terms of the time in minutes (t) the phone is used.

    b. If the grandmother uses her phone for 20 minutes during the first month, what is her bill?

120. Concessions A concessionaire at a football game pays a vendor $60 per game for selling hot dogs at $2.50 each.

    a. Write a function that describes the income I the vendor earns for the concessionaire during the game if the vendor sells h hot dogs.

    b. Find the income if the vendor sells 175 hot dogs.

Discovery and Writing

121. Using words, state three real-life correspondences that represent functions.

122. Write a brief paragraph explaining how to find the domain of a function, given an equation.

123. Write a brief paragraph explaining how to find the range of a function, given an equation.

124. Explain why all functions are relations, but not all relations are functions.

125. Explain why some equations represent y as a function of x and some do not.

126. Write a brief paragraph explaining how to find the domain and range of a function, given a graph.

127. Are the functions $f(x) = x + 3$ and $f(x) = \dfrac{x^2 - 9}{x - 3}$ the same? Explain why or why not.

128. Are the domains of the functions $f(x) = \sqrt{x - 4}$ and $g(x) = \sqrt{4 - x}$ the same or different? What about their ranges?

1.3 Graphs of Functions

In this section, we will learn to

1. Identify the graph of a function.
2. Determine function values from a graph.
3. Determine the domain and range of a function from its graph.
4. Use a graph to find the open intervals where a function is increasing, decreasing, or constant.
5. Identify local maxima and minima on the graph of a function.
6. Classify a function as even, odd, or neither.
7. Use the graph of a function to summarize valuable information about the function's behavior.

As we learned in the last section, a relation is a set of ordered pairs. The ordered pairs can be plotted on a rectangular coordinate system, resulting in a graph.

The graph in Figure 1-25 is the graph of a function. It shows the average retail price of regular gasoline per gallon in the state of Mississippi during the first few months of 2010. Ordered pairs of the form (Date, Price per gallon) are plotted and a smooth curve is then drawn through them, producing the graph of the function. Note that for each day shown, there corresponds exactly one price per gallon. We can see that the price per gallon was approximately $2.69 on April 1, 2010.

FIGURE 1-25 Mississippi Gasoline Prices (*Source:* www.gasbuddy.com; Reprinted by permission.)

Graphs of functions convey very important information, and many applications involve functions. In this section we will continue our study of functions by learning to identify the graph of a function and exploring properties of the graphs of functions.

1. Identify the Graph of a Function

Some graphs represent functions and some do not.

Graph of a Function   The **graph** of a function f in the xy plane is the set of all points (x, y) where x is in the domain of f, y is in the range of f, and $y = f(x)$.

To satisfy the definition of a function, the graph of a function can have only one y value for each x value. We can use this fact to determine whether or not a graph is of a function. We will apply the **vertical-line test.** Graphs that pass the vertical-line test are functions.

Vertical-Line Test

If every vertical line that can be drawn intersects the graph in no more than one point, the graph represents a function [Figure 1-26(a)].

If a vertical line can be drawn that intersects the graph at more than one point, the graph does not represent a function [Figure 1-26(b)].

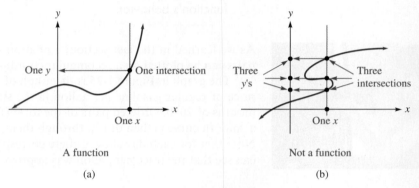

(a)   (b)

FIGURE 1-26

The vertical-line test is exactly what the definition of a function means in a graphical representation.

EXAMPLE 1   **Identifying Graphs of Functions Using the Vertical-Line Test**

Apply the vertical-line test and determine which of the following graphs represent functions:

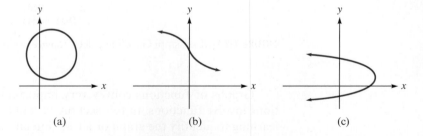

(a)   (b)   (c)

SOLUTION   **a.** This graph fails the vertical-line test, so it does not represent a function. The vertical line drawn crosses the graph at two points.

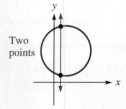

b. This graph passes the vertical-line test, so it does represent a function. A vertical line will never cross the graph at more than one point.

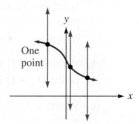

c. This graph fails the vertical-line test, so it does not represent a function. The vertical line drawn crosses the graph at two points.

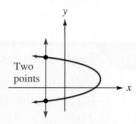

Self Check 1   Which graphs represent functions?

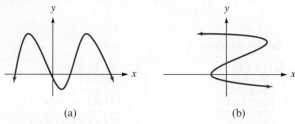

         (a)   (b)

Now Try Exercise 13.

2. Determine Function Values from a Graph

In many applications that involve functions, we may not be given the equation of the function, but will instead have only data. It is important that we learn how to read the value of a function from the graph of that function. Remember, the value of the function is y. It is not a point; it is a y value.

EXAMPLE 2   **Reading the Value of a Function from a Graph**

The graph of a function $f(x)$ is shown here. Find each of the following:

a. $f(-3)$ **b.** $f(1)$ **c.** $f(3)$

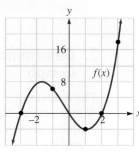

FIGURE 1-27

SOLUTION   **a.** When $x = -3$, we see from the graph that the value of y is 0. This is because the point $(-3, 0)$ lies on the graph of the function. Therefore, we write $f(-3) = 0$. It is important to note that the value of the function at -3 is 0 and not the ordered pair $(-3, 0)$. The ordered pair $(-3, 0)$ is simply a point on the graph of the function.

b. When $x = 1$, we see from the graph that the value of y is -4. This is because the point $(1, -4)$ lies on the graph of the function. Therefore, we write $f(1) = -4$.

c. When $x = 3$, the y value of the function is 18, so we write, $f(3) = 18$.   ■

Self Check 2   Use Figure 1-27 and find $f(-1)$ and $f(2)$.

Now Try Exercise 47e.

ACCENT ON TECHNOLOGY   **Evaluating Functions**

The value of a function can be determined several different ways by a graphing calculator. The value shown may be an approximation. For the function $f(x) = x^3 - 2x + 2$, find $f(-2)$ by reading the value from the graph and from a table. The advantage to using a table is the ease of entering several values at once, and the ability to evaluate the function at any value. When you use the graph, the values that can be evaluated must lie within the values of x in the WINDOW.

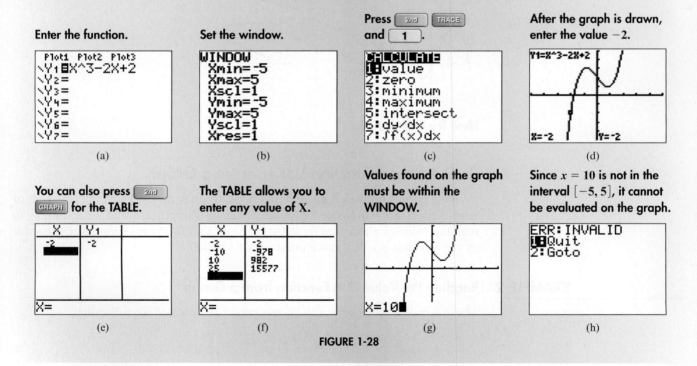

Enter the function.   Set the window.   Press [2nd] [TRACE] and [1].   After the graph is drawn, enter the value -2.

(a)   (b)   (c)   (d)

You can also press [2nd] [GRAPH] for the TABLE.   The TABLE allows you to enter any value of X.   Values found on the graph must be within the WINDOW.   Since $x = 10$ is not in the interval $[-5, 5]$, it cannot be evaluated on the graph.

(e)   (f)   (g)   (h)

FIGURE 1-28

3. Determine the Domain and Range of a Function from Its Graph

The **domain** of a function is the set of **input** values x. Therefore, we can read the domain from the graph of a function by determining the values of x that are included in the graph. Likewise, the **range** of a function is the set of **output** values y, and we determine the range of the function by reading the values of y that are included in the graph (Figure 1-29).

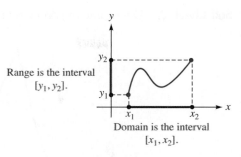

FIGURE 1-29

As we can see, the domain is determined by using the x-axis and the range by using the y-axis. We trace the graph of the function from left to right, noting the x values, and identify the domain. We can follow the graph from its lowest point to its highest point, noting the y values, and determine the range.

EXAMPLE 3   **Determining the Domain and Range of a Function from the Graph**

Determine the domain and range of each function whose graph is shown.

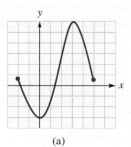   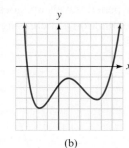   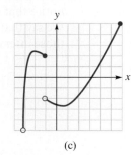   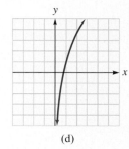

(a)   (b)   (c)   (d)

FIGURE 1-30

SOLUTION   **a.** Trace the graph from left to right. The values of x that are included in the domain are in the interval $-2 \leq x \leq 5$. We can write the domain as $D = \{x | -2 \leq x \leq 5\}$ or we can use interval notation and write $D = [-2, 5]$. Trace the graph from low to high. The values of y that are included in the range are in the interval $-3 \leq y \leq 6$. Using interval notation we can write $R = [-3, 6]$.

b. As we look at this graph, from left to right, we note that there is no "starting point" or "ending point" for the x values. The domain of this function is all real numbers, and we can write that domain as $D = (-\infty, \infty)$. If we look at the graph from its lowest point to its highest point, we see that the graph has a minimum y value of -4, but no maximum value. The graph goes upward without bound. The range is $R = [-4, \infty)$.

c. Trace the graph from left to right. Because we begin with an open circle at $x = -3$, -3 will not be included in the domain. The x values to the right of -3 all the way to and including -1 will be in the domain. Note the closed circle at $x = -1$. Even though there is a break in the graph at $x = -1$, there is no gap in the values of x that are included. The graph continues on to the right, ending at a point with an x value of 6. We write the domain as the interval $D = (-3, 6]$. The range does not include the value $y = -5$ but all values to and including 5. We write the range as $R = (-5, 5]$.

d. This function has no endpoints; however, the value of x is never negative or 0. The graph continues to the right without bound and its domain is $D = (0, \infty)$. Note that the graph goes downward without bound and upward without bound. The range of this function is all the real numbers, $R = (-\infty, \infty)$.

Self Check 3   Determine the domain and range for each function.

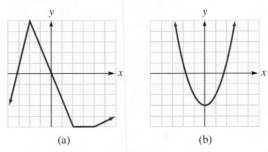

(a)   (b)

FIGURE 1-31

Now Try Exercise 25a.

4. Use a Graph to Find the Open Intervals Where a Function Is Increasing, Decreasing, or Constant

If we take a pencil and trace the graph of the function shown in Figure 1-32, we notice that our pencil travels up, goes down, stays level, and then goes down again. We say that the function is increasing, decreasing, constant, and then decreasing on those intervals.

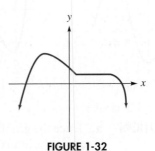

FIGURE 1-32

Increasing, Decreasing, and Constant Functions

- A function f is **increasing** on an open interval if, for any x_1 and x_2 in the interval, $x_1 < x_2$ and $f(x_1) < f(x_2)$.
- A function f is **decreasing** on an open interval if, for any x_1 and x_2 in the interval, $x_1 < x_2$ and $f(x_1) > f(x_2)$.
- A function f is **constant** on an open interval if $f(x_1) = f(x_2)$ for any x_1 and x_2 in the interval.

To determine the open intervals where a function increases, decreases, or remains constant, we will trace the graph from left to right and then write the open intervals, using x values, where we traveled upward or downward or remained constant.

EXAMPLE 4   **Determining the Open Intervals Where a Function Is Increasing, Decreasing, or Constant**

For the graph of the function shown in Figure 1-33, determine the open interval(s) on which the function is

a. Increasing   **b.** Decreasing   **c.** Constant

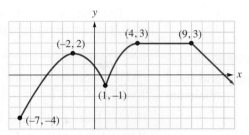

FIGURE 1-33

SOLUTION   **a.** To determine the open intervals where the function increases, we will locate the intervals on the graph where the function values (y values) increase. To do this, we trace the graph with a pencil from the left and note where we travel up. The function shown is increasing on the interval $(-7, -2)$, or $-7 < x < -2$. Notice that we do not use the coordinates $(-7, -4)$ and $(-2, 2)$. Only the values of x are used. The function is also increasing on the interval $(1, 4)$, or $1 < x < 4$.

b. The function is decreasing on the intervals $(-2, 1)$, or $-2 < x < 1$, and $(9, \infty)$, or $x > 9$.

c. The function is constant on the interval $(4, 9)$, or $4 < x < 9$. ■

Self Check 4   For the graph of the function in Figure 1-34, determine the open interval(s) where the function is **a.** Increasing   **b.** Decreasing   **c.** Constant

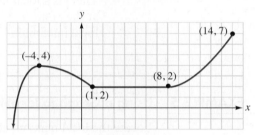

FIGURE 1-34

Now Try Exercise 25d.

5. Identify Local Maxima and Minima on the Graph of a Function

If we ride a roller coaster, we know when we have reached a high point or low point on the ride. We also know there can be more than one high point or low point. Look at the photo of a roller coaster in Figure 1-35, and then look at the path shown as a graph. This is the graph of a function. We call the y value of a high point a **local maximum**, and we call the y value of a low point a **local minimum.** We use the word "local" to indicate that the function reaches a maximum value or a minimum value over an interval. Collectively we use the term **extrema** when we are referring to local maxima and local minima. Sometimes these are also called **relative extrema.**

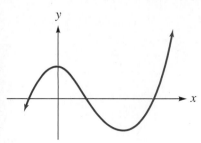

FIGURE 1-35

The local maximum or local minimum is the value of the function, y, at a value of x. It is not an ordered pair.

Local Maximum and Local Minimum	$f(c)$ is a **local maximum** if there exists an interval (a, b) with $a < c < b$ such that $f(x) \leq f(c)$ for all x in (a, b).

$f(c)$ is a **local minimum** if there exists an interval (a, b) with $a < c < b$ such that $f(x) \geq f(c)$ for all x in (a, b).

This definition is not as complicated as it may seem. **A local maximum occurs where the function changes from *increasing* to *decreasing*. Similarly, a local minimum occurs where the function changes from *decreasing* to *increasing*.** There can be more than one extremum over a given interval.

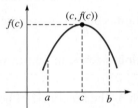

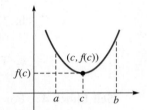

There is a local maximum at the point $(c, f(c))$. The local maximum is $y = f(c)$.

There is a local minimum at the point $(c, f(c))$. The local minimum is $y = f(c)$.

FIGURE 1-36

EXAMPLE 5 **Using a Graph to Identify Local Maxima and Local Minima**

Locate the local maxima and local minima using the graph of the function in Figure 1-37.

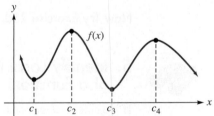

FIGURE 1-37

SOLUTION The local minima are $f(c_1)$ and $f(c_3)$. Notice that we did not say that a local minimum is $(c_1, f(c_1))$. The point $(c_1, f(c_1))$ is *where* a local minimum occurs, but the local minimum is the **value of the function**. The local maxima are $f(c_2)$ and $f(c_4)$. ∎

Self Check 5 Use the graph and identify all local extrema.

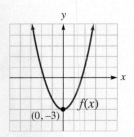

Now Try Exercise 47d.

We can find local extrema by analyzing the graph of a function. However, if we do not have the graph, how do we find the extrema? We will see that for some functions, called *quadratic functions*, there is an algebraic way to do this. For other functions, it may require calculus techniques.

We can find the value of extrema, or perhaps approximations, using a graphing calculator or a computer. To find a maximum or minimum using a graphing calculator, we need to be careful to select a proper window. In spite of finding what seems to be a good window, we may not be able to "see" all the extrema of a function. That is where calculus techniques are necessary.

EXAMPLE 6   **Finding Local Extrema of a Function from a Graph Using a Graphing Calculator**

Graph the function $f(x) = x^3 - 2x^2 - 3x + 2$ on a graphing calculator and find the value of any local extrema.

SOLUTION   We will use our graphing calculator and follow the six steps shown in Figure 1-38.

1. Type the function into the graph menu.   2. Set a window that shows the function.

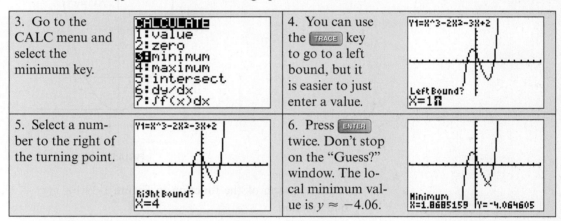

| 3. Go to the CALC menu and select the minimum key. | 4. You can use the TRACE key to go to a left bound, but it is easier to just enter a value. |
| 5. Select a number to the right of the turning point. | 6. Press ENTER twice. Don't stop on the "Guess?" window. The local minimum value is $y \approx -4.06$. |

The maximum is found using the same set of steps with the CALC menu item 4.

FIGURE 1-38

Self Check 6   Graph the function $f(x) = -x^3 + 2x^2 + x - 1$ on a graphing calculator and find the value of any local extrema.

Now Try Exercise 53.

6. Classify a Function as Even, Odd, or Neither

Functions can be classified as being even, odd, or neither. Look at the graphs of the functions in Figure 1-39. Do you notice anything about the symmetry?

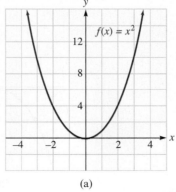

(a)

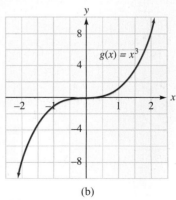

(b)

FIGURE 1-39

The graph of the function in Figure 1-39(a), $f(x) = x^2$, is symmetric about the y-axis and is an example of an **even function**.

Even Function   The function f is an **even** function if it is symmetric about the y-axis. Algebraically, this means that $f(x) = f(-x)$ for all x in the domain.

For example, for the even function $f(x) = x^2$, if $x = 2$, we know that $f(2) = f(-2) = 4$; therefore, both $(2, 4)$ and $(-2, 4)$ are points on the graph. In general, for an even function we can say that if (x, y) is a point on the graph, then $(-x, y)$ is also a point on the graph, as shown in Figure 1-40. Because the graph is symmetric about the y-axis, it is classified as being an even function.

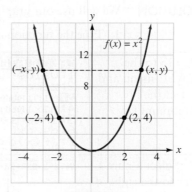

FIGURE 1-40

The graph of the function in Figure 1-39(b), $g(x) = x^3$, is an example of an **odd function**.

Odd Function   The function f is an **odd** function if it is symmetric about the origin. Algebraically, this means that $-f(x) = f(-x)$ for all x in the domain.

For example, for the odd function $g(x) = x^3$, if $x = 2$, we know that $-g(2) = g(-2) = -8$, and both $(2, 8)$ and $(-2, -8)$ are points on the graph. In general, for an odd function you can say that if (x, y) is a point on the graph, then $(-x, -y)$ is also a point on the graph, as shown in Figure 1-41. Because the graph is symmetric about the origin, it is classified as being an odd function.

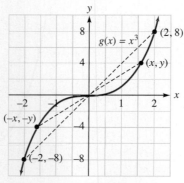

FIGURE 1-41

Another way to determine whether a function is odd is to rotate it one half turn about the origin. If the original graph is obtained, then the function is an odd function.

If a function is not symmetric about the y-axis or the origin, then it is not even or odd.

EXAMPLE 7   **Classifying a Function as Even, Odd, or Neither Using a Graph**

Determine whether each function is even, odd, or neither.

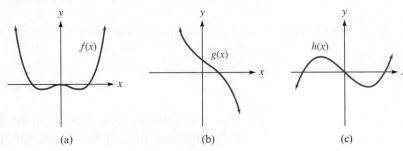

(a)   (b)   (c)

FIGURE 1-42

SOLUTION   **a.** This graph is symmetric about the y-axis, so it is an even function.

b. This graph is neither odd nor even.

c. This graph is symmetric about the origin, so it is an odd function. ∎

Self Check 7   Classify the function as even, odd, or neither.

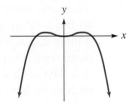

Now Try Exercise 23c.

Next we will use the definition of even or odd functions stated earlier to determine algebraically whether a function is even, odd, or neither.

EXAMPLE 8   **Determining Even and Odd Functions Algebraically**

Determine whether each function is even, odd, or neither.
a. $f(x) = x^4 - x^2$ **b.** $f(x) = x^3 - 2x$ **c.** $f(x) = x^2 - x + 1$

SOLUTION   **a.** To check whether a function is an even function, we find $f(-x)$ and determine whether $f(-x) = f(x)$.

$$f(-x) = (-x)^4 - (-x)^2 = x^4 - x^2$$

Since $f(-x) = f(x)$, the function is even.

b. First we will find $f(-x)$.

$$f(-x) = (-x)^3 - 2(-x) = -x^3 + 2x$$

Since $f(-x) \neq f(x)$, the function is not even. So now we will find out whether the function is odd, by determining whether $f(-x) = -f(x)$.

$$-f(x) = -(x^3 - 2x) = -x^3 + 2x$$

We see that $f(-x) = -f(x)$, and the function is odd.

c. First we will find $f(-x)$.

$$f(-x) = (-x)^2 - (-x) + 1 = x^2 + x + 1$$

Since $f(-x) \neq f(x)$, the function is not even. So now we will find out whether the function is odd, by determining whether $f(-x) = -f(x)$.

$$-f(x) = -(x^2 - x + 1) = -x^2 + x - 1$$

We see that $f(-x) \neq -f(x)$, so the function is neither even nor odd. ∎

Self Check 8   Determine whether the each function is even, odd, or neither.
a. $f(x) = 2x^2 + 1$ **b.** $f(x) = x^3 - 2$ **c.** $f(x) = x^3 + x$

Now Try Exercise 37.

7. Use the Graph of a Function to Summarize Valuable Information about the Function's Behavior

In Section 1.1, we learned about the x- and y-intercepts of the graph of an equation. We can easily apply these concepts to the graphs of functions.

An **x-intercept** of a function is the x-coordinate of any point where the graph of the function crosses or touches the x-axis, if there are any such points. The value of the function at these intercepts is 0, so an x-intercept occurs at a point $(a, 0)$, where $f(a) = 0$. The x-intercept $x = a$ is also called a **zero** of the function. (We will learn more about the **zeros** of a function later.)

The **y-intercept** of a function is the y-coordinate of the point on the graph where the function crosses or touches the y-axis. The value of x at this point is 0, so the y-intercept occurs at the point $(0, b)$, where $f(0) = b$. If a function has a y-intercept, it will have only one. That is because a graph that crosses the y-axis at more than one point fails the vertical-line test and is not the graph of a function.

We conclude this section with an example that summarizes the valuable information we have learned in this section.

EXAMPLE 9   **Determining Intercepts and Other Valuable Information from a Graph**

Use the graph of the function f shown in Figure 1-43 to find each of the following:

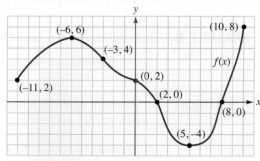

FIGURE 1-43

a. x-intercept(s) and y-intercept

b. Domain and range

c. Open interval(s) where the function is increasing, decreasing, or constant

d. Local extrema

e. $f(-3)$

f. x such that $f(x) = -4$

SOLUTION a. The x-intercepts are 2 and 8, and the y-intercept is 2.

b. $D = [-11, 10]$ and $R = [-4, 8]$.

c. The open intervals where the function is increasing are $(-11, -6) \cup (5, 10)$ and the open interval where the function is decreasing is $(-6, 5)$.

d. Local maximum $= f(-6) = 6$ and local minimum $= f(5) = -4$.

e. $f(-3) = 4$.

f. $f(5) = -4$, so $x = 5$.

Self Check 9 Use the graph of the function f shown to find each of the following:
a. x-intercept(s) and y-intercept   b. Domain and range
c. Open interval(s) where the function is increasing, decreasing, or constant
d. Local extrema   e. $f(-3)$ f. x such that $f(x) = -16$

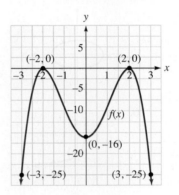

Now Try Exercise 49.

Self Check Answers 1. (a)   2. 6, 0   3. a. $D = (-\infty, \infty); R = (-\infty, \infty)$
b. $D = (-\infty, \infty); R = [-3, \infty)$ 4. a. $(-\infty, -4) \cup (8, 14)$ b. $(-4, 1)$
c. $(1, 8)$ 5. local minimum is $f(0) = -3$; no local maximum
6. local maximum ≈ 1.63; local minimum ≈ -1.11 7. even   8. a. even
b. neither   c. odd   9. a. x-intercepts are -2 and 2; y-intercept is -16
b. $D = (-\infty, \infty); R = (-\infty, 0]$ c. increasing on $(-\infty, -2) \cup (0, 2)$;
decreasing on $(-2, 0) \cup (2, \infty)$ d. local minimum is -16; local maximum is 0
e. -25 f. $x = 0$

Exercises **1.3**

Getting Ready
You should be able to complete these vocabulary and concept statements before you proceed to the practice exercises.

Fill in the blanks.

1. If a _____ line intersects a graph at more than one point, the graph is not of a function.

2. The _____ of a function is the set of input values of x.

3. The _____ of a function is the set of output values.

4. If the graph of a function is symmetric about the _____, it is called an even function.

5. If the graph of a function is symmetric about the _____, it is called an odd function.

6. If a function is even, then $f(-x) =$ _____.

7. If a function is odd, then $f(-x) =$ _____.

8. If the values of $f(x)$ get larger as x increases on an interval, we say that the function is _____ on the interval.

9. If the values of $f(x)$ get smaller as x increases on an interval, we say that the function is ________ on the interval.

10. A function can have at most ________ y-intercept.

Practice

Determine whether each graph represents a function.

11.

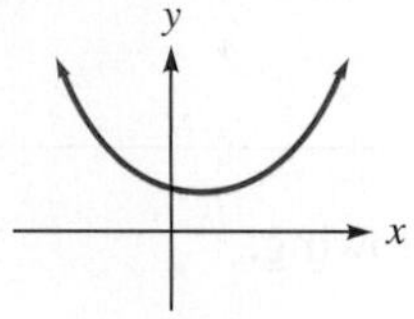

12.

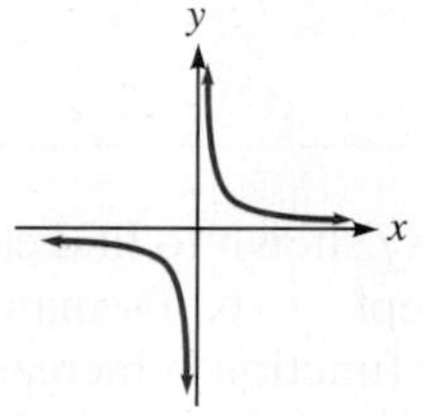

13.

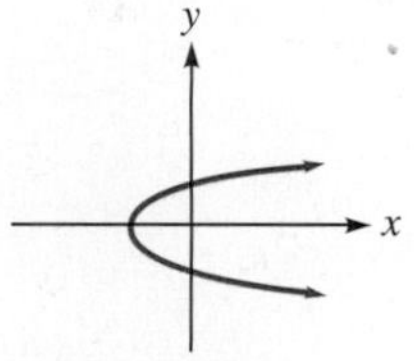

14.

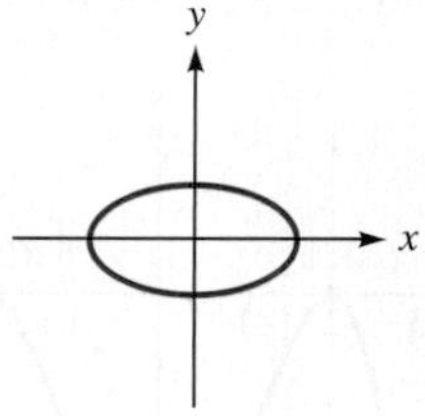

15.

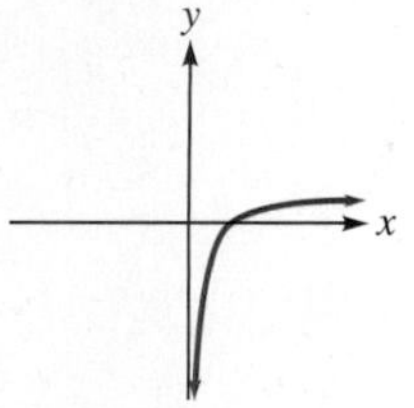

16.

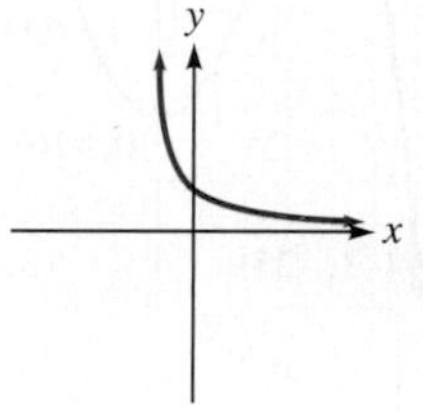

Draw lines to indicate the domain and range of each function as intervals on the x- and y-axes.

17.

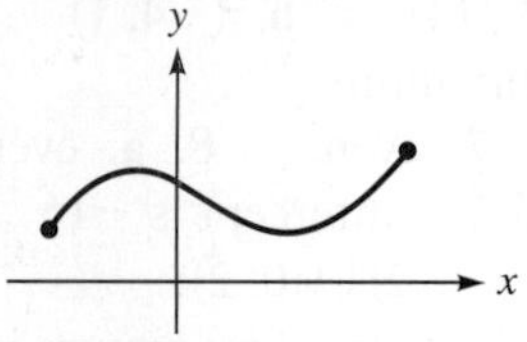

18.

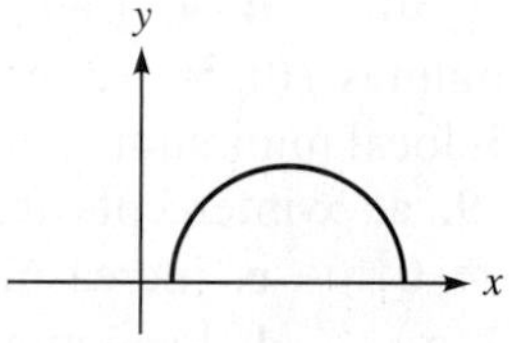

For the graph of each function shown, (a) determine the domain and range, (b) state the values of any intercepts, (c) determine whether the function is odd, even, or neither, and (d) state the open intervals where the function is increasing, decreasing, or constant.

19.

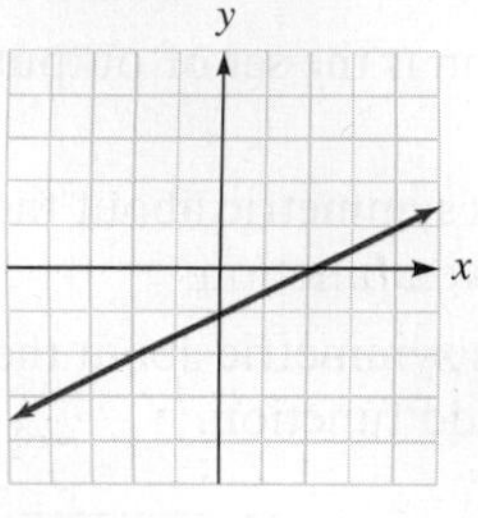

20.

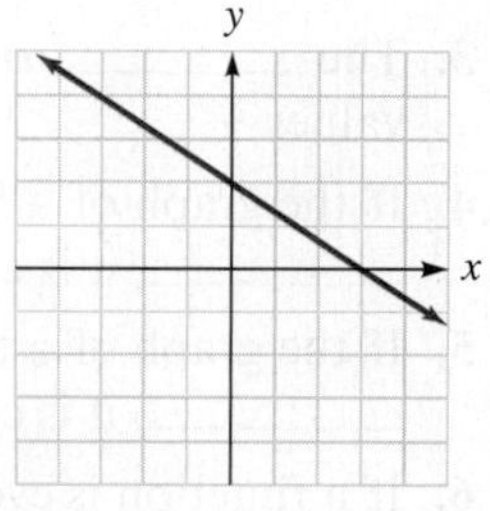

21.

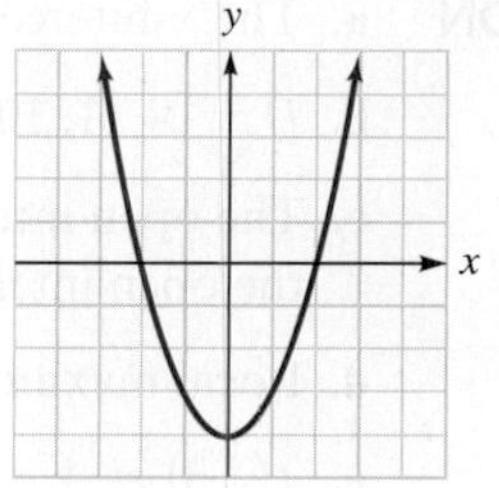

22.

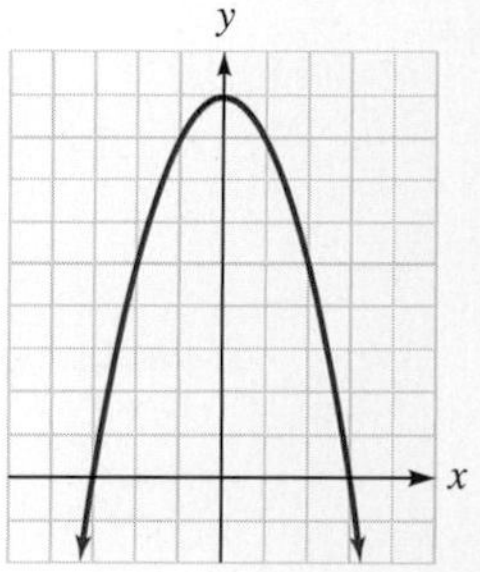

23.

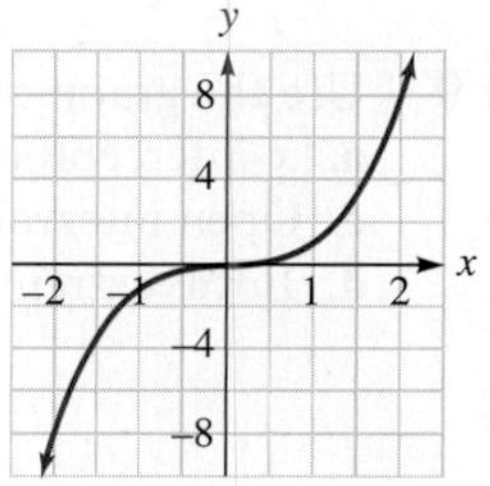

24.

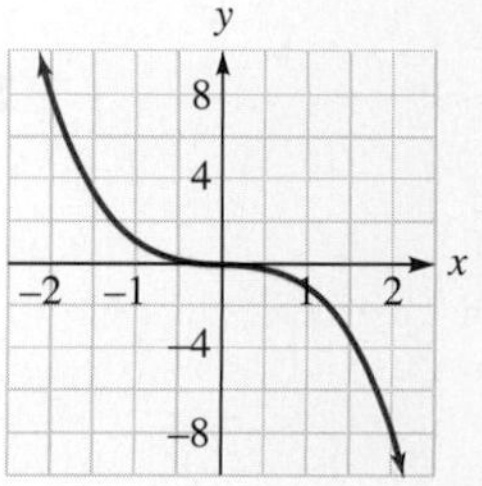

25.

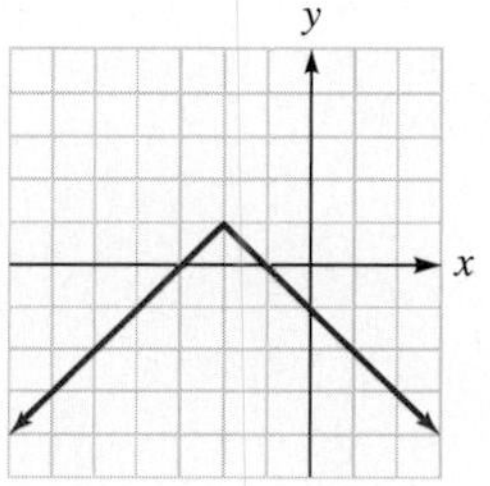

26.

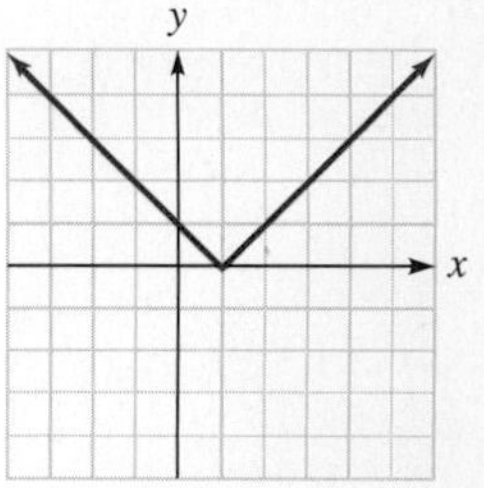

27.

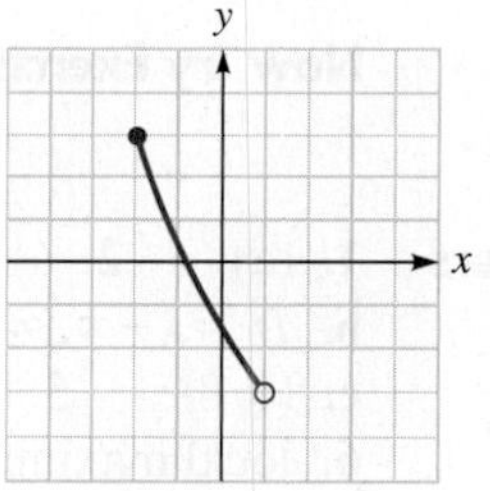

28.

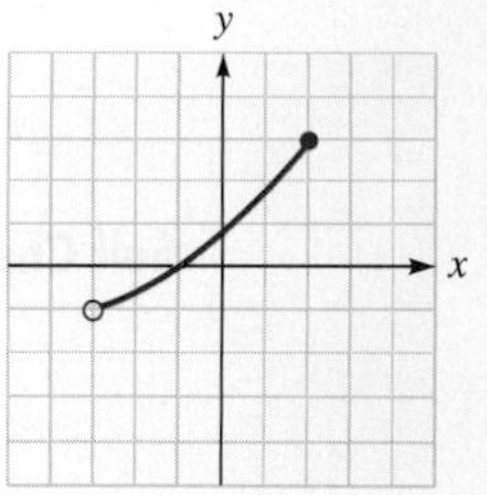

29.

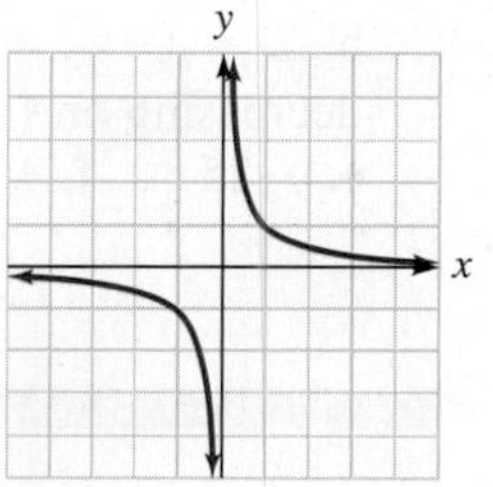

30.

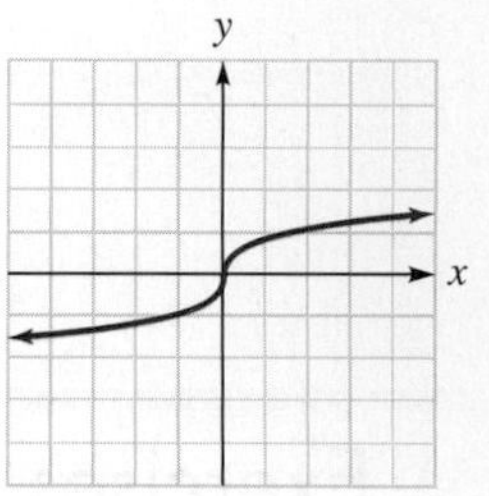

31.

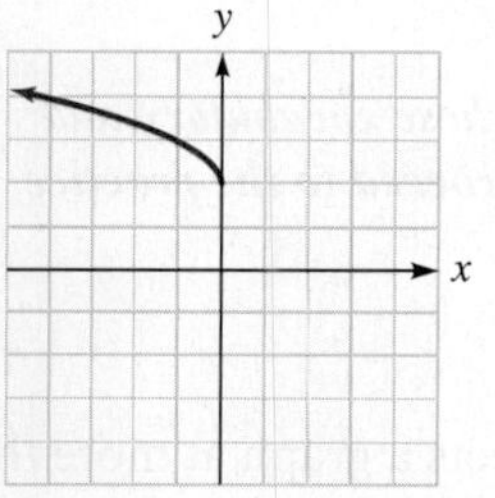

32.

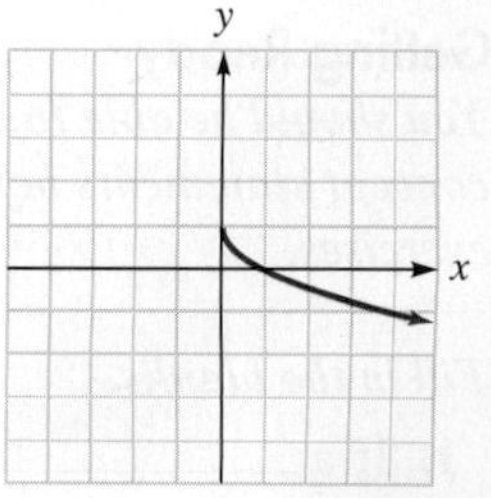

33.
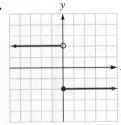

34.
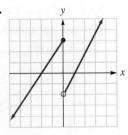

Determine algebraically whether each function is even, odd, or neither.

35. $f(x) = x^4 + x^2$

36. $f(x) = x^6 - x^2$

37. $f(x) = \dfrac{x}{x^2 - 1}$

38. $f(x) = \dfrac{2x}{x^2 - 9}$

39. $f(x) = 2x^3 - 3x$

40. $f(x) = -x^3 + 2x$

41. $f(x) = \dfrac{1}{x^4}$

42. $f(x) = -\dfrac{2}{x^2}$

43. $f(x) = \sqrt{x + 1}$

44. $f(x) = \sqrt{2x - 4}$

45. $f(x) = \dfrac{|x|}{x}$

46. $f(x) = 2x - |x|$

Use the given graph of the function to answer parts (a)–(h).

47. $f(x) = 2x^2 - 2x$

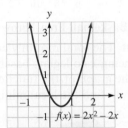

(a) State the domain and the range.

(b) State the value(s) of any intercepts.

(c) State the open intervals where the function is increasing, decreasing, or constant.

(d) State the value(s) of any local extrema.

(e) Find $f\left(-\dfrac{1}{2}\right)$.

(f) Find the values of x for which $f(x) = \dfrac{3}{2}$.

(g) State the value(s) of any zeros of the function.

(h) Is the function odd, even, or neither?

48. $f(x) = -x^2 + 4x$

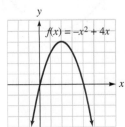

(a) State the domain and range.

(b) State the value(s) of any intercepts.

(c) State the open intervals where the function is increasing, decreasing, or constant.

(d) State the value(s) of any local extrema.

(e) Find $f(1)$.

(f) Find the values of x for which $f(x) = 3$.

(g) State the value(s) of any zeros of the function.

(h) Is the function odd, even, or neither?

49. $f(x) = x(x^2 - 4)$

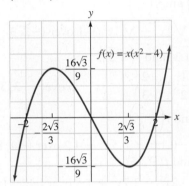

(a) State the domain and range.

(b) State the value(s) of any intercepts.

(c) State the open intervals on which the function is increasing, decreasing, or constant.

(d) State the value(s) of any local extrema.

(e) Find $f(2)$.

(f) Find the value(s) of x for which $f(x) = \dfrac{16\sqrt{3}}{9}$.

(g) State the value(s) of any zeros of the function.

(h) Is the function odd, even, or neither?

50. $f(x) = -x^2(x^2 - 1)$

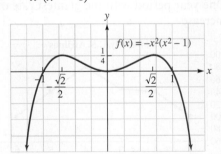

(a) State the domain and range.

(b) State the value(s) of any intercepts.

(c) State the open intervals on which the function is increasing, decreasing, or constant.

(d) State the value(s) of any local extrema.

(e) Find $f(-1)$.

(f) Find the value(s) of x for which $f(x) = \dfrac{1}{4}$.

(g) State the value(s) of any zeros of the function.

(h) Is the function odd, even, or neither?

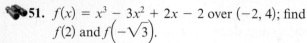

Approximate any local extrema and find the indicated values of the function over the given interval.

51. $f(x) = x^3 - 3x^2 + 2x - 2$ over $(-2, 4)$; find $f(2)$ and $f\left(-\sqrt{3}\right)$.

52. $f(x) = x^3 + 2x^2 - 5x - 3$ over $(-4, 3)$; find $f(-1.3)$ and $f\left(\sqrt{2}\right)$.

53. $f(x) = x^4 - 3x^2 + 2x - 1$ over $(-3, 2)$; find $f(-1.7)$ and $f(1)$.

54. $f(x) = x^6 - 3x^3 - 4$ over $(-2, 2)$; find $f\left(-\dfrac{2}{3}\right)$ and $f\left(\dfrac{5}{7}\right)$.

55. $f(x) = \sqrt{2}x^3 - \sqrt{3}x^2 + 1$ over $(-2, 3)$; find $f\left(-\sqrt{1.3}\right)$ and $f\left(\sqrt{3}\right)$.

56. $f(x) = \dfrac{1}{2}x^4 + \dfrac{2}{3}x^2 + \dfrac{1}{2}x - 5$ over $(-3, 3)$; find $f\left(\dfrac{4}{3}\right)$ and $f\left(-\dfrac{5}{9}\right)$.

57. $f(x) = -0.1x^3 + 0.5x^2 - 4.2$ over $(-4, 8)$; find $f(2.3)$ and $f(-3.5)$.

58. $f(x) = 3.2x^5 - 2.7x^2 + 1.5$ over $(-2, 2)$; find $f(1.8)$ and $f(-0.75)$.

Applications

59. U.S. airlines The graph shows the number of domestic and international passengers, in millions, flying U.S. airlines in 2004 and through 2009. Use the graph to approximate the following:

(a) Domain and range

(b) Open intervals where the graph increases, decreases, or remains constant

(c) Local extrema

(d) One-year period with the greatest rate of increase

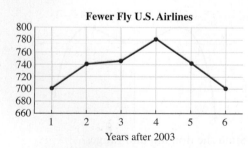

Fewer Fly U.S. Airlines

Years after 2003

Source: *USA Today*

60. Rain in Dallas, Texas The graph shows the average number of inches of rain, per month, in Dallas, Texas for the month of May and through October. Use the graph to approximate the following:

(a) Domain and range

(b) Open intervals where the graph increases, decreases, or remains constant

(c) Local extrema

(d) Average number of inches of rain in May

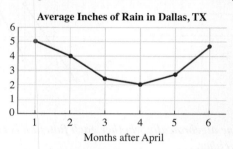

Average Inches of Rain in Dallas, TX

Months after April

Source: http://www.weather.com

61. Maximum height The path of a basketball thrown from a free-throw line can be modeled by the graph shown here. If x represents the horizontal distance, in feet, from the free-throw line and $f(x)$ represents the height, in feet, of the basketball, use the graph to approximate the local maximum.

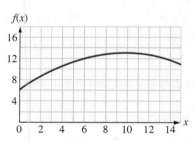

Basketball Free Throw

62. Minimum cost The total monthly cost $C(x)$, in dollars, of producing x collegiate baseball caps can be modeled by the graph of the function shown here. Use the graph to find the local minimum value.

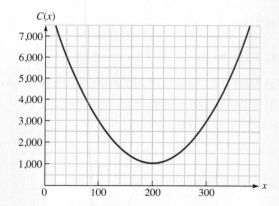

Cost of Producing Baseball Caps

63. Area Using 120 feet of fencing, we wish to maximize the area of a rectangular plot of land that has one side along the bank of a river, as shown here.

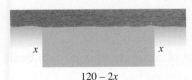

$120 - 2x$

(a) Express the area A as a function of x.

(b) What is the area if $x = 20$ feet?

(c) What is the area if $x = 40$ feet?

 (d) Use a calculator to graph the function A and determine the value of x that will yield the maximum area. What is the maximum area?

64. Volume From a piece of cardboard 60 centimeters by 60 centimeters, square corners are cut out so that the sides can be folded up to form a box.

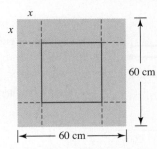

60 cm

60 cm

(a) Express the volume V as a function of x.

(b) What is the volume if $x = 5$ centimeters?

(c) What is the volume if $x = 10$ centimeters?

(d) Use a calculator to graph the function V and determine the value of x that will yield the maximum volume. What is the maximum volume?

65. Maximizing profit A company makes and sells purses. The profit function is given by $P(x) = -0.1x^3 + 1.65x^2 + 24x - 80$. Use a graphing calculator to determine the number of purses x that will produce the maximum profit.

66. Maximum number of mosquitoes Assume that the number of mosquitoes $N(x)$ depends on the rainfall, in inches, according to the function $N(x) = 12,000 + 1,055.25x - 72.6x^2 + x^3$, $0 \le x \le 50$. Use a graphing calculator to determine the following:

(a) The amount of rainfall that will produce the minimum number of mosquitoes

(b) The amount of rainfall that will produce the maximum number of mosquitoes

Discovery and Writing

67. Is there a limit on the number of x-intercepts that a function can have?

68. Why can a function have at most one y-intercept?

69. Describe what happens at the point where the graph of a function changes from increasing to decreasing.

70. Describe what happens at the point where the graph of a function changes from decreasing to increasing.

71. In this section, we discussed maximum and minimum values; state three real-life situations where a maximum or minimum value is important.

72. Give two examples of real-life functions that are increasing.

73. Give two examples of real-life functions that are decreasing.

74. Draw the graph of a function with the following characteristics:

Domain = $[-5, 10]$

Range = $[-10, 10)$

Increasing on $(-5, 0)$

Decreasing on $(0, 10)$

Local minimum = -5

Local maximum = 5

Yuri Arcurs/Shutterstock.com

1.4 Basic Graphs

In this section, we will learn to

1. Graph a function by plotting points.
2. Graph and interpret the behavior of a variety of important basic functions.
3. Evaluate and graph the greatest integer function.
4. Evaluate and graph piecewise-defined functions.
5. Solve applications involving piecewise-defined functions.

Graphs convey very important information. Consider the graph in Figure 1-44. It shows a comparison of the weekly earnings of full-time wage and salary workers who are 25 years old.

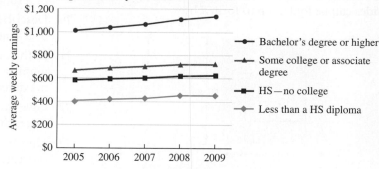

FIGURE 1-44

Source: U.S. Bureau of Labor Statistics

We see from the graph that the weekly earnings in 2009 for workers 25 years old with a bachelor's degree or a higher degree were approximately $1,100. This weekly salary is approximately $400 more than the salary for a worker of the same age who attended some college or who had an associate degree. The graph clearly indicates a salary advantage for people obtaining a bachelor's degree or an advanced degree. The visual representation of data that a graph provides can be quite useful.

We will continue our study of functions by learning basic functions that are used extensively in mathematics. It is important that we learn how to graph these functions, be knowledgeable of their domain and range, and understand how to use the functions in calculations.

1. Graph a Function by Plotting Points

In Section 1.1, we graphed equations by making a table of solutions and plotting the points given in the table. Then we connected the points by drawing a smooth curve through them. We will use that same strategy to graph functions.

EXAMPLE 1   **Graphing Functions**

Graph the functions. **a.** $f(x) = -2|x| + 3$ **b.** $f(x) = \sqrt{x-2}$

SOLUTION   In each case, we will make a table of solutions and plot the points given by the table. Then we will connect the points by drawing a smooth curve through them to obtain the graph of the function.

a. $f(x) = -2|x| + 3$

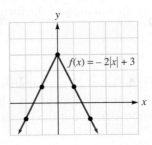

$f(x) = -2\|x\| + 3$		
x	$f(x)$	$(x, f(x))$
-2	-1	$(-2, -1)$
-1	1	$(-1, 1)$
0	3	$(0, 3)$
1	1	$(1, 1)$
2	-1	$(2, -1)$

FIGURE 1-45

b. $f(x) = \sqrt{x - 2}$

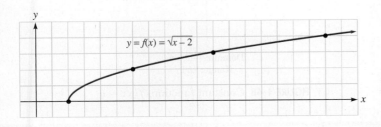

$f(x) = \sqrt{x - 2}$		
x	$f(x)$	$(x, f(x))$
2	0	$(2, 0)$
6	2	$(6, 2)$
11	3	$(11, 3)$
18	4	$(18, 4)$

FIGURE 1-46

Self Check 1   Graph $f(x) = |x + 3|$.

Now Try Exercise 31.

From Figure 1-45, we can see that the domain of the function $f(x) = -2|x| + 3$ is $D = (-\infty, \infty)$, and that the range is $R = (-\infty, 3]$.

From Figure 1-46, we can see that the domain of the function $f(x) = \sqrt{x - 2}$ is $D = [2, \infty)$, and that the range is $R = [0, \infty)$.

2. Graph and Interpret the Behavior of a Variety of Important Basic Functions

Linear Function   A **linear function** (Figure 1-47) is a function determined by an equation of the form $f(x) = mx + b$, or $y = mx + b$.

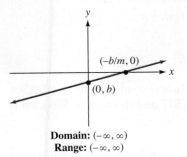

Domain: $(-\infty, \infty)$
Range: $(-\infty, \infty)$

FIGURE 1-47 Linear function, $f(x) = mx + b$

The graph is a nonvertical line. The domain is the set of all real numbers. For $m \neq 0$, the range is all real numbers. The slope of the line is m; the y-intercept is b. If $m > 0$, the line increases. If $m < 0$, the line decreases. If $m = 0$, the line is horizontal. (See "constant function.")

There are two types of linear functions that we will examine: **constant functions** and the **identity function**.

Constant Function   A **constant function** (Figure 1-48) is a linear function with slope $m = 0$. It is determined by an equation of the form $f(x) = b$, or $y = b$, where b is any real number.

The domain is the set of all real numbers, $D = (-\infty, \infty)$, and the range is b, $R = \{b\}$. The graph of a constant function is a horizontal line with y-intercept b. A constant function is symmetric about the y-axis and is an even function: $f(x) = f(-x)$.

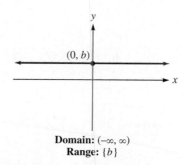

Domain: $(-\infty, \infty)$
Range: $\{b\}$

FIGURE 1-48 Constant function, $f(x) = b$

Identity Function   The **identity function** (Figure 1-49) is a linear function with $m = 1$ and $b = 0$. It is determined by an equation in the form $f(x) = x$, or $y = x$.

The domain and range are both the set of all real numbers, $D = (-\infty, \infty)$ and $R = (-\infty, \infty)$. This function maps every value of x to itself, so for every point on the graph, the x- and y-coordinates are equal. The x- and y-intercepts are both 0. This function is symmetric about the origin and is an odd function: $f(-x) = -f(x)$. This function increases throughout the domain.

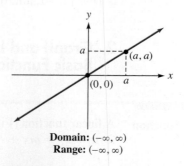

Domain: $(-\infty, \infty)$
Range: $(-\infty, \infty)$

FIGURE 1-49 Identity function, $f(x) = x$

EXAMPLE 2   Finding a Linear Function

The cost of electricity in Eagle River is a linear function of x, the number of kilowatt-hours (kwh) used. If the cost of 100 kwh is $17 and the cost of 500 kwh is $57, find an equation that expresses the function.

SOLUTION   Since C (the cost of electricity) is given to be a linear function of x, there are constants m and b such that $C = mx + b$. We have $C = 17$ when $x = 100$, and the point $(x_1, C_1) = (100, 17)$ lies on the straight-line graph of this function. We also

know that $C = 57$ when $x = 500$, and the point $(x_2, C_2) = (500, 57)$ also lies on that line. The slope of the line is

$$m = \frac{C_2 - C_1}{x_2 - x_1}$$
$$= \frac{57 - 17}{500 - 100}$$
$$= \frac{40}{400}$$
$$= 0.10$$

Now we know that $m = 0.10$. To determine b, we can substitute 0.10 for m and the coordinates of point $P(100, 17)$ into the equation $C = mx + b$ and solve for b.

$$C = mx + b$$
$$17 = 0.10(100) + b$$
$$17 = 10 + b$$
$$7 = b$$

From this we determine that $C = 0.10x + 7$. We can write it using functional notation as $C(x) = 0.10x + 7$. The electric company charges $7 plus 10¢ per kilowatt-hour used.

Self Check 2   Find the cost of using 400 kwh of electricity.

Now Try Exercise 69.

Squaring Function   The **squaring function** (Figure 1-50) is the function determined by the equation $f(x) = x^2$, or $y = x^2$.

The domain is the set of all real numbers $D = (-\infty, \infty)$, and the range is $R = [0, \infty)$. This graph is called a parabola. The x- and y-intercepts are both 0. This function is symmetric about the y-axis and is an even function: $f(x) = f(-x)$. This function decreases over the interval $(-\infty, 0)$ and increases over the interval $(0, \infty)$.

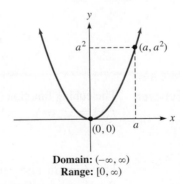

Domain: $(-\infty, \infty)$
Range: $[0, \infty)$

FIGURE 1-50 Squaring function, $f(x) = x^2$

Absolute Value Function   The **absolute value function** (Figure 1-51) is the function determined by the equation $f(x) = |x|$, or $y = |x|$.

The domain is the set of all real numbers $D = (-\infty, \infty)$, and the range is $R = [0, \infty)$. The x- and y-intercepts are both 0. This function is symmetric about the y-axis and is an even function: $f(x) = f(-x)$. This function decreases over the interval $(-\infty, 0)$ and increases over the interval $(0, \infty)$.

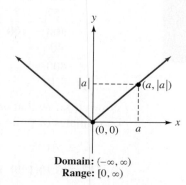

Domain: $(-\infty, \infty)$
Range: $[0, \infty)$

FIGURE 1-51 Absolute value function, $f(x) = |x|$

Square Root Function

The **square root function** (Figure 1-52) is the function determined by the equation $f(x) = \sqrt{x}$, or $y = \sqrt{x}$.

The domain is $D = [0, \infty)$, and the range is $R = [0, \infty)$. The x- and y-intercepts are both 0. The function increases over the interval $(0, \infty)$. Testing for symmetry, we see that $f(x) \neq f(-x)$ and $f(-x) \neq -f(x)$, so this function is neither even nor odd.

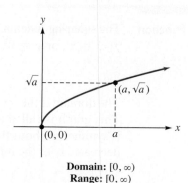

Domain: $[0, \infty)$
Range: $[0, \infty)$

FIGURE 1-52 Square root function, $f(x) = \sqrt{x}$

Cubing Function

The **cubing function** (Figure 1-53) is the function determined by the equation $f(x) = x^3$, or $y = x^3$.

The domain and range are both the set of all real numbers, $D = (-\infty, \infty)$ and $R = (-\infty, \infty)$. The x- and y-intercepts are both 0. Testing for symmetry, we see that

$$f(-x) = (-x)^3 = -x^3$$

$$-f(x) = -x^3$$

Since $f(-x) = -f(x)$, the function is symmetric about the origin and is an odd function. This function increases throughout its domain.

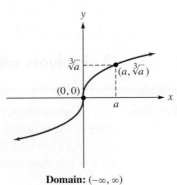

Domain: $(-\infty, \infty)$
Range: $(-\infty, \infty)$

FIGURE 1-53 Cubing function, $f(x) = x^3$

Cube Root Function

The **cube root function** (Figure 1-54) is the function determined by the equation $f(x) = \sqrt[3]{x}$, or $y = \sqrt[3]{x}$.

The domain and range are both the set of all real numbers, $D = (-\infty, \infty)$ and $R = (-\infty, \infty)$. The x- and y-intercepts are both 0. Since $f(-x) = -f(x)$, the function is symmetric about the origin and is an odd function. This function increases throughout its domain.

Domain: $(-\infty, \infty)$
Range: $(-\infty, \infty)$

FIGURE 1-54 Cube root function, $f(x) = \sqrt[3]{x}$

ACCENT ON TECHNOLOGY Graphing the Cube Root Function

We can graph this function on a graphing calculator using the cube root key by selecting MATH 4 from the graph window, as shown in Figure 1-55.

To graph $f(x) = \sqrt[3]{x}$ on the graphing calculator, go to the graph menu and then enter MATH 4 . Graph using the ZOOM 4 menu.

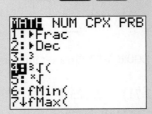

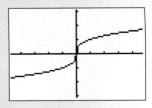

FIGURE 1-55

Reciprocal Function   The **reciprocal function** (Figure 1-56) is the function determined by the equation

$$f(x) = \frac{1}{x}, \text{ or } y = \frac{1}{x}.$$

The domain and range are both the set of all nonzero real numbers, $D = (-\infty, 0) \cup (0, \infty)$ and $R = (-\infty, 0) \cup (0, \infty)$. There are no intercepts. The function decreases throughout its domain. Since $f(-x) = -f(x)$, the function is symmetric about the origin and is an odd function.

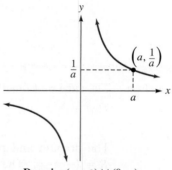

Domain: $(-\infty, 0) \cup (0, \infty)$
Range: $(-\infty, 0) \cup (0, \infty)$

FIGURE 1-56 Reciprocal function, $f(x) = \dfrac{1}{x}$

3. Evaluate and Graph the Greatest Integer Function

Greatest Integer Function   The **greatest integer function** (Figure 1-57), sometimes called the "floor" function, is the function defined by the equation $f(x) = [\![x]\!]$ [also $f(x) = \text{int}(x)$], or $y = [\![x]\!]$, where the value of y that corresponds to x is the greatest integer that is less than or equal to x.

The domain is the set of all real numbers $D = (-\infty, \infty)$, and the range is the set of all integers $R = \{y \,|\, y \text{ is an integer}\}$. The x-intercepts are all the numbers in the interval $[0, 1)$, and the y-intercept is 0.

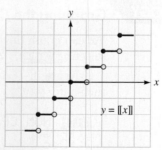

FIGURE 1-57 Greatest integer function, $f(x) = [\![x]\!]$

For example, $[\![2.71]\!] = 2$, $[\![23\frac{1}{2}]\!] = 23$, $[\![10]\!] = 10$, $[\![\pi]\!] = 3$, $[\![-2.5]\!] = -3$. Since the greatest integer function is made up of a series of horizontal line segments, it is an example of a group of functions called **step functions**.

To graph the greatest integer function, we list several intervals and the corresponding values of the greatest integer function:

$[0, 1): y = [\![x]\!] = 0$ For numbers from 0 to 1 (not including 1), the greatest integer in the interval is 0.

$[1, 2): y = [\![x]\!] = 1$ For numbers from 1 to 2 (not including 2), the greatest integer in the interval is 1.

$[2, 3): y = [\![x]\!] = 2$ For numbers from 2 to 3 (not including 3), the greatest integer in the interval is 2.

Within each interval, the values of y are constant, but they jump by 1 at integer values of x.

EXAMPLE 3   **Using the Greatest Integer Function**

To print a business form, a printer charges $10 for the order, plus $20 for each box containing 500 forms. The printer counts any portion of a box as a full box. Graph this step function.

SOLUTION   If we order the forms and then change our minds before the forms are printed, the cost will be $10. Thus, the ordered pair (0, 10) will be on the graph. If we purchase up to one full box, the cost will be $10 for the order and $20 for the printing, for a total of $30. Thus, the ordered pair (1, 30) will be on the graph. The cost for $1\frac{1}{2}$ boxes will be the same as the cost for 2 full boxes, or $50. Thus, the ordered pairs (1.5, 50) and (2, 50) are on the graph. The complete graph is shown in Figure 1-58.

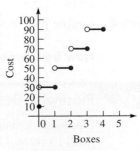

FIGURE 1-58

Self Check 3   Find the cost of $4\frac{1}{2}$ boxes.

Now Try Exercise 73.

4. Evaluate and Graph Piecewise-Defined Functions

Some functions, called **piecewise-defined functions,** are defined by using different equations for different intervals in their domains. To illustrate, we will graph the piecewise-defined function f given by

$$f(x) = \begin{cases} -2 & \text{if } x \le 0, \\ x + 1 & \text{if } x > 0. \end{cases}$$

For each number x, we decide which part of the definition to use. If $x \le 0$, the corresponding value of $f(x)$ is -2. Over the interval $(-\infty, 0)$, the function is constant. If $x > 0$, the corresponding value of $f(x)$ is $x + 1$. Over the interval $(0, \infty)$, the function is increasing. The graph appears in Figure 1-59.

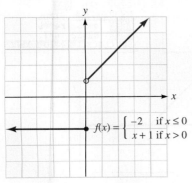

$f(x) = \begin{cases} -2 & \text{if } x \le 0 \\ x + 1 & \text{if } x > 0 \end{cases}$

FIGURE 1-59

EXAMPLE 4   Graphing a Piecewise-Defined Function

Graph the function: $f(x) = \begin{cases} -x & \text{if } x < 0, \\ x^2 & \text{if } 0 \le x \le 1, \\ 1 & \text{if } x > 1. \end{cases}$

SOLUTION   For each number x, we decide which part of the definition to use. If $x < 0$, the value of $f(x)$ is determined by the equation $f(x) = -x$. Over the interval $(-\infty, 0)$, the function is decreasing.

If $0 \le x \le 1$, the value of $f(x)$ is x^2. Over the interval $(0, 1)$, the function is increasing.

If $x > 1$, the value of $f(x)$ is 1. Over the interval $(1, \infty)$, the function is constant. The graph appears in Figure 1-60.

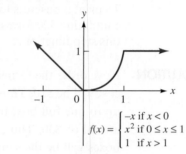

$$f(x) = \begin{cases} -x & \text{if } x < 0 \\ x^2 & \text{if } 0 \le x \le 1 \\ 1 & \text{if } x > 1 \end{cases}$$

FIGURE 1-60

Self Check 4   Graph: $f(x) = \begin{cases} 2x & \text{if } x \le 0, \\ x - 1 & \text{if } x > 0. \end{cases}$

Now Try Exercise 61.

ACCENT ON TECHNOLOGY   **Piecewise-Defined Functions**

Piecewise-defined functions can be graphed on a calculator, but they can be difficult to enter. The inequality symbols are found by pressing [2nd] [MATH] (TEST), shown in Figure 1-61(a). The "and" command is found in the LOGIC menu, which is accessed via the TEST menu [Figure 1-61(b)]. Also, when you are graphing these functions, the calculator should be in DOT mode [Figure 1-61(c)], so that the parts will only be connected if the function defines it that way. The way the function should be entered is shown in Figure 1-61(d). Note that the use of parentheses is essential. The graph of the function is shown in Figure 1-61(e).

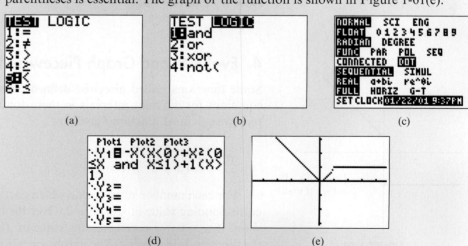

(a)   (b)   (c)

(d)   (e)

FIGURE 1-61

5. Solve Applications Involving Piecewise-Defined Functions

EXAMPLE 5   **Solving an Application of a Piecewise-Defined Function**

The 2007 federal income-tax marginal rates for a single taxpayer are shown in the table.

a. Construct a piecewise-defined function for this table.

b. How much tax will a single person earning $45,000 in 2007 pay?

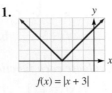

2007 Federal Income-Tax Rates for a Single Person	
Income	**Marginal Tax Rate**
$0−7,825	10%
$7,826−31,850	15%
$31,851−77,100	25%
$77,101−160,850	28%
$160,851−349,700	33%
$349,701+	35%

SOLUTION   **a.** Each piece of the equation is linear. Note, however, that the beginning value for each tax bracket is one more than the ending value for the previous tax bracket. We can construct the function as follows:

$$f(x) = \begin{cases} 0.10x & \text{if } 0 \le x \le 7,825, \\ 0.15x & \text{if } 7,826 \le x \le 31,850, \\ 0.25x & \text{if } 31,851 \le x \le 77,100, \\ 0.28x & \text{if } 77,101 \le x \le 160,850, \\ 0.33x & \text{if } 160,851 \le x \le 349,700, \\ 0.35x & \text{if } x \ge 349,701. \end{cases}$$

b. A person earning $45,000 taxable income falls into the 25% tax bracket and will pay $11,250 in federal taxes. We calculate this by finding $f(45,000)$. $f(45,000) = 0.25(45,000) = 11,250$. ∎

Self Check 5   How much money in federal taxes would a person earning $200,000 a year pay?

Now Try Exercise 81.

Self Check Answers   **1.**

$f(x) = |x + 3|$

2. $47   **3.** $110   **4.**

$$f(x) = \begin{cases} 2x & \text{if } x \le 0 \\ x - 1 & \text{if } x > 0 \end{cases}$$

5. $66,000

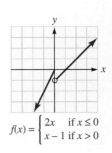

Exercises **1.4**

Getting Ready
You should be able to complete these vocabulary and concept statements before you proceed to the practice exercises.

Fill in the blanks.

1. The graph of the function $f(x) = mx + b$ is a _____ function.

2. The graph of the function $f(x) = x^2$ _____ (increases or decreases) on the interval $(-\infty, 0)$.

3. The identity function is an _____ (even or odd) function.

4. The function $f(x) = x^3$ _____ (increases or decreases) throughout its domain.

5. The reciprocal function $f(x) = \dfrac{1}{x}$ is not defined for $x =$ _____.

6. The greatest integer function is also called a _____ function.

Practice
Sketch the graph of each basic function.

7. $f(x) = -2$
8. $f(x) = x$
9. $f(x) = |x|$
10. $f(x) = \sqrt{x}$
11. $f(x) = \dfrac{1}{x}$
12. $f(x) = x^3$
13. $f(x) = x^2$
14. $f(x) = \sqrt[3]{x}$

Graph each function by plotting points.

15. $f(x) = 2x + 3$
16. $f(x) = -\dfrac{3}{4}x + 4$
17. $f(x) = x^2 - 4$
18. $f(x) = \dfrac{1}{2}x^2 + 3$
19. $f(x) = -x^2 + 2$
20. $f(x) = -x^3 + 1$
21. $f(x) = -x^3 + 2$
22. $f(x) = -(x + 2)^3$
23. $f(x) = (x + 1)^3$
24. $f(x) = -|x|$
25. $f(x) = |x - 2|$
26. $f(x) = -|x - 2|$
27. $f(x) = \left|\dfrac{1}{2}x + 3\right|$
28. $f(x) = -\left|\dfrac{1}{2}x + 3\right|$
29. $f(x) = -\sqrt{x}$
30. $f(x) = \sqrt{x + 1}$
31. $f(x) = -\sqrt{x + 1}$
32. $f(x) = \sqrt{x + 2}$
33. $f(x) = \sqrt{2x - 4}$
34. $f(x) = -\sqrt{2x - 4}$
35. $f(x) = -\sqrt[3]{x}$
36. $f(x) = \sqrt[3]{x} - 2$
37. $f(x) = \sqrt[3]{x} + 2$
38. $f(x) = \sqrt[3]{x} - 1$
39. $f(x) = \dfrac{2}{x}$
40. $f(x) = -\dfrac{3}{x}$

Evaluate each piecewise-defined function.

41. $f(x) = \begin{cases} 2x + 2 & \text{if } x < 0 \\ 3 & \text{if } x \geq 0 \end{cases}$
 a. $f(-2)$ b. $f(0)$

42. $f(x) = \begin{cases} x - 2 & \text{if } x < 1 \\ x^2 & \text{if } x \geq 1 \end{cases}$
 a. $f(1)$ b. $f(5)$

43. $f(x) = \begin{cases} 2x - 5 & \text{if } x < 0 \\ -3x & \text{if } x \geq 0 \end{cases}$
 a. $f(-3)$ b. $f(6)$

44. $f(x) = \begin{cases} 4x + 5 & \text{if } x < 1 \\ x^2 - 1 & \text{if } x \geq 1 \end{cases}$
 a. $f(-1)$ b. $f\left(\dfrac{3}{2}\right)$

45. $f(x) = \begin{cases} 2 & \text{if } x < 0 \\ 2 - x & \text{if } 0 \leq x < 2 \\ x + 1 & \text{if } x \geq 2 \end{cases}$
 a. $f(-1)$ b. $f(1)$ c. $f(2)$

46. $f(x) = \begin{cases} 2x & \text{if } x < 0 \\ 3 - x & \text{if } 0 \leq x < 2 \\ |x| & \text{if } x \geq 2 \end{cases}$
 a. $f(-0.5)$ b. $f(0)$ c. $f(2)$

47. $f(x) = \begin{cases} 2x - 3 & \text{if } x < 0 \\ x^2 & \text{if } 0 \leq x \leq 4 \\ \dfrac{1}{x} & \text{if } x > 4 \end{cases}$
 a. $f(-3)$ b. $f(4)$ c. $f(6)$

48. $f(x) = \begin{cases} |x| & \text{if } x < 0 \\ \sqrt{x} & \text{if } 0 \leq x \leq 9 \\ -\dfrac{1}{x} & \text{if } x > 9 \end{cases}$
 a. $f(-3)$ b. $f(4)$ c. $f(10)$

Evaluate each function at the indicated x values.

49. $f(x) = [x]$ a. $f(3)$ b. $f(-4)$ c. $f(-2.3)$
50. $f(x) = [3x]$ a. $f(4)$ b. $f(-2)$ c. $f(-1.2)$
51. $f(x) = [x + 3]$ a. $f(-1)$ b. $f\left(\dfrac{2}{3}\right)$ c. $f(1.3)$
52. $f(x) = [4x] - 1$ a. $f(-3)$ b. $f(\pi)$ c. $f(0)$

Graph each piecewise-defined function.

53. $f(x) = \begin{cases} x + 2 & \text{if } x < 0 \\ 2 & \text{if } x \geq 0 \end{cases}$

54. $f(x) = \begin{cases} 2x & \text{if } x < 0 \\ -2x & \text{if } x \geq 0 \end{cases}$

55. $f(x) = \begin{cases} x - 2 & \text{if } x \leq 1 \\ 3 & \text{if } x > 1 \end{cases}$

56. $f(x) = \begin{cases} -x & \text{if } x < -4 \\ \frac{1}{2}x & \text{if } x > -4 \end{cases}$

57. $f(x) = \begin{cases} -x & \text{if } x < 0 \\ x^2 & \text{if } x \geq 0 \end{cases}$

58. $f(x) = \begin{cases} |x| & \text{if } x < 0 \\ \sqrt{x} & \text{if } x \geq 0 \end{cases}$

59. $f(x) = \begin{cases} 5 & \text{if } x < 0 \\ x + 5 & \text{if } 0 \leq x \leq 2 \\ 4 - 2x & \text{if } x > 2 \end{cases}$

60. $f(x) = \begin{cases} 2 & \text{if } x < 0 \\ 2 - x & \text{if } 0 \leq x < 2 \\ x & \text{if } x \geq 2 \end{cases}$

61. $f(x) = \begin{cases} x + 3 & \text{if } x < 0 \\ x^2 & \text{if } 0 \leq x \leq 3 \\ -|x| & \text{if } x > 3 \end{cases}$

62. $f(x) = \begin{cases} 5 & \text{if } x < 1 \\ -x + 1 & \text{if } 1 \leq x < 4 \\ |x| & \text{if } x \geq 4 \end{cases}$

Graph each function.

63. $y = [\![2x]\!]$

64. $y = \left[\!\left[\frac{1}{3}x + 3\right]\!\right]$

65. $y = [\![x]\!] - 1$

66. $y = [\![x + 2]\!]$

67. $y = \left[\!\left[\frac{x}{2}\right]\!\right]$

68. $y = \left[\!\left[\frac{x}{3}\right]\!\right]$

Applications

69. Cost of electricity The cost C of electricity in Baltimore is a linear function of x, the number of kilowatt-hours (kwh) used. If the cost of 100 kwh is \$19 and the cost of 500 kwh is \$67, find a linear function that expresses C in terms of x.

70. Temperature conversion The Fahrenheit temperature reading F is a linear function of the Celsius reading C. If $C = 0$ when $F = 32$ and the readings are the same at -40, write F as a function of C.

71. Grading scales A mathematics instructor assigns letter grades according to the following scale:

From	Up to but less than	Grade
60%	70%	D
70%	80%	C
80%	90%	B
90%	100% (including 100%)	A

Find the final semester grade of a student who has test scores of 67%, 73%, 84%, 87%, and 93%. To determine the answer, it might be helpful to graph the ordered pairs (p, g), where p represents the percentage and g represents the grade.

72. Calculating grades See Exercise 71. Find the final semester grade of a student who has test scores of 53%, 65%, 64%, 73%, 89%, and 82%.

73. Renting a jeep A rental company charges \$20 to rent a Jeep Wrangler for one day, plus \$4 for every 100 miles (or portion of 100 miles) that it is driven. Find the cost if the jeep is driven 275 miles. To determine the answer, it might be helpful to graph the ordered pairs (m, C), where m represents the miles driven and C represents the cost.

iStockPhoto/© byltwill

74. Riding in a taxi A taxicab company charges \$3 for a trip up to 1 mile, and \$2 for every extra mile (or portion of a mile). Find the cost to ride $10\frac{1}{4}$ miles. To determine the answer, it might be helpful to graph the ordered pairs (m, C), where m represents the miles traveled and C represents the cost.

75. Computer communications An online information service charges for connect time at a rate of \$12 per hour, computed for every minute or fraction of a minute. Find the cost of $7\frac{1}{2}$ minutes. To determine the answer, it might be helpful to graph the points (t, C), where C is the cost of t minutes of connect time.

76. Plumbing repairs A plumber charges \$30, plus \$40 per hour (or fraction of an hour), to install a new bathtub. If the job took 4 hours, how much did it cost? To determine the answer, it might be helpful to graph the points (t, C), where t is the time it takes to do the job and C is the cost.

77. Rounding numbers Measurements are rarely exact; they are often *rounded* to an appropriate precision. Graph the points (x, y), where y is the result of rounding the number x to the nearest ten.

78. **Signum function** Computer programmers often use the following function, denoted $y = \text{sgn } x$. Graph this function and find its domain and range.

$$y = \begin{cases} -1 & \text{if } x < 0 \\ 0 & \text{if } x = 0 \\ 1 & \text{if } x > 0 \end{cases}$$

79. Graph the function defined by $y = \dfrac{|x|}{x}$ and compare it to the graph in Exercise 78. Are the graphs the same?

80. Graph the function defined by $y = x + |x|$.

81. In 1996, presidential candidate Steve Forbes proposed a flat tax for the United States. He proposed that a person would be taxed at the rate of 17% on all income earned above \$13,300. The first \$13,300 would not be taxed.

    (a) Find an equation for the flat tax using a piecewise-defined function. Use income as the input and the federal tax as the output.

    (b) How much in taxes would a person earning \$25,000 per year pay? \$45,000?

    (c) Noting that the first \$13,300 of income is not taxed, what is the actual percentage of a person's income used to pay federal income tax for an income of \$25,000? \$45,000?

    (d) From part (c) you can see that the actual percentage of the income tax is not the same for everyone, since the first \$13,300 is not taxed. Write a formula for the function where the input is the individual's income and the output is the percentage of that income that pays the federal income tax.

82. Repeat Exercise 81 using a tax rate of 15%, where the first \$12,000 of income is tax free.

Discovery and Writing

83. How many x-intercepts does the function $f(x) = 0$ have? What is the domain and range?

84. Use postal rates for first-class postage to create a step function for calculating costs of mailing a first-class letter.

85. Six Flags White Water in Atlanta, Georgia, charges the following prices for daily admission to the park: general admission, \$36.99; child under 48 inches, \$26.99. Describe how this information can be represented by a piecewise-defined function.

86. Construct a piecewise-defined function that occurs in everyday life.

1.5 Transformations of Graphs

In this section, we will learn to

1. Graph a function by applying a vertical or horizontal shift to the graph of a basic function.
2. Use a combination of vertical and horizontal shifts to graph a function.
3. Graph a function by reflecting the graph of a basic function about the x- or y-axis.
4. Write an equation of a function represented by a given graph or a given transformation.
5. Graph a function by stretching or shrinking the graph of a basic function.
6. Graph a function by using a combination of transformations.

We can often transform the graph of a function into the graph of another function by shifting the graph vertically or horizontally. Also, we can reflect a graph about the x- or y-axis, and stretch or shrink a graph horizontally or vertically to transform the graph of a function into the graph of another function. In this section, we will graph new functions from known ones using these methods.

Consider a white-water rafting trip on the Ocoee River in Tennessee. Suppose one company charges a group of students $20 for each hour on the river, plus $100 for a guide and equipment. The cost of the rafting trip can be represented by the function

$$C_1(t) = 20t + 100,$$

where $C_1(t)$ represents the cost in dollars to raft t hours on the river.

If the company increases its charge for the guide and equipment to $150, the new cost function can be represented by

$$C_2(t) = 20t + 150.$$

The graphs of the two cost functions are shown in Figure 1-62.

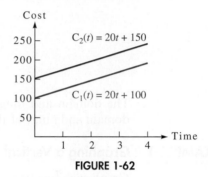

FIGURE 1-62

Note that if we shift the graph of $C_1(t)$ 50 units vertically, we obtain the graph of $C_2(t)$. This shift is called a *translation*.

In Section 1.4, we learned how to graph several important basic functions. Many other functions can be graphed by performing simple transformations of these graphs. In this section, the graphs of $f(x) = x^2$, $f(x) = x^3$, $f(x) = |x|$, $f(x) = \sqrt{x}$, and $f(x) = \sqrt[3]{x}$ will be transformed—that is, translated and stretched in various ways.

1. Graph a Function by Applying a Vertical or Horizontal Shift to the Graph of a Basic Function

The graphs of functions can be identical except for their position in the xy plane. For example, Figure 1-63 shows the graph of the function $f(x) = x^2 + k$ for $k = 0$, $k = 2$, and $k = -3$. The graph of $f(x) = x^2 + 2$ is identical to the graph of $f(x) = x^2$ except that it is shifted 2 units up. The graph of $f(x) = x^2 - 3$ is identical to the graph of $f(x) = x^2$ except that it is shifted 3 units down. These transformations are called **vertical shifts.**

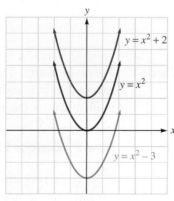

FIGURE 1-63

In general, we can make the following observations:

Vertical Shifts   If f is a function and k is a positive real number, then

- The graph of $g(x) = f(x) + k$ is identical to the graph of $f(x)$ except that it is shifted k units upward.
- The graph of $g(x) = f(x) - k$ is identical to the graph of $f(x)$ except that it is shifted k units downward.

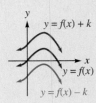

The domain and range of the transformed function may be different from the domain and range of the original function, as we will see in the first example.

EXAMPLE 1   Graphing a Vertical Shift

Graph each function and state the domain and range: **a.** $g(x) = |x| - 2$
b. $h(x) = |x| + 3$

SOLUTION   **a.** The graph of $g(x) = |x| - 2$ is identical to the graph of $f(x) = |x|$ except that it is shifted 2 units down, as shown in Figure 1-64(a). The domain and range are $D = (-\infty, \infty)$ and $R = [-2, \infty)$.

b. The graph of $h(x) = |x| + 3$ is identical to the graph of $f(x) = |x|$ except that it is shifted 3 units up, as shown in Figure 1-64(b). The domain and range are $D = (-\infty, \infty)$ and $R = [3, \infty)$.

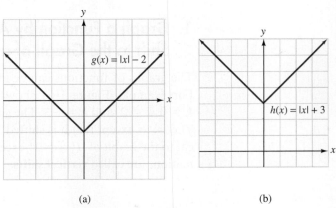

(a)   (b)

FIGURE 1-64

Self Check 1   *Fill in the blanks:* The graph of $g(x) = x^2 + 3$ is identical to the graph of $f(x) = x^2$ except that it is shifted _____ units _____. The graph of $h(x) = x^2 - 4$ is identical to the graph of $f(x) = x^2$ except that it is shifted _____ units _____.

Now Try Exercise 23.

Figure 1-65 shows the graph of $f(x) = (x + h)^2$ for three values of h. If $h = 0$, we have the graph of $f(x) = x^2$. The graph of $f(x) = (x - 2)^2$ is identical to the graph of $f(x) = x^2$ except that it is shifted 2 units to the right. The graph of

$f(x) = (x + 3)^2$ is identical to the graph of $f(x) = x^2$ except that it is shifted 3 units to the left. These transformations are called **horizontal shifts.**

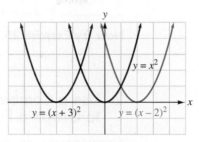

FIGURE 1-65

In general, we can make the following observations:

Horizontal Shifts   If f is a function and k is a positive real number, then

- The graph of $g(x) = f(x - k)$ is identical to the graph of $f(x)$ except that it is shifted k units to the right.
- The graph of $g(x) = f(x + k)$ is identical to the graph of $f(x)$ except that it is shifted k units to the left.

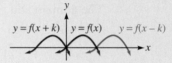

EXAMPLE 2   **Graphing a Horizontal Shift**

Graph each function and state the domain and range: **a.** $g(x) = |x - 4|$
b. $h(x) = |x + 2|$

SOLUTION   **a.** The graph of $g(x) = |x - 4|$ is identical to the graph of $f(x) = |x|$ except that it is shifted 4 units to the right, as shown in Figure 1-66(a). The domain and range are $D = (-\infty, \infty)$ and $R = [0, \infty)$.

b. The graph of $h(x) = |x + 2|$ is identical to the graph of $f(x) = |x|$ except that it is shifted 2 units to the left, as shown in Figure 1-66(b). The domain and range are $D = (-\infty, \infty)$ and $R = [0, \infty)$.

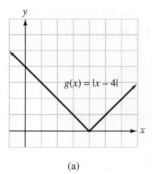

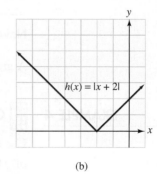

(a)   (b)

FIGURE 1-66

Self Check 2   *Fill in the blanks:* The graph of $g(x) = (x - 3)^2$ is identical to the graph of $f(x) = x^2$ except that it is shifted _____ units to the _____. The graph of $h(x) = (x + 2)^2$ is identical to the graph of $f(x) = x^2$ except that it is shifted _____ units to the _____.

Now Try Exercise 25.

2. Use a Combination of Vertical and Horizontal Shifts to Graph a Function

Sometimes we can obtain a graph by using both a horizontal shift and a vertical shift.

EXAMPLE 3   **Graphing a Function Involving Two Shifts**

Graph each function and state the domain and range: **a.** $g(x) = (x - 5)^3 + 4$ **b.** $h(x) = (x + 2)^2 - 2$

SOLUTION   **a.** The graph of $g(x) = (x - 5)^3 + 4$ is identical to the graph of $f(x) = x^3$ except that it is shifted 5 units to the right and 4 units up, as shown in Figure 1-67(a). The domain and range are $D = (-\infty, \infty)$ and $R = (-\infty, \infty)$.

b. The graph of $h(x) = (x + 2)^2 - 2$ is identical to the graph of $f(x) = x^2$ except that it is shifted 2 units to the left and 2 units down, as shown in Figure 1-67(b). The domain and range are $D = (-\infty, \infty)$ and $R = [-2, \infty)$.

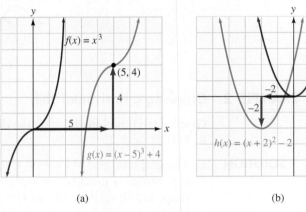

(a)   (b)

FIGURE 1-67

Self Check 3   *Fill in the blanks:* The graph of $g(x) = |x - 4| + 5$ is identical to the graph of $f(x) = |x|$ except that it is shifted _____ units to the _____ and _____ units _____.

Now Try Exercise 31.

A graphing calculator can be used to graph and compare functions.

EXAMPLE 4   **Graphing with a Graphing Calculator**

Use a graphing calculator to compare the graph of each function listed to the graph of $f(x) = \sqrt{x}$. State the domain and range of each function.

a. $g(x) = \sqrt{x} - 4$ **b.** $h(x) = \sqrt{x + 2}$ **c.** $k(x) = \sqrt{x - 2} + 3$

SOLUTION   **a.** Graph the functions $f(x) = \sqrt{x}$ and $g(x) = \sqrt{x} - 4$ on the same screen using the window $X:[-2, 6]$ and $Y:[-6, 3]$. The graphs in Figure 1-68(a) illustrate that the graph of g is obtained by shifting the graph of f down 4 units. The domain and range of g are $D = [0, \infty)$ and $R = [-4, \infty)$.

b. Graph the functions $f(x) = \sqrt{x}$ and $h(x) = \sqrt{x + 2}$ on the same screen using the window $X:[-3, 4]$ and $Y:[-2, 3]$. The graphs in Figure 1-68(b) illustrate that the graph of h is obtained by shifting the graph of f to the left 2 units. The domain and range of h are $D = [-2, \infty)$ and $R = [0, \infty)$.

c. Graph the functions $f(x) = \sqrt{x}$ and $k(x) = \sqrt{x - 2} + 3$ on the same screen using the window $X:[-1, 6]$ and $Y:[-1, 6]$. The graphs in Figure 1-68(c) illustrate that the graph of k is obtained by shifting the graph of f right 2 units and up 3 units. The domain and range of k are $D = [2, \infty)$ and $R = [3, \infty)$.

(a) $f(x) = \sqrt{x}$ is the darker graph, and $g(x) = \sqrt{x} - 4$ is the lighter graph.

(b) $f(x) = \sqrt{x}$ is the darker graph, and $h(x) = \sqrt{x + 2}$ is the lighter graph.

(c) $f(x) = \sqrt{x}$ is the darker graph, and $k(x) = \sqrt{x - 2} + 3$ is the lighter graph.

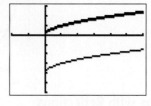 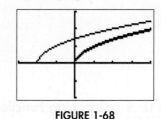 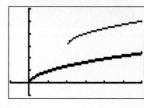

FIGURE 1-68

Self Check 4   Use a graphing calculator to graph these functions in the same graph window: $f(x) = x^3$, $g(x) = (x - 2)^3 + 3$, and $h(x) = (x + 1)^3 - 2$. Describe the transformations performed on the graph of f to produce the graphs of g and h.

Now Try Exercise 101.

3. Graph a Function by Reflecting the Graph of a Basic Function about the x- or y-Axis

Figure 1-69(a) shows that the graph of $f(x) = -\sqrt{x}$ is identical to the graph of $f(x) = \sqrt{x}$ except that it is reflected about the x-axis. Figure 1-69(b) shows that the graph of $f(x) = \sqrt{-x}$ is identical to the graph of $f(x) = \sqrt{x}$ except that it is reflected about the y-axis.

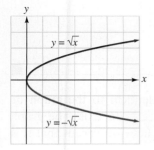

$y = -\sqrt{x}$		
x	y	(x, y)
0	0	$(0, 0)$
1	-1	$(1, -1)$
4	-2	$(4, -2)$

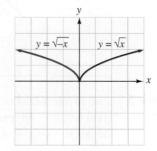

$y = \sqrt{-x}$		
x	y	(x, y)
0	0	$(0, 0)$
-1	1	$(-1, 1)$
-4	2	$(-4, 2)$

(a)   (b)

FIGURE 1-69

In general, we can make the following observations:

Reflections   If f is a function, then

- The graph of $g(x) = -f(x)$ is identical to the graph of $f(x)$ except that it is reflected about the x-axis.

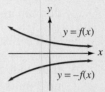

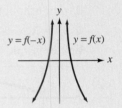

- The graph of $g(x) = f(-x)$ is identical to the graph of $f(x)$ except that it is reflected about the y-axis.

Comment

It is often helpful to think of a reflection as a mirror image of the graph about the x- or y-axis.

EXAMPLE 5   **Graphing Functions with Reflections**

Graph each function and state the domain and range: **a.** $g(x) = -|x + 1|$
b. $h(x) = |-x + 1|$

SOLUTION   **a.** The graph of $g(x) = -|x + 1|$ is identical to the graph of $f(x) = |x + 1|$ except that it is reflected about the x-axis, as shown in Figure 1-70(a). The domain and range of g are $D = (-\infty, \infty)$ and $R = (-\infty, 0]$.

b. The graph of $h(x) = |-x + 1|$ is identical to the graph of $f(x) = |x + 1|$ except that it is reflected about the y-axis, as shown in Figure 1-70(b). The domain and range of h are $D = (-\infty, \infty)$ and $R = [0, \infty)$.

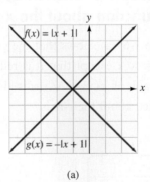

(a)

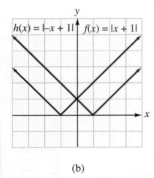

(b)

FIGURE 1-70

Self Check 5   *Fill in the blanks:* The graph of $g(x) = -(x - 4)^2$ is identical to the graph of $f(x) = (x - 4)^2$ except that it is reflected about the _____ axis. The graph of $h(x) = (-x - 4)^2$ is identical to the graph of $f(x) = (x - 4)^2$ except that it is reflected about the _____ axis.

Now Try Exercise 43.

4. Write an Equation of a Function Represented by a Given Graph or a Given Transformation

EXAMPLE 6   **Writing an Equation of a Function Represented by a Given Graph**

The graph shown in Figure 1-71 is a transformation of the graph of $f(x) = \sqrt[3]{x}$. Write an equation of the function represented by the graph.

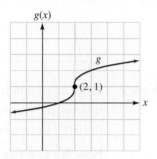

$g(x)$

g

$(2, 1)$

x

FIGURE 1-71

SOLUTION   The graph of g is a horizontal shift to the right 2 units and a vertical shift up 1 unit. Thus, an equation of g is $g(x) = \sqrt[3]{x - 2} + 1$.

Self Check 6   Write an equation of the function represented by the graph:

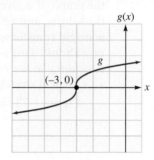

$g(x)$

g

$(-3, 0)$

x

Now Try Exercise 87.

EXAMPLE 7   **Writing an Equation of a Function from Given Transformations**

Write an equation of the function that would result from the following transformations of the graph of the function $f(x) = \sqrt{x}$: y-axis reflection, horizontal shift to the right 2 units, and vertical shift down 3 units. State the domain and range of the new function and draw the graph.

SOLUTION   We will use three steps and apply the transformations stated. We will represent the new function as $g(x)$.

Step 1: Reflection about the y-axis requires that we replace x with $-x$.

$$g(x) = \sqrt{-x}.$$

Step 2: The horizontal shift to the right 2 units requires that we replace x with $x - 2$.

$$g(x) = \sqrt{-x} = \sqrt{-(x - 2)} = \sqrt{-x + 2} = \sqrt{2 - x}$$

Step 3: The vertical shift down 3 units requires that we subtract 3.

$$g(x) = \sqrt{2 - x} - 3$$

The graph is shown in Figure 1-72. The domain and range are $D = (-\infty, 2]$, and $R = [-3, \infty)$.

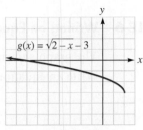

FIGURE 1-72

Self Check 7   Write an equation of the function that would result from the following transformations of the graph of the function $f(x) = \sqrt{x}$: y-axis reflection, horizontal shift to the left 2 units, and vertical shift up 3 units.

Now Try Exercise 95.

5. Graph a Function by Stretching or Shrinking the Graph of a Basic Function

Up until this point, all of the transformations have simply changed the position of the graph of a given function, but have not altered the shape. Horizontal shifts, vertical shifts, and reflections are known as **rigid transformations**. Now we will look at **nonrigid transformations**, which change the basic shape of the graph with a **stretch** or a **shrink**. This is accomplished by multiplying the function or the argument by a constant. To see two nonrigid transformations, look at Figure 1-73. One is known as a **vertical stretch** and the other as a **vertical shrink**.

Figure 1-73 shows the graphs of $y = x^2$, $y = 3x^2$, and $y = \frac{1}{3}x^2$.

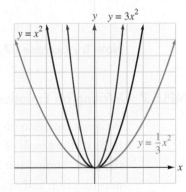

FIGURE 1-73

Because each value of $y = 3x^2$ is 3 times the corresponding value of $y = x^2$, the graph is stretched vertically by a factor of 3. Because each value of $y = \frac{1}{3}x^2$ is one-third of the corresponding value of $y = x^2$, the graph shrinks vertically by a factor of $\frac{1}{3}$.

EXAMPLE 8   **Graphing Using a Transformation**

Graph $g(x) = 3(x^2 - 1)$ using the graph of $f(x) = x^2 - 1$.

SOLUTION   First, we complete the table of solutions for $f(x) = x^2 - 1$, shown in Figure 1-74. Next we multiply each y value in the table by 3 to obtain the table of solutions for

$g(x) = 3(x^2 - 1)$, also shown in the figure. Because each value of $g(x) = 3(x^2 - 1)$ is 3 times the corresponding value of $f(x) = x^2 - 1$, the graph is stretched vertically by a factor of 3. The x-intercepts of both graphs are 1 and -1.

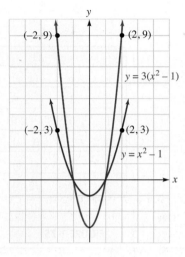

$y = x^2 - 1$			$y = 3(x^2 - 1)$		
x	y	(x, y)	x	y	(x, y)
-2	3	$(-2, 3)$	-2	9	$(-2, 9)$
-1	0	$(-1, 0)$	-1	0	$(-1, 0)$
0	-1	$(0, -1)$	0	-3	$(0, -3)$
1	0	$(1, 0)$	1	0	$(1, 0)$
2	3	$(2, 3)$	2	9	$(2, 9)$

FIGURE 1-74

Self Check 8   *Fill in the blanks:* The graph of $g(x) = 5x^3$ is identical to the graph of $f(x) = x^3$, except that it is vertically _____ by a factor of 5. The graph of $h(x) = \frac{1}{5}x^3$ is identical to the graph $f(x) = x^3$, except that it is vertically _____ by a factor of $\frac{1}{5}$.

Now Try Exercise 41.

In general, we can make the following observations:

Vertical Stretch or Shrink

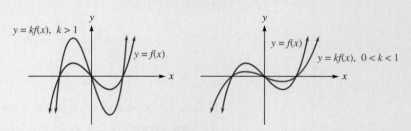

If f is a function and k is a positive real number, then

- If $k > 1$, the graph of $g(x) = kf(x)$ can be obtained by **vertically stretching** the graph of $f(x)$ by a factor of k.
- If $0 < k < 1$, the graph of $g(x) = kf(x)$ can be obtained by **vertically shrinking** the graph of $f(x)$ by a factor of k.

Vertical stretches or shrinks are accomplished by multiplying each y-coordinate of f by k.

We now consider stretching and shrinking a graph in the horizontal direction.

EXAMPLE 9   **Graphing Using a Transformation**

Graph $g(x) = (3x)^2 - 1$ using the graph of $f(x) = x^2 - 1$.

SOLUTION   First, we complete the table of solutions for $f(x) = x^2 - 1$, shown in Figure 1-75. Next we divide each x value in the table by 3 to obtain the table of solutions for

$g(x) = (3x)^2 - 1$. We do this because each value of x in $g(x) = (3x)^2 - 1$ is one-third of the corresponding value of x in $f(x) = x^2 - 1$. The graph of $g(x) = (3x)^2 - 1$ is shrunk horizontally by a factor of $\frac{1}{3}$. The y-intercept of both graphs is -1.

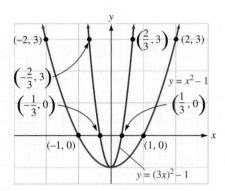

$y = x^2 - 1$		
x	y	(x, y)
-2	3	$(-2, 3)$
-1	0	$(-1, 0)$
0	-1	$(0, -1)$
1	0	$(1, 0)$
2	3	$(2, 3)$

$y = (3x)^2 - 1$		
x	y	(x, y)
$-\frac{2}{3}$	3	$\left(-\frac{2}{3}, 3\right)$
$-\frac{1}{3}$	0	$\left(-\frac{1}{3}, 0\right)$
0	-1	$(0, -1)$
$\frac{1}{3}$	0	$\left(\frac{1}{3}, 0\right)$
$\frac{2}{3}$	3	$\left(\frac{2}{3}, 3\right)$

FIGURE 1-75

Self Check 9   *Fill in the blanks:* The graph of $g(x) = \left(\frac{1}{3}x\right)^2 - 1$ is identical to the graph of $f(x) = x^2 - 1$, except that it is horizontally _____ by a factor of _____.

Now Try Exercise 61.

In general, we can make the following observations:

Horizontal Stretch or Shrink

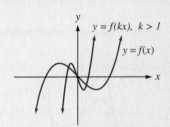
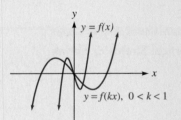

If f is a function and k is a positive real number, then

- If $k > 1$, the graph of $h(x) = f(kx)$ can be obtained by **horizontally shrinking** the graph of $f(x)$.
- If $0 < k < 1$, the graph of $h(x) = f(kx)$ can be obtained by **horizontally stretching** the graph of $f(x)$.

Horizontal stretches or shrinks are accomplished by dividing each x-coordinate by k.

6. Graph a Function by Using a Combination of Transformations

EXAMPLE 10   **Graphing by Using a Sequence of Transformations**

Use the function $f(x) = |x|$ to graph $g(x) = 3|x - 2| + 4$.

SOLUTION   We will graph $g(x) = 3|x - 2| + 4$ by applying three transformations to the basic function $f(x) = |x|$:

Step 1: Translate $f(x) = |x|$ horizontally to the right 2 units

Step 2: Vertically stretch the graph by a factor of 3

Step 3: Translate the graph vertically up 4 units

Step 1: Translate $f(x) = |x|$ horizontally to the right 2 units to obtain the graph of $y = |x - 2|$.

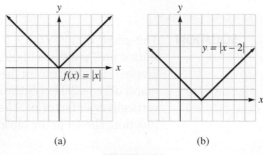

(a)   (b)

FIGURE 1-76

Step 2: Vertically stretch the graph of $y = |x - 2|$ by a factor of 3 to obtain the graph of $y = 3|x - 2|$.

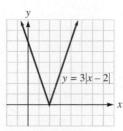

FIGURE 1-77

Step 3: Translate the graph of $y = 3|x - 2|$ vertically up 4 units to obtain the graph of $g(x) = 3|x - 2| + 4$.

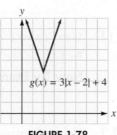

FIGURE 1-78

Self Check 10   Use the graph of the function $f(x) = x^3$ to graph $g(x) = \frac{1}{3}(x + 1)^3 - 2$.

Now Try Exercise 65.

EXAMPLE 11   **Graphing Using Transformations**

Figure 1-79 shows the graph of a function $y = f(x)$. Use this graph and a transformation to find the graph of: **a.** $y = f(x) + 2$ **b.** $y = f(x - 2)$ **c.** $y = 2f(x)$

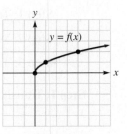

FIGURE 1-79

SOLUTION   The graph of $y = f(x) + 2$ is identical to the graph of $y = f(x)$ except that it is shifted up 2 units [see Figure 1-80(a)]. The graph of $y = f(x - 2)$ is identical to the graph of $y = f(x)$ except that it is shifted to the right 2 units [see Figure 1-80(b)]. The graph of $y = 2f(x)$ is identical to the graph of $y = f(x)$ except that it is stretched vertically by a factor of 2 [see Figure 1-80(c)]. Note the domain and range of each function, shown in the table:

Function	Domain	Range
$y = f(x)$	$[0, \infty)$	$[0, \infty)$
$y = f(x) + 2$	$[0, \infty)$	$[2, \infty)$
$y = f(x - 2)$	$[2, \infty)$	$[0, \infty)$
$y = 2f(x)$	$[0, \infty)$	$[0, \infty)$

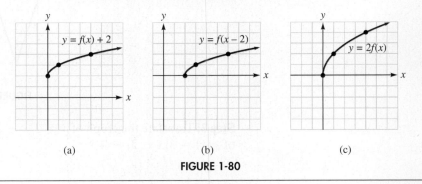

(a)   (b)   (c)

FIGURE 1-80

Self Check 11   Use Figure 1-79 and a reflection to find the graph of:
a. $y = -f(x)$ **b.** $y = f(-x)$

Now Try Exercise 13.

EXAMPLE 12   Using Sequences of Transformations

The graph of $y = f(x)$ is given in Figure 1-81. Use this graph to sketch the indicated functions.

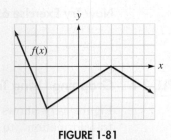

FIGURE 1-81

a. $y = 3f(x)$ **b.** $y = \dfrac{1}{3}f(x)$ **c.** $y = f(3x)$ **d.** $y = f\left(\dfrac{1}{3}x\right)$

e. $y = -2f(x + 1) - 2$

SOLUTION   **a.** The graph of $y = 3f(x)$ is a vertical stretch obtained by multiplying each y-coordinate by 3. Thus, we see the following shifts in the points:

$$(-5, 1) \rightarrow (-5, \mathbf{3}); \ (-3, -4) \rightarrow (-3, \mathbf{-12}); \ (0, -2) \rightarrow (0, \mathbf{-6}); \ (3, 0) \rightarrow (3, \mathbf{0}); \ (6, -2) \rightarrow (6, \mathbf{-6})$$

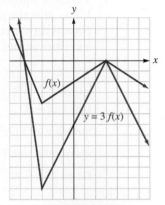

FIGURE 1-82

b. The graph of $y = \tfrac{1}{3}f(x)$ is a vertical shrink obtained by multiplying each y-coordinate by $\tfrac{1}{3}$. For this we have the following shifts in the points:

$$(-5, 1) \rightarrow \left(-5, \frac{1}{3}\right); \ (-3, -4) \rightarrow \left(-3, -\frac{4}{3}\right); \ (0, -2) \rightarrow \left(0, -\frac{2}{3}\right); \ (3, 0) \rightarrow (3, \mathbf{0}); \ (6, -2) \rightarrow \left(6, -\frac{2}{3}\right)$$

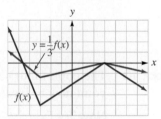

FIGURE 1-83

c. The graph of $y = f(3x)$ is a horizontal shrink obtained by dividing each x-coordinate by 3. For this we have the following shifts in the points:

$$(-5, 1) \rightarrow \left(-\frac{5}{3}, 1\right); \ (-3, -4) \rightarrow (\mathbf{-1}, -4); \ (0, -2) \rightarrow (\mathbf{0}, -2); \ (3, 0) \rightarrow (\mathbf{1}, 0); \ (6, -2) \rightarrow (\mathbf{2}, -2)$$

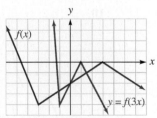

FIGURE 1-84

d. The graph of $y = f\left(\tfrac{1}{3}x\right)$ is a horizontal stretch obtained by dividing each x-coordinate by $\tfrac{1}{3}$, which is the same as multiplying by 3.

$$(-5, 1) \rightarrow (\mathbf{-15}, 1); \ (-3, -4) \rightarrow (\mathbf{-9}, -4); \ (0, -2) \rightarrow (\mathbf{0}, -2); \ (3, 0) \rightarrow (\mathbf{9}, 0); \ (6, -2) \rightarrow (\mathbf{18}, -2)$$

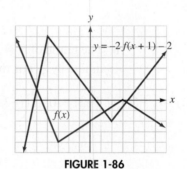

FIGURE 1-85

e. This is a more involved series of transformations, and the order is important. First, shift the graph to the left 1 unit; then stretch the graph by a factor of 2. Next is an x-axis reflection. Finally, shift the graph down vertically 2 units. The transformations are shown in the table.

Original points	$(-5, 1)$	$(-3, -4)$	$(0, -2)$	$(3, 0)$	$(6, -2)$
Left 1 $(x - 1)$	$(-6, 1)$	$(-4, -4)$	$(-1, -2)$	$(2, 0)$	$(5, -2)$
Stretch y by 2 $(2y)$	$(-6, 2)$	$(-4, -8)$	$(-1, -4)$	$(2, 0)$	$(5, -4)$
x-axis reflection $(-y)$	$(-6, -2)$	$(-4, 8)$	$(-1, 4)$	$(2, 0)$	$(5, 4)$
Down 2 $(y - 2)$	$(-6, -4)$	$(-4, 6)$	$(-1, 2)$	$(2, -2)$	$(5, 2)$

$$y = -2f(x + 1) - 2$$

$f(x)$

FIGURE 1-86

Self Check 12 The point $(3, 0)$ lies on the graph of $f(x)$ given in Example 12. Where would this point now be located after the transformation $y = -3f(x - 5) + 2$?

Now Try Exercise 15.

Self Check Answers **1.** 3, up, 4, down **2.** 3, right, 2, left **3.** 4, right, 5, up **4.** to produce the graph of g, shift f 2 units right and 3 units up; to produce the graph of h, shift f to the left 1 unit and down 2 units **5.** x-, y- **6.** $g(x) = \sqrt[3]{x} + 3$ **7.** $g(x) = \sqrt{-2 - x} + 3$ **8.** stretched, shrunk **9.** stretched, 3 **10.**

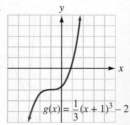

$g(x) = \frac{1}{3}(x + 1)^3 - 2$

11. a.

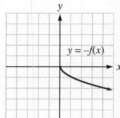

$y = -f(x)$

b.

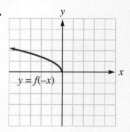

$y = f(-x)$

12. $(8, 2)$

Exercises **1.5**

Getting Ready

You should be able to complete these vocabulary and concept statements before you proceed to the practice exercises.

Fill in the blanks.

1. The graph of $y = f(x) + 5$ is identical to the graph of $y = f(x)$ except that it is translated 5 units _____.

2. The graph of $y =$ _____ is identical to the graph of $y = f(x)$ except that it is translated 7 units down.

3. The graph of $y = f(x - 3)$ is identical to the graph of $y = f(x)$ except that it is translated 3 units _____.

4. The graph of $y = f(x + 2)$ is identical to the graph of $y = f(x)$ except that it is translated 2 units _____.

5. To draw the graph of $y = (x + 2)^2 - 3$, translate the graph of $y = x^2$ _____ units to the left and 3 units _____.

6. To draw the graph of $y = (x - 3)^3 + 1$, translate the graph of $y = x^3$ 3 units to the _____ and 1 unit _____.

7. The graph of $y = f(-x)$ is a reflection of the graph of $y = f(x)$ about the _____.

8. The graph of _____ is a reflection of the graph of $y = f(x)$ about the x-axis.

9. The graph of $y = f(4x)$ shrinks the graph of $y = f(x)$_____ by a factor of $\frac{1}{4}$.

10. The graph of $y = 8f(x)$ stretches the graph of $y = f(x)$_____ by a factor of 8.

Practice

Use the figure showing the graph of $y = f(x)$ and transformations to graph g.

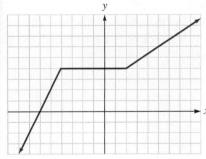

11. $g(x) = f(x) + 2$
12. $g(x) = f(x) - 2$
13. $g(x) = f(x + 2)$
14. $g(x) = f(x - 2)$

15. $g(x) = f(x - 4) - 2$
16. $g(x) = f(x + 4) + 2$
17. $g(x) = f(-x)$
18. $g(x) = -f(x)$
19. $g(x) = 4f(x)$
20. $g(x) = \frac{1}{4}f(x)$
21. $g(x) = f(4x)$
22. $g(x) = f\left(\frac{1}{4}x\right)$

The graph of each function is a transformation of the graph of $f(x) = x^2$. Graph each function.

23. $h(x) = x^2 - 2$
24. $h(x) = x^2 + 2$
25. $h(x) = (x + 3)^2$
26. $h(x) = (x - 3)^2$
27. $h(x) = 2x^2$
28. $h(x) = \frac{1}{2}x^2$
29. $h(x) = -x^2 - 2$
30. $h(x) = -x^2 + 2$
31. $h(x) = (x + 1)^2 + 2$
32. $h(x) = (x - 3)^2 - 1$
33. $h(x) = -(x + 2)^2 - 2$
34. $h(x) = -(x - 1)^2 + 3$
35. $h(x) = 2(x - 1)^2$
36. $h(x) = \frac{1}{2}(x - 1)^2$

The graph of each function is a transformation of the graph of $f(x) = x^3$. Graph each function.

37. $g(x) = (x - 2)^3$
38. $g(x) = (x + 2)^3$
39. $g(x) = (-x)^3$
40. $g(x) = -x^3$
41. $g(x) = \frac{1}{3}x^3$
42. $g(x) = 3x^3$
43. $g(x) = -x^3 + 4$
44. $g(x) = -x^3 - 4$
45. $g(x) = (x - 2)^3 - 1$
46. $g(x) = (x + 2)^3 + 1$
47. $g(x) = -(x - 1)^3 + 2$
48. $g(x) = -(x + 2)^3 - 2$
49. $g(x) = -(2x)^3$
50. $g(x) = (-2x)^3$

The graph of each function is a transformation of the graph of $f(x) = \sqrt{x}$. Graph each function.

51. $g(x) = -\sqrt{x}$
52. $g(x) = \sqrt{-x}$
53. $g(x) = 2\sqrt{x} + 1$
54. $g(x) = \frac{1}{2}\sqrt{x} - 1$
55. $g(x) = \sqrt{4x} - 1$
56. $g(x) = \sqrt{3x} + 2$
57. $g(x) = \sqrt{x - 2} + 1$
58. $g(x) = \sqrt{x + 5} - 2$
59. $g(x) = -2\sqrt{x} + 3$
60. $g(x) = -4\sqrt{x} - 4$

The graph of each function is a transformation of the graph of $f(x) = |x|$. Graph each function.

61. $h(x) = |3x|$
62. $h(x) = 3|x|$
63. $h(x) = |x - 2| + 1$
64. $h(x) = |x + 5| - 2$
65. $h(x) = -2|x| + 3$
66. $h(x) = -2|x + 3|$
67. $h(x) = |3 - x| + 2$
68. $h(x) = |2 - x| - 1$

The graph of each function is a transformation of the graph of $f(x) = \dfrac{1}{x}$. Graph each function.

69. $g(x) = 2\left(\dfrac{1}{x}\right)$

70. $g(x) = \dfrac{1}{2}\left(\dfrac{1}{x}\right)$

71. $g(x) = -\dfrac{1}{x}$

72. $g(x) = -3\left(\dfrac{1}{x}\right)$

73. $g(x) = \dfrac{1}{x-1}$

74. $g(x) = \dfrac{1}{x+1}$

75. $g(x) = \dfrac{1}{x} + 1$

76. $g(x) = \dfrac{1}{x} - 1$

The graph of each function is a transformation of the graph of $f(x) = \sqrt[3]{x}$. Graph each function.

77. $h(x) = -2\sqrt[3]{x}$

78. $h(x) = -\dfrac{1}{2}\sqrt[3]{x}$

79. $h(x) = \sqrt[3]{x} - 3$

80. $h(x) = \sqrt[3]{x} + 1$

81. $h(x) = \sqrt[3]{x-2}$

82. $h(x) = \sqrt[3]{x+3}$

83. $h(x) = -\sqrt[3]{x-1} + 2$

84. $h(x) = \sqrt[3]{2-x} - 1$

85. $h(x) = \sqrt[3]{2x}$

86. $h(x) = \sqrt[3]{\dfrac{1}{2}x}$

The graph of a transformation of one of the basic functions is shown. Identify the basic function, $f(x)$, and write an equation for the graphed function, $g(x)$.

87.
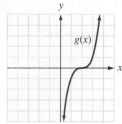

88.
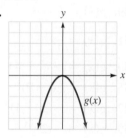

89.
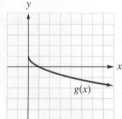

90.
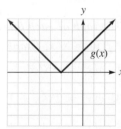

91.
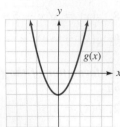

92.
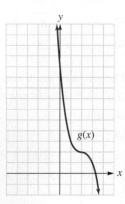

93.
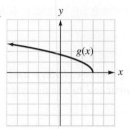

94.
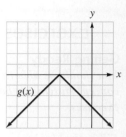

List the transformations that will produce the graph of the given function from the graph of the function $f(x) = x^3$.

95. $g(x) = 2 - x^3$

96. $g(x) = -(x+2)^3$

97. $g(x) = 2(x-1)^3$

98. $g(x) = (2x)^3 - 4$

99. $g(x) = -\dfrac{1}{2}(x+1)^3 - 3$

100. $g(x) = \left(\dfrac{1}{2}x\right)^3 - 1$

Discovery and Writing

Use a graphing calculator to perform each experiment. Write a brief paragraph describing your findings.

101. Investigate the translations of the graph of a function by graphing the parabola $y = (x - k)^2 + k$ for several values of k. What do you observe about successive positions of the vertex?

102. Investigate the translations of the graph of a function by graphing the parabola $y = (x - k)^2 + k^2$ for several values of k. What do you observe about successive positions of the vertex?

103. Investigate the horizontal stretching of the graph of a function by graphing $y = \sqrt{ax}$ for several values of a. What do you observe?

104. Investigate the vertical stretching of the graph of a function by graphing $y = b\sqrt{x}$ for several values of b. What do you observe? Are these graphs different from the graphs in Exercise 103?

Write a paragraph using your own words.

105. Explain why the effect of vertically stretching a graph by a factor of -1 is to reflect the graph across the x-axis.

106. Explain why the effect of horizontally stretching a graph by a factor of -1 is to reflect the graph across the y-axis.

1.6 Operations on Functions

In this section, we will learn to

1. Add, subtract, multiply, and divide functions, specifying domains.

2. Write functions as sums, differences, products, or quotients of other functions.

3. Use composition of functions and determine the domain of the new function.

4. Use composition of functions to solve real-world problems.

Functions can be combined by addition, subtraction, multiplication, and division. In this section, we will explore these operations on functions and give careful attention to their domains and ranges.

Real numbers can be added, subtracted, multiplied, and divided. We can also perform these operations on algebraic expressions. Likewise, these operations can be performed on functions, thus creating a new function.

Suppose that the functions $R(x) = 140x$ and $C(x) = 120{,}000 + 40x$ model a company's yearly revenue and cost for producing and selling surfboards. By subtracting the functions, $R(x) - C(x)$, we would arrive at a new function represented by

$$(R - C)(x) = 140x - (120{,}000 + 40x)$$

$$= 140x - 120{,}000 - 40x \qquad \text{Remove parentheses.}$$

$$= 100x - 120{,}000$$

This function represents the profit made by the company when it sells x surfboards. We will now discuss, in depth, how to add, subtract, multiply, and divide functions.

1. Add, Subtract, Multiply, and Divide Functions, Specifying Domains

Consider the functions $f(x) = 3x + 5$ and $g(x) = x^2 - 4$. We can form new functions using the sum, difference, product, and quotient of f and g as shown here.

$$f(x) + g(x) = (3x + 5) + (x^2 - 4) = x^2 + 3x + 1 \qquad \text{Sum}$$

$$f(x) - g(x) = (3x + 5) - (x^2 - 4) = -x^2 + 3x + 9 \qquad \text{Difference}$$

$$f(x) \cdot g(x) = (3x + 5) \cdot (x^2 - 4) = 3x^3 + 5x^2 - 12x - 20 \qquad \text{Product}$$

$$\frac{f(x)}{g(x)} = \frac{3x + 5}{x^2 - 4}, \quad x \neq \pm 2 \qquad \text{Quotient}$$

We now formally define the arithmetic operations on algebraic functions.

Adding, Subtracting, Multiplying, and Dividing Functions

If the ranges of the functions f and g are subsets of the real numbers, then

1. The **sum** of f and g, denoted as $f + g$, is defined by

$$(f + g)(x) = f(x) + g(x).$$

2. The **difference** of f and g, denoted as $f - g$, is defined by

$$(f - g)(x) = f(x) - g(x).$$

3. The **product** of f and g, denoted as $f \cdot g$, is defined by

$$(f \cdot g) = f(x)g(x).$$

4. The **quotient** of f and g, denoted as f/g, is defined by

$$(f/g)(x) = \frac{f(x)}{g(x)} \quad (g(x) \neq 0).$$

The domain of each function, unless otherwise restricted, is the set of real numbers x that are in the domains of both f and g. In the case of the quotient f/g, there is the further restriction that $g(x) \neq 0$.

EXAMPLE 1 **Finding the Sum and Difference of Functions**

Let $f(x) = 3x + 1$ and $g(x) = 2x - 3$. Find each of the following functions and its domain: **a.** $f + g$ **b.** $f - g$

SOLUTION **a.** Find $f + g$.

$$(f + g)(x) = f(x) + g(x)$$

$$= (3x + 1) + (2x - 3)$$

$$= 5x - 2$$

Since the domain of both f and g is the set of real numbers, the domain of $f + g$ is $D_{f+g} = (-\infty, \infty)$.

b. Find $f - g$.

$$(f - g)(x) = f(x) - g(x)$$

$$= (3x + 1) - (2x - 3)$$

$$= x + 4$$

Since the domain of both f and g is the set of all real numbers, $D_{f-g} = (-\infty, \infty)$. ∎

Self Check 1 Find $g - f$.

Now Try Exercise 9.

EXAMPLE 2 **Finding the Product and Quotient of Functions**

Let $f(x) = 3x + 1$ and $g(x) = 2x - 3$. Find each of the following functions and its domain: **a.** $f \cdot g$ **b.** f/g

SOLUTION **a.** Find $f \cdot g$.

$$(f \cdot g)(x) = f(x) \cdot g(x)$$

$$= (3x + 1)(2x - 3)$$

$$= 6x^2 - 7x - 3$$

Since the domain of both f and g is the set of all real numbers, $D_{f \cdot g} = (-\infty, \infty)$.

b. Find f/g.

$$(f/g)(x) = \frac{f(x)}{g(x)} \qquad (g(x) \neq 0)$$

$$= \frac{3x + 1}{2x - 3} \qquad (2x - 3 \neq 0)$$

Since $2x - 3 \neq 0$, the domain of f/g is the set of all real numbers except $\frac{3}{2}$. This is given by $D_{f/g} = \left(-\infty, \frac{3}{2}\right) \cup \left(\frac{3}{2}, \infty\right)$.

Self Check 2   Find g/f and its domain.

Now Try Exercise 15.

EXAMPLE 3   **Performing Operations on Functions**

Let $f(x) = x^2 - 4$ and $g(x) = \sqrt{x}$. Find each function and its domain.
a. $f + g$ **b.** $f \cdot g$ **c.** f/g **d.** g/f

SOLUTION   Because all real numbers can be squared, the domain of f is the set of all real numbers. Because $\sqrt{x}$ is to be a real number, the domain of g is given by $D = [0, \infty)$.

a. Find $f + g$.

$$(f + g)(x) = f(x) + g(x)$$

$$= x^2 - 4 + \sqrt{x}$$

The domain consists of the numbers x that are in the domain of both f and g. This is $(-\infty, \infty) \cap [0, \infty)$, which is $[0, \infty)$. The domain of $f + g$ is given by $D_{f+g} = [0, \infty)$.

b. Find $f \cdot g$.

$$(f \cdot g)(x) = f(x)g(x)$$

$$= (x^2 - 4)\sqrt{x}$$

$$= x^2\sqrt{x} - 4\sqrt{x}$$

The domain consists of the numbers x that are in the domain of both f and g. The domain of $f \cdot g$ is given by $D_{f \cdot g} = [0, \infty)$.

c. Find f/g.

$$(f/g)(x) = \frac{f(x)}{g(x)} \qquad g(x) \neq 0$$

$$= \frac{x^2 - 4}{\sqrt{x}}$$

The domain consists of the numbers x that are in the domain of both f and g, except 0 (because division by 0 is undefined). The domain of f/g is $D_{f/g} = (0, \infty)$.

d. Find g/f.

$$(g/f)(x) = \frac{g(x)}{f(x)} \qquad f(x) \neq 0$$

$$= \frac{\sqrt{x}}{x^2 - 4}$$

The domain consists of the numbers x that are in the domain of both f and g, except 2 (because division by 0 is undefined). The domain of g/f is $D_{g/f} = [0, 2) \cup (2, \infty)$.

Self Check 3   Find $g - f$ and its domain.

Now Try Exercise 13.

EXAMPLE 4   **Evaluating Combined Functions**

Find $(f + g)(3)$ when $f(x) = x^2 + 1$ and $g(x) = 2x + 1$.

SOLUTION   The first step is to find the combined function $(f + g)(x)$.

$$(f + g)(x) = f(x) + g(x)$$
$$= x^2 + 1 + 2x + 1$$
$$= x^2 + 2x + 2$$

Comment

It is not necessary to find the combined function to evaluate $(f + g)(3)$.
You could also evaluate it by finding
$f(3) + g(3) = (3^2 + 1) + (2(3) + 1) = 17$.
However, if you have to evaluate the combined function at more than one value, then it is more efficient to find the combined function first.

We then find $(f + g)(3)$.

$$(f + g)(x) = x^2 + 2(x) + 2$$
$$(f + g)(3) = 3^2 + 2(3) + 2 \qquad \text{Substitute 3 for } x.$$
$$= 9 + 6 + 2$$
$$= 17$$

Self Check 4   Find $(f \cdot g)(-2)$.

Now Try Exercise 21.

2. Write Functions as Sums, Differences, Products, or Quotients of Other Functions

EXAMPLE 5   **Finding the Components of a Combined Function**

Let $h(x) = x^2 + 3x + 2$. Find two functions f and g such that **a.** $f + g = h$
b. $f \cdot g = h$

SOLUTION   **a.** There is no unique solution to this problem. One possible solution is $f(x) = x^2$ and $g(x) = 3x + 2$, for then

$$(f + g)(x) = f(x) + g(x)$$
$$= (x^2) + (3x + 2)$$
$$= x^2 + 3x + 2$$
$$= h(x)$$

Another solution is $f(x) = x^2 + 2x$ and $g(x) = x + 2$.

b. Again, there are many possibilities. One is suggested by factoring
$h(x) = x^2 + 3x + 2$:

$$h(x) = x^2 + 3x + 2 = (x + 1)(x + 2).$$

If we let $f(x) = x + 1$ and $g(x) = x + 2$, then

$$(f \cdot g)(x) = f(x) \cdot g(x)$$
$$= (x + 1)(x + 2)$$
$$= x^2 + 3x + 2$$
$$= h(x)$$

Another possibility is $f(x) = 3$ and $g(x) = \dfrac{x^2}{3} + x + \dfrac{2}{3}$.

Self Check 5   Find two functions f and g such that $f - g = h$.

Now Try Exercise 39.

3. Use Composition of Functions and Determine the Domain of the New Function

Composition of functions is another way of combining functions. Often one quantity is a function of a second quantity that depends, in turn, on a third quantity. For example, the cost of a car trip is a function of the gasoline consumed. The amount of gasoline consumed, in turn, is a function of the number of miles driven. Such chains of dependence are analyzed mathematically as *composition of functions.*

Suppose that $y = f(x)$ and $y = g(x)$ define two functions. Any number x in the domain of g will produce a corresponding value $g(x)$ in the range of g. If $g(x)$ is in the domain of function f, then $g(x)$ can be substituted into f, and a corresponding value $f(g(x))$ will be determined. This two-step process defines a new function, called a **composite function,** denoted by $f \circ g$. (See Figure 1-87.)

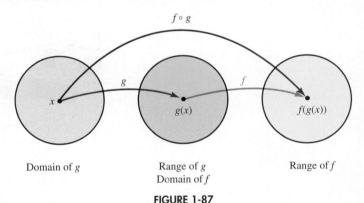

FIGURE 1-87

Composite Functions   The **composite function** $f \circ g$ is defined by $(f \circ g)(x) = f(g(x))$.

The **domain** of $f \circ g$ consists of all those numbers in the domain of g for which $g(x)$ is in the domain of f.

For example, if $f(x) = 5x + 1$ and $g(x) = 4x - 3$, then

$$
\begin{aligned}
(f \circ g)(x) &= f(g(x)) & (g \circ f)(x) &= g(f(x)) \\
&= f(4x - 3) & &= g(5x + 1) \\
&= 5(4x - 3) + 1 & &= 4(5x + 1) - 3 \\
&= 20x - 14 & &= 20x + 1
\end{aligned}
$$

Caution   In the previous example, $(f \circ g)(x) \neq (g \circ f)(x)$. This shows that the composition of functions is not commutative. However, there are functions for which this relationship is true. We will look at those in the next section.

We have seen that a function can be represented by a machine. If we put a number from the domain into the machine (the input), a number from the range comes out (the output). For example, if we put 2 into the machine shown in

Figure 1-88(a), the number $f(2) = 5(2) - 2 = 8$ comes out. In general, if we put x into the machine shown in Figure 1-88(b), the value $f(x)$ comes out.

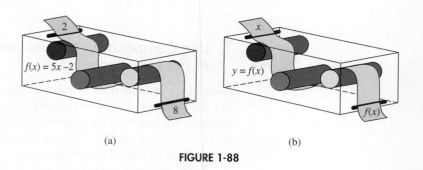

(a)   (b)

FIGURE 1-88

The function machines shown in Figure 1-89 illustrate the composition $f \circ g$. When we put a number x into the function g, the value $g(x)$ comes out. The value $g(x)$ then goes into function f, and $f(g(x))$ comes out.

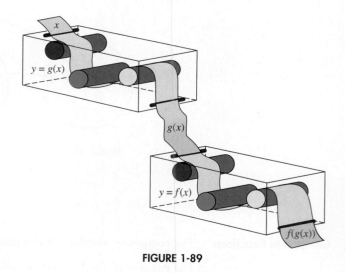

FIGURE 1-89

To further illustrate these ideas, suppose we let $f(x) = 2x + 1$ and $g(x) = x - 4$.

- $(f \circ g)(9)$ means $f(g(9))$. In Figure 1-90(a), function g receives the number 9 and subtracts 4, and the number $g(9) = 5$ comes out. The 5 goes into function f, which doubles it and adds 1. The final result, 11, is the output of the composite function $f \circ g$.

$$(f \circ g)(9) = f(g(9)) = f(5) = 2(5) + 1 = 11$$

- $(f \circ g)(x)$ means $f(g(x))$. In Figure 1-90(a), function g receives the number x and subtracts 4, and the number $(x - 4)$ comes out. The $(x - 4)$ goes into function f, which doubles it and adds 1. The final result, $(2x - 7)$, is the output of the composite function $f \circ g$.

$$(f \circ g)(x) = f(g(x)) = f(x - 4) = 2(x - 4) + 1 = 2x - 7$$

- $(g \circ f)(-2)$ means $g(f(-2))$. In Figure 1-90(b), function f receives the number -2, doubles it and adds 1, and releases -3 into function g. Function g subtracts 4 from -3 and releases a final output of -7.

$$(g \circ f)(x) = g(f(x)) = g(-3) = -3 - 4 = -7$$

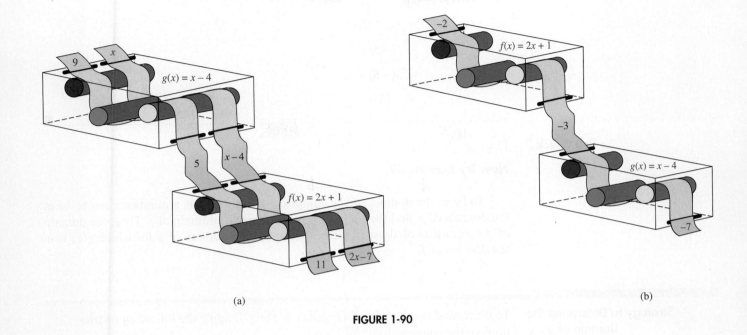

(a)   (b)

FIGURE 1-90

EXAMPLE 6   **Finding Composite Functions**

If $f(x) = 2x + 7$ and $g(x) = 4x + 1$, find **a.** $(f \circ g)(x)$ **b.** $(g \circ f)(x)$

SOLUTION   **a.** Find $(f \circ g)(x)$.

$$(f \circ g)(x) = f(g(x))$$
$$= f(4x + 1)$$
$$= 2(4x + 1) + 7$$
$$= 8x + 9$$

b. Find $(g \circ f)(x)$.

$$(g \circ f)(x) = g(f(x))$$
$$= g(2x + 7)$$
$$= 4(2x + 7) + 1$$
$$= 8x + 29$$

Self Check 6   If $h(x) = x + 1$, find $(f \circ h)(x)$.

Now Try Exercise 53.

EXAMPLE 7   **Evaluating a Composite Function**

If $f(x) = 3x - 2$ and $g(x) = 3x^2 + 6x - 5$, find $(f \circ g)(-2)$.

SOLUTION   Because $(f \circ g)(-2)$ means $f(g(-2))$, we first find $g(-2)$. We then find $f(g(-2))$.

$$g(x) = 3x^2 + 6x - 5$$
$$g(-2) = 3(-2)^2 + 6(-2) - 5$$
$$= 3(4) - 12 - 5$$
$$= -5$$

$$(f \circ g)(-2) = f(g(-2))$$
$$= f(-5)$$
$$= 3(-5) - 2$$
$$= -17$$

Self Check 7   Find $(g \circ f)(-1)$.

Now Try Exercise 49.

To be in the domain of the composite function $f \circ g$, a number x has to be in the domain of g, and the output of g must be in the domain of f. Thus, the domain of $f \circ g$ consists of those inputs x that are in the domain of g for which $g(x)$ is in the domain of f.

Strategy to Determine the Domain of $f \circ g$ To determine the domain of $(f \circ g)(x) = f(g(x))$, apply the following restrictions to the composition:

1. If x is not in the domain of g, it will not be in the domain of $f \circ g$.
2. Any x that has an output $g(x)$ that is not in the domain of f will not be in the domain of $f \circ g$.

EXAMPLE 8   **Finding the Domain of Composite Functions**

Let $f(x) = \sqrt{x}$ and $g(x) = x - 3$. Find the domain of **a.** $(f \circ g)(x)$ **b.** $(g \circ f)(x)$

SOLUTION   For $\sqrt{x}$ to be a real number, x must be a nonnegative real number. Thus, the domain of f is $D_f = [0, \infty)$. Since any real number x can be an input into g, $D_g = (-\infty, \infty)$.

a. First, we will find the function $(f \circ g)(x)$.

$$(f \circ g)(x) = f(g(x)) = f(x - 3) = \sqrt{x - 3}$$

Since the domain and range of g are the set of all real numbers, the only restrictions on the domain of $f \circ g$ can be found by looking at the composite function. Since $x - 3$ must be nonnegative, we can see that $x \geq 3$. Therefore, the domain of $f \circ g$ is $D_{f \circ g} = [3, \infty)$.

ACCENT ON TECHNOLOGY   **Composition of Functions**

We can use a graphing calculator to graph the composite function as shown in Figure 1-91. We can make the graph of the composite function appear bold by scrolling left of Y3 and choosing bold.

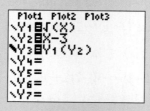
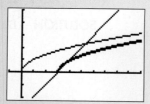

FIGURE 1-91

b. The domain of $(g \circ f)(x)$ is the set of real numbers x such that x is in the domain of f and $f(x)$ is in the domain of g. We also know that the domain of g is the set of all real numbers. So we will find $(g \circ f)(x)$.

$$(g \circ f)(x) = g(f(x)) = g\left(\sqrt{x}\right) = \sqrt{x} - 3$$

From this we can determine $D_{g \circ f} = [0, \infty)$.

Self Check 8 Find the domain of $(f \circ f)(x)$.

Now Try Exercise 61.

Sometimes, the domain of a composite function cannot be determined by merely looking at the definition of the function.

EXAMPLE 9 **Finding the Domain of a Composite Function**

For $f(x) = x^2 - 4$ and $g(x) = \sqrt{4 - x^2}$, find the composition function $(f \circ g)(x)$ and its domain.

SOLUTION The composition function is

$$(f \circ g)(x) = f(g(x))$$

$$= f\left(\sqrt{4 - x^2}\right)$$

$$= \left(\sqrt{4 - x^2}\right)^2 - 4$$

$$= 4 - x^2 - 4$$

$$= -x^2$$

Just by looking at the function $(f \circ g)(x) = -x^2$, it might seem that the domain is the set of all real numbers. However, this is not the case, because the domain of g is $D_g = [-2, 2]$. Any real number can be put into f, but only numbers in the interval $[-2, 2]$ can be put into g. Therefore, we see that $D_{f \circ g} = [-2, 2]$.

Use your graphing calculator to convince yourself that this is the domain. Have the calculator graph only the composite function by turning the first two functions off.

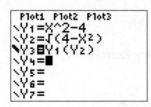 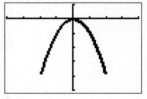

FIGURE 1-92

Self Check 9 For $f(x) = x^2 + 5$ and $g(x) = \sqrt{3 - x}$, find $(f \circ g)(x)$ and its domain.

Now Try Exercise 65.

EXAMPLE 10 **Finding the Domain of a Composite Function**

Let $f(x) = \dfrac{x + 3}{x - 2}$ and $g(x) = \dfrac{1}{x}$. **a.** Find the domain of $(f \circ g)(x)$.
b. Find $(f \circ g)(x)$.

SOLUTION   For $\frac{1}{x}$ to be a real number, x cannot be 0. Thus, the domain of g is $D_g = (-\infty, 0) \cup (0, \infty)$. Since any real number except 2 can be an input into f, the domain of f is $D_f = (-\infty, 2) \cup (2, \infty)$.

a. The domain of $(f \circ g)(x)$ is the set of real numbers x such that x is in the domain of g and $g(x)$ is in the domain of f. We have seen that all values of x but 0 are in the domain of g and that all values of $g(x)$ but 2 are in the domain of f. So we must exclude 0 from the domain of $(f \circ g)(x)$ and all values of x where $g(x) = 2$.

$$g(x) = 2$$

$$\frac{1}{x} = 2 \qquad \text{Substitute } \frac{1}{x} \text{ for } g(x).$$

$$1 = 2x \qquad \text{Since } x \neq 0, \text{ we can multiply both sides by } x.$$

$$x = \frac{1}{2} \qquad \text{Divide both sides by 2.}$$

The domain of $(f \circ g)(x)$ is the set of all real numbers except 0 and $\frac{1}{2}$, which is $(-\infty, 0) \cup \left(0, \frac{1}{2}\right) \cup \left(\frac{1}{2}, \infty\right)$.

b. Find $(f \circ g)(x)$.

$$(f \circ g)(x) = f(g(x))$$

$$= f\left(\frac{1}{x}\right) \qquad \text{Substitute } \frac{1}{x} \text{ for } g(x).$$

$$= \frac{\dfrac{1}{x} + 3}{\dfrac{1}{x} - 2} \qquad \text{Substitute } \frac{1}{x} \text{ for } x \text{ in } f.$$

$$= \frac{1 + 3x}{1 - 2x} \qquad \text{Multiply numerator and denominator by } x.$$

Thus, $(f \circ g)(x) = \dfrac{1 + 3x}{1 - 2x}$.

Self Check 10   Let $f(x) = \dfrac{x}{x - 1}$ and $g(x) = \dfrac{1}{x}$. Find the domain of $(f \circ g)(x)$, and then find $(f \circ g)(x)$.

Now Try Exercise 69.

4. Use Composition of Functions to Solve Real-World Problems

EXAMPLE 11   **Using Composite Functions**

A laboratory sample is removed from a cooler at a temperature of 15°F. Technicians then warm the sample at the rate of 3°F per hour. Express the sample's temperature in degrees Celsius as a function of the time t (in hours) since it was removed from the cooler.

SOLUTION   The temperature of the sample is 15°F when $t = 0$. Because it warms at 3°F per hour, it warms $3t°$ after t hours. Thus, the Fahrenheit temperature after t hours is given by the function

$$F(t) = 3t + 15 \qquad F(t) \text{ is the Fahrenheit temperature, and } t \text{ represents the time in hours.}$$

The Celsius temperature is a function of the Fahrenheit temperature $F(t)$, given by the formula

$$C(F(t)) = \frac{5}{9}(F(t) - 32).$$

To express the sample's Celsius temperature as a function of time, we find the composition function $(C \circ F)(t)$.

$$(C \circ F)(t) = C(F(t))$$

$$= C(3t + 15) \qquad \text{Substitute for } F(t).$$

$$= \frac{5}{9}[(3t + 15) - 32] \qquad \text{Substitute } 3t + 15 \text{ for } F(t) \text{ in } C(F(t)).$$

$$= \frac{5}{9}(3t - 17) \qquad \text{Simplify.}$$

$$= \frac{5}{3}t - \frac{85}{9}$$

Self Check 11   Find $(C \circ F)(10)$.

Now Try Exercise 93.

Self Check Answers   **1.** $(g - f)(x) = -x - 4$ **2.** $(g/f)(x) = \dfrac{2x - 3}{3x + 1}$; $D_{g/f} = \left(-\infty, -\dfrac{1}{3}\right) \cup \left(-\dfrac{1}{3}, \infty\right)$

3. $(g - f)(x) = \sqrt{x} - x^2 + 4$; $D_{g-f} = [0, \infty)$ **4.** -15

5. One possibility is $f(x) = 2x^2$ and $g(x) = x^2 - 3x - 2$.

6. $(f \circ h)(x) = 2x + 9$ **7.** 40

8. $D_{f \circ f} = [0, \infty)$ **9.** $(f \circ g)(x) = 8 - x$; $D_{f \circ g} = (-\infty, 3]$

10. $D_{f \circ g} = (-\infty, 0) \cup (0, 1) \cup (1, \infty)$; $(f \circ g)(x) = \dfrac{1}{1 - x}$ **11.** approximately 7.2

Exercises **1.6**

Getting Ready

You should be able to complete these vocabulary and concept statements before you proceed to the practice exercises.

Fill in the blanks.

1. $(f + g)(x) = $ _____.

2. $(f - g)(x) = $ _____.

3. $(f \cdot g)(x) = $ _____.

4. $(f/g)(x) = $ _____, where $g(x) \neq 0$.

5. The domain of $f + g$ is the _____ of the domain of f and g.

6. $(f \circ g)(x) = $ _____.

7. $(g \circ f)(x) = $ _____.

8. To be in the domain of the composite function $f \circ g$, a number x has to be in the _____ of g, and the output of g must be in the _____ of f.

Practice

Let $f(x) = 2x + 1$ and $g(x) = 3x - 2$. Find each function and its domain.

9. $f + g$ **10.** $f - g$

11. $f \cdot g$ **12.** f/g

Let $f(x) = x^2 + x$ and $g(x) = x^2 - 1$. Find each function and its domain.

13. $f - g$ **14.** $f + g$

15. f/g **16.** $f \cdot g$

Let $f(x) = x^2 - 1$ and $g(x) = 3x - 2$. Find each value, if possible.

17. $(f + g)(2)$ **18.** $(f + g)(-3)$

19. $(f - g)(0)$ **20.** $(f-g)(-5)$

21. $(f \cdot g)(2)$ **22.** $(f \cdot g)(-1)$

23. $(f/g)\left(\dfrac{2}{3}\right)$ **24.** $(f/g)(t)$

Given the graphs of functions f and g, find each value and answer each question.

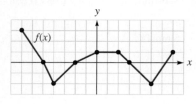

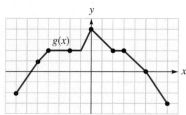

25. $(f + g)(-4)$ **26.** $(f + g)(-2)$

27. $(f - g)(1)$ **28.** $(f - g)(3)$

29. $(f \cdot g)(5)$ **30.** $(f \cdot g)(-5)$

31. $(f/g)(-1)$ **32.** $(f/g)(2)$

33. What is the domain of $(f + g)(x)$?

34. What is the domain of $(f/g)(x)$?

35. Sketch the graph of $(f + g)(x)$.

36. Sketch the graph of $(f - g)(x)$.

Find two functions f and g such that h(x) can be expressed as the function indicated. Several answers are possible.

37. $h(x) = 3x^2 + 2x$; $f + g$

38. $h(x) = 3x^2$; $f \cdot g$

39. $h(x) = \dfrac{3x^2}{x^2 - 1}$; f/g

40. $h(x) = 5x + x^2$; $f - g$

41. $h(x) = x(3x^2 + 1)$; $f - g$

42. $h(x) = (3x - 2)(3x + 2)$; $f + g$

43. $h(x) = x^2 + 7x - 18$; $f \cdot g$

44. $h(x) = 5x^5$; f/g

Let $f(x) = 2x - 5$ and $g(x) = 5x - 2$. Find each value.

45. $(f \circ g)(2)$ **46.** $(g \circ f)(-3)$

47. $(f \circ f)\left(-\dfrac{1}{2}\right)$ **48.** $(g \circ g)\left(\dfrac{3}{5}\right)$

Let $f(x) = 3x^2 - 2$ and $g(x) = 4x + 4$. Find each value.

49. $(f \circ g)(-3)$ **50.** $(g \circ f)(3)$

51. $(f \circ f)\left(\sqrt{3}\right)$ **52.** $(g \circ g)(-4)$

Let $f(x) = 3x$ and $g(x) = x + 1$. Determine the domain of each composite function and then find the composite function.

53. $f \circ g$ **54.** $g \circ f$

55. $f \circ f$ **56.** $g \circ g$

Let $f(x) = x^2$ and $g(x) = 2x$. Determine the domain of each composite function and then find the composite function.

57. $g \circ f$ **58.** $f \circ g$

59. $g \circ g$ **60.** $f \circ f$

Let $f(x) = \sqrt{x}$ and $g(x) = x + 1$. Determine the domain of each composite function and then find the composite function.

61. $f \circ g$ **62.** $g \circ f$

63. $f \circ f$ **64.** $g \circ g$

Let $f(x) = \sqrt{x + 1}$ and $g(x) = x^2 - 1$. Determine the domain of each composite function and then find the composite function.

65. $g \circ f$ **66.** $f \circ g$

67. $g \circ g$ **68.** $f \circ f$

Let $f(x) = \dfrac{1}{x - 1}$ and $g(x) = \dfrac{1}{x - 2}$. Determine the domain of each composite function and then find the composite function.

69. $f \circ g$ **70.** $g \circ f$

71. $f \circ f$ **72.** $g \circ g$

Find two functions f and g such that the composition $f \circ g$ expresses the given correspondence. Several answers are possible.

73. $y = 3x - 2$ **74.** $y = 7x - 5$

75. $y = x^2 - 2$ **76.** $y = x^3 - 3$

77. $y = (x - 2)^2$ **78.** $y = (x - 3)^3$

79. $y = \sqrt{x + 2}$ **80.** $y = \dfrac{1}{x - 5}$

81. $y = \sqrt{x} + 2$ **82.** $y = \dfrac{1}{x} - 5$

83. $y = x$ **84.** $y = 3$

Use the table to evaluate each composite function.

x	−2	−1	0	1	2	3	4
f(x)	3	1	−2	1	2	0	2
g(x)	−2	4	3	−1	−1	2	4

85. $(f \circ g)(1)$

86. $(f \circ g)(2)$

87. $(g \circ f)(0)$

88. $(g \circ f)(-1)$

89. $(f \circ f)(2)$

90. $(g \circ g)(3)$

Applications

91. DVD camcorder Suppose that the functions $R(x) = 300x$ and $C(x) = 60{,}000 + 40x$ model a company's monthly revenue and cost in dollars for producing and selling DVD camcorders.

  a. Find $(R - C)(x)$, the function that models the monthly profit $P(x)$.

  b. Find the company's profit if 500 camcorders are produced and sold in one month.

© Istockphoto.com/Wittelsbach bernd

92. TV screen The height of the television screen shown is 13 inches.

  a. Write a formula to find the area A of the viewing screen.

  b. Use the Pythagorean Theorem to write a formula to find the width w of the screen.

  c. Write a formula to find the area A of the screen as a function of the diagonal d.

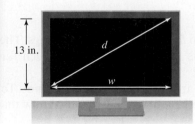

93. Area of an oil spill Suppose an oil spill from a tanker is spreading in the shape of a circular ripple. If the function $d(t) = 3t$ represents the diameter of the spill in inches at time t minutes, express the area A of the oil spill as a function of time. Find the area of the oil spill after 48 hours.

94. Area of a square Write a formula for the area A of a square in terms of its perimeter P.

95. Perimeter of a square Write a formula for the perimeter P of a square in terms of its area A.

96. Ceramics When the temperature of a pot in a kiln is $1{,}200°$F, an artist turns off the heat and leaves the pot to cool at a controlled rate of $81°$F per hour. Express the temperature of the pot in degrees Celsius as a function of the time t (in hours) since the kiln was turned off.

Discovery and Writing

97. Let $f(x) = 3x$. Show that $(f + f)(x) = f(x + x)$.

98. Let $g(x) = x^2$. Show that $(g + g)(x) \neq g(x + x)$.

99. Let $f(x) = \dfrac{x - 1}{x + 1}$. Find $(f \circ f)(x)$.

100. Let $g(x) = \dfrac{x}{x - 1}$. Find $(g \circ g)(x)$.

 Let $f(x) = x^2 - x$, $g(x) = x - 3$, and $h(x) = 3x$. Use a graphing calculator to graph both functions on the same axis. Write a brief paragraph summarizing your observations.

101. f and $f \circ g$

102. f and $g \circ f$

103. f and $f \circ h$

104. f and $h \circ f$

105. What are some practical applications of composition of functions?

1.7 Inverse Functions

In this section, we will learn to

1. Use algebra to determine whether a function is one-to-one.
2. Use the horizontal-line test to determine whether a function is one-to-one.
3. Define the inverse of a one-to-one function and verify inverse functions.
4. Find the inverse of a one-to-one function.
5. Understand the graphical relationship between inverse functions and determine the domain and range of these functions.

In this section, we will discuss inverse functions. A function and its inverse do opposite things.

Suppose we climb the Great Wall of China on a summer day when the temperature reaches a high of 35°C.

The linear function defined by $F = \frac{9}{5}C + 32$ gives a formula to convert degrees Celsius to degrees Fahrenheit. If we substitute a Celsius reading into the formula, a Fahrenheit reading comes out. For example, if we substitute 35° for C, we obtain a Fahrenheit reading of 95°.

$$F = \frac{9}{5}C + 32$$
$$= \frac{9}{5}(35) + 32$$
$$= 63 + 32$$
$$= 95$$

If we want to find a Celsius reading from a Fahrenheit reading, we need a formula into which we can substitute a Fahrenheit reading and have a Celsius reading come out. Such a formula is $C = \frac{5}{9}(F - 32)$, which takes the Fahrenheit reading of 95° and turns it back into a Celsius reading of 35°.

$$C = \frac{5}{9}(F - 32)$$
$$= \frac{5}{9}(95 - 32)$$
$$= \frac{5}{9}(63)$$
$$= 35$$

The functions defined by these two formulas do opposite things. The first turns 35°C into 95°F, and the second turns 95°F back into 35°C. Such functions are called *inverse functions*.

Some functions have inverses that are functions and some do not. To guarantee that the inverse of a function will also be a function, we must know that the function is *one-to-one*.

1. Use Algebra to Determine Whether a Function Is One-to-One

In this section, we will find inverses of functions that are one-to-one. *One-to-one functions* are functions whose inverses are also functions. We now examine what it means for a function to be one-to-one. Consider the following two functions:

Function 1: To each student, there corresponds exactly one eye color.

Function 2: To each student, there corresponds exactly one college identification number.

Function 1 **is not a one-to-one function** because two different students can have the same eye color.

Function 2 **is a one-to-one function** because two different students will always have two different ID numbers.

Recall that each element x in the domain of a function has a single output y. For some functions, different numbers x in the domain can have the same output. (See Figure 1-93(a).) For other functions, called **one-to-one functions**, different numbers x have different outputs. (See Figure 1-93(b).)

One-to-One Function   A function f is called **one-to-one** if and only if for x_1 and x_2 in the domain of f, we have $f(x_1) \neq f(x_2)$ for $x_1 \neq x_2$.

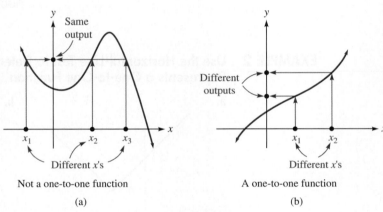

FIGURE 1-93

Equivalently, we can say if $f(x_1) = f(x_2)$ then $x_1 = x_2$.

EXAMPLE 1   **Determine Whether Functions Are One-to-One**

Determine whether the following functions are one-to-one:
a. $f(x) = x^4 + x^2$ **b.** $f(x) = x^3$

SOLUTION   **a.** The function $f(x) = x^4 + x^2$ is not one-to-one, because different numbers in the domain have the same output. For example, 2 and -2 have the same output: $f(2) = f(-2) = 20$.

b. The function $f(x) = x^3$ is one-to-one, because different numbers x produce different outputs $f(x)$. This is because different numbers have different cubes. ∎

Self Check 1   Determine whether $f(x) = \sqrt{x}$ is one-to-one.

Now Try Exercise 7.

2. Use the Horizontal-Line Test to Determine Whether a Function is One-to-One

If a function is one-to-one, it is either always increasing or always decreasing throughout its domain. A **horizontal-line test** can be used to determine whether the graph of a function represents a one-to-one function. If every horizontal line that intersects the graph of a function does so only once, the function is one-to-one. (See Figure 1-94(a).) If any horizontal line intersects the graph of a function more than once, the function is not one-to-one. (See Figure 1-94(b).)

Comment

A one-to-one function satisfies both the horizontal- and vertical-line tests.

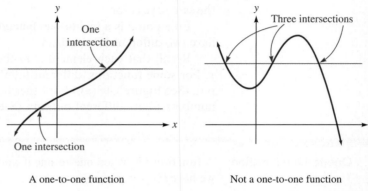

A one-to-one function

(a)

Not a one-to-one function

(b)

FIGURE 1-94

EXAMPLE 2 **Use the Horizontal-Line Test to Determine Whether Each Graph Represents a One-to-One Function**

a.

b.

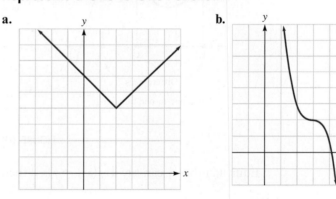

SOLUTION We will use the horizontal-line test and draw many horizontal lines. If every horizontal line that intersects the graph does so exactly once, the function is one-to-one. If any horizontal line intersects the graph more than once, the function is not one-to-one.

a. Because the horizontal line we draw intersects the graph in two places, we know that the function fails the horizontal-line test and is not a one-to-one function. (See Figure 1-95.)

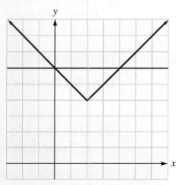

FIGURE 1-95

b. We draw several horizontal lines and note that each one intersects the graph exactly once. We conclude that the graph passes the horizontal-line test and represents a one-to-one function. (See Figure 1-96.)

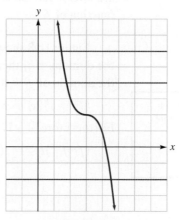

FIGURE 1-96

Self Check 2   Determine whether the graph at the left represents a one-to-one function.

Now Try Exercise 19.

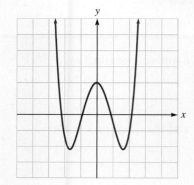

3. Define the Inverse of a One-to-One Function and Verify Inverse Functions

Figure 1-97(a) illustrates a function f from set **X** to set **Y**. Since three arrows point to a single y, the function f is not one-to-one. If the arrows in Figure 1-97(a) were reversed, the diagram would not represent a function, because three values of x in set **X** would correspond to a value y in set **Y**.

If the arrows of the one-to-one function f in Figure 1-97(b) were reversed, as in Figure 1-97(c), the diagram would represent a function. This function is called the **inverse of function f** and is denoted by the symbol f^{-1}. We say that f is **invertible**.

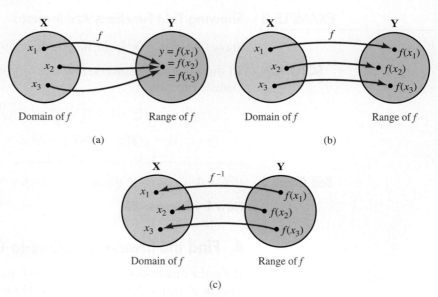

FIGURE 1-97

Inverse Functions

If f and g are two one-to-one functions such that $(f \circ g)(x) = x$ for every x in the domain of g, and $(g \circ f)(x) = x$ for every x in the domain of f, then f and g are **inverse functions**. Function g can be denoted as f^{-1} and is called the **inverse function of f**.

Figure 1-98 shows a one-to-one function f and its inverse f^{-1}. To the number x in the domain of f, there corresponds an output $f(x)$ in the range of f. Since $f(x)$ is in the domain of f^{-1}, the output for $f(x)$ under the function f^{-1} is $f^{-1}(f(x)) = x$. Thus,

$$(f^{-1} \circ f)(x) = f^{-1}(f(x)) = x.$$

Furthermore, if y is any number in the domain of f^{-1}, then

$$(f \circ f^{-1})(y) = f(f^{-1}(y)) = y.$$

Caution

The -1 in the notation for an inverse function is not an exponent. Remember that

$$f^{-1}(x) \neq \frac{1}{f(x)}.$$

Comment

It may be helpful to remember the relationship between the domain and range of a function and its inverse as $D_f = R_{f^{-1}}$ and $R_f = D_{f^{-1}}$.

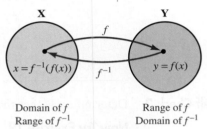

Domain of f Range of f
Range of f^{-1} Domain of f^{-1}

FIGURE 1-98

Properties of a One-to-One Function

Property 1: If f is a one-to-one function, there is a one-to-one function $f^{-1}(x)$ such that

$$(f^{-1} \circ f)(x) = x \qquad \text{and} \qquad (f \circ f^{-1})(x) = x.$$

Property 2: The domain of f is the range of f^{-1}, and the range of f is the domain of f^{-1}.

$$D_f = R_{f^{-1}} \qquad \text{and} \qquad R_f = D_{f^{-1}}$$

To show that one function is the inverse of another, we must show that their composition is the *identity function*, $f(x) = x$.

EXAMPLE 3   **Showing That Functions Are Inverses**

Show that $f(x) = x^3$ is the inverse function of $g(x) = \sqrt[3]{x}$.

SOLUTION   To show that f is the inverse function of g, we must show that $f \circ g$ and $g \circ f$ are the identity function.

$$(f \circ g)(x) = f(g(x)) = f\left(\sqrt[3]{x}\right) = \left(\sqrt[3]{x}\right)^3 = x$$
$$(g \circ f)(x) = g(f(x)) = g(x^3) = \sqrt[3]{x^3} = x$$

Self Check 3   If $x \geq 0$, is $f(x) = x^2$ the inverse of $g(x) = \sqrt{x}$? Note the restrictions on the domain.

Now Try Exercise 23.

4. Find the Inverse of a One-to-One Function

If f is the one-to-one function $y = f(x)$, then f^{-1} reverses the correspondence of f. That is, if $f(a) = b$, then $f^{-1}(b) = a$. For a one-to-one function, if we can uniquely solve the equation $y = f(x)$ for x, then we can use that equation to find the inverse function. To determine f^{-1}, we follow these steps.

Strategy for Finding f^{-1}

Step 1: Replace $f(x)$ with y.

Step 2: Interchange the variables x and y in the equation $y = f(x)$.

Step 3: Solve the resulting equation for y if possible.

Step 4: Replace y with $f^{-1}(x)$.

EXAMPLE 4   **Finding the Inverse of a Function**

Find the inverse of $y = f(x) = \dfrac{3}{2}x + 2$ and verify the result.

SOLUTION   We will use the strategy given above to find f^{-1}. We will then verify the result by showing that $(f \circ f^{-1})(x) = x$ and $(f^{-1} \circ f)(x) = x$.
To find f^{-1}, we use the following steps:

Step 1: Replace $f(x)$ with y.

$$f(x) = \frac{3}{2}x + 2$$

$$y = \frac{3}{2}x + 2$$

Step 2: Interchange the variables x and y.

$$x = \frac{3}{2}y + 2$$

Step 3: Solve the equation for y.

$$x = \frac{3}{2}y + 2$$

$$2x = 3y + 4 \qquad \text{Multiply both sides by 2.}$$

$$2x - 4 = 3y \qquad \text{Subtract 4 from both sides.}$$

$$y = \frac{2x - 4}{3} \qquad \text{Divide both sides by 3.}$$

Step 4: Replace y with $f^{-1}(x)$.

$$y = \frac{2x - 4}{3}$$

$$f^{-1}(x) = \frac{2x - 4}{3}$$

The inverse of $f(x) = \dfrac{3}{2}x + 2$ is $f^{-1}(x) = \dfrac{2x - 4}{3}$.

To verify that $f(x) = \dfrac{3}{2}x + 2$ and $f^{-1}(x) = \dfrac{2x - 4}{3}$ are inverses, we must show that $(f \circ f^{-1})(x) = x$ and $(f^{-1} \circ f)(x) = x$.

$$(f \circ f^{-1})(x) = f(f^{-1}(x)) \qquad\qquad (f^{-1} \circ f)(x) = f^{-1}(f(x))$$

$$= f\left(\frac{2x - 4}{3}\right) \qquad\qquad = f^{-1}\left(\frac{3}{2}x + 2\right)$$

$$= \frac{3}{2}\left(\frac{2x - 4}{3}\right) + 2 \qquad\qquad = \frac{2\left(\frac{3}{2}x + 2\right) - 4}{3}$$

$$= x - 2 + 2 \qquad\qquad = \frac{3x + 4 - 4}{3}$$

$$= x \qquad\qquad\qquad = x$$

Self Check 4   Find $f(2)$. Then find $f^{-1}(5)$. Explain the significance of the results.

Now Try Exercise 29.

5. Understand the Graphical Relationship between Inverse Functions and Determine the Domain and Range of These Functions

Because we interchange the positions of x and y to find the inverse of a function, the point (b, a) lies on the graph of $y = f^{-1}(x)$ whenever the point (a, b) lies on the graph of $y = f(x)$. Thus, the graph of a function and its inverse are reflections of each other about the line $y = x$ (see Figure 1-99).

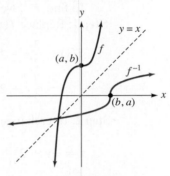

FIGURE 1-99

EXAMPLE 5   **Finding the Inverse of a Function**

Find the inverse of $y = f(x) = x^3 + 3$. Graph the function and its inverse on the same set of coordinate axes.

SOLUTION   We find the inverse of the function $y = f(x) = x^3 + 3$ by interchanging x and y and solving for y.

$$y = x^3 + 3 \qquad \text{The original function.}$$

$$x = y^3 + 3 \qquad \text{Interchange } x \text{ and } y.$$

$$y^3 = x - 3 \qquad \text{Solve for } y.$$

$$y = \sqrt[3]{x - 3} \qquad \text{This is the inverse function.}$$

Thus, $f^{-1}(x) = \sqrt[3]{x - 3}$. The functions are graphed in Figure 1-100. Notice that they are symmetric about the line $y = x$.

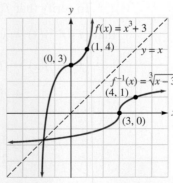

FIGURE 1-100

Self Check 5   Find $f(2)$. Then find $f^{-1}(11)$. Explain the significance of the result.

Now Try Exercise 49.

Example 5 demonstrates a procedure for finding an inverse. This method will always work when we can solve for a unique y. However, for a function such as $y = f(x) = x^5 + x + 3$, we cannot use the above method because the equation $x = y^5 + y + 3$ cannot be solved uniquely for y.

If a function is not one-to-one, it does not have an inverse. We say that it is not *invertible*. However, it is sometimes possible to restrict a domain so that an inverse can be found.

EXAMPLE 6   **Finding the Inverse of a Function and Determining the Domain and Range**

The function $y = f(x) = x^2 + 3$ is not one-to-one. However, it becomes one-to-one when we restrict its domain to the interval $(-\infty, 0]$. Under this restriction:

a. Find the range of f.

b. Find the inverse of f along with its domain and range.

c. Graph each function.

SOLUTION   **a.** The function f is defined by $y = f(x) = x^2 + 3$, with $D_f = (-\infty, 0]$. If x is replaced with numbers in this interval, y ranges over the values 3 and above. (See Figure 1-101.) Thus, the range of f is $R_f = [3, \infty)$.

b. To find the inverse of f, we interchange x and y in the equation that defines f in the restricted domain. Then solve for y.

$$y = x^2 + 3 \quad (x \le 0)$$

$$x = y^2 + 3 \quad (y \le 0) \qquad \text{Interchange } x \text{ and } y.$$

$$x - 3 = y^2 \quad (y \le 0) \qquad \text{Subtract 3 from both sides.}$$

To solve this equation for y, we take the square root of both sides. Since $y \le 0$, we have $-\sqrt{x - 3} = y$. The inverse of f is defined by $y = f^{-1}(x) = -\sqrt{x - 3}$. It has $D_{f^{-1}} = [3, \infty)$ and $R_{f^{-1}} = (-\infty, 0]$. (See Figure 1-101.)

c. The graphs of the functions appear in Figure 1-101. Note that the line of symmetry is $y = x$.

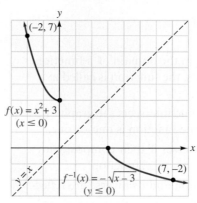

FIGURE 1-101

Self Check 6   Find the inverse of f when its domain is restricted to the interval $[0, \infty)$.

Now Try Exercise 57.

EXAMPLE 7   **Using an Inverse to Find the Range of a Function**

Find the domain and range of $y = f(x) = \dfrac{2}{x} + 3$.

SOLUTION   Since x cannot be 0, $D_f = (-\infty, 0) \cup (0, \infty)$. To find the range of f, we find $D_{f^{-1}}$ as follows (remember that no denominator can be 0):

$$y = \frac{2}{x} + 3$$

$$x = \frac{2}{y} + 3 \qquad \text{Interchange } x \text{ and } y.$$

$$xy = 2 + 3y \qquad \text{Multiply both sides by } y.$$

$$xy - 3y = 2 \qquad \text{Subtract } 3y \text{ from both sides.}$$

$$y(x - 3) = 2 \qquad \text{Factor out } y.$$

$$y = \frac{2}{x - 3} \qquad \text{Divide both sides by } x - 3.$$

The final equation defines f^{-1}, with $D_{f^{-1}} = (-\infty, 3) \cup (3, \infty)$. Because $R_f = D_{f^{-1}}$, $R_f = (-\infty, 3) \cup (3, \infty)$. ∎

Self Check 7   Find the range of $y = f(x) = \dfrac{3}{x} - 1$.

Now Try Exercise 63.

Self Check Answers   1. yes   2. no   3. yes   4. 5, 2   5. 11, 2   6. $y = f^{-1}(x) = \sqrt{x - 3}$
7. $(-\infty, -1) \cup (-1, \infty)$

Exercises **1.7**

Getting Ready
You should be able to complete these vocabulary and concept statements before you proceed to the practice exercises.

Fill in the blanks.

1. If different numbers in the domain of a function have different outputs, the function is called a _____ function.

2. If every _____ line intersects the graph of a function only once, the function is one-to-one.

3. Two functions are inverses if their composite in either order is the _____ function.

4. The graph of a function and its inverse are reflections of each other about the line _____.

Practice
Determine whether each function is one-to-one.

5. $y = 3x$

6. $y = \dfrac{1}{2}x$

7. $y = x^2 + 3$

8. $y = x^4 - x^2$

9. $y = x^3 - x$

10. $y = x^2 - x$

11. $y = |x|$

12. $y = |x - 3|$

13. $y = 5$

14. $y = \sqrt{x - 5}$

15. $y = (x - 2)^2;\ x \geq 2$

16. $y = \dfrac{1}{x}$

Use the horizontal-line test to determine whether each graph represents a one-to-one function.

17.

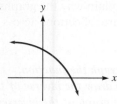

18.

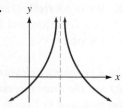

19.

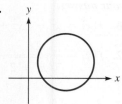

20.

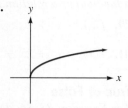

Verify that the functions are inverses by showing that $f \circ g$ and $g \circ f$ are identity functions. That is, $(f \circ g)(x) = x$ and $(g \circ f)(x) = x$.

21. $f(x) = 5x;\ g(x) = \dfrac{1}{5}x$

22. $f(x) = 4x + 5;\ g(x) = \dfrac{x - 5}{4}$

23. $f(x) = x^3 - 4;\ g(x) = \sqrt[3]{x + 4}$

24. $f(x) = \sqrt[3]{x} + 5;\ g(x) = (x - 5)^3$

25. $f(x) = \dfrac{x + 1}{x};\ g(x) = \dfrac{1}{x - 1}$

26. $f(x) = \dfrac{x + 1}{x - 1};\ g(x) = \dfrac{x + 1}{x - 1}$

Each equation defines a one-to-one function f. Determine f^{-1} and verify that $f \circ f^{-1}$ and $f^{-1} \circ f$ are the identity function.

27. $f(x) = 3x$

28. $f(x) = \dfrac{1}{3}x$

29. $f(x) = 3x + 2$

30. $f(x) = 2x - 5$

31. $f(x) = (x - 1)^3$

32. $f(x) = x^3 - 7$

33. $f(x) = \sqrt[3]{x} - 8$

34. $f(x) = \sqrt[3]{x + 4}$

35. $f(x) = \dfrac{1}{x + 3}$

36. $f(x) = \dfrac{1}{x - 2}$

37. $f(x) = \dfrac{1}{2x}$

38. $f(x) = \dfrac{1}{x^3}$

Find the inverse of each one-to-one function and graph both the function and its inverse on the same set of coordinate axes.

39. $f(x) = 5x$

40. $f(x) = \dfrac{3}{2}x$

41. $f(x) = 2x - 4$

42. $f(x) = \dfrac{3}{2}x - 2$

43. $x - y = 2$

44. $x + y = 0$

45. $2x + y = 4$

46. $3x + 2y = 6$

47. $f(x) = x^3 - 2$

48. $f(x) = (x - 5)^3$

49. $f(x) = \sqrt[3]{x - 4}$

50. $f(x) = \sqrt[3]{x} + 6$

51. $f(x) = \dfrac{1}{2x}$

52. $f(x) = \dfrac{1}{x - 3}$

53. $f(x) = \dfrac{x + 1}{x - 1}$

54. $f(x) = \dfrac{x - 1}{x}$

The function f defined by the given equation is one-to-one on the given domain. Find $f^{-1}(x)$ and state the domain.

55. $f(x) = x^2 - 3 \quad (x \le 0)$

56. $f(x) = \dfrac{1}{x^2} \quad (x > 0)$

57. $f(x) = x^4 - 8 \quad (x \ge 0)$

58. $f(x) = -\dfrac{1}{x^4} \quad (x < 0)$

59. $f(x) = \sqrt{4 - x^2} \quad (0 \le x \le 2)$

60. $f(x) = \sqrt{x^2 - 1} \quad (x \le -1)$

Find the domain of f and then find the range of f by finding the domain of f^{-1}.

61. $f(x) = \dfrac{x}{x - 2}$

62. $f(x) = \dfrac{x - 2}{x + 3}$

63. $f(x) = \dfrac{1}{x} - 2$

64. $f(x) = \dfrac{3}{x} - \dfrac{1}{2}$

The given function is not a one-to-one function. Restrict the domain in such a way that the resulting function is one-to-one. Find the inverse of the function and state the domain and range of each. (There is more than one correct answer.)

65. $f(x) = 9 - x^2$

66. $f(x) = x^2 - 4$

67. $f(x) = (x + 2)^2$

68. $f(x) = (x - 3)^2$

69. $f(x) = |x| - 4$

70. $f(x) = |x + 2|$

Use the given graph of f to sketch the graph of f^{-1}.

71.

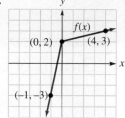

72.

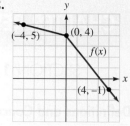

73.

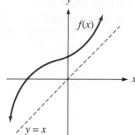

74.

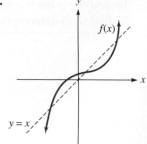

Applications

75. Buying pizza A pizzeria charges $8.50 plus 75¢ per topping for a large pizza.

a. Find a linear function that expresses the cost y of a large pizza in terms of the number of toppings x.

b. Find the cost of a pizza that has four toppings.

c. Find the inverse of the function found in part (a), to find a formula that gives the number of toppings y in terms of the cost x.

d. If Josh has $10, how many toppings can he afford?

76. Cell-phone bills A phone company charges $11 per month plus a nickel per call.

a. Find a rational function that expresses the average cost y of a call in a month when x calls were made.

b. To the nearest tenth of a cent, find the average cost of a call in a month when 68 calls were made.

c. Find the inverse of the function found in part (a), to find a formula that gives the number of calls y that can be made for an average cost x.

d. How many calls need to be made for an average cost of 15¢ per call?

Discovery and Writing

77. Write a brief paragraph to explain why $R_f = D_{f^{-1}}$.

78. Write a brief paragraph to explain why the graphs of a function and its inverse are reflections across the line $y = x$.

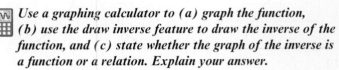

Use a graphing calculator to (a) graph the function, (b) use the draw inverse feature to draw the inverse of the function, and (c) state whether the graph of the inverse is a function or a relation. Explain your answer.

79. $f(x) = x^2 + 2x - 1$ **80.** $f(x) = x^3 - 2x^2 + 3$

81. $f(x) = x^5 + 2x + 2$ **82.** $f(x) = \dfrac{2x}{\sqrt{x^2 + 2}}$

True or False

For Exercises 83–86, determine whether the statement is true or false. If it is false, give an example to demonstrate why.

83. Even functions are *invertible* (have an inverse function).

84. If the point (a, b) is on the graph of f, then $(a, -b)$ is on the graph of the inverse.

85. The domain of a function is the same as the range of the function's inverse.

86. For $f(x) = x^n$, where n is an odd integer, the function will always have an inverse.

87. Give five examples of formulas for a given quantity for which we may want to know the inverse. Think of formulas such as $A = \pi r^2$.

CHAPTER REVIEW

SECTION 1.1   Graphs of Equations

Definitions and Concepts	Examples
To find the *x*-intercepts, let $y = 0$ and solve for x. To find the *y*-intercepts, let $x = 0$ and solve for y.	See Example 1, pages 108–109

Definitions and Concepts	Examples
1. If the points (x, y) and $(-x, y)$ both lie on the graph of an equation, the graph is **symmetric about the y-axis.** 2. If the points (x, y) and $(-x, -y)$ both lie on the graph of an equation, the graph is **symmetric about the origin.** 3. If the points (x, y) and $(x, -y)$ both lie on the graph of an equation, the graph is **symmetric about the x-axis.**	See Example 2, page 111
Standard equation of a circle: Center at (h, k) and radius r: $$(x - h)^2 + (y - k)^2 = r^2$$ Center at $(0, 0)$ and radius r: $$x^2 + y^2 = r^2$$	See Example 7, page 119

EXERCISES

Graph each equation. Find all intercepts and symmetries.

1. $y = x^2 + 2$

2. $y = x^3 - 2$

3. $y = \dfrac{1}{2}|x|$

4. $y = -\sqrt{x - 4}$

5. $y = \sqrt{x} + 2$

6. $y = |x + 1| + 2$

 Use a graphing calculator to graph each equation.

7. $y = |x - 4| + 2$

8. $y = -\sqrt{x + 2} + 3$

9. $y = x + 2|x|$

10. $y^2 = x - 3$

Identify the center and radius of each circle written in standard form.

11. $x^2 + y^2 = 64$

12. $x^2 + (y - 6)^2 = 100$

13. $(x + 7)^2 + y^2 = \dfrac{1}{4}$

14. $(x - 5)^2 + (y + 1)^2 = 9$

Write the equation of each circle in standard form.

15. Center at the origin; $r = 7$

16. Center at $(3, 0)$; $r = \dfrac{1}{5}$

17. Center at $(-2, 12)$; $r = 5$

18. Center at $\left(\dfrac{2}{7}, 5\right)$; $r = 9$

Write the equation of each circle in standard form and general form.

19. Center at $(-3, 4)$; radius 12

20. Ends of diameter at $(-6, -3)$ and $(5, 8)$

Convert the general form of each circle given into standard form.

21. $x^2 + y^2 + 6x - 4y + 4 = 0$

22. $2x^2 + 2y^2 - 8x - 16y - 10 = 0$

Graph each equation.

23. $x^2 + y^2 - 16 = 0$

24. $x^2 + y^2 - 4x = 5$

25. $x^2 + y^2 - 2y = 15$

26. $x^2 + y^2 - 4x + 2y = 4$

 Use a graphing calculator to solve each equation. If an answer is not exact, round to the nearest hundredth.

27. $x^2 - 11 = 0$

28. $x^3 - x = 0$

29. $|x^2 - 2| - 1 = 0$

30. $x^2 - 3x = 5$

SECTION **1.2** Functions and Function Notation

Definitions and Concepts	Examples
Relation: A set of ordered pairs. The set created by the first components in the ordered pairs is called the **domain** (the inputs). The set created by the second components in the ordered pairs is the **range** (the outputs).	See Example 1, page 125

Definitions and Concepts	Examples
A **function** f is a correspondence between a set of input values x (called the **domain**) and a set of output values y (called the **range**), where to each x value in the domain, there corresponds exactly one y value in the range.	See Example 4, pages 129–130
To indicate that x is a function of y, we use **function notation** and write $y = f(x)$, which is read as "$y = f$ of x."	See Example 6, page 131
Difference quotient: $$\frac{f(x + h) - f(x)}{h}, h \neq 0$$	See Example 8 , page 133

EXERCISES

Determine whether the equation defines y as a function of x. Assume that all variables represent real numbers.

31. $y = 3$

32. $y + 5x^2 = 2$

33. $y^2 - x = 5$

34. $y = |x| + x$

Find the domain and the range of each function.

35. $f(x) = 3x^2 - 5$

36. $f(x) = \dfrac{3x}{x - 5}$

37. $f(x) = \sqrt{x - 1}$

38. $f(x) = \sqrt{x^2 + 1}$

Find $f(2)$, $f(-3)$, and $f(k)$.

39. $f(x) = 5x - 2$

40. $f(x) = \dfrac{6}{x - 5}$

41. $f(x) = |x - 2|$

42. $f(x) = \dfrac{x^2 - 3}{x^2 + 3}$

Evaluate the difference quotient.

43. $f(x) = 2x + 3$

44. $f(x) = 5x - 6$

45. $f(x) = 3x^2 - 6x + 7$

46. $f(x) = 2x^2 - 7x + 3$

SECTION 1.3   Graphs of Functions

Definitions and Concepts	Examples
• If every vertical line that can be drawn intersects the graph at no more than one point, the graph represents a function. • If a vertical line can be drawn that intersects the graph at more than one point, the graph does not represent a function.	See Example 1, pages 138–139
It is important that we learn how to read the value of a function from the graph of that function.	See Example 2, pages 139–140
• The **domain** of a function is the set of input values x. Therefore, we can read the domain from the graph of a function by determining the values of x that are included in the graph. • The **range** of a function is the set of output values y, and we determine the range of the function by reading the values of y that are included in the graph.	See Example 4, pages 142–143
• A function f is **increasing** on an open interval if, for any x_1 and x_2 in the interval, $x_1 < x_2$ and $f(x_1) < f(x_2)$. • A function f is **decreasing** on an open interval if, for any x_1 and x_2 in the interval, $x_1 < x_2$ and $f(x_1) > f(x_2)$. • A function f is **constant** on an open interval if $f(x_1) = f(x_2)$ for all x_1 and x_2 in the interval.	See Example 5, page 144

Definitions and Concepts	Examples
• $f(c)$ is a **local maximum** if there exists an interval (a, b) with $a < c < b$ such that $f(x) \leq f(c)$ for all x in (a, b). • $f(c)$ is a **local minimum** if there exists an interval (a, b) with $a < c < b$ such that $f(x) \geq f(c)$ for all x in (a, b).	See Example 6, page 145
• The function f is an **even** function if it is symmetric about the y-axis. Algebraically, this means that $f(x) = f(-x)$ for all x in the domain. • The function f is an **odd** function if it is symmetric about the origin. Algebraically, this means that $-f(x) = f(-x)$ for all x in the domain.	See Example 8, pages 147–148
• An **x-intercept** of a function is the x-coordinate of any point where the graph of the function crosses or touches the x-axis. • The **y-intercept** of a function is the y-coordinate of the point on the graph where the function crosses or touches the y-axis.	See Example 9, pages 148–149

EXERCISES

For the graph of each function shown, (a) determine the domain and range, (b) state the value of any intercepts, (c) determine whether the function is odd, even, or neither, and (d) state the open intervals where the function is increasing, where it is decreasing, and where it is constant.

47.

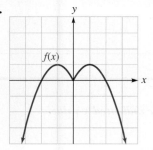

48.

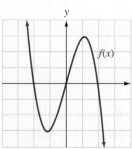

49.

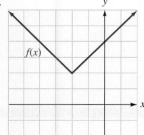

50.

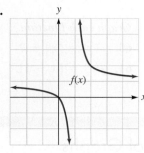

Determine algebraically whether each function is even, odd, or neither.

51. $f(x) = x^4 - x^2 - 2$ **52.** $f(x) = x^3 - 3x$

53. $f(x) = \sqrt{x - 2}$ **54.** $f(x) = \dfrac{4}{x^2}$

Use the graph of the function or the equation of the function to answer each question.

55. $f(x) = -x^2 + 4x - 1$

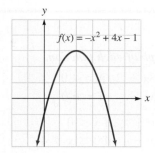

(a) State the value(s) of any extrema.

(b) State the value(s) of any intercepts.

(c) State the domain and the range.

(d) $f(2) = ?$ $f(-3) = ?$

(e) State the value(s) of any zeros of the function.

(f) Is the function even, odd, or neither?

SECTION **1.4**   Basic Graphs

Definitions and Concepts	Examples
Some graphs of functions can be approximated by plotting points.	See Example 1, pages 154–155

Definitions and Concepts	Examples
• A **linear function** is a function determined by an equation of the form $f(x) = mx + b$, or $y = mx + b$. • A **constant function** is a linear function with slope $m = 0$. It is determined by an equation of the form $f(x) = b$, or $y = b$, where b is any real number. • The **identity function** is the linear function with $m = 1$ and $b = 0$. It is determined by the equation in the form $f(x) = x$, or $y = x$.	See Example 2, pages 156–157
• The **squaring function** is the function determined by the equation $f(x) = x^2$, or $y = x^2$. • The **absolute value function** is the function determined by the equation $f(x) = |x|$, or $y = |x|$. • The **square root function** is the function determined by the equation $f(x) = \sqrt{x}$, or $y = \sqrt{x}$. • The **cubing function** is the function determined by the equation $f(x) = x^3$, or $y = x^3$. • The **cube root function** is the function determined by the equation $f(x) = \sqrt[3]{x}$, or $y = \sqrt[3]{x}$. • The **reciprocal function** is the function determined by the equation $f(x) = \dfrac{1}{x}$, or $y = \dfrac{1}{x}$. • The **greatest integer function,** sometimes called the "floor" function, is the function defined by the equation $f(x) = [x]$ [also $f(x) = \text{int}(x)$], or $y = [x]$, where the value of y that corresponds to x is the greatest integer that is less than or equal to x.	See Example 3, page 161

EXERCISES

Sketch the graph of each function.

56. $f(x) = \begin{cases} 3x - 2 & \text{if } x < 0 \\ x^2 & \text{if } x \geq 0 \end{cases}$

57. $f(x) = \begin{cases} -3 & \text{if } x < 0 \\ x - 1 & \text{if } 0 \leq x < 4 \\ x & \text{if } x \geq 4 \end{cases}$

Evaluate each function at the indicated x value.

58. $f(x) = [2x]$; $f(1.7)$

59. $f(x) = [x - 5]$; $f(4.99)$

Graph each function.

60. $f(x) = [x] + 2$

61. $f(x) = [x - 1]$

SECTION **1.5**   Transformations of Graphs

Definitions and Concepts	Examples
Vertical shifts: $y = f(x) + k$ shifts the graph up k units. $y = f(x) - k$ shifts the graph down k units.	See Example 1, page 168
Horizontal shifts: $y = f(x - k)$ shifts the graph right k units. $y = f(x + k)$ shifts the graph left k units.	See Example 2, page 169

Definitions and Concepts	Examples
Reflections:	See Example 5, page 172
$\quad y = -f(x)$ reflects the graph about the x-axis.	
$\quad y = f(-x)$ reflects the graph about the y-axis.	
Vertical stretches or shrinks (multiply each y-coordinate by k):	See Example 8, pages 174–175
$\quad y = kf(x), k > 1,$ is a vertical stretch.	
$\quad y = kf(x), 0 < k < 1,$ is a vertical shrink.	
Horizontal stretches or shrinks (divide each x-coordinate by k):	See Example 11, pages 177–178
$\quad y = f(kx), k > 1,$ is a horizontal shrink.	
$\quad y = f(kx), 0 < k < 1,$ is a horizontal stretch.	

EXERCISES

Use the figure showing the graph of $y = f(x)$ and transformations to graph g.

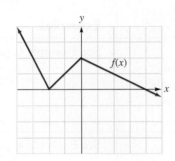

62. $g(x) = f(x) - 3$ **63.** $g(x) = f(x - 3)$
64. $g(x) = f(-x)$ **65.** $g(x) = -f(x)$
66. $g(x) = 2f(x)$ **67.** $g(x) = f(2x)$

The graph of each function is a transformation of the graph of $f(x) = x^2$. Graph both on one set of coordinate axes.

68. $g(x) = x^2 + 1$ **69.** $g(x) = (x + 2)^2$
70. $g(x) = -3x^2$ **71.** $g(x) = (x - 2)^2 - 2$
72. $g(x) = -\dfrac{1}{2}(x - 2)^2 - 2$

Each function is a translation of a basic function, $f(x)$. Graph both on one set of coordinate axes.

73. $g(x) = x^2 + 5$ **74.** $g(x) = (x - 7)^3$
75. $g(x) = \sqrt{x + 2} + 3$ **76.** $g(x) = |x - 4| + 2$

Each function is a stretching of $f(x) = x^3$. Graph both on one set of coordinate axes.

77. $g(x) = \dfrac{1}{3}x^3$ **78.** $g(x) = (-5x)^3$

Graph each function using a combination of translations and stretchings.

79. $g(x) = -|x - 4| + 3$ **80.** $g(x) = \dfrac{1}{4}|x - 4| + 1$
81. $g(x) = 3\sqrt{x + 3} + 2$ **82.** $g(x) = \dfrac{1}{3}(x + 3)^3 + 2$
83. $f(x) = \sqrt{-x} + 3$ **84.** $g(x) = 2\sqrt[3]{x} - 5$

SECTION **1.6**   Operations on Functions

Definitions and Concepts	Examples
The **sum** of f and g, denoted as $f + g$, is defined by $(f + g)(x) = f(x) + g(x)$. The **difference** of f and g, denoted as $f - g$, is defined by $(f - g)(x) = f(x) - g(x)$.	See Example 1, page 184

Definitions and Concepts	Examples
The **product** of f and g, denoted as $f \cdot g$, is defined by $(f \cdot g) = f(x)g(x)$.	See Example 2, pages 184–185
The **quotient** of f and g, denoted as f/g, is defined by $(f/g)(x) = \dfrac{f(x)}{g(x)}$ $(g(x) \neq 0)$.	
The **composite function** $f \circ g$ is defined by $(f \circ g)(x) = f(g(x))$. The **domain** of $f \circ g$ consists of all those numbers in the domain of g for which $g(x)$ is in the domain of f.	See Example 6, page 189

EXERCISES

Let $f(x) = x^2 - 1$ and $g(x) = 2x + 1$. Find each function and its domain.

85. $(f + g)(x)$

86. $(f \cdot g)(x)$

87. $(f - g)(x)$

88. $(f/g)(x)$

89. $(f \circ g)(x)$

90. $(g \circ f)(x)$

SECTION **1.7**   Inverse Functions

Definitions and Concepts	Examples
Inverse functions: If f and g are two one-to-one functions such that $(f \circ g)(x) = x$ for every x in the domain of g, and $(g \circ f)(x) = x$ for every x in the domain of f, then f and g are **inverse functions.** Function g can be denoted as f^{-1} and is called the **inverse function of** f.	See Example 1, page 197
Strategy for finding $f^{-1}(x)$: **Step 1:** Replace $f(x)$ with y. **Step 2:** Interchange the variables x and y in the equation $y = f(x)$. **Step 3:** Solve the resulting equation for y if possible. **Step 4:** Replace y with $f^{-1}(x)$.	See Example 4, page 201
The graph of a function is symmetric to the graph of its inverse. The axis of symmetry is the line $y = x$.	See Example 5, page 202

EXERCISES

Each equation defines a one-to-one function. Find $f^{-1}(x)$.

91. $y = 7x - 1$

92. $y = \dfrac{2}{x}$

93. $y = x^3 - 6$

94. $y = \sqrt[3]{x + 4}$

95. $y = \dfrac{1}{2 - x}$

96. $y = \dfrac{x}{1 - x}$

97. $y = \dfrac{3}{x^3}$

98. Find the inverse of the one-to-one function $f(x) = 2x - 5$ and graph both the function and its inverse on the same set of coordinate axes.

99. Find the range of $y = \dfrac{2x + 3}{5x - 10}$ by finding the domain of f^{-1}.

CHAPTER TEST

Find the x- and y-intercepts of each graph.

1. $y = x^3 - 16x$ **2.** $y = |x - 4|$

Find the symmetries of each graph.

3. $y^2 = x - 1$ **4.** $y = x^4 + 1$

Graph each equation.

5. $y = x^2 - 9$ **6.** $x = |y|$

7. $y = 2\sqrt{x}$ **8.** $x = y^3$

Write the equation of each circle.

9. Center at $(5, 7)$; radius of 8

10. Center at $(2, 4)$; passing through $(6, 8)$

Graph each equation.

11. $x^2 + y^2 = 9$ **12.** $x^2 - 4x + y^2 + 3 = 0$

Find the domain and range of each function.

13. $f(x) = \dfrac{3}{x - 5}$ **14.** $f(x) = \sqrt{x + 3}$

Find $f(-1)$ and $f(2)$.

15. $f(x) = \dfrac{x}{x - 1}$ **16.** $f(x) = \sqrt{x + 7}$

For the graph of each function $f(x)$ shown, (a) determine the domain and range, (b) state the value of any intercepts, (c) determine whether the function is odd, even, or neither, and (d) state the intervals where the function is increasing, where it is decreasing, and where it is constant.

17.

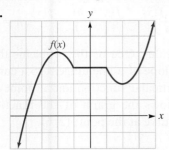

18.

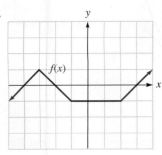

Determine algebraically whether each function is even, odd, or neither.

19. $f(x) = 2x^2 - x - 1$ **20.** $f(x) = \dfrac{1}{x}$

Use the graph or the equation of the function to answer each question.

21. $f(x) = 2x^4 - 2x^2$

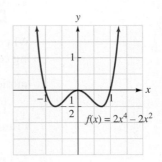

(a) State the value(s) of any extrema.

(b) State the value(s) of any intercepts.

(c) State the domain and the range.

(d) $f(2) = ?$ $f(-3) = ?$

(e) State the value(s) of any zeros of the function.

(f) Is the function odd, even, or neither?

Sketch the graph of each function.

22. $f(x) = \begin{cases} |x| & \text{if } x \le 0 \\ \sqrt{x} & \text{if } x > 0 \end{cases}$ **23.** $f(x) = 2x^2 - 1$

24. $f(x) = -(x - 1)^3$ **25.** $f(x) = \sqrt{x - 1} + 5$

Let $f(x) = 3x$ and $g(x) = x^2 + 2$. Find each function.

26. $f + g$ **27.** $g \circ f$

28. f/g **29.** $f \circ g$

Assume that $f(x)$ is one-to-one. Find f^{-1}.

30. $f(x) = \dfrac{x + 1}{x - 1}$ **31.** $f(x) = x^3 - 3$

Find the range of f by finding the domain of f^{-1}.

32. $y = \dfrac{3}{x} - 2$ **33.** $y = \dfrac{3x - 1}{x - 3}$

LOOKING AHEAD TO **CALCULUS**

1. **Using the Difference Quotient to Find a Tangent Line and Define the Derivative**
   - Connect the difference quotient to tangents to a curve—Section 1.2

2. **Derivative Rules—Using the Algebra and Composition of Functions**
   - Connect the algebra and composition of functions to the derivative—Section 1.6

1. Using the Difference Quotient to Find a Tangent Line and Define the Derivative

When we study algebra and trigonometry, we often have no idea why we are learning certain concepts. In the second section of this chapter, we defined the **difference quotient,** $\dfrac{f(x + h) - f(x)}{h}$, $h \neq 0$. This quotient is very important in calculus.

Let's take a look at an illustration of the difference quotient using the graph of a function.

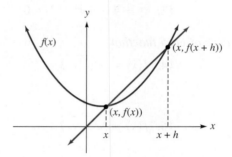

Notice that two points are identified on the graph of the function, $(x, f(x))$ and $(x + h, f(x + h))$. The distance between the two x values is represented by h. Recall that the slope of a line is $m = \dfrac{y_2 - y_1}{x_2 - x_1}$. The secant line drawn through the two points in the graph has as its slope $m = \dfrac{f(x + h) - f(x)}{(x + h) - x} = \dfrac{f(x + h) - f(x)}{h}$, $h \neq 0$. This is the difference quotient.

Now we will see just how this is used in calculus. One of the problems that we will study in calculus is how to determine the slope of a line tangent to a curve at a point. To illustrate this, we will look at a particular function and a point on the graph of that function. The figure below shows the graph of the function $f(x) = 2x^2 - 3x + 4$ and the line tangent to the curve at the point $(2, 6)$.

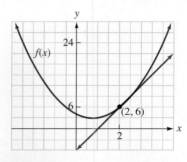

We know that to find the slope of a line, we need two points on the line, but the tangent line and the function only share one point. So all we know right now is that the tangent line contains the point $(2, 6)$. Since the difference quotient is the slope of a secant line, let's draw a secant line connecting the points $(2, 6)$ and $(4, 24)$. The slope of the secant line is $m = \dfrac{f(4) - f(2)}{4 - 2} = \dfrac{24 - 6}{4 - 2} = 9$. This is shown in the figure.

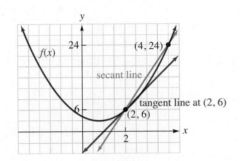

Obviously, the slope of the secant line we drew is not very close to the slope of the tangent line, so how can we get a better approximation? Suppose we were to find the slope of the secant line connecting the points $(2, 6)$ and $(2.5, 9)$? That slope would be $m = \dfrac{9 - 6}{2.5 - 2} = 6$. Do you think that slope would be close to the slope of the tangent? It is certainly closer than our first approximation, but still not what we need.

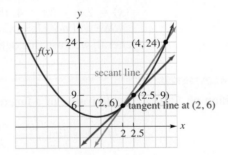

In this table, h values are shown in the first row and the corresponding slopes of the secant lines are shown in the second row. Notice that we calculated the slopes of the secant lines as the values of h got closer and closer to 0, but not equal to 0. (Why can't $h = 0$?)

h	0.5	0.05	0.005	0.0005
slope of secant line, $\dfrac{f(2 + h) - f(2)}{h}$	$\dfrac{9 - 6}{0.5} = 6$	$\dfrac{6.255 - 6}{0.05} = 5.1$	$\dfrac{6.02505 - 6}{0.005} = 5.01$	$\dfrac{6.0025005 - 6}{0.0005} = 5.001$

These slopes are getting closer and closer to 5, so it would appear that the slope of the tangent line is going to be 5, but how can we be certain of that? In the "Looking Ahead to Calculus" section of Chapter R, we developed the concept of the limit, so let's apply that to this problem. The values in the table indicate that the slope of the tangent line is $\displaystyle\lim_{h \to 0} \dfrac{f(2 + h) - f(2)}{h} = 5$. While the table may give us an indication of the value of the limit, it is not a proof of the limit. To prove that the limit is 5, we will evaluate the limit as shown. Pay careful attention to all the algebraic operations that are involved.

$$\lim_{h\to 0}\frac{f(2+h)-f(2)}{h}$$

$$=\lim_{h\to 0}\frac{\overbrace{2(2+h)^2-3(2+h)+4}^{f(2+h)}-\overbrace{(2(2)^2-3(2)+4)}^{f(2)}}{h}$$

$$=\lim_{h\to 0}\frac{(2(4+4h+h^2)-3(2)-3(h)+4)-(2(4)-6+4)}{h}$$

$$=\lim_{h\to 0}\frac{8+8h+2h^2-6-3h+4-8+6-4}{h}$$

$$=\lim_{h\to 0}\frac{5h+2h^2}{h}$$

$$=\lim_{h\to 0}\frac{h(5+2h)}{h}$$

$$=\lim_{h\to 0}(5+2h)=5$$

When we take calculus, we will spend a lot of time analyzing functions. One of the tools that we use to analyze a function is called the **derivative.** The derivative of a function at a point is the *slope of the tangent line* at that point, provided the limit exists. It is also called the *slope of the curve* or the *rate of change* at that point. There is a special notation used for the derivative of a function; it is $f'(x)$, read "f prime of x." Looking at the function we have been examining, we can say that for $f(x)=2x^2-3x+4$, the derivative is $f'(2)=5$.

We will not elaborate here on the uses of this derivative, but there are many! In general, we have this definition from calculus:

Definition of $f'(a)$ The **derivative of a function f at a number a,** denoted by $f'(a)$, is

$$f'(a)=\lim_{h\to 0}\frac{f(a+h)-f(a)}{h},\text{ provided the limit exists.}$$

What if we wanted to find the value of the derivative at a lot of points on the curve? Do we have to find the derivative at each point separately? That would seem to be very tedious and time consuming.

Can we find the derivative for any value of x? The answer is yes—by evaluating the limit using the function rather than a numerical value of x. As we did with the evaluation of the other limit, we are using a lot of algebraic techniques to simplify the expression. One of the main problems that many students have in calculus is a weakness in their algebra skills. If you have weak algebra skills, you may be able to understand the calculus concepts, but you will not be able to perform the actual problems.

Definition of $f'(x)$ The definition of the derivative for a function is given by

$$f'(x)=\lim_{h\to 0}\frac{f(x+h)-f(x)}{h}.$$

Now, you can see a use of the difference quotient.

To find the derivative of $f(x) = 2x^2 - 3x + 4$, our first step is to simplify the difference quotient for the function.

$$f(x) = 2x^2 - 3x + 4$$

$$\frac{f(x+h) - f(x)}{h} = \frac{\overbrace{(2(x+h)^2 - 3(x+h) + 4)}^{f(x+h)} - \overbrace{(2x^2 - 3x + 4)}^{f(x)}}{h}$$

$$= \frac{(2(x^2 + 2xh + h^2) - 3x - 3h + 4) - 2x^2 + 3x - 4}{h}$$

$$= \frac{2x^2 + 4xh + 2h^2 - 3x - 3h + 4 - 2x^2 + 3x - 4}{h}$$

$$= \frac{4xh + 2h^2 - 3h}{h}$$

$$= \frac{h(4x + 2h - 3)}{h}$$

$$= 4x + 2h - 3$$

Now that we have simplified the difference quotient, we will evaluate the limit. Remember, when we evaluate this limit, we can just substitute the value $h = 0$ into the term.

$$f'(x) = \lim_{h \to 0} \frac{f(x+h) - f(x)}{h}$$

$$= \lim_{h \to 0} (4x + 2h - 3)$$

$$= 4x + 2(0) - 3$$

$$= 4x - 3$$

Let's check what we determined earlier by using the function $f'(x) = 4x - 3$. $f'(2) = 4(2) - 3 = 5$, which confirms that the slope of the tangent line at $x = 2$ is 5.

Connect to Calculus:
Using the Difference Quotient to Find a Tangent Line

1. Find an equation of the tangent line for $f(x) = 2x^2 - 3x + 4$ at the values $x = -2$ and $x = 2$.
2. At the vertex of a parabola, what would the value of the derivative equal?
3. For the function $f(x) = \dfrac{1}{x}$, can you find $f'(0)$? Why or why not?
4. If $f(x) = 3x^2 + 5x$, find $f'(-1)$ and use it to find an equation of the tangent line to the parabola $f(x) = 3x^2 + 5x$ at the point $(-1, -2)$.

2. Derivative Rules—Using the Algebra and Composition of Functions

There are rules for finding the derivative of a function that can be developed by using the limit definition for functions of a general form. Some of these rules use the algebra of functions you studied in Section 6 of this chapter, and are listed here for two functions. $(f + g)' = f' + g'$ means that you can add two functions together and find the derivative of the new function, or you can find the derivative of each function separately and then add them together.

These are some rules you will use in calculus:

Derivative Rules

Sum Rule: $(f + g)' = f' + g'$ Difference Rule: $(f - g)' = f' - g'$

Product Rule: $(fg)' = fg' + gf'$ Quotient Rule: $\left(\dfrac{f}{g}\right)' = \dfrac{gf' - fg'}{g^2}, g \neq 0$

A very powerful rule of derivatives is called the **Chain Rule**. Of course, you are not yet expected to understand the mathematics of the Chain Rule, but you should notice that the Chain Rule makes use of the composition of functions and the product of functions, which you have also learned about in this chapter. Here is the definition of the Chain Rule as it appears in calculus texts:

Chain Rule

If f and g are both differentiable and $F = f \circ g$ is the **composite function** defined by $F(x) = f(g(x))$, then F is differentiable and F' is given by the **product** $F'(x) = f'(g(x))g'(x)$.

Perhaps now you can see how important it is to learn the algebra of functions, because you will be using them extensively in calculus.

Connect to Calculus:
Derivative Rules—Using the Algebra and Composition of Functions

5. Suppose that $f(3) = 4, f'(3) = 2, g(3) = -5$, and $g'(3) = 5$. Find the following values:

   a. $(fg)'(3)$ b. $(f/g)'(3)$ c. $(g/f)'(3)$

6. Suppose that $F(x) = f(g(x))$, $g(5) = 7, g'(5) = 3, f'(5) = -2$, and $f'(7) = 2$. Find $F'(5)$.

Polynomial and Rational Functions

2

CAREERS AND MATHEMATICS: Applied Mathematician

Applied mathematicians use theories and techniques, such as computational methods and mathematical modeling, to formulate and solve practical problems that arise in engineering, business, and government, as well as social, life, and physical sciences. This can include such diverse things as analyzing the effects and safety of new drugs and efficiently scheduling airline routes between cities.

Education and Mathematics Required
• Other than with employment in the federal government, generally a PhD in mathematics is required.
• College Algebra; Trigonometry; Calculus I, II, and III; Linear Algebra; Ordinary Differential Equations; Real Analysis; Abstract Algebra; Theory of Analysis; and Complex Analysis form a basic list. Most mathematicians will study additional topics in their area of specialty.

How Applied Mathematicians Use Math and Who Employs Them
• Mathematical theories, algorithms, computational techniques, and computer technology can be used to solve economic, scientific, engineering, and business problems.
• A large number of mathematicians work for the federal government, with about 80% of those working for the U.S. Department of Defense. There are also positions in NASA. Many mathematicians are employed by universities as faculty and divide their time between teaching and research. Mathematicians are also employed by research-and-development laboratories as part of a technical team.

Career Outlook and Earnings
• Much faster-than-average employment growth is expected for mathematicians. PhD holders with a strong background in mathematics and a related field, such as computer science or engineering, will have better employment opportunities in related occupations. Employment of mathematicians is expected to increase by 22% during the 2008–2018 decade.
• The median annual salary is $93,580, with the top 10% of earners having a salary of $142,460.

For more information see http://www.bls.gov/oco.

In this chapter, we continue our study of functions. Polynomial and rational functions are very important. We will develop methods to solve polynomial equations and polynomial and rational inequalities.

2.1 Quadratic Functions

In this section, we will learn to

1. Recognize the general and standard forms of a quadratic function.
2. Graph a quadratic function using the components of its formula.
3. Solve real-life problems using quadratic functions.
4. Find a quadratic function that fits real-world data using regression.

Quadratic functions are important because we can use them to model many real-life problems. For example, the path of a basketball jump shot by Shaquille O'Neal and the path of a guided missile can be modeled with quadratic functions. Businesses like Coca-Cola and Best Buy can use quadratic functions to help maximize the profit and revenue for the items they produce and sell. In this section, we will study quadratic functions.

1. Recognize the General and Standard Forms of a Quadratic Function

A quadratic function is one type of polynomial function.

Polynomial Function

A **polynomial function of x with degree n** is a function of the form

$$P(x) = a_n x^n + a_{n-1} x^{n-1} + \cdots + a_1 x + a_0,$$

where a_n, a_{n-1}, ..., a_1, and a_0 are real numbers with $a_n \neq 0$ and n is a nonnegative integer.

Here are a few examples of polynomial functions:

$f(x) = 6$	$f(x) = \sqrt{5}$	Constant functions
$f(x) = 3x - 5$	$f(x) = \dfrac{1}{4}x + 8$	Linear functions
$f(x) = x^2 - 5x + 4$	$f(x) = -\sqrt{2}x^2 - 3x + \dfrac{1}{2}$	Quadratic functions
$f(x) = 4x^3 - 2x^2 - 5x + 1$	$f(x) = \dfrac{\sqrt{3}}{5}x^3 + 9$	Cubic functions

A function that is defined by a polynomial of the second degree is called a **quadratic function**. Note that the function $f(x) = x^2 - 5x + 4$ listed above is a quadratic function.

Quadratic Function

A **quadratic function** is a second-degree polynomial function in one variable, defined by an equation of the form $f(x) = ax^2 + bx + c$ $(a \neq 0)$ or $y = ax^2 + bx + c$ $(a \neq 0)$, where a, b, and c are real numbers.

A quadratic function written in the form stated in the definition box is said to be in **general form**.

General Form of a Quadratic Function

The **general form** of an equation of a quadratic function is

$$f(x) = ax^2 + bx + c, a \neq 0.$$

The graph of a quadratic function is a curve called a **parabola**, a cup-shaped curve that opens either upward ($\cup$) or downward ($\cap$). It is important that we can recognize the characteristics of a parabola:

- The parabola opens upward if $a > 0$ and opens downward if $a < 0$.
- The **vertex** is the point at which the parabola turns from increasing to decreasing or from decreasing to increasing.
- The y value of the vertex is the **maximum** or **minimum value** of the parabola.
- The **axis of symmetry** of a parabola is the vertical line that intersects the parabola at the vertex. The parabola is symmetric about this vertical line.

The important characteristics of a parabola are shown in Figure 2-1.

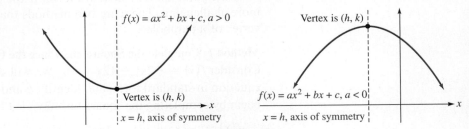

FIGURE 2-1

In Section 1.4 we looked at the squaring function $f(x) = x^2$, which is the simplest quadratic function. Its graph is shown in Figure 2-2.

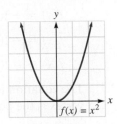

FIGURE 2-2

We also graphed transformations of the squaring function like $g(x) = (x + 2)^2 - 2$ by shifting the graph of our squaring function two units to the left and two units down (see Figure 2-3). The vertex of this parabola is $(-2, -2)$.

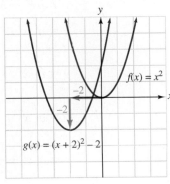

FIGURE 2-3

The equation of a parabola that is written in this form is said to be in **standard form**; this is a convenient form for identifying the vertex of a parabola.

Standard Form of a Quadratic Function	The **standard form** of an equation of a quadratic function is $$y = f(x) = a(x - h)^2 + k \quad (a \neq 0).$$

We now have two forms of a quadratic function: the general form, $f(x) = ax^2 + bx + c$, and the standard form, $f(x) = a(x - h)^2 + k$. The vertex is the point (h, k).

2. Graph a Quadratic Function Using the Components of Its Formula

To graph a parabola, it is very important that we identify its vertex. This is easy to do if the parabola is in standard form. If the parabola is in general form, it can be more challenging. There are two methods that are commonly used to identify the vertex of a parabola.

Method 1: Complete the Square and Place the Quadratic Function in Standard Form
Consider $f(x) = -2x^2 + 12x - 16$. We will complete the square on x, write the equation in standard form, and identify h and k, the coordinates of the vertex. We begin by completing the square on $-2x^2 + 12x$.

$f(x) = -2x^2 + 12x - 16$	Identify a: $a = -2$
$f(x) = -2(x^2 - 6x) - 16$	Factor $a = -2$ from $-2x^2 + 12x$.
$f(x) = -2(x^2 - 6x + \mathbf{9 - 9}) - 16$	One-half of -6 is -3; $(-3)^2 = 9$. Add and subtract 9 within the parentheses.
$f(x) = -2(x^2 - 6x + 9) \mathbf{- 2(-9)} - 16$	Distribute the multiplication by -2.
$f(x) = -2(x - 3)^2 + 18 - 16$	Factor $x^2 - 6x + 9$ and multiply.
$f(x) = -2(x - 3)^2 + 2$	Simplify.

The equation is now in standard form with $h = 3$ and $k = 2$. Therefore, the vertex is the point with coordinates $(h, k) = (3, 2)$.

Method 2: Use a Formula to Find the Vertex of a Parabola
We now derive a formula to use to find the vertex of a parabola that is in general form. To accomplish this we will start with $y = ax^2 + bx + c \ (a \neq 0)$ and complete the square on $ax^2 + bx$ to change the equation into the form $y = a(x - h)^2 + k$. In this form, we can read the coordinates (h, k) of the vertex from the equation.

$y = ax^2 + bx + c$	
$y = a\left(x^2 + \dfrac{b}{a}x\right) + c$	Factor a from $ax^2 + bx$.
$y = a\left(x^2 + \dfrac{b}{a}x + \dfrac{b^2}{4a^2} - \dfrac{b^2}{4a^2}\right) + c$	Add and subtract $\dfrac{b^2}{4a^2}$ within the parentheses.
$y = a\left(x^2 + \dfrac{b}{a}x + \dfrac{b^2}{4a^2}\right) - \dfrac{ab^2}{4a^2} + c$	Distribute the multiplication of a.
$y = a\left(x + \dfrac{b}{2a}\right)^2 + c - \dfrac{b^2}{4a}$	Factor $x^2 + \dfrac{b}{a}x + \dfrac{b^2}{4a^2}$ and simplify $\dfrac{ab^2}{4a^2}$.
(1) $y = a\left[x - \left(-\dfrac{b}{2a}\right)\right]^2 + c - \dfrac{b^2}{4a}$	$-\left(-\dfrac{b}{2a}\right) = \dfrac{b}{2a}$

If we compare Equation (1) to the form $y = a(x - h)^2 + k$, we see that $h = -\frac{b}{2a}$ and $k = c - \frac{b^2}{4a}$. This proves the following theorem:

Vertex Theorem   The graph of the function $y = f(x) = ax^2 + bx + c$ $(a \neq 0)$ is a parabola with vertex at $\left(-\dfrac{b}{2a}, c - \dfrac{b^2}{4a}\right)$ or $\left(-\dfrac{b}{2a}, f\left(-\dfrac{b}{2a}\right)\right)$.

From the theorem, we note that with the form $y = f(x) = a(x - h)^2 + k$, we have $h = -\frac{b}{2a}$ and $k = f(h)$. The theorem provides a very effective method to find the vertex of a parabola, and it will be used extensively in this section.

To graph a quadratic function $y = f(x) = ax^2 + bx + c$, follow these steps:

Strategy for Graphing a Quadratic Function

Step 1: Find the vertex (h, k), where $h = -\dfrac{b}{2a}$ and $k = f(h)$.

Step 2: Determine whether the parabola opens upward or downward. If $a > 0$, the parabola opens upward. If $a < 0$, the parabola opens downward.

Step 3: Find the y-intercept by evaluating $f(0)$. There will always be one y-intercept.

Step 4: Find any x-intercepts by finding the solution to the equation $f(x) = 0$. The number of x-intercepts will be 0, 1, or 2.

Step 5: Plot the vertex, the y-intercept, x-intercept(s) if any, and additional points.

Making use of symmetry is very helpful when graphing a parabola. The **axis of symmetry** is the line $x = h$. Please note that it is not part of the graph of the parabola.

EXAMPLE 1   Graphing a Quadratic Function

Write the function $f(x) = 3x^2 - 6x + 5$ in standard form and then sketch the graph.

SOLUTION   We will apply the five steps outlined in the "Strategy for Graphing a Quadratic Function" box.

Step 1: Find the vertex (h, k), where $h = -\dfrac{b}{2a}$ and $k = f(h)$.

We note that $a = 3$ and $b = -6$. To find h, we substitute 3 in for a and -6 in for b.

$$h = -\frac{b}{2a} = -\frac{-6}{2(3)} = 1$$

The x-coordinate of the vertex is 1.

To find k, the y-coordinate, we find $f(h)$. That is, we substitute 1 for x in the equation and solve for y.

$$k = f(1) = 3(1)^2 - 6(1) + 5 = 2$$

We now have $(h, k) = (1, 2)$, which is the vertex, and we can write the quadratic function in standard form:

$$f(x) = 3x^2 - 6x + 5$$

$$f(x) = 3(x - 1)^2 + 2$$

We plot the vertex on the coordinate system in Figure 2-4 and draw the axis of symmetry.

Step 2: Since $a = 3$, the parabola will open upward.

Step 3: The y-intercept is $f(0) = 3(0)^2 - 6(0) + 5 = 5$, so we plot the point $(0, 5)$.

Step 4: After plotting the vertex and the y-intercept and determining that the parabola opens upward, we can see that there will not be any x-intercepts.

Step 5: Complete the graph. Since there are no x-intercepts, we will plot one additional point to complete the graph. The function is symmetric about the axis of symmetry $x = 1$. Therefore, since the point $(0, 5)$ is on the graph, we can determine that the point $(2, 5)$ is also on the graph.

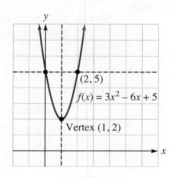

$(2, 5)$

$f(x) = 3x^2 - 6x + 5$

Vertex $(1, 2)$

FIGURE 2-4

Self Check 1   Write the function $f(x) = 2x^2 - 8x + 9$ in standard form and sketch its graph.

Now Try Exercise 25.

EXAMPLE 2   Graphing a Quadratic Function

Write the function $f(x) = -2x^2 - 5x + 3$ in standard form and then sketch the graph.

SOLUTION   We will apply the five steps outlined in the "Strategy for Graphing a Quadratic Function" box.

Step 1: Find the vertex (h, k), where $h = -\dfrac{b}{2a}$ and $k = f(h)$.

We note that $a = -2$ and $b = -5$. To find h, we substitute -2 in for a and -5 in for b.

$$h = -\frac{b}{2a} = -\frac{-5}{2(-2)} = -\frac{5}{4}$$

The x-coordinate of the vertex is $-\frac{5}{4}$.

To find k, the y-coordinate, we find $f(h)$. That is, we substitute $-\frac{5}{4}$ for x in the equation and solve for y.

$$k = f\left(\frac{-5}{4}\right) = -2\left(\frac{-5}{4}\right)^2 - 5\left(\frac{-5}{4}\right) + 3$$

$$= -2\left(\frac{25}{16}\right) + \frac{25}{4} + 3$$

$$= -\frac{25}{8} + \frac{50}{8} + \frac{24}{8}$$

$$= \frac{49}{8}$$

The vertex is the point $\left(-\frac{5}{4}, \frac{49}{8}\right)$. We can now write

$$f(x) = -2x^2 - 5x + 3 = -2\left(x + \frac{5}{4}\right)^2 + \frac{49}{8}.$$

We plot the vertex on the coordinate system in Figure 2-5(a) and draw the axis of symmetry.

Step 2: Since $a = -2$, the parabola will open downward.

Step 3: The y-intercept is $f(0) = -2(0)^2 - 5(0) + 3 = 3$, so we plot the point $(0, 3)$.

Step 4: After plotting the vertex and the y-intercept and determining that the parabola opens downward, we can see that there will be x-intercepts. We will solve the equation $y = f(x) = 0$.

$$0 = -2x^2 - 5x + 3$$

$$0 = 2x^2 + 5x - 3 \qquad \text{Multiply by } -1.$$

$$0 = (2x - 1)(x + 3) \qquad \text{Factor.}$$

$$2x - 1 = 0 \quad \text{or} \quad x + 3 = 0 \qquad \text{Set each factor equal to 0.}$$

$$x = \frac{1}{2} \qquad\qquad x = -3$$

Step 5: Complete the graph. Because of symmetry, we know that the point $\left(-\frac{5}{2}, 3\right)$ is on the graph. We plot this point on the coordinate system in Figure 2-5(a). We can now draw the graph of the function, as shown in Figure 2-5(b).

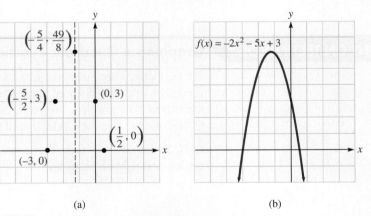

(a)   (b)

FIGURE 2-5

Self Check 2   Write the function $f(x) = -3x^2 + 7x - 2$ in standard form and draw the graph.

Now Try Exercise 27.

EXAMPLE 3   **Graphing a Quadratic Function**

For the function $f(x) = 3x^2 - 3x - 5$, find the vertex and all intercepts. Graph the function.

SOLUTION   Since the vertex is the point (h, k), we find those values as follows:

$$h = -\frac{b}{2a} = -\frac{-3}{2(3)} = \frac{1}{2}$$

$$k = f(h) = f\left(\frac{1}{2}\right) = 3\left(\frac{1}{2}\right)^2 - 3\left(\frac{1}{2}\right) - 5 = -\frac{23}{4}.$$

Thus, the vertex is the point $\left(\frac{1}{2}, -\frac{23}{4}\right)$.

The y-intercept is $f(0) = 3(0)^2 - 3(0) - 5 = -5$.

By plotting the vertex and the y-intercept [Figure 2-6(a)], we see that there will be x-intercepts. To find the x-intercepts, we use the quadratic formula.

$$x = \frac{-(-3) \pm \sqrt{(-3)^2 - 4(3)(-5)}}{2(3)} = \frac{3 \pm \sqrt{69}}{6}$$

We can now draw the graph of the function with the intercepts labeled [Figure 2-6(b)].

Caution

We cannot reduce the expression $\frac{3 \pm \sqrt{69}}{6}$ as shown below.

$$\frac{3 \pm \sqrt{69}}{6} = \frac{1 \pm \sqrt{69}}{2}$$

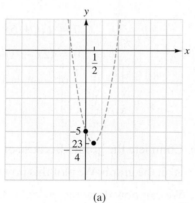

(a)

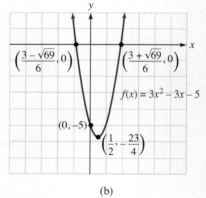

(b)

FIGURE 2-6

ACCENT ON TECHNOLOGY   **Graphing a Quadratic Function**

If exact values are not necessary, then a graphing calculator can be used to draw the graph and find decimal approximations for the values of the intercepts and the vertex. The graph is shown in the figure below. Press 2nd TRACE to access the CALC menu. Use the CALC menu to locate the vertex and the intercepts.

CALC 3:minimum will locate the vertex. To enter the bounds, enter values that you know are left and right of the vertex. Hit ENTER until the minimum is shown.

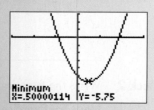

CALC 2:zero will locate the intercepts. Select left- and right-bound values that are obviously left and right of the zero, and hit ENTER until the zero is shown.

 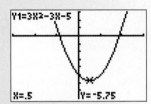

FIGURE 2-7

A graphing calculator can also be used to verify the values found algebraically. For this, we will use the CALC 1:value function.

First we will find the vertex. We can see that the values shown for the function are $\left(\frac{1}{2}, -\frac{23}{4}\right) = (.5, -5.75)$.

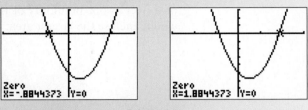 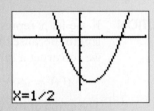 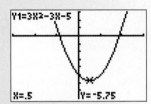

Now we will check the values of the intercepts. We can see that when we enter the exact value of the intercept we found, the point is the same as the intercept found by the calculator.

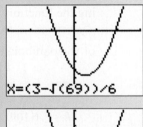 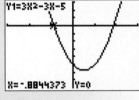

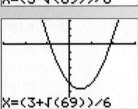 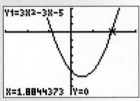

FIGURE 2-8

Self Check 3   Find the vertex and all intercepts of the function $f(x) = 3x^2 + 6x - 1$. Graph the function.

Now Try Exercise 51.

3. Solve Real-Life Problems Using Quadratic Functions

Now we will examine some of the ways a quadratic function can be used in applications.

EXAMPLE 4   Maximizing Area

The Montana Dude Rancher's Association has 400 feet of fencing to enclose a rectangular corral. To save money and fencing, the association intends to use the bank of a river as one boundary of the corral, as in Figure 2-9. Find the dimensions that would enclose the largest area.

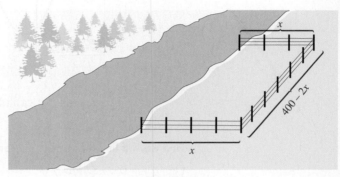

FIGURE 2-9

SOLUTION   If we let x represent the width of the fenced area, then $400 - 2x$ represents the length. Because the area A of a rectangle is the product of the length and the width, we construct a function for the area of the rectangle with width x:

$$A(x) = (400 - 2x)x = -2x^2 + 400x, \ 0 \le x \le 200$$

The graph of this equation is a parabola, and since the coefficient of $-2x^2$ is negative, the parabola opens downward. The vertex is its highest point, and the x-coordinate represents the width of the corral that will give the maximum area $A(x)$. Comparing the function $A(x) = -2x^2 + 400x$ to the general form of a quadratic function, $f(x) = ax^2 + bx + c$, we determine that $a = -2$, $b = 400$, and $c = 0$. The vertex of the parabola is the point $(h, k) = \left(-\frac{b}{2a}, A\left(-\frac{b}{2a}\right)\right)$.

$$h = -\frac{b}{2a} = -\frac{400}{2(-2)} = 100$$

$$k = A(100) = -2(100)^2 + 400(100) = 20{,}000$$

The vertex for this function is (100, 20,000), which tells us that the association's fence should run 100 feet out from the river, 200 feet parallel to the river, and 100 feet back to the river to attain the maximum possible area of 20,000 square feet.

Self Check 4   Find the dimensions yielding the largest area if the amount of fencing available is 600 feet.

Now Try Exercise 61.

To determine a selling price that will maximize revenue, manufacturers must consider the economic principle of supply and demand: Increasing the number of units manufactured decreases the price that can be charged for each unit.

EXAMPLE 5   Maximizing Revenue

A manufacturer of automobile airbags has determined that x units can be manufactured and sold each week for $\$(384 - 0.1x)$ each. Find the weekly production level that will maximize the revenue from sales, and find that revenue.

SOLUTION   We let $y = R(x)$ represent the revenue from sales. We can define **revenue** as the number of items sold times the cost per unit. For this application we have

$$y = R(x) = x(384 - 0.1x) = -0.1x^2 + 384x.$$

The graph of this equation is a parabola. Since the coefficient of $-0.1x^2$ is negative, the parabola opens downward and its vertex represents its highest point. The x-coordinate of the vertex is the production level that will maximize revenue, and the y-coordinate is that maximum revenue $R(x)$. We compare the equations $y = R(x) = -0.1x^2 + 384x$ and $y = ax^2 + bx + c$ to see that $a = -0.1$, $b = 384$, and $c = 0$. The vertex of the parabola is the point $(h, k) = \left(-\frac{b}{2a}, R\left(-\frac{b}{2a}\right)\right)$.

$$h = -\frac{b}{2a} = -\frac{384}{2(-0.1)} = 1{,}920$$

$$k = R(1{,}920) = -0.1(1{,}920)^2 + 384(1{,}920) = 368{,}640$$

The greatest possible revenue from sales is \$368,640, which is attained at a weekly production level of 1,920 units.

Self Check 5   What is the maximum revenue if the airbags can be manufactured and sold each week according to the function $\$(420 - 0.2x)$?

Now Try Exercise 79.

The position of an object at time t fired at an initial velocity of v_0 from initial height h_0 is given by the position function $s(t) = -gt^2 + v_0t + h_0$. The leading coefficient g is the force of gravity. It is negative because gravity pulls downward. If the units are measured in meters, then $g = 4.9$ meters per second per second; if the units are in feet, $g = 16$ feet per second per second. It is important to know how to interpret points on the graph of the function. Each ordered pair on the graph of the function is $(t, s(t)) = $ (time, height at that time).

EXAMPLE 6   **Finding Maximum Height Using Projectile Motion**

An object is launched upward at an initial velocity of 64 feet per second from a height of 80 feet. What is the maximum height the object attains and when does it reach that height? When does the object hit the ground?

SOLUTION   Since the velocity is given in feet per second, we have $g = 16$ feet per second per second. The initial velocity is $v_0 = 64$ and the initial height is $h_0 = 80$. We can form the position function $s(t) = -16t^2 + 64t + 80$. We know that the graph of this position function is a parabola opening downward; we need to find the vertex (h, k) to determine the maximum height and the time the object attains that height.

$$h = -\frac{b}{2a} = -\frac{64}{2(-16)} = -\frac{64}{-32} = 2$$

$$k = s(2) = -16(2)^2 + 64(2) + 80 = -16(4) + 128 + 80 = 144$$

The vertex of the graph of this function is (2, 144) shown in Figure 2-10. From this we can determine that the object will reach a maximum height of 144 feet in 2 seconds.

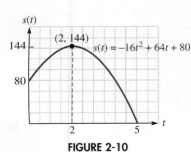

FIGURE 2-10

To determine when the projectile hits the ground, we must solve the quadratic equation $s(t) = -16t^2 + 64t + 80 = 0$.

$$-16t^2 + 64t + 80 = 0 \qquad \text{Set the function equal to 0.}$$

$$-16(t^2 - 4t - 5) = 0 \qquad \text{Factor out } -16.$$

$$t^2 - 4t - 5 = 0 \qquad \text{Divide both sides by } -16.$$

$$(t - 5)(t + 1) = 0 \qquad \text{Factor.}$$

$$t - 5 = 0 \quad \text{or} \quad t + 1 = 0$$

$$t = 5 \qquad\qquad t = -1$$

Since the value of t is a time, we reject the solution $t = -1$. The object hits the ground at $t = 5$ seconds.

Self Check 6   Find the maximum height of the object if the initial velocity is 96 feet per second.

Now Try Exercise 71.

EXAMPLE 7   **Writing an Equation of a Quadratic Function**

Write an equation in standard form for the quadratic function whose graph has the vertex $(3, 2)$ and passes through the point $(2, 0)$.

SOLUTION   The standard form of a quadratic function is $f(x) = a(x - h)^2 + k$. Given that the vertex is the point $(h, k) = (3, 2)$ we will substitute 3 for h and 2 for k in our standard-form equation. This gives us $f(x) = a(x - 3)^2 + 2$. Next, we find the value of a. Since we are given that the parabola passes through the point $(2, 0)$, we can replace $f(x)$ with 0 on the left side and substitute 2 in for x on the right side to then solve for a.

$$f(x) = a(x - 3)^2 + 2$$

$$0 = a(2 - 3)^2 + 2$$

$$0 = a(-1)^2 + 2$$

$$-2 = a$$

$$f(x) = -2(x - 3)^2 + 2$$

The graph is shown in Figure 2-11.

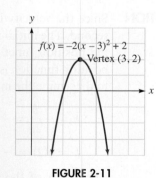

$f(x) = -2(x - 3)^2 + 2$

Vertex $(3, 2)$

FIGURE 2-11

Self Check 7   Write an equation in standard form for the quadratic function whose graph has the vertex $(-1, -2)$ and passes through the point $(0, 2)$.

Now Try Exercise 57.

Karla Neal

EXAMPLE 8   **Finding Hotel Rates**

A 500-room hotel is 80% filled when the nightly room rate is $160. Experience has shown that each $5 increase in the rate results in 10 fewer occupied rooms. Find the nightly rate that will maximize the nightly revenue.

SOLUTION   The nightly revenue is found by multiplying the rate per room by the number of rooms rented. At the rate of $160, the revenue is $R(x) = (\text{number of rooms rented})(\text{room rate}) = (0.80 \cdot 500)(160) = 400 \cdot 160 = 64{,}000$. We will let $x =$ the number of $5 increases. So the room rate will increase by $5x$ and the number of rooms rented will decrease by $10x$. The revenue function is $R(x) = (400 - 10x)(160 + 5x) = 64{,}000 + 400x - 5x^2$. This is a quadratic function; finding the vertex of the parabola will lead us to the maximum value.

$$h = -\frac{b}{2a} = -\frac{400}{2(-50)} = 4$$

$$R(4) = -50(4)^2 + 400(4) + 64{,}000 = 64{,}800$$

From this we can determine that with an increase of four increments of $5, or $20, to a room rate of $180, the revenue will increase to a maximum of $64,800. ∎

Self Check 8   A 500-room hotel is 90% filled when the nightly room rate is $170. Experience has shown that each $10 increase in the rate results in 20 fewer occupied rooms. Find the nightly rate that will maximize the nightly revenue.

Now Try Exercise 81.

4. Find a Quadratic Function That Fits Real-World Data Using Regression

Quadratic regression is a process used to find an equation of a quadratic function that is a "best fit" for a set of data. These models have many uses, such as in economics, where they are used for forecasting and cost-benefit analysis. In order to solve problems of this type, we use a graphing calculator which has built-in functions that analyze the entered data.

EXAMPLE 9   **Finding a Quadratic Function Using Regression**

A study investigated the efficiency of photosynthesis in an Antarctic species of grass. The table lists results for various temperatures. The temperature x is in degrees Celsius and the efficiency y is given as a percentage. The purpose of the research was to determine the temperature at which photosynthesis is most efficient. (Source: D. Brown and P. Rothery, *Models in Biology: Mathematics, Statistics and Computing*.)

X(°C)	−1.5	0	2.5	5	7	10	12	15	17	20	22	25	27	30
Y(%)	33	46	55	80	87	93	95	91	89	77	72	54	46	34

Use a graphing calculator to find the quadratic function that best fits this data.

SOLUTION   **Step 1:** Enter the data into LIST in a calculator. Press 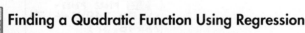 to access the list window and then enter the values (Figure 2-12). Enter the x values into L1 and the y values into L2.

FIGURE 2-12

Step 2: Set up a window for graphing. Use the values in the table to help you with this (Figure 2-13).

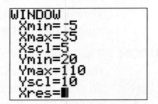

FIGURE 2-13

Step 3: Create a scatter plot and graph as shown in Figure 2-14.

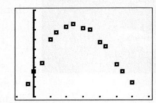

FIGURE 2-14

Step 4: Press STAT and go to the CALC menu. Scroll to 5:QuadReg and press ENTER. Then press ENTER once again. This gives the equation $y = -0.249x^2 + 6.767x + 46.371$ (rounding to three decimals), shown in Figure 2-15.

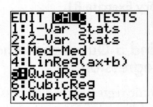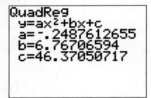

FIGURE 2-15

Step 5: Graph the quadratic function (Figure 2-16).

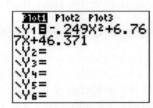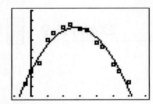

FIGURE 2-16

Step 6: Find the maximum value using the techniques shown in Example 7 in Section 1.3. At this point, you can turn off the plot graph.

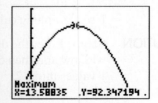

FIGURE 2-17

From Figure 2-17 we see that the vertex of the parabola is the point (13.588, 92.347). From this we determine that the temperature (x) that will produce the most efficient photosynthesis (y) is approximately 13.6°C.

Self Check 9   Using the quadratic regression model obtained, what approximate efficiency would correspond to a temperature of 11°C?

Now Try Exercise 85.

Self Check Answers   **1.** $f(x) = 2(x - 2)^2 + 1$ **2.** $f(x) = -3\left(x - \dfrac{7}{6}\right)^2 + \dfrac{25}{12}$

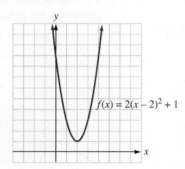

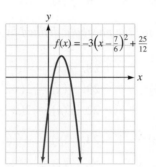

3. Vertex $= (-1, -4)$; y-intercept $= -1$; x-intercepts $= \dfrac{-3 \pm 2\sqrt{3}}{3}$

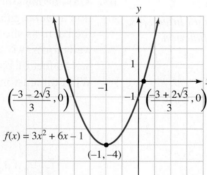

4. 150 ft by 300 ft   **5.** \$220,500   **6.** 224 ft   **7.** $f(x) = 4(x + 1)^2 - 2$
8. \$197.50   **9.** 90.7%

Exercises **2.1**

Getting Ready
You should be able to complete these vocabulary and concept statements before you proceed to the practice exercises.

Fill in the blanks.

1. A quadratic function is defined by the equation
_____ $(a \neq 0)$.

2. The standard form for the equation of a parabola is
_____ $(a \neq 0)$.

3. The graph of a quadratic function opens
_____ if $a < 0$.

4. The graph of a quadratic function opens
_____ if $a > 0$.

5. The point (h, k) is the _____ of the parabola.

6. The x-coordinate of the vertex of the parabolic graph of $y = ax^2 + bx + c$ is _____.

7. The y-coordinate of the vertex of the parabolic graph of $y = ax^2 + bx + c$ is _____.

8. The vertical line that intersects the parabola at its vertex is the _____.

9. If the parabola opens _____, the vertex will be a minimum point.

10. If the parabola opens downward, the vertex will be a _____ point.

Practice

Determine whether the graph of each quadratic function opens upward or downward. State whether a maximum or minimum value occurs at the vertex of the parabola.

11. $f(x) = \frac{1}{2}x^2 + 3$ **12.** $f(x) = 2x^2 - 3x$

13. $f(x) = 3(x + 1)^2 + 2$ **14.** $f(x) = -5(x - 1)^2 - 1$

15. $f(x) = -2x^2 + 5x - 1$ **16.** $f(x) = 2x^2 - 3x + 1$

Find the vertex (h, k) for the parabola defined by each quadratic function.

17. $f(x) = x^2 - 1$ **18.** $f(x) = -x^2 + 2$

19. $f(x) = -5(x - 6)^2$ **20.** $f(x) = 2(x + 1)^2$

21. $f(x) = 3(x + 1)^2 - 1$ **22.** $f(x) = 2(x - 3)^2 - 4$

23. $f(x) = -\frac{1}{2}(x - 3)^2 + 1$ **24.** $f(x) = -\frac{2}{3}(x + 4)^2 + 5$

25. $f(x) = x^2 - 4x + 4$ **26.** $f(x) = x^2 - 10x + 25$

27. $f(x) = -2x^2 + 12x - 17$ **28.** $f(x) = -x^2 + 9x - 2$

29. $f(x) = 3x^2 - 4x + 5$ **30.** $f(x) = -4x^2 + 3x + 4$

For each function: (a) Find the vertex. (b) Find all intercepts. (c) State the domain and range. (d) Graph the function.

31. $f(x) = (x - 3)^2 - 1$ **32.** $f(x) = (x + 3)^2 - 1$

33. $f(x) = -(x + 2)^2 + 1$ **34.** $f(x) = -(x - 3)^2 + 4$

35. $f(x) = 2(x - 4)^2 - 2$ **36.** $f(x) = 3(x - 1)^2 - 3$

37. $f(x) = -\frac{1}{2}(x - 1)^2 + 1$ **38.** $f(x) = \frac{2}{3}(x - 3)^2 + 2$

39. $f(x) = x^2 - x$ **40.** $f(x) = x^2 + 2x$

41. $f(x) = -\frac{1}{2}x^2 + 3$ **42.** $f(x) = \frac{1}{2}x^2 - 2$

43. $f(x) = x^2 - 4x - 5$ **44.** $f(x) = x^2 - 3x - 4$

45. $f(x) = -x^2 + 6x - 9$ **46.** $f(x) = -x^2 - 10x - 25$

47. $f(x) = 2x^2 - 12x + 10$ **48.** $f(x) = -3x^2 - 3x + 18$

49. $f(x) = -3x^2 + 2x + 4$ **50.** $f(x) = -4x^2 + 3x + 2$

51. $f(x) = x^2 - 4x + 1$ **52.** $f(x) = -x^2 - 4x + 1$

Write an equation in standard form of the quadratic function with the given vertex and containing the given point.

53. Vertex $(0, 0)$; point $(-2, 2)$

54. Vertex $(0, 0)$; point $(2, 1)$

55. Vertex $(0, -4)$; point $(-1, 1)$

56. Vertex $(0, 3)$; point $(1, 5)$

57. Vertex $(-1, -3)$; point $(2, 15)$

58. Vertex $(-2, 1)$; point $(-1, 4)$

59. Vertex $(2, 0)$; point $(1, -3)$

60. Vertex $(-3, 0)$; point $(-2, -2)$

Applications

61. Police investigations A police officer seals off the scene of an accident using a roll of yellow tape that is 300 feet long. What dimensions should be used to seal off the maximum rectangular area around the collision? What is that maximum area?

62. Maximizing area The rectangular garden shown has a width of x feet and a perimeter of 100 feet. Find x such that the area of the rectangle is maximized.

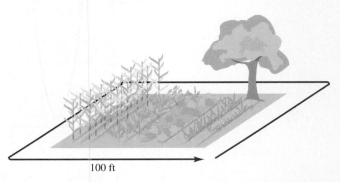

100 ft

63. Maximizing storage area A farmer wants to partition a rectangular feed-storage area in a corner of the barn, as shown in the illustration. The barn walls form two sides of the stall, and the farmer has 50 feet of partition for the remaining two sides. What dimensions will maximize the area?

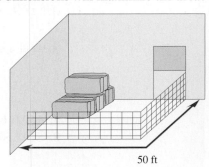

50 ft

64. Maximizing grazing area A rancher wishes to enclose a rectangular partitioned corral with 1,800 feet of fencing. (See the illustration.) What dimensions of the corral would enclose the largest possible area? What is that area?

65. **Sheet-metal fabrication** A 24-inch-wide sheet of metal is to be bent into a rectangular trough with the cross section shown in the illustration. Find the dimensions that will maximize the amount of water the trough can hold—that is, find the dimensions that will maximize the cross-sectional area.

66. **Landscape design** A gardener will use D feet of edging to border a rectangular plot of ground. Show that the maximum area will be enclosed if the rectangle is a square.

67. **Area** A picture has a height that is $\frac{4}{3}$ its width. It is to be enlarged to an area of 192 square inches. What will be the dimensions of the enlargement?

68. **Area** A garden measuring 12 meters by 16 meters is to have a sidewalk around it, increasing the total area to 285 square meters. What will be the width of the sidewalk?

69. **Dimensions** You are to make a square-bottomed box with no lid. The height is to be 3 inches and the volume 192 cubic inches. You will be taking a piece of cardboard, cutting 3-inch squares from each corner, scoring between the corners, and folding up the edges. What should be the dimensions of the cardboard?

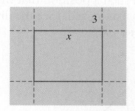

70. **Architecture** A parabolic arch has an equation of $x^2 + 20y - 400 = 0$, where x is measured in feet. Find the maximum height of the arch.

71. **Height of a basketball** The path of a basketball thrown from the free-throw line can be modeled by the quadratic function $f(x) = -0.06x^2 + 1.5x + 6$, where x is the horizontal distance (in feet) from the free-throw line and $f(x)$ is the height (in feet) of the ball. Find the maximum height of the basketball.

72. **Ballistics** A guided missile is projected from the origin of a coordinate system with the x-axis along the ground and the y-axis vertical. Its path, or **trajectory,** is given by the equation $y = 400x - 16x^2$. Find the missile's maximum height.

73. **Ballistics** A child throws a ball up a hill that makes an angle of 45° with the horizontal. The ball lands 100 feet up the hill. Its trajectory is a parabola with equation $y = -x^2 + ax$ for some number a. Find a.

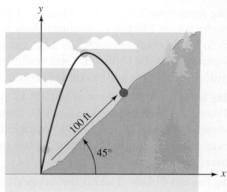

74. **Projectile motion** A ball is thrown straight up from the top of a building 144 feet tall with an initial velocity of 64 feet per second. The distance s (in feet) of the ball from the ground is given by $s = 144 + 64t - 16t^2$. Find the maximum height attained by the ball and the time it takes the ball to hit the ground.

75. **Projectile motion** An object is fired vertically upward from a stand 30 meters high with an initial speed of 75 meters per second. Use the formula $s(t) = -gt^2 + v_0t + h_0$ to find the maximum height the object attains, to the nearest tenth of a meter. When will the object reach the ground, to the nearest second? (Acceleration due to gravity on Earth is 9.8 meters per second per second.)

76. **Digital cameras** A company that produces and sells digital cameras has determined that the total weekly cost C of producing x digital cameras is given by the function $C(x) = 1.5x^2 - 144x + 5,856$. Determine the production level that minimizes the weekly cost for producing the digital cameras and find that minimum weekly cost.

77. **Webcams** A company that makes and sells webcams has found that the total weekly cost C of producing x webcams is given by the function $C(x) = 0.5x^2 - 160x + 26,250$. Find the production level that minimizes the weekly cost and find that minimum weekly cost.

78. **Baseball caps** A company that makes and sells baseball caps has found that the total monthy cost C of producing x caps is given by the function $C(x) = 0.2x^2 - 80x + 9,000$. Find the production level that will minimize the monthly cost and find that minimum cost.

79. Selling television sets A wholesaler of appliances finds that she can sell $2{,}400 - p$ television sets each week when the price is p dollars. What price will maximize revenue?

80. Finding mass-transit fares The Municipal Transit Authority serves 150,000 commuters daily when the fare is $1.80. Market research has determined that every penny decrease in the fare will result in 1,000 new riders. What fare will maximize revenue?

81. Finding hotel rates A 300-room hotel is two-thirds filled when the nightly room rate is $90. Experience has shown that each $5 increase in the rate results in 10 fewer occupied rooms. Find the nightly rate that will maximize income.

82. Selling concert tickets Tickets for a concert are cheaper when purchased in quantity. The first 100 tickets are priced at $10 each, but each additional block of 100 tickets purchased decreases the cost of each ticket by $0.50. How many blocks of tickets should be sold to maximize the revenue?

83. Numbers The sum of two numbers is 6, and the sum of the squares of those two numbers is as small as possible. What are the numbers?

84. Numbers Find two consecutive positive odd integers whose product is 143.

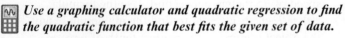 *Use a graphing calculator and quadratic regression to find the quadratic function that best fits the given set of data.*

85. Use the points $(-1, 6)$, $(0, -1)$, $(1, -3)$, $(2, -1.5)$, $(3, 5)$, and $(4, 10)$.

86. Use the points $(-3, 4)$, $(-1, 5)$, $(0, 7)$, $(2, 9)$, $(5, 7)$, and $(6, 5)$.

87. Alligators The length (in inches) and weight (in pounds) of alligators is shown in the table. Find the formula for the best-fit quadratic function and estimate the weight of an alligator that is 130 inches long. (Data from the *AP Statistics Teacher Handbook* from the College Board.)

Length	Weight	Length	Weight
94	130	86	83
74	51	88	70
147	640	72	61
58	28	74	54
86	80	61	44
94	110	90	106
63	33	89	84
86	90	68	39
69	36	76	42
72	38	114	197
128	366	90	102
85	84	78	57
82	80		

88. Alligators If an alligator weighs 125 pounds, what is its approximate length?

Discovery and Writing

89. Find the dimensions of the largest rectangle that can be inscribed in the right triangle ABC shown in the illustration.

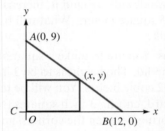

90. Point P lies in the first quadrant and on the line $x + y = 1$ in such a position that the area of triangle OPA is maximized. (See the illustration.) Find the coordinates of P.

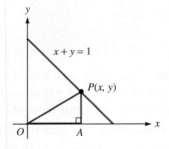

2.2 Polynomial Functions

In this section, we will learn to

1. Understand the characteristics of the graphs of polynomial functions.
2. Find the zeros of a polynomial function.
3. Apply the theory in the section and graph a polynomial function.
4. Determine the multiplicity of zeros and use them to factor a polynomial function.
5. Use the Intermediate Value Theorem to find an interval that contains a root of a polynomial function.

So far, we have discussed two types of polynomial functions—first-degree (or linear) functions and second-degree (or quadratic) functions. In this section, we will discuss polynomial functions of a higher degree.

Polynomial functions of a higher degree can be used to model the path of a roller coaster or to model the fluctuation of gasoline prices over the past few months. Goliath, a hypercoaster, opened a few years ago at Six Flags over Georgia. It climbs to a height of 200 feet and reaches speeds of nearly 70 mph. It has over 4,400 feet of steel track. Portions of Goliath's track can be modeled with a polynomial function.

Recall from Section 2.1 the definition of a polynomial function:

Polynomial Function   A **polynomial function of x with degree n** is a function of the form

$$P(x) = a_n x^n + a_{n-1} x^{n-1} + \cdots + a_1 x + a_0,$$

where $a_n, a_{n-1}, \ldots, a_1$, and a_0 are real numbers with $a_n \neq 0$ and n is a nonnegative integer.

In Chapter 1 we learned how to graph several basic functions. Some of these are polynomial functions. The **degree** of a polynomial function is the largest power of x that appears in the polynomial. Four basic polynomial functions which we have already studied are shown in the table:

Name	Function	Degree	Graph
Zero function	$f(x) = 0$	None	x-axis
Constant function	$f(x) = a_0$ $(a_0 \neq 0)$	0	Horizontal line, y-intercept $= a_0$ at $(0, a_0)$
Linear function	$f(x) = a_1 x + a_0$ $(a_1 \neq 0)$	1	Nonvertical line, slope of a_1, y-intercept $= a_0$ at $(0, a_0)$
Quadratic function	$f(x) = a_2 x^2 + a_1 x + a_0$ $(a_2 \neq 0)$	2	Parabola, opens upward when $a_2 > 0$ and downward when $a_2 < 0$

Now we turn our attention to graphing polynomials of a higher degree.

1. Understand the Characteristics of the Graphs of Polynomial Functions

Look at the graph in Figure 2-18. This is the graph of the polynomial function $f(x) = x^4 - x^3 - 12x^2 - 4x + 16$.

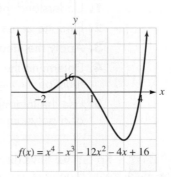

$$f(x) = x^4 - x^3 - 12x^2 - 4x + 16$$

FIGURE 2-18

If we look at the way this function is defined, it may seem that it would be impossible to graph it. In this section and the sections to follow, we are going to learn about the behavior of polynomials and how that can be used to draw graphs. Several characteristics will be covered: end behavior, zeros, and turning points or extrema.

End Behavior

As we look at the graph, we notice the **end behavior** of the function. By that, we mean what happens as x increases *without bound* and as x decreases *without bound* (i.e. $x \to \infty$ and $x \to -\infty$). We see that the function's graph rises on the left and rises on the right. We will learn that polynomials behave in certain predictable ways.

Zeros or x-Intercepts

Looking at the graph, we can note that the function has three **zeros.** A zero is a value of x for which $f(x) = 0$; those values of x are also the x-intercepts. We are going to learn how to find these values of x. Notice that the function touches the x-axis and turns at the zero of $x = -2$. At the zeros of $x = 1$ and $x = 2$ notice that the function crosses the x-axis. We will examine why that happens.

Turning Points or Extrema

The graph also has three **extrema.** Recall from Section 1.3 that an extremum is a maximum or minimum value that the function attains over an interval. We also call these **turning points.** We will not always be able to find the exact value of these extrema, but we will have a good idea of where they are. Finding the exact values of extrema often requires calculus techniques.

The graphs of two other polynomial functions are shown in Figure 2-19.

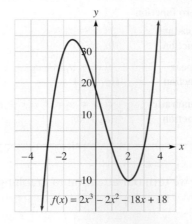

$$f(x) = 2x^3 - 2x^2 - 18x + 18$$

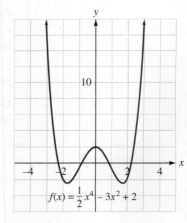

$$f(x) = \frac{1}{2}x^4 - 3x^2 + 2$$

FIGURE 2-19

Note that the graphs of polynomial functions are smooth and continuous curves. Because the graphs are smooth, there are no cusps or corners. Because they are continuous, there are no breaks or holes. The graph of a polynomial function can be drawn without lifting the pencil from the paper. (See Figure 2-20.)

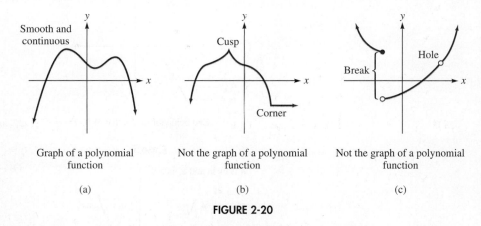

Graph of a polynomial function

(a)

Not the graph of a polynomial function

(b)

Not the graph of a polynomial function

(c)

FIGURE 2-20

A basic polynomial function called a **power function** is a polynomial function of the form $f(x) = ax^n$, where a is a nonzero real number and n is a nonnegative integer. Functions in the form $f(x) = x^n$ have simple graphs. Look at the graphs shown in Figure 2-21.

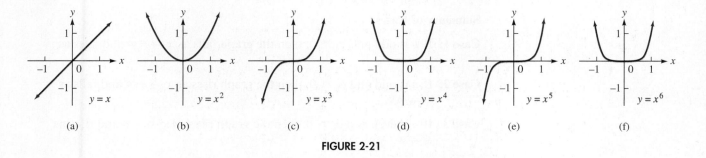

$y = x$ (a) $y = x^2$ (b) $y = x^3$ (c) $y = x^4$ (d) $y = x^5$ (e) $y = x^6$ (f)

FIGURE 2-21

Notice that when n is even, the function is an even function, and the graph has the same general shape as $f(x) = x^2$. The graph is symmetric with respect to the y-axis. When n is odd and greater than 1, the function is an odd function, and the graph has the same general shape as $f(x) = x^3$. The graph is symmetric with respect to the origin. The graphs are flatter at the origin and the functions increase more rapidly as n increases.

Now let's look at other polynomial functions and what characteristics their graphs will have. The end behavior of the graph of a polynomial function is the same as the end behavior of the graph of the term with the highest degree. The **leading coefficient** a_n and the **degree** n of a polynomial will be our starting point. These will determine the end behavior. This is summarized in the Leading-Coefficient Test.

Leading-Coefficient Test and End Behavior

For $f(x) = a_n x^n + a_{n-1} x^{n-1} + \cdots + a_1 x + a_0$ $(a_n \neq 0)$, as x increases without bound $(x \to \infty)$ or decreases without bound $(x \to -\infty)$, the function will eventually increase without bound $(f(x) \to \infty)$ or decrease without bound $(f(x) \to -\infty)$ as shown in Figure 2-22.

Case 1

n is odd and a_n is > 0

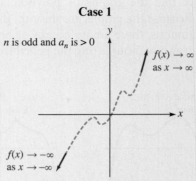

$f(x) \to \infty$
as $x \to \infty$

$f(x) \to -\infty$
as $x \to -\infty$

End behavior is like that of $f(x) = x^3$

Case 2

n is odd and a_n is < 0

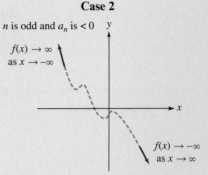

$f(x) \to \infty$
as $x \to -\infty$

$f(x) \to -\infty$
as $x \to \infty$

End behavior is like that of $f(x) = -x^3$

Case 3

n is even and a_n is > 0

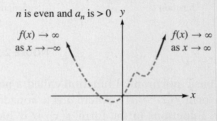

$f(x) \to \infty$
as $x \to -\infty$

$f(x) \to \infty$
as $x \to \infty$

End behavior is like that of $f(x) = x^2$

Case 4

n is even and a_n is < 0

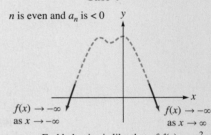

$f(x) \to -\infty$
as $x \to -\infty$

$f(x) \to -\infty$
as $x \to \infty$

End behavior is like that of $f(x) = -x^2$

FIGURE 2-22

Summary of Cases

Case 1: If n is odd and $a_n > 0$, then the graph falls as $x \to -\infty$ and rises as $x \to \infty$.

Case 2: If n is odd and $a_n < 0$, then the graph rises as $x \to -\infty$ and falls as $x \to \infty$.

Case 3: If n is even and $a_n > 0$, then the graph rises as $x \to -\infty$ and rises as $x \to \infty$.

Case 4: If n is even and $a_n < 0$, then the graph falls as $x \to -\infty$ and falls as $x \to \infty$.

EXAMPLE 1 **Using the Leading-Coefficient Test to Determine End Behavior**

Use the Leading-Coefficient Test to describe the end behavior of each function.

a. $f(x) = 3x^3 - 2x + 1$ **b.** $f(x) = 4x^4 - 2x^3 + 3x + 4$

c. $f(x) = -5x^6 - x^3 + 2$ **d.** $f(x) = -2x^5 + 3$

SOLUTION **a.** $f(x) = 3x^3 - 2x + 1$ has degree 3, which is odd, and $a_n = 3 > 0$. The end behavior of the function will be like that of $f(x) = x^3$, which means $f(x) \to -\infty$ as $x \to -\infty$ and $f(x) \to \infty$ as $x \to \infty$. The graph falls on the left and rises on the right. See Figure 2-22 (Case 1).

b. $f(x) = 4x^4 - 2x^3 + 3x + 4$ has degree 4, which is even, and $a_n = 4 > 0$. The end behavior of the function will be like that of $f(x) = x^2$, which means $f(x) \to \infty$ as $x \to -\infty$ and $f(x) \to \infty$ as $x \to \infty$. The graph rises on the left and rises on the right. See Figure 2-22 (Case 3).

c. $f(x) = -5x^6 - x^3 + 2$ has degree 6, which is even, and $a_n = -5 < 0$. The end behavior of the function will be like that of $f(x) = -x^2$, which means $f(x) \to -\infty$ as $x \to -\infty$ and $f(x) \to -\infty$ as $x \to \infty$. The graph falls on the left and falls on the right. See Figure 2-22 (Case 4).

d. $f(x) = -2x^5 + 3$ has degree 5, which is odd, and $a_n = -2 < 0$. The end behavior of the function will be like that of $f(x) = -x^3$, which means $f(x) \to \infty$ as $x \to -\infty$ and $f(x) \to -\infty$ as $x \to \infty$. The graph rises on the left and falls on the right. See Figure 2-22 (Case 2).   ∎

Self Check 1   Describe the end behavior of each polynomial.

a. $f(x) = 2x^7 - 3x^4$ **b.** $f(x) = -\dfrac{2}{3}x^8 + 5x^3 - 3$

Now Try Exercise 7.

ACCENT ON TECHNOLOGY   ## Graphing Polynomial Functions

We can use a graphing calculator to graph some of the functions in Example 1. Setting the proper window for the graph of a polynomial may take several tries, since the value of a polynomial function can become quite large very quickly, especially if the degree of the polynomial is greater than 3. We can use the end behavior to help us come up with a good view of the graph.

(a) Enter the function $f(x) = 3x^3 - 2x + 1$ and press ⁅ZOOM⁆ 4:ZDecimal. We can see that this window is a good window for this function. Notice that the end behavior is like the end behavior of $f(x) = x^3$.

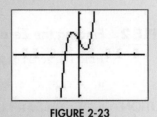

FIGURE 2-23

(b) Enter the function $f(x) = 4x^4 - 2x^3 + 3x + 4$. If we try to use the same window that we used in part (a), practically nothing shows up! Look at Figure 2-24(a). So we have to find a better window. Let's try to use the ⁅ZOOM⁆ 6:Standard window, Figure 2-24(b). This gives us a better view, but it's still not great. It appears that the function never crosses the x-axis. (We will study ways of being sure of this later in this chapter.) If we use the window shown in Figure 2-24(c), then we get a better view of the graph, Figure 2-24 (d).

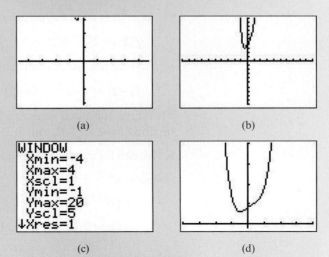

(a)   (b)

```
WINDOW
 Xmin=-4
 Xmax=4
 Xscl=1
 Ymin=-1
 Ymax=20
 Yscl=5
↓Xres=1
```

(c)   (d)

FIGURE 2-24

2. Find the Zeros of a Polynomial Function

Now that we know how to determine the end behavior of a function, our next step is to be able to locate the zeros. Finding the zeros of a polynomial function is very important in algebra because the zeros are the x-intercepts. In this section we will primarily look at polynomial functions that can easily be factored. However, not all polynomials can be factored with just factoring techniques; we will study those types of polynomials and how to find their zeros in later sections. Once we have found the zeros and we know the end behavior, we can begin to graph the function.

Here are two important facts related to the graph of a polynomial function:

Fact 1: The graph of a polynomial function of degree n will have at most n real zeros. (We will discuss this in more detail later in the chapter.)

Fact 2: The graph of a polynomial function of degree n will have at most $n - 1$ turning points (i.e., relative extrema).

Here are three important facts about a zero of a polynomial function $f(x)$. If $x = a$ is a zero of a polynomial function $f(x)$, then

Fact 1: $f(a) = 0$.

Fact 2: $x - a$ is a factor of the polynomial function $f(x)$.

Fact 3: $x = a$ is an x-intercept.

EXAMPLE 2 **Finding the Zeros of a Polynomial Function**

Find the zeros of each polynomial function.

a. $f(x) = x^3 - 4x$ **b.** $f(x) = x^4 - 5x^2 + 4$

SOLUTION **a.** We can factor the function as follows:

$$f(x) = x^3 - 4x$$

$$= x(x^2 - 4) \qquad \text{Factor out } x.$$

$$= x(x - 2)(x + 2) \qquad \text{Factor the difference of two squares.}$$

From this we can determine the values of x that are zeros of the function.

- x is a factor; therefore $x = 0$ is a zero, and we also note that

$$f(0) = 0^3 - 4(0) = 0.$$

- $(x - 2)$ is a factor; therefore $x = 2$ is a zero, and we also note that

$$f(2) = 2^3 - 4(2) = 8 - 8 = 0.$$

- $(x + 2)$ is a factor; therefore $x = -2$ is a zero, and we also note that

$$f(-2) = (-2)^3 - 4(-2) = -8 + 8 = 0.$$

Therefore, the three zeros, and hence the three x-intercepts, of the polynomial function are $x = 0, 2,$ and -2.

b. We can factor the function as follows:

$$f(x) = x^4 - 5x^2 + 4$$

$$= (x^2 - 4)(x^2 - 1) \qquad \text{Factor using } x^2.$$

$$= (x - 2)(x + 2)(x - 1)(x + 1) \qquad \text{Factor the difference of two squares.}$$

From this we can determine the values of x that are zeros of the function.

- $x - 2$ is a factor; therefore $x = 2$ is a zero, and we also note that

$$f(2) = 2^4 - 5(2)^2 + 4 = 16 - 20 + 4 = 0.$$

- $x + 2$ is a factor; therefore $x = -2$ is a zero, and we also note that

$$f(-2) = (-2)^4 - 5(-2)^2 + 4 = 16 - 20 + 4 = 0.$$

- $x - 1$ is a factor; therefore $x = 1$ is a zero, and we also note that

$$f(1) = (1)^4 - 5(1)^2 + 4 = 1 - 5 + 4 = 0.$$

- $x + 1$ is a factor; therefore $x = -1$ is a zero, and we also note that

$$f(-1) = (-1)^4 - 5(-1)^2 + 4 = 1 - 5 + 4 = 0.$$

Therefore, the four zeros, and hence the four x-intercepts, of the polynomial are $x = 1, -1, 2,$ and -2. ∎

Self Check 2   Find the zeros of each polynomial.
a. $f(x) = x^3 - x^2 - 2x$ **b.** $f(x) = -2x^4 + 2x^2$

Now Try Exercise 25(a).

EXAMPLE 3   **Finding a Polynomial, Given the Zeros**

Find a polynomial function with the given zeros. (There is no unique answer.)

a. $x = -2, -1, 0,$ and 3 **b.** $x = -\dfrac{2}{3}, 1,$ and 1

SOLUTION   **a.** Given a zero, we can form a factor of the polynomial function. If $x = -2$ is a zero, then $x + 2$ is a factor. Given the zeros $x = -2, -1, 0,$ and 3, we can form a polynomial function as shown:

$$f(x) = x(x + 2)(x + 1)(x - 3)$$
$$= (x^2 + 2x)(x^2 - 2x - 3)$$
$$= x^4 - 7x^2 - 6x$$

b. With the given zero $x = -\frac{2}{3}$, we can form the factor $x + \frac{2}{3}$. To obtain a factor containing no fractions, we choose to multiply $x + \frac{2}{3}$ by 3 and obtain $3x + 2$. Since $x = 1$ is a zero occurring twice, we can write the factors as $(x - 1)^2$. Now we form the polynomial.

$$f(x) = (3x + 2)(x - 1)^2$$
$$= (3x + 2)(x^2 - 2x + 1)$$
$$= 3x^3 - 4x^2 - x + 2$$ ∎

Self Check 3   Find a polynomial function with the given zeros. (There is no unique answer.)
a. $x = -1, 2,$ and 2 **b.** $x = \dfrac{1}{2}, \dfrac{2}{5}, -1,$ and 1

Now Try Exercise 37.

3. Apply the Theory in the Section and Graph a Polynomial Function

Now that we know how to determine the end behavior of a polynomial function and can identify zeros, we are prepared to graph a polynomial function. Remember that the graph of a polynomial is a smooth curve with no sharp points (cusps) or breaks. We also know that the degree n is the maximum number of zeros and that $n - 1$ is the maximum number of turns. We know that other than at the zeros, the value of a function will be either positive or negative. We will use a sign chart to determine where the graph is above and below the x-axis.

Graphing Polynomial Functions

Step 1: Find the zeros or x-intercepts of the polynomial function.

Step 2: Determine the end behavior of the graph.

Step 3: Make a sign chart to determine where the graph is above and below the x-axis.

The three steps given in the "Graphing Polynomial Functions" box will help us graph a polynomial function. Keep in mind that we can also determine the y-intercept and make use of any symmetries of the graph. These are skills we learned earlier. Also, we can complete a table of values and plot a few additional points, if necessary, to graph the function.

EXAMPLE 4   **Graphing a Polynomial Function Using Zeros, End Behavior, and a Sign Chart**

Draw the graph of each polynomial function using zeros, end behavior, and a sign chart.

a. $f(x) = x^3 - 4x$ **b.** $f(x) = x^4 - 5x^2 + 4$

SOLUTION   **a.** We will use the three steps outlined in "Graphing Polynomial Functions."

Step 1: Find the zeros or x-intercepts of the polynomial function.
In Example 2(a) we determined that the zeros for $f(x) = x^3 - 4x$ are $x = 0, 2$, and -2. Factoring the function, we have $f(x) = x(x - 2)(x + 2)$.

Step 2: Determine the end behavior of the graph.
Since the degree of the function is 3, and the leading coefficient is positive, the end behavior of the function is like that of the function $f(x) = x^3$. This means that $f(x) \to -\infty$ as $x \to -\infty$ and $f(x) \to \infty$ as $x \to \infty$. The graph falls on the left and rises on the right. We also know that the function can turn no more than two times, since the degree is three. So far we have as much of the graph as is shown in Figure 2-25.

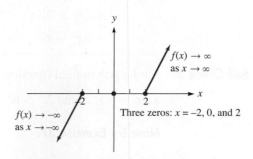

FIGURE 2-25

Step 3: Make a sign chart to determine where the graph is above and below the x-axis.
We will make a sign chart to determine the way the function behaves between the zeros, and then we can complete the sketch of the graph. The zeros divide

the x-axis into four intervals: $(-\infty, -2), (-2, 0), (0, 2),$ and $(2, \infty)$. If we select a value of x in each of those intervals and evaluate the function, we will know whether the function lies above or below the x-axis in that interval.

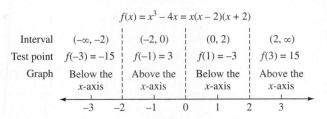

$$f(x) = x^3 - 4x = x(x - 2)(x + 2)$$

Interval	$(-\infty, -2)$	$(-2, 0)$	$(0, 2)$	$(2, \infty)$
Test point	$f(-3) = -15$	$f(-1) = 3$	$f(1) = -3$	$f(3) = 15$
Graph	Below the x-axis	Above the x-axis	Below the x-axis	Above the x-axis

FIGURE 2-26

We now have enough information to complete the graph, as shown in Figure 2-27. We are not able to use algebra to find the extrema, but we can approximate the shape.

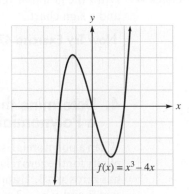

$$f(x) = x^3 - 4x$$

FIGURE 2-27

b. We will use the three steps outlined in the "Graphing Polynomial Functions" box.

Step 1: Find the zeros or x-intercepts of the polynomial function.
From Example 2(b), we have $f(x) = (x - 2)(x + 2)(x - 1)(x + 1)$. The zeros are $x = 1, -1, 2,$ and -2.

Step 2: Determine the end behavior of the graph.
The leading coefficient is positive and the degree is 4. This tells us that the end behavior is like that of $f(x) = x^2$. This means that $f(x) \to \infty$ as $x \to -\infty$ and $f(x) \to \infty$ as $x \to \infty$. The graph rises on the left and rises on the right.

Step 3: Make a sign chart to determine where the graph is above and below the x-axis.
Now we make a sign chart as follows to analyze the behavior of the function in the intervals created by the zeros:

$$f(x) = x^4 - 5x^2 + 4 = (x - 2)(x + 2)(x - 1)(x +1)$$

Interval	$(-\infty, -2)$	$(-2, -1)$	$(-1, 1)$	$(1, 2)$	$(2, \infty)$
Test point	$f(-3) = 40$	$f\left(-\frac{3}{2}\right) = -2.2$	$f(0) = 4$	$f\left(\frac{3}{2}\right) = -2.2$	$f(3) = 40$
Graph	Above the x-axis	Below the x-axis	Above the x-axis	Below the x-axis	Above the x-axis

FIGURE 2-28

From the sign chart we are now able to draw the graph.

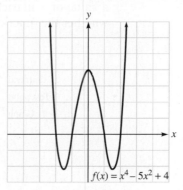

$$f(x) = x^4 - 5x^2 + 4$$

FIGURE 2-29

Self Check 4   Draw the graph of the polynomial $f(x) = x^4 - x^3 - 2x^2$ using zeros, end behavior, and a sign chart.

Now Try Exercise 47.

4. Determine the Multiplicity of Zeros and Use Them to Factor a Polynomial Function

When a polynomial is factored, the number of times a factor occurs gives us the **multiplicity** for the zero related to that factor. For example, $f(x) = (x - 3)(x - 1)^2(x + 2)$ has zeros $x = 3$, 1, and -2. The factors $x - 3$ and $x + 2$ each occur once, so the zeros $x = 3$ and $x = -2$ each have multiplicity one. The factor $x - 1$ occurs twice, so the zero $x = 1$ has multiplicity two.

The multiplicity of a zero affects the behavior of the function at that zero. Look at the graph of $f(x) = (x - 3)(x - 1)^2(x + 2)$ in Figure 2-30.

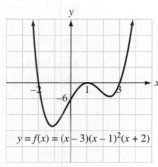

$$y = f(x) = (x - 3)(x - 1)^2(x + 2)$$

FIGURE 2-30

The graph crosses the x-axis where the multiplicity is one and touches the x-axis and turns where the multiplicity is two. The multiplicity of a zero determines whether the graph touches the x-axis and turns or crosses the x-axis.

Multiplicity   If a is a zero with an **odd** multiplicity, the graph **crosses** the x-axis at $x = a$.

If a is a zero with an **even** multiplicity, the graph **touches** the x-axis and turns at $x = a$.

EXAMPLE 5   **Finding a Polynomial Function, Given the Zeros**

Find a polynomial function that has the given zeros. (There is no unique answer.) Do not multiply out.

a. $-2, 0, 1$ (multiplicity three), and 2   **b.** $-3, 1, 3,$ and 4   **c.** $-\dfrac{2}{3}, 3,$ and $\dfrac{3}{5}$

SOLUTION   We will use the given zeros and form factors of the polynomial function.

a. If $x = -2$ is a zero, then $x + 2$ is a factor. If $x = 0$ is a zero then x is a factor. If $x = 1$ is a zero with multiplicity three, then $x - 1$ is a factor three times, and we write $(x - 1)^3$. If $x = 2$ is a zero, then $x - 2$ is a factor. One polynomial that has these zeros is $f(x) = x(x + 2)(x - 1)^3(x - 2)$.

b. Each zero has multiplicity one, so we can form the polynomial
$f(x) = (x + 3)(x - 1)(x - 3)(x - 4)$.

c. Taking $x = -\frac{2}{3}$, we form the factor $x + \frac{2}{3}$, but we also note that we can write the factor as $3x + 2$, which is a form that is more familiar. Likewise, for the zero $x = \frac{3}{5}$, we can write the factor as $x - \frac{3}{5}$ or the more common form $5x - 3$. We now form the polynomial $f(x) = (3x + 2)(x - 3)(5x - 3)$.

Comment

Because any constant multiple of the polynomial functions formed in Example 5 would have exactly the same zeros, there is not a unique solution.

Self Check 5   Find a polynomial function that has the given zeros. (There is no unique answer.)

a. $x = -3, 0$ (multiplicity two), 2, and 3   **b.** $x = -\dfrac{1}{2}, \dfrac{3}{4},$ and 5

Now Try Exercise 33.

EXAMPLE 6   **Using a Graph and Multiplicity to Factor a Polynomial Function**

The graph of $f(x) = x^4 - x^3 - 12x^2 - 4x + 16$ is shown. Use the graph to determine the zeros of the polynomial function and factor the function.

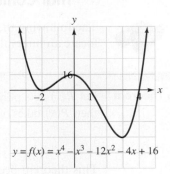

$y = f(x) = x^4 - x^3 - 12x^2 - 4x + 16$

FIGURE 2-31

SOLUTION   The graph seems to indicate that the zeros are $x = -2, 1,$ and 4. If these are the zeros, then $f(-2) = 0$, $f(1) = 0$, and $f(4) = 0$.

$$f(-2) = (-2)^4 - (-2)^3 - 12(-2)^2 - 4(-2) + 16$$

$$= 16 + 8 - 48 + 8 + 16$$

$$= 0$$

$$f(1) = (1)^4 - (1)^3 - 12(1)^2 - 4(1) + 16$$

$$= 1 - 1 - 12 - 4 + 16$$

$$= 0$$

$$f(4) = (4)^4 - (4)^3 - 12(4)^2 - 4(4) + 16$$
$$= 256 - 64 - 192 - 16 + 16$$
$$= 0$$

We note that the degree of the polynomial function is 4 and there can be no more than four zeros. Since the graph touches the x-axis and turns at $x = -2$, that factor will have even multiplicity. The factor $x + 2$ occurs twice. The graph crosses the x-axis at $x = 1$ and $x = 4$, so the zeros $x = 1$ and $x = 4$ each have multiplicity one, and the factors $x - 1$ and $x - 4$ each occur once. We can now factor the polynomial: $f(x) = (x + 2)^2(x - 1)(x - 4)$.

■

Self Check 6   Use the graph in Figure 2-32 to factor the polynomial
$f(x) = x^5 + 5x^4 + 3x^3 - 9x^2$.

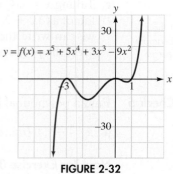

FIGURE 2-32

Now Try Exercise 39.

5. Use the Intermediate Value Theorem to Find an Interval That Contains a Root of a Polynomial Function

The following theorem is called the **Intermediate Value Theorem**. It can be used to locate an interval that contains a zero.

Intermediate Value Theorem   Let $P(x)$ be a polynomial function with real coefficients. If $P(a) \neq P(b)$ for $a < b$, then $P(x)$ takes on all values between $P(a)$ and $P(b)$ on the closed interval $[a, b]$.

This theorem becomes clear when we consider the graph of the polynomial function $y = P(x)$, shown in Figure 2-33.

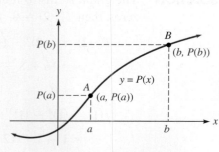

FIGURE 2-33

We have seen that graphs of polynomials are *continuous* curves. There are no holes or gaps in the graph. If $P(a) \neq P(b)$, then the continuous curve joining the points $A(a, P(a))$ and $B(b, P(b))$ must take on all values between $P(a)$ and $P(b)$ in the interval $[a, b]$, because there are no gaps in the curve. If $P(a)$ and $P(b)$ have opposite signs, then there is at least one number r in the interval (a, b) for which $P(r) = 0$.

This number r is a zero of $P(x)$, and a root of the equation $P(x) = 0$.

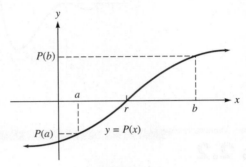

FIGURE 2-34

EXAMPLE 7   **Using the Intermediate Value Theorem**

Show that the equation $3x^3 + 2x^2 - x + 1 = 0$ has at least one root in the interval $[-2, -1]$.

SOLUTION   For $P(x) = 3x^3 + 2x^2 - x + 1$, we find that

$$P(-2) = 3(-2)^3 + 2(-2)^2 - (-2) + 1 = -13$$

and

$$P(-1) = 3(-1)^3 + 2(-1)^2 - (-1) + 1 = 1.$$

Since $P(-2) = -13$ and $P(-1) = 1$, by the Intermediate Value Theorem, $P(x)$ must take on every value between -13 and 1 over the interval $[-2, -1]$. Therefore, there is at least one value of x such that $-2 < x < -1$ and $P(x) = 0$.

ACCENT ON TECHNOLOGY   **Intermediate Value Theorem**

If you have a calculator, draw the graph of $P(x) = 3x^3 + 2x^2 - x + 1$. Use the `TRACE` key to evaluate the function at the two values, or use TABLE to verify that there is a root in the interval $[-2, -1]$.

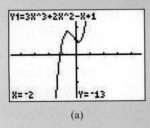

(a)

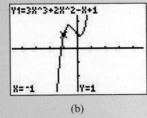

(b)

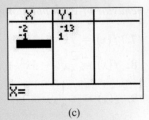

(c)

FIGURE 2-35

Self Check 7   Show that the equation $-2x^4 + 3x^2 + 4 = 0$ has at least one root in each interval. **a** $[-2, -1]$ **b.** $[1, 2]$

Now Try Exercise 67.

Self Check Answers **1. a.** The graph falls on the left and rises on the right.   **b.** The graph falls on the left and falls on the right.   **2. a.** $x = 0, 2,$ and -1 **b.** $x = 0, 1,$ and -1
3. a. $f(x) = x^3 - 3x^2 + 4$ **b.** $f(x) = 10x^4 - 9x^3 - 8x^2 + 9x - 2$
4.

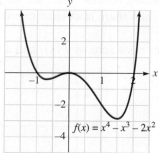

5. a. $f(x) = x^2(x + 3)(x - 2)(x - 3)$
b. $f(x) = (2x + 1)(4x - 3)(x - 5)$
6. $f(x) = x^2(x - 1)(x + 3)^2$
7. a. $P(-2) = -16$ and $P(-1) = 5$
b. $P(1) = 5$ and $P(2) = -16$

Exercises **2.2**

Getting Ready

You should be able to complete these vocabulary and concept statements before you proceed to the practice exercises.

Fill in the blanks.

1. If the degree of a polynomial is odd, the end behavior will be like that of _____.

2. If the degree of a polynomial is even, the end behavior will be like that of _____.

3. A polynomial of degree n can have at most _____ zeros.

4. A polynomial of degree n can have at most _____ turning points.

5. If $f(a) = 0$ for the polynomial $f(x)$, then _____ is a factor of the polynomial.

6. The degree of a polynomial is the _____ power of x that appears in the polynomial.

Practice

Use the Leading-Coefficient Test to determine the end behavior of each polynomial.

7. $f(x) = 7x^4 - 2x^2 + 1$

8. $f(x) = \dfrac{2}{3}x^6 + 3x^3 + 2x$

9. $g(x) = -\dfrac{1}{2}x^5 + 3x^4 + 2x^2 - 4$

10. $g(x) = -3x^7 + 2x^4 - 5x + 2$

11. $h(x) = -3x^4 - 5x - 1$

12. $h(x) = -x^2 + 3x + 2$

13. $s(x) = \sqrt{5}x^7 + 10x^3 - 2x$

14. $s(x) = 4x^9 - 7x^2 + 5x - 12$

Match the graph with the polynomial that would produce the graph.

15. $f(x) = (x - a)(x - b)^2(x - c)$

16. $f(x) = -(x - a)(x - b)(x - c)^3$

17. $g(x) = (x - a)^2(x - b)$

18. $g(x) = -(x - a)^2(x - b)(x - c)$

(a) 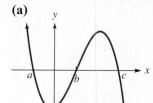   **(b)**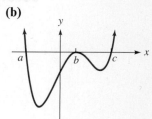

(c) 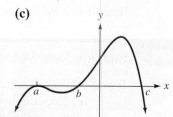   **(d)**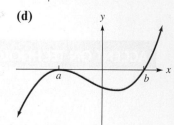

19. The graph of a polynomial function $f(x)$ is shown in the figure.

(a) What is the least degree of $f(x)$ possible?

(b) Is the leading coefficient of $f(x)$ positive or negative?

(c) What are the zeros of $f(x)$?

(d) Give a possible factored form of $f(x)$.

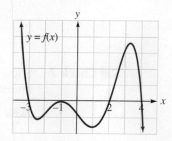

20. Repeat the questions from Exercise 19 for the graph of the polynomial function $g(x)$ shown in the figure.

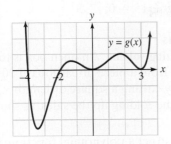

For Exercises 21–32, (a) Find the zeros of each polynomial function and (b) state the multiplicity of each.
(c) For each zero, state whether the graph touches the x-axis and turns or crosses the x-axis.

21. $f(x) = 4x^2 - 25$

22. $f(x) = 64 - 9x^2$

23. $g(x) = 2x^2 + 7x - 15$

24. $g(x) = 6x^2 - x - 2$

25. $h(x) = x^4 + 2x^3 - 3x^2$

26. $h(x) = x^3 - 8x^2 + 16x$

27. $s(x) = \frac{1}{2}x^4 + x^2 - \frac{15}{2}$

28. $s(x) = \frac{2}{3}x^4 - \frac{2}{3}x^2 - \frac{4}{3}$

29. $f(x) = 3x^2(x + 4)^2(x - 5)$

30. $f(x) = 2x(x - 3)^2(x + 1)^2$

31. $g(x) = (2x - 5)(x + 3)(x - 1)^2$

32. $g(x) = (3x + 1)^3(x + 1)^2$

Find a polynomial function that has the given zeros. (There is no unique answer.) Do not multiply out.

33. $x = -1, 0$ (multiplicity 3), 2, and 4

34. $x = -3, -2$ (multiplicity 2), 0, and 3

35. $x = -5$ (multiplicity 2), -3, and -1

36. $x = -3, -1$ (multiplicity 4), 2, and 10

37. $x = -3, -\frac{3}{2}$, and 4   **38.** $x = -\frac{5}{3}, \frac{1}{2}$, and 6

Use the given graph and the multiplicity to factor the given polynomial function.

39. $f(x) = 3x^4 - 9x^3 - 9x^2 + 33x - 18$

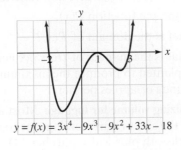

$y = f(x) = 3x^4 - 9x^3 - 9x^2 + 33x - 18$

40. $f(x) = 2x^4 - 12x^3 + 8x^2 + 48x - 64$

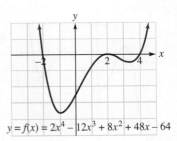

$y = f(x) = 2x^4 - 12x^3 + 8x^2 + 48x - 64$

41. $g(x) = -3x^4 - x^3 + 10x^2$

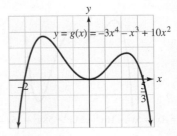

$y = g(x) = -3x^4 - x^3 + 10x^2$

42. $g(x) = -2x^4 + 5x^3 + 4x^2 - 12x$

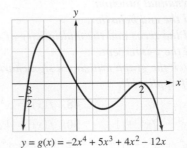

$y = g(x) = -2x^4 + 5x^3 + 4x^2 - 12x$

43. $h(x) = x^5 - 12x^3 - 16x^2$

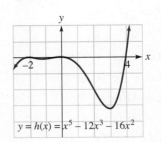

$y = h(x) = x^5 - 12x^3 - 16x^2$

44. $h(x) = 2x^5 + 4x^4 - 22x^3 - 24x^2 + 72x$

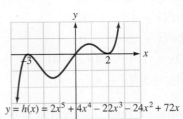

$y = h(x) = 2x^5 + 4x^4 - 22x^3 - 24x^2 + 72x$

45. $f(x) = 4x^5 - 9x^4 - 5x^3 + 15x^2 + x - 6$

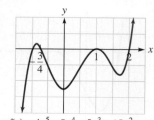

$y = f(x) = 4x^5 - 9x^4 - 5x^3 + 15x^2 + x - 6$

46. $f(x) = 5x^5 - 16x^4 - 31x^3 + 58x^2 + 16x - 32$

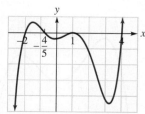

$y = f(x) = 5x^5 - 16x^4 - 31x^3 + 58x^2 + 16x - 32$

For each polynomial function,
(a) Find the end behavior.
(b) Find the zeros and the multiplicity of each zero.
(c) Find the y-intercept.
(d) Sketch the graph utilizing a sign chart and symmetry if applicable.

47. $f(x) = x^3 - x$ **48.** $f(x) = x^3 + x^2$

49. $f(x) = x^4 - 2x^2 + 1$ **50.** $f(x) = 4x^4 - x^2$

51. $g(x) = x^4 + x^3 - 6x^2$ **52.** $g(x) = x^4 - x^3 - 6x^2$

53. $g(x) = -3x^4 + 3x^3$ **54.** $g(x) = -2x^6 + 4x^5$

55. $g(x) = 6x^3 - 4x^2 - 10x$

56. $g(x) = 6x^3 - 9x^2 - 6x$

57. $h(x) = -3x^3 + 5x^2 + 12x$

58. $h(x) = -8x^3 + 6x^2 + 2x$

59. $h(x) = 3(x + 1)^2(x^2 - 9)$

60. $h(x) = 2(x + 2)^2(x^2 - 4)$

61. $h(x) = -x^2(x - 4)^3(x + 1)$

62. $h(x) = -x^3(x - 2)^2(x + 2)$

63. $f(x) = 2x^4(x - 1)(x + 3)^2$

64. $f(x) = 2x^5(x + 1)^2(x - 3)$

65. $f(x) = (2x - 1)^2(x - 4)(x + 4)$

66. $f(x) = (3x - 5)^3(x - 4)(x + 2)$

Use the Intermediate Value Theorem to show that each equation has at least one real root in the given interval.

67. $2x^2 + x - 3 = 0; [-2, -1]$

68. $2x^3 + 17x^2 + 31x - 20 = 0; [-1, 2]$

69. $3x^3 - 11x^2 - 14x = 0; [4, 5]$

70. $2x^3 - 3x^2 + 2x - 3 = 0; [1, 2]$

71. $x^4 - 8x^2 + 15 = 0; [1, 2]$

72. $x^4 - 8x^2 + 15 = 0; [2, 3]$

Applications

73. Maximizing volume An open box is to be constructed from a piece of cardboard 20 inches by 24 inches by cutting a square of length x from each corner and folding up the sides, as shown in the figure.

(a) Construct a polynomial function $V(x)$ that expresses the volume of the constructed box as a function of x.

(b) Use what you have learned about graphing polynomials to sketch the graph of $V(x)$.

(c) Using your graph as a guide, state the domain of this function as it relates to the problem.

(d) Use a calculator to graph the function and estimate the value of x that gives the maximum volume; estimate that maximum volume.

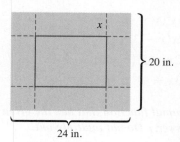

74. Maximizing volume Repeat Exercise 73 using a piece of cardboard with dimensions of 30 inches by 36 inches.

75. Maximizing production If 270 apple trees are planted per acre, the production per tree is 840 pounds. For every tree over 270 planted per acre, the production per tree decreases to $840 - 0.1x^2$ pounds per tree, where x is the number of trees over 270.

(a) Write a production function $P(x)$ for the number of pounds per acre as a function of x.

(b) Graph the function $P(x)$.

(c) What is the domain of $P(x)$?

(d) What is the number of trees per acre that should be planted to produce the maximum number of pounds of apples per acre?

76. Maximizing volume for luggage Most airlines restrict the size of carry-on luggage to a total 45 inches (length plus width plus height). The height of a piece of luggage is to be 4 inches less than the width.

(a) Write a function for the volume V as a function of x.

(b) Graph the function.

(c) What is the domain of the function?

(d) What dimensions will create a piece of luggage with the maximum volume?

77. Online DVD sales A web-based company produces and sells DVDs. The monthly profit $P(x)$ in hundreds of dollars can be modeled by the polynomial function $P(x) = 10x^3 - 100x^2 + 210x$, where $1 \leq x \leq 12$ and x represents the month of the year ($x = 1$ corresponds to January).

(a) What is the profit of the company in December?

(b) Identify the month(s) when the company's profit is $0.

78. Roller coaster A portion of a roller coaster's tracks can be modeled by the polynomial function $f(x) = 0.0001(x^3 - 600x^2 + 90,000x)$, where $0 \leq x \leq 400$. $f(x)$ represents the height of the roller coaster in feet and x represents the horizontal distance in feet.

(a) Find the height of the roller coaster when $x = 100$.

(b) Find the value(s) of x on the interval $[0, 400]$ at which the height of the roller coaster is 0 feet.

79. Containers A box has a length of 16 inches, a width of 10 inches, and a height of between 4 inches and 8 inches. Can it have a volume of 1,000 cubic inches? *Hint:* Model the volume of the box with a function of h and use the Intermediate Value Theorem.

80. Lollipops If a candy company makes spherical lollipops with radii between 1 and 4 centimeters, use the Intermediate Value Theorem to determine whether a lollipop can be made with a volume of 200 cubic centimeters. *Hint:* The volume formula for a sphere is $V = \frac{4}{3}\pi r^3$.

Discovery and Writing

81. Is it possible for a polynomial to have no x-intercept? Explain and give an example.

82. Is it possible for a polynomial to have no y-intercept? Explain.

83. Explain why a polynomial of odd degree must have at least one zero.

84. Explain why the polynomial function $f(x) = -ax^4 + 3x^3 + 2x + 8$, where a is a positive real number, must have at least one positive and one negative zero.

2.3 Rational Zeros of Polynomial Functions

In this section, we will learn to

1. Divide polynomials using the Division Algorithm.
2. Evaluate and factor polynomials using the Remainder and Factor Theorems.
3. Determine all rational zeros of a polynomial function using the Rational-Zeros Theorem.
4. Use Descartes' Rule of Signs to estimate the number of positive and negative zeros of a polynomial function.
5. Solve real-world applications that utilize polynomial functions.

The cheetah is the fastest land animal and is known for its speed and stealth. Cheetahs can run at speeds between 70 mph and 75 mph. They have the ability to accelerate from a standstill to 68 mph in three seconds—faster than most sports cars.

Although we will never run as fast as a cheetah, we do live in a society that wants convenience and speed. We demand fast service, fast food, fast communication, fast technology, and fast vehicles.

In this section, we will use synthetic division, a fast way to divide certain polynomials. Synthetic division can help us find the roots of many polynomial equations.

In Section 2.2 we examined the graphs of polynomial functions and we found zeros of the polynomial functions by factoring. For example, to find the zeros of the third-degree polynomial function $P(x) = x^3 - 3x^2 + 2x$, we proceed as follows:

$$x^3 - 3x^2 + 2x = 0$$

$$x(x^2 - 3x + 2) = 0 \qquad \text{Factor out } x.$$

$$x(x - 1)(x - 2) = 0 \qquad \text{Factor } x^2 - 3x + 2.$$

$$x = 0 \quad \text{or} \quad x - 1 = 0 \quad \text{or} \quad x - 2 = 0 \qquad \text{Set each factor equal to 0.}$$

$$x = 1 \qquad\qquad x = 2$$

The zeros are $x = 0$, 1, and 2.

Of course we know that not all polynomial functions can be factored by conventional factoring techniques. In fact, there is no general formula for finding zeros of polynomial functions of degree 5 or higher. There are some tools, however, that we can use to help us find zeros of polynomial functions. In this section we will study methods of finding rational zeros of polynomial functions. In Section 2.4, we will find real and complex zeros of polynomial functions.

1. Divide Polynomials Using the Division Algorithm

In online Appendix A.4, division of polynomials using long division is discussed. For example, we can divide $P(x) = 3x^3 + 6x + 10$, the **dividend**, by $d(x) = x - 1$, the **divisor**, as shown:

$$
\begin{array}{r}
3x^2 + 3x\ + 9 \\
x - 1\overline{)3x^3 \qquad\quad + 6x + 10} \\
\underline{3x^3 - 3x^2} \\
3x^2 + 6x \\
\underline{3x^2 - 3x} \\
9x + 10 \\
\underline{9x - \ 9} \\
19
\end{array}
$$

From this division we can write $\dfrac{3x^3 + 6x + 10}{x - 1} = 3x^2 + 3x + 9 + \dfrac{19}{x - 1}$ or $3x^3 + 6x + 10 = (x - 1)(3x^3 + 3x + 9) + 19$. This is summarized in the **Division Algorithm.**

Division Algorithm If $P(x)$ and $d(x)$ are polynomials, with $d(x) \neq 0$, there exist unique polynomials $q(x)$ and $r(x)$ such that

$$P(x) = d(x) \cdot q(x) + r(x),$$

where $r(x) = 0$ or is of lesser degree than $d(x)$.

$P(x)$ is the **dividend**; $d(x)$ is the **divisor**; $q(x)$ is the **quotient**; and $r(x)$ is the **remainder**.

Also in online Appendix A.4, **synthetic division** is shown to be an easy way to divide higher-degree polynomials by binomials of the form $x - c$. This will be a very useful tool as we discover techniques that allow us to find the rational zeros of a polynomial function.

Comment If you are not very familiar with synthetic division, you should review it before proceeding in this section. See Example 11 in online Appendix A.4.

2. Evaluate and Factor Polynomials Using the Remainder and Factor Theorems

If a polynomial $P(x)$ is divided by a polynomial in the form $x - c$, there is a relationship between the remainder and the value $P(c)$. In Example 11 in online Appendix A.4, we let $P(x) = 3x^4 - 8x^3 + 10x + 3$ and divided by $x - 2$. Note that $c = 2$. Our remainder when we performed the synthetic division was 7. If we evaluate $P(x)$ at an x value of 2, we will see an important relationship: $P(2) = 3(2)^4 - 8(2)^3 + 10(2) + 3 = 7$. So from this we see that the remainder 7 is the same as $P(2)$. This demonstrates what is known as the **Remainder Theorem**.

Remainder Theorem　　If $P(x)$ is a polynomial, c is any number, and $P(x)$ is divided by $x - c$, the remainder is $P(c)$.

EXAMPLE 1　**Using the Remainder Theorem**

Find the remainder that will occur when $P(x) = 2x^4 - 10x^3 + 17x^2 - 14x - 3$ is divided by

a. $x - 3$　　**b.** $x + 4$

SOLUTION　We will use the Remainder Theorem on both parts.

a. By the Remainder Theorem, the remainder will be $P(3)$.

$$P(x) = 2x^4 - 10x^3 + 17x^2 - 14x - 3$$

$$P(3) = 2(3)^4 - 10(3)^3 + 17(3)^2 - 14(3) - 3 \qquad \text{Substitute 3 for } x.$$

$$= 0$$

Comment

We can find this value without having to perform this tedious calculation by hand or using a calculator. We can use synthetic division, which makes the calculation quite simple and fast.

The remainder will be 0.

$$
\begin{array}{r|rrrrr}
3 & 2 & -10 & 17 & -14 & -3 \\
  &   & 6 & -12 & 15 & 3 \\
\hline
  & 2 & -4 & 5 & 1 & 0
\end{array}
$$

By using synthetic division, we have quickly evaluated $P(3) = 0$.

b. By the Remainder Theorem, the remainder will be $P(-4)$.

$$P(x) = 2x^4 - 10x^3 + 17x^2 - 14x - 3$$

$$P(-4) = 2(-4)^4 - 10(-4)^3 + 17(-4)^2 - 14(-4) - 3 \qquad \text{Substitute } -4 \text{ for } x.$$

$$= 1{,}477$$

We can also use synthetic division to find the remainder if we prefer. ∎

Self Check 1　Find the remainder when $P(x)$ is divided by $x - 2$.

Now Try Exercise 23.

In Section 2.2, we used the fact that if $x = a$ is a zero of the polynomial $P(x)$, then $P(a) = 0$ and $x - a$ is a factor of the polynomial. This is expressed in the **Factor Theorem**.

Factor Theorem   If $P(x)$ is a polynomial function and c is any real number, then

- If $P(c) = 0$, then $x - c$ is a factor of $P(x)$.
- If $x - c$ is a factor of $P(x)$, then $P(c) = 0$.

In order to factor polynomials, we can use the Factor Theorem and synthetic division together to factor polynomials. In Example 1(a), we found $P(3) = 0$ for $P(x) = 2x^4 - 10x^3 + 17x^2 - 14x - 3$, hence $x - 3$ is a factor. Let's look at the synthetic division we did for this problem.

$$\begin{array}{r|rrrrr} 3 & 2 & -10 & 17 & -14 & -3 \\ & & 6 & -12 & 15 & 3 \\ \hline & 2 & -4 & 5 & 1 & 0 \end{array}$$

The numbers in the last line are the coefficients of the quotient and the remainder. Thus we can write $P(x) = (x - 3)(2x^3 - 4x^2 + 5x + 1)$.

EXAMPLE 2   **Using the Factor Theorem and Synthetic Division to Factor a Polynomial Function**

Show that $x + 2$ is a factor of $P(x) = x^4 - 7x^2 - 6x$ using synthetic division. Use that result to factor $P(x)$ and state the zeros.

SOLUTION   We have $(x - c) = (x - (-2))$, so $c = -2$; now we will use synthetic division to find the quotient. The numbers in the bottom line are the coefficients of the quotient.

$$\begin{array}{r|rrrrr} -2 & 1 & 0 & -7 & -6 & 0 \\ & & -2 & 4 & 6 & 0 \\ \hline & 1 & -2 & -3 & 0 & 0 \end{array}$$

These are the coefficients of the quotient.

From this division we can now factor the polynomial.

$$\begin{aligned} P(x) &= (x + 2)(x^3 - 2x^2 - 3x) &&\text{Factor the polynomial.} \\ &= (x + 2)(x)(x^2 - 2x - 3) &&\text{Factor out the } x. \\ &= x(x + 2)(x - 3)(x + 1) &&\text{Factor the quadratic.} \end{aligned}$$

We can now see that the zeros of this polynomial are $x = -2, -1, 0,$ and 3.

Self Check 2   Show that $x - 2$ is a factor of $P(x) = x^3 - 3x^2 - 10x + 24$ using synthetic division. Use that result to factor $P(x)$ and state the zeros.

Now Try Exercise 29.

3. Determine All Rational Zeros of a Polynomial Function Using the Rational-Zeros Theorem

Sometimes we find ourselves in a car wondering how to get to somewhere. Fortunately, many cars are now equipped with a navigation system that can give us directions to a specific address or point of interest. A navigation system uses satellite signals to determine the exact location of our car. Once the system knows the location of the car, it can guide us to our destination.

When finding the zeros of more complicated functions, there are theorems that act like a navigation system and can help us locate and identify the zeros of the function. One of these theorems is the **Rational-Zeros Theorem**.

Recall that a rational number is any number that can be written in the form $\frac{p}{q}$, where p and q are integers and $q \neq 0$. We will now discuss how to find rational zeros of polynomial functions with integer coefficients. The following theorem enables us to list the possible rational zeros of such functions.

Rational-Zeros Theorem

Given a polynomial function

$$P(x) = a_n x^n + a_{n-1} x^{n-1} + a_{n-2} x^{n-2} + \cdots + a_1 x + a_0$$

with integer coefficients, every rational zero of $P(x)$ can be written as $\frac{p}{q}$ (written in lowest terms), where p is a factor of the constant a_0 and q is a factor of the lead coefficient a_n. Each rational zero of $P(x)$ is a solution to the polynomial equation $P(x) = 0$.

EXAMPLE 3   Using the Rational-Zeros Theorem

Find the rational zeros of the polynomial $P(x) = 2x^3 + 3x^2 - 8x + 3$.

SOLUTION   This is a third-degree polynomial and will have at most three zeros. Any rational zeros will have the form $\dfrac{p}{q} = \dfrac{\text{factor of the constant 3}}{\text{factor of the lead coefficient 2}}$; the possible rational zeros are therefore

$$\pm \frac{3}{1}, \quad \pm \frac{1}{1}, \quad \pm \frac{3}{2}, \quad \text{and} \quad \pm \frac{1}{2},$$

or, written in order of increasing size,

$$-3, \quad -\frac{3}{2}, \quad -1, \quad -\frac{1}{2}, \quad \frac{1}{2}, \quad 1, \quad \frac{3}{2}, \quad \text{and} \quad 3.$$

We check each possibility to see whether it is a zero. Using synthetic division, we can start with $\frac{3}{2}$.

$$
\begin{array}{r|rrr}
\frac{3}{2} & 2 & 3 & -8 & 3 \\
 & & 3 & 9 & \frac{3}{2} \\
\hline
 & 2 & 6 & 1 & \frac{9}{2}
\end{array}
$$

Since the remainder is not 0, $\frac{3}{2}$ is not a zero and can be crossed off the list.
We now try $\frac{1}{2}$:

$$
\begin{array}{r|rrr}
\frac{1}{2} & 2 & 3 & -8 & 3 \\
 & & 1 & 2 & -3 \\
\hline
 & 2 & 4 & -6 & 0
\end{array}
$$

Since the remainder is 0, $\frac{1}{2}$ is a zero. The binomial $x - \frac{1}{2}$ is a factor of $P(x)$, and any remaining zeros must be supplied by the remaining factor, which is the quotient $2x^2 + 4x - 6$. Since $2x^2 + 4x - 6$ is a quadratic and can be factored, we can find the other zeros.

$$P(x) = \left(x - \frac{1}{2}\right)(2x^2 + 4x - 6)$$

$$= \left(x - \frac{1}{2}\right)(2)(x^2 + 2x - 3) \qquad \text{Factor out the 2.}$$

$$= (2x - 1)(x^2 + 2x - 3) \qquad \text{Multiply the first factor by 2.}$$

$$= (2x - 1)(x + 3)(x - 1) \qquad \text{Factor the quadratic.}$$

The zeros of the polynomial function are $x = \frac{1}{2}$, -3, and 1.

■

Self Check 3   Find the zeros of $P(x) = 3x^3 - 10x^2 + 9x - 2$.

Now Try Exercise 37.

ACCENT ON TECHNOLOGY   **Using a Graphing Calculator to Find Factors**

If we have a graphing calculator, we can reduce the amount of work involved in trying to find the rational zeros of a polynomial function. With a list of the complete pool of possible rational zeros, we can utilize the TABLE function to find the zeros. Then we use synthetic division to factor the polynomial function. This is demonstrated in Example 4. A scientific calculator can also be used to check for zeros.

EXAMPLE 4   **Using the Rational-Zeros Theorem, Synthetic Division, and a Calculator**

Use the Rational-Zeros Theorem, a calculator, and synthetic division to find all the rational zeros of each polynomial function; factor; and draw the graph.

a. $P(x) = 2x^4 + x^3 - 9x^2 - 4x + 4$ **b.** $P(x) = x^4 + 3x^3 - x - 3$

SOLUTION   **a.** The first step is to find all the possible rational zeros. For the polynomial $P(x) = 2x^4 + x^3 - 9x^2 - 4x + 4$, we see that $p = 4$ and $q = 2$ so $\dfrac{p}{q} = \dfrac{\text{all integer factors of 4}}{\text{all integer factors of 2}} = \dfrac{\pm 1, \pm 2, \pm 4}{\pm 1, \pm 2} = \pm 1, \pm 2, \pm 4$, and $\pm \dfrac{1}{2}$.

The next step is to evaluate the function at these values. The table in Figure 2-36 was created with a graphing calculator using the TABLE function.

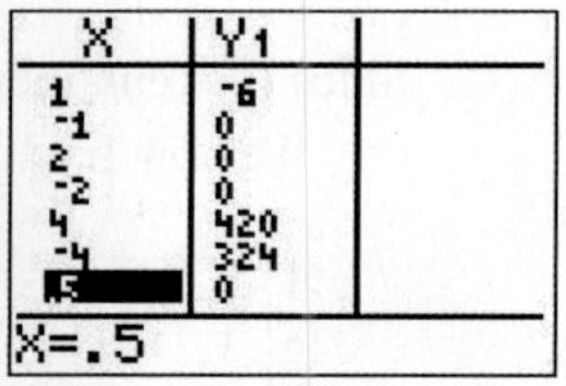

X	Y1
1	-6
-1	0
2	0
-2	0
4	420
-4	324
	0

X=.5

FIGURE 2-36

FIGURE 2-37

Since this is a fourth-degree polynomial, we know that there are at most four real zeros. In this example, all of the zeros are rational numbers, and we have found all four of them: $x = -1, 2, -2$, and $\frac{1}{2}$. Now that we know what the rational zeros are, we create the function $P(x) = (x + 1)(x - 2)(2x - 1)(x + 2)$. The graph is shown in Figure 2-37.

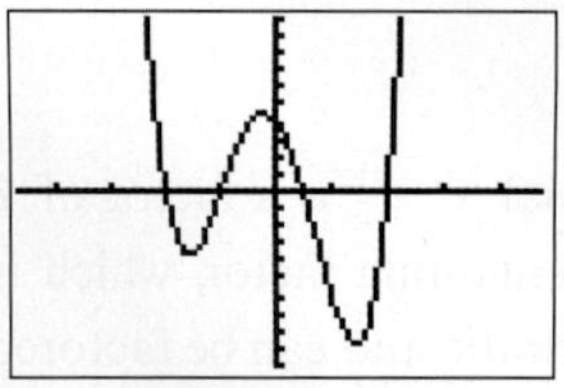

b. For $P(x) = x^4 + 3x^3 - x - 3$, we will first find all possible rational zeros. Since $p = -3$ and $q = 1$, $\dfrac{p}{q} = \dfrac{\text{all integer factors of } -3}{\text{all integer factors of } 1} = \dfrac{\pm 1, \pm 3}{\pm 1} = \pm 1$, and ± 3.

We enter the function in the graph editor and evaluate the polynomial at each of the possible rational zeros by using the TABLE function.

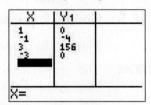

FIGURE 2-38

We see in Figure 2-38 that the only rational zeros are $x = 1$ and $x = -3$. Starting with $x = 1$, we perform synthetic division.

$$\begin{array}{r|rrrrr} 1 & 1 & 3 & 0 & -1 & -3 \\ & & 1 & 4 & 4 & 3 \\ \hline & 1 & 4 & 4 & 3 & 0 \end{array} \longrightarrow P(x) = (x-1)(x^3 + 4x^2 + 4x + 3)$$

Now we will factor the quotient $(x^3 + 4x^2 + 4x + 3)$. We know that $x = -3$ is a zero.

$$\begin{array}{r|rrrr} -3 & 1 & 4 & 4 & 3 \\ & & -3 & -3 & -3 \\ \hline & 1 & 1 & 1 & 0 \end{array} \longrightarrow P(x) = (x-1)(x+3)(x^2 + x + 1)$$

The term $x^2 + x + 1$ does not factor, so we have found all of the rational zeros of the polynomial function. If we look at the graph in Figure 2-39, we see that the function has only two x-intercepts (i.e., zeros).

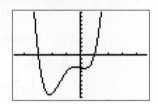

FIGURE 2-39

Self Check 4 Use the Rational-Zeros Theorem, a calculator, and synthetic division to find all the rational zeros and factors of the polynomial function $P(x) = 6x^4 + 5x^3 - 25x^2 - 10x + 24$.

Now Try Exercise 47.

Consider the polynomial function graphed in Figure 2-40. The graph is of a fifth-degree polynomial, but it appears to have only three real zeros.

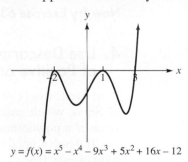

$$y = f(x) = x^5 - x^4 - 9x^3 + 5x^2 + 16x - 12$$

FIGURE 2-40

It looks as though, at two of the zeros, the graph only touches the x-axis and turns. When we use synthetic division to find zeros, we must be mindful of the fact that a zero can have a multiplicity greater than 1. So once we have identified a number as a zero, we cannot eliminate the zero from our list until we examine the new values of p and q. Those values may change and reduce the number of possible rational zeros.

EXAMPLE 5 **Finding Repeating Zeros**

Find all the rational zeros of $P(x) = x^5 - x^4 - 9x^3 + 5x^2 + 16x - 12$.

SOLUTION The graph shown in Figure 2-40 shows us what we are looking for, but keep in mind that we may not always have a graph to help us get started. Our first step is to find all the possible rational zeros. Since $p = -12$ and $q = 1$,

$$\frac{p}{q} = \frac{\text{all integer factors of } -12}{\text{all integer factors of } 1} = \frac{\pm 1, \pm 2, \pm 3, \pm 4, \pm 6, \pm 12}{\pm 1} = \pm 1, \pm 2, \pm 3, \pm 4,$$

± 6, and ± 12. We can see that this is a fairly large list of possible rational zeros, but we may be able to eliminate some of these after we find the first zero.

You can use a scientific or a graphing calculator to evaluate the function and find a rational zero to use for your first synthetic division.

$$
\begin{array}{r|rrrrr}
3 & 1 & -1 & -9 & 5 & 16 & -12 \\
  &   & 3 & 6 & -9 & -12 & 12 \\
\hline
  & 1 & 2 & -3 & -4 & 4 & 0
\end{array}
\longrightarrow P(x) = (x - 3)(x^4 + 2x^3 - 3x^2 - 4x + 4)
$$

With this first factorization, the value of $p = 4$ in the quotient allows us to eliminate some of the values of $\frac{p}{q}$ that were in the original list. We now have $\frac{p}{q} = \pm 1, \pm 2$, and ± 4, and we continue to find factors of the polynomial.

$$
\begin{array}{r|rrrrr}
1 & 1 & 2 & -3 & -4 & 4 \\
  &   & 1 & 3 & 0 & -4 \\
\hline
  & 1 & 3 & 0 & -4 & 0
\end{array}
\longrightarrow P(x) = (x - 3)(x - 1)(x^3 + 3x^2 - 4)
$$

Using $c = 1$ again gives another factor.

$$
\begin{array}{r|rrrr}
1 & 1 & 3 & 0 & -4 \\
  &   & 1 & 4 & 4 \\
\hline
  & 1 & 4 & 4 & 0
\end{array}
\longrightarrow P(x) = (x - 3)(x - 1)^2(x^2 + 4x + 4)
$$

At this point we can factor the quadratic term to get to the complete factorization of the polynomial:

$$P(x) = (x - 3)(x - 1)^2(x + 2)^2$$

The zeros are $x = 3$ (multiplicity 1), 1 (multiplicity 2), and -2 (multiplicity 2). If you were to begin with another zero instead of 1, the final answer would still be the same.

Self Check 5 Find all the rational zeros of $P(x) = 4x^5 - 28x^4 + 49x^3 + 8x^2 - 44x - 16$.

Now Try Exercise 63.

4. Use Descartes' Rule of Signs to Estimate the Number of Positive and Negative Zeros of a Polynomial Function

René Descartes (1596–1650) is credited with a theorem known as **Descartes' Rule of Signs,** which enables us to estimate the number of positive and negative real zeros of a polynomial function.

If a polynomial is written in descending powers of x, and we scan it from left to right, we say that a *variation in sign* occurs whenever successive terms have opposite signs. For example, the polynomial

$$P(x) = \overset{+\text{ to }-}{\overbrace{3x^5 - 2x^4}} \overset{-\text{ to }+}{\overbrace{-5x^3 +}} \overset{+\text{ to }-}{\overbrace{x^2 - x}} - 9$$

has three variations in sign, and the polynomial

$$P(-x) = 3(-x)^5 - 2(-x)^4 - 5(-x)^3 + (-x)^2 - (-x) - 9$$

$$= \overset{-\text{ to }+}{\overbrace{-3x^5 - 2x^4 +}} \overset{+\text{ to }-}{\overbrace{5x^3 + x^2 + x}} - 9$$

has two variations in sign.

Descartes' Rule of Signs

- If $P(x)$ is a polynomial function with real coefficients, the number of positive real zeros of $P(x)$ is either equal to the number of variations in sign of $P(x)$ or less than that number by an even number.
- The number of negative real zeros of $P(x)$ is either equal to the number of variations in sign of $P(-x)$ or less than that number by an even number.

EXAMPLE 6   **Using Descartes' Rule of Signs**

Discuss the possibilities for the real zeros of $P(x) = 3x^3 - 2x^2 + x - 5$.

SOLUTION   Since there are three variations of sign in $P(x) = 3x^3 - 2x^2 + x - 5$, there can be either three positive real zeros or only one (1 is less than 3 by the even number 2). Because

$$P(-x) = 3(-x)^3 - 2(-x)^2 + (-x) - 5$$

$$= -3x^3 - 2x^2 - x - 5$$

has no variations in sign, there are no negative real zeros. Furthermore, 0 is not a zero, because the terms of the polynomial do not have a common factor of x.

The table shows these possibilities, with a total of three zeros.

Number of Positive Zeros	Number of Negative Zeros	Number of Nonreal Zeros
3	0	0
1	0	2

■

Self Check 6   Discuss the possibilities for the real zeros of $P(x) = 5x^3 + 2x^2 - x + 3$.

Now Try Exercise 73.

EXAMPLE 7   **Using Descartes' Rule of Signs**

Discuss the possibilities for the real zeros of $P(x) = 5x^5 - 3x^3 - 2x^2 + x - 1$.

SOLUTION   Since there are three variations of sign in $P(x)$, there are either three or one positive real zeros. Because $P(-x) = -5x^5 + 3x^3 - 2x^2 - x - 1$ has two variations in sign, there are two or zero negative real zeros. The possibilities are shown here, with five total zeros.

Number of Positive Zeros	Number of Negative Zeros	Number of Nonreal Zeros
1	0	4
3	0	2
1	2	2
3	2	0

Self Check 7　Discuss the possibilities for the real zeros of $P(x) = 5x^5 - 2x^2 - x - 1$.

Now Try Exercise 79.

Englishman David Beckham currently plays major-league soccer for the Los Angeles Galaxy. Beckham is one of the highest paid players in the world, and he has twice been runner-up for the International Federation of Association Football's player of the year award.

Brandon_Parry/Shutterstock.com

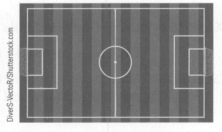

DiverS-VectoR/Shutterstock.com

In soccer, it is important for players to know the boundaries of the playing field. They need to know when it is advantageous to kick the ball out of bounds and when it is advantageous to keep the ball in bounds.

Knowing where the boundaries are can also be helpful in mathematics. Establishing integer **bounds** on zeros can be an important aid in locating and finding the zeros of polynomial functions. We will learn how to find integer bounds for the roots.

The Remainder Theorem and synthetic division provide a way of verifying that a particular number is a zero of a polynomial function, but another theorem provides a way to find bounds on the zeros of a polynomial, enabling us to focus on just where the zeros may be.

Upper and Lower Bounds on Zeros

- If the lead coefficient of a polynomial function $P(x)$ with real coefficients is positive, perform synthetic division of the coefficients by a positive number c. If each term in the last row of the division is nonnegative, no number greater than c can be a zero of $P(x)$—c is an **upper bound** of the real roots.
- If $P(x)$ is synthetically divided by a negative number d and the signs in the last row alternate, no value less than d can be a zero of $P(x)$—d is a **lower bound** of the real roots. (If 0 appears in the third row, that 0 can be assigned either a plus or a minus sign to help the signs alternate.)

EXAMPLE 8　**Finding Bounds on the Zeros of a Polynomial Function**

Establish integer bounds for the zeros of $P(x) = 2x^3 + 3x^2 - 5x - 7$.
(Answers are not unique.)

SOLUTION   The possible rational zeros are $\dfrac{p}{q} = \dfrac{\pm 1, \pm 7}{\pm 1, \pm 2} = \pm 1, \pm 7, \pm\dfrac{1}{2}$, and $\pm\dfrac{7}{2}$. By Descartes' Rule of Signs, there can be only one positive real zero, and there will be two or zero negative real zeros.

We will do several synthetic divisions by positive integers, looking for nonnegative values in the last row. Then we will divide by several negative integers, looking for alternating signs in the last row. Trying 1 first gives

$$
\begin{array}{r|rrr}
\underline{1} & 2 & 3 & -5 & -7 \\
& & 2 & 5 & 0 \\
\hline
& +2 & +5 & 0 & -7
\end{array}
$$

Because one of the signs in the last row is negative, we cannot claim that 1 is an upper bound. We now try 2.

$$
\begin{array}{r|rrr}
\underline{2} & 2 & 3 & -5 & -7 \\
& & 4 & 14 & 18 \\
\hline
& +2 & +7 & +9 & +11
\end{array}
$$

Because the last row is entirely nonnegative, we can claim that 2 is an upper bound. That is, no number greater than 2 can be a zero of the polynomial function.

Perform several divisions by negative integers, looking for alternating signs in the last row. We begin with -3.

$$
\begin{array}{r|rrr}
\underline{-3} & 2 & 3 & -5 & -7 \\
& & -6 & 9 & -12 \\
\hline
& +2 & -3 & +4 & -19
\end{array}
$$

Since the signs in the last row alternate, -3 is a lower bound.

Since -3 is a lower bound and 2 is an upper bound, we have established integer bounds for the real zeros. ∎

Self Check 8   Establish integer bounds for the zeros of $P(x) = 2x^3 + 3x^2 - 11x - 7$. (Answers are not unique.)

Now Try Exercise 87.

5. Solve Real-World Applications That Utilize Polynomial Functions

Many applications involve polynomial functions.

EXAMPLE 9   **Solving an Application of Polynomial Functions**

To protect cranberry crops from the damage of early freezes, growers flood the cranberry bogs. Three irrigation sources, used together, can flood a cranberry bog in one day. If the sources are used one at a time, the second source requires one day longer to flood the bog than the first, and the third requires four days longer than the first. If the bog must be flooded before a freeze that is predicted in three days, can the water in the last two sources be diverted to other bogs?

SOLUTION   Let x represent the number of days it would take the first irrigation source to flood the bog. Then $x + 1$ and $x + 4$ represent the number of days it would take the second and third sources, respectively, to flood the bog. Because the first source alone requires x days to flood the bog, that source could fill $\frac{1}{x}$ of the bog in one day. In one day, the remaining sources could flood $\frac{1}{x+1}$ and $\frac{1}{x+4}$ of the bog.

The part of a bog the first source can flood in one day	plus	the part the second source can flood in one day	plus	the part the third source can flood in one day	equals	one bog
$\dfrac{1}{x}$	$+$	$\dfrac{1}{x+1}$	$+$	$\dfrac{1}{x+4}$	$=$	1

We multiply both sides of the equation by $x(x+1)(x+4)$ to clear it of fractions and then simplify to get

$$x(x+1)(x+4)\left(\frac{1}{x} + \frac{1}{x+1} + \frac{1}{x+4}\right) = 1 \cdot x(x+1)(x+4)$$

$$(x+1)(x+4) + x(x+4) + x(x+1) = x(x+1)(x+4)$$

$$x^2 + 5x + 4 + x^2 + 4x + x^2 + x = x^3 + 5x^2 + 4x$$

$$0 = x^3 + 2x^2 - 6x - 4$$

To solve the equation $x^3 + 2x^2 - 6x - 4 = 0$, we first list the possible rational zeros. We see that $p = -4$ and $q = 1$, resulting in the possible rational zeros $\frac{p}{q} = \pm1, \pm2,$ and ±4. One solution of this equation is $x = 2$, because when we synthetically divide by 2, the remainder is 0:

$$\underline{2|}\ \begin{array}{rrrr} 1 & 2 & -6 & -4 \\ & 2 & 8 & 4 \\ \hline 1 & 4 & 2 & 0 \end{array}$$

We find the other solutions by using the quadratic formula to find solutions to the equation $x^2 + 4x + 2 = 0$. The two solutions are $-2 + \sqrt{2}$ and $-2 - \sqrt{2}$. Since these numbers are negative and the time it takes to flood the bog cannot be negative, these roots must be discarded. The only meaningful solution is $x = 2$.

Since the first source alone can flood the bog in two days, and it is three days until the freeze, the other two water sources can be diverted to flood other bogs.

Self Check 9 If a freeze is predicted in one day, can the water in the last two sources be diverted?

Now Try Exercise 97.

Self Check Answers
1. -11 **2.** $P(x) = (x-2)(x+3)(x-4);\ x = -3, 2,$ and 4
3. $x = 1, 2,$ and $\dfrac{1}{3}$ **4.** $x = -2, \dfrac{3}{2}, -\dfrac{4}{3},$ and 1;

$P(x) = (x+2)(2x-3)(3x+4)(x-1)$ **5.** $x = 2$ (multiplicity 2), 4, and

$-\dfrac{1}{2}$ (multiplicity 2)   **6.** two or zero positive zeros, one negative zero, two or

zero nonreal zeros   **7.** one positive zero, two or zero negative zeros, four or two
or zero nonreal zeros   **8.** $-4, 3$ **9.** no

Exercises **2.3**

Getting Ready

You should be able to complete these vocalulary and concept statements before you proceed to the practice exercises.

Fill in the blanks.

1. The variables in a polynomial have _____-number exponents.

2. A zero of $P(x)$ is any number c for which _____.

3. The Remainder Theorem holds when c is _____ number.

4. If $P(x)$ is a polynomial and $P(x)$ is divided by _____, the remainder will be $P(c)$.

5. If $P(x)$ is a polynomial, then $P(c) = 0$ if and only if $x - c$ is a _____ of $P(x)$.

6. A shortcut method for dividing a polynomial by a binomial of the form $x - c$ is called _____ division.

7. The polynomial $6x^4 + 5x^3 - 2x^2 + 3$ has _____ variations in sign.

8. The polynomial $(-x)^3 - (-x)^2 - 4$ has _____ variations in sign.

9. The equation $7x^4 + 5x^3 - 2x + 1 = 0$ can have at most _____ positive roots.

10. The equation $7x^4 + 5x^3 - 2x + 1 = 0$ can have at most _____ negative roots.

11. If no number less than d can be a root of $P(x) = 0$, then d is called a(n) _____.

12. If no number greater than c can be a root of $P(x) = 0$, then c is called a(n) _____.

13. The rational zeros of the polynomial $P(x) = 3x^3 + 4x - 7$ will have the form $\frac{p}{q}$, where p is a factor of _____ and q is a factor of 3.

14. The rational zeros of the equation $P(x) = 5x^3 + 3x^2 - 4$ will have the form $\frac{p}{q}$, where p is a factor of -4 and q is a factor of _____.

Practice

Use the Division Algorithm and synthetic division to express $P(x) = 3x^3 - 2x^2 - 6x - 4$ in the form $P(x) = d(x) \cdot q(x) + r(x)$.

15. $d(x) = x - 1$ 16. $d(x) = x - 2$

17. $d(x) = x - 3$ 18. $d(x) = x - 4$

19. $d(x) = x + 1$ 20. $d(x) = x + 2$

21. $d(x) = x + 3$ 22. $d(x) = x + 4$

Find each value by substituting the given value of x into the polynomial function and simplifying. Then find the value by doing a synthetic division and finding the remainder.

23. $P(x) = 3x^3 - 2x^2 - 5x - 7;\ P(2)$

24. $P(x) = 5x^3 + 4x^2 + x - 1;\ P(-2)$

25. $P(x) = 7x^4 + 2x^3 + 5x^2 - 1;\ P(-1)$

26. $P(x) = 2x^4 - 2x^3 + 5x^2 - 1;\ P(2)$

27. $P(x) = 2x^5 + x^4 - x^3 - 2x + 3;\ P(1)$

28. $P(x) = 3x^5 + x^4 - 3x^2 + 5x + 7;\ P(-2)$

Use the Factor Theorem to decide whether each statement is true or false.

29. $x - 1$ is a factor of $x^7 - 1$.

30. $x - 2$ is a factor of $x^3 - x^2 + 2x - 8$.

31. $x - 1$ is a factor of $3x^5 + 4x^2 - 7$.

32. $x + 1$ is a factor of $3x^5 + 4x^2 - 7$.

33. $x + 3$ is a factor of $2x^3 - 2x^2 + 1$.

34. $x - 3$ is a factor of $3x^5 - 3x^4 + 5x^2 - 13x - 6$.

35. $x - 1$ is a factor of $x^{1,984} - x^{1,776} + x^{1,492} - x^{1,066}$.

36. $x + 1$ is a factor of $x^{1,984} + x^{1,776} - x^{1,492} - x^{1,066}$.

Find all rational zeros of each polynomial function and state any multiplicity greater than one.

37. $P(x) = x^3 - 5x^2 - x + 5$

38. $P(x) = x^3 + 7x^2 - x - 7$

39. $P(x) = x^3 - 2x^2 - x + 2$

40. $P(x) = x^3 + x^2 - 4x - 4$

41. $P(x) = x^3 - x^2 - 4x + 4$

42. $P(x) = x^3 + 2x^2 - x - 2$

43. $P(x) = x^3 - 2x^2 - 9x + 18$

44. $P(x) = x^3 + 3x^2 - 4x - 12$

45. $P(x) = 2x^3 - x^2 - 2x + 1$

46. $P(x) = 3x^3 + x^2 - 3x - 1$

47. $P(x) = 3x^3 + 5x^2 + x - 1$

48. $P(x) = 2x^3 - 3x^2 + 1$

49. $P(x) = x^4 - 10x^3 + 35x^2 - 50x + 24$

50. $P(x) = x^4 + 4x^3 + 6x^2 + 4x + 1$

51. $P(x) = x^4 + 3x^3 - 13x^2 - 9x + 30$

52. $P(x) = x^4 - 8x^3 + 14x^2 + 8x - 15$

53. $P(x) = x^5 + 3x^4 - 5x^3 - 15x^2 + 4x + 12$

54. $P(x) = x^5 - 3x^4 - 5x^3 + 15x^2 + 4x - 12$

55. $P(x) = x^7 - 12x^5 + 48x^3 - 64x$

56. $P(x) = x^7 + 7x^6 + 21x^5 + 35x^4 + 35x^3 + 21x^2 + 7x + 1$

57. $P(x) = 3x^3 - 2x^2 + 12x - 8$

58. $P(x) = 4x^4 - 8x^3 - x^2 + 8x - 3$

59. $P(x) = 3x^4 - 14x^3 + 11x^2 + 16x - 12$

60. $P(x) = 2x^4 - x^3 - 2x^2 - 4x - 40$

61. $P(x) = 12x^4 + 20x^3 - 41x^2 + 20x - 3$

62. $P(x) = 4x^5 - 12x^4 + 15x^3 - 45x^2 - 4x + 12$

63. $P(x) = 6x^5 - 7x^4 - 48x^3 + 81x^2 - 4x - 12$

64. $P(x) = 36x^4 - x^2 + 2x - 1$

65. $P(x) = 30x^3 - 47x^2 - 9x + 18$

66. $P(x) = 20x^3 - 53x^2 - 27x + 18$

67. $P(x) = 15x^3 - 61x^2 - 2x + 24$

68. $P(x) = 12x^4 + x^3 + 42x^2 + 4x - 24$

69. $P(x) = 20x^3 - 44x^2 + 9x + 18$

70. $P(x) = 24x^3 - 82x^2 + 89x - 30$

71. $P(x) = x^3 - \frac{4}{3}x^2 - \frac{13}{3}x - 2$

72. $P(x) = x^3 - \frac{19}{6}x^2 + \frac{1}{6}x + 1$

Use Descartes' Rule of Signs to find the number of possible positive and negative zeros of each polynomial function.

73. $P(x) = 3x^3 + 5x^2 - 4x + 3$

74. $P(x) = 3x^3 - 5x^2 - 4x - 3$

75. $P(x) = 2x^3 + 7x^2 + 5x + 5$

76. $P(x) = -2x^3 - 7x^2 - 5x - 4$

77. $P(x) = 8x^4 + 5$

78. $P(x) = -3x^3 + 5$

79. $P(x) = x^4 + 8x^2 - 5x - 10$

80. $P(x) = 5x^7 + 3x^6 - 2x^5 + 3x^4 + 9x^3 + x^2 + 1$

81. $P(x) = -x^{10} - x^8 - x^6 - x^4 - x^2 - 1$

82. $P(x) = x^{10} + x^8 + x^6 + x^4 + x^2 + 1$

83. $P(x) = x^9 + x^7 + x^5 + x^3 + x$

84. $P(x) = -x^9 - x^7 - x^5 - x^3 - x$

85. $P(x) = -2x^4 - 3x^2 + 2x + 3$

86. $P(x) = -7x^5 - 6x^4 + 3x^3 - 2x^2 + 7x - 4$

Find integer bounds for the zeros of each polynomial function. Answers can vary.

87. $P(x) = x^2 - 2x - 4$

88. $P(x) = 9x^2 - 6x - 1$

89. $P(x) = 18x^2 - 6x - 1$

90. $P(x) = 2x^2 - 10x - 9$

91. $P(x) = 6x^3 - 13x^2 - 110x$

92. $P(x) = 12x^3 + 20x^2 - x - 6$

93. $P(x) = x^5 + x^4 - 8x^3 - 8x^2 + 15x + 15$

94. $P(x) = 3x^4 - 5x^3 - 9x^2 + 15x$

95. $P(x) = 3x^5 - 11x^4 - 2x^3 + 38x^2 - 21x - 15$

96. $P(x) = 3x^6 - 4x^5 - 21x^4 + 4x^3 + 8x^2 + 8x + 32$

Applications

97. Parallel resistance If three resistors with resistances R_1, R_2, and R_3 are wired in parallel, their combined resistance R is given by the formula $\frac{1}{R} = \frac{1}{R_1} + \frac{1}{R_2} + \frac{1}{R_3}$. The design of a voltmeter requires that the resistance R_2 be 10 ohms greater than the resistance R_1, the resistance R_3 be 50 ohms greater than R_1, and the combined resistance be 6 ohms. Find the value of each resistance.

98. Fabricating sheet metal The open tray shown in the illustration is to be manufactured from a 12-inch-by-14-inch rectangular sheet of metal by cutting squares from each corner and folding up the sides. If the volume of the tray is to be 160 cubic inches and x is to be an integer, what size squares should be cut from each corner?

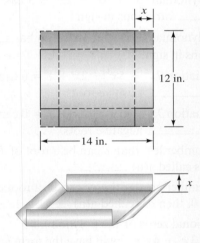

99. Packaging The length of a 25-kilogram FedEx box is 7 inches greater than its height. The width of the box is 4 inches greater than its height. If the volume of the box is 4,420 cubic inches, find the height of the box.

100. Soda cans An aluminum Mountain Dew can is approximately the shape of a cylinder. If the height of the can is 9 centimeters greater than its radius and the volume of the can is approximately 108π cubic centimeters, find the radius of the can. (The formula for the volume of a cylinder is $V = \pi r^2 h$.)

101. Area A rectangle is inscribed in the parabola $y = 16 - x^2$, as shown in the illustration. Find the point (x, y) if the area of the rectangle is 42 square units.

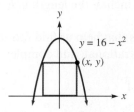

$y = 16 - x^2$

(x, y)

102. Area One corner of the rectangle shown is at the origin, and the opposite corner (x, y) lies in the first quadrant on the curve $y = x^3 - 2x^2$. Find the point (x, y) if the area of the rectangle is 27 square units.

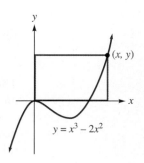

$y = x^3 - 2x^2$

Discovery and Writing

103. If 0 is a zero of
$$P(x) = a_n x^n + a_{n-1} x^{n-1} + \cdots + a_1 x + a_0, \text{ find } a_0.$$

104. If 0 occurs twice as a zero of
$$P(x) = a_n x^n + a_{n-1} x^{n-1} + \cdots + a_1 x + a_0, \text{ find } a_1.$$

105. If $P(2) = 0$, and $P(-2) = 0$, explain why $x^2 - 4$ is a factor of $P(x)$.

106. If $P(x) = x^4 - 3x^3 + kx^2 + 4x - 1$ and $P(2) = 11$, find k.

107. If n is an even integer and c is a positive constant, show that $x^n + c = 0$ has no real roots.

108. If n is an even positive integer and c is a positive constant, show that $x^n - c = 0$ has two real roots.

2.4 Roots of Polynomial Equations

In this section, we will learn to

1. Use the Rational-Zeros Theorem to solve a polynomial equation.

2. Determine the complex roots of a polynomial equation and create a polynomial equation.

3. Understand the Fundamental Theorem of Algebra and its significance.

restyler/Shutterstock.com

FedEx is a company that offers reliable shipping services across town and across the globe. A wide range of envelopes and boxes are available to accommodate the shipping needs of most individuals and companies.

One FedEx box is a 10-kilogram box with the following characteristics:

- The length of the box is 6 inches greater than its height.
- The width of the box is 3 inches greater than its height.
- The volume of the box is 2,080 cubic inches.

To find the dimensions of the box, we can let h represent the height (in inches). Then $h + 6$ will represent the length and $h + 3$ will represent the width. Since the volume of the box is given to be 2,080, we have

$$V = l \cdot w \cdot h$$

$$2{,}080 = (h + 6)(h + 3)h \qquad \text{Substitute.}$$

$$2{,}080 = h^3 + 9h^2 + 18h \qquad \text{Multiply.}$$

$$0 = h^3 + 9h^2 + 18h - 2{,}080 \qquad \text{Subtract 2,080 from both sides.}$$

To find the height h, we must find the roots of the polynomial equation. Note that one root of the equation is $x = 10$, because 10 satisfies the equation:

$$(10)^3 + 9(10)^2 + 18(10) - 2,080 = 0.$$

Therefore, the height of the box is $h = 10$ inches, the length is $h + 6 = 16$ inches, and the width is $h + 3 = 13$ inches.

In Section 2.3, we introduced theorems that help us find rational zeros. We will now discover other ways to find rational, irrational, and complex roots of polynomial equations of the form $P(x) = 0$.

1. Use the Rational-Zeros Theorem to Solve a Polynomial Equation

To solve a polynomial equation, factoring is usually the best starting point. We can use the Rational-Zeros Theorem to help us accomplish this.

EXAMPLE 1   **Solving a Polynomial Equation and Graphing a Polynomial Function**

For $P(x) = x^3 + x^2 - 3x - 3$, solve the equation $P(x) = 0$. Use the solutions to graph the polynomial function.

SOLUTION   Since this is a third-degree equation, we will use the Rational-Zeros Theorem. We know that there can be at most three real zeros of the equation, so our first step is to find any rational zeros. The possible rational zeros are $\dfrac{p}{q} = \dfrac{\text{factors of } -3}{\text{factors of } 1} = \pm 1$ and ± 3. Trying $x = -1$, we find that it is a zero, so we will use synthetic division to factor the polynomial.

$$\begin{array}{r|rrrr} -1 & 1 & 1 & -3 & -3 \\ & & -1 & 0 & 3 \\ \hline & 1 & 0 & -3 & 0 \end{array}$$

We now have $(x + 1)$ as a factor and write the equation as $(x + 1)(x^2 - 3) = 0$. The next step is to set each factor to equal 0 and solve the equations.

$$x + 1 = 0 \quad \text{or} \quad x^2 - 3 = 0$$
$$x = -1 \qquad\qquad x^2 = 3$$
$$x = \pm\sqrt{3}$$

The solution set is $\left\{-1, \sqrt{3}, -\sqrt{3}\right\}$.

Using these solutions, we can now construct a sign chart for the polynomial function.

$$P(x) = x^3 + x^2 - 3x - 3$$

Interval	$(-\infty, -\sqrt{3})$	$(-\sqrt{3}, -1)$	$(-1, \sqrt{3})$	$(\sqrt{3}, \infty)$
Test point	$f(-2) = -1$	$f\left(-\frac{3}{2}\right) = 0.375$	$f(0) = -3$	$f(2) = 3$
Graph	Below the x-axis	Above the x-axis	Below the x-axis	Above the x-axis

$$\longleftarrow \quad -2 \quad -\sqrt{3} \quad -\tfrac{3}{2} \quad -1 \quad 0 \quad \sqrt{3} \quad 2 \quad \longrightarrow$$

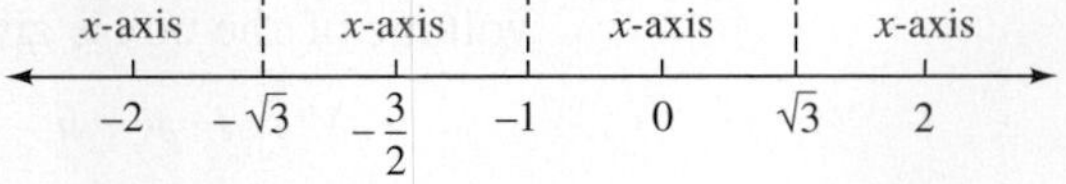

FIGURE 2-41

Using the sign chart, we can graph $P(x) = x^3 + x^2 - 3x - 3$.

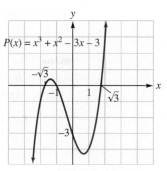

FIGURE 2-42

Self Check 1   For $P(x) = x^3 - 2x^2 - 5x + 10$, solve the equation $P(x) = 0$. Use the solutions to graph the polynomial function.

Now Try Exercise 5.

The Quadratic Formula can be used to find irrational roots for quadratic factors.

EXAMPLE 2   **Using Synthetic Division and the Quadratic Formula to Solve an Equation; Graphing a Polynomial Function**

For $P(x) = x^4 - x^3 - 9x^2 - x + 10$, solve the equation $P(x) = 0$. Use the solutions to graph the polynomial.

SOLUTION   We will begin with the Rational-Zeros Theorem to determine whether there are any rational zeros: $\dfrac{p}{q} = \dfrac{\text{factors of } 10}{\text{factors of } 1} = \dfrac{\pm 1, \pm 2, \pm 5, \pm 10}{\pm 1} = \pm 1, \pm 2, \pm 5, \text{ and } \pm 10.$
Use synthetic division to find zeros and factor until we have a quadratic factor.

$$x^4 - x^3 - 9x^2 - x + 10 = 0$$

$$\underline{1|}\ \begin{array}{rrrrr} 1 & -1 & -9 & -1 & 10 \\ & 1 & 0 & -9 & -10 \\ \hline 1 & 0 & -9 & -10 & 0 \end{array}$$

Use synthetic division again.

$$(x - 1)(x^3 - 9x - 10) = 0$$

$$\underline{-2|}\ \begin{array}{rrrr} 1 & 0 & -9 & -10 \\ & -2 & 4 & 10 \\ \hline 1 & -2 & -5 & 0 \end{array}$$

At this point, we have $(x - 1)(x + 2)(x^2 - 2x - 5) = 0$. The quadratic equation $x^2 - 2x - 5 = 0$ will not factor, so we will use the Quadratic Formula.

$$x - 1 = 0 \quad \text{or} \quad x + 2 = 0 \quad \text{or} \quad x^2 - 2x - 5 = 0$$

$$x = 1 \qquad\qquad x = -2 \qquad\qquad x = \dfrac{-(-2) \pm \sqrt{(-2)^2 - 4(1)(-5)}}{2(1)}$$

$$x = \dfrac{2 \pm \sqrt{24}}{2}$$

$$x = \dfrac{2 \pm 2\sqrt{6}}{2}$$

$$x = 1 \pm \sqrt{6}$$

The solution set is $\left\{1, -2, 1 + \sqrt{6}, 1 - \sqrt{6}\right\}$.

These solutions allow us to construct the sign chart shown in Figure 2-43.

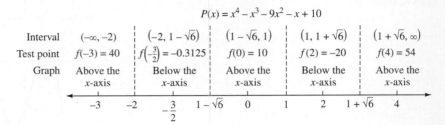

$$P(x) = x^4 - x^3 - 9x^2 - x + 10$$

Interval	$(-\infty, -2)$	$(-2, 1 - \sqrt{6})$	$(1 - \sqrt{6}, 1)$	$(1, 1 + \sqrt{6})$	$(1 + \sqrt{6}, \infty)$
Test point	$f(-3) = 40$	$f\left(-\frac{3}{2}\right) = -0.3125$	$f(0) = 10$	$f(2) = -20$	$f(4) = 54$
Graph	Above the x-axis	Below the x-axis	Above the x-axis	Below the x-axis	Above the x-axis

FIGURE 2-43

The sign chart leads us to the graph shown in Figure 2-44.

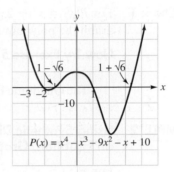

FIGURE 2-44

ACCENT ON TECHNOLOGY   Finding Approximations of Irrational Roots

We have already seen how a calculator can be used to find rational roots of an equation (i.e., the zeros of a polynomial function). A graphing calculator can be used to find approximations of the irrational roots of an equation. In Section 1.1, we used a graphing calculator to find the x-intercepts of a graph. This is equivalent to finding the roots of an equation. If we want to use our calculator to check our solutions, enter the polynomial function $P(x) = x^4 - x^3 - 9x^2 - x + 10$ into the graph editor. Go to the table and enter the roots that were found. We can see that each one gives a value of 0 for the function, which verifies that these are the roots of the equation.

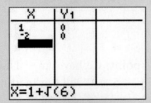

Enter the root in exact form.   The table will give an approximation of x, but a 0 for y.

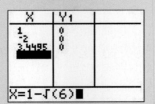

FIGURE 2-45

Self Check 2   For $P(x) = x^4 - 5x^3 + 2x^2 + 11x + 3$, solve the equation $P(x) = 0$. Use the solutions to graph the polynomial.

Now Try Exercise 9.

EXAMPLE 3   **Proving that a Number Is Irrational**

Prove that $\sqrt{2}$ is an irrational number.

SOLUTION   We know that $\sqrt{2}$ is a real root of the equation $x^2 - 2 = 0$. Since any rational root of this equation will have the form $\dfrac{p}{q} = \dfrac{\text{factors of } -2}{\text{factors of } 1} = \dfrac{\pm 1, \pm 2}{\pm 1} = \pm 1 \text{ and } \pm 2$, and none of those possible rational solutions satisfies the equation, $\sqrt{2}$ must be irrational.   ∎

Self Check 3   Prove that $\sqrt{3}$ is irrational.

Now Try Exercise 13.

2. Determine the Complex Roots of a Polynomial Equation and Create a Polynomial Equation

In Section R.2, complex numbers were defined. A complex number is a number that can be written in the form $a + bi$, where a and b are real numbers. Recall that $i = \sqrt{-1}$. Up until now, we have found only real roots of polynomial equations. The complex number system consists of both real numbers and imaginary numbers. We will now find complex roots of polynomial equations.

EXAMPLE 4   **Finding Complex Roots of Polynomial Equations**

Solve: $3x^3 - 5x^2 + 3x - 10 = 0$.

SOLUTION   Our first step is to determine any rational roots. The list of possible rational roots is $\dfrac{p}{q} = \dfrac{\pm 1, \pm 2, \pm 5, \pm 10}{\pm 1, \pm 3} = \pm 1, \pm 2, \pm 5, \pm 10, \pm\dfrac{1}{3}, \pm\dfrac{2}{3}, \pm\dfrac{5}{3}, \text{ and } \pm\dfrac{10}{3}$. We find that $x = 2$ is a root, and we use synthetic division to factor the polynomial.

$$
\begin{array}{r|rrrr}
2 & 3 & -5 & 3 & -10 \\
  &   & 6 & 2 & 10 \\
\hline
  & 3 & 1 & 5 & 0
\end{array}
$$

We have

$$3x^3 - 5x^2 + 3x - 10 = 0$$

$$(x - 2)(3x^2 + x + 5) = 0$$

$$x - 2 = 0 \quad \text{or} \quad 3x^2 + 1x + 5 = 0$$

$$x = 2 \qquad\qquad x = \frac{-1 \pm \sqrt{1^2 - 4(3)(5)}}{2(3)}$$

$$x = \frac{-1 \pm i\sqrt{59}}{6}$$

The solution set is $\left\{ 2, -\dfrac{1}{6} + i\dfrac{\sqrt{59}}{6}, -\dfrac{1}{6} - i\dfrac{\sqrt{59}}{6} \right\}$.

Self Check 4   Solve: $x^3 - 1 = 0$.

Now Try Exercise 21.

EXAMPLE 5   **Finding Complex Roots of Polynomial Equations**

Solve: $x^3 - x^2 + x - 1 = 0$.

SOLUTION   Since $\frac{p}{q} = \pm 1$, we can see that 1 is a root of the equation. We will now factor.

$$
\begin{array}{r|rrrr}
1 & 1 & -1 & 1 & -1 \\
  &   & 1 & 0 & 1 \\
\hline
  & 1 & 0 & 1 & 0
\end{array}
\longrightarrow (x - 1)(x^2 + 1) = 0
$$

Setting each factor to 0, we find the solutions.

$$
\begin{array}{ll}
x - 1 = 0 \quad \text{or} \quad & x^2 + 1 = 0 \\
 \quad x = 1 & x^2 = -1 \\
 & x = \pm\sqrt{-1} \\
 & x = \pm i
\end{array}
$$

The solution set is $\{1, i, -i\}$.

Self Check 5   Solve: $x^3 - 2x^2 + 5x - 10 = 0$.

Now Try Exercise 27.

If we look closely at the answers for Examples 4 and 5, we will notice something special about the complex roots. In each case there were two complex roots, and they were in the form $a + bi$ and $a - bi$. These are conjugates. This leads us to the following theorem:

Conjugate-Pairs Theorem   If a polynomial equation $P(x) = 0$ with real-number coefficients has a complex root $a + bi$ with $b \neq 0$, then the conjugate $a - bi$ is also a root.

EXAMPLE 6   **Using the Conjugate-Pairs Theorem**

Find a second-degree polynomial equation with real coefficients that has a root of $2 + i$.

SOLUTION   Because $2 + i$ is a root, its complex conjugate $2 - i$ is also a root. The equation is

$$
[x - (2 + i)][x - (2 - i)] = 0
$$

$$
(x - 2 - i)(x - 2 + i) = 0
$$

$$
x^2 - 4x + 5 = 0 \qquad \text{Multiply.}
$$

The equation $x^2 - 4x + 5 = 0$ will have roots of $x = 2 + i$ and $2 - i$.

Self Check 6   Find a second-degree polynomial equation with real coefficients that has a root of $1 - i$.

Now Try Exercise 43.

EXAMPLE 7   **Using the Conjugate-Pairs Theorem**

Find a fourth-degree polynomial equation with real coefficients and i as a root with multiplicity two.

SOLUTION   Because i is a root twice and a fourth-degree polynomial equation has four roots, we must find the other two roots. According to the Conjugate-Pairs Theorem, the missing roots are the conjugates of the given roots. Thus, the complete solution set is $\{i, i, -i, -i\}$. The equation is

$$(x - i)(x - i)[x - (-i)][x - (-i)] = 0$$
$$(x - i)(x + i)(x - i)(x + i) = 0$$
$$(x^2 + 1)(x^2 + 1) = 0$$
$$x^4 + 2x^2 + 1 = 0$$

Caution

In order for the Conjugate-Pairs Theorem to apply, the coefficients must be real numbers. If that is not the case, then the theorem does not apply.

Self Check 7   Find a fourth-degree polynomial equation with real coefficients and $-i$ as a root with multiplicity two.

Now Try Exercise 53.

EXAMPLE 8   **Finding Coefficients That Are Not Real Numbers**

Find a second-degree equation with a double root of i.

SOLUTION   If i is a root twice in a second-degree equation, the equation is

$$(x - i)(x - i) = 0$$
$$x^2 - 2ix - 1 = 0$$

In this equation, the coefficient of x is $-2i$, which is not a real number. Therefore, the roots are not complex-conjugate pairs.

Self Check 8   Find a second-degree equation with a double root of $-i$.

Now Try Exercise 65.

3. Understand the Fundamental Theorem of Algebra and Its Significance

Before attempting to find the roots of a polynomial equation, it would be useful to know whether any roots exist. This question was answered by Carl Friedrich Gauss (1777−1855) when he proved the **Fundamental Theorem of Algebra.**

Fundamental Theorem of Algebra   If $P(x)$ is a polynomial function with positive degree, then $P(x)$ has at least one zero.

The Fundamental Theorem of Algebra guarantees that polynomial functions such as $P(x) = 2x^3 + 3$ and $P(x) = 32.75x^{1,984} + \sqrt{2}x^3 - x + 5$ all have at least one zero. Since all polynomial functions with positive degree have zeros, their corresponding polynomial equations all have roots; those are the zeros of the polynomial function.

The next theorem will help us show that every nth-degree polynomial equation has exactly n roots.

Polynomial Factorization Theorem	If $n > 0$ and $P(x)$ is an nth-degree polynomial function, then $P(x)$ has exactly n linear factors: $P(x) = a_n(x - r_1)(x - r_2)(x - r_3) \cdots (x - r_n)$, where $r_1, r_2, r_3, \cdots, r_n$ are numbers and a_n is the lead coefficient of $P(x)$.

EXAMPLE 9   **Using the Polynomial Factorization Theorem**

Given that $-2i$ is a zero of $P(x) = x^4 + x^3 - 2x^2 + 4x - 24$, find the remaining zeros.

SOLUTION   By the Conjugate-Pairs Theorem, if $-2i$ is a zero, then its conjugate $2i$ is also a zero. Therefore, we have two factors of the polynomial function, $x + 2i$ and $x - 2i$. If we multiply these factors, then we will have a quadratic factor of $P(x)$.

$$(x - 2i)(x + 2i) = x^2 - (2i)^2$$
$$= x^2 - 4i^2 \qquad \text{Recall that } i^2 = -1.$$
$$= x^2 + 4$$

Our next step is to use the division algorithm and long division to find another factor. Since we are dividing by a factor, the remainder will be zero.

$$
\begin{array}{r}
x^2 + x - 6 \\
x^2 + 4 \overline{)x^4 + x^3 - 2x^2 + 4x - 24} \\
\underline{x^4 \qquad\quad + 4x^2} \\
x^3 - 6x^2 + 4x \\
\underline{x^3 \qquad\quad + 4x} \\
-6x^2 \qquad\quad - 24 \\
\underline{-6x^2 \qquad\quad - 24} \\
0
\end{array}
$$

At this point, we can factor the polynomial function as

$$P(x) = (x^2 + 4)(x^2 + x - 6) = (x^2 + 4)(x + 3)(x - 2).$$

The four zeros are $x = 2i, -2i, -3,$ and 2.

Self Check 9   Given that $3i$ is a zero of $P(x) = x^4 + 2x^3 + 6x^2 + 18x - 27$, find the remaining zeros.

Now Try Exercise 55.

The values of r from the Polynomial Factorization Theorem need not be distinct. Any number r_i that occurs k times as a root of a polynomial equation is called a **root of multiplicity k.**

The following theorem summarizes the previous discussion:

Roots Theorem	If multiple roots are counted individually, the polynomial equation $P(x) = 0$ with degree n $(n > 0)$ has exactly n roots among the complex numbers.

From the Polynomial Factorization Theorem, we know that a polynomial factors into linear factors if we use complex numbers. If we use only real-number coefficients, a polynomial can always be factored into linear and quadratic factors. If a quadratic cannot be factored over the real numbers, then it is considered irreducible.

EXAMPLE 10   **Factoring a Polynomial Function and Finding the Roots**

a. Factor the polynomial $P(x) = x^6 - 64$ completely into linear and irreducible quadratic factors with real coefficients.

b. Use the factorization in part (a) to find all the roots of the equation $x^6 - 64 = 0$.

SOLUTION   **a.** We begin by factoring the polynomial function as the difference of two squares and then the sum and difference of two cubes.

$P(x) = x^6 - 64$

$\qquad = (x^3 + 8)(x^3 - 8)$ Factor using the difference of two squares.

$\qquad = (x + 2)(x^2 - 2x + 4)(x - 2)(x^2 + 2x + 4)$ Factor using the sum and difference of cubes.

b. We will use these factors to find all six roots of the equation.

$(x + 2)(x - 2)(x^2 - 2x + 4)(x^2 - 2x + 4) = 0$

$$x + 2 = 0 \qquad\qquad\qquad\qquad\qquad\qquad x - 2 = 0$$

$$x = -2 \qquad\qquad\qquad\qquad\qquad\qquad x = 2$$

$$x^2 - 2x + 4 = 0 \qquad\qquad\qquad\qquad\qquad x^2 + 2x + 4 = 0$$

$$x = \frac{-(-2) \pm \sqrt{(-2)^2 - 4(1)(4)}}{2(1)} \qquad\qquad x = \frac{-2 \pm \sqrt{(2)^2 - 4(1)(4)}}{2(1)}$$

$$= \frac{2 \pm \sqrt{-12}}{2} \qquad\qquad\qquad\qquad\qquad = \frac{-2 \pm \sqrt{-12}}{2}$$

$$= \frac{2 \pm 2\sqrt{3}i}{2} \qquad\qquad\qquad\qquad\qquad = \frac{-2 \pm 2\sqrt{3}i}{2}$$

$$= 1 \pm \sqrt{3}i \qquad\qquad\qquad\qquad\qquad\qquad = -1 \pm \sqrt{3}i$$

The solution set is $\left\{2, -2, -1 + i\sqrt{3}, -1 - i\sqrt{3}, 1 + i\sqrt{3}, 1 - i\sqrt{3}\right\}$.

The graph of the equation is shown in Figure 2-46.

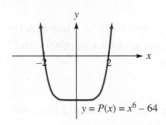

$y = P(x) = x^6 - 64$

FIGURE 2-46

Notice that there are only two real roots. In Section 1.5, we studied the transformations of graphs of functions. Recall that the graph of a function in the form $f(x) = x^n$, when n is even, will have the shape of the graph of $f(x) = x^2$. As the value of n increases, the graph is flatter at the turn.

Self Check 10   **a.** Factor the polynomial function $P(x) = x^4 - 25$ completely into linear and irreducible quadratic factors with real coefficients.

b. Use the factorization in part (a) to find all the roots of the equation $x^4 - 25 = 0$.

Now Try Exercise 77.

Self Check Answers   1. $\left\{2, \sqrt{5}, -\sqrt{5}\right\}$ 2. $\left\{-1, 3, \dfrac{3 + \sqrt{13}}{2}, \dfrac{3 - \sqrt{13}}{2}\right\}$

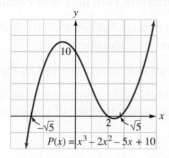

$P(x) = x^3 - 2x^2 - 5x + 10$

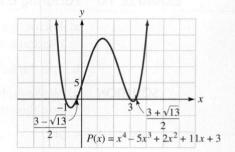

$P(x) = x^4 - 5x^3 + 2x^2 + 11x + 3$

3. The possible rational zeros $x = 1, -1, 3,$ and -3 do not satisfy $x^2 - 3 = 0$.

4. $\left\{1, -\dfrac{1}{2} - i\dfrac{\sqrt{3}}{2}, -\dfrac{1}{2} + i\dfrac{\sqrt{3}}{2}\right\}$ 5. $\left\{2, i\sqrt{5}, -i\sqrt{5}\right\}$ 6. $x^2 - 2x + 2 = 0$

7. $x^4 + 2x^2 + 1 = 0$ 8. $x^2 + 2ix - 1 = 0$ 9. $x = -3i, 1,$ and -3

10. $P(x) = (x^2 - 5)(x^2 + 5); \left\{\sqrt{5}, -\sqrt{5}, i\sqrt{5}, -i\sqrt{5}\right\}$

Exercises 2.4

Getting Ready

You should be able to complete these vocabulary and concept statements before you proceed to the practice exercises.

Fill in the blanks.

1. The conjugate of $a + bi$ is _____.

2. If $2 - 3i$ is a root of a polynomial equation, then _____ is also a root.

3. A fifth-degree polynomial equation can have at most _____ complex roots.

4. The polynomial equation
$3x^5 + x^4 + 21x^3 + 7x^2 + 36x + 12 = 0$ has exactly _____ roots among the complex numbers and _____ linear factors.

Practice

Solve the polynomial equation $P(x) = 0$ and use the solutions to graph the polynomial function.

5. $P(x) = 2x^3 - 3x^2 - 11x + 6$

6. $P(x) = 2x^3 - 9x^2 + x + 12$

7. $P(x) = x^3 - 5x$

8. $P(x) = x^4 - 2x^2$

9. $P(x) = 4x^4 - 12x^3 + x^2 + 15x - 2$

10. $P(x) = 9x^4 - 42x^3 + 50x^2 - 14x - 3$

11. $P(x) = 16x^3 - 48x^2 - 3x + 9$

12. $P(x) = 6x^3 + 3x^2 - 8x - 4$

Prove the given number is an irrational number.

13. $\sqrt{7}$

14. $-\sqrt{5}$

A partial solution set is given for each equation. Find the complete solution set.

15. $x^3 + 3x^2 - 13x - 15 = 0; \{-1\}$

16. $x^3 + 6x^2 + 5x - 12 = 0; \{1\}$

17. $2x^3 + x^2 - 18x - 9 = 0; \left\{-\dfrac{1}{2}\right\}$

18. $2x^3 - 3x^2 - 11x + 6 = 0; \left\{\dfrac{1}{2}\right\}$

19. $x^4 - 2x^3 - 2x^2 + 6x - 3 = 0; \{1, 1\}$

20. $x^4 + 2x^3 - 3x^2 - 4x + 4 = 0; \{1, -2\}$

Solve each polynomial equation.

21. $x^3 - x^2 + 4x - 4 = 0$

22. $3x^3 - 4x^2 + 27x - 36 = 0$

23. $x^3 - 2\sqrt{3}x^2 + 3x = 0$

24. $x^4 - 2\sqrt{5}x^3 + 5x^2 = 0$

25. $x^3 - 5x^2 + 17x - 13 = 0$

26. $x^4 - 2x^3 + 3x^2 - 6x = 0$

27. $3x^3 - 5x^2 + 3x - 1 = 0$

28. $x^3 + x^2 - x + 2 = 0$

29. $x^5 + x^3 + 2x^2 - 12x + 8 = 0$

30. $2x^5 - 9x^4 + 16x^3 - 22x^2 + 24x - 8 = 0$

31. $x^5 + x^4 + 16x^3 + 16x^2 = 0$

32. $x^5 - 4x^4 + 8x^3 - 8x^2 = 0$

Write a polynomial equation of the least possible degree with real coefficients and the given roots.

33. 4, 5

34. −3, 5

35. 1, 1, 1

36. 1, 0, −1

37. 2, 4, 5

38. 7, 6, 3

39. $1, -1, \sqrt{2}, -\sqrt{2}$

40. $0, 0, 0, \sqrt{3}, -\sqrt{3}$

41. $\sqrt{2}, i$

42. $3, -i$

43. $1, i$

44. $2, 2 + i$

45. $-2, 3 - i$

46. $2, 3, i$

47. $1, 2, 1 + i$

48. $i, 1 - i$

49. $i, 2 - i$

50. $0, 1 + i$

51. $2, 3i$

52. $0, 2 - 3i$

53. $i, 2$

54. $i, 2 + i$

Use the given zero to find the remaining zeros of each polynomial function.

55. $P(x) = x^4 - 4x^3 + 4x^2 - 4x + 3; -i$

56. $P(x) = 2x^4 + x^3 + 17x^2 + 9x - 9; -3i$

57. $P(x) = x^5 - 5x^3 + 10x^2 - 6x; 1 + i$

58. $P(x) = x^5 - 10x^4 + 38x^3 - 66x^2 + 45x; 2 - i$

59. $P(x) = x^4 - x^3 + 2x^2 - 4x - 8; 2i$

60. $P(x) = x^5 + 5x^4 + 4x^3 - 6x^2 + 8x + 24; 1 - i$

Given that $1 + i$ is a root of each equation, find the other roots.

61. $x^3 - 5x^2 + 8x - 6 = 0$

62. $x^3 - 2x + 4 = 0$

63. $x^4 - 2x^3 - 7x^2 + 18x - 18 = 0$

64. $x^4 - 2x^3 - 2x^2 + 8x - 8 = 0$

65. Find a second-degree equation with a double root of $2i$.

66. Find a second-degree equation with a double root of $-3i$.

The graph and degree of a polynomial function $P(x)$ is shown. State the number of real zeros and the number of complex zeros, assuming all zeros have multiplicity one.

67. degree 2

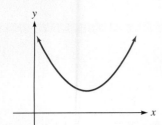

68. degree 2

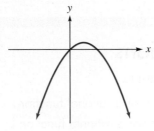

69. degree 3

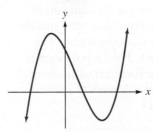

70. degree 3

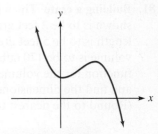

71. degree 4

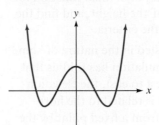

72. degree 4

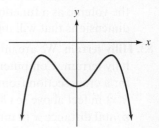

73. degree 4

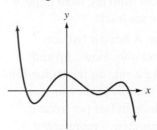

74. degree 4

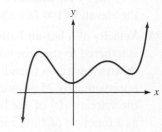

75. degree 5

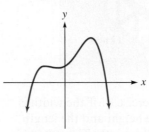

76. degree 5

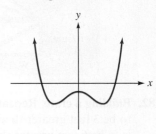

For each polynomial function $P(x)$,

(a) Factor the polynomial completely into linear and irreducible quadratic factors with real coefficients.

(b) Use the factorization in part (a) to find all the roots of the equation $P(x) = 0$.

77. $P(x) = x^4 - 49$

78. $P(x) = x^4 - 144$

79. $P(x) = x^6 - 1$

80. $P(x) = x^6 - 729$

Applications

 81. Building a crate The width of the shipping crate shown is to be 2 feet greater than its height and the length is to be 5 feet greater than the width. The volume is to be 170 cubic feet. Find a polynomial function for the volume as a function of the height, and find the dimensions that will meet the criteria. (Round to the nearest tenth.)

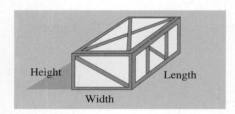

Height

Length

Width

82. Building a crate Repeat Exercise 81 if the width is to be 3 feet greater than the height and the length is to be 6 feet greater than the width. The volume is to be 216 cubic feet. Find a polynomial function for the volume as a function of the height, and find the dimensions that will meet the criteria.

83. Hilly terrain We are interested in the nature of some hilly terrain. Computer simulation has told us that for a cross section from west to east, the elevation $h(x)$ in feet above sea level is related to the horizontal distance x in miles from a fixed point by the function $h(x) = -x^4 + 5x^3 + 91x^2 - 545x + 550$, $x \in [0, 9]$. At what distances from the fixed point is the elevation 100 feet above sea level?

 84. Velocity of a hot-air balloon A hot-air balloon is tethered to the ground and only moves up and down. You and a friend take a ride on the ballon for approximately 25 minutes. On this particular ride, the velocity $v(t)$ of the balloon in feet per minute, as a function of time t in minutes, is represented by the function $v(t) = -t^3 + 34t^2 - 320t + 850$. At what times is the velocity of the balloon 50 feet per minute?

85. Velocity of a robot The velocity of a robotic manufacturing aide that delivers parts along a conveyor belt is represented by the function $v(t) = -t^3 + 18t^2 - 94t + 140$, where velocity is in tens of feet and time is in minutes. It takes the robot 10 minutes to make one trip from the supply point. At what times on the time interval $[0,10]$ is the velocity of the robot 0 feet per minute?

86. Velocity of a robot Using a graphing calculator, approximate (to the nearest tenth of a minute) the times during the trip when the robot in Exercise 85 will have a velocity of 50 feet per minute?

Discovery and Writing

87. Explain why the Fundamental Theorem of Algebra guarantees that every polynomial equation of positive degree has at least one root.

88. Explain why the Fundamental Theorem of Algebra and the Factor Theorem guarantee that an nth-degree polynomial equation has n roots.

89. Prove that any odd-degree polynomial equation with real coefficients must have at least one real root.

90. If a, b, c, and d are positive numbers, prove that $ax^4 + bx^2 + cx - d = 0$ has exactly two nonreal roots.

2.5 Rational Functions

In this section, we will learn to

1. Determine the domain of a rational function.
2. Examine the behavior of a rational function by finding its vertical, horizontal, and oblique asymptotes.
3. Graph a rational function.
4. Analyze applications that involve rational functions.

We have discussed polynomial functions and now focus on another class of functions, called **rational functions**. Rational functions are defined by rational expressions that are quotients of polynomials. For example, consider the time t it takes a NASCAR driver to drive the 500 miles in the Daytona 500 race. The time t can be defined as a function of the average rate of the driver's speed r. That is, $t = f(r) = \frac{500}{r}$. If a driver averages a speed of 170 mph, we can evaluate $f(170)$ to determine the time in hours driven.

$$f(170) = \frac{500}{170} = \frac{50}{17}$$

This is approximately 2.94 hours. We will discuss this type of function in this section.

1. Determine the Domain of a Rational Function

In Section 1.2, we defined a function and the domain of a function. We should recall that the domain of a function is the set of numbers we can input into a function. There are times when the domain of a function is defined by the application of the function; however, unless specific restrictions are stated, we find the domain to be the set of all numbers for which the function is defined. For functions such as the polynomial functions we have studied in the previous sections of this chapter, the domain is the set of all real numbers. For other functions, we have to analyze the function to determine the domain. In Section 1.4, the **reciprocal function**, $f(x) = \frac{1}{x}$, was defined. The domain and range for this function are both the set of all nonzero real numbers, $(-\infty, 0) \cup (0, \infty)$. The graph is shown in Figure 2-47.

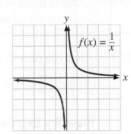

FIGURE 2-47

This function is an example of a **rational function.**

Rational Function   A function of the form $R(x) = \frac{P(x)}{Q(x)}$, where $P(x)$ and $Q(x)$ are polynomial functions and $Q(x) \neq 0$, is a **rational function.**

The **domain of a rational function** is the set of all real numbers for which $Q(x) \neq 0$.

EXAMPLE 1   **Determining the Domain of a Rational Function**

Find the domain of each function.

a. $f(x) = \dfrac{3}{x - 2}$ **b.** $f(x) = \dfrac{5x + 2}{x^2 - 4}$ **c.** $f(x) = \dfrac{3x^2 - 2}{(2x + 1)(x - 3)}$

d. $f(x) = \dfrac{3x + 2}{x^2 - 7x + 12}$ **e.** $f(x) = \dfrac{2x - 1}{x^2 + 25}$

SOLUTION   We can examine the denominator of each function and determine whether there are any values of x that will result in the denominator being equal to 0. These values are not in the domain.

a. Since $x = 2$ is the only value that will make the denominator 0, the domain is all real numbers except 2. It can be expressed using set-builder notation, $D = \{x \mid x \ne 2\}$, or interval notation, $D = (-\infty, 2) \cup (2, \infty)$.

b. To determine the domain, we will factor the denominator.

$$f(x) = \frac{5x + 2}{x^2 - 4} = \frac{5x + 2}{(x - 2)(x + 2)}$$

The denominator is 0 if $x = 2$ or $x = -2$, so those values of x cannot be part of the domain. The domain is $D = \{x \mid x \ne -2, x \ne 2\}$ or $D = (-\infty, -2) \cup (-2, 2) \cup (2, \infty)$.

c. For $x = -\dfrac{1}{2}$ and $x = 3$, the denominator is 0. The domain of the function is

$$D = \left\{ x \,\middle|\, x \ne -\frac{1}{2}, x \ne 3 \right\} \text{ or } D = \left(-\infty, -\frac{1}{2} \right) \cup \left(-\frac{1}{2}, 3 \right) \cup (3, \infty).$$

d. Factoring the denominator, we have $f(x) = \dfrac{3x + 2}{x^2 - 7x + 12} = \dfrac{3x + 2}{(x - 4)(x - 3)},$ which indicates that the domain is $D = \{x \mid x \ne 3, x \ne 4\}$ or $D = (-\infty, 3) \cup (3, 4) \cup (4, \infty)$.

e. The denominator of this function does not factor over the real numbers, and has no real zeros. Therefore there are no restrictions on the domain, and the domain is the set of all real numbers, $D = (-\infty, \infty)$.

Comment

If there are numerous restrictions on the domain, it may be easier for you to use set-builder notation rather than interval notation.

Caution

Do not exclude values of the variable from the domain that make the numerator equal to zero, unless the denominator has the same factor. We will discuss this situation later in this section.

Self Check 1   Find the domain of each function.

a. $f(x) = \dfrac{2x - 3}{x^2 - x - 2}$ **b.** $f(x) = \dfrac{x}{x^2 + 1}$

Now Try Exercise 13.

2. Examine the Behavior of a Rational Function by Finding Its Vertical, Horizontal, and Oblique Asymptotes

Consider the function $t = f(r) = \frac{500}{r}$ from the beginning of the section. A graph of this function is shown in Figure 2-48.

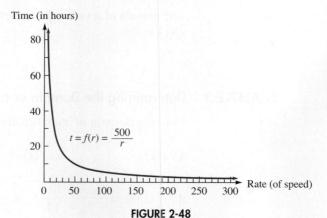

FIGURE 2-48

We see from the graph that as the rate of speed increases, the time it takes to complete the race decreases. In fact, if we drive at rocket speed, we will arrive in almost no time at all. We can express this by saying that as the rate increases without bound (or approaches infinity), the time it takes to complete the race approaches 0 hours. When a graph approaches a line as shown in the figure, we call the line an **asymptote**. The horizontal line representing the rate axis shown in the graph is a **horizontal asymptote**.

We also see from the graph that as the rate of speed decreases, the time it takes to complete the race increases. In fact, if the car goes at turtle speed, it will take almost forever to finish the race. We can express this by saying that as the rate gets slower and slower (or approaches 0 mph), the time approaches infinity. The vertical line representing the time axis shown on the graph is a **vertical asymptote.** A vertical asymptote is a vertical line that the graph approaches, but never touches.

While polynomial functions have continuous graphs with no holes, gaps, or jumps, the same cannot be said of rational functions. Rational functions often exhibit **asymptotic behavior**. At this time, we will consider only rational functions that are reduced, so there are no common factors in the numerator and denominator.

First we will examine how a function behaves as it approaches a value of x for which the function is undefined, and then we will look at how the function behaves as x increases without bound, $x \to \infty$, and decreases without bound, $x \to -\infty$.

Consider again the reciprocal function, $f(x) = \frac{1}{x}$, paying attention to how the numerical values of the function change as the variable approaches 0. Since the domain does not include $x = 0$, we have to answer the question, what happens to the value of the function as x approaches 0 from both the left and from the right? The notation $x \to 0^+$ is equivalent to saying "as x approaches 0 from the right." The notation $x \to 0^-$ is equivalent to saying "as x approaches 0 from the left." Look at the tables in Figure 2-49.

$x \to 0^+$									
x	1,000	100	10	2	1	$\frac{1}{2}$	$\frac{1}{10}$	$\frac{1}{100}$	$\frac{1}{1,000}$
$f(x) = \frac{1}{x}$	$\frac{1}{1,000}$	$\frac{1}{100}$	$\frac{1}{10}$	$\frac{1}{2}$	1	2	10	100	1,000

$x \to 0^-$									
x	$-1,000$	-100	-10	-2	-1	$-\frac{1}{2}$	$-\frac{1}{10}$	$-\frac{1}{100}$	$-\frac{1}{1,000}$
$f(x) = \frac{1}{x}$	$-\frac{1}{1,000}$	$-\frac{1}{100}$	$-\frac{1}{10}$	$-\frac{1}{2}$	-1	-2	-10	-100	$-1,000$

FIGURE 2-49

We see in the first table that as the value of x gets closer and closer to 0 from the right, the value of the function increases without bound. We see in the second table that as the value of x gets closer and closer to 0 from the left, the value of the function decreases without bound. We can see this graphically too (see Figure 2-50). We conclude that the line $x = 0$ is a **vertical asymptote.**

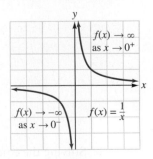

FIGURE 2-50

Vertical Asymptote   The line $x = a$ is a **vertical asymptote** of the graph of a function if the value of the function increases without bound or decreases without bound as x approaches a.

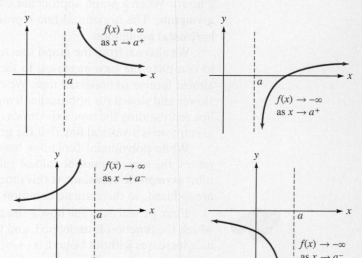

If $R(x) = \frac{P(x)}{Q(x)}$ is a rational function (quotient of polynomial functions), with $Q(x) \neq 0$, and if $P(x)$ and $Q(x)$ have no common factors, there will be a vertical asymptote at all zeros of the polynomial $Q(x)$.

Comment   The graph of a function will never cross a vertical asymptote. The definition of the rational function will determine how many vertical asymptotes (if any) the graph of the function has. Later in the course, we will look at functions that have an infinite number of vertical asymptotes.

EXAMPLE 2   Finding Vertical Asymptotes

Find the vertical asymptotes, if any, of each function.

a. $f(x) = \dfrac{2x - 1}{x + 3}$ **b.** $f(x) = \dfrac{x^2}{x^2 - 25}$ **c.** $f(x) = \dfrac{x + 1}{x^2 + 5}$

SOLUTION   All of the functions shown are reduced, having no common factors. To find vertical asymptotes, we can find the zeros of each denominator.

a. Solving $x + 3 = 0$, we find $x = -3$. The only vertical asymptote is $x = -3$.

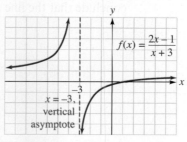

FIGURE 2-51

b. Solving the equation below, we find that there are two vertical asymptotes, $x = 5$ and $x = -5$.

$$x^2 - 25 = 0$$
$$(x - 5)(x + 5) = 0$$
$$x = 5 \quad \text{or} \quad x = -5$$

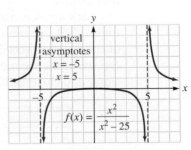

FIGURE 2-52

c. There is no real solution to the equation $x^2 + 5 = 0$; therefore, there are no vertical asymptotes.

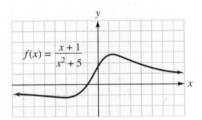

FIGURE 2-53

Self Check 2   Find the vertical asymptote, if any, of each function.

a. $f(x) = \dfrac{x - 1}{x^2 - 4x}$ **b.** $f(x) = \dfrac{3x + 2}{5x - 4}$ **c.** $f(x) = \dfrac{x - 1}{x^2 + 6}$

Now Try Exercise 23.

Caution

When we find vertical asymptotes, it is always important that we factor $P(x)$ and $Q(x)$ completely and remove any common factors. We then identify the zeros of the denominator.

Consider $g(x) = \dfrac{x - 4}{x^2 - 16}$.

$$g(x) = \frac{x - 4}{(x + 4)(x - 4)} \qquad \text{Factor the denominator completely.}$$

We can now identify $x = 4$ and $x = -4$ as zeros of the denominator. However, the only vertical asymptote is $x = -4$. Since the numerator and the denominator have the common factor $x - 4$, there is a "hole" in the graph of the function at $x = 4$, but not a vertical asymptote. If we look only at the given function, and forget to factor, we may incorrectly identify $x = 4$ as a vertical asymptote. See the graph in Figure 2-54.

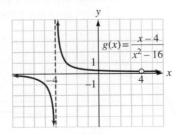

FIGURE 2-54

We will now examine the behavior of rational functions as the value of the variable x increases without bound, $x \to \infty$, and decreases without bound, $x \to -\infty$. We will again restrict this discussion to rational functions that are reduced. This is similar to our study of polynomial functions and determining the end behavior they exhibit.

Let's again consider the function $f(x) = \frac{1}{x}$. Looking at the tables of values in Figure 2-55, we can see in the first table that as the values of x increases, the value of the function gets closer and closer to 0. We can say that as $x \to \infty$, $f(x) \to 0$. We can see in the second table that as x decreases, the value of the function also gets closer and closer to 0: As $x \to -\infty$, $f(x) \to 0$.

$x \to \infty$									
x	$\frac{1}{1,000}$	$\frac{1}{100}$	$\frac{1}{10}$	$\frac{1}{2}$	1	2	10	100	1,000
$f(x) = \frac{1}{x}$	1,000	100	10	2	1	$\frac{1}{2}$	$\frac{1}{10}$	$\frac{1}{100}$	$\frac{1}{1,000}$

$x \to -\infty$									
x	$-\frac{1}{1,000}$	$-\frac{1}{100}$	$-\frac{1}{10}$	$-\frac{1}{2}$	-1	-2	-10	-100	$-1,000$
$f(x) = \frac{1}{x}$	$-1,000$	-100	-10	-2	-1	$-\frac{1}{2}$	$-\frac{1}{10}$	$-\frac{1}{100}$	$-\frac{1}{1,000}$

FIGURE 2-55

We can see these same results graphically in Figure 2-56. We conclude that the line $y = 0$ is a **horizontal asymptote**.

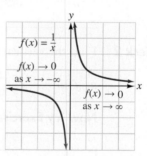

FIGURE 2-56

Comment

In calculus, this behavior is described using **limits.** For example, we would write $\lim\limits_{x \to \infty} \frac{1}{x} = 0$. This means that as x increases without bound, the function gets closer and closer to the limiting value of 0. We could also write $\lim\limits_{x \to 0^+} \frac{1}{x} = \infty$. This is the same as saying that as x gets closer and closer to 0 from the right, the function increases without bound. This is explained further at the end of the chapter in the "Looking Ahead to Calculus" section.

Horizontal Asymptote   The line $y = b$ is a **horizontal asymptote** of the graph of a function if the value of the function approaches b, $f(x) \to b$, when x increases without bound, $x \to \infty$, or x decreases without bound, $x \to -\infty$.

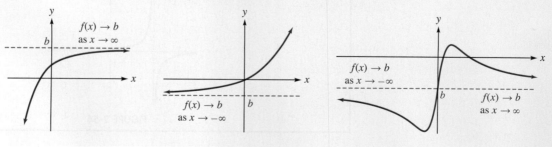

Comment

The graph of a function may cross a horizontal asymptote. A rational function will have at most one horizontal asymptote.

To find whether a rational function has a horizontal asymptote, we can use the degree of the numerator and the degree of the denominator as our starting point and then use the following guidelines. We use these stated guidelines because analyzing a function's behavior and determining what value $f(x)$ is approaching as x increases or decreases without bound is often very complicated. The guidelines simplify this for us.

Strategy for Finding Horizontal Asymptotes

For a rational function $R(x) = \dfrac{P(x)}{Q(x)} = \dfrac{a_m x^m + a_{m-1} x^{m-1} + \cdots + a_1 x + a_0}{b_n x^n + b_{n-1} x^{n-1} + \cdots + b_1 x + b_0}$,

$a_m \neq 0$, $b_n \neq 0$, horizontal asymptotes can be found by comparing the degree of $P(x)$, which is m, and the degree of $Q(x)$, which is n, using these guidelines:

- If $m < n$, the line $y = 0$, which is the x-axis, is the horizontal asymptote of the graph of the function.
- If $m = n$, the line $y = \dfrac{a_m}{b_n}$ is the horizontal asymptote of the graph of the function.
- If $m > n$, there is no horizontal asymptote.

EXAMPLE 3   **Finding Horizontal Asymptotes**

Find the horizontal asymptote, if any, of each function.

a. $f(x) = \dfrac{2x - 5}{x + 3}$ **b.** $f(x) = \dfrac{1 - x}{x^2 + 2}$ **c.** $f(x) = \dfrac{x^2 - 4}{x - 1}$

SOLUTION   We will compare the degree of the numerator to that of the denominator to find any horizontal asymptote that exists.

a. For $f(x) = \frac{2x-5}{x+3}$, the degree of the numerator is $m = 1$ and the degree of the denominator is $n = 1$. Since $m = n$, the horizontal asymptote is the line $y = \frac{a_m}{b_n} = \frac{2}{1} = 2$. See the values in the table in Figure 2-57 that demonstrate the behavior $f(x) \to 2$ as $x \to -\infty$, and as $x \to \infty$. Notice that there is a vertical asymptote at $x = -3$.

$x \to \infty$	$f(x) \to 2$	$x \to -\infty$	$f(x) \to 2$
1,000	1.989	−1,000	2.011
10,000	1.9989	−10,000	2.0011
100,000	1.99989	−100,000	2.00011
1,000,000	1.999989	−1,000,000	2.000011

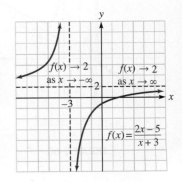

FIGURE 2-57

b. For $f(x) = \frac{1-x}{x^2+2}$, the degree of the numerator is $m = 1$ and the degree of the denominator is $n = 2$. Since $m < n$, the horizontal asymptote is the line $y = 0$. See the values in the table in Figure 2-58 that demonstrate the behavior $f(x) \to 0$ as $x \to -\infty$ and as $x \to \infty$. Notice that there is no vertical asymptote.

$x \to \infty$	$f(x) \to 0$	$x \to -\infty$	$f(x) \to 0$
1,000	-0.001	$-1,000$	0.001
10,000	-0.0001	$-10,000$	0.0001
100,000	-0.00001	$-100,000$	0.00001
1,000,000	-0.000001	$-1,000,000$	0.000001

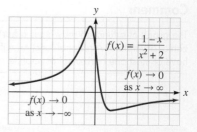

$f(x) = \dfrac{1-x}{x^2+2}$

$f(x) \to 0$ as $x \to \infty$

$f(x) \to 0$ as $x \to -\infty$

FIGURE 2-58

c. For $f(x) = \frac{x^2 - 4}{x - 1}$, the degree of the numerator is $m = 2$ and the degree of the denominator is $n = 1$. Since $m > n$, there is no horizontal asymptote. Notice that there is a vertical asymptote at $x = 1$.

$x \to \infty$	$f(x) \to \infty$	$x \to -\infty$	$f(x) \to -\infty$
1,000	1,001	$-1,000$	-999
10,000	10,001	$-10,000$	$-9,999$
100,000	100,001	$-100,000$	$-99,999$
1,000,000	1,000,001	$-1,000,000$	$-999,999$

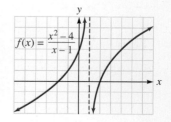

$f(x) = \dfrac{x^2-4}{x-1}$

FIGURE 2-59

Self Check 3 Find the horizontal asymptote, if any, of each function.

a. $f(x) = \dfrac{x^2 - 4}{2x^2 - 1}$ **b.** $f(x) = \dfrac{x^3}{x + 2}$ **c.** $f(x) = \dfrac{x - 1}{x^2 + 1}$

Now Try Exercise 31.

For a reduced rational function $R(x) = \frac{P(x)}{Q(x)}$, if the degree of the numerator is greater than the degree of the denominator, there will be no horizontal asymptote. If the degree of the numerator is one greater than the degree of the denominator, the graph of the function will have an **oblique (slant) asymptote**. There will be at most one oblique asymptote for the graph of any rational function. To determine the equation of the asymptote, we will use long division and the division algorithm to rewrite the function. For example, if $f(x) = \dfrac{3x^3 + 2x^2 + 2}{x^2 - 1}$, we perform long division to rewrite the function using the division algorithm.

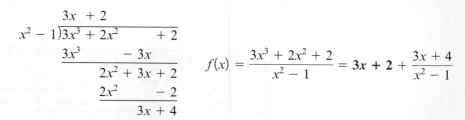

$$f(x) = \frac{3x^3 + 2x^2 + 2}{x^2 - 1} = 3x + 2 + \frac{3x + 4}{x^2 - 1}$$

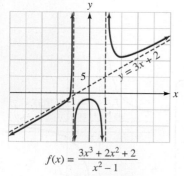

$f(x) = \dfrac{3x^3 + 2x^2 + 2}{x^2 - 1}$

FIGURE 2-60

As $x \to \infty$, the term $\frac{3x + 4}{x^2 - 1} \to 0$, and the graph of the function approaches the **oblique asymptote** $y = 3x + 2$. The graph is shown in Figure 2-60. Notice that there are vertical asymptotes at $x = 1$ and $x = -1$.

EXAMPLE 4 Finding Oblique Asymptotes

Find the oblique asymptote of each function.

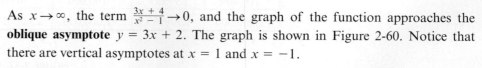

a. $f(x) = \dfrac{2x^3 - x^2 - 7x + 7}{x^2 - 4}$ **b.** $f(x) = \dfrac{3x^2 - 2x - 4}{x - 2}$

SOLUTION   **a.** Since the degree of the numerator is one greater than that of the denominator, we will use long division and the division algorithm to find the oblique asymptote.

$$x^2 - 4\overline{)2x^3 - x^2 - 7x + 7}$$

$$\begin{array}{r} 2x \;\; - \;\; 1 \\ \hline 2x^3 \qquad\quad - 8x \\ \hline -x^2 + \;\; x + 7 \\ -x^2 \qquad\quad + 4 \\ \hline x + 3 \end{array}$$

$$f(x) = \frac{2x^3 - x^2 - 7x + 7}{x^2 - 4} = 2x - 1 + \frac{x + 3}{x^2 - 4}$$

The graph of the function has the oblique asymptote $y = 2x - 1$. The graph is shown in Figure 2-61. Notice that there are vertical asymptotes at $x = 2$ and $x = -2$.

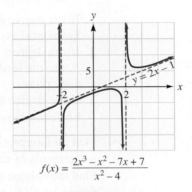

$$f(x) = \frac{2x^3 - x^2 - 7x + 7}{x^2 - 4}$$

FIGURE 2-61

b. We can use synthetic division for the division because we are dividing by $x - 2$, a linear polynomial.

$$\begin{array}{r|rrr} 2 & 3 & -2 & -4 \\ & & 6 & 8 \\ \hline & 3 & 4 & 4 \end{array}$$

$$f(x) = \frac{3x^2 - 2x - 4}{x - 2} = 3x + 4 + \frac{4}{x - 2}$$

The oblique asymptote is $y = 3x + 4$, and the graph is shown in Figure 2-62. Note that there is a vertical asymptote at $x = 2$.

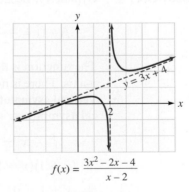

$$f(x) = \frac{3x^2 - 2x - 4}{x - 2}$$

FIGURE 2-62

■

Self Check 4   Find the oblique asymptote of $f(x) = \dfrac{3x^4 - 2x^2 + x - 1}{x^3 - x - 1}$.

Now Try Exercise 59.

3. Graph a Rational Function

Now that we have learned strategies for finding asymptotes, we will combine those strategies with others to develop a strategy for graphing rational functions that are reduced. (Rational functions that have common factors in the numerator and denominator will be covered later in the section.)

Strategy for Graphing a Reduced Rational Function $R(x) = \dfrac{P(x)}{Q(x)}$

- **Check for symmetry.** If $P(x)$ and $Q(x)$ involve only even powers of x, or if $R(-x) = R(x)$, the function is an even function and the graph is symmetric about the y-axis. Check for symmetry about the origin by determining whether $R(-x) = -R(x)$.
- **Find vertical asymptotes.** The real roots of $Q(x) = 0$, if any, determine the vertical asymptotes of the graph.
- **Find the horizontal asymptote.**

  1. If the degree of $P(x)$ is less than the degree of $Q(x)$, the line $y = 0$ is a horizontal asymptote.
  2. If the degrees of $P(x)$ and $Q(x)$ are equal, the line $y = \frac{a_m}{b_n}$, where a_m and b_n are the lead coefficients of $P(x)$ and $Q(x)$, respectively, is a horizontal asymptote.
  3. If the degree of $P(x)$ is greater than the degree of $Q(x)$, there is no horizontal asymptote.

- **Find all intercepts.** Solutions to the equation $P(x) = 0$, if any, are the x-intercepts of the graph. $R(0)$, if defined, is the y-intercept of the graph.
- **Find the oblique asymptote.** If the degree of $P(x)$ is one greater than the degree of $Q(x)$, there is an oblique asymptote. Divide $P(x)$ by $Q(x)$ using long division or synthetic division if possible. The quotient is the oblique asymptote. Ignore the remainder.
- **Plot other points.** Plot at least one point on each side of vertical asymptotes and on each side of any x-intercepts.

EXAMPLE 5 Graphing a Rational Function

Graph: $f(x) = \dfrac{x^2 - 4}{x^2 - 1}$.

SOLUTION
- **Symmetry:** Because all the exponents of x are even, $f(-x) = f(x)$.

$$f(-x) = \frac{(-x)^2 - 4}{(-x)^2 - 1} = \frac{x^2 - 4}{x^2 - 1} = f(x)$$

The function is therefore an even function and its graph is symmetric about the y-axis. There is no symmetry about the origin, because $-f(x) \neq f(-x)$.

- **Vertical asymptotes:** The rational function is in reduced form. To find the vertical asymptotes, we set the denominator equal to 0 and solve for x.

$$x^2 - 1 = 0$$
$$(x + 1)(x - 1) = 0$$
$$x + 1 = 0 \quad \text{or} \quad x - 1 = 0$$
$$x = -1 \quad \mid \quad x = 1$$

There will be vertical asymptotes at $x = -1$ and $x = 1$.

- **Horizontal asymptote:** Since the degrees of the numerator and denominator polynomials are both 2 and the coefficients are both equal to 1, the line $y = 1$ is a horizontal asymptote.

- **Intercepts:** We can find the y-intercept by evaluating $f(0)$.

$$f(0) = \frac{x^2 - 4}{x^2 - 1} = \frac{0^2 - 4}{0^2 - 1} = \frac{-4}{-1} = 4$$

The y-intercept is 4.

We can find any x-intercepts by setting the numerator equal to 0 and solving for x.

$$x^2 - 4 = 0$$
$$(x + 2)(x - 2) = 0$$
$$x + 2 = 0 \quad \text{or} \quad x - 2 = 0$$
$$x = -2 \quad \bigg| \quad x = 2$$

The x-intercepts are 2 and -2.

- **Plot other points:** First we will sketch the information that we have so far. This will allow us to see what other points to plot to allow us to complete the graph of the function. From the information shown in Figure 2-63, we can determine where we need to plot other points to complete the graph. Remember that asymptotes are not part of the graph, but they do indicate the behavior of the function as it approaches them.

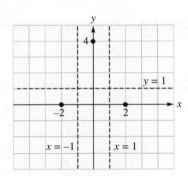

FIGURE 2-63

Using the table in Figure 2-64, we evaluate the function at several points so that we are able to complete the graph. Since the function is even, we know that $f(-x) = f(x)$, so we need only to calculate the function for positive values of x. Plotting these points and the corresponding symmetric points across the y-axis will show us how we should complete the graph.

x	$\dfrac{1}{2}$	$\dfrac{3}{2}$	3
$f(x)$	5	$-\dfrac{7}{5}$	$\dfrac{5}{8}$

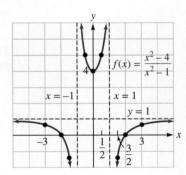

FIGURE 2-64

Self Check 5 Graph: $f(x) = \dfrac{2x^2}{x^2 - 4}$.

Now Try Exercise 75.

ACCENT ON TECHNOLOGY | **Graphing a Rational Function**

A graphing calculator can be very useful in helping us to draw the graph of a rational function. However, certain precautions must be used or we will get a graph that is not going to give us a good idea of how the function looks. For example, Figure 2-65(a) shows the graph of the function from Example 5, $f(x) = \dfrac{x^2 - 4}{x^2 - 1}$. Using the [ZOOM] Standard window, a good representation of the graph is shown. In Figure 2-65(b), the [ZOOM] Decimal window is used, and part of the graph is not shown.

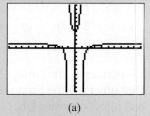

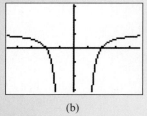

(a)   (b)

FIGURE 2-65

Therefore, while a calculator is a great tool, it should be used to verify what we have already discovered, and not to just draw the graph.

EXAMPLE 6   **Graphing a Rational Function**

Graph: $f(x) = \dfrac{3x}{x - 2}$.

SOLUTION • **Symmetry:** Because $f(-x) = \dfrac{3(-x)}{-x - 2} = \dfrac{3x}{x + 2} \neq f(x)$, there is no symmetry about the y-axis. Neither is there symmetry about the origin, because $f(-x) \neq -f(x)$.

$$f(-x) = \frac{3(-x)}{-x - 2} = \frac{3x}{x + 2} \quad \text{and} \quad -f(x) = -\frac{3x}{x - 2}$$

• **Vertical asymptotes:** To find the vertical asymptotes, we set the denominator equal to 0 and solve for x. Since the solution is 2, there will be a vertical asymptote at $x = 2$.

• **Horizontal asymptote:** Since the degrees of the numerator and denominator polynomials are the same, the line $y = \frac{3}{1} = 3$ is a horizontal asymptote.

• **Intercepts:** We can find the y-intercept by evaluating $f(0) = 0$. The y-intercept is 0, and the graph passes through the origin, $(0, 0)$. We can find any x-intercepts by setting the numerator equal to 0 and solving the resulting equation.

$$3x = 0 \rightarrow x = 0$$

The only x-intercept is 0. Both the x- and y-intercepts are 0.

• **Plot other points:** We will select some values of x to help us complete the graph.

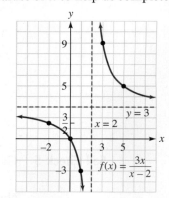

x	-2	1	3	5
$f(x)$	$\dfrac{3}{2}$	-3	9	5

FIGURE 2-66

Self Check 6   Graph: $f(x) = \dfrac{-2x}{x + 3}$.

Now Try Exercise 73.

Listed here are a few important details worth noting about asymptotic behavior:

- As x increases or decreases without bound, the function will approach the horizontal asymptote, if there is one.
- As x gets closer and closer to a vertical asymptote, the function will either increase without bound or decrease without bound.
- A function will never cross a vertical asymptote, but can cross a horizontal asymptote.

To determine whether a function crosses a horizontal asymptote $y = b$, we solve the equation $f(x) = b$. If there is a solution, then the graph crosses or touches the horizontal asymptote.

EXAMPLE 7   Graphing a Rational Function That Crosses a Horizontal Asymptote

Graph: $f(x) = \dfrac{x}{x^2 - 9}$.

SOLUTION
- **Symmetry:** Note that $f(-x) = \dfrac{-x}{(-x)^2 - 9} = \dfrac{-x}{x^2 - 9} = -\dfrac{x}{x^2 - 9} = -f(x)$.
  Therefore, the function is odd, and the graph will be symmetric about the origin.
- **Vertical asymptotes:** To find the vertical asymptotes, we set the denominator equal to 0 and solve for x. We see that there will be vertical asymptotes at $x = 3$ and $x = -3$.

$$x^2 - 9 = 0$$
$$(x - 3)(x + 3) = 0$$
$$x = 3 \quad \text{or} \quad x = -3$$

- **Horizontal asymptote:** Since the degree of the numerator is less than the degree of the denominator, $y = 0$ is the horizontal asymptote.
- **Intercepts:** We can find the y-intercept by evaluating $f(0) = 0$. The y-intercept is 0, and the graph passes through the origin, $(0, 0)$. Setting the numerator equal to 0 yields an x-intercept of 0. Both intercepts are 0.
- **Plot other points:** We will select some values of x to help us complete the graph. Notice that the graph of this function crosses the horizontal asymptote, but it cannot cross a vertical asymptote.

x	-5	-2	2	5
$f(x)$	$-\dfrac{5}{16}$	$\dfrac{2}{5}$	$-\dfrac{2}{5}$	$\dfrac{5}{16}$

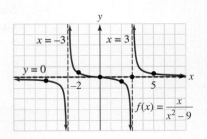

FIGURE 2-67

Self Check 7   Graph: $f(x) = \dfrac{-2x}{x^2 - 1}$.

Now Try Exercise 83.

ACCENT ON TECHNOLOGY   **Graphing a Rational Function; Additional Insights**

There are times when a graphing calculator may show a graph that looks as though it has a break in it, when it doesn't. Look at Figure 2-68(a), which is the graph of the function from Example 7 using a ⟨zoom⟩ Standard window. The graph looks as though it has a gap around the origin, but we know that is not the case. Also take note of the function as it approaches the x-axis, which is the horizontal asymptote. Sometimes students will look at a calculator graph and just copy it as it is without understanding that a calculator can only fill in pixels and cannot always draw a complete graph. The graph in Figure 2-68(b) uses the ⟨zoom⟩ Decimal window, which shows the behavior around the origin a little better but not the asymptotic behavior around the horizontal asymptote. You must use care when using a calculator to graph a rational function.

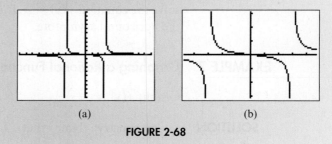

(a)   (b)

FIGURE 2-68

EXAMPLE 8   **Graphing a Rational Function with No Vertical Asymptotes**

Graph: $f(x) = \dfrac{1}{x^2 + 1}$.

SOLUTION   •   **Symmetry:** Since $f(-x) = \dfrac{1}{(-x)^2 + 1} = \dfrac{1}{x^2 + 1} = f(x)$, the function is even, and the graph will be symmetric about the y-axis.

•   **Vertical asymptotes:** The equation $x^2 + 1 = 0$ has no real solutions, and the denominator is never 0. There are no vertical asymptotes.

•   **Horizontal asymptote:** The degree of the numerator is less than the degree of the denominator, and $y = 0$ and is the horizontal asymptote.

•   **Intercepts:** We can find the y-intercept by evaluating $f(0) = 1$. The y-intercept is 1, and the graph passes through $(0, 1)$. Since the numerator cannot be equal to 0, there is no x-intercept.

•   **Plot other points:** We will select some values of x to help us complete the graph. Since the numerator is 1 and the denominator will always be greater than or equal to 1, we can see that 1 is the maximum value the function attains.

x	-3	-2	-1	0	1	2	3
$f(x)$	$\dfrac{1}{10}$	$\dfrac{1}{5}$	$\dfrac{1}{2}$	1	$\dfrac{1}{2}$	$\dfrac{1}{5}$	$\dfrac{1}{10}$

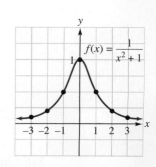

FIGURE 2-69

Self Check 8   Graph: $f(x) = \dfrac{-2}{x^2 + 2}$.

Now Try Exercise 87.

EXAMPLE 9   **Graphing a Rational Function with an Oblique Asymptote**

Graph: $f(x) = \dfrac{x^2 + x - 2}{x - 3}$.

SOLUTION   •   **Symmetry:** We evaluate $f(-x)$.

$$f(x) = \frac{x^2 + x - 2}{x - 3}$$

$$f(-x) = \frac{(-x)^2 + (-x) - 2}{(-x) - 3} \qquad \text{Replace } x \text{ with } -x.$$

$$= \frac{x^2 - x - 2}{-x - 3} \qquad \text{Simplify.}$$

$$= \frac{x^2 - x - 2}{(-1)(x + 3)} \qquad \text{Factor } -1 \text{ out of the denominator.}$$

$$= \frac{-x^2 + x + 2}{x + 3} \qquad \text{Divide by } -1.$$

Because $f(-x) \neq f(x)$ and $f(-x) \neq -f(x)$, there is no symmetry about the y-axis or origin. The function is not odd or even.

•   **Vertical asymptotes:** The vertical asymptote is the line $x = 3$.

•   **Intercepts:** $f(0) = \frac{2}{3}$, so the y-intercept is $\frac{2}{3}$. To find the x-intercepts, we set the numerator equal to 0 and solve. We see that the x-intercepts are $x = 1$ and $x = -2$.

$$x^2 + x - 2 = 0$$

$$(x + 2)(x - 1) = 0$$

$$x = -2 \quad \text{or} \quad x = 1$$

•   **Oblique asymptote:** Since the degree of the numerator is one greater than the degree of the denominator, there is an oblique asymptote. We can use synthetic division to find an equation of the asymptote.

$$\underline{3|}\ \ 1\ \ \ 1\ \ -2$$
$$\underline{\quad\ \ 3\ \ 12}$$
$$1\ \ \ 4\ \ \ 10$$

$$f(x) = \frac{x^2 + x - 2}{x - 3} = x + 4 + \frac{10}{x - 3}$$

The line $y = x + 4$ is the oblique asymptote.

•   **Plot other points:** We will select some values of x to help us complete the graph.

x	-8	-4	-2	2	4	8
$f(x)$	$-\dfrac{54}{11}$	$-\dfrac{10}{7}$	0	-4	18	14

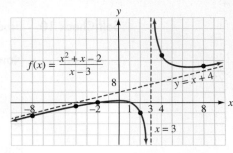

FIGURE 2-70

Self Check 9   Graph: $f(x) = \dfrac{x^2 - 4x - 5}{x - 2}$.

Now Try Exercise 89.

We have discussed rational functions in simplified form. We now consider a rational function $R(x) = \dfrac{P(x)}{Q(x)}$ where $P(x)$ and $Q(x)$ have a common factor. Graphs of such functions have gaps or missing points that are not the result of vertical asymptotes.

EXAMPLE 10   **Graphing a Rational Function That Is Not Reduced**

Find the domain of the function $f(x) = \dfrac{x^2 - x - 12}{x - 4}$ and graph the function.

SOLUTION   Since a denominator cannot be 0, $x \neq 4$. The domain is the set of all real numbers except 4. When we factor the numerator of the expression, we see that the numerator and denominator have a common factor of $x - 4$.

$$f(x) = \frac{x^2 - x - 12}{x - 4}$$

$$= \frac{(x + 3)(x - 4)}{x - 4}$$

If $x \neq 4$, the common factor of $x - 4$ can be divided out. The resulting function is equivalent to the original function only when we keep the restriction that $x \neq 4$. Thus,

$$y = \frac{(x + 3)(x - 4)}{x - 4} = x + 3 \qquad \text{(provided that } x \neq 4\text{)}.$$

When $x = 4$, the function is not defined. The graph of the function appears in Figure 2-71. It is a line with the point that has the x-coordinate of 4 missing.

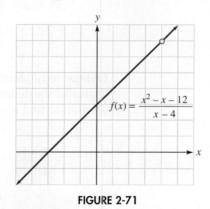

$$f(x) = \frac{x^2 - x - 12}{x - 4}$$

FIGURE 2-71

Self Check 10   Find the domain of the function $f(x) = \dfrac{x^2 + x - 6}{x + 3}$ and graph the function.

Now Try Exercise 91.

4. Analyze Applications That Involve Rational Functions

Rational expressions often define functions. For example, if the cost of a no-contract cell phone is $20 per month plus $0.15 per minute of use, the monthly cost

will be $C(n) = 20 + 0.15n$, where n is the number of minutes used in a month. The average (mean) monthly cost per minute is the total monthly cost divided by the number of minutes of air time. We express the function for the average monthly cost per minute as

$$\overline{C}(n) = \frac{C(n)}{n} = \frac{20 + 0.15n}{n}, \; n > 0.$$

Since $n > 0$, the domain of the function is the interval $(0, \infty)$. Because the numerator and denominator each have degree one, the horizontal asymptote is $y = 0.15$, and there is a vertical asymptote, $x = 0$. The graph of the function is shown in Figure 2-72. As the number of minutes used in a month increases, the average cost per minute gets closer and closer to \$0.15.

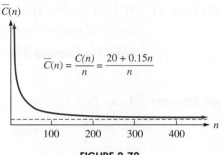

FIGURE 2-72

In business, there are three functions that are often used. The **cost function** is found by adding fixed costs to variable costs. The **revenue function** is the price per item times the number of items sold. The **profit function** is the revenue function minus the cost function. The **average-cost function, average-revenue function,** and **average-profit function** are examples of rational functions.

EXAMPLE 11   **Average-Cost Function**

The daily-cost function for a company that makes and sells sandals is
$C(x) = 0.01x^2 + 5x + 100$.

a. Find the average-cost function.

b. Graph the average-cost function.

c. Use a graphing calculator to find the level of production that yields the minimum average cost.

SOLUTION   **a.** The average-cost function is

$$\overline{C}(x) = \frac{C(x)}{x} = \frac{0.01x^2 + 5x + 100}{x} = 0.01x + 5 + \frac{100}{x}, \; x > 0.$$

b. This function has a vertical asymptote at $x = 0$. There is an oblique asymptote at $y = 0.01x + 5$.

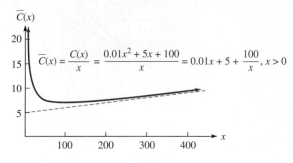

FIGURE 2-73

c. Using the techniques we've already developed, we find the minimum of the average-cost function.

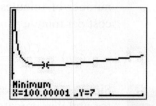

FIGURE 2-74

A production level of 100 units will produce the minimum average cost of $7. ∎

Self Check 11   If $C(x) = x^2 + 160x + 6,400$, use a graphing calculator to find the value of x that gives the minimum average cost.

Now Try Exercise 97.

Self Check Answers   **1. a.** $D = \{x \mid x \neq -1, x \neq 2\}$ **b.** $D = (-\infty, \infty)$ **2. a.** $x = 0, x = 4$

b. $x = \dfrac{4}{5}$ **c.** none   **3. a.** $y = \dfrac{1}{2}$ **b.** none   **c.** $y = 0$ **4.** $y = 3x$

5.

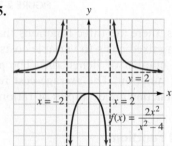

6.

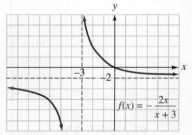

7.

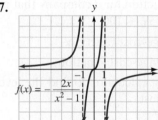

8.

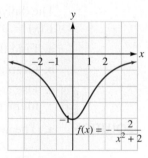

9.

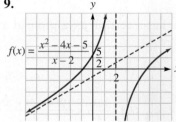

10. $D = (-\infty, -3) \cup (-3, \infty)$

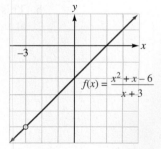

11. $x = 80$

Exercises **2.5**

Getting Ready

You should be able to complete these vocabulary and concept statements before you proceed to the practice exercises.

Fill in the blanks.

1. When a graph approaches a vertical line but never touches it, we call the line a vertical _____.

2. A rational function is a function with a polynomial numerator and a polynomial _____.

3. To find a _____ asymptote of a rational function in reduced form, set the denominator polynomial equal to 0 and solve the equation.

4. To find the ____-intercept of a rational function $R(x)$, find $R(0)$.

5. To find the ____-intercept of a rational function, set the numerator equal to 0 and solve the equation.

6. Given the rational function $R(x) = \frac{P(x)}{Q(x)}$, if the degree of $P(x)$ is less than the degree of $Q(x)$, the horizontal asymptote is _____.

7. Given the rational function $R(x) = \frac{P(x)}{Q(x)}$, if the degree of $P(x)$ and the degree of $Q(x)$ are _____, the horizontal asymptote is $y = \frac{\text{the lead coefficient of the numerator}}{\text{the lead coefficient of the denominator}}$.

8. Given the rational function $R(x) = \frac{P(x)}{Q(x)}$, if the degree of the numerator is one greater than the degree of the denominator, the graph will have an _____ asymptote.

9. The graph of a function may cross a _____ asymptote but will never cross a _____ asymptote.

10. The graph of $f(x) = \frac{x^2 - 4}{x + 2}$ will have a(n) _____ at $x = -2$.

Practice

Find the domain of each rational function and write in interval notation. Do not graph the function.

11. $f(x) = \dfrac{x^2}{x - 2}$

12. $f(x) = \dfrac{x^3 - 3x^2 + 1}{x + 3}$

13. $f(x) = \dfrac{2x^2 + 7x - 2}{x^2 - 25}$

14. $f(x) = \dfrac{5x^2 + 1}{x^2 + 5}$

15. $f(x) = \dfrac{x - 1}{x^3 - x}$

16. $f(x) = \dfrac{x + 2}{2x^2 - 9x + 9}$

17. $f(x) = \dfrac{3x^2 + 5}{x^2 + 1}$

18. $f(x) = \dfrac{7x^2 - x + 2}{x^4 + 4}$

19. $f(x) = \dfrac{x + 1}{x^2 + 5x + 4}$

20. $f(x) = \dfrac{x}{x^3 - x^2}$

Find the vertical asymptotes, if any, of each rational function. Do not graph the function.

21. $f(x) = \dfrac{x}{x - 3}$

22. $f(x) = \dfrac{2x}{2x + 5}$

23. $f(x) = \dfrac{x + 2}{x^2 - 1}$

24. $f(x) = \dfrac{x - 4}{x^2 - 16}$

25. $f(x) = \dfrac{1}{x^2 - x - 6}$

26. $f(x) = \dfrac{x + 2}{2x^2 - 6x - 8}$

27. $f(x) = \dfrac{x^2}{x^2 + 5}$

28. $f(x) = \dfrac{x^3 - 3x^2 + 1}{2x^2 + 3}$

Find the horizontal asymptote, if any, of each rational function. Do not graph the function.

29. $f(x) = \dfrac{2x - 1}{x}$

30. $f(x) = \dfrac{x^2 + 1}{3x^2 - 5}$

31. $f(x) = \dfrac{x^2 + x - 2}{2x^2 - 4}$

32. $f(x) = \dfrac{5x^2 + 1}{5 - x^2}$

33. $f(x) = \dfrac{x + 1}{x^3 - 4x}$

34. $f(x) = \dfrac{x}{2x^2 - x + 11}$

35. $f(x) = \dfrac{x^2}{x - 2}$

36. $f(x) = \dfrac{x^4 + 1}{x - 3}$

Complete each of the following exercises regarding the function $f(x) = \dfrac{1 - x}{x^2 - 2x - 3}$ using the graph for reference.

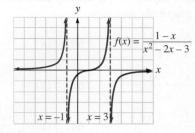

37. State the domain of the function.

38. State the range of the function.

39. State the equation of each asymptote.

40. Identify the y-intercept.

41. Identify the x-intercept.

42. As $x \to \infty$, $f(x) \to$ ____

43. As $x \to -1^-$, $f(x) \to$ ____

44. As $x \to -1^+$, $f(x) \to$ ____

45. As $x \to 3^-$, $f(x) \to$ ____

46. As $x \to 3^+$, $f(x) \to$ ____

Complete each of the following exercises regarding the function $f(x) = \dfrac{1}{x(x-1)^2}$, *using the graph for reference.*

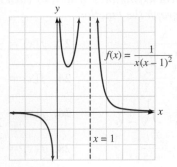

$f(x) = \dfrac{1}{x(x-1)^2}$

$x = 1$

47. State the domain of the function.

48. State the range.

49. Why is there no *x*-intercept?

50. Why is there no *y*-intercept?

51. State the equation of each vertical asymptote.

52. State the equation of the horizontal asymptote.

53. As $x \to \infty$, $f(x) \to$ _____

54. As $x \to -\infty$, $f(x) \to$ _____

55. As $x \to 0^-$, $f(x) \to$ _____

56. As $x \to 0^+$, $f(x) \to$ _____

57. As $x \to 1^-$, $f(x) \to$ _____

58. As $x \to 1^+$, $f(x) \to$ _____

Find the oblique asymptote, if any, of each rational function. Do not graph the function.

59. $f(x) = \dfrac{x^2 + 1}{x - 1}$

60. $f(x) = \dfrac{x^3 - 1}{x^2}$

61. $f(x) = \dfrac{x^2 - 5x - 6}{x - 2}$

62. $f(x) = \dfrac{x^2 - 2x + 11}{x + 3}$

63. $f(x) = \dfrac{2x^2 - 5x + 1}{x - 4}$

64. $f(x) = \dfrac{5x^3 + 1}{x + 5}$

65. $f(x) = \dfrac{x^3 + 2x^2 - x - 1}{x^2 - 1}$

66. $f(x) = \dfrac{-x^3 + 3x^2 - x + 1}{x^2 + 1}$

Use the guidelines of the section to graph each rational function, including vertical, horizontal, and oblique asymptotes; x- and y-intercepts; and symmetry. Check your work with a graphing calculator.

67. $f(x) = \dfrac{1}{x - 2}$

68. $f(x) = \dfrac{3}{x + 3}$

69. $f(x) = \dfrac{x}{x - 1}$

70. $f(x) = \dfrac{x}{x + 2}$

71. $f(x) = \dfrac{x + 1}{x + 2}$

72. $f(x) = \dfrac{x - 1}{x - 2}$

73. $f(x) = \dfrac{2x - 1}{x - 1}$

74. $f(x) = \dfrac{3x + 2}{x^2 - 4}$

75. $g(x) = \dfrac{x^2 - 9}{x^2 - 4}$

76. $g(x) = \dfrac{x^2 - 4}{x^2 - 9}$

77. $g(x) = \dfrac{x^2 - x - 2}{x^2 - 4x + 3}$

78. $g(x) = \dfrac{x^2 + 7x + 12}{x^2 - 7x + 12}$

79. $h(x) = \dfrac{x^2 + 2x - 3}{x^3 - 4x}$

80. $h(x) = \dfrac{x^2 - 9}{x^2}$

81. $h(x) = \dfrac{3x^2 - 12}{x^2}$

82. $f(x) = \dfrac{x}{(x + 3)^2}$

83. $f(x) = \dfrac{x}{(x - 1)^2}$

84. $f(x) = \dfrac{x + 1}{x^2(x - 2)}$

85. $f(x) = \dfrac{x}{x^2 + 1}$

86. $f(x) = \dfrac{x - 1}{x^2 + 2}$

87. $g(x) = \dfrac{3x^2}{x^2 + 1}$

88. $f(x) = \dfrac{x^2 - 9}{2x^2 + 1}$

89. $h(x) = \dfrac{x^2 - 2x - 8}{x - 1}$

90. $f(x) = \dfrac{x^3 + x^2 + 6x}{x^2 - 1}$

Graph each rational function. Note that the numerator and denominator of the fraction share a common factor.

91. $f(x) = \dfrac{x^2}{x}$

92. $f(x) = \dfrac{x^2 - 2x + 1}{x - 1}$

93. $f(x) = \dfrac{x^3 + x}{x}$

94. $f(x) = \dfrac{x^3 - x^2}{x - 1}$

95. $f(x) = \dfrac{x^3 - 1}{x - 1}$

96. $f(x) = \dfrac{x^2 - x}{x^2}$

Applications

97. Cost of membership directory A service club wants to publish a directory of its members. Some investigation shows that the cost of typesetting and photography will be $700, and the cost of printing each directory will be $1.25.

   a. Find a function that gives the total cost *C* of printing *x* directories.

   b. Find a function that gives the mean cost per directory $\overline{C}$ of printing *x* directories.

   c. Find the mean cost per directory if 500 directories are printed.

   d. Find the mean cost per directory if 1,000 directories are printed.

   e. Find the total cost of printing 500 directories.

   f. Find the mean cost per directory if 2,000 directories are printed.

98. Electricity charges An electric company charges $7.50 per month plus $0.09 for each kilowatt-hour (kwh) of electricity used.

   a. Find a function that gives the total cost *C* of *n* kwh of electricity.

   b. Find a function that gives the mean cost per kwh $\overline{C}$ when using *n* kwh.

c. Find the total cost for using 775 kwh.

d. Find the mean cost per kwh when 775 kwh are used.

e. Find the mean cost per kwh when 1,000 kwh are used.

f. Find the mean cost per kwh when 1,200 kwh are used.

99. Utility costs An electric company charges $8.50 per month plus $0.095 for each kilowatt-hour (kwh) of electricity used.

a. Find a linear function that gives the total cost of n kwh of electricity.

b. Find a rational function that gives the average cost per kwh when using n kwh.

c. Find the average cost per kwh when 850 kwh are used.

100. Scheduling work crews The rational function $f(t) = \frac{t^2 + 3t}{2t + 3}$ gives the number of days it would take two construction crews, working together, to frame a house that crew 1 (working alone) could complete in t days and crew 2 (working alone) could complete in $t + 3$ days.

a. If crew 1 could frame a certain house in 21 days, how long would it take both crews working together?

b. If crew 2 could frame a certain house in 25 days, how long would it take both crews working together?

101. Employee bonus A company decides to annually reward employees whose performance is excellent, based on criteria published in their company newsletter. The performance bonus B consists of a fixed amount A plus an equal share of a large bonus b. The number of employees qualifying for the performance bonus is n. The function $B(n) = A + \frac{b}{n}$ is used to calculate the performance bonus.

a. If the company decides to give an individual bonus of $300 to each qualifying employee and let them share a large bonus of $12,000, write the function that represents the performance bonus.

b. How much will each of 30 qualifying employees get?

c. If the number of employees n qualifying for the bonus increases without bound, what will the amount of each employee's performance bonus approach?

102. Employee bonus Use a graphing calculator and the information given in Exercise 101 to answer the following questions.

a. If the company decides *not* to give the fixed amount A, how does the graph of the function change as n increases?

b. How does the graph of the function change if the company also increases the fixed amount A by $100?

103. Oxygen content of a lake Suppose the oxygen content of a lake, expressed as a percentage of the lake's usual oxygen content, at t days after organic waste is dumped into the lake, is given by the function

$$O(t) = \frac{100t^2 + 500t + 10,000}{t^2 + 15t + 100}.$$

a. What is the oxygen content of the lake 20 days after organic waste is dumped into the lake?

b. What is the oxygen content of the lake 50 days after organic waste is dumped into the lake?

c. What can you say about $O(t)$ when t is extremely large?

d. Over time, will nature restore the oxygen content to its natural level?

Discovery and Writing

104. Can a rational function have two horizontal asymptotes? Explain.

105. Can a rational function have two slant asymptotes? Explain.

In Exercises 106–108, a, b, c, and d are nonzero constants.

106. Show that the graph of $f(x) = \frac{ax + b}{cx^2 + d}$ has the horizontal asymptote $y = 0$.

107. Show that the graph of $f(x) = \frac{ax^3 + b}{cx^2 + d}$ has the slant asymptote $y = \frac{a}{c}x$.

108. Show that the graph of $f(x) = \frac{ax^2 + b}{cx^2 + d}$ has the horizontal asymptote $y = \frac{a}{c}$.

109. Graph the rational function $f(x) = \frac{x^3 + 1}{x}$ and explain why the curve is said to have a *parabolic asymptote*.

Use a graphing calculator to perform each experiment. Write a brief paragraph describing your findings.

110. Investigate the positioning of the vertical asymptotes of a rational function by graphing $f(x) = \frac{x}{x - k}$ for several values of k. What do you observe?

111. Investigate the positioning of the vertical asymptotes of a rational function by graphing $f(x) = \frac{x}{x^2 - k}$ for $k = 4, 1, -1,$ and 0. What do you observe?

112. Find the range of the rational function $f(x) = \frac{kx^2}{x^2 + 1}$ for several values of k. What do you observe?

113. Investigate the positioning of the x-intercepts of a rational function by graphing $f(x) = \frac{x^2 - k}{x}$ for $k = 1, -1,$ and 0. What do you observe?

2.6 Polynomial and Rational Inequalities

In this section, we will learn to

1. Solve and graph a polynomial inequality.
2. Solve and graph a rational inequality.
3. Solve a real-world problem involving inequalities.

China was the host of the summer Olympics in 2008. The swimming and diving competitions were held at the Beijing National Aquatics Center, nicknamed the Water Cube. The blue-colored design is not an actual cube, but a cuboid, a rectangular box. To symbolize water, the outside of the center was covered in bubbles.

If a diver performs a simple forward dive from a 10-foot springboard, the height of the diver $s(t)$ in feet at time t in seconds can be modeled by a quadratic function. If the quadratic function $s(t) = -16t^2 + 12t + 10$ represents the height of a diver at time t seconds, how would we determine the time interval in which the diver's height is greater than 6 feet?

To solve this problem we would need to find t such that $s(t) > 6$. That is, we would need to solve the inequality $-16t^2 + 12t + 10 > 6$. You will be asked to solve this problem in Exercise 81.

This type of inequality is known as a **quadratic inequality**. We will learn to solve three types of inequalities in this section: quadratic, polynomial (of degree 3 and higher), and rational.

We can use what we have learned in this chapter about polynomial and rational functions to solve polynomial and rational inequalities. In Section R.5, we solved linear, compound, and absolute-value inequalities. Recall that our solution in most cases was an interval, or intervals. Occasionally a single value was our solution or part of the solution. There were also cases where we had no solution and those where all real numbers satisfied the inequality. These were not encountered very often. These same types of solutions will be seen for polynomial and rational inequalities.

Polynomial Inequality

For a polynomial function $f(x)$, a **polynomial inequality** is an inequality that can be written in the form

$$f(x) > 0, f(x) \geq 0, f(x) < 0, \text{ or } f(x) \leq 0.$$

Note that there are four inequalities possible: $f(x) > 0$, $f(x) \geq 0$, $f(x) < 0$, or $f(x) \leq 0$. There are two approaches that can easily be used to solve a polynomial inequality: graphing and making a sign chart.

When we graphed polynomial functions, we determined the point or points where the function either crossed the x-axis or touched it and turned; these were the zeros of the function. If the function changed signs from positive (above the x-axis) to negative (below the x-axis), or vice versa, then it did so at a zero.

Look at Figure 2-75. Note the intervals listed in the table in Figure 2-76 where the function satisfies each of the inequalities.

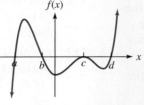

Inequality	Interval
$f(x) > 0$	$(a, b) \cup (d, \infty)$
$f(x) \geq 0$	$[a, b] \cup \{c\} \cup [d, \infty)$
$f(x) < 0$	$(-\infty, a) \cup (b, c) \cup (c, d)$
$f(x) \leq 0$	$(-\infty, a] \cup [b, d]$

FIGURE 2-75   **FIGURE 2-76**

Caution

> Pay special attention to the use of parentheses and brackets. A parenthesis is used when there is a strict inequality, $f(x) < 0$ or $f(x) > 0$. Brackets are used for closed intervals. Never use a bracket with the infinity symbol, ∞; this symbol represents unbounded behavior. Notice that the inequality $f(x) \geq 0$ has two intervals and the value $x = c$ as part of the solution. Since the function is equal to 0 at that point, it is part of the solution set.

In addition to using a graph to solve an inequality, we can also use a sign chart. Since the zeros are the only values of the variable where the function can change signs, the sign of the function is either positive or negative in the interval between two zeros. It cannot be both positive and negative in that interval.

To determine the sign of the function, we will use a sign chart much like we did in Section 2.2, when we discussed graphing polynomials. The sign chart that fits the function in Figure 2-75 is shown in Figure 2-77.

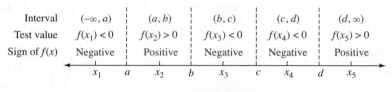

Interval	$(-\infty, a)$	(a, b)	(b, c)	(c, d)	(d, ∞)
Test value	$f(x_1) < 0$	$f(x_2) > 0$	$f(x_3) < 0$	$f(x_4) < 0$	$f(x_5) > 0$
Sign of $f(x)$	Negative	Positive	Negative	Negative	Positive

$x_1 \quad a \quad x_2 \quad b \quad x_3 \quad c \quad x_4 \quad d \quad x_5$

FIGURE 2-77

Strategy for Solving a Polynomial Inequality Using a Sign Chart

Step 1: Express the inequality in one of these forms:

$$f(x) > 0, \ f(x) \geq 0, \ f(x) < 0, \ \text{or} \ f(x) \leq 0.$$

Step 2: Determine the zeros, if any, of the function by solving the equation $f(x) = 0$.

Step 3: Make a sign chart. Mark the zeros on an x-axis, and test a number in each interval to determine the sign of the function in that interval.

Step 4: Select the interval or intervals that satisfy the inequality from the sign chart and write the solution.

1. Solve and Graph a Polynomial Inequality

If $a \neq 0$, inequalities such as $ax^2 + bx + c < 0$ and $ax^2 + bx + c > 0$ are called **quadratic inequalities.** We have graphed quadratic functions by finding the vertex and determining any zeros of the function. We will now use this knowledge to solve quadratic inequalities.

EXAMPLE 1   **Solving a Quadratic Inequality**

Determine the interval(s), if any, where $x^2 - x > 6$.

SOLUTION   We rewrite the inequality as $x^2 - x - 6 > 0$. We define the quadratic function to be $f(x) = x^2 - x - 6$, and we solve the inequality $f(x) > 0$. Quadratic functions can have at most two zeros, and we can find the zeros by using the quadratic formula or factoring, if the function is factorable. This function can be factored.

$$x^2 - x - 6 = 0 \qquad \text{Set } f(x) = 0.$$

$$(x - 3)(x + 2) = 0 \qquad \text{Factor.}$$

$$x - 3 = 0 \quad \text{or} \quad x + 2 = 0$$

$$x = 3 \qquad \qquad x = -2 \qquad \text{Find the zeros}$$

At this point, there are two ways to approach the problem. We will demonstrate both methods.

1. We can find the vertex and graph the function.
2. We can use a sign chart.

Method 1: Find the Vertex and Graph the Function

To graph the quadratic function, we will find the vertex and then plot the vertex and the two zeros. Since the leading coefficient is positive, the parabola opens upward. Recall from Section 2.1 that a quadratic function $f(x) = ax^2 + bx + c$ has its vertex at the point $(h, k) = \left(-\frac{b}{2a}, f\left(-\frac{b}{2a}\right)\right)$.

$$h = -\frac{b}{2a} = -\frac{(-1)}{2(1)} = \frac{1}{2} \qquad \text{From } f(x) = x^2 - x - 6, \text{ we have } a = 1 \text{ and } b = -1.$$

$$k = f\left(-\frac{b}{2a}\right) = f\left(\frac{1}{2}\right) = \left(\frac{1}{2}\right)^2 - \frac{1}{2} - 6 = \frac{1}{4} - \frac{1}{2} - 6 = -\frac{25}{4}$$

The vertex is $\left(\frac{1}{2}, -\frac{25}{4}\right)$.

Now we graph the function, as shown in Figure 2-78, and determine the intervals where the graph is above the x-axis. The solution to the inequality $x^2 - x - 6 > 0$ is $(-\infty, -2) \cup (3, \infty)$.

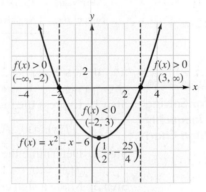

FIGURE 2-78

Method 2: Use a Sign Chart

Draw an x-axis. Mark the zeros, $x = -2$ and $x = 3$. We now evaluate the function at one test value in each interval. The results are shown in Figure 2-79. The results from the sign chart are the same as those found by using the graph.

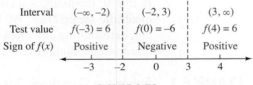

Interval	$(-\infty, -2)$	$(-2, 3)$	$(3, \infty)$
Test value	$f(-3) = 6$	$f(0) = -6$	$f(4) = 6$
Sign of $f(x)$	Positive	Negative	Positive

FIGURE 2-79

The solution set as indicated on the sign chart is $(-\infty, -2) \cup (3, \infty)$. ∎

Self Check 1   Solve: $2x^2 + 5x \leq 12$.

Now Try Exercise 7.

ACCENT ON TECHNOLOGY   **Solving an Inequality Using a Calculator**

A graphing calculator can be used to solve inequalities. We can use the graph or the table to determine where the polynomial is positive and where it is negative. The amount of work we do with the calculator can vary. We can find the roots and then graph the function, or we can graph the function and use the calculator to find the zeros. For higher-degree polynomials, we need to make sure that we have found all the zeros if we decide to use only the calculator. The calculator screens in Figure 2-80 show the solution of the inequality $x^2 - x - 6 > 0$ to be $(-\infty, -2) \cup (3, \infty)$.

Comment

When we create a sign chart, remember that we are using a number line. Therefore, we must make sure the values of x are placed in the correct order. Students often write the values down in any order, which will not yield a correct answer.

Graph and evaluate
the function.

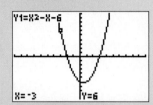

Use the table to evaluate
the function.

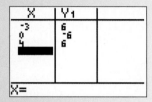

FIGURE 2-80

Polynomial inequalities of degree 3 and higher can be solved by the methods used in Example 1. If the polynomial function can be factored completely using the Rational-Zeros Theorem or other techniques, then either a graph or a sign chart can be used to solve the inequality. In this section, we will restrict the polynomial functions to those that can be factored. If the polynomial function doesn't factor, and has irrational zeros, then techniques discussed earlier in the chapter will be required.

EXAMPLE 2   **Solving a Polynomial Inequality**

Solve the inequality $x^3 + 3x^2 \le 6x + 8$ using a sign chart.

SOLUTION   We will follow the steps indicated in the strategy box to solve this inequality with a sign chart.

Step 1: Rewrite the inequality as $x^3 + 3x^2 - 6x - 8 \le 0$. We can now define the polynomial function as $f(x) = x^3 + 3x^2 - 6x - 8$.

Step 2: Find the zeros of the polynomial function. We use the Rational-Zeros Theorem that was discussed in Section 2.3 to find that the set of possible rational zeros is $\dfrac{p}{q} = \dfrac{\pm 1, \pm 2, \pm 4, \pm 8}{\pm 1} = \pm 1, \pm 2, \pm 4,$ and ± 8. Evaluating the function at these values, we find $f(-1) = 0$, which indicates that $x = -1$ is a zero. At this point, we can use synthetic division to factor the function or we can continue to test the values in the function as shown in the table. Since this is a third-degree polynomial, we know that there can be at most three zeros.

x	1	-1	2	-2	4	-4	8	-8
$f(x)$	-10	0	0	8	80	0	648	-280

Looking at the table, we can determine that the three zeros of this function are $x = -1, 2,$ and -4. Thus $f(x) = x^3 + 3x^2 - 6x - 8 = (x + 1)(x - 2)(x + 4)$.

Step 3: Create the sign chart and evaluate the function at test values to determine the sign of the function.

Interval	$(-\infty, -4)$	$(-4, -1)$	$(-1, 2)$	$(2, \infty)$
Test value	$f(-5) = -28$	$f(-2) = 8$	$f(0) = -8$	$f(3) = 28$
Sign of $f(x)$	Negative	Positive	Negative	Positive

$$\xleftarrow{\hspace{1cm}}\;\underset{-5}{\bullet}\;\underset{-4}{|}\;\underset{-2}{\bullet}\;\underset{-1}{|}\;\underset{0}{\bullet}\;\underset{2}{|}\;\underset{3}{\bullet}\;\xrightarrow{\hspace{1cm}}$$

FIGURE 2-81

Step 4: Solve the inequality using the sign chart. The sign chart indicates that $f(x) \leq 0$ over the intervals $(-\infty, -4] \cup [-1, 2]$. The graph of the function is shown in Figure 2-82, which confirms the answer.

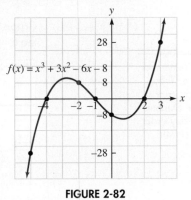

FIGURE 2-82

Self Check 2   Solve the inequality $13x \geq x^3 + 12$ using a sign chart.

Now Try Exercise 37.

2. Solve and Graph a Rational Inequality

A rational inequality can be solved in the same way that a polynomial inequality is solved. Although graphing is a possible method, the sign chart is the most efficient method; we will use it in the examples to follow. A rational function may have a restricted domain, so we must be mindful of that when we state the solution set. A value of the variable that would yield a zero denominator cannot be included in an interval.

As with polynomial inequalities, our first step is to rewrite the inequality in one of the forms $f(x) > 0$, $f(x) \geq 0$, $f(x) < 0$, or $f(x) \leq 0$, where $f(x)$ is a rational function. Some rational inequalities will require algebraic simplification.

EXAMPLE 3   **Solving a Rational Inequality**

Solve the inequality $\dfrac{6}{x} \geq 2$ using a sign chart.

SOLUTION   To get a 0 on the right-hand side, we add -2 to both sides. We then combine like terms on the left-hand side.

$$\frac{6}{x} \geq 2$$

$$\frac{6}{x} - 2 \geq 0$$

$$\frac{6}{x} - \frac{2(x)}{x} \geq 0$$

$$\frac{6 - 2x}{x} \geq 0$$

We define the rational function to be $f(x) = \frac{6 - 2x}{x}$, and we note that the domain is $D = \{x \mid x \in \mathbb{R}, x \neq 0\}$. Even though $x = 0$ is not in the domain of the function, we will include that value in the sign chart, as well as $x = 3$, since $f(3) = 0$. Recall that graphically there will be a vertical asymptote at $x = 0$, but a rational function can change from positive to negative or from negative to positive on either side of an asymptote.

Interval	$(-\infty, 0)$	$(0, 3)$	$(3, \infty)$
Test value	$f(-1) = -8$	$f(1) = 4$	$f(4) = -\frac{1}{2}$
Sign of $f(x)$	Negative	Positive	Negative

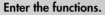

FIGURE 2-83

The solution to the inequality $\dfrac{6 - 2x}{x} \geq 0$ is the interval $(0, 3]$.

Caution

Pay special attention to the fact that the interval does not include $x = 0$, because $f(0)$ is not defined. The graph of the function confirms this solution.

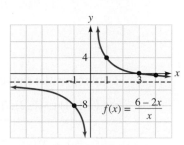

FIGURE 2-84

Self Check 3   Solve: $\dfrac{3}{x - 1} \leq 4$.

Now Try Exercise 61.

ACCENT ON TECHNOLOGY   **Another Way to Solve Inequalities Using a Graphing Calculator**

We can solve an inequality with a graphing calculator without simplifying the inequality. For Example 3, $\frac{6}{x} \geq 2$, we graph two functions, $y = \frac{6}{x}$ and $y = 2$. Once we graph the functions, we then determine the point of intersection and the interval where $\frac{6}{x} \geq 2$. We exclude the value $x = 0$ from the solution set.

Enter the functions.   Graph the functions.   Find the intersection.

 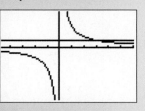 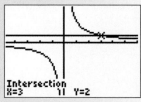

FIGURE 2-85

The interval over which $\frac{6}{x} \geq 2$ is the interval $(0, 3]$.

EXAMPLE 4   **Solving a Rational Inequality**

Solve: $\dfrac{(x - 3)(x - 1)^2}{(x + 1)} < 0$.

SOLUTION   The rational function is $f(x) = \dfrac{(x-3)(x-1)^2}{(x+1)}$. Looking at the numerator, we note that $f(3) = 0$ and $f(1) = 0$, so these are the only zeros of the function. We also note that $f(-1)$ is not defined, because the denominator cannot equal 0. The graph of the function will have a vertical asymptote at $x = -1$. These are the three values that we will use in the sign chart. Recall that when the multiplicity of a zero is even, the graph only touches the x-axis at that value. We note that the zero $x = 1$ has multiplicity two. This will result in there not being a sign change at that value. The sign chart for this function is shown in Figure 2-86.

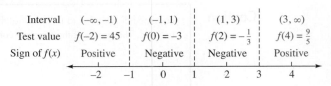

Interval	$(-\infty, -1)$	$(-1, 1)$	$(1, 3)$	$(3, \infty)$
Test value	$f(-2) = 45$	$f(0) = -3$	$f(2) = -\frac{1}{3}$	$f(4) = \frac{9}{5}$
Sign of $f(x)$	Positive	Negative	Negative	Positive

FIGURE 2-86

Looking at the sign chart, we determine that $\dfrac{(x-3)(x-1)^2}{(x+1)} < 0$ has the solution set $(-1, 1) \cup (1, 3)$. The value $x = 1$ is not included because $f(1) = 0$. ∎

Self Check 4   Solve: $\dfrac{(x+3)(x-3)}{x^2} \le 0$.

Now Try Exercise 53.

3. Solve a Real-World Problem Involving Inequalities

Recall from Section 2.1 that we defined the position function $s(t) = -gt^2 + v_0 t + h_0$, which yields the position of a propelled object at time t in seconds propelled at an initial velocity of v_0 from an initial height h_0. If the height is in meters, then $g = 4.9$ meters per second per second; if it is in feet, $g = 16$ feet per second per second.

EXAMPLE 5   **An Application of the Position Function**

A baseball is thrown vertically upward from the top of the Sears Tower in Chicago. The tower is 1,353 feet tall. If the initial velocity of the ball is 200 feet per second, for what interval of time will the ball have a height greater than the tower's?

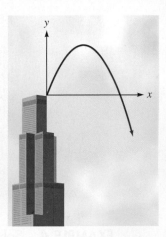

SOLUTION   Using the position function $s(t) = -gt^2 + v_0 t + h_0$, we substitute $g = 16$, $v_0 = 200$, and $h = 1{,}353$. Our height function is

$$s(t) = -16t^2 + 200t + 1{,}353.$$

To determine the time interval in which the ball's height is greater than the building's (1,353 feet), we must solve the following quadratic inequality:

$$s(t) > 1{,}353$$

$$-16t^2 + 200t + 1{,}353 > 1{,}353$$

$$-16t^2 + 200t > 0$$

$$-4t(4t - 50) > 0$$

Since t is time, we know that $t \geq 0$. If we look at the factored inequality, we can determine that the two zeros are $t = 0$ and $4t - 50 = 0 \rightarrow t = 12.5$. These are the values we will use in our sign chart. Notice that no value less than 0 is evaluated, because that is meaningless in this problem. We determine that the ball will stay higher than the tower over the interval $(0, 12.5)$ seconds.

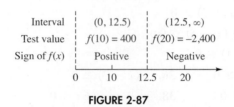

Interval	$(0, 12.5)$	$(12.5, \infty)$
Test value	$f(10) = 400$	$f(20) = -2{,}400$
Sign of $f(x)$	Positive	Negative

FIGURE 2-87

Self Check 5   During what time interval is the baseball's height lower than the top of the Sears Tower? (*Hint:* Refer to Example 5 and find the time the ball hits the ground. Round to the nearest tenth of a second.)

Now Try Exercise 75.

Self Check Answers   1. $\left[-4, \dfrac{3}{2}\right]$ 2. $(-\infty, -4] \cup [1, 3]$ 3. $(-\infty, 1) \cup \left[\dfrac{7}{4}, \infty\right)$

4. $[-3, 0) \cup (0, 3]$ 5. $(12.5, 17.4)$

Exercises **2.6**

Getting Ready
You should be able to complete these vocabulary and concept statements before you proceed to the practice exercises.

Fill in the blanks.

1. For a polynomial or rational function $f(x)$, four inequalities are possible: _____.

2. Inequalities of the form $ax^2 + bx + c < 0$ and $ax^2 + bx + c > 0$, where $a \neq 0$, are called _____ inequalities.

3. Two convenient methods to solve a quadratic inequality are _____ and _____.

4. When solving polynomial inequalities of the form $f(x) \geq 0$ or $f(x) \leq 0$, the zeros are always _____ in the solution set.

5. When solving a rational inequality, the values of x that make the denominator 0 are always _____ from the solution set.

6. When solving a rational inequality, the zeros of the numerator and the _____ asymptotes are always marked off on the sign chart.

Practice
Solve each quadratic inequality and write the answer in interval notation. If there is no solution, state so.

7. $x^2 + x - 12 \leq 0$ 8. $x^2 + x - 20 \leq 0$

9. $x^2 + 7x + 12 < 0$ 10. $x^2 - 3x + 2 < 0$

11. $x^2 - 13x + 12 \leq 0$ 12. $x^2 - 10x + 9 < 0$

13. $x^2 + 2x + 1 < 0$ 14. $x^2 - 4x + 4 < 0$

15. $6x^2 + 5x \geq 4$ 16. $9x^2 - 3x \geq 3$

17. $x^2 + 9 \leq 6x$ 18. $x^2 + 25 \leq -10x$

19. $6x^2 + 5x \geq -1$

20. $6x^2 - 5x < -1$

21. $x^2 + 9x + 20 \geq 0$

22. $x^2 + 9x < -20$

23. $x^2 + x + 1 > 0$

24. $x^2 + x + 2 > 0$

25. $-x^2 + 3x > 10$

26. $-x^2 - 15 > 5x$

27. $9x^2 + 24x > -16$

28. $9x^2 \leq 24x - 16$

Solve each polynomial inequality of degree 3 or 4 and write the answer in interval notation. If there is no solution, state so.

29. $(x + 2)(x - 1)(x - 5) \geq 0$

30. $(x - 9)(x + 3)(x + 2) \leq 0$

31. $x^2(x - 2)(x + 3) > 0$

32. $(x - 1)^2(x - 4)(x + 5) > 0$

33. $-(x + 3)(x - 2)^2(x - 5) \leq 0$

34. $-x^2(x - 2)(x + 1) \leq 0$

35. $(x + 2)(x - 2)^2 \geq 0$

36. $x^2(x - 3) \geq 0$

37. $x^3 - 2x^2 - 3x < 0$

38. $x^3 + x^2 - 6x < 0$

39. $x^3 + 2x^2 \geq x + 2$

40. $x^3 + 6 \geq 4x^2 - x$

41. $4x^3 + 8x^2 - x > 2$

42. $6x^3 + 5x^2 > 8x + 3$

43. $x^3 - 3x + 2 < 0$

44. $x^3 + 7x^2 + 15x + 9 < 0$

45. $4x^3 + 2 \geq 7x - 4x^2$

46. $9x^3 + 3x^2 \geq 8x + 4$

47. $x^4 + x^3 + 2 < 3x^2 + x$

48. $x^4 + x + 2 < x^3 + 3x^2$

49. $2x^4 + x^3 - 26x^2 + 11x + 12 \geq 0$

50. $3x^4 - 8x^3 - 19x^2 + 8x + 16 \leq 0$

Solve each rational inequality and write the answer in interval notation. If there is no solution, state so.

51. $\dfrac{x + 3}{x - 2} < 0$

52. $\dfrac{x - 1}{x + 2} < 0$

53. $\dfrac{x^2 + 2x}{x^2 - 1} \geq 0$

54. $\dfrac{x^2 - 16}{x^2 + x} \geq 0$

55. $\dfrac{x^2 - 4}{x^2 - 9} < 0$

56. $\dfrac{x^2 - 25}{x^2 - 36} < 0$

57. $\dfrac{x^2 + 5x + 6}{x^2 + x - 6} \geq 0$

58. $\dfrac{x^2 + 10x + 25}{x^2 - x - 12} \geq 0$

59. $\dfrac{6x^2 - x - 1}{x^2 + 4x + 4} < 0$

60. $\dfrac{6x^2 - 3x - 3}{x^2 - 2x - 8} < 0$

61. $\dfrac{3}{x} > 2$

62. $\dfrac{5}{x} < 1$

63. $\dfrac{3}{x - 2} \leq 5$

64. $\dfrac{3}{x + 2} \leq 4$

65. $\dfrac{6}{x^2 - 1} < 1$

66. $\dfrac{4}{x^2 - 1} > 2$

67. $\dfrac{1}{x} + \dfrac{3}{x + 2} \geq 2$

68. $\dfrac{1}{x - 4} + \dfrac{1}{x + 2} \geq -\dfrac{1}{4}$

Use a graphing calculator to solve each inequality. Round your answers to three decimal places if necessary.

69. $3x^3 - 4x^2 \leq 5x - 2$

70. $5x^4 - 3x^3 > 2x^2 + 5$

71. $|2x^2 - 5x - 2| \geq 2$

72. $|3x^3 - 2x - 5| > 5$

73. $\dfrac{3x^2}{x^2 + 1} \geq 2$

74. $\dfrac{x - 5}{x^3 - 1} \leq 4$

Applications

For Exercises 75–80, use the position function $s(t) = -gt^2 + v_0t + h_0$. For measurements in meters, use $g = 4.9$; for measurements in feet, use $g = 16$.

75. Water balloon A water balloon is launched at 19.6 meters per second from a platform that is 58.8 meters tall. Over what period of time will the balloon be climbing?

76. Propelled arrow An arrow is propelled directly upward at 64 feet per second from a platform 80 feet high. Over what interval of time will the arrow be falling?

77. Toy rocket A toy rocket is launched from ground level directly upward at 39.2 meters per second. Over what interval of time is the rocket at or above a height of 34.3 meters?

78. Diet Coke and Mentos Mentos are placed into a Diet Coke, resulting in the Diet Coke being propelled upward from ground level at 32 feet per second. Over what time interval is the Diet Coke's height above 12 feet?

Ted Kinsman/Photo Researchers, Inc.

79. Tennis ball A tennis ball is thrown straight up from the top of a building 144 feet tall with an initial velocity of 64 feet per second. Over what interval of time will the ball be more than 144 feet above the ground?

80. Police pistol If a police officer fires a Glock 22 pistol and the bullet is shot straight up from an initial height of 5 feet with a muzzle velocity of 800 feet per second, over what time interval will the height of the bullet be less than 10,005 feet?

81. Height of a diver Refer to the problem introduced in the section opener. Determine the interval of time over which the height of the diver is more than 6 feet above the water.

82. **Area** You have 100 feet of fence and wish to enclose a rectangular garden area. Determine the interval of the values of the width that will yield an area of at least 400 square feet.

83. **Area** You have 1,200 feet of fence to enclose a rectangular pen for your horses. Determine the interval of the values of the width that will yield an area of at least 80,000 square feet.

84. **Yearly bonus** Each year, a small business budgets $50,000 to reward qualifying members of its workforce with a bonus. If $50,000 is split among the number of employees x qualifying for the bonus, the function $A(x) = \frac{50,000}{x}$ represents the amount in dollars that each receives. Detemine the number of employees that can qualify for the bonus if the amount each receives will exceed $5,000.

85. **Cake sales** Isabelle runs a home business and sells coconut, caramel, and red velvet cakes. If it costs her $5 to make each cake and she has fixed weekly

cost of $200, then $\overline{C}(x) = \frac{5x + 600}{x}$ represents the average cost in dollars for each of x cakes sold. How many cakes does Isabelle need to make and sell each week so that her weekly average cost per cake is less than $15?

Discovery and Writing

86. Describe two ways to solve a polynomial or rational inequality using a graphing calculator.

87. What is wrong with the following solution?

$$\frac{x - 1}{x + 2} \le 4$$

$$x - 1 \le 4(x + 2)$$

$$x - 1 \le 4x + 8$$

$$-3x \le 9$$

$$x \ge -3$$

CHAPTER REVIEW

SECTION 2.1   Quadratic Functions

Definitions and Concepts	Examples
A **quadratic function** is a second-degree polynomial function in one variable. It is defined by an equation of the form $f(x) = ax^2 + bx + c \ (a \ne 0)$ or $y = ax^2 + bx + c \ (a \ne 0)$, where a, b, and c are constants.	See complete explanation, pages 220–221
The graph of a quadratic function is a curve called a **parabola**.	See Example 1, pages 223–224
Standard form of a quadratic function: $y = f(x) = a(x - h)^2 + k \ (a \ne 0)$. The vertex is the point (h, k). The parabola opens upward if $a > 0$ and opens downward if $a < 0$. $h = -\frac{b}{2a}$, $k = f(h)$.	See Example 2, pages 224–225

EXERCISES

Determine whether the graph of each quadratic function opens upward or downward. State whether a maximum or minimum value occurs at the vertex of the parabola.

1. $f(x) = \frac{1}{2}x^2 + 4$
2. $f(x) = -4(x + 1)^2 + 5$

Find the vertex of each parabola.

3. $f(x) = 2(x - 1)^2 + 6$
4. $f(x) = -2(x + 4)^2 - 5$
5. $f(x) = x^2 + 6x - 4$
6. $f(x) = -4x^2 + 4x - 9$

Graph each quadratic function. Label the vertex and all intercepts. State the domain and range and the equation for the axis of symmetry.

7. $f(x) = (x - 2)^2 - 3$
8. $f(x) = -(x - 4)^2 + 4$
9. $f(x) = x^2 - x$
10. $f(x) = x - x^2$
11. $f(x) = x^2 - 3x - 4$
12. $f(x) = 3x^2 - 8x - 3$
13. $f(x) = -x^2 + 2x - 3$
14. $f(x) = -x^2 - 2x + 3$

Applications

15. **Architecture** A parabolic arch has an equation of $3x^2 + y - 300 = 0$. Find the maximum height of the arch in feet.

16. Numbers The sum of two numbers is 1, and their product is as large as possible. Find the numbers.

17. Maximizing area A rancher wishes to enclose a rectangular corral with 1,400 feet of fencing. Find the dimensions of the corral that will maximize the area and find that maximum area.

18. Digital cameras A company that produces and sells digital cameras has determined that the total weekly cost C of producing x digital cameras is given by the function $C(x) = 1.5x^2 - 150x + 4{,}850$. Determine the production level that minimizes the weekly cost for producing the digital cameras and find that minimum weekly cost.

SECTION 2.2   Polynomial Functions

Definitions and Concepts	Examples
Leading-Coefficient Test and end behavior: For $f(x) = a_n x^n + a_{n-1} x^{n-1} + \cdots + a_1 x + a_0$, $a_n \neq 0$, as x increases without bound $(x \to \infty)$ or decreases without bound $(x \to -\infty)$, the function will eventually increase without bound $(f(x) \to \infty)$ or decrease without bound $(f(x) \to -\infty)$.	See Example 1, pages 240–241
The graph of a polynomial function of degree n will have at most n real zeros and at most $n - 1$ turning points (i.e., relative extrema).	See Example 2, pages 242–243
If $x = a$ is a zero of a polynomial function $f(x)$, then • $f(a) = 0$. • $(x - a)$ is a factor of the polynomial f. • $x = a$ is an x-intercept.	See Example 3, page 243

EXERCISES

Use the Leading-Coefficient Test to determine the end behavior of each polynomial.

19. $f(x) = 5x^4 + 3x^3 + 2$

20. $f(x) = \dfrac{3}{5}x^5 - 3x^3 + 1$

21. $g(x) = -x^5 - 2x^4 + 5x^2 - 3$

22. $g(x) = -3x^6 + 2x^2 + 3$

For each polynomial function,

(a) Find the end behavior.
(b) Find the zeros and the multiplicity of each zero.
(c) Find the y-intercept.
(d) Sketch the graph, utilizing a sign chart and symmetry if applicable.

23. $f(x) = 2x^3 - 7x^2 + 3x$

24. $g(x) = 3x(x + 2)^2(x - 2)$

25. $f(x) = 2x^3 + 11x^2 - x - 30$

26. $f(x) = -x^5 + 4x^4 + 3x^3 - 18x^2$

27. $h(x) = 4x^4 - 15x^2 - 4$

28. $h(x) = -x^4 + 34x^2 - 225$

SECTION 2.3   Rational Zeros of Polynomial Functions

Definitions and Concepts	Examples
Remainder Theorem: If $P(x)$ is a polynomial function, c is any number, and $P(x)$ is divided by $x - c$, the remainder is $P(c)$.	See Example 1, page 255

Definitions and Concepts	Examples
Factor Theorem: If $P(x)$ is a polynomial function and c is any real number, then • If $P(c) = 0$, then $x - c$ is a factor of $P(x)$. • If $x - c$ is a factor of $P(x)$, then $P(c) = 0$.	See Example 2, page 256
Rational-Zeros Theorem: Given a polynomial function $$P(x) = a_x x^n + a_{n-1}x^{n-1} + a_{n-2}x^{n-2} + \cdots + a_1 x + a_0$$ with integer coefficients, every rational zero of $P(x)$ can be written as $\frac{p}{q}$ (written in lowest terms), where p is a factor of the constant a_0 and q is a factor of the lead coefficient a_n. Each rational zero of $P(x)$ is a solution to the polynomial equation $P(x) = 0$.	See Example 3, pages 257–258
Descartes' Rule of Signs: • If $P(x)$ is a polynomial function with real coefficients, the number of positive real zeros of $P(x)$ is either equal to the number of variations in sign of $P(x)$ or less than that number by an even number. • The number of negative real zeros of $P(x)$ is either equal to the number of variations in sign of $P(-x)$ or less than that number by an even number.	See Example 6, page 261

EXERCISES

Use synthetic division to perform each division and write the result as $P(x) = d(x) \cdot q(x) + r(x)$.

29. $P(x) = 3x^4 + 2x^2 + 3x + 7$; $q(x) = x - 3$

30. $P(x) = 2x^4 - 3x^2 + 3x - 1$; $q(x) = x - 2$

31. $P(x) = 5x^5 - 4x^4 + 3x^3 - 2x^2 + x - 1$; $q(x) = x + 2$

32. $P(x) = 4x^5 + 2x^4 - x^3 + 3x^2 + 2x + 1$; $q(x) = x + 1$

Let $P(x) = 4x^4 + 2x^3 - 3x^2 - 2$. Find the remainder when $P(x)$ is divided by each binomial.

33. $x - 1$ **34.** $x - 2$ **35.** $x + 3$ **36.** $x + 2$

Use the Factor Theorem to determine whether each statement is true or false.

37. $x - 2$ is a factor of $x^3 + 4x^2 - 2x + 4$.

38. $x + 3$ is a factor of $2x^4 + 10x^3 + 4x^2 + 7x + 21$.

39. $x - 5$ is a factor of $x^5 - 3{,}125$.

40. $x - 6$ is a factor of $x^5 - 6x^4 - 4x + 24$.

Let $P(x) = 5x^3 + 2x^2 - x + 1$. Use synthetic division to find each value.

41. $P(3)$ **42.** $P(-3)$

43. $P\left(\dfrac{1}{2}\right)$ **44.** $P(i)$

A partial solution set is given for each equation. Find the complete solution set.

45. $2x^3 - 3x^2 - 11x + 6 = 0$; $\{3\}$

46. $x^4 + 4x^3 - x^2 - 20x - 20 = 0$; $\{-2, -2\}$

Find all rational zeros of each polynomial function.

47. $P(x) = 6x^3 - 11x^2 - 3x + 2$

48. $P(x) = 10x^3 - 11x^2 - 51x - 18$

49. $P(x) = 2x^4 + x^3 - 11x^2 + 11x - 3$

50. $P(x) = 3x^5 - 11x^4 - 22x^3 + 6x^2 + 19x + 5$

Determine how many real roots each equation has.

51. $3x^6 - 4x^5 + 3x + 2 = 0$

52. $2x^6 - 5x^4 + 5x^3 - 4x^2 + x - 12 = 0$

53. $3x^{65} - 4x^{50} + 3x^{17} + 2x = 0$

54. $x^{1{,}984} - 12 = 0$

Use Descartes' Rule of Signs to find the number of possible positive and negative real zeros of each polynomial function.

55. $P(x) = 4x^3 - 3x^2 - 5x + 1$

56. $P(x) = -5x^5 + 2x^3 - 4x - 3$

SECTION 2.4   Roots of Polynomial Equations

Definitions and Concepts	Examples
Conjugate-Pairs Theorem: If a polynomial equation $P(x) = 0$ with real-number coefficients has a complex root $a + bi$ with $b \neq 0$, then the conjugate $a - bi$ is also a root.	See Example 6, page 272
Fundamental Theorem of Algebra: If $P(x)$ is a polynomial function with positive degree, then $P(x)$ has at least one zero.	See Example 9, page 274
Polynomial Factorization Theorem: If $n > 0$ and $P(x)$ is an nth-degree polynomial function, then $P(x)$ has exactly n linear factors: $P(x) = a_n(x - r_1)(x - r_2)(x - r_3) \cdots (x - r_n)$, where $r_1, r_2, r_3, \ldots, r_n$ are numbers and a_n is the lead coefficient of $P(x)$.	See Example 9, page 274
Roots Theorem: If multiple roots are counted individually, the polynomial equation $P(x) = 0$ with degree n ($n > 0$) has exactly n roots among the complex numbers.	See Example 10, page 275

EXERCISES

Write a polynomial equation of the least possible degree with real coefficients and the given roots.

57. $-2, 1 - i$ **58.** $3, 2i$

Use the given zero to find the remaining zeros of each polynomial function.

59. $P(x) = x^4 - x^3 + 7x^2 - 9x - 18; -3i$

60. $P(x) = 2x^3 - 7x^2 + 6x + 5; 2 + i$

Solve each polynomial equation.

61. $x^3 - 3\sqrt{5}x^2 - 4x = 0$

62. $6x^4 - x^3 + 95x^2 - 16x - 16 = 0$

63. $2x^3 - 5x^2 - x + 6 = 0$

64. $2x^3 + 3x^2 - 8x + 3 = 0$

65. $x^4 + 3x^3 - 9x^2 + 3x - 10 = 0$

66. $x^4 + x^3 - 5x^2 + x - 6 = 0$

Applications

67. Designing solar collectors The space available for the installation of three solar collecting panels requires that their lengths differ by the amounts shown in the illustration, and that the total of their widths be 15 meters. To be equally effective, each panel must measure exactly 60 square meters. Find the dimensions of each panel.

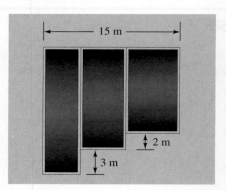

68. Designing a storage tank The design specifications for the cylindrical storage tank shown require that its height be 3 feet greater than the radius of its circular base and that the volume of the tank be 19,000 cubic feet. Use a graphing calculator to find the radius of the tank to the nearest hundredth of a foot.

SECTION 2.5   Rational Functions

Definitions and Concepts	Examples
Rational function: A function of the form $R(x) = \frac{P(x)}{Q(x)}$, where $P(x)$ and $Q(x)$ are polynomial functions and $Q(x) \neq 0$, is a **rational function.** The **domain of a rational function** is the set of all real numbers for which $Q(x) \neq 0$.	See Example 1, pages 279–280
Vertical asymptote: The line $x = a$ is a **vertical asymptote** of the graph of a function if the value of the function increases without bound or decreases without bound as x approaches a. If $R(x) = \frac{P(x)}{Q(x)}$ is a rational function (quotient of polynomial functions), with $Q(x) \neq 0$, and if $P(x)$ and $Q(x)$ have no common factors, there will be a vertical asymptote at all zeros of the polynomial $Q(x)$.	See Example 2, pages 282–283
Horizontal asymptote: The line $y = b$ is a **horizontal asymptote** of the graph of a function if the value of the function approaches b, $f(x) \to b$, when x increases without bound, $x \to \infty$, or x decreases without bound, $x \to -\infty$.	See Example 3, pages 285–286
Oblique asymptote: For a reduced rational function, $R(x) = \frac{P(x)}{Q(x)}$, if the degree of the numerator is one greater than the degree of the denominator, the graph of the function will have an **oblique (slant) asymptote.**	See Example 4, pages 286–287

EXERCISES

Complete each of the following exercises regarding the function $f(x) = \dfrac{x-1}{x(x+2)^2}$, *using the graph for reference.*

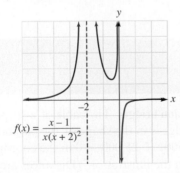

$$f(x) = \frac{x-1}{x(x+2)^2}$$

69. State the domain.   **70.** State the range.

71. Identify the x-intercept.

72. Why is there no y-intercept?

73. State the equation of any vertical asymptotes.

74. State the equation(s) of any horizontal asymptotes.

75. As $x \to -\infty$, $f(x) \to$ ___   **76.** As $x \to \infty$, $f(x) \to$ ___

77. As $x \to 0^-$, $f(x) \to$ ___   **78.** As $x \to 0^+$, $f(x) \to$ ___

79. As $x \to -2^-$, $f(x) \to$ ___

80. As $x \to -2^+$, $f(x) \to$ ___

Find the domain of each rational function.

81. $f(x) = \dfrac{3x^2 + x - 2}{x^2 - 25}$ **82.** $f(x) = \dfrac{2x^2 + 1}{x^2 + 7}$

Find the vertical asymptotes, if any, of each rational function.

83. $f(x) = \dfrac{x + 5}{x^2 - 1}$ **84.** $f(x) = \dfrac{x - 7}{x^2 - 49}$

85. $f(x) = \dfrac{x}{x^2 - x - 6}$ **86.** $f(x) = \dfrac{5x + 2}{2x^2 - 6x - 8}$

Find the horizontal asymptote, if any, of each rational function.

87. $f(x) = \dfrac{2x^2 + x - 2}{4x^2 - 4}$ **88.** $f(x) = \dfrac{5x^2 + 4}{4 - x^2}$

89. $f(x) = \dfrac{x + 1}{x^3 - 4x}$ **90.** $f(x) = \dfrac{x^3}{2x^2 - x + 11}$

Find the slant asymptote, if any, of each rational function.

91. $f(x) = \dfrac{2x^2 - 5x + 1}{x - 4}$ **92.** $f(x) = \dfrac{5x^3 + 1}{x + 5}$

Use the guidelines of the section to graph each rational function, including vertical, horizontal, and oblique asymptotes; x- and y-intercepts; and symmetry. Check your work with a graphing calculator.

93. $f(x) = \dfrac{x}{(x-1)^2}$

94. $f(x) = \dfrac{x-3}{x^2 - 3x - 4}$

95. $f(x) = \dfrac{3x-5}{x+4}$

96. $f(x) = \dfrac{2x^2 - x - 1}{x^2 + x - 6}$

97. $f(x) = \dfrac{x^2}{x-3}$

98. $f(x) = \dfrac{x^2 - 2x + 1}{x+2}$

Graph each rational function. Note that the numerator and denominator of the fraction share a common factor.

99. $f(x) = \dfrac{x^2 + x - 6}{x - 2}$

100. $f(x) = \dfrac{x^3 + 8}{x + 2}$

SECTION **2.6**   Polynomial and Rational Inequalities

Definitions and Concepts	**Examples**
Steps to solving an inequality using a sign chart: **Step 1:** Express the inequality in one of the forms: $f(x) > 0$, $f(x) \geq 0$, $f(x) < 0$, $f(x) \leq 0$. **Step 2:** Determine the zeros, if any, of the function by solving the equation $f(x) = 0$. **Step 3:** Make a sign chart. Mark the zeros on an x-axis, and test a number in each interval to determine the sign of the function in that interval. If you are solving a rational inequality, mark all values where a vertical asymptote will be located. **Step 4:** Use the sign chart to solve the inequality.	Quadratic inequality: see Example 1, pages 301–302 Polynomial inequality of degree 3 or higher: see Example 2, pages 303–304 Rational inequality: see Example 3, pages 304–305

EXERCISES

Solve each inequality, writing the solution in interval notation, or state that there is no solution.

101. $7x - 15 \geq 2x^2$

102. $6x^2 + 13x < 5$

103. $2x^3 + 7x^2 + 2x - 3 \leq 0$

104. $3x^3 - 4x^2 - 9x + 10 > 0$

105. $\dfrac{4}{x} > 6$

106. $\dfrac{x^2 - 2x - 3}{x^2 - 4} \leq 0$

CHAPTER TEST

Find the vertex of each quadratic function.

1. $f(x) = 3x^2 - 24x + 38$

2. $f(x) = 5 - 4x - x^2$

Assume that an object tossed vertically upward reaches a height of h feet after t seconds, where $h(t) = 100t - 16t^2$.

3. In how many seconds does the object reach its maximum height?

4. What is that maximum height?

5. Suspension bridges The cable of a suspension bridge is in the shape of the parabola $x^2 - 2,500y + 25,000 = 0$ in the coordinate system shown in the illustration. (Distances are in feet.) How far above the roadway is the cable's lowest point?

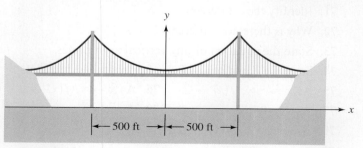

6. Refer to Exercise 5. How far above the roadway does the cable attach to the vertical pillars?

7. For the polynomial function $f(x) = x^5 - 3x^3 + 2x^2$,

     (a) Find the end behavior.

     (b) Find the zeros and the multiplicity of each zero.

     (c) Find the y-intercept.

     (d) Sketch the graph, utilizing a sign chart and symmetry if applicable.

Find all rational zeros of each polynomial function.

8. $P(x) = 2x^3 + 3x^2 - 11x - 6$

9. $P(x) = 2x^5 - 9x^4 + 7x^3 + 6x^2$

Use the given zero to find the remaining zeros of each polynomial function.

10. $P(x) = 3x^4 + 7x^3 + 6x^2 + 28x - 24$; $2i$

11. $P(x) = 2x^4 - 3x^3 + 7x^2 + 7x - 5$; $1 - 2i$

Solve each polynomial equation.

12. $x^3 - 8x^2 + 25x - 26 = 0$

13. $x^4 - x^3 - 2x^2 - 4x - 24 = 0$

Use the guidelines of Section 2.5 to graph each rational function, including vertical, horizontal, and oblique asymptotes; x- and y-intercepts; and symmetry. Check your work with a graphing calculator.

14. $f(x) = \dfrac{2x}{(x + 2)^2}$ **15.** $f(x) = \dfrac{x^2 - 3x + 2}{x^2 - 9}$

16. $f(x) = \dfrac{2x^2 + x - 1}{x^2 - 3x + 2}$ **17.** $f(x) = \dfrac{x^2 + x - 1}{x + 2}$

Graph each rational function. Note that the numerator and denominator of the fraction share a common factor.

18. $f(x) = \dfrac{3x^2 - x - 2}{x - 1}$ **19.** $f(x) = \dfrac{x^3 - 27}{x - 3}$

Solve each inequality, writing the solution in interval notation, or state that there is no solution.

20. $2x^2 + 15x + 25 > 0$ **21.** $x^3 - 5x^2 \geq 4x - 20$

22. $\dfrac{5}{x - 1} < 6$

 # LOOKING AHEAD TO **CALCULUS**

1. Using the Derivative to Analyze the Graph of a Function
- Connect derivatives to analyzing graphs of functions—Section 2.2

2. Using End Behavior to Find Limits to Infinity and Infinite Limits
- Connect limits to the end behavior of a function—Sections 2.2 and 2.5

3. Using the Intermediate Value Theorem to Apply Newton's Method for Finding Roots
- Connect the Intermediate Value Theorem to Newton's Method—Section 2.2

1. Using the Derivative to Analyze the Graph of a Function

In the Looking Ahead to Calculus section of Chapter 1, the derivative of a function was defined. In that section, we learned that the derivative of a function at a point is the slope of the tangent line to the curve at the point. The slope of the tangent line conveys very important characteristics about the graph of a function, which we will now examine. We will restrict our comments and observations to polynomial functions. When we take calculus, we will learn that this topic can be expanded to all types of functions.

What can the derivative or slope of the tangent line tell us about the graph of a function?

- Intervals where the function is increasing or decreasing
- Points where local maxima and local minima occur

Look at the figure.

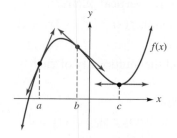

From the graph, we note the following:

1. At the point $(a, f(a))$, the tangent line slants upward from left to right. Its slope is therefore a positive number. At the same time, we see that the graph is increasing on the open interval about this point.
2. At the point $(b, f(b))$, the tangent line slants downward from left to right. Its slope is therefore a negative number. At the same time, we see that the graph is decreasing on the open interval about this point.
3. At the point $(c, f(c))$, the tangent line is a horizontal line. Its slope is zero. At the same time, we see that the graph has a local minimum at that point.

Following these observations, we can conclude the following:

Summary of Derivative Information

- A polynomial function $f(x)$ increases on the open intervals where the slope or derivative is positive, $f'(x) > 0$.
- A polynomial function $f(x)$ decreases on the open intervals where the slope or derivative is negative, $f'(x) < 0$.
- A polynomial function $f(x)$ has a turning point, or a local extremum, where the slope of derivative is zero, $f'(x) = 0$.
- If $f'(x)$ changes from positive to negative at a point then the polynomial function $f(x)$ has a local maximum at that point. If $f'(x)$ changes from negative to positive at a point, then the polynomial function $f(x)$ has a local minimum at that point.

So if we are given a polynomial function and its derivative, we can complete a sign chart for the derivative and use that information to determine many important characteristics about the graph of the function.

Connect to Calculus:
Using the Derivative to Analyze the Graph of a Function

For each polynomial, the derivative is also given. Complete a sign chart for the derivative and use the information to determine the open intervals where the function is increasing and where it is decreasing. Also, identify local maximum and minimum values of the function. You can sketch a graph to help you with these problems.

1. $f(x) = 2x^3 + 3x^2 - 72x; f'(x) = 6(x - 3)(x + 4)$
2. $f(x) = 3x^5 - 5x^3 + 3; f'(x) = 15x^2(x - 1)(x + 1)$
3. $f(x) = x^4 - 4x^2 + 6; f'(x) = 4x(x^2 - 2)$

2. Using End Behavior to Find Limits to Infinity and Infinite Limits

In Section 2.2, we studied the end behavior of a polynomial, i.e., how the function behaves as the value of x goes to positive or negative infinity. In Section 2.5, we examined the behavior of rational functions as the values of x approach asymptotes. Having looked at limits in Chapter R, we can now use limit notation to describe this behavior. The figure shows how these limits are indicated.

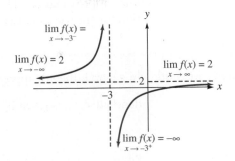

As the value of x goes toward negative infinity, the graph of the function gets closer and closer to the horizontal asymptote $y = 2$; we use the notation $\lim\limits_{x \to -\infty} f(x) = 2$. The same is true as the value of x goes toward positive infinity; we use the notation $\lim\limits_{x \to \infty} f(x) = 2$. These are examples of limits to infinity.

As the value of x approaches $x = -3$ from the left, we use the notation $\lim\limits_{x \to -3^-} f(x) = \infty$ because the value of the function increases without bound. As the value of x approaches $x = -3$ from the right, we use the notation $\lim\limits_{x \to 3^+} f(x) = -\infty$ because the value of the function decreases without bound. These are examples of infinite limits.

Connect to Calculus:
Using End Behavior to Find Limits to Infinity and Infinite Limits

Sketch the graph of an example of a function $f(x)$ that satisfies all of the given conditions.

4. $f(2) = 0$; $\lim\limits_{x \to 1^-} f(x) = -\infty$; $\lim\limits_{x \to 1^+} f(x) = \infty$

5. $\lim\limits_{x \to 2^+} f(x) = \infty$; $\lim\limits_{x \to 2^-} f(x) = -\infty$; $\lim\limits_{x \to \infty} f(x) = 0$; $\lim\limits_{x \to -\infty} f(x) = 0$; $\lim\limits_{x \to 0^+} f(x) = -\infty$; $\lim\limits_{x \to 0^-} f(x) = -\infty$; $f(-2) = 0$

3. Using the Intermediate Value Theorem to Apply Newton's Method for Finding Roots

In this chapter, we studied the Rational-Zeros Theorem for finding rational zeros of a polynomial. For a quadratic equation, we can always use the Quadratic Formula. There are formulas for finding the roots of third- and fourth-degree equations, but they are quite complicated to use. There are many equations for which we have no formula or theorem to help us solve the equation. Graphing calculators have made this much easier, but have you ever wondered how the calculator can find an approximation of a zero? While there are a number of methods that can be used by the calculator, many of them make use of **Newton's Method**, which itself makes use of the derivative. We will not go into the theory of how this method works, but we will look at the basic formula.

We start the process with a guess. Of course the guess must be somewhat close to a root. Can you think of a way to find an interval in which a root lies? Think back to the **Intermediate Value Theorem** we studied in Section 2.2. This theorem tells us that if the value of a function changes signs when evaluated at two different numbers, then at some point between those two values, the function must equal 0. Of course, the function cannot have any gaps or holes in the interval. So we make a number in the interval we obtain from using the Intermediate Value Theorem our first guess.

Here is Newton's Method for finding a root. After we make our initial guess, we apply the method to find our next estimate. Then we apply this estimate to find the third approximation. The process continues until we have found a good approximation for the root, because each estimate gets closer and closer to the root.

Newton's Method

In general, the nth approximation is denoted x_n, and the next approximation, denoted x_{n+1}, is given by $x_{n+1} = x_n - \dfrac{f(x_n)}{f'(x_n)}$.

So we see that once again the derivative is being used.

Let's see how it actually is applied. Suppose we want to find a solution to the equation $x^5 - 3x^2 - 4 = 0$. We know that there is at least one solution to this equation. How? Let's see. If we use function notion, we can write $f(x) = x^5 - 3x^2 - 4$. Since $f(1) = -6$ and $f(2) = 16$, we know that there is a root in the interval $(1, 2)$. So we will make the initial guess of $x_1 = 1.5$. The derivative is $f'(x) = 5x^4 - 6x$. Now we will proceed with the method.

As we calculate successive approximations, the question becomes, "How do we know when we're finished?" For this problem, we are rounding to five decimal places. Notice that the last two values (guesses) are the same. That is how we know we can stop. The approximate root is $x \approx 1.64770$.

$$x_2 = 1.5 - \frac{f(1.5)}{f'(1.5)} \approx 1.69349$$

$$x_3 = 1.69349 - \frac{f(1.69349)}{f'(1.69349)} \approx 1.65070$$

$$x_4 = 1.65070 - \frac{f(1.65070)}{f'(1.65070)} \approx 1.64771$$

$$x_5 = 1.64771 - \frac{f(1.64771)}{f'(1.64771)} \approx 1.64770$$

$$x_6 = 1.64770 - \frac{f(1.64770)}{f'(1.64770)} \approx 1.64770$$

This result is confirmed in the graph.

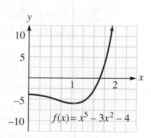

Connect to Calculus:
Using the Intermediate Value Theorem to Apply Newton's Method to Find Roots

Given the function and its derivative, use Newton's Method to find a zero in the given interval, rounding to five decimal places.

6. $f(x) = x^4 - 3x^3 - 4$; $f'(x) = 4x^3 - 9x^2$; $(3, 4)$

7. $f(x) = x^5 - 4x^3 - 3$; $f'(x) = 5x^4 - 12x^2$; $(1, 3)$

Exponential and Logarithmic Functions

3

CAREERS AND MATHEMATICS: Epidemiologist

Epidemiologists investigate and describe the determinants and distribution of disease, disability, and other health outcomes. They also develop means of prevention and control. Applied epidemiologists typically work for state health agencies and are responsible for responding to disease outbreaks and determining the cause and method of containment. Research epidemiologists work in laboratories, studying ways to prevent future outbreaks. This career can be quite rewarding, both mentally and financially. Epidemiologists spend a lot of time saving lives and finding solutions for better health.

Education and Mathematics Required
- Applied epidemiologists are generally required to have a master's degree from a school of public health. Research epidemiologists may need a PhD or medical degree, depending on their area of work.
- College Algebra, Trigonometry, a Calculus sequence, Applied Data Analysis, Survey and Research Methods, Mathematical Statistics, and Biostatistics are required courses.

How Epidemiologists Use Math and Who Employs Them
- Epidemiologists use mathematical models when they are tracking the progress of an infectious disease. The SIR model consists of three variables: S (for susceptible), I (for infectious), and R (for recovered). It is used for infectious diseases such as measles, mumps, and rubella.
- Government agencies employ 57%; hospitals employ 12%; colleges and universities employ 11%; and 9% are employed in scientific research and developmental services, like the American Cancer Society.

Career Outlook and Salary
- Employment growth is projected to be 15% over the decade 2008–2018, which is above average. This is due to an increased threat of bioterrorism and rare but infectious diseases, such as West Nile virus and avian flu.
- The median annual income is $63,000, with the top 10% of salaries at or above $98,380.

For more information, see www.bls.gov/oco.

In this chapter, we will discuss exponential functions, which are often used in banking, ecology, and science. We also will discuss logarithmic functions, which are applied in chemistry, geology, and environmental science.

3.1 Exponential Functions and Their Graphs

In this section, we will learn to

1. Approximate the value of an exponential expression.
2. Approximate an exponential function.
3. Use the properties of exponential functions.
4. Investigate the graphs of exponential functions.
5. Approximate and graph the natural exponential function.
6. Solve compound-interest problems.

Extreme waterslides, called "plunge" or "plummet" slides, are fearsome because of their heights. With near-vertical drops, the slides are designed to allow riders to reach the greatest possible speeds. Summit Plummet at Blizzard Beach, a part of the Walt Disney World Resort in Florida, stands 120 feet tall. On this slide, riders can achieve speeds of up to 55 mph.

The shapes of extreme waterslides can be modeled using *exponential functions*, the topic of this section. Exponential functions are also important in business. Anyone who has ever had a savings account and received interest on that money received that interest according to a mathematical formula. What does it mean to say we are earning 8% compounded monthly? If we never withdraw any funds, our account will grow exponentially as the result of evaluating an *exponential function*.

The graph in Figure 3-1 shows the balance in a mutual fund in which $5,000 was invested in 1990 at 8%, compounded monthly. The graph shows that in the year 2015, the value of the account will be approximately $38,000, and in the year 2030, the value will be approximately $121,000. From the graph, we can see that the longer the money is kept on deposit, the more rapidly it will grow. This is an example of *exponential growth*.

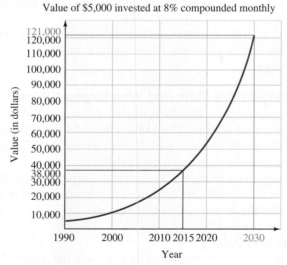

Value of $5,000 invested at 8% compounded monthly

FIGURE 3-1

Before we can discuss exponential functions, we must define irrational exponents.

1. Approximate the Value of an Exponential Expression

We have discussed expressions of the form b^x, where x is a rational number.

- 5^2 means "the square of 5."

- $4^{1/3}$ means "the cube root of 4."

- $6^{-2/5} = \dfrac{1}{6^{2/5}}$ means "the reciprocal of the fifth root of the square of 6."

To understand exponential functions and their graphs requires that we give meaning to b^x when x is an irrational number. We consider the expression $3^{\sqrt{2}}$, where $\sqrt{2}$ is the irrational number $1.414213562. \ldots$. We can use closer and closer approximations, as shown. Since $\sqrt{2}$ is an irrational number, we will use a calculator to find the approximations.

$$3^{\sqrt{2}} \approx 3^{1.4} \approx 4.655536722$$

$$3^{\sqrt{2}} \approx 3^{1.41} \approx 4.706965002$$

$$3^{\sqrt{2}} \approx 3^{1.414} \approx 4.727695035$$

$$3^{\sqrt{2}} \approx 3^{1.4142} \approx 4.72873393$$

Since the exponents of the expressions in the list are getting closer to $\sqrt{2}$, the values of the expressions are getting closer to the value of $3^{\sqrt{2}}$.

On a scientific calculator, there is an exponential key, usually y^x. On a graphing calculator, we can use the ⌃ key.

EXAMPLE 1   Approximating Exponential Expressions

Approximate each expression to six decimal places.

a. $4^{2/3}$ **b.** $5^{-\sqrt{3}}$ **c.** $\left(\dfrac{4}{7}\right)^{\pi}$

SOLUTION   We will use a calculator.

a. $4_{\wedge}^{(2/3)} \approx 2.519842$ Enter 2/3 in parentheses.

b. $5_{\wedge}^{\left(-\sqrt{3}\right)} \approx 0.061567$ Enter $-\sqrt{3}$ in parentheses.

c. $(4/7)_{\wedge}^{\pi} \approx 0.172375$ Enter the base, 4/7, in parentheses.

Figure 3-2 shows the graphing-calculator screen that was used to find the values.

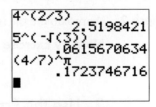

```
4^(2/3)
          2.5198421
5^(-√(3))
          .0615670634
(4/7)^π
          .1723746716
■
```

FIGURE 3-2

Self Check 1   Approximate each expression to six decimal places.

a. $5^{3/5}$ **b.** $-3^{\sqrt{6}}$ **c.** $7^{-2.356}$

Now Try Exercise 17.

If b is a positive number and x is a real number, the expression b^x always represents a positive number. It is also true that the familiar properties of exponents hold for irrational exponents.

2. Approximate an Exponential Function

Caution

Don't confuse an exponential function with a power function. In a power function, such as $f(x) = x^3$, the exponent is a constant and the base is the variable. In an exponential function, such as $f(x) = 3^x$, the base is a constant and the exponent is the variable.

If $b > 0$ and $b \neq 1$, the equation $y = b^x$ defines a function, because for each input x, there is exactly one output y. Since x can be any real number, the domain of the function is the set of real numbers. Since the base b of the expression b^x is positive, y is always positive, and the range is the set of positive numbers. Since b^x is an exponential expression, the function $f(x) = b^x$ is called an **exponential function.**

We make the restriction $b > 0$ to exclude any imaginary numbers that might result from taking even roots of negative numbers. The restriction that $b \neq 1$ excludes the constant function $f(x) = 1^x$, in which $f(x) = 1$ for every real number x.

Exponential Function

An **exponential function with base b** is defined by the equation

$$f(x) = b^x \text{ or } y = b^x \ (b > 0, \ b \neq 1, \text{ and } x \text{ is a real number}).$$

The **domain of this exponential function** is all real numbers, $D = (-\infty, \infty)$.
The **range** is $R = (0, \infty)$.

Exponential functions are useful in numerous applications, such as modeling populations, carbon-dating ancient artifacts, and computing interest earned in an account. A calculator is often required to approximate exponential functions.

EXAMPLE 2   **Approximating an Exponential Function**

Given the exponential function $f(x) = 12.5(2.32)^x - 3$, approximate each value to six decimal places.

a. $f(2)$ **b.** $f\left(\sqrt{5}\right)$ **c.** $f(-7.89)$

SOLUTION   Using a calculator, we find the following:

a. $f(2) = 12.5(2.32)^2 - 3 = 64.28$ (Note that this is an exact value.)

b. $f\left(\sqrt{5}\right) = 12.5(2.32)^{\sqrt{5}} - 3 \approx 79.066515$

c. $f(-7.89) = 12.5(2.32)^{-7.89} - 3 \approx -2.983662$

Self Check 2   Approximate the function $f(x) = 5.89(4.17)^x + 1$ to six decimal places at $x = \sqrt{3}$.

Now Try Exercise 27.

3. Use the Properties of Exponential Functions

Graphs of exponential functions of the form $f(x) = b^x$ will all have the same basic shape, and will all have a horizontal asymptote of $y = 0$. They will be smooth curves with no holes, jumps, gaps, or cusps.

EXAMPLE 3   **Graphing an Exponential Function**

Graph: $f(x) = 2^x$.

SOLUTION   We will evaluate the function at several x values, plot the points obtained, and join them with a smooth curve.

$f(x) = 2^x$		
x	$f(x)$	$(x, f(x))$
-2	$2^{-2} = \dfrac{1}{4}$	$\left(-2, \dfrac{1}{4}\right)$
-1	$2^{-1} = \dfrac{1}{2}$	$\left(-1, \dfrac{1}{2}\right)$
0	$2^0 = 1$	$(0, 1)$
1	$2^1 = 2$	$(1, 2)$
2	$2^2 = 4$	$(2, 4)$

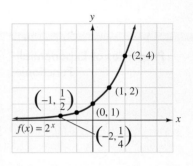

FIGURE 3-3

Notice that the function never touches or crosses the x-axis. The line $y = 0$ is a horizontal asymptote. We can also see that the function is increasing throughout its domain. The domain is $D = (-\infty, \infty)$ and the range is $R = (0, \infty)$.   ∎

Self Check 3   Graph: $f(x) = 3^x$.

Now Try Exercise 31.

EXAMPLE 4   **Graphing an Exponential Function**

Graph: $f(x) = \left(\dfrac{1}{2}\right)^x$.

SOLUTION   We will evaluate the function at several x values, plot the points obtained, and draw the graph.

$f(x) = \left(\dfrac{1}{2}\right)^x$		
x	$f(x)$	$(x, f(x))$
-2	$\left(\dfrac{1}{2}\right)^{-2} = 4$	$(-2, 4)$
-1	$\left(\dfrac{1}{2}\right)^{-1} = 2$	$(-1, 2)$
0	$\left(\dfrac{1}{2}\right)^0 = 1$	$(0, 1)$
1	$\left(\dfrac{1}{2}\right)^1 = \dfrac{1}{2}$	$\left(1, \dfrac{1}{2}\right)$
2	$\left(\dfrac{1}{2}\right)^2 = \dfrac{1}{4}$	$\left(2, \dfrac{1}{4}\right)$

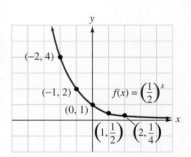

FIGURE 3-4

Notice that the function is decreasing throughout its domain. Also, there is a horizontal asymptote at $y = 0$. The domain is $D = (-\infty, \infty)$ and the range is $R = (0, \infty)$.   ∎

Self Check 4   Graph: $f(x) = \left(\dfrac{1}{3}\right)^x$.

Now Try Exercise 33.

These graphs demonstrate the following properties of exponential functions:

Properties of Exponential Functions

- The **domain of the exponential function** $f(x) = b^x$ is $D = (-\infty, \infty)$, the set of real numbers.
- The **range** is $R = (0, \infty)$, the set of positive real numbers.
- The graph has a y-intercept of 1 and contains the point $(0, 1)$.
- There is no x-intercept.
- The x-axis is an asymptote of the graph.
- The graph of $f(x) = b^x$ passes through the point $(1, b)$.

Two additional properties can be stated:

Additional Properties of Exponential Functions

- If $b > 1$, then $f(x) = b^x$ is an **increasing function**. It models **exponential growth.**
- If $0 < b < 1$, then $f(x) = b^x$ is a **decreasing function**. It models **exponential decay.**

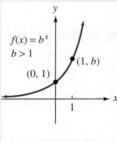

Consider the graphs of the functions $f(x) = 4^x$ and $g(x) = \left(\frac{1}{4}\right)^x$, shown in Figure 3-5. When the base is greater than 1, the function is increasing, which we see demonstrated by the graph of the function $f(x) = 4^x$. When the base is in the interval $(0, 1)$, the function is decreasing, which is demonstrated by the graph of the function $g(x) = \left(\frac{1}{4}\right)^x$.

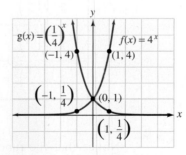

FIGURE 3-5

Exponential growth and decay have extensive and important applications in many areas, which we will discover throughout this chapter.

Comment   Recall that $b^{-x} = \dfrac{1}{b^x} = \left(\dfrac{1}{b}\right)^x$. If $b > 1$, any function of the form $f(x) = b^{-x}$ models exponential

decay, because $0 < \dfrac{1}{b} < 1$.

EXAMPLE 5   **Finding an Exponential Function Using a Graph**

The graph of an exponential function of the form $f(x) = b^x$ is shown in Figure 3-6. Find the value of b.

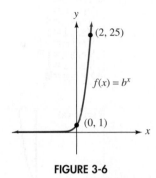

FIGURE 3-6

SOLUTION   We note that the graph passes through the point $(0, 1)$, a property of exponential functions of this form. Since the graph also passes through the point $(2, 25)$, we can find b by substituting 2 for x, then 25 for $f(2)$ in the equation $f(x) = b^x$. Finally we solve for b.

$$f(x) = b^x$$
$$f(2) = b^2$$
$$25 = b^2$$
$$5 = b \qquad b \text{ must be positive.}$$

The base b is 5. Note that the points $(0, 1)$ and $(2, 25)$ satisfy the equation $f(x) = 5^x$.

Self Check 5   Can a graph passing through $(0, 2)$ and $\left(1, \frac{3}{2}\right)$ be the graph of $f(x) = b^x$?

Now Try Exercise 43.

An exponential function $f(x) = b^x$ is either increasing (for $b > 1$) or decreasing (for $0 < b < 1$), and its graph will always pass the horizontal-line test. Also, different real numbers x always determine different values of b^x. As a result, we know that exponential functions are one-to-one.

One-to-One Property of Exponential Functions   An exponential function defined by $f(x) = b^x$ or $y = b^x$, where $b > 0$ and $b \neq 1$, is one-to-one. This implies that

- If $b^r = b^s$, then $r = s$.
- If $r \neq s$, then $b^r \neq b^s$.

The fact that exponential functions are one-to-one means that they are invertible. We will study the inverses of exponential functions in Section 3.3.

ACCENT ON TECHNOLOGY   **Graphing Exponential Functions**

When using a graphing calculator to draw the graphs of exponential functions, use care entering the function. Proper use of parentheses is critical to obtaining the correct graph. Since we know the shape of an exponential function, it is usually easy to find a good window for the graph. Notice that as the value of b, in $f(x) = b^x$, changes, the steepness of the graph increases or decreases. Several graphs are shown in Figure 3-7, along with a window that works well with these basic functions.

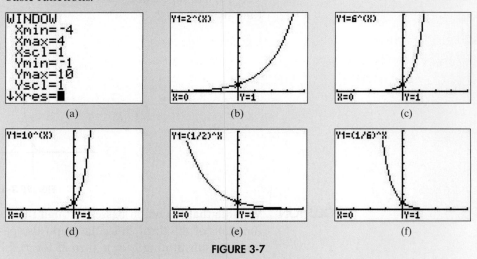

FIGURE 3-7

4. Investigate the Graphs of Exponential Functions

In Chapter 1, we saw how transformations can be used to graph functions. We can apply this to the graphs of exponential functions. These transformations are summarized here:

Transformations of Exponential Functions

If $f(x) = b^x$ is a function and k represents a real number, then we have the following transformations:

Vertical Shift

$y = b^x + k$ shifts the graph of $f(x) = b^x$ upward k units if $k > 0$ and downward k units if $k < 0$.

Horizontal Shift

$y = b^{x-k}$ shifts the graph of $f(x) = b^x$ right k units if $k > 0$ and left k units if $k < 0$.

Reflection

$y = -b^x$ reflects the graph of $f(x) = b^x$ about the x-axis.

$y = b^{-x}$ reflects the graph of $f(x) = b^x$ about the y-axis.

Vertical Stretch or Shrink (Multiply each y-coordinate by k)

$y = kb^x$ is a vertical stretch of the graph of $f(x) = b^x$ if $k > 1$ and a vertical shrink if $0 < k < 1$.

Horizontal Shrink or Stretch (Divide each x-coordinate by k)

$y = b^{kx}$ is a horizontal shrink of the graph of $f(x) = b^x$ if $k > 1$ and a horizontal stretch if $0 < k < 1$.

EXAMPLE 6   **Graphing an Exponential Function Using a Transformation**

Graph $g(x) = 2^x + 3$ using a transformation. State the domain and range and an equation of the horizontal asymptote.

SOLUTION   The graph of this function is identical to the graph of $f(x) = 2^x$ except that it is translated upward 3 units. The graphs of both $f(x) = 2^x$ and $g(x) = 2^x + 3$ are shown in Figure 3-8(a). The graph of $g(x) = 2^x + 3$ is shown in Figure 3-8(b). The domain is $D = (-\infty, \infty)$ and the range is $R = (3, \infty)$. The horizontal asymptote is $y = 3$.

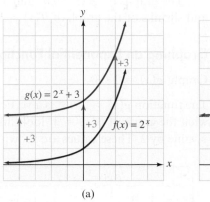 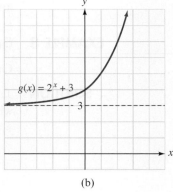

(a)   (b)

FIGURE 3-8

Self Check 6   Graph $g(x) = 2^x - 2$ using a transformation. State the domain and range and an equation of the horizontal asymptote.

Now Try Exercise 49.

EXAMPLE 7   **Graphing an Exponential Function Using a Transformation**

Graph $g(x) = 3^{x-2}$ using a transformation. State the domain and range and an equation of the horizontal asymptote.

SOLUTION   The graph of this function is identical to the graph of $f(x) = 3^x$ except that it is shifted to the right 2 units. The graphs of both $f(x) = 3^x$ and $g(x) = 3^{x-2}$ are shown in Figure 3-9.

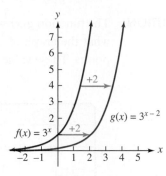

FIGURE 3-9

The domain is $D = (-\infty, \infty)$ and the range is $R = (0, \infty)$. The horizontal asymptote is $y = 0$.

Self Check 7   Graph $g(x) = 3^{x+1}$ using a transformation. State the domain and range and an equation of the horizontal asymptote.

Now Try Exercise 53.

| ACCENT ON TECHNOLOGY | Graphing Exponential Functions |

When we enter a function such as $g(x) = 3^{x-2}$ into a graphing calculator, we must put parentheses around the exponent.

The graphs of $g(x) = kb^x$ and $f(x) = b^{kx}$ are vertical and horizontal stretches and shrinks of the graph of $f(x) = b^x$.

EXAMPLE 8   Graphing an Exponential Function in the Form $g(x) = kb^x$

Graph: $g(x) = 2 \cdot 3^x$.

SOLUTION   The function $g(x) = 2 \cdot 3^x$ is a vertical stretch of the function $f(x) = 3^x$. We know what the graph of $f(x) = 3^x$ looks like, so to graph $g(x) = 2 \cdot 3^x$, we multiply each y value by 2. These values are shown in the table in Figure 3-10.

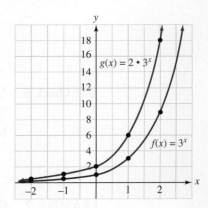

$g(x) = 2 \cdot 3^x$			
x	3^x	$2 \cdot 3^x$	(x, y)
-2	$\dfrac{1}{9}$	$\dfrac{2}{9}$	$\left(-2, \dfrac{2}{9}\right)$
-1	$\dfrac{1}{3}$	$\dfrac{2}{3}$	$\left(-1, \dfrac{2}{3}\right)$
0	1	2	$(0, 2)$
1	3	6	$(1, 6)$
2	9	18	$(2, 18)$

FIGURE 3-10

Self Check 8   Graph: $g(x) = 2 \cdot 4^x$.

Now Try Exercise 71.

EXAMPLE 9   Graphing an Exponential Function in the Form $g(x) = b^{kx}$

Graph: $g(x) = 3^{x/2}$.

SOLUTION   The function $g(x) = 3^{x/2}$ is a horizontal stretch of the function $f(x) = 3^x$. We know what the graph of $f(x) = 3^x$ looks like, so to graph $g(x) = 3^{x/2}$, we will plot some points. These values are shown in the table in Figure 3-11.

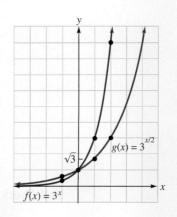

$g(x) = 3^{x/2}$			
x	3^x	$3^{x/2}$	(x, y)
-1	$\dfrac{1}{3}$	$3^{-1/2} = \dfrac{1}{\sqrt{3}}$	$\left(-1, \dfrac{1}{\sqrt{3}}\right)$
0	1	1	$(0, 1)$
1	3	$3^{1/2} = \sqrt{3}$	$\left(1, \sqrt{3}\right)$
2	9	3	$(2, 3)$

FIGURE 3-11

Self Check 9 Graph: $g(x) = 4^{x/2}$.

Now Try Exercise 67.

EXAMPLE 10 **Graphing an Exponential Function Using Reflections and Shifts**

Graph: $g(x) = -2^x + 1$.

SOLUTION The graph of $g(x) = -2^x + 1$ is obtained by transformations of the graph of $f(x) = 2^x$. There is an x-axis reflection and a vertical shift 1 unit up.

$g(x) = -2^x + 1$		
x	$-2^x + 1$	(x, y)
-2	$-2^{-2} + 1 = \dfrac{3}{4}$	$\left(-2, \dfrac{3}{4}\right)$
-1	$-2^{-1} + 1 = \dfrac{1}{2}$	$\left(-1, \dfrac{1}{2}\right)$
0	$-2^0 + 1 = 0$	$(0, 0)$
1	$-2^1 + 1 = -1$	$(1, -1)$
2	$-2^2 + 1 = -3$	$(2, -3)$

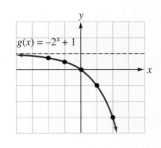

FIGURE 3-12

Self Check 10 Graph: $g(x) = -3^x + 2$.

Now Try Exercise 61.

5. Approximate and Graph the Natural Exponential Function

There are two numbers that are of great importance in mathematics: π and e. While π is fairly well known to many—at least they have heard of it—the number e is not as well known. The number e is much "newer" than π, in that it was discovered much later, and in a less specific way. As far as we know, the first time the number e appeared as a special number was in 1690. Mathematician Gottfried Leibniz wrote of the number, although he used the notation b. Much of the mathematical notation we use today is due to Leonhard Euler, and in fact the notation e is his.

In mathematical models of natural events, the number e often appears as the base of an exponential function.

Natural Exponential Function The function $f(x) = e^x$ is called the **natural exponential function.**

There are several ways that the number e is defined in mathematics. Some require the use of calculus. In 1748, Euler showed that $e = 1 + \frac{1}{1!} + \frac{1}{2!} + \frac{1}{3!} + \frac{1}{4!} + \cdots$. The symbol ! is read as "factorial," and to understand this definition of e we will need to understand **factorial notation.**

Factorial Notation If n is a natural number, the symbol $n!$, read as "n **factorial**," is defined as

$$n! = n(n - 1)(n - 2)(n - 3) \cdots (3)(2)(1).$$

From the definition we see that:

- $1! = 1$.
- $2! = 2(2 - 1) = 2 \cdot 1 = 2$.
- $3! = 3(3 - 1)(3 - 2) = 3 \cdot 2 \cdot 1 = 6$.
- $4! = 4(4 - 1)(4 - 2)(4 - 3) = 4 \cdot 3 \cdot 2 \cdot 1 = 24$.

Comment

Most students memorize the approximation 2.718 for e. This is similar to the approximation 3.14 that many use for π.

If we find $e = 1 + \frac{1}{1!} + \frac{1}{2!} + \frac{1}{3!} + \frac{1}{4!} + \cdots$ to 14 decimal places, we get the approximation $e \approx 2.71828182845904$.

Another way to define e is as the number the exponential expression $\left(1 + \frac{1}{x}\right)^x$ approaches as x increases without bound. This is shown in the table in Figure 3-13. The graph of $f(x) = \left(1 + \frac{1}{x}\right)^x$ is also shown.

x	$\left(1 + \dfrac{1}{x}\right)^x$
1	2
10	2.5937425
100	2.7048138
1,000	2.7169239
1,000,000	2.7182805
1,000,000,000	2.7182818
\multicolumn{2}{c}{As $x \to \infty$, $\left(1 + \dfrac{1}{x}\right)^x \to e$}

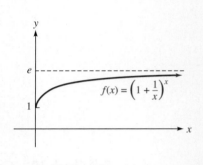

FIGURE 3-13

EXAMPLE 11   **Approximating $f(x) = e^x$**

Use a calculator to approximate each value to six decimal places.

a. $f(-2)$ **b.** $f(2)$ **c.** $f(e)$

SOLUTION   We will need to use either a scientific or a graphing calculator to approximate these values. Look for the e^x key.

a. $f(-2) = e^{-2} \approx 0.135335$

b. $f(2) = e^2 \approx 7.389056$

c. $f(e) = e^e \approx 15.154262$

Self Check 11   Approximate to six decimal places: $f(-1)$.

Now Try Exercise 81.

To graph $f(x) = e^x$, we will approximate $f(x)$ at several x values, plot these points, and join them with a smooth curve. The table and graph are shown in Figure 3-14.

$f(x) = e^x$		
x	$f(x)$	$(x, f(x))$
-1	≈ 0.37	$(-1, 0.37)$
0	1	$(0, 1)$
1	≈ 2.72	$(1, 2.72)$
2	≈ 7.39	$(2, 7.39)$

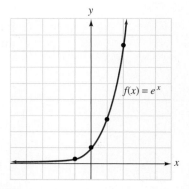

FIGURE 3-14

6. Solve Compound-Interest Problems

Earlier in this section, we stated that if $b > 1$, then $f(x) = b^x$ is an **increasing function.** This function models **exponential growth.** One application of exponential growth involves interest earned in savings accounts or other interest-bearing accounts. Since everyone likes to see their money grow, this is an application that should be understood by anyone who invests money.

Banks pay **interest** for using their customers' money. Interest is calculated as a percentage of the amount on deposit in an account and is paid **annually** (once a year), **quarterly** (four times a year), **monthly** (12 times a year), or **daily** (365 times a year). Interest left on deposit in a bank account will also earn interest. Such accounts are said to earn **compound interest.**

Compound-Interest Formula

If P dollars are deposited in an account earning interest at an annual rate r, compounded n times each year, the amount A in the account after t years is given by the formula

$$A = P\left(1 + \frac{r}{n}\right)^{nt}.$$

The more times interest is compounded per year, the more money we make. Ideally we would like for our bank to compound interest infinitely many times. When this happens, the amount invested grows exponentially according to the formula $A = Pe^{rt}$. Interest calculated this way is said to be **compounded continuously.**

Continuous Compound-Interest Formula

If P dollars are deposited in an account earning interest at an annual rate r, compounded continuously, the amount A after t years is given by the formula

$$A = Pe^{rt}.$$

We will now see how $A = Pe^{rt}$ is derived.

We begin with the compound-interest formula and investigate what happens to it as n becomes infinitely large. To see what happens, we first let $n = rx$, where x is another variable.

$$A = P\left(1 + \frac{r}{n}\right)^{nt}$$

$$A = P\left(1 + \frac{r}{rx}\right)^{rxt} \qquad \text{Substitute } rx \text{ for } n.$$

$$A = P\left(1 + \frac{1}{x}\right)^{rxt} \qquad \text{Simplify } \frac{r}{rx}.$$

$$A = P\left[\left(1 + \frac{1}{x}\right)^{x}\right]^{rt} \qquad \text{Remember that } (a^m)^n = a^{mn}.$$

As x increases to infinity, the value of $\left(1 + \frac{1}{x}\right)^x$ approaches the value of e, and the formula

$$A = P\left[\left(1 + \frac{1}{x}\right)^{x}\right]^{rt}$$

becomes

$$A = Pe^{rt}. \qquad \text{Substitute } e \text{ for } \left(1 + \frac{1}{x}\right)^{x}.$$

As we are beginning to see, the number e is very important and will help us solve many real-life applications.

EXAMPLE 12 **Finding Compound Interest**

Suppose the 5-year interest rate on an individual retirement account (IRA) certificate of deposit (CD) is 3.10%. (a) How much will a $5,000 CD be worth at the end of the 5 years if the interest is compounded daily? (b) If the interest is compounded continuously, what will the value be?

SOLUTION We will substitute the values given into our compound-interest formulas.

a. We will substitute $5,000 for P, 0.031 (3.10% coverted to a decimal) for r, 365 (compounded daily) for n, and 5 for t in our compound-interest formula.

$$A = P\left(1 + \frac{r}{n}\right)^{nt}$$

$$A = 5{,}000\left(1 + \frac{0.031}{365}\right)^{(365)(5)} \approx 5{,}838.25$$

The CD will be worth $5,838.25.

b. We will substitute $5,000 for P, 0.031 for r, and 5 for t in our continuous compound-interest formula.

$$A = Pe^{rt}$$

$$A = 5{,}000e^{(0.031)(5)} \approx 5{,}838.29$$

The CD will be worth $5,838.29 if the interest is compounded continuously.

Self Check 12 For the CD in Example 12, how much will it be worth if $8,000 is the initial amount and interest is compounded daily?

Now Try Exercise 83.

In financial calculations, the initial amount deposited is often called the **present value,** denoted PV. The amount to which the account will grow is called the **future value,** denoted FV. The interest rate for each compounding period is called the **periodic interest rate,** i, and the number of times interest is compounded is the **number of compounding periods,** n. Using these definitions, we have an alternate formula for compound interest:

$$FV = PV(1 + i)^n.$$

Reworking Example 12 using this formula, we have $PV = 5{,}000$, $i = \frac{0.031}{365}$, and $n = 365 \cdot 5 = 1{,}825$.

$$FV = PV(1 + i)^n$$

$$FV = 5{,}000\left(1 + \frac{0.031}{365}\right)^{1{,}825} \approx 5{,}838.25$$

EXAMPLE 13   **Finding Compound Interest**

When Jacob is 20 years old, he receives $25,000 from his grandparents and decides that he will invest part of it in a mutual fund paying 10% interest, compounded monthly. If he wants the account to be worth $1,000,000 when he turns 65, how much should he put into the account now?

SOLUTION   We will use the compound-interest formula $A = P\left(1 + \frac{r}{n}\right)^{nt}$ to solve this problem. Note that in this scenario, we are given A, r, n, and t, and we must find P. To find P, we substitute 1,000,000 for A, 0.10 for r, 12 for n, and 45 for t into our formula and then solve algebraically for P.

$$A = P\left(1 + \frac{r}{n}\right)^{nt}$$

$$1{,}000{,}000 = P\left(1 + \frac{0.10}{12}\right)^{(12)(45)}$$

$$P = \frac{1{,}000{,}000}{\left(1 + \dfrac{0.10}{12}\right)^{(12)(45)}}$$

$$= 1{,}000{,}000\left(1 + \frac{0.10}{12}\right)^{-(12)(45)}$$

$$\approx 11{,}318.08$$

Comment

We could also use the alternate formula $FV = PV(1 + i)^n$ for this problem and solve for PV.

If Jacob invests $11,318.08 at age 20, he will have $1,000,000 in his account at age 65.

Self Check 13   How much should Jacob invest now if he wants to have $500,000 in the account at age 65?

Now Try Exercise 97.

Self Check Answers   **1. a.** 2.626528   **b.** -14.746998 **c.** 0.010208   **2.** 70.859014
3.   **4.**

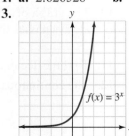

$f(x) = 3^x$

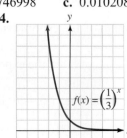

$f(x) = \left(\dfrac{1}{3}\right)^x$

5. no   **6.** $D = (-\infty, \infty); R = (-2, \infty); y = -2$

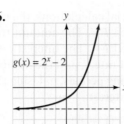

$g(x) = 2^x - 2$

7. $D = (-\infty, \infty); R = (0, \infty); y = 0$

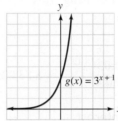

$g(x) = 3^{x+1}$

8.   **9.**

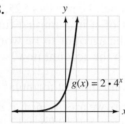   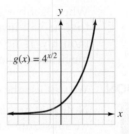

$g(x) = 2 \cdot 4^x$ $g(x) = 4^{x/2}$

10.   **11.** 0.367879   **12.** \$9,341.20   **13.** \$5,659.04

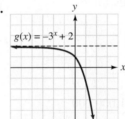

$g(x) = -3^x + 2$

Exercises **3.1**

Getting Ready
You should be able to complete these vocabulary and concept statements before you proceed to the practice exercises.

Fill in the blanks.

1. If $b > 0$ and $b \neq 1$, $y = b^x$ represents a(n) _____ function.

2. If $f(x) = b^x$ represents an increasing function, then $b >$ ____.

3. In interval notation, the domain of the exponential function $f(x) = b^x$ is _____.

4. The number b is called the _____ of the exponential function $f(x) = b^x$.

5. The range of the exponential function $f(x) = b^x$ is _____.

6. The graphs of all exponential functions $f(x) = b^x$ have the same ___-intercept of _____.

7. If $b > 0$ and $b \neq 1$, the graph of $f(x) = b^x$ approaches the x-axis, which is called a(n) _____ of the curve.

8. If $f(x) = b^x$ represents a decreasing function, then _____ $< b <$ _____.

9. If $b > 1$, then $f(x) = b^x$ defines a(n) _____ function.

10. The graph of an exponential function $f(x) = b^x$ always passes through the points $(0, 1)$ and _____.

11. To three decimal places, the value of e is _____.

12. The continuous compound-interest formula is $A =$ _____.

13. Since $e > 1$, the natural exponential function $f(x) = e^x$ is a(n) _____ function.

14. The graph of the natural exponential function $f(x) = e^x$ passes through the points $(0, 1)$ and _____.

Practice

Approximate each value to six decimal places.

15. $4^{\sqrt{3}}$ **16.** $5^{\sqrt{2}}$ **17.** 7^π **18.** $3^{-\pi}$

19. $3^{2.1}$ **20.** $2^{5.6}$ **21.** $5^{-1.3}$ **22.** $6^{-2.1}$

23. $e^{2.3}$ **24.** $e^{1.5}$

Given $f(x) = 15.2(3.19)^x$, approximate each value to six decimal places.

25. $f(3.9)$ **26.** $f(2.2)$ **27.** $f(\pi)$

28. $f(-\pi)$ **29.** $f\left(-\sqrt{2}\right)$ **30.** $f\left(\sqrt{3}\right)$

Graph each exponential function. State the domain and range and the equation of the horizontal asymptote.

31. $f(x) = 4^x$ **32.** $f(x) = 5^x$

33. $f(x) = \left(\dfrac{1}{5}\right)^x$ **34.** $f(x) = \left(\dfrac{1}{3}\right)^x$

35. $f(x) = \left(\dfrac{3}{4}\right)^x$ **36.** $f(x) = \left(\dfrac{4}{5}\right)^x$

37. $f(x) = (1.5)^x$ **38.** $f(x) = (0.3)^x$

Determine whether the graph could represent an exponential function of the form $f(x) = b^x$.

39.

40.

41.

42.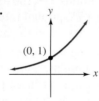

Find the value of b, if any, that would cause the graph of $y = b^x$ to look like the graph indicated.

43.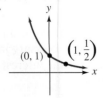

44.

45.

46.

47.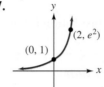

48.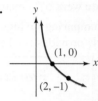

Graph each function using transformations. State the domain, range, equation of the horizontal asymptote, and y-intercept.

49. $f(x) = 3^x - 1$ **50.** $f(x) = 2^x + 3$

51. $h(x) = 3^{x-1}$ **52.** $h(x) = 2^{x+2}$

53. $f(x) = 4^{x+1}$ **54.** $f(x) = 2^{x-3}$

55. $f(x) = e^x - 2$ **56.** $f(x) = e^x + 2$

57. $g(x) = e^{x-2}$ **58.** $g(x) = e^{x+1}$

59. $f(x) = 2^{-x}$ **60.** $f(x) = 3^{-x}$

61. $h(x) = -3^x + 1$ **62.** $h(x) = -2^x - 2$

63. $g(x) = 2^{x+2} - 1$ **64.** $g(x) = 3^{x-1} + 2$

65. $f(x) = 3e^x$ **66.** $f(x) = -2e^x$

67. $g(x) = 3^{2x}$ **68.** $g(x) = 2^{3x}$

69. $h(x) = 2^{2x} - 1$ **70.** $h(x) = 3^{2x} + 2$

71. $g(x) = 3 \cdot 2^x$ **72.** $f(x) = -2 \cdot 3^x$

Use the calculator functions to find all intercepts of each function to six decimal places.

73. $f(x) = 5(2^x)$ **74.** $f(x) = 2(5^x)$

75. $f(x) = 3^{-x}$ **76.** $f(x) = 2^{-x}$

77. $f(x) = 2e^x - 1$ **78.** $f(x) = -3e^{-x} + 2$

79. $f(x) = -5(3^{-0.5x}) + 2$ **80.** $f(x) = 2^{2x} - 3$

81. Given the function $f(x) = 3e^x$, approximate each value to six decimal places.

    (a) $f(2.1)$ (b) $f(-3)$ (c) $f(e)$ (d) $f(\pi)$

82. Given the function $f(x) = e^{2x}$, approximate each value to six decimal places.

    (a) $f(-2)$ (b) $f(3.1)$ (c) $f(-\pi)$ (d) $f\left(\sqrt{3}\right)$

Applications

In Exercises 83–86, assume that there are no additional deposits or withdrawals.

83. **Compound interest** An initial deposit of \$10,000 earns 8% interest, compounded quarterly. How much will be in the account in 10 years?

84. **Compound interest** An initial deposit of $1,000 earns 9% interest, compounded monthly. How much will be in the account in $4\frac{1}{2}$ years?

85. **Comparing interest rates** How much more interest could $500 earn in 5 years, compounded semi-annually (two times a year), if the annual interest rate were $5\frac{1}{2}\%$ instead of 5%?

86. **Comparing savings plans** Which institution in the ads provides the better investment?

> ### *Fidelity Savings & Loan*
> Earn 5.25%
> compounded monthly

> ### **Union Trust**
> Money Market Account
> paying 5.35%
> compounded annually

87. **Compound interest** If $1 had been invested on July 4, 1776, at 5% interest, compounded annually, what would it be worth on July 4, 2076?

88. **360/365 method** Some financial institutions pay daily interest, compounded by the 360/365 method, using the formula $A = P\left(1 + \frac{r}{360}\right)^{365t}$ (t is in years). Using this method, what will an initial investment of $1,000 be worth in 5 years, assuming a 7% annual interest rate?

89. **Carrying charges** A college student takes advantage of the ad shown and buys a bedroom set for $1,100. He plans to pay the $1,100 plus interest when his income-tax refund comes in 8 months. At that time, what will he need to pay?

> ## BUY NOW, PAY LATER!
> Only $1\frac{3}{4}\%$ interest per month.

90. **Credit-card interest** A bank credit card charges interest at the rate of 21% per year, compounded monthly. If a senior in college charges her last tuition bill of $1,500 and intends to pay it in 1 year, what will she have to pay?

Mark Stout Photography/Shutterstock.com

91. **Continuous compound interest** An initial investment of $5,000 earns 8.2% interest, compounded continuously. What will the investment be worth in 12 years?

92. **Continuous compound interest** An initial investment of $2,000 earns 8% interest, compounded continuously. What will the investment be worth in 15 years?

93. **Comparison of compounding methods** An initial deposit of $5,000 grows at an annual rate of 8.5% for 5 years. Compare the final balances resulting from continuous compounding and annual compounding.

94. **Comparison of compounding methods** An initial deposit of $30,000 grows at an annual rate of 8% for 20 years. Compare the final balances resulting from continuous compounding and annual compounding.

95. **Frequency of compounding** $10,000 is invested in each of two accounts, both paying 6% annual interest. In the first account, interest compounds quarterly, and in the second account, interest compounds daily. Find the difference between the accounts after 20 years.

96. **Determining an initial deposit** An account now contains $11,180 and has been accumulating interest at a 7% annual rate, compounded continuously, for 7 years. Find the initial deposit.

97. **Saving for college** In 20 years, a father wants to accumulate $40,000 to pay for his daughter's college expenses. If he can get 6% interest, compounded quarterly, how much must he invest now to achieve his goal?

98. **Saving for college** In Exercise 97, how much should the father invest to achieve his goal if he can get 6% interest, compounded continuously?

99. **Do you want to be a millionaire?** How much of an investment in a mutual fund would be necessary in order to have $2,000,000 in 30 years if the interest rate is 4.25%, compounded monthly?

100. Do you want to be a millionaire? How much of an investment in a mutual fund would be necessary in order to have $3,500,000 in 40 years if the interest rate is 3.9%, compounded monthly?

Discovery and Writing

101. To have P available in n years, A can be invested now in an account paying interest at an annual rate r, compounded annually. Show that $A = P(1 + r)^{-n}$.

102. If $2^{t+4} = k2^t$, find k.

103. If $5^{3t} = k^t$, find k.

104. a. If $e^{t+3} = ke^t$, find k.   **b.** If $e^{3t} = k^t$, find k.

3.2 Applications of Exponential Functions

In this section, we will learn to

1. Investigate real-life situations using exponential and logistic functions.
2. Use a graphing utility to apply the Malthusian model of population growth.
3. Determine an exponential-regression model to fit a given data set.

Flu kills an estimated 36,000 Americans each year and results in a much larger number of hospitalizations. The influenza virus replicates quickly and can rapidly infect a population. The most effective method of preventing the viral infection and its severe complications is a flu vaccination.

An event that changes with time, such as the spread of the influenza virus, can by modeled by an exponential function. In this section, we will see several important applications of these functions: radioactive decay, oceanography, population growth, and epidemiology.

A mathematical description of an observed event is called a **model** of that event. Many events that change with time can be modeled by exponential functions of the form

$$y = f(t) = ab^{kt}, \qquad \text{Remember that } ab^{kt} \text{ means } a(b^{kt}).$$

where a, b, and k are constants and t represents time. If f is an increasing function, we say that y *grows exponentially.* If f is a decreasing function, we say that y *decays exponentially.*

1. Investigate Real-Life Situations Using Exponential and Logistic Functions

The atomic structure of a radioactive material changes as the material emits radiation. Uranium, for example, changes (decays) into thorium, then into radium, and eventually into lead.

Experiments have determined the time it takes for one-half of a sample of a given radioactive element to decompose. That time is a constant, called the element's **half-life.** The amount present decays exponentially according to the following formula:

Radioactive-Decay Formula   The amount A of radioactive material present at time t is given by

$$A = A_0 2^{-t/h},$$

where A_0 is the amount that was present initially (at $t = 0$) and h is the material's half-life.

EXAMPLE 1   **Solving a Radioactive-Decay Problem**

The half-life of radium is approximately 1,600 years. How much of a 1-gram sample will remain after 1,000 years?

SOLUTION   In this example, $A_0 = 1$, $h = 1,600$, and $t = 1,000$. We substitute these values into the formula for radioactive decay and simplify.

$$A = A_0 2^{-t/h}$$

$$A = 1 \cdot 2^{-1,000/1,600}$$

$$\approx 0.648419777 \qquad \text{Use a calculator.}$$

After 1,000 years, approximately 0.65 gram of radium will remain.

Self Check 1   After 800 years, how much radium will remain?

Now Try Exercise 3.

Intensity-of-Light Formula   The intensity I of light (in lumens) at a distance x meters below the surface of a body of water decreases exponentially according to the formula

$$I = I_0 k^x,$$

where I_0 is the intensity of light above the water and k is a constant that depends on the clarity of the water.

Comment   A lumen is a unit of standard measurement that decribes how much light is contained in a certain area. The lumen is part of the photometry group that measures different aspects of light.

EXAMPLE 2   **Solving an Intensity-of-Light Problem**

At one location in the Atlantic Ocean, the intensity of light above the water I_0 is 12 lumens and $k = 0.6$. Find the intensity of light at a depth of 5 meters.

SOLUTION   We will substitute 12 for I_0, 0.6 for k, and 5 for x into the formula for light intensity and then simplify.

$$I = I_0 k^x$$

$$I = 12(0.6)^5$$

$$I \approx 0.93312$$

At a depth of 5 meters, the intensity of the light is slightly less than 1 lumen.

Self Check 2   Find the intensity of light at a depth of 10 meters.

Now Try Exercise 11.

An equation based on the exponential function provides a model for **population growth.** One such model, called the **Malthusian model of population growth,** assumes a constant birth rate and a constant death rate. In this model, the population P grows exponentially according to the following formula:

| **Malthusian Model of Population Growth** | If b is the annual birth rate, d is the annual death rate, t is the time (in years), P_0 is the initial population at $t = 0$, and P is the current population, then |

$$P = P_0 e^{kt},$$

where $k = b - d$ is the **annual growth rate,** the difference between the annual birth rate and annual death rate.

EXAMPLE 3   Solving a Population-Growth Problem

The population of the United States is approximately 300 million people. Assuming that the annual birth rate is 19 per 1,000 and the annual death rate is 7 per 1,000, what does the Malthusian model predict the U.S. population will be in 50 years?

SOLUTION   We can use the stated information to write the Malthusian model for the U.S. population. We will then substitute values into the model to predict the population in 50 years. Since k is the difference between the birth and death rates, we have

Petr Nad/Shutterstock.com

$$k = \textbf{\textit{b}} - \textit{d}$$

$$k = \frac{\mathbf{19}}{\mathbf{1,000}} - \frac{7}{1,000} \qquad \text{Substitute } \frac{19}{1,000} \text{ for } b \text{ and } \frac{7}{1,000} \text{ for } d.$$

$$k = 0.019 - 0.007$$

$$= 0.012$$

We can now substitute 300,000,000 for P_0, 50 for t, and 0.012 for k in the formula for the Malthusian model of population growth and simplify.

$$P = \textbf{\textit{P}}_0 e^{kt}$$

$$P = (\mathbf{300{,}000{,}000}) e^{(0.012)(50)}$$

$$= (300{,}000{,}000) e^{0.6}$$

$$\approx 546{,}635{,}640.1 \qquad \text{Use a calculator.}$$

After 50 years, the U.S. population will exceed 546 million people.

Self Check 3   Find the population in 100 years.

Now Try Exercise 15.

Many infectious diseases, including some caused by viruses, spread most rapidly when they first infect a population but then more slowly as the number of uninfected individuals decreases. These situations are often modeled by a function called a **logistic function.**

| **Logistic Epidemiology Model** | The size P of an infected population at any time t in years is given by the logistic function |

$$P = \frac{M}{1 + \left(\dfrac{M}{P_0} - 1\right) e^{-kt}},$$

where P_0 is the infected population size at time $t = 0$, k is a constant determined by how contagious the virus is in a given environment, and M is the theoretical maximum size of the population P.

EXAMPLE 4 **Calculating Infections**

In a city with a population of 1,200,000, there are currently 1,000 cases of infection with HIV. If the spread of the disease is projected by the logistic function $P = \dfrac{1,200,000}{1 + (1,200 - 1)e^{-0.4t}}$, how many people will be infected in 3 years?

SOLUTION We can substitute 3 for t in the given formula and calculate P.

$$P = \frac{1,200,000}{1 + (1,200 - 1)e^{-0.4t}}$$

$$P = \frac{1,200,000}{1 + (1,199)e^{-0.4(3)}}$$

$$\approx 3,313.710094$$

In 3 years, approximately 3,300 people are expected to be infected.

Self Check 4 How many people will be infected in 10 years?

Now Try Exercise 35.

2. Use a Graphing Utility to Apply the Malthusian Model of Population Growth

The English economist Thomas Robert Malthus (1766–1834) was a pioneer in population study. He believed that poverty and starvation were unavoidable, because the human population tends to grow exponentially, whereas the food supply tends to grow linearly.

Graphing calculators can be used to solve some application problems by finding the intersection of two graphs.

EXAMPLE 5 Using a Calculator to Solve a Population Problem

Suppose that a country with a population of 1,000 people is growing exponentially according to the formula $P = 1,000e^{0.02t}$, where t is in years. Furthermore, assume that the food supply, measured in adequate food per day per person, is growing linearly according to the formula $y = 30.625x + 2,000$. In how many years will the population outstrip the food supply?

SOLUTION We can use a graphing calculator with window settings of $[0, 100]$ for x and $[0, 10,000]$ for y. After graphing the functions, we find the point where the two graphs intersect. We can see that the food supply will be adequate for almost 72 years. At that time, the population of approximately 4,200 people will have problems, assuming all conditions remain static.

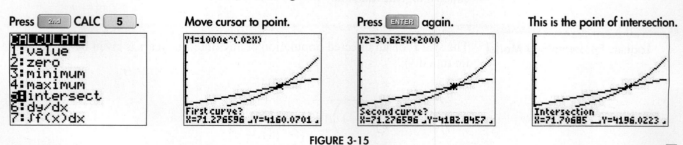

Press [2nd] CALC [5].   Move cursor to point.   Press [ENTER] again.   This is the point of intersection.

FIGURE 3-15

Self Check 5 In 80 years, what is the approximate number of people per day who will not have adequate food?

Now Try Exercise 43.

3. Determine an Exponential-Regression Model to Fit a Given Data Set

In earlier sections, we have studied linear curve fitting and quadratic regression using a graphing calculator. In Example 6, we are going to learn about exponential regression.

EXAMPLE 6   **Using Exponential Regression to Model the Temperature of Coffee**

The table in Figure 3-16 shows the cooling temperatures of a hot cup of coffee after it is made.

a. Determine an exponential-regression model function to represent the given data.

b. If room temperature is 75°F, when will the coffee reach room temperature?

c. In 1992, McDonald's was sued for serving coffee at 180° that spilled and burned a customer. The customer was awarded over $2.5 million. Experts stated that if the coffee had been served at 155°, it would not have seriously burned the customer. How long will it take the coffee in Figure 3-16 to reach 155°?

SOLUTION   **a.** Detailed directions for regression on a graphing calculator were given when we studied quadratic functions (Section 2.1, Example 9). For this example, follow the same steps, but choose ExpReg under the CALC list. The function $f(t) = 171.46172829065 \cdot (0.98824695767991)^x$ is shown to fit the data. Notice, however, that $f(0) \approx 171$ rather than the 179.5° shown in the table. Remember that the function we found is the one that best fits the data; it is not necessarily going to give us every value in the table at the exact time shown.

Time in minutes	Temperature in °F
0	179.5
5	168.7
8	158.1
11	149.2
15	141.7
18	134.6
22	125.4
25	123.5
30	116.3
34	113.2
38	109.1
42	105.7
45	102.2
50	100.5

FIGURE 3-16

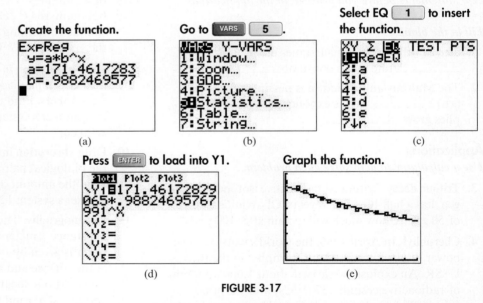

FIGURE 3-17

b. We can find the time when the temperature is 75° by graphing the line $y = 75$ and finding the point of intersection. The window will have to be extended so that the maximum of x is 100. The point of intersection shows that in about 70 minutes the coffee will reach room temperature.

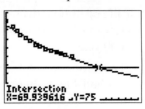

FIGURE 3-18

c. We can find the time when the temperature is 155° by graphing the line $y = 155$ and finding the point of intersection. It will take about 8.5 minutes for the coffee to cool to 155°.

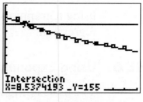

FIGURE 3-19

Self Check 6   According to the model, when does the coffee reach 100°F?

Now Try Exercise 47.

Self Check Answers   **1.** about 0.71 g   **2.** 0.073 lm   **3.** over 996 million   **4.** about 52,000   **5.** 503   **6.** 45.6 min

Exercises **3.2**

Getting Ready
You should be able to complete these vocabulary and concept statements before you proceed to the applications.

Fill in the blanks.

1. The Malthusian model assumes a constant _____ rate and a constant _____ rate.
2. The Malthusian prediction is pessimistic, because a(n) _____ grows exponentially but food supplies grow _____.

Applications
Use a calculator to help solve each problem.

3. **Tritium decay** Tritium, a radioactive isotope of hydrogen, has a half-life of 12.4 years. Of an initial sample of 50 grams, how much will remain after 100 years?

4. **Chernobyl** In April 1986, the world's worst nuclear power disaster occurred at Chernobyl in the then USSR. An explosion released about 1,000 kilograms of radioactive cesium-137 (^{137}Cs) into the atmosphere. If the half-life of ^{137}Cs is 30.17 years, how much will remain in the atmosphere in 100 years?

5. **Chernobyl** Refer to Exercise 4. How much ^{137}Cs will remain in 200 years?

6. **Carbon-14 decay** The half-life of radioactive carbon-14 is 5,700 years. How much of an initial sample will remain after 3,000 years?

7. **Plutonium decay** One of the isotopes of plutonium, plutonium-237, decays with a half-life of 40 days. How much of an initial sample will remain after 60 days?

8. **Comparing radioactive decay** One isotope of holmium, holmium-162 (^{162}Ho), has a half-life of 22 minutes. The half-life of a second isotope, holmium-164 (^{164}Ho), is 37 minutes. Starting with a sample containing equal amounts, find the ratio of the amounts of ^{162}Ho to ^{164}Ho after 1 hour.

9. **Drug absorption in smokers** The biological half-life of the asthma medication theophylline is 4.5 hours for smokers. Find the amount of the drug retained in a smoker's system 12 hours after a dose of 1 unit is taken.

10. **Drug absorption in nonsmokers** For a nonsmoker, the biological half-life of theophylline is 8 hours. Find the amount of the drug retained in a nonsmoker's system 12 hours after a 1-unit dose.

11. **Oceanography** The intensity I of light (in lumens) at a distance x meters below the surface of a body of water is given by $I = I_0 k^x$, where I_0 is the intensity at the surface and k depends on the clarity of the water. At one location in the Arctic Ocean, $I_0 = 8$ and $k = 0.5$. Find the intensity at that location at a depth of 2 meters.

12. **Oceanography** At one location in the Atlantic Ocean, $I_0 = 14$ and $k = 0.7$. Find the intensity of light at that location at a depth of 12 meters. (See Exercise 11.)

13. **Oceanography** At a depth of 3 meters at one location in the Pacific Ocean, the intensity I of light is 1 lumen and $k = 0.5$. Find the intensity I_0 of light at the surface.

14. **Oceanography** At a depth of 2 meters at one location off the coast of Belize, the intensity I of light is 2 lumens and $k = 0.2$. Find the intensity I_0 of light at the surface.

15. **Bluegill population** A Wisconsin lake is stocked with 10,000 bluegills. The population is expected to grow exponentially according to the model $P = P_0 2^{t/2}$. How many bluegills will be in the lake in 5 years?

© istockphoto.com/Dirk Nelson

16. **Community growth** The population of Eagle River is growing exponentially according to the model $P = 375(1.3)^t$, where t is measured in years from the present date. Find the population in 3 years.

17. **Bacterial cultures** A colony of 6 million bacteria is growing in a culture medium. The population P after t hours is given by the formula $P = (6 \times 10^6)(2.3)^t$. Find the population after 4 hours.

18. **Population growth** The growth of a town's population is modeled by the formula $P = 173e^{0.03t}$. How large will the population be when $t = 20$?

19. **Population decline** The decline of a city's population is modeled by the formula $P = 1.2 \times 10^6 e^{-0.008t}$. How large will the population be when $t = 30$?

20. **Epidemics** The spread of hoof-and-mouth disease through a herd of cattle can be modeled by the formula $P = P_0 e^{0.27t}$, where P is the size of the infected population, P_0 is the infected population size at $t = 0$, and t is the time in days. If a rancher does not act quickly to treat two cases, how many cattle will have the disease in 1 week?

21. **World population growth** The population of the Earth is approximately 6 billion people and is growing at an annual rate of 1.9%. Assuming a Malthusian growth model, find the world population in 30 years.

22. **World population growth** See Exercise 21. Assuming a Malthusian growth model, find the world population in 40 years.

23. **World population growth** See Exercise 21. By what factor will the current population of the Earth increase in 50 years?

24. **World population growth** See Exercise 21. By what factor will the current population of the Earth increase in 100 years?

25. **Alcohol absorption** In one individual, the percentage of alcohol absorbed into the bloodstream after drinking two glasses of wine is given by the formula $P = 0.3(1 - e^{-0.05t})$ where t is time in minutes. Find the percentage of alcohol absorbed into the blood after $\frac{1}{2}$ hour.

26. **Drug absorption** The percentage P of the drug triazolam (a drug for treating insomnia) remaining in a person's bloodstream after t hours is given by $P = e^{-0.3t}$. What percentage will remain in the bloodstream after 24 hours?

27. **Medicine** The concentration x of a certain drug in an organ after t minutes is given by $x = 0.08(1 - e^{-0.1t})$. Find the concentration of the drug after $\frac{1}{2}$ hour.

28. **Medicine** Refer to Exercise 27. Find the initial concentration of the drug. (*Hint:* When $t = 0$.)

29. **Spreading the news** Suppose the function $N = P(1 - e^{-0.1t})$ is used to model the length of time t (in hours) it takes for N people living in a town with population P to hear a news flash. How many people in a town of 50,000 will hear the news between 1 and 2 hours after it happens?

30. **Spreading the news** How many people in the town described in Exercise 29 will not have heard the news after 10 hours?

31. **Newton's law of cooling** Some hot water, initially at 100°C, is placed in a room with a temperature of 40°C. The temperature T of the water after t hours is given by $T = 40 + 60(0.75)^t$. Find the temperature in $3\frac{1}{2}$ hours.

32. **Epidemics** Refer to Example 4. How many people will have HIV in 5 years?

33. **Epidemics** Refer to Example 4. How many people will have HIV in 8 years?

34. **Epidemics** In a city with a population of 450,000, there are currently 1,000 cases of hepatitis. If the spread of the disease is projected by the logistic function $P = \dfrac{450{,}000}{1 + (450 - 1)e^{-0.2t}}$, how many people will have contracted hepatitis after 6 years?

35. **Epidemics** In an Indonesian city with a population of 55,000, there are currently 100 cases of the bird flu. If the spread of the disease is projected by the formula $P = \dfrac{55{,}000}{1 + (550 - 1)e^{-0.8t}}$, how many people will have contracted the bird flu after 2 years?

36. Life expectancy The life expectancy of white females can be estimated by using the function $l = 78.5(1.001)^x$, where x is the person's current age. Find the life expectancy of a white female who is currently 50 years old. Give the answer to the nearest tenth of a year.

37. Oceanography The width w (in millimeters) of successive growth spirals of the *Catapulus voluto* shell, shown in the illustration, is given by the function $w = 1.54e^{0.503n}$, where n is the spiral number. To the nearest tenth of a millimeter, find the width of the fifth spiral.

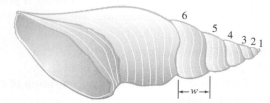

38. Skydiving Before the parachute opens, the velocity v (in meters per second) of a skydiver is given by $v = 50(1 - e^{-0.2t})$. Find the initial velocity.

39. Skydiving Refer to Exercise 38. Find the velocity after 20 seconds.

40. Free-falling objects After t seconds, a certain falling object has a velocity v given by $v = 50(1 - e^{-0.3t})$. Which is falling faster after 2 seconds, the object or the skydiver in Exercise 38?

41. Population growth In 1999, the male population of the United States was about 133 million and the female population was about 139 million. Assuming a Malthusian growth model with a 1% annual growth rate, how many more females than males will there be in 2019?

42. Population growth See Exercise 41. How many more females than males will there be in 2049?

Use a graphing calculator to solve each problem.

43. In Example 5, suppose that better farming methods change the formula for food-supply growth to $y = 31x + 2,000$. How long will the food supply be adequate?

44. In Example 5, suppose that a birth-control program changes the formula for population growth to $P = 1,000e^{0.01t}$. How long will the food supply be adequate?

45. Exponential regression A population of *Escherichia coli* bacteria doubles every 20 minutes. Construct a table that shows the growth of a single *E. coli* bacterium for a 2-hour period. Then use a graphing calculator to plot the data and create an exponential-regression equation to model this growth.

46. Refer to Exercise 45. At what point will the population reach 200 cells?

47. Wireless subscriber connections The table shows the number of wireless subscriber connections according to the Cellular Telecommunications Industry Association for the years 1990, 1995, 2000, 2005, and 2009. Determine the exponential-regression equation that models the growth. Let $x = 0$ correspond to 1990.

Year	Number (in millions)
1990	3.1
1995	33.8
2000	109.5
2005	207.9
2009	285.6

48. Using the model found in Exercise 47, in which year did the number of wireless subscriber connections first exceed 250 million?

Discovery and Writing

49. Graph the function defined by the equation $f(x) = \dfrac{e^x + e^{-x}}{2}$ from $x = -2$ to $x = 2$. The graph will look like a parabola, but it is not. The graph, called a **catenary,** is important in the design of power distribution networks, because it represents the shape of a uniform flexible cable whose ends are suspended from the same height. The function is called the **hyperbolic cosine function.** The hyperbolic cosine function was used in the design and construction of the St. Louis Gateway Arch.

50. Graph the function defined by the equation $f(x) = \dfrac{e^x - e^{-x}}{2}$ from $x = -2$ to $x = 2$. The function is called the **hyperbolic sine function.**

51. Graph the logistic function $P = \dfrac{1,200,000}{1 + (1,199)e^{-0.4t}}$ from Example 4. Use window settings of $[0, 30]$ for x and $[0, 1,500,000]$ for y. Use the capabilities of your graphing calculator to explore the function. As time passes, what value does P approach? How many years does it take for 20% of the population to become infected? For 80%?

52. The tank in the illustration initially contains 20 gallons of pure water. A brine solution containing 0.5 pound of salt per gallon is pumped into the tank, and the well-stirred mixture leaves at the same rate. The amount A in pounds of salt in the tank after t minutes is given by $A = 10(1 - e^{-0.03t})$.

a. Graph this function.

b. What is A when $t = 0$? Explain why that value is expected.

c. What is A after 2 minutes? After 10 minutes?

d. What value does A approach after a long time (as t becomes large)? Explain why this is the value you would expect.

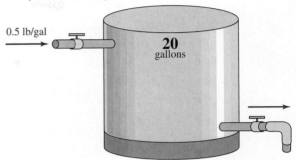

3.3 Logarithmic Functions and Their Graphs

In this section, we will learn to

1. Solve logarithmic equations with base 10, base e, and other bases.

2. Graph logarithmic functions.

3. Determine the domain of a logarithmic function.

4. Use transformations to graph logarithmic functions.

Guests aboard the Royal Caribbean's cruise ship *Freedom of the Seas* can now "hang 10" while out to sea. The FlowRider surf simulator allows riders to surf a wavelike water flow of 34,000 gallons per minute.

It is important that water in the FlowRider have the proper pH value. For example, if the pH is too low, the water will be acidic and will make riders' eyes and noses burn. It will also make their skin dry and itchy. The pH of the water is one of the most important factors in pool-water balance and should be tested frequently. To calculate pH, we need to understand *logarithms*, the topic of this section. As we continue through this chapter, we will see several real-life applications of logarithms.

Since exponential functions are one-to-one functions, each one has an inverse. For example, to find the inverse of the function $y = 3^x$, we interchange the positions of x and y and obtain $x = 3^y$. The graphs of these two functions are shown in Figure 3-20(a).

To find the inverse of the function $y = \left(\frac{1}{3}\right)^x$, we again interchange the positions of x and y and obtain $x = \left(\frac{1}{3}\right)^y$. The graphs of these two functions are shown in Figure 3-20(b).

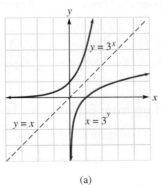

(a)

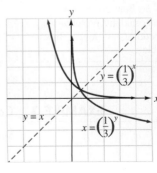

(b)

FIGURE 3-20

In general, the inverse of the function $y = b^x$ is $x = b^y$. When $b > 1$, their graphs appear as shown in Figure 3-21(a). When $0 < b < 1$, their graphs appear as shown in Figure 3-21(b).

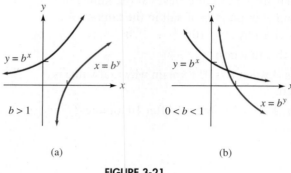

(a)   (b)

FIGURE 3-21

1. Solve Logarithmic Equations with Base 10, Base e, and Other Bases

To express the inverse function of $y = b^x$ in the form $y = f^{-1}(x)$, we must solve the equation $x = b^y$ for y. To do this, we need the following definition:

Logarithmic Function   If $b > 0$ and $b \neq 1$, the **logarithmic function with base b** is defined by

$$y = \log_b x \qquad \text{if and only if} \qquad x = b^y.$$

The **domain of the logarithmic function** is $D = (0, \infty)$.

The **range** is $R = (-\infty, \infty)$.

The logarithmic function is also written as $f(x) = \log_b x$.

Comment

Since the domain of the logarithmic function is the set of positive numbers, the logarithm of 0 and the logarithm of a negative number are undefined in the set of real numbers.

Comment

In general, we usually write $b^y = x$ instead of $x = b^y$. Both are acceptable. Also, many students prefer writing $\log_b x = y$ instead of $y = \log_b x$. This too is acceptable.

The range of the logarithmic function is the set of real numbers, because the value of y in the equation $x = b^y$ can be any real number. The domain is the set of positive numbers, because the value of x in the equation $x = b^y$ ($b > 0$) is always positive.

Since the function $y = \log_b x$ is the inverse of the one-to-one exponential function $y = b^x$, the logarithmic function is also one-to-one.

The expression $x = b^y$ is said to be written in *exponential form*. The equivalent expression $y = \log_b x$ is said to be written in *logarithmic form*. To translate from one form to the other, it is helpful to keep track of the base and the exponent.

      Exponential form   ***Logarithmic form***

        Base   Exponent   Exponent   Base

EXAMPLE 1   **Converting from Exponential Form to Logarithmic Form**

Write each equation in logarithmic form.

a. $2^6 = 64$ **b.** $27^{1/3} = \sqrt[3]{27} = 3$ **c.** $6^{-2} = \dfrac{1}{36}$

SOLUTION   a. $2^6 = 64$ is equivalent to $\log_2 64 = 6$.

b. $27^{1/3} = \sqrt[3]{27} = 3$ is equivalent to $\log_{27} 3 = \dfrac{1}{3}$.

c. $6^{-2} = \dfrac{1}{36}$ is equivalent to $\log_6 \dfrac{1}{36} = -2$.

Self Check 1   Write $13^{-2} = \dfrac{1}{169}$ in logarithmic form.

Now Try Exercise 17.

EXAMPLE 2   **Converting from Logarithmic Form to Exponential Form**

Write each equation in exponential form.

a. $\log_5 125 = 3$ b. $\log_{64} 8 = \dfrac{1}{2}$ c. $\log_{1/4} 16 = -2$

SOLUTION   a. $\log_5 125 = 3$ is equivalent to $5^3 = 125$.

b. $\log_{64} 8 = \dfrac{1}{2}$ is equivalent to $64^{1/2} = 8$.

c. $\log_{1/4} 16 = -2$ is equivalent to $\left(\dfrac{1}{4}\right)^{-2} = 16$.

Self Check 2   Write $\log_5 625 = 4$ in exponential form.

Now Try Exercise 25.

The definition of a logarithm can be used to find the values of many logarithms. Several examples are shown in the table.

$\log_5 25 = \mathbf{2}$	because **2** is the exponent to which 5 is raised to get 25: $5^2 = 25$.
$\log_7 1 = \mathbf{0}$	because **0** is the exponent to which 7 is raised to get 1: $7^0 = 1$.
$\log_{16} 4 = \dfrac{1}{2}$	because $\frac{1}{2}$ is the exponent to which 16 is raised to get 4: $16^{1/2} = \sqrt{16} = 4$.
$\log_2 \dfrac{1}{8} = -3$	because **−3** is the exponent to which 2 is raised to get $\frac{1}{8}$: $2^{-3} = \dfrac{1}{8}$.

In each of these examples, the logarithm of a number is an exponent. In fact,

$\log_b x$ is the exponent to which b is raised to get x.

To express this as an equation, we write

$b^{\log_b x} = x$.

EXAMPLE 3   **Evaluating Logarithms**

Find the value of each logarithm.   a. $\log_2 128$ b. $\log_{11} 1$ c. $\log_9 \dfrac{1}{81}$
d. $\log_{144} 12$

SOLUTION   We will use the definition of a logarithm to find each value.

a. $\log_2 128 = 6$. Why? Because 6 is the exponent we place on 2 to get 128: $2^6 = 128$.

b. $\log_{11} 1 = 0$. Why? Because 0 is the exponent we place on 11 to get 1: $11^0 = 1$.

c. $\log_9 \frac{1}{81} = -2$. Why? Because -2 is the exponent we place on 9 to get $\frac{1}{81}$:

$9^{-2} = \frac{1}{81}$.

d. $\log_{144} 12 = \frac{1}{2}$. Why? Because $\frac{1}{2}$ is the exponent we place on 144 to get 12:

$144^{1/2} = \sqrt{144} = 12$.

Self Check 3 Find the value of each logarithm. **a.** $\log_3 243$ **b.** $\log_7 \sqrt{7}$ **c.** $\log_5 \dfrac{1}{25}$

Now Try Exercise 33.

EXAMPLE 4 **Finding an Unknown Term in a Logarithmic Equation**

Find a in each equation. **a.** $\log_a 32 = 5$ **b.** $\log_9 a = -\dfrac{1}{2}$ **c.** $\log_9 3 = a$

SOLUTION We will write each equation in exponential form and solve for a.

a. $\log_a 32 = 5$ is equivalent to $a^5 = 32$. Since $2^5 = 32$, we have $a^5 = 2^5$ and $a = 2$.

b. $\log_9 a = -\frac{1}{2}$ is equivalent to $9^{-1/2} = a$. Since $9^{-1/2} = \frac{1}{3}$, it follows that $a = \frac{1}{3}$.

c. $\log_9 3 = a$ is equivalent to $9^a = 3$. Since $3 = 9^{1/2}$, we have $9^a = 9^{1/2}$ and $a = \frac{1}{2}$.

Self Check 4 Find d in each equation. **a.** $\log_4 \dfrac{1}{16} = d$ **b.** $\log_d 36 = 2$ **c.** $\log_8 d = -\dfrac{1}{3}$

Now Try Exercise 61.

Many applications use base-10 logarithms (also called **common logarithms**). When the base b is not indicated in the notation $\log x$, we assume that $b = 10$:

$\log x$ means $\log_{10} x$.

Because base-10 logarithms appear so often, we should become familiar with the following base-10 logarithms:

$\log_{10} \dfrac{1}{100} = -2$	because	$10^{-2} = \dfrac{1}{100}$.
$\log_{10} \dfrac{1}{10} = -1$	because	$10^{-1} = \dfrac{1}{10}$.
$\log_{10} 1 = 0$	because	$10^0 = 1$.
$\log_{10} 10 = 1$	because	$10^1 = 10$.
$\log_{10} 100 = 2$	because	$10^2 = 100$.
$\log_{10} 1,000 = 3$	because	$10^3 = 1,000$.

In general, we have $\log_{10} 10^x = x$.

ACCENT ON TECHNOLOGY Evaluating Logarithms on a Calculator

Scientific and graphing calculators will evaluate base-10 logarithms using the
[LOG] key. This takes the place of extensive tables that had to be used before calculators were available. Figure 3-22 shows the evaluation of log 2.34 and demonstrates that **a logarithm is an exponent**. Notice that the logarithm is evaluated, and
then the answer is used as the exponent of 10 to get 2.34. Remember: log 2.34 is
the exponent of 10 that will yield 2.34.

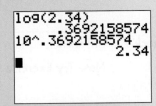

```
log(2.34)
         .3692158574
10^.3692158574
                2.34
```

FIGURE 3-22

EXAMPLE 5 Using a Calculator to Solve a Logarithmic Equation

Find x in the equation log $x = 0.7482$ to four decimal places.

SOLUTION The equation log $x = 0.7482$ is equivalent to $10^{0.7482} = x$. Therefore we have
$x = 10^{0.7482} \approx 5.6002$.

Self Check 5 Solve: log $x = 1.87737$. Give the result to four decimal places.

Now Try Exercise 87.

We have seen the importance of the number e in mathematical models of events
in nature. Base-e logarithms are just as important. They are called **natural logarithms** or **Napierian logarithms,** after John Napier (1550–1617). They are usually
written as ln x, rather than $\log_e x$:

ln x means $\log_e x$.

The domain of $f(x) = \ln x$ is the interval $(0, \infty)$, and the range is the interval
$(-\infty, \infty)$.

To estimate the base-e logarithms of numbers, we can use a calculator. Scientific and graphing calculators have a natural-logarithm key [LN], which is used like
the [LOG] key.

EXAMPLE 6 Calculating Natural Logarithms

Use a calculator to find each value. **a.** ln 17.32   **b.** ln(log 0.05)

SOLUTION **a.** ln 17.32 ≈ 2.851861903, which means $e^{2.851861903} = 17.32$.

b. ln(log 0.05) has no value, because log (0.05) < 0. Our calculator will give an
error message.

Self Check 6 Find each value to four decimal places. **a.** ln π **b.** $\ln\left(\log \dfrac{1}{3}\right)$

Now Try Exercise 83.

EXAMPLE 7 **Solving Equations with Natural Logarithms**

Solve each equation to four decimal places. **a.** $\ln x = 1.335$ **b.** $\ln x = \log 5.5$

SOLUTION   We will write the equation in exponential form and use our calculator to simplify.

a. The equation $\ln x = 1.335$ is equivalent to $e^{1.335} = x$. We will have to use a calculator to evaluate the expression: $x = e^{1.335} \approx 3.8000$.

b. The equation $\ln x = \log 5.5$ is equivalent to $e^{\log 5.5} = x$. Using a calculator, we find $x = e^{\log 5.5} \approx 2.0967$.

Self Check 7   Solve each equation to four decimal places. **a.** $\ln x = 1.9344$ **b.** $\log x = \ln 3.2$

Now Try Exercise 91.

2. Graph Logarithmic Functions

To draw the graph of a logarithmic function, we will use the fact that a logarithmic function is the inverse of an exponential function. Recall the characteristics of inverse functions: If a function contains the point (a, b), then the inverse contains the point (b, a). Because of this, we know that since an exponential function $f(x) = b^x$ has domain $D = (-\infty, \infty)$ and range $R = (0, \infty)$, the inverse of that function, $g(x) = \log_b x$, has domain $D = (0, \infty)$ and range $R = (-\infty, \infty)$.

The exponential function $f(x) = b^x$ has the horizontal asymptote $y = 0$, and the logarithmic function $g(x) = \log_b x$ has the vertical asymptote $x = 0$. We also know that inverse functions are symmetric about the line $y = x$.

This information is summarized in the table in Figure 3-23 and illustrated in Figure 3-24.

$f(x) = b^x$	$g(x) = \log_b x$
$D = (-\infty, \infty)$	$D = (0, \infty)$
$R = (0, \infty)$	$R = (-\infty, \infty)$
H.A. $y = 0$	V.A. $x = 0$
Contains $(0, 1)$	Contains $(1, 0)$
Contains $(1, b)$	Contains $(b, 1)$

FIGURE 3-23

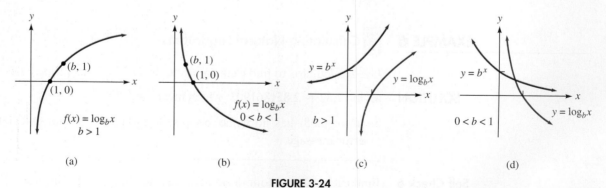

(a)   (b)   (c)   (d)

FIGURE 3-24

EXAMPLE 8   Graphing Logarithmic Functions

Graph each function. **a.** $f(x) = \log_2 x$ **b.** $g(x) = \log_{1/2} x$

SOLUTION   We will complete a table of values, plot the points, and draw a smooth curve through them.

a. Since the base 2 is greater than 1, we know the general shape of the graph of $f(x) = \log_2 x$ will resemble the graph shown in Figure 3-24(a). To graph the logarithmic function $f(x) = \log_2 x$, we will calculate and plot several points and then draw the graph shown in Figure 3-25(b).

$f(x) = \log_2 x$		
x	$f(x)$	$(x, f(x))$
$\frac{1}{4}$	-2	$\left(\frac{1}{4}, -2\right)$
$\frac{1}{2}$	-1	$\left(\frac{1}{2}, -1\right)$
1	0	$(1, 0)$
2	1	$(2, 1)$
4	2	$(4, 2)$
8	3	$(8, 3)$

(a)

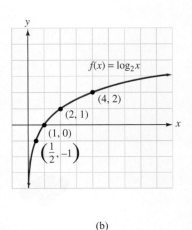

(b)

FIGURE 3-25

b. Since the base $\frac{1}{2}$ is less than 1, we know the general shape of the graph of $g(x) = \log_{1/2} x$ will resemble the graph shown in Figure 3-24(b). To graph $y = g(x) = \log_{1/2} x$, we calculate and plot several points with coordinates (x, y) that satisfy the equation $x = \left(\frac{1}{2}\right)^y$. After joining these points with a smooth curve, we have the graph shown in Figure 3-26(b).

$g(x) = \log_{1/2} x$		
x	$f(x)$	$(x, f(x))$
$\frac{1}{4}$	2	$\left(\frac{1}{4}, 2\right)$
$\frac{1}{2}$	1	$\left(\frac{1}{2}, 1\right)$
1	0	$(1, 0)$
2	-1	$(2, 1)$
4	-2	$(4, 2)$
8	-3	$(8, 3)$

(a)

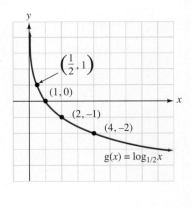

(b)

FIGURE 3-26

Self Check 8   Graph each function. **a.** $g(x) = \log_3 x$ **b.** $g(x) = \log_{1/3} x$

Now Try Exercise 107.

The graph of the common logarithmic function $f(x) = \log x$ is shown in Figure 3-27. Its graph is very important and should be memorized.

Graph of the Common Logarithmic Function
$f(x) = \log x$

$f(x) = \log x$		
x	$f(x)$	$(x, f(x))$
$\dfrac{1}{100}$	-2	$\left(\dfrac{1}{100}, -2\right)$
$\dfrac{1}{10}$	-1	$\left(\dfrac{1}{10}, -1\right)$
1	0	$(1, 0)$
10	1	$(10, 1)$
100	2	$(100, 2)$

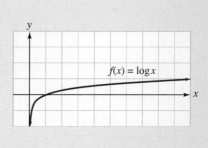

FIGURE 3-27

The graph of the natural logarithmic function $f(x) = \ln x$ is shown in Figure 3-28. Its graph is very important in calculus and should be memorized.

Graph of the Natural Logarithmic Function $f(x) = \ln x$

$f(x) = \ln x$		
x	$f(x)$	$(x, f(x))$
$e^{-1} \approx 0.4$	≈ -1	$(0.4, -1)$
1	0	$(1, 0)$
$e \approx 2.7$	≈ 1	$(2.7, 1)$
$e^2 \approx 7.4$	≈ 2	$(7.4, 2)$

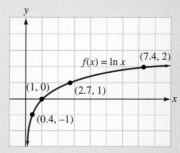

FIGURE 3-28

There are several important things to note about logarithmic functions.

1. (a) $f(x) = \ln x$ and $g(x) = e^x$ are inverses because

$y = \ln x$ if and only if $x = e^y$.   Remember that $\ln x = \log_e x$.

(b) $f(x) = \log_b x$ and $g(x) = b^x$ are inverses because

$y = \log_b x$ if and only if $x = b^y$.

2. (a) $\ln e = 1$ because $\ln e = \log_e e = 1$. $e^1 = e$

(b) $\log_b b = 1$. $b^1 = b$

3. (a) Because $f(x) = \ln x$ and $g(x) = e^x$ are inverses, their graphs are symmetric about the line $y = x$ [Figure 3-29(a)].

(b) Because $f(x) = \log_b x$ and $g(x) = b^x$ are inverses, their graphs are symmetric about the line $y = x$ [Figure 3-29(b)].

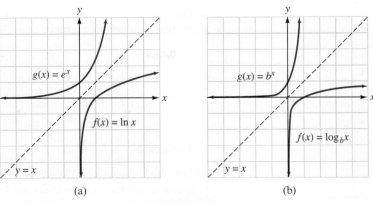

FIGURE 3-29

Two other properties follow from the fact that $f(x) = \ln x$ and $f(x) = e^x$ are inverses. The properties and examples are shown in the table.

Property	Examples
$\ln e^x = x$	$\ln e^8 = 8$ $\ln e^{-5x} = -5x$ $\ln e^{3x+2} = 3x + 2$
$e^{\ln x} = x$	$e^{\ln 4} = 4$ $e^{\ln (7x)} = 7x$ $e^{\ln (9x-2)} = 9x - 2$

3. Determine the Domain of a Logarithmic Function

We can easily find the domain of a logarithmic function if we are looking at its graph. We need a strategy for finding the domain if we are given an equation. We have seen that the domain of the function $f(x) = \log_b x$ is $(0, \infty)$. Recall that the base b is always a positive number and cannot equal 1. So when b is raised to a power, the resulting answer is also positive. Therefore, $x > 0$ and this is the domain for $f(x) = \log_b x$. We can extend this to other logarithmic functions. We refer to x as the *argument* of the logarithmic function; the argument must always be positive.

Domain of Logarithmic Functions   If $b > 1$ and $b \neq 0$, the domain of $f(x) = \log_b[h(x)]$ is all real values of x for which $h(x) > 0$.

Note that this also applies to natural logarithmic functions as well.

EXAMPLE 9   **Finding the Domains of Logarithmic Functions**

Find the domain of each logarithmic function. Write the answer in interval notation.

a. $f(x) = \log_3(2x - 3)$ **b.** $f(x) = \log(5 - x)$

c. $f(x) = \ln(x + 1)^2$ **d.** $f(x) = \ln(x^2 + 1)$

SOLUTION   Each logarithmic function is of the form $f(x) = \log_b[h(x)]$. To determine the domain, we will identify the argument $h(x)$, set $h(x) > 0$, and solve the inequality for x.

a. For $f(x) = \log_3(2x - 3)$, we note that $h(x) = 2x - 3$. The domain is the set of all real numbers for which $2x - 3 > 0$.

$$2x - 3 > 0$$
$$2x > 3$$
$$x > \frac{2}{3}$$

The domain is $D = \left(\frac{2}{3}, \infty\right)$.

b. For $f(x) = \log(5 - x)$, we note that the argument $h(x) = 5 - x$. The domain is the set of real numbers for which $5 - x > 0$, which is the set of real numbers $x < 5$. So the domain is $D = (-\infty, 5)$.

c. To find the domain of $f(x) = \ln(x + 1)^2$, we first note that $h(x) = (x + 1)^2$. We also note that the expression $(x + 1)^2 > 0$ for all real values of x except for $x = -1$. Therefore, the domain is $D = (-\infty, -1) \cup (-1, \infty)$.

d. Since $h(x) = x^2 + 1$, the term $x^2 + 1 > 0$ for all values of x. The domain of $f(x) = \ln(x^2 + 1)$ is therefore all real numbers, $D = (-\infty, \infty)$.

Self Check 9   Find the domain of the function $f(x) = \log_3(3x + 5)$.

Now Try Exercise 97.

4. Use Transformations to Graph a Logarithmic Function

We can graph logarithmic functions using transformations if we use the same concepts that we have employed in prior sections. While logarithmic functions usually have a vertical asymptote, that is not always the case. It is essential that we determine the domain of a logarithmic function prior to sketching the graph. A summary of transformations is shown here. These also hold for the natural logarithmic functions.

Transformations of Logarithmic Functions   If $f(x) = \log_b x$ is a function and k represents a real number, then we have the following transformations:

Vertical Shift

$y = \log_b x + k$ shifts the graph of $f(x) = \log_b x$ upward k units if $k > 0$ and downward k units if $k < 0$.

Horizontal Shift

$y = \log_b(x - k)$ shifts the graph of $f(x) = \log_b x$ right k units if $k > 0$ and left k units if $k < 0$.

Reflection

$y = -\log_b x$ reflects the graph of $f(x) = \log_b x$ about the x-axis.

$y = \log_b(-x)$ reflects the graph of $f(x) = \log_b x$ about the y-axis.

Vertical Stretch or Shrink (Multiply each y-coordinate by k)

$y = k \log_b x$ is a vertical stretch of the graph of $f(x) = \log_b x$ if $k > 1$ and a vertical shrink if $0 < k < 1$.

Horizontal Shrink or Stretch (Divide each x-coordinate by k)

$y = \log_b(kx)$ is a horizontal shrink of the graph of $f(x) = \log_b x$ if $k > 1$ and a horizontal stretch if $0 < k < 1$.

EXAMPLE 10   **Graphing a Logarithmic Function Using a Vertical Translation**

Graph the function $g(x) = 3 + \log_2 x$. State the domain and range, an equation of vertical asymptote, and x-intercept.

SOLUTION   The function $g(x) = 3 + \log_2 x$ is a vertical translation 3 units up of the function $f(x) = \log_2 x$. We will draw the graph of $f(x) = \log_2 x$ [shown in Example 8(a)] and then shift it up 3 units. The domain is $D = (0, \infty)$, the range is $R = (-\infty, \infty)$, and the vertical asymptote is $x = 0$. To find the x-intercept, we set the function to 0 and solve the resulting equation.

$$3 + \log_2 x = 0$$

$$\log_2 x = -3$$

$$2^{-3} = x \qquad \text{Convert to exponential form.}$$

$$\frac{1}{8} = x \qquad \text{This is the } x\text{-intercept.}$$

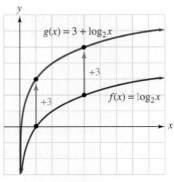

FIGURE 3-30

Self Check 10   Graph the function $g(x) = -2 + \log_2 x$.

Now Try Exercise 111.

EXAMPLE 11   **Graphing a Logarithmic Function Using a Horizontal Translation**

Graph the function $g(x) = \log_{1/2}(x - 1)$. State the domain and range, an equation of vertical asymptote, and x-intercept.

SOLUTION   The function $g(x) = \log_{1/2}(x - 1)$ is a horizontal translation 1 unit to the right of the function $f(x) = \log_{1/2} x$ [shown in Example 8(b)]. The domain is $D = (1, \infty)$; the range is $R = (-\infty, \infty)$; the vertical asymptote is $x = 1$; and the x-intercept is 2.

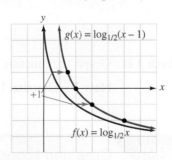

FIGURE 3-31

Self Check 11   Graph the function $g(x) = \log_3(x + 2)$.

Now Try Exercise 113.

EXAMPLE 12 **Graphing Logarithmic Functions with Reflections**

Graph each function.   **a.** $g(x) = \ln(-x)$ **b.** $h(x) = \ln(1 - x)$

SOLUTION   **a.** The function $g(x) = \ln(-x)$ is a y-axis reflection of the function $f(x) = \ln(x)$ (shown in Figure 3-28). The domain is $D = (-\infty, 0)$, the range is $R = (-\infty, \infty)$, and the vertical asymptote is $x = 0$. The x-intercept is $x = -1$.

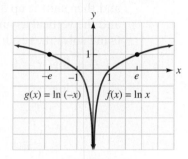

FIGURE 3-32

b. First we find the domain of the function $h(x) = \ln(1 - x)$ by setting the argument $1 - x > 0$ and solving the inequality.

$$1 - x > 0$$

$$-x > -1$$

$$x < 1$$

The domain is $D = (-\infty, 1)$ and there will be a vertical asymptote at $x = 1$.

Next, we note that

$$h(x) = \ln(1 - x)$$

$$= \ln(-x + 1)$$

$$= \ln[-(x - 1)]$$

To graph this function, we reflect the graph of $f(x) = \ln x$ about the y-axis as shown in Figure 3-32, and then apply a horizontal shift 1 unit right. The final graph is shown in Figure 3-33. The range is $R = (-\infty, \infty)$ and the x-intercept is 0.

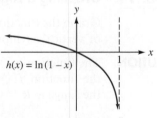

FIGURE 3-33

Self Check 12   Graph the function $g(x) = \log_3 (2 - x)$.

Now Try Exercise 121.

ACCENT ON TECHNOLOGY   **Graphing Logarithmic Functions**

Graphing calculators can be used to graph logarithmic functions. However, the only bases that are built in are 10 and e. There is a way to graph logarithmic functions in other bases, but it involves changing the base. We will cover that in

Section 3.5. In most applications, either common logarithms or natural logarithms are used. When we are graphing logarithmic functions on a graphing calculator, we need to be aware of the domain so that we can set a proper window. Notice that the graph in Figure 3-34 appears to stop on the left and not continue toward negative infinity. That is a result of how the calculator plots points as it approaches the vertical asymptote.

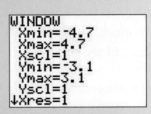
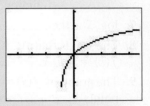
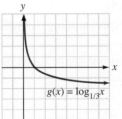

FIGURE 3-34

Self Check Answers

1. $\log_{13} \dfrac{1}{169} = -2$ 2. $5^4 = 625$ 3. **a.** 5   **b.** $\dfrac{1}{2}$ **c.** -2 4. **a.** -2

b. 6   **c.** $\dfrac{1}{2}$ 5. 75.3998   6. **a.** 1.1447   **b.** no value   7. **a.** 6.9199

b. 14.5596   8. **a.** 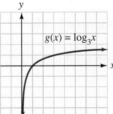   **b.**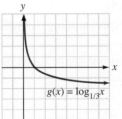

9. $\left(-\dfrac{5}{3}, \infty \right)$ 10. 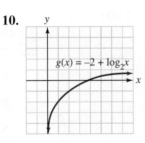   11.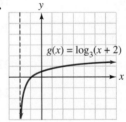

12.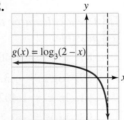

Exercises **3.3**

Getting Ready

You should be able to complete these vocabulary and concept statements before you proceed to the practice exercises.

Fill in the blanks.

1. The equation $y = \log_b x$ is equivalent to _____.

2. The domain of the logarithmic function $f(x) = \log_b x$ is the interval _____.

3. The _____ of the logarithmic function $f(x) = \log_b x$ is the interval $(-\infty, \infty)$.

4. $b^{\log_b x} =$ _____.

5. Because the exponential function is one-to-one, it has a(n) _____ function.

6. The inverse of an exponential function is called a(n) _____ function.

7. $\log_b x$ is the _____ to which b is raised to get x.

8. The y-axis is a(n) _____ of the graph of $f(x) = \log_b x$.

9. The graph of $f(x) = \log_b x$ passes through the points _____ and _____.

10. $\log_{10} 10^x =$ _____.

11. The expression $\ln x$ means _____.

12. The domain of the function $f(x) = \ln x$ is the interval _____.

13. The range of the function $f(x) = \ln x$ is the interval _____.

14. The graph of $f(x) = \ln x$ has the _____ as an asymptote.

15. In the expression $\log x$, the base is understood to be _____.

16. In the expression $\ln x$, the base is understood to be _____.

Practice

Write each equation in logarithmic form.

17. $8^2 = 64$

18. $10^3 = 1,000$

19. $4^{-2} = \dfrac{1}{16}$

20. $3^{-4} = \dfrac{1}{81}$

21. $\left(\dfrac{1}{2}\right)^{-5} = 32$

22. $\left(\dfrac{1}{3}\right)^{-3} = 27$

23. $x^y = z$

24. $m^n = p$

Write each equation in exponential form.

25. $\log_3 81 = 4$

26. $\log_7 7 = 1$

27. $\log_{1/2} \dfrac{1}{8} = 3$

28. $\log_{1/5} 1 = 0$

29. $\log_4 \dfrac{1}{64} = -3$

30. $\log_6 \dfrac{1}{36} = -2$

31. $\log_\pi \pi = 1$

32. $\log_7 \dfrac{1}{49} = -2$

Find the value of each logarithm.

33. $\log_2 8$

34. $\log_3 9$

35. $\log_4 64$

36. $\log_6 216$

37. $\log_{1/2} \dfrac{1}{8}$

38. $\log_{1/3} \dfrac{1}{81}$

39. $\log_9 3$

40. $\log_{125} 5$

41. $\log_{1/2} 8$

42. $\log_{1/2} 16$

Find the value of x.

43. $\log_2 256 = x$

44. $\log_3 729 = x$

45. $\log_4 \dfrac{1}{4} = x$

46. $\log_6 \dfrac{1}{6} = x$

47. $\log_{1/2} \dfrac{1}{32} = x$

48. $\log_{1/9} \dfrac{1}{81} = x$

49. $\log_{121} 11 = x$

50. $\log_{25} 5 = x$

51. $\log_{1/3} 27 = x$

52. $\log_{1/2} 32 = x$

53. $\log_8 x = 2$

54. $\log_7 x = 0$

55. $\log_{25} x = \dfrac{1}{2}$

56. $\log_4 x = \dfrac{1}{2}$

57. $\log_5 x = -2$

58. $\log_3 x = -4$

59. $\log_x 5^3 = 3$

60. $\log_x 5 = 1$

61. $\log_x \dfrac{9}{4} = 2$

62. $\log_x \dfrac{\sqrt{3}}{3} = \dfrac{1}{2}$

63. $\log_x \dfrac{9}{4} = -2$

64. $\log_x \dfrac{\sqrt{3}}{3} = -\dfrac{1}{2}$

65. $2^{\log_2 5} = x$

66. $3^{\log_3 4} = x$

67. $x^{\log_4 6} = 6$

68. $x^{\log_3 8} = 8$

Find each value without using a calculator.

69. $\log 10,000$

70. $\log 1,000,000$

71. $\log 0.001$

72. $\log \dfrac{1}{100,000}$

73. $e^{\ln 7}$

74. $e^{\ln 9}$

75. $\ln e^4$

76. $\ln e^{-6}$

Use a calculator to find each value to four decimal places.

77. $\log 3.25$

78. $\log 0.57$

79. $\log 0.00467$

80. $\log 375.876$

81. $\ln \dfrac{2}{3}$

82. $\ln \dfrac{12}{7}$

83. $\log(\ln 1.7)$

84. $\ln(\log 9.8)$

85. $\ln(\log 0.1)$

86. $\log(\ln 0.01)$

Use a calculator to find x to four decimal places, if possible.

87. $\log x = 1.4023$

88. $\log x = 0.926$

89. $\log x = -3.71$

90. $\log x = \log \pi$

91. $\ln x = 1.4023$

92. $\ln x = 2.6490$

93. $\ln x = -3.71$

94. $\ln x = -0.28$

95. $\log x = \ln 8$

96. $\ln x = \log 7$

Find the domain of each logarithmic function. Write your answer using interval notation.

97. $f(x) = \log(x - 4)$

98. $f(x) = \ln(x + 5)$

99. $f(x) = \ln(2 - x)$

100. $f(x) = \log(x + 3)^2$

101. $f(x) = \ln(x^2 + 4)$

102. $f(x) = \log(x^2 + 1)$

103. $f(x) = \log(x^2 - 25)$ **104.** $f(x) = \ln(16 - x^2)$

105. $f(x) = \ln\left(\dfrac{1}{x}\right)$ **106.** $f(x) = \log_5\left(\dfrac{1}{x - 1}\right)$

Graph each function. State the equation of the vertical asymptote and the x-intercept. Label one other point on the graph.

107. $f(x) = \log_4 x$ **108.** $f(x) = \log_3 x$

109. $f(x) = \log_{1/4} x$ **110.** $f(x) = \log_{1/3} x$

111. $f(x) = 2 + \log_2 x$ **112.** $f(x) = -3 + \log_3 x$

113. $f(x) = \log_3(x - 2)$ **114.** $f(x) = \log_{1/2}(x + 4)$

115. $f(x) = 2 + \log_2(x + 1)$

116. $f(x) = -2 + \log_4(x - 1)$

117. $f(x) = -\log_5 x$ **118.** $f(x) = -\log_4 x$

119. $f(x) = \log(-x)$ **120.** $f(x) = \log_2(-x)$

121. $f(x) = \log_2(3 - x)$ **122.** $f(x) = \log_3(2 - x)$

123. $f(x) = \log(3x)$ **124.** $f(x) = \log\left(\dfrac{x}{3}\right)$

125. $f(x) = \ln(x + 2)$ **126.** $f(x) = \ln(x - 1)$

127. $f(x) = -\ln(x)$ **128.** $f(x) = \ln(-x)$

State the equation of the vertical asymptote and find the values of any intercepts.

129. $f(x) = \log(\pi + x)$

130. $f(x) = -\log(e - x)$

131. $f(x) = -\ln(x^2 + 1) - 2$

132. $f(x) = \ln(2 - x^2) + 2$

133. $f(x) = \log(x + 3)$

134. $f(x) = \log(4 - x)$

135. $f(x) = -3\ln(x + \sqrt{2})$

136. $f(x) = \ln(x^2 - 2x)$

Discovery and Writing

137. Consider the following graphs. Which is greater, a or b, and why?

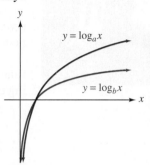

138. Consider the following graphs. Which is greater, a or b, and why?

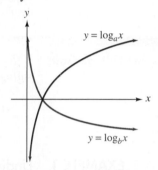

139. Choose two numbers and add their logarithms. Then find the logarithm of the product of those two numbers. What do you observe? Does it work for three numbers?

140. If $\log_a b = 7$, find $\log_b a$.

141. Use a graphing calculator to graph the function $y = \ln e^x$. Explain why the graph is a line. What is a simpler form of the equation of that line?

142. Use a graphing calculator to graph the function $y = e^{\ln x}$. Is its graph a line, or just part of a line? Explain.

143. As x increases to infinity, each of the y-values in the functions $f(x) = x^2$, $g(x) = e^x$, and $h(x) = \ln x$ increases to infinity too, but at different rates. Examine each function numerically and determine which one increases at the highest rate.

3.4 Applications of Logarithmic Functions

In this section, we will learn to

1. Apply our understanding of logarithmic functions to various fields of study.

On December 24, 2004, one of the deadliest natural disasters in history occurred when an earthquake in the Indian Ocean triggered a series of tsunamis, killing more than 225,000 people in 11 countries. Indonesia, Sri Lanka, Thailand, and India

were hit the hardest. The devastation prompted worldwide aid: More than \$7 billion was donated to help with relief efforts. The earthquake was the second largest ever recorded, with a magnitude between 9.1 and 9.3 on the Richter scale.

In 2010, Haiti experienced a catastrophic earthquake, with an epicenter approximately 16 miles west of Port-au-Prince, Haiti's capital. An estimated 3 million people were affected by the earthquake.

The measurement of the intensity of an earthquake is based on a logarithmic function. Logarithmic functions have extensive applications in many fields, including engineering, geology, social sciences, and physics.

1. Apply Our Understanding of Logarithmic Functions to Various Fields of Study

Electrical engineers use common logarithms to measure the voltage gain of devices such as amplifiers or the length of a transmission line. The unit of gain, called the **decibel,** is defined by a logarithmic function.

Decibel Voltage Gain   If E_O is the output voltage of a device and E_I is the input voltage, the **decibel voltage gain** (dB gain) is given by

$$\text{dB gain} = 20 \log \frac{E_O}{E_I}.$$

EXAMPLE 1   **Finding Decibel Voltage Gain**

Find the dB gain of an amplifier if its input is 0.5 volt and its output is 40 volts.

SOLUTION   The decibel voltage gain is found by substituting values into the formula for dB gain.

$$\text{dB gain} = 20 \log \frac{E_O}{E_I}$$

$$\text{dB gain} = 20 \log \frac{\mathbf{40}}{\mathbf{0.5}} \qquad \text{Substitute 40 for } E_O \text{ and 0.5 for } E_I.$$

$$= 20 \log 80$$

$$\approx 38.06179974 \qquad \text{Use a calculator.}$$

To the nearest decibel, the dB gain is 38 decibels.

Self Check 1   Find the dB gain if the input is 0.7 volt.

Now Try Exercise 7.

Seismologists measure the intensity of earthquakes on the **Richter scale,** which is based on a logarithmic function. The earthquake that hit Haiti in 2010 measured 7.0 on this scale.

Haiti's presidential palace after 2010 earthquake

Richter Scale   If R is the intensity of an earthquake, A is the amplitude (measured in micrometers), and P is the period (the time of one oscillation of the Earth's surface, measured in seconds), then

$$R = \log \frac{A}{P}.$$

EXAMPLE 2   **Finding the Intensity of an Earthquake**

Find the intensity of an earthquake with an amplitude of 5,000 micrometers $\left(\frac{1}{2} \text{ centimeter}\right)$ and a period of 0.07 second.

SOLUTION   We substitute 5,000 for A and 0.07 for P in the Richter scale formula and simplify.

$$R = \log \frac{A}{P}$$

$$R = \log \frac{\mathbf{5,000}}{\mathbf{0.07}}$$

$$\approx \log 71,428.57143 \qquad \text{Use a calculator.}$$

$$\approx 4.853871964$$

To the nearest tenth, the earthquake measures 4.9 on the Richter scale.

Self Check 2   Find the intensity of an aftershock with the same period but one-half the amplitude.

Now Try Exercise 13.

A battery charges at a rate that depends on how close it is to being fully charged—it charges fastest when it is most discharged. The formula that determines the time required to charge a battery to a certain level is based on a natural logarithmic function.

Hywit Dimyadi/Shutterstock.com

Charging Batteries   If M is the theoretical maximum charge that a battery can hold and k is a positive constant that depends on the battery and the charger, the length of time t (in minutes) required to charge the battery to a given level C is given by

$$t = -\frac{1}{k} \ln\left(1 - \frac{C}{M}\right).$$

EXAMPLE 3   **Finding Charging Time of a Battery**

How long will it take to bring a fully discharged battery to 90% of full charge? Assume that $k = 0.025$ and that time is measured in minutes.

SOLUTION The condition 90% of full charge means 90% of M. We substitute $0.90M$ for C and 0.025 for k in the formula for charging batteries.

$$t = -\frac{1}{k}\ln\left(1 - \frac{C}{M}\right)$$

$$t = -\frac{1}{0.025}\ln\left(1 - \frac{0.90M}{M}\right)$$

$$= -40\ln(1 - 0.9)$$

$$= -40\ln(0.1)$$

$$\approx 92.10340372 \qquad \text{Use a calculator.}$$

The battery will reach 90% charge in about 92 minutes.

Self Check 3 How long will it take this battery to reach 80% of full charge?

Now Try Exercise 19.

If a population grows exponentially at a certain annual rate, the time required for the population to double is called the **doubling time** and is given by the following formula:

Population Doubling Time If r is the annual growth rate and t is the time (in years) required for a population to double, then

$$t = \frac{\ln 2}{r}.$$

EXAMPLE 4 **Finding Population Doubling Time**

The population of the earth is growing at the approximate rate of 2% per year. If this rate continues, how long will it take the population to double?

SOLUTION Because the population is growing at the rate of 2% per year, we substitute 0.02 for r in the formula for doubling time and simplify.

$$t = \frac{\ln 2}{r}$$

$$t = \frac{\ln 2}{0.02}$$

$$\approx 34.65735903$$

It will take about 35 years for the earth's population to double.

Self Check 4 If the world population's annual growth rate could be reduced to 1.5% per year, what would be the doubling time?

Now Try Exercise 21.

When energy is added to a gas, its temperature and volume can increase. In **isothermal expansion,** the temperature remains constant—only the volume changes. The energy required is calculated as follows:

Isothermal Expansion   If the temperature T is constant, the energy E required to increase the volume of 1 mole of gas from an initial volume V_i to a final volume V_f is given by

$$E = RT \ln\left(\frac{V_f}{V_i}\right).$$

E is measured in joules and T in kelvins. R is the universal gas constant, which is 8.314 joules per mole per kelvin.

EXAMPLE 5   **Finding Energy Required for Isothermal Expansion**

Find the amount of energy that must be supplied to triple the volume of 1 mole of gas at a constant temperature of 300 kelvins.

SOLUTION   We substitute 8.314 for R and 300 for T in the formula. Since the final volume is to be three times the initial volume, we also substitute $3V_i$ for V_f.

$$E = RT \ln\left(\frac{V_f}{V_i}\right)$$

$$E = (8.314)(300) \ln\left(\frac{3V_i}{V_i}\right)$$

$$= 2{,}492.2 \ln 3$$

$$\approx 2{,}740.15877$$

Approximately 2,740 joules of energy must be added to triple the volume of the gas. ∎

Self Check 5   What energy is required to double the volume of the gas?

Now Try Exercise 25.

Self Check Answers   1. about 35 dB   2. about 4.6   3. about 64 min   4. about 46 yr
5. 1,729 J

Exercises **3.4**

Getting Ready
You should be able to complete these vocabulary and concept statements before you proceed to the applications.

Fill in the blanks.

1. The formula for decibel gain is _____.

2. The intensity of an earthquake is measured by the formula $R =$ _____.

3. The formula for charging batteries is _____.

4. If a population grows exponentially at a rate r, the time it will take for the population to double is given by the formula $t =$ _____.

5. The formula for isothermal expansion is _____.

6. The logarithm of a negative number is _____.

Applications
Use a calculator to solve each problem.

7. **Gain of an amplifier** An amplifier produces an output of 17 volts when the input signal is 0.03 volt. Find the decibel voltage gain.

8. **Transmission lines** A 4.9-volt input to a long transmission line decreases to 4.7 volts at the other end. Find the decibel voltage loss.

9. **Gain of an amplifier** Find the dB gain of an amplifier whose input voltage is 0.71 volt and whose output voltage is 20 volts.

10. **Gain of an amplifier** Find the dB gain of an amplifier whose output voltage is 2.8 volts and whose input voltage is 0.05 volt.

11. **Gain of an amplifier** Find the dB gain of the amplifier shown.

12. **Gain of an amplifier** Find the dB gain of the amplifier shown.

13. **Earthquakes** An earthquake has an amplitude of 5,000 micrometers and a period of 0.2 second. Find its measure on the Richter scale.

14. **Earthquakes** An earthquake has an amplitude of 8,000 micrometers and a period of 0.008 second. Find its measure on the Richter scale.

15. **Earthquakes** An earthquake with a period of $\frac{1}{4}$ second has an amplitude of 2,500 micrometers. Find its measure on the Richter scale.

16. **Earthquakes** An earthquake has a period of $\frac{1}{2}$ second and an amplitude of 5 cm. Find its measure on the Richter scale. (*Hint:* 1 cm = 10,000 micrometers.)

17. **Earthquakes** Earthquakes measuring between 3.5 and 5.4 on the Richter scale are often felt but rarely cause damage. Suppose an earthquake in Northern California has an amplitude of 6,000 micrometers and a period of 0.3 second. Is it likely to cause damage?

18. **Earthquakes** An earthquake measuring between 7.0 and 7.9 on the Richter scale is a major earthquake and can cause serious damage over larger areas. Suppose an earthquake in Chile has an amplitude of 198.5 cm and a period of 0.1 second. Would it cause serious damage over large areas? (*Hint:* 1 cm = 10,000 micrometers.)

19. **Charging batteries** If $k = 0.116$, how long will it take a battery to reach a 90% charge? Assume that the battery was fully discharged when it began charging.

20. **Charging batteries** If $k = 0.201$, how long will it take a battery to reach a 40% charge? Assume that the battery was fully discharged when it began charging.

21. **Population growth** A town's population grows at the rate of 12% per year. If this growth rate remains constant, how long will it take the population to double?

22. **Fish population growth** One thousand bass were stocked in Catfish Lake in Eagle River, Wisconsin, which had no bass population. If the population of bass is expected to grow at a rate of 25% per year, how long will it take the population to double?

23. **Population growth** A population growing at an annual rate r will triple in a time t (in years) given by the formula $t = \frac{\ln 3}{r}$. How long will it take the population of the town in Exercise 21 to triple?

24. **Fish population growth** How long will it take the fish population in Exercise 22 to triple?

25. **Isothermal expansion** One mole of gas expands isothermally to triple its volume. If the gas temperature is 400 kelvins, what energy is absorbed?

26. **Isothermal expansion** One mole of gas expands isothermally to double its volume. If the gas temperature is 300 kelvins, what energy is absorbed?

If an investment is growing continuously for t years, its annual growth rate r is given by the formula $r = \dfrac{1}{t} \ln \dfrac{P}{P_0}$, *where P is the current value and P_0 is the amount originally invested.*

27. **Investing** Investment in America Online (AOL) experienced continuous growth for several years in the 1990s. An investment of $10,400 in AOL in 1992 was worth $10,400,000 in 1999. Find AOL's average annual growth rate during this period.

28. **Investing** Dell experienced continuous growth for several years in the 1990s. A $5,000 investment in Dell in 1995 was worth $237,000 in 1999. Find the average annual growth rate of the stock.

29. **Depreciation** In business, equipment is often depreciated using the double declining-balance method. In this method, a piece of equipment with a life expectancy of N years, costing $\$C$, will depreciate to a value of $\$V$ in n years, where n is given by the formula $n = \dfrac{\log V - \log C}{\log\left(1 - \dfrac{2}{N}\right)}$. If a computer that cost $37,000 has a life expectancy of 5 years and has depreciated to a value of $8,000, how old is it?

30. **Depreciation** A machine worth $470 when new had a life expectancy of 12 years. If it is now worth $189, how old is it? (See Exercise 29.)

31. **Annuities** If $\$P$ is invested at the end of each year in an annuity earning interest at an annual rate r, the amount in the account will be $\$A$ after n years, where

$$n = \dfrac{\log\left(\dfrac{Ar}{P} + 1\right)}{\log(1 + r)}.$$

If $1,000 is invested each year in an annuity earning 12% annual interest, when will the account be worth $20,000?

32. Annuities If $5,000 is invested each year in an annuity earning 8% annual interest, when will the account be worth $50,000? (See Exercise 31.)

33. Breakdown voltage The coaxial power cable shown has a central wire with radius $R_1 = 0.25$ cm. It is insulated by a surrounding shield with inside radius $R_2 = 2$ cm. The maximum voltage the cable can withstand is called the **breakdown voltage** V of the insulation. V is given by the formula $V = ER_1 \ln \frac{R_2}{R_1}$ where E is the **dielectric strength** of the insulation. If $E = 400,000$ volts/cm, find V.

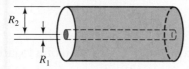

34. Breakdown voltage In Exercise 33, if the inside diameter of the shield were doubled, what voltage could the cable withstand?

Discovery and Writing

35. Suppose you graph the function $f(x) = \ln x$ on a coordinate grid with a unit distance of 1 cm on the x- and y-axes. How far out must you go on the x-axis so that $f(x) = 12$? Give your result to the nearest mile.

36. Suppose you graph the function $f(x) = \log x$ on a coordinate grid with a unit distance of 1 cm on the x- and y-axes. How far out must you go on the x-axis so that $f(x) = 12$? Give the result to the nearest mile. Why is this result so much larger than the result in Exercise 35?

37. One form of the logistic function is given by the equation $y = \dfrac{1}{1 + e^{-2x}}$. Explain how you would find the y-intercept of its graph.

38. Graph the function $y = \ln|x|$. Explain why the graph looks the way it does.

3.5 Properties of Logarithms

In this section, we will learn to

1. Evaluate logarithmic expressions using the properties of logarithms and using a graphing utility.

2. Simplify exponential and logarithmic expressions using the properties of logarithms.

3. Rewrite more complex logarithmic expressions and combine logarithmic expressions using a combination of properties.

4. Use the Change-of-Base Formula to evaluate logarithmic expressions and graph logarithmic functions.

5. Solve real-world problems using logarithms.

Rolling Stone magazine lists the rock band U2 at number 22 in their list of the top 100 artists of all time. The Irish band is noted for lead singer Bono's vocals and its anthemic sound.

Attending a U2 concert or cranking up a car stereo and playing the band's song "With or Without You" is an enjoyable activity. Music is an important part of our pop culture. Since many students prefer their music loud, the loudness of sound and its intensity are interesting concepts. In this section, we will see that loudness and intensity of sound are related by a formula involving the natural logarithmic function.

The properties that we will learn in this section have many uses in algebra, but they have importance in many more areas as well, especially in calculus. It is important that we become proficient in using these properties. Logarithms have many practical uses. Before calculators were available, mathematicians would reduce large numbers to logarithms, and then calculate the logarithms using tables. While this use is almost nonexistent today, logarithms have many other uses, as we will discover in this and the next section.

1. Evaluate Logarithmic Expressions Using the Properties of Logarithms and Using a Graphing Utility

Since logarithms are exponents, the properties of exponents have counterparts in the theory of logarithms. We begin with four basic properties.

Basic Properties of Logarithms   If $b > 0$ and $b \neq 1$, then

1. $\log_b 1 = 0$.   **2.** $\log_b b = 1$.

3. $\log_b b^x = x$.   **4.** $b^{\log_b x} = x$ $(x > 0)$.

Properties (1) through (4) follow directly from the definition of a logarithm.

1. $\log_b 1 = 0$, because $b^0 = 1$.
2. $\log_b b = 1$, because $b^1 = b$.
3. $\log_b b^x = x$, because $b^x = b^x$.
4. $b^{\log_b x} = x$, because $\log_b x$ is the exponent to which b is raised to get x.

EXAMPLE 1   **Using the Basic Properties of Logarithms**

Simplify each expression.   **a.** $\log_3 1$ **b.** $\log_4 4$ **c.** $\log_7 7^3$ **d.** $b^{\log_b 3}$

SOLUTION   **a.** By Property (1), $\log_3 1 = \mathbf{0}$, because $3^0 = 1$.

b. By Property (2), $\log_4 4 = \mathbf{1}$, because $4^1 = 4$.

c. By Property (3), $\log_7 7^3 = \mathbf{3}$, because $7^3 = 7^3$.

d. By Property (4), $b^{\log_b 3} = 3$, because $\log_b 3$ is the power to which b is raised to get 3. ∎

Self Check 1   Simplify each expression.   **a.** $\log_4 1$ **b.** $\log_3 3$ **c.** $\log_2 2^4$ **d.** $5^{\log_5 2}$

Now Try Exercise 11.

EXAMPLE 2   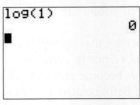 **Verifying the Basic Properties of Logarithms on a Calculator**

Simplify each expression and verify using a calculator.
a. $\log 1$ **b.** $\log 10$ **c.** $\ln e^3$ **d.** $e^{\ln 3}$

SOLUTION   We can use a scientific or a graphing calculator to verify the answers found. These are shown in Figure 3-35.

$\log 1 = 0$, **because** $10^0 = 1$. $\log 10 = 1$, **because** $10^1 = 10$.

```
log(1)
              0
■
```

```
log(10)
              1
■
```

(a)   (b)

FIGURE 3-35

$\ln e^3 = 3$ **because** $e^3 = e^3$.

$e^{\ln 3} = 3$ **because ln 3 is the power to which** e **is raised to get 3.**

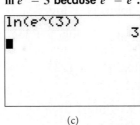

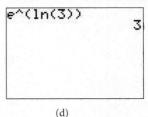

(c)

(d)

FIGURE 3-35 (Continued)

Self Check 2   Simplify each expression and verify using a calculator.

a. $\ln 1$ **b.** $\ln e$ **c.** $\log 10^5$ **d.** $10^{\log 5}$

Now Try Exercise 15.

2. Simplify Exponential and Logarithmic Expressions Using the Properties of Logarithms

Comment

It is important to keep in mind that the properties stated for logarithms also hold for natural logarithms.

The next two properties state that

the logarithm of a product is the sum of the logarithms and

the logarithm of a quotient is the difference of the logarithms.

Product and Quotient Properties of Logarithms

If M, N, and b are positive numbers and $b \neq 1$, then

5. $\log_b MN = \log_b M + \log_b N$ **6.** $\log_b \dfrac{M}{N} = \log_b M - \log_b N$

Property (5) is known as the **Product Rule**. Property (6) is known as the **Quotient Rule**.

The proof is available in online Appendix B.

Caution

By Property (5) of logarithms, the logarithm of a *product* is equal to the *sum* of the logarithms. The logarithm of a sum or a difference usually does not simplify. In general,

$\log_b(M + N) \neq \log_b M + \log_b N$ and $\log_b(M - N) \neq \log_b M - \log_b N$.

By Property (6), the logarithm of a *quotient* is equal to the *difference* of the logarithms. The logarithm of a quotient is not the quotient of the logarithms:

$\log_b \dfrac{M}{N} \neq \dfrac{\log_b M}{\log_b N}$.

These are common mistakes that students make.

EXAMPLE 3   **Using the Product and Quotient Rules**

Assume that x, y, and z are positive numbers and $x > 1$. Use the properties of logarithms to expand each logarithmic expression as much as possible.

a. $\log_b xyz$ **b.** $\log_b \dfrac{x}{25}$ **c.** $\ln \dfrac{x}{yz}$ **d.** $\log_b \dfrac{x(x + 1)}{(x - 1)}$

SOLUTION   **a.** $\log_b xyz = \log_b(xy)z$

$$= \log_b(xy) + \log_b z \qquad \text{The log of a product is the sum of the logs.}$$

$$= \log_b x + \log_b y + \log_b z \qquad \text{The log of a product is the sum of the logs.}$$

b. $\log_b \dfrac{x}{25} = \log_b x - \log_b 25 \qquad \text{The log of a quotient is the difference of the logs.}$

c. $\ln \dfrac{x}{yz} = \ln x - \ln(yz)$

$$= \ln x - (\ln y + \ln z) \qquad \text{The ln of a product is the sum of the lns.}$$

$$= \ln x - \ln y - \ln z \qquad \text{Remove parentheses.}$$

d. $\log_b \dfrac{x(x+1)}{(x-1)} = \log_b x(x+1) - \log_b(x-1) \qquad \text{First use the Quotient Rule.}$

$$= \log_b x + \log_b(x+1) - \log_b(x-1) \qquad \text{Apply the Product Rule.} \ \blacksquare$$

Self Check 3   Use the properties of logarithms to expand each logarithmic expression as much as possible, $x > 4$.   **a.** $\ln(x+3)(x-4)$ **b.** $\log \dfrac{75}{x}$

Now Try Exercise 25.

EXAMPLE 4   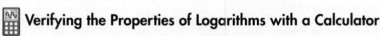 **Verifying the Properties of Logarithms with a Calculator**

Calculate each logarithm by using a graphing calculator. First, use the Product or Quotient Rule; second, use the expression as written.

a. $\ln(75 \cdot 20)$ **b.** $\log \dfrac{240}{13}$

SOLUTION   **a.** Using the Product Rule, we can write the expression as

$$\ln(75 \cdot 20) = \ln 75 + \ln 20.$$

Using a calculator, we verify that these two are equal, as shown in Figure 3-36.

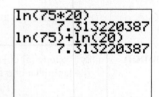

FIGURE 3-36

b. Using the Quotient Rule, we can write the expression as

$$\log \dfrac{240}{13} = \log 240 - \log 13.$$

Using a calculator, we verify that these two are equal, as shown in Figure 3-37.

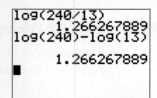

FIGURE 3-37

Self Check 4 Calculate each logarithm by using a graphing calculator. First, use the Product or Quotient Rule; second, use the expression as written.

a. $\log(15 \cdot 7)$ **b.** $\ln \dfrac{140}{11}$

Now Try Exercise 19.

Two more properties state that

the logarithm of a power is the power times the logarithm and

if the logarithms of two numbers are equal, the numbers are equal.

Power Rule and One-to-One Property of Logarithms

If M and b are positive numbers and $b \neq 1$, then

7. $\log_b M^p = p \log_b M$.   **8.** if $\log_b x = \log_b y$, then $x = y$.

Property (7) is known as the **Power Rule**. Property (8) is known as the **One-to-One Property**.

The proof is available in online Appendix B.

EXAMPLE 5 **Using the Power Rule**

Assume that x is a positive number. Use the Power Rule to simplify each expresssion.

a. $\ln x^{10}$ **b.** $\log_5 \sqrt{x}$

SOLUTION **a.** $\ln x^{10} = 10 \ln x$ The log of a power is the power times the log.

b. We will first write the radical expression as an exponential expression and then apply the Power Rule.

$$\log_5 \sqrt{x} = \log_5 x^{1/2} \qquad \text{Rewrite } \sqrt{x} \text{ in exponential form.}$$

$$= \frac{1}{2} \log_5 x \qquad \text{Apply the Power Rule.}$$

Caution

$$\log_b x^2 \neq (\log_b x)^2$$

Self Check 5 Simplify: $\log_4 \sqrt[3]{x^2}$.

Now Try Exercise 29.

ACCENT ON TECHNOLOGY   **Applying the Power Rule**

The development of calculators has made computations extremely easy and accurate, but handheld calculators have a limit to the size of the numbers they can display and evaluate. For example, if we were to try to calculate $\log 12^{1,000}$ [Figure 3-38(a)], we would most likely get an overflow error message as shown in Figure 3-38(b). However, by applying the Power Rule, we can evaluate the number with our calculator as is shown in Figure 3-38(c).

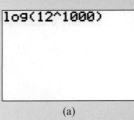

| (a) | (b) | (c) |

FIGURE 3-38

3. Rewrite More Complex Logarithmic Expressions and Combine Logarithmic Expressions Using a Combination of Properties

We can use the properties of logarithms to write a logarithm as the sum or difference of several logarithms. The simplification will often require the use of a combination of the properties and may require several steps.

EXAMPLE 6   Expanding a Logarithmic Expression

Assume that x, y, z, and b are positive numbers and $b \neq 1$. Write each expression in terms of the logarithms of x, y, and z.

a. $\ln(x^3 y^2 z)$ **b.** $\log_b \dfrac{y^2 \sqrt{z}}{x}$

SOLUTION   **a.** $\ln(x^3 y^2 z) = \ln x^3 + \ln y^2 + \ln z$ The ln of a product is the sum of the lns.

$\qquad\qquad\qquad = 3 \ln x + 2 \ln y + \ln z$ The ln of a power is the power times the ln.

b. $\log_b \dfrac{y^2 \sqrt{z}}{x} = \log_b \left(y^2 \sqrt{z} \right) - \log_b x$ The log of a quotient is the difference of the logs.

$\qquad\qquad\qquad = \log_b y^2 + \log_b z^{1/2} - \log_b x$ The log of a product is the sum of the logs.

$\qquad\qquad\qquad = 2 \log_b y + \dfrac{1}{2} \log_b z - \log_b x$ The log of a power is the power times the log.

Self Check 6   Write $\log_b \sqrt[3]{\dfrac{x^2 y}{z}}$ in terms of the logarithms of x, y, and z.

Now Try Exercise 51.

There are problems in calculus in which it is advantageous to expand a logarithmic expression. One such application is called logarithmic differentiation. We can use the laws of logarithms to transform an extremely difficult calculus problem into one that is solved with relative ease. Find out more at the end of the chapter.

We can use the properties of logarithms to combine several logarithms into one.

EXAMPLE 7   Combining Logarithmic Expressions

Assume x, y, z, and b are positive numbers and $b \neq 1$, $x > 2$. Write each expression as one logarithm.

a. $2 \log_b x + \dfrac{1}{3} \log_b y$ **b.** $\dfrac{1}{2} \log_b(x - 2) - \log_b y + 3 \log_b z$

SOLUTION   **a.** $2 \log_b x + \dfrac{1}{3} \log_b y = \log_b x^2 + \log_b y^{1/3}$ A power times a log is the log of the power.

$\qquad\qquad\qquad\qquad\quad = \log_b (x^2 y^{1/3})$ The sum of two logs is the log of the product.

b. $\dfrac{1}{2}\log_b(x-2) - \log_b y + 3\log_b z$

$= \log_b(x-2)^{1/2} - \log_b y + \log_b z^3$ A power times a log is the log of the power.

$= \log_b \dfrac{(x-2)^{1/2}}{y} + \log_b z^3$ The difference of two logs is the log of the quotient.

$= \log_b \dfrac{z^3\sqrt{x-2}}{y}$ The sum of two logs is the log of the product.

Self Check 7   Write as one logarithm: $2\ln x + \dfrac{1}{2}\ln y - 3\ln(x-y)$.

Now Try Exercise 57.

Summary of the Properties of Logarithms

If b, M, and N are positive numbers and $b \neq 1$, then

1. $\log_b 1 = 0$.

2. $\log_b b = 1$.

3. $\log_b b^x = x$.

4. $b^{\log_b x} = x$.

5. $\log_b MN = \log_b M + \log_b N$.

6. $\log_b \dfrac{M}{N} = \log_b M - \log_b N$.

7. $\log_b M^p = p\log_b M$.

8. if $\log_b x = \log_b y$, then $x = y$.

1. $\ln 1 = 0$.

2. $\ln e = 1$.

3. $\ln e^x = x$.

4. $e^{\ln x} = x$.

5. $\ln MN = \ln M + \ln N$.

6. $\ln \dfrac{M}{N} = \ln M - \ln N$.

7. $\ln M^p = p\ln M$.

8. if $\ln x = \ln y$, then $x = y$.

Note that the properties of logarithms are stated along with the corresponding properties of natural logarithms.

EXAMPLE 8   **Using Properties of Logarithms**

Given $\log_b 2 \approx 0.3$, $\log_b 3 \approx 0.5$, and $\log_b 5 \approx 0.7$, find approximations for each value. **a.** $\log_b 18$ **b.** $\log_b 2.5$ **c.** $\log_b \sqrt{\dfrac{5}{6}}$

SOLUTION   **a.** $\log_b 18 = \log_b(2 \cdot 3^2)$

$= \log_b 2 + \log_b 3^2$ Apply the Product Rule.

$= \log_b 2 + 2\log_b 3$ Apply the Power Rule.

$\approx 0.3 + 2(.5)$

≈ 1.3

b. $\log_b 2.5 = \log_b\left(\dfrac{5}{2}\right)$

$= \log_b 5 - \log_b 2$ Apply the Quotient Rule.

$\approx 0.7 - 0.3$

≈ 0.4

c. $\log_b \sqrt{\dfrac{5}{6}} = \log_b \left(\dfrac{5}{2 \cdot 3}\right)^{1/2}$

$\qquad\qquad = \dfrac{1}{2}\left(\log_b \dfrac{5}{2 \cdot 3}\right)$ Apply the Power Rule.

$\qquad\qquad = \dfrac{1}{2}(\log_b 5 - \log_b 2 \cdot 3)$ Apply the Quotient Rule.

$\qquad\qquad = \dfrac{1}{2}(\log_b 5 - (\log_b 2 + \log_b 3))$ Apply the Product Rule.

$\qquad\qquad = \dfrac{1}{2}(\log_b 5 - \log_b 2 - \log_b 3)$ Simplify.

$\qquad\qquad \approx \dfrac{1}{2}(0.7 - 0.3 - 0.5)$

$\qquad\qquad \approx -0.05$

Self Check 8   Use the given values in Example 8 to approximate $\log_b 0.75$.

Now Try Exercise 83.

4. Use the Change-of-Base Formula to Evaluate Logarithmic Expressions and Graph Logarithmic Functions

We have seen how to use a calculator to find base-10 and base-e logarithms. To use a calculator to find logarithms with different bases, such as $\log_7 63$, we can divide the base-10 (or base-e) logarithm of 63 by the base-10 (or base-e) logarithm of 7.

$$\log_7 63 = \frac{\log 63}{\log 7} \qquad\qquad \log_7 63 = \frac{\ln 63}{\ln 7}$$

$$\approx 2.129150068 \qquad\qquad\qquad \approx 2.129150068$$

Caution

The expression $\dfrac{\log_a x}{\log_a b}$ means that one logarithm is to be divided by the other. They are not to be subtracted.

To check the result, we verify that $7^{2.129150068} \approx 63$. This example suggests that if we know the base-a logarithm of a number, we can find its logarithm with some other base b.

Change-of-Base Formulas   If a, b, and x are positive numbers, $a \neq 1$, and $b \neq 1$, then

$$\log_b x = \frac{\log_a x}{\log_a b}.$$

If we know logarithms with base a (for example, $a = 10$), we can find the logarithm of x with a new base b by dividing the base-a logarithm of x by the base-a logarithm of b.

EXAMPLE 9   **Using the Change-of-Base Formula**

Find: $\log_3 5$.

SOLUTION   We can substitute 3 for b, 10 for a, and 5 for x into the Change-of-Base Formula.

$$\log_b x = \frac{\log_a x}{\log_a b}$$

$$\log_3 5 = \frac{\log_{10} 5}{\log_{10} 3}$$ Divide the base-10 logarithm of 5 by the base-10 logarithm of 3.

$$\approx 1.464973521$$

To four decimal places, $\log_3 5 \approx 1.4650$.

Self Check 9   Find $\log_5 3$ to four decimal places.

Now Try Exercise 95.

ACCENT ON TECHNOLOGY   **Using the Change-of-Base Formula to Graph a Logarithmic Function**

We can use the Change-of-Base Formula to graph logarithmic functions that do not have a base of 10 or e. Consider the function

$$f(x) = \log_4(x + 1).$$

Since the base of the function $f(x) = \log_4(x + 1)$ is 4, to graph this function on a calculator, we can rewrite the function as

$$f(x) = \log_4(x + 1) = \frac{\log(x + 1)}{\log 4}.$$

We can also use the natural logarithm function if we prefer, $f(x) = \log_4(x + 1) = \frac{\ln(x + 1)}{\ln 4}$. The domain of this function is $D = (-1, \infty)$, so there will be a vertical asymptote at $x = -1$. We set the graphing window to $[-4.7, 4.7]$ by $[-3.1, 3.1]$. Notice that the graph appears to stop at $x = -1$, but that is not the case. We can have the calculator evaluate the function at a value very close to $x = -1$ to verify this. Notice in Figure 3-39(b), that the point on the graph is not shown, but the value is indicated.

Graph the function

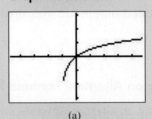

(a)

Evaluate close to $x = -1$

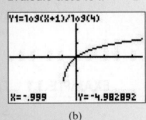

(b)

FIGURE 3-39

5. Solve Real-World Problems Using Logarithms

We must use properties of logarithms to solve many problems.

pH Scale
The more acidic a chemical solution, the greater the concentration of hydrogen ions in the solution. Chemists measure this concentration indirectly by the **pH scale**, or the **hydrogen-ion index.**

---

pH of a Solution   If $[H^+]$ is the hydrogen-ion concentration in gram-ions per liter, then

$$pH = -\log[H^+].$$

Since pure water has approximately 10^{-7} gram-ions of hydrogen per liter, its pH is

$$pH = -\log[\mathbf{H}^+]$$

$$pH = -\log \mathbf{10^{-7}}$$

$$= -(-7) \log 10 \qquad \text{The log of a power is the power times the log.}$$

$$= -(-7) \cdot 1 \qquad \text{Use Property (2) of logarithms: } \log_b b = 1.$$

$$= 7$$

EXAMPLE 10   **Finding the pH of Seawater**

Seawater has a pH of approximately 8.5. Find its hydrogen-ion concentration.

SOLUTION   We can substitute 8.5 for pH and solve the equation $pH = -\log[\mathbf{H}^+]$ for $[\mathbf{H}^+]$.

$$8.5 = -\log[\mathbf{H}^+]$$

$$-8.5 = \log[\mathbf{H}^+]$$

$$[\mathbf{H}^+] = 10^{-8.5} \qquad \text{Change the equation from logarithmic form to exponential form.}$$

We use a calculator to find that $[\mathbf{H}^+] \approx 3.2 \times 10^{-9}$ gram-ions per liter. ∎

Self Check 10   The pH of a solution is 5.7. Find the hydrogen-ion concentration.

Now Try Exercise 107.

Electronics

Recall that if E_O is the output voltage of a device and E_I is the input voltage, the decibel voltage gain is given by

$$\mathbf{dB\ gain = 20 \log \frac{E_O}{E_I}}.$$

If input and output are measured in watts instead of volts, a different formula is needed.

EXAMPLE 11   **Finding an Alternate Formula for Voltage Gain**

Show that an alternate formula for dB voltage gain is

$$\mathbf{dB\ gain = 10 \log \frac{P_O}{P_I}},$$

where P_I is the power input and P_O is the power output.

SOLUTION   Power is directly proportional to the square of the voltage. So for some constant k,

$$P_I = k(E_I)^2 \qquad \text{and} \qquad P_O = k(E_O)^2,$$

and

$$\frac{P_O}{P_I} = \frac{k(E_O)^2}{k(E_I)^2} = \left(\frac{E_O}{E_I}\right)^2.$$

We raise both sides to the $\frac{1}{2}$ power to get

$$\frac{E_O}{E_I} = \left(\frac{P_O}{P_I}\right)^{1/2},$$

which we substitute into the formula for dB gain.

$$\text{dB gain} = 20 \log \frac{E_O}{E_I}$$

$$= 20 \log \left(\frac{P_O}{P_I} \right)^{1/2}$$

$$= 20 \cdot \frac{1}{2} \log \frac{P_O}{P_I} \qquad \text{The log of a power is the power times the log.}$$

$$= 10 \log \frac{P_O}{P_I} \qquad \text{Simplify.}$$

Self Check 11   Find the dB gain of a device to the nearest hundredth when $P_O = 30$ watts and $P_I = 2$ watts.

Now Try Exercise 111.

Physiology
In physiology, experiments suggest that the relationship between the loudness and the intensity of sound is a logarithmic one known as the Weber–Fechner Law.

Weber–Fechner Law   If L is the apparent loudness of a sound and I is the intensity, then

$L = k \ln I$, where k is a constant to be determined experimentally.

EXAMPLE 12   **Using the Weber–Fechner Law**

What increase in the intensity of a sound is necessary to cause a doubling of the apparent loudness?

SOLUTION   We use the formula $L = k \ln I$. To double the apparent loudness, we multiply both sides of the equation by 2 and use Property (7) of logarithms.

$$L = k \ln I$$

$$2L = 2k \ln I$$

$$= k \ln I^2$$

To double the apparent loudness, we must square the intensity.

Self Check 12   What increase is necessary to triple the apparent loudness?

Now Try Exercise 113.

Self Check Answers   **1. a.** 0   **b.** 1   **c.** 4   **d.** 2   **2. a.** 0   **b.** 1   **c.** 5   **d.** 5
3. a. $\ln(x + 3) + \ln(x - 4)$ **b.** $\log 75 - \log x$ **4. a.** 2.021189299

b. 2.54374715   **5.** $\frac{2}{3} \log_4 x$ **6.** $\frac{1}{3}(2 \log_b x + \log_b y - \log_b z)$

7. $\ln \dfrac{x^2 \sqrt{y}}{(x - y)^3}$ **8.** -0.1 **9.** 0.6826   **10.** 2×10^{-6} g ion/L

11. 11.76 dB   **12.** Cube the intensity.

Exercises **3.5**

Getting Ready

You should be able to complete these vocabulary and concept statements before you proceed to the practice exercises.

Fill in blanks.

1. $\log_b 1 = $ _____
2. $\log_b b = $ _____
3. $\log_b MN = \log_b$ _____ $+ \log_b$ _____
4. $b^{\log_b x} = $ _____
5. If $\log_b x = \log_b y$, then _____ $= $ _____.
6. $\log_b \dfrac{M}{N} = \log_b M$ _____ $\log_b N$
7. $\log_b x^p = p \log_b$ _____
8. $\log_b b^x = $ _____
9. $\log_b(A + B)$ _____ $\log_b A + \log_b B$
10. $\log_b A + \log_b B$ _____ $\log_b AB$

Simplify each expression.

11. $\log_6 1$
12. $\log_4 4$
13. $\log_4 4^7$
14. $4^{\log_4 8}$
15. $5^{\log_5 10}$
16. $\log_5 5^2$
17. $\log_5 5$
18. $\log_5 1$

Practice

Use a calculator to verify each equation.

19. $\log[(3.7)(2.9)] = \log 3.7 + \log 2.9$
20. $\ln \dfrac{9.3}{2.1} = \ln 9.3 - \ln 2.1$
21. $\ln(3.7)^3 = 3 \ln 3.7$
22. $\log \sqrt{14.1} = \dfrac{1}{2} \log 14.1$
23. $\log 3.2 = \dfrac{\ln 3.2}{\ln 10}$
24. $\ln 9.7 = \dfrac{\log 9.7}{\log e}$

Assume that x, y, z, and b are positive numbers, x > 3 and b ≠ 1. Use the properties of logarithms to expand each logarithmic expression and simplify.

25. $\log_b 2xy$
26. $\log_b 3xz$
27. $\log_b \dfrac{2x}{y}$
28. $\log_b \dfrac{x}{yz}$
29. $\log_b x^2 y^3$
30. $\log_b x^3 y^2 z$
31. $\log_b(xy)^{1/3}$
32. $\log_b x^{1/2} y^3$
33. $\log_b x\sqrt{z}$
34. $\log_b \sqrt{xy}$
35. $\log_b \dfrac{\sqrt[3]{x}}{\sqrt[3]{yz}}$
36. $\log_b \sqrt[4]{\dfrac{x^3 y^2}{z^4}}$

37. $\ln 5x^{-3}$
38. $\log_b \sqrt{xy^{-4}}$
39. $\log \dfrac{3x}{10,000}$
40. $\log \dfrac{1,000}{x^2}$
41. $\ln \dfrac{11}{e^3}$
42. $\ln \dfrac{e^7}{12}$
43. $\log_2 32x$
44. $\log_3 \dfrac{81}{x}$
45. $\log_4 \dfrac{64}{\sqrt{x-3}}$
46. $\log_5 \dfrac{125}{\sqrt[3]{x+1}}$
47. $\ln 3e$
48. $\ln e\sqrt{x}$
49. $\log 10^5$
50. $\log 10x$
51. $\log_2 \sqrt{\dfrac{x-1}{x^3(x+1)}}$
52. $\log_2 \sqrt[3]{\dfrac{3x}{(x-1)^4}}$
53. $\ln\left[\dfrac{x^4\sqrt{2-x}}{(x+3)^4}\right], x < 2$
54. $\ln\left[\dfrac{(x+2)^4(x-1)}{\sqrt{x+3}}\right]$
55. $\log\left[\dfrac{9(x+1)^3}{100x\sqrt[3]{x+3}}\right]$
56. $\log\left[\dfrac{1,000x^4}{\sqrt[4]{x-3}}\right]$

Assume that x, y, z, and b are positive numbers, x > 5 and b ≠ 1. Use the properties of logarithms to combine logarithmic expressions and write as the logarithm of one quantity.

57. $\log_b(x+1) - \log_b x$
58. $\log_b x + \log_b(x+2) - \log_b 8$
59. $2 \log_b x + \dfrac{1}{3} \log_b y$
60. $-2 \log_b x - 3 \log_b y + \log_b z$
61. $-3 \log_b x - 2 \log_b y + \dfrac{1}{2} \log_b z$
62. $3 \log_b(x+1) - 2 \log_b(x+2) + \log_b x$
63. $\log_b\left(\dfrac{x}{z}+x\right) - \log_b\left(\dfrac{y}{z}+y\right)$
64. $\log_b(xy+y^2) - \log_b(xz+yz) + \log_b z$
65. $3 \ln x + \ln 5$
66. $4 \ln x + \ln 12$
67. $\log_3 486 - \log_3 2$
68. $\log_2 320 - \log_2 5$
69. $\dfrac{1}{4}(2 \ln x - \ln y)$
70. $\dfrac{1}{2}(5 \log x + \log y)$
71. $\ln x + \ln(x-1) - \ln 5 - \ln(x^2-1)$
72. $\log(x^2-25) + \log 12 - \log(x+5)$

Determine whether each statement is true or false.

73. $\log_b ab = \log_b a + 1$
74. $\log_b 0 = 1$

75. $\ln(x + y) \neq \ln x + \ln y$

76. If $\log_a b = c$, then $\log_b a = c$.

77. $\log_7 7^7 = 7$

78. $\ln(-x) = -\ln x$

79. $\dfrac{\log_b A}{\log_b B} = \log_b A - \log_b B$

80. $\log_b \dfrac{1}{5} = -\log_b 5$

81. $\dfrac{1}{3}\log_b a^3 = \log_b a$

82. $\log_b y + \log_{1/b} y = 0$

Assume that $\log_{10} 4 = 0.6021$, $\log_{10} 7 = 0.8451$, *and* $\log_{10} 9 = 0.9542$. *Use these values and the properties of logarithms to find each value. Do not use a calculator.*

83. $\log_{10} 28$

84. $\log_{10} \dfrac{7}{4}$

85. $\log_{10} 2.25$

86. $\log_{10} 36$

87. $\log_{10} \dfrac{63}{4}$

88. $\log_{10} \dfrac{4}{63}$

89. $\log_{10} 252$

90. $\log_{10} 49$

91. $\log_{10} 112$

92. $\log_{10} 324$

93. $\log_{10} \dfrac{144}{49}$

94. $\log_{10} \dfrac{324}{63}$

Use a calculator and the Change-of-Base Formula to find each logarithm to four decimal places.

95. $\log_3 7$

96. $\log_7 3$

97. $\log_\pi 3$

98. $\log_3 \pi$

99. $\log_3 8$

100. $\log_5 10$

101. $\log_{\sqrt{2}} \sqrt{5}$

102. $\log_\pi e$

Use a calculator and the Change-of-Base Formula to graph each function.

103. $f(x) = \log_3(x + 2)$

104. $f(x) = \log_5(x - 4)$

105. $f(x) = \log_{1/2}(3 - x)$

106. $f(x) = \log_{3/4}(2 - x)$

Applications

107. **pH of a solution** Find the pH of a solution with a hydrogen-ion concentration of 1.7×10^{-5} gram-ions per liter.

108. **pH of calcium hydroxide** Find the hydrogen-ion concentration of a saturated solution of calcium hydroxide whose pH is 13.2.

109. **pH of apples** The pH of apples can range from 2.9 to 3.3. Find the range of the hydrogen-ion concentration.

110. **pH of sour pickles** The hydrogen-ion concentration of sour pickles is 6.31×10^{-4} gram-ions per liter. Find the pH.

111. **dB gain** An amplifier produces a 40-watt output with a $\frac{1}{2}$-watt input. Find the dB gain.

112. **dB loss** Losses in a long telephone line reduce a 12-watt input signal to an output of 3 watts. Find the dB gain. (Because it is a loss, the "gain" will be negative.)

113. **Weber–Fechner Law** What increase in intensity is necessary to quadruple the apparent loudness of a sound?

114. **Weber–Fechner Law** What decrease in intensity is necessary to make a sound appear half as loud?

115. **Isothermal expansion** If a certain amount E of energy is added to 1 mole of a gas, the gas expands from an initial volume of 1 liter to a final volume V without changing its temperature according to the formula $E = 8,300 \ln V$. Find the volume if twice that energy is added to the gas.

116. **Richter scale** By what factor must the amplitude of an earthquake change to increase its severity by 1 point on the Richter scale? Assume that the period remains constant. The Richter scale is given by $R = \log \frac{A}{P}$, where A is the amplitude and P the period of the tremor.

Discovery and Writing

117. Simplify: $3^{4 \log_3 2} + 5^{\frac{1}{2} \log_5 25}$.

118. Find the value of $a - b$ if
$5 \log x + \frac{1}{3} \log y - \frac{1}{2} \log x - \frac{5}{6} \log y = \log(x^a y^b)$.

119. Prove Property (6) of logarithms:
$\log_b \frac{M}{N} = \log_b M - \log_b N$.

120. Show that $-\log_b x = \log_{1/b} x$.

121. Show that $e^{x \ln a} = a^x$.

122. Show that $e^{\ln x} = x$.

123. Show that $\ln(e^x) = x$.

124. If $\log_b 3x = 1 + \log_b x$, find b.

125. Explain why $\ln(\log 0.9)$ is undefined.

126. Explain why $\log_b(\ln 1)$ is undefined.

In Exercises 127 and 128, A and B are both negative. Thus, AB and $\frac{A}{B}$ are positive, and log AB and log $\frac{A}{B}$ are defined.

127. Is it still true that $\log AB = \log A + \log B$? Explain.

128. Is it still true that $\log \frac{A}{B} = \log A - \log B$? Explain.

3.6 Exponential and Logarithmic Equations

In this section, we will learn to

1. Solve exponential equations.
2. Solve logarithmic equations.
3. Model real-world problems using exponential and logarithmic equations.

elena moiseeva/Shutterstock.com

Coffee is one of the most popular beverages worldwide. Suppose we visit a coffee shop and order a white-chocolate mocha. After smooth white chocolate is blended with rich espresso and steamed milk and topped with whipped cream, the coffee is served to us at a temperature of 180°F.

Suppose the exponential function $T = 70 + 110e^{-0.2t}$ models the temperature T of the mocha after t minutes. If we were interested in determining how long it will take for the temperature of the coffee to reach 80°F, we would substitute 80 in for T and solve the resulting equation $80 = 70 + 110e^{-0.2t}$ for t. This equation is called an exponential equation because t occurs as an exponent. You will be asked to solve this equation in Exercise 114.

Interestingly, we have to use logarithms to solve this exponential equation. In fact, we have to use logarithms to solve many exponential equations, and we often have to use exponents to solve logarithmic equations. In this section, we will learn to solve both exponential equations and logarithmic equations.

An **exponential equation** is an equation with a variable in one of its exponents. Some examples of exponential equations are

$$3^x = 5, \quad e^{2x} = 7, \quad 6^{x-3} = 2^x, \quad \text{and} \quad 3^{2x+1} - 10(3^x) + 3 = 0.$$

A **logarithmic equation** is an equation with at least one logarithmic expression that contains a variable. Some examples of logarithmic equations are

$$\log 2x = 25, \quad \ln x - \ln(x - 12) = 24, \quad \text{and} \quad \log x = \log \frac{1}{x} + 4.$$

1. Solve Exponential Equations

Some exponential equations can be solved using like bases. If we look at the exponential equation $2^x = 2^4$, we can see that the solution is $x = 4$. Because the function $f(x) = 2^x$ is one-to-one, this is the only solution. Suppose we want to solve the equation $2^{x+1} = 16$. How is this equation similar to $2^x = 2^4$? Since $16 = 2^4$, we can rewrite the equation as $2^{x+1} = 2^4$. When we do, we note that the bases are equal. If the bases are equal, the exponents are too. We can solve the equation as follows:

$$2^{x+1} = 16$$
$$2^{x+1} = 2^4$$
$$x + 1 = 4$$
$$x = 3$$

One-to-One Property of Exponents   If $a^x = a^y$, $a \neq 1$, and $a \neq 0$, then $x = y$.

That is, equal quantities with like bases have equal exponents.

If we can express each side of an exponential equation as a power of the same base, we can solve the equation.

EXAMPLE 1   **Solving Exponential Equations Using Like Bases**

Solve each equation.   **a.** $2^{2x+1} = 32$ **b.** $9^{x+1} = 27^{x-2}$

SOLUTION   We will rewrite each exponential equation with the same base and solve.

a. $2^{2x+1} = 32$

$2^{2x+1} = 2^5$ Rewrite 32 as 2^5.

$2x + 1 = 5$ Exponents are equal.

$2x = 4$

$x = 2$

Check: $2^{2(2)+1} = 2^5 = 32$. The solution checks.

b. $9^{x+1} = 27^{x-2}$

$(3^2)^{x+1} = (3^3)^{x-2}$ $9 = 3^2$ and $27 = 3^3$.

$3^{2x+2} = 3^{3x-6}$ Simplify exponents.

$2x + 2 = 3x - 6$ Exponents are equal.

$x = 8$ Solve for x.

To check this solution, we will use a calculator and find that the solution is $x = 8$.

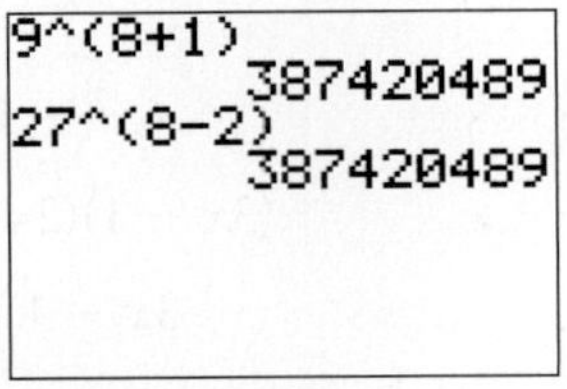

```
9^(8+1)
        387420489
27^(8-2)
        387420489
```

FIGURE 3-40

Self Check 1   Solve each equation.   **a.** $125^{x-1} = 5^{2x+3}$ **b.** $16^{x-1} = 4^{x+3}$

Now Try Exercise 7.

EXAMPLE 2   **Solving Exponential Equations Using Like Bases**

Solve: $2^{x^2+2x} = \dfrac{1}{2}$.

SOLUTION   Since $\frac{1}{2} = 2^{-1}$, we can write the equation in the form $2^{x^2+2x} = 2^{-1}$ and use like bases to solve.

$$2^{x^2+2x} = \frac{1}{2}$$

$2^{x^2+2x} = 2^{-1}$ Bases are equal.

$x^2 + 2x = -1$ Exponents are equal.

$x^2 + 2x + 1 = 0$ Solve the quadratic equation.

$(x + 1)(x + 1) = 0$

$x + 1 = 0$

$x = -1$

We can verify that -1 satisfies the equation by using the TABLE feature on a graphing calculator. Enter each side of the equation in the graph editor, and then go to the TABLE and enter $x = -1$.

FIGURE 3-41

Self Check 2   Solve: $3^{x^2 + 2x} = 27$.

Now Try Exercise 17.

EXAMPLE 3   **Solving Exponential Equations with Like Bases**

Solve: $e^{6x^2} = e^{-x+1}$.

SOLUTION   We can use like bases to solve the exponential equation. Because equal quantities with like bases have equal exponents, we have

$$6x^2 = -x + 1$$

$$6x^2 + x - 1 = 0 \qquad \text{Add } x - 1 \text{ to both sides.}$$

$$(3x - 1)(2x + 1) = 0 \qquad \text{Factor the trinomial.}$$

$$3x - 1 = 0 \quad \text{or} \quad 2x + 1 = 0 \qquad \text{Set each factor equal to 0.}$$

$$x = \frac{1}{3} \qquad\qquad x = -\frac{1}{2} \qquad \text{Solve each equation.}$$

Self Check 3   Solve: $e^{4x^2} = e^{24}$.

Now Try Exercise 23.

If the bases of the terms in an exponential equation are not equal, as is often the case, we use logarithms to solve the equation. Logarithmic functions are one-to-one. Because of that, we know that the following property holds:

One-to-One Property of Logarithms   If $\log_b x = \log_b y$, $b > 0$, $b \ne 1$, $x > 0$, and $y > 0$, then $x = y$.

That is, logarithms of equal numbers are equal.

Also, since logarithmic functions are one-to-one, we can take the logarithm of both sides of an equation. This allows us to use the properties of logarithms to help solve the equation. A logarithm of any base may be used, but it is most efficient to use natural logarithms or common logarithms, because a calculator is often needed to get an approximate answer.

Strategy for Solving Exponential Equations with Different Bases

To solve an exponential equation, we can follow these steps:

Step 1: Isolate the exponential expression.

Step 2: Take the same logarithm of both sides.

Step 3: Simplify using the rules of logarithms.

Step 4: Solve for the variable.

- If the equation contains base e, it is most efficient to use natural logarithms, because $\ln e^x = x$.
- If the equation contains base 10, it is most efficient to use common logarithms, because $\log 10^x = x$.
- If the equation contains exponential terms with bases other than base e or 10, use either natural logarithms or common logarithms.

EXAMPLE 4   **Solving an Exponential Equation with Base 10**

Solve the equation $10^{3x-1} = 53$. Give the answer in logarithmic form. Then use a calculator and obtain a decimal approximation correct to four decimal places.

SOLUTION   We note that the exponential expression 10^{3x-1} is isolated. Since the base is 10, we will use common logarithms to solve the equation.

$$10^{3x-1} = 53$$

$$\log 10^{3x-1} = \log 53 \qquad \text{Take the common log of both sides.}$$

$$(3x - 1)\log 10 = \log 53 \qquad \text{Use the Power Rule of logarithms.}$$

$$(3x - 1)(1) = \log 53 \qquad \text{Since } \log 10 = 1, \text{ substitute 1 for } \log 10.$$

$$3x - 1 = \log 53 \qquad \text{Remove parentheses.}$$

$$3x = \log 53 + 1 \qquad \text{Add 1 to both sides.}$$

$$x = \frac{\log 53 + 1}{3} \qquad \text{Solve for } x. \text{ This is the exact answer.}$$

$$x \approx 0.9081 \qquad \text{This is a decimal approximation.}$$

Self Check 4   Solve the equation $10^{5-x} = 75$. Give the answer in logarithmic form. Then use a calculator and obtain a decimal approximation correct to four decimal places.

Now Try Exercise 35.

EXAMPLE 5   **Solving an Exponential Equation Using Common Logarithms**

Solve the exponential equation $3^x - 5 = 0$. Give the answer in logarithmic form and then use a calculator to obtain a decimal approximation. Check the approximation. Finally, round the approximation to four decimal places.

SOLUTION   First we will isolate the exponential expression 3^x by adding 5 to both sides. Next, we will take the common logarithm of both sides of the equation. We will then use the Power Rule and solve for x.

$$3^x - 5 = 0$$

$$3^x = 5 \qquad \text{Add 5 to both sides.}$$

$$\log 3^x = \log 5 \qquad \text{Take the common logarithm of each side.}$$

$$x \log 3 = \log 5 \qquad \text{Apply the Power Rule for logarithms.}$$

$$x = \frac{\log 5}{\log 3} \qquad \text{Solve for } x. \text{ This is the exact solution.}$$

$$x \approx 1.464973521 \qquad \text{This is an approximate solution.}$$

We can check the solution using a calculator.

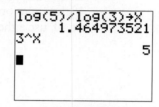

FIGURE 3-42

The solution, rounded to four decimal places, is $x \approx 1.4650$.

Self Check 5   Solve the equation $5^x = 3$. Give the answer in logarithmic form. Then use a calculator and obtain a decimal approximation rounded to four decimal places.

Now Try Exercise 25.

Caution

A careless reading of $x = \frac{\log 5}{\log 3}$ leads to a common error. The right-hand side of the equation calls for a division, not a subtraction.

$$\frac{\log 5}{\log 3} = (\log 5) \div (\log 3)$$

$$\frac{\log 5}{\log 3} \neq \log 5 - \log 3$$

It is the expression $\log \frac{5}{3}$ that is equivalent to $\log 5 - \log 3$.

EXAMPLE 6   **Solving an Exponential Equation Using Natural Logarithms**

Solve the exponential equation $6^{x-3} = 2^x$. Give the answer in logarithmic form and then obtain a decimal approximation rounded to four decimal places.

SOLUTION   First we note that each exponential expression is isolated on one side. We will begin by taking the natural logarithm of each side and then apply the Power Rule of logarithms.

$$6^{x-3} = 2^x$$

$$\ln 6^{x-3} = \ln 2^x \qquad \text{Take the natural logarithm of each side.}$$

$$(x - 3) \ln 6 = x \ln 2 \qquad \text{Use the Power Rule to simplify.}$$

$$x \ln 6 - 3 \ln 6 = x \ln 2 \qquad \text{Multiply and eliminate parentheses.}$$

$$x \ln 6 - x \ln 2 = 3 \ln 6 \qquad \text{Add } 3 \ln 6 - x \ln 2 \text{ to both sides.}$$

$$x(\ln 6 - \ln 2) = 3 \ln 6 \qquad \text{Factor } x \text{ from the left side.}$$

Comment

As we see in Examples 5 and 6, either common logarithms or natural logarithms can be used to solve exponential equations. As a result, the exact logarithmic form of our solution can contain common or natural logarithms.

$$x = \frac{3 \ln 6}{\ln 6 - \ln 2}$$ Solve for x. This is the exact solution.

$$x = \frac{\ln 216}{\ln 3}$$ Use the Power and Quotient Rules of logarithms.

$$x \approx 4.8928$$ This is an approximate solution. ∎

Self Check 6   Solve the equation $5^{x+3} = 3^x$. Give the answer in logarithmic form and then obtain a decimal approximation rounded to four decimal places.

Now Try Exercise 29.

EXAMPLE 7   **Solving Exponential Equations with Base e**

Solve the equation $5e^{2x} = 40$. Give the answer in natural logarithmic form and then obtain a decimal approximation rounded to four decimal places.

SOLUTION   We will first isolate the exponential expression e^{2x}. Because the base is the natural exponent e, we will take the natural logarithm of both sides.

$$5e^{2x} = 40$$

$$e^{2x} = 8$$ Divide both sides by 5.

$$\ln e^{2x} = \ln 8$$ Take the natural log of both sides.

$$2x = \ln 8$$ Use Property (3): $\ln e^x = x$.

$$x = \frac{\ln 8}{2}$$ Solve for x. This is the exact answer.

$$x \approx 1.039720771$$

We can check our answer using our calculator. Store the answer as X and evaluate.

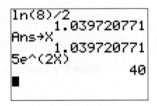

```
ln(8)/2
            1.039720771
Ans→X
            1.039720771
5e^(2X)
                     40
■
```

FIGURE 3-43

Comment

In Example 7, Property (3) was used to write $\ln e^{2x}$ as $2x$. It is also correct to simplify this expression by using the Power Rule:

$$\ln e^{2x} = 2x \ln e = 2x(1) = 2x.$$

The decimal approximation rounded to four places is $x \approx 1.0397$. ∎

Self Check 7   Solve the equation $2e^{-3x} = 17$. Give the answer in natural logarithmic form and then obtain a decimal approximation rounded to four decimal places.

Now Try Exercise 39.

EXAMPLE 8   **Solving an Exponential Equation by Factoring**

Solve the equation $5^{2x} - 2 \cdot 5^x - 3 = 0$. Give the answer in natural logarithmic form and then obtain a decimal approximation rounded to four decimal places.

SOLUTION   The given equation is quadratic in form and can be solved by factoring and logarithms.

$$5^{2x} - 2 \cdot 5^x - 3 = 0$$

$$(5^x - 3)(5^x + 1) = 0 \qquad \text{Factor.}$$

$$5^x - 3 = 0 \qquad \text{or} \qquad 5^x + 1 = 0$$

Comment

If only a decimal approximation of a solution is needed, a graphing calculator can be used to solve exponential equations. Our strategy is to graph each side of the equation and then find the point of intersection.

$$5^x = 3 \qquad\qquad\qquad 5^x = -1$$

$$\ln 5^x = \ln 3 \qquad\qquad \ln 5 = \ln(-1)$$

$$x \ln 5 = \ln 3 \qquad\qquad \ln(-1) \text{ is not defined.}$$

$$x = \frac{\ln 3}{\ln 5} \approx 0.6826$$

Self Check 8   Solve the equation $3^{2x} - 4 \cdot 3^x + 4 = 0$. Give the answer in natural logarithmic form and then obtain a decimal approximation rounded to four decimal places.

Now Try Exercise 45.

EXAMPLE 9   **Solving an Exponential Equation Using a Calculator**

Use a graphing calculator to solve the equation $\left(\sqrt{3}\right)^x = 2^{\pi x - 3}$.

SOLUTION   Enter each side of the equation in the [GRAPH] menu and find the point of intersection. From Figure 3-44, we see that the solution is $x \approx 1.2770786$.

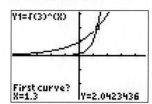 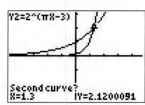 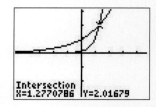

FIGURE 3-44

Self Check 9   Use a graphing calculator to solve the equation $2^{x - \sqrt{3}} = e^x - 2$.

Now Try Exercise 97.

2. Solve Logarithmic Equations

To solve a logarithmic equation, there are several strategies. The strategy we use will be dependent upon the nature of the equation. Four strategies are summarized here:

Strategy for Solving Logarithmic Equations

1. For logarithmic equations containing one logarithm, we can use the definition of the logarithm. We write the logarithm in exponential form and then solve for the variable.
   Note: This strategy can be used to solve the following logarithmic equations:

   $$\log_3(x + 5) = 2, \ \log(x^2 - 4) = 5, \text{ and } 3 \ln 4x = 7.$$

2. For more complicated logarithmic equations, we can combine and isolate the logarithmic expressions. We can then use the definition of the logarithm and write the logarithm in exponential form to solve for the variable.

Note: This strategy can be used to solve the following logarithmic equations:

$$\log x + \log(x - 1) = 3, \log_5(3x + 2) - \log_5(2x - 3) = 2, \text{ and}$$
$$\ln x + \ln(x + 3) = 6.$$

3. We can use the One-to-One Property of logarithms for logarithmic equations of the form $\log_b x = \log_b y$.
   Note: This strategy can be used to solve the following logarithmic equations:

$$\log(5x - 6) = \log x^2 \text{ and } \ln(2x - 7) = \ln(3x + 5).$$

4. For other types of logarithmic equations, we sometimes can combine logarithms on each side and then use the One-to-One Property of logarithms to solve the equations.
   Note: This strategy can be used to solve the following logarithmic equations:

$$\ln x^2 + \ln(x - 3) = \ln(x + 4) \text{ and } \log_4 x - \log_4(x + 2) = \log_4 5 + \log_4 3x.$$

Comment

It is important that we check our answers when solving logarithmic equations. Logarithms have restricted domains, so we must exclude any possible solution that results in the logarithmic expression's being undefined.

EXAMPLE 10 **Solving Logarithmic Equations Using the Defintion of a Logarithm**

Solve each equation and check all solutions.

a. $\log_3(x - 4) = 4$ **b.** $\log(x^2 - 1) = 2$ **c.** $-3 \ln 4x = 9$

SOLUTION Because each of these equations contains one logarithmic expression, we will use the first strategy outlined in the box. We will use the definition of the logarithm, convert to exponential form, and solve for x.

a. $\log_3(x - 4) = 4$

$$3^4 = x - 4 \qquad \text{Convert to exponential form.}$$

$$81 = x - 4 \qquad \text{Simplify.}$$

$$x = 85 \qquad \text{Solve for } x. \text{ This is a possible solution.}$$

Check: $\log_3(\mathbf{85} - 4) = \log_3 81 = 4$ because $3^4 = 81$. The solution checks.

b. $\log(x^2 - 1) = 2$

$$10^2 = x^2 - 1 \qquad \text{Convert to exponential form.}$$

$$100 = x^2 - 1 \qquad \text{Simplify.}$$

$$101 = x^2 \qquad \text{Add 1 to both sides.}$$

$$x = \pm\sqrt{101} \qquad \text{These are possible solutions.}$$

Check: Both solutions check.

$$x = \sqrt{101}: \log\left(\left(\sqrt{\mathbf{101}}\right)^2 - 1\right) = \log(101 - 1) = \log 100 = 2$$

$$x = -\sqrt{101}: \log\left(\left(-\sqrt{\mathbf{101}}\right)^2 - 1\right) = \log(101 - 1) = \log 100 = 2$$

Comment

In part (c), we can also use an alternate strategy to eliminate the natural logarithm in the expression $\ln 4x = -3$. If we take the exponential form of both sides, we can solve for x by using Property (3).

$$e^{\ln(4x)} = e^{-3}$$

$$4x = e^{-3}$$

$$x = \frac{e^{-3}}{4} = \frac{1}{4e^3}$$

c. $-3 \ln 4x = 9$

$$\ln 4x = -3 \qquad \text{Divide both sides by } -3 \text{ to isolate } \ln 4x.$$

$$e^{-3} = 4x \qquad \text{Convert to exponential form.}$$

$$x = \frac{e^{-3}}{4} = \frac{1}{4e^3} \qquad \text{Solve for } x \text{ to find the possible solution.}$$

Check: $-3 \ln \left(4\left(\dfrac{1}{4e^3}\right)\right) = -3 \ln e^{-3} = \ln e^9 = 9$. The solution checks.

Self Check 10   Solve each equation. **a.** $\log_2(x + 3) = 5$ **b.** $2 \ln 3x = 12$

Now Try Exercise 61.

EXAMPLE 11   **Solving a Logarithmic Equation Using the Properties of Logarithms**

Solve: $\log x + \log(x - 3) = 1$.

SOLUTION   We will use the second strategy outlined in the box. We will combine the two logarithmic expressions on the left, then use the definition of the logarithm. Once we convert to exponential form, we can solve for x.

$$\log x + \log(x - 3) = 1$$

$$\log x(x - 3) = 1 \qquad \text{Combine using the Product Rule.}$$

$$x(x - 3) = 10^1 \qquad \text{Convert to exponential form.}$$

$$x^2 - 3x - 10 = 0 \qquad \text{Remove parentheses and subtract 10 from both sides.}$$

$$(x + 2)(x - 5) = 0 \qquad \text{Factor.}$$

$$x + 2 = 0 \quad \text{or} \quad x - 5 = 0$$

$$x = -2 \qquad \qquad x = 5$$

Check: The number -2 can be eliminated as a solution, because $\log(-2)$ is not defined. We check the remaining number, 5.

$$\log x + \log(x - 3) = 1$$

$$\log 5 + \log(5 - 3) \overset{?}{=} 1$$

$$\log 5 + \log 2 \overset{?}{=} 1$$

$$\log 10 = 1 \qquad \text{Use the Product Rule.}$$

Therefore, $x = 5$ is the only solution to the equation.   ∎

Self Check 11   Solve: $\log x + \log(x - 15) = 2$.

Now Try Exercise 73.

EXAMPLE 12   **Solving a Logarithmic Equation Using the One-to-One Property of Logarithms**

Solve: $\log_7(3x + 2) = \log_7(2x - 3)$.

SOLUTION   Because the logarithmic equation is of the form $\log_b x = \log_b y$, we can use the One-to-One Property to solve the equation. This is the third strategy outlined in the box. If $\log_b x = \log_b y$, then $x = y$.

$$\log_7(3x + 2) = \log_7(2x - 3)$$

$$3x + 2 = 2x - 3 \qquad \text{Use the One-to-One Property of logarithms.}$$

$$x = -5 \qquad \text{Solve for } x.$$

Check: It is essential that the possible solution be checked.

$$\log_7(3x + 2) = \log_7(2x - 3)$$

$$\log_7(3(-5) + 2) \overset{?}{=} \log_b(2(-5) - 3)$$

$$\log_b(-13) \overset{?}{=} \log_b(-13)$$

Comment

It is incorrect to state that the logarithms in Example 12 are canceled. The logarithms aren't canceled; the One-to-One Property is used.

Since $\log_b(-13)$ is not defined, there is no solution to the equation.   ∎

Self Check 12   Solve $\log_8(5x + 2) = \log_8(6x + 1)$ using the One-to-One Property.

Now Try Exercise 79.

EXAMPLE 13   **Solving a Logarithmic Equation**

Solve: $\ln(2x + 5) - \ln(x + 5) = \ln\dfrac{5}{3}$.

SOLUTION   We will use the fourth strategy outlined in the box and first combine the two logarithms on the left. Then we will use the One-to-One Property and solve for x.

$$\ln(2x + 5) - \ln(x + 5) = \ln\frac{5}{3}$$

$$\ln\frac{2x + 5}{x + 5} = \ln\frac{5}{3} \qquad \text{Use the Quotient Rule and combine lns.}$$

$$\frac{2x + 5}{x + 5} = \frac{5}{3} \qquad \text{Use the One-to-One Property.}$$

$$3(2x + 5) = 5(x + 5) \qquad \text{Multiply both sides by } 3(x + 5).$$

$$6x + 15 = 5x + 25 \qquad \text{Remove parentheses.}$$

$$x = 10 \qquad \text{Solve for } x.$$

Check: The solution $x = 10$ checks, as we see here:

$$\ln(2x + 5) - \ln(x + 5) \overset{?}{=} \ln\frac{5}{3}$$

$$\ln[2(\mathbf{10}) + 5] - \ln[(\mathbf{10}) + 5] \overset{?}{=} \ln\frac{5}{3} \qquad \text{Substitute 10 in for } x.$$

$$\ln 25 - \ln 15 \overset{?}{=} \ln\frac{5}{3}$$

$$\ln\frac{25}{15} \overset{?}{=} \ln\frac{5}{3} \qquad \text{Use the Quotient Rule.}$$

$$\ln\frac{5}{3} = \ln\frac{5}{3} \qquad \text{Simplify.}$$

Self Check 13   Solve: $\log_7(2x + 2) - \log_7(x + 6) = \log_7\dfrac{4}{3}$.

Now Try Exercise 81.

A calculator can be used to solve a logarithmic equation if the equation does not simplify easily, or if, when simplified, the resulting equation cannot be factored.

EXAMPLE 14   **Solving a Logarithmic Equation Using a Calculator**

Solve the equation $\ln x^2 + \ln(x - 3) = \ln(x + 4)$ using a graphing calculator.

SOLUTION   Enter each side of the equation in the graph editor, and then find the point of intersection. The solution is $x \approx 3.5891327$ (Figure 3-45).

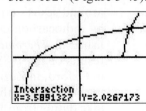

FIGURE 3-45

Self Check 14   Solve: $\log(2x + 3) - \log(x + 2) = \log(x^3)$.

Now Try Exercise 99.

3. Model Real-World Problems Using Exponential and Logarithmic Equations

Carbon-14 Dating

When a living organism dies, its cycle of oxygen and carbon dioxide (common to all living things) ceases; then carbon-14, a radioactive isotope with a half-life of 5,700 years, is no longer absorbed. By measuring the amount of carbon-14 present in an ancient object, archaeologists can estimate the object's age.

The amount A of radioactive material present at time t is given by the model

$$A = A_0 2^{-t/h},$$

where A_0 is the amount present initially and h is the half-life of the material.

EXAMPLE 15   **Using Carbon-14 Dating**

How old is a wooden statue that has two-thirds of its original carbon-14 content?

SOLUTION   To find the time t when $A = \frac{2}{3}A_0$, we substitute $\frac{2A_0}{3}$ for A and 5,700 for h in the radioactive-decay formula and solve for t.

$$A = A_0 2^{-t/h}$$

$$\frac{2A_0}{3} = A_0 2^{-t/5,700}$$

$$\frac{2}{3} = 2^{-t/5,700} \qquad \text{Divide by } A_0$$

$$\ln \frac{2}{3} = \ln 2^{-t/5,700} \qquad \text{Take the ln of each side.}$$

$$\ln \frac{2}{3} = -\frac{t}{5,700}\ln 2 \qquad \text{Use the Power Rule.}$$

$$t = -5,700\left(\frac{\ln \frac{2}{3}}{\ln 2}\right) \qquad \text{Solve for } t.$$

$$t \approx 3,334.286254$$

The wooden statue is approximately 3,334 years old.

Self-Check 15   How old is an artifact that has 60% of its original carbon-14 content?

Now Try Exercise 105.

The Shroud of Turin, a piece of linen long reputed to have been wrapped around Jesus's body after the crucifixion, is much older than the date suggested by radiocarbon dating, according to new microchemical research. A study published in *Thermochimica Acta*, a peer-reviewed chemistry journal, dismisses the results of the 1988 carbon-14 dating of the shroud. Three reputable laboratories in Oxford, in Zurich, and in Tucson, Arizona, concluded that the cloth on which the smudged outline of the body of a man is indelibly impressed was a medieval fake dating from 1260 to 1390, and not the burial cloth wrapped around the body of Christ.

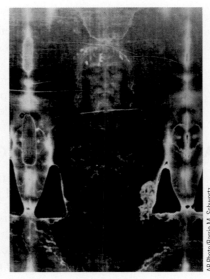

FIGURE 3-46

EXAMPLE 16   **Using Carbon-14 Dating**

Suppose a fossil sample contains 20% of the amount of carbon-14 that was present originally. Given that the half-life of carbon-14 is 5,700 years, answer the following questions:

a. What is the approximate age of the fossil sample?

b. What will the approximate percentage of carbon-14 remaining be when the fossil sample is 35,000 years old?

SOLUTION   **a.** We substitute $A = 0.20A_0$ and $h = 5{,}700$ into the formula $A = A_0 2^{-t/h}$ and solve for t.

$$0.20A_0 = A_0 2^{-t/5{,}700}$$

$$0.20 = 2^{-t/5{,}700} \qquad \text{Divide by } A_0.$$

$$\ln 0.20 = \ln 2^{-t/5{,}700} \qquad \text{Take the ln of each side.}$$

$$\ln 0.20 = -\frac{t}{5{,}700} \ln 2 \qquad \text{Use the Power Rule.}$$

$$t = \frac{-5{,}700 \ln 0.20}{\ln 2} \qquad \text{Solve for } t.$$

$$t \approx 13{,}234.99014$$

The fossil sample is approximately 13,235 years old.

b. The percentage of carbon-14 remaining after 35,000 years does not depend on the original amount. Therefore, we can let $A_0 = 1$ and solve the equation.

$$A = 1 \cdot 2^{-35{,}000/5{,}700}$$

$$A = 2^{-35{,}000/5{,}700}$$

$$A \approx 0.0141765$$

Approximately 1.42% remains after 35,000 years.

Self Check 16   Approximate the percentage of carbon-14 remaining when the fossil sample is 40,000 years old.

Now Try Exercise 106.

Population Growth

When there is sufficient food and space, populations of living organisms tend to increase exponentially according to the Malthusian growth model

$$P = P_0 e^{kt},$$

where P_0 is the initial population at $t = 0$ and k depends on the rate of growth. The value of k is often called the growth constant.

EXAMPLE 17   **Finding Population Growth**

The bacteria in a laboratory culture increased from an initial population of 500 to 1,500 in 3 hours. Find the total time it will take for the population to reach 10,000.

SOLUTION   For this problem, we have $P_0 = 500$, $P = 1,500$, and $t = 3$. We will substitute these values into the growth model and solve for the growth constant k. After we have found the value of the growth constant, we can project the population for a given time t or find the time necessary to reach a given population.

$$P = P_0 e^{kt}$$

$$1,500 = 500(e^{3k}) \qquad \text{Divide by 500.}$$

$$3 = e^{3k} \qquad \text{Take the ln of each side. Remember that } \ln e^x = x.$$

$$\ln 3 = 3k$$

$$k = \frac{\ln 3}{3} \qquad \text{Solve for } k.$$

To find out when the population will reach 10,000, we substitute 10,000 for P, 500 for P_0, and $\frac{\ln 3}{3}$ for k in the equation $P = P_0 e^{kt}$ and solve for t:

$$P = P_0 e^{kt}$$

$$10,000 = 500 e^{\left(\frac{\ln 3}{3}\right)t}$$

$$20 = e^{\left(\frac{\ln 3}{3}\right)t} \qquad \text{Divide by 500.}$$

$$\ln 20 = \frac{\ln 3}{3} t \qquad \text{Take the ln of each side.}$$

$$t = \frac{3 \ln 20}{\ln 3} \qquad \text{Solve for } t.$$

$$t \approx 8.180499084$$

The culture will reach 10,000 bacteria in a total time of little more than 8 hours. ∎

Self Check 17   If the population increases from 1,000 to 3,000 in 3 hours, how long will it take to reach 20,000?

Now Try Exercise 111.

Self Check Answers   **1. a.** 6   **b.** 5   **2.** 1, −3   **3.** $\pm\sqrt{6}$ **4.** $5 - \log 75 \approx 3.1249$

5. $\dfrac{\log 3}{\log 5} \approx 0.6826$ **6.** $\dfrac{-3 \ln 5}{\ln 5 - \ln 3} = \dfrac{\ln 5^{-3}}{\ln \dfrac{5}{3}} \approx -9.4520$

7. $-\dfrac{\ln 8.5}{3} \approx -0.7134$ **8.** $\dfrac{\ln 2}{\ln 3} \approx 0.6309$ **9.** 0.94808746   **10. a.** 29

b. $\dfrac{e^6}{3}$ **11.** 20   **12.** 1   **13.** 9   **14.** 1.1903279   **15.** about 4,200 yr

16. 0.77%   **17.** about 8 hr

Exercises **3.6**

Getting Ready
You should be able to complete these vocabulary and concept statements before you proceed to the practice exercises.

Fill in the blanks.

1. An equation with a variable in its exponent is called a(n) _____ equation.

2. An equation with a logarithmic expression that contains a variable is a(n) _____ equation.

3. The formula for radiocarbon decay is
   $A =$ _____.

4. The formula for population growth is
   $P =$ _____.

Practice
Solve each exponential equation by expressing each term as a power of the same base.

5. $3^{-x} = 81$ 6. $2^{-x} = \dfrac{1}{64}$

7. $2^{3x-1} = 64$ 8. $3^{2x-1} = 27$

9. $5^{2x-1} = 125^x$ 10. $4^{3x} = 64^{2x-3}$

11. $8^x = 16^{x-1}$ 12. $125^x = 25^{x-2}$

13. $6^{2-x} = \dfrac{1}{216}$ 14. $7^{1-x} = \dfrac{1}{49}$

15. $2^{x^2-2x} = 8$ 16. $5^{x^2-3x} = 625$

17. $3^{x^2+4x} = \dfrac{1}{81}$ 18. $7^{x^2+3x} = \dfrac{1}{49}$

19. $10^{x-1/2} = \sqrt[3]{10}$ 20. $5^{x+2/3} = \dfrac{1}{\sqrt{5}}$

21. $e^{x-1} = \dfrac{1}{e^2}$ 22. $e^{x+2} = \dfrac{1}{e^{x-1}}$

23. $e^{x^2-1} = e^{24}$ 24. $e^{x^2+7x} = \dfrac{1}{e^{12}}$

Solve each exponential equation. Give the answer in exact form. If the exact form involves logarithms, also give an approximation rounded to four decimal places.

25. $4^x = 5$ 26. $7^x = 12$

27. $13^{x-1} = 2$ 28. $5^{x+1} = 3$

29. $2^{x+1} = 3^x$ 30. $5^{x-3} = 3^{2x}$

31. $2^x = 3^x$ 32. $3^{2x} = 4^x$

33. $4^{x+2} = 8^x$ 34. $27^{x+1} = 3^{2x+1}$

35. $7^{x^2} = 10$ 36. $5^{x^2} = 2^{5x}$

37. $e^x = 10$ 38. $8e^x = 16$

39. $4e^{2x} = 24$ 40. $2e^{5x} = 18$

Solve each exponential equation. Approximate to four decimal places when necessary.

41. $4^{x+2} - 4^x = 15$ 42. $3^{x+3} + 3^x = 84$

43. $2(3^x) = 6^{2x}$ 44. $2(3^{x+1}) = 3(2^{x-1})$

45. $2^{2x} - 10(2^x) + 16 = 0$ 46. $3^{2x} - 10(3^x) + 9 = 0$

47. $2^{2x+1} - 2^x = 1$ 48. $3^{2x+1} - 10(3^x) + 3 = 0$

49. $e^{2x} + 2e^x - 3 = 0$ 50. $e^{2x} - 3e^x - 4 = 0$

51. $e^{4x} - e^{2x} - 12 = 0$ 52. $e^{4x} - 5e^{2x} + 6 = 0$

Solve each logarithmic equation. Give the answer in exact form.

53. $\log_2 x = 5$ 54. $\log_3 x = -2$

55. $\log x = -2$ 56. $\log x = 3$

57. $\ln x = 4$ 58. $\ln x = -2$

59. $\log_{1/3}(x + 3) = -3$ 60. $\log_{1/2}(x + 2) = -4$

61. $\log_4(2x - 3) = 3$ 62. $\log_6(3x - 1) = 2$

63. $4 \ln 3x = 24$ 64. $2 \ln 2x = 16$

65. $\log_2 \sqrt{x - 1} = 3$ 66. $\log_3 \sqrt{x + 5} = 2$

67. $\log \dfrac{4x + 1}{2x + 9} = 0$ 68. $\log \dfrac{5x + 2}{2(x + 7)} = 0$

69. $\log x^2 = 2$ 70. $\log x^3 = 3$

Solve each logarithmic equation using the strategies from the section. Be sure to check and exclude any possible solution that results in the logarithmic expression's being undefined.

71. $\log x + \log(x - 3) = 1$

72. $\log_5(x + 1) - \log_5(x - 1) = 2$

73. $\log x + \log(x - 48) = 2$

74. $\log x + \log(x - 15) = 2$

75. $\log(x - 3) - \log 6 = 2$

76. $\log 5{,}000 - \log(x - 2) = 3$

77. $\log(2x - 3) - \log(x - 1) = 0$

78. $\log(x + 90) = 3 - \log x$

79. $\log_9(2x - 3) = \log_9(x + 4)$

80. $\log_7(3x + 5) = \log_7(2x + 6)$

81. $\log_7 x + \log_7(x - 5) = \log_7 6$

82. $\ln x + \ln(x - 2) = \ln 120$

83. $\ln 15 - \ln(x - 2) = \ln x$

84. $\ln 10 - \ln(x - 3) = \ln x$

85. $\log(x - 1) - \log 6 = \log(x - 2) - \log x$

86. $\log(x - 6) - \log(x - 2) = \log \dfrac{5}{x}$

87. $\log_5(7 + x) + \log_5(8 - x) - \log_5 2 = 2$

88. $\log_3 x = \log_3 \dfrac{1}{x} + 4$

89. $\log x^2 = (\log x)^2$

90. $\log(\log x) = 1$

91. $\dfrac{\log(3x - 4)}{\log x} = 2$

92. $\dfrac{\log(5x + 6)}{2} = \log x$

93. $\dfrac{1}{2} \log (4x + 5) = \log x$

94. $2 \log_3 x - \log_3(x - 4) = 2 + \log_3 2$

95. $\log(7y + 1) = 2 \log(y + 3) - \log 2$

96. $2 \log(y + 2) = \log(y + 2) - \log 12$

⊞ *Use a graphing calculator to solve each equation, rounding the answer to six decimal places.*

97. $\ln(x^2 - 1) = 4$

98. $\ln(2x + 1) = 3$

99. $\log \sqrt{x - 1} = \dfrac{1}{3}$

100. $\log(x^2 + 1) = \sqrt{2}$

Applications

⊞ *Use a calculator to help solve each problem.*

101. Tritium decay The half-life of tritium is 12.4 years. How long will it take for 25% of a sample of tritium to decompose?

102. Radioactive decay In 2 years, 20% of a radioactive element decays. Find its half-life.

103. Thorium decay An isotope of thorium, thorium-227, has a half-life of 18.4 days. How long will it take 80% of a sample to decompose?

104. Lead decay An isotope of lead, lead-201, has a half-life of 8.4 hours. How many hours ago was there 30% more of the substance?

105. Carbon-14 dating A cloth fragment is found in an ancient tomb. It contains 70% of the carbon-14 that it is assumed to have had initially. How old is the cloth?

106. Carbon-14 dating Only 25% of the carbon-14 in a wooden bowl remains. How old is the bowl?

107. Compound interest If $500 is deposited in an account paying 8.5% annual interest, compounded semiannually, how long will it take for the account to increase to $800?

108. Continuous compound interest In Exercise 107, how long will it take for the account to increase to $800 if the interest is compounded continuously?

109. Compound interest If $1,300 is deposited in a savings account paying 9% interest, compounded quarterly, how long will it take the account to increase to $2,100?

110. Compound interest A sum of $5,000 deposited in an account grows to $7,000 in 5 years. Assuming annual compounding, what interest rate is being paid?

111. Bacterial growth A staphylococcus culture grows according to the formula $P = P_0 a^t$. If it takes 5 days for the culture to triple in size, how long did it take to double in size?

112. Oceanography The intensity I of a light a distance x meters beneath the surface of a lake decreases exponentially. If the light intensity at 6 meters is 70% of the intensity at the surface, at what depth will the intensity be 20%?

113. Rodent control The rodent population in a city is currently estimated at 30,000. If it is expected to double every 5 years, how long will it take for the population to reach 1 million?

114. Temperature of coffee Refer to the section opener and find the time it takes for the white-chocolate mocha to reach a temperature of 80°F if the function $T = 70 + 110e^{-0.2t}$ models the temperature T after t minutes.

115. Newton's law of cooling Water at 100°C is left to cool in a room where the temperature is 60°C. After 3 minutes, the water temperature is 90°C. If the water temperature T is a function of time t given by $T = 60 + 40e^{kt}$, find k.

116. Newton's law of cooling Refer to Exercise 115 and find the time required for the water temperature to reach 70°C.

Discovery and Writing

117. Explain why it is necessary to check the solutions of a logarithmic equation.

118. Use the population-growth formula to show that the doubling time for a population is given by $T = \frac{\ln 2}{r}$.

119. Use the population-growth formula to show that the tripling time for a population is given by $T = \frac{\ln 3}{r}$.

120. Can you solve $x = \log x$ algebraically? Can you find an approximate solution? Find x.

121. Solve: $\log_2(\log_5(\log_7 x)) = 2$.

122. Solve: $\log_8\left(16\sqrt[3]{4{,}096}\right)^{\frac{1}{6}} = x$.

CHAPTER REVIEW

SECTION **3.1**   Exponential Functions and Their Graphs

Definitions and Concepts	Examples
An **exponential function with base b** is defined by the equation $\quad f(x) = b^x$ or $y = b^x$ ($b > 0$, $b \neq 1$, and x is a real number).	See Example 2, page 322
Properties of an exponential function, $f(x) = b^x$: • The **domain** is $D = (-\infty, \infty)$, the set of real numbers. • The **range** is $R = (0, \infty)$, the set of positive real numbers. • The graph has y-intercept 1. • There is no x-intercept. • The x-axis is a horizontal asymptote of the graph. • The graph passes through the points $(1, b)$ and $(0, 1)$. • If $b > 1$, then $f(x) = b^x$ is an **increasing function.** This function models **exponential growth.** • If $0 < b < 1$, then $f(x) = b^x$ is a **decreasing function.** This function models **exponential decay.**	See Example 3, pages 322–323

$f(x) = b^x$

$b > 1$

$(0, 1)$ $(1, b)$

Increasing function

$f(x) = b^x$

$0 < b < 1$

$(0, 1)$ $(1, b)$

Decreasing function

Definitions and Concepts	Examples
Transformations of functions: If f is a function and k represents a real number, then we have the following transformations. • **Vertical shift:** $y = b^x + k$ shifts the graph of f upward k units if $k > 0$ and downward k units if $k < 0$. • **Horizontal shift:** $y = b^{x-k}$ shifts the graph of f right k units if $k > 0$ and left k units if $k < 0$. • **Reflection:** $y = -b^x$ reflects the graph of f about the x-axis and $y = b^{-x}$ reflects the graph of f about the y-axis. • **Vertical stretch or shrink (multiply each y-coordinate by k):** $y = kb^x$ is a vertical stretch of the graph of f if $k > 1$ or a vertical shrink if $0 < k < 1$. • **Horizontal shrink or stretch (divide each x-coordinate by k):** $y = b^{kx}$ is a horizontal shrink of the graph of f if $k > 1$ and a horizontal stretch if $0 < k < 1$.	See Example 6, pages 326–327 See Example 7, page 327
The function $f(x) = e^x$ is called the **natural exponential function.** The number $e \approx 2.718281828$.	See Example 11, page 330

Definitions and Concepts	Examples
Compound-interest formula: • If P dollars are deposited in an account earning interest at an annual rate r, compounded n times each year, the amount A in the account after t years is given by the formula $A = P\left(1 + \dfrac{r}{n}\right)^{nt}$. • If interest is compounded continuously, the formula is $A = Pe^{rt}$.	See Example 12, page 332

EXERCISES

Use properties of exponents to simplify each expression.

1. $5^{\sqrt{2}} \cdot 5^{\sqrt{2}}$

2. $\left(2^{\sqrt{5}}\right)^{\sqrt{2}}$

Graph the function defined by each equation.

3. $f(x) = 3^x$

4. $f(x) = \left(\dfrac{1}{3}\right)^x$

5. The graph of $f(x) = 6^x$ will pass through the points $(0, p)$ and $(1, q)$. Find p and q.

6. Give the domain and range of the function $f(x) = b^x$, with $b > 0$ and $b \neq 1$.

Graph each function by using a transformation.

7. $f(x) = \left(\dfrac{1}{2}\right)^x - 2$

8. $f(x) = \left(\dfrac{1}{2}\right)^{x+2}$

9. $f(x) = -5^x$

10. $f(x) = -5^x + 4$

11. $f(x) = e^x + 1$

12. $f(x) = e^{x-3}$

Applications

13. Compound interest How much will $10,500 become if it earns 9% per year for 60 years, compounded quarterly?

14. Compound interest If $10,500 accumulates interest at an annual rate of 9%, compounded continuously, how much will be in the account in 60 years?

SECTION 3.2   Applications of Exponential Functions

Definitions and Concepts	Examples
Radioactive-decay formula: The amount A of radioactive material present at time t is given by $A = A_0 2^{-t/h}$, where A_0 is the amount that was present initially (at $t = 0$) and h is the material's half-life.	See Example 1, page 338
Intensity-of-light formula: The intensity I of light (in lumens) at a distance x meters below the surface of a body of water decreases exponentially according to the formula $I = I_0 k^x$, where I_0 is the intensity of light above the water and k is a constant that depends on the clarity of the water.	See Example 2, page 338
Malthusian model of population growth: If b is the annual birth rate, d is the annual death rate, t is the time (in years), P_0 is the initial population at $t = 0$, and P is the current population, then $P = P_0 e^{kt}$, where $k = b - d$ is the **annual growth rate,** the difference between the annual birth rate and annual death rate.	See Example 3, page 339

Applications

15. Radioactive decay The half-life of a radioactive material is about 34.2 years. How much of the material is left after 20 years?

16. Oceanography Find the intensity of light at a depth of 12 meters if $I_0 = 14$ and $k = 0.7$.

17. Population growth The population of the United States is approximately 300 million people. Find the population in 50 years if $k = 0.015$.

18. Epidemics In a city with a population of 450,000, there are currently 1,000 cases of hepatitis. If the spread of the disease is projected by the logistic function

$$P = \frac{450,000}{1 + (450 - 1)e^{-0.2t}},$$

how many people will have contracted the hepatitis virus after 5 years?

SECTION **3.3**   Logarithmic Functions and Their Graphs

Definitions and Concepts	Examples
If $b > 0$ and $b \neq 1$, the **logarithmic function with base b** is defined by $y = \log_b x$ if and only if $x = b^y$. The notation $f(x) = \log_b x$ is also used.	See Example 1, pages 346–347
Base-e logarithms are called **natural logarithms.** The notation $\ln x$ means $\log_e x$.	See Example 6, page 349
Properties of a logarithmic function: • The **Domain** is $D = (0, \infty)$. • The **Range** is all real numbers, $R = (-\infty, \infty)$. • The y-axis is a vertical asymptote. • The graph of the function contains the points $(1, 0)$ and $(b, 1)$.	See Example 8, pages 350–351
For any function $h(x)$, the **domain** of $f(x) = \log_b(h(x))$ is all real values of x for which $h(x) > 0$.	See Example 9, pages 353–354
Many logarithmic functions can be graphed using transformations.	See Example 10, page 355

EXERCISES

19. Give the domain and range of the logarithmic function $f(x) = \log_3 x$.

20. Give the domain and range of the logarithmic function $f(x) = \ln x$.

Find each value.

21. $\log_3 9$

22. $\log_9 \dfrac{1}{3}$

23. $\log_x 1$

24. $\log_5 0.04$

25. $\log_a \sqrt{a}$

26. $\log_a \sqrt[3]{a}$

Find x.

27. $\log_2 x = 5$

28. $\log_{\sqrt{3}} x = 4$

29. $\log_{\sqrt{2}} x = 6$

30. $\log_{0.1} 10 = x$

31. $\log_x 2 = -\dfrac{1}{3}$

32. $\log_x 32 = 5$

33. $\log_{0.25} x = -1$

34. $\log_{0.125} x = -\dfrac{1}{3}$

35. $\log_{\sqrt{2}} 32 = x$

36. $\log_{\sqrt{5}} x = -4$

37. $\log_{\sqrt{3}} 9\sqrt{3} = x$

38. $\log_{\sqrt{5}} 5\sqrt{5} = x$

Graph each function. Use transformations.

39. $f(x) = \log(x - 2)$ **40.** $f(x) = 3 + \log x$

Graph each pair of equations on one set of coordinate axes.

41. $y = 4^x$ and $y = \log_4 x$

42. $y = \left(\dfrac{1}{3}\right)^x$ and $y = \log_{1/3} x$

Use a calculator to find each value to four decimal places.

43. $\ln 452$ **44.** $\ln(\log 7.85)$

Use a calculator to solve each equation. Round each answer to four decimal places.

45. $\ln x = 2.336$ **46.** $\ln x = \log 8.8$

Graph each function. Use transformations.

47. $f(x) = 1 + \ln x$ **48.** $f(x) = \ln(x + 1)$

Simplify each expression.

49. $\ln(e^{12})$ **50.** $e^{\ln 14x}$

SECTION **3.4**   Applications of Logarithmic Funcitons

Definitions and Concepts	Examples
Decibel voltage gain: If E_O is the output voltage of a device and E_I is the input voltage, the **decibel voltage gain** (dB gain) is given by dB gain $= 20 \log \frac{E_O}{E_I}$.	See Example 1, page 360
Richter scale: If R is the intensity of an earthquake, A is the amplitude (measured in micrometers), and P is the period (the time of one oscillation of the earth's surface, measured in seconds), then $R = \log \frac{A}{P}$.	See Example 2, page 361
Charging batteries : If M is the theoretical maximum charge that a battery can hold and k is a positive constant that depends on the battery and the charger, the length of time t (in minutes) required to charge the battery to a given level C is given by $t = -\frac{1}{k} \ln\left(1 - \frac{C}{M}\right)$.	See Example 3, pages 361–362
Population doubling time: If r is the annual growth rate and t is the time (in years) required for a population to double, then $t = \frac{\ln 2}{r}$.	See Example 4, page 362
Isothermal expansion: If the temperature T is constant, the energy E required to increase the volume of 1 mole of gas from an initial volume V_i to a final volume V_f is given by $E = RT \ln\left(\frac{V_f}{V_i}\right)$. E is measured in joules and T in kelvins. R is the universal gas constant, which is 8.314 joules per mole per kelvin.	See Example 5, page 363

Applications

51. Gain of an amplifier An amplifier has an output of 18 volts when the input is 0.04 volt. Find the dB gain.

52. Earthquakes An earthquake had a period of 0.3 second and an amplitude of 7,500 micrometers. Find its measure on the Richter scale.

53. Charging batteries How long will it take a dead battery to reach an 80% charge? (Assume $k = 0.17$.)

54. Population growth How long will it take the population of the United States to double if the growth rate is 3% per year?

55. Isothermal expansion Find the amount of energy that must be supplied to double the volume of 1 mole of gas at a constant temperature of 350 kelvins. (*Hint:* $R = 8.314$.)

SECTION **3.5**   Properties of Logarithms

Definitions and Concepts	Examples
If b, M, and N are positive numbers and $b \neq 1$, then **1.** $\log_b 1 = 0$. **2.** $\log_b b = 1$. **3.** $\log_b b^x = x$. **4.** $b^{\log_b x} = x$. **5.** $\log_b MN = \log_b M + \log_b N$. **6.** $\log_b \dfrac{M}{N} = \log_b M - \log_b N$. **7.** $\log_b M^p = p \log_b M$. **8.** if $\log_b x = \log_b y$, then $x = y$.	See Example 1, page 366 See Example 3, pages 367–368 See Example 5, page 369

Definitions and Concepts	Examples
Change-of-base formula: If a, b, and x are positive numbers, $a \neq 1$, and $b \neq 1$, then $\log_b x = \dfrac{\log_a x}{\log_a b}$.	See Example 9, pages 372–373
pH of a solution: If $[\text{H}^+]$ is the hydrogen-ion concentration in gram-ions per liter, then $\text{pH} = -\log[\text{H}^+]$.	See Example 10, page 374
Weber–Fechner Law: If L is the apparent loudness of a sound and I is the intensity, then $L = k \ln I$.	See Example 12, page 375

EXERCISES

Simplify each expression.

56. $\log_7 1$

57. $\log_7 7$

58. $\log_7 7^3$

59. $7^{\log_7 4}$

60. $\ln e^4$

61. $\ln 1$

62. $10^{\log_{10} 7}$

63. $e^{\ln 3}$

64. $\log_b b^4$

65. $\ln e^9$

Write each expression in terms of the logarithms of x, y, and z.

66. $\log_b \dfrac{x^2 y^3}{z^4}$

67. $\log_b \sqrt{\dfrac{x}{yz^2}}$

68. $\ln \dfrac{x^4}{y^5 z^6}$

69. $\ln \sqrt[3]{xyz}$

Write each expression as the logarithm of one quantity.

70. $3 \log_b x - 5 \log_b y + 7 \log_b z$

71. $\dfrac{1}{2}(\log_b x + 3 \log_b y) - 7 \log_b z$

72. $4 \ln x - 5 \ln y - 6 \ln z$

73. $\dfrac{1}{2} \ln x + 3 \ln y - \dfrac{1}{3} \ln z$

Assume that $\log a = 0.6$, $\log b = 0.36$, and $\log c = 2.4$. Find the value of each expression.

74. $\log abc$

75. $\log a^2 b$

76. $\log \dfrac{ac}{b}$

77. $\log \dfrac{a^2}{c^3 b^2}$

78. To four decimal places, find $\log_5 17$.

Applications

79. **pH of grapefruit juice** The pH of grapefruit juice is about 3.1. Find its hydrogen-ion concentration.

80. **Weber–Fechner Law** Find the decrease in the apparent loudness of a sound if the intensity of the sound is cut in half.

SECTION 3.6   Exponential and Logarithmic Equations

Definitions and Concepts	Examples
Solving exponential equations: **1.** If the bases are equal, solve the equation using the One-to-One Property of exponential functions: If $a^x = a^y$, $a \neq 1$, and $a \neq 0$, then $x = y$. **2.** If the equation contains base 10, use common logarithms, because $\log 10^x = x$. **3.** If the bases are not equal, take the logarithm or natural logarithm of both sides and use the Power Rule of logarithms. **4.** If the equation contains base e, use natural logarithms, because $\ln e^x = x$.	See Example 1, page 379 See Example 4, page 381 See Example 6, pages 382–383 See Example 7, page 383

Definitions and Concepts	Examples
Solving logarithmic equations: Solving logarithmic equations can employ many strategies. The one we should use is dependent upon the nature of the equation. 1. For logarithmic equations containing one logarithm, we can use the definition of the logarithm. We write the logarithm in exponential form and then solve for the variable. 2. For more complicated logarithmic equations, we can combine and isolate the logarithmic expressions. We can then use the definition of the logarithm and write the logarithm in exponential form to solve for the variable. 3. We can use the One-to-One Property of logarithms for logarithmic equations of the form $\log_b x = \log_b y$. 4. For other types of logarithmic equations, we sometimes can combine logarithms on each side and then use the One-to-One Property of logarithms to solve the equations. It is important to **check your answers** when solving logarithmic equations.	See Example 10, page 385 See Example 11, page 386 See Example 12, page 386 See Example 13, page 387

EXERCISES

Solve each equation. Give the answer in exact form. If the exact form involves logarithms, also give an approximation rounded to four decimal places.

81. $5^{x+2} = 625$

82. $2^{x^2+4x} = \dfrac{1}{8}$

83. $e^x = e^{-6x+14}$

84. $e^{2x^2} = e^{18}$

85. $3^x = 7$

86. $2^x = 3^{x-1}$

87. $2e^x = 16$

88. $-5e^x = -35$

89. $\log_7(-7x + 2) = \log_7(3x + 32)$

90. $\ln(x + 3) = \ln(-5x + 51)$

91. $\log x + \log(29 - x) = 2$

92. $\log_2 x + \log_2(x - 2) = 3$

93. $\log_2(x + 2) + \log_2(x - 1) = 2$

94. $\log 3 - \log(x - 1) = -1$

95. $\dfrac{\log(7x - 12)}{\log x} = 2$

96. $\ln x + \ln(x - 5) = \ln 6$

97. $e^{x \ln 2} = 9$

98. $\ln x = \ln(x - 1)$

99. $\ln x = \ln(x - 1) + 1$

100. $\ln x = \log_{10} x$ (*Hint:* Use the change-of-base formula.)

Applications

101. Carbon-14 dating A wooden statue found in Egypt has a carbon-14 content that is two-thirds of that found in living wood. If the half-life of carbon-14 is 5,700 years, how old is the statue?

CHAPTER TEST

Match each equation with its graph.

_____ **1.** $y = 3^x$

_____ **2.** $y = 3^x - 1$

_____ **3.** $y = 3^{x-1}$

_____ **4.** $y = \log_3 x$

_____ **5.** $y = \log_3(x - 1)$

_____ **6.** $y = \log_{1/3} x$

a.

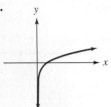

b.

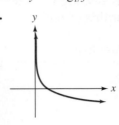

c.

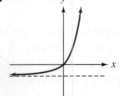

d.

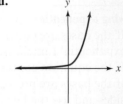

e.

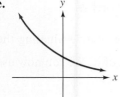

f.

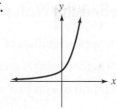

g.

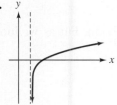

h.

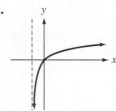

Graph each function.

7. $f(x) = 2^x + 1$

8. $f(x) = e^{x-2}$

Solve each problem.

9. A radioactive material decays according to the formula $A = A_0 2^{-t}$. How much of a 3-gram sample will be left in 6 years?

10. An initial deposit of \$1,000 earns 6% interest, compounded twice a year. How much will be in the account in 1 year?

11. An account contains \$2,000 and earns 8% interest, compounded continuously. How much will be in the account in 10 years?

Find each value.

12. $\log_7 343$

13. $\log_3 \dfrac{1}{27}$

14. $\log_{10} 10^{12} + 10^{\log_{10} 5}$

15. $\log_{3/2} \dfrac{9}{4}$

16. $\log_{2/3} \dfrac{27}{8}$

Graph each function.

17. $f(x) = \log(x - 1)$

18. $f(x) = 2 + \ln x$

Write each expression in terms of the logarithms of a, b, and c.

19. $\log a^2 b c^3$

20. $\ln \sqrt{\dfrac{a}{b^2 c}}$

Write each expression as a logarithm of a single quantity.

21. $\dfrac{1}{2} \log(a + 2) + \log b - 2 \log c$

22. $\dfrac{1}{3}(\log a - 2 \log b) - \log c$

Assume that $\log 2 = 0.3010$ *and* $\log 3 = 0.4771$. *Find each value. Do not use a calculator.*

23. $\log 24$

24. $\log \dfrac{8}{3}$

Use the Change-of-Base Formula to find each logarithm. Do not attempt to simplify the answer.

25. $\log_7 3$

26. $\log_\pi e$

Determine whether each statement is true or false.

27. $\log_a ab = 1 + \log_a b$

28. $\dfrac{\log a}{\log b} = \log a - \log b$

29. Find the pH of a solution with a hydrogen-ion concentration of 3.7×10^{-7} gram-ions per liter. (*Hint:* pH $= -\log[H^+]$.)

30. Find the dB gain of an amplifier when $E_O = 60$ volts and $E_I = 0.3$ volt. (*Hint:* dB gain $= 20 \log \frac{E_O}{E_I}$.)

Solve each equation. Give the answer in exact form. If the exact form involves logarithms, also give an approximation rounded to four decimal places.

31. $3^{x^2 - 2x} = 27$

32. $e^{2x^2} = e^{50}$

33. $3^{x-1} = 100^x$

34. $5e^x = 45$

35. $\ln(5x + 2) = \ln(2x + 5)$

36. $\log x + \log(x - 9) = 1$

37. $\log_6 18 - \log_6(x - 3) = \log_6 3$

 LOOKING AHEAD TO **CALCULUS**

1. **Using Calculus to Define the Natural Exponent *e***
   - Connect the definition of the natural exponent *e* to limits in order to define it—Section 3.1.

2. **Using the Properties of Logarithms to Simplify Derivatives**
   - Connect the properties of logarithms to the process of logarithmic differentiation—Section 3.5.

1. Using Calculus to Define the Natural Exponent *e*

In Section 3.1 we calculated an approximation of the number *e* by evaluating the exponential expression $\left(1 + \frac{1}{x}\right)^x$ at increasingly greater values of x. We can now use limit notation and state the definition of *e* in two ways:

$$e = \lim_{x \to 0} (1 + x)^{1/x} \text{ and } e = \lim_{n \to \infty} \left(1 + \frac{1}{n}\right)^n.$$

At first glance we may think that these are the same expression, but as we look closely we do see the differences.

Connect to Calculus:
Defining the Natural Exponent *e*

Use the approximation $e \approx 2.718281828$ for these problems.

1. Use a calculator to approximate $e = \lim_{x \to 0} (1 + x)^{1/x}$ to nine decimal places.

   To do this, choose values of x so that x gets closer and closer to 0. Be sure to enter the function $y = (1 + x)^{1/x}$ into the graph editor on the calculator and use the table feature. What value of x produces the desired level of accuracy?

2. Now use the definition $e = \lim_{n \to \infty} \left(1 + \frac{1}{n}\right)^n$ to approximate e to nine decimal places. What value of n produces the desired level of accuracy?

2. Using the Properties of Logarithms to Simplify Derivatives

The properties of logarithms can be used to simplify an expression to facilitate taking the derivative of a complicated function that involves products, quotients, or powers. When we solve an equation, we often perform a mathematical operation on each side in order to simplify the equation. For example, we may multiply both sides by the least common denominator to eliminate fractions from the equation. Well, we can also take the logarithm of both sides of an equation, and use the properties of logarithms to help take the derivative. We will introduce the Derivative Rule here because it uses logarithms. In calculus, this rule will be covered in great detail. We will look at some functions we may encounter in calculus to see just how useful the properties of logarithms can be.

Consider the function $f(x) = \dfrac{x^{3/4}\sqrt{x^3 + 2x}}{(4x - 5)^3}$. At first glance, we see that it is a very complicated algebraic function. Suppose we were asked to find the derivative of this function in calculus. We could use various rules of differentiation, but our work would be very challenging and probably take us a long time. If we simplify first by using the properties of logarithms, our work in calculus turns out to be much easier. This process is call logarithmic differentiation.

Let's use y for $f(x)$ and discover the initial strategy used in finding the derivative by using logarithms.

$$y = \frac{x^{3/4}\sqrt{x^3 + 2x}}{(4x - 5)^3}$$

Step 1: Take the logarithm of both sides. Use log or ln.

$$\ln y = \ln \frac{x^{3/4}\sqrt{x^3 + 2x}}{(4x - 5)^3}$$

Step 2: Rewrite the right side of the equation using properties.

$$\ln y = \ln \frac{x^{3/4}\sqrt{x^3 + 2x}}{(4x - 5)^3}$$

$$\ln y = \ln\left(x^{3/4}\sqrt{x^3 + 2x}\right) - \ln(4x - 5)^3 \qquad \text{Use the Quotient Rule.}$$

$$\ln y = \ln(x^{3/4}(x^3 + 2x)^{1/2}) - \ln(4x - 5)^3 \qquad \text{Rewrite } \sqrt{x^3 + 2x} \text{ as } (x^3 + 2x)^{1/2}.$$

$$\ln y = \ln x^{3/4} + \ln(x^3 + 2x)^{1/2} - \ln(4x - 5)^3 \qquad \text{Use the Product Rule.}$$

$$\ln y = \frac{3}{4}\ln x + \frac{1}{2}\ln(x^3 + 2x) + 3\ln(4x - 5) \qquad \text{Use the Power Rule.}$$

It may appear that this is a tremendous amount of work to do before taking the derivative, but rest assured—the last expression is quite easy to differentiate.

While the example function could be differentiated without using our properties of logarithms, there are functions for which there is no other way to find the derivative than to introduce logarithms: for example, $y = x^{\sqrt{x}}$. The function doesn't look very complicated, but we have no rules that would allow us to find the derivative. If we use logarithms, we can easily solve the problem in calculus.

$$y = x^{\sqrt{x}}$$

$$\ln y = \ln x^{\sqrt{x}}$$

$$\ln y = \sqrt{x}\ln x \qquad \text{Use the Power Rule.}$$

$$\ln y = x^{1/2}\ln x$$

After these steps are complete, it is easy to find the derivative.

Connect to Calculus:
Using the Properties of Logarithms to Simplify Derivatives

Use the properties of logarithms to simplify each function so that it can be differentiated.

3. $y = \dfrac{(x^3 - 2)^{3/4}\sqrt{x}}{\sqrt[3]{3x - 5}}$

4. $y = x^{\ln x}$

Trigonometric Functions

4

CAREERS AND MATHEMATICS: Urban Planner

© Simon Rawles/Alamy

Urban planners develop plans for the best use of a community's land and resources for residential, commercial, institutional, and recreational purposes. They help local officials alleviate social, economic, and environmental problems by recommending locations for roads, schools, and other infrastructure. Their work requires forecasting the future needs of the population. Some planners are involved in environmental issues, including pollution control, wetland preservation, and forest conservation.

Education and Mathematics Required
- A master's degree from an accredited urban-planning program provides the best training. Students are admitted to master's-degree programs in planning with a wide range of backgrounds, such as a bachelor's degree in economics, geography, political science, or environmental design.
- College Algebra, Trigonometry, Geometry, Calculus I and II, Number Theory, Analysis, and Statistics are required courses.

How Urban Planners Use Math and Who Employs Them
- Urban planners use math as they design the arrangement, appearance, and functionality of towns and cities. Mathematical models are used to predict the future needs of a group of people.
- As many as 68% are employed by local governments. Companies involved with architectural or consulting services employ an increasing proportion of planners. Others are employed in state government agencies. A small number work for the federal government.

Career Outlook and Salary
- Employment for urban planners is expected to grow 19% from 2008 to 2018, faster than the average for all occupations.
- The median annual salary of urban and regional planners was $59,810 in May 2008. Urban planners can make as much as $91,520 a year.

For more information, see http://www.bls.gov/oco.

In this chapter, we will introduce the unit circle and use it to define trigonometric functions. We will also learn to graph the trigonometric functions.

4.1 Angles and Their Measurement

In this section, we will learn to

1. Describe angles and place angles in standard position.
2. Use degree and radian measures.
3. Convert between degrees and radians.
4. Draw angles in standard position.
5. Understand characteristics of coterminal angles.
6. Solve arc-length problems using radian measure.
7. Find the area of a sector of a circle using radian measure.

Trigonometry permeates many aspects of life. Most of the time, people are not even aware of it. The concept of an angle is familiar to most students, but the use of angles in the world around us is far more widespread than you may realize. Angles play a part in everything from architecture and engineering to billiards and golf.

For example, lie angle is the angle between the bottom, or sole, of a golf club and the shaft. Different golf clubs have different lie angles, which determine the flight of the ball. Having the proper lie angle is critical if we want the opportunity to make accurate golf shots. A lie angle that is too upright will cause the heel of the club to hit the ground first and produce a pull or hook. A lie angle that is too flat causes a push or slice. Approximately 90% of all golfers have a lie angle that is off by at least 2°.

Since angles are basic to the study of trigonometry, we begin our study by defining angles and how they are measured.

1. Describe Angles and Place Angles in Standard Position

An **angle** is two rays (half-lines) originating from a common point called the **vertex.** One ray, the **initial side,** is rotated to the terminal side, forming the angle.

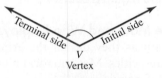

FIGURE 4-1

The arrow indicates the direction of the angle. Angles that are formed by rotations in a counterclockwise direction are considered to be **positive**. Angles that are formed by rotations in a clockwise direction are considered to be **negative**.

Any letter or symbol can be used to name an angle, but typically Greek letters such as α, β, γ, and θ are used. Capital letters such as A, B, and C are often used, especially when labeling the angles in a triangle. Figure 4-2 illustrates several angles. Notice that the position of the two rays does not change, but the direction and amount of rotation indicate different angles.

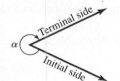
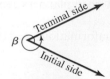

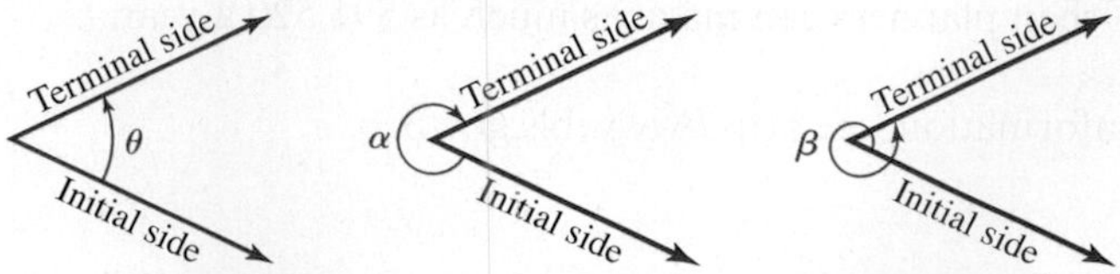

Positive Angle: counterclockwise rotation

Negative Angle: clockwise rotation

Positive Angle: more than 1 rotation

FIGURE 4-2

While an angle can be drawn in any position, we generally use what is known as **standard position**. An angle is said to be in **standard position** when it is drawn on a rectangular coordinate system with its vertex at the origin and its initial side along the positive *x*-axis. The terminal side of the angle will lie in one of the four quadrants, or it will lie along one of the axes. If the terminal side lies on the *x*- or *y*-axis, the angle is called a **quadrantal angle.**

Because each angle in Figure 4-3 has its initial side along the positive *x*-axis and its vertex at the origin, each one is in standard position.

Angles in Standard Position

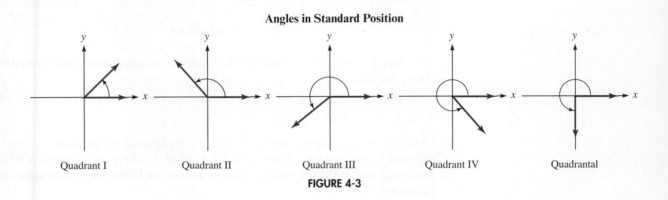

Quadrant I   Quadrant II   Quadrant III   Quadrant IV   Quadrantal

FIGURE 4-3

2. Use Degree and Radian Measures

One way to measure the amount of rotation used to form an angle is **degree measure.** In ancient times, people noticed that the annual progression of the seasons appeared to repeat in 360-day cycles. Because they divided the year's circle into 360 days, it was reasonable to divide the circle itself into 360 equal parts, and the degree measure of angles was born. Although calendars were revised when closer observation showed that a year contains about $365\frac{1}{4}$ days, we still use the 360-degree circle.

Angles are measured according to the amount of rotation from the initial side to the terminal side. One complete revolution is 360 degrees, denoted by 360°. An angle has a measure of 1° if it represents $\frac{1}{360}$ of a complete revolution.

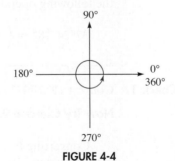

FIGURE 4-4

A positive angle that is less than 90° is called an **acute angle** [see Figure 4-5(a)]. A 90° angle is a **right angle** [see Figure 4-5(b)]. Notice that a small square is used to designate a right angle. An **obtuse angle** is an angle that measures between 90° and 180° [see Figure 4-5(c)]. One-half of a rotation is an angle of measure 180°, called a **straight angle** [see Figure 4-5(d)].

Types of Angles

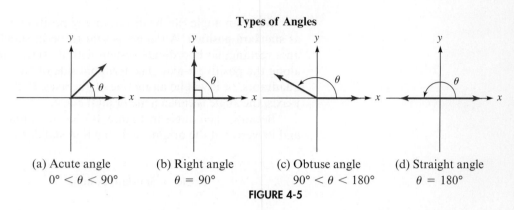

(a) Acute angle
$0° < \theta < 90°$

(b) Right angle
$\theta = 90°$

(c) Obtuse angle
$90° < \theta < 180°$

(d) Straight angle
$\theta = 180°$

FIGURE 4-5

We can indicate fractional parts of a degree by using **degree decimal (DD)** notation. For example,

$$\frac{1}{2}° = 0.5° \quad \text{and} \quad 33\frac{3}{4}° = 33.75°.$$

Another way to express parts of a degree makes use of **minutes,** denoted by the symbol ′, and **seconds,** denoted by the symbol ″, where each degree is 60 minutes and each minute is 60 seconds. This is known as **DMS (degrees-minutes-seconds)** notation:

$$1° = 60' \quad \text{or} \quad \frac{1}{60}° = 1', \quad 1' = 60'' \quad \text{or} \quad \frac{1'}{60} = 1'', \quad \text{and} \quad 1'' = \frac{1}{60}' = \frac{1}{3,600}°.$$

Comment   It can often be helpful for students to think of time when initially learning to work with degrees, minutes, and seconds. With time, we know that 1 hour is 60 minutes and 1 minute is 60 seconds. In trigonometry, we note that one degree is 60 minutes and 1 minute is 60 seconds.

It is sometimes necessary to convert from DMS to DD notation or from DD to DMS. While most calculators have conversion keys that accomplish this, we can also do it without using those keys.

EXAMPLE 1   **Converting from DMS to DD Notation**

Convert 37°24′18″ to degree decimal notation.

SOLUTION   Because $24' = \frac{24}{60}(1°)$ and $18'' = \frac{18}{3600}(1°)$, we can convert from DMS to DD using the following operation and a calculator:

$$37°24'18'' = \left(37 + \frac{24}{60} + \frac{18}{3,600} \right)° = 37.405°$$

■

Self Check 1   Convert 12°37′42″ to degree decimal notation.

Now Try Exercise 9.

Converting from DD to DMS involves more than one step, if it is not done using conversion keys on a calculator.

EXAMPLE 2   **Converting from DD to DMS Notation**

Convert 83.41° to DMS (degrees-minutes-seconds) notation.

SOLUTION   To convert this measurement to DMS, we will first convert to degrees and minutes, and then to degrees, minutes, and seconds.

$$83.41° = 83° + 0.41°$$
$$= 83° + (0.41)(60')$$ Convert 0.41° to minutes; 1° = 60'.
$$= 83° + 24.6'$$
$$= 83° + 24' + 0.6(60'')$$ Convert 0.6' to seconds; 1' = 60''.
$$= 83° + 24' + 36''$$
$$= 83°24'36''$$

Self Check 2   Convert 76.27° to DMS (degrees-minutes-seconds) notation.

Now Try Exercise 15.

ACCENT ON TECHNOLOGY   **Converting from DD to DMS Notation**

Calculators can easily convert from DD to DMS notation. The conversion is shown in Figure 4-6 using a graphing calculator. Generally, for any calculator, we enter the DD notation, then press the DMS key for the conversion.

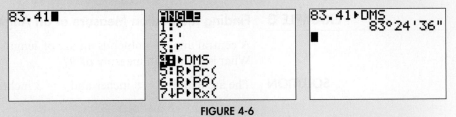

FIGURE 4-6

Degree measurement is used in applications such as navigation and surveying. However, in most areas of mathematics and science, another unit of measure called the **radian** is primarily used for measuring angles.

We know that 1° represents $\frac{1}{360}$ of a complete revolution in a circle. Now we will look at how **1 radian** is defined in relation to a circle.

Radian   The measure of a central angle θ is **1 radian** if and only if θ subtends (intercepts) an arc whose length is equal to the radius of the circle.

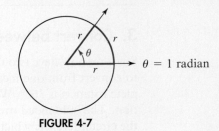

FIGURE 4-7

If the length of the subtended arc is twice that of the radius, the measure of the central angle is 2 radians. This is illustrated in Figure 4-8.

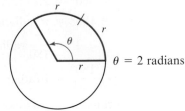

FIGURE 4-8

Notice that the angle formed in Figure 4-8 can be found by dividing the length of the intercepted arc by the radius. This is how we define the radian measure of an angle.

Radian Measure of an Angle   In a circle of radius r, the measure of the central angle θ that subtends an arc of length s is $\theta = \dfrac{s}{r}$ radians.

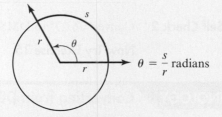

FIGURE 4-9

EXAMPLE 3   Finding the Radian Measure of an Angle

A central angle θ subtends an arc of length 8 inches in a circle of radius 3 inches. What is the radian measure of θ?

SOLUTION   The arc length $s = 8$ inches and $r = 3$ inches, so we have

$$\theta = \frac{s}{r} = \frac{8 \text{ inches}}{3 \text{ inches}} = \frac{8}{3}.$$

Self Check 3   A central angle θ subtends an arc of length 11 inches in a circle of radius 5 inches. What is the radian measure of θ?

Now Try Exercise 21.

Comment   Notice in Example 3 that the inches units cancel. Whereas with degree measure it is *essential* to use the degree symbol, that is not the case with radian measure. We can include the word *radian*, but since a radian measure is a real number, the designation is not required.

3. Convert between Degrees and Radians

Since we now have two types of angular measurement, it is necessary to know how to convert from one measurement to the other. In degree measurement, one complete rotation is 360°. We need to define the radian measure of one complete rotation. The subtended arc of a complete revolution for any circle is equivalent to the circumference, which is $2\pi r$. Substituting this into the definition of the radian measure of a central angle, we have $\theta = \dfrac{s}{r} = \dfrac{2\pi r}{r} = 2\pi$. We can now compare degree and radian measure, beginning with the measure of one complete revolution.

$$360° = 2\pi \text{ radians}$$
$$180° = \pi \text{ radians}$$
$$1° = \frac{\pi}{180} \text{ radians}$$
$$\frac{180°}{\pi} = 1 \text{ radian}$$

From this we can ascertain how to convert from degree to radian measure and from radian to degree measure.

Strategy for Converting between Degree and Radian Measures	**1.** To convert from degree measure to radian measure, multiply the degree measure by $\dfrac{\pi \text{ radians}}{180°}$. **2.** To convert from radian measure to degree measure, multiply the radian measure by $\dfrac{180°}{\pi \text{ radians}}$.

Comment   Sometimes students will use the wrong conversion because they do not have a good idea of the measurement of a radian compared to that of a degree. Since one complete revolution of a circle in radian measure is 2π radians, and $2\pi \approx 6.283$, we know that about 6.3 radians is equivalent to 360°. So we can see that radian measure is going to be a number less than degree measure for the same angle. When we do our conversions, it is a good idea to keep these relationships in mind:

$$1 \text{ radian} = \frac{180°}{\pi \text{ radians}} \approx 57.3°$$

$$1° = \frac{\pi \text{ radians}}{180°} \approx 0.017 \text{ radian}$$

The radian measure of an angle is often given as a multiple of π rather than a decimal approximation. We use the measure $\frac{2\pi}{3}$ rather than the decimal approximation $\frac{2\pi}{3} \approx 2.094395$. The measurement $\frac{2\pi}{3}$ is called **exact form**.

EXAMPLE 4   **Converting from Degree to Radian Measure**

Convert each degree measure to exact radian measure.

a. 15°   **b.** 200°   **c.** −40°

SOLUTION   Because we are changing to radian measure, we multiply by $\dfrac{\pi \text{ radians}}{180°}$.

a. $15° = 15° \cdot \dfrac{\pi \text{ radians}}{180°} = \dfrac{\pi}{12} \text{ radian}$

b. $200° = 200° \cdot \dfrac{\pi \text{ radians}}{180°} = \dfrac{10\pi}{9} \text{ radians}$

c. $-40° = -40° \cdot \dfrac{\pi \text{ radians}}{180°} = -\dfrac{2\pi}{9} \text{ radian}$

Self Check 4   Convert 135° to exact radian measure.

Now Try Exercise 27.

EXAMPLE 5   **Converting from Radian to Degree Decimal Measure**

Convert each radian measure to degree decimal measure. Round to the nearest tenth when necessary.

a. $\dfrac{\pi}{20}$ radian   **b.** $\dfrac{5\pi}{18}$ radian   **c.** $\dfrac{3\pi}{7}$ radians

SOLUTION   Because we are changing to degree measure, we multiply by $\dfrac{180°}{\pi \text{ radians}}$.

$$\textbf{a.} \;\; \frac{\pi}{20} \text{ radian} = \frac{\pi}{20} \text{ radian} \cdot \frac{180°}{\pi \text{ radians}} = 9°$$

$$\textbf{b.} \;\; \frac{5\pi}{18} \text{ radian} = \frac{5\pi}{18} \text{ radian} \cdot \frac{180°}{\pi \text{ radians}} = 50°$$

$$\textbf{c.} \;\; \frac{3\pi}{7} \text{ radian} = \frac{3\pi}{7} \text{ radian} \cdot \frac{180°}{\pi \text{ radians}} = \frac{540°}{7} \approx 77.1°$$

■

Self Check 5   Convert $-\dfrac{\pi}{45}$ radians to degree measure.

Now Try Exercise 45.

4. Draw Angles in Standard Position

An angle is in **standard position** when it is drawn on a rectangular coordinate system with its vertex at the origin and its initial side along the positive x-axis. It is important to learn the equivalence of certain angles. Table 1 gives the degree measure and the corresponding radian measure of eight common angles.

TABLE 1

Degree Measure	0°	30°	45°	60°	90°	180°	270°	360°
Radian Measure	0	$\frac{\pi}{6}$	$\frac{\pi}{4}$	$\frac{\pi}{3}$	$\frac{\pi}{2}$	π	$\frac{3\pi}{2}$	2π

We can use this table to find the radian measure for multiples of these angles. For example, since $60° = \frac{\pi}{3}$, we determine that $120° = 2 \cdot 60° = \frac{2\pi}{3}$.

EXAMPLE 6   **Drawing Angles in Standard Position**

Draw and label each angle in standard position.

   a. $\dfrac{4\pi}{3}$ **b.** $\dfrac{11\pi}{6}$ **c.** $-135°$

SOLUTION   **a.** $\frac{\pi}{3}$ is shown in Figure 4-10(a), so $\frac{4\pi}{3}$ is 4 times that angle and terminates in quadrant III. Note that $\frac{4\pi}{3} = 4 \cdot \frac{\pi}{3} = 4(60°) = 240°$.

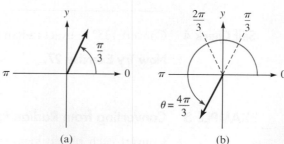

(a)   (b)

FIGURE 4-10

b. $\frac{11\pi}{6}$ terminates in quadrant IV and is 11 times the angle $\frac{\pi}{6}$. Note that $\frac{11\pi}{6} = 11 \cdot \frac{\pi}{6} = 11(30°) = 330°$.

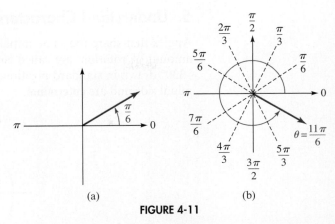

(a)   (b)

FIGURE 4-11

c. $-135° = -3 \cdot 45°$, so we begin with $-45°$. Since the angle is negative, we draw it in a clockwise direction.

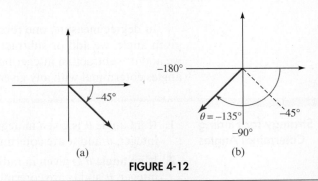

(a)   (b)

FIGURE 4-12

Self Check 6   Draw the angle 210° in standard position.

Now Try Exercise 67.

Figure 4-13 shows common angles in both radian and degree measure.

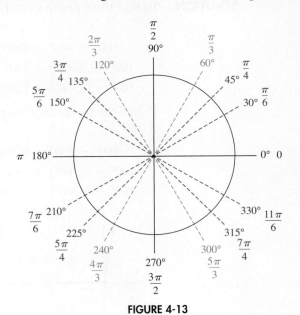

FIGURE 4-13

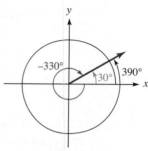
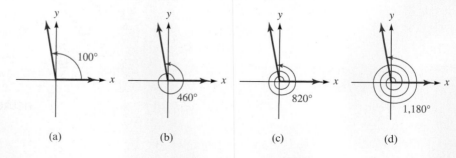

5. Understand Characteristics of Coterminal Angles

Angles that share the same initial and terminal side, but perhaps have a different amount of rotation, are called **coterminal angles**. The three angles 30°, 390°, and −330° drawn in standard position, as shown in Figure 4-14, all share the same terminal side and are **coterminal**.

FIGURE 4-14

In degree measure, one revolution is 360°. To find an angle coterminal with a given angle, we add or subtract an integer multiple of 360°. For radian measure, we add or subtract an integer multiple of 2π. We see that there are infinitely many angles coterminal with any given angle.

Strategy for Finding Coterminal Angles

1. If an angle θ is given in degree measure and $\alpha = \theta \pm 360°k$, where k is an integer, θ and α are coterminal.

2. If an angle θ is given in radian measure and $\alpha = \theta \pm 2\pi k$, where k is an integer, θ and α are coterminal.

EXAMPLE 7   **Finding Coterminal Angles**

Find three positive and three negative angles that are coterminal with an angle in standard position that measures 100°.

SOLUTION   To find three positive coterminal angles to $\theta = 100°$, we add integer multiples of 360°.

$$100° + 360° = 460°$$ 100° plus one revolution; Figure 4-15(b)
$$100° + 2(360°) = 820°$$ 100° plus two revolutions; Figure 4-15(c)
$$100° + 3(360°) = 1,180°$$ 100° plus three revolutions; Figure 4-15(d)

To find three negative coterminal angles to $\theta = 100°$, we subtract integer multiples of 360°.

$$100° - 360° = -260°$$ 100° minus one revolution; Figure 4-15(e)
$$100° - 2(360°) = -620°$$ 100° minus two revolutions; Figure 4-15(f)
$$100° - 3(360°) = -980°$$ 100° minus three revolutions; Figure 4-15(g)

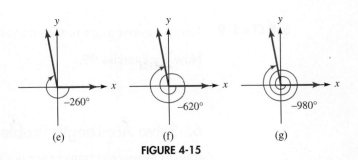

FIGURE 4-15

Self Check 7 Find three positive and three negative angles that are coterminal with an angle in standard position that measures –120°.

Now Try Exercise 93.

EXAMPLE 8 **Finding Coterminal Angles**

Find a positive and a negative angle coterminal to each angle.

a. $\dfrac{5\pi}{6}$ b. $-\dfrac{3\pi}{4}$

SOLUTION a. $\dfrac{5\pi}{6} + 2\pi = \dfrac{17\pi}{6}$ is coterminal to $\dfrac{5\pi}{6}$.

$\dfrac{5\pi}{6} - 2\pi = -\dfrac{7\pi}{6}$ is coterminal to $\dfrac{5\pi}{6}$.

b. $-\dfrac{3\pi}{4} + 2\pi = \dfrac{5\pi}{4}$ is coterminal to $-\dfrac{3\pi}{4}$.

$-\dfrac{3\pi}{4} - 2\pi = -\dfrac{11\pi}{4}$ is coterminal to $-\dfrac{3\pi}{4}$.

Self Check 8 Find a positive and a negative angle coterminal to $\dfrac{2\pi}{3}$.

Now Try Exercise 97.

EXAMPLE 9 **Finding Coterminal Angles**

Find a positive angle in the interval $[0, 2\pi)$ that is coterminal to each angle.

a. $\dfrac{20\pi}{3}$ b. $-\dfrac{29\pi}{12}$

SOLUTION a. Since $\frac{20\pi}{3}$ is greater than 2π, we will subtract a multiple of 2π to find a coterminal angle in the interval $[0, 2\pi)$.

$$\alpha = \dfrac{20\pi}{3} - 6\pi = \dfrac{20\pi}{3} - \dfrac{18\pi}{3} = \dfrac{2\pi}{3}$$

b. Since $-\frac{29\pi}{12}$ is less than 2π, we will add a multiple of 2π to find a coterminal angle in the interval $[0, 2\pi)$.

$$\alpha = -\dfrac{29\pi}{12} + 4\pi = -\dfrac{29\pi}{12} + \dfrac{48\pi}{12} = \dfrac{19\pi}{12}$$

Self Check 9   Find a positive angle in the interval $[0, 2\pi]$ that is coterminal to $\dfrac{19\pi}{6}$.

Now Try Exercise 99.

6. Solve Arc-Length Problems Using Radian Measure

Radian measure permits easy calculation of **arc length** in a circle. When we defined the measure of a radian angle, we used the equation $\theta = \frac{s}{r}$ radians. From this we can solve for the arc length s.

Arc Length   If a central angle θ in a circle with a radius r is in radians, the length s of the subtended arc is given by $s = r\theta$.

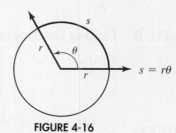

FIGURE 4-16

Caution   If the angle θ is in degrees, then we must convert to radian measure to use the formula.

EXAMPLE 10   **Finding Arc Length**

If a central angle θ of a circle with radius 6.2 cm is 80°, find the length of the subtended arc. Round to the nearest hundredth.

SOLUTION   First we will change θ to radians, and then we can find the arc length.

$$80° = 80° \cdot \frac{\pi \text{ radians}}{180°} = \frac{4\pi}{9} \text{ radians}$$

Now we can find the arc length using $r = 6.2$ cm and $\theta = \dfrac{4\pi}{9}$ radians.

$$s = r\theta = 6.2\left(\frac{4\pi}{9}\right) \approx 8.66 \text{ cm}$$

Self Check 10   If a central angle θ of a circle with radius 7.5 cm is $\frac{5\pi}{3}$, find the length of the subtended arc. Round to the nearest hundredth.

Now Try Exercise 103.

Latitude and longitude are angles that form a coordinate system used to define a point on the earth's surface. Latitude is defined with respect to the equator. If a ray is drawn from the center of the earth to the surface of the equator and another ray is drawn to another point, the angle formed is the latitude of that point. While the surface of the earth is certainly not smooth, we can still use the circular shape to approximate the distance between two points on the surface if we know their latitudes and their longitudes are approximately the same.

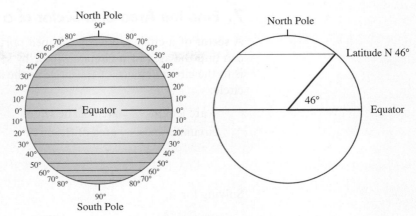

FIGURE 4-17

EXAMPLE 11   Solving an Application Using Arc Length

Baton Rouge, Louisiana, is located at latitude N 30°32′ and La Crosse, Wisconsin, is due north at latitude N 43°48′. Using 3,960 miles as the earth's radius, approximate the distance between the cities to the nearest mile.

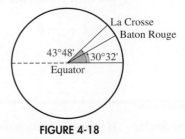

FIGURE 4-18

SOLUTION   Since both cities are north of the equator, the angle formed by the difference of the two latitudes will be the angle used to find the arc length.

$$\begin{array}{r} 43°48' \\ - \ 30°32' \\ \hline 13°16' \end{array}$$

The angle we found is in degree measure; therefore, we need to convert it to radian measure in order to use the formula for arc length.

$$\theta = 13°16' = \left(13 + \frac{16}{60}\right)^{°}\left(\frac{\pi}{180°}\right)$$

$$s = r\theta = (3{,}960)\left(13 + \frac{16}{60}\right)^{°}\left(\frac{\pi}{180°}\right) \approx 917 \text{ miles}$$

It is approximately 917 miles from Baton Rouge, Louisiana, to La Crosse, Wisconsin.

Self Check 11   Find the distance between Austin, Texas, located at latitude N 30°16′; and Oklahoma City, Oklahoma, located at latitude N 35°28′. Use 3,960 miles as the radius of the earth and round your answer to the nearest mile.

Now Try Exercise 125.

7. Find the Area of a Sector of a Circle Using Radian Measure

A **sector** of a circle is a wedge-shaped portion like that shaded in Figure 4-19. We can find the area of a circular sector by setting up a proportion. The arc length s is to the circumference of the circle as the area A of the sector is to the area of the circle:

$$\frac{\text{arc length}}{\text{circumference}} = \frac{\text{area of the sector}}{\text{area of the circle}}$$

$$\frac{s}{2\pi r} = \frac{A}{\pi r^2}$$

Solving for A:

$$\frac{s}{2\pi r} = \frac{A}{\pi r^2}$$

$$A = \frac{s\pi r^2}{2\pi r}$$

$$A = \frac{1}{2}rs$$

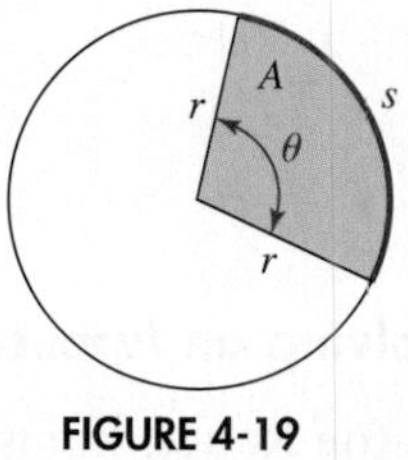

FIGURE 4-19

If θ is measured in radians, we can use the formula $s = r\theta$ and substitute $r\theta$ for s. We then obtain the formula for the area of a sector.

$$A = \frac{1}{2}rr\theta$$

$$A = \frac{1}{2}r^2\theta$$

Area of a Circular Sector   If a sector of a circle with radius r has a central angle θ in radians, the area of the sector is given by the formula

$$A = \frac{1}{2}r^2\theta.$$

EXAMPLE 12   Finding the Area of a Circular Sector

A piece of land is in the shape of a circular sector formed by an arc of length 100 meters and a central angle measuring 150°. Find the area of the sector to the nearest tenth of a square meter.

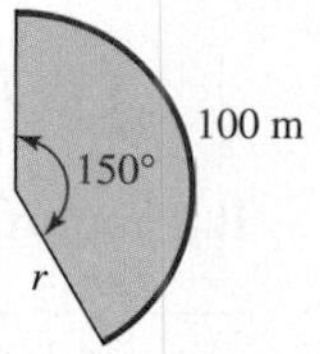

100 m

150°

r

FIGURE 4-20

SOLUTION   First, convert $\theta = 150°$ to radian measure.

$$\theta = 150° = 150°\left(\frac{\pi}{180°}\right) = \frac{150\pi}{180} = \frac{5\pi}{6}$$

Since $s = r\theta$, we have $r = \dfrac{s}{\theta} = \dfrac{100 \text{ m}}{\frac{5\pi}{6}} = \dfrac{600 \text{ m}}{5\pi} = \dfrac{120 \text{ m}}{\pi}$. Now we can find the area

of the circular sector.

$$A = \frac{1}{2}r^2\theta$$

$$= \frac{1}{2}\left(\frac{120 \text{ m}}{\pi}\right)^2\left(\frac{5\pi}{6}\right)$$

$$\approx 1,909.9 \text{ square meters}$$

Self Check 12   Rework Example 12 with an arc of length 120 meters and a central angle of 140°.

Now Try Exercise 115.

Self Check Answers   1. $\approx 12.628°$ 2. $76°16'12''$ 3. $\dfrac{11}{5} = 2.2$ 4. $\dfrac{3\pi}{4}$ 5. $-4°$

6. 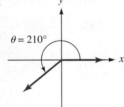   7. $240°, 600°, 960°, -480°, -840°,$ and $-1,200°$

8. $\dfrac{8\pi}{3}$ and $-\dfrac{4\pi}{3}$ 9. $\dfrac{7\pi}{6}$ 10. 39.27 cm   11. 359 mi   12. 2,946.6 m²

Exercises **4.1**

Getting Ready
You should be able to complete these vocabulary and concept statements before you proceed to the practice exercises.

Fill in the blanks.

1. A(n) _____ is defined as a figure formed by two rays originating from a common point called the ____.

2. Two ways to measure an angle are _____ measure and _____ measure.

3. A positive angle that is less than 90° is called a(n) _____ angle.

4. A(n) _____ angle is an angle that measures between 90° and 180°.

5. To convert from degree measure to radian measure, multiply the degree measure by _____.

6. To convert from radian measure to degree measure, multiply the radian measure by _____.

7. If a central angle θ in a circle with radius r is in radians, the length s of the subtended arc is given by ____.

8. If a sector of a circle with radius r has a central angle θ in radians, the area of the sector is given by the formula ____.

Practice
Convert from DMS form to DD form, rounding to three decimal places.

9. $42°14'37''$ 10. $125°35'12''$

11. $321°21''$ 12. $275°15''$

13. $12°14'49''$ 14. $9°6'24''$

Convert from DD form to DMS, rounding the seconds to the nearest tenth when necessary.

15. $14.625°$ 16. $135.71°$

17. $-217.413°$ 18. $-129.526°$

19. $47.08°$ 20. $124.85°$

A central angle θ subtends an arc of length s in a circle of radius r. What is the radian measure of θ?

21. $s = 4$ feet; $r = 2$ feet

22. $s = 6$ yards; $r = 2.5$ yards

23. $s = 12$ inches; $r = 19$ inches

24. $s = 8$ inches; $r = 14$ inches

Convert each degree measurement to radians, expressing the answer in terms of π.

25. 0°   **26.** 30°

27. 270°   **28.** 360°

29. 225°   **30.** 150°

31. 210°   **32.** 240°

33. 330°   **34.** 135°

35. −240°   **36.** −120°

37. −60°   **38.** −315°

Convert each degree measurement to radians, rounding to three decimal places.

39. 39°   **40.** 178°

41. −217.35°   **42.** −59.42°

43. 121°15′29″   **44.** 245°35′54″

Convert each radian measurement to degrees.

45. $\dfrac{\pi}{6}$ **46.** $\dfrac{\pi}{3}$

47. $-\dfrac{5\pi}{2}$ **48.** $-\dfrac{7\pi}{4}$

49. $\dfrac{5\pi}{6}$ **50.** $\dfrac{7\pi}{2}$

51. $\dfrac{9\pi}{4}$ **52.** $\dfrac{5\pi}{4}$

53. $-\dfrac{13\pi}{6}$ **54.** $\dfrac{3\pi}{4}$

55. $\dfrac{5\pi}{3}$ **56.** $\dfrac{11\pi}{3}$

57. $\dfrac{\pi}{9}$ **58.** $\dfrac{5\pi}{18}$

59. $\dfrac{7\pi}{12}$ **60.** $\dfrac{3\pi}{10}$

Convert each radian measurement to degrees, rounding to the nearest hundredth.

61. 3.5   **62.** −2.7

63. $-\dfrac{1}{2}$ **64.** $\dfrac{3}{20}$

65. $\sqrt{7}$ **66.** 4

For each angle:
(a) Draw the angle in standard position.
(b) State the quadrant in which the angle lies, or state that it is quadrantal.
(c) Identify each angle as acute, right, obtuse, straight, or none of these.

67. 35°   **68.** 75°

69. −59°   **70.** −145°

71. $\dfrac{5\pi}{6}$ **72.** $\dfrac{2\pi}{3}$

73. $\dfrac{\pi}{2}$ **74.** π

75. 170°   **76.** 95°

77. 225°   **78.** 350°

79. −270°   **80.** −90°

81. $\dfrac{\pi}{6}$ **82.** $\dfrac{\pi}{4}$

83. $-\dfrac{5\pi}{6}$ **84.** $-\dfrac{11\pi}{6}$

Determine whether the given angles are coterminal.

85. 40° and 400°   **86.** 90° and −270°

87. 135° and 270°   **88.** 135° and −135°

89. $\dfrac{11\pi}{6}$ and $-\dfrac{\pi}{6}$ **90.** $\dfrac{3\pi}{4}$ and $-\dfrac{5\pi}{4}$

91. $\dfrac{16\pi}{3}$ and $\dfrac{29\pi}{3}$ **92.** $\dfrac{19\pi}{5}$ and $\dfrac{31\pi}{5}$

Find a positive angle less than 360° or less than 2π that is coterminal to the given angle.

93. 490°   **94.** 430°

95. −120°   **96.** −210°

97. $\dfrac{17\pi}{4}$ **98.** $\dfrac{21\pi}{4}$

99. $-\dfrac{22\pi}{3}$ **100.** $-\dfrac{19\pi}{3}$

101. $\dfrac{23\pi}{9}$ **102.** $\dfrac{20\pi}{9}$

In each problem, s is the length of the arc subtended by the central angle θ in a circle with radius r. Find the missing value, rounding to three decimal places.

103. $r = 15$ inches; $\theta = \dfrac{\pi}{3}$; find s.

104. $r = 22$ inches; $\theta = \dfrac{3\pi}{4}$; find s.

105. $\theta = 60°$; $r = 30$ meters; find s.

106. $\theta = 210°$; $r = 40$ meters; find s.

107. $\theta = \dfrac{3\pi}{7}$; $s = 27$ inches; find r.

108. $\theta = \dfrac{6\pi}{11}$; $s = 14$ inches; find r.

109. $\theta = 150°$; $s = 15$ centimeters; find r.

110. $\theta = 12°$; $s = 28$ centimeters; find r.

111. $r = 17$ inches; $s = 30$ inches; find θ.

112. $r = 23$ inches; $s = 50$ inches; find θ.

113. $r = 22$ meters; $s = 15$ meters; find θ.
114. $r = 34$ meters; $s = 20$ meters; find θ.

In each problem below, A is the area of a circular sector formed by the central angle θ in a circle with radius r. Find the missing value, rounding to three decimal places.

115. $r = 15$ inches; $\theta = \frac{2\pi}{3}$; find A.
116. $r = 20$ inches; $\theta = \frac{3\pi}{5}$; find A.
117. $r = 12$ centimeters; $\theta = 45°$; find A.
118. $r = 19$ centimeters; $\theta = 210°$; find A.
119. $r = 13$ inches; $A = 140$ square inches; find θ.
120. $r = 10$ inches; $A = 350$ square inches; find θ.
121. $\theta = 3.6$ radians; $A = 500$ square inches; find r.
122. $\theta = 5.2$ radians; $A = 700$ square inches; find r.

Applications

123. **Arc of Sea Dragon** A Sea Dragon ride travels along a 150° arc of a circle. If the radius of the circle is 50 feet, how long is the Sea Dragon's path? Round to one decimal place.

124. **Length of a railroad track** A railroad track curves along a 17° arc of a circle. If the radius of the circle is 255 meters, how long is the track? Round to one decimal place.

125. **Distance between cities** Birmingham, Alabama, located at latitude N 33°31′, is due south of Nashville, Tennessee, located at N 36°10′. Using 3,960 miles as the radius of the earth, determine the distance between the two cities to the nearest mile.

126. **Distance between cities** Detroit, Michigan, located at latitude N 42°20′, is due north of Columbus, Ohio, located at N 39°57′. Using 3,960 miles as the radius of the earth, determine the distance between the two cities to the nearest mile.

127. **Distance to the moon** The diameter of the moon is approximately 2,160 miles. How far is the center of the moon from the earth if the central angle from a point on the earth intercepting a diameter on the moon is 0.56°? Round to the nearest mile.

128. **Angular diameter of the sun** The sun is approximately 93 million miles from the earth, and its diameter is about 864,000 miles. In degrees, find the central angle from a point on the the earth to the sun.

129. **Geometry** Find the exact area of a sector of a circle if the sector has a central angle of exactly 30° and the circle has a radius of exactly 20 units.

130. **Geometry** A circle contains a sector with a central angle of $\frac{2\pi}{3}$ and an area of exactly 30 square meters. Find the diameter of the circle.

131. **Area of a sector** A sprinkler on a football field sprays water over a distance of 20 feet and is set to rotate through an angle of 305°. How many square feet of grass does the sprinkler water? Round to the nearest square foot.

132. **Buying pizza** If one-quarter of a 12-inch-diameter pepperoni pizza sells for $3, what would be a fair price for one-sixth of an 18-inch-diameter pepperoni pizza?

133. **Geometry** A regular hexagon (a six-sided polygon) is inscribed in a circle 24 meters in diameter. How long is an arc intercepted by one of the sides of the hexagon?

134. **Geometry** A regular octagon (an eight-sided polygon) is inscribed in a circle 30 centimeters in diameter. How long is the arc intercepted by one of the sides of the octagon?

135. **Windshield wiper** The rear windshield-wiper blade on a Honda Accord Crosstour is 21 inches long and rotates 95°. If the rubber part of the blade is 16 inches long, what is the area of the glass cleaned by the blade? Round to the nearest square inch.

136. **Silk fan** An Asian silk fan, when opened completely, rotates through an angle of 140° and has an area of approximately 83 square inches. What is the radius of the fan? Round to the nearest inch.

137. **Spotlight** A spotlight used for hunting rotates 160° horizontally and shines a distance of 100 feet. Find the area in square yards that is illuminated by the spotlight. Round to the nearest square yard.

138. **ATV** Jonathan's Polaris Sportsman ATV has tires with a diameter of 26 inches. What distance in feet does Jonathan travel if his tires rotate 1,000 times? Round to the nearest foot.

139. Designing pulleys Three pulleys have the radii shown. If pulley A turns through one revolution, through what angle does pulley B turn?

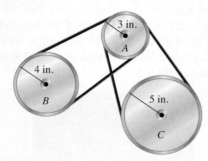

140. Designing pulleys If pulley C in Exercise 139 turns through one revolution, through what angle does pulley A turn?

Discovery and Writing

141. What are the differences in the way a degree measure and a radian measure is used?

142. Why is it more useful for a surveyor to use degree measure than radian measure?

143. Find the latitude of the city where you live, then find the distance from there along the earth's surface to London, England, located at N 51°30'.

144. Explain how you would find the distance between a city located north of the equator and a city located located south of the equator.

145. If the central angle of a circle is held constant and the radius of the circle is tripled, what effect does this have on the arc length?

146. If the central angle of a circle is held constant and the radius of the circle is tripled, what effect does this have on the area of the sector?

4.2 The Unit Circle and Trigonometric Functions

In this section, we will learn to

1. Understand the unit circle.
2. Evaluate quadrantal angles using the unit circle.
3. Evaluate special angles using the unit circle.
4. Use symmetry and the unit circle to evaluate multiples of special angles.
5. Determine the algebraic signs of trigonometric functions.
6. Understand the periodic properties of trigonometric functions.
7. Determine the reference angle of a non-acute angle.
8. Find exact values of trigonometric functions using reference angles.

In algebra we use exponential functions to measure the growth of money in a savings account and we use logarithms to measure the intensity of an earthquake. Trigonometric functions are used in the study of triangles and the modeling of periodic phenomena. The phases of the moon and the ebb and flow of the tide are just two everyday examples of occurrences in nature that follow a predictable pattern. In this section, we will discover why trigonometric functions are used in many applications to predict or model this type of behavior.

Imagine what a big job it is to build the International Space Station. The astronauts have to attach modules that weigh tons and work solar panels that are longer than a bus. How can this be done safely and efficiently? To make this giant task easier, the Canadian Space Agency has developed a new robotic arm called Canadarm2, which is operated by controlling the angles at each of its seven joints. With an astronaut at the end of the arm, it is important that the final position be done very accurately. To accomplish this, repeated use of trigonometric functions of the angles are employed. In this section, we are going to study these trigonometric functions.

1. Understand the Unit Circle

The word *trigonometry* comes from the Greek words *trigonon* (triangle) and *metron* (measure). Most of the work in trigonometry is based on six trigonometric functions. A unit circle is often used to introduce these functions. **A unit circle** is a circle with radius 1, with the center at the origin. We will base these definitions using the same ideas used to define the radian measure of angles. Since most of the problems in calculus involve real numbers, this will allow us to be in conformity with other functions.

The names of the six trigonometric functions and their abbreviations are:

Function Name	Abbreviation
sine	sin
cosine	cos
tangent	tan
cosecant	csc
secant	sec
cotangent	cot

The equation of the unit circle is $x^2 + y^2 = 1$. Our goal is to define the trigonometric functions for any real number. If you look at Figure 4-21, we have drawn a unit circle with an arc of length s marked. The central angle is t, and the radius is 1. Recall that the formula for arc length in radian measure is $s = r\theta$. Substituting the information from the circle, we now have the following:

$$s = r\theta = 1 \cdot t = t.$$

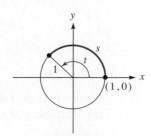

FIGURE 4-21

Therefore, **in a unit circle, the length of the intercepted arc is the same as the radian measure of the central angle**; both are designated by the real number t. We note that any real number t can be associated with exactly one point, $P(x, y)$, on the unit circle. If t is positive, the distance is measured in a counterclockwise direction. If t is a negative number, the distance is measured in a clockwise direction.

$$t > 0 \qquad x^2 + y^2 = 1 \qquad t < 0$$

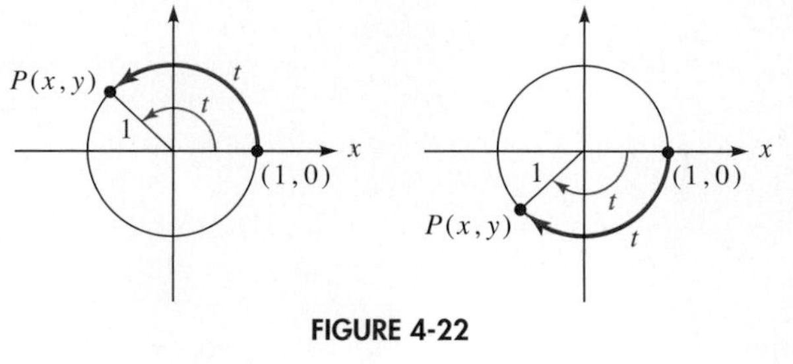

FIGURE 4-22

The tie between the trigonometric functions and the unit circle is so close that the trigonometric functions are often called the **circular functions**. We now define the six trigonometric functions for any real number t.

Trigonometric Functions Defined for the Unit Circle (Circular Functions)

Let t be a real number and $P(x, y)$ be the point on the unit circle that corresponds to t. The six trigonometric functions of t are:

$$\sin t = y \qquad\qquad \csc t = \frac{1}{y}, y \neq 0$$

$$\cos t = x \qquad\qquad \sec t = \frac{1}{x}, x \neq 0$$

$$\tan t = \frac{y}{x}, x \neq 0 \qquad\qquad \cot t = \frac{x}{y}, y \neq 0$$

Since t is the radian measure of an angle and a real number, we state that if $P(x, y)$ is a point on the unit circle, we have

$$y = \sin \,(\text{an angle of } t \text{ radians}) = \sin \,(\text{a real number } t) = \sin t$$

and

$$x = \cos \,(\text{an angle of } t \text{ radians}) = \cos \,(\text{a real number } t) = \cos t.$$

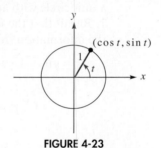

FIGURE 4-23

Notice that $\csc t$ is the reciprocal of $\sin t$, $\sec t$ is the reciprocal of $\cos t$, and $\cot t$ is the reciprocal of $\tan t$.

EXAMPLE 1   **Finding Trigonometric Function Values When Given a Point on the Unit Circle**

The point $P = \left(\frac{1}{3}, \frac{2\sqrt{2}}{3}\right)$ is a point on the unit circle that corresponds to a real number t. Find each of the following:

a. $\sin t$ **b.** $\cos t$ **c.** $\tan t$ **d.** $\csc t$ **e.** $\sec t$ **f.** $\cot t$

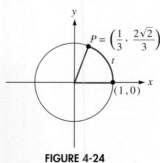

FIGURE 4-24

SOLUTION   **a.** $\sin t = y = \dfrac{2\sqrt{2}}{3}$

b. $\cos t = x = \dfrac{1}{3}$

c. $\tan t = \dfrac{y}{x} = \dfrac{\frac{2\sqrt{2}}{3}}{\frac{1}{3}} = 2\sqrt{2}$

d. $\csc t = \dfrac{1}{y} = \dfrac{1}{\frac{2\sqrt{2}}{3}} = \dfrac{3}{2\sqrt{2}} = \dfrac{3}{2\sqrt{2}} \cdot \dfrac{\sqrt{2}}{\sqrt{2}} = \dfrac{3\sqrt{2}}{4}$ Rationalize the denominator.

e. $\sec t = \dfrac{1}{x} = \dfrac{1}{\frac{1}{3}} = 3$

f. $\cot t = \dfrac{x}{y} = \dfrac{\frac{1}{3}}{\frac{2\sqrt{2}}{3}} = \dfrac{1}{2\sqrt{2}} = \dfrac{1}{2\sqrt{2}} \cdot \dfrac{\sqrt{2}}{\sqrt{2}} = \dfrac{\sqrt{2}}{4}$ Rationalize the denominator.   ∎

Self Check 1   The point $P = \left(\frac{2}{5}, \frac{\sqrt{21}}{5}\right)$ is a point on the unit circle that corresponds to a real number t. Find each of the following:

a. $\sin t$ **b.** $\cos t$ **c.** $\tan t$ **d.** $\csc t$ **e.** $\sec t$ **f.** $\cot t$

Now Try Exercise 9.

In the last section, we stated that Greek letters are often used to indicate an angle. We can extend that notation to match what we have defined here. Typically, we use the letter t to indicate a real number and a Greek letter to indicate the angle in radian measure that has the same value (Figure 4-25).

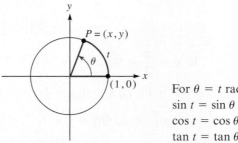

For $\theta = t$ radians,
$\sin t = \sin \theta$ $\csc t = \csc \theta$
$\cos t = \cos \theta$ $\sec t = \sec \theta$
$\tan t = \tan \theta$ $\cot t = \cot \theta$

FIGURE 4-25

Comment   The trigonometric functions are functions of real numbers and functions of angles. We can also use the same definitions if the angle is given in degrees; however, if the angle is in degrees, the degree notation must be used.

2. Evaluate Quadrantal Angles Using the Unit Circle

Recall from the previous section the equivalent measures in degrees and radians for several angles shown in Table 2.

TABLE 2

Degree Measure	0°	30°	45°	60°	90°	180°	270°	360°
Radian Measure	0	$\dfrac{\pi}{6}$	$\dfrac{\pi}{4}$	$\dfrac{\pi}{3}$	$\dfrac{\pi}{2}$	π	$\dfrac{3\pi}{2}$	2π

The angles $30° = \frac{\pi}{6}$, $45° = \frac{\pi}{4}$, and $60° = \frac{\pi}{3}$ are often called **special angles**. These are angles that can have the values of the six trigonometric functions calculated exactly. The other angles in the table $0° = 0$, $90° = \frac{\pi}{2}$, $180° = \pi$, $270° = \frac{3\pi}{2}$, $360° = 2\pi$ are **quadrantal angles**, because the terminal side is one of the axes. We can use the unit circle to find the trigonometric functions of quadrantal angles.

EXAMPLE 2 **Finding the Values of the Trigonometric Functions of Quadrantal Angles**

Find the values of the six trigonometric functions for the quadrantal angles:

a. $0° = 0$ **b.** $90° = \dfrac{\pi}{2}$ **c.** $180° = \pi$ **d.** $270° = \dfrac{3\pi}{2}$

SOLUTION We will use the unit circle and the definitions to find these values.

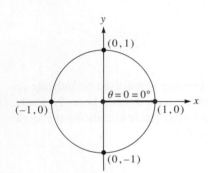

FIGURE 4-26

a. $\theta = 0 = 0°$

$P = (1, 0), x = 1, y = 0$

$\sin 0 = \sin 0° = y = 0$ $\qquad \csc 0 = \csc 0° = \dfrac{1}{y} = \dfrac{1}{0}$ undefined

$\cos 0 = \cos 0° = x = 1$ $\qquad \sec 0 = \sec 0° = \dfrac{1}{x} = \dfrac{1}{1} = 1$

$\tan 0 = \tan 0° = \dfrac{y}{x} = \dfrac{0}{1} = 0$ $\qquad \cot 0 = \cot 0° = \dfrac{x}{y} = \dfrac{1}{0}$ undefined

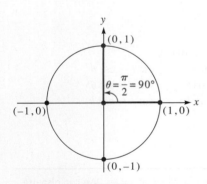

FIGURE 4-27

b. $\theta = \dfrac{\pi}{2} = 90°$

$P = (0, 1), x = 0, y = 1$

$\sin \dfrac{\pi}{2} = \sin 90° = y = 1$ $\qquad \csc \dfrac{\pi}{2} = \csc 90° = \dfrac{1}{y} = \dfrac{1}{1} = 1$

$\cos \dfrac{\pi}{2} = \cos 90° = x = 0$ $\qquad \sec \dfrac{\pi}{2} = \sec 90° = \dfrac{1}{x} = \dfrac{1}{0}$ undefined

$\tan \dfrac{\pi}{2} = \tan 90° = \dfrac{y}{x} = \dfrac{1}{0}$ undefined $\qquad \cot \dfrac{\pi}{2} = \cot 90° = \dfrac{x}{y} = \dfrac{0}{1} = 0$

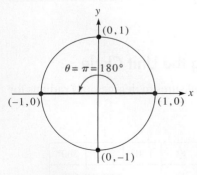

FIGURE 4-28

c. $\theta = \pi = 180°$

$P = (-1, 0), x = -1, y = 0$

$\sin \pi = \sin 180° = y = 0$ $\qquad \csc \pi = \csc 180° = \dfrac{1}{y} = \dfrac{1}{0}$ undefined

$\cos \pi = \cos 180° = x = -1$ $\qquad \sec \pi = \sec 180° = \dfrac{1}{x} = \dfrac{1}{-1} = -1$

$\tan \pi = \tan 180° = \dfrac{y}{x} = \dfrac{0}{-1} = 0$ $\qquad \cot \pi = \cot 180° = \dfrac{x}{y} = \dfrac{-1}{0}$ undefined

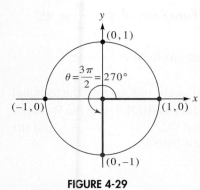

FIGURE 4-29

d. $\theta = \dfrac{3\pi}{2} = 270°$

$P = (0, -1),\ x = 0,\ y = -1$

$\sin \dfrac{3\pi}{2} = \sin 270° = y = -1$ $\qquad$ $\csc \dfrac{3\pi}{2} = \csc 270° = \dfrac{1}{y} = \dfrac{1}{-1} = -1$

$\cos \dfrac{3\pi}{2} = \cos 270° = x = 0$ $\qquad$ $\sec \dfrac{3\pi}{2} = \sec 270° = \dfrac{1}{x} = \dfrac{1}{0}$ undefined

$\tan \dfrac{3\pi}{2} = \tan 270° = \dfrac{y}{x} = \dfrac{-1}{0}$ undefined $\quad$ $\cot \dfrac{3\pi}{2} = \cot 270° = \dfrac{x}{y} = \dfrac{0}{-1} = 0$

Self Check 2   Find the values of the six trigonometric functions for $\theta = -\pi = -180°$.

Now Try Exercise 17.

We summarize these values in Table 3.

TABLE 3

Trigonometric-Function Values for Quadrantal Angles						
θ	$\sin \theta$	$\cos \theta$	$\tan \theta$	$\csc \theta$	$\sec \theta$	$\cot \theta$
$\theta = 0 = 0°$	0	1	0	Undefined	1	Undefined
$\theta = \frac{\pi}{2} = 90°$	1	0	Undefined	1	Undefined	0
$\theta = \pi = 180°$	0	-1	0	Undefined	-1	Undefined
$\theta = \frac{3\pi}{2} = 270°$	-1	0	Undefined	-1	Undefined	0

3. Evaluate Special Angles Using the Unit Circle

We now turn our attention to the special angles, $30° = \frac{\pi}{6}$, $45° = \frac{\pi}{4}$, and $60° = \frac{\pi}{3}$. The unit circle in Figure 4-30 is divided into parts that correspond to the special angles and multiples of those angles. If we can find the coordinates on the unit circle that correspond to the point where the terminal side of each angle intercepts the unit circle, then we can determine the exact values of the trigonometric functions for the special angles. We can then find the point that corresponds to each multiple of the special angles.

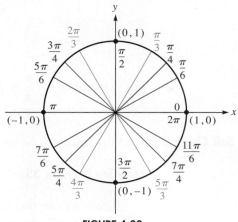

FIGURE 4-30

The first special angle we will examine is $\theta = \frac{\pi}{4} = 45°$.

EXAMPLE 3   Determining the Values of the Trigonometric Functions of $\theta = \dfrac{\pi}{4} = 45°$

Find the exact values of the six trigonometric functions for $\theta = \dfrac{\pi}{4} = 45°$.

SOLUTION   Since $\frac{\pi}{4} = \frac{1}{2} \cdot \frac{\pi}{2}$, we can draw the angle on the unit circle and see that the terminal side of the angle corresponds to the line $y = x$ [Figure 4-31(a)]. Using the Pythagorean Theorem to find the values of x and y, we find that $x = \frac{\sqrt{2}}{2}$, $y = \frac{\sqrt{2}}{2}$, and the coordinates of the point are $P = \left(\frac{\sqrt{2}}{2}, \frac{\sqrt{2}}{2} \right)$ [Figure 4-31(b)].

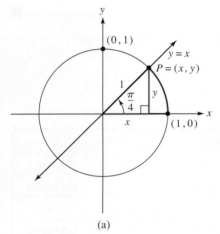

$$x^2 + y^2 = 1$$
$$x^2 + x^2 = 1 \qquad \text{Use } y = x.$$
$$2x^2 = 1$$
$$x^2 = \frac{1}{2}$$
$$x = \sqrt{\frac{1}{2}} = \frac{1}{\sqrt{2}} = \frac{1}{\sqrt{2}} \cdot \frac{\sqrt{2}}{\sqrt{2}} = \frac{\sqrt{2}}{2}$$
$$y = \frac{\sqrt{2}}{2}$$
$$P = \left(\frac{\sqrt{2}}{2}, \frac{\sqrt{2}}{2} \right)$$

(a)   (b)

FIGURE 4-31

Now that we have found the coordinates on the unit circle that correspond to the angle $\theta = \frac{\pi}{4} = 45°$, we can find the values of the six trigonometric functions for the angle. Note that we first found $x = \frac{1}{\sqrt{2}}$, and then we rationalized the denominator to find $x = \frac{\sqrt{2}}{2}$. Likewise, we have $y = \frac{1}{\sqrt{2}} = \frac{\sqrt{2}}{2}$. When we are finding the secant, cosecant, tangent, and cotangent, the form $x = y = \frac{1}{\sqrt{2}}$ will be used to make the calculations easier.

$$\sin \frac{\pi}{4} = \sin 45° = y = \frac{\sqrt{2}}{2} \qquad\qquad \csc \frac{\pi}{4} = \csc 45° = \frac{1}{y} = \frac{1}{\frac{1}{\sqrt{2}}} = \sqrt{2}$$

$$\cos \frac{\pi}{4} = \cos 45° = x = \frac{\sqrt{2}}{2} \qquad\qquad \sec \frac{\pi}{4} = \sec 45° = \frac{1}{x} = \frac{1}{\frac{1}{\sqrt{2}}} = \sqrt{2}$$

$$\tan \frac{\pi}{4} = \tan 45° = \frac{y}{x} = \frac{\frac{1}{\sqrt{2}}}{\frac{1}{\sqrt{2}}} = 1 \qquad\qquad \cot \frac{\pi}{4} = \cot 45° = \frac{x}{y} = \frac{\frac{1}{\sqrt{2}}}{\frac{1}{\sqrt{2}}} = 1$$

Self Check 3   Find the values of the six trigonometric functions for $\theta = -\dfrac{\pi}{4} = -45°$.

Now Try Exercise 35.

Before we can find the values of the trigonometric functions for $\theta = \frac{\pi}{6} = 30°$ and $\theta = \frac{\pi}{3} = 60°$, we need to find the coordinates for the corresponding points on the unit circle. We will use a right triangle to find these points. The sum of the angles in any triangle is 180°; therefore, in a right triangle, if one of the angles is 60°, the other angle is 30°. In Figure 4-32(a), a right triangle is created using 60° and

$30°$ as the two acute angles, with sides of length x and y. If the triangle is reflected as in Figure 4-32(b), an equilateral triangle (all angles are $60°$) is created. Therefore, all the sides in the equilateral triangle have a length of 1. Since the line segment of length y bisects the base of the triangle, each of the shorter sides is $x = \frac{1}{2}$ in length. Using the Pythagorean Theorem, we have

$$x^2 + y^2 = 1$$
$$y^2 = 1 - x^2 = 1 - \left(\frac{1}{2}\right)^2 = 1 - \frac{1}{4} = \frac{3}{4}$$
$$y = \frac{\sqrt{3}}{2}$$

Therefore, the coordinates of point P are $P = \left(\frac{1}{2}, \frac{\sqrt{3}}{2}\right)$ as shown in Figure 4-32(c).

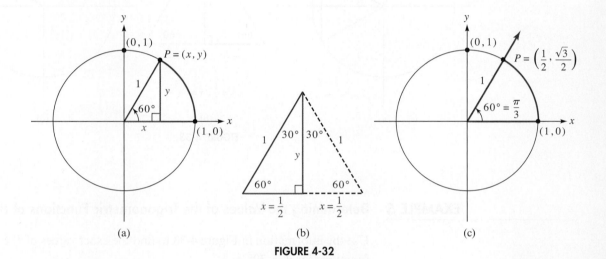

(a)   (b)   (c)

FIGURE 4-32

EXAMPLE 4   **Determining the Values of the Trigonometric Functions of $\theta = 60° = \dfrac{\pi}{3}$**

Use the information in Figure 4-32 to find the exact values of the six trigonometric functions for $\theta = 60° = \frac{\pi}{3}$.

SOLUTION   From Figure 4-32(c), we see that the point on the unit circle corresponding to the angle $\theta = 60° = \frac{\pi}{3}$ is $P = \left(\frac{1}{2}, \frac{\sqrt{3}}{2}\right)$. Therefore, we know that $x = \frac{1}{2}$, and $y = \frac{\sqrt{3}}{2}$, and we can apply the definitions of the six trigonometric functions.

$$\sin 60° = \sin \frac{\pi}{3} = y = \frac{\sqrt{3}}{2} \qquad \csc 60° = \csc \frac{\pi}{3} = \frac{1}{y} = \frac{1}{\frac{\sqrt{3}}{2}} = \frac{2}{\sqrt{3}} = \frac{2}{\sqrt{3}} \cdot \frac{\sqrt{3}}{\sqrt{3}} = \frac{2\sqrt{3}}{3}$$

$$\cos 60° = \cos \frac{\pi}{3} = x = \frac{1}{2} \qquad \sec 60° = \sec \frac{\pi}{3} = \frac{1}{x} = \frac{1}{\frac{1}{2}} = 2$$

$$\tan 60° = \tan \frac{\pi}{3} = \frac{y}{x} = \frac{\frac{\sqrt{3}}{2}}{\frac{1}{2}} = \sqrt{3} \qquad \cot 60° = \cot \frac{\pi}{3} = \frac{x}{y} = \frac{\frac{1}{2}}{\frac{\sqrt{3}}{2}} = \frac{1}{\sqrt{3}} = \frac{1}{\sqrt{3}} \cdot \frac{\sqrt{3}}{\sqrt{3}} = \frac{\sqrt{3}}{3}$$

Self Check 4   Find the values of the six trigonometric functions for $\theta = -60° = -\dfrac{\pi}{3}$.

Now Try Exercise 37.

To find the coordinates for the point on the unit circle that corresponds to $\theta = 30° = \frac{\pi}{6}$, we can use the same steps we used for $\theta = 60° = \frac{\pi}{3}$. We draw the angle $\theta = 30° = \frac{\pi}{6}$ and construct a triangle as before. You can see that the coordinates of x and y are reversed.

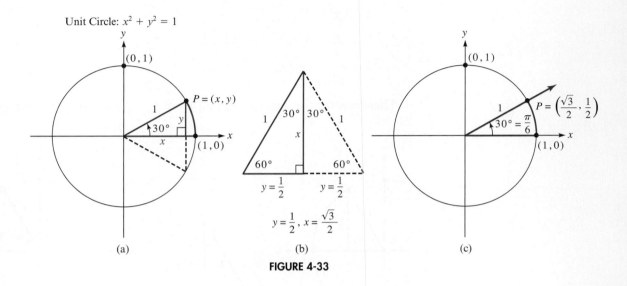

FIGURE 4-33

EXAMPLE 5   Determining the Values of the Trigonometric Functions of $\theta = 30° = \dfrac{\pi}{6}$

Use the information in Figure 4-33 to find the exact values of the six trigonometric functions for $\theta = 30° = \frac{\pi}{6}$.

SOLUTION   From Figure 4-33(c), we see that the point on the unit circle corresponding to the angle $\theta = 30° = \frac{\pi}{6}$ is $P = \left(\frac{\sqrt{3}}{2}, \frac{1}{2} \right)$. Therefore, we know that $x = \frac{\sqrt{3}}{2}$ and $y = \frac{1}{2}$, and we can apply the definitions of the six trigonometric functions.

$$\sin 30° = \sin \frac{\pi}{6} = y = \frac{1}{2}$$

$$\csc 30° = \csc \frac{\pi}{6} = \frac{1}{y} = \frac{1}{\frac{1}{2}} = 2$$

$$\cos 30° = \cos \frac{\pi}{6} = x = \frac{\sqrt{3}}{2}$$

$$\sec 30° = \sec \frac{\pi}{6} = \frac{1}{x} = \frac{1}{\frac{\sqrt{3}}{2}} = \frac{2}{\sqrt{3}} \cdot \frac{\sqrt{3}}{\sqrt{3}} = \frac{2\sqrt{3}}{3}$$

$$\tan 30° = \tan \frac{\pi}{6} = \frac{y}{x} = \frac{\frac{1}{2}}{\frac{\sqrt{3}}{2}} = \frac{1}{\sqrt{3}} \cdot \frac{\sqrt{3}}{\sqrt{3}} = \frac{\sqrt{3}}{3}$$

$$\cot 30° = \cot \frac{\pi}{6} = \frac{x}{y} = \frac{\frac{\sqrt{3}}{2}}{\frac{1}{2}} = \sqrt{3}$$

Self Check 5   Find the values of the six trigonometric functions for $\theta = -30° = -\frac{\pi}{6}$.

Now Try Exercise 39.

We summarize the values that we have found for the special and quadrantal angles in Table 4.

TABLE 4

Trigonometric-Function Values for Special and Quadrantal Angles						
θ	$\sin\theta$	$\cos\theta$	$\tan\theta$	$\csc\theta$	$\sec\theta$	$\cot\theta$
$\theta = 0 = 0°$	0	1	0	Undefined	1	Undefined
$\theta = \dfrac{\pi}{6} = 30°$	$\dfrac{1}{2}$	$\dfrac{\sqrt{3}}{2}$	$\dfrac{\sqrt{3}}{3}$	2	$\dfrac{2\sqrt{3}}{3}$	$\sqrt{3}$
$\theta = \dfrac{\pi}{4} = 45°$	$\dfrac{\sqrt{2}}{2}$	$\dfrac{\sqrt{2}}{2}$	1	$\sqrt{2}$	$\sqrt{2}$	1
$\theta = \dfrac{\pi}{3} = 60°$	$\dfrac{\sqrt{3}}{2}$	$\dfrac{1}{2}$	$\sqrt{3}$	$\dfrac{2\sqrt{3}}{3}$	2	$\dfrac{\sqrt{3}}{3}$
$\theta = \dfrac{\pi}{2} = 90°$	1	0	Undefined	1	Undefined	0
$\theta = \pi = 180°$	0	-1	0	Undefined	-1	Undefined
$\theta = \dfrac{3\pi}{2} = 270°$	-1	0	Undefined	-1	Undefined	0

It is important that you learn these values, because they are used throughout the course.

4. Use Symmetry and the Unit Circle to Evaluate Multiples of Special Angles

Using symmetry, the other coordinates of the points on the unit circle can be found, as shown in Figure 4-34. This allows us to find the values of the six trigonometric functions for multiples of the special angles.

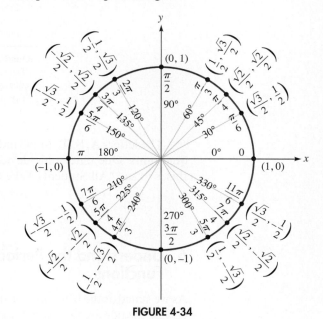

FIGURE 4-34

EXAMPLE 6   **Finding the Values of Trigonometric Functions of Special Angles**

Use the unit circle to find the exact value of each of the following:

a. $\sin\dfrac{3\pi}{4}$ **b.** $\cos 210°$ **c.** $\tan\dfrac{4\pi}{3}$ **d.** $\csc\dfrac{11\pi}{6}$

e. $\sec 225°$ **f.** $\cot 120°$

SOLUTION **a.** $\sin \dfrac{3\pi}{4} = \dfrac{\sqrt{2}}{2}$ **b.** $\cos 210° = -\dfrac{\sqrt{3}}{2}$ **c.** $\tan \dfrac{4\pi}{3} = \sqrt{3}$

d. $\csc \dfrac{11\pi}{6} = -2$ **e.** $\sec 225° = -\sqrt{2}$ **f.** $\cot 120° = -\dfrac{\sqrt{3}}{3}$

Self Check 6 Use the unit circle to find the exact value of each of the following:

a. $\sin \dfrac{4\pi}{3}$ **b.** $\cos 135°$ **c.** $\tan \dfrac{7\pi}{6}$ **d.** $\csc \dfrac{7\pi}{4}$

e. $\sec 330°$ **f.** $\cot 300°$

Now Try Exercise 41.

5. Determine the Algebraic Signs of Trigonometric Functions

If a particular angle is not a quadrantal angle, the sign of a trigonometric function for that angle is determined by the quadrant in which the terminal side of the angle lies. The signs of x and y are the determining factors. The sine and cosecant functions are positive when y is positive, which is in quadrant I and quadrant II (QI and QII). Since x is positive in QI and QIV, the cosine and secant functions are positive in those quadrants. The tangent and cotangent functions are positive when both x and y are positive or both are negative, which is in QI and QIII. Figure 4-35 will help you remember this.

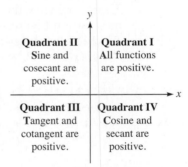

FIGURE 4-35

The letters **A, S, T,** and **C** indicate the functions that are positive in the various quadrants. One mnemonic device that some students find useful in remembering this is to say "**A**ll **S**tudents **T**ake **C**alculus."

6. Understand the Periodic Properties of Trigonometric Functions

As we stated at the beginning of the section, the phases of the moon and the ebb and flow of the tide are two everyday examples of occurrences in nature that follow a predictable pattern or cycle. Tides go from low tide to high tide every 12 hours.

If you look at the unit circle, as the value of $t = \theta$ moves from 0 to 2π, the sine function moves from 0 to 1, then from 1 back to 0, then from 0 to -1, and then back to 0. If you continue to move the value of $t = \theta$ around the unit circle, this cycle of values repeats every 2π units. The cosine function follows a similar repetition, starting at 1, moving to 0, then to -1, back to 0, and ending back at 1 before it begins to repeat the cycle. The sine and cosine functions are called **periodic functions** because as the value of $t = \theta$ increases, the values of $\sin \theta$ and $\cos \theta$ repeat predictably.

Periodic Function   A function f is said to be **periodic** with period p if p is the smallest positive number for which $f(t) = f(t + p)$ for all t in the domain of f.

Since the sine and cosine functions repeat every 2π units, we say that the **period** of the sine and cosine functions is 2π.

Period of the Sine and Cosine Functions   The sine and cosine functions are periodic functions because

$$\sin\theta = \sin(\theta + 2\pi n) \text{ and } \cos\theta = \cos(\theta + 2\pi n)$$

for any integer n and any angle θ. The period is 2π.

This periodic behavior of the sine and cosine function can be used to evaluate trigonometric functions. Assuming that θ is reduced to lowest terms and is not a quadrantal angle, then we know:

- If θ is a multiple of $\dfrac{\pi}{4}$, then θ will be coterminal with one of the angles $\dfrac{\pi}{4}, \dfrac{3\pi}{4}, \dfrac{5\pi}{4}$, or $\dfrac{7\pi}{4}$.

- If θ is a multiple of $\dfrac{\pi}{6}$, then θ will be coterminal with one of the angles $\dfrac{\pi}{6}, \dfrac{5\pi}{6}, \dfrac{7\pi}{6}$, or $\dfrac{11\pi}{6}$.

- If θ is a multiple of $\dfrac{\pi}{3}$, then θ will be coterminal with one of the angles $\dfrac{\pi}{3}, \dfrac{2\pi}{3}, \dfrac{4\pi}{3}$, or $\dfrac{5\pi}{3}$.

EXAMPLE 7   **Using the Period to Evaluate the Sine and Cosine Functions**

Evaluate each of the following by using the unit circle and the period of the function to find a coterminal angle that is in the interval $0 \le \theta < 2\pi$:

a. $\sin\dfrac{13\pi}{4}$ **b.** $\cos\left(-\dfrac{23\pi}{6}\right)$ **c.** $\sin\dfrac{23\pi}{3}$

SOLUTION   **a.** We will use the period of the sine function to obtain an angle coterminal to $\frac{13\pi}{4}$ that is less than 2π. We find $\theta = \frac{13\pi}{4} - 2\pi = \frac{5\pi}{4}$.

$$\sin\frac{13\pi}{4} = \sin\frac{5\pi}{4} = -\frac{\sqrt{2}}{2}$$

b. You may need to add a multiple of 2π to find the coterminal angle in the interval $[0, 2\pi)$.

$$\cos\left(-\frac{23\pi}{6}\right) = \cos\left(-\frac{23\pi}{6} + 2 \cdot 2\pi\right) = \cos\frac{\pi}{6} = \frac{\sqrt{3}}{2}$$

c. $\sin\dfrac{23\pi}{3} = \sin\left(\dfrac{23\pi}{3} - 6\pi\right) = \sin\dfrac{5\pi}{3} = -\dfrac{\sqrt{3}}{2}$

Self Check 7   Evaluate each of the following: **a.** $\sin\left(-\dfrac{27\pi}{4}\right)$ **b.** $\cos\dfrac{29\pi}{3}$

Now Try Exercise 49.

The period of the secant and cosecant functions is also 2π. (Do you know why?) What about the tangent and cotangent functions? To determine the period of the tangent and cotangent functions, we need to find the smallest positive number p for which $\tan \theta = \tan(\theta + p)$. Look at Figure 4-36.

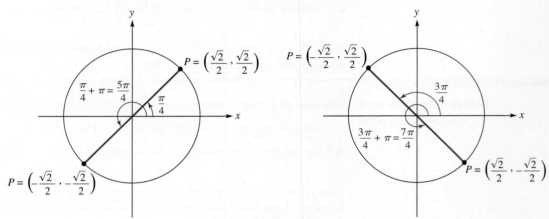

FIGURE 4-36

In quadrants I and III, the sign of $\tan \theta = \frac{y}{x}$ will be positive. Likewise, in quadrants II and IV, the sign of $\tan \theta = \frac{y}{x}$ will be negative. We can see that $\tan \theta = \tan(\theta + \pi)$ for all values of θ. Furthermore, the number π is the smallest positive number p for which $\tan \theta = \tan(\theta + p)$. Therefore, the period of the tangent function, and hence the cotangent function, is π.

Periodic Behavior of the Trigonometric Functions   The period for the sine, cosine, secant, and cosecant functions is 2π. The period for the tangent and cotangent functions is π. For any angle θ and any integer n, the following are true:

$$\sin \theta = \sin(\theta + 2\pi n) \qquad \csc \theta = \csc(\theta + 2\pi n)$$
$$\cos \theta = \cos(\theta + 2\pi n) \qquad \sec \theta = \sec(\theta + 2\pi n)$$
$$\tan \theta = \tan(\theta + \pi n) \qquad \cot \theta = \cot(\theta + \pi n)$$

EXAMPLE 8   **Using the Periodic Behavior of Trigonometric Functions**

Evaluate each of the following by using the period of the function:

a. $\tan \dfrac{16\pi}{3}$ **b.** $\sec\left(-\dfrac{11\pi}{4}\right)$ **c.** $\cot \dfrac{9\pi}{2}$ **d.** $\csc(-7\pi)$

SOLUTION   **a.** The period of the tangent function is π, so we subtract multiples of π until we have an angle in the interval $[0, 2\pi)$.

$$\tan \frac{16\pi}{3} = \tan\left(\frac{16\pi}{3} - 5\pi\right) = \tan \frac{\pi}{3} = \sqrt{3}$$

b. The period of the secant function is 2π, so we add a multiple of 2π.

$$\sec\left(-\frac{11\pi}{4}\right) = \sec\left(-\frac{11\pi}{4} + 2 \cdot 2\pi\right) = \sec \frac{5\pi}{4} = -\sqrt{2}$$

c. The period of the cotangent function is π, so we subtract multiples of π.

$$\cot \frac{9\pi}{2} = \cot\left(\frac{9\pi}{2} - 4\pi\right) = \cot \frac{\pi}{2} = 0$$

d. The period of the cosecant function is 2π, so we add a multiple of 2π.

$$\csc(-7\pi) = \csc(-7\pi + 8\pi) = \csc \pi$$

However, $\csc \pi$ is undefined; therefore, $\csc(-7\pi)$ is undefined. ∎

Self Check 8   Evaluate each of the following by using the period of the function:

a. $\tan\left(-\dfrac{11\pi}{3}\right)$ **b.** $\sec\dfrac{13\pi}{2}$

Now Try Exercise 55.

7. Determine the Reference Angle of a Non-acute Angle

We have seen how to find the exact values of the trigonometric functions for special angles and multiples of those angles using the periodic properties of the functions. Now we will look at another way to find the exact value of a trigonometric function of an angle that is a multiple of a special angle. This involves the use of a **reference angle**.

Reference Angle   The **reference angle** for angle θ is the *acute* angle θ' formed by the terminal side of θ and the nearer part of the x-axis.

The values of the six trigonometric functions of any angle θ are equal to those of the reference angle θ', except possibly for sign.

In Figure 4-37, you can see how the reference angle is drawn. The procedure shown is in degree measure, but the same procedure is used if the angle measure is in radians.

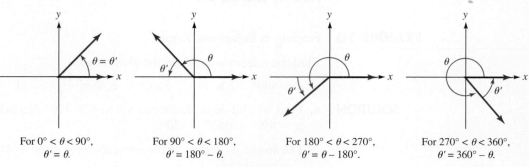

For $0° < \theta < 90°$,   For $90° < \theta < 180°$,   For $180° < \theta < 270°$,   For $270° < \theta < 360°$,
$\theta' = \theta$. $\theta' = 180° - \theta$. $\theta' = \theta - 180°$. $\theta' = 360° - \theta$.

FIGURE 4-37

Reference angles are **acute angles**, and are of the same type of measure as the given angle. If an angle θ is in radian measure, then the reference angle θ' is in radian measure, $0 < \theta' < \frac{\pi}{2}$. If an angle θ is in degree measure, then the reference angle θ' is in degree measure, $0° < \theta' < 90°$.

If the given angle is greater than $360°$ (2π radians) or less than $-360°$ (-2π radians), you should first find a positive coterminal angle less than $360°$ or 2π. Then you proceed to find the reference angle as shown in Figure 4-37.

EXAMPLE 9   Finding a Reference Angle

Find the reference angle θ' for the given angle. Draw the angle in standard position to help you find the reference angle.

a. $\theta = 290°$ **b.** $\theta = -150°$

SOLUTION   **a.** $\theta = 290°$ lies in QIII, as shown in standard position in Figure 4-38. Find θ' as shown.

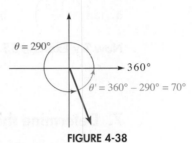

$\theta = 290°$

$360°$

$\theta' = 360° - 290° = 70°$

FIGURE 4-38

b. The angle $\theta = -150°$ lies in QII, as shown in Figure 4-39. Even though this angle is negative, if we look at the figure, it is clear what we need to do to find the reference angle.

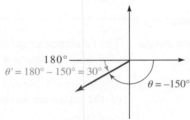

$180°$

$\theta' = 180° - 150° = 30°$

$\theta = -150°$

FIGURE 4-39

Self Check 9   Find the reference angle for the given angle **a.** $\theta = 165°$ **b.** $\theta = -\dfrac{2\pi}{3}$

Now Try Exercise 59.

EXAMPLE 10   Finding a Reference Angle

Find the reference angle for the given angle.

a. $\theta = 490°$ **b.** $\theta = -820°$ **c.** $\theta = \dfrac{17\pi}{3}$ **d.** $\theta = -\dfrac{19\pi}{4}$

SOLUTION   **a.** First we find an angle coterminal to $\theta = 490°$ that is less than $360°$:
$\beta = 490° - 360° = 130°$.

The angles lie in QII, so the reference angle is $\beta' = 180° - 130° = 50°$.

b. To find a positive angle coterminal to $\theta = -820°$ that is less than $360°$, we will add $3 \cdot 360° = 1{,}080°$ to $\theta = -820°$ and get the coterminal angle $\beta = -820° + 3 \cdot 360° = 260°$. β is in QIII, so the reference angle is $\beta' = 260° - 180° = 80°$.

c. Since $\theta = \frac{17\pi}{3}$ is in radian measure, we subtract a multiple of 2π to get the coterminal angle $\beta = \frac{17\pi}{3} - 2 \cdot 2\pi = \frac{17\pi}{3} - 4\pi = \frac{5\pi}{3}$. These angles lie in QIV, so the reference angle is $\beta' = 2\pi - \frac{5\pi}{3} = \frac{\pi}{3}$.

d. To find an angle less than 2π and coterminal to $\theta = -\frac{19\pi}{4}$, we add a multiple of 2π. The coterminal angle is $\beta = -\frac{19\pi}{4} + 6\pi = \frac{5\pi}{4}$, which is in QIII. The reference angle is $\beta' = \frac{5\pi}{4} - \pi = \frac{\pi}{4}$.

Self Check 10   Find the reference angle θ' for the given angle.

a. $\theta = -650°$ **b.** $\theta = \dfrac{27\pi}{8}$

Now Try Exercise 67.

If you look at the angles in Examples 9 and 10, do you notice a pattern for the angles that are multiples of the special angles? In Example 9(b), $\theta = -150°$, which is a multiple of 30°, has as its reference angle 30°. The pattern continues if you look at Examples 10(c) and 10(d). This is not a coincidence. We summarize this in Table 5. Assume that angle θ is not a quadrantal angle and that, if it is a fraction, it is reduced to lowest terms.

TABLE 5

Summary of the Reference Angles	
Angle θ	**Reference Angle θ'**
θ = multiple of 30°	$\theta' = 30°$
θ = multiple of $\dfrac{\pi}{6}$	$\theta' = \dfrac{\pi}{6}$
θ = multiple of 60°	$\theta' = 60°$
θ = multiple of $\dfrac{\pi}{3}$	$\theta' = \dfrac{\pi}{3}$
θ = multiple of 45°	$\theta' = 45°$
θ = multiple of $\dfrac{\pi}{4}$	$\theta' = \dfrac{\pi}{4}$

Using these reference angles will allow us to find the values of the trigonometric functions for any multiple of a special angle.

8. Find Exact Values of Trigonometric Functions Using Reference Angles

Since the six trigonometric functions of any angle θ are equal to those of the reference angle θ', except possibly for sign, we can use the reference angle to find the values of the six trigonometric functions of any multiple of any angle. We specifically use this fact for the special angles, since they are used so often and we can find exact values for the trigonometric functions of special angles. Notice in Example 11 that if we draw the angle, it is clear that if the angle is negative, we can find the reference angle using the absolute value of the angle.

EXAMPLE 11   **Using Reference Angles to Find Exact Values of Trigonometric Functions**

Find the exact value of each trigonometric function using a reference angle.

a. $\sin 210°$ **b.** $\cos(-225°)$ **c.** $\tan 420°$

d. $\csc\left(-\dfrac{4\pi}{3}\right)$ **e.** $\sec\dfrac{11\pi}{6}$ **f.** $\cot\left(-\dfrac{3\pi}{4}\right)$

SOLUTION   **a.** 210° lies in QIII and is a multiple of the special angle 30°, which is the reference angle. We know that $\sin 30° = \frac{1}{2}$, but the sine function is negative in QIII. Therefore, we determine that $\sin 210° = -\sin 30° = -\frac{1}{2}$.

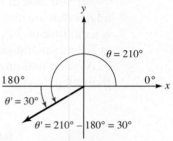

FIGURE 4-40(a)

b. –225° lies in QII and is a multiple of the special angle 45°. The cosine function is negative in QII. Since $\cos 45° = \frac{\sqrt{2}}{2}$, we can state that $\cos(-225°) = -\cos 45° = -\frac{\sqrt{2}}{2}$.

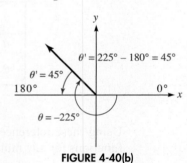

FIGURE 4-40(b)

c. 420° is a multiple of the special angle 60° and lies in QI. All trigonometric functions are positive in QI, so we can evaluate the function as $\tan 420° = \tan 60° = \sqrt{3}$.

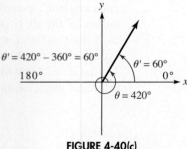

FIGURE 4-40(c)

d. $-\frac{4\pi}{3}$ lies in QII, where the cosecant function is positive. We know that the reference angle is the special angle $\frac{\pi}{3}$; therefore, $\csc\left(-\frac{4\pi}{3}\right) = \csc\frac{\pi}{3} = \frac{2\sqrt{3}}{3}$.

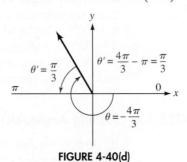

FIGURE 4-40(d)

e. $\frac{11\pi}{6}$ lies in QIV, where the secant function is positive. Therefore, using the reference angle $\frac{\pi}{6}$, we have $\sec\frac{11\pi}{6} = \sec\frac{\pi}{6} = \frac{2\sqrt{3}}{3}$.

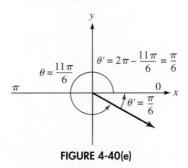

FIGURE 4-40(e)

f. The angle $-\frac{3\pi}{4}$ lies in QIII, where the cotangent function is positive. Using the reference angle $\frac{\pi}{4}$, we have $\cot\left(-\frac{3\pi}{4}\right) = \cot\frac{\pi}{4} = 1$.

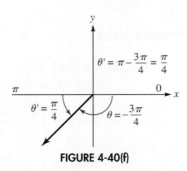

FIGURE 4-40(f)

Self Check 11   Find the exact value of each trigonometric function.

a. $\sin\left(-\frac{5\pi}{6}\right)$ **b.** $\cos\frac{7\pi}{4}$ **c.** $\tan 300°$

Now Try Exercise 83.

We have used special angles as reference angles to find values of trigonometric functions of specific angles, such as in Example 11. There are infinitely many angles with the same reference angle. For example, suppose we say that $\sin\theta = \frac{1}{2}$.

What is the value of θ? Looking at the special angles, we can say that $\theta = 30°$ or $\theta = \frac{\pi}{6}$. However, the sine function is also positive in QII, and, using $\theta = 30°$ as a reference angle, we could also say that $\theta = 150°$ because $\sin 150° = \frac{1}{2}$. In fact, there are infinitely many angles that would answer the question. We could say that $\theta = 30° + 360°n$, where n is any integer. This would be all the angles coterminal with $\theta = 30°$. In addition, we could say that $\sin (30° + 360°n) = \frac{1}{2}$, where n is any integer.

Whether you select one answer or all answers depends on the particular application. Often, the range of answers will be indicated.

EXAMPLE 12 **Finding Angles Using Reference Angles**

If $\sin \theta = -\frac{1}{2}$ and $\cos \theta = \frac{\sqrt{3}}{2}$, find one negative value for θ, $-360° < \theta \le 0°$, and one positive value for θ, $0° \le \theta < 360°$.

SOLUTION Because $\sin \theta$ is negative and $\cos \theta$ is positive, θ lies in QIV. Using what we have learned about special angles, we know that $\sin 30° = \frac{1}{2}$, so $\theta' = 30°$ is our reference angle. Looking at Figure 4-41, we can see that the two values for θ that fit the stated restrictions are $\theta = -30°$ and $\theta = 330°$.

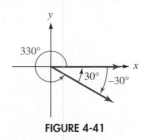

FIGURE 4-41

Self Check 12 If $\cos \theta = -\frac{1}{2}$ and $\tan \theta = \sqrt{3}$, find one negative value for θ, $-360° < \theta \le 0°$ and one positive value for θ, $0° \le \theta < 360°$.

Now Try Exercise 115.

Self Check Answers 1. a. $\dfrac{\sqrt{21}}{5}$ b. $\dfrac{2}{5}$ c. $\dfrac{\sqrt{21}}{2}$ d. $\dfrac{5\sqrt{21}}{21}$ e. $\dfrac{5}{2}$ f. $\dfrac{2\sqrt{21}}{21}$

2. $\sin (-\pi) = 0$ $\qquad\qquad$ $\cos (-\pi) = -1$ $\qquad\qquad$ $\tan (-\pi) = 0$
$\csc (-\pi)$ is undefined $\qquad$ $\sec (-\pi) = -1$ $\qquad$ $\cot (-\pi)$ is undefined

3. $\sin \left(-\dfrac{\pi}{4}\right) = -\dfrac{\sqrt{2}}{2}$ $\qquad$ $\cos \left(-\dfrac{\pi}{4}\right) = \dfrac{\sqrt{2}}{2}$ $\qquad$ $\tan \left(-\dfrac{\pi}{4}\right) = -1$

$\csc \left(-\dfrac{\pi}{4}\right) = -\sqrt{2}$ $\qquad$ $\sec \left(-\dfrac{\pi}{4}\right) = \sqrt{2}$ $\qquad$ $\cot \left(-\dfrac{\pi}{4}\right) = -1$

4. $\sin \left(-\dfrac{\pi}{3}\right) = -\dfrac{\sqrt{3}}{2}$ $\qquad$ $\cos \left(-\dfrac{\pi}{3}\right) = \dfrac{1}{2}$ $\qquad$ $\tan \left(-\dfrac{\pi}{3}\right) = -\sqrt{3}$

$\csc \left(-\dfrac{\pi}{3}\right) = -\dfrac{2\sqrt{3}}{3}$ $\qquad$ $\sec \left(-\dfrac{\pi}{3}\right) = 2$ $\qquad$ $\cot \left(-\dfrac{\pi}{3}\right) = -\dfrac{\sqrt{3}}{3}$

5. $\sin \left(-\dfrac{\pi}{6}\right) = -\dfrac{1}{2}$ $\qquad$ $\cos \left(-\dfrac{\pi}{6}\right) = \dfrac{\sqrt{3}}{2}$ $\qquad$ $\tan \left(-\dfrac{\pi}{6}\right) = -\dfrac{\sqrt{3}}{3}$

$\csc \left(-\dfrac{\pi}{6}\right) = -2$ $\qquad$ $\sec \left(-\dfrac{\pi}{6}\right) = \dfrac{2\sqrt{3}}{3}$ $\qquad$ $\cot \left(-\dfrac{\pi}{6}\right) = -\sqrt{3}$

6. a. $-\dfrac{\sqrt{3}}{2}$ b. $-\dfrac{\sqrt{2}}{2}$ c. $\dfrac{\sqrt{3}}{3}$ d. $-\sqrt{2}$ e. $\dfrac{2\sqrt{3}}{3}$ f. $-\dfrac{\sqrt{3}}{3}$

7. a. $-\dfrac{\sqrt{2}}{2}$ **b.** $\dfrac{1}{2}$ **8. a.** $\sqrt{3}$ **b.** undefined   **9. a.** $15°$ **b.** $\dfrac{\pi}{3}$

10. a. $70°$ **b.** $\dfrac{3\pi}{8}$ **11. a.** $-\dfrac{1}{2}$ **b.** $\dfrac{\sqrt{2}}{2}$ **c.** $-\sqrt{3}$

12. $-120°$ and $240°$

Exercises **4.2**

Getting Ready
You should be able to complete these vocabulary and concept statements before you proceed to the practice exercises.

Fill in the blanks.

1. The equation of the unit circle is ___.

2. If θ is a multiple of $30°$ and lies in QIII, then $\sin\theta =$ ___.

3. In QIV, the two trigonometric functions that are positive are ___ and ___.

4. If the point $P(x, y)$ is a point on the unit circle that corresponds to a real number t, with $x \neq 0$ and $y \neq 0$, then

   a. $\sin t =$ ___   **b.** $\cos t =$ ___   **c.** $\tan t =$ ___

   d. $\csc t =$ ___   **e.** $\sec t =$ ___   **f.** $\cot t =$ ___

5. Three angles that are often referred to as special angles are ___ , ___ , and ___.

6. The period of the sine and cosine functions is ___.

7. The period of the tangent and cotangent functions is ___.

8. A reference angle is always a(n) ___ angle.

Practice
The point $P = (x, y)$ is a point on the unit circle that corresponds to a real number t. Find each of the following:

   (a) $\sin t$ *(b)* $\cos t$ *(c)* $\tan t$
   (d) $\csc t$ *(e)* $\sec t$ *(f)* $\cot t$

9. $\left(\dfrac{1}{2}, \dfrac{\sqrt{3}}{2}\right)$ **10.** $\left(\dfrac{1}{4}, \dfrac{\sqrt{15}}{4}\right)$

11. $\left(-\dfrac{1}{3}, -\dfrac{2\sqrt{2}}{3}\right)$ **12.** $\left(-\dfrac{1}{5}, -\dfrac{2\sqrt{6}}{5}\right)$

13. $\left(\dfrac{\sqrt{3}}{3}, -\dfrac{\sqrt{6}}{3}\right)$ **14.** $\left(\dfrac{\sqrt{2}}{2}, -\dfrac{\sqrt{2}}{2}\right)$

15. $\left(\dfrac{\sqrt{33}}{7}, -\dfrac{4}{7}\right)$ **16.** $\left(-\dfrac{5}{6}, \dfrac{\sqrt{11}}{6}\right)$

Use the coordinates of the point P on the unit circle that corresponds to the real number t to find: (a) $\sin t$, (b) $\cos t$, (c) $\tan t$ or state that the value does not exist.

17. $t = \dfrac{3\pi}{2}$ **18.** $t = -\dfrac{\pi}{2}$

19. $t = -2\pi$ **20.** $t = 2\pi$

21. $t = 3\pi$ **22.** $t = 4\pi$

23. $t = -\dfrac{7\pi}{2}$ **24.** $t = \dfrac{9\pi}{2}$

25. $t = \dfrac{\pi}{4}$ **26.** $t = -\dfrac{\pi}{4}$

27. $t = -\dfrac{3\pi}{4}$ **28.** $t = \dfrac{\pi}{3}$

29. $t = \dfrac{4\pi}{3}$ **30.** $t = -\dfrac{2\pi}{3}$

31. $t = -\dfrac{\pi}{6}$ **32.** $t = \dfrac{5\pi}{6}$

33. $t = \dfrac{11\pi}{3}$ **34.** $t = \dfrac{17\pi}{6}$

Use the unit circle to find the values of the six trigonometric functions for each angle.

35. $\theta = \dfrac{3\pi}{4}$ **36.** $\theta = \dfrac{5\pi}{4}$

37. $\theta = \dfrac{2\pi}{3}$ **38.** $\theta = \dfrac{4\pi}{3}$

39. $\theta = \dfrac{5\pi}{6}$ **40.** $\theta = \dfrac{7\pi}{6}$

Use the unit circle to find the exact value of each function.

41. $\cos\dfrac{7\pi}{4}$ **42.** $\sin\dfrac{11\pi}{6}$

43. $\tan 210°$ **44.** $\sec 135°$

45. $\cot\dfrac{5\pi}{3}$ **46.** $\csc 240°$

Use the periodic properties of the functions to find the exact value of each trigonometric function. If a function is undefined, state so.

47. $\sin \dfrac{11\pi}{6}$

48. $\cos \dfrac{5\pi}{6}$

49. $\cot \left(-\dfrac{3\pi}{4}\right)$

50. $\sin \left(-\dfrac{11\pi}{4}\right)$

51. $\sin 150°$

52. $\cos 120°$

53. $\tan (-135°)$

54. $\cot (-270°)$

55. $\tan \dfrac{5\pi}{3}$

56. $\csc \dfrac{4\pi}{3}$

57. $\cos 210°$

58. $\csc 300°$

Find the reference angle for each angle. Round to three decimal places when necessary.

59. $240°$

60. $135°$

61. $\dfrac{7\pi}{4}$

62. $\dfrac{5\pi}{4}$

63. $-141°$

64. $-215°$

65. $-\dfrac{4\pi}{9}$

66. $-\dfrac{3\pi}{4}$

67. $\dfrac{31\pi}{6}$

68. $\dfrac{19\pi}{6}$

69. $-\dfrac{22\pi}{3}$

70. $-\dfrac{41\pi}{3}$

71. $619°$

72. $421°$

73. $-470°$

74. $-530°$

75. 3.4

76. 6.9

77. -2.6

78. -4.3

Use a reference angle to find the exact value of each trigonometric function. If a function is undefined, state so.

79. $\sin (-510°)$

80. $\cos (-600°)$

81. $\tan \dfrac{5\pi}{2}$

82. $\cot 3\pi$

83. $\cos \left(-\dfrac{4\pi}{3}\right)$

84. $\sin \left(-\dfrac{7\pi}{3}\right)$

85. $\cot 750°$

86. $\sec 480°$

87. $\csc \dfrac{3\pi}{2}$

88. $\sec 5\pi$

89. $\sec 315°$

90. $\csc 630°$

91. $\sin (-3\pi)$

92. $\cos 5\pi$

93. $\cos \left(-\dfrac{7\pi}{6}\right)$

94. $\sin \left(-\dfrac{13\pi}{6}\right)$

95. $\sec \dfrac{7\pi}{4}$

96. $\csc \dfrac{13\pi}{4}$

Use the given information to find all exact values of α, where possible, if $0 \le \alpha < 2\pi$.

97. $\sin \alpha = \dfrac{1}{2}$

98. $\cos \alpha = -\dfrac{\sqrt{3}}{2}$

99. $\tan \alpha = \dfrac{\sqrt{3}}{3}$

100. $\cot \alpha = -\sqrt{3}$

101. $\sec \alpha = -2$

102. $\csc \alpha = \dfrac{2\sqrt{3}}{3}$

103. $\cos \alpha = \dfrac{1}{2}$

104. $\sin \alpha = -\dfrac{\sqrt{3}}{2}$

105. $\cot \alpha = -1$

106. $\tan \alpha = 1$

107. $\sin \alpha \cos \alpha = \dfrac{\sqrt{3}}{4}$

108. $\sin \alpha \cos \alpha = -\dfrac{\sqrt{3}}{4}$

Name the quadrant in which angle θ lies. Assume θ is positive and in standard position.

109. $\sin \theta > 0; \tan \theta < 0$

110. $\sin \theta < 0; \sec \theta < 0$

111. $\cos \theta > 0; \sin \theta > 0$

112. $\tan \theta < 0; \cos \theta > 0$

113. $\tan \theta < 0; \csc \theta < 0$

114. $\cos \theta < 0; \cot \theta > 0$

Use the given information to find one negative value for θ, $-360° < \theta \le 0°$, and one positive value for θ, $0° \le \theta < 360°$.

115. $\tan \theta = \dfrac{\sqrt{3}}{3}; \sin \theta = \dfrac{1}{2}$

116. $\tan \theta = -1; \cos \theta = -\dfrac{\sqrt{2}}{2}$

117. $\tan \theta = -\sqrt{3}; \cos \theta = \dfrac{1}{2}$

118. $\cot \theta = \sqrt{3}; \cos \theta = -\dfrac{\sqrt{3}}{2}$

119. $\sec \theta = -\dfrac{2\sqrt{3}}{3}; \sin \theta = -\dfrac{1}{2}$

120. $\cos \theta = \dfrac{\sqrt{3}}{2}; \csc \theta = 2$

121. $\tan \theta = -1; \sec \theta = \sqrt{2}$

122. $\tan \theta = -\sqrt{3}; \cos \theta = -\dfrac{1}{2}$

123. $\sec \theta = -\sqrt{2}; \cot \theta = -1$

124. $\csc \theta = -\sqrt{2}; \cot \theta = -1$

125. $\sin \theta = -1$

126. $\cos \theta = -1$

127. $\tan \theta$ is undefined; $\sin \theta = 1$

128. $\sin \theta = -\dfrac{\sqrt{3}}{2}; \cos \theta = \dfrac{1}{2}$

Applications

129. Temperature in Baton Rouge The monthly average high temperature in Baton Rouge is modeled by the circular function

$$T(x) = 17 \sin\left(\frac{\pi}{7}(x - 4)\right) + 76,$$

where $T(x)$ is the temperature in degrees Fahrenheit in month x and $x = 1$ corresponds to January and $x = 12$ corresponds to December. Find the average high temperatures in Baton Rouge for the following months. Round to the nearest degree.

  a. February   **b.** June   **c.** November

130. Temperature in Boston The monthly average low temperature in Boston is modeled by the circular function

$$T(x) = 21 \sin\frac{\pi}{6}(x - 4) + 43,$$

where $T(x)$ is the temperature in degrees Fahrenheit in month x and $x = 1$ corresponds to January and $x = 12$ corresponds to December. If you plan to visit Boston and attend a Red Sox game in May, what average low temperature can you expect the night of the game? Round to the nearest degree.

131. Oscillating spring The motion of a weight suspended by a spring is given by

$y(t) = \dfrac{1}{2}\cos\left(\dfrac{\pi}{4}t\right)$, where y represents the

displacement of the weight in inches from equilibrium and t represents the time in seconds. Find the position of the weight after 7 seconds.

132. Oscillating spring The motion of a weight suspended by a spring is given by

$y(t) = \dfrac{1}{8}\cos\left(\dfrac{\pi}{2}t\right)$, where y represents the

displacement of the weight in inches from equilibrium and t represents the time in seconds. At what times will the weight be at equilibrium?

133. Straight-line distance Find the straight-line distance between the two points on the unit circle that correspond to the real numbers $\frac{\pi}{3}$ and $\frac{\pi}{2}$. Round to three decimal places.

134. Distance along a circle Find the distance measured along the unit circle between the two points that correspond to the real numbers $\frac{\pi}{3}$ and $\frac{\pi}{2}$. Round to three decimal places.

Discovery and Writing

135. Show that the area of a sector of the unit circle is given by the formula $A = \frac{1}{2}\theta$, where θ is the radian measure of the central angle of the sector.

Use this figure to complete Exercises 136–138.

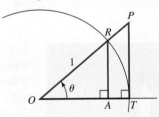

136. Show that the area of triangle OAR is given by the formula $A = \frac{1}{2}\cos\theta\sin\theta$.

137. Show that the area of triangle OTP is given by the formula $A = \frac{1}{2}\tan\theta$.

138. Use the inequality

  area of $\triangle OAR \le$ area of sector $OTR \le$ area of $\triangle OTP$

to show that $\displaystyle\lim_{\theta \to 0}\frac{\sin\theta}{\theta} = 1$.

139. List examples of real-life data that repeat over time and therefore can be modeled using circular functions.

140. Find all angles θ for which $\tan\theta$ and $\cot\theta$ are equivalent.

4.3 Trigonometric Functions of Any Angle; Fundamental Identities

In this section, we will learn to

1. Define the domain and range of the trigonometric functions.
2. Use the Even–Odd Properties to simplify and evaluate a trigonometric function.
3. Determine the trigonometric-function values of any angle in standard position.
4. Use the fundamental trigonometric identities to determine trigonometric-function values and simplify expressions.
5. Approximate trigonometric-function values with a calculator.
6. Solve right triangles using the right-triangle definitions.
7. Understand the relationship between complementary angles and cofunctions.

The Washington Monument is located on the west end of the National Mall in Washington, DC, and was built to commemorate the first U.S. president, George Washington. It is the tallest structure in Washington, DC, at approximately 556 feet tall.

Trigonometry can be used to determine the heights of tall structures. In this section we will learn the right-triangle definitions of the trigonometric functions and apply the definitions to solve right-triangle problems.

In our work with trigonometric functions, it is often necessary to know the domain and range, so we will now define the domain and range of the trigonometric functions.

1. Define the Domain and Range of the Trigonometric Functions

The domain of a function defined by the equation $y = f(x)$ is the set of all values of x for which the function is defined, and the range is the set of all values of y. For virtually all of the functions previously studied, both the domain and the range have been subsets of the real numbers. The domain of the sine and cosine functions is the set of all real numbers, because any real number t can be measured by moving around the unit circle, and for any point P on the unit circle, $P(x, y) = P(\cos t, \sin t)$. To determine the range of these two functions, we look at the unit circle. The values of x and y are in the interval $[-1, 1]$; since $x = \cos t$ and $y = \sin t$, the values of these functions lie in the interval $[-1, 1]$. We can see this illustrated in Figure 4-42.

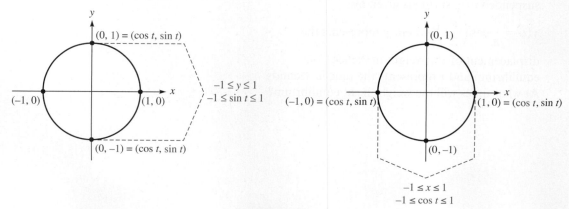

FIGURE 4-42

Domain and Range of the Sine and Cosine Functions

The **domain of the sine and cosine functions** is the set of all real numbers, $D = (-\infty, \infty)$.

The **range of the sine and cosine functions** is all real numbers in the interval $R = [-1, 1]$:

$$-1 \leq \sin t \leq 1 \text{ and } -1 \leq \cos t \leq 1.$$

The definition of the cosecant function is $\csc t = \frac{1}{y}, y \neq 0$. Since $y = 0$ occurs when the terminal side of the angle is on the x-axis, the cosecant function is not defined for integer multiples of π. These are also the points where $\sin t = 0$.

The secant function is defined as $\sec t = \frac{1}{x}, x \neq 0$. The value of x is 0 on the y-axis; therefore, when the terminal side of the angle is on the y-axis, $\sec t$ is not defined.

The range of the cosecant function is found by these calculations:

$-1 \leq \sin t \leq 1$	Range of the sine function
$\|\sin t\| \leq 1$	Absolute-value inequality
$\dfrac{1}{\|\sin t\|} \geq 1$	Algebraic Rule: $a \leq b \Leftrightarrow \dfrac{1}{a} \geq \dfrac{1}{b}$
$\|\csc t\| \geq 1$	$\dfrac{1}{\|\sin t\|} = \left\|\dfrac{1}{y}\right\| = \|\csc t\|$
$\csc t \leq -1 \text{ or } \csc t \geq 1$	Definition of an absolute-value inequality

From this we see that the range of the cosecant function is $R = (-\infty, -1] \cup [1, \infty)$. The range of the secant function can be found the same way.

Domain and Range of the Secant and Cosecant Functions

The **domain of the secant function** is the set of all real numbers except for odd-integer multiples of $\dfrac{\pi}{2}$.

The **domain of the cosecant function** is the set of all real numbers except for integer multiples of π.

The **range of the secant and cosecant functions** is $R = (-\infty, -1] \cup [1, \infty)$:

$$\csc t \leq -1 \text{ or } \csc t \geq 1 \text{ and } \sec t \leq -1 \text{ or } \sec t \geq 1.$$

The tangent function is defined as $\tan t = \frac{y}{x}, x \neq 0$; therefore, the domain of the tangent function is the same as that of the secant function. The cotangent function has the same domain as the cosecant function, because $\cot t = \frac{x}{y}, y \neq 0$.

The range of both the tangent and cotangent functions is the set of all real numbers. Recall the asymptotic behavior demonstrated by rational functions such as $f(x) = \frac{1}{x}$. As the value of the denominator moves to 0, the function demonstrates unbounded behavior. Therefore, the range for the tangent and cotangent functions is the set of all real numbers.

Domain and Range of the Tangent and Cotangent Functions

The **domain of the tangent function** is the set of all real numbers except for odd-integer multiples of $\dfrac{\pi}{2}$.

The **domain of the cotangent function** is the set of all real numbers except for integer multiples of π.

The **range of the tangent and cotangent functions** is the set of all real numbers, $R = (-\infty, \infty)$.

2. Use the Even–Odd Properties to Simplify and Evaluate a Trigonometric Function

In the last section, we evaluated the values of trigonometric functions of multiples of special angles. Several of the functions that were evaluated involved negative numbers. We will now discover the properties that allow us to rewrite the function without having to use a negative angle.

In algebra, an **even function** is defined as a function for which $f(-x) = f(x)$ for all values of x in the domain. An **odd function** is a function for which $f(-x) = -f(x)$ for all values of x in the domain. Look at Figure 4-43.

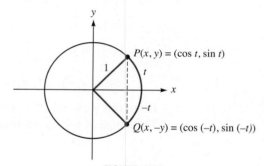

FIGURE 4-43

Figure 4-43 demonstrates the following:

$$\left.\begin{array}{l} \sin(-t) = -y \\ \sin t = y \end{array}\right\} \rightarrow \sin(-t) = -\sin t \qquad \text{Sine is an odd function.}$$

$$\left.\begin{array}{l} \cos(-t) = x \\ \cos t = x \end{array}\right\} \rightarrow \cos(-t) = \cos t \qquad \text{Cosine is an even function.}$$

$$\left.\begin{array}{l} \tan(-t) = -\dfrac{y}{x} \\ \tan t = \dfrac{y}{x} \end{array}\right\} \rightarrow \tan(-t) = -\tan t \qquad \text{Tangent is an odd function.}$$

The reciprocal functions have the same properties.

Cosecant is an odd function.

Secant is an even function.

Cotangent is an odd function.

Even–Odd Properties of Trigonometric Functions

$\sin(-t) = -\sin t$	$\csc(-t) = -\csc t$
$\cos(-t) = \cos t$	$\sec(-t) = \sec t$
$\tan(-t) = -\tan t$	$\cot(-t) = -\cot t$

These properties can be very useful in numerous applications, including graphing trigonometric functions and solving trigonometric equations.

EXAMPLE 1   **Using the Even–Odd Properties**

Use the Even–Odd Properties to simplify and then evaluate each function.

a. $\sin\left(-\dfrac{5\pi}{3}\right)$ b. $\cos\left(-\dfrac{9\pi}{4}\right)$ c. $\tan\left(-\dfrac{17\pi}{6}\right)$

SOLUTION   a. The reference angle is $\dfrac{\pi}{3}$, and $\dfrac{5\pi}{3}$ is in QIV, where $\sin\theta < 0$.

$$\sin\left(-\frac{5\pi}{3}\right) = -\sin\left(\frac{5\pi}{3}\right) = -\left(-\frac{\sqrt{3}}{2}\right) = \frac{\sqrt{3}}{2}$$

b. The reference angle is $\dfrac{\pi}{4}$, and $\dfrac{9\pi}{4}$ is in QI, where $\cos\theta > 0$.

$$\cos\left(-\frac{9\pi}{4}\right) = \cos\left(\frac{9\pi}{4}\right) = \frac{\sqrt{2}}{2}$$

c. The reference angle is $\dfrac{\pi}{6}$, and $\dfrac{17\pi}{6}$ is in QII, where $\tan\theta < 0$.

$$\tan\left(-\frac{17\pi}{6}\right) = -\tan\left(\frac{17\pi}{6}\right) = -\left(-\frac{\sqrt{3}}{3}\right) = \frac{\sqrt{3}}{3}$$

Self Check 1   Use the Even–Odd Properties to simplify and then evaluate each function.

a. $\sin\left(-\dfrac{7\pi}{6}\right)$ b. $\cos\left(-\dfrac{10\pi}{3}\right)$ c. $\tan\left(-\dfrac{15\pi}{4}\right)$

Now Try Exercise 7.

3. Determine the Trigonometric-Function Values of Any Angle in Standard Position

Suppose we know the coordinates of a point on the terminal side of angle θ in standard position, but that point is not on the unit circle. Can we define the trigonometric functions for angle θ? In Figure 4-44, the inner circle is the unit circle. The outer circle has center at the origin, and the equation $x^2 + y^2 = r^2$.

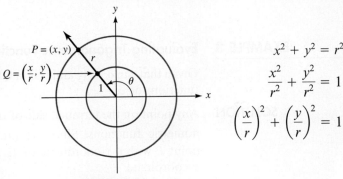

FIGURE 4-44

Since the point $Q = \left(\dfrac{x}{r}, \dfrac{y}{r}\right)$ is a point on the unit circle, and is a point on the terminal side of the angle, we can define the six trigonometric functions for angle θ using the coordinates of any point $P = (x, y)$ on the terminal side of the angle.

Trigonometric Functions of Any Angle

Let θ be an angle in standard position, as shown in Figure 4-44, and $P = (x, y)$ be any point (except the origin) on the terminal side of angle θ and $r = \sqrt{x^2 + y^2}$. The **six trigonometric functions of θ** are:

$$\sin \theta = \frac{y}{r} \qquad\qquad \csc \theta = \frac{r}{y}, y \neq 0$$

$$\cos \theta = \frac{x}{r} \qquad\qquad \sec \theta = \frac{r}{x}, x \neq 0$$

$$\tan \theta = \frac{y}{x}, x \neq 0 \qquad \cot \theta = \frac{x}{y}, y \neq 0$$

EXAMPLE 2   Finding the Values of the Trigonometric Functions of an Angle

The point $P(-3, 4)$ is on the terminal side of an angle θ. Draw the angle in standard position and calculate the exact values of the six trigonometric functions of θ.

SOLUTION   From $P(-3, 4)$ we determine that $x = -3$, $y = 4$, and $r = \sqrt{(-3)^2 + 4^2} = \sqrt{25} = 5$.

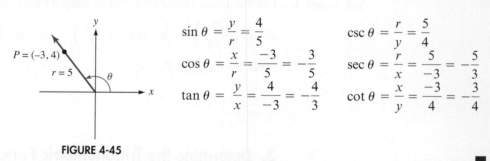

$$\sin \theta = \frac{y}{r} = \frac{4}{5} \qquad\qquad \csc \theta = \frac{r}{y} = \frac{5}{4}$$

$$\cos \theta = \frac{x}{r} = \frac{-3}{5} = -\frac{3}{5} \qquad \sec \theta = \frac{r}{x} = \frac{5}{-3} = -\frac{5}{3}$$

$$\tan \theta = \frac{y}{x} = \frac{4}{-3} = -\frac{4}{3} \qquad \cot \theta = \frac{x}{y} = \frac{-3}{4} = -\frac{3}{4}$$

FIGURE 4-45

Self Check 2   The point $P(4, -5)$ is on the terminal side of an angle θ. Calculate the exact values of the six trigonometric functions of θ.

Now Try Exercise 21.

EXAMPLE 3   Evaluating Trigonometric Functions of an Angle

Given that $\sin \alpha = \frac{12}{13}$ and α is in QII, find the values of the other five trigonometric functions of α.

SOLUTION   Any point on the terminal side of α can be used to compute the value of the trigonometric functions. Since we are given $\sin \alpha = \frac{12}{13}$ and $\sin \alpha = \frac{y}{r}$, we choose a point P with a y-coordinate of 12 and $r = 13$. Now we will find the value of the x-coordinate.

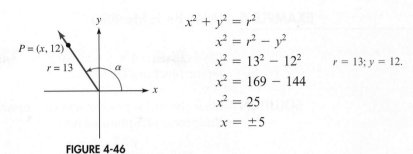

$$x^2 + y^2 = r^2$$
$$x^2 = r^2 - y^2$$
$$x^2 = 13^2 - 12^2 \qquad r = 13; y = 12.$$
$$x^2 = 169 - 144$$
$$x^2 = 25$$
$$x = \pm 5$$

FIGURE 4-46

Since the angle terminates in QII, the x-coordinate of point P is negative, so $x = -5$, and we determine the point is $P(-5, 12)$. We find the values of the other trigonometric functions by substituting the appropriate values $x = -5$, $y = 12$, and $r = 13$.

$$\cos \alpha = \frac{x}{r} = \frac{-5}{13} = -\frac{5}{13} \qquad \tan \alpha = \frac{y}{x} = \frac{12}{-5} = -\frac{12}{5} \qquad \csc \alpha = \frac{r}{y} = \frac{13}{12}$$

$$\sec \alpha = \frac{r}{x} = \frac{13}{-5} = -\frac{13}{5} \qquad \cot \alpha = \frac{x}{y} = \frac{-5}{12} = -\frac{5}{12}$$

Self Check 3   Given that $\cos \beta = -\dfrac{3}{4}$ and β is not in QII, find $\sin \beta$ and $\tan \beta$.

Now Try Exercise 31.

4. Use the Fundamental Trigonometric Identities to Determine Trigonometric-Function Values and Simplify Expressions

If a point $P = (x, y)$ is on the terminal side of an angle θ, we have defined $\sin \theta = \frac{y}{r}$, $r = \sqrt{x^2 + y^2}$. We also define $\csc \theta = \frac{r}{y}$, $y \neq 0$. Since these two functions are reciprocals of one another, as are cosine and secant, and tangent and cotangent, we can define the **Reciprocal Identities.**

Reciprocal Identities   For any angle θ for which the function is defined,

$$\sin \theta = \frac{1}{\csc \theta} \qquad \csc \theta = \frac{1}{\sin \theta} \qquad \sin \theta \cdot \csc \theta = 1$$

$$\cos \theta = \frac{1}{\sec \theta} \qquad \sec \theta = \frac{1}{\cos \theta} \qquad \cos \theta \cdot \sec \theta = 1$$

$$\tan \theta = \frac{1}{\cot \theta} \qquad \cot \theta = \frac{1}{\tan \theta} \qquad \tan \theta \cdot \cot \theta = 1$$

From definitions, we can also develop two other very useful identities, the Quotient Identities.

Quotient Identities   For any angle θ for which the function is defined,

$$\tan \theta = \frac{\sin \theta}{\cos \theta} \quad \text{and} \quad \cot \theta = \frac{\cos \theta}{\sin \theta}.$$

Note that the tangent, cotangent, cosecant, and secant functions can be written in terms of sine and cosine. If you know the value of the sine and cosine for an angle, then you can determine the values of the remaining trigonometric functions.

EXAMPLE 4 **Using Basic Identities**

Given that $\sin\theta = \frac{2}{3}$ and $\cos\theta = -\frac{\sqrt{5}}{3}$, find the values of the remaining trigonometric functions for θ.

SOLUTION Since the sine is positive and the cosine is negative, the angle lies in QII. Using the Reciprocal Identities, we have

$$\csc\theta = \frac{1}{\sin\theta} = \frac{1}{\dfrac{2}{3}} = \frac{3}{2}$$

and

$$\sec\theta = \frac{1}{\cos\theta} = \frac{1}{-\dfrac{\sqrt{5}}{3}} = -\frac{3}{\sqrt{5}} = -\frac{3\sqrt{5}}{5}.$$

Now we will use the Quotient Identities to find $\tan\theta$.

$$\tan\theta = \frac{\sin\theta}{\cos\theta} = \frac{\dfrac{2}{3}}{-\dfrac{\sqrt{5}}{3}} = -\frac{2}{\sqrt{5}} = -\frac{2\sqrt{5}}{5}$$

Finally, we will use the Reciprocal Identity to find $\cot\theta$.

$$\cot\theta = \frac{1}{\tan\theta} = -\frac{\sqrt{5}}{2}$$

Self Check 4 Given that $\sin\theta = \frac{\sqrt{7}}{4}$ and $\cos\theta = \frac{3}{4}$, find the values of the remaining trigonometric functions for θ.

Now Try Exercise 47.

For any point on the unit circle, with $x^2 + y^2 = 1$, $\sin\theta = y$ and $\cos\theta = x$. If we substitute these values into the equation of the unit circle, we have $(\sin\theta)^2 + (\cos\theta)^2 = 1$. However, in trigonometry, $(\sin\theta)^2 = \sin^2\theta$. The same is true for all of the trigonometric functions. Thus we rewrite the identity, known as a **Pythagorean Identity**, as $\sin^2\theta + \cos^2\theta = 1$.

From this identity, we will now develop two more Pythagorean Identities.

$$\sin^2\theta + \cos^2\theta = 1$$

$$\frac{\sin^2\theta}{\cos^2\theta} + \frac{\cos^2\theta}{\cos^2\theta} = \frac{1}{\cos^2\theta} \qquad \text{Divide both sides by } \cos^2\theta \text{ and simplify using the Reciprocal and Quotient Identities.}$$

$$\tan^2\theta + 1 = \sec^2\theta$$

Likewise,

$$\sin^2\theta + \cos^2\theta = 1$$

$$\frac{\sin^2\theta}{\sin^2\theta} + \frac{\cos^2\theta}{\sin^2\theta} = \frac{1}{\sin^2\theta}$$ Divide both sides by $\sin^2\theta$ and simplify using the Reciprocal and Quotient Identities.

$$\mathbf{1 + \cot^2\theta = \csc^2\theta}$$

These identities are summarized as follows:

Pythagorean Identities $\sin^2\theta + \cos^2\theta = 1$ $\tan^2\theta + 1 = \sec^2\theta$ $1 + \cot^2\theta = \csc^2\theta$

Comment   These identities are used extensively in many types of problems. They establish relationships between the functions, and, in other forms, can be modified for substitutions and used in other applications. For example, $\sin^2\theta + \cos^2\theta = 1$ can be written as $\cos^2\theta = 1 - \sin^2\theta = (1 - \sin\theta)(1 + \sin\theta)$. It is important that you learn these identities.

EXAMPLE 5   **Using Identities to Simplify Expressions**

Simplify the expression $\tan^2\theta\csc^2\theta - \tan^2\theta$.

SOLUTION   Using the Pythagorean Identities, we proceed as follows:

$$\tan^2\theta\csc^2\theta - \tan^2\theta = \tan^2\theta(\csc^2\theta - 1)$$ Factor out $\tan^2\theta$.

$$= \tan^2\theta(\cot^2\theta)$$ $\cot^2\theta + 1 = \csc^2\theta \Rightarrow \cot^2\theta = \csc^2\theta - 1$

$$= \tan^2\theta\left(\frac{1}{\tan^2\theta}\right)$$ Use the Reciprocal Identity: $\cot^2\theta = \frac{1}{\tan^2\theta}$.

$$= 1$$

Self Check 5   Simplify the expression $\tan^2\theta(1 - \sin\theta)(1 + \sin\theta)$.

Now Try Exercise 53.

EXAMPLE 6   **Using Identities to Simplify Expressions**

Given that $\sin\theta = \frac{4}{5}$ for an acute angle θ, use identities to find the values of the remaining five trigonometric functions for θ.

SOLUTION   Since θ is an acute angle, it lies in Q1. Because the sine and cosecant functions are reciprocals,

$$\csc\theta = \frac{1}{\sin\theta} = \frac{1}{\frac{4}{5}} = \frac{5}{4}.$$

Since $\sin^2\theta + \cos^2\theta = 1$,

$$\cos^2\theta = 1 - \sin^2\theta$$

$$\cos\theta = \sqrt{1 - \sin^2\theta} \qquad \text{Since } \theta \text{ is acute, } \cos\theta > 0.$$

$$= \sqrt{1 - \left(\frac{4}{5}\right)^2} = \sqrt{\frac{9}{25}} = \frac{3}{5}$$

Now that we have values for both the sine and cosine functions, we can easily find the values of the secant, tangent, and cotangent functions as follows:

$$\sec\theta = \frac{1}{\cos\theta} = \frac{1}{\dfrac{3}{5}} = \frac{5}{3}$$

$$\tan\theta = \frac{\sin\theta}{\cos\theta} = \frac{\dfrac{4}{5}}{\dfrac{3}{5}} = \frac{4}{3}$$

$$\cot\theta = \frac{1}{\tan\theta} = \frac{1}{\dfrac{4}{3}} = \frac{3}{4}$$

Self Check 6 Given, that $\cos\theta = \frac{5}{13}$ for an acute angle θ, use identities to find the values of the remaining five trigonometric functions for θ.

Now Try Exercise 67.

5. Approximate Trigonometric-Function Values with a Calculator

Before scientific and graphing calculators became available, trigonometric tables were used to approximate trigonometric-function values such as sin 39° and tan $\frac{\pi}{5}$. Calculators make it possible to approximate these values very quickly. Your calculator has three trigonometric function keys, SIN, COS, and TAN. For the cosecant, secant, and cotangent functions, use the Reciprocal Identities: for example, sec 40° is found by taking $\frac{1}{\cos 40°}$.

Additionally, your calculator has at least two **modes,** degree and radian, that must be set correctly when approximating the function. (Some scientific calculators have a gradian mode, which is sometimes used in surveying and in a few countries.)

EXAMPLE 7 **Approximating Trigonometric Functions with a Calculator**

Use a calculator to approximate each of the following functions, rounding to three decimals:

a. $\sin 23°$ **b.** $\cos 54.75°$ **c.** $\tan\dfrac{2\pi}{5}$ **d.** $\csc 1.3$

SOLUTION For parts (a) and (b), set the calculator to degree mode. A graphing-calculator screen is shown in Figure 4-47, with the mode setting and the steps to follow.

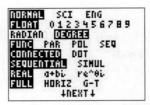
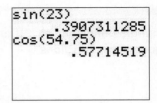

FIGURE 4-47

$\sin 23° \approx 0.391$ and $\cos 54.75° \approx 0.577$.

The functions in parts (c) and (d) are to be approximated in radian mode, as shown in Figure 4-48. Since there is no key for the cosecant function, the reciprocal function $\csc \theta = \dfrac{1}{\sin \theta}$ is used.

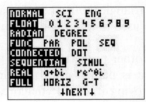
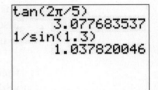

FIGURE 4-48

$\tan \dfrac{2\pi}{5} \approx 3.078$ and $\csc 1.3 \approx 1.038$.

Self Check 7   Approximate each function, rounding to three decimal places.
a. $\sin 0.75$ **b.** $\cot 59°$

Now Try Exercise 73.

6. Solve Right Triangles Using the Right-Triangle Definitions

Because trigonometry gives us the ability to determine all sides and all angles of a right triangle when only some are known, it is important in astronomy, navigation, and surveying.

Many applications involve right triangles. In addition, we will see in future sections that right triangles can be very useful in a wide array of problems. Therefore, we will now view the definitions of the trigonometric functions in a different way by using a right triangle. To develop these new definitions, we place right triangle ABC on a coordinate system, as in Figure 4-49. We call this the reference triangle for the acute angle A. We let a be the length of BC, the side opposite angle A. Then we let b be the length of AC, the side adjacent to angle A, and we let the hypotenuse have length c. The six trigonometric functions of the acute angle A can be defined as ratios involving sides of the right triangle ABC.

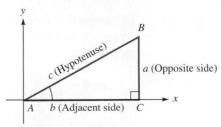

FIGURE 4-49

Right-Triangle Definitions of
Trigonometric Functions

If angle A is an acute angle in the right triangle ABC (Figure 4-49), then

$$\sin A = \frac{\text{opposite side}}{\text{hypotenuse}} = \frac{a}{c} \qquad \csc A = \frac{\text{hypotenuse}}{\text{opposite side}} = \frac{c}{a}$$

$$\cos A = \frac{\text{adjacent side}}{\text{hypotenuse}} = \frac{b}{c} \qquad \sec A = \frac{\text{hypotenuse}}{\text{adjacent side}} = \frac{c}{b}$$

$$\tan A = \frac{\text{opposite side}}{\text{adjacent side}} = \frac{a}{b} \qquad \cot A = \frac{\text{adjacent side}}{\text{opposite side}} = \frac{b}{a}$$

The same identities defined earlier in this section apply to these functions.

The total measure of the three angles in any triangle is 180°. Since one angle in a right triangle is 90°, the other angles are acute and total 90°. Radian measure can also be used in a triangle, although in many triangle applications, degree measure is used.

Another property of right triangles is called the **Pythagorean Theorem.** This theorem is named after the Greek mathematician Pythagoras, who is credited with proving it. The Pythagorean Theorem states that in any right triangle, the square of the hypotenuse is equal to the sum of the squares of the other two sides: $a^2 + b^2 = c^2$. From this we have

$$a^2 = c^2 - b^2 \rightarrow a = \sqrt{c^2 - b^2},$$
$$b^2 = c^2 - a^2 \rightarrow b = \sqrt{c^2 - a^2}, \text{ and}$$
$$c^2 = a^2 + b^2 \rightarrow c = \sqrt{a^2 + b^2}.$$

Caution

> $c = \sqrt{a^2 + b^2} \neq a + b$. Note that because a, b, and c represent distances, there is no need for $\pm$ to precede each radical. This is a common algebraic mistake that students make.

EXAMPLE 8   Finding the Values of Trigonometric Functions

The two legs in a right triangle have measure 8 inches and 15 inches. Find the values of the six trigonometric functions for angle θ, indicated in Figure 4-50.

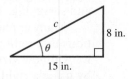

FIGURE 4-50

SOLUTION   Our first step is to find the length of the hypotenuse. We will use the Pythagorean Theorem.

$$c^2 = 8^2 + 15^2 = 64 + 225 = 289$$
$$c = \sqrt{289} = 17$$

For angle θ, the opposite side has measure 8 inches, the adjacent side has measure 15 inches, and the hypotenuse is 17 inches. Using the definitions of the six trigonometric functions, we have the following function values:

$$\sin \theta = \frac{\text{opposite}}{\text{hypotenuse}} = \frac{8}{17} \qquad \csc \theta = \frac{\text{hypotenuse}}{\text{opposite}} = \frac{17}{8}$$

$$\cos \theta = \frac{\text{adjacent}}{\text{hypotenuse}} = \frac{15}{17} \qquad \sec \theta = \frac{\text{hypotenuse}}{\text{adjacent}} = \frac{17}{15}$$

$$\tan \theta = \frac{\text{opposite}}{\text{adjacent}} = \frac{8}{15} \qquad \cot \theta = \frac{\text{adjacent}}{\text{opposite}} = \frac{15}{8}$$

Self Check 8   Find the value of the six trigonometric functions for the angle θ indicated in the figure.

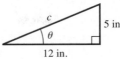

Now Try Exercise 101.

EXAMPLE 9   **Finding the Length of a Side of a Right Triangle, Given an Angle and a Side**

Using the right triangle formed in Figure 4-51, find the length of the side that corresponds to the height of the tree.

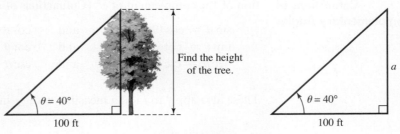

Find the height of the tree.

$\theta = 40°$

100 ft

$\theta = 40°$

100 ft

a

FIGURE 4-51

SOLUTION   To find the length of side a, we can use the definition $\tan \theta = \dfrac{\text{opposite}}{\text{adjacent}}$.

$$\tan 40° = \frac{a}{100}$$

$$a = 100 \tan 40° \qquad \text{Solve for } a.$$

$$a \approx 83.91$$

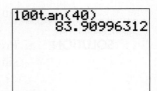

The tree is approximately 84 feet tall.

Self Check 9   Find the height of the tree if you are 75 feet from the tree, and the angle of elevation is 50°. Round to the nearest foot.

Now Try Exercise 121.

7. Understand the Relationship between Complementary Angles and Cofunctions

Since the two acute angles in a right triangle total 90°, they are what is known as **complementary angles.**

Complementary Angles   If the sum of two acute angles is 90° or $\dfrac{\pi}{2}$, the angles are **complementary angles.**

Looking at Figure 4-52, we can see a relationship between the values of the sine and cosine of the two acute angles in the right triangle.

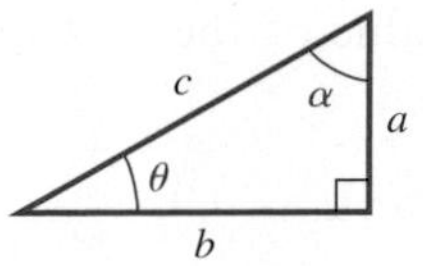

$$\sin \theta = \frac{a}{c} \qquad \cos \alpha = \frac{a}{c}$$

$$\cos \theta = \frac{b}{c} \qquad \sin \alpha = \frac{b}{c}$$

FIGURE 4-52

From this we can derive the following:

$$\theta + \alpha = 90° \qquad\qquad \sin \theta = \cos \alpha$$

$$\alpha = 90° - \theta \qquad\qquad \sin \theta = \cos (90° - \theta)$$

Because of this relationship, the functions sine and cosine are called **cofunctions.** In addition, tangent and cotangent are cofunctions, and secant and cosecant are cofunctions.

Cofunction Theorem: Cofunctions of Complementary Angles

If θ is any acute angle, any trigonometric function of θ is equal to the cofunction of the complement of θ. (**Cofunctions of complementary angles are equal.**)

$$\sin \theta = \cos (90° - \theta) \qquad \text{and} \qquad \cos \theta = \sin (90° - \theta)$$
$$\tan \theta = \cot (90° - \theta) \qquad \text{and} \qquad \cot \theta = \tan (90° - \theta)$$
$$\sec \theta = \csc (90° - \theta) \qquad \text{and} \qquad \csc \theta = \sec (90° - \theta)$$

These also apply to radian measure, with $\frac{\pi}{2}$ in place of $90°$.

EXAMPLE 10   Using Cofunctions

Find an equivalent expression for each of the following, using cofunctions:

a. $\sin 39.5°$ **b.** $\tan \dfrac{\pi}{3}$ **c.** $\csc 42°$

SOLUTION   a. From the Cofunction Theorem, we know that $\sin \theta = \cos (90° - \theta)$, so $\sin 39.5° = \cos (90° - 39.5°) = \cos 50.5°$.

b. $\tan \dfrac{\pi}{3} = \cot \left(\dfrac{\pi}{2} - \dfrac{\pi}{3} \right) = \cot \dfrac{\pi}{6}$

c. $\csc 42° = \sec (90° - 42°) = \sec 48°$ ∎

Self Check 10   Find an equivalent expression for each of the following, use cofunctions:

a. $\cos 14.75°$ **b.** $\sec \dfrac{2\pi}{5}$

Now Try Exercise 109.

Self Check Answers   1. a. $\dfrac{1}{2}$ **b.** $-\dfrac{1}{2}$ **c.** 1 **2.** $\sin \theta = -\dfrac{5\sqrt{41}}{41}$, $\cos \theta = \dfrac{4\sqrt{41}}{41}$, $\tan \theta = -\dfrac{5}{4}$,

$\csc \theta = -\dfrac{\sqrt{41}}{5}$, $\sec \theta = \dfrac{\sqrt{41}}{4}$, $\cot \theta = -\dfrac{4}{5}$ **3.** $\sin \beta = -\dfrac{\sqrt{7}}{4}$; $\tan \beta = \dfrac{\sqrt{7}}{3}$

4. $\csc \theta = \dfrac{4\sqrt{7}}{7}$, $\sec \theta = \dfrac{4}{3}$, $\tan \theta = \dfrac{\sqrt{7}}{3}$, $\cot \theta = \dfrac{3\sqrt{7}}{7}$ **5.** $\sin^2 \theta$

6. $\sin \theta = \dfrac{12}{13}$, $\csc \theta = \dfrac{13}{12}$, $\sec \theta = \dfrac{13}{5}$, $\tan \theta = \dfrac{12}{5}$, $\cot \theta = \dfrac{5}{12}$

7. a. 0.682 **b.** 0.601

8. $\sin \theta = \dfrac{5}{13}, \cos \theta = \dfrac{12}{13}, \tan \theta = \dfrac{5}{12}, \csc \theta = \dfrac{13}{5}, \sec \theta = \dfrac{13}{12}, \cot \theta = \dfrac{12}{5}$

9. 89 ft   **10. a.** $\sin 75.25°$ **b.** $\csc \dfrac{\pi}{10}$

Exercises **4.3**

Getting Ready

You should be able to complete the vocabulary and concept statements before you proceed to the practice exercises.

Fill in the blanks.

1. The domain of $y = \sin x$ is ___.
2. The range of $y = \cos x$ is ___.
3. When evaluating a trigonometric function on a calculator, it is important to select the proper ___.
4. If $P(x, y)$ is a point on the terminal side of angle θ, then $\tan \theta =$ _____.
5. Because $\sin(-x) = -\sin x$, the sine function is an ___ function.
6. One of the Pythagorean Identities states $\sin^2 x + \underline{} = 1$.

Practice

Use the Even–Odd Properties to simplify and find the exact value of each function. Do not use a calculator.

7. $\cos(-120°)$ **8.** $\sin(-120°)$

9. $\sin\left(-\dfrac{\pi}{3}\right)$ **10.** $\cos\left(-\dfrac{5\pi}{3}\right)$

11. $\tan\left(-\dfrac{5\pi}{4}\right)$ **12.** $\cot\left(-\dfrac{5\pi}{3}\right)$

13. $\sin(-5\pi)$ **14.** $\cos\left(-\dfrac{7\pi}{2}\right)$

15. $\sec(-210°)$ **16.** $\csc(-135°)$

17. $\cot(-315°)$ **18.** $\tan(-315°)$

19. $\csc\left(-\dfrac{11\pi}{6}\right)$ **20.** $\sec\left(-\dfrac{7\pi}{6}\right)$

In the following exercises, point P is on the terminal side of angle θ, which is in standard position. Calculate the exact values of sin θ, cos θ, tan θ, csc θ, sec θ, and cot θ.

21. $P(-5, 12)$ **22.** $P(3, -4)$

23. $P(2, 5)$ **24.** $P(3, 6)$

25. $P(1, 1)$ **26.** $P(-3, -3)$

27. $P(-3, -5)$ **28.** $P(-2, -7)$

29. $P(4, -2)$ **30.** $P(5, -6)$

Find the remaining trigonometric functions of the angle θ.

31. $\sin \theta = \dfrac{4}{5}$; θ is in QII

32. $\cot \theta = \dfrac{40}{9}$; θ is in QI

33. $\cos \theta = -\dfrac{3}{4}$; θ is in QIII

34. $\sec \theta = -\dfrac{\sqrt{5}}{2}$; θ is in QII

35. $\tan \theta = -1$; θ is in QIV

36. $\sin \theta = -\dfrac{2}{5}$; θ is in QIII

37. $\csc \theta = \dfrac{7}{5}$; $0 < \theta < \dfrac{\pi}{2}$

38. $\sec \theta = \dfrac{\sqrt{10}}{3}$; $0 < \theta < \dfrac{\pi}{2}$

39. $\cot \theta = -\dfrac{5}{12}$; $\sin \theta > 0$

40. $\cos \theta = \dfrac{\sqrt{5}}{5}$; $\tan \theta < 0$

41. $\sin \theta = -\dfrac{4}{7}$; $270° < \theta < 360°$

42. $\tan \theta = -\dfrac{2}{3}$; $90° < \theta < 180°$

43. $\cos \theta = \dfrac{20}{21}$; $\cot \theta > 0$

44. $\cot \theta = -\dfrac{15}{8}$; $\sin \theta < 0$

45. $\tan \theta = \dfrac{24}{7}$; $\pi < \theta < \dfrac{3\pi}{2}$

46. $\csc \theta = -\dfrac{61}{11}$; $\pi < \theta < \dfrac{3\pi}{2}$

Use the given information and identities to find the value of each of the remaining trigonometric functions for the acute angle θ.

47. $\cot \theta = \dfrac{5}{12}$; $\cos \theta = \dfrac{5}{13}$

48. $\cos \theta = \dfrac{\sqrt{5}}{5}$; $\csc \theta = \dfrac{\sqrt{5}}{2}$

49. $\sec\theta = \dfrac{5}{3}$; $\csc\theta = \dfrac{5}{4}$

50. $\tan\theta = \dfrac{3}{5}$; $\sin\theta = \dfrac{3\sqrt{34}}{34}$

51. $\tan\theta = \dfrac{40}{9}$; $\cos\theta = \dfrac{9}{41}$

52. $\sin\theta = \dfrac{\sqrt{7}}{4}$; $\cos\theta = \dfrac{3}{4}$

Use identities to simplify each expression.

53. $\csc^2\theta - \cos^2\theta\csc^2\theta$ **54.** $\cos^2\theta\tan^2\theta + \cos^2\theta$

55. $\sin^4\theta + \sin^2\theta\cos^2\theta$ **56.** $\cos^4\theta + \sin^2\theta\cos^2\theta$

57. $(1 - \sin\theta)(1 + \sin\theta)$

58. $(1 - \cos\theta)(1 + \cos\theta)$

59. $(\csc\theta + 1)(\csc\theta - 1)$

60. $(\sec\theta - 1)(\sec\theta + 1)$

61. $(\sec\theta + \tan\theta)(\sec\theta - \tan\theta)$

62. $(\csc\theta + \cot\theta)(\csc\theta - \cot\theta)$

63. $\cot^2\theta(1 + \cos\theta)(1 - \cos\theta)$

64. $\sin^2\theta(\csc^2\theta - 1)$

65. $(\cos\theta\sec\theta + \cos\theta)(\cos\theta\sec\theta - \cos\theta)$

66. $(\tan\theta\csc\theta + 1)(\tan\theta\csc\theta - 1)$

Use the given information and identities to find the value of each of the five remaining trigonometric functions for the acute angle θ.

67. $\sin\theta = \dfrac{5}{13}$ **68.** $\csc\theta = \dfrac{11}{8}$

69. $\cos\theta = \dfrac{\sqrt{2}}{5}$ **70.** $\sec\theta = \dfrac{7}{5}$

71. $\tan\theta = \dfrac{9}{4}$ **72.** $\cot\theta = \dfrac{8}{5}$

Use a calculator to approximate each value to four decimal places.

73. $\sin 23.1°$ **74.** $\sin 57.8°$

75. $\cos 41°12'$ **76.** $\cos 17°19'$

77. $\tan 59.34°$ **78.** $\tan 82.6°$

79. $\sec\dfrac{5\pi}{12}$ **80.** $\sec 0.75$

81. $\csc 1.3$ **82.** $\csc\dfrac{3\pi}{7}$

83. $\cot\dfrac{\pi}{3}$ **84.** $\cot 1.05$

Find the exact value of each expression. Do not use a calculator.

85. $\sin 0° + \cos 0°\tan 45°$

86. $\sin^2 90° + \cos 180°\tan 0°$

87. $\cos^2 90° + \cos 90°\sin^2 180°$

88. $\cos^2 0° + \sin^2 90° + \cot^2 90°$

89. $\sin^2\dfrac{3\pi}{2} + \csc^2\dfrac{3\pi}{2} + \cot^2\dfrac{3\pi}{2}$

90. $\cos\pi\sin\pi - \tan^2\pi$

91. $\sin\dfrac{\pi}{6}\cos\dfrac{\pi}{3} - 2\tan^2\dfrac{\pi}{3}$

92. $\sin^2\left(\dfrac{2\pi}{3}\right)\cos^2\left(\dfrac{\pi}{4}\right) + \tan\left(\dfrac{\pi}{4}\right)\sin\left(\dfrac{\pi}{2}\right)$

93. $\sin 45°\cos 315°\tan 150°\tan 60°$

94. $\cos 30°\tan 60° + \cos^2 45°\tan 45°$

95. $\csc^2 210°\sec 30° - \sec 315°\cot 60°$

96. $\csc 90°\csc 210° + \csc 45°\sin 135°$

97. $\sec 0\tan 0 - \cot\dfrac{\pi}{6}\cot\dfrac{\pi}{4}$

98. $\cot^2\dfrac{3\pi}{2}\csc\dfrac{\pi}{2} + \sec\dfrac{\pi}{3}\cot\dfrac{\pi}{6}$

99. $\csc^2\dfrac{\pi}{3}\sin^2\dfrac{4\pi}{3}$ **100.** $\sec^2\dfrac{\pi}{4}\cos^2\dfrac{7\pi}{4}$

Find the value of the six trigonometric functions for the acute angle θ.

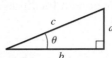

101. $a = 8$, $b = 15$ **102.** $a = 11$, $b = 60$

103. $a = 5$, $c = 11$ **104.** $a = 9$, $c = 13$

105. $b = 24$, $c = 25$ **106.** $b = 35$, $c = 37$

107. $a = 12$, $b = 15$ **108.** $b = 14$, $c = 24$

Find the cofunction with the same value as the given function. Verify your answer with a calculator.

109. $\sin 48.2°$ **110.** $\cos 19.5°$

111. $\tan\dfrac{\pi}{5}$ **112.** $\sec\dfrac{2\pi}{7}$

113. $\cos\dfrac{3\pi}{8}$ **114.** $\cot\dfrac{2\pi}{5}$

115. $\csc 52°$ **116.** $\tan 39°$

State whether each statement is true for all acute angles θ.

117. $\sin^2\theta + \sin^2(90° - \theta) = 1$

118. $\cos\theta = \dfrac{1}{\sec(90° - \theta)}$

119. $\cos(90° - \theta) + \cos\theta = 1$

120. $\cot\theta = \dfrac{\sin(90° - \theta)}{\cos(90° - \theta)}$

Applications

121. Height of a redwood tree If the angle of elevation to the top of a California redwood tree from a point 60 feet from its base is 70°, find the height of the tree to the nearest foot. (The angle that an observer's line of sight makes with the horizontal is called the **angle of elevation.**)

122. Height of the Washington Monument If the angle of elevation to the top of the Washington Monument from a point 100 feet from its base is 79.8°, find the height of the monument. Round to the nearest foot.

123. Angle of depression A person on the edge of a cliff looks down at a boat on a lake. The angle of depression of the person's line of sight is 10°, and the line-of-sight distance from the person to the boat is 555 feet. (If an observer looks down to an object, the angle made with the horizontal is called the **angle of depression.**) How high, to the nearest foot, is the cliff? (Ignore the height of the person.)

124. Height of a lighthouse Kristen waves from the top of the Cape Hatteras Lighthouse in North Carolina to her grandmother standing below on the sidewalk. If the angle of depression of Kristen's line of sight is 63.4° and the line-of-sight distance from the lighthouse to her grandmother is 100 feet, how tall is the lighthouse? (Ignore the height of Kristen's grandmother.)

Audrey Snider-Bell/Shutterstock.com

125. Distance traveled A Mazda RX-8 drives up a hill on a road that makes an angle of 7°30′ with the horizontal. How far will the car travel to reach the top if the horizontal distance traveled is 715 feet? Round to the nearest foot.

126. Length of a ski lift A Colorado ski lift travels up a mountainside and makes an angle of 68.2° with the horizontal. How far does the ski lift travel to reach the top if the horizontal distance traveled is 8,000 feet?

127. Altitude lost On an approach to an airport, a plane is descending at an angle of 3°30′. How much altitude is lost as the plane travels a horizontal distance of 22.3 miles? Round to the nearest tenth of a mile.

128. Altitude lost On an approach to the Kennedy Space Center, a shuttle is descending at an angle of 5°. How much altitude is lost as the shuttle travels a horizontal distance of 15 miles? Round to the nearest tenth of a mile.

129. Angle of descent A plane loses 2,750 feet in altitude as it travels a horizontal distance of 39,300 feet. Find its angle of descent to the nearest tenth of a degree.

130. Angle of ascent A train rises 230 feet as it travels 1 mile up a steep grade. To the nearest tenth of a degree, find its angle of ascent. (*Hint:* 5,280 feet = 1 mile.)

The maximum height of a projectile fired at an angle of inclination θ is given by the equation $h(\theta) = \dfrac{v_0^2 \sin^2 \theta}{2g}$, where v_0 is the initial velocity and gravitational acceleration $g \approx 32.2$ feet per second per second or $g \approx 9.8$ meters per second per second. The range of the projectile is given by $r(\theta) = \dfrac{v_0^2 \sin(2\theta)}{g}$.

131. Projectile motion A football is kicked by Thomas Morstead of the New Orleans Saints with an initial velocity of 25 meters per second at an angle of 45° with the horizontal. To the nearest tenth of a meter, determine the range and maximum height of the football.

J. Meric/Getty Images

132. Projectile motion A long jumper leaves the ground with an initial velocity of 24.8 feet per second at an angle of 28° with the horizontal. How far did he jump? What was his maximum height? Round answers to one decimal place.

Discovery and Writing

One reason for preferring radian measure to degree measure is that for small angles θ, the radian measure of θ, $\sin \theta$, and $\tan \theta$ are approximately equal. In Exercises 133 and 134, let θ be the small angle $\theta = 0.01234$.

133. Set a calculator to radian mode. To how many decimal places do θ and $\sin \theta$ agree?

134. Set a calculator to radian mode. To how many decimal places do θ and $\tan \theta$ agree?

135. What happens to the value of tan θ as θ approaches 90° from the left? Complete a table of values and use the following angles: 88°, 89°, 89.9°, 89.99°, and 89.999°.

136. What happens to the value of cot θ as θ approaches 0° from the right. Complete a table of values and use the following angles: 3°, 2°, 1°, 0.1°, 0.01°, and 0.001°.

4.4 Graphs of the Sine and Cosine Functions

In this section, we will learn to

1. Graph $y = \sin x$ using the unit circle.

2. Graph $y = \cos x$ using the unit circle.

3. Use amplitude to graph sine and cosine functions.

4. Use amplitude and period to graph sine and cosine functions.

5. Use horizontal shifts to graph sine and cosine functions.

6. Use vertical shifts to graph sine and cosine functions.

7. Solve applied problems involving simple harmonic motion and alternating current.

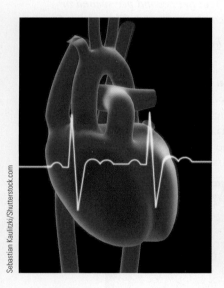

We encounter many things daily that follow a pattern and are often predictable. Examples are our heartbeat, the temperature in Aruba for a given time of year, the monthly sales at Macy's department store, the time of high tide and time of low tide in Acapulco, the time the sun rises in Paris and sets in London, and the daily attendance at Walt Disney World Resort. All of these things are periodic in nature.

Because the sine and cosine functions repeat their values, they are called periodic functions. We will discuss their graphs in this section. To do this, we will use the unit circle and the properties that it reveals about these functions. To graph the functions, we will use the more common function notation, $y = f(x) = \sin x$ and $y = f(x) = \cos x$. To draw the graphs, we could evaluate the functions at several values and plot the points; however, we will first go to the unit circle to discover its relation to the graphs of these functions as x moves from 0 to 2π. In Section 4.2, the periodic properties of these functions were discussed. The period of $y = \sin x$ and $y = \cos x$ is 2π. Because of this periodic property, the graphs complete one cycle every 2π units.

1. Graph $y = \sin x$ Using the Unit Circle

Imagine a unit circle with a central angle x that is steadily increasing from $x = 0$ to $x = 2\pi$. This increasing central angle determines the point $P(\cos x, \sin x)$, which is steadily moving around the circle in a counterclockwise direction. The y-coordinate of P is $\sin x$. We can graph $y = \sin x$ by plotting the values of x on the horizontal axis and the y-coordinate of P on the y-axis, as in Figure 4-53.

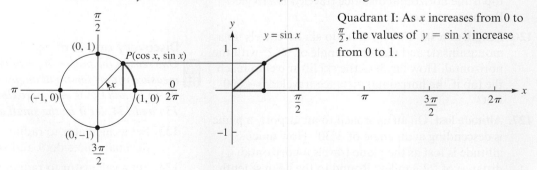

Quadrant I: As x increases from 0 to $\frac{\pi}{2}$, the values of $y = \sin x$ increase from 0 to 1.

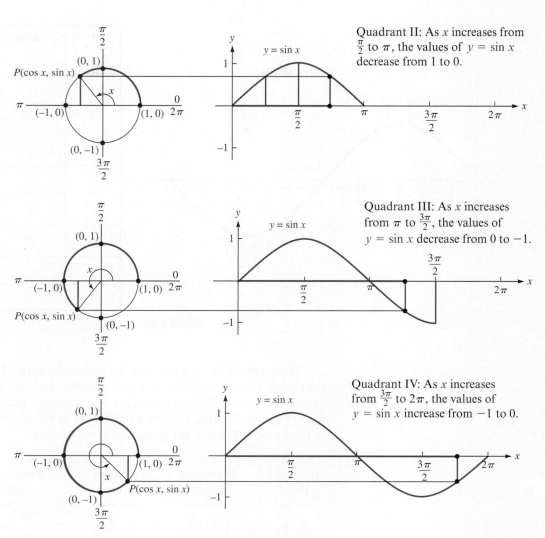

Quadrant II: As x increases from $\frac{\pi}{2}$ to π, the values of $y = \sin x$ decrease from 1 to 0.

Quadrant III: As x increases from π to $\frac{3\pi}{2}$, the values of $y = \sin x$ decrease from 0 to -1.

Quadrant IV: As x increases from $\frac{3\pi}{2}$ to 2π, the values of $y = \sin x$ increase from -1 to 0.

FIGURE 4-53

As P moves once around the unit circle, x increases from 0 to 2π and the graph of $y = \sin x$ completes one cycle over a period of 2π. The graph does not start at $x = 0$, nor does it stop at $x = 2\pi$. It continues to oscillate in identical, repeating cycles of length 2π in both directions, as shown in Figure 4-54. A cycle can begin at any point. When drawing a sine curve, it is not necessary to start at $x = 0$, but that is often a convenient place to begin.

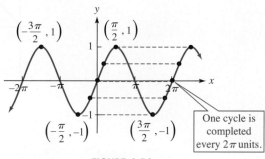

FIGURE 4-54

We could also graph the sine function by plotting points, as shown in Figure 4-55. The points that are plotted are those found by evaluating the sine function for the special angles and multiples of those angles.

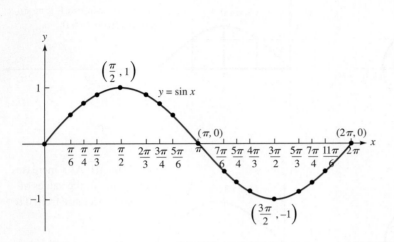

x	$\sin x$	x	$\sin x$
0	0		
$\frac{\pi}{6}$	$\frac{1}{2}$	$\frac{7\pi}{6}$	$-\frac{1}{2}$
$\frac{\pi}{4}$	$\frac{\sqrt{2}}{2}$	$\frac{5\pi}{4}$	$-\frac{\sqrt{2}}{2}$
$\frac{\pi}{3}$	$\frac{\sqrt{3}}{2}$	$\frac{4\pi}{3}$	$-\frac{\sqrt{3}}{2}$
$\frac{\pi}{2}$	1	$\frac{3\pi}{2}$	-1
$\frac{2\pi}{3}$	$\frac{\sqrt{3}}{2}$	$\frac{5\pi}{3}$	$-\frac{\sqrt{3}}{2}$
$\frac{3\pi}{4}$	$\frac{\sqrt{2}}{2}$	$\frac{7\pi}{4}$	$-\frac{\sqrt{2}}{2}$
$\frac{5\pi}{6}$	$\frac{1}{2}$	$\frac{11\pi}{6}$	$-\frac{1}{2}$
π	0	2π	0

FIGURE 4-55

The graph of the sine function is a sinusoidal curve. The shape will not change even when transformations such as horizontal and vertical stretches and shrinks occur. Because of this, students often find it easy to graph the sine function using **five key points.** For this, we start a basic cycle at the origin. These points are shown in Figure 4-56. The five key points for $y = \sin x$ are

Intercept Point	Maximum Point	Intercept Point	Minimum Point		Intercept Point
$(0, 0),$	$\left(\frac{\pi}{2}, 1\right),$	$(\pi, 0),$	$\left(\frac{3\pi}{2}, -1\right),$	and	$(2\pi, 0).$
starting point of cycle	one-quarter point of cycle	one-half point of cycle	three-quarters point of cycle		ending point of cycle

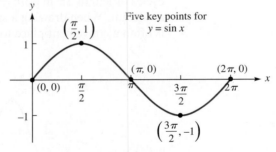

FIGURE 4-56

2. Graph $y = \cos x$ Using the Unit Circle

The graph of the cosine function can be derived from the unit circle in the same manner as that of the sine function. Because the first coordinate of point P in Figure 4-57(a) is $\cos x$, point P's bouncing from side to side will help us generate the graph of $y = \cos x$. At $x = 0$, the first coordinate of P is $\cos 0 = 1$. We begin the graph of $y = \cos x$ at $x = 0$, $y = 1$ [Figure 4-57(b)].

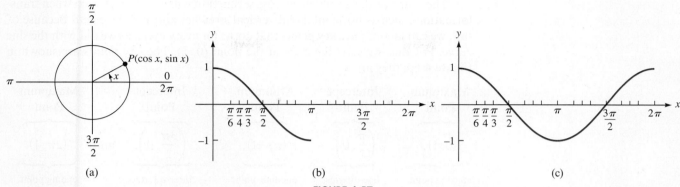

FIGURE 4-57

As x increases from 0 to π, point P moves to the left, with its first coordinate (the value of $\cos x$) decreasing from $\cos 0 = 1$ down to $\cos \pi = -1$. At $x = \frac{\pi}{2}$, the graph intercepts the x-axis, because $\cos \frac{\pi}{2} = 0$. This creates the portion of the graph shown in Figure 4-57(b).

As x continues to increase from π to 2π, point P moves to the right, returning to its starting position at $(1, 0)$. The values of $\cos x$ increase from $\cos \pi = -1$ to $\cos \frac{3\pi}{2} = 0$ and then up to a peak at $\cos 2\pi = 1$. This completes one cycle of the graph of $y = \cos x$, as shown in Figure 4-57(c).

As with the sine function, the cosine function does not start or stop with one cycle, but continues to repeat endlessly. Likewise, a cycle can begin at any point, but $x = 0$, $y = 1$ is a convenient place to begin the graph.

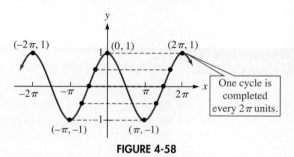

FIGURE 4-58

We can graph the cosine function by plotting points, as shown in Figure 4-59. The points that are plotted are those found by evaluating the cosine function for the special angles and multiples of those angles.

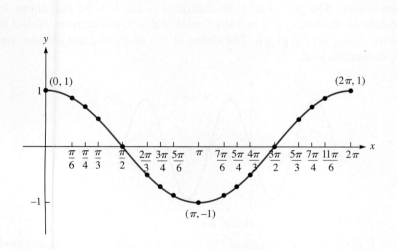

x	$\cos x$	x	$\cos x$
0	1		
$\frac{\pi}{6}$	$\frac{\sqrt{3}}{2}$	$\frac{7\pi}{6}$	$-\frac{\sqrt{3}}{2}$
$\frac{\pi}{4}$	$\frac{\sqrt{2}}{2}$	$\frac{5\pi}{4}$	$-\frac{\sqrt{2}}{2}$
$\frac{\pi}{3}$	$\frac{1}{2}$	$\frac{4\pi}{3}$	$-\frac{1}{2}$
$\frac{\pi}{2}$	0	$\frac{3\pi}{2}$	0
$\frac{2\pi}{3}$	$-\frac{1}{2}$	$\frac{5\pi}{3}$	$\frac{1}{2}$
$\frac{3\pi}{4}$	$-\frac{\sqrt{2}}{2}$	$\frac{7\pi}{4}$	$\frac{\sqrt{2}}{2}$
$\frac{5\pi}{6}$	$-\frac{\sqrt{3}}{2}$	$\frac{11\pi}{6}$	$\frac{\sqrt{3}}{2}$
π	-1	2π	1

FIGURE 4-59

The shape of the graph of the cosine function will not change, even when transformations such as horizontal and vertical stretches and shrinks occur. Because of this, we can identify **five key points** that occur for every cycle, as we did with the sine curve. For this, we start the cycle at the point $(0, 1)$. The key points are shown in Figure 4-60; they are

Maximum Point	Intercept Point	Minimum Point	Intercept Point		Maximum Point
$(0, 1)$,	$\left(\dfrac{\pi}{2}, 0\right)$,	$(\pi, -1)$,	$\left(\dfrac{3\pi}{2}, 0\right)$,	and	$(2\pi, 1)$.
starting point of cycle	one-quarter point of cycle	one-half point of cycle	three-quarters point of cycle		ending point of cycle

The x-intercepts occur at $x = \dfrac{\pi}{2}$ and $x = \dfrac{3\pi}{2}$. The maximum value of 1 occurs when $x = 0$ and $x = 2\pi$. The minimum value of -1 occurs when $x = \pi$.

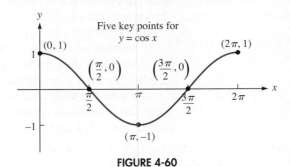

FIGURE 4-60

ACCENT ON TECHNOLOGY   **Comparing Functions**

Use a graphing calculator to graph both $Y_1 = \sin x$ and $Y_2 = \cos x$ and turn off the axes. Do you see any difference between these two? As we progress through this chapter, you will see that both of these functions are called *sinusoidal functions*.

3. Use Amplitude to Graph Sine and Cosine Functions

Now that we have seen how to graph the sine and cosine functions, we will examine vertical shrinks and stretches. For example, how does the graph of the function $y = 3 \sin x$ compare to the graph of $y = \sin x$? In Figure 4-61, the graphs of both functions are shown. The period of both functions is 2π, but the maximum and minimum values of the function $y = 3 \sin x$ are 3 and -3, respectively. Notice the five key points shown on the graph. The shape of the graph did not change; only the extreme values changed.

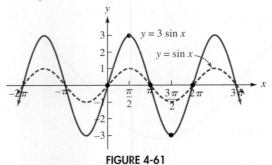

FIGURE 4-61

The change in the extreme values of the function is a change in the **amplitude.** The values of A in the functions $y = A \sin x$ and $y = A \cos x$ affect the maximum and minimum values of these functions. Because $-1 \le \sin x \le 1$, the greatest value that can be attained by $y = A \sin x$ is $|A| \cdot 1 = |A|$. The same is true for the cosine function.

Amplitude   The value $|A|$ is called the **amplitude** of the graph of $y = A \sin x$ and $y = A \cos x$, with $A \neq 0$.

$|A|$ is the maximum value the function attains.

$-|A|$ is the minimum value the function attains.

EXAMPLE 1   **Graphing Functions in the Forms of $y = A \sin x$ and $y = A \cos x$**

Graph each function using the five key points.

a. $y = \dfrac{1}{2} \sin x$ **b.** $y = -2 \cos x$

SOLUTION   **a.** The function $y = \frac{1}{2} \sin x$ has x-intercepts at $x = 0$, $x = \pi$, and $x = 2\pi$. The amplitude is $\frac{1}{2}$. The extreme values occur at $x = \frac{\pi}{2}$ and $x = \frac{3\pi}{2}$. The maximum value of the function is $y = \frac{1}{2} \sin \frac{\pi}{2} = \frac{1}{2}$, and the minimum value is $y = \frac{1}{2} \sin \frac{3\pi}{2} = -\frac{1}{2}$.

x	0	$\dfrac{\pi}{2}$	π	$\dfrac{3\pi}{2}$	2π
$y = \dfrac{1}{2} \sin x$	0	$\dfrac{1}{2}$	0	$-\dfrac{1}{2}$	0
(x, y)	$(0, 0)$	$\left(\dfrac{\pi}{2}, \dfrac{1}{2}\right)$	$(\pi, 0)$	$\left(\dfrac{3\pi}{2}, -\dfrac{1}{2}\right)$	$(2\pi, 0)$

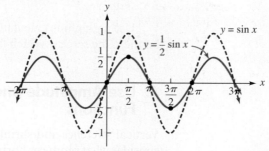

FIGURE 4-62

b. For $y = -2 \cos x$, the value of A is -2, so the amplitude is $|A| = |-2| = 2$. However, since $A < 0$, the graph is reflected across the x-axis. The x-intercepts occur at $x = \frac{\pi}{2}$ and $x = \frac{3\pi}{2}$. The extreme values occur at $x = 0$, $x = \pi$, and $x = 2\pi$. The maximum value of $y = -2 \cos x$ is 2, occurring at $x = \pi$; the minimum value of -2 occurs at $x = 0$ and $x = 2\pi$.

x	0	$\dfrac{\pi}{2}$	π	$\dfrac{3\pi}{2}$	2π
$y = -2 \cos x$	-2	0	2	0	-2
(x, y)	$(0, -2)$	$\left(\dfrac{\pi}{2}, 0\right)$	$(\pi, 2)$	$\left(\dfrac{3\pi}{2}, 0\right)$	$(2\pi, -2)$

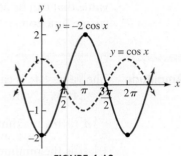

FIGURE 4-63

Self Check 1   Graph each function. **a.** $y = -\dfrac{5}{2}\sin x$ **b.** $y = 5\cos x$

Now Try Exercise 11.

ACCENT ON TECHNOLOGY   **Graphing Trigonometric Functions**

Graphing calculators can graph the sine and cosine functions. While the capability to graph in degree mode is available, we use radian mode because we are graphing functions and use real numbers. When we set the window, we can use a built-in trig window, or we can set the window to fit the graph. Typically, we will use increments of π for functions where only the amplitude is changed. The built-in trig window uses the intervals X: $[-2\pi, 2\pi]$, Y: $[-4, 4]$, with a scale of $\frac{\pi}{2}$ on the x-axis.

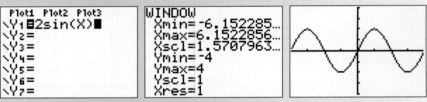

FIGURE 4-64

Use a graphing calculator to graph $Y_1 = \sin x$, $Y_2 = \frac{1}{2}\sin x$, and $Y_3 = 2\sin x$ on the same screen. Which function shows a vertical stretch of $y = \sin x$?

4. Use Amplitude and Period to Graph Sine and Cosine Functions

Vertical stretches and shrinks are determined by the amplitude $|A|$. Now we will consider what produces horizontal stretches and shrinks. The graph of $y = A \sin Bx$ completes one cycle as Bx increases from 0 to 2π or as x itself increases from 0 to $\frac{2\pi}{B}$. We can use an inequality to illustrate this.

$$0 \le Bx \le 2\pi, \qquad B \ne 0$$
$$0 \le x \le \frac{2\pi}{B}$$

The period of the graph of $y = A \sin Bx$ is $\frac{2\pi}{B}$. A change in the period of the function results in a horizontal shrink or stretch.

Amplitude and Period   For the functions $y = A \sin Bx$ and $y = A \cos Bx$, with $A \ne 0$ and $B \ne 0$, $|A|$ is the **amplitude** and $\dfrac{2\pi}{B}$ is the **period.**

For $A > 0$, Figure 4-65(a) shows the five key points of one cycle for $y = A \sin Bx$. If $A < 0$, the graph will be reflected across the x-axis.

- **x-intercepts**   The three x-intercepts of $y = A \sin Bx$ occur at the start of the cycle (at $x = 0$), at the halfway point (at $x = \frac{\pi}{B}$), and at the end of the cycle (at $x = \frac{2\pi}{B}$).
- **Extreme values of y** . The highest point of $y = A \sin Bx$ is at $y = |A|$, which occurs at one-quarter of the period (when $x = \frac{\pi}{2B}$). The lowest point is at $y = -|A|$, which occurs at three-quarters of the period (when $x = \frac{3\pi}{2B}$).

For $A > 0$, Figure 4-65(b) shows the five key points of one cycle of $y = A \cos Bx$. If $A < 0$, the graph will be reflected across the x-axis.

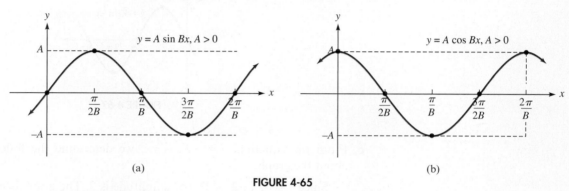

(a)   (b)

FIGURE 4-65

EXAMPLE 2   **Graphing Functions in the Forms of $y = A \sin Bx$ and $y = A \cos Bx$**

Determine the amplitude and the period of each function. Then use the five key points to graph at least one complete cycle.

a. $y = \sin 2x$ **b.** $y = 4 \sin \frac{1}{2}x$ **c.** $y = -2 \cos \pi x$ **d.** $y = 3 \cos\left(-\frac{2}{3}x\right)$

SOLUTION   **a.** From the equation $y = \sin 2x$, we determine the following information about the graph:

- Since $|A| = |1| = 1$, the amplitude is 1.
- The period is $\frac{2\pi}{B} = \frac{2\pi}{2} = \pi$.
- The x-intercepts occur at $x = 0$, $x = \frac{\pi}{2}$ (at one-half of the cycle), and $x = \pi$ (the end of the cycle).
- The maximum value of 1 occurs at $x = \frac{\pi}{4}$ (the one-quarter point of the cycle).
- The minimum value of -1 occurs at $x = \frac{3\pi}{4}$ (the three-quarters point of the cycle).

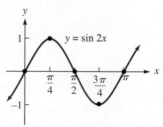

FIGURE 4-66

b. From the equation $y = 4 \sin \frac{1}{2}x$, we determine the following information about the graph:

- Since $|A| = |4| = 4$, the amplitude is 4.
- The period is $\dfrac{2\pi}{B} = \dfrac{2\pi}{\frac{1}{2}} = 4\pi$.
- The x-intercepts occur at $x = 0$, $x = 2\pi$ (at one-half of the cycle), and $x = 4\pi$ (the end of the cycle).

- The maximum value of 4 occurs at $x = \dfrac{\pi}{2B} = \dfrac{\pi}{2\left(\frac{1}{2}\right)} = \pi$ (the one-quarter point of the cycle).
- The minimum value of -4 occurs at $x = \dfrac{3\pi}{2B} = \dfrac{3\pi}{2\left(\frac{1}{2}\right)} = 3\pi$ (the three-quarters point of the cycle).

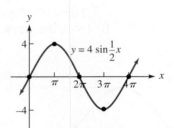

FIGURE 4-67

c. From the equation $y = -2\cos \pi x$, we determine the following information about the graph:

- Since $|A| = |-2| = 2$, the amplitude is 2. The graph is reflected across the x-axis, because $A < 0$.
- The period is $\dfrac{2\pi}{B} = \dfrac{2\pi}{\pi} = 2$.
- The x-intercepts occur at $x = \frac{1}{2}$ (the one-quarter point of the cycle), and at $x = \frac{3}{2}$ (the three-quarters point of the cycle).
- The maximum value of 2 occurs at $x = 1$ (the one-half point of the cycle, because graph is reflected).
- The minimum value of -2 occurs at $x = 0$ (the beginning of the cycle) and $x = 2$ (the end of the cycle).

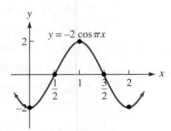

FIGURE 4-68

d. Before we gather information for the graph of the function $y = 3\cos\left(-\frac{2}{3}x\right)$, we will simplify the equation by using the Even–Odd Property $\cos(-x) = \cos x$. We rewrite the equation as $y = 3\cos\left(-\frac{2}{3}x\right) = 3\cos\frac{2}{3}x$. From the equation $y = 3\cos\frac{2}{3}x$, we determine the following information about the graph:

- Since $|A| = |3| = 3$, the amplitude is 3.
- The period is $\dfrac{2\pi}{B} = \dfrac{2\pi}{\frac{2}{3}} = 3\pi$.
- The x-intercepts occur at $x = \frac{1}{4} \cdot 3\pi = \frac{3\pi}{4}$ (the one-quarter point of the cycle), and $x = \frac{3}{4} \cdot 3\pi = \frac{9\pi}{4}$ (the three-quarters point of the cycle).
- The minimum value of -3 occurs at $x = \frac{1}{2} \cdot 3\pi = \frac{3\pi}{2}$ (the one-half point of the cycle).

- The maximum value of 3 occurs at $x = 0$ and at $x = 3\pi$ (the beginning and end of the cycle).

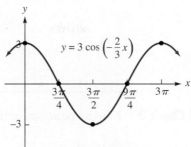

$$y = 3 \cos\left(-\frac{2}{3}x\right)$$

FIGURE 4-69

Self Check 2   Graph each function. **a.** $y = \dfrac{1}{2}\sin\left(-\dfrac{\pi}{2}x\right)$ **b.** $y = 5\cos 4x$

Now Try Exercise 17.

Sinusoidal functions can be used to represent repetitive behaviors and patterns, such as ocean waves, sound waves, and light waves. A sine curve or a cosine curve can often be used to plot average daily temperatures. (A cosine curve is also considered to be sinusoidal.) Later in this chapter, we will see how to use sinusoidal regression to find a curve of best fit for data. Right now, we will discover how to derive a function from a graph using what we know about amplitude and period.

EXAMPLE 3   **Finding an Equation to Fit a Curve**

Find an equation in the form $y = A \sin Bx$ that will produce the graph shown in Figure 4-70.

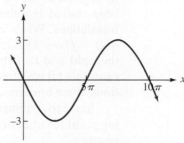

FIGURE 4-70

SOLUTION   To find an equation that fits the graph, we first find the value of A. From the graph, we determine that the amplitude is $|A| = 3$. However, since the graph is reflected across the x-axis, $A = -3$.

One cycle of the function is completed over the interval $[0, 10\pi]$. The period of this function is 10π. We know that period of a sine function is given by $\frac{2\pi}{B}$; therefore, we find B by solving the following equation:

$$\frac{2\pi}{B} = 10\pi$$

$$2\pi = 10\pi(B)$$

$$\frac{2\pi}{10\pi} = B$$

$$B = \frac{2}{10} = \frac{1}{5}$$

Thus we have the equation $y = -3 \sin\frac{1}{5}x$.

This answer can be checked using a graphing calculator. Enter the function and set the window to $[0, 10\pi]$ for X and $[-3, 3]$ for Y as shown in Figure 4-71.

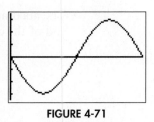

FIGURE 4-71

Self Check 3   Find an equation in the form $y = A \cos Bx$ that will produce the graph shown here.

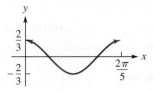

Now Try Exercise 79.

ACCENT ON TECHNOLOGY   **Comparing Functions**

On a graphing calculator, graph the functions $Y_1 = \cos x$, $Y_2 = \cos \frac{1}{2}x$, and $Y_3 = \cos 2x$. Which of the graphs shows a horizontal stretch of $y = \cos x$?

5. Use Horizontal Shifts to Graph Sine and Cosine Functions

In applied work, we often encounter graphs of trigonometric functions that have been shifted in either a vertical or a horizontal direction. Such shifts are called **translations**. When we studied algebraic functions, we learned that the graph of $g(x) = f(x - k)$ is identical to the graph of $f(x)$ except that it is shifted k units to the right and that the graph of $g(x) = f(x + k)$ is identical to the graph of $f(x)$ except that it is shifted k units to the left. When we are graphing trigonometric functions, these horizontal shifts are called **phase shifts**.

Since trigonometric functions are periodic, if we find a starting point and ending point for one complete cycle, then we can graph the function using our five key points.

We begin the graphing process for $y = A \sin (Bx - C)$ with the basic inequality $0 \le Bx - C \le 2\pi$. The starting point for one cycle is where $Bx - C = 0$ or $x = \frac{C}{B}$. The ending point will be where $Bx - C = 2\pi$ or $x = \frac{2\pi}{B} + \frac{C}{B}$. The five key points maintain the same position within the cycle, as shown in Figure 4-72. These can be found by dividing the interval $\left[\frac{C}{B}, \frac{2\pi}{B} + \frac{C}{B}\right]$ into four equal intervals. Each interval will have length $\frac{2\pi}{B} \div 4 = \frac{2\pi}{4B} = \frac{\pi}{2B}$.

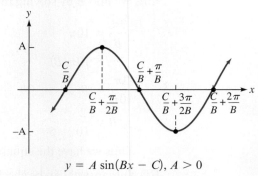

$$y = A \sin(Bx - C), A > 0$$

FIGURE 4-72

The same process is used to graph $y = A \cos(Bx - C)$, as shown in Figure 4-73.

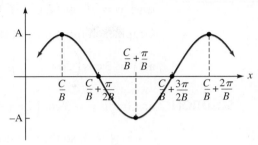

$$y = A\cos(Bx - C), A > 0$$

FIGURE 4-73

Summary of Strategies for Graphing $y = A \sin(Bx - C)$ and $y = A \cos(Bx - C)$

- The graph of $y = A \sin(Bx - C)$ is identical to the graph of $y = A \sin Bx$ except that it is translated $\dfrac{C}{B}$ units to the right if $C > 0$ or $\dfrac{C}{B}$ units to the left if $C < 0$.

- The same holds true for the graph of the function $y = A \cos(Bx - C)$.

- $|A| =$ the **amplitude**; $\dfrac{2\pi}{B} =$ the **period**; $\dfrac{C}{B} =$ the **phase shift**.

Finding Key Points for $y = A \sin(Bx - C)$ and $y = A \cos(Bx - C)$

Step 1: Find one cycle by solving the inequality $0 \le Bx - C \le 2\pi$ to find the interval $\left[\dfrac{C}{B}, \dfrac{C}{B} + \dfrac{2\pi}{B}\right]$.

Step 2: Divide the interval into four equal parts, each having length
$$\frac{2\pi}{B} \div 4 = \frac{2\pi}{4B} = \frac{\pi}{2B}.$$
The values of x to use for this are $x = \dfrac{C}{B}$, $x = \dfrac{C}{B} + \dfrac{\pi}{2B}$, $x = \dfrac{C}{B} + \dfrac{\pi}{B}$,
$x = \dfrac{C}{B} + \dfrac{3\pi}{2B}$, and $x = \dfrac{C}{B} + \dfrac{2\pi}{B}$.

If $B = 1$, then the graph is a horizontal translation with no change in the period.

EXAMPLE 4   **Graphing a Function in the Form $y = A \sin(Bx - C)$ with $B = 1$**

Graph the function $y = \sin\left(x - \frac{\pi}{4}\right)$. Compare it to the graph of $y = \sin x$ on the same set of axes.

SOLUTION   Given the function $y = \sin\left(x - \frac{\pi}{4}\right)$, we note that $A = 1$ and $B = 1$. Therefore, there is no change in the period. Since $C = \frac{\pi}{4}$, the graph is a translation to the right by $\frac{\pi}{4}$ units. It is shown in Figure 4-74, along with the graph of $y = \sin x$.

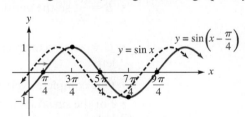

FIGURE 4-74

Self Check 4   Graph the function $y = 2 \cos\left(x + \dfrac{\pi}{6}\right)$.

Now Try Exercise 45.

EXAMPLE 5   **Graphing Functions in the Forms of** $y = A \sin(Bx - C)$
and $y = A \cos(Bx - C)$

Graph each function and label at least one complete cycle.

a. $y = 3 \cos\left(2\pi x - \dfrac{\pi}{2}\right)$ **b.** $y = -2 \sin\left(\dfrac{1}{2}x + \pi\right)$

SOLUTION   **a.** Graph $y = 3 \cos\left(2\pi x - \dfrac{\pi}{2}\right)$.

- $A = 3$, so the amplitude is $|A| = |3| = 3$.
- $B = 2\pi$, so the period is $\dfrac{2\pi}{B} = \dfrac{2\pi}{2\pi} = 1$.

- $C = \dfrac{\pi}{2}$, so the phase shift is $\dfrac{C}{B} = \dfrac{\frac{\pi}{2}}{2\pi} = \dfrac{1}{4}$ unit to the right.

- One complete cycle is over the interval $\left[\dfrac{C}{B}, \dfrac{2\pi}{B} + \dfrac{C}{B}\right] = \left[\dfrac{1}{4}, \dfrac{5}{4}\right]$, which can also
be found by solving the inequality $0 \le 2\pi x - \dfrac{\pi}{2} \le 2\pi$.
- For the cosine function, the maximum points are at the ends of the
interval; at $x = \dfrac{1}{4}$, $y = 3 \cos\left[2\pi\left(\dfrac{1}{4}\right) - \dfrac{\pi}{2}\right] = 3 \cos(0) = 3$, and at $x = \dfrac{5}{4}$,
$y = 3 \cos\left[2\pi\left(\dfrac{5}{4}\right) - \dfrac{\pi}{2}\right] = 3 \cos(2\pi) = 3$.
- The minimum is at the one-half point. At $x = \dfrac{3}{4}$, $y = 3 \cos\left[2\pi\left(\dfrac{3}{4}\right) - \dfrac{\pi}{2}\right] = 3 \cos(\pi) = -3$.
- The intercepts are at the one-quarter point of the interval, $x = \dfrac{1}{2}$, and the
three-quarters point of the interval, $x = 1$.

We can now label the five key points and graph the function.

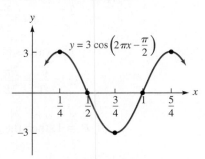

FIGURE 4-75

b. Graph $y = -2 \sin\left(\dfrac{1}{2}x + \pi\right)$.

- $A = -2$, so the amplitude is $|-2| = 2$. Since $A < 0$, the graph is reflected
across the x-axis.
- $B = \dfrac{1}{2}$, so the period is $\dfrac{2\pi}{\frac{1}{2}} = 4\pi$.

- $C = -\pi$, so the phase shift is $\dfrac{C}{B} = \dfrac{-\pi}{\frac{1}{2}} = -2\pi$ units, which is 2π units to the
left.
- One complete cycle is over the interval $\left[\dfrac{C}{B}, \dfrac{2\pi}{B} + \dfrac{C}{B}\right] = [-2\pi, 2\pi]$.
- For the sine function, the x-intercepts are at the ends and middle of the inter-
val, at $x = -2\pi$, $x = 0$, and $x = 2\pi$.
- Since the graph is reflected, the minimum value occurs at the one-quarter
point; at $x = -2\pi + \pi = -\pi$, $y = -2 \sin\left[\dfrac{1}{2}(-\pi) + \pi\right] = -2 \sin\dfrac{\pi}{2} = -2$.
- The maximum value occurs at the three-quarters point; at
$x = -2\pi + 3\pi = \pi$, $y = -2 \sin\left[\dfrac{1}{2}(\pi) + \pi\right] = -2 \sin\dfrac{3\pi}{2} = 2$.

We can now label the five key points and graph the function.

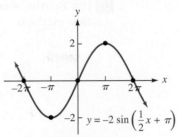

$$y = -2 \sin\left(\tfrac{1}{2}x + \pi\right)$$

FIGURE 4-76

Self Check 5   Graph the function $y = 3 \sin\left(2x - \dfrac{\pi}{6}\right)$.

Now Try Exercise 47.

EXAMPLE 6   **Finding a Function to Fit a Graph**

Find a function in the form $y = A \sin(Bx - C)$ that will produce the graph shown in Figure 4-77.

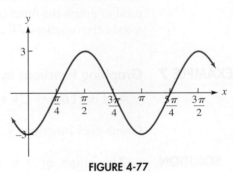

FIGURE 4-77

SOLUTION   To find an equation in the form $y = A \sin(Bx - C)$ that will produce this graph, we need to find the values of A, B, and C. Looking at the graph, the interval $\left[\dfrac{\pi}{4}, \dfrac{5\pi}{4}\right]$ shows one cycle of a sine curve, so we will use that as our starting point.

- The graph shows us that $A = 3$.
- The length of the interval, and the period of the function, is π. Using the fact that the period is $\dfrac{2\pi}{B}$, we solve for B.

$$\frac{2\pi}{B} = \pi$$
$$B = 2$$

- From the graph, we see that the phase shift is $\dfrac{\pi}{4}$. Using the fact that the phase shift is given by $\dfrac{C}{B}$, we can find C.

$$\frac{C}{2} = \frac{\pi}{4}$$
$$C = \frac{\pi}{2}$$

We now have enough information to form the function $y = 3 \sin\left(2x - \dfrac{\pi}{2}\right)$.

You can check your answer by graphing the function on a graphing calculator. Set the window to exactly the same as the graph in Figure 4-77, and the tick marks to the same unit, so that you can easily compare the two graphs.

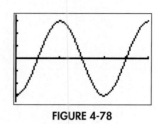

FIGURE 4-78

Self Check 6   For the graph in Figure 4-77, find an equation in the form $y = A\cos(Bx)$ or $y = A\cos(Bx - C)$.

Now Try Exercise 81.

6. Use Vertical Shifts to Graph Sine and Cosine Functions

The graph of $y = f(x) + k$ is obtained by a **vertical translation** of the function $f(x)$ up k units. If $k < 0$, the translation is down k units. The same is true for $y = A\sin(Bx - C) + D$ and $y = A\cos(Bx - C) + D$. The value of D is a vertical translation of the sine and cosine functions. The five key points can still be used to graph the functions. However, instead of the function oscillating along the x-axis, the function will oscillate along the line $y = D$.

EXAMPLE 7   **Graphing Functions in the Forms of $y = A\sin(Bx - C) + D$ and $y = A\cos(Bx - C) + D$**

Graph each function.   **a.** $y = 2\sin \pi x + 3$ **b.** $y = 3\cos\left(x - \dfrac{\pi}{4}\right) - 1$

SOLUTION   **a.** The graph of $y = 2\sin \pi x + 3$ is a vertical translation of the function $y = 2\sin \pi x$ up 3 units.

- $B = \pi$, so the period of the function is $\frac{2\pi}{B} = \frac{2\pi}{\pi} = 2$.
- The function $y = 2\sin \pi x$ has maximum value of 2 and a minimum value of -2. Therefore, the maximum value of the function $y = 2\sin \pi x + 3$ is $y = 2 + 3 = 5$ and the minimum value is $y = -2 + 3 = 1$.
- The function intercepts the line $y = 3$ at the beginning, middle, and end of the cycle that covers the interval $[0, 2]$, at $x = 0$, $x = 1$, and $x = 2$.

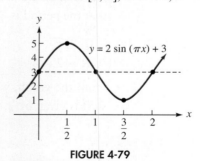

FIGURE 4-79

b. The graph of $y = 3\cos\left(x - \frac{\pi}{4}\right) - 1$ is a vertical translation of the function $y = 3\cos\left(x - \frac{\pi}{4}\right)$ down 1 unit.

- We see that $B = 1$ and $C = \frac{\pi}{4}$. Therefore, the period is $\frac{2\pi}{B} = \frac{2\pi}{1} = 2\pi$ and the phase shift is $\frac{C}{B} = \frac{\pi}{4}$ units to the right. The graph completes one cycle over the interval $\left[\frac{\pi}{4}, \frac{\pi}{4} + 2\pi\right] = \left[\frac{\pi}{4}, \frac{9\pi}{4}\right]$.
- The function $y = 3\cos\left(x - \frac{\pi}{4}\right)$ has a maximum value of 3 and a minimum value of -3. Therefore, the maximum value of the function $y = 3\cos\left(x - \frac{\pi}{4}\right) - 1$ is $y = 3 - 1 = 2$, and the minimum value is $y = -3 - 1 = -4$.
- The graph is not reflected; therefore, the maximum value occurs at the beginning of the period, at $x = \frac{\pi}{4}$, and at the end of the period, at $x = \frac{9\pi}{4}$. The minimum value occurs at the one-half point of the interval, at $x = \frac{5\pi}{4}$.
- The graph intercepts the line $y = -1$ at the one-quarter point, $x = \frac{3\pi}{4}$, and at the three-quarters point, $x = \frac{7\pi}{4}$.

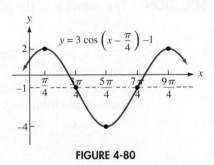

$$y = 3\cos\left(x - \frac{\pi}{4}\right) - 1$$

FIGURE 4-80

Self Check 7   Graph the function $y = 4\sin(2x - \pi) + 2$.

Now Try Exercise 75.

7. Solve Applied Problems Involving Simple Harmonic Motion and Alternating Current

Simple Harmonic Motion

Simple harmonic motion is a motion that is periodic and sinusoidal. Any oscillation that can be described by formulas involving a sine or cosine function is simple harmonic motion. It can be typified by an object suspended from a spring that can be made to bounce, or **oscillate**. In Figure 4-81, the position above or below the equilibrium position is given by the formula $y = A\cos\sqrt{\frac{k}{m}}\,t$.

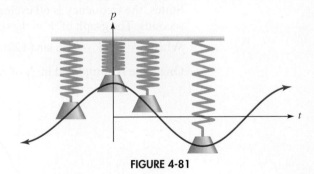

FIGURE 4-81

The amplitude of the oscillation is A. The constant k, called the **spring constant**, measures the stiffness of the spring, and m is the mass of the object. The variable t represents the time in seconds since the object was released. The period of the oscillation,

$$p = \frac{2\pi}{\sqrt{\dfrac{k}{m}}} = 2\pi\sqrt{\frac{m}{k}},$$

indicates the number of seconds per complete cycle. The reciprocal of the period, called the **frequency of the oscillation**, indicates the number of cycles per second.

EXAMPLE 8   **Solving a Simple Harmonic Motion Problem**

A mass of 4 grams, attached to a spring with a spring constant of 9 dynes per centimeter, is raised above the equilibrium position and released. Find the period and frequency of the oscillation.

SOLUTION   The period p of the oscillation is given by $p = 2\pi\sqrt{\frac{m}{k}}$. Substituting 4 for m and 9 for k gives

$$p = 2\pi\sqrt{\frac{m}{k}} = 2\pi\sqrt{\frac{4}{9}} = 2\pi \cdot \frac{2}{3} = \frac{4\pi}{3}.$$

It takes the mass $\frac{4\pi}{3}$ seconds, or about 4.2 seconds, to return to its starting point. Because the frequency, f, of the oscillation is the reciprocal of the period, we have

$$f = \frac{1}{p} = \frac{3}{4\pi}.$$

The frequency is $\frac{3}{4\pi}$, or approximately 0.24 hertz (cycles per second).

Self Check 8   Repeat Example 8 with a mass of 6 grams attached to a spring with a spring constant of 10 dynes per centimeter.

Now Try Exercise 89.

Alternating Current

The humming sound of a fluorescent light fixture or a toy train's transformer is caused by electric current reversing itself 120 times each second. The voltage available at a wall outlet is described by the formula

$$V = V_0 \sin{(2\pi ft)},$$

where V_0 is approximately 155 volts, f is the frequency, and t is time. In the United States, the frequency is 60 cycles per second (or 60 hertz), and time is measured in seconds. The graph of $V = V_0 \sin{(2\pi ft)} = 155 \sin{(120\pi t)}$ is shown in Figure 4-82. When $t = \frac{1}{60}$, $V = 155 \sin\left(120\pi \cdot \frac{1}{60}\right) = 155 \sin 2\pi = 0$.

One cycle is completed in $\frac{1}{60}$ of a second, so there are 60 cycles a second.

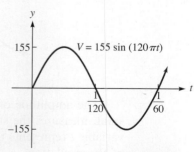

FIGURE 4-82

EXAMPLE 9   **Solving an Alternating Current Problem**

If the voltage is given by $V = V_0 \sin (120\pi t)$, the **root mean square (RMS) value** of the voltage is $\dfrac{\sqrt{2}V_0}{2}$. Show that V is equal to the RMS value when $t = \dfrac{1}{480}$ second.

SOLUTION   We substitute $\frac{1}{480}$ for t into the equation $V = V_0 \sin (120\pi t)$ and find V.

$$V = V_0 \sin (120\pi t) = V_0 \sin \left(120\pi \frac{1}{480} \right) = V_0 \sin \left(\frac{\pi}{4} \right) = V_0 \frac{\sqrt{2}}{2}$$

$$V = \frac{\sqrt{2}V_0}{2}$$

Therefore, V is equal to the RMS value when $t = \frac{1}{480}$ second.   ■

Self Check 9   What is the voltage if $V_0 = 155$ volts and $t = \dfrac{1}{360}$ second?

Now Try Exercise 93.

Self Check Answers   1. a.

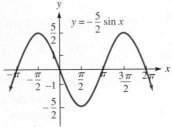

b.

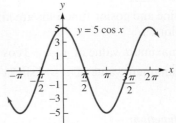

2. a.

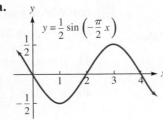

b.

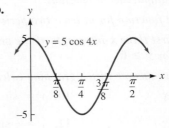

3. $y = \frac{2}{3} \cos 5x$

4.

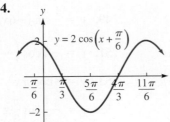

5.

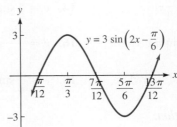

6. $y = -3 \cos 2x$ or $y = 3 \cos (2x - \pi)$

7.

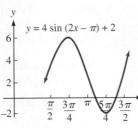

8. $p = 2\pi \sqrt{\dfrac{3}{5}} \approx 4.81$ sec; $f = \dfrac{1}{p} \approx 0.21$ hertz

9. $\dfrac{155\sqrt{3}}{2} V$

Exercises **4.4**

Getting Ready
You should be able to complete these vocabulary and concept statements before you proceed to the practice exercises.

Fill in the blanks.

1. The graph of $y = \sin x$ completes one cycle over a period of ___ units.
2. Over the interval $[-2\pi, 2\pi]$, $y = \sin x = 1$ at $x =$ ___ and ___.
3. The sine and cosine functions can be graphed by first plotting five _____.
4. Over the interval $[-2\pi, 2\pi]$, $y = \cos x = -1$ at $x =$ ___ and ___.
5. For $0 \le x \le 2\pi$, the x-intercepts of $y = \cos x$ are ____ and ___.
6. The sine and cosine functions are known as _____ functions.
7. The maximum value of $y = -3 \cos \pi x$ is ___.
8. The minimum value of $y = 2 \sin 2x$ is ___.

Practice
For each function:
(a) State the amplitude and period.
(b) Graph the function for at least two periods.
(c) Find the first three x-intercepts of the graph of the function for $x \ge 0$.

(d) Check your answer with a graphing calculator.

9. $y = \sin 2x$

10. $y = \sin 4x$

11. $y = 2 \sin x$

12. $y = \dfrac{1}{2} \sin x$

13. $y = 4 \sin \dfrac{1}{2}x$

14. $y = 5 \sin \dfrac{1}{4}x$

15. $y = -3 \sin \pi x$

16. $y = -\dfrac{1}{4} \sin \dfrac{3\pi}{2}x$

17. $y = \dfrac{1}{3} \sin (-2\pi x)$

18. $y = \dfrac{1}{4} \sin \left(-\dfrac{\pi}{3}x\right)$

For each function:
(a) State the amplitude and period.
(b) Graph the function for at least two periods.
(c) Find the first three x-intercepts of the graph of the function for $x \ge 0$.

(d) Check your answer with a graphing calculator.

19. $y = 3 \cos x$

20. $y = -\dfrac{1}{2} \cos x$

21. $y = \cos 2x$

22. $y = \cos 4x$

23. $y = -2 \cos \dfrac{1}{4}x$

24. $y = -\dfrac{1}{4} \cos \dfrac{2}{3}x$

25. $y = \cos \dfrac{\pi}{2}x$

26. $y = \dfrac{1}{2} \cos \dfrac{3\pi}{2}x$

27. $y = \dfrac{1}{2} \cos (-4\pi x)$

28. $y = 2 \cos (-2\pi x)$

For each function:
(a) Graph the function for at least two periods.
(b) Find the first value of x at which the graph of the function reaches its maximum value for $x \ge 0$.
(c) Find the first value of x at which the graph of the function reaches its minimum value for $x \ge 0$.
(d) Check your answer with a graphing calculator.

29. $y = 5 \sin 3x$

30. $y = 5 \sin 2x$

31. $y = 4 \cos 2x$

32. $y = 6 \cos 3x$

33. $y = -3 \cos \pi x$

34. $y = -\pi \cos x$

35. $y = -4 \sin \pi x$

36. $y = -2 \sin 2\pi x$

37. $y = 3 \sin \left(x - \dfrac{\pi}{2}\right)$

38. $y = 4 \sin \left(x + \dfrac{\pi}{2}\right)$

39. $y = -2 \cos (x + \pi)$

40. $y = -3 \cos (x - \pi)$

For each function:
(a) Find the amplitude, period, and phase shift of the function.
(b) Graph the function for at least two periods.
(c) Check your answer with a graphing calculator.

41. $y = 2 \sin 2x$

42. $y = 3 \cos 3x$

43. $y = \cos \dfrac{1}{4}x$

44. $y = \sin \dfrac{1}{3}x$

45. $y = 3 \sin \left(\dfrac{1}{2}x - \dfrac{\pi}{4}\right)$

46. $y = \dfrac{1}{2} \cos \left(\dfrac{1}{3}x - \pi\right)$

47. $y = -\dfrac{1}{2} \cos \left(\pi x - \dfrac{\pi}{3}\right)$

48. $y = 7 \sin \left(2\pi x - \dfrac{\pi}{2}\right)$

49. $y = \dfrac{1}{2} \sin (-6x)$

50. $y = -3 \sin \dfrac{1}{6}x$

51. $y = -4 \cos \left(-\dfrac{1}{3}x\right)$

52. $y = 6 \cos \dfrac{2}{3}x$

53. $y = 8 \cos \pi x$

54. $y = 3 \sin 2\pi x$

55. $y = -\dfrac{1}{3} \sin \dfrac{\pi}{3}x$

56. $y = -\dfrac{5}{3} \cos \dfrac{\pi}{4}x$

57. $y = \sin \left(x - \dfrac{\pi}{3}\right)$

58. $y = \cos \left(x + \dfrac{\pi}{4}\right)$

59. $y = -3 \cos \left(x + \dfrac{\pi}{6}\right)$

60. $y = -\sin \left(x - \dfrac{\pi}{2}\right)$

61. $y = 5 \sin (2x + \pi)$

62. $y = 4 \cos (2\pi x + \pi)$

Write an equation of a sine function in the form $y = A \sin (Bx - C)$ with the stated characteristics.

63. amplitude $= 4$; period $= \pi$

64. amplitude $= \dfrac{1}{2}$; period $= \dfrac{\pi}{2}$

65. amplitude $= \dfrac{\pi}{2}$; period $= 4$

66. amplitude $= 3\pi$; period $= \dfrac{1}{2}$

67. amplitude $= 3$; period $= \pi$; phase shift $= \dfrac{\pi}{2}$

68. amplitude $= \dfrac{4}{5}$; period $= \dfrac{\pi}{4}$; phase shift $= \pi$

69. amplitude $= 2$; period $= \dfrac{1}{3}$; phase shift $= -\dfrac{1}{2}$

70. amplitude $= 4$; period $= \dfrac{3}{5}$; phase shift $= -3$

Graph each function. Label at least two periods.
Check your answer with a graphing calculator.

71. $y = 4 + \sin x$ **72.** $y = -3 + 2\sin x$

73. $y = 2 - \cos\left(x - \dfrac{\pi}{2}\right)$

74. $y = 3 + \cos\left(2x + \pi\right)$

75. $y = -\dfrac{1}{2}\sin\left(x + \dfrac{\pi}{2}\right) + 1$

76. $y = -2 + \sin\left(2x - \pi\right)$

77. $y = 4\cos 2\pi x + 2$

78. $y = -2\cos\dfrac{\pi}{2}x + 4$

Write an equation of a function in the form
$y = A\sin\left(Bx - C\right) + D$ *or* $y = A\cos\left(Bx - C\right) + D$,
with $A \neq 0$ *and* $B \neq 0$, *that will produce the graph shown.*
(Answers are not unique.)
Check your answer with a graphing calculator.

79.
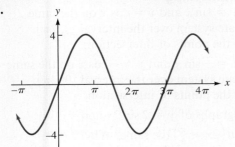

80.
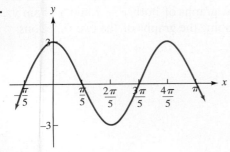

81.
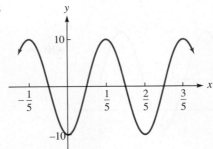

82.
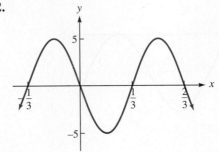

83.
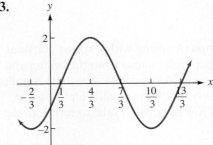

84.
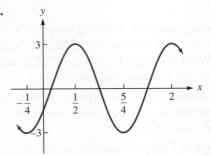

85.
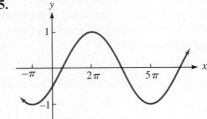

86.
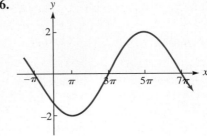

87.

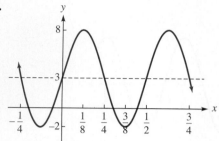

88.

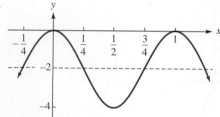

Applications

89. Harmonic motion A spring with a spring constant of 6 newtons per meter hangs from the ceiling and supports a mass of 24 kilograms. The mass is pulled a few centimeters below the equilibrium position and released. How long will it take to return to the starting point?

90. Frequency What is the frequency of the oscillation in Exercise 89?

91. Finding the spring constant A spring supports a mass of 12 grams. The oscillation is $\frac{1}{\pi}$ hertz (cycles per second). What is the spring constant? (The units of your answer will be dynes per centimeter.)

92. Period of oscillation For a given spring, what is the effect on the period of its oscillation if the mass of the suspended object is doubled?

93. Electrical power in Europe Electrical power in Europe has a frequency of 50 cycles per second (or 50 hertz), and the peak voltage available at a power outlet is 340 volts. What formula determines the available voltage at any time t?

94. RMS voltage Show that when $t = \frac{1}{400}$ second, the voltage in Exercise 93 is equal to the RMS value of the voltage.

95. Motion of a pendulum A pendulum swinging through a small arc makes an angle θ with the vertical that is given by $\theta = A \cos\sqrt{\dfrac{9.8}{l}}\,t$. The coefficient A is the amplitude of the swing, and l is the length of the pendulum. If the pendulum is 1 meter long, what are the period and frequency of the oscillation?

96. Length of a pendulum What is the effect on the period of the oscillation of the pendulum in Exercise 95 if the length of the pendulum is doubled?

97. Frequency of a tuning fork The vibration of a tuning fork and the surrounding air is simple harmonic motion governed by the equation $y = A \sin(2\pi f t)$, where f is the frequency, t the time, and A the amplitude of the vibration. The vibrations of one particular tuning fork take 0.0035 second to complete one cycle. What is the frequency of the tone?

Tatiana Popova/Shutterstock.com

98. Wavelength The distance that sound travels in the time it takes to complete one cycle is called the **wavelength** of the sound. The speed of sound in air is approximately 1,100 feet per second. What is the wavelength of the sound in Exercise 97?

Discovery and Writing

99. Graph $y = \sin x$ and $y = \cos x$ on the same coordinate system over the interval $[0, 2\pi]$. Identify the points of intersection.

100. Graph $y = -\sin x$ and $y = -\cos x$ on the same coordinate system over the interval $[0, 2\pi]$. Identify the points of intersection.

101. Are the graphs of $y = 2 \sin 2x$ and $y = 2 \sin\left(2x - \frac{\pi}{4}\right)$ the same? Why?

102. Sketch the graph of $y = x + \sin x$. To do so, first sketch the graphs of both $y = x$ and $y = \sin x$; then combine the graphs of the two functions.

4.5 Graphs of the Tangent, Cotangent, Secant, and Cosecant Functions

In this section, we will learn to

1. Sketch graphs of tangent functions.
2. Sketch graphs of cotangent functions.
3. Sketch graphs of cosecant functions.
4. Sketch graphs of secant functions.

The music genre of hip-hop is very popular today. Stanley Kirk Burrell, best known as MC Hammer, is considered a forefather of pop rap. His signature song "U Can't Touch This" appeared on his first pop rap album, which sold over 10 million copies. We will discuss the graphs of the remaining four trigonometric functions, $y = \tan x$, $y = \cot x$, $y = \csc x$, and $y = \sec x$ in this section. These graphs have vertical asymptotes. When we sketch the graphs of these trigonometric functions, we must remember that our graphs "can't touch" these vertical lines.

In Section 4.2, we discussed the periodic properties of the trigonometric functions. The functions $y = \sin x$, $y = \cos x$, $y = \csc x$, and $y = \sec x$ each have a period of 2π. The period of $y = \tan x$ and $y = \cot x$ is π.

In Section 4.4, we used the interval $[0, 2\pi]$ as the basic period for graphing the sine and cosine functions. Since the graphs of $y = \sin x$ and $y = \cos x$ complete one cycle every 2π units, the origin makes a convenient starting point, especially when transformations are involved. However, the sine and cosine functions are defined for all real numbers. The same is not true for the remaining four trigonometric functions. To graph these functions, we must address the asymptotic behavior of the functions at the values of x where the functions are undefined.

1. Sketch Graphs of Tangent Functions

To graph the function $y = \tan x$, we begin with what we know about the function. Since $y = \tan x = \frac{\sin x}{\cos x}$, $\tan x = 0$, where $\sin x = 0$. These x-intercepts occur at all integer multiples of π—i.e., $x = k\pi$, where k is any integer. The function $y = \tan x$ is not defined where $\cos x = 0$. Since $\cos x = 0$ for all odd-integer multiples of $\frac{\pi}{2}$, the domain of the tangent function is all real numbers except odd-integer multiples of $\frac{\pi}{2}$. Therefore, the graph of the tangent function will have vertical asymptotes at $x = \frac{\pi}{2}n$, where n is an odd integer. Figure 4-83 shows some of the asymptotes and x-intercepts of the tangent function.

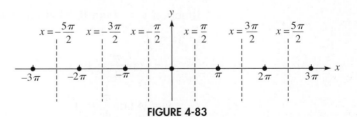

FIGURE 4-83

The graph of a function will not intersect with a vertical asymptote. The function $y = \tan x$ has only one x-intercept in each interval formed by the vertical asymptotes; therefore, the graph can cross the x-axis only once in each interval. We must determine the behavior of the graph as the value of the variable approaches each asymptote. The function will either increase without bound (go to positive infinity) or decrease without bound (go to negative infinity). We will use the interval $\left(-\frac{\pi}{2}, \frac{\pi}{2}\right)$ as our basic interval. Each interval will look the same because the function is periodic. The values in Table 6 show us that as x gets closer to $x = \frac{\pi}{2}$ from the left, the function increases without bound $\left(\lim\limits_{x \to \frac{\pi}{2}^-} \tan x = \infty \right)$. As x gets closer to $-\frac{\pi}{2}$ from the right, the function decreases without bound $\left(\lim\limits_{x \to -\frac{\pi}{2}^+} \tan x = -\infty \right)$.

TABLE 6

$x \to \frac{\pi}{2}^-$	$y = \tan x$	$x \to -\frac{\pi}{2}^+$	$y = \tan x$
$\frac{\pi}{4}$	1	$-\frac{\pi}{4}$	-1
$\frac{\pi}{3}$	$\frac{\sqrt{3}}{3} \approx 1.73$	$-\frac{\pi}{3}$	$-\frac{\sqrt{3}}{3} \approx -1.73$
$\frac{\pi}{2} - .5 \approx 1.07$	≈ 1.83	$-\frac{\pi}{2} + .5 \approx -1.07$	≈ -1.83
$\frac{\pi}{2} - .1 \approx 1.47$	≈ 9.97	$-\frac{\pi}{2} + .1 \approx -1.47$	≈ -9.97
$\frac{\pi}{2} - .01 \approx 1.56$	≈ 100	$-\frac{\pi}{2} + .01 \approx -1.56$	≈ -100
$\frac{\pi}{2} - .001 \approx 1.57$	$\approx 1{,}000$	$-\frac{\pi}{2} + .001 \approx -1.57$	$\approx -1{,}000$

The graph of $y = \tan x$ is shown in Figure 4-84. Note that we can use three key points in the basic interval $\left(-\frac{\pi}{2}, \frac{\pi}{2} \right)$. These points are $(0, 0)$, $\left(\frac{\pi}{4}, 1 \right)$, and $\left(-\frac{\pi}{4}, -1 \right)$. Notice that the x-intercept is at the one-half point of the interval. The other two key points are at the one-quarter and three-quarters points.

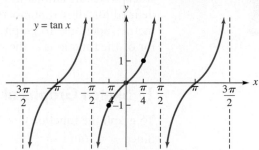

The domain is all real numbers except for odd-integer multiples of $\frac{\pi}{2}$. The range is all real numbers.

FIGURE 4-84

For the sine and cosine functions, the value of A determines the amplitude. The tangent function does not have amplitude, because the range of the function is all real numbers, $R = (-\infty, \infty)$. How do the values of A and B affect the graph of the function $y = A \tan Bx$? Let's look at the function for our two key points $\left(\frac{\pi}{4}, 1 \right)$ and $\left(-\frac{\pi}{4}, -1 \right)$.

$$y = \tan \frac{\pi}{4} = 1 \qquad\qquad y = \tan \left(-\frac{\pi}{4} \right) = -1$$

$$y = A \tan \frac{\pi}{4} = A \qquad \text{and} \qquad y = A \tan \left(-\frac{\pi}{4} \right) = -A$$

Therefore, multiplying the function by A results in a vertical stretch or shrink of A units. If $A < 0$, the graph is reflected across the x-axis.

Just as with the graph of $y = \sin Bx$, a change in the value of B results in a horizontal stretch or shrink by changing the period of the function. The period of $y = A \tan x$ is π; therefore, the period of $y = A \tan Bx$ is $\frac{\pi}{B}$. Using our basic period, we can find the asymptotes by solving the inequality $-\frac{\pi}{2} < Bx < \frac{\pi}{2}$, which results in the inequality $-\frac{\pi}{2B} < x < \frac{\pi}{2B}$. This is summarized in Figure 4-85.

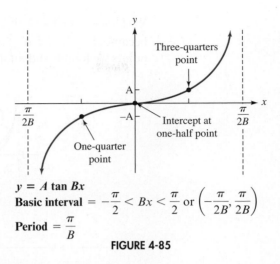

$y = A \tan Bx$

Basic interval $= -\dfrac{\pi}{2} < Bx < \dfrac{\pi}{2}$ or $\left(-\dfrac{\pi}{2B}, \dfrac{\pi}{2B}\right)$

Period $= \dfrac{\pi}{B}$

FIGURE 4-85

EXAMPLE 1   Graphing Functions in the Form $y = A \tan Bx$

Graph two periods of each function.   **a.** $y = 3 \tan \pi x$ **b.** $y = -\tan \dfrac{1}{2}x$

SOLUTION   **a.** The period of the function $y = 3 \tan \pi x$ is $\frac{\pi}{B} = \frac{\pi}{\pi} = 1$. You can find the basic period by using the interval $\left(-\frac{\pi}{2B}, \frac{\pi}{2B}\right) = \left(-\frac{\pi}{2\pi}, \frac{\pi}{2\pi}\right) = \left(-\frac{1}{2}, \frac{1}{2}\right)$. (Or you can solve the inequality $-\frac{\pi}{2} < Bx < \frac{\pi}{2}$ to find the interval.) There will be vertical asymptotes at $x = -\frac{1}{2}$ and $x = \frac{1}{2}$. The x-intercept in this interval will be at one-half of the interval, $x = 0$, and the point is $(0, 0)$. We then plot the two other key points at the one-quarter and three-quarters points. The value of the function at the one-quarter point, at $x = -\frac{1}{4}$, is $y = 3 \tan \pi\left(-\frac{1}{4}\right) = 3 \tan\left(-\frac{\pi}{4}\right) = -3$. The value of the function at the three-quarters point, at $x = \frac{1}{4}$, is $y = 3 \tan\left(\frac{\pi}{4}\right) = 3$. The graph in Figure 4-86 shows the basic period as well as another period, and the points we found, $\left(-\frac{1}{4}, -3\right)$, $(0, 0)$, and $\left(\frac{1}{4}, 3\right)$.

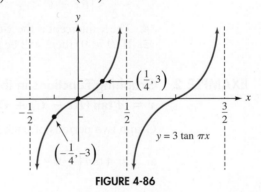

FIGURE 4-86

b. The period of the function $y = -\tan \dfrac{1}{2}x$ is $\dfrac{\pi}{\frac{1}{2}} = 2\pi$. You can find the basic period

by using the interval $\left(-\dfrac{\pi}{2B}, \dfrac{\pi}{2B}\right) = \left(-\dfrac{\pi}{2\left(\frac{1}{2}\right)}, \dfrac{\pi}{2\left(\frac{1}{2}\right)}\right) = (-\pi, \pi)$. (Or you can

solve the inequality $-\frac{\pi}{2} < Bx < \frac{\pi}{2}$ to find the interval.) There will be vertical asymptotes at $x = -\pi$ and $x = \pi$. The x-intercept in this interval will be $x = 0$, and the point is $(0, 0)$. We then plot the two other key points at the one-quarter and three-quarters points. The value of the function at the one-quarter point, at $x = -\frac{\pi}{2}$, is $y = -\tan\frac{1}{2}\left(-\frac{\pi}{2}\right) = -\tan\left(-\frac{\pi}{4}\right) = -(-1) = 1$. The value of the function at the three-quarters point, at $x = \frac{\pi}{2}$, is $y = -\tan\frac{1}{2}\left(\frac{\pi}{2}\right) = -\tan\left(\frac{\pi}{4}\right) = -1$.

We plot the points $(0, 0)$, $\left(-\frac{\pi}{2}, 1\right)$, and $\left(\frac{\pi}{2}, -1\right)$. Note that the graph is reflected, because $A = -1 < 0$. The graph in Figure 4-87 shows the basic period, the points we found, and another period.

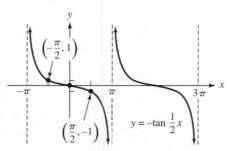

FIGURE 4-87

Self Check 1   Graph the function $y = \frac{1}{3}\tan 3x$.

Now Try Exercise 15.

When the tangent function is in the form $y = A\tan(Bx - C)$, there will be a phase shift (horizontal shift). The phase shift is $\frac{C}{B}$; however, since we have asymptotes at the beginning and end of the basic period we are using, we solve the inequality $-\frac{\pi}{2} < Bx - C < \frac{\pi}{2}$ to find the asymptotes. Adding a constant to the function, $y = A\tan(Bx - C) + D$, produces a vertical shift.

Summary of Strategy for Graphing
$y = A\tan(Bx - C) + D$

1. Determine the period. Period $= \dfrac{\pi}{B}$.

2. Use the basic period $\left(-\dfrac{\pi}{2}, \dfrac{\pi}{2}\right)$.

3. Find the asymptotes of one period by solving the inequality
$-\dfrac{\pi}{2} < Bx - C < \dfrac{\pi}{2}$.

4. The x-intercept is the x-coordinate of the one-half point of the interval if $D = 0$.

5. If $D \neq 0$, there will be a vertical shift of D units.

EXAMPLE 2   **Graphing Functions in the Forms $y = A\tan(Bx - C)$ and $y = A\tan(Bx - C) + D$**

Graph two periods of each function.

a. $y = 4\tan\left(\dfrac{\pi}{2}x - \dfrac{\pi}{2}\right)$ **b.** $y = \tan\left(x + \dfrac{\pi}{2}\right) + 2$

SOLUTION   **a.** The period of the function $y = 4\tan\left(\dfrac{\pi}{2}x - \dfrac{\pi}{2}\right)$ is $\dfrac{\pi}{B} = \dfrac{\pi}{\frac{\pi}{2}} = 2$. We will find a basic period by solving the inequality $-\dfrac{\pi}{2} < Bx - C < \dfrac{\pi}{2}$.

$$-\dfrac{\pi}{2} < \dfrac{\pi}{2}x - \dfrac{\pi}{2} < \dfrac{\pi}{2}$$

$$0 < \dfrac{\pi}{2}x < \pi$$

$$0 < x < \dfrac{\pi}{\left(\dfrac{\pi}{2}\right)}$$

$$0 < x < 2$$

There will be vertical asymptotes at $x = 2k$, where k is an integer. The x-intercepts will be in the middle of each interval bounded by vertical asymptotes.

The one-quarter point is at $x = \dfrac{1}{2}$, yielding

$$y = 4 \tan\left(\frac{\pi}{2} \cdot \frac{1}{2} - \frac{\pi}{2}\right) = 4 \tan\left(-\frac{\pi}{4}\right) = -4.$$

The three-quarters point is at $x = \dfrac{3}{2}$, yielding

$$y = 4 \tan\left(\frac{\pi}{2} \cdot \frac{3}{2} - \frac{\pi}{2}\right) = 4 \tan\frac{\pi}{4} = 4.$$

We plot the points $\left(\dfrac{1}{2}, -4\right)$, $(1, 0)$, and $\left(\dfrac{3}{2}, 4\right)$.

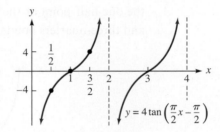

FIGURE 4-88

b. For the function $y = \tan\left(x + \frac{\pi}{2}\right) + 2$, $B = 1$; therefore, the period is π. Since $C = -\frac{\pi}{2}$, the phase shift is $\frac{C}{B} = -\frac{\pi}{2}$. We find the interval by solving the inequality $-\frac{\pi}{2} < Bx - C < \frac{\pi}{2}$.

$$-\frac{\pi}{2} < x + \frac{\pi}{2} < \frac{\pi}{2}$$
$$-\pi < x < 0$$

The basic period we will graph is $(-\pi, 0)$. There is a vertical shift up 2 units. The graph is shown in Figure 4-89.

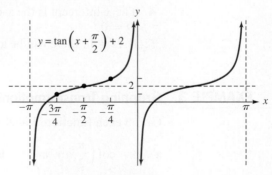

FIGURE 4-89

Self Check 2   Graph each function. **a.** $y = -2\tan\left(x + \dfrac{\pi}{4}\right)$ **b.** $y = \tan 2\pi x - 3$

Now Try Exercise 23.

2. Sketch Graphs of Cotangent Functions

Since $y = \cot x = \frac{\cos x}{\sin x}$, the x-intercepts and vertical asymptotes will be swapped, relative to $y = \tan x = \frac{\sin x}{\cos x}$. Therefore, the x-intercepts will be at odd-integer multiples of $\frac{\pi}{2}$ and the vertical asymptotes will be at integer multiples of π. The period of $y = \cot x$ is π, and the graph completes a cycle every π units. For our basic period, we will use the interval $(0, \pi)$.

The graph of $y = \cot x$ is shown in Figure 4-90. Over the basic interval $(0, \pi)$, we use three key points: $\left(\frac{\pi}{4}, 1\right), \left(\frac{\pi}{2}, 0\right)$, and $\left(\frac{3\pi}{4}, -1\right)$. Notice that the x-intercept is at the one-half point of the interval. The other two key points are at the one-quarter and three-quarters points.

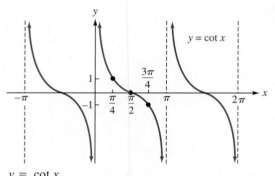

$y = \cot x$
The domain is all real numbers except for integer
multiples of π. The range is all real numbers.
FIGURE 4-90

Transformations of the function $y = \cot x$ are like those we discovered when graphing the tangent function.

Summary of Strategy for Graphing
$y = A\cot(Bx - C) + D$

1. Determine the period. Period $= \dfrac{\pi}{B}$.
2. Use the basic period $(0, \pi)$.
3. Find the asymptotes of one period by solving the inequality
   $0 < Bx - C < \pi$.
4. The x-intercept is the x-coordinate of the one-half point of the interval if $D = 0$.
5. If $D \neq 0$, there will be a vertical shift of D units.

EXAMPLE 3   **Graphing Transformations of** $y = \cot x$

Graph each function. Label at least one complete period.

a. $y = \cot\left(\dfrac{\pi}{2}x + \pi\right)$ **b.** $y = -2\cot 3x + 1$

SOLUTION   **a.** $y = \cot\left(\dfrac{\pi}{2}x + \pi\right)$ has period $= \dfrac{\pi}{B} = \dfrac{\pi}{\frac{\pi}{2}} = 2$. The basic period is found by

solving the inequality $0 < Bx - C < \pi$.

$$0 < \frac{\pi}{2}x + \pi < \pi$$

$$-\pi < \frac{\pi}{2}x < 0$$

$$\frac{-\pi}{\left(\frac{\pi}{2}\right)} < x < 0$$

$$-2 < x < 0$$

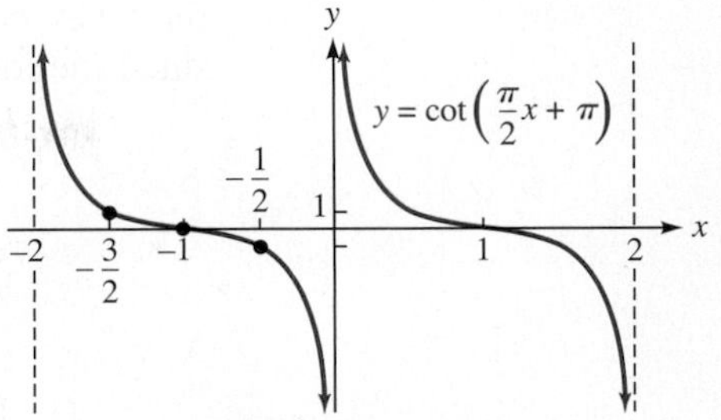

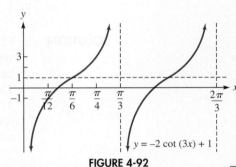

FIGURE 4-91

The x-intercept in the basic interval is at the one-half point, at $x = -1$. The one-quarter point is at $x = -\frac{3}{2}$. We evaluate $y = \cot\left[\frac{\pi}{2} \cdot \left(-\frac{3}{2}\right) + \pi\right] = \cot\frac{\pi}{4} = 1$ and plot the point $\left(-\frac{3}{2}, 1\right)$. The three-quarters point is at $x = -\frac{1}{2}$. We evaluate $y = \cot\left[\frac{\pi}{2} \cdot \left(-\frac{1}{2}\right) + \pi\right] = \cot\frac{3\pi}{4} = -1$ and plot the point $\left(-\frac{1}{2}, -1\right)$.

b. $y = -2\cot 3x + 1$ has period $\frac{\pi}{B} = \frac{\pi}{3}$. The basic period is found by solving the inequality $0 < 3x < \pi \Rightarrow 0 < x < \frac{\pi}{3}$. There is a vertical shift up 1 unit. The function is reflected across the x-axis and there is a vertical stretch of 2. To find the three key points using the period $\left(0, \frac{\pi}{3}\right)$, we evaluate the function at the one-quarter point, at $x = \frac{\pi}{12}$; the one-half point, at $x = \frac{\pi}{6}$; and the three-quarters point, at $x = \frac{\pi}{4}$.

$$-2\cot\left(3 \cdot \frac{\pi}{12}\right) + 1 = -2\cot\frac{\pi}{4} + 1 = -2(1) + 1 = -1$$

$$-2\cot\left(3 \cdot \frac{\pi}{6}\right) + 1 = -2\cot\frac{\pi}{2} + 1 = -2(0) + 1 = 1$$

$$-2\cot\left(3 \cdot \frac{\pi}{4}\right) + 1 = -2\cot\frac{3\pi}{4} + 1 = -2(-1) + 1 = 3$$

FIGURE 4-92

Self Check 3   Graph each function. **a.** $y = 3\cot\left(\pi x - \frac{\pi}{2}\right)$ **b.** $y = \cot\frac{1}{4}x - 2$

Now Try Exercise 29.

3. Sketch Graphs of Cosecant Functions

Graphing the function $y = \csc x$ and its variations can be accomplished by using the definition $y = \csc x = \frac{1}{\sin x}$. The period of both functions is 2π. To graph $y = \csc x$, first graph $y = \sin x$. Because $y = \csc x$ is undefined whenever $y = \sin x$ is 0, the graph will have vertical asymptotes at the x-intercepts of the sine curve, which are multiples of π. Since the function is periodic, we will use $(0, 2\pi)$ as the basic period. The graph can then be extended as far as desired in either direction. An open interval is used because $y = \csc x$ is not defined at the endpoints of the period. When $x = \frac{\pi}{2}$, both functions are 1; their graphs intersect at the point $\left(\frac{\pi}{2}, 1\right)$. Similarly, the graphs intersect at $\left(\frac{3\pi}{2}, -1\right)$. We also know that $\csc x > 0$ when $\sin x > 0$, and $\csc x < 0$ when $\sin x < 0$. We can use this to determine the asymptotic behavior of $y = \csc x$. As $\sin x$ gets closer to 0, the reciprocal $\csc x$ increases without bound toward positive infinity or decreases without bound toward negative infinity. Figure 4-93 shows the graphs of both $y = \csc x$ and $y = \sin x$. Note that the 5 key points of the sine curve are included for reference. If you can graph the sine function, you can graph the cosecant function.

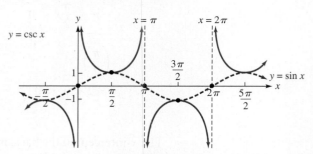

The domain is all real numbers except for integer multiples of π. The range is $(-\infty, -1] \cup [1, \infty)$.

FIGURE 4-93

The function $y = \csc x$ does not have amplitude.

EXAMPLE 4   Graphing Transformations of $y = \csc x$

Graph each function. Label at least one complete period.

a. $y = 2 \csc \pi x$ **b.** $y = -\csc\left(x - \dfrac{\pi}{4}\right)$ **c.** $y = 3 \csc \dfrac{1}{2}x + 2$

SOLUTION   a. First we will graph $y = 2 \sin \pi x$. We will then be able to graph $y = 2 \csc \pi x$ very quickly. The period is $\frac{2\pi}{B} = \frac{2\pi}{\pi} = 2$. To begin the graph, we select the basic period $(0, 2)$. The function $y = 2 \sin \pi x$ is 0 at $x = 0$, $x = 1$, and $x = 2$. These values of vertical asymptotes of $y = 2 \csc \pi x$ occur over the period $(0, 2)$. The amplitude of $y = 2 \sin \pi x$ is 2. We now have enough information to graph the functions. The sine function $y = 2 \sin \pi x$ is shown in the graph, but is not actually part of the graph. It is shown for reference only.

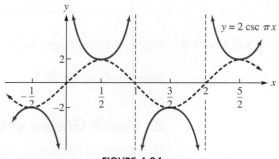

FIGURE 4-94

b. To graph $y = -\csc\left(x - \frac{\pi}{4}\right)$, we begin with the graph of $y = -\sin\left(x - \frac{\pi}{4}\right)$. The period is 2π. To find the basic period of $y = -\sin\left(x - \frac{\pi}{4}\right)$, we solve the inequality

$$0 < x - \frac{\pi}{4} < 2\pi$$

$$\frac{\pi}{4} < x < \frac{9\pi}{4}$$

The basic period that we will use is $\left(\frac{\pi}{4}, \frac{9\pi}{4}\right)$. $y = -\sin\left(x - \frac{\pi}{4}\right)$ will intercept the x-axis, and $y = -\csc\left(x - \frac{\pi}{4}\right)$ will have vertical asymptotes, at $x = \frac{\pi}{4}$, $x = \frac{5\pi}{4}$, and $x = \frac{9\pi}{4}$. The function is reflected across the x-axis.

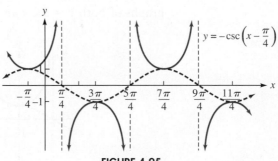

FIGURE 4-95

c. Graphing $y = 3 \csc \frac{1}{2}x + 2$ will involve a vertical shift up 2 units. We will continue the strategy of using the sine function as the guide. The function $y = 3 \sin \frac{1}{2}x + 2$ has period $\dfrac{2\pi}{B} = \dfrac{2\pi}{\frac{1}{2}} = 4\pi$. There is no phase shift. The function $y = 3 \sin \frac{1}{2}x$ oscillates between a maximum value of 3 and a minimum value of -3; therefore, $y = 3 \sin \frac{1}{2}x + 2$ will oscillate between a maximum value of $3 + 2 = 5$ and a minimum value of $-3 + 2 = -1$. The period is 4π, so the basic period we will use for our graph is $(0, 4\pi)$. The function $y = 3 \sin \frac{1}{2}x + 2$ intercepts the horizontal line $y = 2$ at $x = 0$, $x = 2\pi$, and $x = 4\pi$. These are locations of vertical asymptotes of $y = 3 \csc \frac{1}{2}x + 2$. The graph is shown in Figure 4-96.

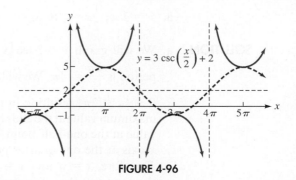

FIGURE 4-96

Self Check 4   Graph each function. **a.** $y = \csc 2x - 3$ **b.** $y = -\dfrac{1}{2}\csc\left(x + \dfrac{\pi}{2}\right)$

Now Try Exercise 43.

4. Sketch Graphs of Secant Functions

We will examine the graph of the function $y = \sec x$ and its variations by using the definition $y = \sec x = \frac{1}{\cos x}$. The period of both functions is 2π. To graph $y = \sec x$, we graph $y = \cos x$; we will use the basic period $(0, 2\pi]$. The function $y = \csc x$ will have vertical asymptotes at the values of x for which $y = \cos x = 0$. These occur at all odd-integer multiples of $x = \frac{\pi}{2}$. Both functions are 1, and the graphs intersect, at the points $(0, 1)$ and $(2\pi, 1)$. Similarly, the graphs intersect at the point $(\pi, -1)$. We also know that $\sec x > 0$ when $\cos x > 0$ and $\sec x < 0$ when $\cos x < 0$. We can use this to determine the asymptotic behavior of $y = \sec x$. As $\cos x$ gets closer to 0, the reciprocal $\sec x$ increases without bound toward positive infinity or decreases without bound toward negative infinity. Figure 4-97 shows the graphs of both $y = \sec x$ and $y = \cos x$.

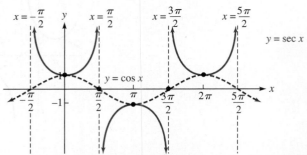

The domain is all real numbers except for odd-integer multiples of $\frac{\pi}{2}$.
The range is $(-\infty, -1] \cup [1, \infty)$.

FIGURE 4-97

The function $y = \sec x$ does not have amplitude.

EXAMPLE 5   **Graphing Transformations of $y = \sec x$**

Graph each function and label at least one period.

a. $y = 3 \sec \dfrac{1}{2}x$ **b.** $y = -\sec\left(x + \dfrac{\pi}{2}\right)$ **c.** $y = 5 \sec \pi x - 3$

SOLUTION   **a.** We will graph $y = 3 \sec \frac{1}{2}x$ by beginning with the graph of $y = 3\cos\frac{1}{2}x$. The period is $\dfrac{2\pi}{\frac{1}{2}} = 4\pi$. We will use the basic period found by solving the inequality $0 \le \frac{1}{2}x \le 2\pi$, resulting in the interval $[0, 4\pi]$. The cosine function reaches its maximum value at the ends of the period, at $x = 0$ and $x = 4\pi$, and the minimum value at the one-half point, at $x = 2\pi$. The function $y = 3\cos\frac{1}{2}x$ intercepts the x-axis at the one-quarter point, $x = \pi$; and the three-quarters point, $x = 3\pi$. Therefore, $x = \pi$ and $x = 3\pi$ will be vertical asymptotes of $y = 3 \sec\frac{1}{2}x$ over the period $[0, 4\pi]$. The graph is shown in Figure 4-98.

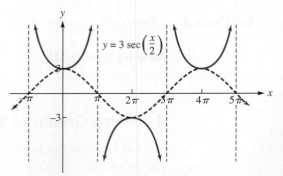

FIGURE 4-98

b. The graph of $y = -\sec\left(x + \frac{\pi}{2}\right)$ will have a phase shift $\frac{\pi}{2}$ units to the left, and will be reflected across the x-axis. So the basic period we will use is $\left[-\frac{\pi}{2}, \frac{3\pi}{2}\right]$. By graphing the function $y = -\cos\left(x + \frac{\pi}{2}\right)$, we can insert the vertical asymptotes where $y = -\cos\left(x + \frac{\pi}{2}\right) = 0$. These points are at the one-quarter point, $x = 0$; and the three-quarters point, $x = \pi$. Since the graph is reflected, the minimum values are at the ends of the period and the maximum value is in the middle of the period. The graph is shown in Figure 4-99.

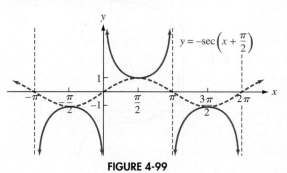

FIGURE 4-99

c. Graphing the function $y = 5\sec \pi x - 3$ will involve a period change and a vertical shift down 3 units. The period is $\frac{2\pi}{B} = \frac{2\pi}{\pi} = 2$. There is no phase shift, so we will use the basic period $[0, 2]$. The function $y = 5\cos \pi x - 3$ will intercept the line $y = -3$ at the one-quarter point, $x = \frac{1}{2}$; and the three-quarters point, $x = \frac{3}{2}$. The function $y = 5\cos \pi x$ has a maximum value of 5, so $y = 5\cos \pi x - 3$ has the maximum value of $5 - 3 = 2$. The minimum value of $y = 5\cos \pi x$ is -5, so $y = 5\cos \pi x - 3$ will have a minimum value of $-5 - 3 = -8$. The maximum value of $y = 5\cos \pi x - 3$ occurs at the ends of the period, and the minimum value occurs at the one-half point. The graph is shown in Figure 4-100.

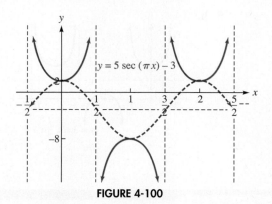

FIGURE 4-100

Self Check 5   Graph each function.   **a.** $y = \sec\left(2x - \frac{\pi}{4}\right)$ **b.** $y = 3\sec\frac{1}{4}x + 4$

Now Try Exercise 45.

Self Check Answers

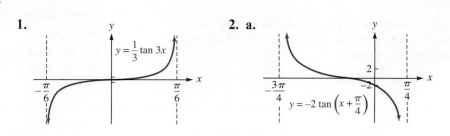

2. b.

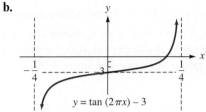

$y = \tan(2\pi x) - 3$

3. a.

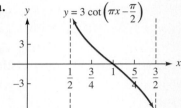

$y = 3\cot\left(\pi x - \dfrac{\pi}{2}\right)$

3. b.

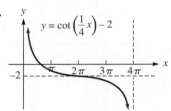

$y = \cot\left(\dfrac{1}{4}x\right) - 2$

4. a.

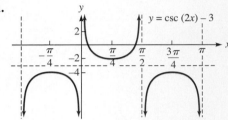

$y = \csc(2x) - 3$

4. b.

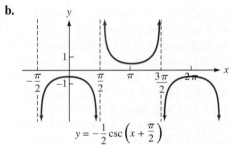

$y = -\dfrac{1}{2}\csc\left(x + \dfrac{\pi}{2}\right)$

5. a.

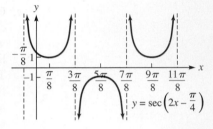

$y = \sec\left(2x - \dfrac{\pi}{4}\right)$

5. b.

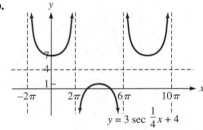

$y = 3\sec\dfrac{1}{4}x + 4$

Exercises 4.5

Getting Ready
You should be able to complete these vocabulary and concept questions before you proceed to the practice exercises.

Fill in the blanks.

1. What is the domain of the tangent function? range?

2. What is the domain of the cotangent function? range?

3. Over the interval $[-2\pi, 2\pi]$, at what values of x does the graph of $y = \tan x$ have vertical asymptotes? x-intercepts?

4. Over the interval $[-2\pi, 2\pi]$, at what values of x does the graph of $y = \cot x$ have vertical asymptotes? x-intercepts?

5. What is the y-intercept of $y = \tan x$?

6. Why does $y = \cot x$ not have a y-intercept?

7. What is the domain of the secant function? range?

8. What is the domain of the cosecant function? range?

9. Over the interval $[-2\pi, 2\pi]$, for what values of x is $\sin x = \csc x$?

10. Over the interval $[-2\pi, 2\pi]$, for what values of x is $\cos x = \sec x$?

Practice

Graph two periods of each function. Label one period, including asymptotes and intercepts.

11. $y = 3 \tan x$

12. $y = \dfrac{1}{2} \tan x$

13. $y = -\dfrac{1}{4} \tan 3x$

14. $y = -5 \tan 2\pi x$

15. $y = 3 \tan \dfrac{\pi}{4} x$

16. $y = 2 \tan 4x$

17. $y = \dfrac{1}{3} \cot x$

18. $y = 4 \cot x$

19. $y = -4 \cot \dfrac{\pi}{2} x$

20. $y = -\dfrac{1}{2} \cot 2x$

21. $y = \tan\left(x - \dfrac{\pi}{4}\right)$

22. $y = -\tan\left(x + \dfrac{\pi}{2}\right)$

23 $y = -3 \tan\left(\dfrac{x}{2} + \dfrac{\pi}{2}\right)$

24. $y = \dfrac{2}{3} \tan\left(\dfrac{1}{2}x - \pi\right)$

25. $y = \tan\left(\pi x - \dfrac{\pi}{2}\right)$

26. $y = 2 \tan(2x - 2\pi)$

27. $y = \cot\left(x + \dfrac{\pi}{4}\right)$

28. $y = -\cot\left(x + \dfrac{\pi}{2}\right)$

29. $y = \dfrac{1}{2} \cot(2x - \pi)$

30. $y = 4 \cot\left(\dfrac{x}{3} - \dfrac{\pi}{2}\right)$

31. $y = 1 - \cot \pi x$

32. $y = -3 + \cot \dfrac{x}{2}$

33. $y = 2 + \tan \dfrac{x}{2}$

34. $y = 4 - \tan \dfrac{\pi x}{2}$

35. $y = 2 \csc x$

36. $y = \dfrac{1}{2} \sec x$

37. $y = 3 \sec \dfrac{x}{3}$

38. $y = 7 \sec \dfrac{\pi x}{4}$

39. $y = 3 - \sec x$

40. $y = 1 + \csc x$

41. $y = -4 \csc \dfrac{2\pi x}{5}$

42. $y = -3 \csc \pi x$

43. $y = \csc\left(x + \dfrac{\pi}{2}\right)$

44. $y = \sec\left(x + \dfrac{\pi}{4}\right)$

45. $y = -2 \sec\left(x - \dfrac{\pi}{3}\right)$

46. $y = 2 \csc\left(\dfrac{x}{3} - \dfrac{\pi}{2}\right)$

Applications

47. Logging A logging company uses a canal to move logs from location A to location B. The canal is 20 feet wide and makes a right-angle turn, becoming 25 feet wide.

a. The length L of the log shown, as a function of the angle θ, is $L(\theta) = 20 \sec \theta + 25 \csc \theta$. Show why this is true.

b. Graph $L(\theta)$ over the interval $\left(0, \dfrac{\pi}{2}\right)$. What is the maximum value of θ? Why?

c. What is the length of the longest log that can be sent down the canal?

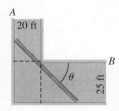

48. Spotlight A prison spotlight is 100 feet from the prison building. The building is 400 feet long. The spotlight rotates at the rate of exactly 1 revolution every 5 seconds, with $\theta = \dfrac{2}{5}\pi t$. The function $d(t) = \left|100 \tan \dfrac{2}{5}\pi t\right|$ determines the distance d in feet of the light beam from point C after t seconds.

a. Graph $d(t)$ over the interval $(0, 1.25)$.

b. What happens to the value of a as t approaches 1.25 seconds?

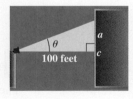

Discovery and Writing

49. Sketch the graphs of $y = \tan x$ and $y = \cot x$. Identify the points of intersection between 0 and 2π.

50. Sketch the graphs of $y = \sec x$ and $y = \csc x$. Identify the points of intersection between 0 and 2π.

4.6 Regression and Other Graphs Involving Trigonometric Functions (optional)

In this section, we will learn to

1. Use sinusoidal regression to model sets of data.
2. Sketch the graphs of trigonometric functions using addition of ordinates.
3. Sketch the graphs of powers of sine functions using a graphing calculator.
4. Use a Fourier series to solve problems.
5. Sketch the graphs of damping functions.

The Sydney Opera House is an Australian landmark that is recognized throughout the world. It was completed in 1973 and is one of the great iconic buildings of the twentieth century. The building houses multiple performance venues and is one of the busiest performing-arts centers in the world, hosting over 1,500 performances each year.

The Sydney Opera House was designed so that each performance venue would have the best acoustics possible. In understanding acoustics and constructing our great concert halls, we apply trigonometry. Many problems in the fields of electrical engineering, optics, and image processing involve trigonometry. We will study applications involving the graphs of the trigonometric functions in this section.

1. Use Sinusoidal Regression to Model Sets of Data

Regression techniques can be tedious and difficult. However, computer software and graphing calculators have greatly enhanced our ability to find an equation of a function that fits a set of data points. The accuracy resulting from technology is, in many cases, far superior to that of regressions done by hand. In Section 4.4, we found a sine function that could fit a given graph. However, when we have data points, there is no complete graph, so it is up to us to find a function that will best fit the data. In the following example, we will work through the problem using a regression program and then do it without a regression program.

EXAMPLE 1   Finding a Sine Function to Fit Data—Sinusoidal Regression

The data in Table 7 show the average monthly temperature (in degrees Fahrenheit) in Baton Rouge, Louisiana. Use the regression capabilities of a graphing calculator to find a sine curve that approximates the function. Then use your knowledge of the sine function to find a sine curve to approximate the data.

TABLE 7

Month	Jan	Feb	Mar	Apr	May	Jun	Jul	Aug	Sep	Oct	Nov	Dec
Avg. Temp.	50	54	61	68	75	81	82	82	78	68	60	54

SOLUTION   Follow the steps we've seen in earlier examples of regression (see, for instance, Example 9 in Section 2.1), but choose SinReg under the CALC list. The resulting equation and graph of the function are shown in Figure 4-101.

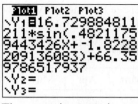
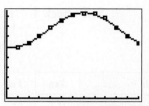

The regression equation will be copied into Y1.

Graph the function and the data. The fit is good.

FIGURE 4-101

Finding an equation without a regression program is not difficult with this set of data. Starting with the equation $y = A \sin (Bx - C) + D$, we will find the value of each variable. The values of A and D are found from the greatest value and least value of the data set.

- $A = \dfrac{\text{greatest value} - \text{least value}}{2} = \dfrac{82 - 50}{2} = 16$, so the amplitude is 16.

- $D = \dfrac{\text{greatest value} + \text{least value}}{2} = \dfrac{82 + 50}{2} = 66$ is the vertical shift.

- The period is 12, so we can find B: $\dfrac{2\pi}{B} = 12 \rightarrow B = \dfrac{\pi}{6}$.

- So far, we have $y = 16 \sin\left(\frac{\pi}{6}x - C\right) + 66$. Taking the first data point $(1, 50)$, we put these values into the function and solve for C.

$$50 = 16 \sin\left(\frac{\pi}{6}(1) - C\right) + 66$$

$$-16 = 16 \sin\left(\frac{\pi}{6} - C\right)$$

$$-1 = \sin\left(\frac{\pi}{6} - C\right)$$

$$\sin\left(-\frac{\pi}{2}\right) = \sin\left(\frac{\pi}{6} - C\right) \qquad \text{Substitute } \sin\left(-\frac{\pi}{2}\right) = -1.$$

$$-\frac{\pi}{2} = \frac{\pi}{6} - C$$

$$C = \frac{\pi}{6} + \frac{\pi}{2}$$

$$C = \frac{2\pi}{3}$$

The function we have found is $y = 16 \sin\left(\frac{\pi}{6}x - \frac{2\pi}{3}\right) + 66$. Both functions and the data points are shown in Figure 4-102.

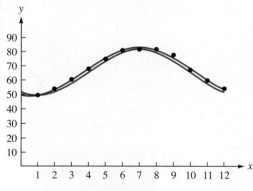

FIGURE 4-102

Self Check 1   The average low monthly temperature (in degrees Fahrenheit) for Anchorage, Alaska, is shown in the table. Use the regression capabilities of a graphing calculator to find a sine curve that approximates the function.

Month	Jan	Feb	Mar	Apr	May	Jun	Jul	Aug	Sep	Oct	Nov	Dec
Avg. Low Temp.	9.3	11.7	18.2	28.7	38.9	47	51.5	49.4	41.4	28.3	15.9	11.4

Now Try Exercise 35.

2. Sketch the Graphs of Trigonometric Functions Using Addition of Ordinates

Trigonometric functions are often used in combinations to produce functions of interest. Addition of ordinates (y values) can be done by hand if the functions are relatively simple, but others will require technology.

EXAMPLE 2   **Using Addition of Ordinates to Sketch the Graph of a Function**

Graph the function $y = x + \cos x$.

SOLUTION   It would be possible to make an extensive table of values, plot the points, and draw the curve. However, it is easier to use our knowledge of the functions $y = x$ and $y = \cos x$ to sketch the function. We will pick some special values of x and add the values of the functions. This is called "addition of ordinates" because we will add the y values (the ordinates).

We know that at even multiples of π, and also at $x = 0$, $y = \cos x = 1$. Therefore, at these values of x, the value of y will be $y = x + \cos x = x + 1$.

Likewise, at odd multiples of π, $y = \cos x = -1$, and the value of y will be $y = x + \cos x = x - 1$ at these values. At odd multiples of $\frac{\pi}{2}$, $y = \cos x = 0$, giving us $y = x + \cos x = x$.

Using these values, we can sketch the graph of $y = x + \cos x$.

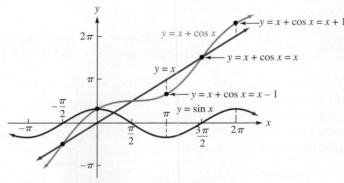

FIGURE 4-103

Graphing calculators can be used to graph combinations of functions. You can graph all three functions, or you can turn off the two functions and graph only the composite function.

Enter all the functions, but turn off the first two.	This is the function using $[-\pi, 2\pi]$ for both axes.	Turning on the function $y = x$ can help the visualization.
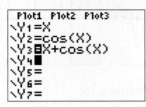	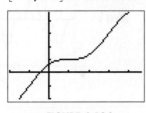	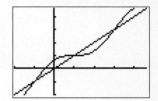

FIGURE 4-104

Self Check 2 Graph the function $y = 2x + \cos x$ over $[0, 2\pi]$.

Now Try Exercise 13.

Some combinations of the trigonometric functions are more difficult to graph by hand, so a graphing calculator is really the most practical way to draw the graph.

EXAMPLE 3   **Using a Calculator to Graph a Composite Function**

Graph the function $y = 3\sin x + 2\sin 2x + \sin 3x$.

SOLUTION The period of $y = \sin x$ is 2π, which is the largest among the three components. Use the window $X: [-\pi, 4\pi]$, $Y: [-7, 7]$. This graph is an approximation of a shape called a **sawtooth waveform**, which has many uses in electronics.

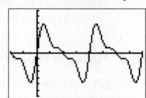

FIGURE 4-105

Self Check 3 Graph the function $y = 3\sin x + 4\sin 2x + 2\sin 3x$.

Now Try Exercise 33.

3. Sketch the Graphs of Powers of Sine Functions Using a Graphing Calculator

In some applications, graphs involving powers of the trigonometric functions appear. Two are shown in Example 4.

EXAMPLE 4   **Graphing Powers of Sine Functions**

Use a graphing calculator to graph the functions $y = \sin^5 x$ and $y = \sin^6 x$.

SOLUTION Use the window $X: [-\pi, 4\pi]$, $Y: [-1.5, 1.5]$. When entering a power of a trigonometric function in a graphing calculator, you need to enter it as in $(\sin (x))^5$, not as in $\sin^5 x$.

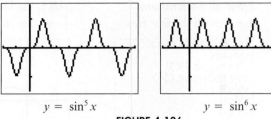

$$y = \sin^5 x \qquad\qquad y = \sin^6 x$$

FIGURE 4-106

Self Check 4 Graph the function $y = \sin^7 x$.

Now Try Exercise 7.

4. Use a Fourier Series to Solve Problems

A thorough knowledge of the nature of sound and of vibrations in both solid material and air is needed for the proper design of musical instruments. For example, the shape of a violin and the materials used in its manufacture (even the varnish) affect the distinctive coloration of the tone.

The analysis of these more complex waveforms still involves the sine function, but in the form $y = A \sin 2\pi ft + A_1 \sin 2\pi f_1 t + A_2 \sin 2\pi f_2 t + \cdots$, where $f_1, f_2, \ldots$ and $A_1, A_2, \ldots$ are the frequencies and amplitudes of the overtones present in the original sound. This function is called a **Fourier sine series**, after French mathematician Jean Baptiste Fourier (1768–1830).

The sum of a Fourier series is also used as an approximation for various square, sawtooth, etc., waveforms that occur in electronics. The approximation is made more accurate by adding more and more terms to the series.

EXAMPLE 5 **Using a Fourier Series**

A stereo system must be of high quality if it is to accurately reproduce a waveform known as a **square wave**. The theory of Fourier explains why: A square wave can be represented as an unending sum of sine functions with frequencies that go beyond audibility. To reproduce a square wave, a stereo system must handle a very wide spectrum of frequencies without distortion. According to Fourier, the function $y = \sin x + \frac{1}{3} \sin 3x + \frac{1}{5} \sin 5x + \frac{1}{7} \sin 7x + \cdots$ represents a square wave. Graph the function through $\sin 7x$, through $\sin 9x$, and through $\sin 11x$. Analyze what happens to the graph as the number of terms increases.

SOLUTION

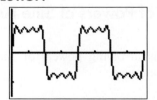

$$y = \sin x + \tfrac{1}{3} \sin 3x + \tfrac{1}{5} \sin 5x + \tfrac{1}{7} \sin 7x$$

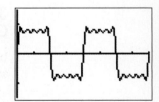

$$y = \sin x + \tfrac{1}{3} \sin 3x + \tfrac{1}{5} \sin 5x + \tfrac{1}{7} \sin 7x + \tfrac{1}{9} \sin 9x$$

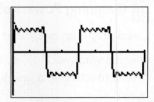

$$y = \sin x + \tfrac{1}{3} \sin 3x + \tfrac{1}{5} \sin 5x + \tfrac{1}{7} \sin 7x + \tfrac{1}{9} \sin 9x + \tfrac{1}{11} \sin 11x$$

FIGURE 4-107

As more terms are added to the function, the graph becomes less bumpy and the rising and falling edges become closer to vertical.   ∎

Self Check 5   Graph the Fourier series

$$f(t) = 2.5 + \frac{10}{\pi} \sin \frac{\pi t}{4} + \frac{2}{\pi} \sin \frac{5\pi t}{4} + \frac{10}{3\pi} \sin \frac{3\pi z}{4} + \frac{10}{7\pi} \sin \frac{7\pi z}{4}.$$

Now Try Exercise 39.

5. Sketch the Graphs of Damping Functions

Functions such as $y = x \sin x$, $y = e^x \cos 2x$, and $y = \ln x \sin \pi x$ are called **damping functions.**

Damping functions have many applications. In music, a WAV file can be approximated by a sine function. When a song fades, the producer decides how a song fades. There are programs that will produce different fades; one offers a linear fade, a logarithmic fade, or an inverse logarithmic fade. These fades all utilize damping functions.

EXAMPLE 6   **Graphing a Damping Function**

Graph the function $y = x \sin x$.

SOLUTION   Just as we did when we added ordinates, we will use our knowledge of the sine function to analyze the function.

$$-1 \le \sin x \le 1$$
$$-x \le x \sin x \le x$$

Therefore, the function $y = x \sin x$ will oscillate between $y = x$ and $y = -x$. If $\sin x = 1$, then $y = x \sin x = x$; this will occur at $x = \frac{\pi}{2} \pm 2\pi k$, where k is any integer. When $\sin x = -1$, which occurs at $x = \frac{3\pi}{2} \pm 2\pi k$, $y = x \sin x = -x$. Also, $\sin x = 0$ yields $y = x \sin x = 0$, when $x = \pi k$. We say that the function is **damped** between the lines $y = x$ and $y = -x$. In this function, x is the damping factor.

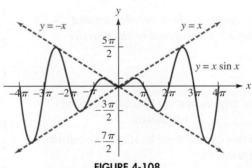

FIGURE 4-108

Self Check 6   Graph the function $y = 2x \sin \dfrac{x}{2}$.

Now Try Exercise 25.

Self Check Answers   1. $y = 21.045295879541 \sin (0.55517100483452x - 2.3243758733976)$
$+ 30.46425268477$

2.

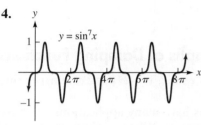

$y = 2x + \cos x$

3.

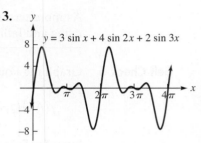

$y = 3 \sin x + 4 \sin 2x + 2 \sin 3x$

4.

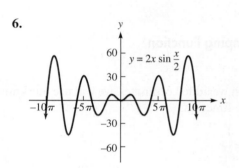

$y = \sin^7 x$

5.

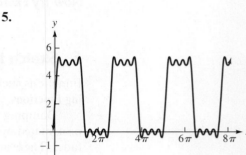

$$f(t) = 2.5 + \frac{10}{\pi} \sin \frac{\pi t}{4} + \frac{2}{\pi} \sin \frac{5\pi t}{4} + \frac{10}{3\pi} \sin \frac{3\pi t}{4} + \frac{10}{7\pi} \sin \frac{7\pi t}{4}$$

6.

$y = 2x \sin \dfrac{x}{2}$

Exercises **4.6**

Getting Ready
You should be able to complete these vocabulary and concept statements before you proceed to the practice exercises.

Fill in the blanks.

1. Combinations of trigonometric functions can be graphed using _____ of ordinates.
2. A(n) _____ series can be used to study the nature of sound and vibrations.
3. In the function $y = x \cos x$, x is the _____ factor.
4. Damping functions can be used to produce _____ files.

Practice
Graph each function.

5. $y = \cos^4 x$
6. $y = -\tan^4 x$
7. $y = -\sin^3 x$
8. $y = \cos^3 x$
9. $y = \sin^{100} x$
10. $y = \cos^{101} x$

11. $y = \cos^2 3x$
12. $y = -\sin^2 \dfrac{x}{2}$
13. $y = \sin x + \cos x$
14. $y = 2 \sin x + 2 \cos \dfrac{x}{2}$
15. $y = \sin \dfrac{x}{2} + \sin x$
16. $y = x + \sin x$
17. $y = 2 \sin x \cos x$
18. $y = \cos^2 x - \sin^2 x$
19. $y = |\sin x|$
20. $y = |\cos x|$
21. $y = \sin x^2$
22. $y = \sin x^3$
23. $y = \sin \sqrt{x}, x \geq 0$
24. $y = \cos \sqrt{x}, x \geq 0$
25. $y = x \cos x$
26. $y = x^2 \cos x$
27. $y = e^{x/8} \sin x$
28. $y = e^{\sqrt{x}} \sin x, x \geq 0$
29. $y = x \sin 10x$
30. $y = 2x \cos 5x$
31. $y = (\ln x)(\sin 15x), x \geq 0$
32. $y = \left(\ln \dfrac{x}{2}\right)(\cos 10x), x \geq 0$
33. $y = 2 \sin x + 3 \sin 2x + \sin 3x$
34. $y = 3 \cos x + 2 \cos 2x + \cos 3x$

⋀ Applications

35. Disneyland attendance The table shows the average monthly attendance (in millions) at Disneyland in Anaheim, California. Use the regression capabilities of a graphing calculator to find a sine curve that approximates the data.

Month	Jan	Feb	Mar	Apr	May	Jun	Jul	Aug	Sep	Oct	Nov	Dec
Avg. Attend.	1.2	1.2	1.2	1.3	1.3	1.4	1.5	1.5	1.5	1.4	1.3	1.2

36. Company sales The table shows the average monthly sales, in thousands of dollars, for a locally owned deli. Use the regression capabilities of a graphing calculator to find a sine curve that approximates the data.

Month	Jan	Feb	Mar	Apr	May	Jun	Jul	Aug	Sep	Oct	Nov	Dec
Avg. Sales	20	18	15	19	23	29	33	37	33	32	28	25

37. Walt Disney World temperature The table shows the average high monthly temperature (in degrees Fahrenheit) at Walt Disney World, in Florida. Use the regression capabilities of a graphing calculator to find a sine curve that approximates the data. Then use your knowledge of the sine function to find a sine curve to approximate the data.

Month	Jan	Feb	Mar	Apr	May	Jun
Avg. Temp.	72	73	77	82	87	90

Month	Jul	Aug	Sep	Oct	Nov	Dec
Avg. Temp.	92	92	90	84	79	73

38. Seattle temperature The table shows the average low monthly temperature (in degrees Fahrenheit) in Seattle, Washington. Use the regression capabilities of a graphing calculator to find a sine curve that approximates the data. Then use your knowledge of the sine function to find a sine curve to approximate the data.

Month	Jan	Feb	Mar	Apr	May	Jun
Avg. Temp.	36	37	39	43	48	53

Month	Jul	Aug	Sep	Oct	Nov	Dec
Avg. Temp.	56	57	53	46	40	36

⋀ Discovery and Writing

39. Graph the first four terms of the Fourier series given by

$$f(x) = \sin x - \frac{1}{3^2} \sin 3x + \frac{1}{5^2} \sin 5x - \frac{1}{7^2} \sin 7x + \cdots .$$

This graph is called a **triangular pulse**. It has applications in timer circuitry in electronics.

40. On the same axes, graph $y = \cos^2 x$ and $y = \frac{1}{2} \cos 2x$. One graph is a vertical translation of the other. Use the TRACE feature to discover the amount of translation, and express this as an equation.

41. Investigate the effect of various values for the exponent n on the graph of $y = \sin^n x$. What values of n cause the graph to lie above the x-axis? What values of n cause the graph to appear both above and below the x-axis?

42. Refer to Exercise 41. What happens to the shape of the peaks of the graph as the value of n increases?

CHAPTER REVIEW

SECTION **4.1**   Angles and Their Measurement

Definitions and Concepts	Examples
To convert from degree measure to radian measure, multiply the degree measure by $\dfrac{\pi \text{ radians}}{180°}$.	See Example 4, page 409
To convert from radian measure to degree measure, multiply the radian measure by $\dfrac{180°}{\pi \text{ radians}}$.	See Example 5, pages 409–410
Angles that share the same initial and terminal side, but perhaps have different amounts of rotation, are called **coterminal angles**. • If an angle θ is given in degree measure and $\alpha = \theta \pm 360°k$, where k is an integer, θ and α are coterminal. • If an angle θ is given in radian measure and $\alpha = \theta \pm 2\pi k$, where k is an integer, θ and α are coterminal.	See Example 7, pages 412–413
Arc length: If a central angle θ in a circle with a radius r is in radians, the length s of the subtended arc is given by $s = r\theta$.	See Example 10, page 414
Area of a circular sector: If a sector of a circle with radius r has a central angle θ in radians, the area of the sector is given by the formula $A = \frac{1}{2}r^2\theta$.	See Example 12, pages 416–417

EXERCISES

For each angle:

(i) State the quadrant in which the angle lies, or state that it is quadrantal.

(ii) Identify each angle as acute, right, obtuse, straight, or none of these.

1. a. $50°$ **b.** $-\dfrac{5\pi}{3}$

Determine whether the given angles are coterminal.

2. a. $150°$ and $510°$ **b.** $\dfrac{4\pi}{7}$ and $\dfrac{10\pi}{7}$

Convert each angle measure from degree decimal (DD) to degrees, minutes, and seconds (DMS). Round the seconds to the nearest tenth when necessary.

3. a. $138.634°$ **b.** $-212.75°$

Convert from DMS form to DD form, rounding to three decimal places.

4. a. $194°24'17''$ **b.** $-225°45'2''$

Convert each degree measurement to radians, expressing the answer in terms of π.

5. a. $-225°$ **b.** $390°$

Convert each degree measurement to radians, rounding your answer to three decimal places.

6. a. $89.36°$ **b.** $-13°10'17''$

Change the radian measure of each angle to degrees.

7. a. $\dfrac{15\pi}{4}$ **b.** $\dfrac{5\pi}{12}$

Convert each radian measure to degree decimal form, rounding your answer to the nearest hundredth.

8. a. 4.6 **b.** $-\sqrt{6}$

In each problem, s is the length of the arc subtended by central angle θ in a circle of radius r. Use $s = r\theta$ to find the missing value, rounding your answer to two decimal places.

9. a. $r = 22$ inches; $\theta = 135°$

     b. $\theta = \dfrac{5\pi}{9}$; $s = 37$ inches

     c. $r = 5$ inches; $s = 12$ inches

In each problem, A is the area of a circular sector formed by central angle θ in a circle of radius r. Use $A = \dfrac{1}{2}r^2\theta$ to find the missing value, rounding your answer to two decimal places.

10. a. $r = 10$ inches; $\theta = \dfrac{4\pi}{5}$

   b. $r = 22$ centimeters; $A = 150$ square centimeters

   c. $\theta = 40°$, $A = 600$ square inches

Applications

11. Distance between cities Dallas, Texas, located at latitude N 32°51′, is due south of Sioux Falls, South Dakota, located at N 43°34′. Using 3,960 miles as the radius of Earth, determine the distance between the two cities to the nearest mile.

12. Geometry A regular pentagon (a five-sided polygon) is inscribed in a circle 3 inches in diameter. How long is the arc intercepted by one of the sides of the pentagon? Round to two decimal places.

SECTION 4.2   The Unit Circle and Trigonometric Functions

Definitions and Concepts	Examples
Trigonometric functions: Let t be a real number and $P(x, y)$ be the point on the unit circle that corresponds to t. The six trigonometric functions of t are: $\sin t = y$ $\csc t = \dfrac{1}{y}, y \neq 0$ $\cos t = x$ $\sec t = \dfrac{1}{x}, x \neq 0$ $\tan t = \dfrac{y}{x}, x \neq 0$ $\cot t = \dfrac{x}{y}, y \neq 0$	See Example 1, pages 422–423
Periods of the trigonometric functions: $\sin \theta = \sin(\theta + 2\pi n)$ $\csc \theta = \csc(\theta + 2\pi n)$ $\cos \theta = \cos(\theta + 2\pi n)$ $\sec \theta = \sec(\theta + 2\pi n)$ $\tan \theta = \tan(\theta + \pi n)$ $\cot \theta = \cot(\theta + \pi n)$	See Example 8, pages 432–433
Reference angle: The **reference angle** for angle θ is the *acute* angle θ' formed by the terminal side of θ and the nearer part of the x-axis. The values of the six trigonometric functions of any angle θ are equal to those of the reference angle θ', except possibly for sign.	See Example 9, page 434

EXERCISES

The given coordinates are of a point on the unit circle that corresponds to a real number t. Find sin t, cos t, tan t, csc t, sec t, and cot t.

13. a. $\left(\dfrac{\sqrt{5}}{3}, \dfrac{2}{3}\right)$ **b.** $\left(-\dfrac{3}{4}, -\dfrac{\sqrt{7}}{4}\right)$

Find the reference angle for each angle.

14. a. $\dfrac{27\pi}{4}$ **b.** $713°$ **c.** $\dfrac{5\pi}{9}$

Find the exact value of each trigonometric function. If a function is undefined, state so.

15. a. $\sin 210°$ **b.** $\tan(-240°)$

   c. $\cos \dfrac{7\pi}{4}$ **d.** $\cot 3\pi$

Use the given information to find all values of α, where possible, if $0 \le \alpha < 2\pi$.

16. a. $\sin \alpha = -\dfrac{\sqrt{3}}{2}$ **b.** $\cos \alpha = \dfrac{\sqrt{2}}{2}$

   c. $\tan \alpha = -\sqrt{3}$

Use the given information to find one negative value for
θ with −360° < θ ≤ 0° and one positive value for θ with
0° ≤ θ < 360°.

17. a. $\cos\theta = -\dfrac{1}{2}; \sin\theta = -\dfrac{\sqrt{3}}{2}$

    b. $\tan\theta = -\dfrac{\sqrt{3}}{3}; \sec\theta = \dfrac{2\sqrt{3}}{3}$

SECTION 4.3   Trigonometric Functions of Any Angle; Fundamental Identities

Definitions and Concepts	**Examples**
Even–Odd Properties: $\sin(-t) = -\sin t$ $\csc(-t) = -\csc t$ $\cos(-t) = \cos t$ $\sec(-t) = \sec t$ $\tan(-t) = -\tan t$ $\cot(-t) = -\cot t$	See Example 1, page 445
Trigonometric functions of any angle: Let θ be an angle in standard position and let $P = (x, y)$ be any point (except the origin) on the terminal side of angle θ. The six **trigonometric functions** of θ are: $\sin\theta = \dfrac{y}{r}$ $\csc\theta = \dfrac{r}{y},\; y \neq 0$ $\cos\theta = \dfrac{x}{r}$ $\sec\theta = \dfrac{r}{x},\; x \neq 0$ $\tan\theta = \dfrac{y}{x},\; x \neq 0$ $\cot\theta = \dfrac{x}{y},\; y \neq 0$	See Example 2, page 446
Reciprocal Identities: For any angle θ for which the function is defined, $\sin\theta = \dfrac{1}{\csc\theta}$ $\csc\theta = \dfrac{1}{\sin\theta}$ $\sin\theta \cdot \csc\theta = 1$ $\cos\theta = \dfrac{1}{\sec\theta}$ $\sec\theta = \dfrac{1}{\cos\theta}$ $\cos\theta \cdot \sec\theta = 1$ $\tan\theta = \dfrac{1}{\cot\theta}$ $\cot\theta = \dfrac{1}{\tan\theta}$ $\tan\theta \cdot \cot\theta = 1$ **Quotient Identities:** For any angle θ for which the function is defined, $\tan\theta = \dfrac{\sin\theta}{\cos\theta}$ and $\cot\theta = \dfrac{\cos\theta}{\sin\theta}$	See Example 4, page 448
Pythagorean Identities: $\sin^2\theta + \cos^2\theta = 1$ $\tan^2\theta + 1 = \sec^2\theta$ $1 + \cot^2\theta = \csc^2\theta$	See Example 5, page 449

Definitions and Concepts	Examples
Right-triangle definitions of trigonometric functions: If angle A is an acute angle in the right triangle ABC, then $$\sin A = \frac{\text{opposite side}}{\text{hypotenuse}} \qquad \csc A = \frac{\text{hypotenuse}}{\text{opposite side}}$$ $$\cos A = \frac{\text{adjacent side}}{\text{hypotenuse}} \qquad \sec A = \frac{\text{hypotenuse}}{\text{adjacent side}}$$ $$\tan A = \frac{\text{opposite side}}{\text{adjacent side}} \qquad \cot A = \frac{\text{adjacent side}}{\text{opposite side}}$$	See Example 8, page 452
Complementary angles: If the sum of two acute angles is 90° or $\frac{\pi}{2}$, the angles are **complementary angles**. **Cofunctions of complementary angles:** If θ is any acute angle, any trigonometric function of θ is equal to the cofunction of the complement of θ. (**Cofunctions of complementary angles are equal.**) $\sin \theta = \cos(90° - \theta)$ and $\cos \theta = \sin(90° - \theta)$ $\tan \theta = \cot(90° - \theta)$ and $\cot \theta = \tan(90° - \theta)$ $\sec \theta = \csc(90° - \theta)$ and $\csc \theta = \sec(90° - \theta)$	See Example 10, page 454

EXERCISES

The point P is on the terminal side of angle θ, which is positive and in standard position. Calculate the exact value of the six trigonometric functions of θ.

18. a. $P(5, -2)$ **b.** $P(-3, 2)$

Use the Even–Odd Properties to simplify and find the exact value of each function. Do not use a calculator.

19. a. $\sin\left(-\dfrac{10\pi}{3}\right)$ **b.** $\cos(-210°)$

   c. $\tan\left(-\dfrac{11\pi}{6}\right)$

Find the remaining trigonometric functions of angle θ. Assume θ is positive and in standard position.

20. a. $\sec \theta = -\dfrac{5}{3}$; θ is in QII

   b. $\tan \theta = -3$; θ is in QIV

Use the given information and identities to find the value of each of the five remaining trigonometric functions for acute angle θ.

21. a. $\cos \theta = \dfrac{2}{5}$ **b.** $\tan \theta = \dfrac{1}{2}$

Use identities to simplify each expression.

22. a. $(\sin \theta \csc \theta + \sin \theta)(\sin \theta \csc \theta - \sin \theta)$

   b. $(\cot \theta \sec \theta + 1)(\cot \theta \sec \theta - 1)$

Use a calculator to compute each value to four decimal places.

23. a. $\sin \dfrac{12\pi}{13}$ **b.** $\sec 39.4°$

Find the exact value of each expression. Do not use a calculator.

24. a. $\sin 30° \cos 240° \cot 150° \cot 60°$

   b. $\sec^2 \dfrac{\pi}{3} \cos^2 \dfrac{5\pi}{3}$

Find the values of the six trigonometric functions for acute angle θ.

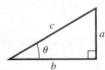

25. a. $a = 4$; $b = 7$ **b.** $b = 5$; $c = 9$

Applications

26. Angle of elevation A man stands 75 feet from a tree. The angle of elevation to the top of the tree is 35°. How tall is the tree? (Round to the nearest tenth of a foot.)

SECTION **4.4**   Graphs of the Sine and Cosine Functions

Definitions and Concepts	Examples
Amplitude and period: For the functions $y = A \sin Bx$ and $y = A \cos Bx$, with $A \neq 0$ and $B \neq 0$, $\|A\|$ is the **amplitude** and $\frac{2\pi}{B}$ is the **period**.	See Example 2, pages 465–467
Graphing $y = A \sin(Bx - C)$ and $y = A \cos(Bx - C)$: The graph of $y = A \sin(Bx - C)$ is identical to the graph of $y = A \sin Bx$ except that it is translated $\frac{C}{B}$ units to the right if $C > 0$ or $\frac{C}{B}$ units to the left if $C < 0$. The same holds true for the graph of the function $y = A \cos(Bx - C)$.	See Example 5, pages 470–471
Graphing $y = A \sin(Bx - C) + D$ and $y = A \cos(Bx - C) + D$: The graphs of $y = A \sin(Bx - C) + D$ and $y = A \cos(Bx - C) + D$ are vertical translations of D units.	See Example 7, pages 472–473

EXERCISES

For each function:

(a) State the amplitude and period.

(b) Graph the function for at least two periods.

(c) Find the first three x-intercepts of the graph for $x \geq 0$.

(d) Check your answer with a graphing calculator.

27. $y = 3 \sin \dfrac{\pi}{4} x$ **28.** $y = -\dfrac{1}{2} \cos 2x$

For each function:

(a) Find the amplitude, period, and phase shift (if any).

(b) Graph the function for at least two periods.

(c) Check your answer with a graphing calculator.

29. $y = \dfrac{3}{4} \sin\left(x + \dfrac{\pi}{4}\right)$

30. $y = -2 \cos\left(\pi x - \dfrac{\pi}{2}\right)$

Write an equation of a sine function with the stated characteristics.

31. amplitude = 4; period = 2

32. amplitude = $\dfrac{3}{5}$; period = $\dfrac{\pi}{2}$; phase shift = $\dfrac{\pi}{4}$

Graph each function. Label at least two periods. Check your answer with a graphing calculator.

33. $y = -3 + \sin\left(x - \dfrac{\pi}{4}\right)$

34. $y = 2 - \cos(2x - \pi)$

35. *Write an equation in the form*
$y = A \sin(Bx - C) + D$ *or*
$y = A \cos(Bx - C) + D$, *with $A \neq 0$ and $B \neq 0$, that will produce the graph shown.*

Check your answer with a graphing calculator.

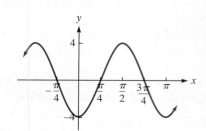

SECTION 4.5   Graphs of the Tangent, Cotangent, Secant, and Cosecant Functions

Definitions and Concepts	Examples
Summary for graphing $y = A\tan(Bx - C) + D$: • Period $= \dfrac{\pi}{B}$. • Use the basic period $\left(-\dfrac{\pi}{2}, \dfrac{\pi}{2}\right)$. • Find the asymptotes of one period by solving the inequality $-\dfrac{\pi}{2} < Bx - C < \dfrac{\pi}{2}$. • The x-intercept is at the one-half point of the interval if $D = 0$. • If $D \neq 0$, there will be a vertical shift of D units.	See Example 2, pages 482–483
Summary for graphing $y = A\cot(Bx - C) + D$: • Period $= \dfrac{\pi}{B}$. • Use the basic period $(0, \pi)$. • Find the asymptotes of one period by solving the inequality $0 < Bx - C < \pi$. • The x-intercept is at the one-half point of the interval if $D = 0$. • If $D \neq 0$, there will be a vertical shift of D units.	See Example 3, pages 484–485
Graphing $y = \csc x$: • The domain is all real numbers except for integer multiples of π. • The range is $(-\infty, -1] \cup [1, \infty)$. • Graph the cosecant function by first graphing the corresponding sine function.	See Example 4, pages 486–487
Graphing $y = \sec x$: • The domain is all real numbers except for odd-integer multiples of $\dfrac{\pi}{2}$. • The range is $(-\infty, -1] \cup [1, \infty)$. • Graph the secant function by first graphing the corresponding cosine function.	See Example 5, pages 488–489

EXERCISES

Graph two periods of each function. Label one period, including asymptotes and intercepts.

36. $y = -4\tan\left(\dfrac{1}{2}x - \pi\right)$ **37.** $y = \dfrac{1}{2}\cot 2\pi x$

38. $y = 2\sec\left(x + \dfrac{\pi}{4}\right)$

39. $y = -\csc(2\pi x - \pi)$

SECTION 4.6   Regression and Other Graphs Involving Trigonometric Functions (optional)

Definitions and Concepts	Examples
Sinusoidal regression is used to find a sine or cosine function that best fits a set of data.	See Example 1, pages 492–493
Addition of ordinates is a technique used to graph functions that involve the sum or difference of trigonometric functions. Simple functions can be done by hand, but more complex functions require technology to be accurate.	See Example 2, pages 494–495

Definitions and Concepts	Examples
The analysis of more complex waveforms, such as in sound and vibrations, can involve the sine function in the form $y = A \sin 2\pi ft + A_1 \sin 2\pi f_1 t + A_2 \sin 2\pi f_2 t + \cdots$, where $f_1, f_2, \ldots$ and $A_1, A_2, \ldots$ are the frequencies and amplitudes of the overtones present in the original sound. This function is called a **Fourier sine series**.	See Example 5, pages 496–497

EXERCISES

Graph each function.

40. $y = -\sin^5 x$

41. $y = 3 \sin \pi x + \cos 2\pi x$

42. $y = x^2 \cos x$

CHAPTER TEST

Unless otherwise indicated, round answers to two decimals.

1. Convert $47°15'37''$ to degree decimal form and radian measure.

2. Covert 3.5 radians to degree decimal form and degrees-minutes-seconds form.

3. Find the arc length s subtended by the central angle $\theta = 60°$ in a circle of radius 10 inches.

4. Find the area A of a circular sector formed by the central angle $\theta = 2.4$ in a circle of radius $r = 12$ centimeters.

5. The point $P = \left(\frac{1}{2}, -\frac{\sqrt{3}}{2}\right)$ is a point on the unit circle that corresponds to a real number t. Find each of the following:

   a. $\sin t$ **b.** $\cos t$ **c.** $\tan t$

   d. $\csc t$ **e.** $\sec t$ **f.** $\cot t$

6. Find the exact value of each trigonometric function.

   a. $\sin \dfrac{3\pi}{4}$ **b.** $\cos\left(-\dfrac{5\pi}{6}\right)$ **c.** $\tan 240°$

7. Find one negative value for θ with $-360° < \theta \le 0°$ and one positive value for θ with $0° \le \theta < 360°$, given that $\sin \theta = -\frac{\sqrt{2}}{2}$ and $\sec \theta = \sqrt{2}$.

8. The point $P(-5, 7)$ is on the terminal side of angle θ, which is positive and in standard position. Calculate the exact values of the six trigonometric functions of angle θ.

9. Given that $\tan \theta = -4$ and θ is in QIV, find the values of the remaining trigonometric functions of θ.

10. In a right triangle, $a = 5$ and $b = 10$. Find the values of the six trigonometric functions of acute angle θ.

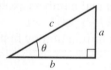

Graph two periods of each function.

11. $y = -3 \sin \pi x$ **12.** $y = \dfrac{1}{2} \cos\left(x - \dfrac{\pi}{3}\right)$

13. $y = \tan(2\pi x + 2\pi)$ **14.** $y = 2 + \csc \dfrac{1}{2}x$

15. $y = \sec \dfrac{\pi x}{2}$ **16.** $y = -2 \cot(2x + \pi)$

17. Write an equation of a sine function with amplitude 3, period $\frac{\pi}{2}$, and phase shift $\frac{\pi}{4}$.

18. Write an equation in the form $y = A \cos(Bx - C)$ with $A \ne 0$ and $B \ne 0$, and an equation in the form $y = A \sin(Bx - C)$ with $A \ne 0$ and $B \ne 0$, that will produce the graph shown.

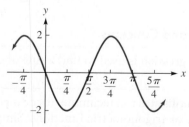

19. Graph the function $y = e^{x/10} \sin x$.

LOOKING AHEAD TO **CALCULUS**

1. **Using the Derivatives of Trigonometric Functions**
   - Connect the definitions of the trigonometric functions to their derivatives—Section 4.3

2. **Using Values of Trigonometric Functions to Define Taylor Polynomials**
   - Connect trigonometric functions to the definition of a Taylor polynomial—Section 4.3

1. Using the Derivatives of Trigonometric Functions

In the Looking Ahead to Calculus section in Chapter 1, we looked at a function called the **derivative**. The derivative of a function is a function that describes the rate of change of the original function. Derivatives have a multitude of uses. The derivatives of the trigonometric functions are shown here.

Derivatives of Trigonometric Functions

Function	Derivative
$f(x) = \sin x$	$f'(x) = \cos x$
$f(x) = \cos x$	$f'(x) = -\sin x$
$f(x) = \tan x$	$f'(x) = \sec^2 x$
$f(x) = \cot x$	$f'(x) = -\csc^2 x$
$f(x) = \csc x$	$f'(x) = -\csc x \cot x$
$f(x) = \sec x$	$f'(x) = \sec x \tan x$

Connect to Calculus:
Derivatives of Trigonometric Functions

1. Find the derivative of each function.
   a. $f(x) = \sin x + \tan x$ **b.** $f(x) = 2 \csc x + 5 \cos x$
2. An object at the end of a vertical spring is stretched 4 cm beyond its rest position and released at time $t = 0$. At time t, its position is $s = f(t) = 4 \cos t$. The velocity of the spring is the derivative of the position function. Find the velocity function.

2. Using Values of Trigonometric Functions to Define Taylor Polynomials

An important application of derivatives is a Taylor polynomial. Essentially, there is a way to write a function, such as $f(x) = \sin x$, as a polynomial. The polynomial can be used to approximate the value of the function and can make other operations in calculus much easier to perform. We will not go through the details of how these polynomials are derived because that is beyond the scope of our work here. However, derivatives are used to find them. We will examine a couple of Taylor polynomials and see how they can be used.

A Taylor polynomial is "centered" around a value. The higher the degree of the polynomial, the more accurately it approximates the function. The third-degree Taylor polynomial centered at $\frac{\pi}{3}$ is

$$\sin x \approx \frac{\sqrt{3}}{2} + \frac{1}{2}\left(x - \frac{\pi}{3}\right) - \frac{\sqrt{3}}{4}\left(x - \frac{\pi}{3}\right)^2 - \frac{1}{12}\left(x - \frac{\pi}{3}\right)^3.$$

The graph shows the sine function in blue and the polynomial as a dashed red line.

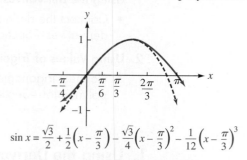

$$\sin x = \frac{\sqrt{3}}{2} + \frac{1}{2}\left(x - \frac{\pi}{3}\right) - \frac{\sqrt{3}}{4}\left(x - \frac{\pi}{3}\right)^2 - \frac{1}{12}\left(x - \frac{\pi}{3}\right)^3$$

Notice that around the value $x = \frac{\pi}{3}$, the two graphs are essentially identical. As you move away from $x = \frac{\pi}{3}$, the graphs begin to separate. This is why we say the polynomial is "centered" around $x = \frac{\pi}{3}$.

Now let's see how we can use this polynomial to approximate the value of $\sin 65°$. The first thing we must do is convert the degree measurement to radian measure. The polynomial is a function, and we use real numbers.

$$x = 65° = 60° + 5° = \frac{\pi}{3} + \frac{\pi}{36}$$

Now we substitute this into the polynomial and evaluate. For $x = 65° = \frac{\pi}{3} + \frac{\pi}{36}$, we have

$$\sin x \approx \frac{\sqrt{3}}{2} + \frac{1}{2}\left(x - \frac{\pi}{3}\right) - \frac{\sqrt{3}}{4}\left(x - \frac{\pi}{3}\right)^2 - \frac{1}{12}\left(x - \frac{\pi}{3}\right)^3$$

$$\sin 65° = \sin\left(\frac{\pi}{3} + \frac{\pi}{36}\right) \approx \frac{\sqrt{3}}{2} + \frac{1}{2}\left(\frac{\pi}{3} + \frac{\pi}{36} - \frac{\pi}{3}\right) - \frac{\sqrt{3}}{4}\left(\frac{\pi}{3} + \frac{\pi}{36} - \frac{\pi}{3}\right)^2 - \frac{1}{12}\left(\frac{\pi}{3} + \frac{\pi}{36} - \frac{\pi}{3}\right)^3$$

$$= \frac{\sqrt{3}}{2} + \frac{1}{2}\left(\frac{\pi}{36}\right) - \frac{\sqrt{3}}{4}\left(\frac{\pi}{36}\right)^2 - \frac{1}{12}\left(\frac{\pi}{36}\right)^3$$

$$\approx 0.90630567$$

If we evaluate using a calculator, we have $\sin 65° \approx 0.90630778$. These values are quite close.

Connect to Calculus: Taylor Polynomials

3. Use the third-degree Taylor polynomial from the section to approximate $\sin 55°$.

4. The third-degree Taylor polynomial for $f(x) = \cos x$ is

$$\cos x \approx \frac{\sqrt{2}}{2} - \frac{\sqrt{2}}{2}\left(x - \frac{\pi}{4}\right) - \frac{\sqrt{2}}{4}\left(x - \frac{\pi}{4}\right)^2 + \frac{\sqrt{2}}{12}\left(x - \frac{\pi}{4}\right)^3.$$

Use this polynomial to approximate $\cos 50°$ and $\cos 25°$. Now use a calculator to approximate each function.

5. What do you see about the approximations you found using the polynomial? Why do you think there is a difference in the accuracy?

Analytic Trigonometry

5

CAREERS AND MATHEMATICS:

Forensic Analyst

Peter Coombs/Alamy

Forensic analysts investigate crimes by collecting and analyzing physical evidence. By performing tests on weapons or on substances such as fiber, hair tissue, and body fluids, they can determine the significance of the substances to the investigation. Some specialize in DNA analysis or firearm examination.

Education and Mathematics Required

- Forensic-analyst positions usually require a bachelor's degree, in either forensic science or another natural science.
- College Algebra, Trigonometry, Geometry, Calculus I and II, and Statistics are required.

How Forensic Analysts Use Math and Who Employs Them

- A forensic analyst uses pattern analysis of blood stains to determine what happened at the scene of a crime. Math principles are used to determine the location of the victim when the blood was shed and the type of weapon or impact that caused the injury.
- Forensic analysts work primarily for state and local governments. Thirty percent of forensic analysts are self-employed. Some work for detective agencies.

Career Outlook and Salary

- Jobs for forensic analysts are expected to increase by 20% from 2008–2018, which is much faster than average. Employment growth in state and local government should be driven by the increasing application of forensic-science techniques, such as DNA analysis, to examining, solving, and preventing crime.
- Median annual wages of forensic analysts were $51,480 in May 2009; 10% of forensic analysts earn more than $84,260.

For more information, see http://www.bls.gov/oco.

In this chapter, we will learn about inverse trigonometric functions and trigonometric identities. Then, we use that knowledge to help us solve trigonometric equations.

5.1 Inverse Trigonometric Functions

In this section, we will learn to

1. Develop and graph the inverse sine function.
2. Develop and graph the inverse cosine function.
3. Develop and graph the inverse tangent function.
4. Find values of inverse trigonometric functions.
5. Convert a trigonometric expression to an algebraic expression.

One of the longest-running kite festivals in the United States is the Zilker Park Kite Festival, held in Austin, Texas. Each year, hundreds of beautiful kites fly in the air. The festival began in 1936.

Suppose we are interested in finding the angle of elevation θ of a kite. To do so, we must be given the value of one of its trigonometric functions. Then we work the problem backwards and determine θ. When we solved application problems involving right triangles, we often were required to use our calculator and $\sin^{-1}$, $\cos^{-1}$, or $\tan^{-1}$ to find θ. These are known as the inverse trigonometric functions. In this section, we will study the inverse trigonometric functions of sine, cosine, and tangent in depth. We will learn different notations and also understand why domain restrictions on the functions are required. These functions are used to solve many real-life problems involving angles.

In Section 1.7 we defined an inverse function. If two functions are inverses, the following statements are true:

- $f(x) = y \Leftrightarrow f^{-1}(y) = x$.
- The functions are one-to-one. That means that they are always increasing or always decreasing throughout the defined domain. (They pass the horizontal-line test.)
- Graphically, the functions are reflections of each other about the line $y = x$.
- The relationship between the domain and range of a function and of its inverse is $D_f = R_{f^{-1}}$ and $R_f = D_{f^{-1}}$.

If a function is not one-to-one, it does not have an inverse. However, as we observed, it is possible to restrict the domain of a function so that it is one-to-one and invertible.

1. Develop and Graph the Inverse Sine Function

The domain of the sine function is all real numbers, and the range is $[-1, 1]$. Figure 5-1 illustrates the fact that the sine function is not one-to-one, failing the horizontal-line test. The horizontal line $y = b$, $-1 \le b \le 1$, will intersect the graph of the sine function at infinitely many points.

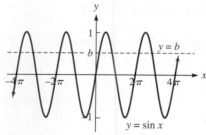

FIGURE 5-1

To define the inverse sine function, it is necessary to restrict the domain of the sine fuction to an interval over which the function is one-to-one and obtains all values in the range $[-1, 1]$. While other intervals are possible, the conventional interval used is $\left[-\frac{\pi}{2}, \frac{\pi}{2}\right]$. This is illustrated in Figure 5-2.

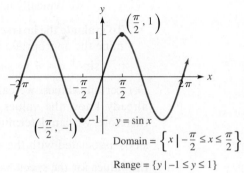

Domain $= \left\{ x \,\middle|\, -\frac{\pi}{2} \le x \le \frac{\pi}{2} \right\}$

Range $= \{ y \,|\, -1 \le y \le 1 \}$

FIGURE 5-2

In Section 1.7, we found the equation of the inverse function of $f(x) = y$ by interchanging the x and y values. If we take the restricted portion of the sine graph shown in Figure 5-2, we follow the same steps to find the inverse sine function. We can now state the definition of the inverse sine function.

Inverse Sine Function

The **inverse sine function**, $y = \sin^{-1} x$, has domain $D = \{ x \,|\, -1 \le x \le 1 \}$ and range $R = \left\{ y \,\middle|\, -\frac{\pi}{2} \le y \le \frac{\pi}{2} \right\}$.

$y = \sin^{-1} x$ if and only if $x = \sin y$ and $-\frac{\pi}{2} \le y \le \frac{\pi}{2}$.

The graph is shown in Figure 5-3.

Caution

$\sin^{-1} x \ne \dfrac{1}{\sin x}$, just as

$f^{-1}(x) \ne \dfrac{1}{f(x)}$. We may occasion-

ally see the inverse sine function written as $y = \arcsin x$.

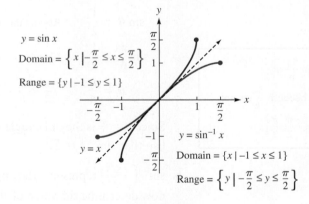

FIGURE 5-3

Understanding the Inverse Sine Function

- The **input of the inverse sine function is the sine of an angle;** therefore, the input must be in the interval $[-1, 1]$.

    The expression $\sin^{-1} \frac{1}{2} = \theta$ is defined because the equation $\sin \theta = \frac{1}{2}$ can be solved with $-\frac{\pi}{2} \le \theta \le \frac{\pi}{2}$.

    The expression $\sin^{-1} 2 = \theta$ is meaningless because the equation $\sin \theta = 2$ cannot be solved.

- The **output of the inverse sine function is an angle, which is a real number** in the interval $\left[-\frac{\pi}{2}, \frac{\pi}{2}\right]$.

  The output is an angle or a real number. In our study of trigonometry, we typically use angles, $\sin^{-1} x = \theta$.

- To **evaluate the inverse sine function, rewrite $\sin^{-1} x = \theta$ as $\sin \theta = x$** and determine the appropriate value of θ.

  $$\sin^{-1} x = \theta \Leftrightarrow \sin \theta = x, \text{ with } -\frac{\pi}{2} \le \theta \le \frac{\pi}{2} \text{ and } -1 \le x \le 1.$$

For certain values, we can find the exact value of the inverse sine function. We already know the values of the trigonometric functions for the special angles; therefore, we can determine the exact value of the inverse sine function for expressions associated with the special angles in the interval $\left[-\frac{\pi}{2}, \frac{\pi}{2}\right]$. Table 1 summarizes the values for the special angles in the interval $\left[-\frac{\pi}{2}, \frac{\pi}{2}\right]$.

TABLE 1

θ	$-\frac{\pi}{2}$	$-\frac{\pi}{3}$	$-\frac{\pi}{4}$	$-\frac{\pi}{6}$	0	$\frac{\pi}{6}$	$\frac{\pi}{4}$	$\frac{\pi}{3}$	$\frac{\pi}{2}$
$\sin \theta$	-1	$-\frac{\sqrt{3}}{2}$	$-\frac{\sqrt{2}}{2}$	$-\frac{1}{2}$	0	$\frac{1}{2}$	$\frac{\sqrt{2}}{2}$	$\frac{\sqrt{3}}{2}$	1

EXAMPLE 1 **Finding the Exact Value of an Inverse Sine Expression**

Find the exact value of each expression or state that it does not exist and why.

a. $\sin^{-1} \frac{1}{2}$ **b.** $\sin^{-1}\left(-\frac{\sqrt{3}}{2}\right)$ **c.** $\sin^{-1} \pi$ **d.** $\sin^{-1} 1$

SOLUTION **a.** $\sin^{-1} \frac{1}{2}$ represents the angle θ with $-\frac{\pi}{2} \le \theta \le \frac{\pi}{2}$ and $\sin \theta = \frac{1}{2}$. We will now determine the value of θ.

$$\sin^{-1} \frac{1}{2} = \theta \qquad -\frac{\pi}{2} \le \theta \le \frac{\pi}{2}$$

$$\sin \theta = \frac{1}{2} \qquad \text{Rewrite the equation.}$$

$$\sin \frac{\pi}{6} = \frac{1}{2} \qquad \text{Find } \theta.$$

$$\theta = \frac{\pi}{6} \qquad -\frac{\pi}{2} \le \frac{\pi}{6} \le \frac{\pi}{2}$$

Caution

Even though $\sin \frac{5\pi}{6} = \frac{1}{2}$,

$\sin^{-1} \frac{1}{2} \ne \frac{5\pi}{6}$, because $\frac{5\pi}{6}$ is not

in the interval $\left[-\frac{\pi}{2}, \frac{\pi}{2}\right]$.

Since $\theta = \frac{\pi}{6}$ is the only angle in the interval $\left[-\frac{\pi}{2}, \frac{\pi}{2}\right]$ such that $\sin \theta = \frac{1}{2}$, we establish that $\sin^{-1} \frac{1}{2} = \frac{\pi}{6}$.

b. $\sin^{-1}\left(-\frac{\sqrt{3}}{2}\right)$ represents the angle θ with $-\frac{\pi}{2} \le \theta \le \frac{\pi}{2}$ and $\sin \theta = -\frac{\sqrt{3}}{2}$. We will now determine the value of θ.

$$\sin^{-1}\left(-\frac{\sqrt{3}}{2}\right) = \theta \qquad -\frac{\pi}{2} \le \theta \le \frac{\pi}{2}$$

$$\sin \theta = -\frac{\sqrt{3}}{2} \qquad \text{Rewrite the equation.}$$

$$\sin\left(-\frac{\pi}{3}\right) = -\frac{\sqrt{3}}{2} \qquad \text{Find } \theta.$$

$$\theta = -\frac{\pi}{3} \qquad -\frac{\pi}{2} \le \left(-\frac{\pi}{3}\right) \le \frac{\pi}{2}$$

Since $-\frac{\pi}{3}$ is the only angle in the interval $\left[-\frac{\pi}{2}, \frac{\pi}{2}\right]$ such that $\sin \theta = -\frac{\sqrt{3}}{2}$, we establish that $\sin^{-1}\left(-\frac{\sqrt{3}}{2}\right) = -\frac{\pi}{3}$.

c. $\sin^{-1} \pi$ does not exist, because π is not in the interval $[-1, 1]$.

d. $\sin^{-1} 1$ represents the angle θ with $-\frac{\pi}{2} \le \theta \le \frac{\pi}{2}$ and $\sin \theta = 1$.

We will now determine the value of θ.

$$\sin^{-1} 1 = \theta \qquad -\frac{\pi}{2} \le \theta \le \frac{\pi}{2}$$

$$\sin \theta = 1 \qquad \text{Rewrite the equation.}$$

$$\sin \frac{\pi}{2} = 1 \qquad \text{Find } \theta.$$

$$\theta = \frac{\pi}{2} \qquad -\frac{\pi}{2} \le \frac{\pi}{2} \le \frac{\pi}{2}$$

Since $\frac{\pi}{2}$ is the only angle in the interval $\left[-\frac{\pi}{2}, \frac{\pi}{2}\right]$ such that $\sin \theta = 1$, we establish that $\sin^{-1} 1 = \frac{\pi}{2}$. ∎

Self Check 1   Find the exact value of each expression or state that it does not exist and why.

a. $\sin^{-1} \dfrac{\sqrt{3}}{2}$ **b.** $\sin^{-1}\left(-\dfrac{\sqrt{2}}{2}\right)$ **c.** $\sin^{-1}(-3)$

Now Try Exercise 9.

ACCENT ON TECHNOLOGY   A calculator is needed to find approximate values for inverse sine expressions that are not equivalent to a special angle. It is important to check the **MODE**. Our calculator should be set to **radian mode**. A calculator will give us an answer in degree mode; however, these are functions, and the input and output are real numbers. Thus we use radian mode, because a radian measure is a real number. The inverse functions are typically the second function assigned to the function key.

FIGURE 5-4

EXAMPLE 2   **Using a Calculator to Evaluate an Inverse Sine**

Find each value in radians rounding to three decimal places. Then check your answer.

a. $\sin^{-1} 0.352$ **b.** $\sin^{-1} \dfrac{\pi}{4}$ **c.** $\sin^{-1}(-0.578)$

SOLUTION   **a.** Set the calculator to radian mode. We find $\sin^{-1} 0.352 \approx 0.360$. To check our answer, and to see the relationship of the sine and inverse sine functions, we evaluate $\sin 0.360 \approx 0.352$.

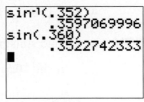

FIGURE 5-5

   b. Since $\frac{\pi}{4} \approx 0.785$, it is in the domain of the inverse sine function.
   $\sin^{-1}\frac{\pi}{4} \approx 0.903$. The calculation and the check are shown in Figure 5-6.

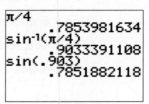

FIGURE 5-6

   c. $\sin^{-1}(-0.578) \approx -0.616$, as shown in Figure 5-7 with the check.

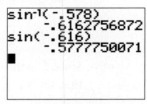

FIGURE 5-7

Self Check 2   Evaluate each expression, rounding to three decimal places.

   a. $\sin^{-1} 0.898$ **b.** $\sin^{-1}\left(-\frac{\pi}{6}\right)$

Now Try Exercise 17.

   Recall from Section 1.7 that $f(f^{-1}(x)) = x$ for all x in the domain of f^{-1} and
that $f^{-1}(f(x)) = x$ for all x in the domain of f. Because $y = \sin x$, $-\frac{\pi}{2} \le x \le \frac{\pi}{2}$,
and $y = \sin^{-1} x$, $-1 \le x \le 1$, are inverse functions, we consider the effect of evaluating the functions in succession.

Composition of Sine and Inverse Sine Functions				
$\sin(\sin^{-1} x) = x$	when	$-1 \le x \le 1$		$[f(f^{-1}(x)) = x]$
$\sin^{-1}(\sin x) = x$	when	$-\dfrac{\pi}{2} \le x \le \dfrac{\pi}{2}$		$[f^{-1}(f(x)) = x]$

   Since $\sin^{-1} x$ is defined only when $-1 \le x \le 1$, $\sin(\sin^{-1} x) = x$ is undefined if x
is not in the interval $[-1, 1]$. The expression $\sin^{-1}(\sin x)$ is always defined, but the
answer will equal x only if $-\frac{\pi}{2} \le x \le \frac{\pi}{2}$.

EXAMPLE 3   **Finding the Composition of the Sine and Inverse Sine Functions**

   Find: **a.** $\sin(\sin^{-1} 0.5)$ **b.** $\sin\left(\sin^{-1}\frac{\pi}{3}\right)$ **c.** $\sin^{-1}\left(\sin\frac{\pi}{4}\right)$

   d. $\sin^{-1}\left(\sin\frac{7\pi}{6}\right)$ **e.** $\sin^{-1}\left(\sin\frac{5\pi}{7}\right)$

SOLUTION   **a.** To find $\sin(\sin^{-1} 0.5)$, use $\sin(\sin^{-1} x) = x$ when $-1 \le x \le 1$.

   $\sin(\sin^{-1} \mathbf{0.5}) = \mathbf{0.5}$, because $-1 \le \mathbf{0.5} \le 1$.

   b. $\sin\left(\sin^{-1}\frac{\pi}{3}\right)$ is undefined, because $\frac{\pi}{3} \approx 1.05 > 1$.

c. To find $\sin^{-1}\left(\sin\frac{\pi}{4}\right)$, use $\sin^{-1}(\sin x) = x$ when $-\frac{\pi}{2} \le x \le \frac{\pi}{2}$.

$$\sin^{-1}\left(\sin\frac{\pi}{4}\right) = \frac{\pi}{4}, \text{ because } -\frac{\pi}{2} \le \frac{\pi}{4} \le \frac{\pi}{2}.$$

d. $\sin^{-1}\left(\sin\frac{7\pi}{6}\right) \ne \frac{7\pi}{6}$, because $\frac{7\pi}{6}$ is not in the interval $\left[-\frac{\pi}{2}, \frac{\pi}{2}\right]$. However, $\frac{7\pi}{6}$ is a multiple of the special angle $\frac{\pi}{6}$, so we can proceed as shown.

$$\sin^{-1}\left(\sin\frac{7\pi}{6}\right) = \sin^{-1}\left(-\frac{1}{2}\right) \qquad \sin\frac{7\pi}{6} = -\frac{1}{2}$$

$$\sin^{-1}\left(-\frac{1}{2}\right) = -\frac{\pi}{6}$$

$$\sin^{-1}\left(\sin\frac{7\pi}{6}\right) = -\frac{\pi}{6}$$

e. $\frac{5\pi}{7}$ is not a multiple of a special angle; therefore, we cannot find the exact value of $\sin\frac{5\pi}{7}$. We know that $\sin^{-1}\left(\sin\frac{5\pi}{7}\right)$ represents an angle in the interval $\left[-\frac{\pi}{2}, \frac{\pi}{2}\right]$, but $\frac{5\pi}{7}$ is not in that interval. Therefore, $\sin^{-1}(\sin x) = x$ cannot be used. We will use the reference angle $\theta' = \frac{2\pi}{7}$ (Figure 5-8) to determine the answer. $\theta = \frac{5\pi}{7}$ lies in QII; therefore, $\sin\frac{5\pi}{7}$ is positive, and $\sin\frac{5\pi}{7} = \sin\frac{2\pi}{7}$.

Since $-\frac{\pi}{2} \le \frac{2\pi}{7} \le \frac{\pi}{2}$,

$$\sin^{-1}\left(\sin\frac{5\pi}{7}\right) = \sin^{-1}\left(\sin\frac{2\pi}{7}\right) = \frac{2\pi}{7}.$$

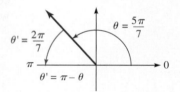

FIGURE 5-8

Self Check 3   Find: **a.** $\sin(\sin^{-1} 0.75)$ **b.** $\sin^{-1}\left(\sin\frac{3\pi}{4}\right)$ **c.** $\sin^{-1}\left[\sin\left(-\frac{4\pi}{5}\right)\right]$

Now Try Exercise 29.

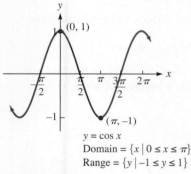

$y = \cos x$
Domain $= \{x \mid 0 \le x \le \pi\}$
Range $= \{y \mid -1 \le y \le 1\}$

FIGURE 5-9

2. Develop and Graph the Inverse Cosine Function

When defining the inverse cosine function, we use similar considerations to those applied when defining the inverse sine function. The function $y = \cos x$ is not a one-to-one function. To define the inverse cosine function, it is necessary to restrict the domain to an interval over which the function is one-to-one and obtains all values in the range $[-1, 1]$. There are many possible intervals, but the conventional interval used is $[0, \pi]$.

Inverse Cosine Function   The **inverse cosine function**, $y = \cos^{-1} x$, has domain $D = \{x \mid -1 \le x \le 1\}$ and range $R = \{y \mid 0 \le y \le \pi\}$.

$$y = \cos^{-1} x \qquad \text{if and only if} \qquad x = \cos y \qquad \text{and} \qquad 0 \le y \le \pi.$$

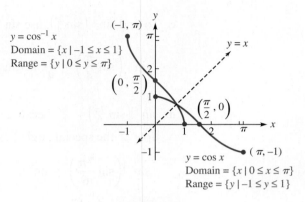

FIGURE 5-10

Understanding the Inverse Cosine Function

- The **input of the inverse cosine function is the cosine of an angle;** therefore, the input must be in the interval $[-1, 1]$.
- The **output of the inverse cosine function is an angle, which is a real number** in the interval $[0, \pi]$.
- To **evaluate the inverse cosine function,** rewrite $\cos^{-1} x = \theta$ as $\cos \theta = x$ and determine the appropriate value of θ.

$$\cos^{-1} x = \theta \Leftrightarrow \cos \theta = x, \text{ with } 0 \le \theta \le \pi \text{ and } -1 \le x \le 1.$$

Just as with the inverse sine function, we can find the exact value of the inverse cosine function for certain values. Table 2 summarizes the values that are associated with the special angles over $[0, \pi]$.

TABLE 2

θ	0	$\dfrac{\pi}{6}$	$\dfrac{\pi}{4}$	$\dfrac{\pi}{3}$	$\dfrac{\pi}{2}$	$\dfrac{2\pi}{3}$	$\dfrac{3\pi}{4}$	$\dfrac{5\pi}{6}$	π
$\cos \theta$	1	$\dfrac{\sqrt{3}}{2}$	$\dfrac{\sqrt{2}}{2}$	$\dfrac{1}{2}$	0	$-\dfrac{1}{2}$	$-\dfrac{\sqrt{2}}{2}$	$-\dfrac{\sqrt{3}}{2}$	-1

EXAMPLE 4   Finding the Exact Value of an Inverse Cosine Expression

Find the exact value of each expression or state that it does not exist and why.

a. $\cos^{-1} \dfrac{\sqrt{2}}{2}$ **b.** $\cos^{-1}\left(-\dfrac{\sqrt{3}}{2}\right)$ **c.** $\cos^{-1} \sqrt{3}$ **d.** $\cos^{-1} 1$

SOLUTION   **a.** $\cos^{-1} \dfrac{\sqrt{2}}{2}$ represents the angle θ in the interval $[0, \pi]$ such that $\cos \theta = \dfrac{\sqrt{2}}{2}$.

$$\cos^{-1} \frac{\sqrt{2}}{2} = \theta \qquad 0 \le \theta \le \pi$$

$$\cos \theta = \frac{\sqrt{2}}{2} \qquad \text{Rewrite the equation.}$$

$$\cos \frac{\pi}{4} = \frac{\sqrt{2}}{2} \qquad \text{Find } \theta.$$

$$\theta = \frac{\pi}{4} \qquad 0 \le \frac{\pi}{4} \le \pi$$

Because $\theta = \frac{\pi}{4}$ is the only number in the interval $[0, \pi]$ such that $\cos \theta = \frac{\sqrt{2}}{2}$, we have established that $\cos^{-1} \frac{\sqrt{2}}{2} = \frac{\pi}{4}$.

b. Refer to Table 2 when solving this problem.

$$\cos^{-1}\left(-\frac{\sqrt{3}}{2}\right) = \theta \qquad\qquad 0 \le \theta \le \pi$$

$$\cos\theta = -\frac{\sqrt{3}}{2} \qquad\qquad \text{Rewrite the equation.}$$

$$\cos\frac{5\pi}{6} = -\frac{\sqrt{3}}{2} \qquad\qquad \text{Find } \theta.$$

$$\theta = \frac{5\pi}{6} \qquad\qquad 0 \le \frac{5\pi}{6} \le \pi$$

c. $\cos^{-1}\sqrt{3}$ is undefined because $\sqrt{3} > 1$ and cannot be the value of the cosine of any angle.

d. $\cos^{-1}1 = 0$ because $\theta = 0$ is the only angle in the interval $[0, \pi]$ that satisfies the equation $\cos\theta = 0$.   ∎

Self Check 4   Find the exact value of each expression or state that it does not exist and why.

a. $\cos^{-1}\dfrac{1}{2}$ **b.** $\cos^{-1}(-1)$ **c.** $\cos^{-1}\sqrt{2}$

Now Try Exercise 11.

Composition of Cosine and Inverse Cosine Functions

$\cos(\cos^{-1}x) = x$	when $-1 \le x \le 1$	$[f(f^{-1}(x)) = x]$
$\cos^{-1}(\cos x) = x$	when $0 \le x \le \pi$	$[f^{-1}(f(x)) = x]$

Since $\cos^{-1}x$ is defined only when $-1 \le x \le 1$, $\cos(\cos^{-1}x) = x$ is undefined if x is not in the interval $[-1, 1]$. The expression $\cos^{-1}(\cos x)$ is always defined, but the answer will equal x only if $0 \le x \le \pi$.

EXAMPLE 5   **Finding the Composition of the Cosine and Inverse Cosine Functions**

Find: **a.** $\cos\left[\cos^{-1}\left(-\dfrac{2}{3}\right)\right]$ **b.** $\cos^{-1}\left(\cos\dfrac{\pi}{2}\right)$ **c.** $\cos^{-1}\left[\cos\left(-\dfrac{\pi}{6}\right)\right]$

SOLUTION   **a.** To find $\cos\left[\cos^{-1}\left(-\frac{2}{3}\right)\right]$, use $\cos(\cos^{-1}x) = x$ when $-1 \le x \le 1$.

$$\cos\left[\cos^{-1}\left(-\frac{2}{3}\right)\right] = -\frac{2}{3}, \text{ because } -1 \le -\frac{2}{3} \le 1.$$

b. To find $\cos^{-1}\left(\cos\frac{\pi}{2}\right)$, use $\cos^{-1}(\cos x) = x$ when $0 \le x \le \pi$.

Since $0 \le \dfrac{\pi}{2} \le \pi$, we know that $\cos^{-1}\left(\cos\dfrac{\pi}{2}\right) = \dfrac{\pi}{2}$.

c. $\cos^{-1}\left[\cos\left(-\frac{\pi}{6}\right)\right] \neq -\frac{\pi}{6}$ because $-\frac{\pi}{6}$ is not in the range of the inverse cosine, $[0, \pi]$. However, we can use the fact that $-\frac{\pi}{6}$ is a multiple of $\frac{\pi}{6}$ to find the solution. In order to find $\cos^{-1}\left[\cos\left(-\frac{\pi}{6}\right)\right]$, we will "work from the inside out."

$$\cos^{-1}\left[\cos\left(-\frac{\pi}{6}\right)\right] = \cos^{-1}\left(\frac{\sqrt{3}}{2}\right) \qquad \cos\left(\frac{-\pi}{6}\right) = \frac{\sqrt{3}}{2}$$

$$\cos^{-1}\left(\frac{\sqrt{3}}{2}\right) = \frac{\pi}{6}$$

$$\cos^{-1}\left[\cos\left(-\frac{\pi}{6}\right)\right] = \frac{\pi}{6}$$

∎

Self Check 5   Find: **a.** $\cos[\cos^{-1}(-0.65)]$ **b.** $\cos^{-1}\left(\cos\frac{4\pi}{3}\right)$

Now Try Exercise 45.

3. Develop and Graph the Inverse Tangent Function

The tangent function is not one-to-one. In order to define the inverse tangent function, we select the interval $\left(-\frac{\pi}{2}, \frac{\pi}{2}\right)$, over which the tangent function is one-to-one. The graphs of the tangent and inverse tangent functions are shown in Figure 5-11.

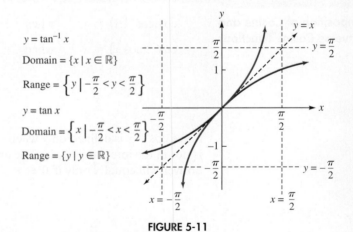

$y = \tan^{-1} x$

Domain $= \{x \mid x \in \mathbb{R}\}$

Range $= \left\{y \mid -\frac{\pi}{2} < y < \frac{\pi}{2}\right\}$

$y = \tan x$

Domain $= \left\{x \mid -\frac{\pi}{2} < x < \frac{\pi}{2}\right\}$

Range $= \{y \mid y \in \mathbb{R}\}$

FIGURE 5-11

Inverse Tangent Function

The **inverse tangent function**, $y = \tan^{-1} x$, has domain $D = $ all real numbers and range $R = \left\{y \mid -\frac{\pi}{2} < y < \frac{\pi}{2}\right\}$.

Understanding the Inverse Tangent Function

- The **input of the inverse tangent function is the tangent of an angle**, which can be any real number. Thus, $y = \tan^{-1} x$ will always be defined.

- The **output of the inverse tangent function is an angle, which is a real number** in the interval $\left(-\frac{\pi}{2}, \frac{\pi}{2}\right)$.

- To **evaluate the inverse tangent function**, rewrite $\tan^{-1} x = \theta$ as $\tan \theta = x$ and determine the appropriate value of θ.

$$\tan^{-1} x = \theta \Leftrightarrow \tan \theta = x, \text{ with } -\frac{\pi}{2} < \theta < \frac{\pi}{2} \text{ and } x \in \mathbb{R}.$$

EXAMPLE 6   **Finding the Exact Value of an Inverse Tangent Expression**

Find the exact value of each expression.   **a.** $\tan^{-1}(-1)$ **b.** $\tan^{-1} \sqrt{3}$

SOLUTION   **a.** $\tan^{-1}(-1)$ represents the angle θ, with $-\frac{\pi}{2} < \theta < \frac{\pi}{2}$, such that $\tan \theta = -1$. We write this as an equation and determine the value of θ.

$$\tan^{-1}(-1) = \theta \qquad -\frac{\pi}{2} < \theta < \frac{\pi}{2}$$

$$\tan \boldsymbol{\theta} = -1 \qquad \text{Rewrite the equation.}$$

$$\tan\left(-\frac{\boldsymbol{\pi}}{\boldsymbol{4}}\right) = -1 \qquad \text{Find } \theta.$$

$$\boldsymbol{\theta} = -\frac{\boldsymbol{\pi}}{\boldsymbol{4}} \qquad -\frac{\pi}{2} < -\frac{\pi}{4} < \frac{\pi}{2}$$

Since $-\frac{\pi}{4}$ is the only number in the interval $\left(-\frac{\pi}{2}, \frac{\pi}{2}\right)$ such that $\tan \theta = -1$, we establish that $\tan^{-1}(-1) = -\frac{\pi}{4}$.

b. Using the definition of the inverse tangent function, we have the following:

$$\tan^{-1} \sqrt{3} = \theta \qquad -\frac{\pi}{2} < \theta < \frac{\pi}{2}$$

$$\tan \boldsymbol{\theta} = \sqrt{3} \qquad \text{Rewrite the equation.}$$

$$\tan \frac{\boldsymbol{\pi}}{\boldsymbol{3}} = \sqrt{3} \qquad \text{Find } \theta.$$

$$\boldsymbol{\theta} = \frac{\boldsymbol{\pi}}{\boldsymbol{3}} \qquad -\frac{\pi}{2} < \frac{\pi}{3} < \frac{\pi}{2}$$

Self Check 6   Find the exact value of $\tan^{-1}\left(-\frac{\sqrt{3}}{3}\right)$.

Now Try Exercise 13.

Composition of Tangent and Inverse Tangent Functions

$\tan(\tan^{-1} x) = x$	for all x	$[f(f^{-1}(x)) = x]$
$\tan^{-1}(\tan x) = x$	when $\quad -\frac{\pi}{2} < x < \frac{\pi}{2}$	$[f^{-1}(f(x)) = x]$

EXAMPLE 7   **Finding the Composition of Tangent and Inverse Tangent Functions**

Find: **a.** $\tan(\tan^{-1} 2.5)$ **b.** $\tan^{-1}\left(\tan \frac{2\pi}{3}\right)$

SOLUTION   **a.** $\tan(\tan^{-1} 2.5) = 2.5$ because $\tan(\tan^{-1} x) = x$ for all x.

b. For $\tan^{-1}\left(\tan \frac{2\pi}{3}\right)$, we note that $\frac{2\pi}{3}$ is not in the interval $\left(-\frac{\pi}{2}, \frac{\pi}{2}\right)$. Therefore, $\tan^{-1}\left(\tan \frac{2\pi}{3}\right) \neq \frac{2\pi}{3}$. Starting with $\tan \frac{2\pi}{3}$, we have the following:

$$\tan^{-1}\left(\tan\frac{2\pi}{3}\right) = \tan^{-1}\left(-\sqrt{3}\right) \qquad \tan\frac{2\pi}{3} = -\sqrt{3}$$

$$\tan^{-1}\left(-\sqrt{3}\right) = -\frac{\pi}{3}$$

$$\tan^{-1}\left(\tan\frac{2\pi}{3}\right) = -\frac{\pi}{3}$$

Self Check 7   Find the exact value of $\tan^{-1}\left(\tan\frac{7\pi}{6}\right)$.

Now Try Exercise 49.

4. Find Values of Inverse Trigonometric Functions

We can use the definitions of the trigonometric functions and the inverse trigono-metric functions to find the exact values of composite expressions.

EXAMPLE 8   **Finding the Exact Values of Trigonometric Expressions**

Find the exact value of each expression.

a. $\cos\left[\sin^{-1}\left(-\frac{1}{2}\right)\right]$ **b.** $\sin\left(\tan^{-1}\frac{5}{12}\right)$ **c.** $\tan\left[\cos^{-1}\left(-\frac{1}{3}\right)\right]$

SOLUTION   **a.** The expression $\sin^{-1}\left(-\frac{1}{2}\right)$ represents the angle $\theta = \sin^{-1}\left(-\frac{1}{2}\right) = -\frac{\pi}{6}$.

$$\cos\left[\sin^{-1}\left(-\frac{1}{2}\right)\right] = \cos\theta = \cos\left(-\frac{\pi}{6}\right) = \frac{\sqrt{3}}{2}$$

b. To find $\sin\left(\tan^{-1}\frac{5}{12}\right)$, we note that $\frac{5}{12} > 0$, which means $\tan^{-1}\frac{5}{12}$ represents an angle θ in QI. The angle is not a special angle. (Why?) It is shown in standard position in Figure 5-12. We determine a point $P(x, y)$ on the termi-nal side of the angle such that $\tan^{-1}\frac{5}{12} = \theta$ or, equivalently, $\tan\theta = \frac{5}{12}$. Since $\tan\theta = \frac{y}{x} = \frac{5}{12}$, we determine that $x = 12$ and $y = 5$. With these values, we can evaluate r.

$$r^2 = x^2 + y^2$$

$$r = \sqrt{x^2 + y^2} = \sqrt{144 + 25} = \sqrt{169} = 13$$

We now have enough information to find the value of the expression.

$$\sin\left(\tan^{-1}\frac{5}{12}\right) = \sin\theta = \frac{y}{r} = \frac{5}{13}$$

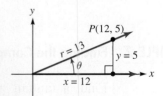

FIGURE 5-12

c. We will use the strategy developed in part (b) to find $\tan\left[\cos^{-1}\left(-\frac{1}{3}\right)\right]$. Because $-\frac{1}{3} < 0$, $\theta = \cos^{-1}\left(-\frac{1}{3}\right)$ lies in QII. The angle is shown in standard position in

Figure 5-13. We determine a point $P(x, y)$ on the terminal side of the angle such that $\cos^{-1}\left(-\frac{1}{3}\right) = \theta$, or equivalently, $\cos\theta = -\frac{1}{3}$. Since $\cos\theta = \frac{x}{r} = -\frac{1}{3}$, we see that $x = -1$ and $r = 3$. (How do we know that $r > 0$?) Now we solve for y.

$$r^2 = x^2 + y^2$$

$$y^2 = r^2 - x^2 = 3^2 - (-1)^2 = 9 - 1 = 8$$

$$y = \pm\sqrt{8} = \pm 2\sqrt{2}$$

However, since the angle lies in QII, $y > 0$; therefore, $y = 2\sqrt{2}$. We now have sufficient information to solve the problem.

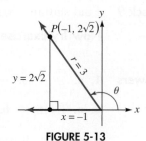

FIGURE 5-13

$$\tan\left[\cos^{-1}\left(-\frac{1}{3}\right)\right] = \tan\theta = \frac{y}{x} = \frac{2\sqrt{2}}{-1} = -2\sqrt{2}$$

Self Check 8   Find the exact value of $\csc\left[\tan^{-1}\left(-\frac{4}{7}\right)\right]$.

Now Try Exercise 65.

5. Convert a Trigonometric Expression to an Algebraic Expression

In calculus, trigonometric expressions are often converted to algebraic expressions, and vice versa. In some problems, algebraic expressions are necessary to complete the problem, while in other problems, the opposite is true. This is explored in more detail in "Looking Ahead to Calculus" at the end of this chapter. However, we will now explore techniques that can be used to convert a trigonometric expression to an algebraic expression.

EXAMPLE 9   **Converting a Trigonometric Expression to an Algebraic Expression**

Find $\cos(\sin^{-1} x)$, given that $0 < x \le 1$.

SOLUTION

Since the value of x is unknown, we don't know what number has a sine of x. Therefore, we will find an algebraic expression in terms of x that is equivalent to $\cos(\sin^{-1} x)$. We do know that $\theta = \sin^{-1} x$ represents an angle in either QI or QIV. (Why?) In Figure 5-14, angle θ has a sine of x ($x > 0$), because we label the side opposite θ with x and the hypotenuse has a value of 1. We can use the Pythagorean Theorem to find that the remaining side of the triangle has length $\sqrt{1 - x^2}$. We can then read the value of $\cos(\sin^{-1} x)$ from Figure 5-14.

$$\theta = \sin^{-1} x$$

$$\sin \theta = \frac{x}{1} \quad \left(\frac{\text{opposite}}{\text{hypotenuse}}\right)$$

$$x^2 + y^2 = 1$$

$$y^2 = 1 - x^2$$

$$y = \sqrt{1 - x^2}$$

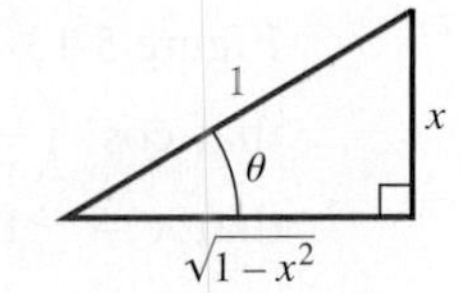

FIGURE 5-14

$$\cos(\sin^{-1} x) = \frac{\text{adjacent side}}{\text{hypotenuse}}$$

$$= \frac{\sqrt{1 - x^2}}{1}$$

$$= \sqrt{1 - x^2}$$

Because $-\frac{\pi}{2} \le \theta \le \frac{\pi}{2}$, the expression $\cos(\sin^{-1} x)$ is never negative, and the positive value of the radical is correct, $\cos(\sin^{-1} x) = \sqrt{1 - x^2}$.

Self Check 9   Find $\sin(\tan^{-1} x)$, given that $0 < x \le 1$.

Now Try Exercise 75.

Self Check Answers   **1. a.** $\dfrac{\pi}{3}$ **b.** $-\dfrac{\pi}{4}$ **c.** It is not defined because -3 is not in the interval $[-1, 1]$.

2. a. 1.115 **b.** -0.551 **3. a.** 0.75 **b.** $\dfrac{\pi}{4}$ **c.** $-\dfrac{\pi}{5}$ **4. a.** $\dfrac{\pi}{3}$

b. π **c.** It is not defined because $\sqrt{2}$ is not in the interval $[-1, 1]$.

5. a. -0.65 **b.** $\dfrac{2\pi}{3}$ **6.** $-\dfrac{\pi}{6}$ **7.** $\dfrac{\pi}{6}$ **8.** $-\dfrac{\sqrt{65}}{4}$ **9.** $\dfrac{x\sqrt{x^2 + 1}}{x^2 + 1}$

Exercises 5.1

Getting Ready
You should be able to complete these vocabulary and concept statements before you proceed to the practice exercises.

Fill in the blanks.

1. The function $f(x) = \sin^{-1} x$ has domain = _____ and range = _____.

2. The function $f(x) = \cos^{-1} x$ has domain = _____ and range = _____.

3. The values of x for which $\sin^{-1}(\sin x) = x$ are _____.

4. The values of x for which $\sin(\sin^{-1} x) = x$ are _____.

5. The values of x for which $\cos^{-1}(\cos x) = x$ are _____.

6. The values of x for which $\cos(\cos^{-1} x) = x$ are _____.

7. True or False: $\cos^{-1}\left(\cos \frac{3\pi}{4}\right) = \frac{3\pi}{4}$. _____

8. True or False: $\sin^{-1}\left(\sin \frac{3\pi}{4}\right) = \frac{3\pi}{4}$. _____

Practice
Find the exact value of each expression, if any. Do not use a calculator. Note that each answer should be a real number.

9. $\sin^{-1}\left(-\dfrac{1}{2}\right)$

10. $\cos^{-1}\left(-\dfrac{1}{2}\right)$

11. $\cos^{-1} 0$

12. $\sin^{-1} 0$

13. $\tan^{-1}(-\sqrt{3})$

14. $\tan^{-1}\left(-\dfrac{\sqrt{3}}{3}\right)$

15. $\tan^{-1} 1$

16. $\sin^{-1}(-1)$

Use a calculator to approximate each value in radians to three decimal places.

17. $\sin^{-1} 0.95$

18. $\sin^{-1} 0.71$

19. $\cos^{-1} \dfrac{\sqrt{3}}{5}$

20. $\cos^{-1} \dfrac{4}{9}$

21. $\tan^{-1} \pi$

22. $\tan^{-1} 34$

23. $\sin^{-1}\left(-\dfrac{2}{3}\right)$

24. $\sin^{-1}\left(-\dfrac{7}{10}\right)$

25. $\cos^{-1}(-0.375)$

26. $\cos^{-1}(-0.924)$

27. $\tan^{-1}(-20)$

28. $\tan^{-1}(-4\pi)$

Find the exact value of each expression or state why the value does not exist. Do not use a calculator.

29. $\sin^{-1}\left(\sin\dfrac{\pi}{6}\right)$

30. $\sin^{-1}\left[\sin\left(-\dfrac{\pi}{3}\right)\right]$

31. $\sin^{-1}\left[\sin\left(-\dfrac{3\pi}{4}\right)\right]$

32. $\sin^{-1}\left[\sin\left(-\dfrac{11\pi}{6}\right)\right]$

33. $\sin(\sin^{-1}0.75)$

34. $\sin[\sin^{-1}(-0.35)]$

35. $\cos^{-1}\left(\cos\dfrac{7\pi}{6}\right)$

36. $\cos^{-1}\left[\cos\left(-\dfrac{5\pi}{6}\right)\right]$

37. $\sin\left(\sin^{-1}\dfrac{1}{2}\right)$

38. $\cos\left(\cos^{-1}\dfrac{1}{2}\right)$

39. $\tan(\tan^{-1}1)$

40. $\sin(\sin^{-1}0)$

41. $\cos(\cos^{-1}1)$

42. $\tan(\tan^{-1}0)$

43. $\tan^{-1}\left[\tan\left(-\dfrac{\pi}{6}\right)\right]$

44. $\tan^{-1}\left[\tan\left(-\dfrac{\pi}{4}\right)\right]$

45. $\cos^{-1}\left(\cos\dfrac{2\pi}{3}\right)$

46. $\cos^{-1}\left(\cos\dfrac{3\pi}{4}\right)$

47. $\sin\left(\sin^{-1}\dfrac{3}{4}\right)$

48. $\cos\left[\cos^{-1}\left(-\dfrac{3}{5}\right)\right]$

49. $\tan(\tan^{-1}5)$

50. $\tan[\tan^{-1}(-6)]$

51. $\cos(\cos^{-1}1.5)$

52. $\sin[\sin^{-1}(-2.1)]$

Find the exact value of each expression. Do not use a calculator.

53. $\sin\left(\cos^{-1}\dfrac{\sqrt{3}}{2}\right)$

54. $\cos\left[\sin^{-1}\left(-\dfrac{1}{2}\right)\right]$

55. $\tan\left[\sin^{-1}\left(-\dfrac{\sqrt{3}}{2}\right)\right]$

56. $\cot\left(\cos^{-1}\dfrac{\sqrt{2}}{2}\right)$

57. $\cos(\tan^{-1}1)$

58. $\sin(\tan^{-1}0)$

59. $\sec\left(\sin^{-1}\dfrac{4}{5}\right)$

60. $\csc\left(\cos^{-1}\dfrac{3}{5}\right)$

61. $\sin\left(\cos^{-1}\dfrac{5}{13}\right)$

62. $\cos\left(\sin^{-1}\dfrac{5}{13}\right)$

63. $\tan\left[\sin^{-1}\left(-\dfrac{4}{5}\right)\right]$

64. $\tan\left[\cos^{-1}\left(-\dfrac{3}{5}\right)\right]$

65. $\cot\left[\cos^{-1}\left(-\dfrac{12}{13}\right)\right]$

66. $\cot\left[\sin^{-1}\left(-\dfrac{12}{13}\right)\right]$

67. $\cos\left[\tan^{-1}\left(\dfrac{5}{6}\right)\right]$

68. $\sin\left(\tan^{-1}\dfrac{4}{9}\right)$

69. $\csc\left(\cos^{-1}\dfrac{2}{3}\right)$

70. $\sec\left(\sin^{-1}\dfrac{3}{7}\right)$

71. $\sin\left[\cos^{-1}\left(-\dfrac{1}{3}\right)\right]$

72. $\cos\left[\sin^{-1}\left(-\dfrac{2}{3}\right)\right]$

73. $\tan\left[\sin^{-1}\left(-\dfrac{3}{4}\right)\right]$

74. $\tan\left[\cos^{-1}\left(-\dfrac{2}{5}\right)\right]$

Rewrite each value as an algebraic expression in the variable x, $0 < x \le 1$. Assume θ is in QI.

75. $\sin(\tan^{-1}x)$

76. $\cos(\tan^{-1}x)$

77. $\tan(\sin^{-1}x)$

78. $\tan(\cos^{-1}x)$

79. $\sin(\cos^{-1}2x)$

80. $\cos(\sin^{-1}4x)$

81. $\cos\left(\sin^{-1}\dfrac{1}{x}\right)$

82. $\sec\left(\tan^{-1}\dfrac{x}{2}\right)$

83. $\csc\left(\cos^{-1}\dfrac{x}{\sqrt{x^2+9}}\right)$

84. $\cot\left(\sin^{-1}\dfrac{\sqrt{x^2-4}}{x}\right)$

Applications

85. Shanghai skyscraper The Shanghai World Financial Center is 1,614 feet tall and is the tallest skyscraper in Shanghai. If the angle of depression from the top of the center to a point on the ground 435 feet away from its base is $\tan^{-1}\dfrac{1,614}{435}$ radians, find the angle of depression. Round to two decimal places.

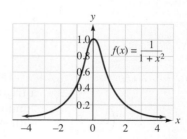

sevenke/Shutterstock.com

86. Sydney Harbor Bridge The angle of elevation to the top of the Sydney Harbor Bridge from your viewpoint on a ferry is $\cos^{-1}\left(\dfrac{2,640}{2,679}\right)$ radians. Find the angle of elevation. Round to two decimal places.

87. Flying a kite A man is flying a kite with a 150 feet of string released. If the height of the kite is y feet, write an expression for θ, the angle of elevation of the kite, as a function of y. Determine θ in degrees if $y = 100$ feet. Round to one decimal place.

88. Hot-air balloon A hot-air balloon is flying at an altitude of 800 feet toward a point directly above you. If you are a horizontal distance of x feet from the balloon, write an expression for θ, the angle of elevation of the balloon, as a function of x. Determine θ in degrees if $x = 300$ feet. Round to one decimal place.

89. Area The graph of $f(x) = \dfrac{1}{1+x^2}$ is shown.

The area bounded by the graph of the function and the x-axis between $x = a$ and $x = b$ can be determined by evaluating the inverse trigonometric expression $\tan^{-1} b - \tan^{-1} a$.

(a) Find the area between $x = -1$ and $x = 1$.

(b) Find the area between $x = -\sqrt{3}$ and $x = \sqrt{3}$.

(c) Find the area between $x = -\frac{\sqrt{3}}{3}$ and $x = \frac{\sqrt{3}}{3}$.

90. Area The graph of $f(x) = \dfrac{1}{\sqrt{1-x^2}}$ is shown.

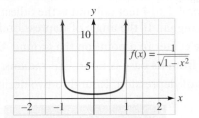

$$f(x) = \frac{1}{\sqrt{1-x^2}}$$

The area bounded by the graph of the function and the x-axis between $x = a$ and $x = b$ can be determined by evaluating the inverse trigonometric expression $\sin^{-1} b - \sin^{-1} a$.

(a) Find the area between $x = -\frac{1}{2}$ and $x = \frac{1}{2}$.

(b) Find the area between $x = -1$ and $x = 1$.

(c) Find the area between $x = -\frac{\sqrt{3}}{2}$ and $x = \frac{\sqrt{3}}{2}$.

Discovery and Writing

91. If $\sin \frac{11\pi}{6} = -\frac{1}{2}$, explain why it is incorrect to write $\sin^{-1}\left(-\frac{1}{2}\right) = \frac{11\pi}{6}$.

92. If $\cos\left(-\frac{\pi}{4}\right) = \frac{\sqrt{2}}{2}$, explain why it is incorrect to write $\cos^{-1}\left(\frac{\sqrt{2}}{2}\right) = -\frac{\pi}{4}$.

5.2 Inverse Cotangent, Secant, and Cosecant Functions (optional)

In this section, we will learn to

1. Develop the inverse cotangent function and evaluate inverse cotangent expressions.

2. Develop the inverse secant and cosecant functions and evaluate inverse secant and cosecant functions.

Word processors are computer applications used to produce printable material. Most word processors have an undo button that reverses the last operation performed. This button is ideal for college students writing assignments and creating laboratory reports.

In trigonometry, we are now learning that we are able to reverse or "undo" a trigonometric-function operation. If we know that $\csc \frac{\pi}{6}$ is 2, we can reverse or "undo" this operation using an inverse trigonometric function. That is, if we begin with the 2, we can evaluate $\csc^{-1} 2$ and obtain $\frac{\pi}{6}$. In this section, we will continue our study of the inverse trigonometric functions and focus our attention on inverse cotangent, inverse secant, and inverse cosecant.

Just as we did in the previous section, we will define each of the remaining inverse functions using an interval over which the function is one-to-one.

1. Develop the Inverse Cotangent Function and Evaluate Inverse Cotangent Expressions

There are at least two conventions for defining the inverse cotangent function, depending upon the period selected for the cotangent function. We have selected the interval $\{x \mid 0 < x < \pi\}$, which is a common selection. Figure 5-15 shows the graph of $y = \cot x$ with its domain restricted. With this restricted domain, the function $y = \cot x$ becomes a one-to-one function and has an inverse.

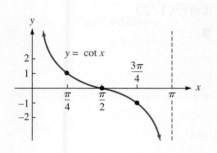

$y = \cot x$	
x	y
$\dfrac{\pi}{4}$	1
$\dfrac{\pi}{2}$	0
$\dfrac{3\pi}{4}$	-1

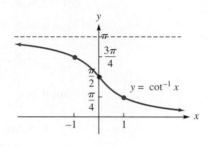

$y = \cot^{-1} x$	
x	y
1	$\dfrac{\pi}{4}$
0	$\dfrac{\pi}{2}$
-1	$\dfrac{3\pi}{4}$

FIGURE 5-15

Inverse Cotangent Function   The **inverse cotangent function,** $y = \cot^{-1} x$, has domain $D =$ all real numbers and range $R = \{y \mid 0 < y < \pi\}$.

$$y = \cot^{-1} x \qquad \text{if and only if} \qquad x = \cot y \qquad \text{and} \qquad 0 < y < \pi.$$

EXAMPLE 1   **Evaluating the Inverse Cotangent Function**

Find the exact value of each expression.

a. $\cot^{-1} 1$ **b.** $\cot^{-1} 0$ **c.** $\cot^{-1}\left(-\sqrt{3}\right)$

SOLUTION   **a.** $\cot^{-1} 1$ represents the angle in the interval $(0, \pi)$ with cotangent 1. Hence, we have $\cot^{-1} 1 = \frac{\pi}{4}$. Note that $\cot \frac{\pi}{4} = 1$ and that $0 < \frac{\pi}{4} < \pi$.

b. The only angle in the interval $(0, \pi)$ with cotangent 0 is $\frac{\pi}{2}$. Therefore,

$$\cot^{-1} 0 = \frac{\pi}{2}.$$

c. Since cotangent is negative in the interval $\left(\frac{\pi}{2}, \pi\right)$, $\cot^{-1}\left(-\sqrt{3}\right)$ represents an angle in that interval with cotangent $-\sqrt{3}$. Since $\cot \frac{5\pi}{6} = -\sqrt{3}$, we have $\cot^{-1}\left(-\sqrt{3}\right) = \frac{5\pi}{6}$. ∎

Self Check 1   Find the exact value of $\cot^{-1}(-1)$.

Now Try Exercise 3.

Using a calculator to evaluate the inverse cotangent function is not straightforward, because the inverse cotangent and inverse tangent do not have the same domain. If the angle is in QI, $\left(0, \frac{\pi}{2}\right)$, then we can use the inverse tangent to find the answer. If the angle is in QII, $\left(\frac{\pi}{2}, \pi\right)$, the use of a reference angle will be required.

EXAMPLE 2  **Evaluating Inverse Cotangent Using a Calculator**

Find each value in radians, and round to three decimal places.

a. $\cot^{-1} 2$ **b.** $\cot^{-1} 1.576$ **c.** $\cot^{-1}(-2.1)$

SOLUTION   **a.** First, make certain that your calculator is in RADIAN mode. Since $\cot^{-1} 2 = \theta$ means $\cot \theta = 2$, we know that θ is in QI, because $\cot \theta > 0 \rightarrow 0 < \theta < \frac{\pi}{2}$. (Recall that we are restricted to the interval $(0, \pi)$, the domain of the inverse cotangent.)

$$\cot^{-1} 2 = \theta \qquad 0 < \theta < \frac{\pi}{2}$$

$$\cot \theta = 2$$

$$\frac{1}{\tan \theta} = 2 \qquad \cot \theta = \frac{1}{\tan \theta}$$

$$\tan \theta = \frac{1}{2}$$

$$\theta = \tan^{-1} \frac{1}{2} \approx 0.463647609 \approx 0.464$$

b. Again, since $1.576 > 0$, we know that $\cot^{-1} 1.576$ represents an angle in QI.

$$\cot^{-1} 1.576 = \theta$$

$$\cot \theta = 1.576$$

$$\frac{1}{\tan \theta} = 1.576$$

$$\tan \theta = \frac{1}{1.576}$$

$$\theta = \tan^{-1} \frac{1}{1.576} \approx 0.565414293 \approx 0.565$$

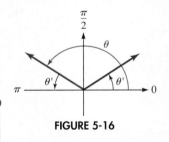

c. To evaluate $\cot^{-1}(-2.1)$ requires the use of a reference angle, because the angle represented by this expression is in QII and that interval is not in the range of the inverse tangent function. We can still use the inverse tangent function, but not directly. In Figure 5-16, notice that we can use the reference angle θ' to find angle θ. So our first step is to use inverse tangent to find θ'.

$$\cot^{-1} 2.1 = \theta'$$

$$\cot \theta' = 2.1$$

$$\tan \theta' = \frac{1}{2.1}$$

$$\boldsymbol{\theta'} = \tan^{-1} \frac{1}{2.1} \approx 0.4444192099$$

$$\theta = \pi - \tan^{-1} \frac{1}{2.1} \approx \mathbf{2.697}$$

FIGURE 5-16

Self Check 2   Approximate each expression to three decimal places.   **a.** $\cot^{-1} 3.5$
b. $\cot^{-1}(-0.57)$

Now Try Exercise 11.

2. Develop the Inverse Secant and Cosecant Functions and Evaluate Inverse Secant and Cosecant Functions

If we restrict the domains of the secant and cosecant functions to appropriate intervals, they too become one-to-one functions and have inverses.

Inverse Secant Function

The **inverse secant function,** $y = \sec^{-1} x$, has domain

$$D = \{x \,|\, x \leq -1 \text{ or } x \geq 1\} \text{ and range } R = \left\{ y \,\middle|\, 0 \leq y \leq \pi \text{ and } y \neq \frac{\pi}{2} \right\}.*$$

$y = \sec^{-1} x$ is equivalent to $x = \sec y$ and $0 \leq y \leq \pi$ and $y \neq \dfrac{\pi}{2}$.

*Some books restrict y to the interval $-\pi \leq y < -\dfrac{\pi}{2}$ or the interval $0 \leq y < \dfrac{\pi}{2}$.

The graphs of the secant function with its domain restricted to $0 \leq x \leq \pi$ and the inverse secant function are shown in Figure 5-17.

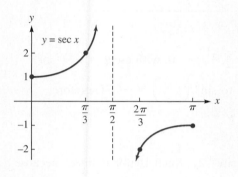

$y = \sec x$	
x	y
0	1
$\dfrac{\pi}{3}$	2
$\dfrac{2\pi}{3}$	-2
π	-1

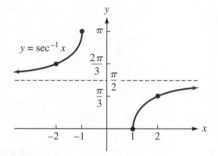

$y = \sec^{-1} x$	
x	y
1	0
2	$\dfrac{\pi}{3}$
-2	$\dfrac{2\pi}{3}$
-1	π

FIGURE 5-17

Inverse Cosecant Function

The **inverse cosecant function,** $y = \csc^{-1} x$, has domain

$$D = \{x \,|\, x \leq -1 \text{ or } x \geq 1\} \text{ and range } R = \left\{ y \,\middle|\, -\frac{\pi}{2} \leq y \leq \frac{\pi}{2} \text{ and } y \neq 0 \right\}.*$$

$y = \csc^{-1} x$ is equivalent to $x = \csc y$ and $-\dfrac{\pi}{2} \leq y \leq \dfrac{\pi}{2}$ and $y \neq 0$.

*Some books restrict y to the interval $-\pi < y \leq -\dfrac{\pi}{2}$ or the interval $0 < y \leq \dfrac{\pi}{2}$.

The graphs of the cosecant function with its domain restricted to $-\dfrac{\pi}{2} \leq x \leq \dfrac{\pi}{2}$ and the inverse cosecant function are shown in Figure 5-18.

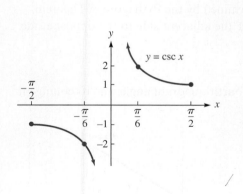

$y = \csc x$	
x	y
$-\dfrac{\pi}{2}$	-1
$-\dfrac{\pi}{6}$	-2
$\dfrac{\pi}{6}$	2
$\dfrac{\pi}{2}$	1

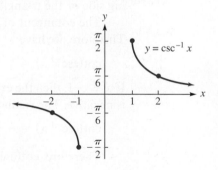

$y = \csc^{-1} x$	
x	y
-1	$-\dfrac{\pi}{2}$
-2	$-\dfrac{\pi}{6}$
2	$\dfrac{\pi}{6}$
1	$\dfrac{\pi}{2}$

FIGURE 5-18

EXAMPLE 3   **Evaluating Inverse Secant and Inverse Cosecant**

Find each value in radians.   **a.** $\sec^{-1} 2$ **b.** $\sec^{-1} 1.75$ **c.** $\csc^{-1}\left(-\dfrac{2\sqrt{3}}{3}\right)$

d. $\csc^{-1} \dfrac{1}{2}$

SOLUTION   **a.** $\sec^{-1} 2$ is the angle in the interval $(0, \pi)$ whose secant is 2 $\left(\text{or whose cosine is } \frac{1}{2}\right)$. Since $\cos \frac{\pi}{3} = \frac{1}{2}$, $\sec \frac{\pi}{3} = 2$ and $\sec^{-1} 2 = \frac{\pi}{3}$.

b. To find $\theta = \sec^{-1} 1.75$, we will use a calculator. Make certain that it is in RADIAN mode. We must use inverse cosine. Following the steps shown, we get to the point where we can use the calculator.

$$\sec^{-1} 1.75 = \theta$$

$$\sec \theta = 1.75$$

$$\cos \theta = \frac{1}{1.75}$$

$$\theta = \cos^{-1}\left(\frac{1}{1.75}\right) \approx 0.963$$

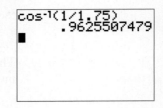

c. $\csc^{-1}\left(-\frac{2\sqrt{3}}{3}\right)$ is an angle in the interval $\left[-\frac{\pi}{2}, \frac{\pi}{2}\right]$, $y \neq 0$, with $\csc \theta = -\frac{2\sqrt{3}}{3}$ or $\sin \theta = -\frac{3}{2\sqrt{3}} = -\frac{\sqrt{3}}{2}$. This is equivalent to $\sin^{-1}\left(-\frac{\sqrt{3}}{2}\right) = -\frac{\pi}{3}$. Therefore,

$$\csc^{-1}\left(-\frac{2\sqrt{3}}{3}\right) = -\frac{\pi}{3}.$$

d. $\csc^{-1} \frac{1}{2}$ is equivalent to the expression $\sin^{-1} 2$, which is not defined, because $\sin \theta = 2$ is not possible.

Self Check 3   Find each value: **a.** $\sec^{-1} 2.15$ **b.** $\csc^{-1} 1$

Now Try Exercise 13.

EXAMPLE 4   **Writing a Trigonometric Expression in Terms of x**

Express $\cot(\csc^{-1} x)$ as an algebraic expression in the variable x.

SOLUTION   The expression $\csc^{-1} x$ represents an angle in the interval $\left[-\frac{\pi}{2}, \frac{\pi}{2}\right]$, $y \neq 0$, whose cosecant is x. If $x \geq 1$, then $\csc^{-1} x$ is the first-quadrant angle shown in Figure 5-19. The ratio of the hypotenuse to the side opposite $\csc^{-1} x$ is $\frac{x}{1} = x$. The remaining side of the triangle is $\sqrt{x^2 - 1}$, as determined by the Pythagorean Theorem.

The cotangent of $\csc^{-1} x$ is the ratio of the adjacent side to the opposite side. Therefore, we have

$$\cot(\csc^{-1} x) = \sqrt{x^2 - 1} \text{ for } x \geq 1.$$

If $x \leq -1$, then the expression $\csc^{-1} x$ is a fourth-quadrant angle, and its cotangent is negative. The result for this case is

$$\cot(\csc^{-1} x) = -\sqrt{x^2 - 1} \text{ for } x \leq -1.$$

Therefore, $\cot(\csc^{-1} x) = \pm\sqrt{x^2 - 1}$.

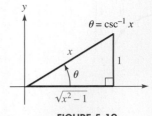

FIGURE 5-19

Self Check 4   Express $\tan(\sec^{-1} x)$ as an algebraic expression in the variable x.

Now Try Exercise 27.

Self Check Answers   1. $\dfrac{3\pi}{4}$ **2. a.** 0.278   **b.** 2.089   **3. a.** 1.087   **b.** $\dfrac{\pi}{2}$ **4.** $\pm\sqrt{x^2-1}$

Exercises **5.2**

Getting Ready
You should be able to complete these vocabulary and concept statements before you proceed to the practice exercises.

Fill in the blanks.

1. The domain of $y = \cot^{-1} x$ is _____ and the range is _____.

2. True or False: An approximation of $y = \cot^{-1}(-2.1)$ can be found just by entering $\dfrac{1}{\tan^{-1}(-2.1)}$ in the calculator. _____

Practice
Evaluate each expression. Do not use a calculator.

3. $\cot^{-1}\sqrt{3}$

4. $\cot^{-1}\left(-\dfrac{\sqrt{3}}{3}\right)$

5. $\sec^{-1}(-1)$

6. $\sec^{-1}(-2)$

7. $\csc^{-1}(-1)$

8. $\csc^{-1} 2$

9. $\cot^{-1}\dfrac{\sqrt{3}}{3}$

10. $\sec^{-1} 0$

Evaluate each expression in radians to three decimal places. If the calculator result is not in the proper range, convert accordingly.

11. $\cot^{-1}(-4)$

12. $\cot^{-1} 4$

13. $\sec^{-1} 2.236$

14. $\sec^{-1}(-2.236)$

15. $\csc^{-1} 1.23$

16. $\csc^{-1}(-1.23)$

17. $\cot^{-1} 7.775$

18. $\cot^{-1}(-7.775)$

Find the exact value of each expression.

19. $\sec(\sec^{-1} 3)$

20. $\csc(\csc^{-1} 5)$

21. $\sec(\csc^{-1} 4)$

22. $\csc(\sec^{-1} 7)$

23. $\cot\left(\sec^{-1}\dfrac{41}{40}\right)$

24. $\tan\left(\csc^{-1}\dfrac{13}{5}\right)$

25. $\sec\left(\csc^{-1}\dfrac{5}{4}\right)$

26. $\csc\left(\sec^{-1}\dfrac{25}{20}\right)$

Rewrite each value as an algebraic expression in the variable x. Make sure that you consider all numbers in the domain of x. Assume θ is in QI.

27. $\sin(\csc^{-1} x)$

28. $\cos(\cot^{-1} x)$

29. $\sec(\cot^{-1} x)$

30. $\csc(\sec^{-1} x)$

31. $\tan(\sec^{-1} x)$

32. $\tan(\cot^{-1} x)$

Applications

33. **Flying a kite** A man is flying a kite with 160 feet of string released. If the height of the kite is y feet, write an expression for θ, the angle of elevation of the kite, as a function of y. (Use inverse cosecant.) Determine θ in degrees if $y = 90$ feet. Round to one decimal place.

34. **Hot-air balloon** A hot-air balloon is flying at an altitude of 700 feet toward a point directly above you. If you are a horizontal distance of x feet from the balloon, write an expression for θ, the angle of elevation of the balloon, as a function of x. Determine θ in degrees if $x = 200$ feet. (Use inverse cotangent.) Round to one decimal place.

35. **Area** The graph of $f(x) = \dfrac{1}{x\sqrt{x^2-1}}$ is shown.

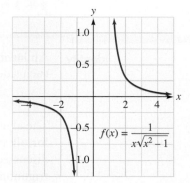
$$f(x) = \dfrac{1}{x\sqrt{x^2-1}}$$

Note that the domain of the function excludes the interval $[-1, 1]$. The area bounded by the graph of the function and the x-axis between $x = a$ and $x = b$ can be determined by evaluating the inverse trigonometric expression $\sec^{-1}|b| - \sec^{-1}|a|$.

(a) Find the area between $x = \sqrt{2}$ and $x = 2$.

(b) Find the area between $x = -\sqrt{2}$ and 2.

36. **Area** Refer to Exercise 35. Use a calculator to find the area between $x = 2$ and $x = 4$. Round to three decimal places.

Discovery and Writing

37. If $\cot\left(-\dfrac{\pi}{4}\right) = -1$, explain why it is incorrect to write $\cot^{-1}(-1) = -\dfrac{\pi}{4}$.

38. If $\sec\left(-\dfrac{\pi}{3}\right) = 2$, explain why it is incorrect to write $\sec^{-1} 2 = -\dfrac{\pi}{3}$.

5.3 Verifying Trigonometric Identities

In this section, we will learn to

1. Verify trigonometric identities.

A puzzle ring is a piece of jewelry made from several separate bands which interlock.

The ring can be disassembled, turning it into a "puzzle" which must be carefully reassembled using a step-by-step procedure. Puzzle rings are worn often for beauty and are often customized and used as wedding, friendship, and promise rings. They are quite popular in the Middle East.

In this section, we are going to use the trigonometric identities that have previously been introduced to verify other trigonometric identities. Identity problems can be viewed as trigonometric "puzzles." As we will see, various strategies can be used to solve these puzzles.

In trigonometry and calculus it is often necessary to convert a trigonometric expression into another, more useful form. We will explore this topic in this section.

1. Verify Trigonometric Identities

In algebra, an equation such as $x^2 - 2x = 5$, for which some numbers are solutions and others are not, is called a **conditional** equation. An equation such as $x + 3 = x + 3$, for which all real numbers are solutions, is called an **identity**.

Identity

An **identity** is an equation that is true for every value of its variable(s) for which each side of the equation is defined.

The algebraic equations $3(x + 2) = 3x + 6$ and $x^2 - y^2 = (x + y)(x - y)$ are true for all values of x and y. Thus, they are algebraic identities. In Section 4.3, we discussed several fundamental trigonometric relationships. These relationships are trigonometric identities. We repeat them here for reference.

Fundamental Trigonometric Identities

These identities are true for any number x for which the functions are defined.

Reciprocal Identities: $\sin x = \dfrac{1}{\csc x}$ $\cos x = \dfrac{1}{\sec x}$ $\tan x = \dfrac{1}{\cot x}$

Quotient Identities: $\tan x = \dfrac{\sin x}{\cos x}$ $\cot x = \dfrac{\cos x}{\sin x}$

Pythagorean Identities: $\sin^2 x + \cos^2 x = 1$ $\tan^2 x + 1 = \sec^2 x$
$$\cot^2 x + 1 = \csc^2 x$$

Even–Odd Identities: $\sin(-x) = -\sin x$ $\cos(-x) = \cos x$ $\tan(-x) = -\tan x$
$$\csc(-x) = -\csc x \quad \sec(-x) = \sec x \quad \cot(-x) = -\cot x$$

To verify other trigonometric identities, we will use these fundamental identities and basic algebraic operations. Algebraic operations are useful because we can use them to add, subtract, multiply, factor, simplify, and combine trigonometric expressions.

EXAMPLE 1 **Simplifying Trigonometric Expressions Using Algebraic Operations**

Combine each expression. **a.** $\dfrac{1}{\tan x} + \dfrac{\sin x}{\cos x}$ **b.** $\dfrac{\cos x}{\sec x + 1} - \dfrac{\cos x}{\sec x - 1}$

SOLUTION **a.** We will use simple addition to add the two expressions.

$$\dfrac{1}{\tan x} + \dfrac{\sin x}{\cos x} = \dfrac{1 \cdot \cos x}{\tan x \cdot \cos x} + \dfrac{\tan x \sin x}{\tan x \cos x} \qquad \text{Find a least common denominator.}$$

$$= \dfrac{\cos x + \tan x \sin x}{\tan x \cos x} \qquad \text{Add the numerators.}$$

$$= \dfrac{\cos x + \tan x \sin x}{\sin x} \qquad \tan x \cos x = \dfrac{\sin x}{\cos x} \cdot \cos x = \sin x$$

b. We can also use other identities to simplify.

$$\dfrac{\cos x}{\sec x + 1} - \dfrac{\cos x}{\sec x - 1} = \dfrac{\cos x(\sec x - 1)}{(\sec x + 1)(\sec x - 1)} - \dfrac{(\sec x + 1)\cos x}{(\sec x + 1)(\sec x - 1)}$$

$$= \dfrac{\cos x \sec x - \cos x - \cos x \sec x - \cos x}{(\sec x + 1)(\sec x - 1)}$$

$$= \dfrac{-2 \cos x}{\sec^2 x - 1} \qquad \text{Combine terms and remove parentheses.}$$

$$= \dfrac{-2 \cos x}{\tan^2 x} \qquad \text{Use Pythagorean Identity.}$$

Self Check 1 Simplify the expression $\dfrac{\sin x + \cos x}{\cos x} + \dfrac{\cos x - \sin x}{\sin x}$.

Now Try Exercise 19.

To verify identities, we will usually work with one side of the equation and use the rules of algebra to transform it into the other side. Here are some suggestions that will make the work easier.

Strategies for Verifying Identities Memorize the fundamental identities.

1. Start with the more complicated side of the equation and try to transform it into the less complicated side.

2. As you work on one side of the identity, keep an eye on the other side. It's easier to hit a target that you can see.

3. It may be helpful to begin by changing all functions into sine and cosine functions. This is usually a good plan when the equation involves more than two of the trigonometric functions.

4. If the numerator or denominator of a fraction contains a factor of $1 + \sin x$ or $1 - \sin x$, consider multiplying both the numerator and the denominator by the conjugate. Because $(1 - \sin x)(1 + \sin x) = 1 - \sin^2 x = \cos^2 x$, we are able to simplify the expression. This same idea applies to factors of $1 \pm \cos x$, $\sec x \pm 1$, and $\sec x \pm \tan x$.

EXAMPLE 2 **Verifying an Identity by Changing to Sine and Cosine**

Verify the identity $\dfrac{\tan x}{\sin x} = \sec x$.

SOLUTION We will begin with the left side because it is more complicated.

$$\frac{\tan x}{\sin x} = \frac{\dfrac{\sin x}{\cos x}}{\sin x} \qquad \text{Rewrite in terms of sine and cosine.}$$

$$= \frac{\cancel{\sin x}}{\cos x} \cdot \frac{1}{\cancel{\sin x}} \qquad \text{Invert the denominator and multiply.}$$

$$= \frac{1}{\cos x} \qquad \text{Simplify.}$$

$$= \sec x \qquad \text{Reciprocal Identity}$$

Self Check 2 Verify the identity $\dfrac{\cot x}{\cos x} = \csc x$.

Now Try Exercise 39.

EXAMPLE 3 **Verifying an Identity Using Sine and Cosine**

Verify the identity $\tan x + \cot x = \csc x \sec x$.

SOLUTION We work with the left-hand side of this equation and change $\tan x$ and $\cot x$ into expressions containing $\sin x$ and $\cos x$.

$$\tan x + \cot x = \frac{\sin x}{\cos x} + \frac{\cos x}{\sin x} \qquad \text{Rewrite with sine and cosine.}$$

$$= \frac{\mathbf{\sin x} \sin x}{\mathbf{\sin x} \cos x} + \frac{\cos x \, \mathbf{\cos x}}{\sin x \, \mathbf{\cos x}} \qquad \text{Common denominator is } \sin x \cos x.$$

$$= \frac{\sin^2 x + \cos^2 x}{\sin x \cos x} \qquad \text{Add the fractions.}$$

$$= \frac{1}{\sin x \cos x} \qquad \text{Identity } \sin^2 x + \cos^2 x = 1$$

$$= \frac{1}{\sin x} \cdot \frac{1}{\cos x} \qquad \text{Rewrite as two fractions.}$$

$$= \csc x \sec x \qquad \text{Reciprocal Identities}$$

Self Check 3 Verify the identity $\csc x - \sin x = \cot x \cos x$.

Now Try Exercise 51.

EXAMPLE 4 **Verifying an Identity Using Factoring**

Verify the identity $\cos^4 x - \sin^4 x = 1 - 2\sin^2 x$.

SOLUTION The left-hand side is the difference of two squares and can be factored. Note that the right-hand side has only the sine function. Our goal is to eliminate the cosine function from the left side. Remember to always keep the target in mind.

$$\cos^4 x - \sin^4 x = (\cos^2 x + \sin^2 x)(\cos^2 x - \sin^2 x) \qquad \text{Factor the difference of squares.}$$

$$= 1 \cdot (\cos^2 x - \sin^2 x) \qquad \sin^2 x + \cos^2 x = 1$$

$$= 1 - \sin^2 x - \sin^2 x \qquad \cos^2 x = 1 - \sin^2 x$$

$$= 1 - 2\sin^2 x$$

Self Check 4   Verify the identity $\sin^4 x - \cos^4 x = 1 - 2\cos^2 x$.

Now Try Exercise 63.

EXAMPLE 5   Verifying an Identity by Combining Fractions

Verify the identity $\dfrac{1}{\sec x - \tan x} - \dfrac{1}{\sec x + \tan x} = 2\tan x$.

SOLUTION   Our strategy is to take the left side, find the common denominator for the fractions, and add them.

$$\frac{1}{\sec x - \tan x} - \frac{1}{\sec x + \tan x} = \frac{(\sec x + \tan x)}{(\sec x - \tan x)(\sec x + \tan x)} - \frac{(\sec x - \tan x)}{(\sec x - \tan x)(\sec x + \tan x)}$$

$$= \frac{\sec x + \tan x - \sec x + \tan x}{\sec^2 x - \tan^2 x} \qquad \text{Combine fractions and multiply denominator.}$$

$$= \frac{2\tan x}{1} \qquad \text{Simplify numerator and use a Pythagorean Identity.}$$

$$= 2\tan x$$

Self Check 5   Verify the identity $\dfrac{\sin x}{1 + \cos x} + \dfrac{1 + \cos x}{\sin x} = 2\csc x$.

Now Try Exercise 73.

EXAMPLE 6   Verifying an Identity Using the Even–Odd Identities

Verify the identity $1 - \dfrac{\cos^2(-x)}{1 + \sin(-x)} = -\sin x$.

SOLUTION   Starting on the left side, we apply the Even–Odd Identities.

$$1 - \frac{\cos^2(-x)}{1 + \sin(-x)} = 1 - \frac{\cos^2 x}{1 - \sin x} \qquad \begin{aligned}\cos(-x) &= \cos x \\ \sin(-x) &= -\sin x\end{aligned}$$

$$= 1 - \frac{1 - \sin^2 x}{1 - \sin x} \qquad \cos^2 x = 1 - \sin^2 x$$

$$= 1 - \frac{(1 - \sin x)(1 + \sin x)}{(1 - \sin x)} \qquad \text{Factor.}$$

$$= 1 - (1 + \sin x) \qquad \text{Reduce.}$$

$$= -\sin x \qquad \text{Simplify.}$$

Self Check 6   Verify the identity $\dfrac{\sin x + \cos(-x)}{\cos(-x)} - \dfrac{\sin(-x) + \cos x}{\sin(-x)} = \sec x \csc x$.

Now Try Exercise 55.

EXAMPLE 7   Verifying an Identity Using Two Different Methods: (1) Splitting a Fraction and (2) Combining Terms

Verify the identity $\dfrac{1 + \cos x}{\sin x} = \csc x + \cot x$.

SOLUTION Often there is more than one way to verify an identity. Typically, one method is clearly the first choice. For this identity, both methods shown are equally effective.

Method 1: Splitting a Fraction

We begin with the left side and split the fraction into two terms.

$$\frac{1 + \cos x}{\sin x} = \frac{1}{\sin x} + \frac{\cos x}{\sin x} \qquad \text{Rewrite as two fractions.}$$

$$= \csc x + \cot x \qquad \text{Identities}$$

Method 2: Combining Terms

$$\csc x + \cot x = \frac{1}{\sin x} + \frac{\cos x}{\sin x} \qquad \text{Rewrite in terms of sine and cosine.}$$

$$= \frac{1 + \cos x}{\sin x} \qquad \text{Combine the terms.}$$

Self Check 7 Verify the identity $\dfrac{1 - \sin x}{\cos x} + \dfrac{\cos x}{1 - \sin x} = 2 \sec x$.

Now Try Exercise 69.

EXAMPLE 8 **Verifying an Identity Using a Conjugate**

Verify the identity $\dfrac{\cos x}{1 + \sin x} = \dfrac{1 - \sin x}{\cos x}$.

SOLUTION Multiplying $1 + \sin x$ by its conjugate $1 - \sin x$ allows us to rewrite the denominator as a single term.

$$\frac{\cos x}{1 + \sin x} = \frac{\cos x}{1 + \sin x} \cdot \frac{\mathbf{1 - \sin x}}{\mathbf{1 - \sin x}} \qquad \begin{array}{l}\text{Multiply numerator and denominator}\\ \text{by the conjugate of } 1 + \sin x.\end{array}$$

$$= \frac{\cos x(1 - \sin x)}{1 - \sin^2 x}$$

$$= \frac{\cos x(1 - \sin x)}{\mathbf{\cos^2 x}} \qquad 1 - \sin^2 x = \cos^2 x$$

$$= \frac{1 - \sin x}{\cos x}$$

Self Check 8 Verify the identity $\dfrac{\sin x}{\cos x + 1} = \dfrac{1 - \cos x}{\sin x}$.

Now Try Exercise 77.

Comment

Do not cross multiply to verify an identity like in Example 8. We cannot treat it as an equation until it has been verified. It is incorrect to multiply or divide both sides of an equation whose truth we are trying to establish by an expression containing a variable, because the resulting equation might not be equivalent to the given expression.

In Examples 2–8, all of the work was performed on one side of the equation. Occasionally, if the equation seems to warrant it, we can work on both sides of the identity independently until each side has been transformed into a common third expression. However, it is essential that each step in the process is reversible, so that either side of the identity can be derived from the other side.

EXAMPLE 9 **Verifying an Identity by Working on Both Sides**

Verify the identity $\dfrac{1 - \cos x}{1 + \cos x} = (\csc x - \cot x)^2$.

SOLUTION   In this example, we will work on each side separately, changing each into a common third expression. Beginning with the left side, we have the following:

$$\frac{1 - \cos x}{1 + \cos x} = \frac{(1 - \cos x)(\mathbf{1 - \cos x})}{(1 + \cos x)(\mathbf{1 - \cos x})}$$

$$= \frac{(1 - \cos x)^2}{1 - \cos^2 x}$$

$$= \frac{(1 - \cos x)^2}{\sin^2 x}$$

$$= \left(\frac{1 - \cos x}{\sin x}\right)^2$$

Now we change the right-hand side.

$$(\csc x - \cot x)^2 = \left(\frac{1}{\sin x} - \frac{\cos x}{\sin x}\right)^2$$

$$= \left(\frac{1 - \cos x}{\sin x}\right)^2$$

Each side has been independently transformed into the expression $\left(\dfrac{1 - \cos x}{\sin x}\right)^2$.

Because each step is reversible, it follows that the given equation is an identity. ∎

Self Check 9   Verify the identity $\dfrac{\cot^2 x}{1 + \csc x} = \dfrac{1 - \sin x}{\sin x}$.

Now Try Exercise 97.

EXAMPLE 10   **Determining Values of the Variable That Create an Identity**

For what values of x is $\sqrt{1 - \cos^2 x} = \sin x$?

SOLUTION   Begin with the identity $1 - \cos^2 x = \sin^2 x$, and take the square root of both sides.

$$\sqrt{1 - \cos^2 x} = \sqrt{\sin^2 x}$$

$$\sqrt{1 - \cos^2 x} = |\sin x| \qquad \text{Recall from algebra that } \sqrt{x^2} = |x|.$$

The equation $\sqrt{1 - \cos^2 x} = |\sin x|$ is an identity, but because $\sin x$ can be negative, $\sqrt{1 - \cos^2 x} = \sin x$ is *not* an identity. If we restrict x to real numbers in the interval $[0, 2\pi]$, $\sin x$ is nonnegative for $0 \leq x \leq \pi$. Hence, $\sqrt{1 - \cos^2 x} = \sin x$ only if x is a number from 0 to π, from 2π to 3π, from 4π to 5π, and so on. ∎

Self Check 10   For what values is $\sqrt{\tan^2 x + 1} = \sec x$?

Now Try Exercise 93.

Self Check Answers   **1.** $\sec x \csc x$ **2–9.** Answers will vary.   **10.** $x = \left(-\dfrac{\pi}{2}, \dfrac{\pi}{2}\right) \cup \left(\dfrac{3\pi}{2}, \dfrac{5\pi}{2}\right)$, etc.

Exercises **5.3**

Getting Ready

You should be able to complete these vocabulary and concept statements before you proceed to the practice exercises.

Fill in the blanks.

1. The Pythagorean Identities are _____.
2. The Even–Odd Identities for sine, cosine, and tangent are _____, _____, and _____.

Practice

Perform the indicated operation, and simplify when possible.

3. $3 \sin x + 4 \sin x$
4. $11 \sec x - 8 \sec x$
5. $4 \csc^2 x - 11 \csc^2 x$
6. $17 \tan^2 x + 12 \tan^2 x$
7. $\sin x(\sin x - 1)$
8. $\cos x(1 + \sin x)$
9. $\tan x(1 - \csc x)$
10. $\cot x(1 + \sin x)$
11. $(1 + \cos x)(1 - \cos x)$
12. $(\sec x + 1)(\sec x - 1)$
13. $(\csc x + 1)(\csc x - 1)$
14. $(1 + \tan x)(1 - \tan x)$
15. $(\cos x + 2)^2$
16. $(\tan x - 3)^2$
17. $(3 \sin x + 2)^2$
18. $(3 \sec x - 5)^2$
19. $\dfrac{1}{\cot x} + \dfrac{\cos x}{\sin x}$
20. $\dfrac{\sec x}{\tan x} + \dfrac{1}{\cot x}$
21. $\dfrac{\sin x}{\cos x + 1} - \dfrac{\sin x}{\cos x - 1}$
22. $\dfrac{\tan x}{\sec x + 1} - \dfrac{\tan x}{\sec x - 1}$

Factor each expression.

23. $\sin^2 x + \sin x \cos x$
24. $\cot x - \cot^2 x \csc x$
25. $\cos^3 x + \cos^2 x$
26. $\csc^4 x - \csc^3 x$
27. $\sec^2 x - 9$
28. $\tan^2 x - 4$
29. $\sec^2 x - \csc^2 x$
30. $\cot^4 x - \tan^2 x$
31. $\sin^2 x + 5 \sin x - 6$
32. $\cos^2 x - 5 \cos x + 6$
33. $2 \tan^2 x + \tan x - 1$
34. $3 \cot^2 x + 2 \cot x - 1$

Verify each identity.

35. $\cos x \tan x = \sin x$
36. $\cot x \sin x = \cos x$
37. $(\tan^2 x + 1)\cos^2 x = 1$
38. $(\cot^2 x + 1)\sin^2 x = 1$
39. $\sin x \sec x = \tan x$
40. $\cos x \csc x = \cot x$
41. $\dfrac{\tan x}{\sec x} = \sin x$
42. $\dfrac{\cot x}{\csc x} = \cos x$
43. $\dfrac{\cot x \sin x}{\cos x} = 1$
44. $\dfrac{\tan x \cos x}{\sin x} = 1$
45. $\dfrac{1 - \cos^2 x}{\sin x} = \sin x$
46. $\dfrac{1 - \sin^2 x}{\cos x} = \cos x$
47. $\tan x \csc x = \sec x$
48. $\cot x \sec x = \csc x$
49. $\dfrac{1 + \tan^2 x}{\sec^2 x} = 1$
50. $\dfrac{1 + \cot^2 x}{\csc^2 x} = 1$
51. $\dfrac{\sin x}{\csc x} + \dfrac{\cos x}{\sec x} = 1$
52. $1 - \dfrac{\cos x}{\sec x} = \sin^2 x$

53. $\dfrac{1}{1 - \sin^2 x} = 1 + \tan^2 x$
54. $\dfrac{1}{1 - \cos^2 x} = 1 + \cot^2 x$
55. $\dfrac{\sin(-x)}{\cos^2(-x) - 1} = -\csc(-x)$
56. $\dfrac{\cos(-x)}{\sin^2(-x) - 1} = -\sec(-x)$
57. $\sin^2 x + \cos^2 x = \sin^2 x \csc^2 x$
58. $\sin^2 x + \cos^2 x = \cos^2 x \sec^2 x$
59. $\dfrac{1 - \sin^2 x}{1 + \tan^2 x} = \cos^4 x$
60. $\dfrac{1 - \sin^2 x}{1 - \cos^2 x} = \cot^2 x$
61. $\dfrac{1 - \cos^2 x}{1 - \sin^2 x} = \tan^2 x$
62. $\tan x \sin x = \dfrac{\csc x}{\cot x + \cot^3 x}$
63. $\sin^2 x \sec x - \sec x = -\cos(-x)$
64. $\cos^2 x \csc x - \csc x = \sin(-x)$
65. $\dfrac{\cos x(\cos x + 1)}{\sin x} = \cos x \cot x + \cot x$
66. $(\sin x - \cos x)(1 + \sin x \cos x) = \sin^3 x - \cos^3 x$
67. $\sin^2 x - \tan^2 x = -\sin^2 x \tan^2 x$
68. $\dfrac{1 + \cot x}{\csc x} = \sin x + \cos x$
69. $\dfrac{1 - \csc x}{\cot x} = \tan x - \sec x$
70. $\dfrac{\cos^2 x - \tan^2 x}{\sin^2 x} = \cot^2 x - \sec^2 x$
71. $\dfrac{1}{\sec x - \tan x} = \tan x + \sec x$
72. $\csc^2 x + \sec^2 x = \csc^2 x \sec^2 x$
73. $\dfrac{\cos x}{\cot x} + \dfrac{\sin x}{\tan x} = \sin x + \cos x$
74. $\dfrac{\cos x}{1 + \sin x} = \sec x - \tan x$
75. $(\sin x + \cos x)^2 + (\sin x - \cos x)^2 = 2$
76. $\cos^2 x + \sin x \cos x = \dfrac{\cos x(\cot x + 1)}{\csc x}$
77. $\dfrac{\cos x}{1 - \sin x} = \dfrac{1 + \sin x}{\cos x}$
78. $\dfrac{\sec x + 1}{\tan x} = \dfrac{\tan x}{\sec x - 1}$
79. $\dfrac{\cos x - \cot x}{\cos x \cot x} = \dfrac{\sin x - 1}{\cos x}$

80. $\cos^4 x - \sin^4 x = 2\cos^2 x - 1$

81. $\dfrac{1}{1 + \sin x} + \dfrac{1}{1 - \sin x} = 2\sec^2 x$

82. $\sqrt{\dfrac{1 - \sin x}{1 + \sin x}} = \sec x - \tan x$ with $0 < x < \dfrac{\pi}{2}$

83. $\sqrt{\dfrac{\csc x - \cot x}{\csc x + \cot x}} = \dfrac{\sin x}{1 + \cos x}$ with $0 < x < \dfrac{\pi}{2}$

84. $-\sqrt{\dfrac{\sec x - 1}{\sec x + 1}} = \dfrac{1 - \sec x}{\tan x}$ with $0 < x < \dfrac{\pi}{2}$

85. $\sqrt{\dfrac{1 - \sin x}{1 + \sin x}} = \dfrac{1 - \sin x}{\cos x}$ with $0 < x < \dfrac{\pi}{2}$

86. $\dfrac{\cos x}{1 - \sin x} - \dfrac{1}{\cos x} = \tan x$

87. $\dfrac{1}{\tan x(\csc x + 1)} = \tan x(\csc x - 1)$

88. $\dfrac{\cot x - \cos x}{\cot x \cos x} = \dfrac{1 - \sin x}{\cos x}$

89. $\dfrac{1}{\sec x(1 + \sin x)} = \sec x(1 - \sin x)$

90. $\dfrac{(\cos x + 1)^2}{\sin^2 x} = 2\csc^2 x + 2\csc x \cot x - 1$

91. $\dfrac{\csc x}{\sec x - \csc x} = \dfrac{\cot^2 x + \csc x \sec x + 1}{\tan^2 x - \cot^2 x}$

92. $(\sin x + \cos x)^2 - (\sin x - \cos x)^2 = 4\sin x \cos x$

Find the values of x that will make each equation an identity.

93. $\sqrt{1 - \sin^2 x} = \cos x$

94. $\sqrt{\sec^2 x - 1} = \tan x$

95. $\sqrt{\csc^2 x - 1} = \cot x$

96. $\sqrt{1 - \cos^2 x} = -\sin x$

Verify each identity.

97. $\dfrac{\cos x + \sin x}{\cos x - \sin x} = \dfrac{\csc x + 2\cos x}{\csc x - 2\sin x}$

98. $\dfrac{\sec x + \tan x}{\csc x + \cot x} = \dfrac{\cot x - \csc x}{\tan x - \sec x}$

99. $\dfrac{\sin x + \cos x + 1}{\sin x + \cos x - 1} = \csc x \sec x + \csc x + \sec x + 1$

100. $\dfrac{\sin^2 x \tan x + \sin^2 x}{\sec x - 2\sin x \tan x} = \dfrac{\sin^2 x}{\cos x - \sin x}$

101. $\dfrac{3\cos^2 x + 11\sin x - 11}{\cos^2 x} = \dfrac{3\sin x - 8}{1 + \sin x}$

102. $\dfrac{3\sin^2 x + 5\cos x - 5}{\sin^2 x} = \dfrac{3\cos x - 2}{1 + \cos x}$

103. $\dfrac{\cos x + \sin x + 1}{\cos x - \sin x - 1} = \dfrac{1 + \cos x}{-\sin x}$

104. $\dfrac{1 + \sin x + \cos x}{1 + \sin x - \cos x} = \dfrac{\sin x}{1 - \cos x}$

105. $\dfrac{1 + \sin x}{\cos x} = \dfrac{\sin x - \cos x + 1}{\sin x + \cos x - 1}$

106. $\dfrac{\tan x + \sec x + 1}{\tan x + \sec x - 1} = \dfrac{1 + \cos x}{\sin x}$

107. $\dfrac{\csc x + 1 + \cot x}{\csc x + 1 - \cot x} = \dfrac{1 + \cos x}{\sin x}$

108. $\dfrac{3\cot x \csc x - 2\csc^2 x}{\csc^2 x + \cot x \csc x} - 3$
$$= -5(\csc^2 x - \csc x \cot x)$$

109. $\dfrac{\sin x + \sin x \cos x - \sin x \cos^2 x - \sin x \cos^3 x}{\cos^4 x - 2\cos^2 x + 1}$
$$= \csc x + \cot x$$

110. $\dfrac{\sin^3 x \cos x + \cos x - \sin^2 x \cos x - \cos x \sin x}{\cos^4 x}$
$$= \sec x - \tan x$$

111. $\dfrac{1}{\sec x + \csc x - \sec x \csc x} = \dfrac{\sin x + \cos x + 1}{2}$

112. $\dfrac{2(\sin x + 1)}{1 + \cot x + \csc x} = \sin x + 1 - \cos x$

113. $\dfrac{2\sin x \cos x + \cos x - 2\sin x - 1}{-\sin^2 x} = \dfrac{2\sin x + 1}{\cos x + 1}$

114. $-4\cos^2 x - 3\sin^2 x = \tan^2 x \cos^2 x - 4$

115. $\sec^2 x(\tan x + 3\sec x)(\tan x - \sec x)$
$$= 2\tan x \sec^3 x - 2\sec^4 x - \sec^2 x$$

116. $\dfrac{\csc x + 1}{\csc x - 1} + \dfrac{\tan x - \sec x}{\tan x + \sec x} = \dfrac{4\tan x}{\cos x}$

Use a graphing calculator to graph both sides of each equation and determine whether the equation is an identity.

117. $\sin^2 x + \cos^2 x = 1$ **118.** $\cos x \tan x = \sin x$

119. $(\sin x + \cos x)^2 = 1$

120. $\dfrac{\cos x}{1 - \sin x} = \dfrac{1 + \sin x}{\cos x}$

121. $\dfrac{1 - \cos^2 x}{\sin x} = \sin x$

122. $\sin^2 x - \cos^2 x = 2\cos^2 x$

Application

123. Rate of change It is known in calculus that the rate of change of the function $f(x) = \tan x - \cot x$, with respect to the variable x, is the trigonometric expression $\sec^2 x + \csc^2 x$. Verify that the trigonometric expression for the rate of change is also $\sec^2 x \csc^2 x$.

124. Rate of change The rate of change of the function $g(x) = \sec^2 x$ is known to be the trigonometric expression $2 \sec^2 x \tan x$. Verify that the trigonometric expression for the rate of change is also $2 \tan x + 2 \tan^3 x$.

 125. Trigonometric substitution In calculus, we often make a trigonometric substitution to eliminate a radical. Consider the radical expression $\sqrt{16 + x^2}$. Substitute $4 \tan \theta$ into the expression for x and simplify.

126. Trigonometric substitution Consider the radical expression $\sqrt{16 - x^2}$. Substitute $4 \sin \theta$ into the expression for x and simplify.

5.4 Sum and Difference Formulas

In this section, we will learn to

1. Find exact values using Sum and Difference Formulas.
2. Establish identities using Sum and Difference Formulas.
3. Use Sum and Difference Formulas involving inverse trigonometric functions.
4. Find the angle between two lines using a difference formula.

Waves are caused by the wind blowing over the surface of the ocean; there is tremendous energy in ocean waves. A variety of technologies have been proposed to capture the energy of waves and to use it as an alternative source of energy.

Sine and cosine are commonly used to model wave motion. To study waves and wave energy, we need an understanding of the rates of change of the trigonometric functions. These are referred to as derivatives in calculus. Initially when we study these derivatives, we will encounter expressions that contain the trigonometric functions of the sum of two angles. It would be nice if $\cos(\alpha + \beta)$ were equivalent to $\cos \alpha + \cos \beta$, but that is generally not the case. For example,

$$\cos(30° + 60°) = \cos 90° = 0$$

$$\cos 30° + \cos 60° = \frac{\sqrt{3}}{2} + \frac{1}{2} = \frac{\sqrt{3} + 1}{2}$$

$$\cos(30° + 60°) \neq \cos 30° + \cos 60°$$

However, it is possible to develop formulas for finding a trigonometric function of the sum or difference of two angles. That is the topic we will learn about in this section. A knowledge of these formulas is very important in calculus.

1. Find Exact Values Using Sum and Difference Formulas

We will use the Distance Formula to derive a formula to evaluate the expression $\cos(\alpha + \beta)$.

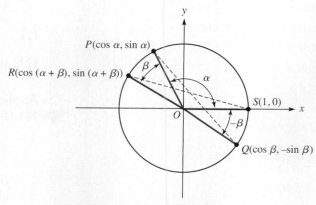

FIGURE 5-20

We draw angles α and $-\beta$ in standard position on the unit circle, as in Figure 5-20; locate point R on the circle so that angle POR is equal to angle β; and form triangles POQ and ROS. Because these two isosceles triangles have equal vertex angles, they are congruent. Hence, $d(RS) = d(PQ)$. The coordinates of the points are:

P: $(\cos \alpha, \sin \alpha)$,

Q: $(\cos(-\beta), \sin(-\beta)) = (\cos \beta, -\sin \beta)$,

R: $(\cos(\alpha + \beta), \sin(\alpha + \beta))$, and

S: $(1, 0)$.

Because $d(RS) = d(PQ)$, their squares are also equal. Therefore, $[d(RS)]^2 = [d(PQ)]^2$. By the Distance Formula, we have

$$[d(RS)]^2 = [d(PQ)]^2$$

$$[\cos(\alpha + \beta) - 1]^2 + [\sin(\alpha + \beta) - 0]^2 = (\cos \alpha - \cos \beta)^2 + (\sin \alpha + \sin \beta)^2$$

$\cos^2(\alpha + \beta) - 2\cos(\alpha + \beta) + 1 + \sin^2(\alpha + \beta) =$ Expand both sides.
$\qquad \cos^2 \alpha - 2\cos \alpha \cos \beta + \cos^2 \beta + \sin^2 \alpha + 2\sin \alpha \sin \beta + \sin^2 \beta$

$1 - 2\cos(\alpha + \beta) + 1 = 1 - 2\cos \alpha \cos \beta + 1 + 2\sin \alpha \sin \beta$ $\sin^2 x + \cos^2 x = 1$
    for any value of x.

$2 - 2\cos(\alpha + \beta) = 2 - 2\cos \alpha \cos \beta + 2\sin \alpha \sin \beta$

$-2\cos(\alpha + \beta) = -2\cos \alpha \cos \beta + 2\sin \alpha \sin \beta$

$\cos(\alpha + \beta) = \cos \alpha \cos \beta - \sin \alpha \sin \beta$ Divide both sides by -2.

The formula for $\cos(\alpha - \beta)$ can be found using the fact that $\alpha - \beta = \alpha + (-\beta)$.

$\cos(\alpha - \beta) = \cos(\alpha + (-\beta))$

$\qquad\qquad = \cos \alpha \cos(-\beta) - \sin \alpha \sin(-\beta)$ Sum Formula

$\qquad\qquad = \cos \alpha \cos \beta - \sin \alpha(-\sin \beta)$ Even–Odd Properties

$\qquad\qquad = \cos \alpha \cos \beta + \sin \alpha \sin \beta$

Sum and Difference Formulas for Cosine

For any angles α and β,

$$\cos(\alpha + \beta) = \cos \alpha \cos \beta - \sin \alpha \sin \beta$$

$$\cos(\alpha - \beta) = \cos \alpha \cos \beta + \sin \alpha \sin \beta$$

The Sum and Difference Formulas can be used to find the exact value of the cosine of an angle that is not a special angle, if it can be written as the sum or difference of special angles or multiples of special angles.

EXAMPLE 1   Using Sum and Difference Formulas for Cosine to Find Exact Values

Find the exact value of each expression. **a.** $\cos 75°$ **b.** $\cos \dfrac{\pi}{12}$

SOLUTION   **a.** We will rewrite $75°$ as the sum or difference of special angles or multiples of special angles, and then use the Sum or Difference Formula to find the exact value. While there are many possibilities, writing $75° = 45° + 30°$ is a good choice.

$$\cos 75° = \cos(45° + 30°)$$

$$= \cos 45° \cos 30° - \sin 45° \sin 30°$$

$$= \frac{\sqrt{2}}{2} \cdot \frac{\sqrt{3}}{2} - \frac{\sqrt{2}}{2} \cdot \frac{1}{2}$$

$$= \frac{\sqrt{6} - \sqrt{2}}{4}$$

We can check our answer with a calculator. Make sure it is in degree mode.

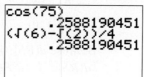

```
cos(75)
            .2588190451
(√(6)-√(2))/4
            .2588190451
```

b. To find $\cos \frac{\pi}{12}$, we use $\frac{\pi}{12} = \frac{\pi}{4} - \frac{\pi}{6}$.

$$\cos \frac{\pi}{12} = \cos\left(\frac{\pi}{4} - \frac{\pi}{6}\right)$$

$$= \cos \frac{\pi}{4} \cos \frac{\pi}{6} + \sin \frac{\pi}{4} \sin \frac{\pi}{6}$$

$$= \frac{\sqrt{2}}{2} \cdot \frac{\sqrt{3}}{2} + \frac{\sqrt{2}}{2} \cdot \frac{1}{2}$$

$$= \frac{\sqrt{6} + \sqrt{2}}{4}$$

Self Check 1   Find the exact value of $\cos 165°$.

Now Try Exercise 7.

In Section 4.3, we discovered that if θ is an acute angle, any trigonometric function of θ is equal to the cofunction of the complement of θ—i.e., **cofunctions of complementary angles are equal**. This property can be extended to any angle. We use the Difference Formula to find $\cos(90° - \theta)$.

$$\cos(90° - \theta) = \cos 90° \cos \theta + \sin 90° \sin \theta$$

$$= 0 \cdot \cos \theta + 1 \cdot \sin \theta$$

$$= \sin \theta$$

We note that $\theta = 90° - (90° - \theta)$ and find a value for $\cos \theta$.

$$\cos \theta = \cos(90° - (90° - \theta))$$

$$= \cos 90° \cos(90° - \theta) + \sin 90° \sin(90° - \theta)$$

$$= \sin(90° - \theta)$$

Therefore, for any angle θ, we have the following identities:

Formulas for Cofunctions of Complementary Angles

For any angle θ,

$$\cos(90° - \theta) = \sin \theta \qquad \text{or} \qquad \cos\left(\frac{\pi}{2} - \theta\right) = \sin \theta$$

$$\sin(90° - \theta) = \cos \theta \qquad \text{or} \qquad \sin\left(\frac{\pi}{2} - \theta\right) = \cos \theta$$

The formulas for cofunctions of complementary angles can be used to develop a formula for $\sin(\alpha + \beta)$. We let $\theta = \alpha + \beta$.

$$\sin \theta = \cos(90° - \theta)$$

$\sin(\alpha + \beta) = \cos(90° - (\alpha + \beta))$	Cofunction Formula; $\theta = \alpha + \beta$
$= \cos((90° - \alpha) - \beta)$	Rewrite.
$= \mathbf{cos(90° - \alpha)}\cos \beta + \sin(90° - \alpha)\sin \beta$	Difference Formula
$= \mathbf{\sin \alpha} \cos \beta + \cos \alpha \sin \beta$	Cofunction Formulas

Thus, we have a Sum Formula for the sine function:
$\sin(\alpha + \beta) = \sin \alpha \cos \beta + \cos \alpha \sin \beta$. To find the formula for $\sin(\alpha - \beta)$, we proceed as follows:

$\sin(\alpha - \beta) = \sin(\alpha + (-\beta))$	
$= \sin \alpha \cos(-\beta) + \cos \alpha \sin(-\beta)$	Sum Formula for Sine
$= \sin \alpha \cos \beta - \cos \alpha \sin \beta$	Even–Odd Identities

We now have both a Sum and Difference Formula for the sine function.

Sum and Difference Formulas for Sine

For any angles α and β,

$$\sin(\alpha + \beta) = \sin \alpha \cos \beta + \cos \alpha \sin \beta$$

$$\sin(\alpha - \beta) = \sin \alpha \cos \beta - \cos \alpha \sin \beta$$

EXAMPLE 2   **Using the Sum and Difference Formulas of the Sine Function**

Find the exact value of each expression.   **a.** $\sin \dfrac{7\pi}{12}$ **b.** $\sin 15°$

SOLUTION   **a.** Find $\sin \frac{7\pi}{12}$ by rewriting $\frac{7\pi}{12}$ as the sum or difference of special angles or multiples of those angles. There are several ways that this can be accomplished, but we will use $\frac{7\pi}{12} = \frac{4\pi}{12} + \frac{3\pi}{12} = \frac{\pi}{3} + \frac{\pi}{4}$ and apply the Sum Formula.

$\sin \dfrac{7\pi}{12} = \sin\left(\dfrac{\pi}{3} + \dfrac{\pi}{4}\right)$	Rewrite the angle as a sum.
$= \sin \dfrac{\pi}{3} \cos \dfrac{\pi}{4} + \cos \dfrac{\pi}{3} \sin \dfrac{\pi}{4}$	Apply the Sum Formula.
$= \dfrac{\sqrt{3}}{2} \cdot \dfrac{\sqrt{2}}{2} + \dfrac{1}{2} \cdot \dfrac{\sqrt{2}}{2}$	Evalulate.
$= \dfrac{\sqrt{6}}{4} + \dfrac{\sqrt{2}}{4}$	Simplify.
$= \dfrac{\sqrt{6} + \sqrt{2}}{4}$	

b. $\sin 15°$ can be rewritten as $\sin 15° = \sin(45° - 30°)$, which allows us to use the Difference Formula.

$$\sin 15° = \sin(45° - 30°) \qquad \text{Rewrite as a difference.}$$

$$= \sin 45° \cos 30° - \cos 45° \sin 30° \qquad \text{Use the Difference Formula.}$$

$$= \frac{\sqrt{2}}{2} \cdot \frac{\sqrt{3}}{2} - \frac{\sqrt{2}}{2} \cdot \frac{1}{2} \qquad \text{Evaluate.}$$

$$= \frac{\sqrt{6}}{4} - \frac{\sqrt{2}}{4} \qquad \text{Simplify.}$$

$$= \frac{\sqrt{6} - \sqrt{2}}{4}$$

Self Check 2   Find the exact value of $\sin 195°$.

Now Try Exercise 11.

EXAMPLE 3   **Using the Sum and Difference Formulas to Find Exact Values**

Find the exact value of each expression. **a.** $\cos 40° \cos 20° - \sin 40° \sin 20°$
b. $\sin 115° \cos 25° - \cos 115° \sin 25°$

SOLUTION   **a.** The key to finding this value is to recognize which formula can be applied. We see that $\cos 40° \cos 20° - \sin 40° \sin 20°$ is the right side of the Sum Formula for Cosine.

$$\cos \mathbf{40°} \cos \mathbf{20°} - \sin \mathbf{40°} \sin \mathbf{20°} = \cos(\mathbf{40°} + \mathbf{20°}) = \cos 60° = \frac{1}{2}$$

b. The expression $\sin 115° \cos 25° - \cos 115° \sin 25°$ is the right side of the Difference Formula for the Sine.

$$\sin \mathbf{115°} \cos \mathbf{25°} - \cos \mathbf{115°} \sin \mathbf{25°} = \sin(\mathbf{115°} - \mathbf{25°}) = \sin 90° = 1$$

Self Check 3   Find the exact value of $\cos 40° \cos 10° + \sin 40° \sin 10°$.

Now Try Exercise 33.

To find formulas for the tangent of the sum and difference of two angles, we use the Quotient Identity $\tan x = \dfrac{\sin x}{\cos x}$ and substitute $x = \alpha + \beta$.

$$\tan(\alpha + \beta) = \frac{\sin(\alpha + \beta)}{\cos(\alpha + \beta)}$$

$$= \frac{\sin \alpha \cos \beta + \cos \alpha \sin \beta}{\cos \alpha \cos \beta - \sin \alpha \sin \beta} \qquad \text{Apply sum formulas.}$$

$$= \frac{\dfrac{\sin \alpha \cos \beta}{\cos \alpha \cos \beta} + \dfrac{\cos \alpha \sin \beta}{\cos \alpha \cos \beta}}{\dfrac{\cos \alpha \cos \beta}{\cos \alpha \cos \beta} - \dfrac{\sin \alpha \sin \beta}{\cos \alpha \cos \beta}} \qquad \text{Divide each term by } \cos \alpha \cos \beta.$$

$$= \frac{\tan \alpha + \tan \beta}{1 - \tan \alpha \tan \beta} \qquad \text{Simplify.}$$

We can substitute $-\beta$ for β in the formula we just found to find the formula for the tangent of the difference of two angles.

Sum and Difference Formulas for Tangent   For any angles α and β,

$$\tan(\alpha + \beta) = \frac{\tan \alpha + \tan \beta}{1 - \tan \alpha \tan \beta}$$

$$\tan(\alpha - \beta) = \frac{\tan \alpha - \tan \beta}{1 + \tan \alpha \tan \beta}$$

EXAMPLE 4   Using Sum and Difference Formulas to Find Exact Values

If $\sin \alpha = \frac{12}{13}$, α in QI, and $\tan \beta = \frac{1}{3}$, β in QIII, find the exact value of each expression.

a. $\cos(\alpha + \beta)$ **b.** $\sin(\alpha - \beta)$ **c.** $\tan(\alpha + \beta)$

SOLUTION   Before we can find $\cos(\alpha + \beta)$, we must find $\sin \alpha$, $\cos \alpha$, $\sin \beta$, $\cos \beta$. We can use the definition of the given trigonometric function to find these values. Since α is in QI, both x and y are positive. Angle β lies in QIII, so both x and y are negative. The values of the missing trigonometric functions are shown in Figure 5-21.

$y = 12, r = 13$

$x^2 + y^2 = r^2$

$x^2 + 12^2 = 13^2$

$x^2 = 169 - 144$

$x^2 = 25$

$x = 5$

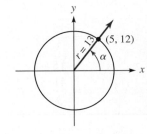

$\sin \alpha = \dfrac{12}{13} = \dfrac{y}{r}, 0 \le \alpha \le \dfrac{\pi}{2}$

$\cos \alpha = \dfrac{x}{r} = \dfrac{5}{13}$

$\tan \alpha = \dfrac{y}{x} = \dfrac{12}{5}$

$x = -3, y = -1$

$r^2 = x^2 + y^2$

$r^2 = (-3)^2 + (-1)^2$

$r^2 = 9 + 1$

$r^2 = 10$

$r = \sqrt{10}$

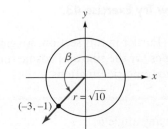

$\tan \beta = \dfrac{1}{3} = \dfrac{y}{x}, \pi \le \beta \le \dfrac{3\pi}{2}$

$\sin \beta = \dfrac{y}{r} = \dfrac{-1}{\sqrt{10}} = -\dfrac{\sqrt{10}}{10}$

$\cos \beta = \dfrac{x}{r} = \dfrac{-3}{\sqrt{10}} = -\dfrac{3\sqrt{10}}{10}$

FIGURE 5-21

a. Evaluate $\cos(\alpha + \beta)$ by applying the Sum Formula for Cosine.

$\cos(\alpha + \beta) = \cos \alpha \cos \beta - \sin \alpha \sin \beta$

$\qquad = \dfrac{5}{13} \cdot \left(-\dfrac{3\sqrt{10}}{10}\right) - \dfrac{12}{13} \cdot \left(-\dfrac{\sqrt{10}}{10}\right)$ Evaluate using the figure.

$\qquad = \dfrac{-15\sqrt{10}}{130} + \dfrac{12\sqrt{10}}{130}$ Simplify.

$\qquad = -\dfrac{3\sqrt{10}}{130}$

b. Evaluate $\sin(\alpha - \beta)$ by applying the Difference Formula for Sine.

$$\sin(\alpha - \beta) = \sin \alpha \cos \beta - \cos \alpha \sin \beta$$

$$= \frac{12}{13} \cdot \left(-\frac{3\sqrt{10}}{10}\right) - \frac{5}{13} \cdot \left(-\frac{\sqrt{10}}{10}\right) \qquad \text{Evaluate using the figure.}$$

$$= \frac{-36\sqrt{10}}{130} + \frac{5\sqrt{10}}{130} \qquad \text{Simplify.}$$

$$= -\frac{31\sqrt{10}}{130}$$

c. Evaluate $\tan(\alpha + \beta)$ by using the Sum Formula for Tangent.

$$\tan(\alpha + \beta) = \frac{\tan \alpha + \tan \beta}{1 - \tan \alpha \tan \beta}$$

$$= \frac{\dfrac{12}{5} + \dfrac{1}{3}}{1 - \left(\dfrac{12}{5}\right)\left(\dfrac{1}{3}\right)} \qquad \text{Evaluate the values using the figure.}$$

$$= \frac{\dfrac{41}{15}}{1 - \dfrac{12}{15}} \qquad \text{Find the common denominator.}$$

$$= \frac{\dfrac{41}{15}}{\dfrac{3}{15}}$$

$$= \frac{41}{3} \qquad \text{Simplify.}$$

Comment

In part (c) of Example 4, we used the Sum Formula. However, if we know the value of $\sin(\alpha + \beta)$ and $\cos(\alpha + \beta)$, we can use the Quotient Identity

$$\tan(\alpha + \beta) = \frac{\sin(\alpha + \beta)}{\cos(\alpha + \beta)}.$$

Self Check 4   Using the information from Example 4, find the exact value of each expression.
a. $\cos(\alpha - \beta)$ **b.** $\sin(\alpha + \beta)$ **c.** $\tan(\alpha - \beta)$

Now Try Exercise 43.

Earlier in this section, we gave the formulas for cofunctions of complementary angles for the sine and cosine functions. These formulas can be extended to all the trigonometric functions.

Formulas for Cofunctions of Complementary Angles

For any angle θ,

$$\cos(90° - \theta) = \sin \theta \qquad \text{or} \qquad \cos\left(\frac{\pi}{2} - \theta\right) = \sin \theta$$

$$\sin(90° - \theta) = \cos \theta \qquad \text{or} \qquad \sin\left(\frac{\pi}{2} - \theta\right) = \cos \theta$$

$$\tan(90° - \theta) = \cot \theta \qquad \text{or} \qquad \tan\left(\frac{\pi}{2} - \theta\right) = \cot \theta$$

$$\cot(90° - \theta) = \tan \theta \qquad \text{or} \qquad \cot\left(\frac{\pi}{2} - \theta\right) = \tan \theta$$

$$\sec(90° - \theta) = \csc \theta \qquad \text{or} \qquad \sec\left(\frac{\pi}{2} - \theta\right) = \csc \theta$$

$$\csc(90° - \theta) = \sec \theta \qquad \text{or} \qquad \csc\left(\frac{\pi}{2} - \theta\right) = \sec \theta$$

2. Establish Identities Using Sum and Difference Formulas

The Sum and Difference Formulas can be used to verify identities.

EXAMPLE 5   **Verifying an Identity**

Verify each identity.

a. $\tan(\theta + 45°) = \dfrac{1 + \tan \theta}{1 - \tan \theta}$ **b.** $\dfrac{\cos(\alpha - \beta)}{\sin \alpha \cos \beta} = \cot \alpha + \tan \beta$

SOLUTION   **a.** Use the formula for the tangent of the sum of two angles.

$$\tan(\theta + 45°) = \frac{\tan \theta + \tan 45°}{1 - \tan \theta \tan 45°}$$

$$= \frac{\tan \theta + 1}{1 - \tan \theta}$$

$$= \frac{1 + \tan \theta}{1 - \tan \theta}$$

b. We can start on either side of the identity. However, the left side seems to be more complicated, so that is where we will begin.

$$\frac{\cos(\alpha - \beta)}{\sin \alpha \cos \beta} = \frac{\cos \alpha \cos \beta + \sin \alpha \sin \beta}{\sin \alpha \cos \beta} \qquad \text{Apply the Difference Formula.}$$

$$= \frac{\cos \alpha \cos \beta}{\sin \alpha \cos \beta} + \frac{\sin \alpha \sin \beta}{\sin \alpha \cos \beta} \qquad \text{Split the fraction.}$$

$$= \frac{\cos \alpha}{\sin \alpha} + \frac{\sin \beta}{\cos \beta} \qquad \text{Simplify.}$$

$$= \cot \alpha + \tan \beta \qquad \text{Quotient Identities}$$

■

Self Check 5   Verify that $\cot(\theta + 45°) = \dfrac{1 - \tan \theta}{1 + \tan \theta}$.

Now Try Exercise 53.

3. Use Sum and Difference Formulas Involving Inverse Trigonometric Functions

When using inverse trigonometric functions, it is important to remember that they are defined over specific intervals. Specifically,

$$y = \sin^{-1} x \qquad \text{if and only if} \qquad x = \sin y \qquad \text{and} \qquad -\frac{\pi}{2} \leq y \leq \frac{\pi}{2},$$

$$y = \cos^{-1} x \qquad \text{if and only if} \qquad x = \cos y \qquad \text{and} \qquad 0 \leq y \leq \pi, \text{ and}$$

$$y = \tan^{-1} x \qquad \text{if and only if} \qquad x = \tan y \qquad \text{and} \qquad -\frac{\pi}{2} < y < \frac{\pi}{2}.$$

These will be used when solving problems such as in the next example.

EXAMPLE 6   **Finding the Exact Value of an Expression with Inverse Trigonometric Functions**

Find the exact value of the expression $\cos\left[\sin^{-1} \dfrac{2}{3} - \cos^{-1}\left(-\dfrac{3}{5}\right)\right]$.

SOLUTION It is important to realize that the expressions $\sin^{-1} \frac{2}{3}$ and $\cos^{-1}\left(-\frac{3}{5}\right)$ each represent an angle. We set $\alpha = \mathbf{\sin^{-1}} \frac{2}{3}$ and $\beta = \mathbf{\cos^{-1}}\left(-\frac{3}{5}\right)$.

$$\cos\left(\underbrace{\sin^{-1}\frac{2}{3}}_{\alpha} - \underbrace{\cos^{-1}\left(-\frac{3}{5}\right)}_{\beta}\right) = \cos(\alpha - \beta) = \cos\alpha \underbrace{\cos\beta}_{-\frac{3}{5}} + \underbrace{\sin\alpha}_{\frac{2}{3}}\sin\beta$$

To find the missing values $\cos\alpha$ and $\sin\beta$, we use a Pythagorean Identity.

• **Find $\cos\alpha$.**

We determine that α lies in QI because sine is positive in QI, and inverse sine is defined only over $\left[-\frac{\pi}{2}, \frac{\pi}{2}\right]$.

$$\sin^2\alpha + \cos^2\alpha = 1$$

$$\mathbf{\cos\alpha} = \sqrt{1 - \sin^2\alpha} \qquad \text{Solve for } \cos\alpha; \alpha \text{ is in QI so } \cos\alpha > 0.$$

$$= \sqrt{1 - \left(\frac{2}{3}\right)^2} \qquad \sin\alpha = \frac{2}{3}$$

$$= \sqrt{\frac{5}{9}} = \frac{\sqrt{5}}{3}$$

• **Find $\sin\beta$.**

We determine that β lies in QII because cosine is negative in QII, and inverse cosine is defined only over $[0, \pi]$.

$$\sin^2\beta + \cos^2\beta = 1$$

$$\mathbf{\sin\beta} = \sqrt{1 - \cos^2\beta} \qquad \text{Solve for } \sin\beta; \beta \text{ is in QII so } \sin\beta > 0.$$

$$= \sqrt{1 - \left(-\frac{3}{5}\right)^2} \qquad \cos\beta = -\frac{3}{5}$$

$$= \sqrt{\frac{16}{25}} = \frac{4}{5}$$

We now have enough information to complete the problem.

$$\cos\left[\sin^{-1}\frac{2}{3} - \cos^{-1}\left(-\frac{3}{5}\right)\right] = \cos(\alpha - \beta) = \cos\alpha\cos\beta + \sin\alpha\sin\beta$$

$$= \frac{\sqrt{5}}{3}\cdot\left(-\frac{3}{5}\right) + \frac{2}{3}\cdot\frac{4}{5} = \frac{-3\sqrt{5}+8}{15} \quad\blacksquare$$

Self Check 6 Find the exact value of $\sin\left[\sin^{-1}\left(-\frac{2}{5}\right) + \cos^{-1}\frac{1}{3}\right]$.

Now Try Exercise 77.

EXAMPLE 7 **Finding the Exact Value of an Expression with Inverse Trigonometric Functions**

Find the exact value of $\sin\left[\tan^{-1}\frac{1}{2} + \sin^{-1}\left(-\frac{3}{4}\right)\right]$.

SOLUTION In the expression $\sin\left[\tan^{-1}\frac{1}{2} + \sin^{-1}\left(-\frac{3}{4}\right)\right]$, we will let $\alpha = \tan^{-1}\frac{1}{2}$ and $\beta = \sin^{-1}\left(-\frac{3}{4}\right)$.

$$\sin\left[\underbrace{\tan^{-1}\left(\frac{1}{2}\right)}_{\alpha} + \underbrace{\mathbf{\sin^{-1}}\left(-\frac{3}{4}\right)}_{\beta}\right] = \sin(\alpha + \beta) = \sin\alpha\cos\beta + \cos\alpha\,\underbrace{\mathbf{\sin\beta}}_{-\frac{3}{4}}$$

• **Find $\sin\alpha$ and $\cos\alpha$.**

Since inverse tangent is defined over the interval $\left(-\frac{\pi}{2}, \frac{\pi}{2}\right)$ and tangent is positive in QI, we determine that $0 < \alpha < \frac{\pi}{2}$. We use the given information, $\tan\alpha = \frac{y}{x} = \frac{1}{2}$; thus, $x = 2$, $y = 1$, and $r = \sqrt{x^2 + y^2} = \sqrt{1 + 4} = \sqrt{5}$.

$$\sin\alpha = \frac{y}{r} = \frac{1}{\sqrt{5}} = \frac{\sqrt{5}}{5} \text{ and } \cos\alpha = \frac{x}{r} = \frac{2}{\sqrt{5}} = \frac{2\sqrt{5}}{5}$$

• **Find $\cos\beta$.**

The inverse sine function is defined over the interval $\left[-\frac{\pi}{2}, \frac{\pi}{2}\right]$, which tells us that β is in QIV because we are given $\beta = \sin^{-1}\left(-\frac{3}{4}\right)$. Consequently, we can state that $-\frac{\pi}{2} < \beta < 0$.

$$\sin\beta = \frac{y}{r} = -\frac{3}{4} \rightarrow y = -3 \text{ and } r = 4.$$

In QIV, $x > 0$. Therefore, we find the value of x to be

$$x = \sqrt{r^2 - y^2} = \sqrt{4^2 - (-3)^2} = \sqrt{16 - 9} = \sqrt{7}, \text{ and } \cos\beta = \frac{\sqrt{7}}{4}.$$

We now have the information needed to complete the problem.

$$\sin\left[\tan^{-1}\frac{1}{2} + \sin^{-1}\left(-\frac{3}{4}\right)\right] = \sin(\alpha + \beta) = \sin\alpha\cos\beta + \cos\alpha\sin\beta$$

$$= \frac{\sqrt{5}}{5}\cdot\frac{\sqrt{7}}{4} + \frac{2\sqrt{5}}{5}\cdot\left(-\frac{3}{4}\right) = \frac{\sqrt{35} - 6\sqrt{5}}{20} \blacksquare$$

Self Check 7   Find the exact value of $\tan\left[\sin^{-1}\left(-\frac{3}{5}\right) + \cos^{-1}\frac{4}{5}\right]$.

Now Try Exercise 71.

Comment   Note that in Examples 6 and 7, we used two different methods for finding the missing trigonometric-function values. In Example 6, we used the Pythagorean Identities, and in Example 7, we used the definitions of the trigonometric functions. Additionally, we could have used reference triangles similar to those used in Example 4. In trigonometry, there is often more than one way to find the solution to a problem.

In Section 5.1, we converted trigonometric expressions into algebraic expressions using inverse trigonometric functions. In the next example, we will expand that concept using the Difference Formula for Sine.

EXAMPLE 8   **Writing a Trigonometric Expression as an Algebraic Expression**

Write the trigonometric expression $\sin(\cos^{-1}x - \sin^{-1}y)$ as an algebraic expression in terms of x and y.

SOLUTION   We will let $\cos^{-1}x = \alpha$, with $0 \le \alpha \le \pi$ and $-1 \le x \le 1$. Likewise, we let

$\sin^{-1}y = \beta$, with $-\frac{\pi}{2} \le \beta \le \frac{\pi}{2}$ and $-1 \le y \le 1$.

Since $0 \le \alpha \le \pi$, $\sin\alpha \ge 0$, and since $-\frac{\pi}{2} \le \beta \le \frac{\pi}{2}$, $\cos\beta \ge 0$, using the Pythagorean Identities, we can determine the following expressions:

$$\sin\alpha = \sqrt{1 - \cos^2\alpha} = \sqrt{1 - x^2} \text{ and } \cos\beta = \sqrt{1 - \sin^2\beta} = \sqrt{1 - y^2}.$$

We can now transform the trigonometric expression into an algebraic expression in terms of x and y.

$$\sin(\cos^{-1} x - \sin^{-1} y) = \sin(\alpha - \beta)$$
$$= \sin \alpha \cos \beta - \cos \alpha \sin \beta$$
$$= \left(\sqrt{1 - x^2}\right)\left(\sqrt{1 - y^2}\right) - xy$$
$$= \sqrt{(1 - x^2)(1 - y^2)} - xy$$
$$= \sqrt{1 - x^2 - y^2 + x^2 y^2} - xy$$

■

Self Check 8   Write the trigonometric expression $\sin(\tan^{-1} x - \sin^{-1} y)$ as an algebraic expression in terms of x and y.

Now Try Exercise 79.

4. Find the Angle between Two Lines Using a Difference Formula

If lines l_1 and l_2, with angles of inclination α_1 and α_2 and slopes m_1 and m_2, intersect at point P (see Figure 5-22), then the angle θ from l_2 to l_1 is the difference of the angles of inclination $\alpha_1 - \alpha_2$: $\theta = \alpha_1 - \alpha_2$.

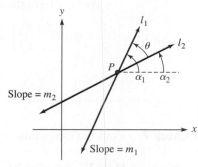

FIGURE 5-22

Angle θ can be found by first finding $\tan \theta$. Using the Difference Formula for Tangent, it follows that

$$\tan \theta = \tan(\alpha_1 - \alpha_2)$$
$$= \frac{\tan \alpha_1 - \tan \alpha_2}{1 + \tan \alpha_1 \tan \alpha_2}$$

The tangent of the angle of inclination of a line is its slope, the rise divided by the run:

$$\tan \alpha = \frac{y}{x} = \frac{\text{rise}}{\text{run}}.$$

From Figure 5-22, we determine

$$\tan \alpha_1 = m_1, \tan \alpha_2 = m_2, \text{ and } \tan \theta = \frac{m_1 - m_2}{1 + m_1 m_2},$$

and θ is the angle from line 2 to line 1.

EXAMPLE 9   **Finding the Angle between Two Lines**

Find the angle from the line $y = 3x + 1$ to the line $y = -2x + 3$.

SOLUTION   Let l_1 be the line $y = -2x + 3$ and l_2 be the line $y = 3x + 1$. Because the slope of a line with equation $y = mx + b$ is m, the slope of $l_1 = -2$ and the slope of $l_2 = 3$ (see Figure 5-23). Since we have $m_1 = -2$ and $m_2 = 3$, we can use the equation

$$\tan \theta = \frac{m_1 - m_2}{1 + m_1 m_2}$$

$$= \frac{-2 - 3}{1 + (-2)(3)}$$

$$= \frac{-5}{-5}$$

$$= 1$$

This leads us to $\theta = \tan^{-1} 1 = 45°$.

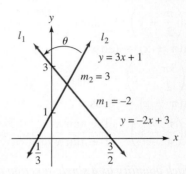

FIGURE 5-23

Self Check 9   Using the information from Example 9, suppose we let l_1 be the line $y = 3x + 1$ and l_2 be the line $y = -2x + 3$. What is the angle from l_1 to l_2?

Now Try Exercise 93.

Self Check Answers   1. $-\dfrac{\sqrt{2} + \sqrt{6}}{4}$ 2. $\dfrac{\sqrt{2} - \sqrt{6}}{4}$ 3. $\dfrac{\sqrt{3}}{2}$ 4. a. $-\dfrac{27\sqrt{10}}{130}$

b. $\dfrac{-41\sqrt{10}}{130}$ c. $\dfrac{31}{27}$ 5. Answers will vary.   6. $\dfrac{-2 + 2\sqrt{42}}{15}$ 7. 0

8. $\dfrac{x\sqrt{x^2 - x^2 y^2 - y^2 + 1} - y\sqrt{x^2 + 1}}{x^2 + 1}$ 9. $-45°$

Exercises **5.4**

Getting Ready
You should be able to complete these vocabulary and concept statements before you proceed to the practice exercises.

Fill in the blanks.

1. $\sin(\alpha + \beta) = $ _____; $\sin(\alpha - \beta) = $ _____;
   $\cos(\alpha + \beta) = $ _____; $\cos(\alpha - \beta) = $ _____.

2. $\tan(\alpha + \beta) = $ _____ and $\tan(\alpha - \beta) = $ _____.

3. Using a sum of special angles or multiples of special angles, rewrite $165° = $ _____ + _____, and now write it as a difference of special angles or multiples of special angles $165° = $ _____ − _____.

4. Using a sum of special angles or multiples of special angles, rewrite $\frac{13\pi}{12} = $ _____ + _____, and now write it as a difference of special angles or multiples of special angles $\frac{13\pi}{15} = $ _____ − _____.

Practice
Use a Sum or Difference Identity to find the exact value of each expression.

5. $\sin 75°$
6. $\sin 195°$
7. $\cos 285°$
8. $\cos 195°$
9. $\tan 105°$
10. $\tan 165°$
11. $\sin 255°$
12. $\sin 105°$
13. $\cos 165°$
14. $\cos 255°$
15. $\sin \dfrac{\pi}{12}$
16. $\sin \dfrac{7\pi}{12}$
17. $\cos \dfrac{13\pi}{12}$
18. $\cos \dfrac{11\pi}{12}$
19. $\tan \dfrac{5\pi}{12}$
20. $\tan \dfrac{23\pi}{12}$
21. $\sec 75°$
22. $\csc 105°$
23. $\cot(-165°)$
24. $\cot(-435°)$
25. $\csc\left(-\dfrac{7\pi}{12}\right)$
26. $\sec\left(-\dfrac{5\pi}{12}\right)$

Verify each statement.

27. $\sin(60° + \theta) = \dfrac{\sqrt{3}}{2}\cos\theta + \dfrac{1}{2}\sin\theta$

28. $\cos\left(\dfrac{\pi}{2} + x\right) = -\sin x$

29. $\tan(\pi + x) = \tan x$

30. $\sin\left(\dfrac{3\pi}{2} - x\right) = -\cos x$

31. $\cos(\pi - x) = -\cos x$

32. $\tan\left(\dfrac{\pi}{4} - x\right) = \dfrac{1 - \tan x}{1 + \tan x}$

Write each quantity as a single function of one angle and evaluate.

33. $\sin 15°\cos 30° + \cos 15° \sin 30°$

34. $\sin 70° \cos 40° - \cos 70° \sin 40°$

35. $\cos 100° \cos 55° + \sin 100° \sin 55°$

36. $\cos 50° \cos 70° - \sin 50° \sin 70°$

37. $\dfrac{\tan 70° + \tan 65°}{1 - \tan 70° \tan 65°}$

38. $\dfrac{\tan 82° - \tan 22°}{1 + \tan 82° \tan 22°}$

39. $\sin\dfrac{\pi}{9}\cos\dfrac{\pi}{18} + \cos\dfrac{\pi}{9}\sin\dfrac{\pi}{18}$

40. $\sin\dfrac{\pi}{9}\cos\dfrac{4\pi}{9} - \cos\dfrac{\pi}{9}\sin\dfrac{4\pi}{9}$

41. $\cos\dfrac{7\pi}{12}\cos\dfrac{5\pi}{12} + \sin\dfrac{7\pi}{12}\sin\dfrac{5\pi}{12}$

42. $\cos\dfrac{7\pi}{18}\cos\dfrac{\pi}{9} - \sin\dfrac{7\pi}{18}\sin\dfrac{\pi}{9}$

For the given set of conditions, find:

(a) $\sin(\alpha + \beta)$ *(b)* $\cos(\alpha - \beta)$ *(c)* $\tan(\alpha + \beta)$

43. $\sin\alpha = \dfrac{3}{5}$, α is in QI; $\cos\beta = -\dfrac{12}{13}$, β is in QIII

44. $\sin\alpha = -\dfrac{7}{25}$, α is in QIII; $\cos\beta = \dfrac{15}{17}$, β is in QIV

45. $\cos\alpha = -\dfrac{9}{41}$, α is in QII; $\tan\beta = -2$, β is in QIV

46. $\sin\alpha = \dfrac{5}{13}$, α is in QII; $\tan\beta = \dfrac{1}{2}$, β is in QIII

47. $\tan\alpha = \dfrac{5}{12}$, α is in QI; $\sin\beta = -\dfrac{\sqrt{5}}{5}$, β is in QIV

48. $\cos\alpha = \dfrac{2\sqrt{5}}{5}$, α is in QI; $\cos\beta = -\dfrac{40}{41}$, β is in QII

49. Given that $\sin\alpha = \frac{12}{13}$, with α in QI, and $\sin(\alpha + \beta) = \frac{24}{25}$, with $\alpha + \beta$ in QII, find $\sin\beta$ and $\cos\beta$.

50. Given that $\cos\beta = -\frac{15}{17}$, with β in QII, and $\sin(\alpha - \beta) = -\frac{24}{25}$, with $\alpha - \beta$ in QIV, find $\sin\alpha$ and $\cos\alpha$.

Verify each identity.

51. $\sin(30° + \theta) - \cos(60° + \theta) = \sqrt{3}\sin\theta$

52. $\sin(60° + \theta) - \cos(30° + \theta) = \sin\theta$

53. $\sin(30° + \theta) + \cos(60° + \theta) = \cos\theta$

54. $\sin(\alpha + \beta) - \sin(\alpha - \beta) = 2\cos\alpha\sin\beta$

55. $\cos(\alpha + \beta) + \cos(\alpha - \beta) = 2\cos\alpha\cos\beta$

56. $\sin(\alpha + \beta)\sin(\alpha - \beta) = \sin^2\alpha - \sin^2\beta$

57. $\cos(\alpha + \beta)\cos(\alpha - \beta) = \cos^2\alpha + \cos^2\beta - 1$

58. $\cot(\alpha + \beta) = \dfrac{\cot\alpha\cot\beta - 1}{\cot\alpha + \cot\beta}$

59. $\cot(\alpha - \beta) = \dfrac{\cot\alpha\cot\beta + 1}{\cot\beta - \cot\alpha}$

60. $\cos(\alpha + \beta) - \cos(\alpha - \beta) = -2\sin\alpha\sin\beta$

61. $\sin(\alpha + \beta) + \sin(\alpha - \beta) = 2\sin\alpha\sin\beta$

62. $-\tan\alpha = \cot\left(\alpha + \dfrac{\pi}{2}\right)$

63. $\dfrac{\tan\alpha + \tan\beta}{1 + \tan\alpha\tan\beta} = \dfrac{1 + \tan\alpha\tan\beta}{\tan\alpha\tan\beta}$

64. $\dfrac{\cos(\alpha + \beta)}{\sin(\alpha - \beta)} = \dfrac{\cot\alpha - \tan\beta}{1 - \cot\alpha\cot\beta}$

65. $\dfrac{\cos(\alpha - \beta)}{\sin(\alpha + \beta)} = \dfrac{1 - \tan\alpha\tan\beta}{\tan\alpha + \tan\beta}$

66. $\dfrac{\sin\alpha\cos\beta + \cos\alpha\sin\beta}{\sin\alpha\cos\beta - \cos\alpha\sin\beta} = \dfrac{1 + \cot\alpha\tan\beta}{1 - \cot\alpha\tan\beta}$

Find the exact value of each expression. Do not use a decimal approximation.

67. $\sin\left[\sin^{-1}\dfrac{1}{2} + \cos^{-1}\dfrac{1}{2}\right]$

68. $\sin\left[\sin^{-1}\left(-\dfrac{1}{2}\right) - \cos^{-1}\dfrac{1}{2}\right]$

69. $\cos\left[\sin^{-1}\left(-\dfrac{\sqrt{2}}{2}\right) - \cos^{-1}\dfrac{\sqrt{3}}{2}\right]$

70. $\cos\left[\tan^{-1}(-1) + \tan^{-1}\sqrt{3}\right]$

71. $\tan\left[\sin^{-1}\dfrac{3}{5} + \tan^{-1}(-2)\right]$

72. $\tan\left[\cos^{-1}\left(-\dfrac{12}{13}\right) - \tan^{-1}\dfrac{1}{3}\right]$

73. $\sin\left[\pi + \sin^{-1}\left(-\dfrac{3}{5}\right)\right]$

74. $\cos\left[\dfrac{\pi}{2} - \cos^{-1}\left(-\dfrac{3}{5}\right)\right]$

75. $\tan\left[\sin^{-1}\dfrac{4}{5} - \dfrac{\pi}{4}\right]$

76. $\tan\left[\tan^{-1}\dfrac{5}{12} + \pi\right]$

77. $\cos\left[\sin^{-1}\dfrac{9}{41} - \cos^{-1}\dfrac{9}{41}\right]$

78. $\sin\left[\cos^{-1}\dfrac{40}{41} + \tan^{-1}\left(-\dfrac{7}{24}\right)\right]$

Write each trigonometric expression as an algebraic expression in x and y.

79. $\sin(\sin^{-1}x + \sin^{-1}y)$ **80.** $\sin(\cos^{-1}x - \cos^{-1}y)$

81. $\tan(\tan^{-1}x + \sin^{-1}y)$ **82.** $\tan\left(\tan^{-1}\dfrac{1}{y} - \sin^{-1}\dfrac{x}{2}\right)$

83. $\cos(\sin^{-1}2x - \cos^{-1}3y)$ **84.** $\cos(\tan^{-1}y + \tan^{-1}x)$

Use a graphing calculator to verify each identity.

85. $\sin x = \cos\left(\dfrac{\pi}{2} - x\right)$ **86.** $\cos x = \sin\left(\dfrac{\pi}{2} - x\right)$

87. $\tan\left(\dfrac{\pi}{2} - x\right) = \cot x$

88. $\tan\left(x + \dfrac{\pi}{4}\right) = \dfrac{1 + \tan x}{1 - \tan x}$

89. $\sin x = -\cos\left(\dfrac{\pi}{2} + x\right)$

90. $\tan(\pi + x) = \tan x$

91. $\sin\left(\dfrac{3\pi}{2} - \theta\right) = -\cos\theta$

92. $\tan\left(x - \dfrac{\pi}{4}\right) = \dfrac{\tan x - 1}{1 + \tan x}$

Applications

Geometry   *In Exercises 93–96, find the angle from line 2 to line 1. Round to one decimal place.*

93. line 1: $y = 3x + 1$; line 2: $y = 2x - 5$

94. line 1: $y = 5x$; line 2: $y = 3x - 8$

95. line 1: $y = 5x$; line 2: $y = -\dfrac{1}{2}x + 3$

96. line 1: $y = -4x$; line 2: $y = \dfrac{2}{3}x$

97. Calculus and rate of change The difference quotient $\dfrac{f(x + h) - f(x)}{h}$ is used in calculus to find the rate of change of a given function $f(x)$. Use the Sum Formula for Sine to show that for $f(x) = \sin x$, the difference quotient simplifies to

$$\cos x\left(\dfrac{\sin h}{h}\right) - \sin x\left(\dfrac{1 - \cos h}{h}\right).$$

98. Calculus and rate of change Use the Sum Formula for Cosine to show that for $f(x) = \cos x$, the difference quotient $\dfrac{f(x + h) - f(x)}{h}$ simplifies to

$$-\sin x\left(\dfrac{\sin h}{h}\right) - \cos x\left(\dfrac{1 - \cos h}{h}\right).$$

Discovery and Writing

99. Derive a formula for $\cos(\alpha + \beta + \delta)$.

100. Derive a formula for $\cos(\alpha - \beta - \delta)$.

101. Derive a formula for $\sin(\alpha - \beta - \delta)$.

102. Derive a formula for $\sin(\alpha + \beta + \delta)$.

103. Derive a formula for $\tan(\alpha + \beta + \delta)$.

104. Show that if lines l_1 and l_2 are perpendicular and each has a nonzero slope, then $m_1 = -\dfrac{1}{m_2}$.

5.5 Double-Angle, Power-Reduction, and Half-Angle Formulas

In this section, we will learn to

1. Find exact values and verify identities using Double-Angle Formulas.

2. Rewrite trigonometric expressions using Power-Reduction Formulas.

3. Use Half-Angle Formulas to find exact values and verify identities.

In Section 5.4 we developed the Sum and Difference Formulas for the sine, cosine, and tangent functions. In this section, we will use these formulas to develop other formulas.

Derek Jeter is an American professional baseball player. He has played his entire career for the New York Yankees and is considered one of the best players of his generation. He is the all-time hits leader among shortstops.

The range r of a baseball hit by Derek Jeter at an angle θ (in degrees) with the horizontal and with an initial velocity of v_0 in feet per second is given by the formula $r = \frac{1}{32} v_0^2 \sin 2\theta$.

Note that this formula involves a double angle. To answer questions related to baseball and other projectile problems, additional knowledge of trigonometry is required. In this section, we will learn to work with trigonometric functions that contain a double angle 2θ. We will also study Power-Reduction Formulas and Half-Angle Formulas. Half-Angle Formulas contain $\frac{\theta}{2}$.

1. Find Exact Values and Verify Identities Using Double-Angle Formulas

Using the fact that we can write $2\theta = \theta + \theta$, we can transform the formulas for the sine, cosine, and tangent of the sum of two angles into the **Double-Angle Formulas**.

$$\sin 2\theta = \sin(\theta + \theta) = \sin \theta \cos \theta + \cos \theta \sin \theta = 2 \sin \theta \cos \theta$$

$$\cos 2\theta = \cos(\theta + \theta) = \cos \theta \cos \theta - \sin \theta \sin \theta = \cos^2 \theta - \sin^2 \theta$$

$$\tan 2\theta = \tan(\theta + \theta) = \frac{\tan \theta + \tan \theta}{1 - \tan \theta \tan \theta} = \frac{2 \tan \theta}{1 - \tan^2 \theta}$$

Because we have the formulas $\cos^2 \theta = 1 - \sin^2 \theta$ and $\sin^2 \theta = 1 - \cos^2 \theta$, we can extend the Double-Angle Formula for cosine:

$$\cos 2\theta = \cos^2 \theta - \mathbf{\sin^2 \theta}$$

$$= \cos^2 \theta - (\mathbf{1 - \cos^2 \theta}) = 2 \cos^2 \theta - 1$$

and

$$\cos 2\theta = \cos^2 \theta - \sin^2 \theta$$

$$= (1 - \sin^2 \theta) - \sin^2 \theta = 1 - 2 \sin^2 \theta$$

We summarize these formulas.

Double-Angle Formulas   For any angle θ for which the function is defined,

$$\sin 2\theta = 2 \sin \theta \cos \theta$$

$$\cos 2\theta = \cos^2 \theta - \sin^2 \theta$$

$$= 2 \cos^2 \theta - 1$$

$$= 1 - 2 \sin^2 \theta$$

$$\tan 2\theta = \frac{2 \tan \theta}{1 - \tan^2 \theta}$$

EXAMPLE 1   **Using Double-Angle Formulas**

Given that $\cos \theta = \dfrac{12}{13}$, θ in QIV, find the exact value of each expression.

a. $\sin 2\theta$ **b.** $\cos 2\theta$ **c.** $\tan 2\theta$

SOLUTION   Figure 5-24 demonstrates how we can find the information we need to determine the desired values.

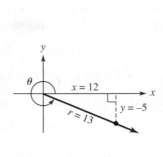

$$\cos \theta = \frac{x}{r} = \frac{12}{13}$$

$$x = 12, r = 13$$

$$x^2 + y^2 = r^2$$

$$12^2 + y^2 = 13^2$$

$$y^2 = 169 - 144$$

$$y^2 = 25$$

$$y = -5 \qquad \text{y is negative in QIV.}$$

FIGURE 5-24

From this information we find $\sin \theta = \frac{y}{r} = -\frac{5}{13}$ and $\tan \theta = \frac{y}{x} = -\frac{5}{12}$.

a. $\sin 2\theta = 2 \sin \theta \cos \theta = 2\left(-\dfrac{5}{13}\right)\left(\dfrac{12}{13}\right) = -\dfrac{120}{169}$

b. $\cos 2\theta = \cos^2 \theta - \sin^2 \theta = \left(\dfrac{12}{13}\right)^2 - \left(-\dfrac{5}{13}\right)^2 = \dfrac{119}{169}$

c. $\tan 2\theta = \dfrac{2 \tan \theta}{1 - \tan^2 \theta} = \dfrac{2\left(-\dfrac{5}{12}\right)}{1 - \left(-\dfrac{5}{12}\right)^2} = \dfrac{-\dfrac{10}{12}}{1 - \dfrac{25}{144}} = \dfrac{-\dfrac{5}{6}}{\dfrac{119}{144}} = -\dfrac{5}{6} \cdot \dfrac{144}{119} = -\dfrac{120}{119}$

Comment   Note that in Example 1, part (c), we could also use the formula

$$\tan 2\theta = \frac{\sin 2\theta}{\cos 2\theta} = \frac{-\dfrac{120}{169}}{\dfrac{119}{169}} = -\frac{120}{119}.$$

Self Check 1   Given that $\sin \theta = -\dfrac{3}{5}$, θ in QIII, find the exact value of each expression.

a. $\sin 2\theta$ **b.** $\cos 2\theta$ **c.** $\tan 2\theta$

Now Try Exercise 41(a)–(c).

EXAMPLE 2   **Verifying Identities**

Verify each identity.
a. $\cos 2\theta = \cos^4 \theta - \sin^4 \theta$ **b.** $\cos \theta \sin 2\theta = 2 \sin \theta - 2 \sin^3 \theta$
c. $\sin 2\theta \sec^2 \theta = 2 \tan \theta$ **d.** $2 \cot 2\theta = \cot \theta - \tan \theta$

SOLUTION   **a.** Because the right-hand side factors, we begin with that side.

$$\cos^4 \theta - \sin^4 = (\cos^2 \theta + \sin^2 \theta)(\cos^2 \theta - \sin^2 \theta) \qquad \text{Factor.}$$

$$= 1 \cdot (\cos^2 \theta - \sin^2 \theta) \qquad \cos^2 \theta + \sin^2 \theta = 1$$

$$= \cos 2\theta \qquad \cos^2 \theta - \sin^2 \theta = \cos 2\theta$$

b. Although the right-hand side looks more complicated, we will begin with the left-hand side because we can use a Double-Angle Formula.

$$\cos\theta\,\sin 2\theta = \cos\theta(2\sin\theta\cos\theta) \qquad \text{Double-Angle Formula for Sine}$$

$$= 2\cos^2\theta\,\sin\theta \qquad\qquad \text{Simplify.}$$

$$= 2(1-\sin^2\theta)\sin\theta \qquad \cos^2\theta = 1-\sin^2\theta$$

$$= 2\sin\theta - 2\sin^3\theta \qquad\quad \text{Simplify.}$$

c. We will begin on the left-hand side of the identity.

$$\sin 2\theta\,\sec^2\theta = 2\sin\theta\cos\theta\,\sec^2\theta \qquad \text{Double-Angle Formula for Sine}$$

$$= 2\sin\theta\cos\theta\,\frac{1}{\cos^2\theta} \qquad \text{Reciprocal Identity}$$

$$= 2\sin\theta\,\frac{1}{\cos\theta} \qquad\qquad \text{Simplify.}$$

$$= 2\tan\theta \qquad\qquad\qquad \text{Quotient Identity}$$

d. We will begin with the left-hand side.

$$2\cot 2\theta = 2\left(\frac{1}{\tan 2\theta}\right) \qquad \cot 2\theta = \frac{1}{\tan 2\theta}$$

$$= 2\left(\frac{1-\tan^2\theta}{2\tan\theta}\right) \qquad \text{Reciprocal of } \tan 2\theta$$

$$= \frac{1-\tan^2\theta}{\tan\theta} \qquad\qquad \text{Simplify.}$$

$$= \frac{1}{\tan\theta} - \frac{\tan^2\theta}{\tan\theta} \qquad \text{Rewrite as two fractions.}$$

$$= \cot\theta - \tan\theta \qquad\quad \text{Simplify using Reciprocal Identity and algebra.}$$

■

Self Check 2 Verify that $\sec 2\theta = \dfrac{\sec^2\theta}{2-\sec^2\theta}$.

Now Try Exercise 53.

EXAMPLE 3   Using Formulas to Find Exact Values

Find the exact value of each expression.

a. $2\sin 15°\cos 15°$ **b.** $\dfrac{2\tan 75°}{1-\tan^2 75°}$

SOLUTION   **a.** $2\sin 15°\cos 15° = \sin(2\cdot 15°) = \sin 30° = \dfrac{1}{2}$

b. $\dfrac{2\tan 75°}{1-\tan^2 75°} = \tan(2\cdot 75°) = \tan 150° = -\dfrac{\sqrt{3}}{3}$

■

Self Check 3 Find the exact value of $\cos^2 105° - \sin^2 105°$.

Now Try Exercise 35.

2. Rewrite Trigonometric Expressions Using Power-Reduction Formulas

There are many instances in calculus when a trigonometric expression involving powers of trigonometric functions will need to be reduced to an expression with only first-degree expressions. We will examine this in the "Looking Ahead to Calculus" section at the end of the chapter.

We can use the Double-Angle Formulas to develop the **Power-Reduction Formulas**.

Power-Reduction Formulas

$$\sin^2 \theta = \frac{1}{2}(1 - \cos 2\theta)$$

$$\cos^2 \theta = \frac{1}{2}(1 + \cos 2\theta)$$

$$\tan^2 \theta = \frac{1 - \cos 2\theta}{1 + \cos 2\theta}$$

To prove these formulas, we use two versions of the Double-Angle Formula for cosine.

$$\cos 2\theta = 1 - 2\sin^2 \theta \qquad\qquad \cos 2\theta = 2\cos^2 \theta - 1$$
$$2\sin^2 \theta = 1 - \cos 2\theta \qquad\qquad 1 + \cos 2\theta = 2\cos^2 \theta$$
$$\sin^2 \theta = \frac{1}{2}(1 - \cos 2\theta) \qquad\qquad \cos^2 \theta = \frac{1}{2}(1 + \cos 2\theta)$$

The Tangent Formula can be verified using these two identities.

$$\tan^2 \theta = \frac{\sin^2 \theta}{\cos^2 \theta} = \frac{\frac{1}{2}(1 - \cos 2\theta)}{\frac{1}{2}(1 + \cos 2\theta)} = \frac{1 - \cos 2\theta}{1 + \cos 2\theta}$$

EXAMPLE 4   Rewriting Trigonometric Expressions Using Power-Reduction Formulas

Rewrite $\sin^4 x$ and simplify.

SOLUTION   This is the type of problem that is common in calculus. We will use both algebraic techniques and Power-Reduction Formulas.

$$\sin^4 x = \sin^2 x \cdot \sin^2 x \qquad\qquad \text{Factor into two terms.}$$

$$= \left[\frac{1}{2}(1 - \cos 2x)\right]\left[\frac{1}{2}(1 - \cos 2x)\right] \qquad\qquad \text{Power-Reduction Formula}$$

$$= \frac{1}{4}(1 - 2\cos 2x + \cos^2 2x) \qquad\qquad \text{Multiply the terms.}$$

$$= \frac{1}{4} - \frac{1}{2}\cos 2x + \frac{1}{4}\cos^2 2x \qquad\qquad \text{Clear parentheses.}$$

$$= \frac{1}{4} - \frac{1}{2}\cos 2x + \frac{1}{4}\left[\frac{1}{2}(1 + \cos 2(2x))\right] \qquad\qquad \begin{array}{l}\text{Power-Reduction Formula} \\ \text{(note: } 2x = \theta \text{ in the formula)}\end{array}$$

$$= \frac{1}{4} - \frac{1}{2}\cos 2x + \frac{1}{8} + \frac{1}{8}\cos 4x \qquad\qquad \text{Clear parentheses.}$$

$$= \frac{3}{8} - \frac{1}{2}\cos 2x + \frac{1}{8}\cos 4x \qquad\qquad \text{Simplify.}$$

Self Check 4   Rewrite $\sin^2 x \cos 2x$ and simplify.

Now Try Exercise 79.

3. Use Half-Angle Formulas to Find Exact Values and Verify Identities

If the trigonometric functions of angle θ are known, the Double-Angle Formulas allow us to find the values of the trigonometric functions of 2θ. Now we will use these formulas to find the values of the trigonometric functions of $\frac{1}{2}\theta$. We will begin by solving the formula $\cos 2\theta = 2\cos^2\theta - 1$ for $\cos\theta$.

$$\cos 2\theta = 2\cos^2\theta - 1$$

$$2\cos^2\theta = 1 + \cos 2\theta$$

$$\cos^2\theta = \frac{1 + \cos 2\theta}{2}$$

$$\cos\theta = \pm\sqrt{\frac{1 + \cos 2\theta}{2}}$$

If we let $\theta = \dfrac{\alpha}{2}$, we have $\cos\dfrac{\alpha}{2} = \pm\sqrt{\dfrac{1 + \cos\alpha}{2}}$.

We can find the Half-Angle Formula for the sine function in a similar way.

$$\cos 2\theta = 1 - 2\sin^2\theta$$

$$2\sin^2\theta = 1 - \cos 2\theta$$

$$\sin^2\theta = \frac{1 - \cos 2\theta}{2}$$

$$\sin\theta = \pm\sqrt{\frac{1 - \cos 2\theta}{2}}$$

Comment

You do not use both signs. You must determine either $+$ or $-$.

Again, we let $\theta = \dfrac{\alpha}{2}$, which gives us the formula $\sin\dfrac{\alpha}{2} = \pm\sqrt{\dfrac{1 - \cos\alpha}{2}}$.

Half-Angle Formulas $\sin\dfrac{\alpha}{2} = \pm\sqrt{\dfrac{1 - \cos\alpha}{2}}$ $\cos\dfrac{\alpha}{2} = \pm\sqrt{\dfrac{1 + \cos\alpha}{2}}$

The sign preceding these radicals is determined by the quadrant in which the angle $\dfrac{\alpha}{2}$ lies.

EXAMPLE 5   **Using a Half-Angle Formula to Find Exact Values of a Trigonometric Function**

Find the exact value of $\sin 15°$.

SOLUTION   First, note that $15° = \frac{1}{2}(30°)$. Since the angle $15°$ is in QI, we use the positive sign in the formula.

$$\sin 15° = \sin \frac{30°}{2} \qquad\qquad \text{Let } \alpha = 30°.$$

$$= +\sqrt{\frac{1 - \cos 30°}{2}} \qquad \text{Use } \sin \frac{\alpha}{2} = +\sqrt{\frac{1 - \cos \alpha}{2}}, \text{ since } \sin 15° > 0.$$

$$= \sqrt{\frac{1 - \dfrac{\sqrt{3}}{2}}{2}} \qquad\qquad \text{The + sign is not necessary.}$$

$$= \sqrt{\frac{\left(1 - \dfrac{\sqrt{3}}{2}\right)}{2} \cdot \frac{2}{2}} \qquad \text{Multiply by } \frac{2}{2}.$$

$$= \sqrt{\frac{2 - \sqrt{3}}{4}}$$

$$= \frac{\sqrt{2 - \sqrt{3}}}{2} \qquad\qquad \text{Simplify.}$$

Self Check 5 Find the exact value of $\sin 22.5°$.

Now Try Exercise 29(a).

EXAMPLE 6 **Using a Half-Angle Formula to Find Exact Values of a Trigonometric Function**

Find the exact value of $\cos \dfrac{7\pi}{12}$.

SOLUTION Because $\frac{7\pi}{12}$ is $\frac{1}{2} \cdot \frac{7\pi}{6}$, and $\frac{7\pi}{6}$ is a multiple of a special angle, we can use the Half-Angle Formula for cosine to evaluate the function. We must determine the quadrant in which the angle lies. The angle $\frac{7\pi}{12}$ lies in QII; therefore, $\cos \frac{7\pi}{12} < 0$.

$$\cos \frac{7\pi}{12} = \cos \frac{1}{2} \cdot \frac{7\pi}{6} \qquad\qquad \text{Let } \alpha = \frac{7\pi}{6}.$$

$$= -\sqrt{\frac{1 + \cos \dfrac{7\pi}{6}}{2}} \qquad \text{Use } \cos \frac{\alpha}{2} = -\sqrt{\frac{1 + \cos \alpha}{2}}, \text{ since } \cos \frac{7\pi}{12} < 0.$$

$$= -\sqrt{\frac{1 - \dfrac{\sqrt{3}}{2}}{2}}$$

$$= -\sqrt{\frac{\left(1 - \dfrac{\sqrt{3}}{2}\right)}{2} \cdot \frac{2}{2}} \qquad \text{Multiply the numerator and denominator by 2.}$$

$$= -\sqrt{\frac{2 - \sqrt{3}}{4}}$$

$$= -\frac{\sqrt{2 - \sqrt{3}}}{2} \qquad\qquad \text{Simplify.}$$

Self Check 6   Find the exact value of $\cos \dfrac{11\pi}{12}$.

Now Try Exercise 31(b).

To derive the Half-Angle Formula for the tangent function, we will show that $\tan \theta = \dfrac{\sin 2\theta}{1 + \cos 2\theta}$ and then substitute $\theta = \dfrac{\alpha}{2}$.

$$\dfrac{\sin 2\theta}{1 + \cos 2\theta} = \dfrac{2 \sin \theta \cos \theta}{1 + (2 \cos^2 \theta - 1)} \qquad \sin 2\theta = 2 \sin \theta \cos \theta; \cos 2\theta = 2 \cos^2 \theta - 1.$$

$$= \dfrac{2 \sin \theta \cos \theta}{2 \cos^2 \theta}$$

$$= \dfrac{\sin \theta}{\cos \theta}$$

$$= \tan \theta$$

Taking $\tan \theta = \dfrac{\sin 2\theta}{1 + \cos 2\theta}$ and substituting $\theta = \dfrac{\alpha}{2}$, we have a Half-Angle Formula for the tangent function:

$$\tan \dfrac{\alpha}{2} = \dfrac{\sin \alpha}{1 + \cos \alpha}.$$

By multiplying both the numerator and the denominator of the formula we just derived by the conjugate of the denominator, we obtain another version of the Half-Angle Tangent Formula.

$$\tan \dfrac{\alpha}{2} = \dfrac{\sin \alpha}{1 + \cos \alpha} \cdot \dfrac{1 - \cos \alpha}{1 - \cos \alpha}$$

$$= \dfrac{\sin \alpha(1 - \cos \alpha)}{1 - \cos^2 \alpha}$$

$$= \dfrac{\sin \alpha(1 - \cos \alpha)}{\sin^2 \alpha}$$

$$= \dfrac{1 - \cos \alpha}{\sin \alpha}$$

We now have two Half-Angle Formulas for the tangent function.

A third Half-Angle Formula, $\tan \dfrac{\alpha}{2} = \pm\sqrt{\dfrac{1 - \cos \alpha}{1 + \cos \alpha}}$, is easily derived.

Half-Angle Formulas for Tangent $\tan \dfrac{\alpha}{2} = \dfrac{\sin \alpha}{1 + \cos \alpha}$, $\tan \dfrac{\alpha}{2} = \dfrac{1 - \cos \alpha}{\sin \alpha}$, $\tan \dfrac{\alpha}{2} = \pm\sqrt{\dfrac{1 - \cos \alpha}{1 + \cos \alpha}}$

EXAMPLE 7   **Using the Tangent Half-Angle Formulas**

Find the exact value of each expression.   **a.** $\tan \dfrac{\pi}{8}$ **b.** $\tan 157.5°$

SOLUTION   **a.** We will use the formula $\tan \dfrac{\alpha}{2} = \dfrac{1 - \cos \alpha}{\sin \alpha}$. Because $\dfrac{\pi}{8} = \dfrac{1}{2} \cdot \dfrac{\pi}{4}$, we let $\alpha = \dfrac{\pi}{4}$.

$$\tan \dfrac{\pi}{8} = \tan \dfrac{\dfrac{\pi}{4}}{2} = \dfrac{1 - \cos \dfrac{\pi}{4}}{\sin \dfrac{\pi}{4}}$$

$$= \frac{1 - \dfrac{\sqrt{2}}{2}}{\dfrac{\sqrt{2}}{2}} \cdot \frac{2}{2}$$ Multiply to simplify the complex fraction.

$$= \frac{2 - \sqrt{2}}{\sqrt{2}} \cdot \frac{\sqrt{2}}{\sqrt{2}}$$ Multiply to rationalize the denominator.

$$= \frac{2\sqrt{2} - 2}{2} = \frac{2(\sqrt{2} - 1)}{2}$$ Factor.

$$= \sqrt{2} - 1$$ Simplify.

b. To find $\tan 157.5°$, we will use the formula $\tan \dfrac{\alpha}{2} = \dfrac{\sin \alpha}{1 + \cos \alpha}$. Since $157.5°$ is $\dfrac{315°}{2}$, we let $\alpha = 315°$.

$$\tan 157.5° = \tan \frac{315°}{2} = \frac{\sin 315°}{1 + \cos 315°}$$

$$= \frac{-\dfrac{\sqrt{2}}{2}}{1 + \dfrac{\sqrt{2}}{2}} \cdot \frac{2}{2}$$ Multiply to simplify the complex fraction.

$$= \frac{-\sqrt{2}}{2 + \sqrt{2}} \cdot \frac{2 - \sqrt{2}}{2 - \sqrt{2}}$$ Multiply by the conjugate of $2 + \sqrt{2}$ to rationalize the denominator.

$$= \frac{-2\sqrt{2} + 2}{2} = \frac{2(-\sqrt{2} + 1)}{2}$$ Factor.

$$= 1 - \sqrt{2}$$ Simplifiy.

In Example 7, we used both of the formulas for the half-angle tangent value. Typically, you can select either formula. There may be times when one form is more useful than the other.

Self Check 7 Find the exact value of $\tan 112.5°$.

Now Try Exercise 27(c).

EXAMPLE 8 **Finding Exact Values of Trigonometric Functions Using the Half-Angle Formulas**

Given that $\sin \theta = -\dfrac{3}{5}$, θ in QIII, find the exact value of each expression.

a. $\cos \dfrac{\theta}{2}$ **b.** $\sin \dfrac{\theta}{2}$ **c.** $\tan \dfrac{\theta}{2}$

SOLUTION We are given that $\sin \theta = -\dfrac{3}{5}$ and that θ is in QIII. Since $\sin \theta = \dfrac{y}{r} = \dfrac{-3}{5}$, we know that $y = -3$ and $r = 5$. Using the Pythagorean Theorem, we can find x as shown in Figure 5-25.

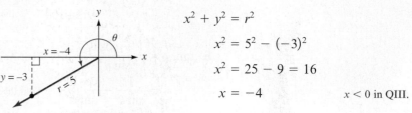

$$x^2 + y^2 = r^2$$
$$x^2 = 5^2 - (-3)^2$$
$$x^2 = 25 - 9 = 16$$
$$x = -4 \qquad x < 0 \text{ in QIII.}$$

FIGURE 5-25

We select $x = -4$ because the angle lies in QIII, where $x < 0$. Therefore, we now have $\cos\theta = -\frac{4}{5}$. Our next step is to determine where the angle $\frac{\theta}{2}$ lies.

$$\pi < \theta < \frac{3\pi}{2}$$

$$\frac{\pi}{2} < \frac{\theta}{2} < \frac{3\pi}{4}$$

This tells us that $\frac{\theta}{2}$ lies in QII; therefore, $\sin\frac{\theta}{2} > 0$ and $\cos\frac{\theta}{2} < 0$. It is important to determine these signs so that we can select the proper signs for the Half-Angle Formulas.

a. $\cos\dfrac{\theta}{2} = -\sqrt{\dfrac{1 + \cos\theta}{2}} = -\sqrt{\dfrac{1 + \left(-\dfrac{4}{5}\right)}{2}} = -\sqrt{\dfrac{\dfrac{1}{5}}{2}} = -\sqrt{\dfrac{1}{10}} = -\dfrac{1}{\sqrt{10}} = -\dfrac{\sqrt{10}}{10}$

b. $\sin\dfrac{\theta}{2} = \sqrt{\dfrac{1 - \cos\theta}{2}} = \sqrt{\dfrac{1 - \left(-\dfrac{4}{5}\right)}{2}} = \sqrt{\dfrac{\dfrac{9}{5}}{2}} = \sqrt{\dfrac{9}{10}} = \dfrac{3}{\sqrt{10}} = \dfrac{3\sqrt{10}}{10}$

c. $\tan\dfrac{\theta}{2} = \dfrac{\sin\dfrac{\theta}{2}}{\cos\dfrac{\theta}{2}} = \dfrac{\dfrac{3\sqrt{10}}{10}}{-\dfrac{\sqrt{10}}{10}} = -3$

■

Self Check 8   Given that $\sin\theta = -\dfrac{5}{13}$, θ in QIII, find the exact value of each expression.

a. $\cos\dfrac{\theta}{2}$ **b.** $\sin\dfrac{\theta}{2}$ **c.** $\tan\dfrac{\theta}{2}$

Now Try Exercise 41(d)–(f).

EXAMPLE 9   **Verifying Identities**

Verify each identity.

a. $\dfrac{\sin(-20\theta)}{-\cos 20\theta - 1} = \tan 10\theta$

b. $\tan\dfrac{\theta}{2} = \csc\theta - \cot\theta$

c. $2\sin^2\dfrac{\theta}{2}\tan\theta = \tan\theta - \sin\theta$

SOLUTION   **a.** $\dfrac{\sin(-20\theta)}{-\cos 20\theta - 1} = \dfrac{-\sin 20\theta}{-(1 + \cos 20\theta)}$ Even–Odd Property, and factor out -1 in the denominator.

$$= \dfrac{\sin 20\theta}{1 + \cos 20\theta}$$ Simplify.

$$= \tan\dfrac{20\theta}{2}$$ Half-Angle Formula

$$= \tan 10\theta$$ Simplify.

b. $\tan\dfrac{\theta}{2} = \dfrac{1 - \cos\theta}{\sin\theta}$ Half-Angle Formula

$$= \dfrac{1}{\sin\theta} - \dfrac{\cos\theta}{\sin\theta}$$ Split the fraction.

$$= \csc\theta - \cot\theta$$ Simplify.

c. $2 \sin^2 \dfrac{\theta}{2} \tan \theta = 2 \cdot \dfrac{1 - \cos \theta}{2} \cdot \dfrac{\sin \theta}{\cos \theta}$ $\left(\sin \dfrac{\theta}{2}\right)^2 = \left(\sqrt{\dfrac{1 - \cos \theta}{2}}\right)^2; \tan \theta = \dfrac{\sin \theta}{\cos \theta}$

$= \dfrac{(1 - \cos \theta)\sin \theta}{\cos \theta}$ Simplify.

$= \dfrac{\sin \theta - \cos \theta \sin \theta}{\cos \theta}$ Clear parentheses.

$= \dfrac{\sin \theta}{\cos \theta} - \dfrac{\cos \theta \sin \theta}{\cos \theta}$ Split the fraction.

$= \tan \theta - \sin \theta$ Simplify.

∎

Self Check 9   Verify that $\sec \dfrac{\theta}{2} = \pm\sqrt{\dfrac{2 \tan \theta}{\tan \theta + \sin \theta}}$.

Now Try Exercise 67.

Self Check Answers   **1. a.** $\dfrac{24}{25}$ **b.** $\dfrac{7}{25}$ **c.** $\dfrac{24}{7}$ **2.** Answers will vary.   **3.** $-\dfrac{\sqrt{3}}{2}$

4. $-\dfrac{1}{4} + \dfrac{1}{2}\cos 2x - \dfrac{1}{4}\cos 4x$ **5.** $\dfrac{\sqrt{2 - \sqrt{2}}}{2}$ **6.** $-\dfrac{\sqrt{2 + \sqrt{3}}}{2}$

7. $-\sqrt{2} - 1$ **8. a.** $-\dfrac{\sqrt{26}}{26}$ **b.** $\dfrac{5\sqrt{26}}{26}$ **c.** -5

9. Answers will vary.

Exercises 5.5

Getting Ready
You should be able to complete these vocabulary and concept statements before you proceed to the practice exercises.

Fill in the blanks.

1. The Double-Angle Formula for the sine function is $\sin 2\theta =$ _____.

2. The three forms of the Double-Angle Formula of the cosine function are $\cos 2\theta =$ _____ = _____ = _____.

3. Assuming that $0 \le \theta < 2\pi$ lies in the given quadrant, state the quadrant in which $\dfrac{\theta}{2}$ lies.
   a. θ in QI means $\dfrac{\theta}{2}$ lies in _____.
   b. θ in QII means $\dfrac{\theta}{2}$ lies in _____.
   c. θ in QIII means $\dfrac{\theta}{2}$ lies in _____.
   d. θ in QIV means $\dfrac{\theta}{2}$ lies in _____.

4. The first step in evaluating $\sin \dfrac{\theta}{2}$ and $\cos \dfrac{\theta}{2}$ is to _____.

Practice
Write each expression in terms of a single trigonometric function of twice the given angle.

5. $2 \sin \alpha \cos \alpha$ 6. $2 \sin 3\theta \cos 3\theta$
7. $2 \cos^2 \alpha - 1$ 8. $2 \cos^2 2\alpha - 1$
9. $\cos^2 \beta - \sin^2 \beta$ 10. $1 - 2 \sin^2 \beta$
11. $2 \cos^2 \dfrac{\beta}{2} - 1$ 12. $\cos^2 \alpha - \dfrac{1}{2}$
13. $4 \sin^2 2\theta \cos^2 2\theta$ 14. $4 \sin \theta \cos \theta$
15. $\cos^2 9\theta - \sin^2 9\theta$ 16. $3 - 6 \sin^2 6\theta$
17. $\dfrac{2 \tan \alpha}{1 - \tan^2 \alpha}$ 18. $\dfrac{2 \tan 4\beta}{1 - \tan^2 4\beta}$

Write each expression as a single trigonometric function of half the given angle.

19. $\sqrt{\dfrac{1 + \cos 30°}{2}}$ 20. $\sqrt{\dfrac{1 - \cos 30°}{2}}$
21. $\dfrac{1 - \cos 200°}{\sin 200°}$ 22. $\dfrac{\sin 50°}{1 + \cos 50°}$

23. $\left(\dfrac{1 - \cos 2x}{\sin 2x}\right)^2$

24. $\left(\dfrac{\sin 4x}{1 + \cos 4x}\right)^2$

For each angle, use the Half-Angle Formulas to find each of the following values. Do not use a calculator.

(a) $\sin \theta$ (b) $\cos \theta$ (c) $\tan \theta$

25. $15°$ **26.** $22.5°$ **27.** $75°$ **28.** $105°$

29. $112.5°$ **30.** $165°$ **31.** $\dfrac{3\pi}{8}$ **32.** $\dfrac{7\pi}{8}$

33. $-\dfrac{9\pi}{8}$ **34.** $-\dfrac{15\pi}{8}$

Find the exact value of each expression.

35. $2 \sin 22.5° \cos 22.5°$

36. $2 \sin \dfrac{5\pi}{12} \cos \dfrac{5\pi}{12}$

37. $\cos^2 157.5° - \sin^2 157.5°$ **38.** $1 - 2 \sin^2 67.5°$

39. $\dfrac{2 \tan \dfrac{7\pi}{12}}{1 - \tan^2 \dfrac{7\pi}{12}}$

40. $\dfrac{2 \tan \dfrac{11\pi}{12}}{1 - \tan^2 \dfrac{11\pi}{12}}$

Use the given information to find the exact value of

(a) $\sin 2\theta$ (b) $\cos 2\theta$ (c) $\tan 2\theta$

(d) $\sin \dfrac{\theta}{2}$ (e) $\cos \dfrac{\theta}{2}$ (f) $\tan \dfrac{\theta}{2}$

41. $\sin \theta = \dfrac{12}{13}, \theta$ is in QI

42. $\sin \theta = \dfrac{8}{17}, \theta$ is in QII

43. $\cos \theta = -\dfrac{9}{41}, \theta$ is in QIII

44. $\cos \theta = -\dfrac{\sqrt{5}}{3}, \theta$ is in QII

45. $\tan \theta = 2, \theta$ is in QIII

46. $\tan \theta = \dfrac{\sqrt{5}}{2}, \theta$ is in QI

47. $\csc \theta = -\dfrac{2\sqrt{7}}{5}, \theta$ is in QIV

48. $\csc \theta = -\dfrac{25}{24}, \theta$ is in QIII

49. $\sec \theta = \dfrac{5}{3}, \theta$ is in QIV

50. $\sec \theta = \dfrac{\sqrt{13}}{3}, \theta$ is in QI

51. $\cot \theta = -\dfrac{4}{3}, \theta$ is in QII

52. $\cot \theta = -\dfrac{2\sqrt{5}}{5}, \theta$ is in QIV

Verify each identity.

53. $\dfrac{\tan 2x}{\sin 2x} = \sec 2x$

54. $\dfrac{\cot 2x}{\cos 2x} = \csc 2x$

55. $2 \csc 2\alpha = \sec \alpha \csc \alpha$

56. $\sin 2\alpha + 2 \sin \alpha = \dfrac{-2 \sin^3 \alpha}{\cos \alpha - 1}$

57. $\cos 2x - \dfrac{\sin 2x}{\cos x} + 2 \sin^2 x = 1 - 2 \sin x$

58. $\sin 2x - 2 \sin x = -\dfrac{2 \sin^3 x}{\cos x + 1}$

59. $\sin^2 \dfrac{\theta}{2} = \dfrac{1}{2}(1 - \cos \theta)$

60. $\cos^2 \dfrac{\theta}{2} = \dfrac{1}{2}(1 + \cos \theta)$

61. $\cot \dfrac{\theta}{2} = \dfrac{1 + \cos \theta}{\sin \theta}$

62. $\cot \dfrac{\theta}{2} = \dfrac{\sin \theta}{1 - \cos \theta}$

63. $2 \cos^4 \theta + 2 \sin^2 \theta \cos^2 \theta - 1 = \cos 2\theta$

64. $2 \sin^4 \theta + 2 \sin^2 \theta \cos^2 \theta - 1 = -\cos 2\theta$

65. $2 \cos \alpha - 1 = \dfrac{\cos 2\alpha + \cos \alpha}{\cos \alpha + 1}$

66. $2 \cos \alpha + 1 = \dfrac{\cos 2\alpha - \cos \alpha}{\cos \alpha - 1}$

67. $\sin^2 \dfrac{\theta}{2} = \dfrac{\sec \theta - 1}{2 \sec \theta}$

68. $\csc^2 \dfrac{\theta}{2} = 2 \csc^2 \theta + 2 \cot \theta \csc \theta$

69. $\tan\left(\dfrac{\pi}{4} + \dfrac{\theta}{2}\right) = \dfrac{1 + \cos \theta + \sin \theta}{1 + \cos \theta - \sin \theta}$

70. $\csc \theta = \dfrac{1}{2} \csc \dfrac{\theta}{2} \sec \dfrac{\theta}{2}$

71. $\cos 4x = 8 \cos^4 x - 8 \cos^2 x + 1$

72. $\dfrac{\sec^2 x}{\sec 2x} = 2 - \sec^2 x$

73. $\dfrac{1 - \tan^2 x}{1 + \tan^2 x} = \cos 2x$

74. $\dfrac{1 + \tan x}{1 - \tan x} = \dfrac{\cos 2x}{1 - \sin 2x}$

75. $\tan \alpha \sin 2\alpha + \cos 2\alpha = 1$

76. $-\tan 2\alpha = \dfrac{2}{\tan \alpha - \cot \alpha}$

77. $\dfrac{1}{2} \sin x \tan \dfrac{x}{2} \csc^2 \dfrac{x}{2} = 1$

78. $\tan \dfrac{\beta}{2} \cos \beta + \tan \dfrac{\beta}{2} = \sin \beta$

Rewrite each expression. Use a Power-Reduction Formula.

79. $\cos^4 x$

80. $\sin^6 x$

81. $\sin^2 x \cos^2 x$

82. $\cos^2 x \sin 2x$

Find each value. Do not use a calculator.

83. $\sin\left(2 \sin^{-1} \dfrac{\sqrt{2}}{2}\right)$

84. $\cos\left(2 \sin^{-1} \dfrac{1}{2}\right)$

85. $\sin\left(2 \sin^{-1} \dfrac{3}{5}\right)$

86. $\cos\left(2 \cos^{-1} \dfrac{4}{5}\right)$

87. $\cos\left(2 \cos^{-1} \dfrac{5}{13}\right)$

88. $\sin\left(2 \sin^{-1} \dfrac{12}{13}\right)$

 Rewrite each value as an algebraic expression in the variable x. Assume all angles are acute.

89. $\sin(2 \sin^{-1} x)$

90. $\cos(2 \cos^{-1} x)$

91. $\cos(2 \sin^{-1} x)$

92. $\sin(2 \cos^{-1} x)$

93. $\tan(2 \tan^{-1} x)$

94. $\tan\left(\dfrac{1}{2} \sin^{-1} x\right)$

95. $\cos\left(\dfrac{1}{2} \cos^{-1} x\right)$

96. $\tan\left(\dfrac{1}{2} \tan^{-1} \dfrac{\sqrt{1 - x^2}}{x}\right)$

97. $\sin\left(\dfrac{1}{2} \cos^{-1} x\right)$

98. $\cos\left(\dfrac{1}{2} \sin^{-1} x\right)$

Applications
The range r of an object projected at an angle θ (in degrees) with the horizontal and with an initial velocity of v_0 in feet per second is given by the formula

$$r = \frac{1}{32} v_0^2 \sin 2\theta.$$

Use this formula for Exercises 99 and 100.

99. Range of a soccer ball If a soccer ball is kicked by Mia Hamm at angle of 27° with an initial velocity of 20 feet per second, find the range of the soccer ball. Round to one decimal place.

100. Range of a golf ball If a golf ball is hit by Jeff Hughes at angle of 48° with an initial velocity of 60 feet per second, find the range of the golf ball. Round to one decimal place.

The ratio of an airplane's speed to the speed of sound is known as the Mach number of the airplane. When an airplane travels faster than the speed of sound, the sound waves form a cone behind the airplane. The Mach number of the airplane is related to the apex angle θ in degrees of the cone by the formula

$$\sin \frac{\theta}{2} = \frac{1}{M}.$$

Use this formula for Exercises 101 and 102.

101. Mach number If an airplane travels faster than the speed of sound so that the sound waves form a cone with an apex angle 60°, find the Mach number of the airplane.

102. Mach number Find the Mach number of a fighter jet if the apex angle of the cone is 75°. Round to one decimal place.

Discovery and Writing

103. To obtain a trigonometric expression for $\sin^2 x$ that is required in calculus, we can square both sides of the Half-Angle Formula $\sin \dfrac{\theta}{2} = \pm\sqrt{\dfrac{1 - \cos \theta}{2}}$ and then let $\dfrac{\theta}{2} = x$. Write the expression for $\sin^2 x$ that results.

104. To obtain a trigonometric expression for $\cos^2 x$ that is required in calculus, we can square both sides of the Half-Angle Formula $\cos \dfrac{\theta}{2} = \pm\sqrt{\dfrac{1 + \cos \theta}{2}}$ and then let $\dfrac{\theta}{2} = x$. Write the expression for $\cos^2 x$ that results.

5.6 Product-to-Sum and Sum-to-Product Formulas and Sums of the Form $A \sin \theta + B \cos \theta$

In this section, we will learn to

1. Express trigonometric products as sums.
2. Express trigonometric sums as products.
3. Write expressions of the form $A \sin \theta + B \sin \theta$ in the form $k \sin(\theta + \varphi)$.
4. Solve application problems.

Have you ever tried to play a simple song using a touch-tone phone? It's possible. A unique sound is produced by each button pressed. The sound we hear is the sum of two tones $y = \sin 2\pi lt$ and $y = \sin 2\pi ht$, where l and h represent the low and high frequencies in cycles per second and t is time in seconds.

Oleksiy Mark/Shutterstock.com
viktordekalb/Shutterstock.com

Using a trigonometry formula, it is possible to write the sound, the sum of the two tones, as a product of sines and cosines. Formulas help us solve many application problems. In this section we will learn how to express trigonometric products as sums and how to express trigonometric sums as products. We will examine an application that involves tuning a piano in this section.

1. Express Trigonometric Products as Sums

If the formulas for $\sin(\alpha + \beta)$ and $\sin(\alpha - \beta)$ are added, some of these new formulas result.

$$\sin(\alpha + \beta) = \sin \alpha \cos \beta + \cancel{\cos \alpha \sin \beta}$$
$$+ \ \underline{\sin(\alpha - \beta) = \sin \alpha \cos \beta - \cancel{\cos \alpha \sin \beta}} \qquad \text{Add the formulas.}$$
$$\sin(\alpha + \beta) + \sin(\alpha - \beta) = 2 \sin \alpha \cos \beta$$

If we divide both sides of the equation $\sin(\alpha + \beta) + \sin(\alpha - \beta) = 2 \sin \alpha \cos \beta$ by 2, we have $\mathbf{\sin \alpha \cos \beta = \frac{1}{2}[\sin(\alpha + \beta) + \sin(\alpha - \beta)]}$.

If the formulas are subtracted, the result is

$$\sin(\alpha + \beta) = \cancel{\sin \alpha \cos \beta} + \cos \alpha \sin \beta$$
$$- \ \underline{\sin(\alpha - \beta) = \cancel{\sin \alpha \cos \beta} + \cos \alpha \sin \beta} \qquad \text{Subtract the formulas.}$$
$$\sin(\alpha + \beta) - \sin(\alpha - \beta) = 2 \cos \alpha \sin \beta$$

Again we divide by 2, with the result **$\cos \alpha \sin \beta = \frac{1}{2}[\sin(\alpha + \beta) - \sin(\alpha - \beta)]$**.
Now we add the cosine formulas for the sum and difference.

$$\cos(\alpha + \beta) = \cos \alpha \cos \beta - \cancel{\sin \alpha \sin \beta}$$
$$+ \underline{\cos(\alpha - \beta) = \cos \alpha \cos \beta + \cancel{\sin \alpha \sin \beta}} \qquad \text{Add the formulas.}$$
$$\cos(\alpha + \beta) + \cos(\alpha - \beta) = 2 \cos \alpha \cos \beta$$

Dividing both sides by 2 yields the formula
$\cos \alpha \cos \beta = \frac{1}{2}[\cos(\alpha + \beta) + \cos(\alpha - \beta)]$.
Now we will subtract the formulas.

$$\cos(\alpha + \beta) = \cancel{\cos \alpha \cos \beta} - \sin \alpha \sin \beta$$
$$- \underline{\cos(\alpha - \beta) = \cancel{\cos \alpha \cos \beta} - \sin \alpha \sin \beta} \qquad \text{Subtract the formulas.}$$
$$\cos(\alpha + \beta) - \cos(\alpha - \beta) = \qquad\qquad -2 \sin \alpha \sin \beta$$
$$\cos(\alpha - \beta) - \cos(\alpha + \beta) = \qquad\qquad 2 \sin \alpha \sin \beta \qquad \text{Multiply both sides by } -1.$$

Now divide both sides by 2, resulting in **$\sin \alpha \sin \beta = \frac{1}{2}[\cos(\alpha - \beta) - \cos(\alpha + \beta)]$**.
These formulas are used in calculus to rewrite expressions for easier use in several applications.

Product-to-Sum Formulas

1. $\sin \alpha \cos \beta = \dfrac{1}{2}[\sin(\alpha + \beta) + \sin(\alpha - \beta)]$

2. $\cos \alpha \sin \beta = \dfrac{1}{2}[\sin(\alpha + \beta) - \sin(\alpha - \beta)]$

3. $\sin \alpha \sin \beta = \dfrac{1}{2}[\cos(\alpha - \beta) - \cos(\alpha + \beta)]$

4. $\cos \alpha \cos \beta = \dfrac{1}{2}[\cos(\alpha + \beta) + \cos(\alpha - \beta)]$

EXAMPLE 1   Using Product-to-Sum Formulas

Calculate the exact value of each expression using Product-to-Sum Formulas.
a. $\sin 67.5° \cos 22.5°$ **b.** $\sin 105° \sin 15°$

SOLUTION   a. Letting $\alpha = 67.5°$ and $\beta = 22.5°$, we use Formula (1).

$$\sin \alpha \cos \beta = \frac{1}{2}[\sin(\alpha + \beta) + \sin(\alpha - \beta)]$$

$$\sin \mathbf{67.5°} \cos \mathbf{22.5°} = \frac{1}{2}[\sin(\mathbf{67.5°} + \mathbf{22.5°}) + \sin(\mathbf{67.5°} - \mathbf{22.5°})]$$

$$= \frac{1}{2}(\sin 90° + \sin 45°)$$

$$= \frac{1}{2}\left(1 + \frac{\sqrt{2}}{2}\right)$$

$$= \frac{1}{2} + \frac{\sqrt{2}}{4}$$

b. Letting $\alpha = 105°$ and $\beta = 15°$, we use Formula (3).

$$\sin \alpha \sin \beta = \frac{1}{2}[\cos(\alpha - \beta) - \cos(\alpha + \beta)]$$

$$\sin 105° \sin 15° = \frac{1}{2}[\cos(105° - 15°) - \cos(105° + 15°)]$$

$$= \frac{1}{2}(\cos 90° - \cos 120°)$$

$$= \frac{1}{2}\left[0 - \left(-\frac{1}{2}\right)\right]$$

$$= \frac{1}{4}$$

Self Check 1   Calculate the exact value of $\cos 75° \sin 15°$.

Now Try Exercise 3.

EXAMPLE 2   **Using Product-to-Sum Formulas**

Write each expression as a sum or difference using Product-to-Sum Formulas.
a. $\cos 5\theta \sin 2\theta$ **b.** $\cos 3\theta \cos 4\theta$

SOLUTION   **a.** We will use Formula (2) by letting $\alpha = 5\theta$ and $\beta = 2\theta$.

$$\cos \alpha \sin \beta = \frac{1}{2}[\sin(\alpha + \beta) - \sin(\alpha - \beta)]$$

$$\cos 5\theta \sin 2\theta = \frac{1}{2}[\sin(5\theta + 2\theta) - \sin(5\theta - 2\theta)]$$

$$= \frac{1}{2}(\sin 7\theta - \sin 3\theta)$$

$$= \frac{1}{2}\sin 7\theta - \frac{1}{2}\sin 3\theta$$

b. We will use Formula (4) by letting $\alpha = 3\theta$ and $\beta = 4\theta$.

$$\cos \alpha \cos \beta = \frac{1}{2}[\cos(\alpha + \beta) + \cos(\alpha - \beta)]$$

$$\cos 3\theta \cos 4\theta = \frac{1}{2}[\cos(3\theta + 4\theta) + \cos(3\theta - 4\theta)]$$

$$= \frac{1}{2}(\cos 7\theta + \cos(-\theta))$$

$$= \frac{1}{2}\cos 7\theta + \frac{1}{2}\cos \theta \qquad \cos(-\theta) = \cos \theta \text{ (Even–Odd Property)}$$

■

Self Check 2   Write $\sin 6\theta \cos 3\theta$ as a sum or difference.

Now Try Exercise 25.

2. Express Trigonometric Sums as Products

We now develop a formula to convert the sum of two sines or the sum of two cosines into a product. To accomplish this, we start with the following equations:

$$\alpha = x + y \qquad\qquad \alpha = x + y$$
$$+ \;\underline{\beta = x - y} \qquad\qquad - \;\underline{\beta = x - y}$$
$$\alpha + \beta = 2x \qquad\qquad \alpha - \beta = 2y$$
$$x = \frac{1}{2}(\alpha + \beta) \qquad\qquad y = \frac{1}{2}(\alpha - \beta)$$

From these equations, we will develop the four Sum-to-Product Formulas.

$$\sin \alpha + \sin \beta = \sin(x + y) + \sin(x - y)$$
$$= \sin x \cos y + \cos x \sin y + \sin x \cos y - \cos x \sin y$$
$$= 2 \sin x \cos y$$
$$= 2 \sin \frac{\alpha + \beta}{2} \cos \frac{\alpha - \beta}{2}$$

$$\sin \alpha - \sin \beta = \sin(x + y) - \sin(x - y)$$
$$= \sin x \cos y + \cos x \sin y - \sin x \cos y + \cos x \sin y$$
$$= 2 \cos x \sin y$$
$$= 2 \cos \frac{\alpha + \beta}{2} \sin \frac{\alpha - \beta}{2}$$

$$\cos \alpha + \cos \beta = \cos(x + y) + \cos(x - y)$$
$$= \cos x \cos y - \sin x \sin y + \cos x \cos y + \sin x \sin y$$
$$= 2 \cos x \cos y$$
$$= 2 \cos \frac{\alpha + \beta}{2} \cos \frac{\alpha - \beta}{2}$$

$$\cos \alpha - \cos \beta = \cos(x + y) - \cos(x - y)$$
$$= \cos x \cos y - \sin x \sin y - \cos x \cos y - \sin x \sin y$$
$$= -2 \sin x \sin y$$
$$= -2 \sin \frac{\alpha + \beta}{2} \sin \frac{\alpha - \beta}{2}$$

Sum-to-Product Formulas

5. $\sin \alpha + \sin \beta = 2 \sin \dfrac{\alpha + \beta}{2} \cos \dfrac{\alpha - \beta}{2}$

6. $\sin \alpha - \sin \beta = 2 \cos \dfrac{\alpha + \beta}{2} \sin \dfrac{\alpha - \beta}{2}$

7. $\cos \alpha + \cos \beta = 2 \cos \dfrac{\alpha + \beta}{2} \cos \dfrac{\alpha - \beta}{2}$

8. $\cos \alpha - \cos \beta = -2 \sin \dfrac{\alpha + \beta}{2} \sin \dfrac{\alpha - \beta}{2}$

EXAMPLE 3   Using Sum-to-Product Formulas

Find the exact value of each expression using Sum-to-Product Formulas.
a. $\sin 75° + \sin 15°$ **b.** $\cos 105° - \cos 15°$

SOLUTION   **a.** For this expression we will use Formula (5). Let $\alpha = 75°$ and $\beta = 15°$.

$$\sin \alpha + \sin \beta = 2 \sin \frac{\alpha + \beta}{2} \cos \frac{\alpha - \beta}{2}$$
$$\sin 75° + \sin 15° = 2 \sin \frac{75° + 15°}{2} \cos \frac{75° - 15°}{2}$$
$$= 2 \sin \frac{90°}{2} \cos \frac{60°}{2}$$
$$= 2 \sin 45° \cos 30°$$
$$= 2 \cdot \frac{\sqrt{2}}{2} \cdot \frac{\sqrt{3}}{2}$$
$$= \frac{\sqrt{6}}{2}$$

b. To evaluate this expression, we use Formula (8) and let $\alpha = 105°$ and $\beta = 15°$.

$$\cos \alpha - \cos \beta = -2 \sin \frac{\alpha + \beta}{2} \sin \frac{\alpha - \beta}{2}$$

$$\cos 105° - \cos 15° = -2 \sin \frac{105° + 15°}{2} \sin \frac{105° - 15°}{2}$$

$$= -2 \sin \frac{120°}{2} \sin \frac{90°}{2}$$

$$= -2 \sin 60° \sin 45°$$

$$= -2 \cdot \frac{\sqrt{3}}{2} \cdot \frac{\sqrt{2}}{2}$$

$$= -\frac{\sqrt{6}}{2}$$

Self Check 3   Evaluate $\cos 285° + \cos 15°$.

Now Try Exercise 13.

EXAMPLE 4   **Using Sum-to-Product Formulas**

Write $\cos 2\theta + \cos 6\theta$ as a product of two functions.

SOLUTION   To perform this operation, we will use Formula (7). We let $\alpha = 2\theta$ and $\beta = 6\theta$.

$$\cos \alpha + \cos \beta = 2 \cos \frac{\alpha + \beta}{2} \cos \frac{\alpha - \beta}{2}$$

$$\cos 2\theta + \cos 6\theta = 2 \cos \frac{2\theta + 6\theta}{2} \cos \frac{2\theta - 6\theta}{2}$$

$$= 2 \cos 4\theta \cos(-2\theta) \qquad \cos(-\theta) = \cos \theta$$

$$= 2 \cos 4\theta \cos 2\theta$$

Self Check 4   Write $\sin 4\theta - \sin 8\theta$ as a product of two functions.

Now Try Exercise 35.

EXAMPLE 5   **Verifying an Identity**

Verify that $\tan 3\theta = \dfrac{\sin 4\theta + \sin 2\theta}{\cos 4\theta + \cos 2\theta}$.

SOLUTION   We will start with the right-hand side because it is the more complicated side.

$$\frac{\sin 4\theta + \sin 2\theta}{\cos 4\theta + \cos 2\theta} = \frac{2 \sin \left(\dfrac{4\theta + 2\theta}{2} \right) \cos \left(\dfrac{4\theta - 2\theta}{2} \right)}{2 \cos \left(\dfrac{4\theta + 2\theta}{2} \right) \cos \left(\dfrac{4\theta - 2\theta}{2} \right)}$$

$$= \frac{2 \sin 3\theta \cancel{\cos \theta}}{2 \cos 3\theta \cancel{\cos \theta}}$$

$$= \tan 3\theta$$

Self Check 5   Verify that $\tan x = \dfrac{\sin 3x - \sin x}{\cos 3x + \cos x}$.

Now Try Exercise 43.

3. Write Expressions of the Form $A \sin \theta + B \sin \theta$ in the Form $k \sin(\theta + \varphi)$

The graph of $y = A \sin \theta + B \cos \theta$ is a sine curve of the form $y = k \sin(\theta + \varphi)$. The amplitude k and the phase shift angle φ are determined by the values of A and B. The following discussion will establish this result and provide the proper values for k and φ.

From the terms of the expression $A \sin \theta + B \cos \theta$, we factor out a common factor of $\sqrt{A^2 + B^2}$. This is not difficult if we realize that the product

$$\sqrt{A^2 + B^2} \cdot \frac{A}{\sqrt{A^2 + B^2}}$$

equals A, and that the product

$$\sqrt{A^2 + B^2} \cdot \frac{B}{\sqrt{A^2 + B^2}}$$

equals B. Thus, we have

(1) $A \sin \theta + B \cos \theta = \sqrt{A^2 + B^2}\left(\dfrac{A}{\sqrt{A^2 + B^2}} \sin \theta + \dfrac{B}{\sqrt{A^2 + B^2}} \cos \theta \right).$

Because the sum of the squares of the coefficients $\dfrac{A}{\sqrt{A^2 + B^2}}$ and $\dfrac{B}{\sqrt{A^2 + B^2}}$ is 1, we can let one of these coefficients be $\sin \varphi$ and the other $\cos \varphi$, for some angle φ. If φ is an angle such that

$$\sin \varphi = \frac{B}{\sqrt{A^2 + B^2}} \text{ and } \cos \varphi = \frac{A}{\sqrt{A^2 + B^2}},$$

then Equation (1) becomes

$$A \sin \theta + B \cos \theta = \sqrt{A^2 + B^2}(\cos \varphi \sin \theta + \sin \varphi \cos \theta)$$

$$= \sqrt{A^2 + B^2}[\sin (\theta + \varphi)]$$

In summary, we have the following theorem:

Theorem $A \sin \theta + B \cos \theta = k \sin(\theta + \varphi)$, where $k = \sqrt{A^2 + B^2}$ and φ is any angle for which $\sin \varphi = \dfrac{B}{\sqrt{A^2 + B^2}} = \dfrac{B}{k}$ and $\cos \varphi = \dfrac{A}{\sqrt{A^2 + B^2}} = \dfrac{A}{k}.$

EXAMPLE 6   **Writing an Expression of the Form $A \sin \theta + B \cos \theta$ in the Form $k \sin(\theta + \varphi)$**

Write the expression $3 \sin \theta + 4 \cos \theta$ in the form $k \sin(\theta + \varphi)$.

SOLUTION   We begin by evaluating k.

$$k = \sqrt{A^2 + B^2} = \sqrt{3^2 + 4^2} = \sqrt{25} = 5$$

Here φ is an angle such that $\sin \varphi = \frac{4}{5} = 0.8$ and $\cos \varphi = \frac{3}{5} = 0.6$. We can use a calculator to determine that $\varphi \approx 53.1°$. Thus, we have

$$3 \sin \theta + 4 \cos \theta \approx 5 \sin(\theta + 53.1°).$$

Self Check 6   Write the expression $5 \sin \theta + 2 \cos \theta$ in the form $k \sin(\theta + \varphi)$.

Now Try Exercise 53.

4. Solve Application Problems

If a weight, hanging by a spring from the ceiling, is pushed upward a small distance and released, it starts to bounce. Its position y at any time t can be found by the formula

$$y = A \cos 2\pi ft.$$

If, as the weight is released, it is given a push upward, it will still continue to bounce. However, its position depends on time in a more complicated way:

$$y = A \sin 2\pi ft + B \cos 2\pi ft,$$

where A, B, and f are dependent on the characteristics of the spring, the upward displacement, and the initial push.

EXAMPLE 7   **Using Harmonic Motion**

A weight is attached to a certain spring and is pushed to start it bouncing. Suppose the position y depends on time t according to the formula $y = 3 \sin 6t + 4 \cos 6t$. What is the amplitude of the oscillation?

SOLUTION   By the theorem just proven, it follows that

$$A \sin \theta + B \cos \theta = \sqrt{A^2 + B^2} \sin(\theta + \varphi)$$

$$y = 3 \sin 6t + 4 \cos 6t = \sqrt{3^2 + 4^2} \sin(6t + \varphi)$$

$$y = 5 \sin(6t + \varphi)$$

The oscillation is described by the sine function. Because the amplitude is 5, the weight is at most 5 units above (or below) the equilibrium position.   ∎

Self Check 7   Work Example 7 using the formula $y = 4 \sin 4t + 5 \cos 4t$.

Now Try Exercise 67.

Self Check Answers   **1.** $\dfrac{2 - \sqrt{3}}{4}$ **2.** $\dfrac{1}{2} \sin 9\theta + \dfrac{1}{2} \sin 3\theta$ **3.** $\dfrac{\sqrt{6}}{2}$ **4.** $-2 \cos 6\theta \sin 2\theta$

5. $\dfrac{\sin 3x - \sin x}{\cos 3x + \cos x} = \dfrac{2 \cos\left(\dfrac{3x + x}{2}\right) \sin\left(\dfrac{3x - x}{2}\right)}{2 \cos\left(\dfrac{3x + x}{2}\right) \cos\left(\dfrac{3x - x}{2}\right)} = \dfrac{2 \cos 2x \sin x}{2 \cos 2x \cos x} = \tan x$

6. $\sqrt{29} \sin(\theta + 21.8°)$ **7.** $\sqrt{41} \approx 6.4$ units above or below the equilibrium position

Exercises **5.6**

Getting Ready

You should be able to complete these vocabulary and concept statements before you proceed to the practice exercises.

Fill in the blanks.

1. The Product-to-Sum Formulas are _____.

2. The Sum-to-Product Formulas are _____.

Practice

Express each product as a sum or a difference and find its value. Do not use a calculator.

3. $\cos 75° \cos 15°$

4. $\sin 15° \cos 75°$

5. $\sin 165° \sin 105°$

6. $\sin 165° \sin 75°$

7. $\cos 22.5° \cos 67.5°$

8. $\cos 105° \sin 15°$

9. $\cos \dfrac{19\pi}{12} \sin \dfrac{5\pi}{12}$

10. $\sin \dfrac{5\pi}{12} \cos \dfrac{13\pi}{12}$

11. $\cos \dfrac{7\pi}{12} \cos \dfrac{5\pi}{12}$

12. $\cos \dfrac{7\pi}{12} \sin \dfrac{13\pi}{12}$

Express each sum as a product and find its value. Do not use a calculator.

13. $\cos 75° + \cos 15°$

14. $\sin 75° + \sin 15°$

15. $\sin 165° - \sin 105°$

16. $\sin 15° - \sin 75°$

17. $\cos 165° - \cos 105°$

18. $\cos 105° - \cos 15°$

19. $\sin \dfrac{\pi}{12} + \sin \dfrac{5\pi}{12}$

20. $\sin \dfrac{5\pi}{12} - \sin \dfrac{13\pi}{12}$

21. $\cos \dfrac{7\pi}{12} + \cos \dfrac{5\pi}{12}$

22. $\sin \dfrac{7\pi}{12} + \sin \dfrac{13\pi}{12}$

Express each product as a sum or a difference.

23. $\sin 40° \sin 30°$

24. $\sin 75° \cos 70°$

25. $\cos \dfrac{5\pi}{7} \cos \dfrac{\pi}{7}$

26. $\sin \dfrac{4\pi}{5} \cos \dfrac{3\pi}{5}$

27. $\cos 5\theta \cos 3\theta$

28. $\sin 3\theta \sin 2\theta$

29. $\sin \alpha \cos \dfrac{\alpha}{2}$

30. $\cos 3\beta \sin \beta$

31. $\sin \dfrac{x}{2} \sin \dfrac{x}{3}$

32. $\cos \dfrac{x}{3} \cos x$

Express each sum as a product.

33. $\sin 50° + \sin 30°$

34. $\sin 75° - \sin 70°$

35. $\cos \dfrac{7\pi}{9} - \cos \dfrac{5\pi}{9}$

36. $\cos \dfrac{3\pi}{5} + \cos \dfrac{\pi}{5}$

37. $\cos 5\theta + \cos 3\theta$

38. $\cos 7\theta - \cos \theta$

39. $\sin \alpha - \sin \dfrac{\alpha}{2}$

40. $\sin 5\beta + \sin \beta$

41. $\cos \dfrac{x}{2} + \cos \dfrac{x}{3}$

42. $\cos x - \cos \dfrac{x}{2}$

Verify each identity.

43. $\dfrac{\sin \alpha + \sin \beta}{\sin \alpha - \sin \beta} = \tan \dfrac{1}{2}(\alpha + \beta) \cot \dfrac{1}{2}(\alpha - \beta)$

44. $\dfrac{\sin \alpha + \sin \beta}{\cos \alpha + \cos \beta} = \tan \dfrac{1}{2}(\alpha + \beta)$

45. $\dfrac{\sin \alpha + \sin \beta}{\cos \alpha - \cos \beta} = -\cot \dfrac{1}{2}(\alpha - \beta)$

46. $\dfrac{\cos \alpha + \cos \beta}{\sin \alpha - \sin \beta} = \cot \dfrac{1}{2}(\alpha - \beta)$

47. $\dfrac{\cos \alpha + \cos 5\alpha}{\cos \alpha - \cos 5\alpha} = \cot 3\alpha \cot 2\alpha$

48. $\sin^2 \alpha - \sin^2 \beta = \sin(\alpha + \beta) \sin(\alpha - \beta)$

49. $\cos^2 \alpha - \cos^2 \beta = \sin(\beta + \alpha) \sin(\beta - \alpha)$

50. $\cos 2\alpha(1 + 2 \cos \alpha) = \cos \alpha + \cos 2\alpha + \cos 3\alpha$

51. $2 \cos 5\alpha \sin 2\alpha = \sin 7\alpha - \sin 3\alpha$

52. $\cot 7\alpha \cot 5\alpha = \dfrac{\cos 12\alpha + \cos 2\alpha}{\cos 2\alpha - \cos 12\alpha}$

Write each $A \sin \theta + B \cos \theta$ expression in the form $k \sin(x + \varphi)$.

53. $6 \sin x + 8 \cos x$

54. $12 \sin x + 5 \cos x$

55. $6 \sin x - 8 \cos x$

56. $12 \sin x - 5 \cos x$

57. $2 \sin x + \cos x$

58. $\sin x - \cos x$

59. $\sin x + \cos x$

60. $2 \sin x - \cos x$

61. $-\sin x + 5 \cos x$

62. $-\sin x - 5 \cos x$

63. $\sqrt{3} \sin x - 3 \cos x$

64. $\sqrt{3} \sin x + 3 \cos x$

Applications

65. Sound A piano tuner counts 50 pulsations of sound in 20 seconds. How far apart are the frequencies of the two piano strings that the tuner has struck?

66. Sound Two tuning forks, cut for frequencies of 2,000 and 2,003 hertz (cycles per second), respectively, are struck and simultaneously touched to a sounding board. How many pulsations of the resulting sound will be heard each second?

© istockphoto.com/CWLawrence

67. Harmonic motion A weight is attached to a spring, pulled down, and released to start it oscillating. Suppose the position y of the weight depends on time t according to the formula $y = 5 \sin 3t + 12 \cos 3t$. What is the maximum number of units the weight will be from its starting position?

68. Harmonic motion A ball is attached to a spring and pushed to start it bouncing. Suppose the position y of the ball depends on time t according to the formula $y = \sqrt{2} \sin 5t + \sqrt{3} \cos 5t$. What is the maximum distance the ball will get from the equilibrium position?

69. Calculus A process known as integration is learned in calculus and is useful for solving many applications. Many trigonometric expressions must be rewritten first before integration can be performed. This is especially true of the products of the trigonometric functions. Use a Product-to-Sum Formula to rewrite $\cos 6\theta \sin 5\theta$ to prepare it for integration in calculus.

70. Calculus Refer to Exercise 69. Use a Product-to-Sum Formula to rewrite $\sin 2\theta \sin 4\theta$ to prepare it for integration in calculus.

Discovery and Writing

Assume that $\alpha + \beta + \delta = 180°$. Verify each identity.

71. $\sin \alpha + \sin \beta + \sin \delta = 4 \cos \dfrac{\alpha}{2} \cos \dfrac{\beta}{2} \cos \dfrac{\delta}{2}$

72. $\cos \alpha + \cos \beta - \cos \delta = 4 \cos \dfrac{\alpha}{2} \cos \dfrac{\beta}{2} \sin \dfrac{\delta}{2} - 1$

73. $\tan \alpha + \tan \beta + \tan \delta = \tan \alpha \tan \beta \tan \delta$

74. $\cos \alpha + \cos \beta + \cos \delta = 1 + 4 \sin \dfrac{\alpha}{2} \sin \dfrac{\beta}{2} \sin \dfrac{\delta}{2}$

5.7 Trigonometric Equations I

In this section, we will learn to

1. Solve equations that have a trigonometric function of a single angle.
2. Solve trigonometric equations involving multiple angles.
3. Use factoring to solve trigonometric equations.
4. Use identities to solve trigonometric equations.

As we've seen, trigonometric functions can be used to model many real-life problems. Suppose the population P of lemurs in a certain region of Madagascar can be modeled by the equation

$$P = 200 \sin \frac{\pi}{12}x + 400,$$

where x is the number of years since 2010.

If we are interested in determining the first year the population of lemurs is expected to reach 600, we would substitute 600 in for P and solve the trigonometric equation for x. By inspection, we can easily see that the solution is $x = 6$.

$$P = 200 \sin \frac{\pi}{12}(6) + 400$$

$$= 200 \sin \frac{\pi}{2} + 400$$

$$= 200(1) + 400$$

$$= 600$$

The year the population will first reach 600 is 2016.

In this section we will study several strategies to solve trigonometric equations. These are equations that involve trigonometric functions.

1. Solve Equations That Have a Trigonometric Function of a Single Angle

Many of the techniques used to solve algebraic equations can be used to solve trigonometric equations. Trigonometric equations require consideration of the interval over which the equation is to be solved. For example, when we solve the equation $x^2 = 25$, we know the only two solutions are $x = 5$ and $x = -5$. If you examine the trigonometric equation $\sin \theta = \frac{1}{2}$, you might immediately give the solution $\theta = \frac{\pi}{6}$. While this is correct, if you think about this simple equation, there are infinitely many solutions. Look at Figure 5-26.

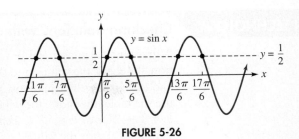

FIGURE 5-26

If we wish to describe the complete solution set to the equation $\sin \theta = \frac{1}{2}$, we use the Periodic Properties of the sine function. Over the interval $[0, 2\pi)$, there are two solutions to the equation, $\theta = \frac{\pi}{6}$ and $\theta = \frac{5\pi}{6}$. Since the period of the sine function is 2π, the function will repeat every 2π units. Therefore, to describe the complete solution set, we use the following notation.

$$\sin \theta = \frac{1}{2}$$

$$\theta = \frac{\pi}{6} + 2k\pi$$

$$\text{or} \quad \theta = \frac{5\pi}{6} + 2k\pi, \quad \text{where } k \text{ is any integer}$$

These are called the **general solutions** of the equation.

When solving equations, you will make extensive use of what you have learned about the special angles and reference angles. Typically, you will determine the special angle that is your reference angle. Then you will be able to determine the solution(s) to the equation.

EXAMPLE 1   Finding General Solutions of a Trigonometric Equation

Find the general solutions of the trigonometric equation $\cos \theta = \frac{1}{2}$.

SOLUTION   Over the interval $[0, 2\pi)$, $\cos \theta > 0$ in QI and QIV. Therefore, over this interval, there are two solutions to the equation $\cos \theta = \frac{1}{2}$. The reference angle is $\theta' = \cos^{-1} \frac{1}{2}$ or $\theta' = \frac{\pi}{3}$. The two angles that solve this equation are $\theta_1 = \frac{\pi}{3}$ and $\theta_2 = \frac{5\pi}{3}$ (see Figure 5-27).

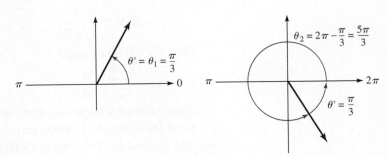

FIGURE 5-27

The period of the cosine function is 2π. Using the Periodic Property of the cosine function, we can write the general solutions to this equation as

$$\theta = \frac{\pi}{3} + 2k\pi \text{ or } \theta = \frac{5\pi}{3} + 2k\pi, \text{ where } k \text{ is any integer.}$$

Self Check 1   Find the general solutions to the trigonometric equation $\cos \theta = -\frac{1}{2}$.

Now Try Exercise 5.

ACCENT ON TECHNOLOGY   Checking Solutions with a Calculator

A graphing calculator can be used to check our solutions to a trigonometric equation. To check the solution to the equation in Example 1 using the graphing editor and table, enter the equations $Y_1 = \cos x$ and $Y_2 = \frac{1}{2}$. The easiest method is to use the table; therefore, the window setting is not important. Enter some values of $\theta = \frac{\pi}{3} + 2k\pi$ or $\theta = \frac{5\pi}{3} + 2k\pi$ for various values of k. The table should indicate that $Y_1 = Y_2$ for the values.

Enter the functions.

Enter $\frac{\pi}{3}$ and $\frac{5\pi}{3}$ into the table. Then enter $\frac{\pi}{3} + 2\pi$ [×] 2

The result from entering $\frac{\pi}{3} + 2\pi$ [×] 2 is shown in this frame.

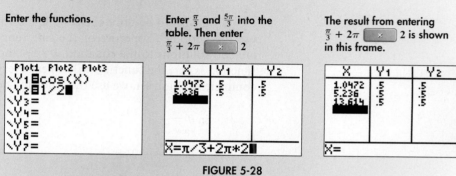

FIGURE 5-28

Once all the general solutions of a trigonometric equation are found, you can select a solution set over a particular interval. Generally, we will find the solution set over the interval $[0, 2\pi)$. To solve some trigonometric equations, we can use the techniques used to solve linear equations.

EXAMPLE 2   Solving a Trigonometric Equation

Solve the equation $\sqrt{2} \sin \theta + 1 = 0$ over the interval $[0, 2\pi)$.

SOLUTION   To solve this equation, we will (1) find the value of $\sin \theta$, (2) find the reference angle, and (3) use the reference angle to find the solution(s) to the equation over the given interval.

Simplify the equation to find the value of $\sin \theta$.

$$\sqrt{2} \sin \theta + 1 = 0$$
$$\sqrt{2} \sin \theta = -1$$
$$\sin \theta = -\frac{1}{\sqrt{2}} = -\frac{\sqrt{2}}{2}$$

Find the reference angle

$$\theta' = \sin^{-1} \frac{\sqrt{2}}{2} = \frac{\pi}{4}.$$

Since a reference angle is an acute angle, we evaluate $\sin^{-1} \frac{\sqrt{2}}{2}$ and not $\sin^{-1}\left(-\frac{\sqrt{2}}{2}\right)$.

Find the solution(s). Since $\sin \theta < 0$, there are two solutions to the equation over the interval $[0, 2\pi)$, one in QIII, and one in QIV. Figure 5-29 illustrates how we use the reference angle to find the solutions.

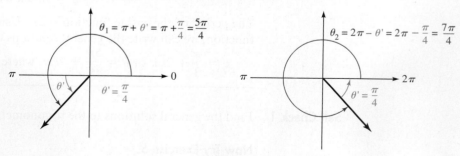

FIGURE 5-29

$$\left.\begin{aligned}\theta = \pi + \frac{\pi}{4} = \frac{5\pi}{4}\\[4pt]\theta = 2\pi - \frac{\pi}{4} = \frac{7\pi}{4}\end{aligned}\right\}\quad \theta = \frac{5\pi}{4} \text{ or } \frac{7\pi}{4}$$

Self Check 2   Solve the equation $\sqrt{2}\sin\theta - 1 = 0$ over the interval $[0, 2\pi)$.

Now Try Exercise 17.

EXAMPLE 3   **Solving a Trigonometric Equation**

Solve the equation $\cos\left(\theta + \frac{\pi}{2}\right) = 1$ over the interval $[0, 2\pi)$. Check your answer graphically.

SOLUTION   Using the strategy utilized in Example 2, our first step is to find the reference angle. However, we note that $\theta' = \cos^{-1} 1 = 0$, which is a quadrantal angle. While this is not truly a reference angle (because a reference angle is an acute angle), we can still use this information. Over the interval $[0, 2\pi)$, the cosine function equals 1 at the endpoints, 0 and 2π. Therefore, we have the following:

$$\cos\left(\theta + \frac{\pi}{2}\right) = 1$$

$$\theta + \frac{\pi}{2} = 0 \quad \text{or} \quad \theta + \frac{\pi}{2} = 2\pi$$

$$\theta = -\frac{\pi}{2} \qquad\qquad \theta = \frac{3\pi}{2}$$

However, the only solution in the designated interval is $\theta = \frac{3\pi}{2}$.

To check this solution graphically, we graph the function $y = \cos\left(\theta + \frac{\pi}{2}\right)$ as shown in Figure 5-30. Note that this function is the graph of the cosine function with a horizontal shift to the left $\frac{\pi}{2}$ units. We see that on the interval $[0, 2\pi)$, $y = 1$ at $\theta = \frac{3\pi}{2}$.

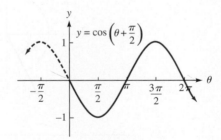

FIGURE 5-30

Self Check 3   Solve the equation $\sin\left(\theta - \frac{\pi}{2}\right) = 1$ over the interval $[0, 2\pi)$.

Now Try Exercise 23.

EXAMPLE 4   **Solving a Trigonometric Equation**

Solve the equation $\tan^2\theta - 3 = 0$ over the interval $[0, 2\pi)$.

SOLUTION   If we were to solve the algebraic equation $x^2 - 3 = 0$, we would isolate the variable and take the square root of each side, resulting in the solutions $x = \pm\sqrt{3}$. We will use the same strategy to solve the given equation.

$$\tan^2 \theta - 3 = 0$$

$$\tan^2 \theta = 3$$

$$\tan \theta = \pm\sqrt{3}$$

Since the function can be either positive or negative, there are solutions in all four quadrants. The reference angle is $\theta' = \tan^{-1}\sqrt{3}$ or $\theta' = \frac{\pi}{3}$.

We can now find the solutions to the equation $\tan^2 \theta - 3 = 0$. Figure 5-31 illustrates how we use the reference angle to find the following solutions.

$$\left.\begin{array}{l} \theta_1 = \dfrac{\pi}{3} \\[2mm] \theta_2 = \pi - \dfrac{\pi}{3} = \dfrac{2\pi}{3} \\[2mm] \theta_3 = \pi + \dfrac{\pi}{3} = \dfrac{4\pi}{3} \\[2mm] \theta_4 = 2\pi - \dfrac{\pi}{3} = \dfrac{5\pi}{3} \end{array}\right\} \theta = \dfrac{\pi}{3}, \dfrac{2\pi}{3}, \dfrac{4\pi}{3}, \text{ and } \dfrac{5\pi}{3}$$

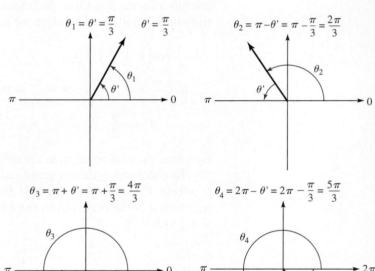

FIGURE 5-31

Self Check 4   Solve the equation $\tan^2 \theta - 1 = 0$ over the interval $[0, 2\pi)$.

Now Try Exercise 29.

2. Solve Trigonometric Equations Involving Multiple Angles

A trigonometric equation may involve a multiple of an angle, such as 2θ or $\frac{\theta}{2}$. How might this affect the solution(s) to equations such as $\sin 2\theta = \frac{1}{2}$ or $\tan \frac{\theta}{2} = -1$? Recall that the period of a trigonometric function in the form $y = A\sin(Bx - C)$ is $\frac{2\pi}{B}$. Therefore, when we are solving an equation involving the multiple of an angle, the change in the period can affect the number of solutions over a particular interval. In Figure 5-32, the graphs of $y = \sin 2x$, $y = \sin \frac{1}{2}x$, and $y = \sin x$ are drawn over the interval $[0, 2\pi]$. The horizontal line $y = \frac{1}{2}$ is also drawn. Notice that as the period is changed, the line intersects the graphs at a different number of points. In the next two examples, we will examine a strategy for solving trigonometric equations with multiple angles.

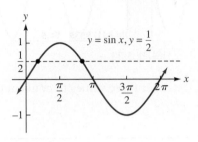
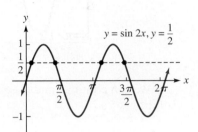
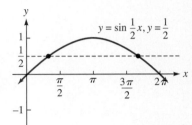

FIGURE 5-32

EXAMPLE 5   **Solving a Trigonometric Equation with Multiple Angles**

Solve the equation $\sin 2\theta = \frac{1}{2}$ over the interval $[0, 2\pi)$.

SOLUTION   The period of $y = \sin 2\theta$ is π, which means there are two complete periods over the interval $[0, 2\pi]$. The sine function equals $\frac{1}{2}$ at $\frac{\pi}{6}$ and $\frac{5\pi}{6}$ over the interval $[0, 2\pi)$. We will find the general solutions to the equation and then find the solutions that fall in the interval.

$$\sin 2\theta = \frac{1}{2}$$

$$2\theta = \frac{\pi}{6} + 2k\pi \qquad \text{or} \qquad 2\theta = \frac{5\pi}{6} + 2k\pi$$

$$\theta = \frac{\pi}{12} + k\pi \qquad\qquad \theta = \frac{5\pi}{12} + k\pi \qquad \text{Divide both sides by 2.}$$

$$\theta = \frac{\pi}{12} + (0)\pi = \frac{\boldsymbol{\pi}}{\boldsymbol{12}} \qquad \theta = \frac{5\pi}{12} + (0)\pi = \frac{\boldsymbol{5\pi}}{\boldsymbol{12}} \qquad \text{Subsitute 0 for } k.$$

$$\theta = \frac{\pi}{12} + (1)\pi = \frac{\boldsymbol{13\pi}}{\boldsymbol{12}} \qquad \theta = \frac{5\pi}{12} + (1)\pi = \frac{\boldsymbol{17\pi}}{\boldsymbol{12}} \qquad k = 1$$

We do not substitute $k = 2$ because that would create an angle out of the interval $[0, 2\pi)$. Therefore the solutions to the equation are

$$\theta = \frac{\pi}{12}, \frac{5\pi}{12}, \frac{13\pi}{12}, \text{ and } \frac{17\pi}{12}.$$

Caution

$\sin 2\theta = \dfrac{1}{2}$	Do NOT divide by 2 at this point.
$\sin \theta = \dfrac{1}{4}$	These are NOT equivalent equations.

ACCENT ON TECHNOLOGY   **Checking Solutions with a Graphing Calculator**

A graphing calculator can be used to check our solutions. First, we graph the equations $y = \sin 2\theta$ and $y = \frac{1}{2}$ using the window $X: [0, 2\pi]$, $Y: [-1, 1]$. Once we have graphed the equations, we can use the VALUE function (2nd , TRACE , 1) to verify that the solutions are correct, or we can use the TABLE to check our solutions.

Use **CALC: VALUE** to enter $\frac{\pi}{12}$.   Press [ENTER] to see that the value is a solution.

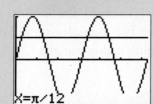
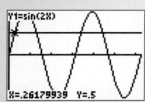

FIGURE 5-33

We can continue and check the other three values to confirm that they are the solutions.

Self Check 5   Solve the equation $\cos 2\theta = \frac{1}{2}$ over the interval $[0, 2\pi)$.

Now Try Exercise 31.

EXAMPLE 6   Solving a Trigonometric Equation with Multiple Angles

Solve the equation $\tan \frac{\theta}{2} = -1$ over the interval $[0, 2\pi)$.

SOLUTION   The period of $y = \tan \theta$ is π; the period of $y = \tan \frac{\theta}{2}$ is 2π. Over the interval $[0, 2\pi)$, the tangent function equals -1 at $\frac{3\pi}{4}$ and $\frac{7\pi}{4}$. We will find the general solution to the equation, and then we will find the solutions over the interval $[0, 2\pi)$.

$$\tan \frac{\theta}{2} = -1$$

$$\frac{\theta}{2} = \frac{3\pi}{4} + k\pi \qquad \text{or} \qquad \frac{\theta}{2} = \frac{7\pi}{4} + k\pi \qquad \text{General solution to the equation.}$$

$$\theta = \frac{6\pi}{4} + 2k\pi \qquad\qquad \theta = \frac{14\pi}{4} + 2k\pi \qquad \text{Multiply by 2.}$$

$$\theta = \frac{3\pi}{2} + 2(0)\pi = \frac{3\pi}{2} \qquad \theta = \frac{7\pi}{2} + 2(0)\pi = \frac{7\pi}{2} > 2\pi \qquad \text{Set } k = 0.$$

Therefore, the only solution to the equation is $\theta = \frac{3\pi}{2}$. A graph used to check the answer is shown in Figure 5-34.

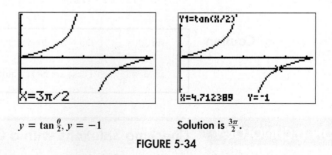

$y = \tan \frac{\theta}{2}, y = -1$ Solution is $\frac{3\pi}{2}$.

FIGURE 5-34

Self Check 6   Solve the equation $\cos \frac{\theta}{2} = -\frac{1}{2}$ over the interval $[0, 2\pi)$.

Now Try Exercise 35.

3. Use Factoring to Solve Trigonometric Equations

In attempting to solve a nonlinear polynomial equation, factoring is often the first technique to be tried. We can solve some trigonometric equations by factoring.

EXAMPLE 7   **Solving a Trigonometric Equation by Factoring**

Solve the equation $2\cos^3\theta = \cos\theta$ over the interval $[0, 2\pi)$.

SOLUTION   We will solve this equation by factoring.

$$2\cos^3\theta = \cos\theta$$

$$2\cos^3\theta - \cos\theta = 0 \qquad \text{Subtract } \cos\theta \text{ from both sides.}$$

$$\cos\theta(2\cos^2\theta - 1) = 0 \qquad \text{Factor out } \cos\theta.$$

$$\cos\theta = 0 \qquad \text{or} \quad 2\cos^2\theta - 1 = 0 \qquad \text{Set each factor to 0 and solve for } \theta.$$

Only values over $[0, 2\pi)$ $\quad \theta = \dfrac{\pi}{2}, \dfrac{3\pi}{2}$

$$\cos^2\theta = \frac{1}{2}$$

$$\cos\theta = \pm\frac{1}{\sqrt{2}} = \pm\frac{\sqrt{2}}{2} \qquad \text{Take the square root of each side.}$$

$$\theta = \frac{\pi}{4}, \frac{3\pi}{4}, \frac{5\pi}{4}, \text{ and } \frac{7\pi}{4} \qquad \text{See Example 4.}$$

$\theta = \dfrac{\pi}{4}, \dfrac{\pi}{2}, \dfrac{3\pi}{4}, \dfrac{5\pi}{4}, \dfrac{3\pi}{2}$, and $\dfrac{7\pi}{4}$ are the solutions over the interval $[0, 2\pi)$. ∎

Self Check 7   Solve the equation $\cot^2\theta = \cot\theta$ over the interval $[0, 2\pi)$.

Now Try Exercise 39.

Caution

In Example 7, it would be incorrect to divide both sides by $\cos\theta$, because a solution to the equation would be lost. This possibility exists whenever we divide both sides of an equation by an expression that might be 0.

$$\cancel{2\cos^3\theta = \cos\theta} \qquad \text{Do not divide by } \cos\theta.$$
$$\cancel{2\cos^2\theta = 1}$$

EXAMPLE 8   **Solving a Trigonometric Equation by Factoring**

Solve the equation $2\sin^2\theta + \sin\theta = 1$ over the interval $[0, 2\pi)$.

SOLUTION   The first step will be to subtract 1 from both sides of the equation. Then we will be able to solve this equation just as a quadratic equation can often be solved by factoring.

$$2\sin^2\theta + \sin\theta = 1$$

$$2\sin^2\theta + \sin\theta - 1 = 0 \qquad \text{Subtract 1 from each side.}$$

$$(2\sin\theta - 1)(\sin\theta + 1) = 0 \qquad \text{Factor.}$$

$$2 \sin \theta - 1 = 0 \qquad \text{or} \quad \sin \theta + 1 = 0 \qquad \text{Set each factor to 0.}$$

Reference angle is $\theta' = \dfrac{\pi}{6}$. $\qquad \sin \theta = \dfrac{1}{2}$ $\qquad\qquad\qquad$ $\sin \theta = -1$ $\qquad$ θ is quadrantal.

$\sin \theta > 0$ in QI and QII. $\qquad \theta = \dfrac{\pi}{6} \text{ and } \dfrac{5\pi}{6}$ $\qquad\qquad$ $\theta = \dfrac{3\pi}{2}$ $\qquad$ Only value in $[0, 2\pi)$

$\theta = \dfrac{\pi}{6}, \dfrac{5\pi}{6},$ and $\dfrac{3\pi}{2}$ are the solutions over the interval $[0, 2\pi)$. ∎

Self Check 8   Solve the equation $2 \cos^2 \theta - \cos \theta = 1$ over the interval $[0, 2\pi)$.

Now Try Exercise 55.

If a trigonometric equation contains functions that are not defined over all the real numbers, it is necessary to check all the possible solutions in the original equation to determine whether they are valid for the given equation.

EXAMPLE 9   **Solving a Trigonometric Equation by Factoring**

Solve the equation $(\sin \theta - 1)(\tan \theta - 1) = 0$ over the interval $[0, 2\pi)$.

SOLUTION   This equation is already in factored form and equal to 0; therefore, our first step will be to set each factor to 0.

$$(\sin \theta - 1)(\tan \theta - 1) = 0$$

$$\sin \theta - 1 = 0 \quad \text{or} \quad \tan \theta - 1 = 0$$

$\sin \theta = 1$ $\qquad\qquad$ $\tan \theta = 1$ $\qquad$ Reference angle is $\theta' = \dfrac{\pi}{4}$.

$\theta = \dfrac{\pi}{2}$ $\qquad\qquad$ $\theta = \dfrac{\pi}{4} \text{ and } \dfrac{5\pi}{4}$ $\qquad$ $\tan \theta > 0$ in QI and QIII

The possible solutions are $\theta = \frac{\pi}{2}, \frac{\pi}{4},$ and $\frac{5\pi}{4}$. However, $\tan \frac{\pi}{2}$ is not defined. Therefore, the only solutions to the equation are $\theta = \dfrac{\pi}{4}$ and $\dfrac{5\pi}{4}$. If you check your solutions using a graphing calculator, it's a good idea to use a scale of $\frac{\pi}{4}$ units when setting the window. ∎

Self Check 9   Solve the equation $(\cot \theta - 1)(\cos \theta + 1) = 0$ over the interval $[0, 2\pi)$.

Now Try Exercise 75.

EXAMPLE 10   **Solving a Trigonometric Equation by Factoring**

Solve the equation $\sqrt{3} \tan \theta \cot \theta + \cot \theta - \sqrt{3} \tan \theta - 1 = 0$ over the interval $[0, 2\pi)$.

SOLUTION   To factor this equation, we will factor by grouping.

$$\sqrt{3} \tan \theta \cot \theta + \cot \theta - \sqrt{3} \tan \theta - 1 = 0$$

$$\cot \theta(\sqrt{3} \tan \theta + 1) - (\sqrt{3} \tan \theta + 1) = 0 \qquad \text{Factor by grouping.}$$

$$(\sqrt{3} \tan \theta + 1)(\cot \theta - 1) = 0 \qquad \text{Factor out the common } \sqrt{3} \tan \theta + 1.$$

$$\sqrt{3}\tan\theta + 1 = 0 \qquad \text{or} \quad \cot\theta - 1 = 0 \qquad \text{Set each factor to 0.}$$

Reference angle is $\theta' = \dfrac{\pi}{6}$. $\tan\theta = -\dfrac{1}{\sqrt{3}} = -\dfrac{\sqrt{3}}{3}$ $\qquad\qquad \cot\theta = 1 \qquad\qquad$ Reference angle is $\theta' = \dfrac{\pi}{4}$.

$\tan\theta < 0$ in QII and QIV $\qquad \theta = \dfrac{5\pi}{6}$ and $\dfrac{11\pi}{6}$ $\qquad\qquad \theta = \dfrac{\pi}{4}$ and $\dfrac{5\pi}{4}$ $\qquad$ $\cot\theta > 0$ in QI and QIII

$$\theta = \dfrac{\pi}{4}, \dfrac{5\pi}{6}, \dfrac{5\pi}{4}, \text{ and } \dfrac{11\pi}{6} \text{ are the solutions over the interval } [0, 2\pi).$$

Self Check 10 Solve the equation $2\sin\theta\cos\theta + \cos\theta - 2\sin\theta - 1 = 0$ over the interval $[0, 2\pi)$.
Now Try Exercise 89.

4. Use Identities to Solve Trigonometric Equations

In addition to using algebraic techniques to solve trigonometric equations, there are times when we will use trigonometric identities to change an equation to a form that can be solved using algebraic techniques.

EXAMPLE 11   **Solving a Trigonometric Equation Using an Identity**

Solve the equation $\sin 2\theta = 2\sin\theta$ over the interval $[0, 2\pi)$.

SOLUTION   We will use the Double-Angle Formula for the sine function.

$$\sin 2\theta = 2\sin\theta$$
$$2\sin\theta\cos\theta = 2\sin\theta \qquad \text{Double-Angle Formula for sine.}$$
$$\sin\theta\cos\theta = \sin\theta \qquad \text{Divide both sides by 2.}$$
$$\sin\theta\cos\theta - \sin\theta = 0 \qquad \text{Subtract } \sin\theta \text{ from both sides.}$$
$$\sin\theta(\cos\theta - 1) = 0 \qquad \text{Factor out } \sin\theta.$$

θ is quadrantal. $\sin\theta = 0 \qquad$ or $\quad \cos\theta - 1 = 0 \qquad$ Set each factor to 0.

Only values in $[0, 2\pi)$ $\boldsymbol{\theta = 0 \text{ and } \pi}$ $\qquad\qquad \cos\theta = 1 \qquad$ θ is quadrantal.

$\qquad\qquad\qquad\qquad\qquad\qquad\qquad\qquad \boldsymbol{\theta = 0} \qquad$ Only value in $[0, 2\pi)$

$\theta = 0$ and π are the solutions over the interval $[0, 2\pi)$.

Self Check 11   Solve the equation $\sin 2\theta = \cos\theta$ over the interval $[0, 2\pi)$.
Now Try Exercise 63.

Caution

$\sin 2\theta = 2\sin\theta$	Do NOT divide by 2 at this point.
~~$\sin\theta = \sin\theta$~~	These are NOT equivalent equations.

Part of successfully solving a trigonometric equation is selecting the proper identity to use, as demonstrated in the next example.

EXAMPLE 12   **Solving a Trigonometric Equation Using an Identity**

Solve the equation $2\cos^2\theta = \sin\theta - 1$ over the interval $[0, 2\pi)$.

SOLUTION   While this equation appears to have the form of a quadratic equation, it cannot be factored. However, if we use the Pythagorean Identity $\sin^2\theta + \cos^2\theta = 1$, and rewrite it as $\cos^2\theta = 1 - \sin^2\theta$, we can substitute this into the equation and solve by factoring.

$$2\cos^2\theta = \sin\theta - 1$$

$$2(1 - \sin^2\theta) = \sin\theta - 1 \qquad \text{Use } \cos^2\theta = 1 - \sin^2\theta.$$

$$2 - 2\sin^2\theta = \sin\theta - 1 \qquad \text{Clear parentheses.}$$

$$2\sin^2\theta + \sin\theta - 3 = 0 \qquad \text{Subtract } \sin\theta - 1 \text{ from both sides and multiply both sides by } -1.$$

$$(2\sin\theta + 3)(\sin\theta - 1) = 0 \qquad \text{Factor.}$$

Solve for $\sin\theta$. $2\sin\theta + 3 = 0$ or $\sin\theta - 1 = 0$ Set each factor to 0.

$$\sin\theta = -\frac{3}{2} \qquad\qquad \sin\theta = 1 \qquad \theta \text{ is quadrantal.}$$

$-\frac{3}{2}$ is not in $[-1, 1]$.   No solution. $\theta = \dfrac{\pi}{2}$ Only value in $[0, 2\pi)$

$\theta = \dfrac{\pi}{2}$ is the only solution over the interval $[0, 2\pi)$. ∎

Self Check 12   Solve the equation $\cos 2\theta = \sin\theta + 1$ over the interval $[0, 2\pi)$.

Now Try Exercise 95.

Self Check Answers   1. $\theta = \dfrac{2\pi}{3} + 2\pi k$ or $\theta = \dfrac{4\pi}{3} + 2\pi k$ 2. $\dfrac{\pi}{4}, \dfrac{3\pi}{4}$ 3. π

4. $\dfrac{\pi}{4}, \dfrac{3\pi}{4}, \dfrac{5\pi}{4}, \dfrac{7\pi}{4}$ 5. $\dfrac{\pi}{6}, \dfrac{5\pi}{6}, \dfrac{7\pi}{6}, \dfrac{11\pi}{6}$ 6. $\dfrac{4\pi}{3}$

7. $\dfrac{\pi}{4}, \dfrac{\pi}{2}, \dfrac{5\pi}{4}, \dfrac{3\pi}{2}$ 8. $0, \dfrac{2\pi}{3}, \dfrac{4\pi}{3}$ 9. $\dfrac{\pi}{4}, \dfrac{5\pi}{4}$

10. $0, \dfrac{7\pi}{6}, \dfrac{11\pi}{6}$ 11. $\dfrac{\pi}{6}, \dfrac{\pi}{2}, \dfrac{5\pi}{6}, \dfrac{3\pi}{2}$ 12. $0, \pi, \dfrac{7\pi}{6}, \dfrac{11\pi}{6}$

Exercises **5.7**

Getting Ready
You should be able to complete these vocabulary and concept statements before you proceed to the practice exercises.

Fill in the blanks.

1. The equation $\sin\theta = 1$ has _____ solutions over all real numbers.

2. True or False: To solve the equation $\sin\theta\cos\theta = \cos^2\theta$, it is OK to divide both sides by $\cos\theta$. _____

3. To solve an equation such as $\cos 2\theta = \frac{1}{2}$ over the interval $[0, 2\pi)$, you should first find the _____ solution.

4. True or False: To solve the equation $2\cos\theta = \cos 2\theta$, you can divide both sides by 2. _____

Practice

Find the general solution(s) for each equation.

5. $\sin \theta = \dfrac{\sqrt{3}}{2}$

6. $\sin \theta = -\dfrac{\sqrt{3}}{2}$

7. $\tan \theta = -1$

8. $\tan \theta = \sqrt{3}$

9. $\cos 2\theta = -\dfrac{\sqrt{3}}{2}$

10. $\cos 2\theta = -\dfrac{\sqrt{2}}{2}$

11. $\sin 2\theta = 1$

12. $\sin 2\theta = -\dfrac{1}{2}$

13. $\tan \dfrac{\theta}{2} = \dfrac{\sqrt{3}}{3}$

14. $\cot \dfrac{\theta}{2} = -\sqrt{3}$

Solve each equation over the interval $[0, 2\pi)$.

15. $5 \sin \theta = 0$

16. $-4 \cos \theta = 0$

17. $\sqrt{2} \cos \theta + 1 = 0$

18. $\sqrt{2} \sin \theta - 1 = 0$

19. $4 \sin^2 \theta = 1$

20. $\cos^2 \theta - 1 = 0$

21. $\tan^2 \theta = \dfrac{1}{3}$

22. $\cos^2 \theta = 3$

23. $\sin\left(\theta + \dfrac{\pi}{4}\right) = \dfrac{1}{2}$

24. $\sin\left(\theta - \dfrac{\pi}{6}\right) = \dfrac{\sqrt{3}}{2}$

25. $\cos\left(2\theta - \dfrac{\pi}{6}\right) = \dfrac{1}{2}$

26. $\cos\left(2\theta - \dfrac{\pi}{4}\right) = 1$

27. $\tan\left(\dfrac{\theta}{2} + \pi\right) = 0$

28. $\cot\left(\dfrac{\theta}{2} + \dfrac{\pi}{2}\right) = 0$

29. $\sec^2 \theta - 2 = 0$

30. $\csc^2 \theta - 4 = 0$

31. $\cos 2\theta = \dfrac{\sqrt{3}}{2}$

32. $\cos 3\theta = \dfrac{1}{2}$

33. $\tan 2\theta = 1$

34. $\tan 2\theta = 0$

35. $\sin \dfrac{\theta}{2} - \dfrac{\sqrt{3}}{2} = 0$

36. $\cos \dfrac{\theta}{4} = \dfrac{1}{2}$

Solve each equation over the interval $[0, 2\pi)$.

37. $\cos^2 \theta - \cos \theta = 0$

38. $\sin^2 \theta + \sin \theta = 0$

39. $\tan^2 \theta = \tan \theta$

40. $\sin \theta \tan \theta = \tan \theta$

41. $\cos \theta \tan \theta = \sqrt{3} \cos \theta$

42. $2 \sin \theta \cos \theta = \sqrt{3} \cos \theta$

43. $\sin 2\theta = \sin \theta$

44. $\cos 2\theta - \cos^2 \theta = 0$

45. $2 \sin^2 \theta + 5 \sin \theta - 3 = 0$

46. $2 \sin^2 \theta - \sin \theta - 1 = 0$

47. $2 \cos^2 \theta + \cos \theta - 1 = 0$

48. $2 \cos^2 \theta - 3 \cos \theta - 2 = 0$

49. $\sin \theta \cos \theta = 0$

50. $\sin \theta \tan \theta = 0$

51. $\cos \theta \sin \theta = \cos \theta$

52. $\cos \theta \sin \theta + \cos \theta = 0$

53. $\tan^2 \theta - \tan \theta = 0$

54. $\tan^2 \theta - \sqrt{3} \tan \theta = 0$

55. $\sin^2 \theta - 3 \sin \theta + 2 = 0$

56. $\cos^2 \theta - 3 \cos \theta + 2 = 0$

57. $2 \cos^2 \theta + 3 \cos \theta + 1 = 0$

58. $4 \sin^2 \theta + 4 \sin \theta + 1 = 0$

59. $2 \sin^2 \theta + \sin \theta = 6$

60. $\sin^2 \theta + \sin \theta = 6$

61. $\sqrt{3} \sin \theta \tan \theta = \sin \theta$

62. $\cos^2 \theta + 4 \cos \theta + 3 = 0$

63. $\sin 2\theta + \cos \theta = 0$

64. $\sin 2\theta - \cos \theta = 0$

65. $\cos \theta = \sin \theta$

66. $\cos \theta = \sqrt{3} \sin \theta$

67. $\cos \theta = \sin \dfrac{\theta}{2}$

68. $\cos \theta = \cos \dfrac{\theta}{2}$

69. $\cos 2\theta = \cos \theta$

70. $\cos 2\theta = \sin \theta$

71. $\tan \theta = -\sin \theta$

72. $\cot \theta = -\cos \theta$

73. $\cos^2 \theta + \sin^2 \theta = \dfrac{1}{2}$

74. $2 \cos^2 \theta + \sin^2 \theta = 3$

75. $(2 \sin \theta - 1)(\cos \theta + 2) = 0$

76. $(\tan \theta - 1)\left(\cot \theta + \sqrt{3}\right) = 0$

77. $(3 \sec \theta + 1)(\csc \theta + 1) = 0$

78. $\left(\sqrt{3} \sec \theta - 2\right)\left(\sqrt{3} \sin \theta - 2\right) = 0$

79. $\cos 2\theta = 1 - \sin \theta$

80. $\cos 2\theta = \cos \theta - 1$

81. $\cos^2 \theta + \sin \theta - 1 = 0$

82. $2 \sin^2 \theta - \cos \theta + 1 = 0$

83. $\sin 4\theta = \sin 2\theta$

84. $\cos 4\theta = \cos 2\theta$

85. $\tan 2\theta + \sec 2\theta = 1$

86. $\tan 2\theta + 2 \sin \theta = 0$

87. $2 \cos^4 \theta = \sin^2 \theta$

88. $6 \cos^2 \theta = -9 \sin \theta$

89. $2 \cos \theta \sin \theta - \sqrt{2} \cos \theta = \sqrt{2} \sin \theta - 1$

90. $4 \sin \theta \cos \theta - 2 \sin \theta = 2 \cos \theta - 1$

91. $2 \sin \theta - \sqrt{2} \tan \theta = \sqrt{2} \sec \theta - 2$

92. $\cot^2 \theta \cos \theta + \cot^2 \theta - 3 \cos \theta - 3 = 0$

93. $2 \sin^4 \theta - 9 \sin^2 \theta + 4 = 0$

94. $2 \sin^3 \theta + \sin^2 \theta = \sin \theta$

95. $4 \cos \theta \sin^2 \theta - \cos \theta = 0$

96. $\tan^2 \theta = 1 + \sec \theta$

97. $\csc^4 \theta = 2 \csc^2 \theta - 1$

98. $\tan^2 \theta - 2 \tan \theta + 1 = 0$

99. $4 \sin^2 \theta + 4 \sin \theta + 1 = 0$

100. $4 \cos^2 2\theta - 4 \cos 2\theta + 1 = 0$

101. $\tan^2 \theta - \tan \theta = 0$

102. $\tan \theta + \cot \theta = -2$

103. $2 \sin \theta \tan \theta + \cos \theta = 1$

104. $\sin^2 \theta - \cos^2 \theta - 3 \sin \theta - 1 = 0$

Solve each equation over the interval $[0, 2\pi)$*. Consider using a Product-to-Sum or Sum-to-Product Formula.*

105. $\sin \theta \cos \theta = \dfrac{1}{2}$

106. $\cos \theta \sin \theta = \dfrac{\sqrt{3}}{4}$

107. $2 \sin \dfrac{3\theta}{2} \cos \dfrac{\theta}{2} = \sin \theta$

108. $2 \cos \dfrac{3\theta}{2} \cos \dfrac{\theta}{2} = \cos 2\theta$

109. $\sin 3\theta \cos \theta = \dfrac{1}{2} \sin 2\theta$

110. $\cos 3\theta \cos \theta = \cos^2 2\theta$

111. $\sin 4\theta = -\sin 2\theta$

112. $\cos 4\theta = -\cos 2\theta$

113. $\cos 9\theta = -\cos 3\theta$

114. $\sin 12\theta = -\sin 4\theta$

Applications

115. Lemur population Suppose the population P of lemurs in a certain region of Madagascar can be modeled by the equation $P = 200 \sin \frac{\pi}{12}x + 400$, where x is the number of years since 2010. Determine the year the lemur population will reach 500.

116. Howler-monkey population Suppose the population P of howler monkeys in a certain region of Central America can be modeled by the equation $P = 600 \sin \frac{\pi}{12}x + 400$, where x is the number of years since 2010. Determine the year the number of howler monkeys in the region will reach 1,000.

Bronson Chang/Shutterstock.com

117. Projectile motion The range r of a baseball hit at an angle θ (in degrees) with the horizontal and with an initial velocity of v_0 in feet per second is given by the formula $r = \frac{1}{32}v_0^2 \sin 2\theta$. If a baseball travels a range of 219.6 feet and is hit with an initial velocity of 90 feet per second, determine the angle θ at which the baseball was hit. Round to the nearest degree.

118. Projectile motion If a football is punted a range of 112.5 feet with an initial velocity of 60 feet per second, use the formula given in Exercise 117 to determine the angle θ at which the football was punted. Round to the nearest degree.

Discovery and Writing

119. Solve $\sin(\sin x) = 0$ over the interval $[0, 2\pi)$.

120. Solve $\tan(\sin x) = 0$ over the interval $[0, 2\pi)$.

121. Write a paragraph explaining the difference between an identity and an equation.

122. In the equation $\cos nx = \frac{\sqrt{3}}{2}$, n is a fixed, positive integer. Find the general solution of this equation.

5.8   Trigonometric Equations II

In this section, we will learn to

1. Use a calculator to approximate solutions of trigonometric equations.

2. Use the Quadratic Formula to solve trigonometric equations.

3. Approximate the solutions of trigonometric equations using a graphing calculator.

4. Solve trigonometric equations in the form $A \sin x + B \cos x = y$.

Sometimes it is necessary to find approximate solutions to trigonometric equations. Consider the Statue of Liberty. The sculpture sits on Liberty Island in New York Harbor; its height from the ground to the tip of the torch is approximately 154 feet.

To determine the approximate angle of elevation in degrees from a point 300 feet away from the base, we would solve the following trigonometric equation:

$$\tan \theta = \frac{305}{300} = \frac{61}{60}:$$

$$\theta = \tan^{-1} \frac{61}{60}$$

$$\theta \approx 45.5°$$

This trigonometric equation actually has an infinite number of solutions. Note that we have a natural restriction placed on θ since we are solving a right-triangle problem. We are finding solutions over the interval $[0°, 90°]$. Therefore we have exactly one solution. We will solve trigonometric equations of this type in this section.

1. Use a Calculator to Approximate Solutions of Trigonometric Equations

Keep in mind that if the exact solution(s) to an equation cannot be found using a special angle, a calculator can be used to find an approximate solution. Even though we may use a calculator, we will continue to make use of a reference angle via the inverse function keys. Remember, a reference angle is an acute angle; therefore, always use a positive number when using the inverse function keys. If the value of the function is negative, you will use that to determine the quadrants in which the solution angles lie.

EXAMPLE 1   **Solving a Trigonometric Equation with a Calculator**

Solve the equation $\sin \theta = 0.6$ over the interval $[0, 2\pi)$.

SOLUTION   Since $\sin \theta > 0$, we know that the solution angles lie in QI and QII, where the sine function is positive. Step one is to find the reference angle, $\theta' = \sin^{-1} 0.6$.

In QI, the solution angle and the reference angle are the same.

$$\theta_1 = \theta' = \sin^{-1} 0.6 \approx 0.64351$$

For the solution in QII, we subtract and find

$$\theta_2 = \pi - \theta' = \pi - \sin^{-1} 0.6 \approx 2.4981.$$

These are demonstrated in Figure 5-35.

$\theta_1 = \theta' = \sin^{-1} 0.6 \approx 0.6435$
$\theta_2 = \pi - \theta' = \pi - \sin^{-1} 0.6 \approx 2.4981$

θ_2
θ_1
π
0
$\theta' = \sin^{-1} 0.6 \approx 0.6435$

FIGURE 5-35

The solutions to the equation $\sin \theta = 0.6$ are $\theta \approx 0.6435$ and 2.4981.

The solutions can be checked with a calculator.

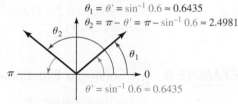

Self Check 1   Solve the equation $\sin \theta = 0.9$ over the interval $[0, 2\pi)$.

Now Try Exercise 5.

EXAMPLE 2 **Solving a Trigonometric Equation with a Calculator**

Solve the equation $\cos\theta = -0.3$ over the interval $[0, 2\pi)$.

SOLUTION The cosine function is negative in QII and QIII. The reference angle is $\theta' = \cos^{-1} 0.3$.

The solution in QII is

$$\theta_1 = \pi - \theta' = \pi - \cos^{-1} 0.3 \approx 1.8755.$$

The solution in QIII is

$$\theta_2 = \pi + \theta' = \pi + \cos^{-1} 0.3 \approx 4.4077.$$

These are illustrated in Figure 5-36. Notice that it is not necessary to write the approximate value of θ', because it is more efficient to let the calculator do that as you find the solution.

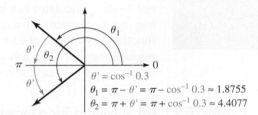

$\theta' = \cos^{-1} 0.3$
$\theta_1 = \pi - \theta' = \pi - \cos^{-1} 0.3 \approx 1.8755$
$\theta_2 = \pi + \theta' = \pi + \cos^{-1} 0.3 \approx 4.4077$

FIGURE 5-36

The solutions are $\theta \approx 1.8755$ and 4.4077. ∎

Self Check 2 Solve the equation $\cos\theta = -0.7$ over the interval $[0, 2\pi)$.

Now Try Exercise 11.

2. Use the Quadratic Formula to Solve Trigonometric Equations

A trigonometric equation that can be written in quadratic form may not be factorable; however, the Quadratic Formula can be used to find approximate values of the solution(s).

EXAMPLE 3 **Solving a Trigonometric Equation Using the Quadratic Formula**

Solve the equation $\sin^2\theta - 2\sin\theta - 2 = 0$ over the interval $[0, 2\pi)$.

SOLUTION Recall that a quadratic equation in the form $ax^2 + bx + c = 0$ can be solved using the Quadratic Formula $x = \dfrac{-b \pm \sqrt{b^2 - 4ac}}{2a}$. Since $\sin^2\theta - 2\sin\theta - 2$ cannot be factored, we will use the Quadratic Formula to approximate any solution(s). Note that when we set up the equation, the unknown value is $\sin\theta$, not x.

For $\sin^2\theta - 2\sin\theta - 2 = 0$, we set $a = 1$, $b = -2$, and $c = -2$, and then use the Quadratic Formula.

$$\sin\theta = \frac{-(-2) \pm \sqrt{(-2)^2 - 4(1)(-2)}}{2(1)}$$

$$\sin\theta = \frac{2 \pm \sqrt{12}}{2} = \frac{2 \pm \sqrt{4\cdot 3}}{2} = \frac{2 \pm 2\sqrt{3}}{2} = \frac{2(1 \pm \sqrt{3})}{2} = 1 \pm \sqrt{3}$$

$$\sin\theta = 1 + \sqrt{3} \quad \text{or} \quad \sin\theta = 1 - \sqrt{3}$$

Now that we have possible values for $\sin \theta$, we will find the values of θ. Since $\left(1 + \sqrt{3}\right) > 1$, that is not a possible value for $\sin \theta$. The equation $\sin \theta = 1 - \sqrt{3} \approx -0.732$ does have two solutions. Recall from Section 5.7 that we used a reference angle to solve equations of this type. The sine function is negative in QIII and QIV.

$$\sin \theta = 1 - \sqrt{3}$$

$$\theta' = \sin^{-1} |1 - \sqrt{3}| \qquad \text{A reference angle is acute; therefore the absolute value is used.}$$

The angle in QIII is

$$\theta_1 = \pi + \sin^{-1} |1 - \sqrt{3}| \approx 3.9629.$$

The angle in QIV is

$$\theta_2 = 2\pi - \sin^{-1} |1 - \sqrt{3}| \approx 5.4619.$$

The solutions over the interval $[0, 2\pi)$ are $\theta \approx 3.9629$ and 5.4619.
Check these solutions with a calculator.

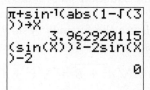 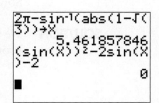

Self Check 3   Solve the equation $2 \cos^2 \theta + \cos \theta - 2 = 0$ over the interval $[0, 2\pi)$.

Now Try Exercise 21.

3. Approximate the Solutions of Trigonometric Equations Using a Graphing Calculator

The technology of a graphing calculator allows us to find approximate solutions of trigonometric equations that cannot be solved using the techniques in the previous examples. Selecting the proper window is essential.

EXAMPLE 4   **Solving a Trigonometric Equation Using a Graphing Calculator**

Solve the equation $5 \cos^2 \theta + 2 \sin \theta + 1 = 0$ over the interval $[0, 2\pi)$. Round to three decimal places.

SOLUTION   The equation is put into a calculator using the window $X \colon [0, 2\pi]$, $Y \colon [-3, 7]$. Be careful when you enter the function, making liberal use of parentheses. After you graph the function, use the ZERO function to find the roots. We see that the solutions to the equation are $\theta \approx 4.2935$ and 5.1312.

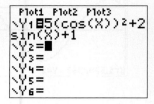  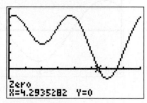 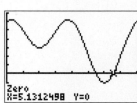

Self Check 4   Solve the equation $3 \sin^2 \theta + 3 \cos \theta - 2 = 0$ over the interval $[0, 2\pi)$. Round to three decimal places.

Now Try Exercise 35.

4. Solve Trigonometric Equations in the Form
$A \sin x + B \cos x = y$

An equation such as $\sin \theta + \cos \theta = 1$ can be solved by squaring both sides. When this is done, there is a possibility of extraneous roots. However, we can also use the theorem from Section 5.6 involving a sum of the form $A \sin \theta + B \cos \theta$.

Theorem $A \sin \theta + B \cos \theta = k \sin(\theta + \varphi)$, where $k = \sqrt{A^2 + B^2}$ and φ is any angle for which $\sin \varphi = \dfrac{B}{\sqrt{A^2 + B^2}} = \dfrac{B}{k}$ and $\cos \varphi = \dfrac{A}{\sqrt{A^2 + B^2}} = \dfrac{A}{k}$.

EXAMPLE 5   **Solving an Equation in the Form $A \sin \theta + B \cos \theta = y$**

Solve the equation $\sin \theta + \cos \theta = 1$ over the interval $[0, 2\pi)$.

SOLUTION   We will use the formula $A \sin \theta + B \cos \theta = k \sin(\theta + \varphi)$ to solve the equation. From $\sin \theta + \cos \theta = 1$, we determine that $A = 1$ and $B = 1$, which leads to

$$k = \sqrt{A^2 + B^2} = \sqrt{1^2 + 1^2} = \sqrt{2},$$

$$\sin \varphi = \frac{B}{k} = \frac{1}{\sqrt{2}}, \text{ and}$$

$$\varphi = \sin^{-1} \frac{1}{\sqrt{2}} = \frac{\pi}{4}.$$

We now can solve the equation.

$$\sin \theta + \cos \theta = 1$$

$$k \sin(\theta + \varphi) = 1 \qquad \text{Substitute } \sin \theta + \cos \theta = k \sin(\theta + \varphi).$$

$$\sqrt{2} \sin\left(\theta + \frac{\pi}{4}\right) = 1 \qquad k = \sqrt{2} \text{ and } \varphi = \frac{\pi}{4}.$$

$$\sin\left(\theta + \frac{\pi}{4}\right) = \frac{1}{\sqrt{2}}$$

$$\theta + \frac{\pi}{4} = \frac{\pi}{4} \quad \text{or} \quad \theta + \frac{\pi}{4} = \frac{3\pi}{4} \qquad \sin \theta > 0 \text{ in QI and QII.}$$

$$\theta = 0 \qquad\qquad\quad \theta = \frac{\pi}{2}$$

We can check the solution with a calculator.

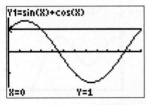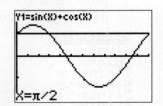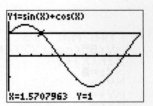

FIGURE 5-37

Self Check 5   Solve the equation $\sin \theta + \sqrt{3} \cos \theta = 1$ over the interval $[0, 2\pi)$.

Now Try Exercise 27.

Self Check Answers   **1.** 1.1198, 2.0218   **2.** 2.3462, 3.9370   **3.** 0.6748, 5.6083

4. 1.8377, 4.4455   **5.** $\dfrac{\pi}{2}, \dfrac{11\pi}{6}$

Exercises 5.8

Getting Ready
You should be able to complete these vocabulary and concept statements before you proceed to the practice exercises.

Fill in the blanks.

1. If a trigonometric equation in quadratic form cannot be factored, use the _____.

2. A _____ can be used to find approximate solutions to trigonometric equations.

3. If you square both sides of trigonometric equation, will all solutions found always work in the original equation? Why or why not?

4. $A \sin \theta + B \cos \theta = k \sin(\theta + \varphi)$, where $k =$ _____.

Practice
Solve each equation over the interval $[0, 2\pi)$. Round to three decimal places.

5. $\sin \theta = 0.4$ 6. $\sin \theta = 0.2$

7. $\sin \theta = -0.34$ 8. $\sin \theta = -0.56$

9. $\cos \theta = 0.25$ 10. $\cos \theta = 0.48$

11. $\cos \theta = -0.8$ 12. $\cos \theta = -0.6$

13. $\tan \theta = 1.57$ 14. $\tan \theta = 1.15$

15. $\tan \theta = -2.65$ 16. $\tan \theta = -3.9$

17. $5 \sin^2 \theta = 4$ 18. $3 \sin^2 \theta - 2 = 0$

19. $9 \cos^2 \theta = 7$ 20. $4 - 5 \cos^2 \theta = 0$

21. $\sin^2 \theta - 4 \sin \theta - 1 = 0$

22. $\cos^2 \theta - 2 \cos \theta - 4 = 0$

23. $\sin^2 \theta + 6 \sin \theta \cos \theta = \cos^2 \theta$

24. $\sin^2 \theta + 2 \sin 2\theta = 3 \cos^2 \theta$

25. $2 \cos^2 \theta - 3 \cos \theta - 1 = 0$

26. $3 \sin^2 \theta + 2 \sin \theta - 3 = 0$

Use the fact that the expression $A \sin \theta + B \cos \theta$ can be written as $k \sin (\theta + \varphi)$ to solve each equation over the interval $[0, 2\pi)$.

27. $\dfrac{\sqrt{3}}{2} \sin \theta + \dfrac{1}{2} \cos \theta = \dfrac{1}{2}$

28. $\sqrt{2} \sin \theta + \sqrt{2} \cos \theta = \sqrt{3}$

29. $\dfrac{1}{2} \sin \theta + \dfrac{\sqrt{3}}{2} \cos \theta = 1$ 30. $\cos \theta - \sin \theta = \sqrt{2}$

31. $\cos \theta - \sqrt{3} \sin \theta = 1$ 32. $\sin \theta + \cos \theta = -\sqrt{2}$

Use a graphing calculator to solve each equation over the interval $[0, 2\pi)$.

33. $3 \cos^2 x - 4 \cos x = 2$ 34. $5 \cos^2 x = 6 - 4 \cos x$

35. $3 \sin^2 x - 4 \cos x = 0$ 36. $2 \sin x = 3 \cos^2 x$

37. $2x - 3 \sin x = 0$ 38. $x = 4 \sin x$

39. $5x + 3 \cos x = 6$ 40. $2x - 3 \cos 2x = 1$

41. $\cos x - \sin x = \dfrac{1}{2}x$ 42. $2 \cos x - 3x = \sin x$

43. $\cos x - 2x^2 = 0$ 44. $\sin^2 x - x^3 = 0$

45. $\ln(x + 2) = 2 \sin 2x, x > -2$

46. $\ln x + \cos \dfrac{1}{2}x = 0, x > 0$

47. $4 - e^{2x} = \cos x$ 48. $\sin 2x + 3 = e^x$

49. $\tan x = 2 \sin x - 1$ 50. $\tan 2x = \ln x, x > 0$

51. $\sin x \cos x = \ln x, x > 0$ 52. $\tan x - 3x = 0$

Applications

53. **Rate of change** In calculus, it is known that the rate of change of the function $f(x) = \tan x$ is $\sec^2 x$. If x is on the interval $[0, 2\pi)$, determine the values of x for which the rate of change is 3. Round to three decimal places. Convert secant to cosine if you prefer.

54. **Rate of change** In calculus, it is known that the rate of change of the function $f(x) = \cot x$ is $-\csc^2 x$. If x is on the interval $[0, 2\pi)$, determine the values of x for which the rate of change is -5. Round to three decimal places. Convert cosecant to sine if you prefer.

55. **Projectile motion** A missile is fired with an initial velocity v_0 of 1,000 feet per second and aimed at a target 5 miles away. Determine the minimum angle of elevation θ (in degrees) of the missile if the range r of the missile in feet is represented by $r = \frac{1}{32}v_0^2 \sin 2\theta$. Round to one decimal place.

trekandshoot/Shutterstock.com

56. **Projectile motion** A football is thrown with an initial velocity v_0 of 55 feet per second to a receiver 30 yards away. Determine the minimum angle of elevation θ (in degrees) of the football if the range r of the football in feet is represented by $r = \frac{1}{32}v_0^2 \sin 2\theta$. Round to one decimal place.

CHAPTER REVIEW

SECTION 5.1   Inverse Trigonometric Functions

Definitions and Concepts	Examples
Inverse sine function: $y = \sin^{-1} x$ if and only if $x = \sin y$ and $-\dfrac{\pi}{2} \le y \le \dfrac{\pi}{2}$. Domain $= \{x \mid -1 \le x \le 1\}$; range $= \left\{ y \mid -\dfrac{\pi}{2} \le y \le \dfrac{\pi}{2} \right\}$.	See Example 1, pages 512–513
Composition of sine and inverse sine functions: $\sin(\sin^{-1} x) = x$ when $-1 \le x \le 1$. $\sin^{-1}(\sin x) = x$ when $-\dfrac{\pi}{2} \le x \le \dfrac{\pi}{2}$.	See Example 3, pages 514–515
Inverse cosine function: $y = \cos^{-1} x$ if and only if $x = \cos y$ and $0 \le y \le \pi$. Domain $= \{x \mid -1 \le x \le 1\}$; range $= \{y \mid 0 \le y \le \pi\}$.	See Example 4, pages 516–517
Composition of cosine and inverse cosine functions: $\cos(\cos^{-1} x) = x$ when $-1 \le x \le 1$. $\cos^{-1}(\cos x) = x$ when $0 \le x \le \pi$.	See Example 5, pages 517–518
Inverse tangent function: $y = \tan^{-1} x$ if and only if $x = \tan y$ and $-\dfrac{\pi}{2} < y < \dfrac{\pi}{2}$. Domain $=$ all real numbers; range $= \left\{ y \mid -\dfrac{\pi}{2} < y < \dfrac{\pi}{2} \right\}$.	See Example 6, page 519

EXERCISES

Find the exact value of each expression, if any. Do not use a calculator. Note that each answer should be a real number.

1. $\sin^{-1} \dfrac{\sqrt{2}}{2}$

2. $\cos^{-1}\left(-\dfrac{\sqrt{3}}{2}\right)$

3. $\tan^{-1} 0$

Use a calculator to find each value in radians to three decimal places.

4. $\sin^{-1} 0.56$

5. $\cos^{-1}(-0.34)$

6. $\tan^{-1}(-25)$

Find the exact value of each expression or state why the value does not exist. Do not use a calculator.

7. $\cos\left(\cos^{-1} \dfrac{4}{5}\right)$

8. $\sin^{-1}\left(\sin \dfrac{7\pi}{4}\right)$

9. $\tan^{-1}\left(\tan \dfrac{5\pi}{6}\right)$

10. $\sin(\sin^{-1} 1.5)$

Find the exact value of each expression. Do not use a calculator.

11. $\tan\left[\cos^{-1}\left(-\dfrac{5}{13}\right)\right]$

12. $\cos\left(\tan^{-1} \dfrac{4}{7}\right)$

13. $\sin\left(\cos^{-1} \dfrac{5}{13}\right)$

14. $\sec\left[\sin^{-1}\left(-\dfrac{2}{3}\right)\right]$

Applications

15. Dubai skyscraper The tallest building in the world is the Burj Khalifa, in Dubai. It is 2,716.5 feet high. If the angle of depression from the top of the building to a point on the ground 800 feet away from its base is represented by $\tan^{-1} \dfrac{2{,}716.5}{800}$ radians, find the angle of depression. Round to two decimal places.

SECTION 5.2   Inverse Cotangent, Secant, and Cosecant Functions (optional)

Definitions and Concepts	Examples
Inverse cotangent function: $y = \cot^{-1} x$ if and only if $x = \cot y$ and $0 < y < \pi$. Domain $=$ all real numbers; range $= \{y \mid 0 < y < \pi\}$.	See Example 1, page 525
Inverse secant function: $y = \sec^{-1} x$ is equivalent to $x = \sec y$ and $0 \le y \le \pi$ and $y \ne \dfrac{\pi}{2}$. Domain $= \{x \mid x \le -1 \text{ or } x \ge 1\}$; range $= \left\{y \,\middle|\, 0 \le y \le \pi \text{ and } y \ne \dfrac{\pi}{2}\right\}$.	See Example 3(a)–(b), pages 527–528
Inverse cosecant function: $y = \csc^{-1} x$ is equivalent to $x = \csc y$ and $-\dfrac{\pi}{2} \le y \le \dfrac{\pi}{2}$ and $y \ne 0$. Domain $= \{x \mid x \le -1 \text{ or } x \ge 1\}$; range $= \left\{y \,\middle|\, -\dfrac{\pi}{2} \le y \le \dfrac{\pi}{2} \text{ and } y \ne 0\right\}$.	See Example 3(c)–(d), pages 527–528

EXERCISES

Evaluate each expression. Do not use a calculator.

16. $\sec^{-1}\left(-\sqrt{2}\right)$ **17.** $\csc^{-1} 2$ **18.** $\cot^{-1} 1$

SECTION 5.3   Verifying Trigonometric Identities

Definitions and Concepts	Examples
Reciprocal Identities: $\sin x = \dfrac{1}{\csc x} \quad \cos x = \dfrac{1}{\sec x} \quad \tan x = \dfrac{1}{\cot x}$ $\csc x = \dfrac{1}{\sin x} \quad \sec x = \dfrac{1}{\cos x} \quad \cot x = \dfrac{1}{\tan x}$	See Example 2, pages 531–532
Quotient Identities: $\tan x = \dfrac{\sin x}{\cos x} \quad \cot x = \dfrac{\cos x}{\sin x}$	See Example 3, page 532
Pythagorean Identities: $\sin^2 x + \cos^2 x = 1 \quad \tan^2 x + 1 = \sec^2 x \quad \cot^2 x + 1 = \csc^2 x$	See Example 5, page 533
Even–Odd Properties: $\sin(-x) = -\sin x \quad \cos(-x) = \cos x \quad \tan(-x) = -\tan x$ $\csc(-x) = -\csc x \quad \sec(-x) = \sec x \quad \cot(-x) = -\cot x$	See Example 6, page 533

EXERCISES

Simplify each expression so that the final answer is not a rational expression.

19. $\dfrac{1}{1 + \cos x} + \dfrac{1}{1 - \cos x}$ **20.** $\dfrac{8}{\sec x + \tan x}$

Verify each identity.

21. $\dfrac{1 + \tan^2 x}{1 - \sin^2 x} = \sec^4 x$

22. $\tan^2 x - \csc^2 x = \dfrac{\sin^2 x - \cot^2 x}{\cos^2 x}$

23. $\dfrac{\sin x + \cos x}{\sin x} - \dfrac{\cos x - \sin x}{\cos x} = \csc x \sec x$

24. $\dfrac{\tan x}{1 + \sec x} + \dfrac{1 + \sec x}{\tan x} = 2 \csc x$

Applications

25. Trigonometric substitution In calculus, we often make a trigonometric substitution to eliminate a radical. Consider the radical expression $\sqrt{9x^2 - 25}$. Substitute $3x = 5 \sec \theta$, $0 \le \theta < \frac{\pi}{2}$, into the expression and simplify.

26. Rate of change The rate of change of the function $f(x) = \sec x + \cos x$ can be given by the expression $\sec x \tan x - \sin x$. Demonstrate how the rate of change can also be written as the expression $\sin x \tan^2 x$.

SECTION 5.4   Sum and Difference Formulas

Definitions and Concepts	Examples
Sum and Difference Formulas for Cosine: For any angles α and β, $\cos(\alpha + \beta) = \cos \alpha \cos \beta - \sin \alpha \sin \beta$ $\cos(\alpha - \beta) = \cos \alpha \cos \beta + \sin \alpha \sin \beta$	See Example 1, pages 539–540
Sum and Difference Formulas for Sine: For any angles α and β, $\sin(\alpha + \beta) = \sin \alpha \cos \beta + \cos \alpha \sin \beta$ $\sin(\alpha - \beta) = \sin \alpha \cos \beta - \cos \alpha \sin \beta$	See Example 2, pages 541–542
Sum and Difference Formulas for Tangent: For any angles α and β, $\tan(\alpha + \beta) = \dfrac{\tan \alpha + \tan \beta}{1 - \tan \alpha \tan \beta}$ $\tan(\alpha - \beta) = \dfrac{\tan \alpha - \tan \beta}{1 + \tan \alpha \tan \beta}$	See Example 4, pages 543–544
The angle θ from line 2, with slope m_2, to line 1, with slope m_1, can be found using $\tan \theta = \dfrac{m_1 - m_2}{1 + m_1 m_2}$.	See Example 9, pages 548–549

EXERCISES

Find the exact value of each expression using a Sum or Difference Formula.

27. $\cos 75°$ **28.** $\tan 255°$

29. $\sin \dfrac{5\pi}{12}$ **30.** $\cos\left(-\dfrac{7\pi}{12}\right)$

Express each quantity as a single function of one angle and evaluate.

31. $\dfrac{\tan 154° - \tan 19°}{1 + \tan 154° \tan 19°}$

32. $\sin \dfrac{\pi}{8} \cos \dfrac{3\pi}{8} + \cos \dfrac{\pi}{8} \sin \dfrac{3\pi}{8}$

For the given set of conditions, find:

(a) $\sin(\alpha + \beta)$ *(b)* $\cos(\alpha - \beta)$ *(c)* $\tan(\alpha + \beta)$

33. $\cos \alpha = -\dfrac{3}{5}$, α is in QIII; $\tan \beta = -3$, β is in QII

34. $\sin \alpha = -\dfrac{5}{13}$, α is in QIV; $\cot \beta = \dfrac{1}{2}$, β is in QI

Evaluate each expression. Do not use a calculator.

35. $\sin\left[\tan^{-1} 2 + \cos^{-1}\left(-\dfrac{1}{2} \right) \right]$

36. $\tan\left(\sin^{-1}\dfrac{2}{3} - \cos^{-1}\dfrac{3}{5} \right)$

Applications

Find the angle from line 2 to line 1. Round to one decimal place.

37. line 1: $y = 2x - 5$; line 2: $y = 3x + 2$

38. line 1: $y = 4x$; line 2: $y = -\dfrac{3}{4}x$

SECTION **5.5**   Double-Angle, Power-Reduction, and Half-Angle Formulas

Definitions and Concepts	Examples
Double-Angle Formulas: For any angle θ for which the function is defined, $\sin 2\theta = 2 \sin \theta \cos \theta$ $\cos 2\theta = \cos^2 \theta - \sin^2 \theta$ $\qquad = 2 \cos^2 \theta - 1$ $\qquad = 1 - 2 \sin^2 \theta$ $\tan 2\theta = \dfrac{2 \tan \theta}{1 - \tan^2 \theta}$	See Example 1, pages 552–553
Power-Reduction Formulas: $\sin^2 \theta = \dfrac{1}{2}(1 - \cos 2\theta)$ $\cos^2 \theta = \dfrac{1}{2}(1 + \cos 2\theta)$ $\tan^2 \theta = \dfrac{1 - \cos 2\theta}{1 + \cos 2\theta}$	See Example 4, page 555
Half-Angle Formulas: $\sin \dfrac{\alpha}{2} = \pm\sqrt{\dfrac{1 - \cos \alpha}{2}}$ $\cos \dfrac{\alpha}{2} = \pm\sqrt{\dfrac{1 + \cos \alpha}{2}}$ $\tan \dfrac{\alpha}{2} = \dfrac{\sin \alpha}{1 + \cos \alpha}$ $\qquad = \dfrac{1 - \cos \alpha}{\sin \alpha}$ $\qquad = \pm\sqrt{\dfrac{1 - \cos \alpha}{1 + \cos \alpha}}$	See Example 7, pages 558–559

EXERCISES

For each angle, use the Half-Angle Formulas to find each of the following values. Do not use a calculator.

(a) $\sin \theta$ (b) $\cos \theta$ (c) $\tan \theta$

39. $\theta = 15°$

40. $\theta = \dfrac{7\pi}{12}$

41. $\theta = \dfrac{5\pi}{8}$

42. $\theta = 22.5°$

Use the given information to find the exact value of

(a) $\sin 2\theta$ (b) $\cos 2\theta$ (c) $\tan 2\theta$

(d) $\sin \dfrac{\theta}{2}$ (e) $\cos \dfrac{\theta}{2}$ (f) $\tan \dfrac{\theta}{2}$

43. $\sin \theta = -\dfrac{4}{5}, \theta$ is in QIII   **44.** $\cos \theta = \dfrac{2}{3}, \theta$ is in QIV

SECTION **5.6**   Product-to-Sum and Sum-to-Product Formulas and Sums of the Form $A \sin \theta + B \cos \theta$

Definitions and Concepts	Examples
Product-to-Sum Formulas:	See Example 1, pages 565–566
1. $\sin \alpha \cos \beta = \dfrac{1}{2}[\sin(\alpha + \beta) + \sin(\alpha - \beta)]$	
2. $\cos \alpha \sin \beta = \dfrac{1}{2}[\sin(\alpha + \beta) - \sin(\alpha - \beta)]$	
3. $\sin \alpha \sin \beta = \dfrac{1}{2}[\cos(\alpha - \beta) - \cos(\alpha + \beta)]$	
4. $\cos \alpha \cos \beta = \dfrac{1}{2}[\cos(\alpha + \beta) + \cos(\alpha - \beta)]$	
Sum-to-Product Formulas:	See Example 3, pages 567–568
5. $\sin \alpha + \sin \beta = 2 \sin \dfrac{\alpha + \beta}{2} \cos \dfrac{\alpha - \beta}{2}$	
6. $\sin \alpha - \sin \beta = 2 \cos \dfrac{\alpha + \beta}{2} \sin \dfrac{\alpha - \beta}{2}$	
7. $\cos \alpha + \cos \beta = 2 \cos \dfrac{\alpha + \beta}{2} \cos \dfrac{\alpha - \beta}{2}$	
8. $\cos \alpha - \cos \beta = -2 \sin \dfrac{\alpha + \beta}{2} \sin \dfrac{\alpha - \beta}{2}$	
Theorem:	See Example 6, page 569
$A \sin \theta + B \cos \theta = k \sin(\theta + \varphi)$, where $k = \sqrt{A^2 + B^2}$ and φ is any angle for which $\sin \varphi = \dfrac{B}{\sqrt{A^2 + B^2}} = \dfrac{B}{k}$ and $\cos \varphi = \dfrac{A}{\sqrt{A^2 + B^2}} = \dfrac{A}{k}$	

EXERCISES

Find the exact value of each quantity using Product-to-Sum or Sum-to-Product Formula. Do not use a calculator.

45. $\sin 75° \sin 15°$

46. $8 \sin \dfrac{\pi}{3} \cos \dfrac{5\pi}{6}$

47. $\sin 195° - \sin 105°$

48. $\cos \dfrac{5\pi}{12} - \cos \dfrac{13\pi}{12}$

Applications

49. Harmonic motion A weight is attached to a spring, pulled down, and released to start it oscillating. Suppose the position y of the weight depends on time t according to the formula $y = 4 \sin 3t + 10 \cos 3t$. What is the maximum number of units the weight will be from starting position?

SECTION **5.7**   Trigonometric Equations I

Definitions and Concepts	Examples
Solving trigonometric equations: When solving equations, we will make extensive use of what we have learned about the special angles and reference angles. Typically, we will determine the special angle that is our reference angle. Then we will be able to determine the solution(s) to the equation.	See Example 2, pages 574–575
Non-linear trigonometric equations: In attempting to solve a nonlinear polynomial equation, factoring is often the first technique to be tried.	See Example 8, pages 579–580

EXERCISES

Find the general solution for each equation.

50. $\tan 2\theta = \dfrac{\sqrt{3}}{3}$

51. $\sin\left(\theta + \dfrac{\pi}{3}\right) = 0$

Solve each equation over the interval $[0, 2\pi)$.

52. $\cos \theta \tan \theta = 0$

53. $6 \cos^2 \theta - 11 \cos \theta = -4$

54. $\sin 2\theta - \sin \theta = 0$

55. $\cot^2 \theta = 1 + \csc \theta$

Applications

56. Population of lemurs Suppose the population P of lemurs in a certain region of Madagascar can be modeled by the equation $P = 200 \sin\left(\frac{\pi}{12}x\right) + 500$, where x is the number of years since 2010. Determine the year the lemur population will reach 700.

SECTION **5.8**   Trigonometric Equations II

Definitions and Concepts	Examples
Approximating solutions: If the solution(s) to an equation cannot be found using a special angle, a calculator can be used.	See Example 1, page 585
Solving trigonometric equations of the form $A \sin x + B \sin x = y$: Use the following theorem from Section 5.6: $A \sin \theta + B \cos \theta = k \sin(\theta + \varphi)$ where $k = \sqrt{A^2 + B^2}$ and φ is any angle for which $\sin \varphi = \dfrac{B}{\sqrt{A^2 + B^2}} = \dfrac{B}{k}$ and $\cos \varphi = \dfrac{A}{\sqrt{A^2 + B^2}} = \dfrac{A}{k}$.	See Example 5, page 588

EXERCISES

Solve each equation over the interval $[0, 2\pi)$. Round to three decimal places.

57. $\sin \theta = -0.58$

58. $5 \cos^2 \theta = 2$

59. $5 \tan^2 \theta - 3 \tan \theta - 2 = 0$

60. $4 - 7 \cos^2 \theta = 0$

Applications

61. Projectile motion The range of a projectile fired at an angle θ with the horizontal and with an initial velocity of v_0 feet per second is given by $r = \frac{1}{32} v_0^2 \sin 2\theta$.

(a) Find the range if the initial velocity is 120 feet per second at an angle of $67.5°$.

(b) Find the initial velocity necessary to fire a projectile 400 feet at an angle of $\theta = 30°$.

CHAPTER TEST

Find the exact value of each expression, if any. Do not use a calculator. Note that each answer should be a real number.

1. $\cos^{-1}\left(-\dfrac{\sqrt{2}}{2}\right)$

2. $\tan^{-1}\dfrac{\sqrt{3}}{3}$

Find the exact value of each expression. Do not use a calculator.

3. $\csc\left(\cos^{-1}\dfrac{4}{5}\right)$

4. $\sin\left[\tan^{-1}\left(-\dfrac{2}{7}\right)\right]$

Verify each identity.

5. $\tan^2 x - \cot^2 x = \sec^2 x - \csc^2 x$

6. $\dfrac{(\sin x + \cos x)^2}{\sin x \cos x} = 2 + \sec x \csc x$

Find the exact value of each expression.

7. $\cos\dfrac{3\pi}{8}$

8. $\tan 15°$

9. Given that $\cos \alpha = \dfrac{5}{13}$ (α is in QIV) and $\cot \beta = 2$ (β in QIII), find the exact value of each of the following:

   (a) $\sin(\alpha + \beta)$ **(b)** $\cos(\alpha - \beta)$ **(c)** $\tan(\alpha + \beta)$

10. Given that $\sin \theta = \dfrac{24}{25}$ (θ is in QII), find the exact value of each of the following:

   (a) $\sin 2\theta$ **(b)** $\cos 2\theta$ **(c)** $\tan 2\theta$

   (d) $\sin\dfrac{\theta}{2}$ **(e)** $\cos\dfrac{\theta}{2}$ **(f)** $\tan\dfrac{\theta}{2}$

Find the exact solutions of each equation over the interval $[0, 2\pi)$.

11. $\sin \theta = \cos 2\theta$

12. $4 \sin^2 \theta + 2 \cos^2 \theta = 3$

13. $2 \cos 3\theta = 1$

14. $2 \sin\dfrac{\theta}{3} - \sqrt{3} = 0$

Solve each equation over the interval $[0, 2\pi)$. Round to three decimal places.

15. $5 \sin^2 \theta - 1 = 0$

16. $3 \sin \theta = 7 \cos \theta$

17. $3 \sin^2 \theta - 7 \sin \theta + 2 = 0$

18. $\tan^4 \theta - 13 \tan^2 \theta + 36 = 0$

19. **Projectile motion** A missile is fired with an initial velocity v_0 of 1,250 feet per second and aimed at a target 8 miles away. Determine the minimum angle of elevation θ (in degrees) of the missile if the range r of the missile in feet is represented by $r = \frac{1}{32}v_0^2 \sin 2\theta$. Round to one decimal place.

20. Find the angle from line 2, $y = -5x + 1$, to line 1, $y = 3x$. Round to one decimal place.

 LOOKING AHEAD TO **CALCULUS**

1. **Using Identities to Simplify Expressions for Trigonometric Integration**
   • Connect the Product-to-Sum Formulas to trigonometric integration—Section 5.6

2. **Using Identities to Perform Trigonometric Substitution**
   • Connect identities to trigonometric substitution—Section 5.3

1. Using Identities to Simplify Expressions for Trigonometric Integration

In calculus, trigonometric identities are used to simplify powers of trigonometric functions to perform an operation known as **integration**. Integration has an almost inexhaustible number of applications. The actual process of integration is beyond the scope of this discussion. However, some of the trigonometric simplifications can be discussed.

Expressions in the form $\sin mx \cos nx$, $\sin mx \sin nx$, or $\cos mx \cos nx$ can be simplified using the following formulas:

1. $\sin A \cos B = \dfrac{1}{2}[\sin(A + B) + \sin(A - B)]$

2. $\sin A \sin B = \dfrac{1}{2}[\cos(A - B) - \cos(A + B)]$

3. $\cos A \cos B = \dfrac{1}{2}[\cos(A + B) + \cos(A - B)]$

EXAMPLE 1   **Rewriting $\sin 4x \cos 5x$ in terms of $\sin x$**

Use Formula (1) and let $A = 4$ and $B = 5$.

$$\sin 4x \cos 5x = \frac{1}{2}[\sin((4 + 5)x) + \sin((4 - 5)x)]$$
$$= \frac{1}{2}[\sin 9x + \sin(-x)]$$
$$= \frac{1}{2}\sin 9x + \frac{1}{2}(-\sin x)$$
$$= \frac{1}{2}\sin 9x - \frac{1}{2}\sin x$$

The resulting expression is much easier to integrate in calculus than the original. ∎

Other identities can be used to rewrite trigonometric expressions.

EXAMPLE 2   **Rewriting $\tan^2 x$ as a quotient of trigonometric functions of degree one**

$$\tan^2 x = \frac{\sin^2 x}{\cos^2 x} = \frac{\frac{1}{2}(1 - \cos 2x)}{\frac{1}{2}(1 + \cos 2x)} = \frac{1 - \cos 2x}{1 + \cos 2x}$$

Connect to Calculus:
Using Identities to Rewrite Expressions

1. Rewrite $\sin 5x \sin 2x$ in terms of $\cos x$.
2. Rewrite $\cos 7x \cos 5x$ in terms of $\cos x$.
3. Rewrite $\csc^2 x$ as a quotient of trigonometric functions of degree one.

2. Using Identities to Perform Trigonometric Substitution

In finding the area of an ellipse or circle, the expression $\sqrt{a^2 - x^2}$ $(a > 0)$ is used in an integration problem. The difficulty with this expression is that it cannot be integrated in the algebraic form. Therefore, a suitable trigonometric substitution is used. With $\sqrt{a^2 - x^2}$, we let $x = a \sin \theta \Rightarrow x^2 = a^2 \sin^2 \theta$. We then substitute and simplify.

$$\sqrt{a^2 - a^2 \sin^2 \theta} = \sqrt{a^2(1 - \sin \theta)} = \sqrt{a^2 \cos^2 \theta} = a|\cos \theta|.$$

If we select a proper interval for the angle, we can eliminate the need for the absolute value. We select $-\frac{\pi}{2} \le \theta \le \frac{\pi}{2}$, which allows us to write the expression as $\sqrt{a^2 - x^2} = a \cos \theta$, $x = a \sin \theta$.

EXAMPLE 3 **Rewriting $\dfrac{\sqrt{9 - x^2}}{x^2}$ as a simplified trigonometric expression**

For $\sqrt{9 - x^2}$, we have $a = 3$ and use the substitution $x = 3 \sin \theta$, $-\dfrac{\pi}{2} \le x \le \dfrac{\pi}{2}$.

$$\sqrt{9 - x^2} = \sqrt{9 - 9 \sin^2 \theta} = \sqrt{9(1 - \sin^2 \theta)} = 3\sqrt{\cos^2 \theta} = 3 \cos \theta$$

$$\frac{\sqrt{9 - x^2}}{x^2} = \frac{3 \cos \theta}{9 \sin^2 \theta} = \frac{\cos \theta}{3 \sin \theta \sin \theta} = \frac{1}{3} \cot \theta \csc \theta$$

For the expression $\sqrt{a^2 + x^2}$, use the substitution $x = a \tan \theta$, $-\dfrac{\pi}{2} \le \theta \le \dfrac{\pi}{2}$. ∎

Connect to Calculus:
Using Identities to Rewrite Functions

Rewrite each expression in terms of trigonometric functions.

4. $x^2\sqrt{4 + x^2}$

5. $\dfrac{x}{\sqrt{x^2 + 4}}$

6. $\dfrac{1}{x^2\sqrt{25 - x^2}}$

Applications; Oblique Triangles

CAREERS AND MATHEMATICS: Architect

© Istockphoto/MarcuzClackson

Without architects, what would the world look like? Every home, church, shop, and restaurant has its beginning in an idea by an architect. Architects are licensed professionals who develop concepts and then turn those concepts into a plan and then a structure. To get licensed requires a degree in architecture and at least 3 years of real experience. Architects do a lot more than just draw plans: They are usually involved in every aspect of the construction, working with engineers, urban planners, interior designers, landscape architects, and other professionals.

Education and Mathematics Required
- An architecture degree is generally attained via a 5-year bachelor's-degree program. A master's degree in architecture will take an additional 1 to 5 years. After attainment of a degree, work experience as an intern is required. The final step to attaining a license is the Architect Registration Examination.
- College Algebra, Trigonometry, Calculus I and II, Probability and Statistics, and Linear Programming are required courses.

How Architects Use Math and Who Employees Them
- Architects use mathematics to create a design that can be used by construction engineers to build the structure. Mathematics is essential to analyzing and calculating structural issues to ensure the safety of the building. Mathematical principles such as the Pythagorean Theorem and trigonometric functions make it possible to describe the sizes and shapes of the design.
- About 21% of architects are self-employed. A small percentage work for residential and nonresidential building construction firms and for government agencies. Others are employed by large architectural firms.

Career Outlook and Salary
- Employment for architects is expected to grow 16% from 2008 to 2018, faster than the average for all occupations.
- Median annual wages of architects were $70,320 in May 2008. The top 10% earned more than $119,200 a year.

For more information see http://www.bls.gov/oco.

The ability to solve triangles has many real-world applications. In this chapter, we will learn to solve right triangles and use the Law of Sines and Law of Cosines to solve oblique triangles. Applications involving bearing, area, and linear and angular velocity of an object will be presented.

6.1 Right Triangles

In this section, we will learn to

1. Use the definitions of the trigonometric functions to solve right triangles.

2. Solve application problems involving right triangles.

The *Black Pearl* is a fictional ship in Disney's Pirates of the Caribbean movies. The *Black Pearl* is easily recognized by its black hulls and sails.

In marine navigation, the direction from one object—say, a ship—to another object is called a bearing. Usually this is the direction to an object from one's own vessel. In this section we will study bearing, as well as other applications of right triangles.

In Section 4.3, the six trigonometric functions were defined in terms of ratios formed by the sides of a right triangle, and we solved some application problems. In this section we will continue to examine several applications that can be solved by utilizing the properties of right triangles and the trigonometric functions.

1. Use the Definitions of the Trigonometric Functions to Solve Right Triangles

A triangle has three sides and three angles. If the length of one of the sides is known, along with any two of the other parts, it is possible to **solve** the triangle. Solving a triangle is the process of finding the measurements of the missing parts of the triangle. Solving a right triangle can be accomplished by using the Pythagorean Theorem and the six trigonometric functions, defined using a right triangle.

Pythagorean Theorem | In any right triangle with sides of lengths a, b, and hypotenuse c, $a^2 + b^2 = c^2$.

Additionally, we have these definitions from Section 4.3.

Right-Triangle Definitions of Trigonometric Functions | If angle A is an acute angle in right triangle ABC (Figure 6-1), then

$$\sin A = \frac{\text{opposite side}}{\text{hypotenuse}} = \frac{a}{c} \qquad \csc A = \frac{\text{hypotenuse}}{\text{opposite side}} = \frac{c}{a}$$

$$\cos A = \frac{\text{adjacent side}}{\text{hypotenuse}} = \frac{b}{c} \qquad \sec A = \frac{\text{hypotenuse}}{\text{adjacent side}} = \frac{c}{b}$$

$$\tan A = \frac{\text{opposite side}}{\text{adjacent side}} = \frac{a}{b} \qquad \cot A = \frac{\text{adjacent side}}{\text{opposite side}} = \frac{b}{a}$$

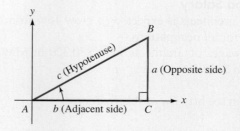

FIGURE 6-1

Throughout this section, we will label any right triangles using the labels in Figure 6-1. When we are asked to solve a right triangle, one angle, C, is the right angle. Consequently, the remaining two angles are complementary angles—i.e., we have $A + B = 90°$. Typically, degree measurements are used when solving right triangles.

Comment   Whenever we solve a triangle, it is always preferable, when possible, to use measurements that are given or exact, rather than approximated. The right angle $C = 90°$ is always a given in a right triangle.

EXAMPLE 1   ## Solving a Right Triangle When Given Two Sides

In a right triangle, $a = 2$ and $b = 5$. Angle C is the right angle. Find the remaining measurements: c, A, and B. Round the side to two decimal places and the angles to one decimal place.

SOLUTION   It's always a good idea to sketch the triangle, making it easier to see the relationships that we are given and the locations of the missing measurements.

• Since we are given two sides, we will use the Pythagorean Theorem to find c.

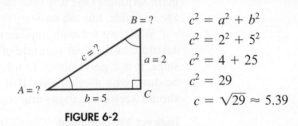

$$c^2 = a^2 + b^2$$
$$c^2 = 2^2 + 5^2$$
$$c^2 = 4 + 25$$
$$c^2 = 29$$
$$c = \sqrt{29} \approx 5.39$$

FIGURE 6-2

• Now we will find angle A by using the given values and the tangent function, then find angle B using the fact that $A + B = 90°$.

$$\tan A = \frac{a}{b}$$

$$\tan A = \frac{2}{5} \qquad \text{Substitute the values.}$$

$$A = \tan^{-1}\frac{2}{5} \qquad \text{Solve for angle } A.$$

$$A \approx 21.8° \qquad \text{Round to one decimal place.}$$

$$B \approx 90° - 21.8° = 68.2°$$

Self Check 1   Solve the right triangle that has $a = 4$ and $c = 7$. Round the side to two decimal places and the angles to one decimal place.

Now Try Exercise 11.

EXAMPLE 2   ## Solving a Right Triangle When Given One Side and One Angle

Solve the right triangle with hypotenuse $c = 10$ and acute angle $A = 33°$. Note that the right angle $C = 90°$ is always a given in a right triangle. Round the lengths of the sides to two decimal places.

SOLUTION   Looking at Figure 6-3, we first determine that angle $B = 90° - A = 90° - 33° = 57°$. This is an exact measurement; should we choose, it would be acceptable to use it to find the lengths of the sides a and b. We demonstrate this as we use the sine function to find both sides.

$$\sin 33° = \frac{a}{10} \qquad\qquad \sin 57° = \frac{b}{10}$$

$$a = 10 \sin 33° \qquad\qquad b = 10 \sin 57°$$

$$a \approx 5.45 \qquad\qquad\quad b \approx 8.39$$

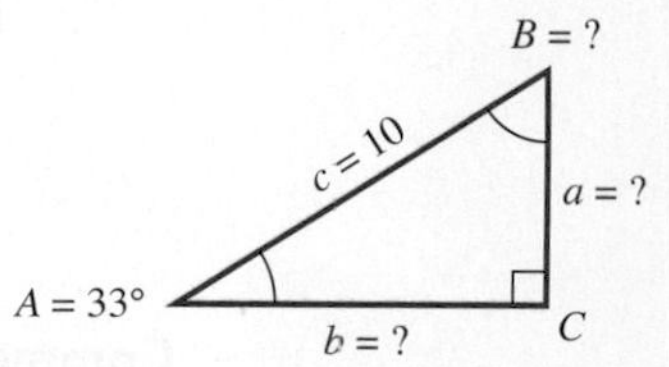

FIGURE 6-3

Self Check 2   Solve the right triangle that has hypotenuse $c = 8.5$ and angle $A = 15.5°$. Round the lengths of the sides to two decimal places.

Now Try Exercise 13.

2. Solve Application Problems Involving Right Triangles

Many applications can be solved using right triangles. Some measurements are very hard to calculate. For example, how do we measure the height of a mountain or a giant sequoia? One way is to use trigonometry. As technology advances, many tools are available, but the basis of their operation is trigonometry.

When we solve an application problem using triangles, answers can only be as accurate as the least accurate of the given data. When a calculation requires several steps, it is a good policy to allow as much of the final computation as possible to be done with a calculator. If it is necessary to find an intermediate value, then we should keep extra digits and round off only when the final answer is found.

Indirect Measurement

If an observer looks up at an object such as a plane, the top of a building, etc., the angle that the observer's line of sight makes with the horizontal is called the **angle of elevation.** If the observer looks down to see the object, the angle made with the horizontal is called the **angle of depression.** The angle of elevation to an object is equal to the angle of depression from the object. Recall from geometry that if two horizontal lines are crossed by a transversal, the alternate interior angles are equal (Figure 6-4).

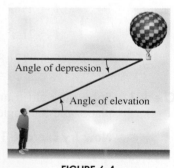

FIGURE 6-4

EXAMPLE 3   Solving an Angle of Elevation Problem

From a location 23.0 meters from the base of a flagpole, the angle of elevation to the top of the flagpole is 37.5°. Find the height of the flagpole to the nearest tenth of a meter.

SOLUTION   Looking at Figure 6-5, we see that a right triangle can be formed using the given measurements and the flagpole. We let h represent the height of the flagpole and use the tangent function to determine h.

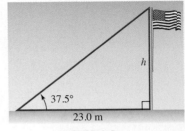

$$\tan 37.5° = \frac{h}{23.0}$$ $\tan \theta = \dfrac{\text{opposite}}{\text{adjacent}}$

$$h = 23.0 \tan 37.5°$$ Solve for h.

$$h \approx 17.6 \text{ m}$$

FIGURE 6-5

The flagpole is approximately 17.6 meters tall.

Self Check 3   Repeat Example 3 if the distance from the base of the pole is 15 meters with an angle of elevation of 47.2°. Round to the nearest tenth of a meter.

Now Try Exercise 21.

EXAMPLE 4   **Solving an Angle of Inclination Problem**

A guy wire that is 92.0 feet long is attached to a point 4 feet from the top of a tower 79 feet high. What is the angle of inclination formed by the ground and the wire? Round to the nearest tenth of a degree.

SOLUTION   Refer to Figure 6-6, which shows the measurements given. Note that the distance from the ground to the point of attachment on the pole is 75 feet ($79 - 4 = 75$). To find angle θ, we will use the sine function.

$$\sin \theta = \frac{75}{92}$$ Definition of sine.

$$\theta = \sin^{-1} \frac{75}{92}$$ Solve the equation for θ.

$$\theta \approx 54.6°$$

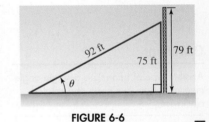

FIGURE 6-6

Self Check 4   A guy wire that is 120.0 feet long is attached to a point 6 feet from the top of a tower 90 feet high. What is the angle of inclination formed by the ground and the wire? Round to the nearest tenth of a degree.

Now Try Exercise 23.

To solve some problems we may need to use two right triangles.

EXAMPLE 5   **Finding the Height of a Statue on Top of a Building**

A statue stands on top of a building. From a point 105.3 feet from the base of the building, the angles of elevation to the top and the base of the statue are 52.1° and 47.4°, respectively (Figure 6-7). Find the height s of the statue. Round to the nearest tenth of a foot.

SOLUTION   There are two right triangles in the figure. We will use both of them. First, we will find the value of h using the tangent ratio.

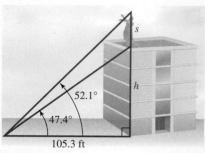

FIGURE 6-7

$$\tan 47.4° = \frac{h}{105.3}$$

$$h = 105.3 \tan 47.4°$$

At this point we will not actually calculate this value, leaving that to be done in the final step. Now we find the value of $s + h$, again by using a tangent ratio. Then we will be able to find the value of s.

$\tan 52.1° = \dfrac{s + h}{105.3}$	Set up tangent ratio.
$s + h = 105.3 \tan 52.1°$	
$s = 105.3 \tan 52.1° - h$	Solve for s.
$s = 105.3 \tan 52.1° - 105.3 \tan 47.4°$	Substitute value of h.
$s \approx 20.8$ ft	Calculator approximation

The height of the statue is approximately 20.8 feet.

Self Check 5   A cellular tower stands on top of a building. From a point 112 feet from the base of the building, the angles of elevation to the top and the base of the tower are 50.5° and 35°, respectively. Find the height of the tower. Round to the nearest tenth of a foot.

Now Try Exercise 31.

Comment

Another type of bearing is a single angle measured from due north and rotated clockwise.

Bearing

Navigators and surveyors use the concept of bearing. Look at Figure 6-8. The **bearing** of point A from point O (the observer) is the acute angle measured from the north–south line to the segment OA. This bearing is denoted as **N 30° E** and is read as "north 30° east." The bearing of point B from O is **N 75° W**, the bearing of point C from O is **S 20° W**, and the bearing of point D from O is **S 80° E**.

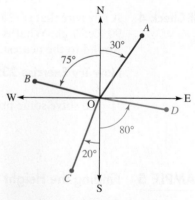

FIGURE 6-8

EXAMPLE 6   Finding Bearing from One City to Another

Rochelle is 25 miles due south of Rockford, and North Chicago is 65 miles due east of Rockford. Find the bearing of North Chicago from Rochelle. Round to the nearest degree.

SOLUTION   Referring to Figure 6-9, we see that a right triangle can be used to solve this problem. To find the bearing, we find angle θ by using the tangent ratio.

$$\tan \theta = \frac{\text{opposite}}{\text{adjacent}}$$

$$\tan \theta = \frac{65}{25} \qquad \text{Tangent ratio}$$

$$\theta = \tan^{-1} \frac{65}{25} \qquad \text{Solve for } \theta.$$

$$\theta \approx 69°$$

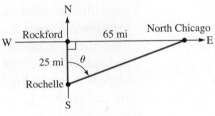

FIGURE 6-9

North Chicago is N 69° E from Rochelle.

Self Check 6   Zachary, Louisiana, is 14 miles due north of Baton Rouge. Hammond is 44 miles due east of Baton Rouge. What is the bearing of Zachary from Hammond? Round to the nearest degree.

Now Try Exercise 37.

EXAMPLE 7   **Using Bearings to Locate a Position**

Two forest-fire lookouts are on a north–south line, 17.2 miles apart. The bearing of a fire from lookout A is S 27.0° W, and the bearing of the fire from lookout B is N 63.0° W. How far is the fire from lookout B? Round to the nearest tenth of a mile.

SOLUTION   See Figure 6-10. Because the sum of the three angles of any triangle is 180°, the angle at point F is 90°, so triangle AFB is a right triangle with hypotenuse AB. We can find the distance of B from the fire by using the cosine ratio.

$$\cos 63.0° = \frac{\text{adjacent side}}{\text{hypotenuse}}$$

$$\cos 63.0° = \frac{a}{17.2} \qquad \text{Set up the ratio.}$$

$$a = 17.2 \cos 63.0° \qquad \text{Solve for } a.$$

$$a \approx 7.8$$

The is fire is approximately 7.8 miles from lookout B.

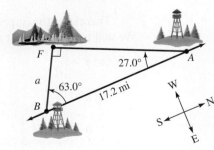

FIGURE 6-10

Self Check 7   Observation tower A is 15.7 miles due west of tower B. A fire is spotted at a bearing of N 70° E from tower A and N 20° W from tower B. How far is the fire from each tower? Round to the nearest tenth of a mile.

Now Try Exercise 39.

Electrical Engineering

Certain components of an electronic circuit behave in ways that can be analyzed with trigonometric functions. For example, a resistor and an inductor connected as a series network will oppose the flow of an alternating current. A measure of this opposition is called the network's **impedance.** The impedance z is determined by the **inductive reactance** X of the inductor and the **resistance** R of the resistor. The three quantities z, R, and X represent the sides of the right triangle shown in Figure 6-11 and are related (by the Pythagorean Theorem) by the equation $z^2 = X^2 + R^2$. The angle θ shown in Figure 6-11 is called the **phase angle,** and from the figure we can see that $\tan \theta = \frac{X}{R}$.

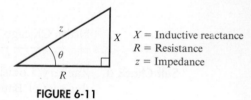

X = Inductive reactance
R = Resistance
z = Impedance

FIGURE 6-11

EXAMPLE 8   Finding Impedance and Phase Angle

An alternating-current generator **G** and an impedance z, consisting of a resistance R and an inductive reactance X, are connected as in Figure 6-12. If $R = 300$ ohms and $X = 400$ ohms, find the impedance z and the phase angle θ, to the nearest tenth of a degree.

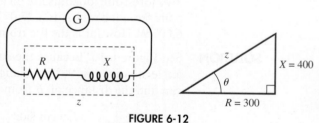

FIGURE 6-12

SOLUTION   Using the circuit and the impedance triangle shown in Figure 6-12, we can determine both z and θ.

$$z^2 = X^2 + R^2 \qquad \text{Pythagorean Theorem}$$
$$z = \sqrt{400^2 + 300^2} \qquad \text{Substitute values for } X \text{ and } R.$$
$$z = \sqrt{250,000}$$
$$z = 500$$

Therefore, the impedance is 500 ohms.
We will use the tangent ratio to find the phase angle θ.

$$\tan \theta = \frac{X}{R} \qquad \text{Tangent ratio}$$
$$\tan \theta = \frac{400}{300} \qquad \text{Substitute values.}$$
$$\theta = \tan^{-1} \frac{4}{3} \qquad \text{Solve for } \theta.$$
$$\theta \approx 53.1°$$

The phase angle θ is approximately 53.1°.

Self Check 8   Repeat Example 8 with $R = 350$ ohms and $X = 450$ ohms.

Now Try Exercise 43.

In calculus, the rate at which a quantity is changing is the basis of problems that are called **related-rates** problems. Some of these problems utilize the properties of right triangles and the trigonometric functions defined by a right triangle in order to find the desired rate of change.

Self Check Answers   1. $b = 5.74$; $A = 34.8°$; $B = 55.2°$ 2. $B = 74.5°$; $a = 2.27$; $b = 8.19$
3. 16.2 m   4. 44.4°   5. 57.4 ft   6. N 72° W
7. 5.4 mi from B; 14.8 mi from A   8. 570.1 Ω; $\theta = 52.1°$

Exercises **6.1**

Getting Ready
You should be able to complete these vocabulary and concept statements before you proceed to the practice exercises.

Fill in the blanks.

1. The Pythagorean Theorem is _____.
2. To solve a triangle means to _____.
3. The longest side in a right triangle is always opposite the _____ angle.
4. State the six definitions of the trigonometric functions of angle θ, using the figure shown.

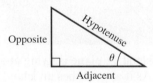

Practice
Solve each triangle by finding the unknown sides and angles. Round each answer to the nearest tenth.

5.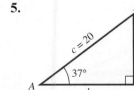

6.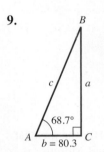

7.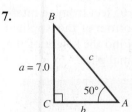

8.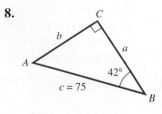

9.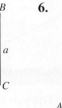

10.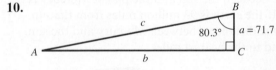

Solve the right triangle with the given sides and given that angle C is a right angle. Round each answer to the nearest tenth.

11. $b = 5$; $c = 11$ 12. $a = 3$; $b = 7$
13. $A = 42°$; $c = 15$ 14. $A = 54°$; $c = 10$
15. $B = 63°$; $a = 4.3$ 16. $B = 71°$; $a = 3.9$
17. $a = 4$; $c = 8$ 18. $b = 11$; $c = 15$
19. $a = 7.2$; $b = 9.6$ 20. $a = 5.3$; $b = 4.1$

Applications

21. **Angle of elevation**   A plane flying at 18,100 feet passes over an observer. Thirty seconds later, the observer notes that the plane's angle of elevation is 31.0°. How fast is the plane moving in miles per hour? (Hint: 5,280 feet = 1 mile.) Round to the nearest unit.

22. **Angle of elevation**   An observer noted that the angle of elevation to a plane over a landmark was 32°40′. If the landmark was 1,530 meters from the observer, find the altitude of the plane. Round to the nearest meter.

23. **Angle of depression** The angle of depression from a point at the top of a tree to a point on the ground is 57°45′. Find the height of the tree if the line-of-sight distance from the top of the tree to the point on the ground is 34.23 feet. Round to two decimal places.

24. **Angle of depression** A plane flying horizontally at 650 miles per hour passes directly over a small city. One minute later, the pilot notes that the angle of depression to that city is 13°. Find the plane's altitude in feet. Round to the nearest foot.

25. **Meteorology** The height h of a cloud can be measured by shining a searchlight vertically, as shown in the figure, and measuring the angle of elevation θ at a distance D from the light. If $D = 557$ feet and $\theta = 78.3°$, find h. Round to the nearest foot.

26. **Astronomy** As viewed from the earth, the greatest angle between the sun and the planet Mercury is 28°. If the earth is 93 million miles from the sun, find the distance d between Mercury and the sun. Round to the nearest mile.

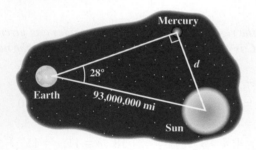

Use the information from the right triangles shown in the figure to find the value of x. Round to two decimal places.

27.
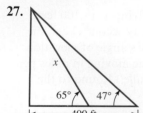

28.
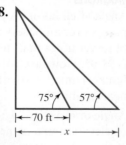

29.
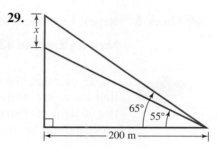

30.
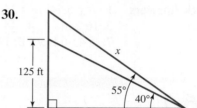

31. **Height on Mount Rushmore** Use the information in the figure to find the height of George Washington's face on Mount Rushmore. Round to one decimal place.

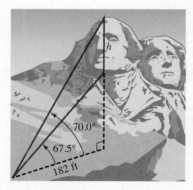

32. **Length of island** A plane is flying at an altitude of 5,120 feet. As it approaches an island, the navigator determines the angles of depression as shown in the figure. Find the length of the island in feet and in miles. Round to the nearest foot and to the nearest tenth of a mile.

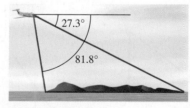

33. **Height of stoplight** The sun visor in Angie's car prevents her from seeing objects more than 17° above the horizon. When her car is 65 feet from an intersection, she barely sees the bottom of the stoplight hanging above the roadway. Find the height of the stoplight. Round to one decimal place.

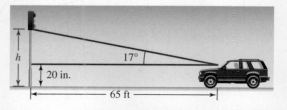

34. Length of shadow When the sun is 75.0° above the horizon, a vertical flagpole casts a shadow of 17.2 ft on the horizontal ground. When the sun sinks to 65.0° above the horizon, how long is the shadow? Round to one decimal place.

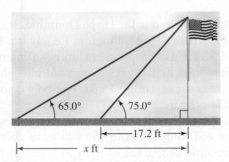

35. Navigation A ship leaves from port on a bearing of S 12.7° E. How far south has the ship traveled during a trip of 327 miles? Round to the nearest mile.

36. Bearing A ship leaves port and sails 8,800 kilometers due west. It then sails 4,500 kilometers due south. To the nearest tenth of a degree, find the ship's bearing from its port.

37. Bearing The bearing of Madison, Wisconsin, from Stevens Point is S 4.1° E. The distance between the cities is 108 miles. By how much is Madison east of Stevens Point? Round to two decimal places.

38. Bearing A ship is 3.3 miles from a lighthouse. It is also due north of a buoy that is 2.5 miles due east of the lighthouse. Find the bearing of the ship from the lighthouse. Round to one decimal place.

39. Bearing Two lighthouses are on an east–west line. The bearing of a ship from one lighthouse is N 59° E, and its bearing from the other lighthouse is N 31° W. How far apart are the lighthouses if the ship is 5.0 miles from the first lighthouse? Round to one decimal place.

40. Flight of a rocket From an observation point 6,500 meters from a launch site, an observer watches the vertical flight of a rocket. At one instant, the angle of elevation of the rocket is 15°. How far will the rocket ascend in the time it takes the angle of elevation to increase to 57°? Round to one decimal place.

41. Bearing Two lookouts are on a north–south line. The bearing of a forest fire from one lookout is S 67° E, and its bearing from the second lookout is N 23° E. If the fire is 3.5 kilometers from the second lookout, how far is it from the first? Round to one decimal place.

42. Bearing A boat is 537 meters from a lighthouse and has a bearing from the lighthouse of N 33.7° W. A second boat is 212 meters from the same lighthouse and has a bearing from the lighthouse of S 20.1° W. How many meters north of the second boat is the first? Round to one decimal place.

43. Impedance An impedance consists of a resistance of 120 ohms and an inductive reactance of 50.0 ohms. Find the impedance. Round to the nearest ohm.

44. Phase angle Use the information in Exercise 43 to find the phase angle. Round to one decimal place.

45. Phase angle An impedance consists of a resistance of 200 ohms in series with an unknown inductance. If the impedance measures 290 ohms, find the phase angle. Round to one decimal place.

46. Inductive reactance Use the information in Exercise 45 to find the inductive reactance. Round to the nearest ohm.

47. Resistance The impedance of a series resistor-inductor network is 240 ohms and the phase angle is 60.0°. Find the resistance R. Round to the nearest ohm.

48. Inductive reactance Use the information in Exercise 47 to find the inductive reactance X. Round to one decimal place.

49. Angle of expansion The top of the stepladder in the figure is 8.0 feet above the floor, and the legs open at an angle θ. The legs of the ladder are 4.5 feet apart. Find the angle θ. Round to one decimal place.

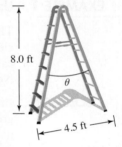

50. Geometry Find the perimeter of a regular pentagon inscribed in a circle with a radius of 12.0 inches. Round to one decimal place.

Discovery and Writing

51. Refer to the figure to find θ. Round to one decimal place.

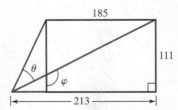

52. Refer to the figure in Exercise 51 to find φ. Round to one decimal place.

6.2 More Right-Triangle Applications

In this section, we will learn to

1. Solve application problems using right triangles.

One of the most famous diamonds in the world is the Hope Diamond. It is housed in the National Museum of Natural History in Washington, DC. It is a beautiful deep-blue diamond that weighs 45.52 carats. The stone exhibits an intense luminescence, and after exposure to ultraviolet light it produces a red phosphorescent glow-in-the-dark color. This color lasts for several seconds. The blue coloration is attributed to very small amounts of boron in the stone.

Optics is the branch of physics that studies the behavior and properties of light. In this section, one of the topics we will study is optics. We will learn about Snell's Law and use trigonometry to understand properties of light as it travels from one medium to another.

There are an inexhaustible number of applications that can use right triangles in the solution. In this section, we continue our study of these applications that utilize right-triangle trigonometry.

1. Solve Application Problems Using Right Triangles

EXAMPLE 1 **Using Two Right Triangles to Solve a Problem**

The Space Needle of Seattle is due north of observer A, who finds the angle of elevation to its top to be 44.4°. A second observer, B, is 706 feet due east of A and calculates the bearing of the Space Needle to be N 48.8° W. How tall is the Space Needle? Round to the nearest foot.

SOLUTION In this problem, we consider two right triangles: triangle PAB, which lies on the ground, and triangle APQ, which sits on the edge with the Space Needle as one of its sides. (See Figure 6-13.) Not enough information about triangle APQ is given for us to use it to determine the height h. However, if we find the value of x first, the value of h can be computed. From triangle APQ, we have

$$\tan 44.4° = \frac{h}{x} \quad \text{or} \quad h = x \tan 44.4°.$$

FIGURE 6-13

The distance x can be found using right triangle PAB. Angle PBA is $90° - 48.8° = 41.2°$. We can now find the value of x, and then the value of h, as shown.

$$\tan 41.2° = \frac{x}{706} \qquad \text{or} \qquad x = 706 \tan 41.2°$$

$$h = \boldsymbol{x} \tan 44.4°$$

$$= \boldsymbol{706 \tan 41.2°} \tan 44.4°$$

$$\approx 605$$

Therefore, we have determined that the Space Needle is approximately 605 feet tall.   ∎

Self Check 1   Find the value of h in the figure. Round to two decimal places.

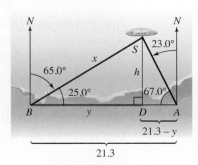

Now Try Exercise 1.

Bearing

EXAMPLE 2   Using Two Right Triangles to Solve a Problem

Two Coast Guard lookouts, A and B, are on an east–west line, 21.3 kilometers apart. The bearing of a ship from lookout A is N 23.0° W, and its bearing from lookout B is N 65.0° E. How far is the ship from lookout B? Round to the nearest tenth.

SOLUTION   In Figure 6-14, we see that triangle BAS is not a right triangle. However, if the perpendicular line segment SD of length h is drawn from point S to side BA, then the two right triangles are formed. If we let y represent the length of BD, then $21.3 - y$ will represent the length of DA.

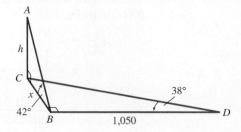

FIGURE 6-14

We now have

$$\tan 25.0° = \frac{h}{y} \qquad \text{or} \qquad \boldsymbol{h = y \tan 25.0°}$$

and

$$\tan 67.0° = \frac{h}{21.3 - y} \qquad \text{or} \qquad \boldsymbol{h = (21.3 - y) \tan 67.0°}.$$

Setting the two values of h equal to each other and solving for y gives

$$y \tan 25.0° = (21.3 - y) \tan 67.0°$$
$$y \tan 25.0° = 21.3 \tan 67.0° - y \tan 67.0°$$
$$y \tan 25.0° + y \tan 67.0° = 21.3 \tan 67.0°$$
$$y(\tan 25.0° + \tan 67.0°) = 21.3 \tan 67.0°$$
$$y = \frac{21.3 \tan 67.0°}{\tan 25.0° + \tan 67.0°}$$

Instead of finding an approximation for y, we will wait until the final step and do it all at once with a calculator. We can now use the cosine ratio from right triangle BSD to find x.

$$\cos 25.0° = \frac{y}{x}$$
$$x = \frac{y}{\cos 25.0°}$$
$$x = \frac{21.3 \tan 67.0°}{\tan 25.0° + \tan 67.0°} \cdot \frac{1}{\cos 25.0°}$$
$$x \approx 19.6$$

The ship is approximately 19.6 kilometers from lookout B. ∎

Self Check 2   Repeat Example 2 if the bearing from lookout A is N 33° W and the bearing from lookout B is N 63° E. Round to the nearest tenth.

Now Try Exercise 3.

Angle of Inclination

If a nonvertical line passes through points $P(x_1, y_1)$ and $Q(x_2, y_2)$, the **slope** of the line is defined as

$$\text{slope of } PQ = \frac{\text{rise}}{\text{run}} = \frac{y_2 - y_1}{x_2 - x_1} \qquad (x_2 \ne x_1).$$

In right triangle PQR in Figure 6-15, $y_2 - y_1$ is the length of the side opposite angle α, and $x_2 - x_1$ is the length of the side adjacent to angle α. Therefore, the slope is the ratio of the opposite side to the adjacent side, and

$$\text{slope of } PQ = \tan \alpha.$$

Angle α, measured counterclockwise from the horizontal line segment, is called the **angle of inclination** of line PQ. When $\alpha = 90°$, $\tan \alpha$ is undefined. This indicates that a vertical line has no slope.

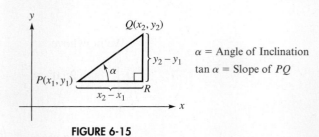

FIGURE 6-15

EXAMPLE 3   Finding the Angle of Inclination

Find the angle of inclination of the line that passes through the points $(-1, -2)$ and $(-5, 2)$.

SOLUTION   Our first step is to find the slope of the line.

$$m = \frac{2-(-2)}{-5-(-1)} = \frac{4}{-4} = -1$$

From this we determine that $\tan \alpha = -1$.
From Figure 6-16, we see that $90° < \alpha < 180°$.
The reference angle is $45°$, since $\tan 45° = 1$,
and we determine that $\alpha = 135°$.

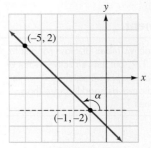

FIGURE 6-16

Self Check 3   Find the angle of inclination of the line that passes through the points $(2, 3)$ and $(0, 1)$.

Now Try Exercise 11.

Refractive Index

Have you ever partially submerged a straight stick, like a ruler, into water and noticed that the stick appears to be bent at the point where it enters the water? Light travels more rapidly in air than in water or glass. When light passes from one substance into another in which it slows down, its path is bent. How much this happens depends on the **refractive indexes** of the two media and the angle between the light ray and the line perpendicular to the surface separating the two media. This is the principle behind camera lenses, eyeglasses, and the sparkle in diamonds.

In 1621, Willebrord Snell (1591–1626), a Dutch physicist, was able to determine the relationship between the different angles of light as it passes from one transparent medium to another. When light passes from one transparent medium to another, it bends according to **Snell's Law:**

$$\frac{\sin i}{\sin r} = n,$$

where i is the **angle of incidence,** r is the **angle of refraction,** and n is a constant called the **refractive index.**

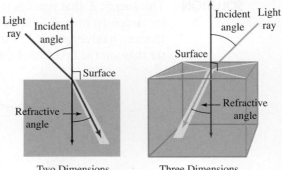

FIGURE 6-17

EXAMPLE 4   **Finding the Index of Refraction**

If light striking a piece of glass at a 10° angle is refracted to an angle of 6°, find the index of refraction of the glass. Round to one decimal place.

SOLUTION   The incident angle is $i = 10°$, and the refractive angle is $r = 6°$, so the index of refraction of the glass is

$$
\begin{aligned}
n &= \frac{\sin i}{\sin r} \\
&= \frac{\sin 10°}{\sin 6°} \\
&\approx 1.7
\end{aligned}
$$

Comment

The refractive index of glass ranges from 1.59 to 1.9, but the refractive index of diamonds is approximately 2.4, which is why diamonds sparkle more than rhinestones.

Self Check 4   If light striking a piece of glass at a 12° angle is refracted to an angle of 7°, find the index of refraction of the glass. Round to one decimal place.

Now Try Exercise 17.

Solar Energy

If we asked you, "When is the earth closest to the sun?" you most likely would answer that it is in the summer (for those who live in the Northern Hemisphere). In fact, the earth is closest to the sun in December. Summers and winters are caused by the angle at which the sun's energy hits the earth. This angle varies as the earth moves along its orbit, because the axis of the earth is tilted by approximately 23°30′. In the summer, sunlight approaches the earth at angles close to 90°; but in the winter, the sunlight is not nearly as direct.

In the Northern Hemisphere at noon on June 21 (the summer solstice), the sun is as far north as it will ever get; this is the first day of summer. Six months later, at noon on December 22, the sun appears 47° farther south in the sky; this is the beginning of winter. The amount of solar heat energy that reaches the earth is proportional to the cosine of the angle between the sun's rays and the vertical. Because this angle is greater during the winter months, less energy reaches us, and the days are colder.

EXAMPLE 5   **Computing Solar Energy**

At noon on the first day of summer in Havana, Cuba, the sun is directly overhead. How much less energy reaches Havana at noon on the first day of winter?

SOLUTION   The energy E that reaches the earth is proportional to the cosine of the angle of incidence. In June, $E = k \cos 0° = k$, whereas in December, $E = k \cos 47° \approx 0.68k$. Havana receives about $1 - 0.68 = 0.32$ or 32% less energy on the first day of winter than on the first day of summer.

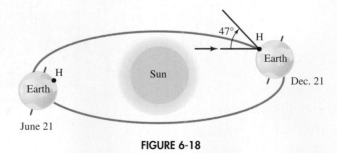

FIGURE 6-18

Self Check 5   Find the amount of the sun's energy that reaches the earth if the angle of incidence is 45°.

Now Try Exercise 19.

Trigonometric functions can be used to relate various sides and angles of geometric figures.

EXAMPLE 6   **Using Geometry**

An isosceles triangle has equal sides of k units and a vertex angle of α degrees. Express the length b of its base in terms of k and α.

SOLUTION   In Figure 6-19, the segment CD is the perpendicular drawn from the vertex of the triangle to the base. Because this perpendicular bisects the vertex angle, we have

$$\sin\frac{\alpha}{2} = \frac{x}{k}$$

$$x = k\sin\frac{\alpha}{2}$$

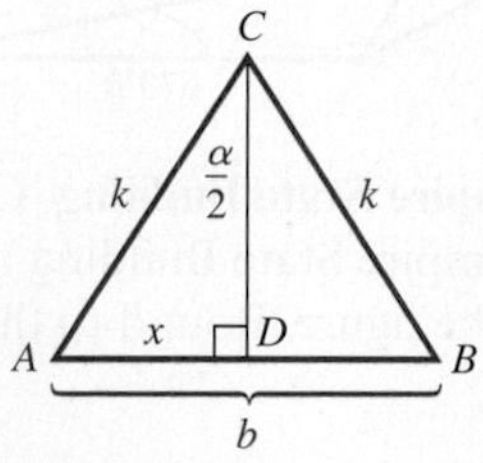

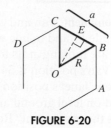

FIGURE 6-19

Because the perpendicular CD also bisects the base, we have $b = 2x$ and we substitute $k\sin\frac{\alpha}{2}$ for x to obtain $b = 2k\sin\frac{\alpha}{2}$.

Self Check 6   A regular polygon has n equal sides, each of length a. The radius of its circumscribed circle is R. Express a as a function of n and R. (See Figure 6-20.)

FIGURE 6-20

Now Try Exercise 23.

Self Check Answers   **1.** 738.65   **2.** 18.0 km   **3.** 45°   **4.** 1.7   **5.** $\dfrac{\sqrt{2}}{2}k$

6. $a = 2R\sin\dfrac{180°}{n}$

Exercises **6.2**

Applications

1. **Height of the Willis Tower** Compute the height h of the Willis Tower, formerly the Sears Tower, using the information given in the figure. Round to the nearest foot.

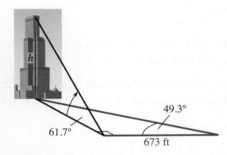

2. **Height of the Empire State Building** Compute the height h of the Empire State Building using the information given in the figure. Round to the nearest foot.

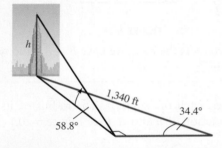

3. **Height of the Washington Monument** Van and Janet, standing on the same side of and in line with the Washington Monument, are looking at its top. The angle of elevation from Van's position is 34.1°, and the angle of elevation from Janet's position is 60.0°. If Van and Janet stand on level ground and are 500 feet apart, how tall is the monument? Round to the nearest foot.

4. **Height of the Statue of Liberty** Use the information given in the figure to compute the height of the figure part of the Statue of Liberty. Round to the nearest foot.

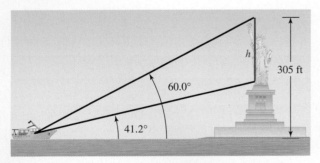

5. **Height of the Gateway Arch** Use the information given in the figure to compute the height h of the Gateway Arch in Saint Louis. Round to the nearest foot.

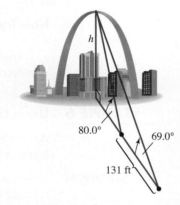

6. **Bearing** The bearing of point C from point A is N 25.0° E, and its bearing from point B is N 40.0° W. If A and B are on an east–west line and 200 kilometers apart, how far is B from C? Round to the nearest kilometer.

7. **Geometry** Find the height h of the triangle shown in the figure. Round to one decimal place.

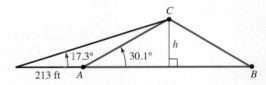

8. **Geometry** Find the length of side AC in the figure in Exercise 7. Round to one decimal place.

9. **Measuring a tower** Cheryl is standing 152 feet due east of a television tower. She then travels along a bearing of S 60.0° W. How tall is the tower if the angle of elevation to its top from her closest position to the tower is 58.2°? Round to the nearest foot.

10. **Height of the Eiffel Tower** Jonathan stands due south of the Eiffel Tower and notes that the angle of elevation to its top is 60.0°, as shown in the figure. His brother Joshua, standing 298 feet due east of Jonathan, calculates an angle of elevation of 56.9°. How tall is the tower? (*Hint*: Consider a triangle on the ground and use the Pythagorean Theorem.) Round to the nearest tenth of a foot.

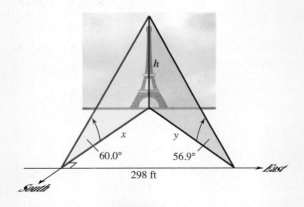

11. **Angle of inclination** Find the angle of inclination of the line that passes through the points $(1, 3)$ and $(3, 5)$.

12. **Angle of inclination** Find the angle of inclination of the line that passes through the points $(0, -1)$ and $(-2\sqrt{3}, 1)$.

13. **Angle of inclination** Find the angle of inclination of the line that passes through the points $(-7, 10)$ and $(-7, -4)$.

14. **Angle of inclination** Find the angle of inclination of the line that passes through the points $(12, -4)$ and $(-7, -4)$.

15. **Optics** A beam of light enters a block of ice at an angle of incidence of $8.0°$ and is refracted to an angle of $6.1°$, as shown in the figure. Find the index of refraction of the ice. Round to two decimal places.

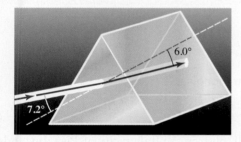

16. **Optics** A beam of light enters a prism, passing through the angle shown in the figure. Find the index of refraction of the glass. Round to two decimal places.

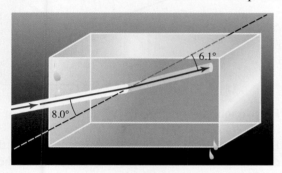

17. **Index of refraction** Tetrachlorethylene, a clear liquid (used in dry cleaning), has an index of refraction equal to that of a certain brand of glass. Could you see a piece of that glass submerged in the liquid?

18. **Finding counterfeits** Refer to Exercise 17 and explain how you could distinguish real diamonds from fake diamonds.

In Exercises 19 and 20, assume that the amount of the sun's energy reaching the earth is directly proportional to the cosine of the angle of incidence.

19. **Solar heating** Find the amount of the sun's energy that reaches the earth if the angle of incidence is $30°$.

20. **Angle of incidence** If the sun's energy reaching the earth is 90% of what it would be if the sun were directly overhead, what is the angle of incidence?

21. **Photographic lighting** A photographer aims his light directly at a painting he wishes to photograph. How much less light will hit the painting if he lowers his light by $20°$? (See the figure.) Assume that the light remains the same distance from the painting. Round to the nearest percent.

22. **Photographic lighting** The light in the figure moves from $30.0°$ below the screen to $10.0°$ above. How much more light hits the screen with the light in the second position? Round to the nearest percent.

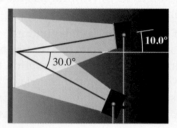

Discovery and Writing

23. **Geometry** Two tangents are drawn from a point P to a circle of radius r, as shown in the figure. The angle between the tangents is θ. In terms of r and θ, find the length of each tangent.

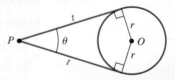

24. **Geometry** A regular polygon has n equal sides, each of length a. The radius of the inscribed circle is r. Express a as a function of n and r.

25. **Geometry** An isosceles triangle has a base angle of α and a side of k centimeters. Express the length b of its base in terms of α and k.

26. **Navigation** At noon, a ship left port on a bearing of N α E and steamed k kilometers. At the same time, another ship left port on a bearing of N α W and steamed k kilometers. Express the distance d between the two ships in terms of α and k.

27. **Height of a building** Point A is k meters from a building. From point A, the angle of elevation to the top of the building is α and the angle of depression to the base of the building is β. Express the height H of the building in terms of $k, \alpha,$ and β.

28. Sliding ladder A ladder k feet long reaches a height of H feet on the side of a building, and the ladder makes an angle of α with the horizontal. The ladder slips so that it reaches a height of only h feet on the side of the building. It then makes an angle of β with the horizontal. Express the distance d that the ladder has come down the building in terms of k, α, and β.

29. Flight of a rocket From an observation point D meters from a launch site, an observer watches the vertical flight of a rocket. At one instant, the angle of elevation of the rocket is θ; at a later instant, it is φ. How far has the rocket traveled during that time?

30. Angle of depression Two helicopters are hovering at an altitude of d feet. The pilot of one helicopter observes a crate being dropped from the other helicopter. At one instant, the pilot measures the angle of depression of the crate to be α. When the crate hits the ground, the pilot measures the angle of depression to be β. How far has the crate fallen in that time?

31. Height of a building Two tall buildings are separated by a distance of d meters. From the top of the shorter building, the angle of elevation to the top of the taller building is α and the angle of depression to the base of the taller building is β. Express the height a of the shorter building in terms of d and β.

32. Height of a building Express the height b of the taller building described in Exercise 31 in terms of d, α, and a.

33. Passive solar heating In summer, when the sun is highest in the sky, the roof overhang shown in the figure prevents the sun from striking the south-facing wall. Find the amount of overhang x in terms of θ and h.

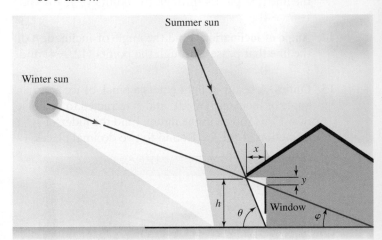

34. Passive solar heating In the winter, when the sun is lowest in the sky, the window in the figure in Exercise 33 is not shaded by the overhang. Find the distance y between the top of the window and the roof in terms of φ, θ, and h.

6.3 Law of Sines

In this section, we will learn to

1. Use the Law of Sines to solve triangles.

2. Solve applications using the Law of Sines.

The Tower of Pisa is a famous bell tower in the Italian city of Pisa. It stands in the cathedral square next to the cathedral. Construction began in 1173 and took place over a period of 177 years. Although the tower was intended to stand vertically, it began to lean shortly after construction started, due to foundation problems.

Suppose we are standing 542 feet away from the base of the Tower of Pisa and the angle of elevation to its top is 17.9°. If the tower makes an angle of 94.5° with the ground, how tall is the tower? Initially, this seems like an easy triangle problem to solve, but because the tower isn't vertical, the triangle isn't a right triangle, and we currently do not know any methods to use to solve the problem.

A triangle that does not have a right angle is called an **oblique triangle.** In this section, we will learn about oblique triangles and the **Law of Sines,** which can be used to solve some oblique triangles. Once we learn the Law of Sines, we can use it to solve the Tower of Pisa problem. This is assigned as Exercise 48 in the problem set.

An oblique triangle will have either three acute angles or two acute angles and one obtuse angle. We use the letters A, B, and C to denote the angles and a, b, and c for the sides, as shown in Figure 6-21.

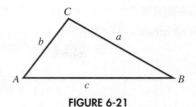

FIGURE 6-21

To solve an oblique triangle, you cannot use the techniques we used when solving right triangles. To solve any triangle, you must have the measurement of one side and any two of the other parts. This leads to four different possibilities:

1. two angles and one side: ASA or SAA.
2. two sides and the angle opposite one of the sides: SSA.
3. two sides and the included angle: SAS.
4. three sides: SSS.

The first two cases utilize the Law of Sines, which we will use in this section, and the last two cases use the Law of Cosines, which we will study in the next section. Note that there is no AAA (angle-angle-angle) possibility, as that would not yield a unique triangle but rather a set of *similar* triangles. For example, all equilateral triangles have three angles of 60° but do not all have the same side lengths.

To develop the Law of Sines, we refer to the triangles in Figures 6-22 and 6-23, in which line segment CD is perpendicular to side AB (or to an extension of AB).

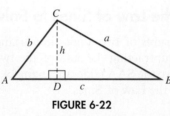

FIGURE 6-22

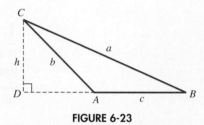

FIGURE 6-23

If h is the length of segment CD in the right triangles in Figure 6-22, the following formulas are true:

$$h = b \sin A \qquad \text{and} \qquad h = a \sin B.$$

Setting both expressions for h equal to each other, we have

$$b \sin A = a \sin B \qquad \text{or} \qquad \frac{a}{\sin A} = \frac{b}{\sin B}.$$

In the two right triangles shown in Figure 6-23, we see that

$$h = a \sin B \qquad \text{and} \qquad \begin{aligned} h &= b \sin (180° - A) \\ &= b \sin A \end{aligned}$$

Use the Sum Formula for Sine.

$\sin (180° - A)$

$= \sin 180° \cos A - \cos 180° \sin A$

$= \sin A.$

From this result, we can now state

$$b \sin A = a \sin B \qquad \text{or} \qquad \frac{a}{\sin A} = \frac{b}{\sin B}.$$

Even though we drew the perpendicular in Figure 6-22 from point C to AB, it could have also been drawn from another vertex, and similar reasoning would give the following results:

$$\frac{a}{\sin A} = \frac{c}{\sin C} \qquad \text{and} \qquad \frac{b}{\sin B} = \frac{c}{\sin C}.$$

Because of the transitive law of equality, it follows that

$$\frac{a}{\sin A} = \frac{b}{\sin B} = \frac{c}{\sin C}.$$

Law of Sines   The sides in any triangle are proportional to the sines of the angles opposite those sides. (See Figure 6-24.)

$$\frac{a}{\sin A} = \frac{b}{\sin B} = \frac{c}{\sin C}$$

or

$$\frac{\sin A}{a} = \frac{\sin B}{b} = \frac{\sin C}{c}$$

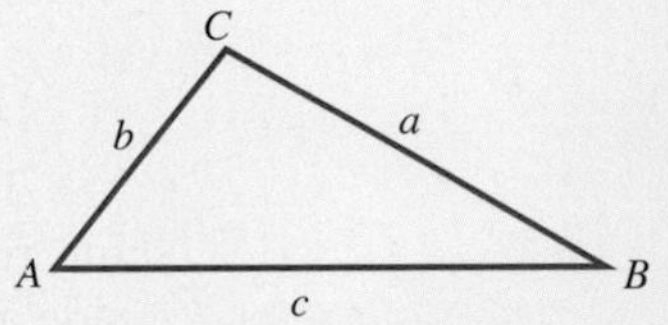

FIGURE 6-24

Since there are several ways that information can be given, it is a good idea to organize the data that we are given in a table. Once that is done, we will be able to select the best way to solve the triangle.

Comment   We always use exact measurements when it is possible. We should only use approximated measurements when necessary.

1. Use the Law of Sines to Solve Triangles

If the measures of two angles in a triangle are given, then the third angle can be found by subtracting the sum of the two given angles from 180°. Therefore, any time we have the SAA (side-angle-angle) or ASA (angle-side-angle) case, we will be able to use the Law of Sines.

EXAMPLE 1   **Solving a SAA Triangle Using the Law of Sines**

Solve the triangle shown in Figure 6-25, where $a = 14$, $A = 21°$, and $B = 35°$. Round sides to two decimal places.

FIGURE 6-25

SOLUTION   To solve the triangle, we must find the measure of angle C and sides b and c. Our first step will be to organize our data in a table, allowing us to get a quick picture of what we need to find and the best way to find it. Since we are given two angles, we know that we will use the Law of Sines to solve this triangle.

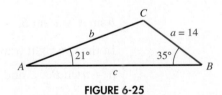

$a = 14$	$A = 21°$
$b = \underline{\ ?\ }$	$B = 35°$
$c = \underline{\ ?\ }$	$C = \underline{\ ?\ }$

From this table, we see that we have enough information to find the measure of side b and angle C. Once we have found angle C, we can then find the measure of side c. Since the sum of the angles in a triangle is 180°, we have

$$C = 180° - (A + B) = 180° - (21° + 35°) = \mathbf{124°}.$$

Now we can find the measures of sides b and c.

$$\frac{a}{\sin A} = \frac{b}{\sin B} \qquad\qquad \frac{a}{\sin A} = \frac{c}{\sin C}$$

$$\frac{14}{\sin 21°} = \frac{b}{\sin 35°} \qquad\qquad \frac{14}{\sin 21°} = \frac{c}{\sin 124°}$$

$$14 \sin 35° = b \sin 21° \qquad\qquad 14 \sin 124° = c \sin 21°$$

$$b = \frac{14 \sin 35°}{\sin 21°} \qquad\qquad c = \frac{14 \sin 124°}{\sin 21°}$$

$$b \approx \mathbf{22.41} \qquad\qquad c \approx \mathbf{32.39}$$

$a = 14$	$A = 21°$
$b \approx 22.41$	$B = 35°$
$c \approx 32.39$	$C = 124°$

Notice that we used the information given for side a and did not use the approximate value of side b to find side c. Always use exact measurements when possible. ∎

Self Check 1   Solve the triangle where $a = 7$, $A = 45°$, and $B = 40°$. Round the sides to two decimal places.

Now Try Exercise 13.

EXAMPLE 2   **Solving an ASA Triangle Using the Law of Sines**

Solve the triangle shown in Figure 6-26, where $a = 29$, $B = 42°$, and $C = 31°$.

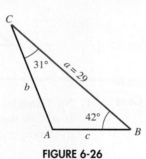

FIGURE 6-26

SOLUTION   To solve the triangle, we must find the measure of angle A and sides b and c. Once again, our first step will be to organize our data in a table, allowing us to get a quick picture of what we need to find and the best way to find it. Since we are given two angles, we know that we will use the Law of Sines to solve this triangle.

$a = 29$	$A = \underline{?}$
$b = \underline{?}$	$B = 42°$
$c = \underline{?}$	$C = 31°$

From this table, we see that we can easily determine the measure of angle A. Since the sum of the angles in a triangle is 180°, we have

$$A = 180° - (B + C) = 180° - (42° + 31°) = \mathbf{107°}.$$

Now we can find the measures of sides b and c.

$$\frac{a}{\sin A} = \frac{b}{\sin B} \qquad\qquad \frac{a}{\sin A} = \frac{c}{\sin C}$$

$$\frac{29}{\sin 107°} = \frac{b}{\sin 42°} \qquad\qquad \frac{29}{\sin 107°} = \frac{c}{\sin 31°}$$

$$29 \sin 42° = b \sin 107° \qquad\qquad 29 \sin 31° = c \sin 107°$$

$$b = \frac{29 \sin 42°}{\sin 107°} \qquad\qquad c = \frac{29 \sin 31°}{\sin 107°}$$

$$b \approx 20.29 \qquad\qquad c \approx 15.62$$

The triangle is now solved for the values given.

$a = 29$	$A = 107°$
$b \approx 20.29$	$B = 42°$
$c \approx 15.62$	$C = 31°$

Self Check 2   Solve the triangle where $a = 15$, $B = 70°$, and $C = 53°$. Round the sides to two decimal places.

Now Try Exercise 17.

If two sides and an angle opposite one of the sides is given, which is the SSA (side-side-angle) case, it is known as the **ambiguous case.** The information in this case may result in one triangle, two triangles, or no triangle.

When side b and angle A are given, the length of side a determines the number of triangles. Figure 6-27 illustrates the fact that $h = b \sin A$. The relationship between the length of a and the length of h determines the number of solutions, if any.

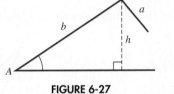

$$\sin A = \frac{h}{b} \Rightarrow h = b \sin A$$

FIGURE 6-27

Laws of Sines—Ambiguous Case

1. **No Triangle**
    If $a < h$, there is no triangle, because a is not long enough to complete the triangle.

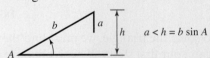

$a < h = b \sin A$

2. **One Right Triangle**
    If $a = h$, then one right triangle can be formed.

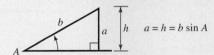

$a = h = b \sin A$

3. **One Oblique Triangle**
    If $a > h$ and $a > b$, then one oblique triangle can be formed.

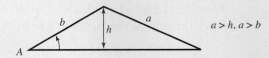

$a > h, a > b$

4. Two Oblique Triangles
If $a > h$ and $a < b$, then two distinct oblique triangles can be formed.

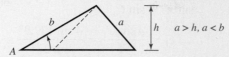

Comment | Even though we have demonstrated the possibilities by using the relationship between the values of a, b, and h, when we actually work the problems, we can use the Law of Sines and the table as we did in the previous examples.

EXAMPLE 3 | **Solving a SSA Triangle Using the Law of Sines—One Triangle**

In a triangle ABC, $a = 4$, $b = 2$, and $A = 27°$. Solve the triangle. Round angles to one decimal place and sides to two decimal places.

SOLUTION | Our first step will be to place the given information in a table. Note that we are using subscripts for the unknown values. Since this is the ambiguous case, we have to consider the possibility that there may be two triangles, which would lead to our having to find c_2, B_2, and C_2.

$a = 4$	$A = 27°$
$b = 2$	$B_1 = ?$
$c_1 = ?$	$C_1 = ?$

The first measurement that we can find is B_1.

$$\frac{\sin A}{a} = \frac{\sin B_1}{b}$$

$$\sin B_1 = \frac{b \sin A}{a} = \frac{2 \sin 27°}{4}$$

$$B_1 = \sin^{-1}\left(\frac{2 \sin 27°}{4}\right)$$

$$B_1 \approx 13.1°$$

Since the sine function is positive over the interval $(0°, 180°)$, there are two angles in the interval with the same sine. The second angle is the supplement of the first.

$$B_2 \approx 180° - B_1 = 180° - 13.1° = 166.9°$$

Note that you can check this with your calculator.

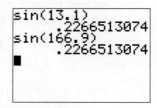

Since $A + B_2 = 27° + 166.9° > 180°$, there cannot be two triangles, because the sum of all three angles is $180°$.

Now that we have determined that there is only one triangle, we will find the values for C_1 and c_1. Our first step will be to find C_1.

$$C_1 \approx 180° - (A + B_1) = 180° - (27° + 13.1°) = 139.9°$$

We can now complete the triangle by finding the value of c_1.

$$\frac{a}{\sin A} = \frac{c_1}{\sin C_1}$$

$$c_1 = \frac{a \sin C_1}{A} = \frac{4 \sin 139.9°}{\sin 27°} \approx \mathbf{5.6752}$$

We have now solved the triangle. It is shown in Figure 6-28.

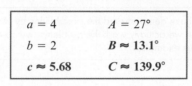

$a = 4$	$A = 27°$
$b = 2$	$B \approx \mathbf{13.1°}$
$c \approx \mathbf{5.68}$	$C \approx \mathbf{139.9°}$

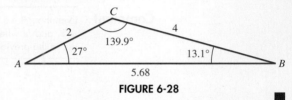

FIGURE 6-28

Self Check 3   In a triangle ABC, $a = 3$, $b = 2$, and $A = 35°$. Solve the triangle. Round angles to one decimal place and the sides to two decimal places.

Now Try Exercise 21.

EXAMPLE 4   **Solving a SSA Triangle Using the Law of Sines—Two Triangles**

In a triangle ABC, $a = 4$, $b = 5$, and $A = 35°$. Determine the number of triangles and then solve the triangle(s). Round angles to one decimal place and sides to two decimal places.

SOLUTION   The ambiguous case requires that we consider the possibility that there may be two triangles.

$a = 4$	$A = 35°$
$b = 5$	$B_1 = ?$
$c_1 = ?$	$C_1 = ?$

Using the Law of Sines, our first step is to find B_1. The value of this angle will determine the number of triangles possible with this data.

$$\frac{\sin A}{a} = \frac{\sin B_1}{b}$$

$$\frac{\sin 35°}{4} = \frac{\sin B_1}{5}$$

$$\sin B_1 = \frac{5 \sin 35°}{4} \quad \Rightarrow B_1 = \sin^{-1}\left(\frac{5 \sin 35°}{4}\right) \approx \mathbf{45.8°}$$

We know there is at least one triangle, and we can determine whether there is a second triangle by finding the value of B_2. If $A + B_2 < 180°$, there will be two triangles possible from the given values.

$$B_2 \approx 180° - B_1 = 180° - 45.8° = \mathbf{134.2°}$$

$$A + B_2 = 35° + 134.2° = 169.2° < 180°$$

Since $A + B_2 < 180°$, there will be two triangles. We add these values to the table.

Triangle 1		Triangle 2	
$a = 4$	$A = 35°$	$a = 4$	$A = 35°$
$b = 5$	$B_1 \approx \mathbf{45.8°}$	$b = 5$	$B_2 \approx \mathbf{134.2°}$
$c_1 = ?$	$C_1 = ?$	$c_2 = ?$	$C_2 = ?$

We now have enough information to complete the triangles. First, we will find the two angles, C_1 and C_2.

$$C_1 \approx 180° - (A + B_1) = 180° - (35° + 45.8°) = \textbf{99.2°}$$
$$C_2 \approx 180° - (A + B_2) = 180° - (35° + 134.2°) = \textbf{10.8°}$$

Using these values, we complete the triangles by finding the remaining side in each triangle.

$$\frac{a}{\sin A} = \frac{c_1}{\sin C_1} \Rightarrow \frac{4}{\sin 35°} = \frac{c_1}{\sin 99.2°} \Rightarrow c_1 = \frac{4 \sin 99.2°}{\sin 35°} \approx \textbf{6.88}$$

$$\frac{a}{\sin A} = \frac{c_2}{\sin C_2} \Rightarrow \frac{4}{\sin 35°} = \frac{c_2}{\sin 10.8°} \Rightarrow c_2 = \frac{4 \sin 10.8°}{\sin 35°} \approx \textbf{1.31}$$

We have now solved both triangles that are possible from the given values.

Triangle 1		Triangle 2	
$a = 4$	$A = 35°$	$a = 4$	$A = 35°$
$b = 5$	$B_1 \approx \textbf{45.8°}$	$b = 5$	$B_2 \approx \textbf{134.2°}$
$c_1 \approx \textbf{6.88}$	$C_1 \approx \textbf{99.2°}$	$c_2 \approx \textbf{1.31}$	$C_2 \approx \textbf{10.8°}$

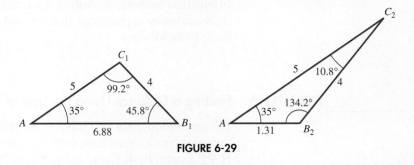

FIGURE 6-29

Self Check 4   In a triangle ABC, $a = 4$, $b = 6$, and $A = 20°$. Determine the number of triangles and then solve the triangle(s). Round angles to one decimal place and sides to two decimal places.

Now Try Exercise 23.

EXAMPLE 5   **Solving a SSA Triangle Using the Law of Sines—No Triangle**

In a triangle ABC, $a = 3$, $b = 7$, and $A = 50°$. Solve the triangle or state that no triangle is possible.

SOLUTION   The ambiguous case requires that we consider the possibility that there may be two triangles.

$a = 3$	$A = 50°$
$b = 7$	$B_1 = ?$
$c_1 = ?$	$C_1 = ?$

Using the Law of Sines, our first step is to find B_1. The value of this angle will determine the number of triangles possible using this data.

$$\frac{\sin A}{a} = \frac{\sin B_1}{b}$$

$$\frac{\sin 50°}{3} = \frac{\sin B_1}{7}$$

$$\sin B_1 = \frac{7 \sin 50°}{3} \approx 1.787$$

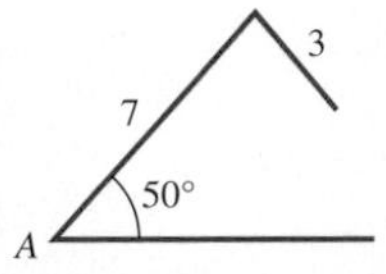

FIGURE 6-30

There is no triangle possible with this set of values, since $1.787 > 1$ and is not a possible value for the sine of an angle.

Self Check 5   In a triangle ABC, $a = 1$, $b = 2$, and $A = 45°$. Solve the triangle or state that no triangle is possible.

Now Try Exercise 25.

2. Solve Applications Using the Law of Sines

In previous sections, we solved applications using right-triangle trigonometry, but there are many applications that can use, or may even require, the Law of Sines for finding the solution.

EXAMPLE 6   **Finding a Distance Using the Law of Sines**

Ranger station Zulu is 7 miles from station Able on an east–west line. Each station observes a signal flare fired into the air. From Able, the bearing of the flare is N 40° E, and from Zulu, its bearing is N 35° W. How far is each station from the location of the flare (find a and z)? Round to two decimal places.

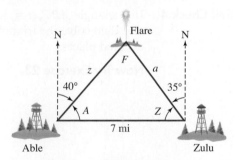

FIGURE 6-31

SOLUTION   Our first step is to find the values of angles A, Z, and F.

$$A = 90° - 40° = 50° \qquad Z = 90° - 35° = 55°$$
$$F = 180° - (A + B) = 180° - (50° + 55°) = 75°$$

Now we can apply the Law of Sines to find the values of a and z.

$$\frac{a}{\sin 50°} = \frac{7}{\sin 75°} \Rightarrow a = \frac{7 \sin 50°}{\sin 75°} \approx 5.55$$

$$\frac{z}{\sin 55°} = \frac{7}{\sin 75°} \Rightarrow z = \frac{7 \sin 55°}{\sin 75°} \approx 5.94$$

Therefore, we have found that station Able is approximately 5.94 miles from the flare and station Zulu is approximately 5.55 miles from the flare.

Self Check 6   Lighthouse B is 10 miles due east of lighthouse A. Offshore, a ship sends a distress signal that is visible from both lighthouses. From lighthouse A, the bearing of the ship is N 35° E, and from lighthouse B, its bearing is N 65° W. How far is the ship from each lighthouse? Round to one decimal place.

Now Try Exercise 51.

Example 7 can be solved using two right triangles, but it is much easier to use a combination of the Law of Sines and a right triangle.

EXAMPLE 7   **Finding the Height of an Airplane Using the Law of Sines and a Right Triangle**

At one end of a runway, the angle of elevation to a plane flying overhead is 33°. Another reading 1,000 feet closer to the plane shows the angle of elevation to be 42°. Find the height of the plane to the nearest foot.

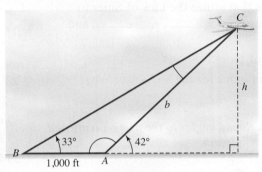

FIGURE 6-32

SOLUTION   Our first step will be to determine the values of angles A and C. Then we will have enough information to use the Law of Sines to find b. Once we have found b, we will use right-triangle trigonometry to find the value of h.

$$A = 180° - 42° = 138°$$ Find angle A.

$$C = 180° - (A + B) = 180° - (138° + 33°) = 9°$$ Find angle C.

$$\frac{b}{\sin 33°} = \frac{1,000}{\sin 9°}$$ Apply the Law of Sines.

$$b = \frac{1,000 \sin 33°}{\sin 9°}$$ Solve for b.

$$\sin 42° = \frac{h}{b}$$ Right-triangle trigonometry.

$$h = b \sin 42° = \frac{1,000 \sin 33°}{\sin 9°} \cdot \sin 42°$$ Solve for h.

$$h \approx 2,330$$

The plane is approximately 2,330 feet high.

Self Check 7   One observer spots a plane with an angle of elevation of 41°. Another observer who is 2,000 feet due east of the first observer spots the plane at an angle of elevation of 52°. What is the height of the plane to the nearest foot?

Now Try Exercise 45.

Self Check Answers   **1.** $b \approx 6.36$; $c \approx 9.86$; $C = 95°$ **2.** $A = 57°$; $b \approx 16.81$; $c \approx 14.28$

3. $B \approx 22.5°$; $C \approx 122.5°$; $c \approx 4.41$
4. $B_1 \approx 30.9°$; $C_1 \approx 129.1°$; $c_1 \approx 9.08$; $B_2 \approx 149.1°$; $C_2 \approx 10.9°$; $c_2 \approx 2.21$
5. no triangle   6. 4.3 mi from lighthouse A; 8.3 mi from lighthouse B
7. 5,419 ft

Exercises **6.3**

Getting Ready

You should be able to complete these vocabulary and concept statements before you proceed to the practice exercises.

Fill in the blanks.

1. A triangle that is not a right triangle is called a(n) _____ triangle.

2. The three cases that utilize the Law of Sines to solve a triangle are _____, _____, and _____.

3. State the Law of Sines.

4. What is another name for the SSA case? Why?

Practice

Solve each triangle. Round angles to the nearest tenth and sides to the nearest hundredth.

5.
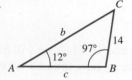

6.
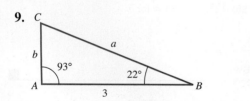

7.   8.
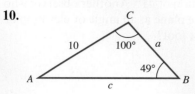

9.
C

10.

11.
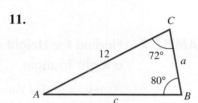

12.
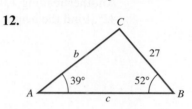

Solve each triangle. Round angles to the nearest tenth and sides to the nearest hundredth.

13. $a = 15$; $A = 72°$; $B = 12°$

14. $a = 40$; $A = 64°$; $B = 31°$

15. $b = 110$; $B = 80°$; $C = 40°$

16. $b = 75$; $B = 48°$; $C = 59°$

17. $c = 74$; $A = 89°$; $B = 43°$

18. $c = 38$; $A = 42°$; $C = 94°$

19. $a = 120$; $A = 140°$; $B = 10°$

20. $a = 200$; $A = 108°$; $B = 24°$

In Exercises 21–36, two sides and an angle are given (SSA). Determine whether one triangle, two triangles, or no triangles are possible. Solve any triangles that do exist. Round angles to the nearest tenth and sides to the nearest hundredth.

21. $a = 123$; $b = 96$; $A = 42°$

22. $a = 102$; $b = 88$; $A = 56°$

23. $a = 35.7$; $c = 45.3$; $A = 12°$

24. $a = 12.5$; $c = 15.4$; $A = 50°$

25. $b = 10.5$; $c = 7.4$; $C = 47.6°$

26. $b = 10.5$; $c = 12$; $B = 85.5°$

27. $a = 5,293$; $b = 3,761$; $B = 9.9°$

28. $a = 2,536$; $b = 5,322$; $A = 25°$

29. $a = 45.4$; $c = 70.4$; $C = 80°$

30. $a = 40$; $c = 12$; $A = 38°$

31. $b = 53.5$; $c = 65$; $B = 98°$

32. $a = 98.5$; $b = 124$; $A = 80°$

33. $a = 200$; $c = 250$; $A = 45°$

34. $b = 3.4$; $c = 4.7$; $B = 7.5°$

35. $a = 120$; $b = 150$; $B = 75°$

36. $a = 54$; $c = 27$; $A = 74°$

In Exercises 37–40, solve for x in the given figure. Round to the nearest tenth.

37.

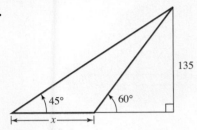

38.

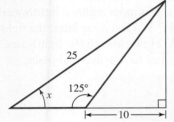

39.

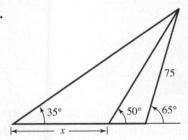

40.

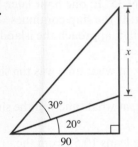

Applications

41. Measuring distance To measure the distance up a steep hill, Josie finds the measurements shown in the figure. Find the length of *BC*. Round to one decimal place.

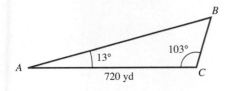

42. Measuring distance Refer to the figure in Exercise 41 and find the length of *AB*. Round to one decimal place.

43. Bearing angle A ship sails 3.2 nautical miles on a bearing of N 33° E. After reaching a lighthouse, the ship turns and sails another 6.7 nautical miles to a position that is due east of its starting point. Find the bearing θ of the lighthouse from the ship's final position. Round to the nearest tenth of a degree.

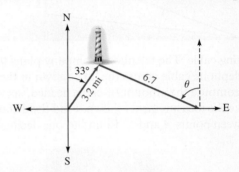

44. Width of a river Points *A* and *B* are on opposite sides of a river, as shown in the figure. A tree at point *C* is 310 feet from point *A*. If angle *A* measures 125° and angle *C* measures 32°, how wide is the river? Round to one decimal place.

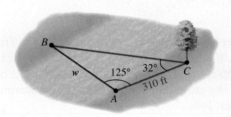

45. Height of a balloon Observers at points *A* and *B* are directly in line with a hot-air balloon and are themselves 215 feet apart. The angles of elevation of the balloon from points *A* and *B* are shown in the figure. Find the height of the balloon. (*Hint:* Find side *b* first.) Round to one decimal place.

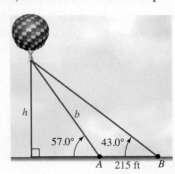

46. Angle of elevation The radio tower shown in the figure is 175 feet tall and is located at the top of a hill. At a point 800 feet down the hill, the angle of elevation to the top of the tower is 19.0°. Find the angle the hill makes with the horizontal. Round to the nearest tenth of a degree.

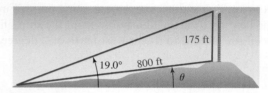

47. Laying cable The telephone company plans to bury a telephone cable across the lake shown in the figure. To compute the amount of cable needed, surveyors make the measurements shown. Find the distance between points *A* and *C*. Round to one decimal place.

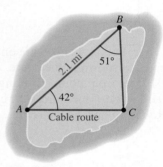

48. Tower of Pisa From a point 542 feet away from the base of the Tower of Pisa, the angle of elevation to its top is 17.9°. If the tower makes an angle of 94.5° with the ground, how tall is the tower? Round to the nearest foot.

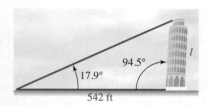

49. Pyramids From a point 312 feet from the base of the great Pyramid of Khufu (Cheops) at Giza, the angle of elevation to its top is 25.5°. If the pyramid makes an angle of 141.8° with the ground, find its slant height. Round to the nearest foot.

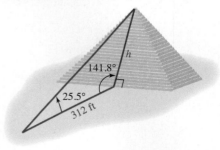

50. Width of a river Two children stand on one side of a river at points *A* and *B*, 120 feet apart. Each one sights the same tree on the opposite side. They determine the measurements shown in the figure. How wide is the river? Round to one decimal place.

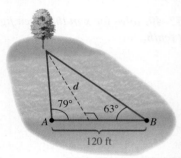

51. Navigation A ship sails due north at 3.2 knots per hour. At 2:00 p.m., the skipper sights a lighthouse in the direction of N 43° W. One hour later, the light-house bears S 78° W. How close to the lighthouse did the ship sail? Round to one decimal place.

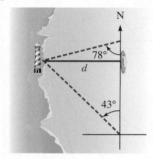

52. Navigation A ship sails on a course bearing N 21° E at a speed of 14 knots. At noon, the first mate sights an island in the direction N 35° E; one hour later, the same island is due east. If the ship continues on its course, how close will it approach the island? Round to one decimal place.

53. Navigation In Exercise 51, at what time was the ship closest to the lighthouse?

54. Navigation In Exercise 52, at what time will the ship be nearest to the island?

55. Flagpole length A flagpole leans 10.5° from the vertical, toward an observer 17.0 feet from the flagpole's base. If the angle of elevation of the top of the flagpole is 75.5°, how long is the flagpole? Round to one decimal place.

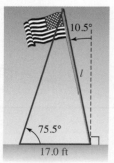

56. Surveying To determine the distance between points *P* and *Q* on opposite sides of a river, a surveyor measures the distances and angles shown in the figure. What is the distance *PQ*? Round to one decimal place.

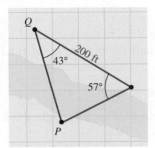

57. Height of a tower A television tower is on top of a building. From point *A*, the angle of elevation of the top of the tower is 40°. From *B*, 150 feet closer to the building, the angle of elevation of the top of the tower is 51° and the angle of elevation of the base of the tower is 45°. Find the height of the tower. Round to one decimal place.

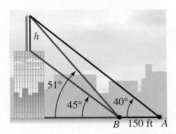

58. Height of a building In Exercise 57, find the height of the building. Round to the nearest foot.

59. Tracking satellites A satellite *S* circles the earth in an orbit 1,230 miles above the earth's surface. When the tracking station *T* sights the satellite at an angle of elevation of 24.1°, how large is angle α? Assume that the radius of the earth is 3,969 miles. Round to one decimal place.

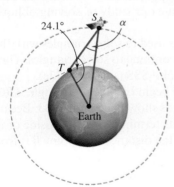

60. Distance Refer to Exercise 59. When the tracking station *T* sights the satellite at an elevation of 24.1°, how far is the satellite from the station? Round to the nearest mile.

61. Solar heating panels A 16-foot solar heating panel is to be installed on the roof shown in the figure. Find the measure of angle β. Round to the nearest tenth of a degree.

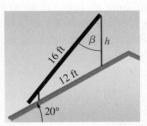

62. Height Refer to Exercise 61. Find the height *h* of the vertical support. Round to one decimal place.

63. Surveying Tiger's Roar Construction Company owns a triangular piece of land that is bounded by three roads. The angle between two of them, Jefferson Highway and Tiger Bend Road, is 43°. The property runs for 2,500 feet along Tiger Bend, and for 2,000 feet along the third road, Airline Highway. How much land might front on Jefferson Highway? Round to the nearest foot.

64. Height of a building A 210-foot television tower stands on the top of an office building. From a point on level ground, the angles of elevation to the top and base of the tower are 25.2° and 21.1°. How tall is the office building? Round to the nearest foot.

65. Height of a tower From the roof of a country inn, the angle of elevation to the top of a cell-phone tower is 15°. From the building's ground level 40 feet below, the angle of elevation of the top of the tower is 25°. How far above ground level is the top of the tower? Round to the nearest foot.

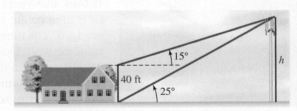

66. Engine design A 10-inch piston rod is connected to a 3-inch crank, as shown in the figure. When the piston rod makes an angle of 7° with its line of motion, what is the piston-to-crankshaft distance *d*? Round to one decimal place.

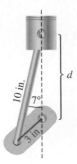

67. Distance A 96.0-foot tower is on the roof of a building, as shown in the figure. Measured from position O, the difference between the angles of elevation of the top and bottom of the tower is 4.0°. If the line-of-sight distance from O to the bottom of the tower is 531 feet, how far is O from the top of the tower? Round to one decimal place.

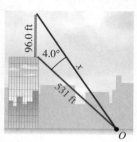

68. Height In Exercise 67, find the height of the building. Round to the nearest foot.

Discovery and Writing

69. Show that in any triangle ABC, the following relation holds:

$$\frac{\sin A - \sin B}{\sin A + \sin B} = \frac{a - b}{a + b}.$$

70. If R is the radius of the circle circumscribed about the triangle ABC in the figure, and s is the **semiperimeter** (half the perimeter) of the triangle, then show that

$$R(\sin A + \sin B + \sin C) = s.$$

(*Hint:* Angle $BOC = 2A$, angle $BOA = 2C$, and angle $COA = 2B$.)

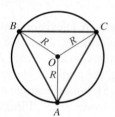

6.4 Law of Cosines

In this section, we will learn to

1. Use the Law of Cosines to solve triangles.
2. Solve applications using the Law of Cosines.

The Lærdal Tunnel in Norway, at 15.2 miles long, is the world's longest road tunnel. It took five years to construct and was completed in 2000. Engineers and architects used bright blue lights and subtle curves in its design in order to keep drivers engaged. The tunnel, one of the most spectacular tunnels in the world, is part of the route between Oslo and Bergen.

To estimate the cost of building a tunnel, engineers must first determine the length of the tunnel. The length can be approximated by using the Law of Cosines, the topic of this section. We will continue our study of solving oblique triangles in this section.

In Section 6.3, the Law of Sines was used to solve triangles with the cases SAA, ASA, and SSA. In this section, we will continue to solve triangles. This time we will look at the cases SAS (side-angle-side) and SSS (side-side-side).

To develop the Law of Cosines, we begin by placing triangle ABC in a coordinate system with vertex A at the origin, as shown in Figure 6-33. Because angle A is in standard position and point B is on its terminal side, the cosine of angle A is the ratio of the x-coordinate of point B to the distance that point B is from the origin. Thus, we have

$$\cos A = \frac{x}{c} \text{ or } x = c \cos A.$$

Furthermore, we have

$$\sin A = \frac{y}{c} \text{ or } y = c \sin A.$$

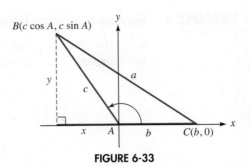

FIGURE 6-33

Thus, the coordinates of point B are $(c \cos A, c \sin A)$, the coordinates of point A are $(0, 0)$, and the coordinates of point C are $(b, 0)$. We can now use the Distance Formula to compute a^2.

$$a = \sqrt{(c \cos A - b)^2 + (c \sin A - 0)^2}$$

$$\begin{aligned} a^2 &= (c \cos A - b)^2 + (c \sin A - 0)^2 && \text{Square both sides.} \\ &= c^2 \cos^2 A - 2bc \cos A + b^2 + c^2 \sin^2 A && \text{Clear parentheses.} \\ &= c^2(\cos^2 A + \sin^2 A) + b^2 - 2bc \cos A && \text{Factor } c^2 \text{ from first and last terms.} \\ &= c^2 + b^2 - 2bc \cos A && \cos^2 A + \sin^2 A = 1. \end{aligned}$$

This is one of the three formulas for the Law of Cosines. The other two cases can be determined in the same manner. Although the triangle in Figure 6-33 is obtuse, this derivation is valid for any triangle.

Law of Cosines The square of a side of a triangle is equal to the sum of the squares of the remaining two sides minus twice the product of those two sides and the cosine of the angle between them. See Figure 6-34.

$$a^2 = b^2 + c^2 - 2bc \cos A$$
$$b^2 = a^2 + c^2 - 2ac \cos B$$
$$c^2 = a^2 + b^2 - 2ab \cos C$$

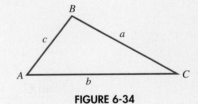

FIGURE 6-34

1. Use the Law of Cosines to Solve Triangles

If you are given the SAS information of an oblique triangle, you will use the Law of Cosines to find the missing side. Once you have found the missing side, you can then use the Law of Sines to find the next angle. However, you have to be very careful because of the ambiguous case. The following guidelines should be followed:

Strategy for Using the Law of Cosines in the SAS Case

Step 1: Find the missing side using the Law of Cosines.

Step 2: Use the Law of Sines to find the smaller angle of the two remaining angles. This will avoid having to deal with the ambiguous case. Remember that the smaller angle will be opposite the shorter side.

Step 3: Find the third angle by subtracting the sum of the other angles from 180°.

EXAMPLE 1   Solving a Triangle Using SAS

Solve the oblique triangle shown in Figure 6-35, with $b = 27$, $c = 14$, and $A = 43°$. Round sides to two decimal places and angles to one decimal place.

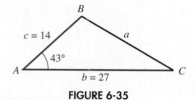

FIGURE 6-35

SOLUTION   It is always a good idea to organize your data in a table so that you can determine the best way to proceed with solving the triangle. From the table, we determine that this requires the use of the Law of Cosines, and our first step will be to find side a.

$a = \underline{?}$	$A = 43°$
$b = 27$	$B = \underline{?}$
$c = 14$	$C = \underline{?}$

$$a^2 = b^2 + c^2 - 2bc \cos A$$

$$a^2 = 27^2 + 14^2 - 2(27)(14) \cos 43° \qquad \text{Substitute the given values.}$$

$$a = \sqrt{27^2 + 14^2 - 2(27)(14) \cos 43°} \qquad \text{Evaluate.}$$

$$a \approx 19.29 \qquad \text{Round to the nearest hundredth.}$$

Inserting this value into the table, we determine that we can now use the Law of Sines. However, we must take care to select the smaller of the two remaining angles. Since the smaller angle is opposite the shorter side, we see that we should find angle C in our next calculation.

$a \approx \mathbf{19.29}$	$A = 43°$
$b = 27$	$B = \underline{?}$
$c = 14$	$C = \underline{?}$

$$\frac{\sin C}{14} = \frac{\sin 43°}{19.29} \qquad \text{Use the Law of Sines.}$$

$$\sin C = \frac{14 \sin 43°}{19.29}$$

$$C = \sin^{-1}\left[\frac{14 \sin 43°}{19.29}\right] \qquad \text{Solve for } C.$$

$$C \approx \mathbf{29.7°} \qquad \text{Round to nearest tenth.}$$

To find the remaining angle B, we subtract the values of angle A and angle C from $180°$.

$$B \approx 180° - 43° - 29.7° = \mathbf{107.3°}$$

We have now solved the triangle.

$a \approx \mathbf{19.29}$	$A = 43°$
$b = 27$	$B \approx \mathbf{107.3°}$
$c = 14$	$C \approx \mathbf{29.7°}$

Self Check 1   Solve the oblique triangle with $b = 15$, $c = 20$, and $A = 62°$. Round angles to one decimal place and sides to two decimal places.

Now Try Exercise 13.

If we are given the three sides of a triangle, the SSS case, then we will use the Law of Cosines. We can follow these guidelines:

Strategy for Using the Law of Cosines in the SSS Case

Step 1: Find the largest angle first, using the Law of Cosines. The largest angle is opposite the longest side.

Step 2: Use the Law of Sines to find either one of the two remaining angles.

Step 3: Find the third angle by subtracting the sum of the other angles from 180°.

Comment   The sum of the lengths of any two sides of a triangle must be greater than the length of the other side. Consequently, the lengths $a = 2$, $b = 4$, and $c = 12$ could not form a triangle.

EXAMPLE 2   **Solving a Triangle Using SSS**

Solve the oblique triangle shown in Figure 6-36, with $a = 3$, $b = 6$, and $c = 7$. Round each angle to one decimal place.

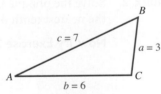

FIGURE 6-36

SOLUTION   The SSS case uses the Law of Cosines to find the largest of the three angles first. By organizing our data in a table, we see that angle C is the largest angle.

$a = 3$	$A = ?$
$b = 6$	$B = ?$
$c = 7$	$C = ?$

We will use the form of the Law of Cosines that involves angle C to find its value.

$$c^2 = a^2 + b^2 - 2ab \cos C$$
$$2ab \cos C = a^2 + b^2 - c^2$$
$$\cos C = \left[\frac{a^2 + b^2 - c^2}{2ab}\right]$$
$$C = \cos^{-1}\left[\frac{a^2 + b^2 - c^2}{2ab}\right]$$
$$C = \cos^{-1}\left[\frac{3^2 + 6^2 - 7^2}{2(3)(6)}\right]$$

$$C \approx 96.4$$

We now have the following information.

$a = 3$	$A = \underline{?}$
$b = 6$	$B = \underline{?}$
$c = 7$	$C \approx 96.4°$

Since we found the largest angle first, we can use the Law of Sines to find angle A and do not have to be concerned with the ambiguous case.

$$\frac{\sin A}{3} = \frac{\sin 96.4°}{7}$$

$$\sin A = \frac{3 \sin 96.4°}{7}$$

$$A = \sin^{-1}\left[\frac{3 \sin 96.4°}{7}\right]$$

$$A \approx 25.2°$$

Finding $B \approx 180° - 25.2° - 96.4° = 58.4°$, we can now complete the chart.

$a = 3$	$A \approx 25.2°$
$b = 6$	$B \approx 58.4°$
$c = 7$	$C \approx 96.4°$

Self Check 2 Solve the oblique triangle with sides $a = 8$, $b = 9$, and $c = 5$. Round each angle to the nearest tenth of a degree.

Now Try Exercise 21.

2. Solve Applications Using the Law of Cosines

EXAMPLE 3 **Finding Distance and Bearing**

An airplane takes off from an airport on a bearing of N 50° W. After flying for 160 miles, the plane develops mechanical problems, and the pilot turns and flies on a bearing of S 30° E. The plane makes a safe emergency landing after flying on this bearing for 145 miles (Figure 6-37).

(a) How far is the plane from the airport? Round to two decimal places.
(b) On what bearing should the rescue helicopter fly to reach the plane? Round to the nearest tenth of a degree.

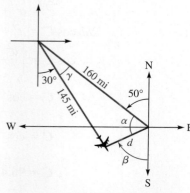

FIGURE 6-37

SOLUTION   **(a)** Since the original angle was N 50° W, we can determine $\gamma = 50° - 30° = 20°$. Since we now have two sides and an included angle, we can use the Law of Cosines to find the value of d.

$$d^2 = 145^2 + 160^2 - 2(145)(160)\cos 20°$$
$$d = \sqrt{145^2 + 160^2 - 2(145)(160)\cos 20°}$$
$$d \approx 54.98$$

The distance is approximately 54.98 miles.

(b) In order to find β, we first determine the value of α. This can be done using the Law of Sines.

$$\frac{\sin \alpha}{145} = \frac{\sin 20°}{54.98}$$
$$\sin \alpha = \frac{145 \sin 20°}{54.98}$$
$$\alpha = \sin^{-1}\left[\frac{145 \sin 20°}{54.98}\right]$$
$$\alpha \approx 64.4°$$

Therefore, $\beta \approx 180° - 50° - 64.4° = 65.6°$. The rescue helicopter should fly on a bearing of S 65.6° W to reach the disabled plane. ∎

Self Check 3   A boat leaves port on a bearing of N 40° W. After traveling for 40 miles, the boat turns and travels on a bearing of S 20° E for 30 miles. At this point, the boat runs out of fuel.

(a) How far is the boat from port? Round to two decimal places.

(b) On what bearing should a rescue boat head to reach the boat? Round to the nearest tenth of a degree.

Now Try Exercise 29.

EXAMPLE 4   **Finding an Angle**

A triangle is formed by the points $A(-4, 2)$, $B(3, 3)$, and $C(0, -3)$. Find angle A to the nearest tenth of a degree.

SOLUTION   We plot the points as shown in Figure 6-38. Our first step is to find the length of each side of the triangle.

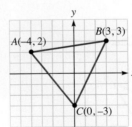

FIGURE 6-38

$$d(AB) = \sqrt{(3 - (-4))^2 + (3 - 2)^2} \approx 7.07$$
$$d(AC) = \sqrt{(0 - (-4))^2 + (-3 - 2)^2} \approx 6.40$$
$$d(BC) = \sqrt{(0 - 3)^2 + (-3 - 3)^2} \approx 6.71$$

Now we can apply the Law of Cosines to find angle A.

$$A = \cos^{-1}\left[\frac{7.07^2 + 6.40^2 - 6.71^2}{2(7.07)(6.40)}\right] \approx 59.5°$$

■

Self Check 4   Use the Law of Cosines to find the value of angle B in Figure 6-38. Round to one decimal place.

Now Try Exercise 39.

Self Check Answers   **1.** $a \approx 18.53$; $B \approx 45.6°$; $C \approx 72.4°$ **2.** $A \approx 62.2°$; $B \approx 84.3°$; $C \approx 33.6°$
3. a. 15.64 mi   **b.** N 81.0° W   **4.** 55.3°

Exercises **6.4**

Getting Ready
You should be able to complete these vocabulary and concept statements before you proceed to the practice exercises.

Fill in the blanks.

1. The two cases that utilize the Law of Cosines to solve a triangle are _____ and _____.

2. The Law of Cosines is _____.

3. $a = 2$, $b = 3$, and $c = 10$ cannot be the sides of a triangle because _____.

4. When given three sides of a triangle, the _____ angle should be found first.

⊪ Practice
Solve each triangle. Round angles to the nearest tenth. Round sides to the nearest hundredth.

5.
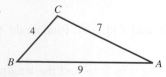

6.
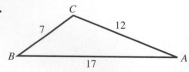

7.
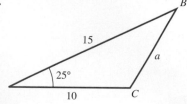

8.
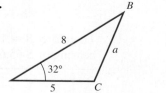

9.
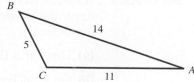

10.
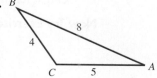

11.
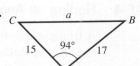

12.
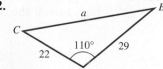

Solve each triangle. Round angles to the nearest tenth. Round sides to the nearest hundredth.

13. $a = 4$; $b = 7$; $C = 50°$

14. $b = 5$; $c = 8$; $A = 45°$

15. $a = 42$; $c = 37$; $B = 99°$

16. $a = 19$; $b = 8$; $C = 21°$

17. $b = 9$; $c = 13$; $A = 85°$

18. $a = 10$; $c = 12$; $B = 100°$

19. $a = 14;\ b = 14;\ C = 80°$
20. $b = 7;\ c = 9;\ A = 90°$
21. $a = 3;\ b = 5;\ c = 7$
22. $a = 4;\ b = 9;\ c = 12$
23. $a = 4;\ b = 4;\ c = 4$
24. $a = 5;\ b = 5;\ c = 8$
25. $a = 6;\ b = 8;\ c = 10$
26. $a = 12;\ b = 5;\ c = 13$
27. $a = 14;\ b = 29;\ c = 21$
28. $a = 30;\ b = 40;\ c = 55$

Applications

29. **Navigation** A ship leaves port and sails 21.2 nautical miles in the direction N 42° W. The captain then turns onto a course of S 15° E. After sailing another 19 nautical miles, how far is the ship from port? Through what angle should the ship turn to return to port? Round to one decimal place.

30. **Navigation** A ship sails 14 nautical miles in the direction S 28° W and then turns onto a course of S 52° W. After sailing another 23 nautical miles, how far is the ship from its starting point? Through what angle should the ship turn to return to port? Round to the nearest degree.

31. **Surveying** To measure the length of a lake, a surveyor finds the measurements shown in the figure. Find the length of the lake. Round to one decimal place.

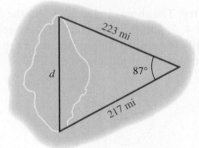

32. **Building tunnels** To estimate the cost of building a tunnel, a surveyor must find the distance through a hill. To do so, the surveyor finds the measurements shown in the figure. How long must the tunnel be to pass through the hill? Round to one decimal place.

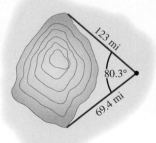

33. **Satellite orbit** A satellite S has a circular orbit around the earth, as shown in the figure. The angle of elevation to the satellite from a tracking station F on the earth is 31.8°. If the tracking station is 721 miles from the satellite, find the height of the satellite's orbit. Assume that the radius of the earth is 3,960 miles. Round to the nearest mile.

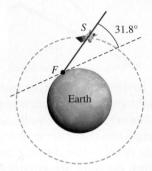

34. **Height of a satellite** The tracking antenna A shown in the figure is aimed at a satellite S at an angle of 42.32° above the horizon. If the distance between the antenna and the satellite is 8,752 miles, how high is the satellite above the earth? Assume the radius of the earth to be 3,960 miles. Round to the nearest mile.

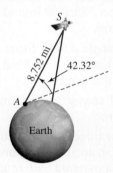

35. **Length of a solar collector** A solar collecting panel is to be installed on a roof that makes an angle of 14° with the horizontal, as shown in the figure. Find the length of the solar panel. Round to one decimal place.

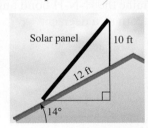

36. Installing solar collectors A solar collecting panel is to be installed on a roof, as shown in the figure. Find the length of the vertical brace. Round to one decimal place.

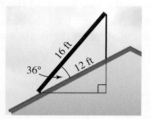

37. Carpentry To determine whether two interior walls meet at a right angle, carpenters often mark a point 3 feet from the corner on one wall and a point (at the same height) on the other wall 4 feet from the corner. If the straight-line distance between those points is 5 feet, the walls are square. At what angle do the walls meet if the distance measures 4 feet 10 inches? Round to the nearest tenth of a degree.

38. Cabinetmaking To build a countertop for a kitchen, a cabinetmaker must determine the angle at which two walls meet. The method of Exercise 37 is used. What is the angle between the walls if the measured distance is 5 feet 3 inches? Round to the nearest tenth of a degree.

39. Geometry Triangle ABC is formed by the points $A(3, 4)$, $B(1, 5)$, and $C(5, 9)$. Find angle A to the nearest tenth of a degree.

40. Geometry In Exercise 39, find angle B to the nearest tenth of a degree.

41. Chemistry The three hydrogen atoms and one nitrogen atom in an ammonia molecule form a tetrahedron—a pyramid with four triangular faces. Three of these faces are isosceles triangles with the nitrogen atom at the vertex. If the distance between the hydrogen atoms is 1.64 angstroms, and the nitrogen atom is 1.02 angstroms from each hydrogen atom, determine the H-N-H bond angle. Round to the nearest tenth of a degree.

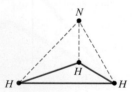

42. Navigation A lighthouse is 15 nautical miles N 23° W of a dock. A ship leaves the dock heading due east at 26.3 knots (nautical miles per hour). How long will it take for the ship to reach a distance of 35 nautical miles from the lighthouse? Round to the nearest hour.

43. Geometry The diagonals of a parallelogram are 12 and 18 feet, and they intersect at an angle of 40°. Find the lengths of the sides of the parallelogram. (*Hint:* The diagonals of a parallelogram bisect each other.) Round to the nearest foot.

44. Geometry The sides of a parallelogram are 95 and 55 meters, and one angle is 110°. Find the length of the longest diagonal. Round to the nearest meter.

45. Geometry In the figure, D is the midpoint of BC. Find angle ADC to the nearest tenth of a degree.

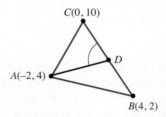

46. Horology The pendulum in the figure is 1.73 meters long, and the horizontal distance between the extremities of its swing is 0.370 meter. Through what angle does the pendulum swing? Round to the nearest tenth of a degree.

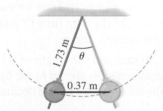

47. Geometry The three circles shown in the figure have radii of 4, 7, and 9 centimeters. If the circles are externally tangent to each other, what are the angles of the triangle that joins their centers?

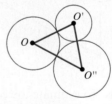

48. Geometry Suppose the three circles in Exercise 47 have radii of 21.2, 19.3, and 31.2 centimeters. If the circles are externally tangent to each other, what are the angles of the triangle that joins their centers? Round to the nearest tenth of a degree.

Discovery and Writing

49. Circles with radii R and r have the same center C. A chord of the larger circle is tangent to the smaller circle and subtends a central angle θ. (See the figure.) Express the length x of the chord as a function of R and θ.

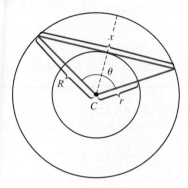

50. From one corner of a cube, diagonals are drawn along two of its faces. (See the figure.) Find the angle θ between the diagonals.

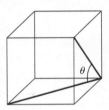

51. Prove that the Pythagorean Theorem is a special case of the Law of Cosines.

6.5 Area of Triangles

In this section, we will learn to

1. Find the area of triangles.

2. Solve application problems involving area.

Bermuda is a British overseas territory in the North Atlantic Ocean. Four hundred years ago, an English vessel was shipwrecked on a cluster of islands, giving birth to this beautiful island nation. It is home to bustling bistros, pale-pink beaches, and kelly-green golf courses.

Bermuda is sometimes associated with the Bermuda Triangle. This triangle, also known as the Devil's Triangle, is a region where aircraft and vessels have allegedly disappeared mysteriously. The triangular region is formed by Bermuda, Miami, and Puerto Rico. In this section, we will study the area of triangles; we will find the area of the Bermuda Triangle in Exercise 45.

1. Find the Area of Triangles

The most well-known formula for the area of a triangle is *one-half of the base times the height* of the triangle,

$$A = \frac{1}{2}bh.$$

Since the height of the triangle must be known for this formula to be used, we need other formulas when the height is unknown. To develop these formulas, we will first consider a triangle for which two sides and the included angle are known (SAS).

In triangle ABC, shown in Figure 6-39, we assume that b, a, and angle C are given. Because $\sin C = \frac{h}{a}$, we can multiply both sides of this equation by a to obtain

$h = a \sin C$.

The area K of triangle ABC is

$$K = \frac{1}{2}bh;$$

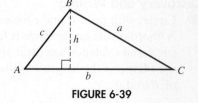

FIGURE 6-39

by substituting $a \sin C$ for h, we now have

$$K = \frac{1}{2}ba \sin C.$$

Similar arguments can be made for any combination of SAS information.

| Area of a Triangle, Given SAS | If two sides and the included angle of a triangle are known, the area of the triangle is one-half the product of the two sides and the sine of the included angle.

$$K = \frac{1}{2}ab \sin C \qquad K = \frac{1}{2}ac \sin B \qquad K = \frac{1}{2}bc \sin A$$

EXAMPLE 1 **Finding the Area of an SAS Triangle**

Find the area of the triangle shown in Figure 6-40.

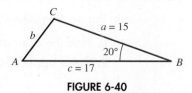

FIGURE 6-40

SOLUTION We are given sides a and c and included angle B, so we use the formula with B.

$$K = \frac{1}{2}ac \sin B$$

$$= \frac{1}{2}(15)(17) \sin 20°$$

$$\approx 43.61 \text{ units}^2$$

Self Check 1 Find the area of the triangle shown in Figure 6-41. Round to two decimal places.

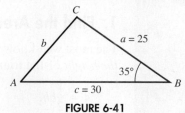

FIGURE 6-41

Now Try Exercise 7.

If three sides of a triangle are given (SSS), the area can be calculated using a formula attributed to Heron (Hero) of Alexandria (circa 250 BC). The proof is given in online Appendix B.

Heron's Formula	If a, b, and c are the three sides of a triangle with semiperimeter

$$s = \frac{a + b + c}{2},$$

then the area of the triangle is given by

$$K = \sqrt{s(s - a)(s - b)(s - c)}.$$

EXAMPLE 2   **Finding the Area of a SSS Triangle—Heron's Formula**

Find the area of a triangle that has sides measuring 5, 7, and 10 inches.

SOLUTION   We let $a = 5$, $b = 7$, and $c = 10$. Then

$$s = \frac{5 + 7 + 10}{2} = 11.$$

We can now apply Heron's Formula.

$$
\begin{aligned}
K &= \sqrt{s(s - a)(s - b)(s - c)} \\
&= \sqrt{11(11 - 5)(11 - 7)(11 - 10)} \qquad \text{Substitute values for } a, b, c, \text{ and } s. \\
&= \sqrt{11(6)(4)(1)} \\
&= \sqrt{264} \\
&= 2\sqrt{66} \\
&\approx 16.25
\end{aligned}
$$

The area of the triangle is 16.25 square inches.   ∎

Self Check 2   Find the area of a triangle that has sides measuring 9, 12, and 17 inches. Round to two decimal places.

Now Try Exercise 9.

If two angles and a side are given (SAA or ASA), we can modify the equation $K = \frac{1}{2}cb \sin A$ to find the area. By the Law of Sines, we have

$$\frac{b}{\sin B} = \frac{c}{\sin C}$$

$$b = \frac{c \sin B}{\sin C}$$

We can now substitute this into Heron's Formula.

$$
\begin{aligned}
K &= \frac{1}{2}bc \sin A \\
&= \frac{1}{2}\frac{c \sin B}{\sin C}c \sin A \\
&= \frac{c^2 \sin A \sin B}{2 \sin C}
\end{aligned}
$$

Using similar arguments, we have the following formulas:

| Area of a Triangle (SAA or ASA) | If two angles and a side of a triangle are given, one of the following formulas can be used to find the area of the triangle: |

$$K = \frac{c^2 \sin A \sin B}{2 \sin C} \qquad K = \frac{a^2 \sin B \sin C}{2 \sin A} \qquad K = \frac{b^2 \sin A \sin C}{2 \sin B}$$

EXAMPLE 3   **Finding the Area of a Triangle Using SAA**

Find the area of a triangle, given $A = 20°$, $C = 15°$, and $c = 23$.

SOLUTION   First, we find angle $B = 180° - 20° - 15° = 145°$. Now we can find the area.

$$K = \frac{c^2 \sin A \sin B}{2 \sin C}$$

$$= \frac{23^2 \sin 20° \sin 145°}{2 \sin 15°}$$

$$\approx 200.48 \text{ units}^2$$

Self Check 3   Find the area of a triangle, given $B = 30°$, $C = 105°$, and $b = 15$. Round to two decimal places.

Now Try Exercise 21.

2. Solve Application Problems Involving Area

EXAMPLE 4   **Finding the Area and Cost of a Triangular Piece of Land**

A triangular lot in Baton Rouge, Louisiana, has sides of 120, 150, and 192 feet. If the property is $25 per square foot, how much does the lot cost? Round the area to the nearest square foot to compute cost.

SOLUTION   Since the three sides of the property form a triangle, and their values are known, we can use Heron's Formula to compute the area.

$$s = \frac{120 + 150 + 192}{2} = 231$$

$$K = \sqrt{231(231 - 120)(231 - 150)(231 - 192)}$$

$$= 8{,}999.9955$$

We will use 9,000 square feet as the area of the triangle, and we multiply this by the cost per square foot.

$$(9{,}000)(\$25) = \$225{,}000$$

Self Check 4   A triangular plot of land has three sides measuring 400, 550, and 450 feet. If the property is $42 per square foot, how much does the lot cost? Round the area to the nearest square foot to calculate cost.

Now Try Exercise 45.

Self Check Answers   **1.** 215.09 units^2 **2.** 51.58 in^2 **3.** 153.68 units^2 **4.** $3,727,122

Exercises **6.5**

Getting Ready

You should be able to complete these vocabulary and concept statements before you proceed to the practice exercises.

Fill in the blanks.

1. The formula to find the area of a triangle if two angles and an included side are known is _____.

2. Heron's Formula is _____.

3. The formula to find the area of a triangle, given SAA or ASA, is _____.

4. The geometric formula for the area of a triangle is _____.

Practice

Find the area of each triangle. Round to two decimal places.

5.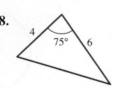

6.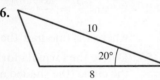

7.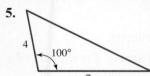

8.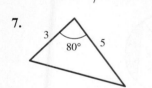

9.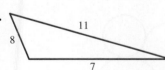

10.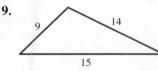

11.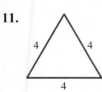

12.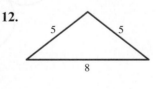

Find the area of each triangle. Round to two decimal places.

13. $a = 17; b = 23; C = 80°$

14. $b = 4; c = 2; A = 60°$

15. $a = 32; c = 22; B = 120°$

16. $B = 33°; a = 101; c = 97$

17. $a = 3; b = 5; c = 7$

18. $a = 2; b = 5; c = 6$

19. $a = 3; b = 4; c = 5$

20. $a = 5; b = 12; c = 13$

21. $A = 55°; B = 45°; c = 12$

22. $A = 102°; C = 47°; b = 8$

23. $A = 70°; B = 15°; b = 23$

24. $B = 62°; C = 41°; c = 17$

25. $a = 6; b = 5; c = 9$

26. $a = 27; b = 32; c = 55$

27. $a = 9; b = 14; c = 20$

28. $a = 14; b = 9; c = 15$

29. $a = 10; b = 12; C = 56°$

30. $b = 8; c = 10; A = 95°$

31. $a = 12; c = 12; B = 80°$

32. $a = 3; b = 4; C = 90°$

33. $A = 44°; B = 51°; c = 10$

34. $B = 19°; C = 49°; a = 7$

35. $A = 24°; B = 51°; c = 10$

36. $A = 17°; B = 80°; c = 12$

Applications

37. **Area of a lot** To find the area of a triangular lot, the owner starts at one corner and walks due west 205 feet to a second corner. After turning through an angle of N 47.3° E, he walks 307 feet to the third corner. What is the area of the lot in square feet? Round to the nearest square foot.

38. **Painting a building** A painter wishes to estimate the area of the gable end of a building. What is its area in square feet if the triangle has the dimensions shown in the figure? Round to nearest square foot.

39. **Making signs** A printer wishes to make a sign in the form of an isosceles triangle with base angles of 70° and the base side of 15 meters. Find the area of the triangle. Round to one decimal place.

40. **Painting a design** Art students at LSU are to draw a figure with an equilateral triangle inside of a larger equilateral triangle, as shown in the figure. If the length of the sides of the larger triangle are 8 feet, what is the total area of the gold triangles formed? Round to two decimal places.

41. **Surveying** Point C has a bearing of N 20° E from point A and a bearing of N 10° E from point B. What is the area of triangle ABC if B is due east of A and 17 kilometers from C? Round to one decimal place.

42. **Surveying** Point C is 40 miles from point A on a bearing of S 30° W. Point B is due south from point A and the bearing of point C from point B is N 50° W. Find the area of triangle ABC. Round to two decimal places.

43. **Hiking** Arnold walks 520 feet, turns and walks 490 feet, and turns again and walks 670 feet, returning to his starting point. What area does his walk encompass? Round to the nearest square foot.

44. **Hiking** Shelly hikes 523 meters, turns and jogs 412 meters, and turns again and runs 375 meters, returning to her starting point. What area does her trip encompass? Round to the nearest square meter.

45. **Bermuda Triangle** The Bermuda Triangle is an area associated with many stories of strange happenings; it is shown in the figure. The distance from Miami, Florida, to Hamilton, Bermuda, is 899 nautical miles. It is 897 nautical miles from Miami to San Juan, Puerto Rico, and 829 nautical miles from San Juan to Hamilton. What is the area of the Bermuda Triangle? Round to the nearest square nautical mile.

46. **Easter Island** Easter Island is a triangular island with three volcanic peaks that form an isosceles triangle. The base angles are 36° and the equal sides are 6.8 miles each. What is the area of the triangle formed by these volcanic peaks? Round to the nearest square mile.

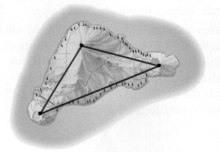

Discussion and Writing

47. Find the area of an isosceles triangle with vertex angle α and base b.

48. Find the area of an isosceles triangle having vertex angle α and equal sides of length a.

49. Find the area of an isosceles triangle if the base b is one-half the length of the one of the equal sides.

50. Prove that the area of a parallelogram is half the product of the diagonals and the sine of the angle between the diagonals.

51. Prove that in any triangle,
$$\cos^2 \frac{A}{2} = \frac{s(s-a)}{bc},$$
where s is one-half the perimeter.

52. Prove that in any triangle,
$$\sin^2 \frac{A}{2} = \frac{(s-b)(s-c)}{bc},$$
where s is one-half the perimeter.

53. Derive the formula for the area of the segment of the circle (the shaded area) shown in the figure.

54. Three externally tangent circles, as shown in the figure, have radii of 5, 7, and 8. What is the area of the shaded, curve-sided "triangle" they enclose? Round to one decimal place.

6.6 Linear and Angular Velocity

In this section, we will learn to

1. Solve linear- and angular-velocity problems.

If we were to watch Chris Johnson, running back for the Tennessee Titans, run during a football game, we might ask, "How fast is he running?" The answer to that question indicates his *linear velocity* (or *linear speed*). He doesn't have to be running a straight line; it is a measure of how far he runs in one unit of time.

Similarly, consider the world's tallest carousel, the Columbia Carousel. It is a double-decker carousel located at California's Great America amusement park and also at Six Flags Great America in Gurnee, Illinois. If we ask the question "How fast is the carousel turning?" the answer to the question would be the *angular velocity.* Angular velocity is a measure of the angle (radians, degrees, or revolutions) through which the Columbia rotates in one unit of time (hour, minute, second, etc.).

In this section, we will study linear and angular velocity in depth.

1. Solve Linear- and Angular-Velocity Problems

Linear and angular velocity are related. The question "How fast are the tires of that car moving?" might be answered in two ways. Because the tire moves with the car, it has linear velocity. Because the tire is spinning, it has angular velocity as well. Since for a given linear velocity, small motorcycle tires spin faster than the larger tires of a semitrailer, linear velocity, angular velocity, and the radius of the wheel are all related.

If a wheel of radius r rolls a distance of s without slipping, any point on the circumference of the wheel also moves a distance s, measured along the arc. (See Figure 6-42.) The wheel rotates through an angle θ (measured in radians), and $s = r\theta$. If this movement is accomplished in a length of time t, then $\frac{s}{t}$ is the **linear velocity** of a point on the circumference of the wheel and $\frac{\theta}{t}$ is the angular velocity of the wheel.

$$\frac{s}{t} = r\left(\frac{\theta}{t}\right)$$

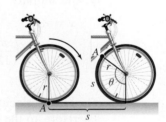

FIGURE 6-42

If we denote the linear velocity $\frac{s}{t}$ by v and the angular velocity $\frac{\theta}{t}$ by ω, the relation between linear and angular velocity is $v = r\omega$.

Linear and Angular Velocity

The **linear velocity** v of an object is the total distance traveled divided by the elapsed time.

$$v = \frac{s}{t}$$

Let θ be the angle swept in time t by an object moving in a circular path. The **angular velocity** ω of the object is defined as the angle θ, measured in radians, divided by the elapsed time t.

$$\omega = \frac{\theta}{t}$$

If the object moves along an arc of length s, then the relationship $s = r\theta$ (in radian measure) yields a corresponding relationship between linear and angular velocity:

$$v = \frac{s}{t} = r\frac{\theta}{t} = r\omega, \text{ thus } v = r\omega.$$

EXAMPLE 1 **Finding the Angular Velocity of the Earth**

What is the angular velocity of the earth about its axis?

SOLUTION There are actually several ways to express the answer. Because the earth rotates once in 24 hours, its angular velocity is one revolution per 24 hours, or $\frac{1}{24}$ revolution per hour. Since one revolution is 2π radians, the earth's angular velocity is also 2π radians per 24 hours, or $\frac{\pi}{12}$ radian per hour. Other possible answers are 1 revolution per day, 15 degrees per hour, and 2π radians per day. ∎

Self Check 1 Mars rotates about its axis once every 24.62 hours. What is the angular velocity of Mars about its axis? State your answer in radians per hour and degrees per hour. Round to two decimal places.

Now Try Exercise 11.

EXAMPLE 2 **Finding Linear Velocity**

Find the linear velocity of a point on the earth's equator. Use 3,960 miles as the radius of the earth and express the answer in miles per hour. Round to the nearest mile.

SOLUTION From Example 1, the angular velocity of the earth is $\omega = \frac{\pi}{12}$ radian per hour. To find the linear velocity, we use the formula $v = r\omega$, substitute 3,960 for r and $\frac{\pi}{12}$ for ω, and simplify.

$$v = r\omega$$
$$= 3,960\left(\frac{\pi}{12}\right)$$
$$\approx 1,037$$

The linear velocity of a point on the equator is approximately 1,037 miles per hour. Therefore, to the passengers in an airplane flying at 1,037 miles per hour from east to west along the equator, the sun would appear to stand still. ∎

Self Check 2 Find the linear velocity of a point on the equator of Mars. Use 2,110 miles for the radius. Round to the nearest unit.

Now Try Exercise 15.

Comment When solving a linear- or angular-velocity problem, we must use care to make sure that the units are correct. Carefully constructing the problem and paying attention to the units will help you solve the problem.

EXAMPLE 3   **Finding Angular Velocity**

A bicycle with 24-inch wheels is traveling at 10 miles per hour. Find the angular velocity of the wheels in revolutions per minute.

SOLUTION   Because the radius of each wheel is 12 inches, or 1 foot, and the angular velocity is to be given in revolutions per minute, we change 10 miles per hour to units of feet per minute.

$$10\,\frac{\text{mi}}{\text{hr}} = 10\,\frac{\cancel{\text{mi}}}{\cancel{\text{hr}}} \cdot 5{,}280\,\frac{\text{ft}}{\cancel{\text{mi}}} \cdot \frac{1}{60}\,\frac{\cancel{\text{hr}}}{\text{min}} = 880\,\frac{\text{ft}}{\text{min}}$$

We substitute 880 for v and 1 for r in the formula $v = r\omega$ and solve for ω.

$$v = r\omega$$
$$880 = 1\omega$$

The angular velocity is 880 radians per minute.

Because there are 2π radians in one revolution, there is $\frac{1}{2\pi}$ revolution in one radian. To find ω in revolutions per minute, we multiply $880\,\dfrac{\text{radians}}{\text{min}}$ by $\dfrac{1}{2\pi}\,\dfrac{\text{rev}}{\text{radian}}$.

$$\omega = 880\,\frac{\text{radians}}{\text{min}} \cdot \frac{1}{2\pi}\,\frac{\text{rev}}{\text{radian}}$$

$$\omega = \frac{440}{\pi}\,\frac{\text{rev}}{\text{min}}$$

$$\omega \approx 140\,\frac{\text{rev}}{\text{min}}$$

Self Check 3   A bicycle with 18-inch wheels is traveling at 6 miles per hour. Find the angular velocity of the wheels in revolutions per minute. Round to the nearest unit.

Now Try Exercise 17.

EXAMPLE 4   **Finding Angular Velocity**

An 8-inch-diameter pulley drives a 6-inch-diameter pulley. If the larger pulley makes 15 revolutions per second, find the angular velocity of the smaller pulley in revolutions per second.

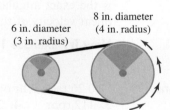

6 in. diameter
(3 in. radius)

8 in. diameter
(4 in. radius)

SOLUTION   First we express the angular velocity of the driving pulley in radians per second.

$$15\,\frac{\text{rev}}{\text{sec}} = 15\,\frac{\text{rev}}{\text{sec}} \cdot 2\pi\,\frac{\text{radians}}{\text{rev}} = 30\pi\,\frac{\text{radians}}{\text{sec}}$$

We assume that the belt connecting the two pulleys does not slip. The linear velocities of points on the edges of the two pulleys must be the same—the product $r_1\omega_1$ for the first pulley (the driving pulley) must equal $r_2\omega_2$ for the second pulley (the driven pulley).

$$r_1\omega_1 = r_2\omega_2$$
$$4(30\pi) = 3\omega_2$$
$$\frac{4(30\pi)}{3} = \omega_2$$
$$40\pi = \omega_2$$

The angular velocity of the smaller pulley is 40π radians per second. To convert to revolutions per second, we multiply by $\frac{1}{2\pi}$ revolution per radian and simplify.

$$40\pi \frac{\text{radians}}{\text{sec}} = 40\pi \frac{\text{radians}}{\text{sec}} \cdot \frac{1}{2\pi} \frac{\text{rev}}{\text{radian}} = 20 \frac{\text{rev}}{\text{sec}}$$

The angular velocity of the smaller pulley is 20 revolutions per second. ∎

Self Check 4 A 10-inch-diameter pulley drives an 8-inch-diameter pulley. If the larger pulley makes 18 revolutions per second, find the angular velocity of the smaller pulley in revolutions per second. Round to the nearest tenth.

Now Try Exercise 21.

EXAMPLE 5 **Finding Angular and Linear Velocity**

A Ferris wheel rotates three times each minute. The diameter of the ride is 50 feet. What is the exact angular velocity of the wheel in radians per minute? What is the linear velocity of those who are on the ride?

SOLUTION $$3 \frac{\text{rev}}{\text{min}} = 3 \frac{\text{rev}}{\text{min}} \cdot 2\pi \frac{\text{radians}}{\text{rev}} = 6\pi \frac{\text{radians}}{\text{min}}$$

Using the equation $v = r\omega$, we calculate the linear velocity.

$$v = 25 \text{ ft} \cdot 6\pi \frac{\text{radians}}{\text{min}} = 150\pi \frac{\text{ft}}{\text{min}}$$

We will convert this linear velocity to miles per hour.

$$v = 150\pi \frac{\text{ft}}{\text{min}} \cdot 60 \frac{\text{min}}{\text{hr}} \cdot \frac{1 \text{ mi}}{5,280 \text{ ft}} = \frac{9,000\pi}{5,280} \frac{\text{mi}}{\text{hr}} \approx 5.35 \text{ mph}$$ ∎

Self Check 5 A carousel rotates four times each minute. The diameter of the ride is 40 feet. What is the exact angular velocity of the carousel in radians per minute? What is the linear velocity of those who are on the ride? Round to two decimal places.

Now Try Exercise 15.

Self Check Answers 1. 0.26 radian/hr or 14.62° per hr   2. 538 mph
3. 112 rpm   4. 22.5 rev/sec   5. 8π radians/min; 5.71 mph

Exercises **6.6**

Getting Ready

You should be able to complete these vocabulary and concept statements before you proceed to the practice exercises.

Fill in the blanks.

1. The _____ velocity of an object is the total distance traveled divided by the elapsed time.

2. The _____ velocity ω of the object is defined as the angle θ, measured in _____, divided by the elapsed time t.

3. The relation between linear and angular velocity is _____.

4. When solving a linear or angular velocity problem, you must use care to make sure that the _____ are correct.

Practice

Find the angular velocity of the object in the specified units.

5. The minute hand of a clock, in radians per hour.

6. The second hand of a clock, in radians per second.

7. The minute hand of a clock, in radians per second.

8. The earth in its orbit, in radians per month.

9. The moon in its orbit, in radians per day. (Assume that the moon circles the earth in 29.5 days.)

10. A wheel turning at 150 revolutions per minute, in radians per second.

Applications

11. Angular velocity of tires A car is traveling 60 miles per hour. How fast are its 30-inch-diameter tires spinning, in revolutions per second?

12. Angular velocity of tires A motorcycle is traveling 30 miles per hour. Find the angular velocity of its 22-inch-diameter tires, in revolutions per minute.

13. Angular velocity of tires A large truck is traveling at 65 miles per hour. How fast are its 20-inch-radius tires spinning, in revolutions per minute?

14. Speed of a van The 30-inch-diameter tires of a van are turning at 400 revolutions per minute. Find the linear velocity of the van in miles per hour.

15. Speed of a bicycle The 27-inch-diameter tires of a bicycle are turning at a rate of 125 revolutions per minute. Find the linear velocity of the bicycle, in feet per minute.

16. Angular and linear velocity of a ceiling fan A ceiling fan rotates 30 times per minute. What is the angular velocity in radians per hour? What is the linear velocity, in miles per hour, of the fan blade if the blade is 2 feet long?

17. Computer hard drive If a hard drive on a computer rotates at 7,200 revolutions per minute, what is the angular velocity in radians per minute? What is the linear velocity in inches per minute of a particle located 2 inches from the center of the hard drive? (Round to the nearest unit.) In miles per hour? (Round to two decimal places.)

18. Angular velocity of a particle A particle moves on a circular path with a linear velocity of 300 feet per second. If the particle makes three revolutions per second, what is its angular velocity? What is the radius of the circle?

19. Pulley design A pulley driven by a belt is making one revolution per second. If the pulley is 6 inches in diameter, find the linear velocity of the belt in feet per second.

20. Pulley design An idler pulley 3 inches in diameter is making 95 revolutions per minute. In inches per second, what is the linear velocity of the belt that is driving the pulley?

21. RPM of two wheels A belt drives two wheels that are 15 and 20 inches in diameter. The 20-inch wheel is turning at 10 revolutions per minute. How fast is the other wheel turning, in revolutions per minute?

22. RPM of two pulleys A belt drives two pulleys. One pulley is 10 inches in diameter, and the other is 12 inches in diameter. How many revolutions per minute is the large pulley making if the small pulley is turning at 20 revolutions per minute?

In Exercises 23–26, assume that the radius of the earth is 3,960 miles.

23. Linear velocity and latitude Find the linear velocity due to the rotation of the earth of Green Bay, Wisconsin, at latitude N 44.5°. (*Hint:* Green Bay moves on a circle with radius *r*.) Round your answer to the nearest unit.

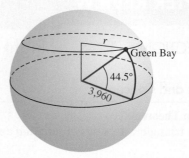

24. Linear velocity and latitude Find the linear velocity due to the rotation of the earth of Miami, Florida, at latitude N 25.8°. Round your answer to the nearest unit.

25. Finding latitude Find the latitude of Baton Rouge, Louisiana, if its linear velocity due to the rotation of the earth is 893 mph. Round your answer to the nearest tenth of a degree.

26. Finding latitude Find the latitude of Jackson, Mississippi, if its linear velocity due to the rotation of the earth is 876 mph. Round your answer to the nearest tenth of a degree.

27. Finding distance If an airplane at the equator flew from east to west at 1,040 mph, the sun would appear to stand still. (See Example 2.) For another plane, flying east to west at 100 mph, the sun also appears to stand still. How far is the second plane from the North Pole? Round to the nearest unit.

28. Dentist's drill A dentist's turbine drill spins at 7,000 revolutions per second. If the radius of the drill tip is 0.5 millimeter, find the linear velocity of the drill's cutting edge, in millimeters per second. Round to the nearest unit.

Discussion and Writing

29. Show that if the gears in the figure are separated by an idler gear of any radius, the angular velocity of the driven gear is $\dfrac{R_1}{R_2}$ times that of the driving gear.

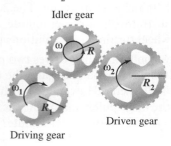

Idler gear

Driven gear

Driving gear

30. If a gear of radius R_1 drives a gear of radius R_2, show that the angular velocity of the driven gear is $\dfrac{R_1}{R_2}$ times that of the driving gear.

Driving gear

Driven gear

CHAPTER REVIEW

SECTION 6.1   Right Triangles

Definitions and Concepts	Examples
Pythagorean Theorem: In any right triangle with sides of length a and b and hypotenuse of length c, $a^2 + b^2 = c^2$.	See Example 1, page 601
Right-triangle definitions of trigonometric functions: If angle A is an acute angle in right triangle ABC, then $\sin A = \dfrac{\text{opposite side}}{\text{hypotenuse}} = \dfrac{a}{c}$ $\csc A = \dfrac{\text{hypotenuse}}{\text{opposite side}} = \dfrac{c}{a}$ $\cos A = \dfrac{\text{adjacent side}}{\text{hypotenuse}} = \dfrac{b}{c}$ $\sec A = \dfrac{\text{hypotenuse}}{\text{adjacent side}} = \dfrac{c}{b}$ $\tan A = \dfrac{\text{opposite side}}{\text{adjacent side}} = \dfrac{a}{b}$ $\cot A = \dfrac{\text{adjacent side}}{\text{opposite side}} = \dfrac{b}{a}$	See Example 2, pages 601–602
Angle of elevation and depression: If an observer looks up to see an object, the angle that the observer's line of sight makes with the horizontal is called the **angle of elevation**; if the observer looks down to see the object, the angle made with the horizontal is called the **angle of depression**.	See Example 3, pages 602–603

EXERCISES
Solve the right triangle with the given dimensions and given that angle C is a right angle. Round angles to the nearest tenth and sides to the nearest hundredth.

1. $a = 5$; $b = 9$ **2.** $b = 7$; $c = 12$

3. $c = 11$; $A = 21°$ **4.** $a = 14$; $B = 47°$

Applications

5. Angle of elevation A hot-air balloon is flying at 2,000 feet when it passes over an observer. Two minutes later, the observer notes that the plane's angle of elevation is 50.0°. How fast is the wind moving the balloon in miles per hour? (*Hint:* 5,280 feet = 1 mile.) Round to the nearest tenth.

SECTION 6.2   More Right-Triangle Applications

Definitions and Concepts	Examples
Angle of inclination: The **angle of inclination** can be found using the slope of the line passing through two points.	See Example 3, page 613

EXERCISES

6. Find the angle of inclination that passes through the points $(1, 1)$ and $(10, 4)$.

7. Use right triangles to find the value of h in the figure. Round to one decimal place.

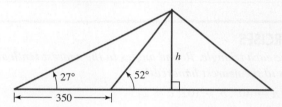

SECTION 6.3   Law of Sines

Definitions and Concepts	Examples
Law of Sines: The sides in any triangle are proportional to the sines of the angles opposite those sides. $$\frac{a}{\sin A} = \frac{b}{\sin B} = \frac{c}{\sin C} \quad \text{or} \quad \frac{\sin A}{a} = \frac{\sin B}{b} = \frac{\sin C}{c}$$	See Example 1, pages 620–621

EXERCISES

Solve each triangle, if it exists. Round angles to the nearest tenth and sides to the nearest hundredth. If two triangles are possible, solve both.

8. $A = 40°$; $B = 82°$; $a = 5$

9. $A = 25°$; $B = 17°$; $c = 12$

10. $a = 18$; $b = 15$; $A = 70°$

11. $a = 11$; $b = 15$; $A = 52°$

12. $a = 10$; $b = 12$; $A = 50°$

13. Solve for x in the figure. Round to two decimal places.

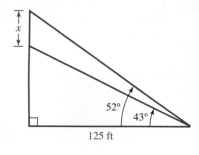

Applications

14. **Canyon width** Two observers stand on one side of a canyon at points A and B, 310 feet apart. Each one sights the same totem pole on the opposite side. They determine the measurements shown in the figure. How wide is the canyon? Round to the nearest tenth of a foot.

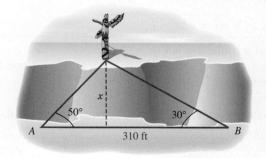

SECTION 6.4   Law of Cosines

Definitions and Concepts	Examples
Law of Cosines: The square of any side of any triangle is equal to the sum of the squares of the remaining two sides minus twice the product of those two sides and the cosine of the included angle. $a^2 = b^2 + c^2 - 2bc \cos A$ $b^2 = a^2 + c^2 - 2ac \cos B$ $c^2 = a^2 + b^2 - 2ab \cos C$	See Example 1, page 634

EXERCISES

Solve each triangle. Round angles to the nearest tenth and sides to the nearest hundredth.

15. $A = 40°$; $b = 102$; $c = 21$

16. $a = 10$; $b = 15$; $c = 22$

Applications

17. Bearing of ship A ship leaves port and sails 210 nautical miles in the direction N 50° E. The captain then turns onto a course of S 12° W. After sailing another 100 nautical miles, how far is the ship from port? (Round to the nearest tenth of a mile.) Through what angle should the ship turn to return to port? (Round to the nearest degree.)

SECTION 6.5   Area of Triangles

Definitions and Concepts	Examples
Area of a triangle, given SAS: If two sides and the included angle of a triangle are known, the area of the triangle is one-half the product of the two sides and the sine of the included angle. $K = \dfrac{1}{2}ab \sin C \qquad K = \dfrac{1}{2}ac \sin B \qquad K = \dfrac{1}{2}bc \sin A$	See Example 1, page 642
Heron's Formula: If a, b, and c are the three sides of a triangle and $s = \dfrac{a + b + c}{2},$ then the area of the triangle is given by $K = \sqrt{s(s - a)(s - b)(s - c)}.$	See Example 2, page 643
Area of a triangle (SAA or ASA): If two angles and a side of a triangle are given, one of the following formulas can be used to find the area of the triangle: $K = \dfrac{c^2 \sin A \sin B}{2 \sin C} \qquad K = \dfrac{a^2 \sin B \sin C}{2 \sin A} \qquad K = \dfrac{b^2 \sin A \sin C}{2 \sin B}$	See Example 3, page 644

EXERCISES

Find the area of each triangle. Round to two decimal places.

18. $a = 7; b = 13; C = 110°$

19. $a = 13; b = 15; c = 17$

20. $A = 104°; B = 29°; c = 12$

SECTION **6.6**   Linear and Angular Velocity

Definitions and Concepts	**Examples**
Linear and angular velocity: The **linear velocity** v of an object is the total distance traveled divided by the elapsed time: $$v = \frac{s}{t}.$$ Let θ be the angle swept in time t by an object moving in a circular path. The **angular velocity** ω of the object is defined as the angle θ, measured in radians, divided by the elapsed time t: $$\omega = \frac{\theta}{t}.$$ If the object moves along an arc of length s, then the relationship $s = r\theta$ (in radian measure) yields a corresponding relationship between linear and angular velocity: $$v = \frac{s}{t} = r\frac{\theta}{t} = r\omega, \text{ thus } v = r\omega.$$	See Example 1, page 648

Applications

21. Angular and linear velocity A 15-inch disk rotates through 12 revolutions per second. What is the angular velocity of the disk in radians per second? What is the linear velocity in inches per second?

22. Blade rotation A 7.5-inch circular saw blade rotates at 2,400 revolutions per minute. Find the angular speed in radians per second and the linear speed in feet per second of the saw as it contacts the lumber.

CHAPTER TEST

Solve the right triangle with the given dimensions and given that angle C is a right angle. Round angles to the nearest tenth and sides to the nearest hundredth.

1. $a = 11; b = 15$

2. $c = 7; A = 31°$

3. Height of building At a distance of 902 feet, two angle-of-elevation measurements are taken. The angle to the observation deck is 22°. The angle to the top of the building is 27°. Use right triangles to find the distance from the observation deck to

the top of the building, the value of h in the figure. What is the total height of the building? Round your answers to the nearest foot.

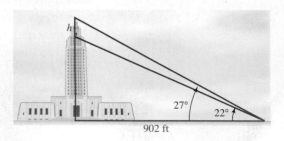

Solve each triangle, if it exists. Round angles to the nearest tenth and sides to the nearest hundredth. If two triangles are possible, solve both.

4. $A = 54°$; $B = 26°$; $a = 15$

5. $a = 15$; $b = 12$; $A = 75°$

6. $a = 9$; $b = 15$; $A = 24°$

7. $A = 54°$; $b = 12$; $c = 19$

8. $a = 12$; $b = 14$; $c = 19$

9. Solve for x in the figure. Round to the nearest tenth of a foot.

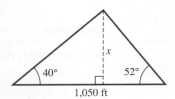

10. **Bearing** A ship leaves port on a course of N 47° E and travels 75 miles. The ship then turns and heads on a bearing of S 27° E for 40 miles. How far from port is the ship, and through what angle must the captain turn to return to port? Round distance to the nearest tenth of a mile. Round the angle to the nearest degree.

Find the area of each triangle. Round to two decimal places.

11. $a = 5$; $b = 11$; $C = 87°$

12. $a = 7$; $b = 12$; $c = 8$

13. $A = 78°$; $B = 41°$; $c = 19$

14. **Angular and linear velocity** A CD spins at 200 revolutions per minute. What is the angular velocity in radians per minute? Radians per second? What is the linear velocity of a piece of dust on the CD 6 centimeters from the center?

 LOOKING AHEAD TO **CALCULUS**

1. **Using Right Triangles to Solve Related-Rates Problems**
   - Connect right triangles to related rates—Section 6.1

1. Using Right Triangles to Solve Related-Rates Problems

Related-rates problems compute the rate of change of one quantity in terms of the rate of change of another quantity at the times the other quantity can be measured more easily. Right triangles are often used in solving certain types of related-rates problems, and many will use trigonometric functions.

EXAMPLE 1   **Finding the Rate of Change of an Angle**

A ladder 12 feet long rests against a wall. If the bottom of the ladder slides away from the wall at 2 feet per second, how fast is the angle between the top of the ladder and the wall changing when the angle is $\frac{\pi}{3}$?

We are given that the change in x is 2 feet per second, which means that the change in x is a function of time t: $\frac{dx}{dt} = 2$. From right triangles, we also know that $\sin \theta = \frac{x}{12}$. This gives us $x = 12 \sin \theta$. Now we use implicit differentiation to differentiate with respect to time t.

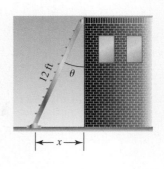

$$x = 12 \sin \theta$$

$$\frac{dx}{dt} = 12 \cos \theta \, \frac{d\theta}{dt}$$

When $\theta = \dfrac{\pi}{3}$, we have

$$2 = 12 \cos \frac{\pi}{3} \cdot \frac{d\theta}{dt}$$

$$\frac{d\theta}{dt} = \frac{2}{12 \cos \dfrac{\pi}{3}} = \frac{2}{12 \cdot \dfrac{1}{2}} = \frac{1}{3}$$

The angle is changing at $\frac{1}{3}$ radian per second.

Connect to Calculus:
Using Right Triangles to Solve Related-Rates Problems

1. A 10-foot ladder leans against the side of a building. If the top of the ladder begins to slide down the wall at the rate of two feet per second, how fast is the bottom of the ladder sliding away from the wall when the top of the ladder is 8 feet from the ground?

2. A lighthouse is located 3 miles from the nearest point on a straight shoreline. The light rotates four times each minute. How fast is the beam of light moving along the shoreline when it is one mile from the point?

Polar Coordinates; Vectors

CAREERS AND MATHEMATICS: Air Traffic Controller

Air traffic controllers work within the National Airspace System to make certain that planes remain a safe distance apart by coordinating the movement of air traffic. In addition to their main goal of safety, controllers also direct planes in an efficient manner so as to minimize delays. Some controllers regulate airport traffic through designated airspaces, and others regulate airport arrivals and departures. Those who work as *terminal controllers* are responsible for watching over all planes traveling in an airport's airspace.

Education and Mathematics Required
- A mathematics degree is a terrific avenue to a career as an air traffic controller. In order to become an air traffic controller, you must also complete a program approved by the Federal Aviation Administration (FAA), pass a pre-employment exam, receive a recommendation from your school, meet basic qualifications in accordance with federal law, and qualify by passing the FAA-authorized pre-employment exam.
- College Algebra, Trigonometry, Geometry, Calculus I and II, and Statistics are required courses.

How Air Traffic Controllers Use Math and Who Employs Them
- Air traffic controllers use math to understand distances and measurements at a moment's notice. Being able to do mental math both quickly and accurately is a necessary part of the job. Correctly directing aircraft to fly at a certain speed and altitude requires a strong math background, knowing that an error in these calculations could be fatal.
- A large majority of air traffic controllers are employed by the FAA. They work in towers and flight-service stations, as well as in air-route traffic control centers. Some work for the U.S. Department of Defense.

Career Outlook and Salary
- Employment for air traffic controllers is expected to grow 13% from 2008 to 2018, which is about average for all occupations.
- Air traffic controllers earn relatively high salaries and enjoy good benefits. Median annual wages were $111,870 in May 2008. The highest 10% earned more than $161,010 a year.

For more information see http://www.bls.gov/oco.

In this chapter, we will learn about the Polar Coordinate System and vectors. The Polar Coordinate System uses an angle and a distance to represent a point in the plane. Many calculations in mathematics are easier to solve using this system. We will also study vectors, a quantity possessing both magnitude and direction. Vectors have many applications in mathematics, physics, and engineering fields.

7.1 Polar Coordinates

In this section, we will learn to

1. Plot points in the polar coordinate system.
2. Convert from polar to rectangular coordinates.
3. Convert from rectangular to polar coordinates.
4. Convert equations between coordinate systems.
5. Use the distance formula with polar coordinates.

While the rectangular coordinate system may be the best-known system used in mathematics, it is not the only coordinate system. Consider the graphs shown in Figure 7-1.

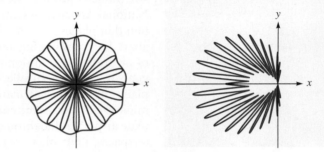

FIGURE 7-1

To easily produce these graphs, another system, the *polar coordinate system,* can be used.

Curves are interesting designs. They catch our eye, inspire art and architecture, and occur in nature. Consider the cutaway of a nautilus shell shown in Figure 7-2. It is one of the best examples of a *logarithmic spiral.* These spirals can easily be produced in the polar coordinate system, as shown on the right in Figure 7-2.

FIGURE 7-2

Many organized patterns appear in nature, such as fractals, and spirals are seen in hurricanes. These are shown in Figure 7-3.

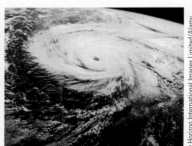

FIGURE 7-3

In this section, we will learn about polar coordinates and the polar coordinate system. This system is used to solve many calculus applications.

1. Plot Points in the Polar Coordinate System

In a rectangular coordinate system, a point is represented by an ordered pair (x, y). The **polar coordinate system** is based on a ray, called the **polar axis**, and its endpoint, called the **pole**. In Figure 7-4, ray OA is the polar axis and O is the pole. Any point P in the plane can be located if two numbers are known: r, the directed distance from O to the point, and θ, an angle in standard position. We label a point in the polar coordinate system as $P(r, \theta)$.

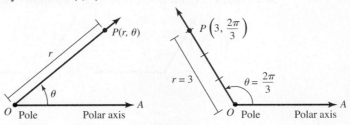

FIGURE 7-4

Angle θ can be measured in degrees or radians, and r can be positive or negative. To plot the point $P(r, \theta)$ when r is positive, we draw angle θ in standard position and count r units along the terminal side of θ. This determines the point $P(r, \theta)$ when $r > 0$. To plot $P(r, \theta)$ when r is negative, we draw angle θ in standard position and count $|r|$ units along the extension of the terminal side of θ, but in the opposite direction. This determines the point $P(r, \theta)$ when $r < 0$ [Figure 7-5 (a)]. For example, to plot the point $\left(-4, \frac{\pi}{6}\right)$, we draw the angle $\theta = \frac{\pi}{6}$ and then go 4 units in the opposite direction along the extension of the terminal side of the angle.

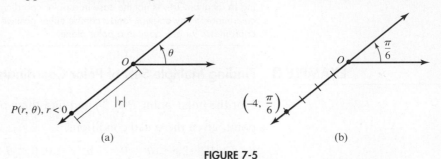

FIGURE 7-5

EXAMPLE 1   **Plotting Points in a Polar Coordinate System**

Plot each point with the given polar coordinates.

a. $\left(2, \frac{3\pi}{4}\right)$ **b.** $\left(-3, \frac{\pi}{3}\right)$ **c.** $(5, 180°)$ **d.** $(-4, -150°)$

SOLUTION   The points are shown in Figure 7-6.

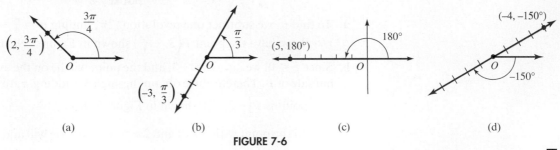

FIGURE 7-6

Self Check 1   Plot each point with the given polar coordinates.

a. $\left(4, \dfrac{7\pi}{6}\right)$ b. $(3, -90°)$ c. $(-2, -\pi)$

Now Try Exercise 7.

EXAMPLE 2   **Plotting Points in a Polar Coordinate System**

Plot the following points to demonstrate that they represent the same polar point:

a. $\left(1, \dfrac{2\pi}{3}\right)$ b. $\left(-1, \dfrac{5\pi}{3}\right)$ c. $\left(-1, -\dfrac{\pi}{3}\right)$ d. $\left(1, -\dfrac{4\pi}{3}\right)$

SOLUTION   Each point is plotted and shown in Figure 7-7.

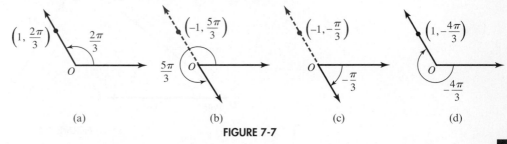

(a)   (b)   (c)   (d)

FIGURE 7-7

Self Check 2   Show that $\left(1, -\dfrac{\pi}{4}\right)$ and $\left(-1, \dfrac{3\pi}{4}\right)$ are polar coordinates for the same polar point.

Now Try Exercise 17.

Comment   In Figure 7-7, we see that the same point is represented by four different sets of polar coordinates. In a rectangular coordinate system, every point has a unique ordered pair of numbers (x, y) denoting its location. This is not the case in a polar coordinate system: Since infinitely many angles are coterminal to any angle θ, and r can be either positive or negative, there are infinitely many sets of coordinates for any point in a polar plane.

EXAMPLE 3   **Finding Multiple Sets of Polar Coordinates for a Point**

Plot the polar point $P\left(3, \dfrac{\pi}{4}\right)$ and find three other sets of polar coordinates for the point, given the stated conditions.

a. $r > 0, -2\pi \le \theta \le 0$ b. $r < 0, 0 \le \theta \le 2\pi$ c. $r > 0, 2\pi \le \theta \le 4\pi$

SOLUTION   Our first step is to plot the point, shown in Figure 7-8.

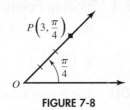

FIGURE 7-8

a. To find θ, we subtract one revolution, 2π, finding $\theta = \dfrac{\pi}{4} - 2\pi = -\dfrac{7\pi}{4}$. Since $r > 0$, we have the point $P\left(3, -\dfrac{7\pi}{4}\right)$ shown in Figure 7-9(a).

b. Since $r < 0$, we have $r = -3$, and the point will lie on the extension of the terminal side of θ. Therefore, we obtain the angle by adding π, to get $\theta = \dfrac{\pi}{4} + \pi = \dfrac{5\pi}{4}$. The point is $P\left(-3, \dfrac{5\pi}{4}\right)$, shown in Figure 7-9(b).

c. To find the point with $r > 0$ and $2\pi \le \theta \le 4\pi$, we will add one revolution to the angle to obtain the point $P\left(3, \dfrac{\pi}{4} + 2\pi\right) = P\left(3, \dfrac{9\pi}{4}\right)$ [Figure 7-9(c)].

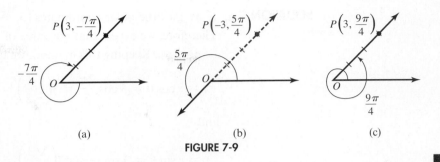

(a)   (b)   (c)

FIGURE 7-9

Self Check 3   Given the polar point $P\left(-2, \frac{\pi}{6}\right)$, find three other sets of coordinates for the point, given the stated conditions. **a.** $r > 0, 0 \le \theta \le 2\pi$ **b.** $r > 0, -2\pi \le \theta \le 0$ **c.** $r < 0, 2\pi \le \theta \le 4\pi$

Now Try Exercise 25.

There is a way to find multiple sets of coordinates for any polar point:

Multiple Sets of Polar Coordinates for a Given Point   For a point with polar coordinates (r, θ) and any integer n, the polar coordinates $(r, \theta + 2\pi n)$ and $(-r, \theta + \pi + 2\pi n)$ represent the same point in the polar coordinate system.

2. Convert from Polar to Rectangular Coordinates

There is a relationship between the rectangular coordinates (x, y) and the polar coordinates (r, θ) of a point. Establishing this relationship is important because we often need to convert polar points to rectangular points, and vice versa. This, in turn, will allow us to convert equations back and forth between rectangular form and polar form.

Suppose that point R in Figure 7-10 has rectangular coordinates (x, y) and polar coordinates (r, θ). We draw PA perpendicular to the x-axis to form the right triangle OAP with $OA = x$, $AP = y$, and $OP = r$. Because angle θ is in standard position, the hypotenuse OP is the terminal side of angle θ. Thus,

$$\cos \theta = \frac{x}{r} \quad \text{and} \quad \sin \theta = \frac{y}{r},$$

or

$$x = r \cos \theta \quad \text{and} \quad y = r \sin \theta.$$

To change from polar to rectangular coordinates, we use the following equations:

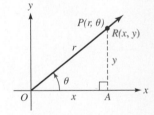

FIGURE 7-10

Converting from Polar to Rectangular Coordinates   Use these formulas to convert $P(r, \theta)$ to $R(x, y)$:

$$x = r \cos \theta \quad \text{and} \quad y = r \sin \theta.$$

EXAMPLE 4   **Converting Polar to Rectangular Coordinates**

For each point with the given polar coordinates, find the corresponding rectangular coordinates.

a. $\left(3, \frac{2\pi}{3}\right)$ **b.** $\left(-5, -\frac{5\pi}{6}\right)$

SOLUTION   **a.** If we plot the polar coordinates $\left(3, \frac{2\pi}{3}\right)$, we see that it lies in QII (Figure 7-11). Therefore, we expect that the value of x will be negative and the value of y will be positive. Keeping this in mind, we use $r = 3$ and $\theta = \frac{2\pi}{3}$ to find

$$x = r \cos \theta = 3 \cos \frac{2\pi}{3} = 3\left(-\frac{1}{2}\right) = -\frac{3}{2}$$

and

$$y = r \sin \theta = 3 \sin \frac{2\pi}{3} = 3\left(\frac{\sqrt{3}}{2}\right) = \frac{3\sqrt{3}}{2},$$

giving us the rectangular coordinates $\left(-\frac{3}{2}, \frac{3\sqrt{3}}{2}\right)$. The point is in QII as we expected it to be.

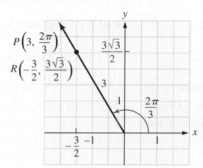

FIGURE 7-11

b. The polar coordinates $\left(-5, -\frac{5\pi}{6}\right)$ lie in QI, as we see in Figure 7-12. Thus, we expect that the values of both x and y will be positive. With $r = -5$ and $\theta = -\frac{5\pi}{6}$, we calculate

$$x = r \cos \theta = -5 \cos\left(-\frac{5\pi}{6}\right) = -5\left(-\frac{\sqrt{3}}{2}\right) = \frac{5\sqrt{3}}{2}$$

and

$$y = r \sin \theta = -5 \sin\left(-\frac{5\pi}{6}\right) = -5\left(-\frac{1}{2}\right) = \frac{5}{2},$$

giving us the rectangular coordinates $\left(\frac{5\sqrt{3}}{2}, \frac{5}{2}\right)$. The point is in QI, which we expected.

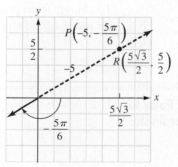

FIGURE 7-12

Self Check 4   For each point with the given polar coordinates, find the corresponding rectangular coordinates. **a.** $\left(-2, \frac{\pi}{4}\right)$ **b.** $\left(3, -\frac{\pi}{3}\right)$

Now Try Exercise 33.

ACCENT ON TECHNOLOGY   ## Converting from Polar to Rectangular Coordinates

A graphing calculator will convert from polar to rectangular coordinates, but will not give an exact value such as $x = \frac{\sqrt{3}}{2}$. The steps to convert the polar point $\left(-5, -\frac{5\pi}{6}\right)$ to rectangular coordinates are shown in Figure 7-13.

Pressing 2nd APPS **takes us to the ANGLE menu. Press** 7 **to find the x-coordinate.**

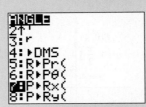

On the home screen, enter the r and θ coordinates of the point; press ENTER.

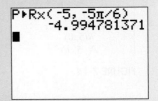

Pressing 2nd APPS **takes us to the ANGLE menu. Press** 8 **to find the y-coordinate.**

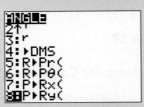

On the home screen, enter the r and θ coordinates of the point; press ENTER. **This is the decimal rectangular approximation of the polar coordinates.**

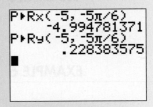

FIGURE 7-13

3. Convert from Rectangular to Polar Coordinates

When we convert from polar to rectangular coordinates, there can be only one representation. Because a point in a polar coordinate system does not have a unique set of coordinates indicating its location, converting from rectangular to polar coordinates is not as straightforward as converting from polar to rectangular coordinates. Referring to Figure 7-10, from the right triangle OAP it follows that

$$r^2 = x^2 + y^2 \qquad \text{and} \qquad \tan \theta = \frac{y}{x}, x \neq 0.$$

To find polar coordinates for a point, we must determine a value for r and a value for θ. We know that $r = \pm\sqrt{x^2 + y^2}$ and $\theta = \tan^{-1}\frac{y}{x}$. However, recall that the inverse tangent function will only return a value in the interval $\left(-\frac{\pi}{2}, \frac{\pi}{2}\right)$. Therefore, we will determine the quadrant in which the point lies, and use a reference angle $\theta' = \tan^{-1}\left|\frac{y}{x}\right|$ to find a value of θ that terminates in the correct quadrant.

Let's start by looking at rectangular points that lie on an axis. These are points that can be converted to polar coordinates without using the inverse tangent function or the definition of r.

EXAMPLE 5   **Converting Rectangular Coordinates on an Axis to Polar Coordinates**

Find two sets of polar coordinates for the given rectangular coordinates.
a. $R(5, 0)$ **b.** $R(0, 4)$

SOLUTION   **a.** The point $R(5, 0)$ lies on the positive x-axis. Therefore, we set $r = 5$ and $\theta = 0$, giving us the point with polar coordinates $P(5, 0)$, in Figure 7-14. Note that these points look exactly alike, which demonstrates why it is important to designate whether a set of coordinates is rectangular or polar. To find another set of polar coordinates for this point, we let $r = -5$ and $\theta = \pi$, and the polar point is $P(-5, \pi)$.

b. The point $R(0, 4)$ lies on the positive y-axis. Thus, we set $r = 4$ and $\theta = \frac{\pi}{2}$, which gives us the polar point $P\left(4, \frac{\pi}{2}\right)$. To find another point, we can use $\theta = -\frac{\pi}{2}$ and $r = -4$, resulting in the point $P\left(-4, -\frac{\pi}{2}\right)$, in Figure 7-15.

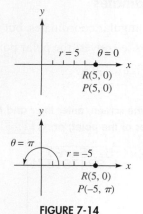

FIGURE 7-14

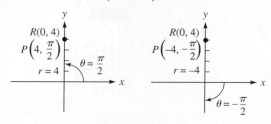

FIGURE 7-15

Self Check 5   Find two sets of polar coordinates for the given rectangular coordinates.
a. $R(-2, 0)$ **b.** $R(0, -6)$

Now Try Exercise 57.

EXAMPLE 6   **Converting Rectangular Coordinates to Polar Coordinates**

For each given rectangular point, find two sets of polar coordinates $P(r, \theta)$ such that for one set $r > 0$ and $0 \leq \theta \leq 2\pi$ and for the other, $r < 0$ and $0 \leq \theta \leq 2\pi$.

a. $R\left(1, \sqrt{3}\right)$ **b.** $R(-3, 3)$ **c.** $R(4, -5)$ (to two decimal places)

SOLUTION   **a.** The point $R\left(1, \sqrt{3}\right)$ lies in QI, and we use $r^2 = x^2 + y^2$ to find the positive and negative values of r. We will also find the reference angle using $\theta' = \tan^{-1}\left|\frac{y}{x}\right|$, and use that to find the two values of θ that meet our criteria (Figure 7-16).

$$r^2 = x^2 + y^2$$
$$r^2 = (1)^2 + \left(\sqrt{3}\right)^2$$
$$r^2 = 1 + 3 = 4 \qquad\qquad \theta' = \tan^{-1}\left|\frac{\sqrt{3}}{1}\right| = \tan^{-1}\sqrt{3} = \frac{\pi}{3}$$
$$r = \pm\sqrt{4} = \pm 2$$

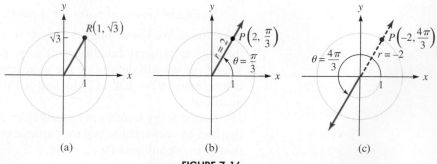

(a)   (b)   (c)

FIGURE 7-16

The polar point with $r > 0$ and $0 \leq \theta \leq 2\pi$ that corresponds to the rectangular point $R\left(1, \sqrt{3}\right)$ has coordinates $P\left(2, \frac{\pi}{3}\right)$. The point with $r < 0$ and $0 \leq \theta \leq 2\pi$ has coordinates $P\left(-2, \frac{4\pi}{3}\right)$.

b. $R(-3, 3)$ lies in QII. We will find the positive and negative values of r and the reference angle θ'.

$$r^2 = x^2 + y^2$$
$$r^2 = (-3)^2 + (3)^2$$
$$r^2 = 9 + 9 = 18$$
$$r = \pm\sqrt{18} = \pm 3\sqrt{2}$$

$$\theta' = \tan^{-1}\left|\frac{y}{x}\right| = \tan^{-1}\left|\frac{3}{-3}\right| = \tan^{-1} 1 = \frac{\pi}{4}$$

Since the point is in QII, we will find θ using a reference angle.

$$\theta = \pi - \theta' = \pi - \frac{\pi}{4} = \frac{3\pi}{4}$$

The polar point with $r > 0$ and $0 \le \theta \le 2\pi$ that corresponds to the rectangular point $R(-3, 3)$ is the point with coordinates $P\left(3\sqrt{2}, \frac{3\pi}{4}\right)$, and the point with $r < 0$ has coordinates $P\left(-3\sqrt{2}, \frac{7\pi}{4}\right)$, as shown in Figure 7-17.

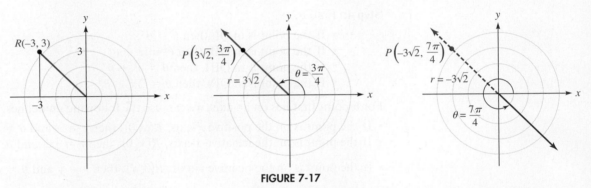

FIGURE 7-17

c. We will use a calculator to find approximate polar coordinates for the point $R(4, -5)$, which lies in QIV. We will find the positive and negative values of r, and then we will use a reference angle to find approximate values of θ.

$$r^2 = x^2 + y^2$$
$$r^2 = (4)^2 + (-5)^2$$
$$r^2 = 16 + 25 = 41$$
$$r = \pm\sqrt{41} \approx \pm 6.40$$

$$\theta' = \tan^{-1}\left|\frac{y}{x}\right| = \tan^{-1}\left|\frac{-5}{4}\right| = \tan^{-1}\frac{5}{4}$$

$$\theta = 2\pi - \tan^{-1}\frac{5}{4} \approx 5.39 \qquad \text{or}$$

$$\theta = \pi - \tan^{-1}\frac{5}{4} \approx 2.25$$

The polar point with $r > 0$ and $0 \le \theta \le 2\pi$ that corresponds to the rectangular point $R(4, -5)$ is approximated by the coordinates $P(6.403, 5.387)$, and the point with $r < 0$ and $0 \le \theta \le 2\pi$ is approximated by the coordinates $P(-6.403, 2.246)$. See Figure 7-18.

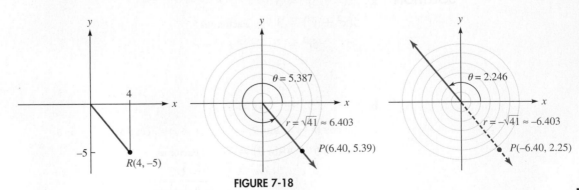

FIGURE 7-18

Self Check 6   Find a set of polar coordinates $P(r, \theta)$, with $r > 0$ and $0 \le \theta < 2\pi$, for each given rectangular point.

a. $R\left(-2\sqrt{3}, -2\right)$ **b.** $R(-2, 5)$ (to three decimal places)

Now Try Exercise 61.

Guidelines for Converting Rectangular to Polar Coordinates with $r > 0$ and $0 \le \theta < 2\pi$

Given a rectangular point $R(x, y)$, we have $r^2 = x^2 + y^2$, $\tan \theta = \dfrac{y}{x}$, and $\theta' = \tan^{-1}\left|\dfrac{y}{x}\right|$, $x \ne 0$. To find a polar point with coordinates $P(r, \theta)$, we can follow these steps:

Step 1: Plot the point $R(x, y)$ to determine the quadrant in which it lies.

Step 2: Find $r > 0$ using $r = \sqrt{x^2 + y^2}$, provided that both x and y are nonzero.

Step 3: Find the reference angle $\theta' = \tan^{-1}\left|\dfrac{y}{x}\right|$, $x \ne 0$.

Step 4: Find θ.

- If the point is in QI, then $\theta = \theta'$.
- If the point is in QII, then $\theta = \pi - \theta'$.
- If the point is in QIII, then $\theta = \pi + \theta'$.
- If the point is in QIV, then $\theta = 2\pi - \theta'$.

For a point that lies on an axis, we can use the following guidelines:

- If the point is on the positive x-axis, $R(x, 0)$, then $r = x$ and $\theta = 0$.
- If the point is on the negative x-axis, $R(x, 0)$, then $r = |x|$ and $\theta = \pi$.
- If the point is on the positive y-axis, $R(0, y)$, then $r = y$ and $\theta = \dfrac{\pi}{2}$.
- If the point is on the negative y-axis, $R(0, y)$, then $r = |y|$ and $\theta = \dfrac{3\pi}{2}$.

---

4. Convert Equations between Coordinate Systems

To convert an equation written in rectangular form to an equation in polar form, you can use the definitions $x = r \cos \theta$, $y = r \sin \theta$, and $x^2 + y^2 = r^2$. Final answers can be in various forms.

EXAMPLE 7   **Converting a Rectangular Equation to a Polar Equation**

Convert each equation written in rectangular form to an equation in polar form.
a. $5x^2 + 5y^2 = 3$ **b.** $3x - 4y = 7$ **c.** $6xy = 5$

SOLUTION   **a.** $5x^2 + 5y^2 = 3$

$$5(x^2 + y^2) = 3 \qquad \text{Factor out 5.}$$

$$5r^2 = 3 \qquad x^2 + y^2 = r^2$$

$$r^2 = \frac{3}{5}$$

b.
$$3x - 4y = 7$$

$$3r \cos \theta - 4r \sin \theta = 7 \qquad \text{Use } x = r \cos \theta \text{ and } y = r \sin \theta.$$

$$r(3 \cos \theta - 4 \sin \theta) = 7 \qquad \text{Factor out } r.$$

$$r = \frac{7}{3 \cos \theta - 4 \sin \theta}$$

c. $6xy = 5$

$6r \cos \theta \cdot r \sin \theta = 5$ Use $x = r \cos \theta$ and $y = r \sin \theta$.

$6r^2 \cos \theta \sin \theta = 5$

This equation can be simplified more if we desire, by using the double-angle formula for sine.

$6r^2 \cos \theta \sin \theta = 5$

$3r^2(2 \sin \theta \cos \theta) = 5$ Factor out $3r^2$.

$3r^2 \sin 2\theta = 5$ Use the identity $\sin 2\theta = 2 \sin \theta \cos \theta$. ∎

Self Check 7   Convert each equation written in rectangular form to an equation in polar form.
a. $4x^2y = 3$ **b.** $x = -6$ **c.** $y^2 = 2x$

Now Try Exercise 77.

Converting from polar equations to rectangular equations is not as straightforward as conversion from rectangular to polar. Final answers can be in different forms. There are general strategies that are often employed in these conversions.

1. Multiply both sides of the equation by r or a power of r.
2. Square both sides of the equation.
3. If of the form $\theta = n$, where n is any real number, take the tangent of both sides.
4. Use the definitions of the trigonometric functions.

EXAMPLE 8   **Converting a Polar Equation to a Rectangular Equation**

Convert each equation written in polar form to an equation in rectangular form.

a. $r = 8 \sin \theta$ **b.** $\theta = \dfrac{\pi}{3}$ **c.** $r = 3 \sec \theta$ **d.** $r = 7$

SOLUTION   **a.** To convert $r = 8 \sin \theta$ to rectangular form, we will multiply both sides by r.

$r = 8 \sin \theta$

$r^2 = 8r \sin \theta$ Multiply both sides by r.

$x^2 + y^2 = 8y$ Substitute $r^2 = x^2 + y^2$ and $r \sin \theta = y$.

$x^2 + y^2 - 8y = 0$ Rewrite.

This is an equation of a circle, and if we complete the square, we can determine the center and radius.

$x^2 + y^2 - 8y + \mathbf{16} = \mathbf{16}$ Complete the square in y.

$x^2 + (y - 4)^2 = 16$ Factor.

This is an equation of a circle with center at $(0, 4)$ and radius $r = 4$.

b. To convert $\theta = \frac{\pi}{3}$ to rectangular form, we will take the tangent of both sides.

$\theta = \dfrac{\pi}{3}$

$\tan \theta = \tan \dfrac{\pi}{3}$ Take the tangent of both sides.

$\dfrac{y}{x} = \sqrt{3}$ Substitute using the definition of tangent and value of $\tan \frac{\pi}{3}$.

$y = \sqrt{3}x$ Solve for y.

This is an equation of a line through the origin with slope $\sqrt{3}$.

c. We will use the reciprocal identity $\sec\theta = \dfrac{1}{\cos\theta}$ to convert $r = 3\sec\theta$ to rectangular form.

$$r = 3\sec\theta$$

$$r = \frac{3}{\cos\theta} \qquad \text{Use the reciprocal identity } \sec\theta = \frac{1}{\cos\theta}.$$

$$r\cos\theta = 3 \qquad \text{Multiply both sides by } \cos\theta.$$

$$x = 3 \qquad \text{Substitute } r\cos\theta = x.$$

This is an equation of the vertical line $x = 3$.

d. We will square both sides to convert $r = 7$ to rectangular form.

$$r = 7$$

$$r^2 = 49 \qquad \text{Square both sides.}$$

$$x^2 + y^2 = 49 \qquad \text{Substitute } r^2 = x^2 + y^2.$$

This is an equation of a circle with center at $(0, 0)$ and radius $r = 7$.

Self Check 8 Convert each equation written in polar form to an equation in rectangular form.

a. $r = -3\cos\theta$ **b.** $\theta = -\dfrac{\pi}{3}$ **c.** $r = -4\csc\theta$ **d.** $r = 6$

Now Try Exercise 89.

5. Use the Distance Formula with Polar Coordinates

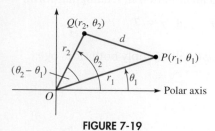

FIGURE 7-19

The Law of Cosines can be used to derive a distance formula for points given in polar coordinates. If $P(r_1, \theta_1)$ and $Q(r_2, \theta_2)$ are two points in Figure 7-19, given in polar coordinates, then the distance d between them is found by applying the Law of Cosines to triangle OPQ.

$$d^2 = r_1{}^2 + r_2{}^2 - 2r_1r_2\cos(\theta_2 - \theta_1)$$

$$d = \sqrt{r_1{}^2 + r_2{}^2 - 2r_1r_2\cos(\theta_2 - \theta_1)}$$

EXAMPLE 9 **Finding Distance with Polar Coordinates**

Find the distance between the polar points $P\left(3, \dfrac{\pi}{6}\right)$ and $Q\left(2, \dfrac{\pi}{2}\right)$.

SOLUTION Let $r_1 = 3$, $\theta_1 = \dfrac{\pi}{6}$, $r_2 = 2$, and $\theta_2 = \dfrac{\pi}{2}$.

$$d = \sqrt{r_1{}^2 + r_2{}^2 - 2r_1r_2\cos(\theta_2 - \theta_1)}$$

$$= \sqrt{3^2 + 2^2 - 2(3)(2)\cos\left(\frac{\pi}{2} - \frac{\pi}{6}\right)}$$

$$= \sqrt{9 + 4 - 12\cos\frac{\pi}{3}}$$

$$= \sqrt{13 - 12\left(\frac{1}{2}\right)}$$

$$= \sqrt{13 - 6}$$

$$= \sqrt{7}$$

The distance PQ is $\sqrt{7}$ units.

Self Check 9   Find the distance between the polar points $P\left(4, \dfrac{5\pi}{6}\right)$ and $Q\left(-5, \dfrac{11\pi}{6}\right)$.

Now Try Exercise 97.

Self Check Answers   **1.**

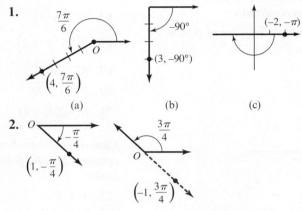

(a)   (b)   (c)

2.

$\left(1, -\dfrac{\pi}{4}\right)$ $\left(-1, \dfrac{3\pi}{4}\right)$

3. a. $\left(2, \dfrac{7\pi}{6}\right)$ **b.** $\left(2, -\dfrac{5\pi}{6}\right)$ **c.** $\left(-2, \dfrac{13\pi}{6}\right)$

4. a. $\left(-\sqrt{2}, -\sqrt{2}\right)$ **b.** $\left(\dfrac{3}{2}, -\dfrac{3\sqrt{3}}{2}\right)$ **5.** Answers will vary, but may include:

a. $(2, \pi), (-2, 0)$ **b.** $\left(6, \dfrac{3\pi}{2}\right), \left(-6, \dfrac{\pi}{2}\right)$ **6. a.** $\left(4, \dfrac{7\pi}{6}\right)$

b. $(5.385, 1.951)$ **7. a.** $2r^3 \cos\theta \sin 2\theta = 3$ **b.** $r\cos\theta = -6$

c. $r\sin^2\theta = 2\cos\theta$ **8. a.** $x^2 + y^2 + 3x = 0$ **b.** $y = -\sqrt{3}x$

c. $y = -4$ **d.** $x^2 + y^2 = 36$ **9.** 1

Exercises **7.1**

Getting Ready
You should be able to complete these vocabulary and concept statements before you proceed to the practice exercises.

Fill in the blanks.

1. The polar coordinate system is based on a ray, called the _____, and its endpoint, called the _____.

2. A point in the polar coordinate system is labeled _____.

3. True or False: The value of r is always positive.

4. True or False: A point in the polar plane has only one set of coordinates in rectangular form.

5. For a point with polar coordinates (r, θ) and any integer n, the polar coordinates $(r, \theta + \underline{\hspace{1cm}})$ and $(-r, \theta + \underline{\hspace{1cm}} + \underline{\hspace{1cm}})$ represent the same point in the polar coordinate system.

6. To convert from polar to rectangular coordinates, use $x = \underline{\hspace{1cm}}$ and $y = \underline{\hspace{1cm}}$.

Practice
Plot each point with the given polar coordinates.

7. $P\left(3, \dfrac{\pi}{6}\right)$

8. $P\left(-3, \dfrac{\pi}{4}\right)$

9. $P\left(-5, -\dfrac{3\pi}{4}\right)$

10. $P\left(5, -\dfrac{\pi}{3}\right)$

11. $P(-4, \pi)$

12. $P\left(3, -\dfrac{3\pi}{2}\right)$

13. $P(-4, 240°)$

14. $P(-5, -210°)$

15. $P(5, -135°)$

16. $P(-6, 120°)$

Determine whether each set of coordinates represent the same point in a polar coordinate system.

17. $P\left(3, \dfrac{4\pi}{3}\right)$ and $Q\left(-3, \dfrac{\pi}{3}\right)$

18. $P(4, \pi)$ and $Q(-4, 0)$

19. $P\left(-2, \dfrac{11\pi}{6}\right)$ and $Q\left(2, -\dfrac{5\pi}{6}\right)$

20. $P\left(-3, \dfrac{3\pi}{4}\right)$ and $Q\left(-3, -\dfrac{\pi}{4}\right)$

21. $P(-2, 225°)$ and $Q(-2, -135°)$

22. $P(-4, 180°)$ and $Q(-4, -180°)$
23. $P(3, 330°)$ and $Q(-3, 210°)$
24. $P(5, 60°)$ and $Q(5, -240°)$

Refer to the figure and give a possible set of polar coordinates for the point with the given conditions.

(a) $r > 0, 0 \leq \theta \leq 2\pi$ *(b)* $r > 0, -2\pi \leq \theta \leq 0$
(c) $r < 0, 0 \leq \theta \leq 2\pi$

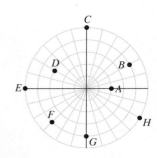

25. A	26. B	27. C	28. D
29. E	30. F	31. G	32. H

The polar coordinates of a point are given. Find the rectangular coordinates of the point. Round to two decimal places when necessary.

33. $P\left(2, \dfrac{\pi}{6}\right)$ 34. $P\left(5, \dfrac{3\pi}{4}\right)$

35. $P(7, 300°)$ 36. $P(20, 225°)$

37. $P\left(-3, -\dfrac{\pi}{3}\right)$ 38. $P\left(-7, -\dfrac{2\pi}{3}\right)$

39. $P\left(2, \dfrac{\pi}{2}\right)$ 40. $P\left(4, \dfrac{3\pi}{2}\right)$

41. $P\left(-2, \dfrac{13\pi}{6}\right)$ 42. $P\left(4, \dfrac{7\pi}{3}\right)$

43. $P\left(5, -\dfrac{9\pi}{4}\right)$ 44. $P\left(6, -\dfrac{5\pi}{3}\right)$

45. $P(-10, -90°)$ 46. $P(-15, -45°)$

47. $P(6, -315°)$ 48. $P(-4, -135°)$

49. $P(4, 5\pi)$ 50. $P(-3, -6\pi)$

51. $P(3.7, 100°)$ 52. $P(5.3, -200°)$

53. $P(-8.5, 4.2)$ 54. $P(-6.3, 5.1)$

Find two sets of polar coordinates for the point with the given rectangular coordinates.

55. $R(7, 0)$ 56. $R(0, 5)$
57. $R(0, -5)$ 58. $R(-4, 0)$
59. $R(0, 0)$ 60. $R(-1, 0)$

The rectangular coordinates of a point are given. Find two sets of polar coordinates for the point, one with $r > 0$ and $0 \leq \theta < 2\pi$, and the other with $r < 0$ and $0 \leq \theta < 2\pi$. Round to two decimal places when necessary.

61. $R(1, 1)$ 62. $R(-3, 3)$

63. $R\left(1, \sqrt{3}\right)$ 64. $R\left(-2, -2\sqrt{3}\right)$

65. $R\left(2\sqrt{3}, -2\right)$ 66. $R\left(-\sqrt{3}, -1\right)$

67. $R\left(1, \sqrt{3}\right)$ 68. $R\left(-\sqrt{3}, 1\right)$

69. $R(3.1, 2.6)$ 70. $R(4.1, -5.3)$

71. $R(-2.6, 5.4)$ 72. $R(-6.1, -3.7)$

Convert each equation written in rectangular form to an equation in polar form. Final answers can be in various forms.

73. $x = 3$ 74. $y = -7$
75. $3x + 2y = 3$ 76. $2x - y = 7$
77. $x^2 + y^2 = 9x$ 78. $x^2 + y^2 = 9$
79. $xy = 12$ 80. $5xy = 14$
81. $y^2 = 2x - x^2$ 82. $x^2 = 2y + 1$
83. $(x^2 + y^2)^3 = 4x^2y^2$ 84. $(x^2 + y^2)^2 = x^2 - y^2$

Convert each equation written in polar form to an equation in rectangular form. Final answers can be in various forms.

85. $r = 3$ 86. $r = -5$
87. $\theta = \pi$ 88. $\theta = 135°$
89. $r^2 = \sin 2\theta$ 90. $r^2 = \cos 2\theta$
91. $3r \cos \theta + 2r \sin \theta = 2$ 92. $4r \sin \theta - 8r \cos \theta = 3$

93. $r = \dfrac{1}{1 + \sin \theta}$ 94. $r = \dfrac{1}{1 - \cos \theta}$

95. $r = 3 \csc \theta + 2 \sec \theta$ 96. $r(2 - \cos \theta) = 2$

Find the distance between the points whose polar coordinates are given.

97. $P\left(5, \dfrac{5\pi}{6}\right), Q\left(3, -\dfrac{\pi}{6}\right)$

98. $P(3, \pi), Q\left(4, -\dfrac{\pi}{2}\right)$

99. $P(-1, \pi), Q\left(2, \dfrac{3\pi}{2}\right)$

100. $P\left(\sqrt{2}, \dfrac{\pi}{2}\right), Q\left(\sqrt{2}, \dfrac{\pi}{6}\right)$

101. $P\left(\sqrt{2}, 12°\right), Q\left(-\sqrt{2}, -78°\right)$

102. $P(2, 40°), Q\left(\sqrt{3}, 70°\right)$

Applications

103. Spiral of Archimedes Draw concentric circles with $r = 1, r = 2, \ldots r = 7$. Plot the polar coordinates listed in the table and draw a smooth curve through them. A spiral of Archimedes is formed.

θ	r
0	0
$\frac{\pi}{6}$	≈ 0.5
$\frac{\pi}{3}$	≈ 1.0
$\frac{\pi}{2}$	≈ 1.6
$\frac{2\pi}{3}$	≈ 2.1
$\frac{5\pi}{6}$	≈ 2.6
π	≈ 3.1

θ	r
$\frac{7\pi}{6}$	≈ 3.7
$\frac{4\pi}{3}$	≈ 4.2
$\frac{3\pi}{2}$	≈ 4.7
$\frac{5\pi}{3}$	≈ 5.2
$\frac{11\pi}{6}$	≈ 5.8
2π	≈ 6.3

104. Spiral of Archimedes Draw concentric circles with $r = 1, r = 2, r = 3$, and $r = 4$. Plot the polar coordinates listed in the table and draw a smooth curve through them. A spiral of Archimedes is formed.

θ	r
0	0
$\frac{\pi}{6}$	≈ 0.3
$\frac{\pi}{3}$	≈ 0.5
$\frac{\pi}{2}$	≈ 0.8
$\frac{2\pi}{3}$	≈ 1.0
$\frac{5\pi}{6}$	≈ 1.3
π	≈ 1.6

θ	r
$\frac{7\pi}{6}$	≈ 1.8
$\frac{4\pi}{3}$	≈ 2.1
$\frac{3\pi}{2}$	≈ 2.4
$\frac{5\pi}{3}$	≈ 2.6
$\frac{11\pi}{6}$	≈ 2.9
2π	≈ 3.1

Discovery and Writing

105. Describe the strategy you use to convert polar coordinates into rectangular coordinates.

106. Describe the strategy you use to convert rectangular coordinates into polar coordinates.

7.2   Polar Equations and Graphs

In this section, we will learn to

1. Graph polar equations using rectangular equations.

2. Test polar equations for symmetry.

3. Graph polar equations.

4. Use rapid sketching.

pasphotography/Shutterstock.com

Carrie Underwood is one of the most popular female singers today. When singing live, it is important that she uses a microphone that picks up the sound of her voice from the front of the microphone while minimizing the noise from the rear and sides. Cardioid microphones are often used by singers for this purpose. These microphones have sensitivity patterns that are similar in shape to a heart when graphed. This heart-shaped design enables the microphone to pick up the sound of the voice extremely well during a live performance. In this section, we will learn about polar equations and their graphs. The cardioid, a heart-shaped curve, is one of the graphs we will study in this section.

In the rectangular coordinate system, the graphs of some functions, written in the form $y = f(x)$, are easily recognized or graphed. For example, we know that the graph of $f(x) = x^2$ is a parabola with vertex at the origin. Other functions, such as some exponential, logarithmic, and trigonometric functions, can be graphed either very quickly or by plotting points. Sometimes symmetry is used as an aid to graph these functions.

Because the graphs of many interesting and useful curves—circles, for example—are not the graphs of functions, they cannot be easily graphed using the form $y = f(x)$. We will learn in this section that many of these curves can be graphed using a polar equation in the form $r = f(\theta)$. Instead of using a rectangular coordinate system to graph these polar equations, we will use a polar grid, which consists of concentric circles centered at the pole, with rays placed at intervals.

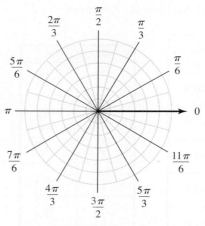

FIGURE 7-20

1. Graph Polar Equations Using Rectangular Equations

In Section 7.1, we converted polar equations to rectangular equations. We will revisit that now with the goal of using these conversions to help us graph polar equations.

EXAMPLE 1 **Graphing a Polar Equation by Converting to a Rectangular Equation**

Convert the equation $r = 4$ to a rectangular equation and then graph the equation.

SOLUTION We will square both sides of the equation and use the equivalent rectangular form.

$$r = 4$$

$$r^2 = 16 \qquad \text{Square both sides.}$$

$$x^2 + y^2 = 16 \qquad \text{Substitute } r^2 = x^2 + y^2.$$

Now we use the fact that the polar equation $r = 4$ is equivalent to the rectangular equation $x^2 + y^2 = 16$, which is an equation of the circle with its center at the origin $(0, 0)$ and a radius of 4, to graph the equation.

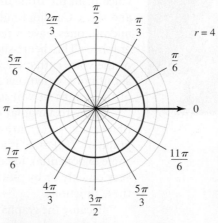

FIGURE 7-21

Self Check 1   Convert the equation $r = -3$ to a rectangular equation and then graph the equation.

Now Try Exercise 7.

EXAMPLE 2   **Graphing a Polar Equation by Converting to a Rectangular Equation**

Convert the equation $r = -4 \cos \theta$ to a rectangular equation and then graph the equation.

SOLUTION   To convert this polar equation to a rectangular equation, we will multiply both sides by r.

$$r = -4 \cos \theta$$

$$r^2 = -4r \cos \theta \qquad \text{Multiply both sides by } r.$$

$$x^2 + y^2 = -4x \qquad \text{Substitute } r^2 = x^2 + y^2 \text{ and } r \cos \theta = x.$$

Now that we have converted the equation to a rectangular form, we recognize this as an equation of a circle. Our next step will be to write the equation in standard form.

$$x^2 + y^2 = -4x$$

$$x^2 + 4x + y^2 = 0 \qquad \text{Add } 4x \text{ to both sides of the equation.}$$

$$(x^2 + 4x + 4) + y^2 = 4 \qquad \text{Complete the square on } x \text{ by adding 4 to both sides.}$$

$$(x + 2)^2 + y^2 = 4 \qquad \text{Standard form for an equation of a circle}$$

Thus, the polar equation $r = -4 \cos \theta$ is equivalent to the rectangular equation $(x + 2)^2 + y^2 = 4$, which is an equation of a circle with center at $(-2, 0)$ and radius 2, and is shown in the graph in Figure 7-22.

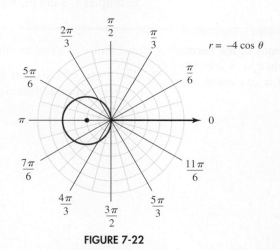

FIGURE 7-22

Self Check 2   Convert the equation $r = 2 \sin \theta$ to a rectangular equation and then graph the equation.

Now Try Exercise 19.

EXAMPLE 3   **Graphing a Polar Equation by Converting to a Rectangular Equation**

Convert the equation $r \sin \theta = 4$ to a rectangular equation and then graph the equation.

SOLUTION To convert this polar equation to a rectangular equation, we use the polar definition $y = r \sin \theta$.

$$r \sin \theta = 4$$

$$y = 4$$

This is an equation of the horizontal line $y = 4$. Therefore, the graph of the polar equation $r \sin \theta = 4$ is a horizontal line.

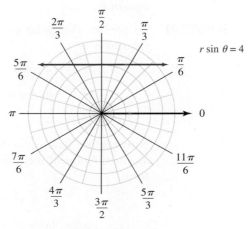

FIGURE 7-23

Self Check 3 Convert the equation $r \cos \theta = -2$ to a rectangular equation and then graph the equation.

Now Try Exercise 23.

Examples 1–3 are examples of the following theorem:

Polar Equations of the Forms
$r = a;\ r = a \sin \theta;\ r = a \cos \theta;$
$r \sin \theta = a;\ r \cos \theta = a$

Assume that a is a real number, with $a \neq 0$.

1. The graph of the polar equation $r = a$ in rectangular coordinates is a circle with its center at the origin $(0, 0)$ and a radius of $|a|$.

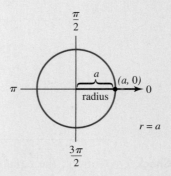

2. The graph of the polar equation $r = a \sin \theta$ in rectangular coordinates is a circle with its center at $\left(0, \dfrac{a}{2}\right)$ and a radius of $\left|\dfrac{a}{2}\right|$. The graph of the polar equation $r = a \cos \theta$ in rectangular coordinates is a circle with its center at $\left(\dfrac{a}{2}, 0\right)$ and a radius of $\left|\dfrac{a}{2}\right|$.

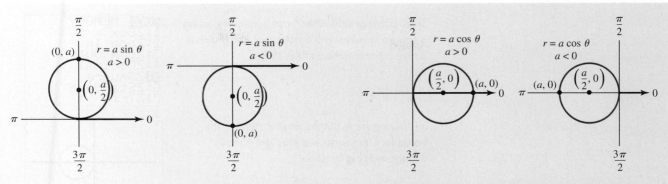

3. The graph of the polar equation $r \sin \theta = a$ in rectangular coordinates is the horizontal line $y = a$. The graph of the polar equation $r \cos \theta = a$ in rectangular coordinates is the vertical line $x = a$.

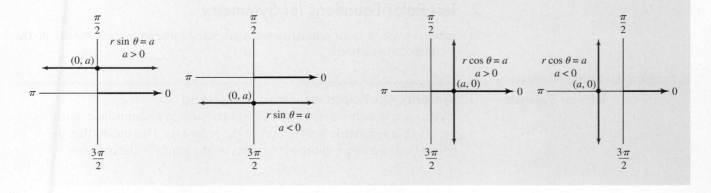

ACCENT ON TECHNOLOGY   **Graphing Polar Equations**

A graphing calculator can be used to graph some polar equations. To graph a polar equation with a graphing calculator, we have to be able to write the equation in terms of r. Graphing a polar equation requires an additional step in setting up the window.

To graph in polar coordinates, we need to set the proper modes. While the graphs can be done in degree mode, we will use the radian mode. We will also need to set the calculator to polar mode, POL.

When we go to the graph menu, we will notice that instead of the functions being entered in terms of y, they are written as functions of r. The variable key will enter θ instead of x.

Setting the window in polar mode involves more than just setting the values for the axes. We must also set the boundaries and increments for the variable. In radian mode, we can set as shown and get a good graph.

Since polar graphs often have symmetry, using a square window will ensure that our graph is a true representation of the shape.

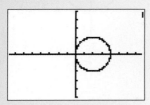

It is important to set the proper step for the variable. Otherwise, we may get a really strange-looking function.

2. Test Polar Equations for Symmetry

In algebra, we use tests for symmetry when graphing some equations. We can do the same with polar equations.

Tests for Symmetry

1. **Symmetry with Respect to the Polar Axis (x-axis)**
   If replacing θ with $-\theta$ in a polar equation results in an equivalent equation, the graph is symmetric with respect to the polar axis. This means that the points (r, θ) and $(r, -\theta)$ are both points on the graph of the equation.

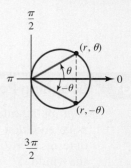

2. **Symmetry with Respect to the Line $\theta = \dfrac{\pi}{2}$ (y-axis)**
   If replacing θ with $\pi - \theta$ in a polar equation results in an equivalent equation, the graph is symmetric with respect to the vertical line $\theta = \dfrac{\pi}{2}$. This means that the points (r, θ) and $(r, \pi - \theta)$ are both points on the graph of the equation.
   An alternate way to test for this symmetry is to determine whether the points (r, θ) and $(-r, -\theta)$ are both points on the graph of the equation.

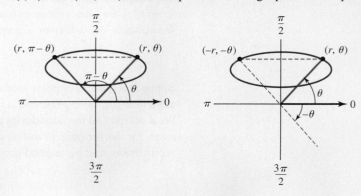

3. **Symmetry with Respect to the Pole (Origin)**
   If replacing r with $-r$ in a polar equation results in an equivalent equation, the graph is symmetric with respect to the pole. This means that the points (r, θ) and $(-r, \theta)$ are both points on the graph of the equation.

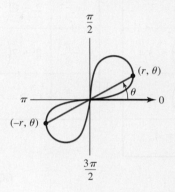

Caution

> Tests for symmetry are useful; if an equation passes one or more of the tests, then we can use that information to draw the graph. However, there are equations that may not pass any of the tests but their graphs may still show symmetry of some type.

3. Graph Polar Equations

When graphing a polar equation, we can use symmetry, if there is any, and then we evaluate the equation at special angles.

EXAMPLE 4   **Graphing a Polar Equation—Cardioid**

Graph the polar equation $r = 2 + 2 \sin \theta$.

SOLUTION   Our first step will be to test for symmetry.

Polar axis	$\theta = \dfrac{\pi}{2}$ axis	Pole
Replace θ with $-\theta$.	Replace θ with $\pi - \theta$.	Replace r with $-r$.
$r = 2 + 2 \sin(-\theta)$	$r = 2 + 2 \sin(\pi - \theta)$	$-r = 2 + 2 \sin \theta$
$\quad = 2 - 2 \sin \theta$	$\quad = 2 + 2(\sin \pi \cos \theta - \cos \pi \sin \theta)$	$\quad r = -2 - 2 \sin \theta$
	$\quad = 2 + 2[0 \cdot \cos \theta - (-1)\sin \theta]$	
	$\quad = 2 + 2 \sin \theta$	
Equations are not equivalent.	Equations are equivalent.	Equations are not equivalent.
There may or may not be symmetry with respect to the polar axis.	There is symmetry with respect to the $\theta = \dfrac{\pi}{2}$ axis.	There may or may not be symmetry with respect to the pole.

Since there is symmetry with respect to the $\theta = \frac{\pi}{2}$ axis, we evaluate the equation at values of special angles over the interval $\left[-\frac{\pi}{2}, \frac{\pi}{2}\right]$ and then complete the graph of the equation.

θ	$r = 2 + 2\sin\theta$
0	$2 + 2(0) = 2$
$\dfrac{\pi}{6}$	$2 + 2\left(\dfrac{1}{2}\right) = 3$
$\dfrac{\pi}{4}$	$2 + 2\left(\dfrac{\sqrt{2}}{2}\right) \approx 3.4$
$\dfrac{\pi}{3}$	$2 + 2\left(\dfrac{\sqrt{3}}{2}\right) \approx 3.7$
$\dfrac{\pi}{2}$	$2 + 2(1) = 4$

θ	$r = 2 + 2\sin\theta$
$-\dfrac{\pi}{6}$	$2 + 2\left(-\dfrac{1}{2}\right) = 1$
$-\dfrac{\pi}{4}$	$2 + 2\left(\dfrac{-\sqrt{2}}{2}\right) \approx 0.6$
$-\dfrac{\pi}{3}$	$2 + 2\left(\dfrac{-\sqrt{3}}{2}\right) \approx 0.3$
$-\dfrac{\pi}{2}$	$2 + 2(-1) = 0$

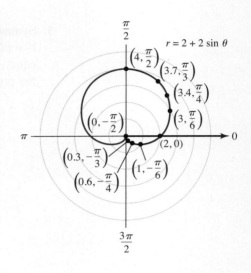

FIGURE 7-24

The graph, shown in Figure 7-24, is an example of a **cardioid**, or heart-shaped curve. ■

Self Check 4   Graph the polar equation $r = 2 - 2\cos\theta$.

Now Try Exercise 27.

EXAMPLE 5   **Graphing a Polar Equation—Limaçon without an Inner Loop**

Graph the polar equation $r = 3 - 2\cos\theta$.

SOLUTION   Our first step will be to test for symmetry.

Polar axis	$\theta = \dfrac{\pi}{2}$ axis	Pole
Replace θ with $-\theta$.	Replace θ with $\pi - \theta$.	Replace r with $-r$.
$r = 3 - 2\cos(-\theta)$ $\quad = 3 - 2\cos\theta$	$r = 3 - 2\cos(\pi - \theta)$ $\quad = 3 - 2(\cos\pi\cos\theta + \sin\pi\sin\theta)$ $\quad = 3 - 2[(-1)\cdot\cos\theta + (0)\sin\theta]$ $\quad = 3 + 2\cos\theta$	$-r = 3 - 2\cos\theta$ $\quad r = -3 + 2\cos\theta$
Equations are equivalent.	Equations are not equivalent.	Equations are not equivalent.
There is symmetry with respect to to the polar axis.	There may or may not be symmetry with respect to the $\theta = \dfrac{\pi}{2}$ axis.	There may or may not be symmetry with respect to the pole.

Since there is symmetry with respect to the polar axis, we evaluate the equation at values of special angles over the interval $[0, \pi]$. After we find the points, we use symmetry to graph the equation.

θ	$r = 3 - 2\cos\theta$
0	$3 - 2(1) = 1$
$\dfrac{\pi}{6}$	$3 - 2\left(\dfrac{\sqrt{3}}{2}\right) \approx 1.3$
$\dfrac{\pi}{4}$	$3 - 2\left(\dfrac{\sqrt{2}}{2}\right) \approx 1.6$
$\dfrac{\pi}{3}$	$3 - 2\left(\dfrac{1}{2}\right) = 2$
$\dfrac{\pi}{2}$	$3 - 2(0) = 3$

θ	$r = 3 - 2\cos\theta$
$\dfrac{2\pi}{3}$	$3 - 2\left(-\dfrac{1}{2}\right) = 4$
$\dfrac{3\pi}{4}$	$3 - 2\left(-\dfrac{\sqrt{2}}{2}\right) \approx 4.4$
$\dfrac{5\pi}{6}$	$3 - 2\left(-\dfrac{\sqrt{3}}{2}\right) \approx 4.7$
π	$3 - 2(-1) = 5$

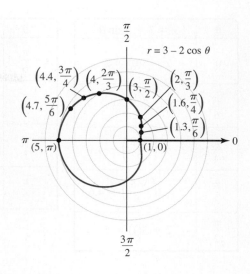

FIGURE 7-25

The curve, shown in Figure 7-25 is an example of a **limaçon without an inner loop**. (The word *limaçon* is the French word for *snail*.)

Self Check 5   Graph the polar equation $r = 4 + 3\sin\theta$.

Now Try Exercise 29.

EXAMPLE 6   **Graphing a Polar Equation—Limaçon with an Inner Loop**

Graph the polar equation $r = 2 + 3\sin\theta$.

SOLUTION   Our first step will be to test for symmetry.

Polar axis	$\theta = \dfrac{\pi}{2}$ axis	Pole
Replace θ with $-\theta$.	Replace θ with $\pi - \theta$.	Replace r with $-r$.
$r = 2 + 3\sin(-\theta)$ $\quad= 2 - 3\sin\theta$	$r = 2 + 3\sin(\pi - \theta)$ $\quad= 2 + 3(\sin\pi\cos\theta - \cos\pi\sin\theta)$ $\quad= 2 + 3[(0)\cdot\cos\theta - (-1)\sin\theta]$ $\quad= 2 + 3\sin\theta$	$-r = 2 + 3\sin\theta$ $\quad r = -2 - 3\sin\theta$
Equations are not equivalent.	Equations are equivalent.	Equations are not equivalent.
There may or may not be symmetry with respect to the polar axis.	There is symmetry with respect to the $\theta = \dfrac{\pi}{2}$ axis.	There may or may not be symmetry with respect to the pole.

Since the symmetry is with respect to the $\theta = \frac{\pi}{2}$ axis, we evaluate the equation at values of special angles over the interval $\left[-\frac{\pi}{2}, \frac{\pi}{2}\right]$ and use symmetry to complete the graph.

θ	$r = 2 + 3 \sin \theta$
0	$2 + 3(0) = 2$
$\dfrac{\pi}{6}$	$2 + 3\left(\dfrac{1}{2}\right) = \dfrac{7}{2}$
$\dfrac{\pi}{4}$	$2 + 3\left(\dfrac{\sqrt{2}}{2}\right) \approx 4.1$
$\dfrac{\pi}{3}$	$2 + 3\left(\dfrac{\sqrt{3}}{2}\right) \approx 4.6$
$\dfrac{\pi}{2}$	$2 + 3(1) = 5$

θ	$r = 2 + 3 \sin \theta$
$-\dfrac{\pi}{6}$	$2 + 3\left(-\dfrac{1}{2}\right) = \dfrac{1}{2}$
$-\dfrac{\pi}{4}$	$2 + 3\left(-\dfrac{\sqrt{2}}{2}\right) \approx -0.1$
$-\dfrac{\pi}{3}$	$2 + 3\left(-\dfrac{\sqrt{3}}{2}\right) \approx -0.6$
$-\dfrac{\pi}{2}$	$2 + 3(-1) = -1$

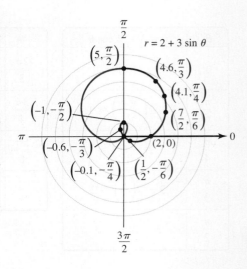

FIGURE 7-26

The graph, in Figure 7-26, is called a **limaçon with an inner loop.**

Self Check 6   Graph the polar equation $r = 1 - 2 \cos \theta$.

Now Try Exercise 33.

EXAMPLE 7   Graphing Polar Equations—Rose

Graph the equation $r = 3 \cos 2\theta$.

SOLUTION   First, we check for symmetry.

Polar axis	$\theta = \dfrac{\pi}{2}$ axis	Pole
Replace θ with $-\theta$.	Replace θ with $\pi - \theta$.	Replace r with $-r$.
$r = 3 \cos(-2\theta)$ $\quad = 3 \cos 2\theta$	$r = 3 \cos[2(\pi - \theta)]$ $\quad = 3 \cos(2\pi - 2\theta)$ $\quad = 3(\cos 2\pi \cos 2\theta + \sin 2\pi \sin 2\theta)$ $\quad = 3(1 \cdot \cos 2\theta + 0 \cdot \sin 2\theta)$ $\quad = 3 \cos 2\theta$	$-r = 3 \cos 2\theta$ $\quad r = -3 \cos 2\theta$
Equations are equivalent.	Equations are equivalent.	Equations are not equivalent.
There is symmetry with respect to to the polar axis.	There is symmetry with respect to the $\theta = \dfrac{\pi}{2}$ axis.	There may or may not be symmetry with respect to the pole.

Even though the test for symmetry about the pole is inconclusive, since there is symmetry about the polar axis and the line $\theta = \frac{\pi}{2}$, there has to be symmetry about the pole.

After we calculate the values in the table and plot the points, we will complete the graph using symmetry with respect to the line $\theta = \frac{\pi}{2}$ [Figure 7-27(a)], and then with respect to the polar axis [Figure 7-27(b)]. This equation produces a graph that is called a four-petal rose, in Figure 7-27(c).

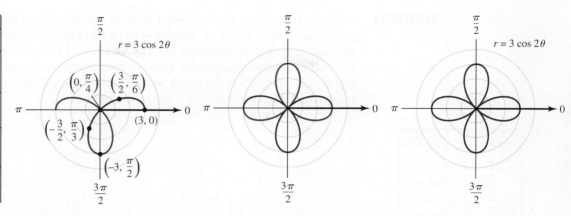

θ	$r = 3 \cos 2\theta$
0	$3(1) = 3$
$\dfrac{\pi}{6}$	$3\left(\dfrac{1}{2}\right) = \dfrac{3}{2}$
$\dfrac{\pi}{4}$	$3 \cdot 0 = 0$
$\dfrac{\pi}{3}$	$3\left(-\dfrac{1}{2}\right) = -\dfrac{3}{2}$
$\dfrac{\pi}{2}$	$3(-1) = -3$

(a) complete symmetry with respect to $\theta = \dfrac{\pi}{2}$

(b) complete symmetry with respect to polar axis

(c) complete graph of the 4-petal rose

FIGURE 7-27

Self Check 7   Graph the polar equation $r = 3 \sin 3\theta$.

Now Try Exercise 37.

EXAMPLE 8   **Graphing Polar Equations—Lemniscate**

Graph the equation $r^2 = 9 \cos 2\theta$.

SOLUTION   We know that this equation is symmetric with respect to the polar axis, since $9 \cos(-2\theta) = 9 \cos 2\theta$. Since $r^2 \geq 0$, $9 \cos 2\theta \geq 0$. Because the cosine function is nonnegative in QI and QIV, $\left[-\dfrac{\pi}{2}, \dfrac{\pi}{2}\right]$, we can find the values of θ that can be used in the equation.

$$-\frac{\pi}{2} \leq 2\theta \leq \frac{\pi}{2} \Rightarrow -\frac{\pi}{4} \leq \theta \leq \frac{\pi}{4}$$

Plotting these points and using symmetry with respect to the polar axis results in the graph shown in Figure 7-28.

θ	$r^2 = 9 \cos 2\theta$	r
0	$9(1)$	± 3
$\dfrac{\pi}{6}$	$9\left(\dfrac{1}{2}\right) = \dfrac{9}{2}$	± 2.12
$\dfrac{\pi}{4}$	$9 \cdot 0 = 0$	0

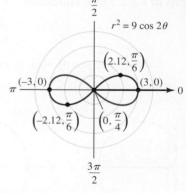

FIGURE 7-28

The curve shown in Figure 7-28 is an example of a lemniscate.

Self Check 8   Graph the equation $r^2 = 4 \sin 2\theta$.

Now Try Exercise 41.

EXAMPLE 9   **Graphing Polar Equations—Spiral**

Graph the equation $r = \theta$.

SOLUTION   We will use a calculator to make a table of values. When we plot the points, the graph produced is called an **Archimedean spiral**. Typically, nonnegative values of θ are used to graph the equation, but any value can be used. If negative values of θ are used to graph the equation, then there is symmetry about the line $\theta = \frac{\pi}{2}$. The graph in Figure 7-29 is the graph using nonnegative values. Figure 7-30 shows the graph when negative values of θ are plotted.

θ	$r = \theta$
0	0
$\frac{\pi}{6}$	≈ 0.52
$\frac{\pi}{3}$	≈ 1.05
$\frac{\pi}{2}$	≈ 1.57
$\frac{2\pi}{3}$	≈ 2.09
$\frac{5\pi}{6}$	≈ 2.62
π	≈ 3.14
$\frac{3\pi}{2}$	≈ 4.71
2π	≈ 6.28

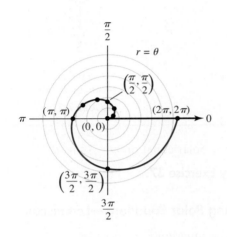

FIGURE 7-29

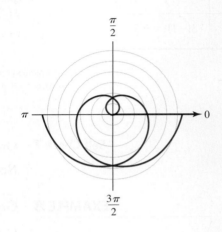

FIGURE 7-30

Self Check 9   Graph the equation $r = \dfrac{\theta}{2}$ over $[0, 2\pi]$.

Now Try Exercise 45.

Graphs of Basic Polar Equations

Cardiod

A cardioid is the curve given by an equation in one of the following forms, with real number $a > 0$:

$$r = a(1 \pm \cos\theta) = a \pm a\cos\theta \qquad \text{or} \qquad r = a(1 \pm \sin\theta) = a \pm a\sin\theta.$$

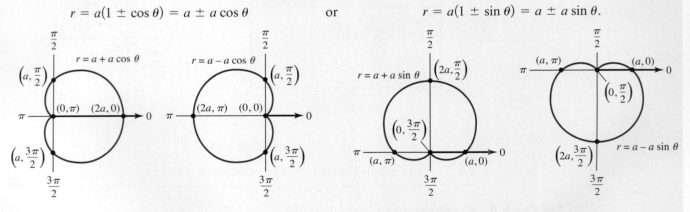

Limaçon

A limaçon is the curve given by an equation in one of the following forms, with real number $a > 0$ and $b > 0$:

$$r = a \pm b\sin\theta \qquad \text{or} \qquad r = a \pm b\cos\theta.$$

If $a > b$, the graph is a limaçon without an inner loop.

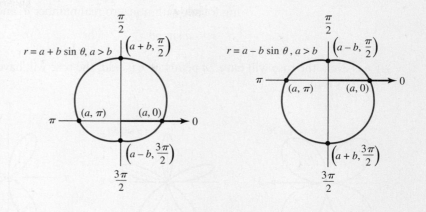

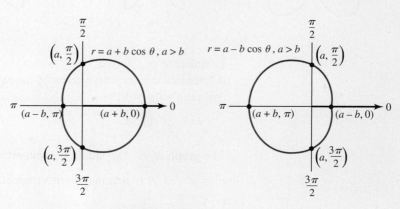

If $a < b$, the graph is a limaçon with an inner loop.

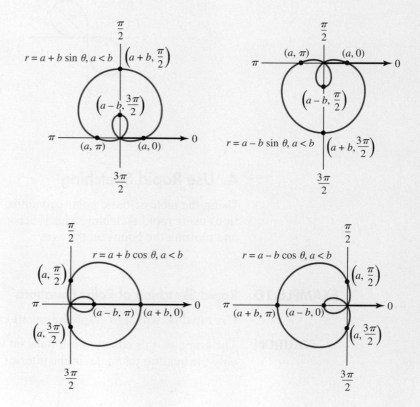

Rose

A rose curve is the shape of the graph given by an equation in one of the following forms, with nonzero real number a, and an integer $n \geq 2$.

$$r = a \sin n\theta \qquad \text{or} \qquad r = a \cos n\theta.$$

If n is even, the rose will have $2n$ petals. If n is odd, the rose will have n petals. The petal of the rose is $|a|$ units long.

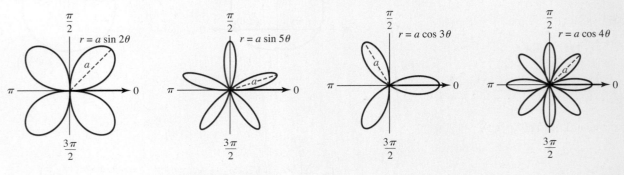

Lemniscate

A lemniscate is the curve given by an equation in one of the following forms, with real number $a \neq 0$:

$$r^2 = a^2 \sin 2\theta \qquad \text{or} \qquad r^2 = a^2 \cos 2\theta.$$

The graph of $r^2 = a^2 \sin 2\theta$ is symmetric with respect to the pole; the graph of $r^2 = a^2 \cos 2\theta$ is symmetric with respect to the polar axis, the pole, and $\theta = \dfrac{\pi}{2}$.

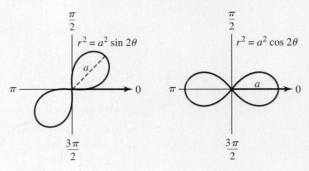

4. Use Rapid Sketching

Using the table of basic polar equations, we can draw the graphs of similar equations using rapid sketching. This is accomplished by recognizing the type of curve and plotting the points on the axes.

EXAMPLE 10 **Rapid Sketching of Polar Equations**

Use rapid sketching to graph the equation $r = 4 + 3 \cos \theta$.

SOLUTION The graph of this equation is a limaçon without a loop. We will plot points using only the quadrantal angles in the interval $[0, 2\pi]$ and then complete the curve.

θ	$r = 4 + 3\cos\theta$
0	$4 + 3(1) = 7$
$\dfrac{\pi}{2}$	$4 + 3(0) = 4$
π	$4 + 3(-1) = 1$
$\dfrac{3\pi}{2}$	$4 + 3(0) = 4$
2π	$4 + 3(1) = 7$

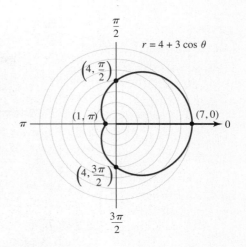

FIGURE 7-31

Self Check 10   Use rapid sketching to graph the equation $r = 2 - 3\sin\theta$.

Now Try Exercise 31.

Self Check Answers   **1.** $x^2 + y^2 = 9$

2. $x^2 + (y - 1)^2 = 1$

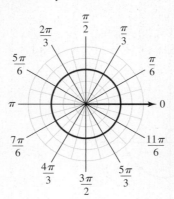

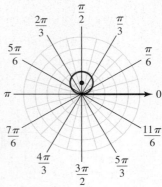

3. $x = -2$

4.

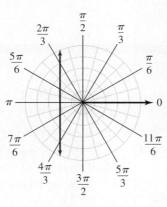

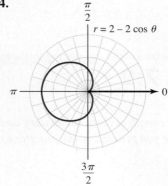

5.

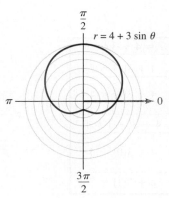

$r = 4 + 3 \sin \theta$

6.

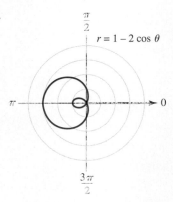

$r = 1 - 2 \cos \theta$

7.

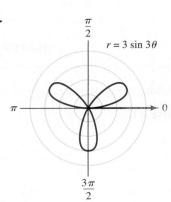

$r = 3 \sin 3\theta$

8.

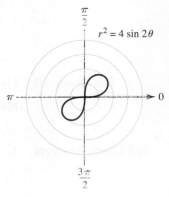

$r^2 = 4 \sin 2\theta$

9.

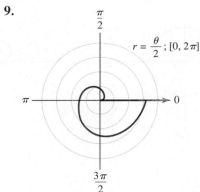

$r = \dfrac{\theta}{2}; [0, 2\pi]$

10.

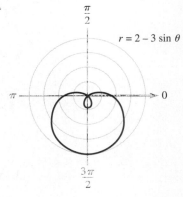

$r = 2 - 3 \sin \theta$

Exercises **7.2**

Getting Ready
You should be able to complete these vocabulary and concept statements before you proceed to the practice exercises.

Fill in the blanks.

1. True or False: If the graph of the equation $r = a + b \cos \theta$ is a limaçon with a loop, then $a > b$.

2. True or False: If the graph of the equation $r = a - b \sin \theta$ is a limaçon without a loop, then $a > b$.

3. The graph of the equation $r = a + a \cos \theta$ is symmetric with respect to the _____ axis.

4. The graph of the equation $r = a - a \sin \theta$ is symmetric with respect to the line _____.

5. If n is odd, the graph of the equation $r = a \sin n\theta$ has _____ petals.

6. The graph of the polar equation $r = a \sin \theta$ is a circle with center at _____.

Practice
Convert each polar equation to an equation using rectangular coordinates and then graph it on a rectangular coordinate system.

7. $r = 8$

8. $r = 6$

9. $\theta = \dfrac{\pi}{6}$

10. $\theta = -\dfrac{\pi}{3}$

11. $3r \cos \theta = -5$ **12.** $5r \sin \theta = 2$

13. $r = 2 \sec \theta$ **14.** $r = -3 \csc \theta$

15. $r^2 = 9$ **16.** $r^2 = 1$

17. $r \cos \theta + r \sin \theta = 1$ **18.** $r \cos \theta - 3r \sin \theta = 3$

Graph each polar equation.

19. $r = 5 \cos \theta$ **20.** $r = -6 \sin \theta$

21. $r = -3 \sin \theta$ **22.** $r = -5 \cos \theta$

23. $r \cos \theta = 5$ **24.** $r \sin \theta = -6$

25. $r = 1 - \sin \theta$ **26.** $r = 1 - \cos \theta$

27. $r = 3 + 3 \cos \theta$ **28.** $r = 4 + 4 \sin \theta$

29. $r = 2 - \cos \theta$ **30.** $r = 3 - 2 \sin \theta$

31. $r = 4 + 2 \sin \theta$ **32.** $r = 5 + 3 \cos \theta$

33. $r = 1 + 2 \sin \theta$ **34.** $r = 2 + 3 \cos \theta$

35. $r = 3 - 4 \cos \theta$ **36.** $r = 4 - 5 \sin \theta$

37. $r = 4 \cos 3\theta$ **38.** $r = 5 \sin 5\theta$

39. $r = 3 \sin 2\theta$ **40.** $r = 2 \cos 4\theta$

41. $r^2 = 16 \cos 2\theta$ **42.** $r^2 = 9 \sin 2\theta$

43. $r^2 = \sin 2\theta$ **44.** $r^2 = \cos 2\theta$

45. $r = -\theta, \theta \geq 0$ **46.** $r = \dfrac{\theta}{\pi}, \theta \geq 0$

Convert each equation to a polar equation and then graph it on a polar coordinate system.

47. $x^2 + y^2 = 16$ **48.** $x^2 + y^2 = 25$

49. $x^2 + y^2 = x$ **50.** $x^2 + y^2 = 2y$

51. $\dfrac{y}{x} = 1$ **52.** $y = \sqrt{3}x$

State a polar equation that will produce the given graph.

53.

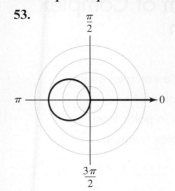

54.

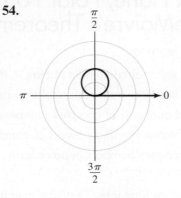

55.

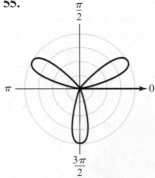

56.

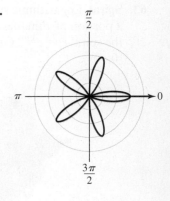

57.

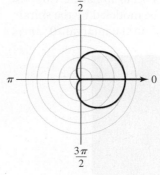

58.

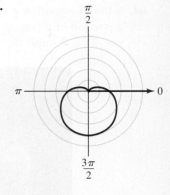

59.

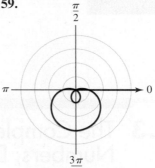

60.

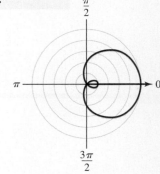

Applications

61. Cardioid microphone A cardioid microphone used by a singer can be modeled by the polar equation $r = -1.5 - 1.5 \sin \theta$. Graph the cardioid.

62. Cardioid microphone A cardioid microphone used by a singer can be modeled by the polar equation $r = 1.5 + 1.5 \cos \theta$. Graph the cardioid.

63. Spiral Logarithmic spirals occur in nature. Suppose a portion of romanesco broccoli can be modeled by the spiral $r = e^{0.1\theta}$. Graph the spiral on the interval $[0, 4\pi]$.

Serhiy Shullye/Shutterstock.com

64. Spiral Logarithmic spirals occur in nature. Suppose a portion of a fractal can be modeled by the spiral $r = 2e^{0.1\theta}$. Graph the spiral on the interval $[0, 2\pi]$.

Christopher S. Howeth/Shutterstock.com

65. Orbit of a planet Suppose the orbit of a planet can be modeled by the polar equation $r = \dfrac{1}{1 - 0.9\cos\theta}$. Graph the path of the planet's orbit.

66. Racetrack A racetrack for toy cars can be modeled by the lemniscate $r^2 = 225\cos 2\theta$. Graph the path of the track.

Discovery and Writing

Use a graphing calculator or utility for Exercises 67 and 68.

67. Graph the four-leaved rose $r = 4\sin 2(\theta + h)$ for several values of h, such as $h = 0, \frac{\pi}{6}, -\frac{\pi}{6}$, and $\frac{\pi}{4}$. What do you observe?

68. Graph the cardioid $r = 3 - 3\sin(\theta + h)$ for several values of h, such as $h = 0, \frac{\pi}{4}, \frac{\pi}{2}$, and π. What do you observe?

7.3 The Complex Plane; Polar Form of Complex Numbers; DeMoivre's Theorem

In this section, we will learn to

1. Find the absolute value of complex numbers in the complex plane.
2. Convert complex numbers from rectangular to polar form.
3. Find products and quotients of complex numbers in polar form.
4. Use DeMoivre's Theorem to find powers of complex numbers.
5. Find *n*th roots of complex numbers in polar form.

Ronen/Shutterstock.com

Electrical engineers have exciting jobs. They design and test new and improved electronics as well as contribute to the development of a wide range of technologies. Many work on the development of telecommunication systems, electric power stations, and the lighting and wiring of buildings. Others work to design and test electrical systems on automobiles and airplanes.

In the fields of electrical engineering and physics, complex numbers are used a great deal. There are many real-life applications that are better understood through the mathematics of complex numbers; electric-circuit theory and electromagnetism are two areas where complex numbers are important. In this section, we will build upon our knowledge of complex numbers and learn about them in greater depth.

1. Find the Absolute Value of Complex Numbers in the Complex Plane

Complex numbers were introduced in Section R.2. Although complex numbers obey most of the properties of real numbers, there is one property that does not carry over: There is no way of ordering the complex numbers that is consistent with the ordering established for real numbers. It makes no sense to say, for example, that $3i$ is greater than 5, that 0 is less than $3 + i$, or that $5 + 5i$ is greater than $-5 + 6i$. Because complex numbers are not linearly ordered, it is impossible to graph them on a number line in any way that preserves the ordering of the real numbers. This inability to graph complex numbers on the number line is not surprising, because real numbers provide coordinates for all points on the number line. However, there is a method for graphing the complex numbers.

We construct two perpendicular axes, as in Figure 7-32, and consider the horizontal axis to be the axis of real numbers and the vertical axis to be the axis of imaginary numbers. These axes determine what is called the **complex plane.** Although these axes resemble those in the rectangular plane, instead of plotting x and y values we plot ordered pairs of real numbers (a, b), where a and b are the parts of the complex number $a + bi$. The real axis is used for plotting a, and the imaginary axis is used for plotting b.

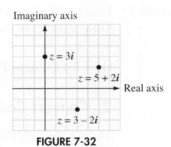

FIGURE 7-32

For any complex number $z = a + bi$, the distance of the point (a, b) from the origin, $0 + 0i = (0, 0)$, is the **absolute value** of the number.

Absolute Value of a Complex Number

If $z = a + bi$ is a complex number, the **absolute value** of z is

$$|z| = \sqrt{a^2 + b^2}.$$

(See Figure 7-33.) This value is also called the **magnitude.**

Recall that the **conjugate** of a complex number $z = a + bi$ is $\bar{z} = a - bi$. Since $z\bar{z} = a^2 + b^2$, we can also define the absolute value as

$$|z| = \sqrt{z\bar{z}}.$$

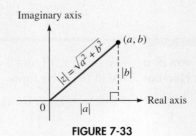

FIGURE 7-33

EXAMPLE 1   **Finding Absolute Value and Plotting Points in the Complex Plane**

Find the absolute value of each complex number and plot the point corresponding to the number in the complex plane.

a. $z = -3 + 2i$ **b.** $z = 4 - i$ **c.** $z = 4i$ **d.** $z = 5 + 6i$

SOLUTION   Points are shown in Figure 7-34.

a. $|-3 + 2i| = \sqrt{(-3)^2 + 2^2} = \sqrt{13}$ **b.** $|4 - i| = \sqrt{4^2 + (-1)^2} = \sqrt{17}$

c. $|4i| = \sqrt{0^2 + 4^2} = 4$ **d.** $|5 + 6i| = \sqrt{5^2 + 6^2} = \sqrt{61}$

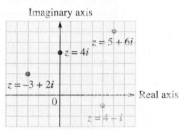

FIGURE 7-34

Self Check 1   Find the absolute value of each complex number and plot the point corresponding to the number in the complex plane.

a. $z = 5 - i$ **b.** $z = -4 + 3i$

Now Try Exercise 7.

2. Convert Complex Numbers from Rectangular to Polar Form

The complex number $z = a + bi$ is said to be written in **rectangular** or **algebraic form.** To determine how to write a complex number in polar form, we refer to Figure 7-35. Since r is the absolute value of the complex number, we have $r = \sqrt{a^2 + b^2}$. Likewise, from the relations $\cos \theta = \frac{a}{r}$ and $\sin \theta = \frac{b}{r}$, we see that $a = r \cos \theta$ and $b = r \sin \theta$. We also note that $\tan \theta = \frac{b}{a}$.

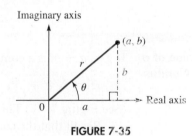

FIGURE 7-35

We now substitute these values into $z = a + bi$ to obtain the polar form of a complex number.

$$z = a + bi = r \cos \theta + (r \sin \theta)i = r(\cos \theta + i \sin \theta)$$

Polar Form of a Complex Number   If $r = |a + bi|$ and $0 \le \theta < 2\pi$, the complex number $z = a + bi$ can be written in the **polar form**

$$z = r(\cos \theta + i \sin \theta).$$

If a complex number is written in polar form $z = r(\cos \theta + i \sin \theta)$, the number r is called the **modulus** and the angle θ is called the **argument.**

EXAMPLE 2   **Converting Complex Numbers to Polar Form**

Write each complex number in polar form.

a. $z = 2 - 2i$ **b.** $z = -1 + 3i$ Round to two decimal places.

SOLUTION   **a.** It is always a good idea to plot the point corresponding to the complex number so that you know the quadrant in which the point lies. This will help you find the correct value of θ. The point is shown in Figure 7-36.

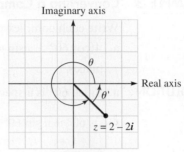

FIGURE 7-36

The complex number $z = 2 - 2i$ lies in QIV. To find θ, we will employ the same technique we used when converting rectangular coordinates to polar coordinates in Section 7.1.

For $z = 2 - 2i$, we have $a = 2$ and $b = -2$. Now we find the value of r.

$$r = \sqrt{a^2 + b^2} = \sqrt{2^2 + (-2)^2} = \sqrt{8} = 2\sqrt{2}$$

Now we will find the value of θ by using $\tan \theta' = \left| \frac{b}{a} \right|$.

$$\tan \theta' = \left| \frac{-2}{2} \right| = 1 \rightarrow \theta' = \tan^{-1} 1 = \frac{\pi}{4}$$

$$\theta = 2\pi - \theta' = 2\pi - \frac{\pi}{4} = \frac{7\pi}{4}$$

We use $r = 2\sqrt{2}$ and $\theta = \frac{7\pi}{4}$ to write $z = 2 - 2i$ in polar form.

$$z = r(\cos \theta + i \sin \theta) = 2\sqrt{2}\left(\cos \frac{7\pi}{4} + i \sin \frac{7\pi}{4} \right)$$

b. The point corresponding to the complex number $z = -1 + 3i$ lies in QII (Figure 7-37).

We find r: $a = -1, b = 3, r = \sqrt{a^2 + b^2} = \sqrt{(-1)^2 + 3^2} = \sqrt{10}$

We find θ': $\tan \theta' = \left| \frac{b}{a} \right| = \left| \frac{3}{-1} \right| = 3 \Rightarrow \theta' = \tan^{-1} 3$

Referring to Figure 7-37, we see that $\theta = \pi - \theta' = \pi - \tan^{-1} 3 \approx 1.89$.

Therefore, we have $z = r(\cos \theta + i \sin \theta) = \sqrt{10} (\cos 1.89 + i \sin 1.89)$.

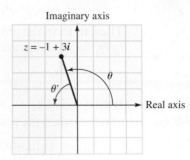

FIGURE 7-37

Self Check 2   Write each complex number in polar form.

     a. $z = \sqrt{3} - i$ **b.** $z = 3 + 4i$ Round to two decimal places.

     Now Try Exercise 19.

EXAMPLE 3   **Converting a Complex Number in Polar Form to Rectangular Form**

     Write each complex number in rectangular form.

     a. $4(\cos 60° + i \sin 60°)$ **b.** $\sqrt{5}(\cos 4.3 + i \sin 4.3)$ Round to two decimal places.

SOLUTION   **a.** Since $60°$ is a special angle, we can find exact values.

$$4(\cos 60° + i \sin 60°) = 4 \cos 60° + 4i \sin 60° = 4\left(\frac{1}{2}\right) + 4i\left(\frac{\sqrt{3}}{2}\right) = 2 + 2i\sqrt{3}$$

     b. It will be necessary to use a calculator to approximate the values. Make sure you have your calculator in radian mode.

$$\sqrt{5}(\cos 4.3 + i \sin 4.3) = \sqrt{5} \cos 4.3 + i\sqrt{5} \sin 4.3 \approx -0.90 - 2.05i$$

     ∎

Self Check 3   Write each complex number in rectangular form.

     a. $4\left(\cos \dfrac{5\pi}{6} + i \sin \dfrac{5\pi}{6}\right)$ **b.** $2.1(\cos 3.2 + i \sin 3.2)$ Round to two decimal places.

     Now Try Exercise 31.

3. Find Products and Quotients of Complex Numbers in Polar Form

It is easy to add complex numbers written in algebraic form, but it is difficult to multiply or divide them. However, it is easy to do so when they are in polar form. To see how, suppose that $z_1 = r_1(\cos \theta_1 + i \sin \theta_1)$ and $z_2 = r_2(\cos \theta_2 + i \sin \theta_2)$. The product $z_1 z_2$ can be found as follows:

$$z_1 z_2 = r_1(\cos \theta_1 + i \sin \theta_1) \cdot r_2(\cos \theta_2 + i \sin \theta_2)$$

$$= r_1 r_2(\cos \theta_1 \cos \theta_2 + i \cos \theta_1 \sin \theta_2 + i \sin \theta_1 \cos \theta_2 - \sin \theta_1 \sin \theta_2)$$

$$= r_1 r_2[(\cos \theta_1 \cos \theta_2 - \sin \theta_1 \sin \theta_2) + i(\cos \theta_1 \sin \theta_2 + \sin \theta_1 \cos \theta_2)]$$

$$= r_1 r_2[\cos(\theta_1 + \theta_2) + i \sin(\theta_1 + \theta_2)]$$

Thus, to form the product of two complex numbers written in polar form, we multiply the moduli r_1 and r_2 and add the arguments θ_1 and θ_2.

Product of Complex Numbers in Polar Form   If $z_1 = r_1(\cos \theta_1 + i \sin \theta_1)$ and $z_2 = r_2(\cos \theta_2 + i \sin \theta_2)$ are two complex numbers, then the **product of the numbers** is given by

$$z_1 z_2 = r_1 r_2[\cos(\theta_1 + \theta_2) + i \sin(\theta_1 + \theta_2)], 0 \le \theta < 2\pi \text{ or } 0° \le \theta < 360°.$$

Quotient of Complex Numbers in Polar Form

If $z_1 = r_1(\cos\theta_1 + i\sin\theta_1)$ and $z_2 = r_2(\cos\theta_2 + i\sin\theta_2)$ are two complex numbers, then the **quotient of the numbers** is given by

$$\frac{z_1}{z_2} = \frac{r_1}{r_2}[\cos(\theta_1 - \theta_2) + i\sin(\theta_1 - \theta_2)], 0 \leq \theta < 2\pi \text{ or } 0° \leq \theta < 360°.$$

The proof of the theorem for the division of complex numbers is similar to the proof for multiplication.

EXAMPLE 4   **Finding the Product and Quotient of Complex Numbers in Polar Form**

Given $z_1 = 4(\cos 50° + i\sin 50°)$ and $z_2 = 3(\cos 110° + i\sin 110°)$, find each of the following in polar form:

a. $z_1 \cdot z_2$ b. $\dfrac{z_1}{z_2}$

SOLUTION   a. $z_1z_2 = 4(\cos 50° + i\sin 50°) \cdot 3(\cos 110° + i\sin 110°)$

$$= (4 \cdot 3)[\cos(50° + 110°) + i\sin(50° + 110°)]$$

$$= 12(\cos 160° + i\sin 160°)$$

b. Now we find the quotient.

$$\frac{z_1}{z_2} = \frac{4(\cos 50° + i\sin 50°)}{3(\cos 110° + i\sin 110°)}$$

$$= \frac{4}{3}[\cos(50° - 110°) + i\sin(50° - 110°)]$$

$$= \frac{4}{3}[\cos(-60)° + i\sin(-60)°]$$

$$= \frac{4}{3}(\cos 300° + i\sin 300°) \qquad \text{Argument must be in } [0°, 360°).$$

Note that the Even-Odd Properties can be used to write the equally acceptable answer $\frac{4}{3}(\cos 60° - i\sin 60°)$.   ∎

Self Check 4   Given $z_1 = 5(\cos 70° + i\sin 70°)$ and $z_2 = 7(\cos 80° + i\sin 80°)$, find each of the following in polar form: **a.** $z_1 \cdot z_2$ **b.** $\dfrac{z_1}{z_2}$

Now Try Exercise 43.

4. Use DeMoivre's Theorem to Find Powers of Complex Numbers

We can use the theorem for multiplying complex numbers in polar form to find powers of complex numbers that are expressed in polar form. For example, we can find the cube of $z = 3(\cos 40° + i\sin 40°)$ as follows:

$$z^3 = [3(\cos 40° + i\sin 40°)]^3$$

$$= [3(\cos 40° + i\sin 40°)][3(\cos 40° + i\sin 40°)][3(\cos 40° + i\sin 40°)]$$

$$= 3^3[\cos(40° + 40° + 40°) + i\sin(40° + 40° + 40°)]$$

$$= 27(\cos 120° + i\sin 120°)$$

The generalization of this example is called **DeMoivre's Theorem.**

DeMoivre's Theorem   If n is a positive integer and $z = r(\cos \theta + i \sin \theta)$ is a complex number in polar form, then

$$z^n = r^n(\cos n\theta + i \sin n\theta).$$

This theorem was first developed around 1730 by the French mathematician Abraham De Moivre. Let's look at some examples of how to use this theorem.

EXAMPLE 5   **Using DeMoivre's Theorem**

Write $[2(\cos 15° + i \sin 15°)]^4$ in the standard $a + bi$ form.

SOLUTION   Using DeMoivre's Theorem, we have the following:

$$[2(\cos 15° + i \sin 15°)]^4 = 2^4[\cos(4 \cdot 15°) + i \sin(4 \cdot 15°)]$$

$$= 16(\cos 60° + i \sin 60°)$$

We can now change to standard form.

$$16(\cos 60° + i \sin 60°) = 16\left(\frac{1}{2} + i\frac{\sqrt{3}}{2}\right) = 8 + 8i\sqrt{3}$$

Self Check 5   Write $[3(\cos 15° + i \sin 15°)]^2$ in the standard $a + bi$ form.

Now Try Exercise 51.

EXAMPLE 6   **Using DeMoivre's Theorem**

Write $(1 + i)^6$ in the standard $a + bi$ form.

SOLUTION   Our first step is to write the complex number $z = 1 + i$ in polar form. We have $a = 1$ and $b = 1$, which means the number would be plotted in QI. We find $r = \sqrt{1^2 + 1^2} = \sqrt{2}$ and $\tan \theta = \frac{1}{1} = 1$, which means $\theta = \tan^{-1} 1 = \frac{\pi}{4}$. Therefore, we have $1 + i = r(\cos \theta + i \sin \theta) = \sqrt{2}\left(\cos \frac{\pi}{4} + i \sin \frac{\pi}{4}\right)$. Now we use DeMoivre's Theorem to find $(1 + i)^6$.

$$(1 + i)^6 = \left[\sqrt{2}\left(\cos \frac{\pi}{4} + i \sin \frac{\pi}{4}\right)\right]^6$$

$$= \left(\sqrt{2}\right)^6\left[\cos\left(6 \cdot \frac{\pi}{4}\right) + i \sin\left(6 \cdot \frac{\pi}{4}\right)\right]$$

$$= 8\left(\cos \frac{3\pi}{2} + i \sin \frac{3\pi}{2}\right)$$

$$= 8(0 - i) = -8i$$

Self Check 6   Write $(1 + i)^5$ in the standard $a + bi$ form.

Now Try Exercise 61.

5. Find nth Roots of Complex Numbers in Polar Form

DeMoivre's Theorem can be used to find all the nth roots of any number. Because both real and complex numbers can be written in the form $a + bi$, they can be written in polar form as well. Thus, one nth root of $a + bi$ is

$$\sqrt[n]{a + bi} = (a + bi)^{1/n}$$

$$= [r(\cos\theta + i\sin\theta)]^{1/n}$$

$$= \sqrt[n]{r}\left(\cos\frac{\theta}{n} + i\sin\frac{\theta}{n}\right)$$

Recall from algebra that the equation $x^2 = 9$ has two distinct roots, 3 and -3, and each qualifies as a square root of 9. Likewise, for a complex number w, the equation $z^n = w$ has n distinct roots, and each qualifies as an nth root of w. It follows that there are n distinct nth roots of any complex number, and DeMoivre's Theorem can be used to find them all.

Finding Complex Roots Using DeMoivre's Theorem

Let $w = r(\cos\theta + i\sin\theta)$ be a nonzero complex number. For $n \geq 2$ and $k = 0, 1, 2, \ldots, n - 1$, w has n distinct complex nth roots found with the formula

$$z_k = \sqrt[n]{r}\left(\cos\frac{\theta + 2\pi k}{n} + i\sin\frac{\theta + 2\pi k}{n}\right) \text{ (radian measure)}$$

or

$$z_k = \sqrt[n]{r}\left(\cos\frac{\theta + 360°k}{n} + i\sin\frac{\theta + 360°k}{n}\right) \text{ (degree measure)}.$$

This tells us that every complex number has two distinct square roots, three distinct cube roots, etc.

EXAMPLE 7   **Finding Complex Roots**

Find the cube roots of 8. Write answers in $a + bi$ form.

SOLUTION   Because 8 can be expressed as $8 + 0i$, we have $a = 8$, $b = 0$, $r = \sqrt{8^2 + 0^2} = 8$ and $\tan\theta = \left|\frac{b}{a}\right| = \left|\frac{0}{8}\right|$. From this we know that $\theta = \tan^{-1}0 = 0°$. Now substitute the values $n = 3$, $r = 8$, and $\theta = 0°$ for $k = 0$, 1, and 2.

$$z_k = \sqrt[3]{8}\left(\cos\frac{0° + 360°k}{3} + i\sin\frac{0° + 360°k}{3}\right)$$

$$z_0 = \sqrt[3]{8}\left(\cos\frac{0° + 360°(0)}{3} + i\sin\frac{0° + 360°(0)}{3}\right) = 2(\cos 0° + i\sin 0°) = 2(1 + 0) = 2$$

$$z_1 = \sqrt[3]{8}\left(\cos\frac{0° + 360°(1)}{3} + i\sin\frac{0° + 360°(1)}{3}\right) = 2(\cos 120° + i\sin 120°) = 2\left(-\frac{1}{2} + i\frac{\sqrt{3}}{2}\right) = -1 + i\sqrt{3}$$

$$z_2 = \sqrt[3]{8}\left(\cos\frac{0° + 360°(2)}{3} + i\sin\frac{0° + 360°(2)}{3}\right) = 2(\cos 240° + i\sin 240°) = 2\left[-\frac{1}{2} + i\left(-\frac{\sqrt{3}}{2}\right)\right] = -1 - i\sqrt{3}$$

The values 2, $-1 + i\sqrt{3}$, and $-1 - i\sqrt{3}$ are the three distinct cube roots of 8. If these three cube roots of 8 are graphed in the complex plane, they are equally spaced at intervals of 120° around a circle of radius 2 [Figure 7-38(a)]. Figure 7-38(b) demonstrates that if the three points are connected by line segments, the segments form an equilateral triangle.

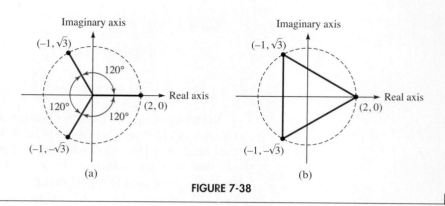

FIGURE 7-38

Self Check 7   Find the cube roots of $8i$.

Now Try Exercise 71.

Self Check Answers   **1. a.** $\sqrt{26}$ **b.** 5   **2. a.** $z = 2\left(\cos \dfrac{11\pi}{6} + i \sin \dfrac{11\pi}{6}\right)$

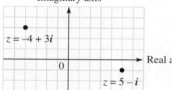

b. $z = 5(\cos 0.93 + i \sin 0.93)$ **3. a.** $-2\sqrt{3} + 2i$ **b.** $\approx -2.10 - 0.12i$

4. a. $35(\cos 150° + i \sin 150°)$ **b.** $\dfrac{5}{7}(\cos 350° + i \sin 350°)$

5. $\dfrac{9\sqrt{3}}{2} + \dfrac{9}{2}i$ **6.** $-4 - 4i$ **7.** $-2i, \sqrt{3} + i,$ and $-\sqrt{3} + i$

Exercises **7.3**

Getting Ready
You should be able to complete these vocabulary and concept statements before you proceed to the practice exercises.

Fill in the blanks.

1. For any complex number $z = a + bi$, the distance of the point (a, b) from the origin is the _____ of the number, and is calculated by _____. This is also called the _____.

2. The conjugate of a complex number $z = a + bi$ is _____.

3. The complex number $z = a + bi$ can be written in the polar form $z = r(\cos \theta + i \sin \theta)$, where $r =$ _____ and is called the _____, and the angle θ is called the _____.

4. If $z_1 = r_1(\cos \theta_1 + i \sin \theta_1)$ and $z_2 = r_2(\cos \theta_2 + i \sin \theta_2)$ are two complex numbers, then the product of the numbers is given by $z_1 z_2 =$ _____ and the quotient is given by $\dfrac{z_1}{z_2} =$ _____.

5. According to DeMoivre's Theorem, if n is a positive integer and $z = r(\cos \theta + i \sin \theta)$ is a complex number in polar form, then $z^n =$ _____.

6. Every complex number has _____ distinct cube roots.

Practice
Find the absolute value of each complex number and plot the point corresponding to the number in the complex plane.

7. $2 + 3i$ 8. $3 - 5i$ 9. $\dfrac{3}{4} + \dfrac{1}{2}i$ 10. $\dfrac{5}{2} - \dfrac{3}{2}i$

11. $-7i$ 12. $3i$ 13. 4 14. -5

Write each complex number in polar form using radian measure for the argument.

15. 6

16. -7

17. $-3i$

18. $4i$

19. $-1 - i$

20. $-1 + i$

21. $3 + 3i\sqrt{3}$

22. $7 - 7i$

23. $-1 - i\sqrt{3}$

24. $-3\sqrt{3} + 3i$

25. $-\sqrt{3} - i$

26. $1 + i\sqrt{3}$

Write each complex number in polar form using degree measure for the argument, rounding to two decimal places.

27. $4 - 3i$

28. $2 + 4i$

29. $-5 + 2i$

30. $-4 - 7i$

Write each complex number in rectangular form. Round to two decimal places when exact values cannot be found.

31. $2(\cos 30° + i \sin 30°)$

32. $5(\cos 45° + i \sin 45°)$

33. $7(\cos 90° + i \sin 90°)$

34. $12(\cos 0° + i \sin 0°)$

35. $-2\left(\cos \dfrac{2\pi}{3} + i \sin \dfrac{2\pi}{3}\right)$

36. $3\left(\cos \dfrac{4\pi}{3} + i \sin \dfrac{4\pi}{3}\right)$

37. $\dfrac{1}{2}(\cos \pi + i \sin \pi)$

38. $-\dfrac{2}{3}(\cos 2\pi + i \sin 2\pi)$

39. $-3(\cos 225° + i \sin 225°)$

40. $-3(\cos 300° + i \sin 300°)$

41. $\sqrt{3}(\cos 1.25 + i \sin 1.25)$

42. $6(\cos 3.1 + i \sin 3.1)$

For the given complex numbers, find $z_1 \cdot z_2$ and $\dfrac{z_1}{z_2}$. Write answers in polar form.

43. $z_1 = 4(\cos 30° + i \sin 30°); z_2 = 2(\cos 60° + i \sin 60°)$

44. $z_1 = 3(\cos 45° + i \sin 45°); z_2 = 2(\cos 120° + i \sin 120°)$

45. $z_1 = \cos 300° + i \sin 300°; z_2 = \cos 0° + i \sin 0°$

46. $z_1 = 5(\cos 85° + i \sin 85°); z_2 = 2(\cos 65° + i \sin 65°)$

47. $z_1 = 2(\cos \pi + i \sin \pi); z_2 = 3(\cos \pi + i \sin \pi)$

48. $z_1 = \cos \dfrac{\pi}{2} + i \sin \dfrac{\pi}{2}; z_2 = \cos \dfrac{3\pi}{2} + i \sin \dfrac{3\pi}{2}$

49. $z_1 = 1 - i\sqrt{3}; z_2 = -\sqrt{2} + i\sqrt{2}$

50. $z_1 = 4 - 4i; z_2 = -\sqrt{3} + i$

Find each power. Write answers in polar form.

51. $[3(\cos 30° + i \sin 30°)]^3$

52. $[4(\cos 15° + i \sin 15°)]^6$

53. $(\cos 15° + i \sin 15°)^{12}$

54. $[2(\cos 120° + i \sin 120°)]^9$

55. $[5(\cos 2° + i \sin 2°)]^5$

56. $[0.5(\cos 100° + i \sin 100°)]^3$

57. $\left[3\left(\cos \dfrac{\pi}{4} + i \sin \dfrac{\pi}{4}\right)\right]^4$

58. $\left[2\left(\cos \dfrac{3\pi}{2} + i \sin \dfrac{3\pi}{2}\right)\right]^6$

59. $[4(\cos 3 + i \sin 3)]^4$

60. $[2(\cos 5 + i \sin 5)]^{20}$

61. $\left(1 - \sqrt{3}i\right)^6$

62. $(-2 - 2i)^8$

63. $(1 + i)^5$

64. $\left(-\sqrt{3} + i\right)^3$

Find the indicated nth root of each expression in $a + bi$ form.

65. A cube root of $8(\cos 180° + i \sin 180°)$

66. A fifth root of $32(\cos 150° + i \sin 150°)$

67. A fifth root of $\cos 300° + i \sin 300°$

68. A fourth root of $256(\cos \pi + i \sin \pi)$

69. A sixth root of $64(\cos 2\pi + i \sin 2\pi)$

70. A sixth root of $3^6(\cos \pi + i \sin \pi)$

Find all the complex roots. Write answers in $a + bi$ form.

71. The cube roots of -8

72. The fourth roots of 16

73. The square roots of i

74. The cube roots of i

75. The fourth roots of $-8 + 8i\sqrt{3}$

76. The eighth roots of 1

Find all the complex roots. Write answers in polar form with the argument in degrees.

77. The fifth roots of $-i$

78. The cube roots of $\dfrac{\sqrt{2}}{2} + \dfrac{\sqrt{2}}{2}i$

79. The cube roots of $4\sqrt{3} + 4i$

80. The fifth roots of $-16 - 16i\sqrt{3}$

Applications

81. Electric circuits The state of a certain circuit is described by two real numbers, the voltage V across it and the current I flowing through it. If $V = 12$ volts and $I = 2$ amperes, describe the real numbers V and I by the single complex number $z = V + Ii$.

82. Electric circuits The state of a certain circuit is described by two real numbers, the voltage C across it and the inductance L flowing through it. If $C = \dfrac{1}{260}$ farad and $L = \dfrac{1}{2}$ henry, describe the real numbers C and L by the single complex number $z = C + Li$.

83. **Electrical impedance** The opposition to alternating current, known as **impedance,** is given by the complex number $Z = R + jX$, where $j = \sqrt{-1}$. Find the magnitude of impedance.

84. **Electrical admittance** In electrical engineering, the **admittance** Y is a measure of how easily a circuit will allow a current to flow. It is defined as the reciprocal of the impedance Z; that is, $Y = \frac{1}{Z}$. If $Z = R + jX$, where $j = \sqrt{-1}$, find a formula for Y.

Discovery

In his study of complex numbers, Leonhard Euler (1707–1783) used the following formula to define imaginary exponents:

$$e^{i\theta} = \cos\theta + i\sin\theta,$$

where e and θ are real numbers and $i = \sqrt{-1}$. Use Euler's formula to prove each statement.

85. $(e^{i\theta})^3 = e^{3i\theta}$

86. $e^{i\theta_1}e^{i\theta_2} = e^{i\theta_1 + i\theta_2}$

87. $\dfrac{e^{i\theta_1}}{e^{i\theta_2}} = e^{i\theta_1 - i\theta_2}$

88. $e^{i\pi} + 1 = 0$

7.4 Vectors

In this section, we will learn to

1. Add and subtract geometric vectors.
2. Determine the components of a position vector.
3. Calculate the magnitude and scalar product of a vector.
4. Add and subtract vectors using components.
5. Use unit coordinate vectors.
6. Find a vector using magnitude and direction.
7. Solve applications that employ vectors.

Carnival Cruise Lines is a leader in family cruising. Its ships are known as fun ships and it has one goal in mind: that every time someone walks up the gangway, they enter a whole new world of fun. Suppose we are aboard a cruise ship traveling at 20 nautical miles per hour due east. If the ocean current is moving at 10 nautical miles per hour due north, what is the actual velocity of the cruise ship? To answer this question, we will have to learn about vectors, the topic of this section. Velocity is one example of a vector. Vectors surface in many applications.

Some physical quantities, called **scalar quantities,** can be described completely by their numerical values, or magnitudes. Some scalar quantities are temperature, distance, area, volume, and elapsed time. Other quantities, called **vector quantities,** have both magnitude and direction and represent a displacement. Some vector quantities are force, velocity, acceleration, and displacement.

1. Add and Subtract Geometric Vectors

A **geometric vector** is a directed line segment. An arrow is used to indicate the direction of the vector. In Figure 7-39, A is the **initial point** and B is the **terminal point** of the vector $\overrightarrow{AB}$. A vector is denoted by a bold letter, such as $\mathbf{v} = \overrightarrow{AB}$. The length of the vector is called the **magnitude** and is denoted by the scalar quantity $|\mathbf{v}|$.

Two vectors are **equal** if they have equal magnitude and the same direction. Remember, a vector represents a displacement, so they are equal if the magnitude and direction are equal. Therefore, all the vectors in Figure 7-40 are equal. The **zero vector 0** has no magnitude or direction.

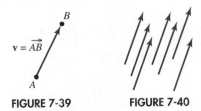

FIGURE 7-39   **FIGURE 7-40**

There are two ways that vectors, **u** and **v**, can be added together geometrically. The first is called the **Triangle Law.** Addition by this method is accomplished by moving the initial point of one vector to the terminal point of the other, as shown in Figure 7-41. Note that a triangle is formed. The vector **u** + **v** is called the **resultant** and is the same as taking one displacement and then the other. The other method used is the **Parallelogram Law.** If two vectors originating at a common point are adjacent sides of a parallelogram, their vector sum is the vector represented by the diagonal of the parallelogram that is drawn from the common point.

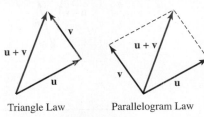

Triangle Law   Parallelogram Law

FIGURE 7-41

The vector $-\mathbf{v}$ has the same magnitude as but the opposite direction of vector **v**. (See Figure 7-42.) Thus $\mathbf{v} + (-\mathbf{v}) = \mathbf{0}$.

FIGURE 7-42

Properties of Addition of Vectors   For vectors **u**, **v**, and **w**, the following properties apply:

1. Commutative: $\mathbf{u} + \mathbf{v} = \mathbf{v} + \mathbf{u}$
2. Associative: $\mathbf{u} + (\mathbf{v} + \mathbf{w}) = (\mathbf{u} + \mathbf{v}) + \mathbf{w}$
3. Zero vector: $\mathbf{u} + \mathbf{0} = \mathbf{0} + \mathbf{u} = \mathbf{u}$

The **difference of two vectors** is defined as $\mathbf{u} - \mathbf{v} = \mathbf{u} + (-\mathbf{v})$. (See Figure 7-43.)

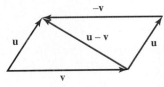

FIGURE 7-43

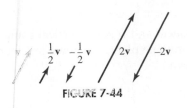

FIGURE 7-44

If k is a real number and **v** is a vector, then the product $k\mathbf{v}$ is also a vector. Its magnitude is $|k||\mathbf{v}|$. When k is a positive real number, $k\mathbf{v}$ has the same direction as **v**; however, when k is negative, $k\mathbf{v}$ has the opposite direction of **v**. (See Figure 7-44.) The vector $k\mathbf{v}$ is called a **scalar multiple** of **v**.

EXAMPLE 1   **Graphing Vectors**

Use the geometric vectors shown in Figure 7-45 to graph each vector indicated.
a. **u − v**   b. **3v + 2w**   c. **2w − u + v**

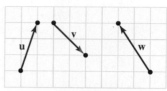

FIGURE 7-45

SOLUTION   Answers are shown in Figure 7-46.

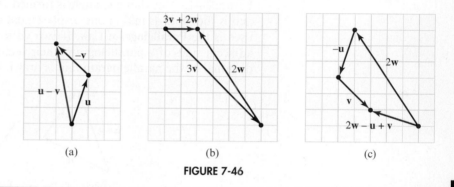

(a)   (b)   (c)

FIGURE 7-46

Self Check 1   Use the geometric vectors in Figure 7-45 to graph the vector **w + 2u + v**.

Now Try Exercise 11.

2. Determine the Components of a Position Vector

Vectors are easier to handle mathematically when they are placed on a rectangular coordinate system. The vector **v** in Figure 7-47, for example, is placed on a rectangular coordinate system so that it starts at the origin and goes to the point $(3, 2)$. If we assume that all vectors start at the origin, then each one is completely determined by its endpoint. Thus, we can denote the vector shown as $\mathbf{v} = \langle 3, 2 \rangle$, where the values 3 and 2 are called the **components** of the vector, and vector **v** is called a **position vector**. We use angle brackets $\langle \, \rangle$ to distinguish the vector $\langle 3, 2 \rangle$ from the ordered pair $(3, 2)$.

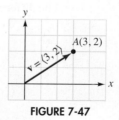

FIGURE 7-47

For a vector $\mathbf{v} = \langle a, b \rangle$, a is the **horizontal component** and b is the **vertical component**. The vector can be placed anywhere in a coordinate plane, as shown in Figure 7-48. Three representations of the vector are shown.

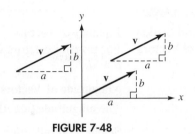

FIGURE 7-48

Position Vector   If a vector **v** has the initial point $P(x_1, y_1)$ and the terminal point $Q(x_2, y_2)$, the components of the vector are $a = x_2 - x_1$ and $b = y_2 - y_1$, so that $\mathbf{v} = \langle x_2 - x_1, y_2 - y_1 \rangle$ and the **position vector** is $\mathbf{v} = \langle a, b \rangle$ (Figure 7-49).

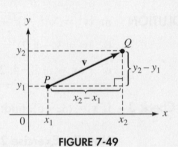

FIGURE 7-49

EXAMPLE 2   **Finding the Component Form of a Vector**

a. Find the position vector of vector **v** if the initial point is $P(-3, 4)$ and the terminal point is $Q(5, 3)$.

b. Draw the vector **v**, and then draw the position vector.

SOLUTION   a. $\mathbf{v} = \langle 5 - (-3), 3 - 4 \rangle = \langle 8, -1 \rangle$

b. Vector $\mathbf{v} = \langle 8, -1 \rangle$ and the position vector are shown in Figure 7-50.

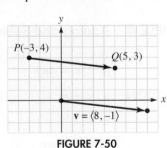

FIGURE 7-50

Self Check 2   a. Find the position vector of vector **v** if the initial point is $P(-3, 1)$, and the terminal point is $Q(2, 4)$.   b. Draw the vector **v**, and then draw the position vector.

Now Try Exercise 17.

3. Calculate the Magnitude and Scalar Product of a Vector

We stated earlier that two vectors are **equal** if they have the same direction and magnitude. Now that we have defined vectors algebraically, we can give a more precise definition of the equality and magnitude of vectors and scalar multiplication.

Equality, Magnitude, and Scalar Multiplication of Vectors

Equality of Vectors
For position vectors $\mathbf{v} = \langle a, b \rangle$ and $\mathbf{w} = \langle c, d \rangle$, $\mathbf{v} = \mathbf{w}$ if and only if $a = c$ and $b = d$.

Magnitude of Vectors
The **magnitude** (length) of a vector $\mathbf{v} = \langle a, b \rangle$ is $|\mathbf{v}| = \sqrt{a^2 + b^2}$.

Scalar Multiplication
If k is a scalar and $\mathbf{v}$ is the vector $\langle a, b \rangle$, then $k\mathbf{v} = k\langle a, b \rangle = \langle ka, kb \rangle$.

EXAMPLE 3   **Finding the Magnitude and Scalar Product of a Vector**

If $\mathbf{v} = \langle -4, 2 \rangle$, find: **a.** $|\mathbf{v}|$ **b.** $3\mathbf{v}$ **c.** $-5\mathbf{v}$

SOLUTION   **a.** $|\mathbf{v}| = \sqrt{(-4)^2 + 2^2} = \sqrt{16 + 4} = \sqrt{20} = 2\sqrt{5}$

b. $3\mathbf{v} = 3\langle -4, 2 \rangle = \langle -12, 6 \rangle$

c. $-5\mathbf{v} = -5\langle -4, 2 \rangle = \langle 20, -10 \rangle$

Self Check 3   If $\mathbf{v} = \langle 3, -5 \rangle$, find: **a.** $|\mathbf{v}|$ **b.** $\dfrac{2}{3}\mathbf{v}$ **c.** $-3\mathbf{v}$

Now Try Exercise 23.

4. Add and Subtract Vectors Using Components

Addition and subtraction of vectors is easy to do when the vectors are defined with components.

Addition and Subtraction of Vectors

If $\mathbf{v} = \langle a, b \rangle$ and $\mathbf{w} = \langle c, d \rangle$, then $\mathbf{v} + \mathbf{w} = \langle a + c, b + d \rangle$ and $\mathbf{v} - \mathbf{w} = \langle a - c, b - d \rangle$.

EXAMPLE 4   **Adding and Subtracting Vectors**

If $\mathbf{v} = \langle 5, -2 \rangle$ and $\mathbf{w} = \langle -3, 4 \rangle$, find: **a.** $\mathbf{v} + \mathbf{w}$ **b.** $\mathbf{w} - \mathbf{v}$ **c.** $3\mathbf{v} + 2\mathbf{w}$
d. $|2\mathbf{v} - 4\mathbf{w}|$

SOLUTION   **a.** $\mathbf{v} + \mathbf{w} = \langle 5, -2 \rangle + \langle -3, 4 \rangle = \langle 5 + (-3), -2 + 4 \rangle = \langle 2, 2 \rangle$

b. $\mathbf{w} - \mathbf{v} = \langle -3, 4 \rangle - \langle 5, -2 \rangle = \langle -3 - 5, 4 - (-2) \rangle = \langle -8, 6 \rangle$

c. $3\mathbf{v} + 2\mathbf{w} = 3\langle 5, -2 \rangle + 2\langle -3, 4 \rangle$
$= \langle 15, -6 \rangle + \langle -6, 8 \rangle = \langle 15 + (-6), -6 + 8 \rangle = \langle 9, 2 \rangle$

d. To find $|2\mathbf{v} - 4\mathbf{w}|$, we will first find vector $2\mathbf{v} - 4\mathbf{w}$.

$2\mathbf{v} - 4\mathbf{w} = 2\langle 5, -2 \rangle - 4\langle -3, 4 \rangle$
$= \langle 10, -4 \rangle - \langle -12, 16 \rangle = \langle 10 - (-12), -4 - 16 \rangle = \langle 22, -20 \rangle$

Now we can find the magnitude.

$|2\mathbf{v} - 4\mathbf{w}| = |\langle 22, -20 \rangle| = \sqrt{22^2 + (-20)^2} = \sqrt{884} = 2\sqrt{221}$

Self Check 4   If $\mathbf{v} = \langle -2, 1 \rangle$ and $\mathbf{w} = \langle 4, 6 \rangle$, find: **a.** $-\mathbf{v} + 3\mathbf{w}$ **b.** $|\mathbf{w} - 2\mathbf{v}|$

Now Try Exercise 29.

Properties of Vectors   For vectors **u**, **v**, and **w**, and real numbers c and d, the following properties apply:

Vector Addition

1) $\mathbf{u} + \mathbf{v} = \mathbf{v} + \mathbf{u}$
2) $\mathbf{u} + (\mathbf{v} + \mathbf{w}) = (\mathbf{u} + \mathbf{v}) + \mathbf{w}$
3) $\mathbf{u} + \mathbf{0} = \mathbf{u}$
4) $\mathbf{u} + (-\mathbf{u}) = \mathbf{0}$

Scalar Multiplication

5) $c(\mathbf{u} + \mathbf{v}) = c\mathbf{u} + c\mathbf{v}$
6) $(c + d)\mathbf{u} = c\mathbf{u} + d\mathbf{u}$
7) $(cd)\mathbf{u} = c(d\mathbf{u}) = d(c\mathbf{u})$
8) $1\mathbf{u} = \mathbf{u}$
9) $0\mathbf{u} = \mathbf{0}$
10) $c\mathbf{0} = \mathbf{0}$

Magnitude of a Vector:
$|c\mathbf{u}| = |c||\mathbf{u}|$

5. Use Unit Coordinate Vectors

If any nonzero vector **v** is multiplied by the reciprocal of its magnitude, the result is $\frac{\mathbf{v}}{|\mathbf{v}|}$, a vector with the same direction as **v** and a length of 1 unit.

Unit Vector   The vector $\mathbf{u} = \dfrac{\mathbf{v}}{|\mathbf{v}|}$ is called the **unit vector in the direction of v.**

If **i** is the vector $\langle 1, 0 \rangle$, then **i** is a unit vector of length 1 unit pointing in the positive x direction. Similarly, if **j** is the vector $\langle 0, 1 \rangle$, then **j** is a vector of length 1 unit pointing in the positive y direction. (See Figure 7-51.)

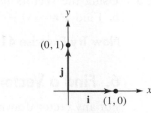

FIGURE 7-51

Any vector can be written as the sum of scalar multiples of the vectors **i** and **j**. For example,

$$\langle 5, 2 \rangle = \langle 5, 0 \rangle + \langle 0, 2 \rangle$$
$$= 5\langle 1, 0 \rangle + 2\langle 0, 1 \rangle$$
$$= 5\mathbf{i} + 2\mathbf{j}$$

The two vectors $5\mathbf{i}$ and $2\mathbf{j}$ are called the x and y **components** of the vector $\langle 5, 2 \rangle$. (See Figure 7-52.)

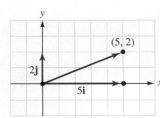

FIGURE 7-52

Unit Coordinate Vectors

The vectors $\mathbf{i} = \langle 1, 0 \rangle$ and $\mathbf{j} = \langle 0, 1 \rangle$ are called **unit coordinate vectors.** Any vector $\mathbf{v} = \langle x, y \rangle$ is **resolved** into its x or **horizontal component** $x\mathbf{i}$ and its y or **vertical component** $y\mathbf{j}$ when it is written in the form $\mathbf{v} = x\mathbf{i} + y\mathbf{j}$.

EXAMPLE 5   **Using the Properties of Vectors**

If $\mathbf{v} = 8\mathbf{i} - 4\mathbf{j}$, $\mathbf{w} = -3\mathbf{i} + 3\mathbf{j}$, and $\mathbf{u} = \mathbf{i} + 6\mathbf{j}$, find each of the following vectors:

a. The unit vector in the direction of $\mathbf{v}$.

b. $2\mathbf{u} - (3\mathbf{v} + \mathbf{w})$.

SOLUTION   **a.** The unit vector is $\frac{\mathbf{v}}{|\mathbf{v}|}$, so our first step is to find the magnitude of $\mathbf{v}$.

$$|\mathbf{v}| = \sqrt{8^2 + (-4)^2} = \sqrt{80} = 4\sqrt{5}$$

Therefore, $\dfrac{\mathbf{v}}{|\mathbf{v}|} = \dfrac{8\mathbf{i} - 4\mathbf{j}}{4\sqrt{5}} = \dfrac{8}{4\sqrt{5}}\mathbf{i} - \dfrac{4}{4\sqrt{5}}\mathbf{j} = \dfrac{2}{\sqrt{5}}\mathbf{i} - \dfrac{1}{\sqrt{5}}\mathbf{j} = \dfrac{2\sqrt{5}}{5}\mathbf{i} - \dfrac{\sqrt{5}}{5}\mathbf{j}$.

b. Clear the parentheses and then evaluate.

$$2\mathbf{u} - (3\mathbf{v} + \mathbf{w}) = 2\mathbf{u} - 3\mathbf{v} - \mathbf{w}$$
$$= 2(\mathbf{i} + 6\mathbf{j}) - 3(8\mathbf{i} - 4\mathbf{j}) - (-3\mathbf{i} + 3\mathbf{j})$$
$$= 2\mathbf{i} + 12\mathbf{j} - 24\mathbf{i} + 12\mathbf{j} + 3\mathbf{i} - 3\mathbf{j}$$
$$= -19\mathbf{i} + 21\mathbf{j}$$

Self Check 5   Using the vectors in Example 5: **a.** Find the unit vector in the direction of $\mathbf{u}$. **b.** Find $4(\mathbf{w} + \mathbf{v}) - 5\mathbf{u}$.

Now Try Exercise 41.

6. Find a Vector Using Magnitude and Direction

For any vector drawn in standard position, the angle formed by the vector and the positive x-axis is the **direction** of the vector (see Figure 7-53). If the magnitude and the direction of the vector are known, we are able to find the horizontal and vertical components of the vector.

Horizontal and Vertical Components of a Vector

For vector $\mathbf{v} = \langle a, b \rangle = a\mathbf{i} + b\mathbf{j}$, with magnitude $|\mathbf{v}|$ and direction θ, the **horizontal component** is $a = |\mathbf{v}|\cos\theta$ and the **vertical component** is $b = |\mathbf{v}|\sin\theta$. Therefore, the vector can be expressed as

$$\mathbf{v} = |\mathbf{v}|(\cos\theta\,\mathbf{i} + \sin\theta\,\mathbf{j}).$$

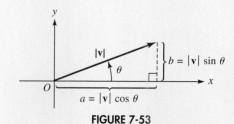

FIGURE 7-53

EXAMPLE 6 **Finding a Vector Using Magnitude and Direction**

A vector has magnitude 12 and a direction of 150°. Find the horizontal and vertical components of the vector and write the vector in terms of **i** and **j**.

SOLUTION First we find $a = 12 \cos 150° = 12\left(-\frac{\sqrt{3}}{2}\right) = -6\sqrt{3}$ and

$b = 12 \sin 150° = 12\left(\frac{1}{2}\right) = 6$. Therefore, the vector is $\mathbf{v} = -6\sqrt{3}\mathbf{i} + 6\mathbf{j}$. ■

Self Check 6 A vector has magnitude 10 and a direction of 240°. Find the horizontal and vertical components of the vector and write the vector in terms of **i** and **j**.

Now Try Exercise 49.

EXAMPLE 7 **Finding the Direction of a Vector**

Find the direction of vector $\mathbf{v} = -3\mathbf{i} + 3\mathbf{j}$.

SOLUTION Looking at Figure 7-54, we see that the vector lies in QII.

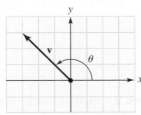

FIGURE 7-54

Using the figure, we can see that $\tan \theta = \frac{3}{-3} = -1$. Since $\tan 45° = 1$, we use the reference angle 45° and find $\theta = 180° - 45° = 135°$ to be the direction of the vector. ■

Self Check 7 Find the direction of vector $\mathbf{v} = -\sqrt{3}\mathbf{i} + \mathbf{j}$.

Now Try Exercise 55.

7. Solve Applications That Employ Vectors

If a **force** is acting on an object, either pushing or pulling, it can be represented by a vector. The **force vector** points in the direction in which it is trying to accelerate the object it is acting on. If more than one force is acting on an object, the sum of the forces is the **resultant force,** which produces the same force on the object as the individual forces together.

EXAMPLE 8 **Finding Resultant Force**

Two forces $\mathbf{F}_1$ and $\mathbf{F}_2$, with magnitudes of 15 and 25 pounds, respectively, act on an object as shown in Figure 7-55. Find the resultant force acting on the object.

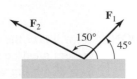

FIGURE 7-55

SOLUTION   Our first step is to write the vectors in composite form.

$$\mathbf{F}_1 = (15 \cos 45°)\mathbf{i} + (15 \sin 45°)\mathbf{j} = \frac{15\sqrt{2}}{2}\mathbf{i} + \frac{15\sqrt{2}}{2}\mathbf{j}$$

$$\mathbf{F}_2 = (25 \cos 150°)\mathbf{i} + (25 \sin 150°)\mathbf{j} = -\frac{25\sqrt{3}}{2}\mathbf{i} + \frac{25}{2}\mathbf{j}$$

The resultant force **F** is the sum of the two vectors. The vectors are shown in Figure 7-56.

$$\mathbf{F} = \mathbf{F}_1 + \mathbf{F}_2$$

$$= \left(\frac{15\sqrt{2}}{2}\mathbf{i} + \frac{15\sqrt{2}}{2}\mathbf{j}\right) + \left(-\frac{25\sqrt{3}}{2}\mathbf{i} + \frac{25}{2}\mathbf{j}\right)$$

$$= \left(\frac{15\sqrt{2} - 25\sqrt{3}}{2}\right)\mathbf{i} + \left(\frac{15\sqrt{2} + 25}{2}\right)\mathbf{j}$$

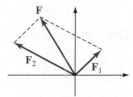

FIGURE 7-56

Self Check 8   Redo Exercise 8 with magnitudes of 30 and 45 pounds for $\mathbf{F}_1$ and $\mathbf{F}_2$, respectively.

Now Try Exercise 65.

Some vector problems involve suspending objects off the ground using cables. There are usually two cables at varying angles used to balance the force of gravity with the forces of tension in the cables. Since the object is not accelerating, the net force must be zero, and the resultant vector is zero. The hanging object is said to be in equilibrium. In the next example, we will see how we can use the Law of Sines to help us solve a vector problem.

EXAMPLE 9   **Finding Tensions**

A large speaker weighing 145 pounds is hung from the ceiling of an auditorium with two cables, as shown in Figure 7-57. What are the tensions on each cable?

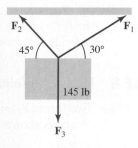

FIGURE 7-57

SOLUTION   We will take the vectors and construct a triangle. Using the fact that the sum of the vectors must be zero for the speaker to be in equilibrium, we can use the Triangle Law of adding vectors to produce the triangle shown in Figure 7-58. Note that the angles have to be drawn carefully. Once the triangle is constructed, the Law of Sines can be used to find the magnitudes of $\mathbf{F}_1$ and $\mathbf{F}_2$.

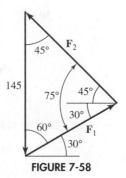

FIGURE 7-58

$$\frac{145}{\sin 75°} = \frac{|\mathbf{F}_1|}{\sin 45°}$$

$$|\mathbf{F}_1| = \frac{145 \sin 45°}{\sin 75°} \approx 106.1 \text{ pounds}$$

$$\frac{145}{\sin 75°} = \frac{|\mathbf{F}_2|}{\sin 60°}$$

$$|\mathbf{F}_2| = \frac{145 \sin 60°}{\sin 75°} \approx 130 \text{ pounds}$$

Self Check 9 Repeat Example 9 using a weight of 150 pounds and angles of 40° for $\mathbf{F}_1$ and 30° for $\mathbf{F}_2$.

Now Try Exercise 67.

A vector can be used to model the **velocity** of a moving object. The direction of the vector is equal to the direction of the moving object, and the magnitude of the vector is the speed of the object. For example, when an airplane flies through wind, both the speed and direction of the airplane are affected. The true velocity of the plane relative to the ground is the sum of a vector representing the velocity of the plane and a vector representing the velocity of the wind.

EXAMPLE 10 **Finding the Speed and Direction of an Airplane**

An airplane is flying due east at 350 mph, and encounters a 30 mph crosswind that is flowing in the direction N 60° E.

a. Express the velocity $\mathbf{u}$ of the airplane relative to the air, and the velocity $\mathbf{v}$ of the wind in component form.

b. What is the true velocity of the airplane expressed as a vector?

c. What is the true speed and direction of the airplane?

SOLUTION **a.** The velocity of the airplane relative to the air can be expressed as $\mathbf{u} = 350\mathbf{i} + 0\mathbf{j} = 350\mathbf{i}$.

   Since the wind is flowing in the direction N 60° E, the angle is $\theta = 30°$. The component form for the wind vector is

$$\mathbf{v} = (30 \cos 30°)\mathbf{i} + (30 \sin 30°)\mathbf{j}$$

$$= 15\sqrt{3}\mathbf{i} + 15\mathbf{j}$$

The vectors are shown in Figure 7-59.

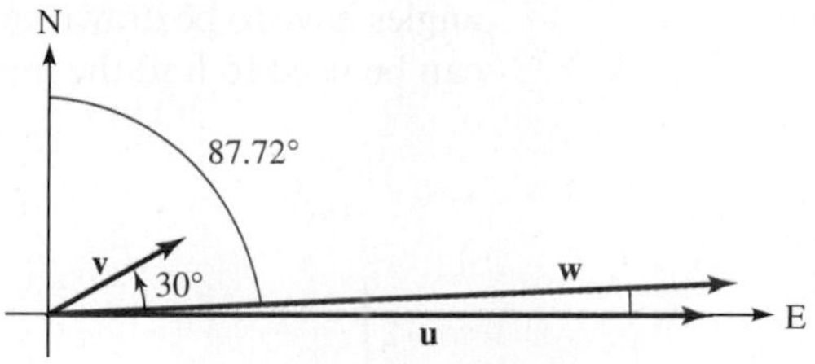

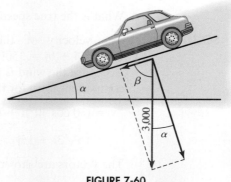

FIGURE 7-59

b. The true velocity of the airplane in vector form is the sum of these two vectors.

$$\mathbf{w} = \mathbf{v} + \mathbf{u}$$
$$= \left(15\sqrt{3}\mathbf{i} + 15\mathbf{j}\right) + 350\mathbf{i}$$
$$= \left(15\sqrt{3} + 350\right)\mathbf{i} + 15\mathbf{j}$$

c. The magnitude of $\mathbf{w}$ is the true speed of the airplane.

$$|\mathbf{w}| = \sqrt{(15\sqrt{3})^2 + 15^2} \approx 376.28$$

The true speed of the plane is 376.28 mph.

The heading of the plane is found by first finding $\alpha = \tan^{-1}\frac{15}{375.98} \approx 2.28°$. Therefore, the direction of the plane is $90° - 2.28° = 87.72°$, which we write as N 87.72° E. (See Figure 7-59.)

■

Self Check 10   Repeat Example 10 with the plane flying due east at 400 mph encountering a crosswind of 25 mph flowing in the direction N 45° E.

Now Try Exercise 77.

It is possible to separate a single vector into several components. Suppose that a car weighing 3,000 pounds is parked on a hill. It is pulled directly downward by gravity with a force of 3,000 pounds. Part of this force appears as a tendency to roll the car down the hill and another part presses the car against the road. Just how the 3,000 pounds is apportioned depends on the angle of the hill. If there were no hill, there would be no tendency to roll. If the hill were very steep, only a small force would hold the car to the road. The weight of the car is said to be **resolved** into two components—one directed down the hill and the other directed into the hill. These vectors obey the Parallelogram Law. (See Figure 7-60.) Note that the angle of the hill α is also the angle between two of the vectors, because both of these angles are complementary to angle β.

FIGURE 7-60

EXAMPLE 11   Finding Force on an Inclined Plane

A 3,000-pound car sits on an incline of 23°. What force is required to prevent the car from rolling down the hill? With what force is it held to the roadway? Round to the nearest pound.

SOLUTION   We must find the magnitudes of vectors **t** and **n** in Figure 7-61. Because $OACB$ is a rectangle, the opposite sides are equal and angle $B = 90°$. Hence, we have

$$\sin 23° = \frac{|\mathbf{t}|}{3,000} \qquad \cos 23° = \frac{|\mathbf{n}|}{3,000}$$

$$|\mathbf{t}| = 3,000 \sin 23° \qquad |\mathbf{n}| = 3,000 \cos 23°$$

$$|\mathbf{t}| \approx 1,172 \qquad\quad |\mathbf{n}| \approx 2,762$$

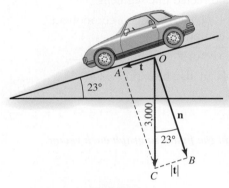

FIGURE 7-61

A force of approximately 1,172 pounds is required to prevent rolling, and a force of approximately 2,762 pounds keeps the car on the hill.

Self Check 11   What force is required to keep a 2,210-pound car from rolling down a ramp that makes a 10.0° angle with the horizontal? Round to the nearest pound.

Now Try Exercise 81.

Self Check Answers   **1.**

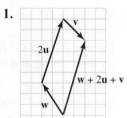

2. a. $\mathbf{v} = \langle 5, 3 \rangle$ **b.**

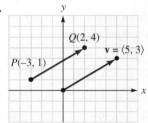

3. a. $\sqrt{34}$ **b.** $\left\langle 2, -\dfrac{10}{3} \right\rangle$ **c.** $\langle -9, 15 \rangle$ **4. a.** $\langle 14, 17 \rangle$ **b.** $4\sqrt{5}$

5. a. $\dfrac{\sqrt{37}}{37}\mathbf{i} + \dfrac{6\sqrt{37}}{37}\mathbf{j}$ **b.** $15\mathbf{i} - 34\mathbf{j}$ **6.** $\mathbf{v} = -5\mathbf{i} - 5\sqrt{3}\mathbf{j}$ The horizontal and vertical components are just used to find **v**.   **7.** $150°$

8. $\left(15\sqrt{2} - \dfrac{45\sqrt{3}}{2} \right)\mathbf{i} + \left(15\sqrt{2} + \dfrac{45}{2} \right)\mathbf{j}$ **9.** $F_1 \approx 138.2$ lb; $F_2 \approx 122.3$ lb

10. a. $\mathbf{u} = 400\mathbf{i}$; $\mathbf{v} = \dfrac{25\sqrt{2}}{2}\mathbf{i} + \dfrac{25\sqrt{2}}{2}\mathbf{j} \approx 17.68\mathbf{i} + 17.68\mathbf{j}$

b. $\mathbf{w} = \mathbf{u} + \mathbf{v} \approx 417.68\mathbf{i} + 17.68\mathbf{j}$ **c.** 418.05 mph; N 87.58° E
11. 384 lb

Exercises **7.4**

Getting Ready
You should be able to complete these vocabulary and concept statements before you proceed to the practice exercises.

Fill in the blanks.

1. Both magnitude and direction can be represented in a _____.

2. The _____ vector has a magnitude of 1.

3. A _____ vector is a directed line segment.

4. For vector $\mathbf{v} = a\mathbf{i} + b\mathbf{j}$, a is the _____ component and b is the _____ component.

5. The magnitude of vector $\mathbf{v} = a\mathbf{i} + b\mathbf{j}$ is _____.

6. Vector $\mathbf{v} = a\mathbf{i} + b\mathbf{j}$ can also be expressed as $\mathbf{v} = |\mathbf{v}|(\underline{\quad} + \underline{\quad})$.

Practice
Use the vectors shown in the figure to graph each vector indicated.

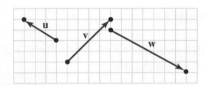

7. $\mathbf{u} + \mathbf{v}$

8. $\mathbf{v} + \mathbf{w}$

9. $3\mathbf{u} - 2\mathbf{v}$

10. $2\mathbf{w} + \mathbf{v}$

11. $\mathbf{u} + 3\mathbf{v} + 2\mathbf{w}$

12. $2\mathbf{u} + 2\mathbf{v} + \mathbf{w}$

13. $2\mathbf{v} + 3\mathbf{u} - \mathbf{w}$

14. $\mathbf{w} + \dfrac{1}{2}\mathbf{v} - 2\mathbf{u}$

Find the position vector of vector $\mathbf{v}$ if the initial point is P and the terminal point is Q.

15. $P(3, 4)$; $Q(3, 5)$

16. $P(0, 7)$; $Q(9, 0)$

17. $P(1, -9)$; $Q(-3, -1)$

18. $P(-3, 8)$; $Q(0, 0)$

19. $P(-5, -5)$; $Q(4, 4)$

20. $P(6, 6)$; $Q(-3, -3)$

21. $P(-2, 9)$; $Q(6, 9)$

22. $P(0, 0)$; $Q(7, -7)$

For each vector $\mathbf{v}$, find $|\mathbf{v}|$.

23. $\mathbf{v} = \langle 3, -1 \rangle$

24. $\mathbf{v} = \langle -5, 2 \rangle$

25. $\mathbf{v} = \langle 0, 7 \rangle$

26. $\mathbf{v} = \langle 5, 0 \rangle$

27. $\mathbf{v} = \left\langle \dfrac{1}{3}, \dfrac{2}{5} \right\rangle$

28. $\mathbf{v} = \left\langle -\dfrac{2}{3}, -\dfrac{3}{4} \right\rangle$

For the given vectors find: (a) $\mathbf{u} + 2\mathbf{v}$ (b) $3\mathbf{u} - 5\mathbf{w}$ (c) $4\mathbf{v} - 2\mathbf{u} - \mathbf{w}$ (d) $3\mathbf{w} - \mathbf{u} + 2\mathbf{v}$

29. $\mathbf{u} = \langle 3, 2 \rangle$; $\mathbf{v} = \langle 4, 4 \rangle$; $\mathbf{w} = \langle -1, -1 \rangle$

30. $\mathbf{u} = \langle 0, -4 \rangle$; $\mathbf{v} = \langle -4, 0 \rangle$; $\mathbf{w} = \langle 2, 2 \rangle$

31. $\mathbf{u} = \langle 5, 0 \rangle$; $\mathbf{v} = \langle 0, -2 \rangle$; $\mathbf{w} = \langle 2, -1 \rangle$

32. $\mathbf{u} = \langle -1, -1 \rangle$; $\mathbf{v} = \langle 7, 2 \rangle$; $\mathbf{w} = \langle 2, -3 \rangle$

33. $\mathbf{u} = \mathbf{i} + 2\mathbf{j}$; $\mathbf{v} = -2\mathbf{i} + \mathbf{j}$; $\mathbf{w} = 3\mathbf{i} + 3\mathbf{j}$

34. $\mathbf{u} = 3\mathbf{i} - 4\mathbf{j}$; $\mathbf{v} = 2\mathbf{i} + 2\mathbf{j}$; $\mathbf{w} = -4\mathbf{i} + 5\mathbf{j}$

35. $\mathbf{u} = -\mathbf{i} - \mathbf{j}$; $\mathbf{v} = 4\mathbf{i} + 4\mathbf{j}$; $\mathbf{w} = 3\mathbf{j}$

36. $\mathbf{u} = -2\mathbf{i} + 4\mathbf{j}$; $\mathbf{v} = 3\mathbf{i} + \mathbf{j}$; $\mathbf{w} = \mathbf{i}$

Find $|\mathbf{u}|$, $|\mathbf{v}|$, $|\mathbf{u} - \mathbf{v}|$, $|2\mathbf{u} + 3\mathbf{v}|$, and $|\mathbf{v} - 3\mathbf{u}|$.

37. $\mathbf{u} = 2\mathbf{i} - 3\mathbf{j}$; $\mathbf{v} = -\mathbf{i} + \mathbf{j}$

38. $\mathbf{u} = -3\mathbf{i} + \mathbf{j}$; $\mathbf{v} = 4\mathbf{i} + 2\mathbf{j}$

39. $\mathbf{u} = \langle 0, 4 \rangle$; $\mathbf{v} = \langle -2, -3 \rangle$

40. $\mathbf{u} = \langle 2, 2 \rangle$; $\mathbf{v} = \langle -2, -2 \rangle$

Find the unit vector in the same direction as $\mathbf{v}$.

41. $\mathbf{v} = 2\mathbf{i} - 3\mathbf{j}$

42. $\mathbf{v} = 4\mathbf{i} + 6\mathbf{j}$

43. $\mathbf{v} = 5\mathbf{j}$

44. $\mathbf{v} = 12\mathbf{i}$

45. $\mathbf{v} = -7\mathbf{i} + 6\mathbf{j}$

46. $\mathbf{v} = -8\mathbf{i} + 7\mathbf{j}$

Resolve each vector into its horizontal and vertical components by writing it in $a\mathbf{i} + b\mathbf{j}$ form. Round to two decimal places when necessary.

47. $\langle 3, 5 \rangle + \langle 5, 3 \rangle$

48. $\langle -2, 7 \rangle + \langle 2, 3 \rangle$

49. $|\mathbf{v}| = 10$; $\theta = 30°$

50. $|\mathbf{v}| = 15$; $\theta = 45°$

51. $|\mathbf{v}| = 23.3$; $\theta = 37.2°$

52. $|\mathbf{v}| = 19.1$; $\theta = 183.7°$

Find the magnitude and direction (in degrees) of the vector. Round to two decimal places, when necessary.

53. $\mathbf{v} = \langle -2, 2 \rangle$

54. $\mathbf{v} = \left\langle 1, \sqrt{3} \right\rangle$

55. $\mathbf{v} = 5\mathbf{i} - 3\mathbf{j}$

56. $\mathbf{v} = -2\sqrt{3}\mathbf{i} - 2\mathbf{j}$

57. $\mathbf{v} = -4\mathbf{i} - 5\mathbf{j}$

58. $\mathbf{v} = \sqrt{7}\mathbf{i} + 3\mathbf{j}$

Use the Law of Cosines to find the angle θ between the two vectors, with $0 \le \theta < 2\pi$. Round to three decimal places.

59. $\mathbf{u} = \langle 3, 7 \rangle$; $\mathbf{v} = \langle -2, 10 \rangle$

60. $\mathbf{u} = \langle -4, -8 \rangle$; $\mathbf{w} = \langle 3, -7 \rangle$

61. $\mathbf{v} = \mathbf{i} - 2\mathbf{j}$; $\mathbf{w} = -3\mathbf{i}$

62. $\mathbf{v} = 4\mathbf{j}$; $\mathbf{w} = 4\mathbf{i} + 6\mathbf{j}$

Applications
63. **Velocity** A rock is shot from a slingshot with an initial velocity of 80 feet per second at an angle of 30° with the horizontal, as shown in the figure. Find the vertical and horizontal components of the velocity vector.

64. **Velocity** The M136 AT4 light antitank weapon has a muzzle velocity of 950 feet per second. If it launches at an angle of 10°, find the vertical and horizontal components of the velocity vector.

65. **Resultant force** Two forces F_1 and F_2, with magnitudes of 70 and 90 pounds, respectively, act on an object as shown in the figure. Find the direction and magnitude of the resultant force acting on the object.

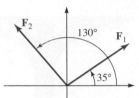

66. **Resultant force** Two forces F_1 and F_2, with magnitudes of 48 and 56 pounds, respectively, act on an object as shown in the figure. Find the direction and magnitude of the resultant force acting on the object.

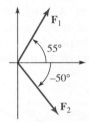

67. **Resultant force** Two men are trying to move a large box. One man is pulling on the box with a force of 50 pounds, and the other with a force of 60 pounds, as shown in the figure. If the box requires a force of at least 110 pounds to move it, will these two men be able to do so?

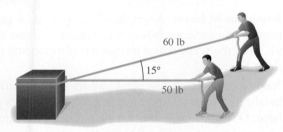

68. **Resultant force** Two mules are needed to pull a wagon. One pulls with a force of 800 pounds and the other with a force of 700 pounds along the line shown in the figure. Find angle θ to the nearest tenth of a degree.

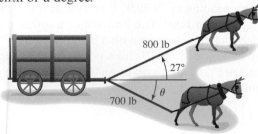

69. **Finding forces** Two men pull on ropes that are attached to the bumper of a car that is stuck. One

man pulls with a force of 114 pounds and the other pulls with a force of 97 pounds. If the angle between the ropes is 13°, what force is exerted on the car?

70. **Finding angles between forces** In Exercise 69, find the angle between the resultant force and the rope pulled by the stronger man.

71. **Finding resultant force and angles between forces** Two forces, one of 75 pounds and one of 90 pounds, are exerted at an angle of 102° from each other. Find the magnitude of the resultant force and the angle between the resultant force and the direction of the 90-pound force.

72. **Moving stones** A donkey and a horse are tied to a large stone. The horse pulls with a force of 950 pounds, and the donkey pulls with a force of 150 pounds. If the angle between their tethers is 19.5°, with what force do they pull on the stone?

73. **Equilibrium** A video screen weighing 500 pounds is held in suspension by two cables as shown in the figure. What is the tension on each cable?

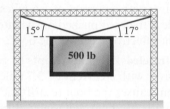

74. **Equilibrium** Use the information given in the figure to determine the tension on each cable supporting the weight.

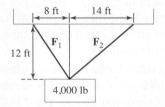

75. **Navigation** A boat that can travel up to 12 knots in still water maintains a course of due west while crossing a river that is flowing due south at 4.0 knots. What is the velocity (magnitude and direction) of the boat relative the bottom of the river?

76. **Navigation** A river is flowing due east at 3 mph. A boat crosses the river heading due south at 4 mph relative to the river. What is the boat's actual speed and direction relative to the bottom of the river?

77. **Speed and direction of an airplane** An airplane traveling north at a speed of 255 mph encounters a crosswind of 46 mph blowing due east from the west. In what direction should the pilot fly in order to maintain a course of due north? What is the ground speed of the plane?

78. **Speed and direction of an airplane** An airplane traveling at 500 mph in the direction N 30° W encounters a crosswind of 70 mph blowing in the direction

N 45° E. What are the resultant speed and direction of the airplane?

79. **Weight of a vehicle** A vehicle presses against a roadway with a force of 1,100 pounds. How much does the vehicle weigh if the roadway has an 18.0° grade?

80. **Angle of a ramp** A garden tractor weighing 351 pounds is being driven up a ramp onto a trailer. If the tractor presses against the ramp with 341 pounds of force, what angle does the ramp make with the horizontal?

81. **Weight of a barrel** A force of 25 pounds is necessary to hold a barrel in place on a ramp that makes an angle of 7° with the horizontal. How much does the barrel weigh?

82. **Inclined planes** If a force of 21.3 pounds is necessary to keep a 50.1-pound barrel from rolling down an inclined plane, what angle does the plane make with the horizontal?

83. **Effective speed** A boat capable of a speed of 11 miles per hour in still water attempts to go directly across a river with a current of 5.6 miles per hour. By what angle is the boat pushed off its intended path? What is the effective speed of the boat?

84. **Distance traveled** A boat attempts to go directly across a river with a current of 3.7 miles per hour. The current causes the boat to drift 23° from its intended path. How far will the boat travel if the trip takes 10 minutes?

85. **Flight path and speed** A plane has an airspeed of 411 miles per hour flying due east. A wind from the north is blowing at 31.0 miles per hour. By how many degrees is the plane blown off its heading? What is the ground speed of the plane (its speed relative to the ground)?

86. **Shooting a rifle** A rifle that fires bullets with a muzzle velocity of 4,100 feet per second is fixed at an angle of elevation of 32°. What is the horizontal component of the bullet's velocity?

87. **Hanging a weight** A 317-pound weight is hanging from the ceiling on a long rope. A man pushes horizontally against the weight to rotate the rope through an angle of 11.2°. What resultant force is being counteracted by the rope?

88. **Speed of a plane** A plane leaves an airport with a heading of S 10° E. At the same time, a second plane leaves the same airport with a heading of S 80° W. One hour later, the first plane is 1,300 miles directly southeast of the second plane. How fast is the first plane going? (Assume no wind.)

89. **Resolution of forces** A 312-pound force is directed due west. What force, directed exactly southeast,

would cause the resultant force to be directed due south?

90. **Resolution of forces** A 50-pound girl is playing in a tire swing hanging from a limb of a large tree, as shown in the figure. To get her started, a friend pulls the swing backward until it makes an angle of 40° with the vertical. What horizontal force **F** is required to hold the swing in this position?

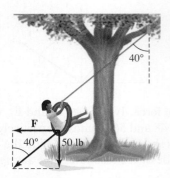

91. **Resolution of forces** A weight of 160 pounds is supported by a wire as in the figure. Find the vertical and horizontal components of force $\mathbf{F}_1$. (*Hint:* The vertical component of $\mathbf{F}_1$ supports half the weight.)

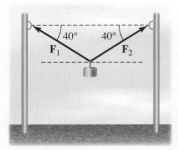

92. **Resolution of forces** A weight of 220 pounds is supported by a cable as shown in the figure in Exercise 91. Find the vertical and horizontal components of force $\mathbf{F}_1$.

93. **Navigation** A plane leaves an airport at noon with a heading of N 60° E and an airspeed of 451 miles per hour. One hour later, another plane leaves the same airport with a heading of N 70° W and an airspeed of 611 miles per hour. What is the bearing of the first plane from the second at 2:00 p.m.? (Assume no wind.)

94. **Navigation** A plane leaves an airport at 3:00 p.m. with a heading of N 70.0° E and an airspeed of 512 miles per hour. Two hours later, another plane leaves the same airport with a heading of S 80° E and an airspeed of 621 miles per hour. How far apart are the two planes at 6:00 p.m.? (Assume no wind.)

Discovery and Writing

95. Prove that $\cos\theta\mathbf{i} + \sin\theta\mathbf{j}$ is a unit vector for all values of θ.

96. The vectors shown form a pentagon. Explain why the sum of these vectors is **0**. Would this be true for any polygon?

97. Two equal forces of f pounds make an angle of θ with each other. What is the magnitude of their resultant?

7.5 Dot Product

In this section, we will learn to

1. Find the dot product of two vectors.

2. Find the angle between two vectors.

3. Find vector and scalar projections.

4. Compute work.

Kasia/Shutterstock.com

Have you ever thought about how to find the distance between two cities or airports? This distance is known as the **great-circle distance.** It is the shortest distance between two points on the surface of a sphere. It is important that airlines know the distance between airports such as O'Hare in Chicago and LAX in Los Angeles, so that they know how much fuel to load onto the airplanes. Using knowledge of longitude and latitude for the location of the cities and a vector operation known as the **dot product,** the great-circle distance can be calculated.

In this chapter, we have already studied two vector operations, addition and scalar multiplication. We will learn about the dot product in this section. The dot product is a scalar; it can be used to find distance, the angle between two vectors, the projection of one vector onto another, and work.

1. Find the Dot Product of Two Vectors

The definition of scalar multiplication provides the way to multiply a vector by a real number. We now define a way, called the **dot product**, to multiply one vector by another.

Dot Product of Two Vectors   Let $\mathbf{v} = \langle a, b\rangle = a\mathbf{i} + b\mathbf{j}$ and $\mathbf{w} = \langle c, d\rangle = c\mathbf{i} + d\mathbf{j}$. The **dot product** of vectors $\mathbf{v}$ and $\mathbf{w}$ is the *scalar*

$$\mathbf{v} \cdot \mathbf{w} = ac + bd.$$

EXAMPLE 1   **Finding a Dot Product**

Find each dot product.

a. $(2\mathbf{i} + 3\mathbf{j}) \cdot (5\mathbf{i} - 4\mathbf{j})$ **b.** $\langle 4, 5\rangle \cdot \langle -2, 7\rangle$

SOLUTION   **a.** $(2\mathbf{i} + 3\mathbf{j}) \cdot (5\mathbf{i} - 4\mathbf{j}) = 2 \cdot 5 + 3 \cdot (-4) = -2$

b. $\langle 4, 5\rangle \cdot \langle -2, 7\rangle = 4 \cdot (-2) + 5 \cdot 7 = 27$

■

Self Check 1   Find the dot product $(-3\mathbf{i} + 6\mathbf{j}) \cdot (5\mathbf{i} - 2\mathbf{j})$.

Now Try Exercise 7.

Properties of real numbers lead to the following properties of the dot product:

Properties of the Dot Product   For all vectors $\mathbf{u}$, $\mathbf{v}$, and $\mathbf{w}$, and real numbers k:

1. $\mathbf{u} \cdot \mathbf{u} = |\mathbf{u}|^2$
2. $\mathbf{u} \cdot \mathbf{v} = \mathbf{v} \cdot \mathbf{u}$
3. $\mathbf{u} \cdot (\mathbf{v} + \mathbf{w}) = \mathbf{u} \cdot \mathbf{v} + \mathbf{u} \cdot \mathbf{w}$
4. $(\mathbf{u} + \mathbf{v}) \cdot \mathbf{w} = \mathbf{u} \cdot \mathbf{w} + \mathbf{v} \cdot \mathbf{w}$
5. $k(\mathbf{u} \cdot \mathbf{v}) = (k\mathbf{u}) \cdot \mathbf{v} = \mathbf{u} \cdot (k\mathbf{v})$
6. $\mathbf{u} \cdot \mathbf{0} = 0$

EXAMPLE 2   **Using the Properties of the Dot Product**

Given the vectors $\mathbf{u} = \langle -2, 4 \rangle$, $\mathbf{v} = \langle 3, 5 \rangle$, and $\mathbf{w} = \langle 1, -2 \rangle$, demonstrate the equality of the property shown.

a. $\mathbf{u} \cdot (\mathbf{v} + \mathbf{w}) = \mathbf{u} \cdot \mathbf{v} + \mathbf{u} \cdot \mathbf{w}$ b. $k(\mathbf{u} \cdot \mathbf{v}) = (k\mathbf{u}) \cdot \mathbf{v} = \mathbf{u} \cdot (k\mathbf{v})$ for $k = 4$

SOLUTION   a. To demonstrate the property $\mathbf{u} \cdot (\mathbf{v} + \mathbf{w}) = \mathbf{u} \cdot \mathbf{v} + \mathbf{u} \cdot \mathbf{w}$, we first find the value of the left-hand side.

$$\mathbf{u} \cdot (\mathbf{v} + \mathbf{w}) = \langle -2, 4 \rangle \cdot (\langle 3, 5 \rangle + \langle 1, -2 \rangle) = \langle -2, 4 \rangle \cdot \langle 4, 3 \rangle = (-2)(4) + 4 \cdot 3 = 4$$

Now we find the value of the right-hand side.

$$\mathbf{u} \cdot \mathbf{v} + \mathbf{u} \cdot \mathbf{w} = \langle -2, 4 \rangle \cdot \langle 3, 5 \rangle + \langle -2, 4 \rangle \cdot \langle 1, -2 \rangle$$
$$= [(-2)(3) + 4 \cdot 5] + [(-2)(1) + 4(-2)] = 14 - 10 = 4$$

b. To demonstrate the property $k(\mathbf{u} \cdot \mathbf{v}) = (k\mathbf{u}) \cdot \mathbf{v} = \mathbf{u} \cdot (k\mathbf{v})$ for $k = 4$, we find the value of each expression.

$$4(\mathbf{u} \cdot \mathbf{v}) = 4[\langle -2, 4 \rangle \cdot \langle 3, 5 \rangle] = 4(-6 + 20) = 56$$
$$(4\mathbf{u}) \cdot \mathbf{v} = [4\langle -2, 4 \rangle] \cdot \langle 3, 5 \rangle = \langle -8, 16 \rangle \cdot \langle 3, 5 \rangle = (-8)(3) + 16 \cdot 5 = 56$$
$$\mathbf{u} \cdot (4\mathbf{v}) = \langle -2, 4 \rangle \cdot [4\langle 3, 5 \rangle] = \langle -2, 4 \rangle \cdot \langle 12, 20 \rangle = (-2)(12) + 4 \cdot 20 = 56$$

Self Check 2   Demonstrate the equality of the property $(\mathbf{u} + \mathbf{v}) \cdot \mathbf{w} = \mathbf{u} \cdot \mathbf{w} + \mathbf{v} \cdot \mathbf{w}$ using $\mathbf{u} = \langle 1, -4 \rangle$, $\mathbf{v} = \langle 2, 3 \rangle$, and $\mathbf{w} = \langle 0, 5 \rangle$.

Now Try Exercise 17.

2. Find the Angle between Two Vectors

The dot product can also be used to find the angle between two vectors. This geometric interpretation of the dot product is useful in applications.

Dot Product and the Angle between Two Vectors   If θ is the angle between $\mathbf{v}$ and $\mathbf{w}$, the dot product of vectors $\mathbf{v}$ and $\mathbf{w}$ is the scalar

$$\mathbf{v} \cdot \mathbf{w} = |\mathbf{v}||\mathbf{w}| \cos \theta, \text{ where } 0° \le \theta \le 180°.$$

To find θ, solve the equation

$$\cos \theta = \frac{\mathbf{v} \cdot \mathbf{w}}{|\mathbf{v}||\mathbf{w}|} \text{ for } \theta.$$

EXAMPLE 3   **Finding the Angle between Two Vectors**

Find the angle between $\mathbf{v} = 3\mathbf{i} + 4\mathbf{j}$ and $\mathbf{w} = 5\mathbf{i} - 12\mathbf{j}$.

SOLUTION   See Figure 7-62. By the property, $\cos \theta = \dfrac{\mathbf{v} \cdot \mathbf{w}}{|\mathbf{v}||\mathbf{w}|}$, so we proceed as follows:

$$\cos \theta = \frac{\mathbf{v} \cdot \mathbf{w}}{|\mathbf{v}||\mathbf{w}|}$$

$$= \frac{(3\mathbf{i} + 4\mathbf{j}) \cdot (5\mathbf{i} - 12\mathbf{j})}{\sqrt{3^2 + 4^2}\sqrt{5^2 + (-12)^2}}$$

$$= \frac{15 - 48}{5 \cdot 13}$$

$$\cos \theta = -\frac{33}{65}$$

$$\theta = \cos^{-1}\left(-\frac{33}{65}\right)$$

$$\theta \approx 120.5°$$

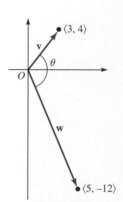

FIGURE 7-62

Self Check 3   Find the angle between $\mathbf{v} = -\mathbf{i} + 3\mathbf{j}$ and $\mathbf{w} = 4\mathbf{i} - \mathbf{j}$.

Now Try Exercise 27.

If the dot product $\mathbf{v} \cdot \mathbf{w} = |\mathbf{v}||\mathbf{w}| \cos \theta$ of two nonzero vectors is 0, then $\cos \theta$ must be 0. If $\cos \theta = 0$, then $\theta = 90°$ and the two vectors are **orthogonal** (perpendicular). Thus, the dot product provides a test to determine whether two vectors are orthogonal. If $\theta = 0°$ or $\theta = 180°$, the vectors are said to be **parallel**.

Orthogonal Vectors   Two nonzero vectors are orthogonal if and only if their dot product is zero.

EXAMPLE 4   **Determining Whether Vectors Are Orthogonal**

Determine whether the given vectors are orthogonal.

a. $\mathbf{v} = 6\mathbf{i} - 2\mathbf{j}$; $\mathbf{w} = \mathbf{i} + 3\mathbf{j}$ **b.** $\mathbf{v} = \langle 2, -1 \rangle$; $\mathbf{w} = \langle 1, -2 \rangle$

SOLUTION   **a.** $\mathbf{v} \cdot \mathbf{w} = (6\mathbf{i} - 2\mathbf{j}) \cdot (\mathbf{i} + 3\mathbf{j}) = 6 \cdot 1 + (-2)(3) = 6 - 6 = 0$. Since the dot product is 0, the vectors are orthogonal.

b. $\mathbf{v} \cdot \mathbf{w} = \langle 2, -1 \rangle \cdot \langle 1, -2 \rangle = 2 \cdot 1 + (-1)(-2) = 4 \neq 0$. Therefore, the vectors are not orthogonal.

Self Check 4   Determine whether the given vectors are orthogonal.

a. $\mathbf{v} = \langle 3, 1 \rangle$; $\mathbf{w} = \langle -3, -3 \rangle$ **b.** $\mathbf{v} = 3\mathbf{i} - 2\mathbf{j}$; $\mathbf{w} = 4\mathbf{i} + 6\mathbf{j}$

Now Try Exercise 33.

EXAMPLE 5 **Finding Components**

Find the horizontal and vertical components of a 2-pound force that makes an angle of 30° with the x-axis. Express the 2-pound force in $a\mathbf{i} + b\mathbf{j}$ form.

SOLUTION The horizontal component of the given force is vector **OA**, which is one leg of the right triangle *OAC*. (See Figure 7-63.) We find the magnitude of **OA** as follows:

$$|\mathbf{OA}| = |\mathbf{OC}| \cos \theta$$
$$= 2 \cos 30°$$
$$= 2\left(\frac{\sqrt{3}}{2}\right)$$
$$= \sqrt{3}$$

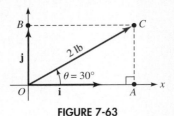

FIGURE 7-63

The horizontal component is $\sqrt{3}$ pounds. We now find the vertical component **OB**.

$$|\mathbf{OB}| = |\mathbf{OC}| \sin \theta$$
$$= 2 \sin 30°$$
$$= 1$$

The vertical component is 1 pound. Thus, $\mathbf{OC} = \sqrt{3}\mathbf{i} + \mathbf{j}$. ∎

Self Check 5 Find the horizontal and vertical components of a 5.0-pound force that makes an angle of 60° with the x-axis. Express the 5.0-pound force in $a\mathbf{i} + b\mathbf{j}$ form.

Now Try Exercise 39.

3. Find Vector and Scalar Projections

In Section 7.4, we worked problems that involved resolving a vector into horizontal and vertical components. It is often necessary to determine a component of a vector in a direction other than that of an axis. If **v** is not the zero vector, then the component of a vector **w** in the direction of **v** is called the **scalar projection of w on v**, denoted **comp$_v$ w**. Figure 7-64 shows that this component is the scalar $|\mathbf{w}| \cos \theta$, the length of vector **OP**. The vector **OP** itself is called the **vector projection of w on v**, denoted as **proj$_v$ w**.

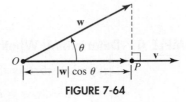

FIGURE 7-64

The scalar projection of **w** on **v** can be expressed by using the dot product, as follows:

$$\mathbf{comp_v\ w} = |\mathbf{w}| \cos \theta$$
$$= \frac{|\mathbf{w}||\mathbf{v}|\cos \theta}{|\mathbf{v}|} \qquad \text{Multiply numerator and denominator by } |\mathbf{v}|.$$
$$= \frac{\mathbf{w} \cdot \mathbf{v}}{|\mathbf{v}|} \qquad \text{Use the property } \mathbf{w} \cdot \mathbf{v} = |\mathbf{w}||\mathbf{v}| \cos \theta.$$

To derive a formula for the vector projection of **w** on **v**, we first form the vector $\frac{\mathbf{v}}{|\mathbf{v}|}$, which is the **unit** vector in the direction of **v**. We then multiply this unit vector by the scalar **comp$_v$ w** to stretch it to the length required for **proj$_v$ w.** Finally, we use the properties of the dot product to write the result in compact form.

$$\textbf{proj}_v\, \textbf{w} = (\textbf{comp}_v\, \textbf{w})(\text{unit vector in direction of } \textbf{v})$$

$$= \frac{\textbf{w} \cdot \textbf{v}}{|\textbf{v}|}\, \frac{\textbf{v}}{|\textbf{v}|}$$

$$= \frac{\textbf{w} \cdot \textbf{v}}{|\textbf{v}|^2}\, \textbf{v}$$

$$= \frac{\textbf{w} \cdot \textbf{v}}{\textbf{v} \cdot \textbf{v}}\, \textbf{v} \qquad \text{For any vector } \textbf{v},\ |\textbf{v}|^2 = \textbf{v} \cdot \textbf{v}.$$

Thus, we have the formulas for the scalar and vector projections of **w** on **v.**

Scalar and Vector Projections of w on v

If $\textbf{v} \neq \textbf{0}$, then

The scalar projection of **w** on $\textbf{v} = \textbf{comp}_v\, \textbf{w} = \dfrac{\textbf{w} \cdot \textbf{v}}{|\textbf{v}|}$.

The vector projection of **w** on $\textbf{v} = \textbf{proj}_v\, \textbf{w} = \dfrac{\textbf{w} \cdot \textbf{v}}{\textbf{v} \cdot \textbf{v}}\, \textbf{v}$.

EXAMPLE 6   **Finding Vector Projection**

A force **w** of exactly 2 pounds makes an angle of exactly 30° with the horizontal.

a. Find the component of this force in the direction of $\textbf{v} = 5\sqrt{3}\textbf{i} + 3\textbf{j}$.

b. Find the vector projection of **w** on **v.**

SOLUTION   **a.** To find the component of the given force **w** in the direction of **v**, we must find **comp$_v$ w.** (See Figure 7-65.)

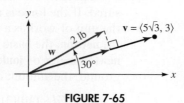

FIGURE 7-65

From Example 5 and from the given information, we know that

$$\textbf{w} = \sqrt{3}\textbf{i} + \textbf{j} \qquad \text{and} \qquad \textbf{v} = 5\sqrt{3}\textbf{i} + 3\textbf{j}.$$

Thus, we have

$$\textbf{comp}_v\, \textbf{w} = \frac{\textbf{w} \cdot \textbf{v}}{|\textbf{v}|} = \frac{\left(\sqrt{3}\textbf{i} + \textbf{j}\right) \cdot \left(5\sqrt{3}\textbf{i} + 3\textbf{j}\right)}{\left|5\sqrt{3}\textbf{i} + 3\textbf{j}\right|}$$

$$= \frac{15 + 3}{\sqrt{75 + 9}} = \frac{18}{\sqrt{84}}$$

$$= \frac{18}{2\sqrt{21}} = \frac{3\sqrt{21}}{7}$$

The component of the 2-pound force in the direction of **v** is $\dfrac{3\sqrt{21}}{7}$ pounds.

b. To find the vector projection of **w** on **v**, we proceed as follows:

$$\textbf{proj}_v\, \textbf{w} = \frac{\textbf{w} \cdot \textbf{v}}{\textbf{v} \cdot \textbf{v}}\textbf{v}$$

$$= \frac{\left(\sqrt{3}\textbf{i} + \textbf{j}\right) \cdot \left(5\sqrt{3}\textbf{i} + 3\textbf{j}\right)}{\left(5\sqrt{3}\textbf{i} + 3\textbf{j}\right) \cdot \left(5\sqrt{3}\textbf{i} + 3\textbf{j}\right)}\left(5\sqrt{3}\textbf{i} + 3\textbf{j}\right)$$

$$= \frac{18}{84}\left(5\sqrt{3}\textbf{i} + 3\textbf{j}\right) = \frac{3}{14}\left(5\sqrt{3}\textbf{i} + 3\textbf{j}\right)$$

$$= \frac{3 \cdot 5\sqrt{3}}{14}\textbf{i} + \frac{3 \cdot 3\textbf{j}}{14} = \frac{15\sqrt{3}}{14}\textbf{i} + \frac{9}{14}\textbf{j}$$

The vector projection of the 2-pound force **w** in the direction of **v** is $\dfrac{15\sqrt{3}}{14}\textbf{i} + \dfrac{9}{14}\textbf{j}$.

Self Check 6   A force **w** of exactly 3 pounds makes an angle of exactly 45° with the horizontal.

a. Find the component of this force in the direction of $\textbf{v} = 3\sqrt{2}\textbf{i} + 2\textbf{j}$.

b. Find the vector projection of **w** on **v**.

Now Try Exercise 45.

4. Compute Work

Work is done when an object is moved by a force over a certain distance. For example, work is done when a horse pulls a wagon, or when a person pushes a lawn mower. The concept of work is an important component in scientific studies dealing with energy.

If a constant force having magnitude F moves an object a total distance d along a straight line, then the **work** that is done is defined as

$W = Fd$ or work = force times distance.

The unit of work is determined by the way the force and distance are measured. If the force is measured in **pounds** and the distance is measured in **feet**, then the unit of work is a **foot-pound** (ft-lb). In scientific work, the basic unit of force is the **newton,** the basic unit of distance is the **meter,** and the basic unit of work is the **newton-meter** or **joule.** For example, if a 10-pound box is lifted three feet off the ground, the amount of work done is

$W = Fd = (10)(3) = 30$ foot-pounds.

This formula does not apply if the force directed is not along the direction of motion, which is the more common scenario. For example, when a person pushes a lawn mower across a lawn, the force exerted is downward along the handle. In this situation, the horizontal component is responsible for the motion and is the only component of force in the direction of motion. If a lawn mower is pushed 40 feet across a level lawn with a constant force of 55 pounds directed down the handle at a constant angle of 34°, then the work done by the force is

$W = $ (Component of force along the horizontal)(Displacement)

$= (55 \cos 34°)(40)$

$\approx 1{,}824$ foot-pounds

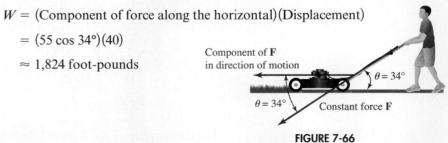

Component of **F** in direction of motion

$\theta = 34°$

$\theta = 34°$ Constant force **F**

FIGURE 7-66

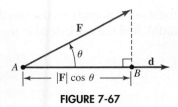

FIGURE 7-67

Suppose a force **F** moves an object from point A to point B, as shown in Figure 7-67. The only force affecting the object is the component of the force in the direction $\mathbf{d} = \overrightarrow{AB}$, which is the displacement vector. For an object to move in the direction of the displacement vector, angle θ must be acute.

Work   The work W done by a force **F** in moving along a vector **d** is

$$W = (\text{comp}_\mathbf{d}\, \mathbf{F})|\mathbf{d}| = \frac{\mathbf{F} \cdot \mathbf{d}}{|\mathbf{d}|}|\mathbf{d}| = \mathbf{F} \cdot \mathbf{d}.$$

EXAMPLE 7   **Calculating Work**

Calculate the work done by a force $\mathbf{F} = \langle 4, 7 \rangle$ that moves an object from the origin to the point $P(8, 11)$.

SOLUTION   The displacement vector is $\mathbf{d} = \langle 8, 11 \rangle$; therefore, the work done is

$$W = \mathbf{F} \cdot \mathbf{d} = \langle 4, 7 \rangle \cdot \langle 8, 11 \rangle = 4 \cdot 8 + 7 \cdot 11 = 109.$$

If the unit of force is pounds and the distance is measured in feet, then the work done is 109 foot-pounds. ∎

Self Check 7   Calculate the work done by a force $\mathbf{F} = \langle -8, 4 \rangle$ that moves an object from the origin to the point $P(-12, -4)$. Use foot-pounds as the unit of work.

Now Try Exercise 51.

EXAMPLE 8   **Calculating Work**

A child pulls on a sled by exerting a force of 20 pounds on the rope. If the rope makes an angle of 30° with the horizontal, find the work done in moving the sled 100 feet. (See Figure 7-68.) Round to the nearest foot-pound.

SOLUTION   By placing the problem on a rectangular coordinate system, we can equate the problem to an object being moved from the origin to the point $P(100, 0)$.

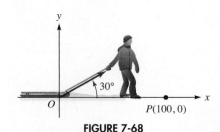

FIGURE 7-68

The displacement vector is $\mathbf{d} = 100\mathbf{i}$. The force on the rope can be written in terms of components as

$$\mathbf{F} = (20 \cos 30°)\mathbf{i} + (20 \sin 30°)\mathbf{j} = 10\sqrt{3}\mathbf{i} + 10\mathbf{j}.$$

Thus, the work done is

$$W = \mathbf{F} \cdot \mathbf{d} = (10\sqrt{3}\mathbf{i} + 10\mathbf{j}) \cdot (100\mathbf{i}) = 1{,}000\sqrt{3} \text{ foot-pounds} \approx 1{,}732 \text{ foot-pounds.} \quad ∎$$

Self Check 8   A man pulls a wagon horizontally by exerting a force of 30 pounds on the handle. If the handle makes an angle of 60° with the horizontal, find the work done in moving the wagon 150 feet.

Now Try Exercise 55.

Self Check Answers   1. -27 2. Both equal -5.   3. $122.5°$

4. a. not orthogonal   b. orthogonal   5. $\dfrac{5}{2}i + \dfrac{5\sqrt{3}}{2}j$

6. a. $\dfrac{9\sqrt{22} + 6\sqrt{11}}{22}$ lb   b. $\dfrac{27\sqrt{2} + 18}{22}i + \dfrac{9 + 3\sqrt{2}}{11}j$

7. 80 ft-lb   8. 2,250 ft-lb

Exercises **7.5**

Getting Ready

You should be able to complete these vocabulary and concept statements before you proceed to the practice exercises.

Fill in the blanks.

1. The dot product is way to ___ two vectors, and the result is a ___.
2. If θ is the angle between two vectors **v** and **w**, then $\cos\theta =$ ___.
3. Two vectors are orthogonal if their dot product is ___.
4. The scalar projection of **w** on **v** is ___.
5. The vector projection of **w** on **v** is ___.
6. The work W done by a force **F** in moving along a vector **d** is ___.

Practice

Find each dot product.

7. $\langle 2, -3 \rangle \cdot \langle 3, -1 \rangle$
8. $\langle 1, -5 \rangle \cdot \langle 5, 1 \rangle$
9. $(2i + 5j) \cdot (i + j)$
10. $(3i - 3j) \cdot (2i + j)$
11. $i \cdot j$
12. $2i \cdot 3i$
13. $\langle 2, 7 \rangle \cdot \langle -1, -4 \rangle$
14. $\langle -5, 3 \rangle \cdot \langle 0, 7 \rangle$

Use the vectors $u = \langle -2, 1 \rangle$, $v = \langle 3, 3 \rangle$, *and* $w = \langle 4, -1 \rangle$ *to find the indicated quantity.*

15. $u \cdot u$
16. $v \cdot w$
17. $u \cdot (v + w)$
18. $(2v + w) \cdot u$
19. $(4w + v) \cdot u$
20. $3w \cdot (2v - u)$

Use the dot product to find the magnitude of **u.**

21. $u = \langle -3, 5 \rangle$
22. $u = \langle -5, 1 \rangle$
23. $u = 3i - j$
24. $u = -4i + 2j$
25. $u = 8i$
26. $u = -10j$

Find the angle between the given vectors in degrees. Round to two decimal places when necessary.

27. $\langle 2, 2 \rangle$; $\langle 5, 0 \rangle$
28. $\langle \sqrt{3}, 1 \rangle$; $\langle 3, 3 \rangle$
29. $\langle \sqrt{3}, -1 \rangle$; $\langle -1, \sqrt{3} \rangle$
30. $\langle 2, 3 \rangle$; $\langle -3, 2 \rangle$
31. $3i - j$; $3i + j$
32. $3i - 4j$; $5i + 12j$

Determine whether the given vectors are orthogonal.

33. $\langle 2, 3 \rangle$; $\langle -3, 2 \rangle$
34. $\langle 2, 3 \rangle$; $\langle 3, 2 \rangle$
35. $5i + j$; $i + j$
36. $6i - 2j$; $i + 3j$
37. i; j
38. $-i$; $3i$

Find the horizontal and vertical components of the given force and angle with the x-axis.

39. 3.5-pound force; $\theta = 30°$
40. 4.5-pound force; $\theta = 30°$
41. 7-pound force; $\theta = 60°$
42. 2-pound force; $\theta = 60°$
43. 6-pound force; $\theta = 45°$
44. 10-pound force; $\theta = 45°$

For each set of vectors,
(a) *Find the component of the first vector in the direction*
of the second vector.
(b) *Find the vector projection of the first vector on the*
second vector.

45. $\langle 3, 4 \rangle$; $\langle 5, 12 \rangle$

46. $\langle 1, 1 \rangle$; $\langle 3, 2 \rangle$

47. $6\mathbf{i} + 8\mathbf{j}$; $4\mathbf{i} - 3\mathbf{j}$

48. $\mathbf{i}$; $\mathbf{i} + \mathbf{j}$

49. $\langle -3, 5 \rangle$; $\left\langle \dfrac{\sqrt{2}}{2}, \dfrac{\sqrt{2}}{2} \right\rangle$

50. $\langle 11, 3 \rangle$; $\langle -3, -2 \rangle$

Find the work done by the force **F** *in moving the object*
from P to Q. Use foot-pounds as the unit of work.

51. $\mathbf{F} = 4\mathbf{i} + 5\mathbf{j}$; $P(0, 0)$; $Q(3, 8)$

52. $\mathbf{F} = 40\mathbf{i} + 50\mathbf{j}$; $P(-1, 1)$; $Q(20, 1)$

53. $\mathbf{F} = 10\mathbf{i} - 3\mathbf{j}$; $P(2, 3)$; $Q(6, 2)$

54. $\mathbf{F} = -4\mathbf{i} + 20\mathbf{j}$; $P(0, 12)$; $Q(5, 25)$

Applications

55. Work A force of 45 pounds in the direction of 60° above the horizontal is required to slide a couch across the floor. Find the work done if the couch is pushed 20 feet.

56. Work A child's wagon is pulled by exerting a force of 26 pounds on a handle that makes a 25° angle with the horizontal. Find the work done to pull the wagon 60 feet.

57. Work The force $\mathbf{F} = 3\mathbf{i} - 8\mathbf{j}$ moves an object 10 feet along the positive x-axis. Find the work done using the pound as the unit of force.

58. Work A man pulls a wagon by exerting a force of 22 pounds on the handle, which is at an angle of 45°. If the wagon is pulled 300 feet, what is the amount of work done?

59. Distance Using the dot product, it has been determined that the angle subtended by two position vectors for Atlanta and Dallas is 0.1819 radian. Use the fact that the radius of the earth is 3,959 miles to obtain the approximate distance between the two cities.

60. Distance Using the dot product, it has determined that the angle subtended by two position vectors for London and New York City is 0.8767 radian. Use the fact that the radius of the earth is 3,959 miles to obtain the approximate distance between the two cities.

61. Vector projection A boat weighing 200 pounds sits on a ramp inclined at 30°. What force is required to keep the boat from sliding down the ramp? (*Hint:* Project the force of the boat onto a unit vector in the direction of the ramp.)

62. Vector projection A sled weighing 22 pounds sits on a snow slope inclined at 30°. What force is required to keep the sled from sliding down the ramp? (*Hint:* Project the force of the sled onto a unit vector in the direction of the snow slope.)

Discovery and Writing

63. Find an example to illustrate that $|\mathbf{u} + \mathbf{v}| \neq |\mathbf{u}| + |\mathbf{v}|$.

64. Find an example to support the Distributive Property $(c + d)\mathbf{v} = a\mathbf{v} + b\mathbf{v}$.

65. Find an example to support the Distributive Property $c(\mathbf{v} + \mathbf{w}) = c\mathbf{v} + c\mathbf{w}$.

66. Find an example to support the Associative Property $k(\mathbf{v} \cdot \mathbf{w}) = (k\mathbf{v}) \cdot \mathbf{w}$.

67. Prove the theorem that if $\mathbf{v}$ is any vector, then $\mathbf{v} \cdot \mathbf{v} = |\mathbf{v}|^2$.

68. Prove the theorem that if $|\mathbf{v}| = |\mathbf{w}|$, then $(\mathbf{v} + \mathbf{w}) \cdot (\mathbf{v} - \mathbf{w}) = 0$.

CHAPTER REVIEW

SECTION 7.1   Polar Coordinates

Definitions and Concepts	Examples
Plotting points: To plot the point $P(r, \theta)$ when $r > 0$: • draw angle θ in standard position and • count r units along the terminal side of θ.	See Example 1(a) and 1(c), page 661

Definitions and Concepts	Examples
Plotting points: To plot $P(r, \theta)$ when $r < 0$: • draw angle θ in standard position and • count $\lvert r \rvert$ units along the extension of the terminal side of θ, but in the opposite direction.	See Example 1(b) and 1(d), page 661
Multiple sets of polar coordinates: For a point with polar coordinates (r, θ) and any integer n, the polar coordinates $(r, \theta + 2\pi n)$ and $(-r, \theta + \pi + 2\pi n)$ represent the same point in the polar coordinate system.	See Example 2, page 662
Converting from polar to rectangular coordinates: $\quad x = r \cos \theta \qquad y = r \sin \theta$	See Example 4, pages 663–664
Converting rectangular to polar coordinates with $r > 0$ and $0 \le \theta < 2\pi$: $\quad r^2 = x^2 + y^2,\ \tan \theta = \dfrac{y}{x},\ \text{and}\ \theta' = \tan^{-1}\left\lvert \dfrac{y}{x} \right\rvert,\ x \ne 0$	See Example 5, pages 665–666
Converting equations from rectangular to polar form: To convert an equation written in rectangular form to an equation in polar form, you can use the definitions $x = r \cos \theta$, $y = r \sin \theta$, and $x^2 + y^2 = r^2$, $\tan \theta = \dfrac{y}{x}$.	See Example 7, pages 668–669
Converting equations from polar form to rectangular form: There are two techniques that are often employed in converting a polar equation to a rectangular equation: **1.** Multiply both sides of the equation by r or a power of r. **2.** Square both sides of the equation. **3.** If of the form $\theta = n$, where n is any real number, take the tangent of both sides. **4.** Use the definitions of the trigonometric functions.	See Example 8, pages 669–670
If $P(r_1, \theta_1)$ and $Q(r_2, \theta_2)$ are two points given in polar coordinates, then the distance between them is $d = \sqrt{r_1{}^2 + r_2{}^2 - 2r_1 r_2 \cos(\theta_2 - \theta_1)}$.	See Example 9, page 670

EXERCISES

Refer to the figure and give a possible set of polar coordinates for the point with the given conditions.

(a) $r > 0, 0 \le \theta \le 2\pi$ *(b)* $r > 0, -2\pi \le \theta \le 0$
(c) $r < 0, 0 \le \theta \le 2\pi$

1. A **2.** B **3.** C **4.** D

The polar coordinates of a point are given. Find the rectangular coordinates of the point. Round to two decimal places when necessary.

5. $P\left(4, \dfrac{7\pi}{6}\right)$ **6.** $P\left(-3, -\dfrac{3\pi}{4}\right)$

7. $P(8, 330°)$ **8.** $P(-5, -240°)$

The rectangular coordinates of a point are given. Find two sets of polar coordinates for the point, one with $r > 0$ and $0 \le \theta < 2\pi$, and the other with $r < 0$ and $0 \le \theta < 2\pi$.

9. $R(-4, 4)$ **10.** $R(4, 0)$

Convert each equation written in rectangular form to an equation in polar form.

11. $3y - 2x = 4$ **12.** $x^2 + y^2 = 49$

13. $7xy = 12$ **14.** $x^2 = y^2 - 3y$

Convert each equation written in polar form to an equation in rectangular form.

15. $r + 2\cos\theta = -2\sin\theta$

16. $r = \dfrac{3}{1 - \sin\theta}$

17. $r(2\cos\theta + \sin\theta) = 2$

18. $r = 2\sec\theta$

Find the distance between the points whose polar coordinates are given. Round to three decimal places.

19. $P\left(3, \dfrac{7\pi}{6}\right)$; $Q\left(-2, \dfrac{\pi}{4}\right)$

20. $P(-1, 135°)$; $Q(3, 330°)$

SECTION **7.2**   Polar Equations and Graphs

Definitions and Concepts	**Examples**
Graph polar equations using rectangular coordinates: The graph of the polar equation $r = a$ in rectangular coordinates is a circle with its center at the origin $(0, 0)$ and a radius of a.	See Example 1, page 674
Graph polar equations using rectangular coordinates: The graph of the polar equation $r = a\sin\theta$ in rectangular coordinates is a circle with its center at $\left(0, \frac{a}{2}\right)$ and a radius of $\frac{a}{2}$. The graph of the polar equation $r = a\cos\theta$ in rectangular coordinates is a circle with its center at $\left(\frac{a}{2}, 0\right)$ and a radius of $\frac{a}{2}$.	See Example 2, page 675
Graph polar equations using rectangular coordinates: The graph of the polar equation $r\sin\theta = a$ in rectangular coordinates is the horizontal line $y = a$. The graph of the polar equation $r\cos\theta = a$ in rectangular coordinates is the vertical line $x = a$.	See Example 3, pages 675–676
Cardioid: A cardioid is the curve given by an equation in one of the following forms, with real number $a > 0$: $\quad r = a(1 \pm \cos\theta) = a \pm a\cos\theta \quad$ or $\quad r = a(1 \pm \sin\theta) = a \pm a\sin\theta$.	See Example 4, pages 679–680
Limaçon: A limaçon is the curve given by an equation in one of the following forms, with real numbers $a > 0$ and $b > 0$: $\quad r = a \pm b\sin\theta \quad$ or $\quad r = a \pm b\cos\theta$. If $a > b$, the graph is a limaçon without an inner loop. If $a < b$, the graph is a limaçon with an inner loop.	See Example 5, pages 680–681
Rose: A rose curve is the shape of the graph given by an equation in one of the following forms, with nonzero real number a and an integer $n \geq 2$. $\quad r = a\sin n\theta \quad$ or $\quad r = a\cos n\theta$. If n is even, the rose will have $2n$ petals. If n is odd, the rose will have n petals. The petal of the rose is a units long.	See Example 7, pages 682–683
Lemniscate: A lemniscate is the curve given by an equation in one of the following forms, with real number $a \neq 0$: $\quad r^2 = a^2\sin 2\theta \quad$ or $\quad r^2 = a^2\cos 2\theta$. The graph of $r^2 = a^2\sin 2\theta$ is symmetric with respect to the pole; the graph of $r^2 = a^2\cos 2\theta$ is symmetric with respect to the polar axis, the pole, and $\theta = \frac{\pi}{2}$.	See Example 8, page 683

EXERCISES

Convert each polar equation to an equation using rect-
angular coordinates and then graph it on a rectangular
coordinate system.

21. $r = -3$

22. $\theta = \dfrac{\pi}{4}$

23. $-2r \cos \theta = 4$

24. $r^2 = 64$

Graph each polar equation.

25. $r = -4 \sin \theta$

26. $r = 6 - 6 \cos \theta$

27. $r = 5 + 4 \cos \theta$

28. $r = 4 \sin 3\theta$

Convert each equation to a polar equation and then graph
it on a polar coordinate system.

29. $x^2 + y^2 = 100$

30. $x^2 + y^2 = -5x$

State a polar equation that will produce the given graph.

31.

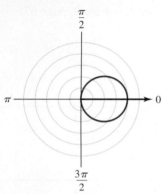

32.

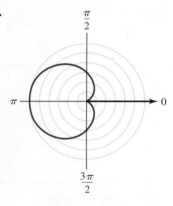

33.

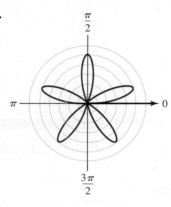

34.

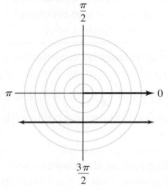

SECTION **7.3**   The Complex Plane; Polar Form of Complex Numbers;
DeMoivre's Theorem

Definitions and Concepts	Examples
Absolute value of a complex number: If $z = a + bi$ is a complex number, the absolute value of z is $\quad \lvert z \rvert = \sqrt{a^2 + b^2}$. This value is also called the **magnitude**.	See Example 1, page 692

Definitions and Concepts	Examples
Polar form of a complex number: If $r = \|a + bi\|$ and $0 \le \theta < 2\pi$, the complex number $z = a + bi$ can be written in the **polar form** $\quad z = r(\cos\theta + i\sin\theta).$	See Example 2, page 693
Product of complex numbers in polar form: If $z_1 = r_1(\cos\theta_1 + i\sin\theta_1)$ and $z_2 = r_2(\cos\theta_2 + i\sin\theta_2)$ are two complex numbers, then the product of the numbers is given by $\quad z_1 z_2 = r_1 r_2 [\cos(\theta_1 + \theta_2) + i\sin(\theta_1 + \theta_2)].$	See Example 4(a), page 695
Quotient of complex numbers in polar form: If $z_1 = r_1(\cos\theta_1 + i\sin\theta_1)$ and $z_2 = r_2(\cos\theta_2 + i\sin\theta_2)$ are two complex numbers, then the quotient of the numbers is given by $\quad \dfrac{z_1}{z_2} = \dfrac{r_1}{r_2}[\cos(\theta_1 - \theta_2) + i\sin(\theta_1 - \theta_2)].$	See Example 4(b), page 695
DeMoivre's Theorem: If n is a positive integer and $z = r(\cos\theta + i\sin\theta)$ is a complex number in polar form, then $\quad z^n = r^n(\cos n\theta + i\sin n\theta).$	See Example 5, page 696
Finding complex roots using DeMoivre's Theorem: Let $w = r(\cos\theta + i\sin\theta)$ be a nonzero complex number. For $n \ge 2$ and $k = 0, 1, 2, \ldots, n - 1$, w has n distinct complex nth roots found with the formula $\quad z_k = \sqrt[n]{r}\left(\cos\dfrac{\theta + 2\pi k}{n} + i\sin\dfrac{\theta + 2\pi k}{n}\right)$ (radian measure) or $\quad z_k = \sqrt[n]{r}\left(\cos\dfrac{\theta + 360°k}{n} + i\sin\dfrac{\theta + 360°k}{n}\right)$ (degree measure).	See Example 7, pages 697–698

EXERCISES

Find the absolute value of each complex number and plot the point corresponding to the number in the complex plane.

35. $4 + 7i$

36. $9i$

Write each complex number in polar form using radian measure for the argument.

37. $5 - 5i$

38. 9

Write each complex number in rectangular form. Round to two decimal places when needed.

39. $4(\cos 135° + i\sin 135°)$

40. $-5(\cos 3.2 + i\sin 3.2)$

For the given values, find the product $z_1 z_2$ and the quotient $\dfrac{z_1}{z_2}$. Leave the number in polar form.

41. $z_1 = 2(\cos 210° + i\sin 210°)$; $z_2 = 3(\cos 0° + i\sin 0°)$

42. $z_1 = 2 - 2i$; $z_2 = 1 - i\sqrt{3}$

Find each power. Leave all answers in standard $a + bi$ form.

43. $[2(\cos 48° + i\sin 48°)]^5$

44. $(2 + i)^4$

Find the indicated nth root of each expression in $a + bi$ form.

45. A cube root of $8(\cos 270° + i\sin 270°)$

46. A fifth root of $\cos 225° + i\sin 225°$

Find all the complex roots. Leave the answers in polar form with the argument in degrees.

47. The cube roots of $-i$ **48.** The fourth roots of 81

SECTION **7.4**   Vectors

Definitions and Concepts	Examples
If k is a real number and $\mathbf{v}$ is a vector, then the product $k\mathbf{v}$ is also a vector.	See Example 1, page 702
Position vector: If a vector $\mathbf{v}$ has the initial point $P = (x_1, y_1)$ and the terminal point $Q = (x_2, y_2)$, the components of the vector are $a = x_2 - x_1$ and $b = y_2 - y_1$, so that $\mathbf{v} = \langle x_2 - x_1, y_2 - y_1 \rangle$ and the **position vector** is $\mathbf{v} = \langle a, b \rangle$.	See Example 2, page 703
Equality of vectors: For position vectors $\mathbf{v} = \langle a, b \rangle$ and $\mathbf{w} = \langle c, d \rangle$, $\mathbf{v} = \mathbf{w}$ if and only if $a = c$ and $b = d$. **Magnitude of vectors:** The **magnitude** (length) of a vector $\mathbf{v} = \langle a, b \rangle$ is $|\mathbf{v}| = \sqrt{a^2 + b^2}$. **Scalar multiplication:** If k is a scalar and $\mathbf{v}$ is the vector $\langle a, b \rangle$, then $k\mathbf{v} = k\langle a, b \rangle = \langle ka, kb \rangle$.	See Example 3, page 704
Addition and subtraction of vectors: If $\mathbf{v} = \langle a, b \rangle$ and $\mathbf{w} = \langle c, d \rangle$, then $\mathbf{v} + \mathbf{w} = \langle a + c, b + d \rangle$ and $\mathbf{v} - \mathbf{w} = \langle a - c, b - d \rangle$.	See Example 4, page 704
Unit vector: The vector $\mathbf{u} = \dfrac{\mathbf{v}}{|\mathbf{v}|}$ is called the **unit vector in the direction of v.** **Unit coordinate vectors:** The vectors $\mathbf{i} = \langle 1, 0 \rangle$ and $\mathbf{j} = \langle 0, 1 \rangle$ are called **unit coordinate vectors.** Any vector $\mathbf{v} = \langle x, y \rangle$ is **resolved** into its x or **horizontal component** $x\mathbf{i}$ and its y or **vertical component** $y\mathbf{j}$ when it is written in the form $\mathbf{v} = x\mathbf{i} + y\mathbf{j}$.	See Example 5, page 706
Horizontal and vertical components of a vector: For vector $\mathbf{v} = \langle a, b \rangle = a\mathbf{i} + b\mathbf{j}$, with magnitude $|\mathbf{v}|$ and direction θ, $a = |\mathbf{v}|\cos\theta$ and $b = |\mathbf{v}|\sin\theta$. Therefore, the vector can be expressed as $\mathbf{v} = |\mathbf{v}|(\cos\theta\,\mathbf{i} + \sin\theta\,\mathbf{j})$.	See Example 6, page 707

EXERCISES

Find the position vector of vector v if the initial point is P and the terminal point is Q.

49. $P(2, 5)$; $Q(6, -1)$ **50.** $P(-3, 2)$; $Q(4, 10)$

Use the given vectors to find:
(a) $|\mathbf{v}|$ *(b)* $-4|\mathbf{u}|$ *(c)* $|\mathbf{v} - 2\mathbf{u}|$ *(d)* $|2\mathbf{u}| - |\mathbf{v}|$

51. $\mathbf{u} = \langle 3, 0 \rangle$; $\mathbf{v} = \langle 2, -8 \rangle$ **52.** $\mathbf{u} = -4\mathbf{i} + 5\mathbf{j}$; $\mathbf{v} = \mathbf{i} - 3\mathbf{j}$

Find $\mathbf{u} - 3\mathbf{v}$, $2\mathbf{u} + 4\mathbf{w}$, $5\mathbf{v} - \mathbf{u} + 6\mathbf{w}$, *and* $5\mathbf{w} + 3\mathbf{u} - \mathbf{v}$ *for the given vectors.*

53. $\mathbf{u} = \langle 4, 0 \rangle$; $\mathbf{v} = \langle -2, 5 \rangle$; $\mathbf{w} = \langle 3, -1 \rangle$

54. $\mathbf{u} = 5\mathbf{i} - 2\mathbf{j}$; $\mathbf{v} = -\mathbf{i} + 3\mathbf{j}$; $\mathbf{w} = 2\mathbf{j}$

Find the unit vector in the same direction as v.

55. $\mathbf{v} = 6\mathbf{i} - 4\mathbf{j}$ **56.** $\mathbf{v} = -7\mathbf{j}$

Resolve each vector into its horizontal and vertical components by writing it in $a\mathbf{i} + b\mathbf{j}$ form.

57. $\langle 4, 6 \rangle + \langle -3, -2 \rangle$ **58.** $|\mathbf{v}| = 12$; $\theta = 60°$

Find the magnitude and direction (in degrees) of the vector. Round to two decimal places when necessary.

59. $\mathbf{v} = \langle 5, -5 \rangle$ **60.** $\mathbf{v} = \mathbf{i} - 7\mathbf{j}$

Use the Law of Cosines to find the angle θ between the two vectors, with $0° \leq \theta \leq 180°$. Round to the nearest tenth.

61. $\mathbf{u} = \langle 5, 1 \rangle$; $\mathbf{v} = \langle -4, 4 \rangle$

62. $\mathbf{u} = -2\mathbf{i} + 3\mathbf{j}$; $\mathbf{v} = 2\mathbf{i} - 5\mathbf{j}$

Applications

63. Force A force of 25 pounds is required to push an 80-pound box up a ramp. What angle does the ramp make with the horizontal?

64. Speed of boat A boat capable of a speed of 6 mph in still water attempts to go directly across a river. As the boat crosses the river, it drifts 30° from its intended path. How strong is the current? What is the effective speed of the boat?

65. Angle of current Laura can row a boat 0.5 mph in still water. She attempts to row straight across a river that has a current of 1 mph. If she must row for 2 hours to cross the river, by what angle is the current pushing her off her intended path?

66. Force A board will break if it is subjected to a force greater than 350 pounds. Will the board hold the 450-pound piano shown in the figure as it slides up the board and into a truck?

67. Force A 201-pound force is directed due east. What force directed due north is needed to produce a resultant force of 301 pounds? What angle is formed by the vectors representing the 201- and 301-pound forces?

68. Navigation An airplane is headed on a bearing of S 6° E at an airspeed of 240 kilometers per hour. A wind of 50 kilometers per hour is blowing in a direction of S 65° W. Find the ground speed and resultant bearing of the plane.

SECTION **7.5**   Dot Product

Definitions and Concepts	**Examples**
Dot product of two vectors: Let $\mathbf{v} = \langle a, b \rangle = a\mathbf{i} + b\mathbf{j}$ and $\mathbf{w} = \langle c, d \rangle = c\mathbf{i} + d\mathbf{j}$. The **dot product** of vectors $\mathbf{v}$ and $\mathbf{w}$ is the *scalar* $\mathbf{v} \cdot \mathbf{w} = ac + bd$.	See Example 1, page 715
Properties of the dot product: For all vectors $\mathbf{u}$, $\mathbf{v}$, and $\mathbf{w}$, and real numbers k: 1. $\mathbf{u} \cdot \mathbf{u} = \|\mathbf{u}\|^2$ 4. $(\mathbf{u} + \mathbf{v}) \cdot \mathbf{w} = \mathbf{u} \cdot \mathbf{w} + \mathbf{v} \cdot \mathbf{w}$ 2. $\mathbf{u} \cdot \mathbf{v} = \mathbf{v} \cdot \mathbf{u}$ 5. $k(\mathbf{u} \cdot \mathbf{v}) = (k\mathbf{u}) \cdot \mathbf{v} = \mathbf{u} \cdot (k\mathbf{v})$ 3. $\mathbf{u} \cdot (\mathbf{v} + \mathbf{w}) = \mathbf{u} \cdot \mathbf{v} + \mathbf{u} \cdot \mathbf{w}$ 6. $\mathbf{u} \cdot \mathbf{0} = 0$	See Example 2, page 716
Dot product and the angle between two vectors: If θ is the angle between $\mathbf{v}$ and $\mathbf{w}$, the dot product of vectors $\mathbf{v}$ and $\mathbf{w}$ is the scalar $$\mathbf{v} \cdot \mathbf{w} = \|\mathbf{v}\|\,\|\mathbf{w}\| \cos \theta, \text{ where } 0° \le \theta \le 180°.$$ To find θ, solve the equation $$\cos \theta = \frac{\mathbf{v} \cdot \mathbf{w}}{\|\mathbf{v}\|\,\|\mathbf{w}\|} \text{ for } \theta.$$	See Example 3, page 717
Orthogonal vectors: Two nonzero vectors are orthogonal if and only if their dot product is zero.	See Example 4, page 717
Scalar and vector projections of w on v: If $\mathbf{v} \neq \mathbf{0}$, then The scalar projection of $\mathbf{w}$ on $\mathbf{v} = \mathbf{comp_v\, w} = \dfrac{\mathbf{w} \cdot \mathbf{v}}{\|\mathbf{v}\|}$. The vector projection of $\mathbf{w}$ on $\mathbf{v} = \mathbf{proj_v\, w} = \dfrac{\mathbf{w} \cdot \mathbf{v}}{\mathbf{v} \cdot \mathbf{v}}\,\mathbf{v}$.	See Example 6, pages 719–720

Definitions and Concepts	Examples
Work: The work W done by a force $\mathbf{F}$ in moving along a vector $\mathbf{d}$ is $W = (\text{comp}_{\mathbf{d}}\mathbf{F})|\mathbf{d}| = \dfrac{\mathbf{F} \cdot \mathbf{d}}{|\mathbf{d}|}|\mathbf{d}| = \mathbf{F} \cdot \mathbf{d}.$	See Example 7, page 721

EXERCISES

Find each dot product.

69. $\langle -4, 3 \rangle \cdot \langle 2, 0 \rangle$ **70.** $(2\mathbf{i} - \mathbf{j}) \cdot (-7\mathbf{i} + 3\mathbf{j})$

Use the vectors $\mathbf{u} = \langle 3, -1 \rangle$, $\mathbf{v} = \langle 2, 2 \rangle$, and $\mathbf{w} = \langle -5, 2 \rangle$ to find the indicated quantity.

71. $\mathbf{u} \cdot (\mathbf{v} + 2\mathbf{w})$ **72.** $2\mathbf{w} \cdot (4\mathbf{v} - \mathbf{u})$

Use the dot product to find the magnitude of $\mathbf{u}$.

73. $\langle 4, -2 \rangle$ **74.** $-5\mathbf{i} + 3\mathbf{j}$

Find the angle between the given vectors in degrees. Round to two decimal places when necessary.

75. $\langle -4, 4 \rangle$; $\langle 0, -5 \rangle$ **76.** $4\mathbf{i} - 2\mathbf{j}$; $-3\mathbf{i} + \mathbf{j}$

Determine whether the given vectors are orthogonal.

77. $\langle 1, 2 \rangle$; $\langle -6, 3 \rangle$ **78.** $3\mathbf{i} + 4\mathbf{j}$; $6\mathbf{i} + 8\mathbf{j}$

Find the horizontal and vertical components of the given force and angle with the x-axis.

79. 8-pound force; $\theta = 45°$

80. 3-pound force; $\theta = 210°$

For each set of vectors,
(a) Find the component of the first vector in the direction of the second vector.
(b) Find the vector projection of the first vector on the second vector.

81. $\langle 2, 5 \rangle$; $\langle -3, 1 \rangle$ **82.** $3\mathbf{i} - \mathbf{j}$; $2\mathbf{j}$

Find the work done by the force $\mathbf{F}$ in moving the object from P to Q. Use foot-pounds as the unit of work.

83. $\mathbf{F} = 2\mathbf{i} + 4\mathbf{j}$; $P(1, 2)$, $Q(5, 12)$

84. $\mathbf{F} = -5\mathbf{i} + 20\mathbf{j}$; $P(15, 0)$, $Q(0, 20)$

Applications

85. Force A 200-pound trailer sits on a ramp inclined at 30°. What force is required to keep the trailer from rolling down the ramp?

86. Work To close a sliding door on a storage building, a person pulls a rope with a constant force of 50 pounds at a constant angle of 60°. Find the work done in moving the door 12 feet.

CHAPTER TEST

1. Find the rectangular coordinates of the point that has polar coordinates $\left(5, \dfrac{3\pi}{4}\right)$.

2. Find two sets of polar coordinates for the point that has rectangular coordinates $\left(-2\sqrt{3}, 2\right)$, one with $r > 0$ and $0 \le \theta < 2\pi$, and the other with $r < 0$ and $0 \le \theta < 2\pi$.

3. Convert the rectangular equation $x^2 = 8y$ to an equation in polar form.

4. Convert the polar equation $r = 4\cos\theta + 3\sin\theta$ to an equation in rectangular form.

5. Find the distance between the points whose polar coordinates are $P\left(-4, \dfrac{5\pi}{4}\right)$ and $Q\left(3, \dfrac{2\pi}{3}\right)$. Round to three decimal places.

6. Graph each polar equation.

  a. $r = 3 - 4\sin\theta$ **b.** $r = 5\cos 3\theta$

  c. $r = 6\cos\theta$ **d.** $-4r\sin\theta = 12$

7. State an equation that will produce the given graph.

  a.

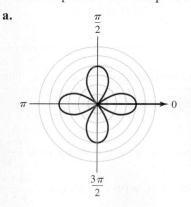

b.

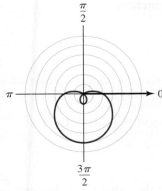

c.

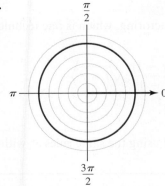

8. Write the complex number $6(\cos 210° + i \sin 210°)$ in rectangular form.

9. Write the complex number $-1 - i\sqrt{3}$ in polar form.

10. If $z_1 = 4\sqrt{3} - 4i$ and $z_2 = 8i$, write z_1 and z_2 in polar form and then find the product $z_1 z_2$ and the quotient $\dfrac{z_1}{z_2}$ in $a + bi$ form.

11. Find $\left(2\sqrt{3} + 2i\right)^5$ using DeMoivre's Theorem. Put your answer in $a + bi$ form.

12. Find the fourth roots of $-81i$.

13. For $\mathbf{u} = 3\mathbf{i} - 2\mathbf{j}$, $\mathbf{v} = 4\mathbf{i}$, and $\mathbf{w} = -2\mathbf{i} + 5\mathbf{j}$, find each of the following:
   a. $|3\mathbf{v}|$ **b.** $\mathbf{u} + 3\mathbf{w}$
   c. $4\mathbf{u} + 2\mathbf{v} - \mathbf{w}$ **d.** $\mathbf{v} \cdot (2\mathbf{w} - 3\mathbf{u})$

14. Find the magnitude and direction (in degrees) of the vector $\mathbf{v} = \mathbf{i} - \sqrt{3}\mathbf{j}$.

15. Find the angle between the vectors $\langle -3, 3 \rangle$ and $\langle 1, -1 \rangle$ in degrees.

16. A plane is headed due east at a speed of 425 mph relative to the air. It encounters a wind blowing due north at 40 mph.
   a. Find the true velocity vector of the plane.
   b. Find the true speed and direction of the plane.

17. Find the magnitude of the force necessary to keep a 3,000-pound car from sliding down a road inclined at 5.6° from the horizontal. Round to the nearest pound.

18. Are the vectors $\langle 2, 1 \rangle$ and $\langle -1, 2 \rangle$ orthogonal?

19. For the vectors $\mathbf{v} = -3\mathbf{i} + 4\mathbf{j}$, and $\mathbf{w} = 2\mathbf{i} + 5\mathbf{j}$ find:
   a. The component of $\mathbf{w}$ on $\mathbf{v}$.
   b. The vector projection of $\mathbf{w}$ on $\mathbf{v}$.

20. Find the work (in foot-pounds) done by the force $\mathbf{F} = 4\mathbf{i} + 5\mathbf{j}$ in moving an object from $P(2, -4)$ to $Q(10, 50)$.

 LOOKING AHEAD TO **CALCULUS**

1. Using Vectors to Define Vector Functions
   • Connect vectors and their properties to functions—Section 7.4

1. Using Vectors to Define Vector Functions

Throughout algebra and trigonometry, functions are an integral part of the courses. Generally, a function takes a value in a domain and assigns it a value in a range. A **vector function** is a function whose domain is the set of real numbers and whose range is a set of vectors. Each part of the vector function $\mathbf{r}(t) = \langle f(t), g(t) \rangle = f(t)\mathbf{i} + g(t)\mathbf{j}$ is called a **component function**.

For example, if $\mathbf{r}(t) = \langle 3t^2 + 1, e^t \rangle$, the **component functions** are $f(t) = 3t^2 + 1$ and $g(t) = e^t$. We have previously studied the topics of limits and derivatives; these operations can be performed on vector functions.

EXAMPLE 1   **Determining the Limit of a Vector Valued Function**

Find each limit.

a. $\lim\limits_{t \to 0} \mathbf{r}(t)$ where $\mathbf{r}(t) = (2t^2 + t + 1)\mathbf{i} + \cos t\mathbf{j}$.

Since both of the component functions are defined over all the real numbers, we can determine the limit by evaluating each function at $t = 0$.

$$\lim_{t \to 0} \langle 2t^2 + t + 1, \cos t \rangle = \langle 2(0)^2 + 0 + 1, \cos 0 \rangle = \langle 1, 1 \rangle = \mathbf{i} + \mathbf{j}$$

b. $\lim\limits_{t \to 1} \mathbf{r}(t)$ where $\mathbf{r}(t) = \dfrac{t^2 - t}{t - 1}\mathbf{i} + \sqrt{t + 8}\mathbf{j}$.

We cannot just evaluate the limit by inserting the value $t = 1$, because $\dfrac{t^2 - t}{t - 1}$ is not defined at that value.

However, the expression can be reduced by factoring, which is one technique of determining limits.

$$\lim_{t \to 1} \mathbf{r}(t) = \lim_{t \to 1} \left\langle \frac{t^2 - t}{t - 1}, \sqrt{t + 8} \right\rangle = \lim_{t \to 1} \left\langle \frac{t(t - 1)}{t - 1}, \sqrt{t + 8} \right\rangle$$

$$= \lim_{t \to 1} \langle t, \sqrt{t + 8} \rangle = \langle 1, 3 \rangle = \mathbf{i} + 3\mathbf{j}$$

Derivatives of vector functions can be found using the same rules as with other functions.

EXAMPLE 2   **Determining the Derivative of a Vector Valued Function**

Find $\mathbf{r}'(t)$ for each vector function.

a. $\mathbf{r}(t) = \langle t^3, \cos t \rangle$

Recalling the derivative rules covered previously, we have

$$\mathbf{r}'(t) = \langle 3t^2, -\sin t \rangle.$$

b. $\mathbf{r}(t) = \langle \sin t, 2\sqrt{t} \rangle$

Again, we can apply simple rules of derivatives.

$$\mathbf{r}'(t) = \langle \cos t, t^{-1/2} \rangle = \left\langle \cos t, \frac{1}{t^{1/2}} \right\rangle$$

Connect to Calculus:
Using Vectors to Define Vector Functions

1. Find each limit.
   a. $\lim\limits_{t \to \frac{\pi}{4}} \langle \sin t, 2 \cos t \rangle$

   b. $\lim\limits_{t \to 2} \langle e^{2t}, \ln e^t \rangle$

2. Find $\mathbf{r}'(t)$.
   a. $\mathbf{r}(t) = 4t^3\mathbf{i} - \sqrt{t}\mathbf{j}$

   b. $\mathbf{r}(t) = 3 \sin 4t\mathbf{i} + 4 \cos t\mathbf{j}$

Linear Systems

8

CAREERS AND MATHEMATICS: Cryptanalyst

Cryptanalysts are essential to the implementation of protection that corporations and private citizens use to keep hackers out of important data systems. They also play an important part in the safety of cities, states, and countries by analyzing and deciphering secret coding systems used by terrorists and other enemies. With increased reliance on the Internet to conduct business, store data, and pass information between groups, the job of a cryptanalyst is becoming increasingly more important for ensuring both private and governmental security of data.

Education and Mathematics Required

- Most cryptanalysts are required to have at least an undergraduate degree in mathematics or computer science. Many will also have a graduate degree in mathematics, with a PhD required for those wishing to be employed by a research facility or university.
- College Algebra; Trigonometry; Calculus I, II, and III; Linear Algebra; Differential and Partial Differential Equations; Elementary Number Theory; Introduction to Real Analysis; Analysis I and II; Methods of Complex Analysis; and Mathematical Cryptography are math courses required.

How Cryptanalysts Use Math and Who Employs Them

- Among the many ways that cryptanalysts use mathematics are encoding and encrypting systems and databases, devising systems to prevent hackers from stealing consumer data, and consulting with business and industry to solve problems related to security.
- Cryptanalysts are employed in all levels of government, including intelligence agencies and special services, universities, financial institutions, insurance companies, telecommunications companies, computer design firms, consulting firms, and science and engineering firms.

Career Outlook and Salary

- Employment of cryptanalysts is expected to grow by 22% through 2018, which is much faster than the average for all occupations.
- The median annual wages of cryptanalysts are approximately $101,645.

For more information see http://www.bls.gov/oco.

In this chapter, we learn to solve systems of equations—sets of several equations, most with more than one variable. One method of solution uses matrices, an important tool in mathematics and its applications.

*Available online at http://www.cengagebrain.com.

8.1 Systems of Linear Equations

In this section, we will learn to

1. Solve systems using the graphing method.
2. Solve systems using the substitution method.
3. Solve systems using the addition method.
4. Solve systems with infinitely many solutions.
5. Solve inconsistent systems.
6. Solve systems involving three equations in three variables.
7. Solve applications involving systems of equations.

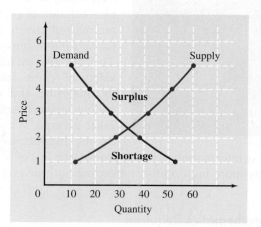

In economics we often talk about the law of supply and demand. As the illustration shows, few units of a product will be in demand when the price is high. However, more units will be in demand when the price is low. As the illustration also shows, few units of product will be supplied when the price is low. However, more product will be supplied when the price is high.

In a free economy, the intersection of the graph of the supply function and graph of the demand function will be the market price of the product. To see that this is true, suppose that the price of a product is $3. At this price, fewer than 30 units of product would be in demand, but over 40 units would be supplied. Since supply is greater than demand, the price would come down.

At a price of $2, somewhat fewer than 40 units would be in demand, but fewer than 30 units would be supplied. Since demand is greater than supply, the price would go up.

Only where the graphs cross are the demand and supply in equilibrium.

In this section, we will begin to discuss how to find the common solution of two equations that occur simultaneously.

Equations with two variables have infinitely many solutions. For example, the tables shown give a few of the solutions of $x + y = 5$ and $x - y = 1$.

$x + y = 5$			$x - y = 1$	
x	y		x	y
1	4		8	7
2	3		3	2
3	2		1	0
5	0		0	−1

Only the pair $x = 3$ and $y = 2$ satisfies both equations. The pair of equations

$$\begin{cases} x + y = 5 \\ x - y = 1 \end{cases}$$

is called a **system of equations,** and the solution $x = 3$, $y = 2$ is called its **simultaneous solution,** or just its **solution.** The process of finding the solution of a system of equations is called **solving the system.**

The graph of an equation in two variables displays the equation's infinitely many solutions. For example, one of the lines in Figure 8-1 represents the infinitely many solutions of the equation $5x - 2y = 1$. The other line in the figure is the graph of the infinitely many solutions of $2x + 3y = 8$.

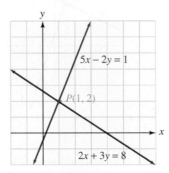

FIGURE 8-1

Because only point $P(1, 2)$ lies on both lines, only its coordinates satisfy both equations. Thus, the simultaneous solution of the system of equations

$$\begin{cases} 5x - 2y = 1 \\ 2x + 3y = 8 \end{cases}$$

is the pair of numbers $x = 1$ and $y = 2$, or simply the pair $(1, 2)$.

This discussion suggests a graphical method of solving systems of equations in two variables.

1. Solve Systems Using the Graphing Method

We can use the following steps to solve a system of two equations in two variables:

Strategy for Using the Graphing Method

Step 1: On one coordinate grid, graph each equation.

Step 2: Find the coordinates of the point or points where all of the graphs intersect. These coordinates give the solutions of the system.

Step 3: If the graphs have no point in common, the system has no solution.

EXAMPLE 1   **Using the Graphing Method to Solve a System of Equations**

Use the graphing method to solve each system.

a. $\begin{cases} 3x + y = 1 \\ -x + 2y = 9 \end{cases}$ **b.** $\begin{cases} 2x - 3y = 4 \\ 4x = -4 + 6y \end{cases}$ **c.** $\begin{cases} y = 4 - x \\ 2x + 2y = 8 \end{cases}$

SOLUTION   **a.** The graphs of the equations are the lines shown in Figure 8-2(a). The solution of this system is given by the coordinates of the point $(-1, 4)$, where the lines intersect. By checking both values in both equations, we can verify that the solution is $x = -1$ and $y = 4$.

b. The graphs of the equations are the parallel lines shown in Figure 8-2(b). Since parallel lines do not intersect, the system has no solution.

c. The graphs of the equations are the lines shown in Figure 8-2(c). Since the lines are the same, they have infinitely many points in common, and the system has infinitely many solutions. All ordered pairs whose coordinates satisfy one of the equations satisfy the other also. To find some solutions, we substitute numbers for x in either equation and solve for y. If $x = 3$, for example, then $y = 1$. One solution is the pair $(3, 1)$. Other solutions are $(0, 4)$ and $(5, -1)$.

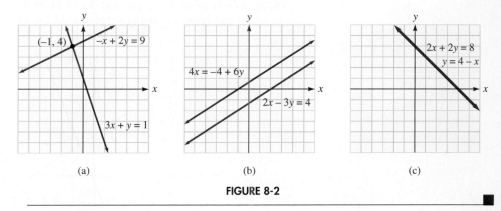

(a)   (b)   (c)

FIGURE 8-2

Self Check 1   Solve: $\begin{cases} y = 2x \\ x + y = 3 \end{cases}$.

Now Try Exercise 13.

Example 1 illustrates three possibilities that can occur when we solve systems of equations. If a system of equations has at least one solution, as in parts (a) and (c) of Example 1, the system is called **consistent.** If it has no solution, as in part (b), it is called **inconsistent.**

If a system of two equations in two variables has exactly one solution, as in part (a), or no solution, as in part (b), the equations in the system are called **independent.** If a system of linear equations has infinitely many solutions, as in part (c), the equations of the system are called **dependent.**

There are three possibilities that can occur when two equations, each with two variables, are graphed:

Possible figure	If ...	Then ...
(graph: two distinct intersecting lines)	the lines are distinct and intersect,	the equations are independent and the system is consistent. The system has one solution.
(graph: two distinct parallel lines)	the lines are distinct and parallel,	the equations are independent and the system is inconsistent. The system has no solution.
(graph: two coincident lines)	the lines coincide,	the equations are dependent and the system is consistent. The system has infinitely many solutions.

We now consider two algebraic methods for finding solutions of systems containing any number of variables.

2. Solve Systems Using the Substitution Method

We use the following steps to solve a system of two equations in two variables by substitution:

Strategy for Using the Substitution Method

Step 1: Solve one equation for a variable—say, y.

Step 2: Substitute the expression obtained for y for every y in the second equation.

Step 3: Solve the equation that results.

Step 4: Substitute the solution found in Step 3 into the equation found in Step 1 and solve for y.

EXAMPLE 2 **Using the Substitution Method to Solve a System of Equations**

Use the substitution method to solve $\begin{cases} 3x + y = 1 \\ -x + 2y = 9 \end{cases}$.

SOLUTION We can solve the first equation for y and then substitute that result for y in the second equation.

$$\begin{cases} 3x + y = 1 \rightarrow y = 1 - 3x \\ -x + 2y = 9 \end{cases}$$

The substitution gives one linear equation with one variable, which we can solve for x:

$$-x + 2(1 - 3x) = 9$$
$$-x + 2 - 6x = 9$$
$$-7x = 7 \qquad \text{Combine like terms and subtract 2 from both sides.}$$
$$x = -1 \qquad \text{Divide both sides by } -7.$$

To find y, we substitute -1 for x in the equation $y = 1 - 3x$ and simplify:

$$y = 1 - 3x$$
$$= 1 - 3(-1)$$
$$= 1 + 3$$
$$= 4$$

Comment

We could solve either equation for x or y. However, it is easier to solve for a variable with a coefficient of 1 or -1; then fractions will be avoided.

The solution is the pair $(-1, 4)$.

Self Check 2 Solve: $\begin{cases} 2x + 3y = 1 \\ x - y = 3 \end{cases}$.

Now Try Exercise 25.

3. Solve Systems Using the Addition Method

As with substitution, the addition method combines the equations of a system to eliminate terms involving one of the variables.

Strategy for Using the Addition Method

Step 1: Write the equations of the system in standard form so that terms with the same variable are aligned vertically.

Step 2: Multiply all the terms of one or both of the equations by constants chosen to make the coefficients of x (or y) differ only in sign.

Step 3: Add the equations. Solve the equation that results, if possible.

Step 4: Substitute the value obtained in Step 3 into either of the original equations and solve for the remaining variable. The results obtained in Step 3 and this step are the solution of the system.

EXAMPLE 3   Using the Addition Method to Solve a System of Equations

Use the addition method to solve $\begin{cases} 3x + 2y = 8 \\ 2x - 5y = 18 \end{cases}$.

SOLUTION   To eliminate y, we multiply the terms of the first equation by 5 and the terms of the second equation by 2, to obtain an equivalent system in which the coefficients of y differ only in sign. Then we add the equations to eliminate y.

$$\begin{cases} 3x + 2y = 8 \rightarrow 15x + 10y = 40 \qquad \text{Multiply by 5.} \\ 2x - 5y = 18 \rightarrow \underline{4x - 10y = 36} \qquad \text{Multiply by 2.} \end{cases}$$

$$19x \qquad\quad = 76 \qquad \text{Add the equations.}$$
$$x = 4 \qquad \text{Divide both sides by 19.}$$

We now substitute 4 for x into either of the original equations and solve for y. We use the first equation.

$$3x + 2y = 8$$
$$3(4) + 2y = 8 \qquad \text{Substitute 4 for } x.$$
$$12 + 2y = 8 \qquad \text{Simplify.}$$
$$2y = -4 \qquad \text{Subtract 12 from both sides.}$$
$$y = -2 \qquad \text{Divide both sides by 2.}$$

The solution is $(4, -2)$.

Comment

To make the coefficients of y differ only in sign and be the smallest numbers possible, we multiply each equation by a number that makes the new coefficients of y the least common multiple of the original coefficients.

Self Check 3   Solve: $\begin{cases} 2x + 5y = 8 \\ 2x - 4y = -10 \end{cases}$.

Now Try Exercise 35.

EXAMPLE 4   Using the Addition Method to Solve a System of Equations

Solve: $\begin{cases} \dfrac{x + 2y}{4} + \dfrac{x - y}{5} = \dfrac{6}{5} \\ \dfrac{x + y}{7} - \dfrac{x - y}{3} = -\dfrac{12}{7} \end{cases}$.

SOLUTION   To clear the first equation of fractions, we multiply both sides by the lowest common denominator, 20. To clear the second equation of fractions, we multiply both sides by the lowest common denominator, which is 21.

$$\begin{cases} 20\left(\dfrac{x + 2y}{4} + \dfrac{x - y}{5}\right) = 20\left(\dfrac{6}{5}\right) \\ 21\left(\dfrac{x + y}{7} - \dfrac{x - y}{3}\right) = 21\left(-\dfrac{12}{7}\right) \end{cases}$$

$$\begin{cases} 5(x + 2y) + 4(x - y) = 24 \\ 3(x + y) - 7(x - y) = -36 \end{cases}$$

$$\begin{cases} 9x + 6y = 24 \\ -4x + 10y = -36 \end{cases} \qquad \text{Remove parentheses and combine like terms.}$$

$$\begin{cases} 3x + 2y = 8 \qquad \text{Divide both sides by 3.} \\ 2x - 5y = 18 \qquad \text{Divide both sides by } -2. \end{cases}$$

In Example 3, we saw that the solution to this final system is $(4, -2)$.

Self Check 4   Solve: $\begin{cases} \dfrac{x + y}{3} + \dfrac{x - y}{4} = \dfrac{3}{4} \\ \dfrac{x - y}{3} - \dfrac{2x - y}{2} = -\dfrac{1}{3} \end{cases}.$

Now Try Exercise 51.

4. Solve Systems with Infinitely Many Solutions

EXAMPLE 5   **Solving a System with Infinitely Many Solutions**

Use the addition method to solve $\begin{cases} x + 2y = 3 \\ 2x + 4y = 6 \end{cases}.$

SOLUTION   To eliminate the variable y, we can multiply both sides of the first equation by -2 and add the result to the second equation to get

$$\begin{array}{r} -2x - 4y = -6 \\ \underline{2x + 4y = 6} \\ 0 = 0 \end{array}$$

Although the result $0 = 0$ is true, it does not give the value of y.

Since the second equation in the original system is twice the first, the equations are equivalent. If we were to solve this system by graphing, the two lines would coincide. The (x, y) coordinates of points on that one line form the infinite set of solutions to the given system. The system is consistent, but the equations are dependent.

To find a general solution, we can solve either equation of the system for y. If we solve the first equation for y, we obtain

$$\begin{aligned} x + 2y &= 3 \\ 2y &= 3 - x \\ y &= \frac{3 - x}{2} \end{aligned}$$

All ordered pairs that are solutions will have the form $\left(x, \dfrac{3 - x}{2}\right)$. For example:

- If $x = 3$, then $y = 0$ and $(3, 0)$ is a solution.

- If $x = 0$, then $y = \frac{3}{2}$ and $\left(0, \frac{3}{2}\right)$ is a solution.

- If $x = -7$, then $y = 5$ and $(-7, 5)$ is a solution.

Self Check 5   Solve: $\begin{cases} 3x - y = 2 \\ 6x - 2y = 4 \end{cases}.$

Now Try Exercise 43.

5. Solve Inconsistent Systems

EXAMPLE 6   **Solving an Inconsistent System**

Solve: $\begin{cases} x + y = 3 \\ x + y = 2 \end{cases}$.

SOLUTION   We can multiply both sides of the second equation by –1 and add the results to the first equation to get

$$\begin{aligned} x + y &= 3 \\ \underline{-x - y} &= \underline{-2} \\ 0 &= 1 \end{aligned}$$

Because $0 \neq 1$, the system has no solution. If we graph each equation in this system, the graphs will be parallel lines. This system is inconsistent, and the equations are independent.

Self Check 6   Solve: $\begin{cases} 3x + 2y = 2 \\ 3x + 2y = 3 \end{cases}$.

Now Try Exercise 45.

6. Solve Systems Involving Three Equations in Three Variables

To solve three equations in three variables, we use addition to eliminate one variable. This will produce a system of two equations in two variables, which we can solve using the methods previously discussed.

EXAMPLE 7   **Solving a System Involving Three Equations in Three Variables**

Solve the system:

$$\begin{array}{ll} (1) \\ (2) \\ (3) \end{array} \begin{cases} x + 2y + z = 8 \\ 2x + y - z = 1 \\ x + y - 2z = -3. \end{cases}$$

SOLUTION   We can add Equations (1) and (2) to eliminate z,

$$\begin{array}{ll} (1) & x + 2y + z = 8 \\ (2) & \underline{2x + y - z = 1} \\ (4) & 3x + 3y = 9, \end{array}$$

and divide both sides of Equation (4) by 3 to get

$$(5) \quad x + y = 3.$$

We now choose a different pair of equations—say, Equations (1) and (3)—and eliminate z again. If we multiply both sides of Equation (1) by 2 and add the result to Equation (3), we get

$$\begin{array}{ll} & 2x + 4y + 2z = 16 \\ (3) & \underline{x + y - 2z = -3} \\ (6) & 3x + 5y = 13. \end{array}$$

Equations (5) and (6) form the system $\begin{cases} x + y = 3 \\ 3x + 5y = 13, \end{cases}$ which we can solve by

substitution or addition. We will use substitution.

(5) $x + y = 3 \rightarrow y = \boxed{3 - x}$
(6) $3x + 5y = 13$

$$3x + 5(\mathbf{3 - x}) = 13 \qquad \text{Substitute } 3 - x \text{ for } y.$$
$$3x + 15 - 5x = 13 \qquad \text{Remove parentheses.}$$
$$-2x = -2 \qquad \text{Combine like terms and subtract 15 from both sides.}$$
$$x = 1 \qquad \text{Divide both sides by } -2.$$

To find y, we substitute 1 for x in the equation $y = 3 - x$.

$$y = 3 - \mathbf{1}$$
$$y = 2$$

To find z, we substitute 1 for x and 2 for y in any one of the original equations that includes z, and we find that $z = 3$. The solution is the triple $(x, y, z) = (1, 2, 3)$. Because there is one solution, the system is consistent and its equations are independent.

Self Check 7   Solve: $\begin{cases} x - y + 2z = 2 \\ 2x + y + z = 4 \\ -x - 2y + 3z = 0 \end{cases}$.

Now Try Exercise 53.

EXAMPLE 8   Solving a System Involving Three Dependent Equations in Three Variables

Solve the system:

(1) $\begin{cases} x + 2y + z = 8 \\ 2x + y - z = 1 \\ x - y - 2z = -7. \end{cases}$
(2)
(3)

SOLUTION   We can add Equations (1) and (2) to eliminate z.

$$\begin{array}{r} x + 2y + z = 8 \\ 2x + y - z = 1 \\ \hline 3x + 3y \quad\quad = 9 \end{array}$$
(4)

We can now multiply Equation (1) by 2 and add it to Equation (3) to eliminate z again.

$$\begin{array}{r} 2x + 4y + 2z = 16 \\ x - y - 2z = -7 \\ \hline 3x + 3y \quad\quad = 9 \end{array}$$
(5)

Since Equations (4) and (5) are the same, the system is consistent but the equations are dependent. There will be infinitely many solutions. To find a general solution, we can solve either Equation (4) or (5) for y to get

(6) $y = 3 - x.$

We can find the value of z in terms of x by substituting the right side of Equation (6) into any of the first three equations—say, Equation (1).

$$x + 2y + z = 8$$
$$x + 2(\mathbf{3 - x}) + z = 8 \qquad \text{Substitute } 3 - x \text{ for } y.$$
$$x + 6 - 2x + z = 8 \qquad \text{Use the Distributive Property to remove parentheses.}$$
$$-x + 6 + z = 8 \qquad \text{Combine terms.}$$
$$z = x + 2 \qquad \text{Solve for } z.$$

A general solution to this system is $(x, y, z) = (x, 3 - x, x + 2)$. To find some specific solutions, we can substitute numbers for x and compute y and z. For example:

• If $x = 1$, then $y = 2$ and $z = 3$. One possible solution is $(1, 2, 3)$.
• If $x = 0$, then $y = 3$ and $z = 2$. Another possible solution is $(0, 3, 2)$.

■

Self Check 8   Solve: $\begin{cases} x + y + z = 3 \\ x - y + 3z = 4 \\ 2x + 2y + 2z = 5 \end{cases}$.

Now Try Exercise 69.

7. Solve Applications Involving Systems of Equations

EXAMPLE 9   **Solving an Application Problem**

An airplane flies 600 miles with the wind for 2 hours and returns against the wind in 3 hours. Find the speed of the wind and the air speed of the plane.

SOLUTION   If a represents the air speed and w represents wind speed, the ground speed of the plane with the wind is the combined speed $a + w$. On the return trip, against the wind, the ground speed is $a - w$. The information in this problem is organized in Figure 8-3 and can be used to give a system of two equations in the variables a and w.

	d	$=$	r	$\cdot$	t
Outbound trip	600		$a + w$		2
Return trip	600		$a - w$		3

FIGURE 8-3

Since $d = rt$, we have $\begin{cases} 600 = 2(a + w) \\ 600 = 3(a - w) \end{cases}$, which can be written as

(1) $\begin{cases} 300 = a + w \\ 200 = a - w \end{cases}$.

We can add these equations to get

$500 = 2a$

$a = 250$ Divide both sides by 2.

To find w, we substitute 250 for a into either of the previous equations [we'll use Equation (1)] and solve for w:

(1) $300 = a + w$

$300 = \mathbf{250} + w$

$w = 50$ Subtract 250 from both sides.

The air speed of the plane is 250 mph. With a 50 mph tailwind, the ground speed is $250 + 50$, or 300 mph. At 300 mph, the 600-mile trip will take 2 hours.

With a 50 mph headwind, the ground speed is $250 - 50$, or 200 mph. At 200 mph, the 600-mile trip will take 3 hours. The answers check.

■

Self Check 9　An airplane flies 500 miles with the wind for 2 hours and returns against the wind in 3 hours. Find the speed of the wind and the air speed of the plane. Round to tenths.

Now Try Exercise 76.

Self Check Answers　**1.** $(1, 2)$　**2.** $(2, -1)$　**3.** $(-1, 2)$　**4.** $(1, 2)$　**5.** $(x, 3x - 2)$
6. The system is inconsistent; no solution.　**7.** $(1, 1, 1)$
8. The system is inconsistent; no solution.
9. wind: 41.7 mph; plane: 208.3 mph

Exercises **8.1**

Getting Ready
You should be able to complete these vocabulary and concept statements before you proceed to the practice exercises.

Fill in the blanks.

1. A set of several equations with several variables is called a _____ of equations.

2. Any set of numbers that satisfies each equation of a system is called a _____ of the system.

3. If a system of equations has a solution, the system is _____.

4. If a system of equations has no solution, the system is _____.

5. If a system of equations has only one solution, the equations of the system are _____.

6. If a system of equations has infinitely many solutions, the equations of the system are _____.

7. The system $\begin{cases} x + y = 5 \\ x - y = 1 \end{cases}$ is _____ (consistent, inconsistent).

8. The system $\begin{cases} x + y = 5 \\ x + y = 1 \end{cases}$ is _____ (consistent, inconsistent).

9. The equations of the system $\begin{cases} x + y = 5 \\ 2x + 2y = 10 \end{cases}$ are _____ (dependent, independent).

10. The equations of the system $\begin{cases} x + y = 5 \\ x - y = 1 \end{cases}$ are _____ (dependent, independent).

11. The pair $(1, 3)$ _____ (is, is not) a solution of the system $\begin{cases} x + 2y = 7 \\ 2x - y = -1 \end{cases}$.

12. The pair $(1, 3)$ _____ (is, is not) a solution of the system $\begin{cases} 3x + y = 6 \\ x - 3y = -8 \end{cases}$.

Practice
Solve each system of equations by graphing. If the system has no solution, state "inconsistent system." If the system has infinitely many solutions, state "dependent equations."

13. $\begin{cases} y = -3x + 5 \\ x - 2y = -3 \end{cases}$

14. $\begin{cases} x - 2y = -3 \\ 3x + y = -9 \end{cases}$

15. $\begin{cases} 3x + 2y = 2 \\ -2x + 3y = 16 \end{cases}$

16. $\begin{cases} x + y = 0 \\ 5x - 2y = 14 \end{cases}$

17. $\begin{cases} y = -x + 5 \\ 3x + 3y = 27 \end{cases}$

18. $\begin{cases} x - 3y = -3 \\ 2x - 6y = 12 \end{cases}$

19. $\begin{cases} y = -x + 6 \\ 5x + 5y = 30 \end{cases}$

20. $\begin{cases} 2x - y = -3 \\ 8x - 4y = -12 \end{cases}$

Use a graphing calculator to approximate the solutions of each system. Give answers to the nearest tenth.

21. $\begin{cases} y = -5.7x + 7.8 \\ y = 37.2 - 19.1x \end{cases}$

22. $\begin{cases} y = 3.4x - 1 \\ y = -7.1x + 3.1 \end{cases}$

23. $\begin{cases} y = \dfrac{5.5 - 2.7x}{3.5} \\ 5.3x - 9.2y = 6.0 \end{cases}$

24. $\begin{cases} 29x + 17y = 7 \\ -17x + 23y = 19 \end{cases}$

Solve each system by substitution, if possible.

25. $\begin{cases} y = x - 1 \\ y = 2x \end{cases}$

26. $\begin{cases} y = 2x - 1 \\ x + y = 5 \end{cases}$

27. $\begin{cases} 2x + 3y = 0 \\ y = 3x - 11 \end{cases}$

28. $\begin{cases} 2x + y = 3 \\ y = 5x - 11 \end{cases}$

29. $\begin{cases} 4x + 3y = 3 \\ 2x - 6y = -1 \end{cases}$

30. $\begin{cases} 4x + 5y = 4 \\ 8x - 15y = 3 \end{cases}$

31. $\begin{cases} x + 3y = 1 \\ 2x + 6y = 3 \end{cases}$

32. $\begin{cases} x - 3y = 14 \\ 3(x - 12) = 9y \end{cases}$

33. $\begin{cases} y = 3x - 6 \\ x = \dfrac{1}{3}y + 2 \end{cases}$

34. $\begin{cases} 3x - y = 12 \\ y = 3x - 12 \end{cases}$

Solve each system by the addition method, if possible.

35. $\begin{cases} 5x - 3y = 12 \\ 2x - 3y = 3 \end{cases}$ **36.** $\begin{cases} 2x + 3y = 8 \\ -5x + y = -3 \end{cases}$

37. $\begin{cases} x - 7y = -11 \\ 8x + 2y = 28 \end{cases}$ **38.** $\begin{cases} 3x + 9y = 9 \\ -x + 5y = -3 \end{cases}$

39. $\begin{cases} 3(x - y) = y - 9 \\ 5(x + y) = -15 \end{cases}$ **40.** $\begin{cases} 2(x + y) = y + 1 \\ 3(x + 1) = y - 3 \end{cases}$

41. $\begin{cases} 2 = \dfrac{1}{x + y} \\ 2 = \dfrac{3}{x - y} \end{cases}$ **42.** $\begin{cases} \dfrac{1}{x + y} = 12 \\ \dfrac{3x}{y} = -4 \end{cases}$

43. $\begin{cases} y + 2x = 5 \\ 0.5y = 2.5 - x \end{cases}$ **44.** $\begin{cases} -0.3x + 0.1y = -0.1 \\ 6x - 2y = 2 \end{cases}$

45. $\begin{cases} x + 2(x - y) = 2 \\ 3(y - x) - y = 5 \end{cases}$ **46.** $\begin{cases} 3x = 4(2 - y) \\ 3(x - 2) + 4y = 0 \end{cases}$

47. $\begin{cases} x + \dfrac{y}{3} = \dfrac{5}{3} \\ \dfrac{x + y}{3} = 3 - x \end{cases}$ **48.** $\begin{cases} 3x - y = 0.25 \\ x + \dfrac{3}{2}y = 2.375 \end{cases}$

49. $\begin{cases} \dfrac{3}{2}x + \dfrac{1}{3}y = 2 \\ \dfrac{2}{3}x + \dfrac{1}{9}y = 1 \end{cases}$ **50.** $\begin{cases} \dfrac{x + y}{2} + \dfrac{x - y}{5} = 2 \\ x = \dfrac{y}{2} + 1 \end{cases}$

51. $\begin{cases} \dfrac{x - y}{5} + \dfrac{x + y}{2} = 6 \\ \dfrac{x - y}{2} - \dfrac{x + y}{4} = 3 \end{cases}$ **52.** $\begin{cases} \dfrac{x - 2}{5} + \dfrac{y + 3}{2} = 5 \\ \dfrac{x + 3}{2} + \dfrac{y - 2}{3} = 6 \end{cases}$

Solve each system, if possible.

53. $\begin{cases} x + y + z = 3 \\ 2x + y + z = 4 \\ 3x + y - z = 5 \end{cases}$ **54.** $\begin{cases} x - y - z = 0 \\ x + y - z = 0 \\ x - y + z = 2 \end{cases}$

55. $\begin{cases} x - y + z = 0 \\ x + y + 2z = -1 \\ -x - y + z = 0 \end{cases}$ **56.** $\begin{cases} 2x + y - z = 7 \\ x - y + z = 2 \\ x + y - 3z = 2 \end{cases}$

57. $\begin{cases} 2x + y = 4 \\ x - z = 2 \\ y + z = 1 \end{cases}$ **58.** $\begin{cases} 3x + y + z = 0 \\ 2x - y + z = 0 \\ 2x + y + z = 0 \end{cases}$

59. $\begin{cases} x + y + z = 6 \\ 2x + y + 3z = 17 \\ x + y + 2z = 11 \end{cases}$ **60.** $\begin{cases} x + y + z = 3 \\ 2x + y + z = 6 \\ x + 2y + 3z = 2 \end{cases}$

61. $\begin{cases} x + y + z = 3 \\ x + z = 2 \\ 2x + 2y + 2z = 3 \end{cases}$ **62.** $\begin{cases} x + y + z = 3 \\ x + z = 2 \\ 2x + y + 2z = 5 \end{cases}$

63. $\begin{cases} x + 2y - z = 2 \\ 2x - y = -1 \\ 3x + y + z = 1 \end{cases}$ **64.** $\begin{cases} x + y = 2 \\ y + z = 2 \\ 3x + 3y = 2 \end{cases}$

65. $\begin{cases} 3x + 4y + 2z = 4 \\ 6x - 2y + z = 4 \\ 3x - 8y - 6z = -3 \end{cases}$ **66.** $\begin{cases} x + y = 2 \\ y + z = 2 \\ x - z = 0 \end{cases}$

67. $\begin{cases} 2x - y - z = 0 \\ x - 2y - z = -1 \\ x - y - 2z = -1 \end{cases}$ **68.** $\begin{cases} x + 3y - z = 5 \\ 3x - y + z = 2 \\ 2x + y = 1 \end{cases}$

69. $\begin{cases} (x + y) + (y + z) + (z + x) = 6 \\ (x - y) + (y - z) + (z - x) = 0 \\ x + y + 2z = 4 \end{cases}$

70. $\begin{cases} (x + y) + (y + z) = 1 \\ (x + z) + (x + z) = 3 \\ (x - y) - (x - z) = -1 \end{cases}$

Applications
Use systems of equations to solve each problem.

71. Price of food items If Jonathan purchases two hamburgers and four orders of french fries for $8 and Hannah purchases 3 hamburgers and two orders of fries for $8, what is the price of each item?

72. Price of tennis equipment Hunter purchases two tennis rackets and four cans of tennis balls for $102. Jana purchases three tennis rackets and two cans of tennis balls for $141. What is the price of each item?

73. Planning for harvest A farmer raises corn and soybeans on 350 acres of land. Because of expected prices at harvest time, he thinks it would be wise to plant 100 acres more of corn than of soybeans. How many acres of each does he plant?

74. Club memberships There is an initiation fee to join the Pine River Country Club, as well as monthly dues. The total cost after 7 months' membership will be $3,025; after $1\frac{1}{2}$ years, $3,850. Find both the initiation fee and the monthly dues.

75. Framing pictures A rectangular picture frame has a perimeter of 1,900 centimeters and a width that is 250 centimeters less than its length. Find the area of the picture.

76. Boating A Mississippi riverboat can travel 30 kilometers downstream in 3 hours and can make the return trip in 5 hours. Find the speed of the boat in still water.

77. **Making an alloy** A metallurgist wants to make 60 grams of an alloy that is to be 34% copper. She has samples that are 9% copper and 84% copper. How many grams of each must she use?

78. **Archimedes' law of the lever** The two weights shown will be in balance if the product of one weight and its distance from the fulcrum is equal to the product of the other weight and its distance from the fulcrum. Two weights are in balance when one is 2 meters and the other 3 meters from the fulcrum. If the fulcrum remained in the same spot and the weights were interchanged, the closer weight would need to be increased by 5 pounds to maintain balance. Find the weights.

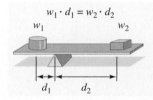

$$w_1 \cdot d_1 = w_2 \cdot d_2$$

79. **Lifting weights** A 112-pound force can lift the 448-pound load shown. If the fulcrum is moved 1 additional foot away from the load, a 192-pound force is required. Find the length of the lever.

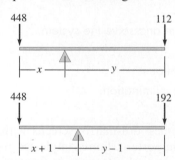

80. **Writing test questions** For a test question, a mathematics teacher wants to find two constants a and b such that the test item "Simplify $a(x + 2y) - b(2x - y)$" will have an answer of $-3x + 9y$. What constants a and b should the teacher use?

81. **Break-even point** Rollowheel Inc. can manufacture a pair of in-line skates for $43.53. Daily fixed costs of manufacturing in-line skates amount to $742.72. A pair of in-line skates can be sold for $89.95. Find equations expressing the expenses E and the revenue R as functions of x, the number of pairs manufactured and sold. At what production level will expenses equal revenues?

82. **Choosing salary options** For its sales staff, a company offers two salary options. One is $326 per week plus a commission of $3\frac{1}{2}\%$ of sales. The other is $200 per week plus $4\frac{1}{4}\%$ of sales. Find equations S_1 and S_2 that express incomes as functions of sales x, and find the weekly sales level that produces equal salaries.

Use systems of three equations in three variables to solve each problem.

83. **Work schedules** A college student earns $198.50 per week working three part-time jobs. Half of his 30-hour workweek is spent cooking hamburgers at a fast-food chain, earning $5.70 per hour. In addition, the student earns $6.30 per hour working at a gas station and $10 per hour doing janitorial work. How many hours per week does the student work at each job?

84. **Investment income** A woman invested $22,000 in three banks paying 5%, 6%, and 7% annual interest. She invested $2,000 more at 6% than at 5%. The total annual interest she earned was $1,370. How much did she invest at each rate?

85. **Age distribution** Approximately 3 million people live in Costa Rica. Of those, 2.61 million are younger than 50 and 1.95 million are older than 14. How many people are in each of the categories 0–14 years, 15–49 years, and 50+ years?

86. **Designing arches** The engineer designing a parabolic arch knows that its equation has the form $y = ax^2 + bx + c$. Use the information in the illustration to find a, b, and c. (*Hint:* The coordinates of points on the parabola satisfy its equation.)

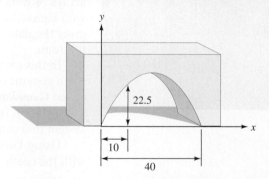

87. **Geometry** The sum of the angles of a triangle is 180°. In a certain triangle, the largest angle is 20° greater than the sum of the other two and is 10° greater than 3 times the smallest. How large is each angle?

88. Ballistics The path of a thrown object is a parabola with the equation $y = ax^2 + bx + c$. Use the information in the illustration to find a, b, and c.

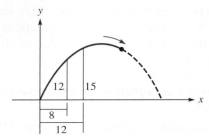

Discovery and Writing

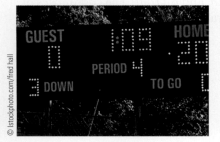

89. Use a graphing calculator to attempt to find the solution of the system $\begin{cases} x - 8y = -51 \\ 3x - 25y = -160 \end{cases}$.

90. Solve the system of Exercise 89 algebraically. Which method is easier, and why?

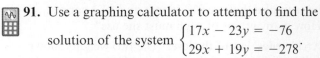

91. Use a graphing calculator to attempt to find the solution of the system $\begin{cases} 17x - 23y = -76 \\ 29x + 19y = -278 \end{cases}$.

92. Solve the system of Exercise 91 algebraically. Which method is easier, and why?

93. Invent a system of two equations with the solution $(1, 2)$ which is difficult to solve with a graphing calculator (as in Exercise 91).

94. Invent a system of two equations with the solution $(1, 2)$ that is easier to solve graphically than algebraically.

8.2 Gaussian Elimination and Matrix Methods

In this section, we will learn to

1. Write a matrix in row echelon form and solve the system.
2. Solve an inconsistent system.
3. Solve a system of dependent equations.
4. Solve systems using Gauss–Jordan elimination.

Since the invention of modern electronics, dot-matrix screens with rectangular arrays of dots (or pixels) have become important. For example, we are all familiar with signs like the football scoreboard shown here, a sign in front of a bank that gives the date and temperature, or a sign that shows which lanes are open on the toll road.

In this section, we will show how rectangular arrays of numbers can be used to solve systems of equations. We begin by introducing a method for solving systems called **Gaussian elimination,** after the German mathematician Carl Friedrich Gauss (1777–1855). In this method, we transform a system of equations into an equivalent system that can be solved by a process called **back substitution.**

Using Gaussian elimination, we solve systems of equations by working only with the coefficients of the variables. From the system of equations

$$\begin{cases} x + 2y + z = 8 \\ 2x + y - z = 1 \\ x + y - 2z = -3 \end{cases},$$

for example, we can form three rectangular arrays of numbers. Each is called a **matrix.** The first is the **coefficient matrix,** which contains the coefficients of the variables of the system. The second matrix contains the constants from the right side of the system. We can write the third matrix, called the **augmented matrix** or the **system matrix,** by joining the two. The coefficient matrix appears to the left of the dashed line, and the matrix of constants appears to the right.

Coefficient matrix   *Constants*   *System matrix*

$$\begin{bmatrix} 1 & 2 & 1 \\ 2 & 1 & -1 \\ 1 & 1 & -2 \end{bmatrix} \qquad \begin{bmatrix} 8 \\ 1 \\ -3 \end{bmatrix} \qquad \left[\begin{array}{ccc|c} 1 & 2 & 1 & 8 \\ 2 & 1 & -1 & 1 \\ 1 & 1 & -2 & -3 \end{array}\right]$$

Each row of the system matrix represents one equation of the system.

- The first row represents the equation $x + 2y + z = 8$.
- The second row represents the equation $2x + y - z = 1$.
- The third row represents the equation $x + y - 2z = -3$.

Because the system matrix has three rows and four columns, it is called a 3×4 matrix (read as "three by four"). The coefficient matrix is 3×3, and the constants form a 3×1 matrix.

To illustrate Gaussian elimination, we solve this system by the addition method of the previous section. In a second column, we keep track of the changes to the system matrix.

EXAMPLE 1   Using Gaussian Elimination to Solve a System of Equations

Use Gaussian elimination to solve

$$(1) \quad \begin{cases} x + 2y + z = 8 \\ (2) \quad 2x + y - z = 1 \\ (3) \quad x + y - 2z = -3. \end{cases}$$

SOLUTION We multiply each term in Equation (1) by –2 and add the result to Equation (2) to obtain Equation (4). Then we multiply each term of Equation (1) by –1 and add the result to Equation (3) to obtain Equation (5). This gives the equivalent system

$$(1) \quad \begin{cases} x + 2y + z = 8 \\ (4) \quad -3y - 3z = -15 \\ (5) \quad -y - 3z = -11. \end{cases} \qquad \left[\begin{array}{ccc|c} 1 & 2 & 1 & 8 \\ 0 & -3 & -3 & -15 \\ 0 & -1 & -3 & -11 \end{array}\right]$$

Now we divide both sides of Equation (4) by –3 to obtain Equation (6),

$$(1) \quad \begin{cases} x + 2y + z = 8 \\ (6) \quad y + z = 5 \\ (5) \quad -y - 3z = -11, \end{cases} \qquad \left[\begin{array}{ccc|c} 1 & 2 & 1 & 8 \\ 0 & 1 & 1 & 5 \\ 0 & -1 & -3 & -11 \end{array}\right]$$

and add Equation (6) to Equation (5) to obtain Equation (7).

$$(1) \quad \begin{cases} x + 2y + z = 8 \\ (6) \quad y + z = 5 \\ (7) \quad -2z = -6 \end{cases} \qquad \left[\begin{array}{ccc|c} 1 & 2 & 1 & 8 \\ 0 & 1 & 1 & 5 \\ 0 & 0 & -2 & -6 \end{array}\right]$$

Finally, we divide both sides of Equation (7) by –2 to obtain the system

$$(1) \quad \begin{cases} x + 2y + z = 8 \\ (6) \quad y + z = 5 \\ \quad z = 3. \end{cases} \qquad \left[\begin{array}{ccc|c} 1 & 2 & 1 & 8 \\ 0 & 1 & 1 & 5 \\ 0 & 0 & 1 & 3 \end{array}\right]$$

The system can now be solved by back substitution. Because $z = 3$, we can substitute 3 for z in Equation (6) and solve for y.

$$(6) \quad y + z = 5$$
$$y + 3 = 5$$
$$y = 2$$

We can now substitute 2 for y and 3 for z in Equation (1) and solve for x.

$$(1) \qquad x + 2y + z = 8$$
$$x + 2(\mathbf{2}) + 3 = 8$$
$$x + 7 = 8$$
$$x = 1$$

The solution is the ordered triple $(1, 2, 3)$.

Self Check 1   Solve: $\begin{cases} x - 2y - z = 1 \\ 2x + y + 2z = 9 \\ -x + 3y - 3z = -14 \end{cases}$.

Now Try Exercise 17.

1. Write a Matrix in Row Echelon Form and Solve the System

In Example 1, we performed operations on the equations as well as on the corresponding matrices. These matrix operations are called **elementary row operations.**

Elementary Row Operations

If any of the following operations are performed on the rows of a system matrix, the matrix of an equivalent system results:

Type 1 row operation: Two rows of a matrix can be interchanged.

Type 2 row operation: The elements of a row of a matrix can be multiplied by a nonzero constant.

Type 3 row operation: Any row can be changed by adding a multiple of another row to it.

- A type 1 row operation is equivalent to writing the equations of a system in a different order.
- A type 2 row operation is equivalent to multiplying both sides of an equation by a nonzero constant.
- A type 3 row operation is equivalent to adding a multiple of one equation to another.

Any matrix that can be obtained from another matrix by a sequence of elementary row operations is **row equivalent** to the original matrix. This means that row equivalent matrices represent equivalent systems of equations.

The process of Example 1 changed a system matrix into a special row equivalent form called **row echelon form.**

Row Echelon Form of a Matrix

A matrix is in **row echelon form** if it has these three properties:

1. The first nonzero entry in each row (called the **lead entry**) is 1.

2. Lead entries appear farther to the right as we move down the rows of the matrix.

3. Any rows containing only 0s are at the bottom of the matrix.

Examples of Matrices in Row Echelon Form:
The matrices

$$\begin{bmatrix} 1 & 3 & 0 & 7 \\ 0 & 0 & 1 & 8 \\ 0 & 0 & 0 & 0 \end{bmatrix}, \quad \begin{bmatrix} 0 & 1 & 2 \\ 0 & 0 & 1 \end{bmatrix}, \quad \text{and} \quad \begin{bmatrix} 1 & 1 \\ 0 & 0 \\ 0 & 0 \\ 0 & 0 \end{bmatrix}$$

are in row echelon form, because the first nonzero entry in each row is 1, the lead entries appear farther to the right as we move down the rows, and any rows that consist entirely of 0s are at the bottom of the matrix.

Examples of Matrices Not in Row Echelon Form:
The next three matrices are not in row echelon form.

$$\begin{bmatrix} 1 & 3 & 0 & 7 \\ 0 & 0 & 1 & 8 \\ 0 & 0 & 1 & 0 \end{bmatrix}$$

The lead entry in the last row is not to the right of the lead entry in the middle row.

$$\begin{bmatrix} 0 & 3 & 0 \\ 0 & 0 & 1 \end{bmatrix}$$

The lead entry of the first row is not 1.

$$\begin{bmatrix} 0 & 0 \\ 1 & 0 \\ 0 & 1 \end{bmatrix}$$

The row of 0s is not last.

EXAMPLE 2   **Writing a System Matrix in Row Echelon Form and Using Back Substitution to Solve the System**

Solve the following system by transforming its system matrix into row echelon form and back substituting:

$$\begin{cases} x + 2y + 3z = 4 \\ 2x - y - 2z = 0 \\ x - 3y - 3z = -2 \end{cases}.$$

SOLUTION   This system is represented by the system matrix

$$\begin{bmatrix} 1 & 2 & 3 & 4 \\ 2 & -1 & -2 & 0 \\ 1 & -3 & -3 & -2 \end{bmatrix}.$$

We will reduce the system matrix to row echelon form. We can use the 1 in the upper left corner to zero out the rest of the first column. To do so, we multiply the first row by –2 and add the result to the second row. We indicate this operation with the notation "$(-2)R1 + R2$."

$$\begin{bmatrix} 1 & 2 & 3 & 4 \\ 2 & -1 & -2 & 0 \\ 1 & -3 & -3 & -2 \end{bmatrix} (-2)R1 + R2 \rightarrow \begin{bmatrix} 1 & 2 & 3 & 4 \\ 0 & -5 & -8 & -8 \\ 1 & -3 & -3 & -2 \end{bmatrix}$$

Next, we multiply row one by –1 and add the result to row three to get a new row three. This fills the rest of the first column with 0s.

$$(-1)R1 + R3 \rightarrow \begin{bmatrix} 1 & 2 & 3 & 4 \\ 0 & -5 & -8 & -8 \\ 0 & -5 & -6 & -6 \end{bmatrix}$$

Comment

When reducing a matrix to row echelon form, we obtain 0s below each lead entry in a column by adding multiples of the row containing the lead entry.

We multiply row two by –1 and add the result to row three to get

$$(-1)R2 + R3 \to \begin{bmatrix} 1 & 2 & 3 & \vdots & 4 \\ 0 & -5 & -8 & \vdots & -8 \\ 0 & 0 & 2 & \vdots & 2 \end{bmatrix}.$$

Finally, we multiply row two by $-\frac{1}{5}$ and row three by $\frac{1}{2}$.

$$\begin{matrix} \left(-\frac{1}{5}\right)R2 \to \\ \left(\frac{1}{2}\right)R3 \to \end{matrix} \begin{bmatrix} 1 & 2 & 3 & \vdots & 4 \\ 0 & 1 & \frac{8}{5} & \vdots & \frac{8}{5} \\ 0 & 0 & 1 & \vdots & 1 \end{bmatrix}$$

The final matrix, now in row echelon form, represents the equivalent system

$$\begin{cases} x + 2y + 3z = 4 \\ y + \dfrac{8}{5}z = \dfrac{8}{5} \\ z = 1 \end{cases}.$$

We can now use back substitution to solve the system. To find y, we substitute 1 for z in the second equation.

$$y + \frac{8}{5}z = \frac{8}{5}$$

$$y + \frac{8}{5}(\mathbf{1}) = \frac{8}{5}$$

$$y = 0$$

To solve for x, we substitute 0 for y and 1 for z in the first equation.

$$x + 2y + 3z = 4$$

$$x + 2(\mathbf{0}) + 3(\mathbf{1}) = 4$$

$$x + 3 = 4$$

$$x = 1$$

The solution of the original system is the triple $(x, y, z) = (1, 0, 1)$. ∎

Self Check 2   Solve: $\begin{cases} x + y + 2z = 5 \\ y - 2z = 0 \\ x - z = 0 \end{cases}$.

Now Try Exercise 35.

ACCENT ON TECHNOLOGY / **Reduce a Matrix to Row Echelon Form**

A matrix can be reduced to row echelon form with a graphing calculator. As an example, we will reduce the matrix

$$\begin{bmatrix} 1 & 2 & 3 & 4 \\ 2 & -1 & -2 & 0 \\ 1 & -3 & -3 & -2 \end{bmatrix}$$

to row echelon form.

The MATRIX menu is accessed by pressing `2nd` `x⁻¹`.

Move to EDIT, select matrix [A], and press `ENTER`. We will see this screen.

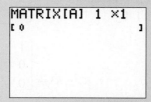

To change the matrix to row echelon form, go to the home screen; access the MATRIX MATH menu; and select A:ref(.

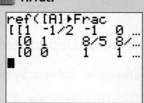

If we want the answers in fraction form, we can add that command by selecting `MATH` 1:Frac.

The window we see will list the matrices that are available to edit.

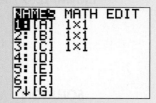

Change the dimensions of the matrix to 3 × 4 and use the cursor to enter the values of the matrix.

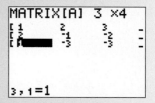

Now select the MATRIX menu again and select [A]. Press `ENTER` and the row echelon form is shown.

FIGURE 8-4

Caution

The row echelon form we obtained using the graphing calculator as shown in Figure 8-4 is not the same as the row echelon form found in Example 2. However, if we multiply row two of the last matrix shown in Figure 8-4 by $\frac{5}{2}$ and add the results to row one, we obtain the equivalent row echelon form found in Example 2.

$$\begin{bmatrix} 1 & -\frac{1}{2} & -1 & 0 \\ 0 & 1 & \frac{8}{5} & \frac{8}{5} \\ 0 & 0 & 1 & 1 \end{bmatrix} \quad \left(\tfrac{5}{2}\right)R2 + R1 \rightarrow \begin{bmatrix} 1 & 2 & 3 & 4 \\ 0 & 1 & \frac{8}{5} & \frac{8}{5} \\ 0 & 0 & 1 & 1 \end{bmatrix}$$

If we back substitute into the equations represented by either row echelon form, we will obtain the same values for x, y, and z.

Since the systems of Examples 1 and 2 have solutions, each system is consistent. The next two examples illustrate a system that is inconsistent and a system whose equations are dependent.

2. Solve an Inconsistent System

EXAMPLE 3   **Using Matrix Methods to Solve an Inconsistent System of Equations**

Use matrix methods to solve $\begin{cases} x + y + z = 3 \\ 2x - y + z = 2. \\ 3y + z = 1 \end{cases}$

SOLUTION   We form the system matrix and use row operations to write it in row echelon form. We use the 1 in the top left position to zero out the rest of the first column.

$$\begin{bmatrix} 1 & 1 & 1 & \vdots & 3 \\ 2 & -1 & 1 & \vdots & 2 \\ 0 & 3 & 1 & \vdots & 1 \end{bmatrix} \begin{matrix} \\ (-2)R1 + R2 \to \\ \\ \end{matrix} \begin{bmatrix} 1 & 1 & 1 & \vdots & 3 \\ 0 & -3 & -1 & \vdots & -4 \\ 0 & 3 & 1 & \vdots & 1 \end{bmatrix}$$

$$\begin{matrix} \\ \\ R2 + R3 \to \end{matrix} \begin{bmatrix} 1 & 1 & 1 & \vdots & 3 \\ 0 & -3 & -1 & \vdots & -4 \\ 0 & 0 & 0 & \vdots & -3 \end{bmatrix}$$

Since the last row of the matrix represents the equation

$$0x + 0y + 0z = -3$$

and no values of x, y, and z could make $0 = -3$, there is no point in continuing. The given system has no solution and is inconsistent. ∎

Self Check 3   Solve: $\begin{cases} x + y + z = 5 \\ y - z = 1 \\ x + 2y = 3 \end{cases}$.

Now Try Exercise 43.

3. Solve a System of Dependent Equations

EXAMPLE 4   **Using Matrix Methods to Solve a System of Dependent Equations**

Use matrices to solve $\begin{cases} x + 2y + z = 8 \\ 2x + y - z = 1 \\ x - y - 2z = -7 \end{cases}$.

SOLUTION   We can set up the system matrix and use row operations to reduce it to row echelon form.

$$\begin{bmatrix} 1 & 2 & 1 & \vdots & 8 \\ 2 & 1 & -1 & \vdots & 1 \\ 1 & -1 & -2 & \vdots & -7 \end{bmatrix} \begin{matrix} \\ (-2)R1 + R2 \to \\ (-1)R1 + R3 \to \end{matrix} \begin{bmatrix} 1 & 2 & 1 & \vdots & 8 \\ 0 & -3 & -3 & \vdots & -15 \\ 0 & -3 & -3 & \vdots & -15 \end{bmatrix}$$

$$\begin{matrix} \\ \left(-\frac{1}{3}\right)R2 \to \\ \left(-\frac{1}{3}\right)R3 \to \end{matrix} \begin{bmatrix} 1 & 2 & 1 & \vdots & 8 \\ 0 & 1 & 1 & \vdots & 5 \\ 0 & 1 & 1 & \vdots & 5 \end{bmatrix}$$

$$\begin{matrix} \\ \\ (-1)R2 + R3 \to \end{matrix} \begin{bmatrix} 1 & 2 & 1 & \vdots & 8 \\ 0 & 1 & 1 & \vdots & 5 \\ 0 & 0 & 0 & \vdots & 0 \end{bmatrix}$$

The final matrix is in echelon form and represents the system

(1) $\begin{cases} x + 2y + z = 8 \\ y + z = 5 \\ 0x + 0y + 0z = 0. \end{cases}$
(2)
(3)

Since all coefficients in Equation (3) are 0, it can be ignored. To solve this system by back substitution, we solve Equation (2) for y in terms of z,

$$y = 5 - z,$$

and then substitute $5 - z$ for y in Equation (1) and solve for x in terms of z.

(1) $\qquad x + 2y + z = 8$

$x + 2(\mathbf{5 - z}) + z = 8$

$x + 10 - 2z + z = 8 \qquad$ Remove parentheses.

$x + 10 - z = 8 \qquad$ Combine like terms.

$x = -2 + z \qquad$ Solve for x.

The solution of this system is $(x, y, z) = (-2 + z, 5 - z, z)$.

There are infinitely many solutions to this system. We can choose any real number for z, but once it is chosen, x and y are determined. For example:

- If $z = 3$, then $x = 1$ and $y = 2$. One possible solution of this system is $x = 1$, $y = 2$, and $z = 3$.
- If $z = 2$, then $x = 0$ and $y = 3$. Another solution is $x = 0$, $y = 3$, and $z = 2$.

Because this system has solutions, it is consistent. Because there are infinitely many solutions, the equations are dependent.

Self Check 4   Solve: $\begin{cases} x + 2y + z = 2 \\ x - z = 2 \\ 2x + 2y = 4 \end{cases}$.

Now Try Exercise 41.

4. Solve Systems Using Gauss–Jordan Elimination

In Gaussian elimination, we perform row operations until the system matrix is in row echelon form. Then we convert the matrix back into equation form and solve by back substitution. A modification of this method, called **Gauss–Jordan elimination,** uses row operations to produce a matrix in *reduced* **row echelon** form. With this method, we can obtain the solution of the system from that matrix directly, and back substitution is not needed.

Reduced Row Echelon Form of a Matrix

A matrix is in **reduced row echelon form** if

1. It is in row echelon form.

2. Entries above each lead entry are zero.

Examples of Matrices in Reduced Row Echelon Form:
The matrices

$$\begin{bmatrix} 1 & 0 & 0 \\ 0 & 1 & 0 \end{bmatrix} \quad \text{and} \quad \begin{bmatrix} 1 & 0 & 0 & 7 \\ 0 & 1 & 0 & 8 \\ 0 & 0 & 1 & 3 \end{bmatrix}$$

are in reduced row echelon form because each matrix is in row echelon form and the entries above each lead entry are 0s.

Examples of Matrices Not in Reduced Row Echelon Form:

$$\begin{bmatrix} 1 & 3 & 0 & 7 \\ 0 & 0 & 1 & 8 \\ 0 & 0 & 1 & 0 \end{bmatrix} \qquad \text{The matrix is not in row echelon form.}$$

$$\begin{bmatrix} 0 & 3 & 5 \\ 0 & 1 & 1 \end{bmatrix} \qquad \text{The number above the lead entry of 1 is not 0.}$$

In the next example, we solve a system by Gauss–Jordan elimination.

EXAMPLE 5 **Using Gauss–Jordan Elimination to Solve a System of Equations**

Use Gauss–Jordan elimination to solve $\begin{cases} w + 2x + 3y + z = 4 \\ x + 4y - z = 0 \\ w - x - y + 2z = 2 \end{cases}$.

SOLUTION We reduce the system matrix as follows. We first use the 1 in the upper left corner to zero out the rest of the first column.

$$\begin{bmatrix} 1 & 2 & 3 & 1 & \vdots & 4 \\ 0 & 1 & 4 & -1 & \vdots & 0 \\ 1 & -1 & -1 & 2 & \vdots & 2 \end{bmatrix} \underset{(-1)R1 + R3 \rightarrow}{} \begin{bmatrix} 1 & 2 & 3 & 1 & \vdots & 4 \\ 0 & 1 & 4 & -1 & \vdots & 0 \\ 0 & -3 & -4 & 1 & \vdots & -2 \end{bmatrix}$$

The lead entry in the second row is already 1. We use it to zero out the rest of the second column.

$$\begin{matrix} (-2)R2 + R1 \rightarrow \\ \\ (3)R2 + R3 \rightarrow \end{matrix} \begin{bmatrix} 1 & 0 & -5 & 3 & \vdots & 4 \\ 0 & 1 & 4 & -1 & \vdots & 0 \\ 0 & 0 & 8 & -2 & \vdots & -2 \end{bmatrix}$$

To make the lead entry in the third row equal to 1, we multiply the third row by $\frac{1}{8}$ and use it to zero out the rest of the third column.

$$\left(\tfrac{1}{8}\right)R3 \rightarrow \begin{bmatrix} 1 & 0 & -5 & 3 & \vdots & 4 \\ 0 & 1 & 4 & -1 & \vdots & 0 \\ 0 & 0 & 1 & -\frac{1}{4} & \vdots & -\frac{1}{4} \end{bmatrix}$$

$$\begin{matrix} (5)R3 + R1 \rightarrow \\ (-4)R3 + R2 \rightarrow \\ \\ \end{matrix} \begin{bmatrix} 1 & 0 & 0 & \frac{7}{4} & \vdots & \frac{11}{4} \\ 0 & 1 & 0 & 0 & \vdots & 1 \\ 0 & 0 & 1 & -\frac{1}{4} & \vdots & -\frac{1}{4} \end{bmatrix}$$

The final matrix is in reduced row echelon form. Note that each lead entry is 1, and each is alone in its column. The matrix represents the system of equations

$$\begin{cases} w + \dfrac{7}{4}z = \dfrac{11}{4} \\ x = 1 \\ y - \dfrac{1}{4}z = -\dfrac{1}{4} \end{cases} \quad \text{or} \quad \begin{cases} w = \dfrac{11}{4} - \dfrac{7}{4}z \\ x = 1 \\ y = -\dfrac{1}{4} + \dfrac{1}{4}z \end{cases}.$$

A general solution is $(w, x, y, z) = \left(\frac{11}{4} - \frac{7}{4}z, 1, -\frac{1}{4} + \frac{1}{4}z, z\right)$. To find some specific solutions, we choose values for z, and the corresponding values of w, x, and y will be determined. For example, if $z = 1$, then $w = 1$, $x = 1$, and $y = 0$. Thus, $(w, x, y, z) = (1, 1, 0, 1)$ is a solution. If $z = -1$, another solution is $(w, x, y, z) = \left(\frac{9}{2}, 1, -\frac{1}{2}, -1\right)$.   ∎

Self Check 5   Solve: $\begin{cases} w + x + y - z = 2 \\ 2w + 2x + 2y - 2z = 3 \\ w + x - y + 3z = 4 \\ w + 3x + 2y - 2z = -1 \end{cases}$.

Now Try Exercise 59.

Self Check Answers   **1.** $(2, -1, 3)$ **2.** $(1, 2, 1)$ **3.** The system is inconsistent; no solution.
4. $(2 + z, -z, z)$ **5.** The system is inconsistent; no solution.

Exercises **8.2**

Getting Ready
You should be able to complete these vocabulary and concept statements before you proceed to the practice exercises.

Fill in the blanks.

1. A rectangular array of numbers is called a
   _____.

2. A 3×5 matrix has _____ rows and
   _____ columns.

3. The matrix containing the coefficients of the variables is called the _____ matrix.

4. The coefficient matrix joined to the column of constants is called the _____ matrix or _____ matrix.

5. Each row of a system matrix represents one
   _____.

6. The rows of the system matrix are changed using elementary _____.

7. If one system matrix is changed to another using row operations, the matrices are _____.

8. If two system matrices are row equivalent, then the systems have the _____.

9. In a type 1 row operation, two rows of a matrix can be _____.

10. In a type 2 row operation, one entire row can be _____ by a nonzero constant.

11. In a type 3 row operation, any row can be changed by _____ to it any _____ of another row.

12. The first nonzero entry in a row is called that row's _____.

Practice
Use Gaussian elimination to solve each system.

13. $\begin{cases} x + y = 7 \\ x - 2y = -1 \end{cases}$ 14. $\begin{cases} x + 3y = 8 \\ 2x - 5y = 5 \end{cases}$

15. $\begin{cases} x - y = 1 \\ 2x - y = 8 \end{cases}$ 16. $\begin{cases} x - 5y = 4 \\ 2x + 3y = 21 \end{cases}$

17. $\begin{cases} x + 2y - z = 2 \\ x - 3y + 2z = 1 \\ x + y - 3z = -6 \end{cases}$ 18. $\begin{cases} x + 5y - z = 2 \\ x + 2y + z = 3 \\ x + y + z = 2 \end{cases}$

19. $\begin{cases} x - y - z = -3 \\ 5x + y = 6 \\ y + z = 4 \end{cases}$ 20. $\begin{cases} x + y = 1 \\ x + z = 3 \\ y + z = 2 \end{cases}$

Determine whether each matrix is in row echelon form, reduced row echelon form, or neither.

21. $\begin{bmatrix} 1 & 3 & 0 & 5 \\ 0 & 1 & 2 & 7 \\ 0 & 0 & 1 & 0 \end{bmatrix}$ 22. $\begin{bmatrix} 1 & 3 & 0 & 5 \\ 0 & 1 & 2 & 7 \\ 0 & 0 & 0 & 0 \end{bmatrix}$

23. $\begin{bmatrix} 1 & 0 & 1 \\ 0 & 1 & 5 \\ 0 & 0 & 0 \\ 0 & 0 & 0 \end{bmatrix}$ 24. $\begin{bmatrix} 1 & 0 & 1 \\ 0 & 1 & 5 \\ 0 & 0 & 1 \\ 0 & 0 & 0 \end{bmatrix}$

Write each system as a matrix and solve it by Gaussian elimination. If a system has infinitely many solutions, show a general solution.

25. $\begin{cases} 2x + y = 3 \\ x - 3y = 5 \end{cases}$ 26. $\begin{cases} x + 2y = -1 \\ 3x - 5y = 19 \end{cases}$

27. $\begin{cases} x - 7y = -2 \\ 5x - 2y = -10 \end{cases}$ 28. $\begin{cases} 3x - y = 3 \\ 2x + y = -3 \end{cases}$

29. $\begin{cases} 2x - y = 5 \\ x + 3y = 6 \end{cases}$ 30. $\begin{cases} 3x - 5y = -25 \\ 2x + y = 5 \end{cases}$

31. $\begin{cases} x - 2y = 3 \\ -2x + 4y = 6 \end{cases}$ 32. $\begin{cases} 2(2y - x) = 6 \\ 4y = 2(x + 3) \end{cases}$

33. $\begin{cases} 2x - y = 7 \\ -x + \dfrac{1}{3}y = -\dfrac{7}{3} \end{cases}$ 34. $\begin{cases} 45x - 6y = 60 \\ 30x + 15y = 63.75 \end{cases}$

35. $\begin{cases} x - y + z = 3 \\ 2x - y + z = 4 \\ x + 2y - z = -1 \end{cases}$ 36. $\begin{cases} 2x + y - z = 1 \\ x + y - z = 0 \\ 3x + y + 2z = 2 \end{cases}$

37. $\begin{cases} x + y - z = -1 \\ 3x + y = 4 \\ y - 2z = -4 \end{cases}$ 38. $\begin{cases} 3x + y = 7 \\ x - z = 0 \\ y - 2z = -8 \end{cases}$

39. $\begin{cases} x - y + z = 2 \\ 2x + y + z = 5 \\ 3x - 4z = -5 \end{cases}$ 40. $\begin{cases} x + z = -1 \\ 3x + y = 2 \\ 2x + y + 5z = 3 \end{cases}$

41. $\begin{cases} x + y - z = 5 \\ x + y + z = 2 \\ 3x + 3y - z = 12 \end{cases}$ 42. $\begin{cases} x + y + 2z = 4 \\ -x - y - 3z = -5 \\ 2x + y + z = 2 \end{cases}$

43. $\begin{cases} 2x - y + z = 6 \\ 3x + y - z = 2 \\ -x + 3y - 3z = 8 \end{cases}$ 44. $\begin{cases} -x + 3y + 2z = -10 \\ 3x - 2y - 2z = 7 \\ -2x + y - z = -10 \end{cases}$

Write each system as a matrix and solve it by Gauss–Jordan elimination.

45. $\begin{cases} x - 2y = 7 \\ y = 3 \end{cases}$ 46. $\begin{cases} x - 2y = 7 \\ y = 8 \end{cases}$

47. $\begin{cases} x + 2y - z = 3 \\ y + 3z = 1 \\ z = -2 \end{cases}$ 48. $\begin{cases} x - 3y + 2z = -1 \\ y - 2z = 3 \\ z = 5 \end{cases}$

49. $\begin{cases} x - y = 7 \\ x + y = 13 \end{cases}$ 50. $\begin{cases} x + 2y = 7 \\ 2x - y = -1 \end{cases}$

51. $\begin{cases} x - \dfrac{1}{2}y = 0 \\ x + 2y = 0 \end{cases}$ 52. $\begin{cases} x - y = 5 \\ -x + \dfrac{1}{5}y = -9 \end{cases}$

53. $\begin{cases} x + y + 2z = 0 \\ x + y + z = 2 \\ x + z = 1 \end{cases}$ 54. $\begin{cases} x + 2y = -3 \\ x + 4y = -2 \\ 2x + z = -8 \end{cases}$

55. $\begin{cases} 2x + y - 2z = 1 \\ -x + y - 3z = 0 \\ 4x + 3y = 4 \end{cases}$ 56. $\begin{cases} 3x + y = 3 \\ 3x + y - z = 2 \\ 6x + z = 5 \end{cases}$

57. $\begin{cases} 2x - 2y + 3z + t = 2 \\ x + y + z + t = 5 \\ -x + 2y - 3z + 2t = 2 \\ x + y + 2z - t = 4 \end{cases}$

58. $\begin{cases} x + y + 2z + t = 1 \\ x + 2y + z + t = 2 \\ 2x + y + z + t = 4 \\ x + y + z + 2t = 3 \end{cases}$

59. $\begin{cases} x + y + t = 4 \\ x + z + t = 2 \\ 2x + 2y + z + 2t = 8 \\ x - y + z - t = -2 \end{cases}$

60. $\begin{cases} x - y + 2z + t = 3 \\ 3x - 2y - z - t = 4 \\ 2x + y + 2z - t = 10 \\ x + 2y + z - 3t = 8 \end{cases}$

61. $\begin{cases} \dfrac{1}{3}x + \dfrac{3}{4}y - \dfrac{2}{3}z = -2 \\ x + \dfrac{1}{2}y + \dfrac{1}{3}z = 1 \\ \dfrac{1}{6}x - \dfrac{1}{8}y - z = 0 \end{cases}$ 62. $\begin{cases} \dfrac{1}{4}x + y + 3z = 1 \\ \dfrac{1}{2}x - 4y + 6z = -1 \\ \dfrac{1}{3}x - 2y - 2z = -1 \end{cases}$

63. $\begin{cases} \dfrac{1}{2}x + \dfrac{1}{4}y - z = 2 \\ \dfrac{2}{3}x + \dfrac{1}{4}y + \dfrac{1}{2}z = \dfrac{3}{2} \\ \dfrac{2}{3}x + z = -\dfrac{1}{3} \end{cases}$ 64. $\begin{cases} \dfrac{5}{7}x - \dfrac{1}{3}y + z = 0 \\ \dfrac{2}{7}x + y + \dfrac{1}{8}z = 9 \\ 6x + 4y - \dfrac{27}{4}z = 20 \end{cases}$

65. $\begin{cases} 3x - 6y + 9z = 18 \\ 2x - 4y + 3z = 12 \\ x - 2y + 3z = 6 \end{cases}$ 66. $\begin{cases} x + 2y - z = 7 \\ 2x - y + z = 2 \\ 3x - 4y + 3z = -3 \end{cases}$

Each system contains a different number of equations than variables. Solve each system using Gauss–Jordan elimination. If a system has infinitely many solutions, show a general solution.

67. $\begin{cases} x + y = -2 \\ 3x - y = 6 \\ 2x + 2y = -4 \\ x - y = 4 \end{cases}$ 68. $\begin{cases} x - y = -3 \\ 2x + y = -3 \\ 3x - y = -7 \\ 4x + y = -7 \end{cases}$

69. $\begin{cases} x + 2y + z = 4 \\ 3x - y - z = 2 \end{cases}$ 70. $\begin{cases} x + 2y - 3z = 5 \\ 5x + y - z = -11 \end{cases}$

71. $\begin{cases} w + x = 1 \\ w + y = 0 \\ x + z = 0 \end{cases}$ **72.** $\begin{cases} w + x - y + z = 2 \\ 2w - x - 2y + z = 0 \\ w - 2x - y + z = -1 \end{cases}$

73. $\begin{cases} x + y = 3 \\ 2x + y = 1 \\ 3x + 2y = 2 \end{cases}$ **74.** $\begin{cases} x + 2y + z = 4 \\ x - y + z = 1 \\ 2x + y + 2z = 2 \\ 3x + 3z = 6 \end{cases}$

Applications
Use matrix methods to solve each problem.

75. Flight range The speed of an airplane with a tail-wind is 300 miles per hour, and with a headwind, 220 miles per hour. On a day with no wind, how far could the plane travel on a 5-hour fuel supply?

76. Resource allocation 120,000 gallons of fuel are to be divided between two airlines. Triple A Airways requires twice as much fuel as UnityAir. How much fuel should be allocated to Triple A?

77. Library shelving To use space effectively, librarians like to fill shelves completely. One 35-inch shelf can hold 3 dictionaries, 5 atlases, and 1 thesaurus; or 6 dictionaries and 2 thesauruses; or 2 dictionaries, 4 atlases, and 3 thesauruses. How wide is one copy of each book?

78. Copying-machine productivity When both copying machines A and B are working, secretaries can make 100 copies in one minute. In one minute, copiers A and C together produce 140 copies. All three working together produce 180 copies in one minute. How many copies per minute can each machine produce separately?

79. Nutritional planning One ounce of each of three foods has the vitamin and mineral content shown in the table. How many ounces of each must be used to provide exactly 22 milligrams of niacin, 12 milligrams of zinc, and 20 milligrams of vitamin C?

Food	Niacin	Zinc	Vitamin C
A	1 mg	1 mg	2 mg
B	2 mg	1 mg	1 mg
C	2 mg	1 mg	2 mg

80. Chainsaw sculpting A wood sculptor carves three types of statues with a chainsaw. The number of hours required for carving, sanding, and painting a totem pole, a bear, and a deer are shown in the table. How many of each should be produced to use all available labor hours?

	Totem Pole	Bear	Deer	Time Available
Carving	2 hr	2 hr	1 hr	14 hr
Sanding	1 hr	2 hr	2 hr	15 hr
Painting	3 hr	2 hr	2 hr	21 hr

Discovery and Writing
81. Explain the difference between the row echelon form and the reduced row echelon form of a matrix.

82. If the upper left entry of a matrix is zero, what row operation might you do first?

83. What characteristics of a row reduced matrix would let you conclude that the system is inconsistent?

84. Explain the differences between Gaussian elimination and Gauss–Jordan elimination.

Use matrix methods to solve each system.

85. $\begin{cases} x^2 + y^2 + z^2 = 14 \\ 2x^2 + 3y^2 - 2z^2 = -7 \\ x^2 - 5y^2 + z^2 = 8 \end{cases}$

(*Hint:* Solve first as a system in x^2, y^2, and z^2.)

86. $\begin{cases} 5\sqrt{x} + 2\sqrt{x} + \sqrt{z} = 22 \\ \sqrt{x} + \sqrt{y} - \sqrt{z} = 5 \\ 3\sqrt{x} - 2\sqrt{y} - 3\sqrt{z} = 10 \end{cases}$

8.3 Matrix Algebra

In this section, we will learn to

1. Add and subtract matrices.
2. Multiply a matrix by a constant.
3. Multiply matrices.
4. Solve applications using matrices.
5. Recognize the identity matrix.

With certain restrictions, we can add, subtract, and multiply matrices. In fact, many problems can be solved using the arithmetic of matrices.

To illustrate the addition of matrices, suppose there are 108 police officers employed at two different locations.

Downtown Station		
	Male	Female
Day Shift	21	18
Night Shift	12	6

Suburban Station		
	Male	Female
Day Shift	14	12
Night Shift	15	10

The employment information about the police officers is contained in two matrices,

$$D = \begin{bmatrix} 21 & 18 \\ 12 & 6 \end{bmatrix} \quad \text{and} \quad S = \begin{bmatrix} 14 & 12 \\ 15 & 10 \end{bmatrix}.$$

The entry 21 in matrix D indicates that 21 male officers work the day shift at the downtown station. The entry 10 in matrix S indicates that 10 female officers work the night shift at the suburban station.

To find the citywide totals, we can add the corresponding entries of matrices D and S.

$$D + S = \begin{bmatrix} 21 & 18 \\ 12 & 6 \end{bmatrix} + \begin{bmatrix} 14 & 12 \\ 15 & 10 \end{bmatrix} = \begin{bmatrix} 35 & 30 \\ 27 & 16 \end{bmatrix}$$

We interpret the total to mean:

	Male	Female
Day Shift	35	30
Night Shift	27	16

To illustrate how to multiply a matrix by a number, suppose that one-third of the officers at the downtown station retire. The downtown staff would then consist of $\frac{2}{3}D$ officers. We can compute $\frac{2}{3}D$ by multiplying each entry of matrix D by $\frac{2}{3}$.

$$\frac{2}{3}D = \frac{2}{3}\begin{bmatrix} 21 & 18 \\ 12 & 6 \end{bmatrix} = \begin{bmatrix} 14 & 12 \\ 8 & 4 \end{bmatrix}$$

After retirements, the downtown staff will be:

	Male	Female
Day Shift	14	12
Night Shift	8	4

These examples show two calculations used in the algebra of matrices, which is the topic of this section. We begin by giving a formal definition of a matrix and defining when two matrices are equal.

Matrix An $m \times n$ **matrix** is a rectangular array of mn numbers arranged in m rows and n columns. We say that the matrix is of **size** (or **order**) $m \times n$.

Matrices are often denoted by letters such as A, B, and C. To denote the entries in an $m \times n$ matrix A, we use double-subscript notation: The entry in the first row, third column is a_{13}, and the entry in the ith row, jth column is a_{ij}. We can use any of the following notations to denote the $m \times n$ matrix A:

$$A, \quad [a_{ij}], \quad \text{or} \quad \left.\begin{bmatrix} a_{11} & a_{12} & a_{13} & \cdots & a_{1n} \\ a_{21} & a_{22} & a_{23} & \cdots & a_{2n} \\ \vdots & \vdots & \vdots & \ddots & \vdots \\ a_{m1} & a_{m2} & a_{m3} & \cdots & a_{mn} \end{bmatrix}\right\} m \text{ rows}$$

$$\underbrace{}_{n \text{ columns}}$$

Two matrices are equal if they are the same size, with the same entries in corresponding positions.

Equality of Matrices If $A = [a_{ij}]$ and $B = [b_{ij}]$ are both $m \times n$ matrices, then $A = B$, provided that each entry a_{ij} in matrix A is equal to the corresponding entry b_{ij} in matrix B.

The following matrices are equal, because they are the same size and corresponding entries are equal:

$$\begin{bmatrix} \sqrt{9} & 0.5 \\ 1 & 4 \end{bmatrix} = \begin{bmatrix} 3 & \frac{1}{2} \\ 1 & 2^2 \end{bmatrix}.$$

The following matrices are not equal, because they are not the same size:

$$\begin{bmatrix} 1 & 2 & 3 \\ 1 & 2 & 3 \end{bmatrix} \neq \begin{bmatrix} 1 & 2 & 3 \\ 1 & 2 & 3 \\ 1 & 2 & 3 \end{bmatrix}. \qquad \text{The first matrix is } 2 \times 3 \text{ and the second is } 3 \times 3.$$

1. Add and Subtract Matrices

We can add matrices of the same size by adding the entries in corresponding positions.

Sum of Two Matrices Let A and B be two $m \times n$ matrices. The **sum** $A + B$ is the $m \times n$ matrix C found by adding the corresponding entries of matrices A and B,

$$A + B = C,$$

where each entry c_{ij} in C is equal to the sum of a_{ij} in A and b_{ij} in B.

EXAMPLE 1 **Adding Matrices**

Add: $\begin{bmatrix} 2 & 1 & 3 \\ 1 & -1 & 0 \end{bmatrix} + \begin{bmatrix} 1 & -1 & 2 \\ -1 & 1 & 5 \end{bmatrix}.$

SOLUTION Since each matrix is 2×3, we can find their sum by adding their corresponding entries.

$$\begin{bmatrix} 2 & 1 & 3 \\ 1 & -1 & 0 \end{bmatrix} + \begin{bmatrix} 1 & -1 & 2 \\ -1 & 1 & 5 \end{bmatrix} = \begin{bmatrix} 2+1 & 1-1 & 3+2 \\ 1-1 & -1+1 & 0+5 \end{bmatrix}$$

$$= \begin{bmatrix} 3 & 0 & 5 \\ 0 & 0 & 5 \end{bmatrix}$$

Self Check 1   Add: $\begin{bmatrix} 3 & -5 \\ 2 & 0 \\ -6 & 5 \end{bmatrix} + \begin{bmatrix} -3 & 4 \\ -1 & -3 \\ 7 & 0 \end{bmatrix}$.

Now Try Exercise 13.

Caution

> Matrices that are not the same size cannot be added. Do not try to add them.

In arithmetic, 0 is the **additive identity,** because $a + 0 = 0 + a = a$ for any real number a. In matrix algebra, the matrix $\mathbf{0} = \begin{bmatrix} 0 & 0 \\ 0 & 0 \end{bmatrix}$ is called the additive identity for 2×2 matrices because $A + \mathbf{0} = \mathbf{0} + A$. For example,

$$\begin{bmatrix} 1 & 2 \\ 3 & 4 \end{bmatrix} + \begin{bmatrix} 0 & 0 \\ 0 & 0 \end{bmatrix} = \begin{bmatrix} 0 & 0 \\ 0 & 0 \end{bmatrix} + \begin{bmatrix} 1 & 2 \\ 3 & 4 \end{bmatrix} = \begin{bmatrix} 1 & 2 \\ 3 & 4 \end{bmatrix}.$$

Additive Identity Matrix   Let A be any $m \times n$ matrix. There is an $m \times n$ matrix $\mathbf{0}$, called the **zero matrix** or the **additive identity matrix,** for which

$$A + \mathbf{0} = \mathbf{0} + A = A.$$

The matrix $\mathbf{0}$ consists of m rows and n columns of 0s.

Every matrix also has an additive inverse.

Additive Inverse of a Matrix   Any $m \times n$ matrix A has an **additive inverse,** an $m \times n$ matrix $-A$ with the property that the sum of A and $-A$ is the zero matrix:

$$A + (-A) = (-A) + A = \mathbf{0}.$$

The entries of $-A$ are the negatives of the corresponding entries of A.

The additive inverse of $A = \begin{bmatrix} 1 & -3 & 2 \\ 0 & 1 & -5 \end{bmatrix}$ is the matrix

$$-A = \begin{bmatrix} -1 & 3 & -2 \\ 0 & -1 & 5 \end{bmatrix} \qquad \begin{array}{l}\text{Each entry of } -A \text{ is the negative of}\\ \text{the corresponding entry of } A.\end{array}$$

because their sum is the zero matrix:

$$A + (-A) = \begin{bmatrix} 1 & -3 & 2 \\ 0 & 1 & -5 \end{bmatrix} + \begin{bmatrix} -1 & 3 & -2 \\ 0 & -1 & 5 \end{bmatrix}$$

$$= \begin{bmatrix} 1-1 & -3+3 & 2-2 \\ 0+0 & 1-1 & -5+5 \end{bmatrix}$$

$$= \begin{bmatrix} 0 & 0 & 0 \\ 0 & 0 & 0 \end{bmatrix}$$

Subtraction of matrices is similar to the subtraction of real numbers.

Difference of Two Matrices   If A and B are $m \times n$ matrices, their **difference** $A - B$ is the sum of A and the additive inverse of B:

$$A - B = A + (-B).$$

For example,

$$\begin{bmatrix} 3 & 7 \\ -4 & 0 \end{bmatrix} - \begin{bmatrix} -1 & 4 \\ -5 & 1 \end{bmatrix} = \begin{bmatrix} 3 & 7 \\ -4 & 0 \end{bmatrix} + \begin{bmatrix} 1 & -4 \\ 5 & -1 \end{bmatrix} = \begin{bmatrix} 4 & 3 \\ 1 & -1 \end{bmatrix}.$$

2. Multiply a Matrix by a Constant

As we have seen in the police example at the beginning of the section, we can multiply a matrix by a constant by multiplying each of its entries by that constant. For example, if $A = \begin{bmatrix} 1 & -2 \\ 3 & 4 \end{bmatrix}$, then

$$5A = 5\begin{bmatrix} 1 & -2 \\ 3 & 4 \end{bmatrix} = \begin{bmatrix} 5 \cdot 1 & 5(-2) \\ 5 \cdot 3 & 5 \cdot 4 \end{bmatrix} = \begin{bmatrix} 5 & -10 \\ 15 & 20 \end{bmatrix}$$

and

$$-A = -1A = -1\begin{bmatrix} 1 & -2 \\ 3 & 4 \end{bmatrix} = \begin{bmatrix} -1 \cdot 1 & -1(-2) \\ -1 \cdot 3 & -1 \cdot 4 \end{bmatrix} = \begin{bmatrix} -1 & 2 \\ -3 & -4 \end{bmatrix}.$$

The second example above illustrates that we can find the additive inverse of a matrix by multiplying the matrix by -1.

If A is a matrix, the real number k in the product kA is called a **scalar.**

Product of a Scalar and a Matrix   If A and B are two $m \times n$ matrices and k is a scalar, then the **scalar multiple of A by k is B,** where each entry b_{ij} in B is equal to k times the corresponding entry a_{ij} in A.

EXAMPLE 2   **Solving a Matrix Equation**

Let $\begin{bmatrix} 5 & y \\ 15 & z \end{bmatrix} = 5\begin{bmatrix} x & 3 \\ 3 & y \end{bmatrix}$. Find y and z.

SOLUTION   We simplify the right side of the equation by multiplying each entry of the matrix by 5.

$$\begin{bmatrix} 5 & y \\ 15 & z \end{bmatrix} = \begin{bmatrix} 5x & 15 \\ 15 & 5y \end{bmatrix}$$

Because the matrices are equal, their corresponding entries are equal. So $y = 15$ and $z = 5y$. We conclude that $y = 15$ and $z = 5 \cdot 15 = 75$. ∎

Self Check 2   Find x in Example 2.

Now Try Exercise 11.

3. Multiply Matrices

To introduce multiplication of matrices, we consider the grading policy at a school where a student's final grade is based on a test-score average and a homework average.

The grades for three students are shown on the next page, along with the weights of the test scores and the homework scores.

	Test Score Average	Homework Average
Ann	83	95
Tonya	72	80
Carlos	85	92

	Weighting	
	System 1	System 2
Test average	0.5	0.3
Homework average	0.5	0.7

To calculate the students' final grades, we must use information from both tables. For example, if their teacher used weighting system 1, their final grades would be

- Ann's grade = 83(0.5) + 95(0.5) = 41.5 + 47.5 = 89;
- Tonya's grade = 72(0.5) + 80(0.5) = 36 + 40 = 76;
- Carlos's grade = 85(0.5) + 92(0.5) = 42.5 + 46 = 88.5.

If we calculate their final grades using weighting system 2, we get

- Ann's grade = 83(0.3) + 95(0.7) = 24.9 + 66.5 = 91.4;
- Tonya's grade = 72(0.3) + 80(0.7) = 21.6 + 56 = 77.6;
- Carlos's grade = 85(0.3) + 92(0.7) = 25.5 + 64.4 = 89.9.

The result of each grade calculation is the sum of the products of the elements in one row of the first matrix and one column of the second matrix. Each calculation is a part of the product of the matrices

$$A = \begin{bmatrix} 83 & 95 \\ 72 & 80 \\ 85 & 92 \end{bmatrix} \quad \text{and} \quad B = \begin{bmatrix} 0.5 & 0.3 \\ 0.5 & 0.7 \end{bmatrix}.$$

To further illustrate how to find the product of two matrices, we will find the product of the 3 × 2 matrix A and the 2 × 2 matrix B. This product will exist because the number of columns in A is equal to the number of rows in B. The result will be a 3 × 2 matrix C.

$$AB = \begin{bmatrix} 83 & 95 \\ 72 & 80 \\ 85 & 92 \end{bmatrix} \begin{bmatrix} 0.5 & 0.3 \\ 0.5 & 0.7 \end{bmatrix} = \begin{bmatrix} ? & ? \\ ? & ? \\ ? & ? \end{bmatrix} = C$$

Each entry of matrix C will be the result of multiplying the entries in one row of A and the entries in one column of B and adding the results. The first-row, first-column entry of matrix C is the sum of the products of entries of the first row of A and the first column of B. The second-row, second-column entry of matrix C is the sum of the products of entries of the second row of A and the second column of B.

$$AB = \begin{bmatrix} 83 & 95 \\ 72 & 80 \\ 85 & 92 \end{bmatrix} \begin{bmatrix} 0.5 & 0.3 \\ 0.5 & 0.7 \end{bmatrix} = \begin{bmatrix} 83(0.5) + 95(0.5) & ? \\ ? & 72(0.3) + 80(0.7) \\ ? & ? \end{bmatrix} = \begin{bmatrix} 89 & ? \\ ? & 77.6 \\ ? & ? \end{bmatrix} = C$$

The first-row, second-column entry of matrix C is the sum of the products of entries of the first row of A and the second column of B. The second-row, first-column entry of matrix C is the sum of the products of entries of the second row of A and the first column of B.

$$AB = \begin{bmatrix} 83 & 95 \\ 72 & 80 \\ 85 & 92 \end{bmatrix} \begin{bmatrix} 0.5 & 0.3 \\ 0.5 & 0.7 \end{bmatrix} = \begin{bmatrix} 83(0.5) + 95(0.5) & 83(0.3) + 95(0.7) \\ 72(0.5) + 80(0.5) & 72(0.3) + 80(0.7) \\ ? & ? \end{bmatrix} = \begin{bmatrix} 89 & 91.4 \\ 76 & 77.6 \\ ? & ? \end{bmatrix} = C$$

The third-row, first-column entry of matrix C is the sum of the products of entries of the third row of A and the first column of B. The third-row, second-column entry of matrix C is the sum of the products of entries of the third row of A and the second column of B.

$$AB = \begin{bmatrix} 83 & 95 \\ 72 & 80 \\ 85 & 92 \end{bmatrix} \begin{bmatrix} 0.5 & 0.3 \\ 0.5 & 0.7 \end{bmatrix} = \begin{bmatrix} 83(0.5) + 95(0.5) & 83(0.3) + 95(0.7) \\ 72(0.5) + 80(0.5) & 72(0.3) + 80(0.7) \\ 85(0.5) + 92(0.5) & 85(0.3) + 92(0.7) \end{bmatrix} = \begin{bmatrix} 89 & 91.4 \\ 76 & 77.6 \\ 88.5 & 89.9 \end{bmatrix} = C$$

The resulting 3×2 matrix C is the product of matrices A and B, and gives the final grades of each student with respect to each weighting system.

	Weighting	
	System 1	System 2
Ann	89	91.4
Tonya	76	77.6
Carlos	88.5	89.9

For the product AB of two matrices to exist, the number of columns of A must equal the number of rows of B. If the product exists, it will have as many rows as A and as many columns as B.

$$\begin{array}{ccccc} A & \cdot & B & = & C \\ m \times n & & n \times p & & m \times p \end{array}$$

These must agree.

The product is of size $m \times p$.

We now formally define the product of two matrices.

Product of Two Matrices   Let A be an $m \times n$ matrix and B be an $n \times p$ matrix. The **product** AB is the $m \times p$ matrix C,

$$AB = C,$$

where each entry c_{ij} in C is the sum of the products of the corresponding entries in the ith row of A and the jth column of B, where $i = 1, 2, 3, \dots, m$, and $j = 1, 2, 3, \dots, p$.

EXAMPLE 3   Multiplying Matrices

Find $C = AB$ if $A = \begin{bmatrix} 1 & 2 & 4 \\ -2 & 1 & -1 \end{bmatrix}$ and $B = \begin{bmatrix} 1 & 5 \\ -2 & 4 \\ 1 & -3 \end{bmatrix}$.

SOLUTION   Because matrix A is 2×3 and B is 3×2, the number of columns of A is the same as the number of rows of B. Therefore, the product C exists and it will be a 2×2 matrix. To find entry c_{11} of C, we find the sum of the products of the entries in the first row of A and the first column of B:

$$c_{11} = 1 \cdot 1 + 2 \cdot (-2) + 4 \cdot 1 = 1.$$

$$\begin{bmatrix} 1 & 2 & 4 \\ -2 & 1 & -1 \end{bmatrix} \begin{bmatrix} 1 & 5 \\ -2 & 4 \\ 1 & -3 \end{bmatrix} = \begin{bmatrix} 1 & ? \\ ? & ? \end{bmatrix}$$

To find entry c_{12}, we move across the first row of A and down the second column of B:

$$c_{12} = 1 \cdot 5 + 2 \cdot 4 + 4 \cdot (-3) = 1.$$

$$\begin{bmatrix} 1 & 2 & 4 \\ -2 & 1 & -1 \end{bmatrix} \begin{bmatrix} 1 & 5 \\ -2 & 4 \\ 1 & -3 \end{bmatrix} = \begin{bmatrix} 1 & 1 \\ ? & ? \end{bmatrix}$$

To find entry c_{21}, we move across the second row of A and down the first column of B:

$$c_{21} = (-2) \cdot 1 + 1 \cdot (-2) + (-1) \cdot 1 = -5.$$

$$\begin{bmatrix} 1 & 2 & 4 \\ -2 & 1 & -1 \end{bmatrix} \begin{bmatrix} 1 & 5 \\ -2 & 4 \\ 1 & -3 \end{bmatrix} = \begin{bmatrix} 1 & 1 \\ -5 & ? \end{bmatrix}$$

To find entry c_{22}, we move across the second row of A and down the second column of B:

Caution

When we multiply matrices of the same size, we must keep in mind that we do not multiply corresponding entries.

$$c_{22} = (-2) \cdot 5 + 1 \cdot 4 + (-1) \cdot (-3) = -3.$$

$$\begin{bmatrix} 1 & 2 & 4 \\ -2 & 1 & -1 \end{bmatrix} \begin{bmatrix} 1 & 5 \\ -2 & 4 \\ 1 & -3 \end{bmatrix} = \begin{bmatrix} 1 & 1 \\ -5 & -3 \end{bmatrix}$$

Self Check 3   Find $D = EF$ if $E = \begin{bmatrix} 1 & -2 \\ 2 & 0 \end{bmatrix}$ and $F = \begin{bmatrix} 2 & 1 \\ 3 & 5 \end{bmatrix}$.

Now Try Exercise 27.

EXAMPLE 4   **Multiplying Matrices**

Find each product.   **a.** $\begin{bmatrix} 1 & 2 & 3 \end{bmatrix} \begin{bmatrix} 4 \\ 5 \\ 6 \end{bmatrix}$ **b.** $\begin{bmatrix} 1 \\ 2 \\ 3 \end{bmatrix} \begin{bmatrix} 4 & 5 & 6 \end{bmatrix}$

SOLUTION   **a.** Since the first matrix is 1×3 and the second matrix is 3×1, the product is a 1×1 matrix.

$$\begin{bmatrix} 1 & 2 & 3 \end{bmatrix} \begin{bmatrix} 4 \\ 5 \\ 6 \end{bmatrix} = \begin{bmatrix} 1 \cdot 4 + 2 \cdot 5 + 3 \cdot 6 \end{bmatrix} = \begin{bmatrix} 32 \end{bmatrix}$$

b. Since the first matrix is 3×1 and the second matrix is 1×3, the product is a 3×3 matrix.

$$\begin{bmatrix} 1 \\ 2 \\ 3 \end{bmatrix} \begin{bmatrix} 4 & 5 & 6 \end{bmatrix} = \begin{bmatrix} 1 \cdot 4 & 1 \cdot 5 & 1 \cdot 6 \\ 2 \cdot 4 & 2 \cdot 5 & 2 \cdot 6 \\ 3 \cdot 4 & 3 \cdot 5 & 3 \cdot 6 \end{bmatrix} = \begin{bmatrix} 4 & 5 & 6 \\ 8 & 10 & 12 \\ 12 & 15 & 18 \end{bmatrix}$$

Self Check 4   Find each product.

a. $\begin{bmatrix} 1 & 3 & 5 \end{bmatrix} \begin{bmatrix} 1 \\ 0 \\ 1 \end{bmatrix}$ **b.** $\begin{bmatrix} 1 \\ 2 \end{bmatrix} \begin{bmatrix} 3 & 4 \end{bmatrix}$

Now Try Exercise 31.

Comment

There are cases where the multiplication of two matrices is commutative, $AB = BA$, but in general, the multiplication of matrices is not commutative.

The multiplication of matrices is not commutative. To show this, we compute AB and BA, where $A = \begin{bmatrix} 1 & 1 \\ 0 & 0 \end{bmatrix}$ and $B = \begin{bmatrix} 0 & 1 \\ 0 & 1 \end{bmatrix}$.

$$AB = \begin{bmatrix} 1 & 1 \\ 0 & 0 \end{bmatrix}\begin{bmatrix} 0 & 1 \\ 0 & 1 \end{bmatrix} = \begin{bmatrix} 0 & 2 \\ 0 & 0 \end{bmatrix}$$

$$BA = \begin{bmatrix} 0 & 1 \\ 0 & 1 \end{bmatrix}\begin{bmatrix} 1 & 1 \\ 0 & 0 \end{bmatrix} = \begin{bmatrix} 0 & 0 \\ 0 & 0 \end{bmatrix}$$

Since the products are not equal, matrix multiplication is not commutative.

4. Solve Applications Using Matrices

EXAMPLE 5   **Solving an Application Problem**

Suppose that supplies must be purchased for the police officers discussed at the beginning of the section. The quantities and prices of each item required for each shift are as follows:

	Quantities				Unit Price (in $)
	Uniforms	Badges	Whistles		
Day Shift	17	13	19	Uniforms	47
Night Shift	14	24	27	Badges	7
				Whistles	5

Find the cost of supplies for each shift.

SOLUTION   We can write the quantities Q and prices P in matrix form and multiply them to get a cost matrix.

$$C = QP$$

$$= \begin{bmatrix} 17 & 13 & 19 \\ 14 & 24 & 27 \end{bmatrix}\begin{bmatrix} 47 \\ 7 \\ 5 \end{bmatrix}$$

$$= \begin{bmatrix} 17 \cdot 47 + 13 \cdot 7 + 19 \cdot 5 \\ 14 \cdot 47 & 24 \cdot 7 & 27 \cdot 5 \end{bmatrix}$$ (17 uniforms)($47) + (13 badges)($7) + (19 whistles)($5)
(14 uniforms)($47) + (24 badges)($7) + (27 whistles)($5)

$$= \begin{bmatrix} 985 \\ 961 \end{bmatrix}$$

It will cost $985 to buy supplies for the day shift and $961 to buy supplies for the night shift.

Slef Check 5   If the unit prices (in dollars) in Example 5 increase to

$$\begin{matrix} \text{Uniforms} \\ \text{Badges} \\ \text{Whistles} \end{matrix}\begin{bmatrix} 50 \\ 10 \\ 6 \end{bmatrix},$$

how much will it cost to buy supplies for the day shift and night shift?

Now Try Exercise 55.

5. Recognize the Identity Matrix

The number 1 is called the *identity for multiplication* because multiplying a number by 1 does not change the number: $a \cdot 1 = 1 \cdot a = a$. There is a **multiplicative identity matrix** with a similar property.

Identity Matrix   Let A be an $n \times n$ matrix. There is an $n \times n$ **identity matrix** I for which

$$AI = IA = A.$$

It is the matrix I consisting of 1s on its diagonal and 0s elsewhere.

$$I = \begin{bmatrix} 1 & 0 & 0 & \cdots & 0 \\ 0 & 1 & 0 & \cdots & 0 \\ 0 & 0 & 1 & \cdots & 0 \\ \vdots & \vdots & \vdots & \ddots & \vdots \\ 0 & 0 & 0 & \cdots & 1 \end{bmatrix}$$

Comment   An identity matrix is always a square matrix—a matrix that has the same number of rows and columns.

We illustrate the previous definition for the 3×3 identity matrix.

$$\begin{bmatrix} 1 & 0 & 0 \\ 0 & 1 & 0 \\ 0 & 0 & 1 \end{bmatrix}\begin{bmatrix} 1 & 2 & 3 \\ 4 & 5 & 6 \\ 7 & 8 & 9 \end{bmatrix} = \begin{bmatrix} 1\cdot 1 + 0\cdot 4 + 0\cdot 7 & 1\cdot 2 + 0\cdot 5 + 0\cdot 8 & 1\cdot 3 + 0\cdot 6 + 0\cdot 9 \\ 0\cdot 1 + 1\cdot 4 + 0\cdot 7 & 0\cdot 2 + 1\cdot 5 + 0\cdot 8 & 0\cdot 3 + 1\cdot 6 + 0\cdot 9 \\ 0\cdot 1 + 0\cdot 4 + 1\cdot 7 & 0\cdot 2 + 0\cdot 5 + 1\cdot 8 & 0\cdot 3 + 0\cdot 6 + 1\cdot 9 \end{bmatrix}$$

$$= \begin{bmatrix} 1 & 2 & 3 \\ 4 & 5 & 6 \\ 7 & 8 & 9 \end{bmatrix}$$

$$\begin{bmatrix} 1 & 2 & 3 \\ 4 & 5 & 6 \\ 7 & 8 & 9 \end{bmatrix}\begin{bmatrix} 1 & 0 & 0 \\ 0 & 1 & 0 \\ 0 & 0 & 1 \end{bmatrix} = \begin{bmatrix} 1 & 2 & 3 \\ 4 & 5 & 6 \\ 7 & 8 & 9 \end{bmatrix}$$

The following properties of real numbers carry over to matrices. In the exercises, you will be asked to illustrate many of these properties.

Properties of Matrices   Let A, B, and C be matrices and a and b be scalars.

Commutative Property of Addition:	$A + B = B + A$
Associative Property of Addition:	$A + (B + C) = (A + B) + C$
Associative Properties of Scalar Multiplication:	$\begin{cases} a(bA) = (ab)A \\ a(AB) = (aA)B \end{cases}$
Distributive Properties of Scalar Multiplication:	$\begin{cases} (a + b)A = aA + bA \\ a(A + b) = aA + aB \end{cases}$
Associative Property of Multiplication:	$A(BC) = (AB)C$
Distributive Properties of Matrix Multiplication:	$\begin{cases} A(B + C) = AB + AC \\ (A + B)C = AC + BC \end{cases}$

EXAMPLE 6   Verifying the Distributive Property of Matrix Multiplication

Verify the Distributive Property $A(B + C) = AB + AC$ using the matrices

$$A = \begin{bmatrix} 2 & 1 \\ 3 & 1 \end{bmatrix}, \qquad B = \begin{bmatrix} 3 & -4 \\ 1 & -1 \end{bmatrix}, \qquad \text{and} \qquad C = \begin{bmatrix} -1 & 3 \\ 0 & 1 \end{bmatrix}.$$

SOLUTION   We do the operations on the left side and the right side separately and compare the results.

$$\begin{bmatrix} 2 & 1 \\ 3 & 1 \end{bmatrix}\left(\begin{bmatrix} 3 & -4 \\ 1 & -1 \end{bmatrix} + \begin{bmatrix} -1 & 3 \\ 0 & 1 \end{bmatrix}\right) = \begin{bmatrix} 2 & 1 \\ 3 & 1 \end{bmatrix}\begin{bmatrix} 2 & -1 \\ 1 & 0 \end{bmatrix}$$ Do the addition within the parentheses.

$$= \begin{bmatrix} 5 & -2 \\ 7 & -3 \end{bmatrix}$$ Multiply.

$$\begin{bmatrix} 2 & 1 \\ 3 & 1 \end{bmatrix}\begin{bmatrix} 3 & -4 \\ 1 & -1 \end{bmatrix} + \begin{bmatrix} 2 & 1 \\ 3 & 1 \end{bmatrix}\begin{bmatrix} -1 & 3 \\ 0 & 1 \end{bmatrix} = \begin{bmatrix} 7 & -9 \\ 10 & -13 \end{bmatrix} + \begin{bmatrix} -2 & 7 \\ -3 & 10 \end{bmatrix}$$ Do the multiplications.

$$= \begin{bmatrix} 5 & -2 \\ 7 & -3 \end{bmatrix}$$ Add.

Because the left and right sides agree, this example illustrates the Distributive Property. ∎

Self Check 6   Verify the Distributive Property for multiplication,

$$(A + B)C = AC + BC,$$

for the matrices given in Example 6.

Now Try Exercise 43.

Self Check Answers   1. $\begin{bmatrix} 0 & -1 \\ 1 & -3 \\ 1 & 5 \end{bmatrix}$ 2. 1   3. $\begin{bmatrix} -4 & -9 \\ 4 & 2 \end{bmatrix}$ 4. **a.** $[6]$ **b.** $\begin{bmatrix} 3 & 4 \\ 6 & 8 \end{bmatrix}$

5. $1,094 for day shift; $1,102 for night shift   6. $\begin{bmatrix} 5 & -2 \\ 7 & -3 \end{bmatrix} = \begin{bmatrix} 5 & -2 \\ 7 & -3 \end{bmatrix}$.

Exercises **8.3**

Getting Ready
You should be able to complete these vocabulary and concept statements before you proceed to the practice exercises.

Fill in the blanks.

1. In a matrix A, the symbol a_{ij} is the entry in row _____ and column _____.

2. For matrices A and B to be equal, they must be the same _____, and corresponding entries must be _____.

3. To find the sum of matrices A and B, we add the _____ entries.

4. To multiply a matrix by a scalar, we multiply _____ by that scalar.

5. The product of a 3×2 matrix A and a 2×4 matrix B will exist because the number of _____ of A is equal to the number of _____ of B.

6. The product C of the matrices in Exercise 5 will be a _____ matrix.

7. Among 2×2 matrices, $\begin{bmatrix} 0 & 0 \\ 0 & 0 \end{bmatrix}$ is the _____ matrix.

8. Among 2×2 matrices, $\begin{bmatrix} 1 & 0 \\ 0 & 1 \end{bmatrix}$ is the _____ matrix.

Practice
Find values of x and y, if any, that will make the matrices equal.

9. $\begin{bmatrix} x & y \\ 1 & 3 \end{bmatrix} = \begin{bmatrix} 2 & 5 \\ 1 & 3 \end{bmatrix}$

10. $\begin{bmatrix} x & 5 \\ 3 & y \end{bmatrix} = \begin{bmatrix} 0 & 5 \\ 3 & 2 \end{bmatrix}$

11. $\begin{bmatrix} x+y & 3+x \\ -2 & 5y \end{bmatrix} = \begin{bmatrix} 3 & 4 \\ -2 & 10 \end{bmatrix}$

12. $\begin{bmatrix} x+y & x-y \\ 2x & 3y \end{bmatrix} = \begin{bmatrix} -x & x-2 \\ -y & 8-y \end{bmatrix}$

Find A + B.

13. $A = \begin{bmatrix} 2 & 1 & -1 \\ -3 & 2 & 5 \end{bmatrix}$; $B = \begin{bmatrix} -3 & 1 & 2 \\ -3 & -2 & -5 \end{bmatrix}$

14. $A = \begin{bmatrix} 3 & 2 & 1 \\ -2 & 3 & -3 \\ -4 & -2 & -1 \end{bmatrix}$; $B = \begin{bmatrix} -2 & 6 & -2 \\ 5 & 7 & -1 \\ -4 & -6 & 7 \end{bmatrix}$

Find the additive inverse of each matrix.

15. $A = \begin{bmatrix} 5 & -2 & 7 \\ -5 & 0 & 3 \\ -2 & 3 & -5 \end{bmatrix}$

16. $A = \begin{bmatrix} 3 & -\frac{2}{3} & -5 & \frac{1}{2} \end{bmatrix}$

Find A − B.

17. $A = \begin{bmatrix} -3 & 2 & -2 \\ -1 & 4 & -5 \end{bmatrix}$; $B = \begin{bmatrix} 3 & -3 & -2 \\ -2 & 5 & -5 \end{bmatrix}$

18. $A = \begin{bmatrix} 2 & 2 & 0 \\ -2 & 8 & 1 \\ 3 & -3 & -8 \end{bmatrix}$; $B = \begin{bmatrix} -4 & 3 & 7 \\ -1 & 2 & 0 \\ 1 & 4 & -1 \end{bmatrix}$

Find 5A.

19. $A = \begin{bmatrix} 3 & -3 \\ 0 & -2 \end{bmatrix}$

20. $A = \begin{bmatrix} 3 & \frac{3}{5} \\ 0 & -1 \end{bmatrix}$

21. $A = \begin{bmatrix} 5 & 15 & -2 \\ -2 & -5 & 1 \end{bmatrix}$

22. $A = \begin{bmatrix} -3 & 1 & 2 \\ -8 & -2 & -5 \end{bmatrix}$

Find 5A + 3B.

23. $A = \begin{bmatrix} 3 & 1 & -2 \\ -4 & 3 & -2 \end{bmatrix}$; $B = \begin{bmatrix} 1 & -2 & 2 \\ -5 & -5 & 3 \end{bmatrix}$

24. $A = \begin{bmatrix} 2 & -5 \\ -5 & 2 \end{bmatrix}$; $B = \begin{bmatrix} 5 & -2 \\ 2 & -5 \end{bmatrix}$

Find each product, if possible.

25. $\begin{bmatrix} 2 & 3 \\ 3 & -2 \end{bmatrix}\begin{bmatrix} 1 & 2 \\ 0 & -2 \end{bmatrix}$

26. $\begin{bmatrix} -2 & 3 \\ 3 & -2 \end{bmatrix}\begin{bmatrix} 2 & 4 \\ -5 & 7 \end{bmatrix}$

27. $\begin{bmatrix} -4 & -2 \\ 21 & 0 \end{bmatrix}\begin{bmatrix} -5 & 6 \\ 21 & -1 \end{bmatrix}$

28. $\begin{bmatrix} -5 & 4 \\ 4 & -5 \end{bmatrix}\begin{bmatrix} 6 & -2 \\ 1 & 3 \end{bmatrix}$

29. $\begin{bmatrix} 2 & 1 & 3 \\ 1 & 2 & -1 \\ 0 & 1 & 0 \end{bmatrix}\begin{bmatrix} 1 & 2 & 3 \\ 2 & -2 & 1 \\ 0 & 0 & 1 \end{bmatrix}$

30. $\begin{bmatrix} 2 & 1 & 1 \\ 1 & 1 & 2 \\ 1 & -2 & -1 \end{bmatrix}\begin{bmatrix} 1 & 2 & 3 \\ 1 & 2 & -3 \\ -1 & -1 & 3 \end{bmatrix}$

31. $\begin{bmatrix} 1 \\ -2 \\ -3 \end{bmatrix}\begin{bmatrix} 4 & -5 & -6 \end{bmatrix}$

32. $\begin{bmatrix} 1 & -2 & -3 \\ 2 & 0 & 1 \end{bmatrix}\begin{bmatrix} 4 \\ -5 \\ -6 \end{bmatrix}$

33. $\begin{bmatrix} 1 & 2 & 3 \end{bmatrix}\begin{bmatrix} 4 & 5 & 6 \\ 7 & 8 & 9 \end{bmatrix}$

34. $\begin{bmatrix} 2 & 5 \\ -1 & 7 \end{bmatrix}\begin{bmatrix} 3 & 5 & -8 \\ -2 & 7 & 5 \\ 3 & -6 & 2 \end{bmatrix}$

35. $\begin{bmatrix} 2 & 3 & 4 \\ 1 & 2 & 3 \\ -2 & 2 & 2 \end{bmatrix}\begin{bmatrix} -1 \\ 2 \\ 3 \end{bmatrix}$

36. $\begin{bmatrix} 2 & 5 \\ -3 & 1 \\ 0 & -2 \\ 1 & -5 \end{bmatrix}\begin{bmatrix} 3 & -2 & 4 \\ -2 & -3 & 1 \end{bmatrix}$

37. $\begin{bmatrix} 1 & 2 & 3 \\ 4 & 5 & 6 \\ 7 & 8 & 9 \end{bmatrix}\begin{bmatrix} 1 & 2 \\ 3 & 4 \end{bmatrix}$

38. $\begin{bmatrix} 1 & 4 & 0 & 0 \\ -4 & 1 & 0 & -2 \\ 0 & 0 & 1 & 0 \\ 0 & 2 & 0 & 1 \end{bmatrix}\begin{bmatrix} 1 \\ 2 \\ -2 \\ -1 \end{bmatrix}$

Let $A = \begin{bmatrix} 2.3 & -1.7 & 3.1 \\ -2 & 3.5 & 1 \\ -8 & 4.7 & 9.1 \end{bmatrix}$, $B = \begin{bmatrix} -2.5 \\ 5.2 \\ -7 \end{bmatrix}$, *and*

$C = \begin{bmatrix} -5.8 \\ 2.9 \\ 4.1 \end{bmatrix}$. *Use a graphing calculator to find each result.*

39. AB
40. $B + C$
41. A^2
42. $AB + C$

Let $A = \begin{bmatrix} 2 & 3 \\ 1 & 3 \end{bmatrix}$, $B = \begin{bmatrix} 2 & 1 & -5 \\ 1 & 1 & 2 \end{bmatrix}$,

$C = \begin{bmatrix} -2 & -1 & 6 \\ 0 & -1 & -1 \end{bmatrix}$, $D = \begin{bmatrix} 1 & 2 \\ 1 & 3 \end{bmatrix}$, *and*

$E = \begin{bmatrix} 1 & -2 \\ 2 & 3 \end{bmatrix}$.

Verify each property by doing the operations on each side of the equation and comparing the results.

43. Distributive Property: $A(B + C) = AB + AC$

44. Associative Property of Scalar Multiplication: $5(6A) = (5 \cdot 6)A$

45. Associative Property of Scalar Multiplication: $3(AB) = (3A)B$

46. Associative Property of Multiplication: $A(DE) = (AD)E$

Let $A = \begin{bmatrix} 1 & 3 \\ 2 & 5 \end{bmatrix}$, $B = \begin{bmatrix} -1 \\ 3 \end{bmatrix}$, and $C = \begin{bmatrix} 3 & 2 \end{bmatrix}$. *Perform the given operations, if possible.*

47. $A - BC$ **48.** $AB + B$

49. $CB - AB$ **50.** CAB

51. ABC **52.** $CA + C$

53. A^2B **54.** $(BC)^2$

Applications
Use a graphing calculator to help solve each problem.

55. Sporting goods Two suppliers manufactured footballs, baseballs, and basketballs in the quantities and at the costs given in the tables. Find matrices Q and C that represent the quantities and costs, find the product QC, and interpret the result.

	Quantity		
	Footballs	Baseballs	Basketballs
Supplier 1	200	300	100
Supplier 2	100	200	200

Unit Cost (in $)	
Football	5
Baseball	2
Basketball	4

56. Retailing Three ice-cream stores sold cones, sundaes, and shakes in the quantities and at the prices given in the tables. Find matrices Q and P that represent the quantities and prices, find the product QP, and interpret the results.

	Quantity		
	Cones	Sundaes	Shakes
Store 1	75	75	32
Store 2	80	69	27
Store 3	62	40	30

Unit Price	
Cone	$1.50
Sundae	$1.75
Shake	$3.00

57. Beverage sales Beverages were sold to parents and children at a school basketball game in the quantities and at the prices given in the tables. Find matrices Q and P that represent the quantities and prices, find the product QP, and interpret the result.

	Quantity		
	Coffee	Milk	Cola
Adult males	217	23	319
Adult females	347	24	340
Children	3	97	750

Price	
Coffee	$0.75
Milk	$1.00
Cola	$1.25

58. Production costs Each of four factories manufactures three products in the daily quantities and at the unit costs given in the tables. Find a suitable matrix product to represent production costs.

Production Quantity			
Factory	Product A	Product B	Product C
Ashtabula	19	23	27
Boston	17	21	22
Chicago	21	18	20
Denver	27	25	22

Unit Production Cost		
	Day shift	Night shift
Product A	$1.20	$1.35
Product B	$0.75	$0.85
Product C	$3.50	$3.70

59. Connectivity matrix An entry of 1 in the following **connectivity matrix** A indicates that the person associated with that row knows the address of the person associated with that column. For example, the 1 in Bill's row and Al's column indicates that Bill can write to Al. The 0 in Bill's row and Carl's column indicates that Bill cannot write to Carl. However, Bill could ask Al to forward his letter to Carl. The matrix A^2 indicates the number of ways that one person can write to another with a letter that is forwarded exactly once. Find A^2.

$$\begin{array}{c} \\ \text{Al} \\ \text{Bill} \\ \text{Carl} \end{array} \begin{array}{ccc} \text{Al} & \text{Bill} & \text{Carl} \end{array}$$

$$\begin{matrix} \text{Al} \\ \text{Bill} \\ \text{Carl} \end{matrix} \begin{bmatrix} 0 & 1 & 1 \\ 1 & 0 & 0 \\ 0 & 1 & 0 \end{bmatrix} = A$$

60. Communication routing Refer to Exercise 59. Find the matrix $A + A^2$. Can everyone receive a letter from everyone else with at most one forwarding?

Discovery and Writing

61. A long-distance telephone carrier has established several direct microwave links among four cities. In the following connectivity matrix, entries a_{ij} and a_{ji} indicate the number of direct links between cities i and j. For example, cities 2 and 4 are not linked directly but could be connected through city 3. Find and interpret matrix A^2.

$$A = \begin{bmatrix} 0 & 2 & 1 & 0 \\ 2 & 0 & 1 & 0 \\ 1 & 1 & 0 & 2 \\ 0 & 0 & 2 & 0 \end{bmatrix}$$

62. Three communication centers are linked as indicated in the illustration, with communication only in the direction of the arrows. Thus, location 1 can send a message directly to location 2 along two paths, but location 2 can return a message directly on only one path. Entry c_{ij} of matrix C indicates the number of channels from i to j. Find and interpret C^2.

$$C = \begin{bmatrix} 0 & 2 & 2 \\ 1 & 0 & 1 \\ 1 & 0 & 0 \end{bmatrix}$$

63. If A and B are 2×2 matrices, is $(AB)^2$ equal to A^2B^2? Support your answer.

64. Let a, b, and c be real numbers. If $ab = ac$ and $a \neq 0$, then $b = c$. Find 2×2 matrices A, B, and C, where $A \neq 0$, to show that such a law does not hold for all matrices.

65. Another property of the real numbers is that if $ab = 0$, then either $a = 0$ or $b = 0$. To show that this property is not true for matrices, find two nonzero 2×2 matrices A and B such that $AB = 0$.

66. Find 2×2 matrices to show that $(A + B)(A - B) \neq A^2 - B^2$.

8.4 Matrix Inversion

In this section, we will learn to

1. Find the inverse of a square matrix using row operations.

2. Solve a system of equations by matrix inversion.

3. Solve applications using matrix inversion.

Since the beginning of civilized society, governments and businesses have been interested in communicating in secret. This required the development of secret codes. Today, credit-card information, bank-account information, and even many e-mail messages are *encrypted*. Encryption is a process of converting ordinary text into unreadable text called **cipher text**.

Codes and code breaking have a long history in warfare. A very important event that helped the United States win World War II was the capture of German U-boat *U-110*, which carried an Enigma encryption machine like the one shown here. By breaking the code, the British and U.S. navies were able to determine the movements of German submarines.

Some of the most sophisticated codes we have today involve matrices. In Exercises 39 and 40, we must find the inverse of a matrix to decode a message.

Two real numbers are called **multiplicative inverses** if their product is the multiplicative identity 1. Some matrices have multiplicative inverses also.

Multiplicative Inverse of a Matrix

If A and B are $n \times n$ matrices, I is the $n \times n$ identity matrix, and

$$AB = BA = I,$$

then A and B are called **multiplicative inverses.** Matrix A is the **inverse** of B, and B is the **inverse** of A.

It can be shown that if a matrix A has an inverse, it has only one inverse. The inverse of A is written as A^{-1}:

$$AA^{-1} = A^{-1}A = I.$$

EXAMPLE 1   Showing That Two Matrices Are Inverses

Show that A and B are inverses if

$$A = \begin{bmatrix} 1 & 1 & 0 \\ 4 & 3 & 0 \\ 2 & 1 & -1 \end{bmatrix} \quad \text{and} \quad B = \begin{bmatrix} -3 & 1 & 0 \\ 4 & -1 & 0 \\ -2 & 1 & -1 \end{bmatrix}.$$

SOLUTION   We must show that both AB and BA are equal to I.

$$AB = \begin{bmatrix} 1 & 1 & 0 \\ 4 & 3 & 0 \\ 2 & 1 & -1 \end{bmatrix} \begin{bmatrix} -3 & 1 & 0 \\ 4 & -1 & 0 \\ -2 & 1 & -1 \end{bmatrix}$$

$$= \begin{bmatrix} -3 + 4 + 0 & 1 - 1 + 0 & 0 + 0 + 0 \\ -12 + 12 + 0 & 4 - 3 + 0 & 0 + 0 + 0 \\ -6 + 4 + 2 & 2 - 1 - 1 & 0 + 0 + 1 \end{bmatrix}$$

$$= \begin{bmatrix} 1 & 0 & 0 \\ 0 & 1 & 0 \\ 0 & 0 & 1 \end{bmatrix}$$

$$BA = \begin{bmatrix} -3 & 1 & 0 \\ 4 & -1 & 0 \\ -2 & 1 & -1 \end{bmatrix} \begin{bmatrix} 1 & 1 & 0 \\ 4 & 3 & 0 \\ 2 & 1 & -1 \end{bmatrix} = \begin{bmatrix} 1 & 0 & 0 \\ 0 & 1 & 0 \\ 0 & 0 & 1 \end{bmatrix}$$

Self Check 1   Are C and D inverses if $C = \begin{bmatrix} 2 & 5 \\ 1 & 3 \end{bmatrix}$ and $D = \begin{bmatrix} 3 & -5 \\ -1 & 2 \end{bmatrix}$?

Now Try Exercise 1.

1. Find the Inverse of a Square Matrix Using Row Operations

A matrix that has an inverse is called a **nonsingular matrix** and is said to be **invertible.** If a matrix does not have an inverse, it is a **singular matrix** and is not invertible. The following method provides a way to find the inverse of an invertible matrix:

Strategy for Finding the Inverse of a Matrix

If a sequence of row operations performed on the $n \times n$ matrix A reduces A to the $n \times n$ identity matrix I, then those same row operations performed in the same order on I will transform I into A^{-1}.

If no sequence of row operations will reduce A to I, then A is not invertible.

Comment

The row operations we use to change the matrix into the identity matrix are the same row operations we use in Gauss–Jordan elimination.

To use this method to find the inverse of an invertible matrix, we perform row operations on matrix A to change it to the identity matrix I. At the same time, we perform the same row operations on I. This changes I into A^{-1}.

A notation for this process uses an n-row-by-$2n$-column matrix, with matrix A as the left half and matrix I as the right half. If A is invertible, the proper row operations performed on $[A|I]$ will transform it into $[I|A^{-1}]$.

EXAMPLE 2 **Finding the Inverse of a 2 × 2 Matrix**

Find the inverse of matrix A if $A = \begin{bmatrix} 2 & -4 \\ 4 & -7 \end{bmatrix}$.

SOLUTION We can set up a 2×4 matrix with A on the left and I on the right of the dashed line:

$$[A|I] = \begin{bmatrix} 2 & -4 & \vdots & 1 & 0 \\ 4 & -7 & \vdots & 0 & 1 \end{bmatrix}.$$

We perform row operations on the entire matrix to transform the left half into I.

$$\begin{bmatrix} 2 & -4 & \vdots & 1 & 0 \\ 4 & -7 & \vdots & 0 & 1 \end{bmatrix} \begin{array}{c} \left(\frac{1}{2}\right)R1 \to \\ (-2)R1 + R2 \to \end{array} \begin{bmatrix} 1 & -2 & \vdots & \frac{1}{2} & 0 \\ 0 & 1 & \vdots & -2 & 1 \end{bmatrix}$$

$$(2)R2 + R1 \to \begin{bmatrix} 1 & 0 & \vdots & -\frac{7}{2} & 2 \\ 0 & 1 & \vdots & -2 & 1 \end{bmatrix}$$

Since matrix A has been transformed into I, the right side of the new matrix is A^{-1}. We can verify this by finding AA^{-1} and $A^{-1}A$ and showing that each product is I:

Caution

We do not find the inverse of a square matrix by finding the multiplicative inverse of each of its entries.

$$AA^{-1} = \begin{bmatrix} 2 & -4 \\ 4 & -7 \end{bmatrix} \begin{bmatrix} -\frac{7}{2} & 2 \\ -2 & 1 \end{bmatrix} = \begin{bmatrix} 1 & 0 \\ 0 & 1 \end{bmatrix}$$

$$A^{-1}A = \begin{bmatrix} -\frac{7}{2} & 2 \\ -2 & 1 \end{bmatrix} \begin{bmatrix} 2 & -4 \\ 4 & -7 \end{bmatrix} = \begin{bmatrix} 1 & 0 \\ 0 & 1 \end{bmatrix}.$$

Self Check 2 Find the inverse of $A = \begin{bmatrix} 3 & 2 \\ 4 & 3 \end{bmatrix}$.

Now Try Exercise 5.

EXAMPLE 3 **Finding the Inverse of a 3 × 3 Matrix**

Find the inverse of matrix A if $A = \begin{bmatrix} 1 & 1 & 0 \\ 1 & 2 & 1 \\ 2 & 3 & 2 \end{bmatrix}$.

SOLUTION We set up a 3×6 matrix with A on the left and I on the right of the dashed line.

$$[A|I] = \begin{bmatrix} 1 & 1 & 0 & \vdots & 1 & 0 & 0 \\ 1 & 2 & 1 & \vdots & 0 & 1 & 0 \\ 2 & 3 & 2 & \vdots & 0 & 0 & 1 \end{bmatrix}$$

We then perform row operations on the matrix to transform the left half into I.

$$\begin{bmatrix} 1 & 1 & 0 & 1 & 0 & 0 \\ 1 & 2 & 1 & 0 & 1 & 0 \\ 2 & 3 & 2 & 0 & 0 & 1 \end{bmatrix} \begin{matrix} \\ (-1)R1 + R2 \to \\ (-2)R1 + R3 \to \end{matrix} \begin{bmatrix} 1 & 1 & 0 & 1 & 0 & 0 \\ 0 & 1 & 1 & -1 & 1 & 0 \\ 0 & 1 & 2 & -2 & 0 & 1 \end{bmatrix}$$

$$\begin{matrix} (-1)R2 + R1 \to \\ \\ (-1)R2 + R3 \to \end{matrix} \begin{bmatrix} 1 & 0 & -1 & 2 & -1 & 0 \\ 0 & 1 & 1 & -1 & 1 & 0 \\ 0 & 0 & 1 & -1 & -1 & 1 \end{bmatrix}$$

$$\begin{matrix} R3 + R1 \to \\ (-1)R3 + R2 \to \\ \\ \end{matrix} \begin{bmatrix} 1 & 0 & 0 & 1 & -2 & 1 \\ 0 & 1 & 0 & 0 & 2 & -1 \\ 0 & 0 & 1 & -1 & -1 & 1 \end{bmatrix}$$

Since the left half of the matrix has been transformed into the identity matrix, the right half has become A^{-1}, and

$$A^{-1} = \begin{bmatrix} 1 & -2 & 1 \\ 0 & 2 & -1 \\ -1 & -1 & 1 \end{bmatrix}.$$

Self Check 3   Find the inverse of $B = \begin{bmatrix} 1 & 1 & 1 \\ 2 & 1 & 4 \\ 2 & 2 & 3 \end{bmatrix}$.

Now Try Exercise 9.

EXAMPLE 4   **Finding the Inverse (If Possible) of a Square Matrix**

If possible, find the inverse of $A = \begin{bmatrix} 1 & 2 \\ 2 & 4 \end{bmatrix}$.

SOLUTION   We form the 2×4 matrix

$$[A|I] = \begin{bmatrix} 1 & 2 & 1 & 0 \\ 2 & 4 & 0 & 1 \end{bmatrix}$$

and begin to transform the left side of the matrix into the identity matrix I.

$$\begin{bmatrix} 1 & 2 & 1 & 0 \\ 2 & 4 & 0 & 1 \end{bmatrix} \begin{matrix} \\ (-2)R1 + R2 \to \end{matrix} \begin{bmatrix} 1 & 2 & 1 & 0 \\ \mathbf{0} & \mathbf{0} & -2 & 1 \end{bmatrix}$$

In obtaining the second-row, first-column position of A, we zero out the entire second row of A. Since we cannot transform matrix A to the identity matrix, A is not invertible.

Self Check 4   If possible, find the inverse of $B = \begin{bmatrix} 1 & -2 \\ -3 & 6 \end{bmatrix}$.

Now Try Exercise 13.

Caution

> Do not attempt to find the inverse of a matrix that isn't square. It does not have an inverse.

The previous examples suggest that some square matrices have inverses and others do not.

2. Solve a System of Equations by Matrix Inversion

If we multiply the matrices on the left side of the equation

$$\begin{bmatrix} 1 & 1 & 0 \\ 1 & 2 & 1 \\ 2 & 3 & 2 \end{bmatrix} \begin{bmatrix} x \\ y \\ z \end{bmatrix} = \begin{bmatrix} 20 \\ 30 \\ 55 \end{bmatrix}$$

and set the corresponding entries equal, we get the following system of equations:

$$\begin{cases} x + y = 20 \\ x + 2y + z = 30 \\ 2x + 3y + 2z = 55 \end{cases}.$$

A system of equations can always be written as a matrix equation $AX = B$, where A is the coefficient matrix of the system, X is a column matrix of variables, and B is the column matrix of constants. If matrix A is invertible, the matrix equation $AX = B$ is easy to solve, as we will see in the next example.

EXAMPLE 5 **Using the Inverse of a Matrix to Solve a System of Equations**

Solve: $\begin{bmatrix} 1 & 1 & 0 \\ 1 & 2 & 1 \\ 2 & 3 & 2 \end{bmatrix} \begin{bmatrix} x \\ y \\ z \end{bmatrix} = \begin{bmatrix} 20 \\ 30 \\ 55 \end{bmatrix}$.

SOLUTION The 3×3 matrix on the left is the matrix whose inverse was found in Example 3. We multiply both sides of the equation by this inverse to obtain an equivalent system of equations.

$$\begin{bmatrix} 1 & -2 & 1 \\ 0 & 2 & -1 \\ -1 & -1 & 1 \end{bmatrix} \begin{bmatrix} 1 & 1 & 0 \\ 1 & 2 & 1 \\ 2 & 3 & 2 \end{bmatrix} \begin{bmatrix} x \\ y \\ z \end{bmatrix} = \begin{bmatrix} 1 & -2 & 1 \\ 0 & 2 & -1 \\ -1 & -1 & 1 \end{bmatrix} \begin{bmatrix} 20 \\ 30 \\ 55 \end{bmatrix}$$

$$\begin{bmatrix} 1 & 0 & 0 \\ 0 & 1 & 0 \\ 0 & 0 & 1 \end{bmatrix} \begin{bmatrix} x \\ y \\ z \end{bmatrix} = \begin{bmatrix} 15 \\ 5 \\ 5 \end{bmatrix}$$ Multiply the matrices. On the left, remember that $A^{-1}A = I$.

$$\begin{bmatrix} x \\ y \\ z \end{bmatrix} = \begin{bmatrix} 15 \\ 5 \\ 5 \end{bmatrix}$$ $IX = X$

The solution of this system can be read directly from the matrix on the right side. Verify that the values $x = 15$, $y = 5$, and $z = 5$ satisfy the original equations. ∎

Self Check 5 Solve: $\begin{bmatrix} 2 & -4 \\ 4 & -7 \end{bmatrix} \begin{bmatrix} x \\ y \end{bmatrix} = \begin{bmatrix} -4 \\ 2 \end{bmatrix}$. (The inverse was found in Example 2.)

Now Try Exercise 29.

Example 5 suggests the following property:

Solving Systems of Equations If A is invertible, the solution of the matrix equation $AX = B$ is

$$X = A^{-1}B.$$

This method is especially useful for finding solutions of several systems of equations that differ from each other only in the column matrix B. If the coefficient matrix A remains unchanged from one system of equations to the next, then A^{-1} needs to be found only once. The solution of each system is found by a single matrix multiplication, $A^{-1}B$. (Always check the results in the original equation.)

3. Solve Applications Using Matrix Inversion

EXAMPLE 6   **Solving an Application Problem Using the Inverse of a Matrix**

A company that manufactures medical equipment spends time on paperwork, manufacture, and testing for each of three versions of a circuit board. The times spent on each and the total time available are given in the following tables. Find the cost of supplies of each shift.

	Hours Required Per Unit		
	Product A	Product B	Product C
Paperwork	1	1	0
Manufacture	1	2	1
Testing	2	3	2

Hours Available	
Paperwork	20
Manufacture	30
Testing	55

SOLUTION   We can let x, y, and z represent the number of units of products A, B, and C to be manufactured, respectively. We can then set up the following system of equations:

Paperwork:	$x + y = 20$	One hour is needed for every A and one hour for every B.
Manufacture:	$x + 2y + z = 30$	One hour is needed for every A and C and two hours for every B.
Testing:	$2x + 3y + 2z = 55$	Two hours are needed for every A and C and three hours for every B.

In matrix form, the system becomes

$$\begin{bmatrix} 1 & 1 & 0 \\ 1 & 2 & 1 \\ 2 & 3 & 2 \end{bmatrix} \begin{bmatrix} x \\ y \\ z \end{bmatrix} = \begin{bmatrix} 20 \\ 30 \\ 55 \end{bmatrix}.$$

We solved this equation in Example 5 to get $x = 15$, $y = 5$, and $z = 15$. To use all of the available time, the company should manufacture 15 units of product A and 5 units each of products B and C. ∎

Self Check 6   In Example 6, if 18 hours are available for paperwork, 32 for manufacture, and 57 for testing, how many of each product should the company manufacture?

Now Try Exercise 37.

Self Check Answers   **1.** yes   **2.** $\begin{bmatrix} 3 & -2 \\ -4 & 3 \end{bmatrix}$ **3.** $\begin{bmatrix} 5 & 1 & -3 \\ -2 & -1 & 2 \\ -2 & 0 & 1 \end{bmatrix}$ **4.** not invertible

5. $x = 18$, $y = 10$ **6.** 11 units of product A and 7 units each of products B and C

Exercises 8.4

Getting Ready

You should be able to complete these vocabulary and concept statements before you proceed to the practice exercises.

Fill in the blanks.

1. Matrices A and B are multiplicative inverses if _____.

2. A nonsingular matrix _____ (is, is not) invertible.

3. If A is invertible, elementary row operations can change $[A|I]$ into _____.

4. If A is invertible, the solution of $AX = B$ is _____.

Practice

Find the inverse of each matrix, if possible.

5. $\begin{bmatrix} 3 & -4 \\ -2 & 3 \end{bmatrix}$

6. $\begin{bmatrix} 2 & 3 \\ 3 & 5 \end{bmatrix}$

7. $\begin{bmatrix} 3 & 7 \\ 2 & 5 \end{bmatrix}$

8. $\begin{bmatrix} 1 & -2 \\ 2 & -5 \end{bmatrix}$

9. $\begin{bmatrix} 1 & 0 & 3 \\ -1 & 1 & 3 \\ -2 & 1 & 1 \end{bmatrix}$

10. $\begin{bmatrix} 2 & 1 & -1 \\ 2 & 2 & -1 \\ -1 & -1 & 1 \end{bmatrix}$

11. $\begin{bmatrix} 3 & 2 & 1 \\ 1 & 1 & -1 \\ 4 & 3 & 1 \end{bmatrix}$

12. $\begin{bmatrix} -2 & 1 & -3 \\ 2 & 3 & 0 \\ 1 & 0 & 1 \end{bmatrix}$

13. $\begin{bmatrix} 1 & 3 & 5 \\ 0 & 1 & 6 \\ 1 & 4 & 11 \end{bmatrix}$

14. $\begin{bmatrix} 1 & 2 & 3 \\ 4 & 5 & 6 \\ 7 & 8 & 9 \end{bmatrix}$

15. $\begin{bmatrix} 1 & 2 & 3 \\ 0 & 1 & 2 \\ 0 & 0 & 1 \end{bmatrix}$

16. $\begin{bmatrix} 1 & 2 & 3 \\ 0 & 1 & 1 \\ 0 & -1 & 0 \end{bmatrix}$

17. $\begin{bmatrix} 1 & 6 & 4 \\ 1 & -2 & -5 \\ 2 & 4 & -1 \end{bmatrix}$

18. $\begin{bmatrix} 1 & 1 & 1 \\ 1 & 0 & -1 \\ 1 & 2 & 3 \end{bmatrix}$

Use a graphing calculator to find the inverse of each matrix.

19. $\begin{bmatrix} 1 & 2 & 3 & 4 \\ 0 & 1 & 2 & 3 \\ 0 & 0 & 1 & 2 \\ 0 & 0 & 0 & 1 \end{bmatrix}$

20. $\begin{bmatrix} 1 & 0 & 0 & 0 \\ 1 & 1 & 0 & 0 \\ 1 & 1 & 1 & 0 \\ 1 & 2 & 2 & 1 \end{bmatrix}$

21. $\begin{bmatrix} 1 & 1 & -1 \\ 0.5 & 1 & 0.5 \\ 1 & 1 & -1.5 \end{bmatrix}$

22. $\begin{bmatrix} -2 & -1 & 1 \\ 0.5 & -1.5 & -0.5 \\ 0 & 1 & 0.5 \end{bmatrix}$

23. $\begin{bmatrix} 3 & 3 & -3 & 2 \\ 1 & -4 & 3 & -5 \\ 3 & 0 & -2 & -1 \\ -1 & 5 & -3 & 6 \end{bmatrix}$

24. $\begin{bmatrix} 1 & 0 & 0 & 0 \\ 2 & 1 & 0 & 0 \\ 3 & 2 & 1 & 0 \\ 4 & 3 & 2 & 1 \end{bmatrix}$

Use matrix inversion to solve each system of equations. Note that several systems have the same coefficient matrix.

25. $\begin{cases} 3x - 4y = 1 \\ -2x + 3y = 5 \end{cases}$

26. $\begin{cases} 3x - 4y = -1 \\ -2x + 3y = 3 \end{cases}$

27. $\begin{cases} 3x - 4y = 0 \\ -2x + 3y = 0 \end{cases}$

28. $\begin{cases} 3x - 4y = -3 \\ -2x + 3y = -2 \end{cases}$

29. $\begin{cases} 2x + y - z = 2 \\ 2x + 2y - z = 4 \\ -x - y + z = -1 \end{cases}$

30. $\begin{cases} 2x + y - z = 3 \\ 2x + 2y - z = -1 \\ -x - y + z = 4 \end{cases}$

31. $\begin{cases} -2x + y - 3z = 2 \\ 2x + 3y = -3 \\ x + z = 5 \end{cases}$

32. $\begin{cases} -2x + y - 3z = 5 \\ 2x + 3y = 1 \\ x + z = -2 \end{cases}$

Use a graphing calculator to solve each system of equations with matrix inversion.

33. $\begin{cases} 5x + 3y = 13 \\ -7x + 5y = -9 \end{cases}$

34. $\begin{cases} 8x - 3y = 7 \\ -3x + 2y = 0 \end{cases}$

35. $\begin{cases} 5x + 2y + 3z = 12 \\ 2x + 5z = 7 \\ 3x + z = 4 \end{cases}$

36. $\begin{cases} 3x + 2y - z = 0 \\ 5x - 2y = 5 \\ 3x + y + z = 6 \end{cases}$

Applications

37. **Manufacturing and testing** The number of hours required to manufacture and test each of two models of heart monitor are given in the first table, and the number of hours available each week for manufacturing and testing are given in the second table.

Hours Required Per Unit		
	Model A	Model B
Manufacturing	23	27
Testing	21	22

Hours Available	
Manufacturing	127
Testing	108

How many of each model can be manufactured each week?

38. Making clothes A clothing manufacturer makes coats, shirts, and slacks. The time required for cutting, sewing, and packaging each item is shown in the table. How many of each should be made to use all available labor hours?

Time Required Per Unit			
	Coats	**Shirts**	**Slacks**
Cutting	20 min	15 min	10 min
Sewing	60 min	30 min	24 min
Packaging	5 min	12 min	6 min

Time Available	
Cutting	115 hr
Sewing	280 hr
Packaging	65 hr

39. Cryptography The letters of a message, called **plain text,** are assigned values 1–26 (for A–Z) and are written in groups of two as 2×1 matrices. To write the message in **cipher text**, each 2×1 matrix B is multiplied by a matrix A, where

$$A = \begin{bmatrix} 1 & 1 \\ 2 & 3 \end{bmatrix}.$$

Find the plain text if the cipher text of one message is

$$AB = \begin{bmatrix} 17 \\ 43 \end{bmatrix}.$$

40. Cryptography The letters of a message, called **plain text,** are assigned values 1–26 (for A–Z) and are written in groups of three as 3×1 matrices. To write the message in **cipher text**, each 3×1 matrix Y is multiplied by matrix A, where

$$A = \begin{bmatrix} 1 & 1 & 0 \\ 2 & 3 & 3 \\ 1 & 1 & 1 \end{bmatrix}.$$

Find the plain text if the cipher text of one message is

$$AY = \begin{bmatrix} 30 \\ 122 \\ 49 \end{bmatrix}.$$

Discovery and Writing
Use examples chosen from 2×2 matrices to support each answer.

41. Does $(AB)^{-1} = A^{-1}B^{-1}$?

42. Does $(AB)^{-1} = B^{-1}A^{-1}$?

Let $A = \begin{bmatrix} -1 & -1 \\ 1 & 1 \end{bmatrix}.$

43. Show that $A^2 = 0$.   **44.** Show that the inverse of $I - A$ is $I + A$.

Let $A = \begin{bmatrix} 3 & 0 & 0 \\ -2 & -1 & -2 \\ 3 & 6 & 3 \end{bmatrix}$ *and* $X = \begin{bmatrix} x \\ y \\ z \end{bmatrix}.$ *Solve each equation. Each solution is called an eigenvector of the matrix A.*

45. $(A - 2I)X = 0$ **46.** $(A - 3I)X = 0$

47. Suppose that A, B, and C are $n \times n$ matrices and A is invertible. If $AB = AC$, prove that $B = C$.

48. Prove that $\begin{bmatrix} a & b \\ c & d \end{bmatrix}$ has an inverse if and only if $ad - bc \neq 0$. (*Hint:* Try to find the inverse and see what happens.)

49. Suppose that B is any matrix for which $B^2 = 0$. Show that $I - B$ is invertible by showing that the inverse of $I - B$ is $I + B$.

50. Suppose that C is any matrix for which $C^3 = 0$. Show that $I - C$ is invertible by showing that the inverse of $I - C$ is $I + C + C^2$.

8.5 Determinants

In this section, we will learn to

1. Evaluate determinants of higher-order matrices.
2. Understand and use properties of determinants.
3. Use determinants to solve systems of equations.

There is a function, called the **determinant function,** that associates a number with every square matrix. The domain of this function is the set of all square matrices, and the range is the set of numbers. Historically, determinants were considered before matrices. They first appeared in the third century BC, when they were considered in a Chinese textbook called *The Nine Chapters on the Mathematical Art*. Gerolamo Cardano, an Italian mathematician, considered two-by-two determinants toward the end of the sixteenth century. However, it was the great Swiss mathematician Gabriel Cramer who made them popular.

In this section we will use determinants to solve systems of linear equations. The determinant function is written as $\det(A)$ or as $|A|$:

Determinant of a 2 × 2 Matrix   If a, b, c, and d are numbers, the **determinant** of $A = \begin{bmatrix} a & b \\ c & d \end{bmatrix}$ is

$$\det(A) = \begin{vmatrix} a & b \\ c & d \end{vmatrix} = ad - bc.$$

Caution   Do not confuse the determinant notation $|A|$ with absolute-value.

EXAMPLE 1   **Finding the Determinant of a 2 × 2 Matrix**

Evaluate: **a.** $\begin{vmatrix} 1 & 2 \\ 3 & 4 \end{vmatrix}$ **b.** $\begin{vmatrix} -2 & 3 \\ -\pi & \frac{1}{2} \end{vmatrix}$

a. $\begin{vmatrix} 1 & 2 \\ 3 & 4 \end{vmatrix} = 1 \cdot 4 - 2 \cdot 3$

$= 4 - 6$

$= -2$

b. $\begin{vmatrix} -2 & 3 \\ -\pi & \frac{1}{2} \end{vmatrix} = (-2)\left(\frac{1}{2}\right) - 3(-\pi)$

$= -1 + 3\pi$

Self Check 1   Evaluate: $\begin{vmatrix} 3 & -2 \\ 5 & -4 \end{vmatrix}$.

Now Try Exercise 7.

1. Evaluate Determinants of Higher-Order Matrices

To evaluate determinants of higher-order matrices, we must define the **minor** and the **cofactor** of an element in a matrix.

Minor and Cofactor of a Matrix

Let $A = [a_{ij}]$ be a square matrix of size (or order) $n \geq 2$.

- The **minor** of a_{ij}, denoted M_{ij}, is the determinant of the $n - 1 \times n - 1$ matrix formed by deleting the ith row and the jth column of A.

- The **cofactor** of a_{ij}, denoted C_{ij}, is $\begin{cases} M_{ij} \text{ when } i + j \text{ is even} \\ -M_{ij} \text{ when } i + j \text{ is odd} \end{cases}$.

EXAMPLE 2 **Finding Minors and Cofactors of a Matrix**

If $A = \begin{bmatrix} 1 & 2 & 3 \\ 4 & 5 & 6 \\ 7 & 8 & 9 \end{bmatrix}$, find the minor and cofactor of each element.

a. a_{31} **b.** a_{12}

SOLUTION **a.** The minor M_{31} is the minor of $a_{31} = 7$ appearing in row 3, column 1. It is found by deleting row 3 and column 1.

$$M_{31} = \begin{vmatrix} 1 & 2 & 3 \\ 4 & 5 & 6 \\ 7 & 8 & 9 \end{vmatrix} = \begin{vmatrix} 2 & 3 \\ 5 & 6 \end{vmatrix} = -3.$$

Because $i + j$ is even $(3 + 1 = 4)$, the cofactor of the minor M_{31} is M_{31}.

$$C_{31} = M_{31} = \begin{vmatrix} 2 & 3 \\ 5 & 6 \end{vmatrix} = 2 \cdot 6 - 3 \cdot 5 = 12 - 15 = -3$$

b. The minor M_{12} is the minor of $a_{12} = 2$ appearing in row 1, column 2. It is found by deleting row 1 and column 2.

$$M_{12} = \begin{vmatrix} 1 & 2 & 3 \\ 4 & 5 & 6 \\ 7 & 8 & 9 \end{vmatrix} = \begin{vmatrix} 4 & 6 \\ 7 & 9 \end{vmatrix} = -6$$

Because $i + j$ is odd $(1 + 2 = 3)$, the cofactor of the minor M_{12} is $-M_{12}$.

$$C_{12} = -M_{12} = -\begin{vmatrix} 4 & 6 \\ 7 & 9 \end{vmatrix} = -(4 \cdot 9 - 6 \cdot 7) = -(36 - 42) = -(-6) = 6$$

Self Check 2 Find the cofactor of a_{23} in Example 2.

Now Try Exercise 13.

We are now ready to evaluate determinants of higher-order matrices.

Determinant of a Matrix

If A is a square matrix of order $n \geq 2$, the **determinant of** A, $|A|$, is the sum of the products of the elements in any row (or column) and the cofactors of those elements.

In Example 3, we evaluate a 3×3 determinant by expanding the determinant in three ways: along two different rows and along a column. This method is called **expanding a determinant by cofactors.**

EXAMPLE 3 **Finding the Determinant of a 3×3 Matrix**

Evaluate $\begin{vmatrix} 1 & 2 & -3 \\ -1 & 0 & 1 \\ -2 & 2 & 1 \end{vmatrix}$ by expanding by cofactors along row 1, row 3, and column 2.

SOLUTION **Expanding on Row 1**

$$\begin{vmatrix} \mathbf{1} & \mathbf{2} & \mathbf{-3} \\ -1 & 0 & 1 \\ -2 & 2 & 1 \end{vmatrix} = a_{11}C_{11} + a_{12}C_{12} + a_{13}C_{13}$$

$$= 1\begin{vmatrix} 0 & 1 \\ 2 & 1 \end{vmatrix} + 2\left(-\begin{vmatrix} -1 & 1 \\ -2 & 1 \end{vmatrix}\right) + (-3)\begin{vmatrix} -1 & 0 \\ -2 & 2 \end{vmatrix}$$

$$= 1(-2) + 2(-1) - 3(-2)$$

$$= -2 - 2 + 6$$

$$= 2$$

Expanding on Row 3

$$\begin{vmatrix} 1 & 2 & -3 \\ -1 & 0 & 1 \\ \mathbf{-2} & \mathbf{2} & \mathbf{1} \end{vmatrix} = a_{31}C_{31} + a_{32}C_{32} + a_{33}C_{33}$$

$$= -2\begin{vmatrix} 2 & -3 \\ 0 & 1 \end{vmatrix} + 2\left(-\begin{vmatrix} 1 & -3 \\ -1 & 1 \end{vmatrix}\right) + 1\begin{vmatrix} 1 & 2 \\ -1 & 0 \end{vmatrix}$$

$$= -2(2) + 2(2) + 1(2)$$

$$= -4 + 4 + 2$$

$$= 2$$

Expanding on Column 2

$$\begin{vmatrix} 1 & \mathbf{2} & -3 \\ -1 & \mathbf{0} & 1 \\ -2 & \mathbf{2} & 1 \end{vmatrix} = a_{12}C_{12} + a_{22}C_{22} + a_{32}C_{32}$$

$$= 2\left(-\begin{vmatrix} -1 & 1 \\ -2 & 1 \end{vmatrix}\right) + 0\begin{vmatrix} 1 & -3 \\ -2 & 1 \end{vmatrix} + 2\left(-\begin{vmatrix} 1 & -3 \\ -1 & 1 \end{vmatrix}\right)$$

$$= 2(-1) + 0(-5) + 2(2)$$

$$= -2 + 4$$

$$= 2$$

In each case, the result is 2. ∎

Comment

Note that we don't need to evaluate the determinant $0\begin{vmatrix} 1 & -3 \\ -2 & 1 \end{vmatrix}$, because it has a coefficient of 0.

When we expand by cofactors along a row or column containing 0s, the work is easier because we have fewer determinants to evaluate.

Self Check 3 Evaluate $\begin{vmatrix} 1 & 0 & 1 \\ 2 & -1 & 0 \\ 3 & 1 & -1 \end{vmatrix}$ by expanding on its first row and second column.

Now Try Exercise 19.

Determinants can be applied in many areas of mathematics in addition to those covered in this section. In the "Looking Ahead to Calculus" section at the end of this chapter, we will explore how determinants are used in calculus.

EXAMPLE 4   **Finding the Determinant of a 4 × 4 Matrix**

Evaluate: $\begin{vmatrix} 0 & 0 & 2 & 0 \\ 1 & 2 & 17 & -3 \\ -1 & 0 & 28 & 1 \\ -2 & 2 & -37 & 1 \end{vmatrix}$.

SOLUTION   Because row 1 contains three 0s, we expand the determinant along row 1. Then only one cofactor needs to be evaluated.

$$\begin{vmatrix} \mathbf{0} & \mathbf{0} & \mathbf{2} & \mathbf{0} \\ 1 & 2 & 17 & -3 \\ -1 & 0 & 28 & 1 \\ -2 & 2 & -37 & 1 \end{vmatrix} = 0|?| - 0|?| + 2\begin{vmatrix} 1 & 2 & -3 \\ -1 & 0 & 1 \\ -2 & 2 & 1 \end{vmatrix} - 0|?|$$

$$= 2(2) \qquad \text{See Example 3.}$$

$$= 4$$

Self Check 4 Evaluate: $\begin{vmatrix} 2 & 0 & 0 & 0 \\ 1 & 0 & 1 & 0 \\ 2 & 2 & 11 & -2 \\ 3 & 1 & 13 & 1 \end{vmatrix}$.

Now Try Exercise 27.

Example 4 suggests the following theorem:

Zero Row or Column Theorem   If every entry in a row or column of a square matrix A is 0, then $|A| = 0$.

2. Understand and Use Properties of Determinants

We have seen that there are row operations for transforming matrices. There are similar row and column operations for transforming determinants.

Row and Column Operations and Determinants   Let A be a square matrix and k be a real number.

1. If a matrix B is obtained from matrix A by interchanging two rows (or columns), then $|B| = -|A|$.

2. If B is obtained from A by multiplying every element in a row (or column) of A by k, then $|B| = k|A|$.

3. If B is obtained from A by adding k times any row (or column) of A to another row (or column) of A, then $|B| = |A|$.

We will illustrate each row operation by showing that the operation holds true for the 2×2 matrix $A = \begin{bmatrix} 3 & 4 \\ 1 & 2 \end{bmatrix}$.

$$|A| = \begin{vmatrix} 3 & 4 \\ 1 & 2 \end{vmatrix} = 3(2) - 4(1) = 6 - 4 = 2.$$

- **Interchanging Two Rows**

  Let B be obtained from A by interchanging its two rows.

  $$|B| = \begin{vmatrix} 1 & 2 \\ 3 & 4 \end{vmatrix} = 1(4) - 2(3) = 4 - 6 = -2$$

  Since $-2 = -1(2)$, we have $|B| = -|A|$.

In general, given $|A|$ and $|B|$ as follows, we have

$$|A| = \begin{vmatrix} a & b \\ c & d \end{vmatrix} = ad - bc$$

and

$$|B| = \begin{vmatrix} c & d \\ a & b \end{vmatrix} = cb - da = -(ad - bc) = -|A|.$$

- **Multiplying Every Element in a Row by k**

  Let B be obtained from A by multiplying its second row by 3.

  $$|B| = \begin{vmatrix} 3 & 4 \\ 3(1) & 3(2) \end{vmatrix} = \begin{vmatrix} 3 & 4 \\ 3 & 6 \end{vmatrix} = 3(6) - 4(3) = 18 - 12 = 6$$

  Since $6 = 3(2)$, we have $|B| = 3|A|$.

In general, given $|A|$ and $|B|$ as follows, we have

$$|A| = \begin{vmatrix} a & b \\ c & d \end{vmatrix} = ad - bc$$

and

$$|B| = \begin{vmatrix} a & b \\ kc & kd \end{vmatrix} = akd - bkc = k(ad - bc) = k|A|.$$

- **Adding k Times Any Row of A to Another Row of A**

  Let B be obtained from A by adding 3 times its first row to its second row.

  $$|B| = \begin{vmatrix} 3 & 4 \\ 3(3) + 1 & 3(4) + 2 \end{vmatrix} = 3(14) - 4(10) = 42 - 40 = 2$$

  Since $2 = 2$, we have $|A| = |B|$.

In general, if B is obtained from A by adding k times its first row to its second row, then

$$|B| = \begin{vmatrix} a & b \\ ka + c & kb + d \end{vmatrix} = a(kb + d) - b(ka + c)$$

$$= akb + ad - bka - bc$$

$$= ad - bc$$

$$= |A|$$

Evaluating higher-order determinants can be very time consuming. However, as we have seen, if we use row and column operations to introduce as many 0s as possible in a row (or column) and then expand the determinant by cofactors along that row (or column), the work is much easier. In the next example, we will use several row and column operations to introduce 0s into the determinant.

EXAMPLE 5 **Using Row and Column Operations to Help Evaluate the Determinant**

Use row and column operations to evaluate $\begin{vmatrix} 10 & 20 & -10 & 20 \\ 2 & 1 & 1 & 1 \\ 1 & 2 & -3 & 2 \\ 2 & -1 & -1 & 1 \end{vmatrix}.$

SOLUTION To get smaller numbers in the first row, we can use a type 2 row operation and multiply each entry in the first row by $\frac{1}{10}$. However, we must then multiply the resulting determinant by 10 to retain its original value.

$$(1) \quad \begin{vmatrix} 10 & 20 & -10 & 20 \\ 2 & 1 & 1 & 1 \\ 1 & 2 & -3 & 2 \\ 2 & -1 & -1 & 1 \end{vmatrix} \begin{array}{c} (\frac{1}{10})R1 \to \\ = 10 \end{array} \begin{vmatrix} 1 & 2 & -1 & 2 \\ 2 & 1 & 1 & 1 \\ 1 & 2 & -3 & 2 \\ 2 & -1 & -1 & 1 \end{vmatrix}$$

To get three 0s in the first row of the second determinant in Equation (1), we perform a type 3 row operation and expand the new determinant along its first row.

$$10 \begin{vmatrix} 1 & 2 & -1 & 2 \\ 2 & 1 & 1 & 1 \\ 1 & 2 & -3 & 2 \\ 2 & -1 & -1 & 1 \end{vmatrix} \begin{array}{c} (-1)R3 + R1 \to \\ = 10 \end{array} \begin{vmatrix} 0 & 0 & 2 & 0 \\ 2 & 1 & 1 & 1 \\ 1 & 2 & -3 & 2 \\ 2 & -1 & -1 & 1 \end{vmatrix}$$

$$(2) \qquad\qquad\qquad\qquad\qquad\qquad = 10(2)\begin{vmatrix} 2 & 1 & 1 \\ 1 & 2 & 2 \\ 2 & -1 & 1 \end{vmatrix}$$

To introduce 0s into the 3×3 determinant in Equation (2), we perform a column operation on the 3×3 determinant and expand the result on its first column.

$$(-2)C3 + C1$$
$$\downarrow$$

$$20 \begin{vmatrix} 2 & 1 & 1 \\ 1 & 2 & 2 \\ 2 & -1 & 1 \end{vmatrix} = 20 \begin{vmatrix} 0 & 1 & 1 \\ -3 & 2 & 2 \\ 0 & -1 & 1 \end{vmatrix}$$

$$= 20\left[-(-3)\begin{vmatrix} 1 & 1 \\ -1 & 1 \end{vmatrix} \right]$$

$$= 20(3)[1 - (-1)]$$

$$= 60(2)$$

$$= 120$$

Self Check 5 Evaluate: $\begin{vmatrix} 15 & 20 & 5 & 10 \\ 1 & 2 & 2 & 3 \\ 0 & 3 & -1 & 1 \\ 1 & 0 & 0 & -1 \end{vmatrix}.$

Now Try Exercise 29.

We will consider one final theorem.

Identical Row or Column Theorem   If A is a square matrix with two identical rows (or columns), then $|A| = 0$.

PROOF   If the square matrix A has two identical rows (or columns), we can apply a type 3 row (or column) operation to zero out one of those rows (or columns). Since the matrix would then have an all-zero row (or column), its determinant would be 0 when we expand by cofactors.   ∎

---

3. Use Determinants to Solve Systems of Equations

We can solve the system $\begin{cases} ax + by = e \\ cx + dy = f \end{cases}$ by multiplying the first equation by d, multiplying the second equation by $-b$, and adding.

$$\begin{aligned} adx + bdy &= ed \\ -bcx - bdy &= -bf \\ \hline adx - bcx &= ed - bf \end{aligned}$$

If $ad \neq bc$, we can solve the resulting equation for x.

$$\begin{aligned} adx - bcx &= ed - bf \\ (ad - bc)x &= ed - bf &&\text{Factor out } x. \end{aligned}$$

$$(3) \qquad x = \frac{ed - bf}{ad - bc} \qquad\qquad \text{Divide both sides by } ad - bc.$$

If $ad \neq bc$, we can also solve the system for y.

$$(4) \qquad y = \frac{af - ec}{ad - bc}$$

We can write the values of x and y in Equations (3) and (4) using determinants.

$$x = \frac{\begin{vmatrix} e & b \\ f & d \end{vmatrix}}{\begin{vmatrix} a & b \\ c & d \end{vmatrix}} = \frac{ed - bf}{ad - bc} \qquad\qquad y = \frac{\begin{vmatrix} a & e \\ c & f \end{vmatrix}}{\begin{vmatrix} a & b \\ c & d \end{vmatrix}} = \frac{af - ec}{ad - bc}$$

If we compare these formulas with the original system

$$\begin{cases} ax + by = e \\ cx + dy = f, \end{cases}$$

we see that the denominators are the determinant of the coefficient matrix:

$$\text{denominator determinant } = \begin{vmatrix} a & b \\ c & d \end{vmatrix}.$$

To find the numerator determinant for x, we replace the a and c in the first column of the denominator determinant with the constants e and f.

To find the numerator determinant for y, we replace the b and d in the second column of the denominator determinant with the constants e and f.

$$x = \frac{\begin{vmatrix} e & b \\ f & d \end{vmatrix}}{\begin{vmatrix} a & b \\ c & d \end{vmatrix}} \qquad\qquad y = \frac{\begin{vmatrix} a & e \\ c & f \end{vmatrix}}{\begin{vmatrix} a & b \\ c & d \end{vmatrix}}$$

This method of using determinants to solve systems of equations is called **Cramer's Rule.**

Cramer's Rule for Two Equations in Two Variables

If the system $\begin{cases} ax + by = e \\ cx + dy = f \end{cases}$ has a single solution, it is given by

$$x = \frac{D_x}{D} \quad \text{and} \quad y = \frac{D_y}{D},$$

where $D = \begin{vmatrix} a & b \\ c & d \end{vmatrix}$, $D_x = \begin{vmatrix} e & b \\ f & d \end{vmatrix}$, and $D_y = \begin{vmatrix} a & e \\ c & f \end{vmatrix}$.

If D, D_x, and D_y are all 0, the system is consistent but the equations are dependent. If $D = 0$ and $D_x \neq 0$ or $D_y \neq 0$, the system is inconsistent.

EXAMPLE 6   **Using Cramer's Rule to Solve a System of Linear Equations**

Use Cramer's Rule to solve the system $\begin{cases} 3x + 2y = 7 \\ -x + 5y = 9 \end{cases}$.

SOLUTION

$$x = \frac{\begin{vmatrix} 7 & 2 \\ 9 & 5 \end{vmatrix}}{\begin{vmatrix} 3 & 2 \\ -1 & 5 \end{vmatrix}} = \frac{7 \cdot 5 - 2 \cdot 9}{3 \cdot 5 - 2(-1)} = \frac{35 - 18}{15 + 2} = \frac{17}{17} = 1$$

$$y = \frac{\begin{vmatrix} 3 & 7 \\ -1 & 9 \end{vmatrix}}{\begin{vmatrix} 3 & 2 \\ -1 & 5 \end{vmatrix}} = \frac{3 \cdot 9 - 7(-1)}{3 \cdot 5 - 2(-1)} = \frac{27 + 7}{15 + 2} = \frac{34}{17} = 2$$

Verify that the pair $(1, 2)$ satisfies both of the equations in the system.

Self Check 6   Solve: $\begin{cases} 2x + 5y = 9 \\ 3x + 7y = 13 \end{cases}$.

Now Try Exercise 39.

We can use Cramer's Rule to solve systems of n equations in n variables where each equation has the form

$$a_1x_1 + a_2x_2 + \cdots + a_nx_n = c.$$

To do so, we let D be the determinant of the coefficient matrix of the system and let D_x be the determinant formed by replacing the ith column of D by the column of constants from the right of the equal signs. If $D \neq 0$, Cramer's Rule provides the following solution:

$$x_1 = \frac{D_{x_1}}{D}, x_2 = \frac{D_{x_2}}{D}, \ldots, x_n = \frac{D_{x_n}}{D}.$$

EXAMPLE 7   **Using Cramer's Rule to Solve a System of Linear Equations**

Use Cramer's Rule to solve the system $\begin{cases} 2x - y + 2z = 3 \\ x - y + z = 2 \\ x + y + 2z = 3 \end{cases}$.

SOLUTION   Each of the values x, y, and z is the quotient of two 3×3 determinants. The denominator of each quotient is the determinant consisting of the nine coefficients of the variables. The numerators for x, y, and z are modified copies of this denominator determinant. We substitute the column of constants for the coefficients of the variable for which we are solving.

$$\begin{cases} 2x - y + 2z = 3 \\ x - y + z = 2 \\ x + y + 2z = 3 \end{cases}$$

$$x = \frac{\begin{vmatrix} 3 & -1 & 2 \\ 2 & -1 & 1 \\ 3 & 1 & 2 \end{vmatrix}}{\begin{vmatrix} 2 & -1 & 2 \\ 1 & -1 & 1 \\ 1 & 1 & 2 \end{vmatrix}} = \frac{3\begin{vmatrix} -1 & 1 \\ 1 & 2 \end{vmatrix} - (-1)\begin{vmatrix} 2 & 1 \\ 3 & 2 \end{vmatrix} + 2\begin{vmatrix} 2 & -1 \\ 3 & 1 \end{vmatrix}}{2\begin{vmatrix} -1 & 1 \\ 1 & 2 \end{vmatrix} - (-1)\begin{vmatrix} 1 & 1 \\ 1 & 2 \end{vmatrix} + 2\begin{vmatrix} 1 & -1 \\ 1 & 1 \end{vmatrix}} = \frac{2}{-1} = -2$$

$$y = \frac{\begin{vmatrix} 2 & 3 & 2 \\ 1 & 2 & 1 \\ 1 & 3 & 2 \end{vmatrix}}{\begin{vmatrix} 2 & -1 & 2 \\ 1 & -1 & 1 \\ 1 & 1 & 2 \end{vmatrix}} = \frac{2\begin{vmatrix} 2 & 1 \\ 3 & 2 \end{vmatrix} - 3\begin{vmatrix} 1 & 1 \\ 1 & 2 \end{vmatrix} + 2\begin{vmatrix} 1 & 2 \\ 1 & 3 \end{vmatrix}}{-1} = \frac{1}{-1} = -1$$

$$z = \frac{\begin{vmatrix} 2 & -1 & 3 \\ 1 & -1 & 2 \\ 1 & 1 & 3 \end{vmatrix}}{\begin{vmatrix} 2 & -1 & 2 \\ 1 & -1 & 1 \\ 1 & 1 & 2 \end{vmatrix}} = \frac{2\begin{vmatrix} -1 & 2 \\ 1 & 3 \end{vmatrix} - (-1)\begin{vmatrix} 1 & 2 \\ 1 & 3 \end{vmatrix} + 3\begin{vmatrix} 1 & -1 \\ 1 & 1 \end{vmatrix}}{-1} = \frac{-3}{-1} = 3$$

Verify that the triple $(-2, -1, 3)$ satisfies each equation in the system.

Self Check 7   Solve: $\begin{cases} 2x - y + 2z = 6 \\ x - y + z = 2 \\ x + y + 2z = 9 \end{cases}$.

Now Try Exercise 43.

Self Check Answers   **1.** -2 **2.** 6 **3.** 6 **4.** -8 **5.** 30 **6.** $(2, 1)$ **7.** $(1, 2, 3)$

Exercises **8.5**

Getting Ready
You should be able to complete these vocabulary and concept statements before you proceed to the practice exercises.

Fill in the blanks.

1. The determinant of a square matrix A is written as _____ or _____.

2. $\begin{vmatrix} a & b \\ c & d \end{vmatrix} =$ _____.

3. If every entry in one row or one column of A is zero, then the determinant of A is _____.

4. If a matrix B is obtained from matrix A by adding one row to another, then $|B| =$ _____.

5. If two columns of A are identical, then the determinant of A is _____.

6. In Cramer's Rule, the denominator is the determinant of the _____.

Practice
Evaluate each determinant.

7. $\begin{vmatrix} 2 & 1 \\ -2 & 3 \end{vmatrix}$

8. $\begin{vmatrix} -3 & -6 \\ 2 & -5 \end{vmatrix}$

9. $\begin{vmatrix} 2 & -3 \\ -3 & 5 \end{vmatrix}$

10. $\begin{vmatrix} 5 & 8 \\ -6 & -2 \end{vmatrix}$

In Exercises 11–18, $A = \begin{vmatrix} 1 & -2 & 3 \\ 4 & 5 & -6 \\ -7 & 8 & 9 \end{vmatrix}$. *Find each minor or cofactor.*

11. M_{21}

12. M_{13}

13. M_{33}

14. M_{32}

15. C_{21}

16. C_{13}

17. C_{33}

18. C_{32}

Evaluate each determinant by expanding by cofactors.

19. $\begin{vmatrix} 2 & -3 & 5 \\ -2 & 1 & 3 \\ 1 & 3 & -2 \end{vmatrix}$

20. $\begin{vmatrix} 1 & 3 & 1 \\ -2 & 5 & 3 \\ 3 & -2 & -2 \end{vmatrix}$

21. $\begin{vmatrix} 1 & -1 & 2 \\ 2 & 1 & 3 \\ 1 & 1 & -1 \end{vmatrix}$

22. $\begin{vmatrix} 1 & 3 & 1 \\ 2 & 1 & -1 \\ 2 & -1 & 1 \end{vmatrix}$

23. $\begin{vmatrix} 2 & 1 & -1 \\ 1 & 3 & 5 \\ 2 & -5 & 3 \end{vmatrix}$

24. $\begin{vmatrix} 3 & 1 & -2 \\ -3 & 2 & 1 \\ 1 & 3 & 0 \end{vmatrix}$

25. $\begin{vmatrix} 0 & 1 & -3 \\ -3 & 5 & 2 \\ 2 & -5 & 3 \end{vmatrix}$

26. $\begin{vmatrix} 1 & -7 & -2 \\ -2 & 0 & 3 \\ -1 & 7 & 1 \end{vmatrix}$

27. $\begin{vmatrix} 0 & 0 & 1 & 0 \\ -2 & 1 & 0 & 1 \\ 1 & 0 & 1 & 2 \\ 2 & 0 & 1 & 2 \end{vmatrix}$

28. $\begin{vmatrix} 1 & 0 & -2 & 1 \\ 0 & 1 & 0 & 1 \\ 0 & 3 & -1 & 2 \\ 0 & -1 & 0 & 1 \end{vmatrix}$

29. $\begin{vmatrix} 1 & 2 & 1 & 3 \\ -2 & 1 & -3 & 1 \\ -1 & 0 & 1 & -2 \\ 2 & -1 & -1 & 3 \end{vmatrix}$

30. $\begin{vmatrix} -1 & 3 & -2 & 5 \\ 2 & 1 & 0 & 1 \\ 1 & 3 & -2 & 5 \\ 2 & -1 & 0 & -1 \end{vmatrix}$

Determine whether each statement is true or false. Do not evaluate the determinants.

31. $\begin{vmatrix} 1 & 3 & -4 \\ -2 & 1 & 3 \\ 1 & 3 & 2 \end{vmatrix} = -\begin{vmatrix} -2 & 1 & 3 \\ 1 & 3 & -4 \\ 1 & 3 & 2 \end{vmatrix}$

32. $\begin{vmatrix} 4 & 6 & 8 \\ 10 & 5 & 15 \\ 20 & 5 & 10 \end{vmatrix} = \begin{vmatrix} 2 & 3 & 4 \\ 10 & 5 & 15 \\ 20 & 5 & 10 \end{vmatrix}$

33. $\begin{vmatrix} -2 & -3 & -4 \\ 5 & -1 & 2 \\ 1 & 2 & 3 \end{vmatrix} = -\begin{vmatrix} 2 & 3 & 4 \\ -5 & 1 & -2 \\ 1 & 2 & 3 \end{vmatrix}$

34. $\begin{vmatrix} 1 & 2 & 3 \\ 4 & 5 & 6 \\ 7 & 8 & 9 \end{vmatrix} = \begin{vmatrix} 5 & 7 & 9 \\ 4 & 5 & 6 \\ 7 & 8 & 9 \end{vmatrix}$

If $\begin{vmatrix} a & b & c \\ d & e & f \\ g & h & i \end{vmatrix} = 3$, *find the value of each determinant.*

35. $\begin{vmatrix} d & e & f \\ a & b & c \\ -g & -h & -i \end{vmatrix}$

36. $\begin{vmatrix} 5a & 5b & 5c \\ -d & -e & -f \\ 3g & 3h & 3i \end{vmatrix}$

37. $\begin{vmatrix} a+g & b+h & c+i \\ d & e & f \\ g & h & i \end{vmatrix}$

38. $\begin{vmatrix} g & h & i \\ a & b & c \\ d & e & f \end{vmatrix}$

Use Cramer's Rule to find the solution of each system, if possible.

39. $\begin{cases} 3x + 2y = 7 \\ 2x - 3y = -4 \end{cases}$

40. $\begin{cases} x - 5y = -6 \\ 3x + 2y = -1 \end{cases}$

41. $\begin{cases} x - y = 3 \\ 3x - 7y = 9 \end{cases}$

42. $\begin{cases} 2x - y = -6 \\ x + y = 0 \end{cases}$

43. $\begin{cases} x + 2y + z = 2 \\ x - y + z = 2 \\ x + y + 3z = 4 \end{cases}$

44. $\begin{cases} x + 2y - z = -1 \\ 2x + y - z = 1 \\ x - 3y - 5z = 17 \end{cases}$

45. $\begin{cases} 2x - y + z = 5 \\ 3x - 3y + 2z = 10 \\ x + 3y + z = 0 \end{cases}$

46. $\begin{cases} x - y - z = 2 \\ x + y + z = 2 \\ -x - y + z = -4 \end{cases}$

47. $\begin{cases} \dfrac{x}{2} + \dfrac{y}{3} + \dfrac{z}{2} = 11 \\ \dfrac{x}{3} + y - \dfrac{z}{6} = 6 \\ \dfrac{x}{2} + \dfrac{y}{6} + z = 16 \end{cases}$

48. $\begin{cases} \dfrac{x}{2} + \dfrac{y}{5} + \dfrac{z}{3} = 17 \\ \dfrac{x}{5} + \dfrac{y}{2} + \dfrac{z}{5} = 32 \\ x + \dfrac{y}{3} + \dfrac{z}{2} = 30 \end{cases}$

49. $\begin{cases} 2p - q + 3r - s = 0 \\ p + q - s = -1 \\ 3p - r = 2 \\ p - 2q + 3s = 7 \end{cases}$

50. $\begin{cases} a + b + c + d = 8 \\ a + b + c + 2d = 7 \\ a + b + 2c + 3d = 3 \\ a + 2b + 3c + 4d = 4 \end{cases}$

In Exercises 51–53, illustrate each column operation by showing that it is true for the determinant $\begin{vmatrix} a & b \\ c & d \end{vmatrix}$.

51. Interchanging two columns

52. Multiplying each element in a column by k

53. Adding k times any column to another column

54. Use the method of addition to solve $\begin{cases} ax + by = e \\ cx + dy = f \end{cases}$
for y and thereby show that $y = \dfrac{af - ec}{ad - bc}$.

Expand the determinants and solve for x.

55. $\begin{vmatrix} 3 & x \\ 1 & 2 \end{vmatrix} = \begin{vmatrix} 2 & -1 \\ x & -5 \end{vmatrix}$ **56.** $\begin{vmatrix} 4 & x^2 \\ 1 & -1 \end{vmatrix} = \begin{vmatrix} x & 4 \\ 2 & 3 \end{vmatrix}$

57. $\begin{vmatrix} 3 & x & 1 \\ x & 0 & -2 \\ 4 & 0 & 1 \end{vmatrix} = \begin{vmatrix} 2 & x \\ x & 4 \end{vmatrix}$

58. $\begin{vmatrix} x & -1 & 2 \\ -2 & x & 3 \\ 4 & -3 & -1 \end{vmatrix} = \begin{vmatrix} 2 & 2 \\ 5 & x \end{vmatrix}$

Applications

59. Investing A student wants to average a 6.6% return by investing $20,000 in the three stocks listed in the table. Because HiTech is a high-risk investment, the student wants to invest three times as much in Save-Tel and OilCo combined as he invests in HiTech. How much should he invest in each stock?

Stock	Rate of Return
HiTech	10%
SaveTel	5%
OilCo	6%

60. Ice skating The illustration shows three circles traced out by a figure skater during her performance. If the centers of the circles are the given distances apart, find the radius of each circle.

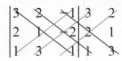

Discovery and Writing
Evaluate each determinant. What do you discover?

61. $\begin{vmatrix} 1 & 3 & 4 \\ 0 & 5 & 2 \\ 0 & 0 & 2 \end{vmatrix}$ **62.** $\begin{vmatrix} 2 & 1 & -2 \\ 0 & 3 & 4 \\ 0 & 0 & -1 \end{vmatrix}$

63. $\begin{vmatrix} 1 & 2 & 4 & 3 \\ 0 & 2 & 2 & 1 \\ 0 & 0 & 3 & 2 \\ 0 & 0 & 0 & 4 \end{vmatrix}$ **64.** $\begin{vmatrix} 2 & 1 & -2 & 1 \\ 0 & 2 & 2 & -1 \\ 0 & 0 & 3 & 1 \\ 0 & 0 & 0 & 2 \end{vmatrix}$

65. Another way to evaluate a 3×3 determinant is to copy its first two columns to the right of the determinant, as shown. Then find the product of the numbers on each red diagonal and find their sum. Then find the product of the numbers on each blue diagonal and find their sum. Finally, subtract the sum of the products on the blue diagonals from the sum of the products on the red diagonals. Find the value of the determinant.

$\begin{vmatrix} 3 & 2 & -1 \\ 2 & 1 & -2 \\ 1 & 3 & -1 \end{vmatrix} \begin{matrix} 3 & 2 \\ 2 & 1 \\ 1 & 3 \end{matrix}$

66. Use the method of Exercise 65 to evaluate the determinant $\begin{vmatrix} 0 & 1 & -3 \\ -3 & 5 & 2 \\ 2 & -5 & 3 \end{vmatrix}$.

67. Use an example chosen from 2×2 matrices to show that the determinant of the product of two matrices is the product of the determinants of those two matrices.

68. Find an example among 2×2 matrices to show that the determinant of a sum of two matrices is not equal to the sum of the determinants of those matrices.

69. A determinant is a function that associates a number with every square matrix. Give the domain and the range of that function.

70. Use an example chosen from 2×2 matrices to show that for $n \times n$ matrices A and B, $AB \neq BA$ but $|AB| = |BA|$.

71. If A and B are matrices and $|AB| = 0$, must either $|A| = 0$ or $|B| = 0$? Explain.

72. If A and B are matrices and $|AB| = 0$, must either $A = 0$ or $B = 0$? Explain.

 Use a graphing calculator to evaluate each determinant.

73. $\begin{vmatrix} 2.3 & 5.7 & 6.1 \\ 3.4 & 6.2 & 8.3 \\ 5.8 & 8.2 & 9.2 \end{vmatrix}$

74. $\begin{vmatrix} 0.32 & -7.4 & -6.7 \\ 3.3 & 5.5 & -0.27 \\ -8 & -0.13 & 5.47 \end{vmatrix}$

8.6 Partial Fractions

In this section, we will learn to

1. Decompose a fraction when the denominator has distinct linear factors.

2. Decompose a fraction when the denominator has distinct quadratic factors.

3. Decompose a fraction when the denominator has repeated linear factors.

4. Decompose a fraction when the denominator has repeated quadratic factors.

5. Decompose a fraction when the degree of the numerator is equal to or greater than the degree of the denominator.

© Istockphoto.com/Jameson Weston

In this section, we will discuss how to write a complicated fraction as a sum of simpler fractions. This skill is used in calculus to solve problems such as modeling a population where there is a maximum population (called the **carrying capacity**) that an environment can sustain.

Suppose there are P gorillas in a specific region in Africa. If there is a carrying capacity of 300 gorillas in that region, the fraction $\dfrac{1}{P^2 - 300P}$ will occur in the model of population growth of gorillas in that region. To solve this modeling problem, we need to rewrite this fraction as a sum of two simpler fractions. In this section, we will learn to do that. The process is called **partial-fraction decomposition.**

You will be asked to decompose this fraction in Exercise 10 in the problem set.

We begin the discussion by reviewing how to add fractions. For example, to find the sum

$$\frac{2}{x} + \frac{6}{x+1} + \frac{-1}{(x+1)^2},$$

we write each fraction with a lowest common denominator of $x(x+1)^2$, add the fractions by adding their numerators and keeping the common denominator, and simplify.

$$\frac{2}{x} + \frac{6}{x+1} + \frac{-1}{(x+1)^2} = \frac{2(x+1)^2}{x(x+1)^2} + \frac{6x(x+1)}{(x+1)x(x+1)} + \frac{-1x}{(x+1)^2 x}$$

$$= \frac{2x^2 + 4x + 2 + 6x^2 + 6x - x}{x(x+1)^2}$$

$$= \frac{8x^2 + 9x + 2}{x(x+1)^2}$$

To reverse the addition process and write a fraction as the sum of simpler fractions with denominators of smallest possible degree, we must **decompose a fraction into partial fractions.** To decompose the fraction

$$\frac{8x^2 + 9x + 2}{x(x + 1)^2}$$

into partial fractions, we will assume that there are constants A, B, and C such that

$$\frac{8x^2 + 9x + 2}{x(x + 1)^2} = \frac{A}{x} + \frac{B}{x + 1} + \frac{C}{(x + 1)^2}.$$

After writing the terms on the right side as fractions with a lowest common denominator of $x(x + 1)^2$, we add the fractions.

$$\frac{8x^2 + 9x + 2}{x(x + 1)^2} = \frac{A(x + 1)^2}{x(x + 1)^2} + \frac{Bx(x + 1)}{(x + 1)x(x + 1)} + \frac{Cx}{(x + 1)^2 x}$$

$$= \frac{Ax^2 + 2Ax + A + Bx^2 + Bx + Cx}{x(x + 1)^2}$$

(1) $\quad \dfrac{8x^2 + 9x + 2}{x(x + 1)^2} = \dfrac{(A + B)x^2 + (2A + B + C)x + A}{x(x + 1)^2}$ $\qquad$ Factor x^2 from $Ax^2 + Bx^2$ and x from $2Ax + Bx + Cx$.

Since the fractions on the left and right sides of Equation (1) are equal and their denominators are equal, the coefficients of their polynomial numerators are equal.

$$\begin{cases} A + B = 8 \\ 2A + B + C = 9 \\ A = 2 \end{cases}$$
$\qquad$ These are the coefficients of x^2.
$\qquad$ These are the coefficients of x.
$\qquad$ These are the constants.

We can solve this system of equations to find that $A = 2$, $B = 6$, and $C = -1$. Then we know that

$$\frac{8x^2 + 9x + 2}{x(x + 1)^2} = \frac{2}{x} + \frac{6}{x + 1} + \frac{-1}{(x + 1)^2}.$$

To check the result, we can add the previous fractions and show that the sum is the original fraction.

To find the partial-fraction decomposition of a fraction, we will follow these steps:

Strategy for Decomposing a Fraction into Partial Fractions When the Denominator Has Distinct Linear Factors

Step 1: Set up the decomposition with unknown constants A, B, C, ... in the numerator of the composition.

Step 2: Write each fraction with a common denominator, add the fractions, and simplify.

Step 3: Set the coefficients of the corresponding terms of each numerator equal to each other.

Step 4: Solve the resulting system of linear equations.

Step 5: Use the values of A, B, C, ... to write the decomposition.

1. Decompose a Fraction When the Denominator Has Distinct Linear Factors

EXAMPLE 1   **Decomposing a Fraction with a Denominator with Distinct Linear Factors**

Decompose $\dfrac{9x + 2}{(x + 2)(3x - 2)}$ into partial fractions.

SOLUTION   **Step 1:**   Since each factor in the denominator is linear, there are constants A and B such that

$$\frac{9x + 2}{(x + 2)(3x - 2)} = \frac{A}{x + 2} + \frac{B}{3x - 2}.$$

Step 2:   After writing the terms on the right side as fractions with a lowest common denominator of $(x + 2)(3x - 2)$, we add the fractions.

(2) $\dfrac{9x + 2}{(x + 2)(3x - 2)} = \dfrac{A(3x - 2)}{(x + 2)(3x - 2)} + \dfrac{B(x + 2)}{(x + 2)(3x - 2)}$

$\dfrac{9x + 2}{(x + 2)(3x - 2)} = \dfrac{3Ax - 2A + Bx + 2B}{(x + 2)(3x - 2)}$

(3) $\dfrac{9x + 2}{(x + 2)(3x - 2)} = \dfrac{(3A + B)x - 2A + 2B}{(x + 2)(3x - 2)}$ Factor x from $3Ax + Bx$.

Step 3:   Since the fractions in Equation (3) are equal, the coefficients of their polynomial numerators are equal, and we have

$$\begin{cases} 9 = 3A + B & \text{These are the coefficients of } x. \\ 2 = -2A + 2B. & \text{These are the constants.} \end{cases}$$

Step 4:   The solution of this system of linear equations is $A = 2$ and $B = 3$.

Step 5:   Finally, we substitute the values of A and B and write the decomposition.

$$\frac{9x + 2}{(x + 2)(3x - 2)} = \frac{2}{x + 2} + \frac{3}{3x - 2}$$

Self Check 1   Decompose $\dfrac{2x + 2}{x(x + 2)}$ into partial fractions.

Now Try Exercise 5.

Comment   The values of A and B in Example 1 can also be found in the following way: From Equation (2), we know that $9x + 2 = A(3x - 2) + B(x + 2)$. In this equation, we can then let $x = -2$ to eliminate the term involving B and solve for A. Then we can let $x = \frac{2}{3}$ to eliminate the term involving A and solve for B. As before, we will find that $A = 2$ and $B = 3$.

2. Decompose a Fraction When the Denominator Has Distinct Quadratic Factors

We can use the following theorem to decompose a fraction with linear and nonlinear factors in the denominator. We begin with an algebraic fraction—like $\dfrac{P(x)}{Q(x)}$, the quotient of two polynomials—and write the fraction as the sum of two or more fractions with simpler denominators. By the theorem, we know that they will be either first-degree or irreducible second-degree polynomials, or powers of those.

Polynomial Factorization Theorem	The factorization of any polynomial $Q(x)$ with real coefficients is the product of polynomials of the forms

$$(ax + b)^n \quad \text{and} \quad (ax^2 + bx + c)^n,$$

where n is a positive integer and $ax^2 + bx + c$ is irreducible over the real numbers.

If $Q(x)$ has a prime quadratic factor of the form $ax^2 + bx + c$, the partial-fraction decomposition of $\frac{P(x)}{Q(x)}$ will have a corresponding term with a linear numerator and a quadratic denominator.

Partial-Fraction Decomposition of $\dfrac{P(x)}{Q(x)}$: $Q(x)$ Has a Prime Quadratic Factor	If $Q(x)$ has a prime quadratic factor of the form $ax^2 + bx + c$, the partial-fraction decomposition of $\frac{P(x)}{Q(x)}$ will have a corresponding term of the form $$\frac{Ax + B}{ax^2 + bx + c}.$$

Caution

> Be sure that the quadratic factor is not factorable over the real numbers—check it with the discriminant. If $b^2 - 4ac$ is negative, the factor is prime.

EXAMPLE 2   **Decomposing a Fraction When One of the Factors in the Denominator Is a Prime Quadratic Factor**

Decompose $\dfrac{2x^2 + x + 1}{x^3 + x}$ into partial fractions.

SOLUTION   **Step 1:**   Since the denominator can be written as a product of a linear factor and a prime quadratic factor, the partial fractions have the form

$$\frac{2x^2 + x + 1}{x(x^2 + 1)} = \frac{A}{x} + \frac{Bx + C}{x^2 + 1}.$$

Step 2:   We add the fractions and simplify.

$$\frac{2x^2 + x + 1}{x(x^2 + 1)} = \frac{A(x^2 + 1)}{x(x^2 + 1)} + \frac{(Bx + C)x}{x(x^2 + 1)}$$

$$= \frac{Ax^2 + A + Bx^2 + Cx}{x(x^2 + 1)}$$

$$= \frac{(A + B)x^2 + Cx + A}{x(x^2 + 1)} \qquad \text{Factor } x^2 \text{ from } Ax^2 + Bx^2.$$

Step 3:   We can equate the corresponding coefficients of the numerators $2x^2 + 1x + 1$ and $(A + B)x^2 + Cx + A$ to get the system

$$\begin{cases} A + B = 2 & \text{These are the coefficients of } x^2. \\ C = 1 & \text{This is the coefficient of } x. \\ A = 1 & \text{This is the constant.} \end{cases}$$

Step 4: This system has solutions of $A = 1$, $B = 1$, $C = 1$.

Step 5: Substituting the values of A, B, and C, we find that the partial-fraction decomposition is

$$\frac{2x^2 + x + 1}{x(x^2 + 1)} = \frac{A}{x} + \frac{Bx + C}{x^2 + 1}$$

$$= \frac{1}{x} + \frac{1x + 1}{x^2 + 1}$$

$$= \frac{1}{x} + \frac{x + 1}{x^2 + 1}$$

■

Self Check 2 Decompose $\dfrac{2x^2 + 1}{x^3 + x}$ into partial fractions.

Now Try Exercise 19.

 Some problems in calculus involving rational expressions cannot be solved easily, or perhaps not at all, without using partial-fraction decomposition. This will be explored in the "Looking Ahead to Calculus" section at the end of the chapter.

3. Decompose a Fraction When the Denominator Has Repeated Linear Factors

If $Q(x)$ has n linear factors of $ax + b$, then $(ax + b)^n$ is a factor of $Q(x)$. When this occurs, the partial-fraction decomposition will contain a sum of n fractions for this term. We will include one fraction with a constant numerator for each power of $ax + b$.

Partial-Fraction Decomposition of $\dfrac{P(x)}{Q(x)}$: $Q(x)$ Has Repeated Linear Factors

If $Q(x)$ has n linear factors of $ax + b$, then $(ax + b)^n$ is a factor of $Q(x)$. Each factor of the form $(ax + b)^n$ generates the following sum of n partial fractions:

$$\frac{A}{ax + b} + \frac{B}{(ax + b)^2} + \frac{C}{(ax + b)^3} + \cdots + \frac{D}{(ax + b)^n}.$$

EXAMPLE 3 **Decomposing a Fraction When the Denominator Has Repeated Linear Factors**

Decompose $\dfrac{3x^2 - x + 1}{x(x - 1)^2}$ into partial fractions.

SOLUTION **Step 1:** Here, each factor in the denominator is linear. The linear factor x appears once and the linear factor $x - 1$ appears twice. Thus, there are constants A, B, and C such that

$$\frac{3x^2 - x + 1}{x(x - 1)^2} = \frac{A}{x} + \frac{B}{x - 1} + \frac{C}{(x - 1)^2}.$$

Caution

Note that $x - 1$ is a repeated factor. It would be incorrect to write

$$\frac{3x^2 - x + 1}{x(x - 1)^2} = \frac{A}{x} + \frac{B}{x - 1} + \frac{C}{x - 1}.$$

Step 2: After writing the terms on the right side as fractions with a lowest common denominator of $x(x-1)^2$, we combine them.

$$\frac{3x^2 - x + 1}{x(x-1)^2} = \frac{A(x-1)^2}{x(x-1)^2} + \frac{Bx(x-1)}{x(x-1)(x-1)} + \frac{Cx}{(x-1)^2 x}$$

$$= \frac{Ax^2 - 2Ax + A + Bx^2 - Bx + Cx}{x(x-1)^2}$$

$$= \frac{(A+B)x^2 + (-2A - B + C)x + A}{x(x-1)^2} \qquad \text{Factor } x^2 \text{ from } Ax^2 + Bx^2 \text{ and } x \text{ from } -2Ax - Bx + Cx.$$

Step 3: Since the fractions are equal, the coefficients of the polynomial numerators are equal, and we have

$$\begin{cases} A + B = 3 & \text{These are the coefficients of } x^2. \\ -2A - B + C = -1 & \text{These are the coefficients of } x. \\ A = 1 & \text{This is the constant.} \end{cases}$$

Step 4: We can solve this system to find that $A = 1$, $B = 2$, and $C = 3$.

Step 5: Substituting the values of A, B, and C, we find that the partial-fraction decomposition is

$$\frac{3x^2 - x + 1}{x(x-1)^2} = \frac{A}{x} + \frac{B}{x-1} + \frac{C}{(x-1)^2}$$

$$= \frac{1}{x} + \frac{2}{x-1} + \frac{3}{(x-1)^2}$$

■

Self Check 3 Decompose $\dfrac{3x^2 + 7x + 1}{x(x+1)^2}$ into partial fractions.

Now Try Exercise 23.

4. Decompose a Fraction When the Denominator Has Repeated Quadratic Factors

If $Q(x)$ has n prime factors of $ax^2 + bx + c$, then $(ax^2 + bx + c)^n$ is a factor. Each factor of the form $(ax^2 + bx + c)^n$ generates a sum of n partial fractions for this term. We will include one fraction with a linear numerator for each power of $ax^2 + bx + c$.

Partial-Fraction Decomposition of $\dfrac{P(x)}{Q(x)}$: $Q(x)$ Has Repeated Quadratic Factors

If $Q(x)$ has n prime factors of $ax^2 + bx + c$, then $(ax^2 + bx + c)^n$ is a factor. Each factor of the form $(ax^2 + bx + c)^n$ generates a sum of n partial fractions of the form

$$\frac{Ax + B}{ax^2 + bx + c} + \frac{Cx + D}{(ax^2 + bx + c)^2} + \cdots + \frac{Ex + F}{(ax^2 + bx + c)^n}.$$

EXAMPLE 4 **Decomposing a Fraction When the Denominator Has Repeated Quadratic Factors**

Decompose $\dfrac{3x^2 + 5x + 5}{(x^2 + 1)^2}$ into partial fractions.

SOLUTION   **Step 1:**   Since the quadratic factor $x^2 + 1$ is used twice, we must find constants A, B, C, and D such that

$$\frac{3x^2 + 5x + 5}{(x^2 + 1)^2} = \frac{Ax + B}{x^2 + 1} + \frac{Cx + D}{(x^2 + 1)^2}.$$

Step 2:   We add the fractions on the right side.

$$\frac{3x^2 + 5x + 5}{(x^2 + 1)^2} = \frac{(Ax + B)(x^2 + 1)}{(x^2 + 1)(x^2 + 1)} + \frac{Cx + D}{(x^2 + 1)^2}$$

$$= \frac{Ax^3 + Ax + Bx^2 + B + Cx + D}{(x^2 + 1)^2}$$

$$= \frac{Ax^3 + Bx^2 + (A + C)x + B + D}{(x^2 + 1)^2} \qquad \text{Factor } x \text{ from } Ax + Cx.$$

Step 3:   If we add the term $0x^3$ to the numerator on the left side, we can equate the corresponding coefficients in the numerators to get

$$\begin{cases} A = 0 & \text{This is the coefficient of } x^3. \\ B = 3 & \text{This is the coefficient of } x^2. \\ A + C = 5 & \text{These are the coefficients of } x. \\ B + D = 5. & \text{These are the constants.} \end{cases}$$

Step 4:   The solution to this system is $A = 0$, $B = 3$, $C = 5$, and $D = 2$.

Step 5:   Substituting the values of A, B, C, and D, we find that the partial-fraction decomposition is

$$\frac{3x^2 + 5x + 5}{(x^2 + 1)^2} = \frac{Ax + B}{x^2 + 1} + \frac{Cx + D}{(x^2 + 1)^2}$$

$$= \frac{0x + 3}{x^2 + 1} + \frac{5x + 2}{(x^2 + 1)^2}$$

$$= \frac{3}{x^2 + 1} + \frac{5x + 2}{(x^2 + 1)^2} \qquad ■$$

Self Check 4   Decompose $\dfrac{x^3 + 2x + 3}{(x^2 + 2)^2}$ into partial fractions.

Now Try Exercise 39.

5. Decompose a Fraction When the Degree of the Numerator Is Equal to or Greater Than the Degree of the Denominator

When the degree of $P(x)$ is equal to or greater than the degree of $Q(x)$ in the fraction $\frac{P(x)}{Q(x)}$, we do a long division before decomposing the fraction into partial fractions.

EXAMPLE 5   **Decomposing a Fraction When the Degree of the Numerator Is Equal to or Greater Than the Degree of the Denominator**

Decompose $\dfrac{x^2 + 4x + 2}{x^2 + x}$ into partial fractions.

SOLUTION   Because the degree of the numerator and denominator are the same, we must do a long division and express the fraction in quotient $+ \frac{\text{remainder}}{\text{divisor}}$ form.

$$
\begin{array}{r}
1 \\
x^2 + x\overline{\smash{\big)}\,x^2 + 4x + 2} \\
\underline{x^2 + x} \\
3x + 2
\end{array}
$$

So we can write

(4) $\dfrac{x^2 + 4x + 2}{x^2 + x} = 1 + \dfrac{3x + 2}{x^2 + x}.$

Because the degree of the numerator of the fraction on the right side of Equation (4) is less than the degree of the denominator, we can find its partial-fraction decomposition.

$$
\frac{3x + 2}{x^2 + x} = \frac{3x + 2}{x(x + 1)}
$$

$$
= \frac{A}{x} + \frac{B}{x + 1}
$$

$$
= \frac{A(x + 1) + Bx}{x(x + 1)}
$$

$$
= \frac{(A + B)x + A}{x(x + 1)}
$$

We equate the corresponding coefficients in the numerator and solve the resulting system of equations to find that the solution is $A = 2$, $B = 1$. So we have

$$
\frac{x^2 + 4x + 2}{x^2 + x} = 1 + \frac{2}{x} + \frac{1}{x + 1}. \quad\blacksquare
$$

Self Check 5   Decompose $\dfrac{x^2 + x - 1}{x^2 - x}$ into partial fractions.

Now Try Exercise 41.

Self Check Answers   **1.** $\dfrac{1}{x} + \dfrac{1}{x + 2}$ **2.** $\dfrac{1}{x} + \dfrac{x}{x^2 + 1}$ **3.** $\dfrac{1}{x} + \dfrac{2}{x + 1} + \dfrac{3}{(x + 1)^2}$

4. $\dfrac{x}{x^2 + 2} + \dfrac{3}{(x^2 + 2)^2}$ **5.** $1 + \dfrac{1}{x} + \dfrac{1}{x - 1}$

Exercises 8.6

Getting Ready
You should be able to complete these vocabulary and concept statements before you proceed to the practice exercises.

Fill in the blanks.

1. A polynomial with real coefficients factors as the product of _____ and _____ factors or powers of those.

2. The second-degree factors of a polynomial with real coefficients are _____, which means they don't factor further over the real numbers.

Practice
Decompose each fraction into partial fractions.

3. $\dfrac{3x - 1}{x(x - 1)}$

4. $\dfrac{4x + 6}{x(x + 2)}$

5. $\dfrac{2x - 15}{x(x - 3)}$

6. $\dfrac{5x + 21}{x(x + 7)}$

7. $\dfrac{3x + 1}{(x + 1)(x - 1)}$

8. $\dfrac{9x - 3}{(x + 1)(x - 2)}$

9. $\dfrac{-4}{x^2 - 2x}$

10. $\dfrac{1}{P^2 - 300P}$

11. $\dfrac{-2x + 11}{x^2 - x - 6}$

12. $\dfrac{7x + 2}{x^2 + x - 2}$

13. $\dfrac{3x - 23}{x^2 + 2x - 3}$

14. $\dfrac{-x - 17}{x^2 - x - 6}$

15. $\dfrac{9x - 31}{2x^2 - 13x + 15}$

16. $\dfrac{-2x - 6}{3x^2 - 7x + 2}$

17. $\dfrac{4x^2 + 4x - 2}{x(x^2 - 1)}$

18. $\dfrac{x^2 - 6x - 13}{(x + 2)(x^2 - 1)}$

19. $\dfrac{x^2 + x + 3}{x(x^2 + 3)}$

20. $\dfrac{5x^2 + 2x + 2}{x^3 + x}$

21. $\dfrac{3x^2 + 8x + 11}{(x + 1)(x^2 + 2x + 3)}$

22. $\dfrac{-3x^2 + x - 5}{(x + 1)(x^2 + 2)}$

23. $\dfrac{5x^2 + 9x + 3}{x(x + 1)^2}$

24. $\dfrac{2x^2 - 7x + 2}{x(x - 1)^2}$

25. $\dfrac{-2x^2 + x - 2}{x^2(x - 1)}$

26. $\dfrac{x^2 + x + 1}{x^3}$

27. $\dfrac{3x^2 - 13x + 18}{x^3 - 6x^2 + 9x}$

28. $\dfrac{3x^2 + 13x + 20}{x^3 + 4x^2 + 4x}$

29. $\dfrac{x^2 - 2x - 3}{(x - 1)^3}$

30. $\dfrac{x^2 + 8x + 18}{(x + 3)^3}$

31. $\dfrac{x^3 + 4x^2 + 2x + 1}{x^4 + x^3 + x^2}$

32. $\dfrac{3x^3 + 5x^2 + 3x + 1}{x^2(x^2 + x + 1)}$

33. $\dfrac{4x^3 + 5x^2 + 3x + 4}{x^2(x^2 + 1)}$

34. $\dfrac{2x^2 + 1}{x^4 + x^2}$

35. $\dfrac{-x^2 - 3x - 5}{x^3 + x^2 + 2x + 2}$

36. $\dfrac{-2x^3 + 7x^2 + 6}{x^2(x^2 + 2)}$

37. $\dfrac{x^3 + 4x^2 + 3x + 6}{(x^2 + 2)(x^2 + x + 2)}$

38. $\dfrac{x^3 + 3x^2 + 2x + 4}{(x^2 + 1)(x^2 + x + 2)}$

39. $\dfrac{2x^4 + 6x^3 + 20x^2 + 22x + 25}{x(x^2 + 2x + 5)^2}$

40. $\dfrac{x^3 + 3x^2 + 6x + 6}{(x^2 + x + 5)(x^2 + 1)}$

41. $\dfrac{x^3}{x^2 + 3x + 2}$

42. $\dfrac{2x^3 + 6x^2 + 3x + 2}{x^3 + x^2}$

43. $\dfrac{3x^3 + 3x^2 + 6x + 4}{3x^3 + x^2 + 3x + 1}$

44. $\dfrac{x^4 + x^3 + 3x^2 + x + 4}{(x^2 + 1)^2}$

45. $\dfrac{x^3 + 3x^2 + 2x + 1}{x^3 + x^2 + x}$

46. $\dfrac{x^4 - x^3 + x^2 - x + 1}{(x^2 + 1)^2}$

47. $\dfrac{2x^4 + 2x^3 + 3x^2 - 1}{(x^2 - x)(x^2 + 1)}$

48. $\dfrac{x^4 - x^3 + 5x^2 + x + 6}{(x^2 + 3)(x^2 + 1)}$

Discovery and Writing

49. Is the polynomial $x^3 + 1$ prime?

50. Decompose $\dfrac{1}{x^3 + 1}$ into partial fractions.

8.7 Graphs of Inequalities

In this section, we will learn to

1. Graph inequalities.

2. Graph the solution set of a system of inequalities.

William Casey/Shutterstock.com

A company builds 84-inch and 120-inch motorized-drive projection screens. If the company needs 2 hours to build an 84-inch screen, it will take $2x$ hours to build x of them. If the company needs 3 hours to build a 120-inch screen, it will take $3y$ hours to build y of them. If the company has 300 hours of labor available each day and cannot build a negative number of either screen, the following system of inequalities provide restrictions on x and y, the number of screens it can build:

$$\begin{cases} 2x + 3y \le 300 \\ x \ge 0 \\ y \ge 0 \end{cases}$$

The time it takes to build x small screens plus the time it takes to build y large screens must be less than or equal to 300. Both x and y must be greater than or equal to 0.

In this section, we will show how to find the solution sets of many types of systems of inequalities. In Example 6, we will solve this system.

1. Graph Inequalities

The **graph of an inequality** in x and y is the graph of all ordered pairs (x, y) that satisfy the inequality. We will start by considering graphs of **linear inequalities**—inequalities that can be written in one of the following forms:

* $Ax + By < C$;
* $Ax + By > C$;
* $Ax + By \le C$;
* $Ax + By \ge C$.

Comment

Note that $y = 3x + 2$ can be written in the form $-3x + y = 2$.

To graph the inequality $y > 3x + 2$, we note that one of the following statements must be true:

$$y = 3x + 2, \ y < 3x + 2, \ \text{or} \ y > 3x + 2.$$

The graph of $y = 3x + 2$ is a line, as shown in Figure 8-5(a). The graphs of the inequalities are half-planes, one on each side of that line. We can think of the graph of $y = 3x + 2$ as a boundary separating the two half-planes. The graph of $y = 3x + 2$ is drawn with a dashed line to show that it is not part of the graph of $y > 3x + 2$.

To find which half-plane is the graph of $y > 3x + 2$, we can substitute the coordinates of any point on one side of the line—say, the origin $(0, 0)$—into the inequality.

$y > 3x + 2$
$0 > 3(0) + 2$
$0 > 2$ False

Since $0 > 2$ is false, the coordinates $(0, 0)$ do not satisfy the inequality, and the origin is not in the half-plane that is the graph of $y > 3x + 2$. Thus the graph is the half-plane located on the other side of the dashed line. The graph of the inequality $y > 3x + 2$ is shown in Figure 8-5(b).

$y = 3x + 2$		
x	y	(x, y)
-1	-1	$(-1, -1)$
0	2	$(0, 2)$
1	5	$(1, 5)$

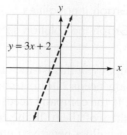

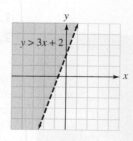

(a)   (b)

FIGURE 8-5

We can use the following steps to graph linear inequalities:

Strategy for Graphing Linear Inequalities in Two Variables	
Step 1:	Write the inequality as an equation by replacing the inequality symbol with an equal sign.
Step 2:	Graph the resulting equation to establish a boundary line. If the inequality is < or >, draw a dashed line. If the inequality is ≤ or ≥, draw a solid line.
Step 3:	Select a test point that is not on the boundary.
Step 4:	Substitute the coordinates of the test point into the inequality.
Step 5:	Shade the region that contains the test point if the coordinates of the test point satisfy the inequality. If they don't, shade the region that does not contain the test point.

EXAMPLE 1   Graphing a Linear Inequality

Graph the inequality $2x - 3y \le 6$.

SOLUTION   **Step 1:**   This inequality is the combination of $2x - 3y < 6$ and $2x - 3y = 6$.

Step 2:   We start by graphing $2x - 3y = 6$ to establish the boundary line that separates the plane into two half-planes. We draw a solid line, because equality is permitted. See Figure 8-6(a).

Step 3:   Because the computations will be easy, we select the origin as a test point.

Step 4:   We will substitute the coordinates of the origin into the inequality and see whether the coordinates satisfy the inequality.

$$2x - 3y \le 6$$
$$2(0) - 3(0) \le 6$$
$$0 \le 6 \quad \text{True.}$$

Step 5:   Because $0 \le 6$ is true, we shade the half-plane that contains the test point. The graph is shown in Figure 8-6(b).

$2x - 3y = 6$		
x	y	(x, y)
0	−2	$(0, -2)$
3	0	$(3, 0)$

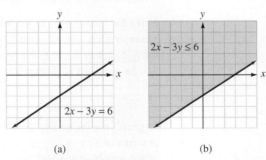

(a)   (b)

FIGURE 8-6

Self Check 1   Graph: $3x + 2y \le 6$.

Now Try Exercise 5.

EXAMPLE 2   **Graphing a Linear Inequality**

Graph the inequality $y < 2x$.

SOLUTION   **Step 1:**   We write the equation $y = 2x$.

Step 2:   Since the graph of $y = 2x$ is not part of the graph of the inequality, we graph the boundary with a dashed line, as in Figure 8-7(a).

Step 3:   We cannot use the origin as a test point, because the boundary line passes through the origin. So we choose some other point—say, $(3, 1)$.

Step 4:   To determine which half-plane represents the graph $y < 2x$, we check whether the coordinates of the test point $(3, 1)$ satisfy the inequality.

$$y < 2x$$
$$1 < 2(3)$$
$$1 < 6 \qquad \text{True}$$

Step 5:   Since $1 < 6$, the point $(3, 1)$ lies in the graph. The graph is shown in Figure 8-7(b).

	$y = 2x$	
x	y	(x, y)
0	0	$(0, 0)$
1	2	$(1, 2)$

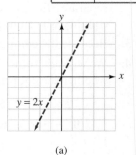 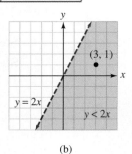

(a)   (b)

FIGURE 8-7

Self Check 2   Graph: $y > 3x$.

Now Try Exercise 11.

ACCENT ON TECHNOLOGY   **Graphing Inequalities**

Many calculators have the capability of graphing inequalities by shading areas either above or below a curve.

Move the cursor over to the left of the equation name, Y_1, Y_2, ..., and press ENTER. Each time the ENTER key is used, the type of curve that will be graphed is changed.

To graph the inequality $2x - 3y \le 6$, solve for y in terms of x; $y \le \frac{2x}{3} - 2$. Enter the expression in the graph editor and select the style that shades the area below the line.

FIGURE 8-8

Use ⌜ZOOM⌝ 6:ZStandard for the WINDOW.

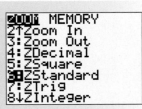

Press ⌜ENTER⌝ and graph the inequality.

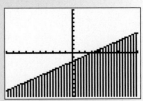

FIGURE 8-8 (Continued)

We can graph many inequalities that are not linear inequalities.

EXAMPLE 3   **Graphing a Nonlinear Inequality**

Graph the inequality $x^2 + y^2 > 25$.

SOLUTION   **Step 1:**   We form the equation $x^2 + y^2 = 25$.

Step 2:   The graph of $x^2 + y^2 = 25$ is a circle. Since the inequality is >, we draw the circle as a dashed circle, as in Figure 8-9(a).

Step 3:   Because the work is easy, we select the origin as the test point.

Step 4:   We substitute the coordinates of the origin into the inequality.

$$x^2 + y^2 > 25$$
$$0^2 + 0^2 > 25$$
$$0 > 25 \qquad \text{False}$$

Step 5:   Since the coordinates of the origin do not satisfy the inequality, we shade the area outside the circle, as shown in Figure 8-9(b).

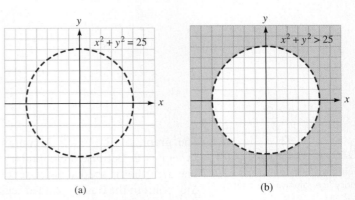

(a)   (b)

FIGURE 8-9

Self Check 3   Graph the inequality $x^2 + y^2 \leq 36$.

Now Try Exercise 15.

2. Graph the Solution Set of a System of Inequalities

We now consider systems of inequalities. To graph the solution set of the system

$$\begin{cases} y < 5 \\ x \le 6, \end{cases}$$

we graph each inequality on the same set of coordinate axes, as in Figure 8-10. The graph of the inequality $y < 5$ is the half-plane that lies below the line $y = 5$. The graph of the inequality $x \le 6$ includes the half-plane that lies to the left of the line $x = 6$ together with the line $x = 6$.

The portion of the xy plane where the two graphs intersect is the graph of the system. Any point that lies in the doubly shaded region has coordinates that satisfy both inequalities in the system.

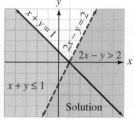

FIGURE 8-10

EXAMPLE 4   **Graphing the Solution Set of a System of Linear Inequalities**

Graph the solution set of $\begin{cases} x + y \le 1 \\ 2x - y > 2 \end{cases}$.

SOLUTION   On the same set of coordinate axes, we graph each inequality, as in Figure 8-11. The graph of $x + y \le 1$ includes the graph of $x + y = 1$ and all points below it. Because the boundary line is included, we draw it as a solid line.

x + y = 1		
x	y	(x, y)
0	1	(0, 1)
1	0	(1, 0)

2x − y = 2		
x	y	(x, y)
0	−2	(0, −2)
1	0	(1, 0)

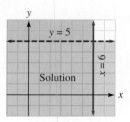

FIGURE 8-11

Comment

To be sure you graph the correct side of the boundary line, always use a test point.

The graph of $2x - y > 2$ contains only those points below the line graph of $2x - y = 2$. Because the boundary line is not included, we draw it as a dashed line. The area that is shaded twice represents the solution of the system of inequalities. Any point in the doubly shaded region has coordinates that satisfy both inequalities.

Self Check 4   Graph the solution set of $\begin{cases} x + y < 2 \\ x - 2y \ge 2 \end{cases}$.

Now Try Exercise 23.

EXAMPLE 5   **Graphing the Solution Set of a System of Inequalities**

Graph the solution set of $\begin{cases} y < x^2 \\ y > \dfrac{x^2}{4} - 2 \end{cases}$.

SOLUTION   The graph of $y = x^2$ is a parabola opening upward with vertex at the origin, as shown in Figure 8-12. The points with coordinates that satisfy the inequality are the points below the parabola.

The graph of $y = \dfrac{x^2}{4} - 2$ is also a parabola opening upward. The points that satisfy the inequality are the points above the parabola. The graph of the solution set is the shaded area between the two parabolas.

$y = x^2$		
x	y	(x, y)
0	0	(0, 0)
1	1	(1, 1)
−1	1	(−1, 1)
2	4	(2, 4)
−2	4	(−2, 4)

$y = \dfrac{x^2}{4} - 2$		
x	y	(x, y)
0	−2	(0, −2)
2	−1	(2, −1)
−2	−1	(−2, −1)
4	2	(4, 2)
−4	2	(−4, 2)

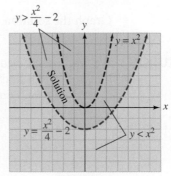

FIGURE 8-12

Self Check 5   Graph the solution set of $\begin{cases} y \le 4 - x^2 \\ y > \dfrac{x^2}{4} \end{cases}$.

Now Try Exercise 29.

EXAMPLE 6   **Solving an Application Problem**

Graph the solution set of the system that began the section:

$$\begin{cases} 2x + 3y \le 300 \\ x \ge 0 \\ y \ge 0 \end{cases}.$$

SOLUTION   We graph each line, as shown in Figure 8-13. The graph of $2x + 3y \le 300$ includes the line $2x + 3y = 300$ and all points below it. Because the boundary line is included, we will draw it as a solid line.

The graph of $x \geq 0$ includes the y-axis and all points to the right of it. The graph of $y \geq 0$ includes the x-axis and all points above it.

The solution set of the system is the set of points in the shaded region.

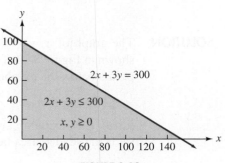

FIGURE 8-13

Self Check 6   Graph the solution set of the system $\begin{cases} x + 2y \leq 50 \\ x \geq 0 \\ y \geq 0 \end{cases}$.

Now Try Exercise 37.

EXAMPLE 7   Graphing the Solution Set of a System of Inequalities

Graph the solution set of $\begin{cases} x + y \leq 4 \\ x - y \leq 6. \\ x \geq 0 \end{cases}$

SOLUTION   We graph each inequality, as in Figure 8-14. The graph of $x + y \leq 4$ includes the line $x + y = 4$ and all points below it. Because the boundary line is included, we draw it as a solid line. The graph of $x - y \leq 6$ contains the line $x - y = 6$ and all points above it.

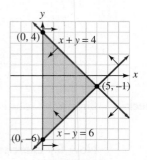

FIGURE 8-14

The graph of the inequality $x \geq 0$ contains the y-axis and all points to the right of the y-axis. The solution of the system of inequalities is the shaded area in the figure.

The coordinates of the corner points of the shaded area are $(0, 4)$, $(0, -6)$, and $(5, -1)$.

Self Check 7   Graph the solution set of $\begin{cases} x + y \leq 5 \\ x - 3y \leq -3. \\ x \geq 0 \end{cases}$

Now Try Exercise 31.

EXAMPLE 8 **Graphing the Solution Set of a System of Inequalities**

Graph the solution set of the system $\begin{cases} x \geq 1 \\ y \geq x \\ 4x + 5y < 20 \end{cases}$.

SOLUTION The graph of the solution set of $x \geq 1$ includes those points on the graph of $x = 1$ and to the right. See Figure 8-15(a).

The graph of the solution set of $y \geq x$ includes those points on the graph of $y = x$ and above it. See Figure 8-15(b).

The graph of the solution set of $4x + 5y < 20$ includes those points below the graph of $4x + 5y = 20$. See Figure 8-15(c).

If these graphs are merged onto a single set of coordinate axes, as in Figure 8-15(d), the graph of the original system of inequalities includes those points within the shaded triangle together with the points on the sides of the triangle drawn as solid lines. The coordinates of the corner points are $\left(1, \frac{16}{5}\right)$, $(1, 1)$, and $\left(\frac{20}{9}, \frac{20}{9}\right)$.

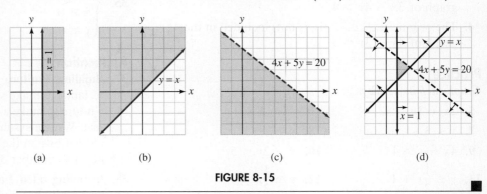

| (a) | (b) | (c) | (d) |

FIGURE 8-15

Self Check 8 Is the point $(2, 2)$ in the solution set of Example 8?

Now Try Exercise 35.

Self Check Answers

1.   2.   3.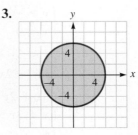

4.   5.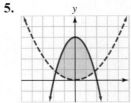

6. 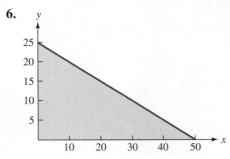 7. 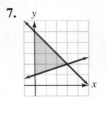 8. yes

Exercises **8.7**

Getting Ready

You should be able to complete these vocabulary and concept statements before you proceed to the practice exercises.

Fill in the blanks.

1. The graph of $Ax + By = C$ is a line. The graph of $Ax + By \le C$ is a _____ with the line as its _____.

2. The boundary of the graph $Ax + By < C$ is _____ (included in, excluded from) the graph.

3. The origin _____ (is, is not) included in the graph of $3x - 4y > 4$.

4. The origin _____ (is, is not) included in the graph of $4x + 3y \le 5$.

Practice

Graph each inequality.

5. $2x + 3y < 12$

6. $4x - 3y > 6$

7. $x < 3$

8. $y > -1$

9. $4x - y > 4$

10. $x - 2y < 5$

11. $y \le \dfrac{1}{2}x + 1$

12. $y \ge \dfrac{1}{3}x - 1$

13. $y < x^2$

14. $y \ge |x|$

15. $x^2 + y^2 \le 4$

16. $x^2 + y^2 > 4$

Graph the solution set of each system.

17. $\begin{cases} y < 3 \\ x \ge 2 \end{cases}$

18. $\begin{cases} y \ge -2 \\ x < 0 \end{cases}$

19. $\begin{cases} y \le x - 2 \\ y \ge 2x + 1 \end{cases}$

20. $\begin{cases} y < 3x + 2 \\ y < -2x + 3 \end{cases}$

21. $\begin{cases} x + y < 2 \\ x + y \le 1 \end{cases}$

22. $\begin{cases} 3x + 2y \ge 6 \\ x + 3y \le 2 \end{cases}$

23. $\begin{cases} x + 2y < 3 \\ 2x - 4y < 8 \end{cases}$

24. $\begin{cases} 3x + y \le 1 \\ -x + 2y \ge 9 \end{cases}$

25. $\begin{cases} 2x - 3y \ge 6 \\ 3x + 2y < 6 \end{cases}$

26. $\begin{cases} 4x + 2y \le 6 \\ 2x - 4y \ge 10 \end{cases}$

27. $\begin{cases} y \ge x^2 - 4 \\ y \le \dfrac{1}{2}x \end{cases}$

28. $\begin{cases} y \le -x^2 + 4 \\ y > -x - 1 \end{cases}$

29. $\begin{cases} y \ge x^2 \\ y < 4 - x^2 \end{cases}$

30. $\begin{cases} x^2 + y \le 1 \\ y - x^2 \ge -1 \end{cases}$

31. $\begin{cases} 2x - y \le 0 \\ x + 2y \le 10 \\ y \ge 0 \end{cases}$

32. $\begin{cases} 3x - 2y \ge 5 \\ 2x + y \ge 8 \\ x \le 5 \end{cases}$

33. $\begin{cases} x - 2y \ge 0 \\ x - y \le 2 \\ x \ge 0 \end{cases}$

34. $\begin{cases} 2x + 3y \le 6 \\ x - y \ge 4 \\ y \ge -4 \end{cases}$

35. $\begin{cases} x + y \le 4 \\ x - y \le 4 \\ x \ge 0 \\ y \ge 0 \end{cases}$

36. $\begin{cases} 2x + 3y \ge 12 \\ 2x - 3y \le 6 \\ x \ge 0 \\ y \le 4 \end{cases}$

Applications

37. **Building furniture** A furniture maker has 60 hours of labor to make sofas (s) and loveseats (l). It takes 6 hours to make a sofa and 4 hours to make a loveseat. Write a system of inequalities that provides the restrictions on the variables. (*Hint*: Remember that a negative number of pieces cannot be made.)

38. **Installing video** Each week, Prime Time Video and Audio has 90 hours of labor to install satellite dishes (d) and home theater systems (t). On average, it takes 5 hours to install a satellite dish and 6 hours to install a home theater system. Write a system of inequalities that provides the restrictions on the variables. (*Hint*: Remember that a negative number of units cannot be installed.)

Discovery and Writing

39. In graphing a linear inequality, explain how to determine the boundary.

40. In graphing a linear inequality, explain how to decide whether the boundary is included.

41. In graphing a linear inequality, explain how to decide which side of the boundary to shade.

42. Does the method you describe in Exercise 41 work for the inequality $3x - 2y \le 0$? Explain.

CHAPTER REVIEW

SECTION 8.1   Systems of Linear Equations

Definitions and Concepts	Examples
Solving a system by graphing: To solve a system of equations by graphing, graph each equation in the system and find the coordinates of the point where all of the graphs intersect.	See Example 1(a), pages 735–736
If the graphs have no point in common, the system has no solution. If the graphs coincide, the system has infinitely many solutions.	See Examples 1(b) and 1(c), pages 735–736
Solving a system by substitution: To solve a system by substitution, solve one equation for one variable, and substitute that result into the other equation. Then solve for the other variable.	See Example 2, page 737
Solving a system by addition: To solve a system by addition, multiply one or both equations by suitable constants so that when the results are added, one variable will be eliminated. Then back substitute to find the other variable.	See Example 3, page 738

EXERCISES

Solve each system by graphing.

1. $\begin{cases} 2x - y = -1 \\ x + y = 7 \end{cases}$ 2. $\begin{cases} 5x + 2y = 1 \\ 2x - y = -5 \end{cases}$

3. $\begin{cases} y = 5x + 7 \\ x = y - 7 \end{cases}$ 4. $\begin{cases} 3x + 2y = 6 \\ y = -\dfrac{3}{2}x + 3 \end{cases}$

5. $\begin{cases} 4x - y = 4 \\ y = 4(x - 2) \end{cases}$

Solve each system by substitution.

6. $\begin{cases} 2y + x = 0 \\ x = y + 3 \end{cases}$ 7. $\begin{cases} 2x + y = -3 \\ x - y = 3 \end{cases}$

8. $\begin{cases} \dfrac{x + y}{2} + \dfrac{x - y}{3} = 1 \\ y = 3x - 2 \end{cases}$ 9. $\begin{cases} y = 3x - 4 \\ 9x - 3y = 12 \end{cases}$

10. $\begin{cases} x = -\dfrac{3}{2}y + 3 \\ 2x + 3y = 4 \end{cases}$

Solve each system by addition.

11. $\begin{cases} x + 5y = 7 \\ 3x + y = -7 \end{cases}$ 12. $\begin{cases} 2x + 3y = 11 \\ 3x - 7y = -41 \end{cases}$

13. $\begin{cases} 2(x + y) - x = 0 \\ 3(x + y) + 2y = 1 \end{cases}$ 14. $\begin{cases} 8x + 12y = 24 \\ 2x + 3y = 4 \end{cases}$

15. $\begin{cases} 3x - y = 4 \\ 9x - 3y = 12 \end{cases}$

Solve each system by any method.

16. $\begin{cases} 3x + 2y - z = 2 \\ x + y - z = 0 \\ 2x + 3y - z = 1 \end{cases}$ 17. $\begin{cases} 5x - y + z = 3 \\ 3x + y + 2z = 2 \\ x + y = 2 \end{cases}$

18. $\begin{cases} 2x - y + z = 1 \\ x - y + 2z = 3 \\ x - y + z = 1 \end{cases}$

Applications

19. **Department store order** The buyer for a large department store must order 40 coats, some fake fur and some leather. He is unsure of the expected sales. He can buy 25 fur coats and the rest leather for $9,300, or 10 fur coats and the rest leather for $12,600. How much does he pay if he decides to split the order evenly?

20. **Ticket sales** Adult tickets for the championship game are usually $5, but on Seniors' Day, seniors paid $4. Children's tickets were $2.50. Sales of 1,800 tickets totaled $7,425, and children and seniors accounted for one-half of the tickets sold. How many of each were sold?

SECTION **8.2** Gaussian Elimination and Matrix Methods

Definitions and Concepts	Examples
There are three matrices associated with a system of equations: a **coefficient matrix**, a **matrix of constants**, and an **augmented** or **system matrix**.	See Section 8.2 opening, pages 746–747
Elementary row operations: • Any two rows of a matrix can be interchanged. • The elements of any row of a matrix can be multiplied by any nonzero constant. • Any row of a matrix can be changed by adding a multiple of another row to it.	See Example 1, pages 747–748
Row echelon form of a matrix: • The first nonzero entry in each row is 1. • Lead entries appear farther to the right as you move down the rows of the matrix. • Any rows containing only 0s are at the bottom of the matrix.	See Example 2, pages 749–750

EXERCISES

Solve by matrix methods, if possible.

21. $\begin{cases} 2x + 5y = 7 \\ 3x - y = 2 \end{cases}$

22. $\begin{cases} 3x - y = -4 \\ -6x + 2y = 8 \end{cases}$

23. $\begin{cases} x + 3y - z = 8 \\ 2x + y - 2z = 11 \\ x - y + 5z = -8 \end{cases}$

24. $\begin{cases} x + 3y + z = 3 \\ 2x - y + z = -11 \\ 3x + 2y + 3z = 2 \end{cases}$

25. $\begin{cases} x + y + z = 4 \\ 3x - 2y - 2z = -3 \\ 4x - y - z = 0 \end{cases}$

SECTION **8.3** Matrix Algebra

Definitions and Concepts	Examples
Equality of matrices: Matrices are equal if and only if they are the same size and have the same corresponding entries.	See Section 8.3 opening, page 758
Sum of two matrices: Two $m \times n$ matrices are added by adding the corresponding elements of those matrices.	See Example 1, page 759
Difference of two matrices: Two $m \times n$ matrices are subtracted by the rule $A - B = A + (-B)$.	See page 760
Multiplication of two matrices: The **product** AB of the $m \times n$ matrix A and the $n \times p$ matrix B is the $m \times p$ matrix C. The ith-row, jth-column entry of C is found by keeping a running total of the products of the elements in the ith row of A and the corresponding elements in the jth column of B.	See Example 3, pages 763–764

EXERCISES

26. Solve for x and y if $\begin{bmatrix} 1 & -4 \\ x & 2 \\ 0 & x+7 \end{bmatrix} = \begin{bmatrix} 1 & x \\ -4 & 2 \\ x+4 & y \end{bmatrix}$.

31. $\begin{bmatrix} 1 & -3 & 2 \end{bmatrix} \begin{bmatrix} 2 \\ 1 \\ 3 \end{bmatrix}$

Perform the matrix operations, if possible.

27. $\begin{bmatrix} 3 & 2 & 1 \\ 3 & 2 & 1 \end{bmatrix} + \begin{bmatrix} -2 & 1 & 3 \\ 1 & -2 & 1 \end{bmatrix}$

32. $\begin{bmatrix} 1 \\ 2 \\ 1 \\ 5 \end{bmatrix} \begin{bmatrix} 2 & -1 & 1 & 3 \end{bmatrix}$

28. $\begin{bmatrix} 2 & 3 & 5 \\ 1 & -2 & 4 \\ 2 & 1 & -2 \end{bmatrix} - \begin{bmatrix} 0 & -2 & 1 \\ 3 & 4 & -2 \\ 6 & -4 & 1 \end{bmatrix}$

33. $\begin{bmatrix} 1 & -5 & 3 \\ 2 & 1 & -1 \end{bmatrix} \begin{bmatrix} 2 \\ -2 \\ 3 \end{bmatrix} \begin{bmatrix} 1 & -1 \\ -1 & 3 \end{bmatrix} \begin{bmatrix} 1 \\ -2 \end{bmatrix}$

29. $\begin{bmatrix} 1 & -2 \\ -3 & 1 \end{bmatrix} \begin{bmatrix} 2 & 3 \\ -1 & 2 \end{bmatrix}$

34. $\begin{bmatrix} 1 & -3 & 2 \end{bmatrix} \begin{bmatrix} 2 \\ 1 \\ -5 \end{bmatrix} + \begin{bmatrix} 1 & -3 \end{bmatrix} \begin{bmatrix} 2 \\ 5 \end{bmatrix}$

30. $\begin{bmatrix} -2 & 3 & 5 \\ 1 & -2 & -3 \end{bmatrix} \begin{bmatrix} 2 & 1 \\ -1 & 2 \\ -2 & 3 \end{bmatrix}$

35. $\left(\begin{bmatrix} 1 & -3 \\ 3 & 1 \end{bmatrix} + \begin{bmatrix} -1 & 3 \\ 1 & 1 \end{bmatrix} \right) \begin{bmatrix} 1 \\ -5 \end{bmatrix}$

SECTION **8.4**   Matrix Inversion

Definitions and Concepts	Examples
Inverse of a matrix: The **inverse** of an $n \times n$ matrix A is A^{-1}, where $AA^{-1} = A^{-1}A = I$.	See Example 1, page 771
Finding the inverse of a matrix: Use elementary row operations to transform $[A\|I]$ into $[I\|A^{-1}]$, where I is an identity matrix. If A cannot be transformed into I, then A is singular.	See Example 3, pages 772–773
Solving a system by matrix inversion: If A is invertible, then the solution of $AX = B$ is $X = A^{-1}B$.	See Example 5, page 774

EXERCISES

Find the inverse of each matrix, if possible.

36. $\begin{bmatrix} 2 & 3 \\ 3 & 5 \end{bmatrix}$

37. $\begin{bmatrix} 1 & 0 & 0 \\ 2 & 0 & -2 \\ 1 & 2 & 2 \end{bmatrix}$

38. $\begin{bmatrix} 1 & 0 & 8 \\ 3 & 7 & 6 \\ 1 & 2 & 3 \end{bmatrix}$

39. $\begin{bmatrix} -6 & 4 \\ -3 & 2 \end{bmatrix}$

40. $\begin{bmatrix} 4 & 4 & 1 \\ 1 & 1 & 1 \\ -1 & -1 & 0 \end{bmatrix}$

Use the inverse of the coefficient matrix to solve each system of equations.

41. $\begin{cases} 4x - y + 2z = 0 \\ x + y + 2z = 1 \\ x + z = 0 \end{cases}$

42. $\begin{cases} w + 3x + y + 3z = 1 \\ w + 4x + y + 3z = 2 \\ x + y = 1 \\ w + 2x - y + 2z = 1 \end{cases}$

SECTION **8.5**   Determinants

Definitions and Concepts	Examples
Determinant: The **determinant** of an $n \times n$ matrix A is the sum of the products of the elements of any row (or column) and the cofactors of those elements. For a 2×2 matrix, $A = \begin{bmatrix} a & b \\ c & d \end{bmatrix}$, $\det(A) = |A| = \begin{vmatrix} a & b \\ c & d \end{vmatrix} = ad - bc.$	See Example 1, page 778
Cramer's Rule: Form quotients of two determinants. The denominator is the determinant of the coefficient matrix A. The numerator is the determinant of a modified coefficient matrix; when solving for the ith variable, replace the ith column of A with a column of constants B.	See Example 6, page 785

EXERCISES

Evaluate each determinant.

43. $\begin{vmatrix} 3 & -2 \\ 1 & -3 \end{vmatrix}$

44. $\begin{vmatrix} 1 & -2 & 3 \\ 2 & -1 & 3 \\ 1 & -1 & 0 \end{vmatrix}$

45. $\begin{vmatrix} 1 & 3 & -1 \\ 1 & 2 & 1 \\ 1 & 0 & 2 \end{vmatrix}$

46. $\begin{vmatrix} 1 & 2 & 3 & 4 \\ -1 & 3 & -3 & 2 \\ 0 & 0 & 0 & -1 \\ 3 & 3 & 4 & 3 \end{vmatrix}$

If $\begin{vmatrix} a & b & c \\ d & e & f \\ g & h & i \end{vmatrix} = 7$, *evaluate each determinant.*

51. $\begin{vmatrix} 3a & 3b & 3c \\ d & e & f \\ g & h & i \end{vmatrix}$

52. $\begin{vmatrix} a & b & c \\ d+g & e+h & f+i \\ g & h & i \end{vmatrix}$

Use Cramer's Rule to solve each system.

47. $\begin{cases} x + 3y = -5 \\ -2x + y = -4 \end{cases}$

48. $\begin{cases} x - y + z = -1 \\ 2x - y + 3z = -4 \\ x - 3y + z = -1 \end{cases}$

49. $\begin{cases} x - 3y + z = 7 \\ x + y - 3z = -9 \\ x + y + z = 3 \end{cases}$

50. $\begin{cases} w + x - y + z = 4 \\ 2w + x + z = 4 \\ x + 2y + z = 0 \\ w + y + z = 2 \end{cases}$

SECTION **8.6**   Partial Fractions

Definitions and Concepts	Examples
Partial-fraction decomposition: The fraction $\frac{P(x)}{Q(x)}$ can be written as the sum of simpler fractions with denominators determined by the prime factors of $Q(x)$.	See Example 1, page 791

EXERCISES

Decompose each fraction into partial fractions.

53. $\dfrac{7x + 3}{x^2 + x}$

54. $\dfrac{4x^3 + 3x + x^2 + 2}{x^4 + x^2}$

55. $\dfrac{x^2 + 5}{x^3 + x^2 + 5x}$

56. $\dfrac{x^2 + 1}{(x + 1)^3}$

SECTION **8.7**   Graphs of Inequalities

Definitions and Concepts	Examples
Systems of inequalities: Systems of inequalities in two variables can be solved by graphing.	See Example 4, page 802
The solution is represented by a plane region with boundaries determined by graphing the inequalities as if they were equations.	See Example 5, page 803

EXERCISES

57. Graph $y \geq -2x - 1$.

58. Graph $x^2 + y^2 > 4$.

Solve each system by graphing.

59. $\begin{cases} 3x + 2y \leq 6 \\ x - y > 3 \end{cases}$

60. $\begin{cases} y \leq x^2 + 1 \\ y \geq x^2 - 1 \end{cases}$

CHAPTER TEST

Solve each system of equations by the graphing method.

1. $\begin{cases} x - 3y = -5 \\ 2x - y = 0 \end{cases}$

2. $\begin{cases} x = 2y + 5 \\ y = 2x - 4 \end{cases}$

Solve each system of equations by the substitution or addition method.

3. $\begin{cases} 3x + y = 0 \\ 2x - 5y = 17 \end{cases}$

4. $\begin{cases} \dfrac{x + y}{2} + x = 7 \\ \dfrac{x - y}{2} - y = -6 \end{cases}$

5. Mixing solutions A chemist has two solutions; one has a 20% concentration and the other a 45% concentration. How many liters of each must she mix to obtain 10 liters of 30% concentration?

6. Wholesale distribution Ace Electronics, Hi-Fi Stereo, and CD World buy a total of 175 DVD players from the same distributor each month. Because CD World buys 25 more units than the other two stores combined, its cost is only $160 per unit. The players cost Hi-Fi $165 each and Ace $170 each. How many players does each retailer buy each month if the distributor receives $28,500 each month from the sale of the players to the three stores?

Write each system of equations as a matrix and solve it by Gaussian elimination.

7. $\begin{cases} 3x - 2y = 4 \\ 2x + 3y = 7 \end{cases}$

8. $\begin{cases} x + 3y - z = 6 \\ 2x - y - 2z = -2 \\ x + 2y + z = 6 \end{cases}$

Write each system of equations as a matrix and solve it by Gauss–Jordan elimination. If the system has infinitely many solutions, show a general solution.

9. $\begin{cases} x + 2y + 3z = -5 \\ 3x + y - 2z = 7 \\ y - z = 2 \end{cases}$

10. $\begin{cases} x + 2y + z = 0 \\ 3x - 2y - 2z = 7 \\ 4x - z = 7 \end{cases}$

Perform the indicated operations.

11. $3\begin{bmatrix} 2 & -3 & 5 \\ 0 & 3 & -1 \end{bmatrix} - 5\begin{bmatrix} -2 & 1 & -1 \\ 0 & 3 & 2 \end{bmatrix}$

12. $\begin{bmatrix} 1 & 2 & 3 \end{bmatrix}\begin{bmatrix} 2 & -2 \\ -2 & 2 \\ 1 & 0 \end{bmatrix}\begin{bmatrix} 3 \\ -2 \end{bmatrix}$

Find the inverse of each matrix, if possible.

13. $\begin{bmatrix} 5 & 19 \\ 2 & 7 \end{bmatrix}$

14. $\begin{bmatrix} -1 & 3 & -2 \\ 4 & 1 & 4 \\ 0 & 3 & -1 \end{bmatrix}$

Use the inverses found in Exercises 13 and 14 to solve each system.

15. $\begin{cases} 5x + 19y = 3 \\ 2x + 7y = 2 \end{cases}$

16. $\begin{cases} -x + 3y - 2z = 1 \\ 4x + y + 4z = 3 \\ 3y - z = -1 \end{cases}$

Evaluate each determinant.

17. $\begin{vmatrix} 3 & -5 \\ -3 & 1 \end{vmatrix}$

18. $\begin{vmatrix} 3 & 5 & -1 \\ -2 & 3 & -2 \\ 1 & 5 & -3 \end{vmatrix}$

Use Cramer's Rule to solve each system for y.

19. $\begin{cases} 3x - 5y = 3 \\ -3x + y = 2 \end{cases}$

20. $\begin{cases} 3x + 5y - z = 2 \\ -2x + 3y - 2z = 1 \\ x + 5y - 3z = 0 \end{cases}$

Graph the solution set of each system.

23. $\begin{cases} x - 3y \geq 3 \\ x + 3y \leq 3 \end{cases}$

24. $\begin{cases} 3x + 4y \leq 12 \\ 3x + 4y \geq 6 \\ x \geq 0 \\ y \geq 0 \end{cases}$

Decompose each fraction into partial fractions.

21. $\dfrac{5x}{2x^2 - x - 3}$

22. $\dfrac{3x^2 + x + 2}{x^3 + 2x}$

LOOKING AHEAD TO **CALCULUS**

1. **Using Determinants to Find the Cross (Vector) Product**
   - Connect the determinant to the definition of the cross (vector) product of vectors—Section 8.5

2. **Using Partial Fractions and Integration of Rational Functions**
   - Connect partial-fraction decomposition to the integration of rational functions—Section 8.6

1. Using Determinants to Find the Cross (Vector) Product

In Section 7.4, we defined a two-dimensional vector $\mathbf{a} = a_1\mathbf{i} + a_2\mathbf{j}$, where a_1 is the horizontal component associated with the x-axis and a_2 is the vertical component associated with the y-axis. A *three-dimensional* coordinate system has wide uses in calculus, especially in the area of vectors. A three-dimensional coordinate system uses x-, y-, and z-axes as shown in the figure. Similarly, we can define two *three-dimensional* vectors as $\mathbf{a} = a_1\mathbf{i} + a_2\mathbf{j} + a_3\mathbf{k} = \langle a_1, a_2, a_3 \rangle$ and $\mathbf{b} = b_1\mathbf{i} + b_2\mathbf{j} + b_3\mathbf{k} = \langle b_1, b_2, b_3 \rangle$.

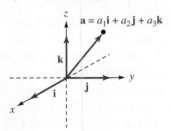

Unlike the dot product of two vectors, which is a scalar (number), the **cross product $\mathbf{a} \times \mathbf{b}$** of two three-dimensional vectors is a vector. In fact, it is a vector that is perpendicular to $\mathbf{a}$ and $\mathbf{b}$, as shown below.

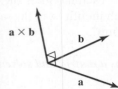

A most important use of calculus was its use in proving Kepler's laws of planetary motion, which approximately describe the motion of any two bodies in orbit around each other. Those proofs are accomplished by using cross products.

Cross products are used in many areas, including finding the equation of a plane, finding the angle between two three-dimensional vectors, and determining whether vectors are parallel.

The cross product is defined as $\mathbf{a} \times \mathbf{b} = \langle a_2b_3 - a_3b_2, a_3b_1 - a_1b_3, a_1b_2 - a_2b_1 \rangle$. In order to make this definition easier to remember, the notation of determinants (covered in Section 8.5) is used. We now define the cross product as

$$\mathbf{a} \times \mathbf{b} = \begin{vmatrix} \mathbf{i} & \mathbf{j} & \mathbf{k} \\ a_1 & a_2 & a_3 \\ b_1 & b_2 & b_3 \end{vmatrix} = \begin{vmatrix} a_2 & a_3 \\ b_2 & b_3 \end{vmatrix}\mathbf{i} - \begin{vmatrix} a_1 & a_3 \\ b_1 & b_3 \end{vmatrix}\mathbf{j} + \begin{vmatrix} a_1 & a_2 \\ b_1 & b_2 \end{vmatrix}\mathbf{k}.$$

EXAMPLE 1   **Finding the Cross Product of Two Vectors**

If $\mathbf{a} = \langle 1, 3, 4 \rangle$ and $\mathbf{b} = \langle 2, 7, -5 \rangle$, then

$$\mathbf{a} \times \mathbf{b} = \begin{vmatrix} \mathbf{i} & \mathbf{j} & \mathbf{k} \\ 1 & 3 & 4 \\ 2 & 7 & -5 \end{vmatrix} = \begin{vmatrix} 3 & 4 \\ 7 & -5 \end{vmatrix}\mathbf{i} - \begin{vmatrix} 1 & 4 \\ 2 & -5 \end{vmatrix}\mathbf{j} + \begin{vmatrix} 1 & 3 \\ 2 & 7 \end{vmatrix}\mathbf{k}$$

$$= (-15 - 28)\mathbf{i} - (-5 - 8)\mathbf{j} + (7 - 6)\mathbf{k} = -43\mathbf{i} + 13\mathbf{j} + \mathbf{k}$$

Three vectors are **coplanar** if $\mathbf{a} \cdot (\mathbf{b} \times \mathbf{c}) = 0$. (Coplanar means that the three vectors are in the same plane.) The value of $\mathbf{a} \cdot (\mathbf{b} \times \mathbf{c})$ is called the **scalar triple.** ■

EXAMPLE 2   **Showing Vectors Are Coplanar**

Use the scalar triple to show that the vectors $\mathbf{a} = \langle 1, 4, -7 \rangle$, $\mathbf{b} = \langle 2, -1, 4 \rangle$, and $\mathbf{c} = \langle 0, -9, 18 \rangle$ are coplanar. The first row is $\mathbf{a}$, the second row is $\mathbf{b}$, and the third row is $\mathbf{c}$.

$$\mathbf{a} \cdot (\mathbf{b} \times \mathbf{c}) = \begin{vmatrix} 1 & 4 & -7 \\ 2 & -1 & 4 \\ 0 & -9 & 18 \end{vmatrix} = 1\begin{vmatrix} -1 & 4 \\ -9 & 18 \end{vmatrix} - 4\begin{vmatrix} 2 & 4 \\ 0 & 18 \end{vmatrix} - 7\begin{vmatrix} 2 & -1 \\ 0 & -9 \end{vmatrix}$$

$$= 1(18) - 4(36) - 7(-18) = 0$$

This shows that the vectors all lie in the same plane. ■

Connect to Calculus:

Using Determinants to Find the Cross (Vector) Product

1. Find the cross product $\mathbf{a} \times \mathbf{b}$.
    (a) $\mathbf{a} = \langle 2, 1, -1 \rangle$; $\mathbf{b} = \langle 0, 1, 2 \rangle$
    (b) $\mathbf{a} = \mathbf{i} - \mathbf{j} + \mathbf{k}$; $\mathbf{b} = \mathbf{i} + \mathbf{j} + \mathbf{k}$
2. Determine whether the vectors are coplanar.
    (a) $\mathbf{a} = \langle 6, 3, -1 \rangle$; $\mathbf{b} = \langle 0, 1, 2 \rangle$; $\mathbf{c} = \langle 4, -2, 5 \rangle$
    (b) $\mathbf{a} = 2\mathbf{i} + 3\mathbf{j} + \mathbf{k}$; $\mathbf{b} = \mathbf{i} - \mathbf{j}$; $\mathbf{c} = 7\mathbf{i} + 3\mathbf{j} + 2\mathbf{k}$

2. Using Partial Fractions and Integration of Rational Functions

In Section 8.6, the process of partial-fraction decomposition was used to rewrite a rational expression as the sum of simpler fractions. This process is used in calculus to decompose rational functions so that the functions can be *integrated*. The actual process of integration is beyond the scope of this discussion, but we will look at how the skill that is learned here will be used in calculus.

For example, suppose we need to integrate the rational function $f(x) = \dfrac{x + 5}{x^2 + x - 2}$. The notation for this is

$$\int \frac{x + 5}{x^2 + x - 2}\,dx.$$

As it is written, this problem cannot be done, except by a computer or calculator. However, if the rational function is decomposed using partial fractions, the problem can be done.

$$\frac{x + 5}{x^2 + x - 2} = \frac{x + 5}{(x + 2)(x - 1)} = \frac{A}{x + 2} + \frac{B}{x - 1} = \frac{A(x - 1) + B(x + 2)}{(x + 2)(x - 1)} = \frac{Ax + Bx - A + 2B}{(x + 2)(x - 1)}$$

Now we can find the values of A and B using the technique learned in Section 8.6.

$$x + 5 = x(A + B) - A + 2B$$

$$\left.\begin{array}{r} A + B = 1 \\ \underline{-A + 2B = 5} \\ 3B = 6 \end{array}\right\} \Rightarrow B = 2,\ A = -1$$

Thus, $\dfrac{x + 5}{x^2 + x - 2} = \dfrac{-1}{x + 2} + \dfrac{2}{x - 1}$, and the problem

$$\int \frac{x + 5}{x^2 + x - 2}\,dx = \int \left(\frac{-1}{x + 2} + \frac{2}{x + 2}\right)dx \text{ becomes one that can be}$$

completed in calculus.

Connect to Calculus:
Using Partial Fractions and Integration of Rational Functions

Use partial-fraction decomposition to simplify each integral.

3. $\displaystyle\int \frac{x - 9}{(x + 5)(x - 2)}\,dx$ **4.** $\displaystyle\int \frac{x - 1}{x^3 + x^2}\,dx$

9

Conic Sections and Quadratic Systems

CAREERS AND MATHEMATICS: Engineer

Nomad_Soul/Shutterstock.com

Engineers are the link between a discovery in science and the practical application of that discovery to meet consumer demands. Engineers will often design and develop new products. They are also supervisors in manufacturing facilities. Most engineers have a specialty. Aerospace engineers, biological engineers, chemical engineers, electrical engineers, mechanical engineers, civil engineers, and petroleum engineers are a few of the specialists in engineering.

Education and Mathematics Required

- Engineers typically have a bachelor's degree in an engineering specialty. Research positions will often require a graduate degree.
- College Algebra, Geometry, Trigonometry, a Calculus sequence, Linear Algebra, Differential Equations, and Statistics are math courses required.

How Engineers Use Math and Who Employs Them

- Mathematics is important for all types of engineers. It is the language of engineering and physical science. Mathematics can develop the intellectual maturity to solve difficult problems. An understanding of mathematical analysis is essential to validate the work of computer programs and to write code that can make programs even more powerful.
- About 37% of engineers are employed in manufacturing industries, and 28% in professional, scientific, and technical services. Federal, state, and local government agencies employ engineers in highway and public-works departments. Potential employers are determined by the engineering specialty.

Career Outlook and Salary

- Employment of engineers is expected to grow about as fast as the average for all occupations, although growth will vary by specialty. Job opportunities are expected to be good.
- Earnings vary significantly among the specialties. Median salaries run from $108,020 for petroleum engineers to $68,730 for agricultural engineers. The highest salaries can be over $165,000.

For more information see http://www.bls.gov/oco.

In this chapter, we will study second-degree equations in x and y. They have many applications in such fields as navigation, astronomy, and satellite communications.

9.1 The Circle and the Parabola

In this section, we will learn to

1. Write an equation of a circle.
2. Graph circles.
3. Write an equation of a parabola in standard form.
4. Graph parabolas.
5. Solve application problems involving parabolas.

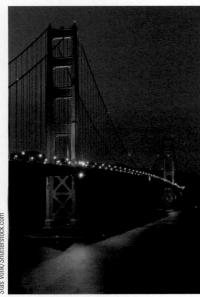

Stas Volik/Shutterstock.com

The suspension bridge is one of the oldest types of bridges. An example is the Golden Gate Bridge, which spans the opening of San Francisco Bay into the Pacific Ocean. Completed in 1937, it is currently the second-longest suspension bridge in the United States. As the illustration shows, the main cables that hold up the bridge hang in the shape of a parabola. A parabola is an example of a *conic section*.

Second-degree equations in x and y have the general form

$$Ax^2 + Bxy + Cy^2 + Dx + Ey + F = 0,$$

where at least one of the coefficients A, B, and C is not zero. The graphs of these equations fall into one of several categories: a point, a pair of lines, a circle, a parabola, an ellipse, a hyperbola, or no graph at all. These graphs are called **conic sections,** because each one is the intersection of a plane and a right-circular cone, as shown in Figure 9-1.

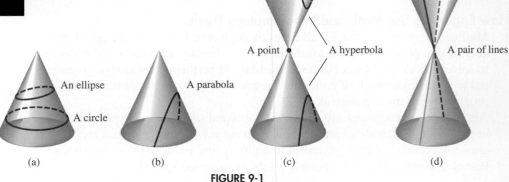

| A point | A hyperbola | A pair of lines |

| An ellipse | A parabola |
| A circle | |

(a)   (b)   (c)   (d)

FIGURE 9-1

Although these shapes have been known since the time of the ancient Greeks, it wasn't until the seventeenth century that René Descartes (1596–1650) and Blaise Pascal (1623–1662) developed the mathematics needed to study them in detail.

In this section, we will consider two conic sections, the circle and the parabola.

1. Write an Equation of a Circle

In Section 1.1, we developed the following standard equations for circles:

Standard Equation of a Circle with Center at (h, k)

The graph of any equation that can be written in the form

$$(x - h)^2 + (y - k)^2 = r^2$$

is a circle with radius r and center at point (h, k).

If $h = 0$ and $k = 0$, we have this result:

Standard Equation of a Circle
with Center at $(0, 0)$

The graph of any equation that can be written in the form

$$x^2 + y^2 = r^2$$

is a circle with radius r and center at the origin.

The graphs of two circles, one with center at (h, k) and one with center at $(0, 0)$, are shown in Figure 9-2.

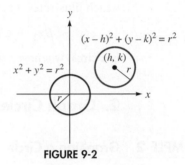

FIGURE 9-2

EXAMPLE 1   **Writing an Equation of a Circle**

Find an equation of the circle shown in Figure 9-3.

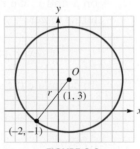

FIGURE 9-3

SOLUTION   We will use the distance formula to find the radius of the circle and then substitute the coordinates of the center and radius into the standard equation of the circle.

To find the radius of the circle, we substitute the coordinates of the points $(1, 3)$ and $(-2, -1)$ into the distance formula and simplify.

$$r = \sqrt{(x_2 - x_1)^2 + (y_2 - y_1)^2}$$
$$r = \sqrt{(-2 - 1)^2 + (-1 - 3)^2} \qquad \text{Substitute 1 for } x_1, \text{3 for } y_1, -2 \text{ for } x_2, \text{ and } -1 \text{ for } y_2.$$
$$= \sqrt{(-3)^2 + (-4)^2}$$
$$= \sqrt{9 + 16}$$
$$= \sqrt{25}$$
$$= 5$$

To find an equation of a circle with radius 5 and center at $(1, 3)$, we substitute 1 for h, 3 for k, and 5 for r in the standard equation of the circle.

$$(x - h)^2 + (y - k)^2 = r^2$$
$$(x - 1)^2 + (y - 3)^2 = 5^2$$

To write the equation in general form, we square the binomials and simplify.

$$x^2 - 2x + 1 + y^2 - 6y + 9 = 25$$
$$x^2 + y^2 - 2x - 6y - 15 = 0 \qquad \text{Subtract 25 from both sides and simplify.}$$ ∎

Self Check 1   Find a general equation of the circle with center at $(-2, 1)$ and radius of 4.

Now Try Exercise 17.

The final equation in Example 1 can be written as

$$1x^2 + 0xy + 1y^2 - 2x - 6y - 15 = 0,$$

which illustrates that the graph of

$$Ax^2 + Bxy + Cy^2 + Dx + Ey + F = 0$$

is a circle whenever $B = 0$ and $A = C$.

2. Graph Circles

EXAMPLE 2   **Graphing a Circle**

Graph the circle whose equation is $2x^2 + 2y^2 + 4x + 3y = 3$.

SOLUTION   We will convert the general form of the circle into standard form so we can identify the center and radius. Then we will graph the circle.

To find the coordinates of the center and the radius, we complete the square on x and y. We begin by dividing both sides of the equation by 2 and rearranging terms to get

$$x^2 + 2x + y^2 + \frac{3}{2}y = \frac{3}{2}.$$

To complete the square on x and y, we add 1 and $\frac{9}{16}$ to both sides.

$$x^2 + 2x + \mathbf{1} + y^2 + \frac{3}{2}y + \frac{9}{16} = \frac{3}{2} + \mathbf{1} + \frac{9}{16}$$

We can factor $x^2 + 2x + 1$ and $y^2 + \frac{3}{2}y + \frac{9}{16}$ and simplify on the right side.

$$(x + 1)^2 + \left(y + \frac{3}{4}\right)^2 = \frac{49}{16}$$

$$[x - (\mathbf{-1})]^2 + \left[y - \left(-\frac{3}{4}\right)\right]^2 = \left(\frac{7}{4}\right)^2$$

From the equation, we see that the coordinates of the center of the circle are $h = -1$ and $k = -\frac{3}{4}$ and that the radius of the circle is $\frac{7}{4}$. The graph is shown in Figure 9-4.

Comment

Since the graphs of circles fail the vertical-line test, their equations do not represent functions. It is somewhat more difficult to use a graphing calculator to graph equations that are not functions. Please see the "Graphing a Circle" Accent on Technology in Section 1.1. In many cases, it is easier to graph the circle by hand.

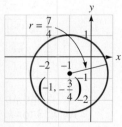

$2x^2 + 2y^2 + 4x + 3y = 3$

FIGURE 9-4

Self Check 2   Graph the circle whose equation is $x^2 + y^2 - 6x - 2y = -6$.

Now Try Exercise 23.

3. Write an Equation of a Parabola in Standard Form

In Chapter 2, we saw that the graphs of quadratic functions of the forms

$$y = ax^2 + bx + c \qquad \text{or} \qquad y = a(x - h)^2 + k \quad (a \neq 0)$$

are parabolas that open up or down. We now discuss parabolas that open to the left or to the right and examine the properties of all parabolas in greater detail.

Parabola	A **parabola** is the set of all points in a plane equidistant from a line l (called the **directrix**) and a fixed point F (called the **focus**) that is not on line l. See Figure 9-5.

The point on the parabola that is closest to the directrix is called the **vertex**, and the line passing through the vertex and the focus is called the **axis of the parabola.**

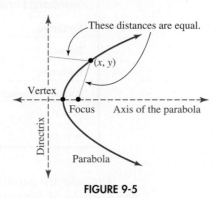

FIGURE 9-5

A basic parabola is one that has its vertex at the origin and opens upward. Consider the parabola shown in Figure 9-6.

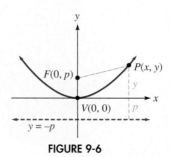

FIGURE 9-6

To find an equation of the parabola, we will find the distance from point $P(x, y)$ on the parabola to the focus and find the distance from point $P(x, y)$ to the directrix and set the two distances equal. Using the Distance Formula, we can find the distance from $P(x, y)$ to $F(0, p)$.

$$d(PF) = \sqrt{x^2 + (y - p)^2}$$

From the figure, we see that the distance from $P(x, y)$ to the directrix $y = -p$ is

$$|y - (-p)| = |y + p|.$$

We can equate these distances and simplify.

$$\sqrt{x^2 + (y - p)^2} = |y + p|$$

$$x^2 + (y - p)^2 = |y + p|^2 \qquad \text{Square both sides.}$$

$$x^2 + (y - p)^2 = (y + p)^2 \qquad |y + p|^2 = (y + p)^2$$

$$x^2 + y^2 - 2yp + p^2 = y^2 + 2py + p^2 \qquad \text{Expand the binomials.}$$

$$x^2 - 2py = 2py \qquad \text{Subtract } y^2 + p^2 \text{ from both sides.}$$

(1) $$x^2 = 4py \qquad \text{Add } 2py \text{ to both sides.}$$

Equation (1) is one of the **standard equations of a parabola** with vertex at the origin. If $p > 0$, as in Equation (1), the graph of the equation will be a parabola that opens upward. If $p < 0$, the graph of the equation will be a parabola that opens downward.

It can be shown that a parabola that has its vertex at the origin and opens to the left or to the right has an equation of the form $y^2 = 4px$.

The standard equations of a parabola with vertex $V(0, 0)$ and some of the characteristics of the parabola are summarized in the table.

Standard Form of an Equation of a Parabola with Vertex $(0, 0)$		
Equation	$x^2 = 4py$	$y^2 = 4px$
Focus	$(0, p)$	$(p, 0)$
Directrix	$y = -p$	$x = -p$
Axis of Symmetry	y-axis	x-axis
$p > 0$	Opens up	Opens right
$p < 0$	Opens down	Opens left

Consider the parabola with equation $x^2 = 8y$ that is shown in Figure 9-7. Since the equation of the parabola is written in standard form, $x^2 = 4py$, we know that $4p = 8$ and $p = 2$. Thus the focus of the parabola is $F(0, p) = F(0, 2)$ and the directrix is $y = -p$, or $y = -2$. Because $p = 2$ and $2 > 0$, the parabola opens upward. The axis of symmetry is a vertical line, which is the y-axis.

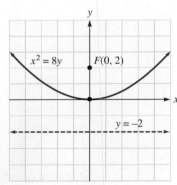

FIGURE 9-7

EXAMPLE 3   Finding an Equation of a Parabola

Find an equation of the parabola with vertex at the origin and focus at $(3, 0)$.

SOLUTION   A sketch of the parabola is shown in Figure 9-8. Because the focus is to the right of the vertex, the parabola opens to the right, and because the vertex is the origin, the standard equation is $y^2 = 4px$. The distance between the focus and the vertex is $p = 3$. We can substitute 3 for p in the standard equation.

$$y^2 = 4px$$
$$y^2 = 4(3)x$$
$$y^2 = 12x$$

An equation of the parabola is $y^2 = 12x$.

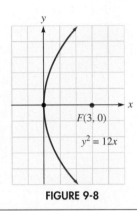

FIGURE 9-8

Self Check 3   Find an equation of the parabola with vertex at the origin and focus at $(-3, 0)$.

Now Try Exercise 31.

In Section 1.5 we used translations to sketch graphs of functions. Translations can also be applied to graphs that are parabolas. The table summarizes the characteristics of parabolas with vertex $V(h, k)$.

Standard Form of an Equation of a Parabola with Vertex (h, k)		
Equation	$(x - h)^2 = 4p(y - k)$	$(y - k)^2 = 4p(x - h)$
Focus	$(h, k + p)$	$(h + p, k)$
Directrix	$y = -p + k$	$x = -p + h$
Axis of Symmetry	$x = h$	$y = k$
$p > 0$	Opens up	Opens right
$p < 0$	Opens down	Opens left

EXAMPLE 4   **Finding an Equation of a Parabola**

Find an equation of the parabola that opens up, has its vertex at the point $(4, 5)$, and passes through the point $(0, 7)$.

SOLUTION   Because the parabola opens upward, we will substitute the coordinates of the given vertex and the point into the equation $(x - h)^2 = 4p(y - k)$ and solve for p. Then we will substitute the coordinates of the vertex and the value of p into the equation.

Since $(h, k) = (4, 5)$ and the point $(0, 7)$ is on the curve, we can substitute 4 for h, 5 for k, 0 for x, and 7 for y in the standard equation and solve for p.

$$(x - h)^2 = 4p(y - k)$$
$$(0 - 4)^2 = 4p(7 - 5)$$
$$16 = 8p$$
$$2 = p$$

To find an equation of the parabola, we substitute 4 for h, 5 for k, and 2 for p in the standard equation and simplify.

$$(x - h)^2 = 4p(y - k)$$

$$(x - 4)^2 = 4(2)(y - 5)$$

$$(x - 4)^2 = 8(y - 5)$$

The graph of the equation appears in Figure 9-9.

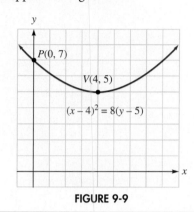

FIGURE 9-9

Self Check 4   Find an equation of the parabola that opens up, has its vertex at $(4, 5)$, and passes through $(0, 9)$.

Now Try Exercise 43.

EXAMPLE 5   **Finding Equations of Two Parabolas**

Find equations of two parabolas with a vertex at $(2, 4)$ that pass through $(0, 0)$.

SOLUTION   The two parabolas are shown in Figure 9-10.

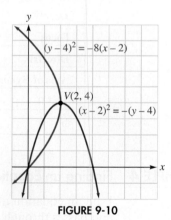

FIGURE 9-10

Part 1:   To find an equation of the parabola that opens to the left, we use the equation $(y - k)^2 = 4p(x - h)$. Since the curve passes through the point $(x, y) = (0, 0)$ and the vertex $(h, k) = (2, 4)$, we substitute 0 for x, 0 for y, 2 for h, and 4 for k in the standard equation and solve for p.

$$(y - k)^2 = 4p(x - h)$$

$$(0 - 4)^2 = 4p(0 - 2)$$

$$16 = -8p$$

$$-2 = p$$

Since $h = 2$, $k = 4$, and $p = -2$, the equation is

$$(y - \mathbf{k})^2 = 4p(x - h)$$
$$(y - \mathbf{4})^2 = 4(-\mathbf{2})(x - \mathbf{2})$$
$$(y - 4)^2 = -8(x - 2)$$

Part 2:   To find an equation of the parabola that opens down, we use the equation $(x - h)^2 = 4p(y - k)$, substitute 2 for h, 4 for k, 0 for x, and 0 for y, and solve for p.

$$(\mathbf{x} - h)^2 = 4p(y - k)$$
$$(\mathbf{0} - \mathbf{2})^2 = 4p(\mathbf{0} - \mathbf{4})$$
$$4 = -16p$$
$$p = -\frac{1}{4}$$

Since $h = 2$, $k = 4$, and $p = -\frac{1}{4}$, the equation is

$$(x - \mathbf{h})^2 = 4p(y - k)$$
$$(x - \mathbf{2})^2 = 4\left(-\frac{1}{4}\right)(y - \mathbf{4}) \qquad \text{Substitute 2 for } h, -\frac{1}{4} \text{ for } p, \text{ and 4 for } k.$$
$$(x - 2)^2 = -(y - 4)$$

Self Check 5   Find equations of two parabolas with vertex at $(2, 4)$ that pass through $(0, 8)$.

Now Try Exercise 45.

4. Graph Parabolas

EXAMPLE 6   **Graphing a Parabola**

Find the vertex, y-intercepts, and graph of the parabola with the equation $y^2 + 8x - 4y = 28$ and graph it.

SOLUTION   To identify the coordinates of the vertex, we will complete the square on y and write the equation in standard form. To find the y-intercepts, we will let $x = 0$ and solve for y. Finally, we will graph the parabola by plotting the vertex and y-intercepts and drawing a smooth curve through the points.

Step 1:   Complete the square and write the equation in standard form.

$$y^2 + 8x - 4y = 28$$
$$y^2 - 4y = -8x + 28 \qquad \text{Subtract } 8x \text{ from both sides.}$$
$$y^2 - 4y + 4 = -8x + 28 + 4 \qquad \text{Add 4 to both sides.}$$
$$(2) \qquad (y - 2)^2 = -8(x - 4) \qquad \text{Factor both sides.}$$

Equation (2) represents a parabola opening to the left with vertex $(4, 2)$.

Step 2:   Find the y-intercepts.

To find the y-intercepts, we substitute 0 for x in Equation (2) and solve for y.

$$(y - 2)^2 = -8(\mathbf{x} - 4)$$
$$(y - 2)^2 = -8(\mathbf{0} - 4) \qquad \text{Substitute 0 for } x.$$
$$y^2 - 4y + 4 = 32 \qquad \text{Remove parentheses.}$$
$$(3) \quad y^2 - 4y - 28 = 0$$

We can use the Quadratic Formula to find that the roots of Equation (3) are $y \approx 7.7$ and $y \approx -3.7$. These are the y-intercepts, and the points with coordinates of approximately $(0, 7.7)$ and $(0, -3.7)$ lie on the graph of the parabola and are the y-intercepts.

Step 3: Graph the parabola.

We can use the information above and the knowledge that the graph opens to the left and has vertex at $(4, 2)$ to draw the graph, shown in Figure 9-11.

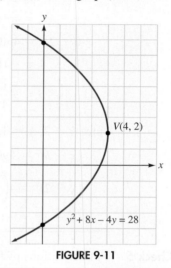

$V(4, 2)$

$y^2 + 8x - 4y = 28$

FIGURE 9-11

Self Check 6   Find the vertex and y-intercepts of the parabola with the equation $y^2 - x + 2y = 3$. Then graph it.

Now Try Exercise 53.

5. Solve Application Problems Involving Parabolas

Parabolas arise in many real-world settings. A few examples are:

- As shown in the illustration, a bouncing ball travels in a series of parabolic paths. Air resistance causes the ball to deviate slightly from a perfect parabola shape.

- A satellite dish is a type of parabolic antenna designed with the specific purpose of transmitting signals to and receiving from satellites.

• Objects extended in space often follow parabolic paths, such as a diver jumping from a diving board or cliff. The diver may follow a complex motion, but the center of mass follows a parabola.

EXAMPLE 7   Solving an Application Problem

The Gateway Arch in Saint Louis has a shape that approximates a parabola. (See Figure 9-12.) The vertex of the parabola is $V(0, 630)$ and the x-intercepts are 315 and −315. Find an equation of the parabola that models the arch.

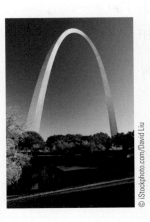

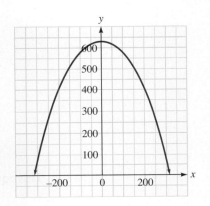

FIGURE 9-12

SOLUTION   To find an equation of the parabola that opens downward, we use the equation $(x - h)^2 = 4p(y - k)$, substitute in the vertex $V(0, 630)$ and the point $(315, 0)$, and solve for p. Then we write an equation of the parabola.

We substitute 0 for h, 630 for k, 315 for x, and 0 for y and solve for p.

$$(\boldsymbol{x} - \boldsymbol{h})^2 = 4p(\boldsymbol{y} - \boldsymbol{k})$$

$$(\boldsymbol{315} - 0)^2 = 4p(0 - \boldsymbol{630})$$

$$99{,}225 = -2{,}520p$$

$$p = -\frac{99{,}225}{2{,}520}$$

$$= -\frac{315}{8}$$

Since $h = 0$, $k = 630$, and $p = -\frac{315}{8}$, the equation is

$$(x - h)^2 = 4p(y - k)$$

$$(x - 0)^2 = 4\left(-\frac{315}{8}\right)(y - 630) \qquad \text{Substitute 0 for } h,\ -\frac{315}{8} \text{ for } p, \text{ and 630 for } k.$$

$$x^2 = -\frac{315}{2}(y - 630) \qquad \text{Simplify.}$$

Self Check 7   What is the approximate height of the arch 115 feet from the base?

Now Try Exercise 81.

EXAMPLE 8   **Solving an Applied Problem**

If water is propelled straight up by a "super nozzle" during a water-fountain show in Las Vegas, the equation $s = 128 - 16t^2$ expresses the water's height s (in feet) at t seconds after it is propelled. Find the maximum height reached by the water.

SOLUTION   The graph of $s = 128t - 16t^2$, which expresses the height s of the water at t seconds after it is propelled, is the parabola shown in Figure 9-13. To find the maximum height reached by the water, we find the s-coordinate k of the vertex of the parabola. To find k, we write an equation of the parabola in standard form by completing the square on t.

$$s = 128t - 16t^2$$

$$16t^2 - 128t = -s \qquad \text{Multiply both sides by } -1.$$

$$t^2 - 8t = -\frac{s}{16} \qquad \text{Divide both sides by 16.}$$

$$t^2 - 8t + 16 = -\frac{s}{16} + 16 \qquad \text{Add 16 to both sides to complete the square.}$$

$$(t - 4)^2 = -\frac{s - 256}{16} \qquad \text{Factor } t^2 - 8t + 16 \text{ and combine terms.}$$

$$(t - 4)^2 = -\frac{1}{16}(s - 256) \qquad \text{Factor out } -\frac{1}{16}.$$

This equation indicates that the maximum height is 256 feet.

Caution

The parabola shown in Figure 9-13 is not the path of the water. The water goes straight up and straight down.

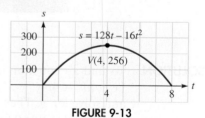

FIGURE 9-13

Self Check 8   At what time will the water strike the ground? (*Hint:* Find the t-intercept.)

Now Try Exercise 77.

Self Check Answers   1. $x^2 + y^2 + 4x - 2y - 11 = 0$ 2.   3. $y^2 = -12x$

4. $(x - 4)^2 = 4(y - 5)$ **5.** $(y - 4)^2 = -8(x - 2)$; $(x - 2)^2 = y - 4$
6. vertex: $(-4, -1)$; y-intercepts: 1 and -3 **7.** 546 ft   **8.** 8 sec

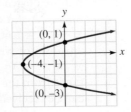

Exercises **9.1**

Getting Ready

You should be able to complete these vocabulary and concept statements before you proceed to the practice exercises.

Fill in the blanks.

1. $(x - 2)^2 + (y + 5)^2 = 9$: center (__ , __); radius __

2. $x^2 + y^2 - 36 = 0$: center (__ , __); radius __

3. $x^2 + y^2 = 5$: center (__ , __); radius __

4. $2(x - 9)^2 + 2y^2 = 7$: center (__ , __); radius __

Determine whether the parabolic graph of each equation opens up, down, left, or right.

5. $y^2 = -4x$ **6.** $y^2 = 10x$

7. $x^2 = -8(y - 3)$ **8.** $(x - 2)^2 = (y + 3)$

Fill in the blanks.

9. A parabola is the set of all points in a plane equidistant from a line, called the ___, and a fixed point not on the line, called the ___.

10. The general form of a second-degree equation in the variables x and y is ___.

Identify the conic as a circle or a parabola.

11. $x^2 - 5x + y^2 = 12$

12. $3x^2 + 3y^2 + 18x + 6y = -24$

13. $x^2 - 8y = 6x - 1$

14. $2y^2 + 4y - 6x = 4$

Practice

Write an equation in standard form of each circle.

15.

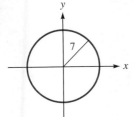

16.

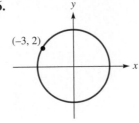

17.

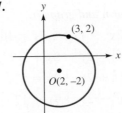

18.

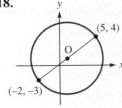

19. Radius of 6; center at the intersection of $3x + y = 1$ and $-2x - 3y = 4$

20. Radius of 8; center at the intersection of $x + 2y = 8$ and $2x - 3y = -5$

Graph each circle.

21. $x^2 + y^2 = 4$ **22.** $x^2 - 2x + y^2 = 15$

23. $3x^2 + 3y^2 - 12x - 6y = 12$

24. $2x^2 + 2y^2 + 4x - 8y + 2 = 0$

Find the vertex, focus, and directrix of each parabola.

25. $x^2 = 12y$ **26.** $y^2 = -12y$

27. $(y - 3)^2 = 20x$ **28.** $x^2 = -\dfrac{1}{2}(y + 5)$

29. $(x + 2)^2 = -24(y - 1)$ **30.** $(y + 1)^2 = 28(x - 2)$

Find an equation in standard form of each parabola.

31. Vertex at $(0, 0)$; focus at $(0, 3)$

32. Vertex at $(0, 0)$; focus at $(0, -3)$

33. Vertex at $(0, 0)$; focus at $(-3, 0)$

34. Vertex at $(0, 0)$; focus at $(3, 0)$

35. Vertex at $(3, 5)$; focus at $(3, 2)$

36. Vertex at $(3, 5)$; focus at $(-3, 5)$

37. Vertex at $(3, 5)$; focus at $(3, -2)$

38. Vertex at $(3, 5)$; focus at $(6, 5)$

39. Vertex at $(0, 2)$; directrix at $y = 3$

40. Vertex at $(-3, 4)$; directrix at $y = 2$

41. Vertex at $(1, -5)$; directrix at $x = -1$

42. Vertex at $(3, 5)$; directrix at $x = 6$

43. Vertex at $(2, 2)$; passes through $(0, 0)$

44. Vertex at $(-2, -2)$; passes through $(0, 0)$

45. Vertex at $(-4, 6)$; passes through $(0, 3)$

46. Vertex at $(-2, 3)$; passes through $(0, -3)$

47. Vertex at $(6, 8)$; passes through $(5, 10)$ and $(5, 6)$

48. Vertex at $(2, 3)$; passes through $\left(1, \dfrac{13}{4}\right)$ and $\left(-1, \dfrac{21}{4}\right)$

49. Vertex at $(3, 1)$; passes through $(4, 3)$ and $(2, 3)$

50. Vertex at $(-4, -2)$; passes through $(-3, 0)$ and $\left(\dfrac{9}{4}, 3\right)$

Write each parabola in standard form and graph it.

51. $y = x^2 + 4x + 5$ **52.** $2x^2 - 12x - 7y = 10$

53. $y^2 + 4x - 6y = -1$ **54.** $x^2 - 2y - 2x = -7$

55. $y^2 - 4y = 4x - 8$ **56.** $y^2 + 2x - 2y = 5$

57. $y^2 - 4y = -8x + 20$ **58.** $y^2 - 2y = 9x + 17$

59. $x^2 - 6y + 22 = -4x$ **60.** $4y^2 - 4y + 16x = 7$

61. $4x^2 - 4x + 32y = 47$ **62.** $4y^2 - 16x + 17 = 20y$

Applications

63. Broadcast range A television tower broadcasts a signal with a circular range, as shown in the illustration. Can a city 50 miles east and 70 miles north of the tower receive the signal?

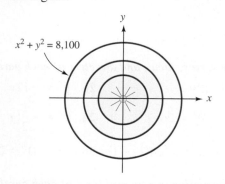

$x^2 + y^2 = 8{,}100$

64. Warning sirens A tornado warning siren can be heard in the circular range shown in the illustration. Can a person 4 miles west and 5 miles south of the siren hear its sound?

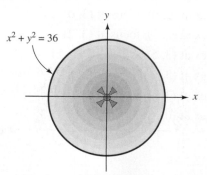

$x^2 + y^2 = 36$

65. Radio translators Some radio stations extend their broadcast range by installing a translator—a remote device that receives the signal and retransmits it. A station with a broadcast range given by $x^2 + y^2 = 1{,}600$, where x and y are in miles, installs a translator with a broadcast area bounded by $x^2 + y^2 - 70y + 600 = 0$. Find the greatest distance from the main transmitter at which the signal can be received.

66. Ripples in a pond When a stone is thrown into the center of a pond, the ripples spread out in a circular pattern, moving at a rate of 3 feet per second. If the stone is dropped at the point $(0, 0)$ in the illustration, when will the ripple reach the seagull floating at the point $(15, 36)$?

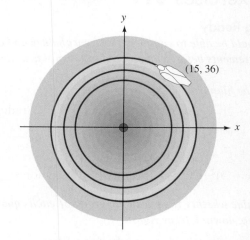

(15, 36)

67. Writing equations of circles Find an equation of the outer rim of the circular arch shown in the illustration.

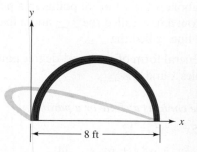

8 ft

68. Writing equations of circles The shape of the window shown is a combination of a rectangle and a semicircle. Find an equation of the circle of which the semicircle is a part.

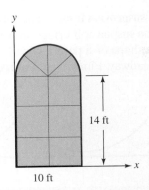

69. Meshing gears For design purposes, the large gear is described by the circle $x^2 + y^2 = 16$. The smaller gear is a circle centered at $(7, 0)$ and tangent to the larger circle. Find an equation of the smaller gear.

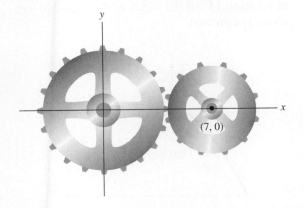

70. Walkways The walkway shown is bounded by the two circles $x^2 + y^2 = 2,500$ and $(x - 10)^2 + y^2 = 900$, measured in feet. Find the largest and the smallest width of the walkway.

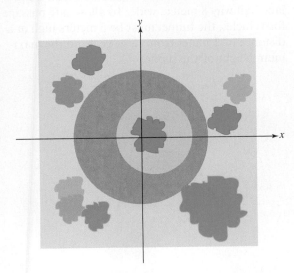

71. Solar furnaces A parabolic mirror collects the rays of the sun and concentrates them at its focus. In the illustration, how far from the vertex of the parabolic mirror will it get the hottest? (All measurements are in feet.)

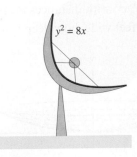

72. Searchlight reflectors A parabolic mirror reflects light in a beam when the light source is placed at its focus. In the illustration, how far from the vertex of the parabolic reflector should the light source be placed? (All measurements are in feet.)

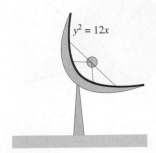

73. Writing equations of parabolas Derive an equation of the parabolic arch shown.

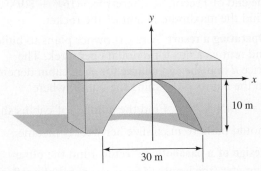

74. Projectiles The cannonball in the illustration follows the parabolic trajectory $y = 30x - x^2$. How far short of the castle does it land?

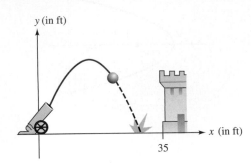

75. Satellite antennas The cross section of the satellite antenna in the illustration is a parabola given by the equation $y = \frac{1}{16}x^2$, with distances measured in feet. If the dish is 8 feet wide, how deep is it?

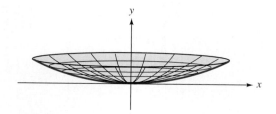

76. Design of a satellite antenna The cross section of the satellite antenna shown is a parabola with the pickup P at its focus. Find the distance d from the pickup to the center of the dish.

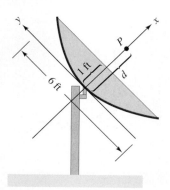

77. Toy rockets A toy rocket is s feet above the earth at the end of t seconds, where $s = -16t^2 + 80t\sqrt{3}$. Find the maximum height of the rocket.

78. Operating a resort A resort owner plans to build and rent n cabins for d dollars per week. The price d that she can charge for each cabin depends on the number of cabins she builds, where $d = -45\left(\frac{n}{32} - \frac{1}{2}\right)$. Find the number of cabins she should build to maximize her weekly income.

79. Design of a parabolic reflector Find the outer diameter (the length $\overline{AB}$) of the parabolic reflector shown.

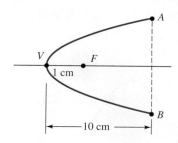

80. Design of a suspension bridge The cable between the towers of the suspension bridge shown in the illustration has the shape of a parabola with vertex 15 feet above the roadway. Find an equation of the parabola.

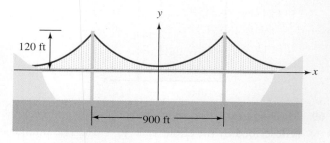

81. Gateway Arch The Gateway Arch in Saint Louis has a shape that approximates a parabola. (See the illustration.) Find the width of the arch 200 feet above the ground.

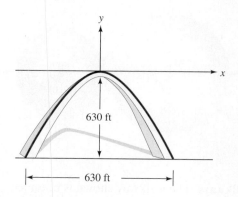

82. Building tunnels A construction firm plans to build a tunnel whose arch is in the shape of a parabola. (See the illustration.) The tunnel will span a two-lane highway 8 meters wide. To allow safe passage for vehicles, the tunnel must be 5 meters high at a distance of 1 meter from its edge. Find the maximum height of the tunnel.

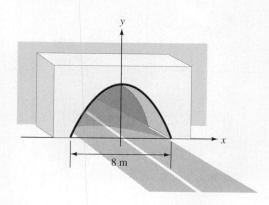

Discovery and Writing

83. Show that the standard form of an equation of a parabola $(y - 2)^2 = 8(x - 1)$ is a special case of the general form of a second-degree equation in two variables.

84. Show that the standard form of an equation of a circle $(x + 2)^2 + (y - 5)^2 = 36$ is a special case of the general form of a second-degree equation in two variables.

Find an equation in the form $(x - h)^2 + (y - k)^2 = r^2$ of the circle passing through the given points.

85. $(0, 8)$, $(5, 3)$, and $(4, 6)$

86. $(-2, 0)$, $(2, 8)$, and $(5, -1)$

Find an equation in the form $y = ax^2 + bx + c$ of the parabola passing through the given points.

87. $(1, 8)$, $(-2, -1)$, and $(2, 15)$

88. $(1, -3)$, $(-2, 12)$, and $(-1, 3)$

89. A stone tossed upward is s feet above the earth after t seconds, where $s = -16t^2 + 128t$. Show that the stone's height x seconds after it is thrown is equal to its height x seconds before it hits the ground.

90. Show that the stone in Exercise 89 reaches its greatest height in one-half of the time it takes until it strikes the ground.

9.2 The Ellipse

In this section, we will learn to

1. Understand the definition of an ellipse.

2. Write an equation of an ellipse.

3. Graph ellipses.

4. Solve application problems using ellipses.

Washington, DC, has many historical sites. Many people touring the United States Capitol visit the National Statuary Hall, originally the chamber of the House of Representatives. It is said that because of the oval shape of the ceiling, politicians on one side of the hall could eavesdrop on politicians talking on the opposite side of the hall. Such rooms are often called *whispering galleries*. Exercise 52 in the problems discusses some properties of whispering galleries.

Today, Statuary Hall houses a collection of statues donated by individual states. Sam Houston, Daniel Webster, and Will Rogers are among those honored with a statue.

A third important conic is an oval-shaped curve called an *ellipse*.

1. Understand the Definition of an Ellipse

Ellipse   An **ellipse** is the set of all points P in a plane such that the sum of the distances from P to two other fixed points F and F' is a positive constant.

We can illustrate this definition by showing how to construct an ellipse. To do so, we place two thumbtacks fairly close together, as in Figure 9-14. We then tie each end of a piece of string to a thumbtack, catch the loop with the point of a pencil, and (while keeping the string stretched tightly) draw the ellipse. Note that $d(F_1P) + d(PF_2)$ will be a positive constant.

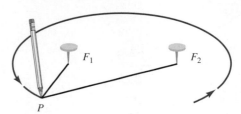

FIGURE 9-14

As we see in the figure, the graph of an ellipse is egg-shaped. The graphs of two ellipses are shown in Figure 9-15.

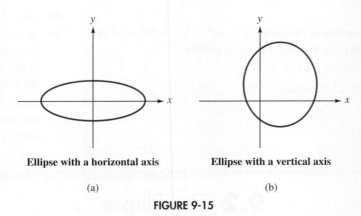

Ellipse with a horizontal axis

(a)

Ellipse with a vertical axis

(b)

FIGURE 9-15

Ellipses occur in many real-life settings:

• Rotating an ellipse about its axis produces a football.

• Planets such as Mars revolve around the sun in an elliptical path.

• A workout machine called an *elliptical trainer* uses elliptical movements for exercise without causing damage to the joints.

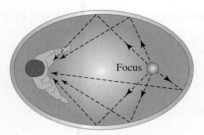

FIGURE 9-16

• An ellipse has an interesting property: Any light or sound that starts at one focus will be reflected through the other. This property is the basis of a medical procedure for treating kidney stones called *lithotripsy*. The patient is placed in an elliptical tank of water with the kidney stone at one focus. Shock waves from a controlled explosion at the other focus are concentrated on the stone, pulverizing it. See Figure 9-16. Though some machines are still in use today, most are being replaced; patients now lie down on a soft water-filled cushion.

2. Write an Equation of an Ellipse

In the ellipse shown in Figure 9-17(a), the fixed points F and F' are called **foci** (each is a **focus**), the midpoint of the chord FF' is called the **center**, and the chord VV' is called the **major axis**. Each of the endpoints V and V' of the major axis is called a **vertex**. The chord BB', perpendicular to the major axis and passing through the center C of the ellipse, is called the **minor axis**.

To derive an equation of the ellipse shown in Figure 9-17(b), we note that point O is the midpoint of chord FF' and let $d(OF) = d(OF') = c$, where $c > 0$. Then the coordinates of point F are $(c, 0)$ and the coordinates of F' are $(-c, 0)$. We also let $P(x, y)$ be any point on the ellipse.

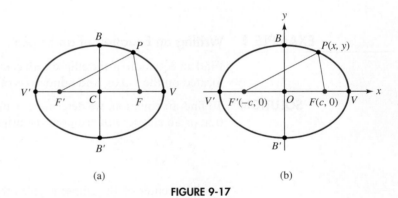

(a)   (b)

FIGURE 9-17

Ellipse: Major Axis on the x-Axis, Center at (0, 0)

The standard equation of an ellipse with its center at the origin and major axis (horizontal) on the x-axis is

$$\frac{x^2}{a^2} + \frac{y^2}{b^2} = 1, \text{ where } a > b > 0.$$

Vertices (ends of the major axis): $V(a, 0)$ and $V'(-a, 0)$

Length of the major axis: $2a$

Ends of the minor axis: $B(0, b)$ and $B'(0, -b)$

Length of the minor axis: $2b$

Foci: $F(c, 0)$ and $F'(-c, 0)$, where $c^2 = a^2 - b^2$

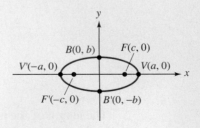

A derivation of the standard form for an equation of an ellipse can be found in online Appendix A.

A similar equation results if the ellipse has its major axis on the *y*-axis and center at the origin:

Ellipse: Major Axis on the y-Axis, Center at (0, 0)

The standard equation of an ellipse with its center at the origin and major axis (vertical) on the *y*-axis is

$$\frac{x^2}{b^2} + \frac{y^2}{a^2} = 1, \text{ where } a > b > 0.$$

Vertices (ends of the major axis): $V(0, a)$ and $V'(0, -a)$

Length of the major axis: $2a$

Ends of the minor axis: $B(b, 0)$ and $B'(-b, 0)$

Length of the minor axis: $2b$

Foci: $F(0, c)$ and $F'(0, -c)$, where $c^2 = a^2 - b^2$

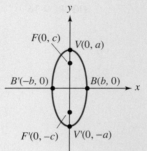

EXAMPLE 1   **Writing an Equation of an Ellipse**

Find an equation of the ellipse with center at the origin, major axis of length 6 units located on the *x*-axis, and minor axis of length 4 units.

SOLUTION   To find an equation, we determine *a* and *b* and substitute into the standard equation of an ellipse with center at the origin and major axis on the *x*-axis,

$$\frac{x^2}{a^2} + \frac{y^2}{b^2} = 1.$$

The center of the ellipse is given to be the origin and the length of the major axis is 6. Since the length of the major axis of an ellipse centered at the origin is $2a$, we have $2a = 6$ or $a = 3$. The coordinates of the vertices are (3, 0) and (–3, 0), as shown in Figure 9-18.

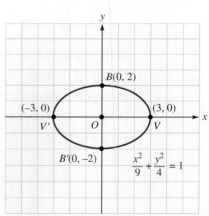

FIGURE 9-18

The length of the minor axis is given to be 4. Since the length of the minor axis of an ellipse centered at the origin is $2b$, we have $2b = 4$ or $b = 2$. The coordinates of *B* and *B'* are (0, 2) and (0, –2).

To find the equation of the ellipse, we substitute 3 for a and 2 for b in the standard equation of an ellipse with center at the origin and major axis on the x-axis.

$$\frac{x^2}{a^2} + \frac{y^2}{b^2} = 1$$

$$\frac{x^2}{3^2} + \frac{y^2}{2^2} = 1$$

$$\frac{x^2}{9} + \frac{y^2}{4} = 1$$

Self Check 1   Find an equation of the ellipse with center at the origin, major axis of length 10 on the y-axis, and minor axis of length 8.

Now Try Exercise 15.

EXAMPLE 2   **Writing an Equation of an Ellipse**

Find an equation of the ellipse with center at the origin, focus at $\left(2\sqrt{3}, 0\right)$, and vertex at $(4, 0)$.

SOLUTION   Because the vertex and focus are on the x-axis, the major axis of the ellipse is on the x-axis. See Figure 9-19. To find an equation of the ellipse, we will determine a^2 and b^2 and substitute into the standard equation

$$\frac{x^2}{a^2} + \frac{y^2}{b^2} = 1.$$

The distance between the center of the ellipse and the vertex is $a = 4$. The distance between the focus and the center is $c = 2\sqrt{3}$.

Since $b^2 = a^2 - c^2$, we can substitute 4 for a and $2\sqrt{3}$ for c and solve for b^2.

$$b^2 = a^2 - c^2$$
$$b^2 = 4^2 - \left(2\sqrt{3}\right)^2$$
$$= 16 - 12$$
$$= 4$$

To find an equation of the ellipse, we substitute 16 for a^2 and 4 for b^2 in the standard equation.

$$\frac{x^2}{a^2} + \frac{y^2}{b^2} = 1$$

$$\frac{x^2}{16} + \frac{y^2}{4} = 1$$

The graph of the ellipse $\dfrac{x^2}{16} + \dfrac{y^2}{4} = 1$ is shown in the figure.

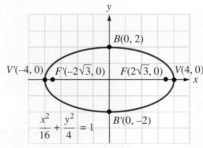

FIGURE 9-19

Self Check 2   Find an equation of the ellipse with center at the origin, focus at $\left(0, 2\sqrt{3}\right)$, and vertex at $(0, 4)$.

Now Try Exercise 17.

To translate an ellipse to a new position centered at the point (h, k) instead of the origin, we replace x and y in the standard equations with $x - h$ and $y - k$, respectively, to get the following results:

Ellipse: Major Axis Horizontal, Center at (h, k)

The standard equation of an ellipse with center at (h, k) and major axis horizontal is

$$\frac{(x - h)^2}{a^2} + \frac{(y - k)^2}{b^2} = 1, \text{ where } a > b > 0.$$

Vertices (ends of the major axis): $V(a + h, k)$ and $V'(-a + h, k)$

Ends of the minor axis: $B(h, b + k)$ and $B'(h, -b + k)$

Foci: $F(h + c, k)$ and $F'(h - c, k)$, where $c^2 = a^2 - b^2$

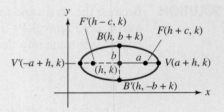

There is a similar result when the major axis is vertical.

Ellipse: Major Axis Vertical, Center at (h, k)

The standard equation of an ellipse with center at (h, k) and major axis vertical is

$$\frac{(x - h)^2}{b^2} + \frac{(y - k)^2}{a^2} = 1, \text{ where } a > b > 0.$$

Vertices (ends of the major axis): $V(h, a + k)$ and $V'(h, -a + k)$

Ends of the minor axis: $B(b + h, k)$ and $B'(-b + h, k)$

Foci: $F(h, k + c)$ and $F'(h, k - c)$, where $c^2 = a^2 - b^2$

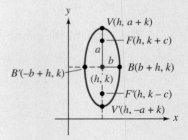

In each case, the length of the major axis is $2a$ and the length of the minor axis is $2b$.

EXAMPLE 3   **Writing an Equation of an Ellipse**

Find an equation of the ellipse with focus at $(-1, 7)$ and vertices at $V(-1, 8)$ and $V'(-1, -2)$.

SOLUTION   We will use the coordinates of the focus and the vertices to determine the center of the ellipse and a and b. We will then substitute these coordinates into the appropriate standard equation of an ellipse. Because the major axis passes through points V and V', we see that the major axis is a vertical line that is parallel to the y-axis. So the standard equation to use is

$$\frac{(x - h)^2}{b^2} + \frac{(y - k)^2}{a^2} = 1, \text{ where } a > b > 0.$$

Since the midpoint of the major axis is the center of the ellipse, the coordinates of the center are $(-1, 3)$, as shown in Figure 9-20.

From Figure 9-20, we see that the distance between the center of the ellipse and either vertex is $a = 5$. We also see that the distance between the focus and the center is $c = 4$.

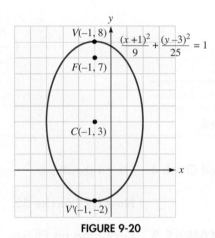

FIGURE 9-20

Since $b^2 = a^2 - c^2$, we can substitute 5 for a and 4 for c and solve for b^2.

$$b^2 = a^2 - c^2$$
$$= 5^2 - 4^2$$
$$= 9$$

To find an equation of the ellipse, we substitute -1 for h, 3 for k, 25 for a^2, and 9 for b^2 in the standard equation and simplify.

$$\frac{(x - h)^2}{b^2} + \frac{(y - k)^2}{a^2} = 1$$

$$\frac{[x - (-1)]^2}{9} + \frac{(y - 3)^2}{25} = 1$$

$$\frac{(x + 1)^2}{9} + \frac{(y - 3)^2}{25} = 1$$

Self Check 3   Find an equation of the ellipse with focus at $(3, 1)$ and vertices at $V(5, 1)$ and $V'(-5, 1)$.

Now Try Exercise 29.

3. Graph Ellipses

EXAMPLE 4   **Graphing an Ellipse**

Graph the ellipse $\dfrac{x^2}{49} + \dfrac{y^2}{4} = 1$.

SOLUTION   To graph the ellipse, we will first determine the coordinates of the vertices and the endpoints of the minor axis. We will then plot these points and draw an ellipse through them.

Because the equation is the standard form of an ellipse centered at the origin with major axis on the x-axis, we know that the center of the ellipse is at $(0, 0)$ and that the vertices lie on the x-axis. Because $a = 7$, the vertices are 7 units to the right and left of the origin, at points $(7, 0)$ and $(-7, 0)$. Because $b^2 = 4$, $b = 2$ and the endpoints of the minor axis are 2 units above and below the origin, at points $(0, 2)$ and $(0, -2)$. Using these points, we can sketch the ellipse shown in Figure 9-21.

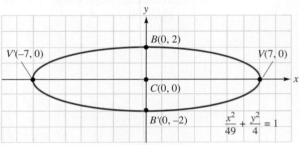

FIGURE 9-21

Comment

In Example 4, note that $49 > 4$. Because 49 is the denominator of the x^2 term, the axis of the ellipse is horizontal.

Self Check 4   Graph the ellipse $\dfrac{x^2}{4} + \dfrac{y^2}{9} = 1$.

Now Try Exercise 35.

EXAMPLE 5   **Graphing an Ellipse**

Graph the ellipse $\dfrac{(x + 2)^2}{4} + \dfrac{(y - 2)^2}{9} = 1$.

SOLUTION   The equation is the standard form of an equation of an ellipse whose major axis is vertical and whose center is at $(-2, 2)$. We will use translations to graph the ellipse.

Because $a^2 = 9$, we have $a = 3$ and the vertices are 3 units above and 3 units below the center, at points $(-2, 5)$ and $(-2, -1)$. Because $b^2 = 4$, $b = 2$ and the endpoints of the minor axis are 2 units to the right and 2 units to the left of the center, at points $(0, 2)$ and $(-4, 2)$. Using these points, we can sketch the ellipse as shown in Figure 9-22.

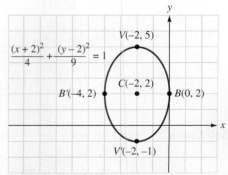

FIGURE 9-22

Comment

In Example 5, note that $9 > 4$. Because 9 is the denominator of the y^2 term, the axis of the ellipse is vertical.

Self Check 5   Graph the ellipse $\dfrac{(x-2)^2}{9} + \dfrac{(y+1)^2}{25} = 1$.

Now Try Exercise 39.

EXAMPLE 6   **Graphing an Ellipse**

Graph: $4x^2 + 9y^2 - 16x - 18y = 11$.

SOLUTION   We write the equation in standard form by completing the square on x and y, and then use translations to graph the ellipse.

$$4x^2 + 9y^2 - 16x - 18y = 11$$

$$4x^2 - 16x + 9y^2 - 18y = 11$$

$$4(x^2 - 4x) + 9(y^2 - 2y) = 11$$

$$4(x^2 - 4x + \mathbf{4}) + 9(y^2 - 2y + \mathbf{1}) = 11 + \mathbf{16} + \mathbf{9}$$

$$4(x - 2)^2 + 9(y - 1)^2 = 36$$

$$\frac{(x-2)^2}{9} + \frac{(y-1)^2}{4} = 1$$

We can now see that the graph is an ellipse with center at $(2, 1)$ and major axis parallel to the x-axis. Because $a^2 = 9$, we have $a = 3$ and the vertices are 3 units to the left and right of the center, at $(-1, 1)$ and $(5, 1)$. Because $b^2 = 4$, we have $b = 2$ and the endpoints of the minor axis are 2 units above and below the center, at $(2, 3)$ and $(2, -1)$. Using these points, we can sketch the ellipse as shown in Figure 9-23.

Caution

> If the coefficients of x^2 and y^2 in the equation of an ellipse are equal, the ellipse is a circle. If the coefficients of x^2 and y^2 are not equal, the ellipse is not a circle.

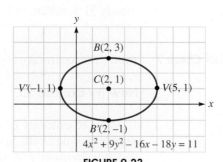

FIGURE 9-23

Self Check 6   Graph: $4x^2 + 9y^2 - 8x + 36y = -4$.

Now Try Exercise 47.

4. Solve Application Problems Using Ellipses

EXAMPLE 7   **Solving an Application Problem**

The orbit of the earth is approximately an ellipse, with the sun at one focus. The ratio of c to a (called the **eccentricity** of the ellipse) is about $\frac{1}{62}$, and the length of the major axis is approximately 186,000,000 miles. How close does the earth get to the sun?

SOLUTION   We will assume that the ellipse has its center at the origin and vertices V' and V at $(-93,000,000, 0)$ and $(93,000,000, 0)$, as shown in Figure 9-24.

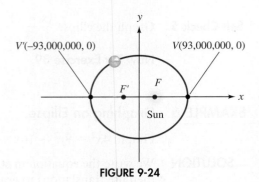

FIGURE 9-24

Because the eccentricity $\frac{c}{a}$ is given to be $\frac{1}{62}$ and $a = 93{,}000{,}000$, we have

$$\frac{c}{a} = \frac{1}{62}$$

$$c = \frac{1}{62}a$$

$$c = \frac{1}{62}(93{,}000{,}000)$$

$$= 1{,}500{,}000$$

The distance $d(FV)$ is the shortest distance between the earth and the sun. (You'll be asked to prove this in the exercises.) Thus,

$$d(FV) = a - c = 93{,}000{,}000 - 1{,}500{,}000 = 91{,}500{,}000 \text{ miles.}$$

The earth's point of closest approach to the sun (called the **perihelion**) is approximately 91.5 million miles.

Self Check 7   Find the eccentricity of the ellipse $\dfrac{(x + 2)^2}{9} + \dfrac{(y - 5)^2}{25} = 1$.

Now Try Exercise 55.

The eccentricity of an ellipse provides a measure of how much the curve resembles a true circle. Specifically, the eccentricity of a true circle equals 0. We can use the eccentricity of an ellipse to judge its shape. If the eccentricity is close to 1, the ellipse is relatively flat, as in the illustration.

If the eccentricity is close to 0, the ellipse is more circular, as shown.

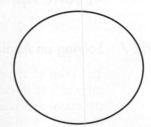

Since the eccentricity of the earth's orbit is $\frac{1}{62}$, the earth's orbit is almost a circle.

Self Check Answers   1. $\dfrac{x^2}{16} + \dfrac{y^2}{25} = 1$ 2. $\dfrac{x^2}{4} + \dfrac{y^2}{16} = 1$ 3. $\dfrac{x^2}{25} + \dfrac{(y-1)^2}{16} = 1$

4.

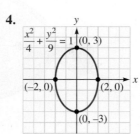

$\dfrac{x^2}{4} + \dfrac{y^2}{9} = 1$ (0, 3)   (−2, 0)   (2, 0)   (0, −3)

5.

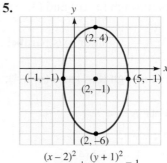

(2, 4)   (−1, −1)   (2, −1)   (5, −1)   (2, −6)

$\dfrac{(x-2)^2}{9} + \dfrac{(y+1)^2}{25} = 1$

6.

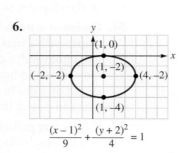

(1, 0)   (1, −2)   (−2, −2)   (4, −2)   (1, −4)

$\dfrac{(x-1)^2}{9} + \dfrac{(y+2)^2}{4} = 1$

7. $\dfrac{4}{5}$

Exercises **9.2**

Getting Ready
You should be able to complete these vocabulary and concept statements before you proceed to the practice exercises.

Fill in the blanks.

1. An ellipse is the set of all points in the plane such that the _____ of the distances from two fixed points is a positive _____.

2. Each of the two fixed points in the definition of an ellipse is called a _____ of the ellipse.

3. The chord that joins the _____ is called the major axis of the ellipse.

4. The chord through the center of an ellipse and perpendicular to the major axis is called the _____ axis.

5. In the ellipse $\dfrac{x^2}{a^2} + \dfrac{y^2}{b^2} = 1$ $(a > b > 0)$, the vertices are $V(\underline{\quad}, \underline{\quad})$ and $V'(\underline{\quad}, \underline{\quad})$.

6. In an ellipse, the relationship between a, b, and c is _____.

7. To draw an ellipse that is 26 inches wide and 10 inches tall, the piece of string should be _____ inches long and the tacks should be _____ inches apart.

8. To draw an ellipse that is 20 centimeters wide and 12 centimeters tall, the string should be _____ inches long, and the tacks should be _____ inches apart.

Identify the conic as a circle, parabola, or ellipse.

9. $x^2 - 6x + y^2 = 7$

10. $5x^2 + 5y^2 + 10y + \dfrac{24}{5} = 0$

11. $x^2 - 4y + 5 = -2x$

12. $y^2 - 8y = -2x - 20$

13. $7x^2 + 5y^2 - 35 = 0$

14. $5x^2 + 2y^2 - 10x + 4y = 13$

Practice
Write an equation of the ellipse with the given characteristics that has its center at the origin.

15. Major axis of length 8 units located on the x-axis; minor axis of length 6 units

16. Major axis of length 14 units located on the y-axis; minor axis of length 10 units

17. Focus at $(3, 0)$; vertex at $(5, 0)$

18. Focus at $(0, 4)$; vertex at $(0, 7)$

19. Focus at $(0, 1)$; $\dfrac{4}{3}$ is one-half the length of the minor axis

20. Focus at $(1, 0)$; $\dfrac{4}{3}$ is one-half the length of the minor axis

21. Focus at $(0, 3)$; major axis of length 8

22. Focus at $(5, 0)$; major axis of length 12

Write an equation in standard form of each ellipse.

23. Center at $(3, 4)$; $a = 3$; $b = 2$; major axis parallel to the y-axis

24. Center at $(3, 4)$; passes through $(3, 10)$ and $(3, -2)$; $b = 2$

25. Center at $(3, 4)$; $a = 3$; $b = 2$; major axis parallel to the x-axis

26. Center at $(3, 4)$; passes through $(8, 4)$ and $(-2, 4)$; $b = 2$

27. Foci at $(-2, 4)$ and $(8, 4)$; $b = 4$

28. Foci at $(8, 5)$ and $(4, 5)$; $b = 3$

29. Vertex at $(6, 4)$; foci at $(-4, 4)$ and $(4, 4)$

30. Center at $(-4, 5)$; $\dfrac{c}{a} = \dfrac{1}{3}$; vertex at $(-4, -1)$

31. Foci at $(6, 0)$ and $(-6, 0)$; $\dfrac{c}{a} = \dfrac{3}{5}$

32. Vertices at $(2, 0)$ and $(-2, 0)$; $\dfrac{2b^2}{a} = 2$

Graph each ellipse.

33. $\dfrac{x^2}{25} + \dfrac{y^2}{9} = 1$ **34.** $\dfrac{x^2}{36} + \dfrac{y^2}{25} = 1$

35. $\dfrac{x^2}{25} + \dfrac{y^2}{49} = 1$ **36.** $4x^2 + y^2 = 4$

37. $\dfrac{x^2}{16} + \dfrac{(y + 2)^2}{36} = 1$ **38.** $(x - 1)^2 + \dfrac{4y^2}{25} = 4$

39. $\dfrac{(x - 4)^2}{49} + \dfrac{(y - 2)^2}{9} = 1$

40. $\dfrac{(x - 1)^2}{25} + \dfrac{y^2}{4} = 1$

Write each ellipse in standard form.

41. $4x^2 + y^2 - 2y = 15$

42. $4x^2 + 25y^2 + 8x - 96 = 0$

43. $9x^2 + 4y^2 + 18x + 16y - 11 = 0$

44. $x^2 + 4y^2 - 10x - 8y = -13$

Graph each ellipse.

45. $x^2 + 4y^2 - 4x + 8y + 4 = 0$

46. $x^2 + 4y^2 - 2x - 16y = -13$

47. $16x^2 + 25y^2 - 160x + 200y + 400 = 0$

48. $3x^2 + 2y^2 + 7x - 6y = -1$

Applications

49. Pool tables Find an equation of the outer edge of the elliptical pool table shown.

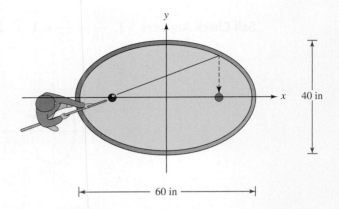

50. Equation of an arch An arch is a semiellipse 12 meters wide and 5 meters high. Write an equation of the ellipse if the ellipse is centered at the origin.

51. Design of a track A track is built in the shape of an ellipse with a maximum length of 100 meters and a maximum width of 60 meters. Write an equation of the ellipse and find its **focal width.** That is, find the length of a chord that is perpendicular to the major axis and passes through either focus of the ellipse.

52. Whispering galleries Any sound from one focus of an ellipse reflects off the ellipse directly back to the other focus. This property explains whispering galleries such as Statuary Hall in Washington, DC. The ceiling of the whispering gallery shown has the shape of a semiellipse. Find the distance sound travels as it leaves focus F and returns to focus F'.

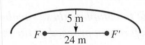

53. Finding the width of a mirror The mirror shown is in the shape of an ellipse. Find the width of the mirror 12 inches above its base.

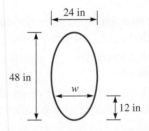

54. Finding the height of a window The window shown has the shape of half an ellipse. Find the height of the window 20 inches from one end.

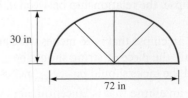

55. Astronomy The moon has an orbit that is an ellipse, with the earth at one focus. If the major axis of the orbit is 378,000 miles and the ratio of c to a is approximately $\frac{11}{200}$, how far does the moon get from the earth? (This farthest point in an orbit is called the **apogee**.)

56. Area of an ellipse The area A of the ellipse

$$\frac{x^2}{a^2} + \frac{y^2}{b^2} = 1$$

is given by $A = \pi ab$. Find the area of the ellipse $9x^2 + 16y^2 = 144$.

Discovery and Writing

57. If F is a focus of the ellipse shown and B is an endpoint of the minor axis, use the distance formula to prove that the length of segment FB is a. (*Hint:* In an ellipse, $c^2 = a^2 - b^2$.)

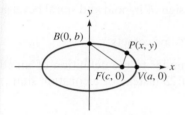

58. If F is a focus of the ellipse shown in Exercise 57 and P is any point on the ellipse, use the distance formula to show that the length of FP is $a - \frac{c}{a}x$. (*Hint:* In an ellipse, $a^2 - c^2 = b^2$.)

59. In the ellipse shown, chord AA' passes through the focus F and is perpendicular to the major axis. Show that the length of AA' (called the **focal width**) is $\frac{2b^2}{a}$.

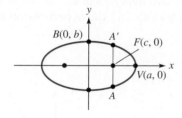

60. Prove that segment FV in Example 7 is the shortest distance between the earth and the sun. (*Hint:* Refer to Exercise 58.)

61. The ends of a piece of string 6 meters long are attached to two thumbtacks that are 2 meters apart. A pencil catches the loop and draws it tight. As the pencil is moved about the thumbtacks (always keeping the tension), an ellipse is produced, with the thumbtacks as foci. Write the equation of the ellipse. (*Hint:* You'll have to establish a coordinate system.)

62. The distance between the point $P(x, y)$ and the point $(0, 2)$ is $\frac{1}{3}$ of the distance from point P to the line $y = 18$. Find the equation of the curve on which point P lies.

63. Prove that $a > b$ in the development of the standard equation of an ellipse.

64. Show that the expansion of the standard equation of an ellipse is a special case of the general second-degree equation in two variables.

65. The eccentricity of an ellipse provides a measure of how much the curve resembles a true circle. Specifically, the eccentricity of a true circle equals 0. When analyzing planetary orbits, astronomers plot the relationship between the length of the orbit's semimajor axis (measured in astronomical units, where 1 AU = 149,598,000 kilometers) and the eccentricity of the orbit. Note that the semimajor axis is perpendicular to the axis containing the foci of an ellipse. Use the given data plot to estimate how many of the 75 planets shown follow orbits that are true circles.

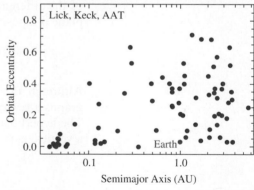

Source: Eccentricity vs. semimajor axis for extrasolar planets. The 75 planets shown were found in a Doppler survey of 1,300 FGKM main sequence stars using the Lick, Keck, and AAT telescopes. The survey was carried out by the California-Carnegie planet search team. http://exoplanets.org/newsframe.html

9.3 The Hyperbola

In this section, we will learn to

1. Write an equation of a hyperbola.

2. Graph hyperbolas.

When a military plane exceeds the speed of sound, it is said to break the sound barrier. When this happens, a visible vapor cloud appears and an explosion-like sound results that we call a *sonic boom*. As the plane moves faster beyond the speed of sound, a cone shape intersects the ground, forming one branch of a curve called a **hyperbola**, a fourth type of conic section. The sonic boom is heard at points inside this hyperbola.

Hyperbolas appear in many real-life applications:

- Hyperbolas are the basis of a navigational system known as LORAN (LOng-RAnge Navigation). It is a radio navigation system that uses time intervals between radio signals to determine the location of a ship or aircraft.
- Hyperbolas are the basis for the design of hypoid and spiral bevel gears that are used in motor vehicles.
- Hyperbolas describe the orbits of some comets.

The definition of a hyperbola is similar to the definition of an ellipse, except that we require a constant difference of $2a$ instead of a constant sum.

Hyperbola   A **hyperbola** is the set of all points P in a plane such that the absolute value of the difference of the distances from point P to two other fixed points in the plane is a positive constant.

Although the definitions of an ellipse and a hyperbola are quite similar, their graphs are very different. The graphs of two hyperbolas are shown in Figure 9-25. The hyperbola on the left has a horizontal axis and the hyperbola on the right has a vertical axis.

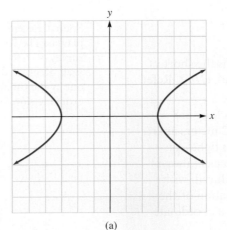

(a)

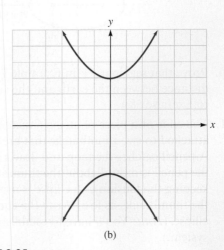

(b)

FIGURE 9-25

1. Write an Equation of a Hyperbola

In the graph of the hyperbola shown in Figure 9-26, points F and F' are called the **foci** of the hyperbola, and the midpoint of chord FF' is called the **center**. The points V and V', where the hyperbola intersects FF', are called **vertices**. The segment VV' is called the **transverse axis**.

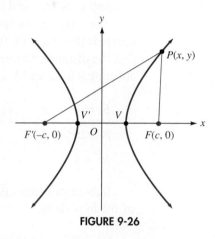

FIGURE 9-26

Hyperbola: Foci on x-Axis, Center at (0, 0)

The standard equation of a hyperbola with center at the origin and foci on the x-axis is

$$\frac{x^2}{a^2} - \frac{y^2}{b^2} = 1, \text{ where } a^2 + b^2 = c^2.$$

Vertices: $V(a, 0)$ and $V'(-a, 0)$
Foci: $F(c, 0)$ and $F'(-c, 0)$

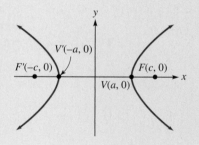

A derivation of the standard form for an equation of a hyperbola can be found in online Appendix A.

If the foci are on the y-axis, a similar equation results:

Hyperbola: Foci on y-Axis, Center at (0, 0)

The standard equation of a hyperbola with center at the origin and foci on the y-axis is

$$\frac{y^2}{a^2} - \frac{x^2}{b^2} = 1, \text{ where } a^2 + b^2 = c^2.$$

Vertices: $V(0, a)$ and $V'(0, -a)$
Foci: $F(0, c)$ and $F'(0, -c)$

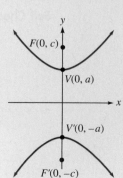

EXAMPLE 1   **Writing an Equation of a Hyperbola**

Write an equation of the hyperbola with vertices $V(4, 0)$ and $V'(-4, 0)$ and a focus at $F(5, 0)$.

SOLUTION   Because the foci lie on the x-axis, we use the standard equation $\dfrac{x^2}{a^2} - \dfrac{y^2}{b^2} = 1$. We will find a^2 and b^2 and substitute the results into the standard equation.

The center of the hyperbola is midway between the vertices V and V'. Thus, the center is the origin $(0, 0)$. The distance between the vertex and the center is $a = 4$, and the distance between the focus and the center is $c = 5$. We can find b^2 by substituting 4 for a and 5 for c in the following equation:

$$b^2 = c^2 - a^2 \qquad \text{In a hyperbola, } b^2 = c^2 - a^2.$$

$$b^2 = 5^2 - 4^2$$

$$b^2 = 9$$

Substituting the values for a^2 and b^2 in the standard equation gives an equation of the hyperbola.

$$\frac{x^2}{a^2} - \frac{y^2}{b^2} = 1$$

$$\frac{x^2}{16} - \frac{y^2}{9} = 1$$

The graph of $\dfrac{x^2}{16} - \dfrac{y^2}{9} = 1$ is shown in Figure 9-27.

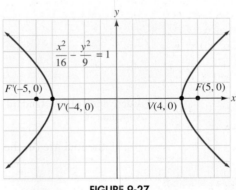

FIGURE 9-27

Self Check 1   Write an equation of the hyperbola with vertices $(0, 4)$ and $(0, -4)$ and a focus at $(0, 5)$.

Now Try Exercise 15.

To translate the hyperbola to a new position centered at the point (h, k) instead of the origin, we replace x and y with $x - h$ and $y - k$, respectively. We get the following results:

Hyperbola: Transverse Axis Horizontal, Center at (h, k)

The standard equation of a hyperbola with center at the point (h, k) and foci on a line parallel to the x-axis is

$$\frac{(x - h)^2}{a^2} - \frac{(y - k)^2}{b^2} = 1, \text{ where } a^2 + b^2 = c^2.$$

Vertices: $V(a + h, k)$ and $V'(-a + h, k)$

Foci: $F(c + h, k)$ and $F'(-c + h, k)$

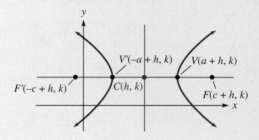

Hyperbola: Transverse Axis Vertical, Center at (h, k)

The standard equation of a hyperbola with center at the point (h, k) and foci on a line parallel to the y-axis is

$$\frac{(y - k)^2}{a^2} - \frac{(x - h)^2}{b^2} = 1, \text{ where } a^2 + b^2 = c^2.$$

Vertices: $V(h, a + k)$ and $V'(h, -a + k)$

Foci: $F(h, c + k)$ and $F'(h, -c + k)$

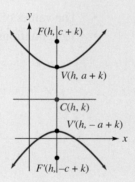

EXAMPLE 2   **Writing an Equation of a Hyperbola**

Write an equation of the hyperbola with vertices $(3, -3)$ and $(3, 3)$ and a focus at $(3, 5)$.

SOLUTION   Because the foci lie on a vertical line, as shown in Figure 9-28, we will use the standard form of the equation

$$\frac{(y - k)^2}{a^2} - \frac{(x - h)^2}{b^2} = 1,$$

determine h, k, a^2, and b^2, and substitute the results into the standard equation.

The center of the hyperbola is midway between the vertices V and V'. Thus, the center is the point $(3, 0)$, $h = 3$, and $k = 0$. The distance between the vertex and the

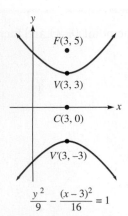

$$\frac{y^2}{9} - \frac{(x-3)^2}{16} = 1$$

FIGURE 9-28

center is $a = 3$, and the distance between the focus and the center is $c = 5$. We can find b^2 by substituting 3 for a and 5 for c in the following equation:

$$b^2 = c^2 - a^2 \qquad \text{In a hyperbola, } b^2 = c^2 - a^2.$$

$$b^2 = 5^2 - 3^2$$

$$b^2 = 16$$

Substituting the values for h, k, a^2, and b^2 in the standard equation gives an equation of the hyperbola.

$$\frac{(y-k)^2}{a^2} - \frac{(x-h)^2}{b^2} = 1$$

$$\frac{(y-0)^2}{9} - \frac{(x-3)^2}{16} = 1$$

$$\frac{y^2}{9} - \frac{(x-3)^2}{16} = 1$$

■

Self Check 2   Write an equation of the hyperbola with vertices $(3, 1)$ and $(-3, 1)$ and a focus at $(5, 1)$.

Now Try Exercise 23.

2. Graph Hyperbolas

Consider the graph of the hyperbola with equation $\dfrac{x^2}{25} - \dfrac{y^2}{4} = 1$ shown in Figure 9-29. From the graph, we can see that

- The values of a and b are $a = 5$ and $b = 2$.
- The equations of the asymptotes of the graph are $y = \frac{2}{5}x$ and $y = -\frac{2}{5}x$.

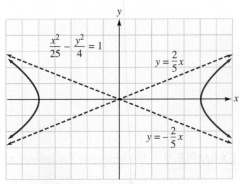

FIGURE 9-29

The values of a and b are important aids in finding the equations of the asymptotes of the hyperbola. In fact, the equations of the asymptotes are

$$y = \frac{b}{a}x \qquad \text{and} \qquad y = -\frac{b}{a}x,$$

in this case $y = \frac{2}{5}x$ and $y = -\frac{2}{5}x$.

The values of a and b are also aids in drawing a special rectangle that can be used to help graph hyperbolas. To show that this is true, we consider the hyperbola with the equation $\dfrac{x^2}{a^2} - \dfrac{y^2}{b^2} = 1$. This hyperbola has its center at the origin and vertices at $V(a, 0)$ and $V'(-a, 0)$. We can plot the points V, V', $B(0, b)$, and $B'(0, -b)$

to form rectangle *RSQP*, called the **fundamental rectangle,** as shown in Figure 9-30. The extended diagonals of this rectangle are the asymptotes of the hyperbola.

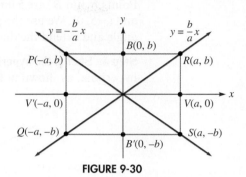

FIGURE 9-30

The method for verifying that the extended diagonals are asymptotes of the hyperbola can be found in online Appendix A.

Using the fundamental rectangle, asymptotes, and vertices as guides, we can sketch the hyperbola shown in Figure 9-31. The segment *BB'* is called the **conjugate axis** of the hyperbola.

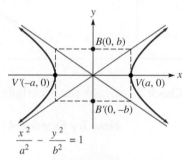

FIGURE 9-31

If the hyperbola is not in standard form, we simply convert it to standard form by completing the square and making use of translations to graph the hyperbola. This information is summarized as follows:

Strategy for Graphing a Hyperbola

Step 1: Write the hyperbola in standard form.
Step 2: Identify the center, *a*, and *b*.
Step 3: Draw the fundamental rectangle and asymptotes.
Step 4: Sketch the hyperbola beginning with the vertices and using the asymptotes as guides.

EXAMPLE 3   **Graphing a Hyperbola**

Graph the hyperbola $\dfrac{y^2}{9} - \dfrac{x^2}{25} = 1$.

SOLUTION   We will apply the four steps to graph the hyperbola.

Step 1: Write the hyperbola in standard form. The equation of the hyperbola $\dfrac{y^2}{9} - \dfrac{x^2}{25} = 1$ is already written in the standard form $\dfrac{y^2}{a^2} - \dfrac{x^2}{b^2} = 1$.

Step 2: Identify the center, *a*, and *b*. From the standard equation of a hyperbola, we see that the center is $(0, 0)$, that $a = 3$ and $b = 5$, and that the vertices are on the *y*-axis.

Step 3: Draw the fundamental rectangle and asymptotes. Vertices V and V' are 3 units up and down from the origin and have coordinates of $(0, 3)$ and $(0, -3)$. Points B and B' are 5 units right and left of the origin and have coordinates of $(5, 0)$ and $(-5, 0)$. We use the points V, V', B, and B' to construct the fundamental rectangle and extend the diagonals to graph the asymptotes, as shown in Figure 9-32.

Step 4: Sketch the hyperbola. We begin with the vertices and sketch the graph of the hyperbola, as shown in Figure 9-32, using the asymptotes as guides.

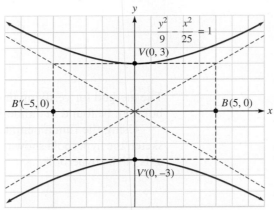

FIGURE 9-32

Caution

The graphs of hyperbolas look very similar to the graphs of parabolas, but hyperbolas are not parabolas.

Self Check 3   Graph the hyperbola $\dfrac{y^2}{9} - \dfrac{(x + 1)^2}{25} = 1$.

Now Try Exercise 41.

EXAMPLE 4   **Graphing a Hyperbola**

Graph the hyperbola $x^2 - y^2 - 2x + 4y = 12$.

SOLUTION   We will use the four steps to graph the hyperbola.

Step 1: Write the hyperbola in standard form. We complete the square on x and y to write the equation in standard form.

$$x^2 - y^2 - 2x + 4y = 12$$
$$x^2 - 2x - (y^2 - 4y) = 12$$
$$x^2 - 2x + 1 - (y^2 - 4y + 4) = 12 + 1 - 4$$
$$(x - 1)^2 - (y - 2)^2 = 9$$
$$\frac{(x - 1)^2}{9} - \frac{(y - 2)^2}{9} = 1$$

Step 2: Identify the center, a, and b. From the standard equation of a hyperbola, we see that the center is $(1, 2)$, that $a = 3$ and $b = 3$, and that the vertices are on a line segment parallel to the x-axis, as shown in Figure 9-33.

Step 3: Draw the fundamental rectangle and asymptotes. The vertices V and V' are 3 units to the right and left of the center and have coordinates of $(4, 2)$ and $(-2, 2)$. Points B and B', 3 units above and below the center, have coordinates of $(1, 5)$ and $(1, -1)$. We use the points V, V', B, and B' to construct the fundamental rectangle and extend the diagonals to draw the asymptotes, as shown in Figure 9-33.

Step 4: Sketch the hyperbola. We begin with the vertices and sketch the graph of the hyperbola, as shown in Figure 9-33, using the asymptotes as guides.

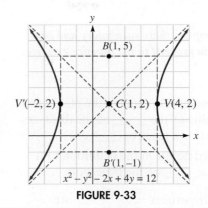

FIGURE 9-33

Self Check 4   Graph the hyperbola $x^2 - y^2 + 6x + 2y = 8$.

Now Try Exercise 47.

We have considered only those hyperbolas with a major axis that is horizontal or vertical. However, some hyperbolas have nonhorizontal or nonvertical major axes. For example, the graph of the equation $xy = 4$ is a hyperbola with vertices at $(2, 2)$ and $(-2, -2)$, as shown in Figure 9-34.

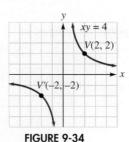

FIGURE 9-34

Self Check Answers   1. $\dfrac{y^2}{16} - \dfrac{x^2}{9} = 1$ 2. $\dfrac{x^2}{9} - \dfrac{(y-1)^2}{16} = 1$

3.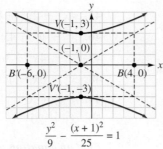

4.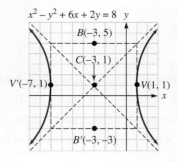

Exercises **9.3**

Getting Ready

You should be able to complete these vocabulary and concept statements before you proceed to the practice exercises.

Fill in the blanks.

1. A hyperbola is the set of all points in the plane such that the absolute value of the _____ of the distances from two fixed points is a positive _____.

2. Each of the two fixed points in the definition of a hyperbola is called a _____ of the hyperbola.

3. The vertices of the hyperbola $\dfrac{x^2}{a^2} - \dfrac{y^2}{b^2} = 1$ are

   $V(\underline{\quad}, \underline{\quad})$ and $V'(\underline{\quad}, \underline{\quad})$.

4. The vertices of the hyperbola $\dfrac{y^2}{a^2} - \dfrac{x^2}{b^2} = 1$ are

   $V(\underline{\quad}, \underline{\quad})$ and $V'(\underline{\quad}, \underline{\quad})$.

5. The chord that joins the vertices is called the _____ of the hyperbola.

6. In a hyperbola, the relationship between a, b, and c is _____.

Identify the conic as a circle, parabola, ellipse, or hyperbola.

7. $x^2 + (y - 4)^2 = 12$
8. $7x^2 + 7y^2 + 70x - 14y = -119$
9. $y^2 - 2x + 23 = 5y$
10. $(x - 8)^2 = 5(y + 6)$
11. $\dfrac{x^2}{35} + \dfrac{y^2}{9} = 1$
12. $2x^2 + 32y^2 + 4x - 30 = 0$
13. $x^2 - 2y^2 - 4y - 6 = 0$
14. $\dfrac{(x - 3)^2}{4} - \dfrac{y^2}{9} = 1$

Practice

Write an equation in standard form of each hyperbola.

15. Vertices $(5, 0)$ and $(-5, 0)$; focus $(7, 0)$
16. Focus $(3, 0)$; vertex $(2, 0)$; center $(0, 0)$
17. Center $(2, 4)$; $a = 2$; $b = 3$; transverse axis is horizontal
18. Center $(-1, 3)$; vertex $(1, 3)$; focus $(2, 3)$
19. Center $(5, 3)$; vertex $(5, 6)$; passes through $(1, 8)$
20. Foci $(0, 10)$ and $(0, -10)$; $\dfrac{c}{a} = \dfrac{5}{4}$
21. Vertices $(0, 3)$ and $(0, -3)$; $\dfrac{c}{a} = \dfrac{5}{3}$
22. Focus $(4, 0)$; vertex $(2, 0)$; center $(0, 0)$
23. Center $(1, 4)$; focus $(7, 4)$; vertex $(3, 4)$

24. Center $(1, -3)$; $a^2 = 4$; $b^2 = 16$
25. Center at the origin; passes through $(4, 2)$ and $(8, -6)$
26. Center $(3, -1)$; y-intercept $= -1$; x-intercept $= 3 + \dfrac{3\sqrt{5}}{2}$

Find the area of the fundamental rectangle of each hyperbola.

27. $4(x - 1)^2 - 9(y + 2)^2 = 36$
28. $x^2 - y^2 - 4x - 6y = 6$
29. $x^2 + 6x - y^2 + 2y = -11$
30. $9x^2 - 4y^2 = 18x + 24y + 63$

Write an equation of each hyperbola.

31. Center $(-2, -4)$; $a = 2$; area of fundamental rectangle is 36 square units
32. Center $(3, -5)$; $b = 6$; area of fundamental rectangle is 24 square units
33. Vertex $(6, 0)$; one end of conjugate axis at $\left(0, \dfrac{5}{4}\right)$
34. Vertex $(3, 0)$; focus $(-5, 0)$; center $(0, 0)$

Graph each hyperbola.

35. $\dfrac{x^2}{9} - \dfrac{y^2}{4} = 1$
36. $\dfrac{y^2}{4} - \dfrac{x^2}{9} = 1$
37. $4x^2 - 3y^2 = 36$
38. $3x^2 - 4y^2 = 36$
39. $y^2 - x^2 = 1$
40. $x^2 - \dfrac{y^2}{4} = 1$
41. $\dfrac{(x + 2)^2}{9} - \dfrac{y^2}{4} = 1$
42. $\dfrac{y^2}{9} - \dfrac{(x - 2)^2}{36} = 1$
43. $4(y - 2)^2 - 9(x + 1)^2 = 36$
44. $9(y + 2)^2 - 4(x - 1)^2 = 36$
45. $4x^2 - 2y^2 + 8x - 8y = 8$
46. $x^2 - y^2 - 4x - 6y = 6$
47. $y^2 - 4x^2 + 6y + 32x = 59$
48. $x^2 + 6x - y^2 + 2y = -11$
49. $-xy = 6$
50. $xy = 20$

Find an equation of the curve on which point P lies.

51. The difference of the distances between $P(x, y)$ and the points $(-2, 1)$ and $(8, 1)$ is 6.
52. The difference of the distances between $P(x, y)$ and the points $(3, -1)$ and $(3, 5)$ is 5.
53. The distance between $P(x, y)$ and the point $(0, 3)$ is $\dfrac{3}{2}$ of the distance between P and the line $y = -2$.
54. The distance between $P(x, y)$ and the point $(5, 4)$ is $\dfrac{5}{3}$ of the distance between P and the line $x = -3$.

Applications

55. Fluids In the illustration, two glass plates in contact at the left and separated by about 5 millimeters on the right are dipped in beet juice, which rises by capillary action to form a hyperbola. The hyperbola is modeled by an equation of the form $xy = k$. If the curve passes through the point $(12, 2)$, what is k?

56. Astronomy Some comets have a hyperbolic orbit, with the sun as one focus. When the comet shown in the illustration is far away from the earth, it appears to be approaching the earth along the line $y = 2x$. Find an equation of its orbit if the comet comes within 100 million miles of the earth.

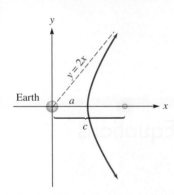

57. Alpha particles The particle in the illustration approaches the nucleus at the origin along the path $9y^2 - x^2 = 81$ in the coordinate system shown. How close does the particle come to the nucleus?

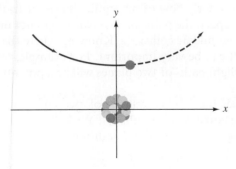

58. Physics Parallel beams of similarly charged particles are shot from two atomic accelerators 20 meters apart, as shown in the illustration. If the particles were not deflected, the beams would be 2.0×10^{-4} meters apart. However, because the charged particles repel each other, the beams follow the hyperbolic path $y = \frac{k}{x}$, for some k. Find k.

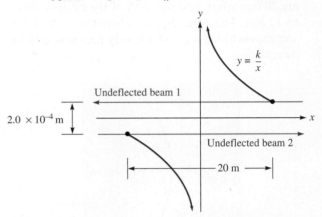

59. Navigation The LORAN (LOng-RAnge Navigation) system in the illustration uses two radio transmitters 26 miles apart to send simultaneous signals. The navigator on a ship at $P(x, y)$ receives the closer signal first, and determines that the difference of the distances between the ship and each transmitter is 24 miles. That places the ship on a certain curve. Identify the curve and find its equation.

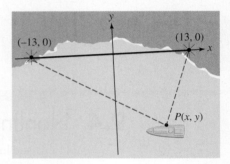

60. Navigation By determining the difference of the distances between the ship in the illustration and two radio transmitters, the LORAN navigation system places the ship on the hyperbola $x^2 - 4y^2 = 576$ in the coordinate system shown. If the ship is 5 miles out to sea, find its coordinates.

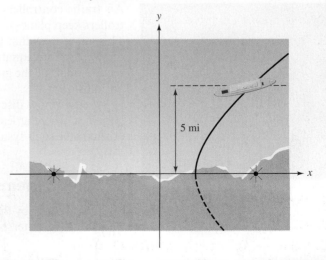

61. Wave propagation Stones dropped into a calm pond at points A and B create ripples that propagate in widening circles. In the illustration, points A and B are 20 feet apart, and the radii of the circles differ by 12 feet. The point $P(x, y)$ where the circles intersect moves along a curve. Identify the curve and find its equation.

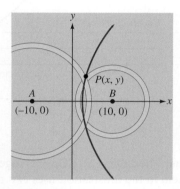

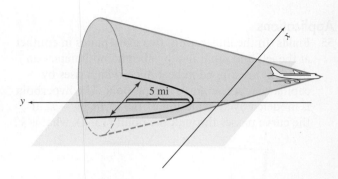

62. Sonic boom The position of a sonic boom caused by the supersonic aircraft is one branch of the hyperbola $y^2 - x^2 = 25$ in the coordinate system shown. How wide is the hyperbola 5 miles from its vertex?

Discovery and Writing

63. Prove that $c > a$ for a hyperbola with center at $(0, 0)$ and line segment FF' on the x-axis.

64. Show that the extended diagonals of the fundamental rectangle of the hyperbola $\dfrac{x^2}{a^2} - \dfrac{y^2}{b^2} = 1$ are $y = \dfrac{b}{a}x$ and $y = -\dfrac{b}{a}x$.

65. Show that the expansion of the standard equation of a hyperbola is a special case of the general equation of second degree with $B = 0$.

66. Write a paragraph describing how you can tell from the equation of a hyperbola whether the transverse axis is vertical or horizontal.

9.4 Nonlinear Systems of Equations

In this section, we will learn to

1. Solve systems by graphing.
2. Solve systems by substitution.
3. Solve systems by addition.
4. Solve problems using systems of nonlinear equations.

Air traffic controllers maintain an orderly flow of air traffic. Since air traffic controllers keep planes a safe distance apart, the paths of two planes will not intersect. This guarantees that there will be no midair collisions. Knowing where the graphs of two or more equations intersect can be very important. For example, the intersection point of the graphs of the flight paths of two planes will be a potential point of collision.

We will now discuss techniques for solving systems of two equations in two variables, where at least one of the equations is nonlinear. We will use three methods to solve such systems: graphing, substitution, and addition.

1. Solve Systems by Graphing

To solve systems by the graphing method, we can graph each equation and identify the point(s) of intersection of the two graphs.

EXAMPLE 1   **Solving a System of Equations by Graphing**

Solve $\begin{cases} x^2 + y^2 = 25 \\ 2x + y = 10 \end{cases}$ by graphing.

SOLUTION   To solve this system, we will graph each equation and identify the point(s) of intersection of the two graphs.

The graph of $x^2 + y^2 = 25$ is a circle with center at the origin and radius of 5. The graph of $2x + y = 10$ is a line with x-intercept at $x = 5$ and y-intercept at $y = 10$.

After graphing the circle and the line, as shown in Figure 9-35, we see that there are two intersection points, $P(3, 4)$ and $P'(5, 0)$. The solutions to the system are $(3, 4)$ and $(5, 0)$.

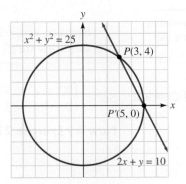

FIGURE 9-35

Comment

The solutions can be verified by substituting into the two equations and checking to see that both are satisfied.

Self Check 1   Solve: $\begin{cases} x^2 + y^2 = 25 \\ 2x - y = 5 \end{cases}$.

Now Try Exercise 9.

A line such as $2x + y = 10$, shown in Example 1, is called a **secant line** because it intersects the graph of the circle at two points. In this case, the system has two solutions. If a line intersects the graph of a circle at one point, the line is called a **tangent line.** In that case, there is one solution. If the line does not intersect the circle, there is no solution.

2. Solve Systems by Substitution

The substitution method can be used to find exact solutions. By making an appropriate substitution, we can convert a nonlinear system of two equations in two variables into one equation in one variable. We can then solve the resulting equation for one variable and back substitute to find the other variable.

EXAMPLE 2   **Solving a System of Equations by Substitution**

Solve $\begin{cases} x^2 + y^2 = 25 \\ 2x + y = 10 \end{cases}$ by substitution.

SOLUTION   We will solve the linear equation for y to obtain an equation called the **substitution equation.** We will then substitute the expression that represents y into the second-degree equation to obtain an equation in the one variable x. After solving the equation for x, we can find y by back substituting the resulting values of x into the substitution equation and simplifying. We begin by solving the linear equation for y.

$$2x + y = 10 \qquad \text{This equation is linear.}$$

$$y = -2x + 10 \qquad \text{This is the substitution equation.}$$

We can now substitute $-2x + 10$ for y in the second-degree equation and solve the resulting quadratic equation for x.

$$x^2 + y^2 = 25 \qquad \text{This equation is nonlinear.}$$

$$x^2 + (-2x + 10)^2 = 25 \qquad \text{Substitute } -2x + 10 \text{ for } y.$$

$$x^2 + 4x^2 - 40x + 100 = 25 \qquad \text{Square the binomial.}$$

$$5x^2 - 40x + 75 = 0 \qquad \text{Subtract 25 from both sides and combine like terms.}$$

$$x^2 - 8x + 15 = 0 \qquad \text{Divide both sides by 5.}$$

$$(x - 5)(x - 3) = 0 \qquad \text{Factor } x^2 - 8x + 15.$$

$$x - 5 = 0 \quad \text{or} \quad x - 3 = 0$$

$$x = 5 \qquad\qquad x = 3$$

Comment

Note that the solutions we obtained by graphing and by substitution are the same.

Because $y = -2x + 10$, if $x = 5$, then $y = 0$; and if $x = 3$, then $y = 4$. The two solutions are $(5, 0)$ and $(3, 4)$.

Self Check 2   Solve $\begin{cases} x^2 + y^2 = 25 \\ 2x - y = 5 \end{cases}$ by substitution.

Now Try Exercise 21.

EXAMPLE 3   **Solving a System of Equations by Substitution**

Solve $\begin{cases} 4x^2 + 9y^2 = 5 \\ y = x^2 \end{cases}$ by substitution.

SOLUTION   In this example, both equations are nonlinear. To solve it, we will substitute y for x^2 in the top equation to obtain one equation in the variable y. Then we will solve the resulting equation for y and use back substitution to find x.

$$4x^2 + 9y^2 = 5$$

$$4y + 9y^2 = 5 \qquad \text{Substitute } y \text{ for } x^2.$$

$$9y^2 + 4y - 5 = 0 \qquad \text{Add } -5 \text{ to both sides.}$$

$$(9y - 5)(y + 1) = 0 \qquad \text{Factor } 9y^2 + 4y - 5.$$

$$9y - 5 = 0 \quad \text{or} \quad y + 1 = 0$$

$$y = \frac{5}{9} \qquad\qquad y = -1$$

Because $y = x^2$, we can find x by solving the equations

$$x^2 = \frac{5}{9} \quad \text{and} \quad x^2 = -1.$$

The solutions of $x^2 = \frac{5}{9}$ are

$$x = \frac{\sqrt{5}}{3} \quad \text{and} \quad x = -\frac{\sqrt{5}}{3}.$$

Since the equation $x^2 = -1$ has no real solutions, the only solutions of the system are

$$\left(\frac{\sqrt{5}}{3}, \frac{5}{9} \right) \quad \text{and} \quad \left(-\frac{\sqrt{5}}{3}, \frac{5}{9} \right).$$

Comment

The substitution method is a good choice when one variable in the system is raised to the first power.

Self Check 3   Solve $\begin{cases} x^2 + 3y^2 = 13 \\ x = y^2 - 1 \end{cases}$ by substitution.

Now Try Exercise 23.

3. Solve Systems by Addition

When we have two second-degree equations of the form $ax^2 + by^2 = c$, we can solve the system by using the addition method and eliminating one of the variables.

EXAMPLE 4   **Solving a System of Equations by Addition**

Solve $\begin{cases} 3x^2 + 2y^2 = 36 \\ 4x^2 - y^2 = 4 \end{cases}$ by addition.

SOLUTION   In this example, we have two second-degree equations of the form $ax^2 + by^2 = c$. In such cases, we can solve the system by eliminating one of the variables by addition, solving the resulting equation, and using back substitution to solve for the second variable.

To eliminate the terms involving y^2, we copy the first equation and multiply the second equation by 2 to obtain the following equivalent system:

$$\begin{cases} 3x^2 + 2y^2 = 36 \\ 8x^2 - 2y^2 = 8 \end{cases}.$$

We can then add the equations and solve the resulting equation for x.

$$11x^2 = 44$$

$$x^2 = 4$$

$$x = 2 \quad \text{or} \quad x = -2$$

To find y, we substitute 2 for x and then -2 for x in the first equation.

$3x^2 + 2y^2 = 36$	$3x^2 + 2y^2 = 36$
$3(2)^2 + 2y^2 = 36$	$3(-2)^2 + 2y^2 = 36$
$12 + 2y^2 = 36$	$12 + 2y^2 = 36$
$2y^2 = 24$	$2y^2 = 24$
$y^2 = 12$	$y^2 = 12$
$y = \sqrt{12} \quad \text{or} \quad y = -\sqrt{12}$	$y = \sqrt{12} \quad \text{or} \quad y = -\sqrt{12}$
$y = 2\sqrt{3} \quad\mid\quad y = -2\sqrt{3}$	$y = 2\sqrt{3} \quad\mid\quad y = -2\sqrt{3}$

The four solutions of this system are

$$\left(2, 2\sqrt{3}\right), \quad \left(2, -2\sqrt{3}\right), \quad \left(-2, 2\sqrt{3}\right), \quad \text{and} \quad \left(-2, -2\sqrt{3}\right).$$

Self Check 4   Solve $\begin{cases} 2x^2 + y^2 = 23 \\ 3x^2 - 2y^2 = 17 \end{cases}$ by addition.

Now Try Exercise 27.

4. Solve Problems Using Systems of Nonlinear Equations

EXAMPLE 5   Solving an Application Problem

The area of a tennis court for singles matches is 2,106 square feet, and the perimeter is 210 feet. Find the dimensions of the court.

Arthur Ashe Tennis Stadium

SOLUTION   We will set up a system of two equations that models the problem and then use the substitution method to solve it.

FIGURE 9-36

Step 1: Write a system of two equations that models the problem. We let x represent the length of the tennis court and y represent its width, as shown in Figure 9-36. Because the area of the tennis court (2,106 square feet) is the product of its length and width, we can form the equation $xy = 2,106$. Because the perimeter of the court (210 feet) is the sum of twice its length and twice its width, we can write the equation $2x + 2y = 210$, or $x + y = 105$. This gives the following system of equations that models the problem:

$$\begin{cases} xy = 2,106 & \text{This equation is nonlinear.} \\ x + y = 105 & \text{This equation is linear.} \end{cases}$$

Step 2: Use substitution to solve the system. First, we solve the linear equation for y.

$$x + y = 105$$

$$y = -x + 105 \qquad \text{This is the substitution equation.}$$

We can now substitute $-x + 105$ for y in the nonlinear equation and solve the resulting equation for x.

$$xy = 2,106$$
$$x(-x + 105) = 2,106 \qquad \text{Substitute } -x + 105 \text{ for } y.$$
$$-x^2 + 105x = 2,106 \qquad \text{Remove parentheses.}$$
$$-x^2 + 105x - 2,106 = 0 \qquad \text{Subtract 2,106 from both sides.}$$
$$x^2 - 105x + 2,106 = 0 \qquad \text{Multiply both sides by } -1.$$
$$(x - 78)(x - 27) = 0 \qquad \text{Factor } x^2 - 105x + 2,106.$$
$$x - 78 = 0 \quad \text{or} \quad x - 27 = 0$$
$$x = 78 \qquad \qquad x = 27$$

Because $y = -x + 105$, if $x = 78$, then $y = 27$; and if $x = 27$, then $y = 78$. The two solutions are (78, 27) and (27, 78).

The dimensions of the tennis court are 78 feet by 27 feet.   ∎

Self Check 5   If the area of a tennis court for doubles matches is 2,808 square feet and the perimeter is 228 feet, find the dimensions of the court.

Now Try Exercise 45.

Self Check Answers   **1.** (4, 3) and (0, −5)   **2.** (4, 3) and (0, −5)   **3.** $\left(2, \sqrt{3}\right)$ and $\left(2, -\sqrt{3}\right)$

4. $\left(3, \sqrt{5}\right), \left(3, -\sqrt{5}\right), \left(-3, \sqrt{5}\right)$, and $\left(-3, -\sqrt{5}\right)$

5. 78 ft by 36 ft

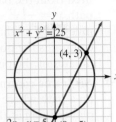

Exercises **9.4**

Getting Ready
You should be able to complete these vocabulary and concept statements before you proceed to the practice exercises.

Fill in the blanks.

1. Solutions of nonlinear systems of equations are the points of intersection of the _____ of conic sections.

2. Approximate solutions of nonlinear systems can be found _____, and exact solutions can be found algebraically using the methods of _____ and _____.

Practice
Solve each system of equations by graphing.

3. $\begin{cases} 8x^2 + 32y^2 = 256 \\ x = 2y \end{cases}$

4. $\begin{cases} x^2 + y^2 = 2 \\ x + y = 2 \end{cases}$

5. $\begin{cases} x^2 + y^2 = 90 \\ y = x^2 \end{cases}$

6. $\begin{cases} x^2 + y^2 = 5 \\ x + y = 3 \end{cases}$

7. $\begin{cases} x^2 + y^2 = 25 \\ 12x^2 + 64y^2 = 768 \end{cases}$

8. $\begin{cases} x^2 + y^2 = 13 \\ y = x^2 - 1 \end{cases}$

9. $\begin{cases} x^2 - 13 = -y^2 \\ y = 2x - 4 \end{cases}$

10. $\begin{cases} x^2 + y^2 = 20 \\ y = x^2 \end{cases}$

11. $\begin{cases} x^2 - 6x - y = -5 \\ x^2 - 6x + y = -5 \end{cases}$

12. $\begin{cases} x^2 - y^2 = -5 \\ 3x^2 + 2y^2 = 30 \end{cases}$

Use a graphing calculator to solve each system of equations.

13. $\begin{cases} y = x + 1 \\ y = x^2 + x \end{cases}$

14. $\begin{cases} y = 6 - x^2 \\ y = x^2 - x \end{cases}$

15. $\begin{cases} 6x^2 + 9y^2 = 10 \\ 3y - 2x = 0 \end{cases}$

16. $\begin{cases} x^2 + y^2 = 68 \\ y^2 - 3x^2 = 4 \end{cases}$

Solve each system of equations for real values of x and y using substitution or addition.

17. $\begin{cases} 25x^2 + 9y^2 = 225 \\ 5x + 3y = 15 \end{cases}$

18. $\begin{cases} x^2 + y^2 = 20 \\ y = x^2 \end{cases}$

19. $\begin{cases} x^2 + y^2 = 2 \\ x + y = 2 \end{cases}$

20. $\begin{cases} x^2 + y^2 = 36 \\ 49x^2 + 36y^2 = 1,764 \end{cases}$

21. $\begin{cases} x^2 + y^2 = 5 \\ x + y = 3 \end{cases}$

22. $\begin{cases} x^2 - x - y = 2 \\ 4x - 3y = 0 \end{cases}$

23. $\begin{cases} x^2 + y^2 = 13 \\ y = x^2 - 1 \end{cases}$

24. $\begin{cases} x^2 + y^2 = 25 \\ 2x^2 - 3y^2 = 5 \end{cases}$

25. $\begin{cases} x^2 + y^2 = 30 \\ y = x^2 \end{cases}$

26. $\begin{cases} 9x^2 - 7y^2 = 81 \\ x^2 + y^2 = 9 \end{cases}$

27. $\begin{cases} 2x^2 + y^2 = 6 \\ x^2 - y^2 = 3 \end{cases}$

28. $\begin{cases} x^2 + y^2 = 13 \\ x^2 - y^2 = 5 \end{cases}$

29. $\begin{cases} x^2 + y^2 = 20 \\ x^2 - y^2 = -12 \end{cases}$

30. $\begin{cases} xy = -\dfrac{9}{2} \\ 3x + 2y = 6 \end{cases}$

31. $\begin{cases} y^2 = 40 - x^2 \\ y = x^2 - 10 \end{cases}$

32. $\begin{cases} x^2 - 6x - y = -5 \\ x^2 - 6x + y = -5 \end{cases}$

33. $\begin{cases} y = x^2 - 4 \\ x^2 - y^2 = -16 \end{cases}$

34. $\begin{cases} 6x^2 + 8y^2 = 182 \\ 8x^2 - 3y^2 = 24 \end{cases}$

35. $\begin{cases} x^2 - y^2 = -5 \\ 3x^2 + 2y^2 = 30 \end{cases}$

36. $\begin{cases} \dfrac{1}{x} + \dfrac{1}{y} = 5 \\ \dfrac{1}{x} - \dfrac{1}{y} = -3 \end{cases}$

37. $\begin{cases} \dfrac{1}{x} + \dfrac{2}{y} = 1 \\ \dfrac{2}{x} - \dfrac{1}{y} = \dfrac{1}{3} \end{cases}$

38. $\begin{cases} \dfrac{1}{x} + \dfrac{3}{y} = 4 \\ \dfrac{2}{x} - \dfrac{1}{y} = 7 \end{cases}$

39. $\begin{cases} 3y^2 = xy \\ 2x^2 + xy - 84 = 0 \end{cases}$

40. $\begin{cases} x^2 + y^2 = 10 \\ 2x^2 - 3y^2 = 5 \end{cases}$

41. $\begin{cases} xy = \dfrac{1}{6} \\ y + x = 5xy \end{cases}$

42. $\begin{cases} xy = \dfrac{1}{12} \\ y + x = 7xy \end{cases}$

Applications

43. **Geometry** The area of a rectangle is 63 square centimeters and its perimeter is 32 centimeters. Find the dimensions of the rectangle.

44. **Dimensions of a whiteboard** The area of a Smart Board interactive whiteboard is 2,880 square inches and its perimeter is 216 inches. Find the dimensions of the whiteboard.

45. **Fencing pastures** The rectangular pasture shown is to be fenced in along a riverbank. If 260 feet of fencing is to enclose an area of 8,000 square feet, find the dimensions of the pasture.

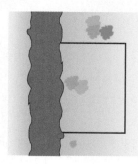

46. **Investments** Grant receives $225 annual income from one investment. Jeff invested $500 more than Grant, but at an annual rate of 1% less. Jeff's annual income is $240. Find the amount and rate of Grant's investment.

47. **Investments** Carol receives $67.50 annual income from one investment. John invested $150 more than Carol at an annual rate of $1\frac{1}{2}$% more. John's annual income is $94.50. Find the amount and rate of Carol's investment. (*Hint:* There are two answers.)

48. **Finding rate and time** Jim drove 306 miles. Jim's brother made the same trip at a speed 17 miles per hour slower than Jim did and required an extra $1\frac{1}{2}$ hours. Find Jim's rate and time.

49. **Paint ball** In the illustration, a liquid-filled paint ball is shot from the base of an incline and follows the parabolic path $y = -\frac{1}{300}x^2 + \frac{1}{5}x$, with distances measured in feet. The incline has a slope of $\frac{1}{10}$. Find the coordinates of the point of intersection of the paint ball and the ground at impact.

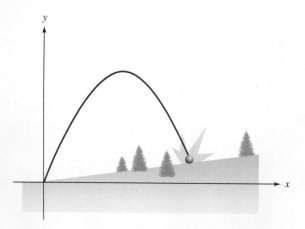

50. **Artillery** See the illustration for Exercise 49. A shell fired from the base of a hill follows the parabolic path $y = -\frac{1}{6}x^2 + 2x$, with distances measured in miles. The hill has a slope of $\frac{1}{3}$. How far from the cannon is the point of impact? (*Hint:* Find the coordinates of the point and then the distance.)

51. **Air traffic control** A plane is flying over an airport on a path whose equation is $y = x^2$. If a second plane, flying at the same altitude, is traveling on a path whose equation is $x + y = 2$, is there any danger of a midair collision?

52. **Ship traffic** One ship is steaming on a path whose equation is $y = x^2 + 1$ and another is steaming on a path whose equation is $x + y = -4$. Is there any danger of collision?

53. **Radio reception** A radio station located 120 miles due east of Collinsville has a listening radius of 100 miles. A straight road joins Collinsville with Harmony, a town 200 miles to the east and 100 miles north. See the illustration. If a driver leaves Collinsville and heads toward Harmony, how far from Collinsville will the driver pick up the station?

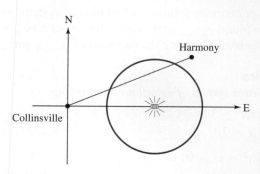

54. **Listening ranges** For how many miles will a driver in Exercise 53 continue to receive the signal?

Discovery and Writing

55. Is it possible for a system of second-degree equations to have no common solution? If so, sketch the graphs of the equations of such a system.

56. Can a system have exactly one solution? If so, sketch the graphs of the equations of such a system.

57. Can a system have exactly two solutions? If so, sketch possible graphs.

58. Can a system have three solutions? If so, sketch the graphs.

59. Can it have exactly four solutions? If so, sketch the graphs.

60. Can a system have more than four solutions? If so, sketch the graphs.

9.5 Plane Curves and Parametric Equations

In this section, we will learn to

1. Graph plane curves defined by parametric equations.
2. Express a curve defined parametrically as a rectangular equation.
3. Find parametric equations for a curve.
4. Explore applications using parametric equations.

Ski jumping is a sport in which skiers go down a takeoff ramp, jump, and attempt to land as far as possible down the hill below. In addition to the length of the jump, judges give points for style. Ski jumping has been part of the Olympic Winter Games since the first games, in Chamonix in 1924.

To describe the motion of an object, such as a ski jumper, mathematicians often use *parametric equations*. These equations are defined using a third variable that we call a **parameter.** By introducing a parameter of, say time, we are able to specify the location of a body in motion and know the time it took to get there.

Up until this point, we have described curves in the plane using rectangular equations, such as $y = x^2 + 1$, or polar equations, such as $r = 3 + 4 \sin \theta$. In both the rectangular and polar coordinate systems, each point on the graph of the curve satisfies the equation. In this section, we will learn more about curves in the plane and study parametric equations.

Look at the curve in Figure 9-37. If a particle moves along this curve, it is impossible to describe the path by an equation in the form $y = f(x)$ because the curve is not that of a function. If we introduce a third variable, $t =$ time, we can define the x-coordinate as $x = f(t)$ and the y-coordinate as $y = g(t)$. This pair of equations is often used as a convenient way to describe a curve.

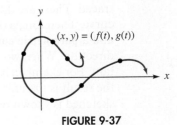

FIGURE 9-37

1. Graph Plane Curves Defined by Parametric Equations

Thinking of the x- and y-coordinates as functions of time leads us to the following definition.

Plane Curves and Parametric Equations	If f and g are continuous functions of t defined on an interval I, the set of ordered pairs $(f(t), g(t))$ is a **plane curve.** The equations $$x = f(t) \qquad \text{and} \qquad y = g(t)$$ are **parametric equations** for the curve, with **parameter** t, $t \in I$.

To sketch a plane curve defined by a pair of parametric equations, we can find several points and then plot them.

EXAMPLE 1   Graphing a Plane Curve

Sketch the curve defined by the parametric equations $x = t^2 - t$ and $y = t + 2$.

SOLUTION   Selecting a value for t and then evaluating the x- and y-coordinates yields a point on the curve. We can find several of these values and plot the points.

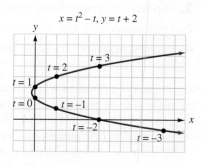

t	x	y
-3	12	-1
-2	6	0
-1	2	1
0	0	2
1	0	3
2	2	4
3	6	5

FIGURE 9-38

If a particle moves along this curve, as the value of t increases in either a positive or negative direction, the particle moves in the direction of the arrows.

Self Check 1   Sketch the curve defined by the parametric equations $x = 2t - 3$ and $y = t - t^2$.

Now Try Exercise 9.

In Example 1, we did not place any restrictions on the value of t, and we assumed that t could be any value. However, sometimes the value of t is restricted to a finite interval. If a specific interval is given, the curve will have an **initial point** and a **terminal point.** We use arrowheads to indicate the direction in which the curve is traced. There are different sets of parametric equations that can represent the same curve. Therefore, a distinction is made between a **curve,** which is just a set of points, and a **parametric curve,** in which the points are traced along a particular path and direction. When working with parametric curves, the *direction* of the path is important. For example, if you graph the equations $x = \sin t$ and $y = \cos t, 0 \le t \le 2\pi$, the result is the unit circle shown in Figure 9-39. The direction in which the curve is sketched is shown by arrows.

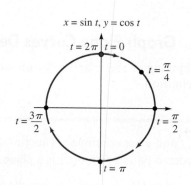

t	x	y
0	0	1
$\dfrac{\pi}{4}$	$\dfrac{\sqrt{2}}{2}$	$\dfrac{\sqrt{2}}{2}$
$\dfrac{\pi}{2}$	1	0
π	0	-1
$\dfrac{3\pi}{2}$	-1	0
2π	0	1

FIGURE 9-39

Likewise, if you graph the equations $x = \cos t$ and $y = \sin t, 0 \le t \le 2\pi$, the graph is also a unit circle, but the curve is traced in a different direction, as shown by the arrows in Figure 9-40.

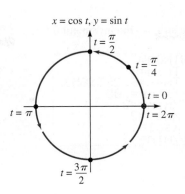

$x = \cos t, \; y = \sin t$

$t = \dfrac{\pi}{2}$

$t = \dfrac{\pi}{4}$

$t = 0$

$t = \pi$

$t = 2\pi$

$t = \dfrac{3\pi}{2}$

t	x	y
0	1	0
$\dfrac{\pi}{4}$	$\dfrac{\sqrt{2}}{2}$	$\dfrac{\sqrt{2}}{2}$
$\dfrac{\pi}{2}$	0	1
π	-1	0
$\dfrac{3\pi}{2}$	0	-1
2π	1	0

FIGURE 9-40

 In calculus, parametric equations are used to analyze the motion of a particle along the defined curve. We will examine this in the Looking Ahead to Calculus section at the end of the chapter by finding the position, velocity, acceleration, and speed of the particle.

ACCENT ON TECHNOLOGY | Graphing Parametric Equations

Graphing calculators are particularly useful when you are trying to sketch a complicated curve, one that would be virtually impossible to sketch by hand. To graph parametric equations on a graphing calculator, we will have to change the mode, enter the functions, and set a proper window. Setting a proper window is very important, as we shall see in the example. Different increments for the variable T can produce dramatically different results. Therefore, it is essential that the proper increments be selected so that the correct curve is shown.

We will graph the equations $x = \sin 3t$ and $y = 2 \cos t$.

Set the calculator to parametric mode. Since the equations we are graphing involve trigonometric equations, we will use radian mode. We could use degree mode, but the window entries would have to reflect that.

```
NORMAL   SCI  ENG
FLOAT    0123456789
RADIAN   DEGREE
FUNC  PAR  POL  SEQ
CONNECTED   DOT
SEQUENTIAL  SIMUL
REAL  a+bi  re^θi
FULL  HORIZ  G-T
SET CLOCK 10/11/24 18:53
```

Access the graph editor and enter the equations. The variable key will automatically enter the T.

```
Plot1 Plot2 Plot3
\X₁ᴛ■sin(3T)
 Y₁ᴛ■2cos(T)■
\X₂ᴛ=
 Y₂ᴛ=
\X₃ᴛ=
 Y₃ᴛ=
\X₄ᴛ=
```

Set the window. We are going to show how the results can vary greatly with changes in the increment for T. For the first graph, we will use an increment of 0.5.

```
WINDOW
 Tmin=0
 Tmax=2π■
 Tstep=.5
 Xmin=-2
 Xmax=2
 Xscl=1
↓Ymin=-3
```

The calculator evaluates the equations at $T = 0, 0.5, 1, 1.5, \ldots, 6$. This is not a good representation of the curve.

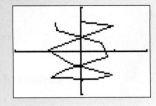

Now let's use the increment 0.05.

This is a true representation of the curve.

```
WINDOW
  Tmin=0
  Tmax=6.2831853…
  Tstep=.05
  Xmin=-2
  Xmax=2
  Xscl=1
↓Ymin=-3
```

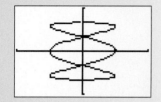

2. Express a Curve Defined Parametrically as a Rectangular Equation

Sometimes a curve given by parametric equations can also be represented by a rectangular equation in x and y. The process of converting the equations to a rectangular equation is called *eliminating the parameter*. When converting the equations, some restriction on the domain of the rectangular equation may be necessary.

Eliminating the Parameter

1. Select the equation that is the simpler to solve for t and then solve for t.
2. Substitute the expression into the other equation.
3. Simplify the equation which is now written in terms of x and y.

We use other strategies and substitutions such as a Pythagorean Identity.

EXAMPLE 2   Eliminating the Parameter

Eliminate the parameter in the parametric equations from Example 1, $x = t^2 - t$ and $y = t + 2$, and write as a rectangular equation.

SOLUTION   Since $y = t + 2$ is the simpler of the two equations, we will begin with this equation and solve for t.

$$y = t + 2$$

$$t = y - 2$$

Now substitute $t = y - 2$ into the equation $x = t^2 - t$ and simplify.

$$x = t^2 - t$$

$$x = (y - 2)^2 - (y - 2)$$

$$x = y^2 - 4y + 4 - y + 2$$

$$x = y^2 - 5y + 6$$

From Section 9.1, we know that the graph of the equation $x = y^2 - 5y + 6$ is a parabola, as shown in Example 1, Figure 9-38. ∎

Self Check 2   Eliminate the parameter in the parametric equations from Self Check 1, $x = 2t - 3$ and $y = t - t^2$, and write as a rectangular equation.

Now Try Exercise 21.

Comment   A curve defined parametrically might be only a portion of the graph of the rectangular equation obtained by eliminating the parameter.

EXAMPLE 3   **Eliminating the Parameter and Graphing Parametric Equations**

Sketch the curve defined by the parametric equations $x = t^2$ and $y = -2t^2 + 1$.

SOLUTION   We eliminate the parameter t by using substitution, as shown.

$$x = t^2$$
$$y = -2t^2 + 1 = -2x + 1$$

The graph of the equation $y = -2x + 1$ is a line. The graph of the parametric equations, however, is not the entire line. Since t^2 is nonnegative and $x = t^2$, it follows that $x \geq 0$. Thus, the graph of the given parametric equations is only part of the line $y = -2x + 1$ as shown in Figure 9-41.

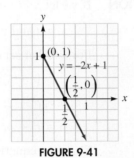

FIGURE 9-41

Self Check 3   Sketch the curve formed by the parametric equations $x = t^2$ and $y = t^2 - 3$.

Now Try Exercise 11.

EXAMPLE 4   **Eliminating the Parameter**

Eliminate the parameter from the parametric equations $x = 3 \cos t$ and $y = 3 \sin t$, for $0 \leq t \leq 2\pi$, and write as a rectangular equation.

SOLUTION   To eliminate the parameter from these equations, we will use the Pythagorean Identity $\sin^2 t + \cos^2 t = 1$. Since $x = 3 \cos t$, we know that $\cos t = \frac{x}{3}$. Likewise, $y = 3 \sin t$ gives us $\sin t = \frac{y}{3}$.

$$\sin^2 t + \cos^2 t = 1$$
$$\left(\frac{y}{3}\right)^2 + \left(\frac{x}{3}\right)^2 = 1$$
$$\frac{y^2}{9} + \frac{x^2}{9} = 1$$
$$x^2 + y^2 = 9$$

We recognize the equation $x^2 + y^2 = 9$ as that of the circle with center at $(0, 0)$ and radius 3.

Self Check 4   Eliminate the parameter from the parametric equations $x = -3 \sin t$ and $y = -3 \cos t$, for $0 \leq t \leq 2\pi$, and write as a rectangular equation.

Now Try Exercise 31.

Both sets of parametric equations in Example 4 and Self Check 4 are equations for the curve that is a circle with the center at $(0, 0)$ and radius 3. If we were to graph the parametric equations in Example 4, the curve would begin and end at the point $(3, 0)$. However, the curve traced by the equations in Self Check 4 would start and end at the point $(0, -3)$. These examples illustrate the versatility of parametric equations that makes them very useful in many applications.

3. Find Parametric Equations for a Curve

We will now look at how to find parametric equations that describes a curve given by a rectangular equation. If y is defined as a function of x, then we can simply set $x = t$ and substitute to obtain the equation for y. If $y = f(x)$ and $x = t$, then we have the equation $y = f(t)$, and t must be in the domain of f. We are not restricted to the substitution $x = t$.

EXAMPLE 5   Finding Parametric Equations for a Curve

Find two sets of parametric equations for the rectangular equation $y = x^2 + 2x - 3$.

SOLUTION   If we let $x = t$, then the equations are

$$x = t \text{ and } y = t^2 + 2t - 3, \, t \in \mathfrak{R}.$$

For another set of equations, we let $x = t + 1$.

$$y = (t + 1)^2 + 2(t + 1) - 3$$
$$y = t^2 + 2t + 1 + 2t + 2 - 3$$
$$y = t^2 + 4t$$

Thus, the equations are

$$x = t + 1 \text{ and } y = t^2 + 4t, \, t \in \mathfrak{R}.$$

Self Check 5   Find two sets of parametric equations for the rectangular equation $y = 1 - x^2$.

Now Try Exercise 39.

Caution

> We can define x as many functions of t, but we must be careful to make certain that the substitution for x yields a function that allows x to be in the domain of f. For example, suppose in Example 5 we had used the substitution $x = t^2$. The resulting equation $y = t^4 + 2t^2 - 3$ does not result in an equivalent parametric equation.

A fixed point on a circle that rolls along a straight line traces out a curve called a **cycloid** (Figure 9-42).

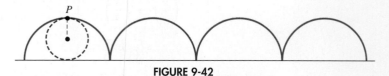

FIGURE 9-42

Finding the rectangular equation of a cycloid would be extremely complicated. However, we can find parametric equations for the cycloid curve.

EXAMPLE 6   Finding Equations of a Cycloid

A point P is on a circle of radius r that traces out a cycloid. Find a parametric description of the cycloid.

SOLUTION   Instead of using t as the parameter, we will choose the angle of rotation θ of the circle as the parameter. ($\theta = 0$ when P is at the origin.) We will assume that a circle with radius r has rotated through θ radians. Since the circle has remained in contact with the line, we can see in Figure 9-43 that the distance it has rolled from the origin is the length of the arc PE. We know from Section 4.1 that the length of the arc is $r\theta$. Also, the length of the segment OE is equal to the length of the arc PE. Thus,

$$|OE| = \text{arc } PE = r\theta.$$

Therefore, the center of the circle is at $C(r\theta, r)$. We let the coordinates of P be (x, y), and we see from Figure 9-43 that

$$x = |OE| - |PD| = r\theta - r\sin\theta = r(\theta - \sin\theta) \quad \text{Since } \sin\theta = \frac{|PD|}{r}, |PD| = r\sin\theta.$$

and

$$y = |CE| - |CD| = r - r\cos\theta = r(1 - \cos\theta). \quad \text{Since } \cos\theta = \frac{|CD|}{r}, |CD| = r\cos\theta.$$

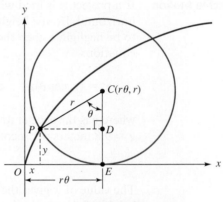

FIGURE 9-43

Self Check 6   Choose a point P on a circle of radius 3 and find a parametric description of the curve that is traced out by P as the circle rolls along the x-axis.

Now Try Exercise 57.

Parametric Equations of a Cycloid

The **parametric equations** of a cycloid are

$$x = r(\theta - \sin\theta) \text{ and } y = r(1 - \cos\theta), \theta \in \mathfrak{R}.$$

Cycloids are associated with many applications, but two of them are quite famous and interesting. These two applications involve one arch of an inverted cycloid.

The **brachistochrone curve**, or curve of fastest descent, is the curve between two points that is covered in the least amount of time by a body that starts at the first point with zero speed and is constrained to move along the curve to the second point, under the action of constant gravity and assuming no friction. The brachistochrone is the cycloid.

FIGURE 9-44   Start the two balls at point A at the same time. The one rolling along the cycloid path travels farther but reaches point B first.

The **tautochrone problem** was solved by the Dutch physicist Huygens, who proved that no matter where a particle is placed on an inverted cycloid, it takes the same amount of time to slide to the bottom. Using this fact, he invented the pendulum clock. Pendulums swing in a cycloidal arc, because they take the same amount of time to make a complete oscillation regardless of the size of the arc through which they swing.

4. Explore Applications Using Parametric Equations

If an object is launched upward, the resulting motion is known as **projectile motion**. Parametric equations are particularly useful when working with problems involving projectile motion. Calculus can be used to verify the following equations:

Projectile Motion   If a projectile is fired with an initial velocity of v_0 at an angle of θ above the horizontal, from a height h above the horizontal, and air resistance is assumed to be negligible, then the projectile's position after t seconds is given by the equations

$$x = (v_0 \cos \theta)t \quad \text{and} \quad y = -\frac{1}{2}gt^2 + (v_0 \sin \theta)t + h,$$

where g is the acceleration due to gravity ($g \approx 32$ feet per second per second or $g \approx 9.8$ meters per second per second).

The value of x gives the horizontal position at time t, and the value of y is the height of the projectile at time t.

EXAMPLE 7   **Solving a Projectile Motion Problem**

A professional golfer hits a wedge. When the club contacts the ball, the angle is 30° and the velocity of the ball is 120 feet per second.

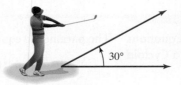

FIGURE 9-45

   a. What are the parametric equations that describe the ball's position as a function of time?
   b. How long does the ball take to hit the ground?
   c. What is the maximum height of the ball, and when does it reach that height?
   d. How far does the ball travel?

SOLUTION   **a.** The velocity $v_0 = 120$ feet per second, the acceleration due to gravity is $g = 32$ feet per second per second, $\theta = 30°$, and $h = 0$ (since the ball is on the ground). Substituting into the parametric equations, we have

$$x = (v_0 \cos \theta)t = (120 \cos 30°)t = 60t\sqrt{3}$$

and

$$y = -\frac{1}{2}gt^2 + (v_0 \sin \theta)t + h = -\frac{1}{2}(32)t^2 + (120 \sin 30°)t + 0 = -16t^2 + 60t.$$

   b. The ball will hit the ground when $y = 0$.

$$-16t^2 + 60t = 0$$

$$-4t(4t - 15) = 0$$

$$t = 0 \quad \text{or} \quad 4t - 15 = 0$$

$$t = 3.75 \text{ seconds}$$

The ball will hit the ground at 3.75 seconds.

c. The value of y indicates the height of the ball. Notice that the equation of y is the equation of a parabola that opens downward. Therefore, the coordinates of the vertex will give us the maximum height and the time when that height is achieved. For the equation $y = -16t^2 + 60t$, the value of t at the vertex is

$$t = \frac{-b}{2a} = \frac{-60}{2(-16)} = 1.875 \text{ seconds.}$$

When $t = 1.875$, we have the maximum height of
$y = -16(1.875)^2 + 60(1.875) = 56.25$ feet.

d. Since the ball is in the air for 3.75 seconds, the horizontal distance traveled is

$$x = 60\sqrt{3}(3.75) \approx 389.7 \text{ feet.}$$ ∎

Self Check 7   Suppose the golfer hits a 15° hybrid. When the club contacts the ball, the velocity of the ball is 220 feet per second. Round to the nearest tenth.

a. What are the parametric equations that describe the ball's position as a function of time?
b. How long does the ball take to hit the ground?
c. What is the maximum height of the ball, and when does it reach that height?
d. How far does the ball travel?

Now Try Exercise 49.

EXAMPLE 8   **Simulating Contextual Horizontal Motion**

Two groups of students drive from Oxford, Mississippi, to Baton Rouge, Louisiana, to attend an Ole Miss–LSU football game. The first group leaves Oxford at 10:00 a.m. and travels at a constant velocity of 45 miles per hour. The second group leaves at 11:30 a.m. and travels at a constant rate of 70 miles per hour. At what time and distance from Oxford will the second group catch the first group?

SOLUTION   We will use a set of parametric equations to describe each group's motion. We let time $t = 0$ be when Group 1 leaves Oxford. We arbitrarily set $y_1 = 1$ as Group 1's path and $y_2 = 2$ as Group 2's path. The horizontal distances traveled at time t are

Group 1: $x_1 = 45t$ and   Group 2: $x_2 = 70(t - 1.5)$.

The groups will meet when $x_1 = x_2$.

$$45t = 70(t - 1.5)$$
$$45t = 70t - 105$$
$$-25t = -105$$
$$t = 4.2$$

Group 2 will catch Group 1 in 4.2 hours—at 2:12 p.m., the groups will meet 189 miles from Oxford. ∎

Self Check 8   A motorist leaves Houston, Texas, at 4:00 a.m. and drives toward El Paso, Texas, at a uniform velocity of 65 miles per hour. At 6:00 a.m., his wife leaves El Paso and drives toward Houston at a uniform velocity of 75 miles per hour. Where and at what time will they meet? The distance between Houston and El Paso is 746 miles.

Now Try Exercise 53.

The development of computers has added to the many applications of parametric equations. Most are beyond the scope of this text, such as the use of parametric curves in computer-aided design (CAD). One such curve is called the Bézier curve, shown below.

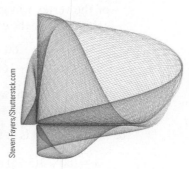

Bézier curves were first used by Pierre Bézier in the late 1960s while working on the design of cars as an engineer at the Renault automobile plant. Now these curves are used in the fields of technical and creative design, computer graphics, animation, and even to model curves in economics.

Self Check Answers   **1.**

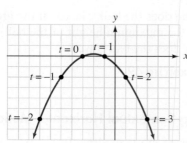

2. $y = -\dfrac{1}{4}(x^2 + 4x + 3)$

3.

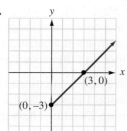

4. $x^2 + y^2 = 9$
5. $x = t$ and $y = 1 - t^2$; $x = 1 - t$, $y = 2t - t^2$
6. $x = 3(\theta - \sin\theta)$ and $y = 3(\theta - \cos\theta)$
7. **a.** $x = 212.5t$ and $y = -16t^2 + 56.9t$
b. 3.55625 sec   **c.** 50.5877 ft at 1.778125 sec
d. 755.7 ft
8. 276.7 mi from Houston, after 4.25 hours, or at 8:15 a.m.

Exercises **9.5**

Getting Ready

You should be able to complete these vocabulary and concept statements before you proceed to the practice exercises.

Fill in the blanks.

1. If f and g are continuous functions of t defined on an interval I, the set of ordered pairs $(f(t), g(t))$ is a _____. The equations $x = f(t)$ and $y = g(t)$ are _____ equations for the curve, with parameter _____.

2. A fixed point on a circle that rolls along a straight line traces out a curve called a _____.

3. The parametric equations used in analyzing projectile motion are $x =$ _____ and $y =$ _____.

4. In order to rewrite parametric equations as a rectangular equation, you must _____ the parameter.

Practice

Sketch the curve determined by the given parametric equations.

5. $x = 2t$; $y = t - 5$

6. $x = 5t$; $y = 2 - t$

7. $x = 2(t - 3)$; $y = 3(t - 2)$

8. $x = 3(t - 3); y = \dfrac{1}{2}(t - 2)$

9. $x = t^2; y = t - 2$ 10. $x = 3t + 1; y = t^2$

The graph of the given parametric equations is only a portion of the graph of the rectangular equation obtained by eliminating the parameter. Sketch the graph of the parametric equations.

11. $x = t^2; y = 2t^2 + 1$ 12. $x = 3t^2; y = 5t^2$

13. $x = 2t^2 - 1; y = 1 - t^2$

14. $x = 2(t^2 + 1); y = 1 - t^2$

15. $x = \sqrt{t}; y = t$ 16. $x = \sqrt{t - 1}; y = \sqrt{t}$

17. $x = 2 \sin t; y = 3 \sin t$ 18. $x = \sin t; y = \cos^2 t$

For each pair of parametric equations, eliminate the parameter and write a rectangular equation for the curve. Adjust the domain of the resulting rectangular equation if necessary. Then sketch the curve.

19. $x = 2t; y = t + 6$ 20. $x = 3t; y = 1 - 4t$

21. $x = t^2; y = t + 3$ 22. $x = t + 2; y = t^2$

23. $x = \sqrt{t}; y = t - 3; t \geq 0$

24. $x = 4t; y = \sqrt{t}; t \geq 0$

25. $x = 2t; y = |t - 2|$ 26. $x = |t - 1|; y = t + 2$

27. $x = 3e^t; y = e^t + 1; 0 \leq t \leq 2$

28. $x = e^t; y = e^{2t}; 0 \leq t \leq 2$

29. $x = \ln 2t; y = 2t^2; 0 < t \leq 2$

30. $x = t^3; y = 3 \ln t; t > 0$

31. $x = 3 \cos t; y = 2 \sin t; 0 \leq t \leq 2\pi$

32. $x = 2 \sin t; y = 4 \cos t; 0 \leq t \leq 2\pi$

33. $x = 3 \cos t; y = 5 \sin t; 0 \leq t \leq 2\pi$

34. $x = \tan t; y = \cot t; 0 < t < \dfrac{\pi}{2}$

35. $x = \cos t; y = \sec t; 0 \leq t < \dfrac{\pi}{2}$

36. $x = \sin t; y = \csc t; 0 < t \leq \dfrac{\pi}{2}$

 37. $x = t^3 - 3t^2 + 2t; y = t - 1; -1 \leq t \leq 3$

38. $x = 2t - 4; y = 8t^3 - 4t^2 + 3; -1 \leq t \leq 1$

Find two sets of parametric equations for each rectangular equation. Adjust the domains of the resulting equations if necessary. Answers can vary.

39. $y = 3x + 7$ 40. $y = 4 - 5x$

41. $x = \dfrac{1}{y}$ 42. $x = \dfrac{1}{y - 1}$

43. $y = 2x^2 - 3$ 44. $y = 1 - 4x^2$

45. $x = y^{5/2}$ 46. $x = \sqrt{y - 1}$

Applications

47. **Projectile motion** Joshua punts a football with an initial velocity of 90 feet per second, from a height of 3 feet, at an angle of 60°.

a. What are the parametric equations that describe the ball's position as a function of time?

b. How long is the hang time for the punt?

c. What is the maximum height of the ball, and when does it reach that height?

d. How far does the ball travel?

 e. Use a graphing calculator to simulate the motion of the ball.

48. **Projectile motion** Jonathan punts a football with an initial velocity of 80 feet per second, from a height of 3 feet, at an angle of 45°.

a. What are the parametric equations that describe the ball's position as a function of time?

b. How long is the hang time for the punt?

c. What is the maximum height of the ball, and when does it reach that height?

d. How far does the ball travel?

 e. Use a graphing calculator to simulate the motion of the ball.

49. **Projectile motion** A quarterback passes a football to a receiver directly downfield at an initial velocity of 66 feet per second. The pass is released at a height of 7 feet from the surface of the field, at an angle of 65° to the horizontal, and the receiver catches the ball at a height of 5 feet.

a. Write a set of parametric equations for the path of the football.

b. How long did it take for the receiver to get to the point of the reception?

c. How far down the field does the receiver catch the ball?

50. Projectile motion A kicker kicks off a football at an initial velocity of 85 feet per second from his team's 30-yard line. The angle of the kick is 50° from the horizontal.

   a. Write a set of parametric equations for the path of the ball.

   b. How long will the kick be in the air if no one catches the ball?

   c. If no one catches the ball, will it fly over the goal line?

   d. Would a receiver 6 feet tall be able to catch the ball at the goal line?

51. Projectile motion A golfer at an indoor driving range is hitting balls from the second level, 10 feet above the ground. The initial velocity of the ball is 120 feet per second at an angle of 25° from the horizontal. From the tee, the distance to the end of the dome is 275 feet; the height of the dome is 70 feet.

   a. How close to the top of the dome will the ball be at its maximum height?

   b. At what height will the ball hit the end wall?

52. Projectile motion An arrow is shot from a height of 5 feet above the ground with an initial velocity of 88 feet per second and an angle of 45°. Will the arrow clear a 20-foot wall that is 220 feet from the archer?

53. Moving vehicles The distance from Orange, Texas, to El Paso, Texas, is about 850 miles via highway. Shelly leaves Orange at 6:00 a.m. and travels at 65 mph. Clara leaves El Paso at 10:00 a.m. and averages 70 mph.

   a. Write a set of parametric equations as functions of time that describes the movement of each driver.

   b. At what time do they meet?

   c. How far will each driver have traveled when the two cars meet?

54. Moving vehicles A train is traveling due east at 70 mph and is 100 miles from an intersection. At the same time, a man on a motorcycle is traveling due north at 60 mph and is 75 miles from the intersection.

   a. Write a set of parametric equations as functions of time that describes the movement of each vehicle.

   b. Write the distance between the two as a function of time.

   c. Use a graphing calculator to graph the distance.

   d. What is the minimum distance between the train and the motorcycle?

   e. Turn off the axes and graph the equations in part (a).

   f. Which will be the first to the intersection?

55. Precalculus Find parametric equations for a circle with radius a and center at the origin. Let P be a point on the circle, and let the parameter t be the angle that OP makes with the x-axis.

56. Precalculus Find parametric equations of the parabola $x^2 = 4py$. Let t be the slope of the line OP.

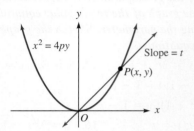

57. Precalculus Circles of radius a and b are centered at the origin. (See the figure.) Line OBA makes an angle t with the x-axis. Line PB is horizontal, and line PA is vertical. Using t as the parameter, determine parametric equations for the curve traced by P.

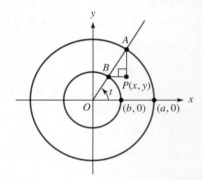

58. Precalculus The line segment AB in the figure is $2k$ units long. Point A lies on the positive x-axis, and B lies on the positive y-axis. Determine parametric equations for the midpoint of segment AB. Use as the parameter the angle t that OM makes with the x-axis.

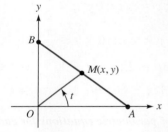

Discovery and Writing

59. A projectile fired at an angle θ with the horizontal and with an initial velocity v_0 has position $P(x, y)$ at any time t given by

$$x = v_0 t \cos \theta \text{ and } y = v_0 t \sin \theta - 16t^2.$$

Eliminate the parameter t and identify the curve.

60. In Exercise 59, determine the range of the projectile and its maximum height.

9.6 Conic Sections in Polar Coordinates

In this section, we will learn to

1. Analyze and sketch the graphs of conic sections.

2. Write a polar equation of a conic in rectangular form.

An important property of the parabola is its reflective property. A light source placed at the focus of a parabola reflects off the parabola, and the light rays emerge parallel to the axis of the parabola. It is this property that is used in flashlights and headlights on automobiles.

Using polar coordinates, we can define conics in an alternate way. This is very helpful in many situations. When we use rectangular coordinates, we place the most importance on the location of the center of the conic. When we use polar coordinates, we place more importance on the location of the focus of a conic. As we have seen, the reflective property of a parabola places importance on the location of the focus.

In this section, we use polar coordinates to write equations for the parabola, ellipse, and hyperbola. The discussion is based on the fact that each of these conics is completely determined by a point, a directrix, and a positive real number called the **eccentricity.**

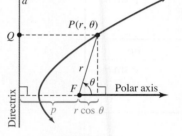

FIGURE 9-46

1. Analyze and Sketch the Graphs of Conic Sections

We begin by identifying the pole as a fixed point F, which is the **focus,** and a fixed line d perpendicular to an extension of the polar axis, called the **directrix,** as in Figure 9-46. Furthermore, we let p $(p > 0)$ be the distance from point F to the directrix d, and we let e be a positive constant representing the **eccentricity.**

Conics   For focus F and directrix d, a **conic** is the set of all points $P(r, \theta)$ that lie on the plane in any location for which the ratio of $d(FP)$ to $d(QP)$ is equal to the constant

$$e = \frac{d(FP)}{d(QP)}.$$

If $e = 1$, the conic is a **parabola**.

If $e < 1$, the conic is an **ellipse**.

If $e > 1$, the conic is a **hyperbola**.

To derive the polar equation for the conic, we note from Figure 9-46 that $d(FP)$ and $d(QP) = p + r \cos \theta$. Then we have

$$e = \frac{r}{p + r \cos \theta}$$

or, by solving for r, we have the standard form of an equation of a conic.

Polar Equation of a Conic   The standard form of the equation of a conic with eccentricity e, focus at the pole, and directrix perpendicular to an extension of the polar axis is

$$r = \frac{ep}{1 - e \cos \theta}.$$

There are three other standard forms for the equation of a conic, determined by rotating the directrix and graph in Figure 9-46 through angles of $\frac{\pi}{2}$, π, and $\frac{3\pi}{2}$. Table 1 summarizes the four possibilities for the standard equation of a conic written in polar coordinates.

TABLE 1

Graph	Equation
1.	$r = \dfrac{ep}{1 - e\cos\theta}$
2.	$r = \dfrac{ep}{1 + e\cos\theta}$
3.	$r = \dfrac{ep}{1 - e\sin\theta}$
4.	$r = \dfrac{ep}{1 + e\sin\theta}$

EXAMPLE 1 **Graphing a Conic**

Analyze and graph the conic determined by $r = \dfrac{3}{4 - 2\cos\theta}$.

SOLUTION By dividing the numerator and denominator by 4, we can write the equation as

$$r = \frac{\dfrac{3}{4}}{1 - \dfrac{1}{2}\cos\theta}.$$

Comparing this to Equation (1) in Table 1, we determine that $e = \frac{1}{2}$ and $ep = \frac{3}{4}$. Now we calculate p as follows:

$$ep = \frac{3}{4}$$

$$\frac{1}{2}p = \frac{3}{4} \qquad \text{Substitute } \frac{1}{2} \text{ for } e.$$

$$p = \frac{3}{2}$$

Since $e = \frac{1}{2} < 1$, the conic is an ellipse. Because $p = \frac{3}{2}$, the directrix is $\frac{3}{2}$ units to the left of the focus.

To graph the ellipse, we first find its vertices by determining r for $\theta = 0$ and for $\theta = \pi$. Then we find its intercepts with the vertical axis by letting $\theta = \frac{\pi}{2}$ and $\theta = \frac{3\pi}{2}$. We plot these points and sketch the graph, as in Figure 9-47.

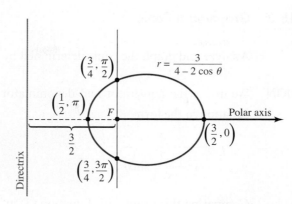

FIGURE 9-47

Self Check 1   Identify, graph, and label the conic determined by $r = \dfrac{4}{3 - 2\cos\theta}$.

Now Try Exercise 9.

EXAMPLE 2   **Graphing a Conic**

Analyze and graph the conic determined by $r = \dfrac{5}{3 + 3\cos\theta}$.

SOLUTION   We divide the numerator and the denominator of the fraction by 3 and write the equation in the form

$$r = \dfrac{\dfrac{5}{3}}{1 + 1\cos\theta}.$$

Comparing this result with Equation (2) in Table 1, we determine that $e = 1$, and the graph of this conic is a parabola opening to the left. Because $p = \frac{5}{3}$, the directrix is $\frac{5}{3}$ units to the right of the pole.

To graph the parabola, we determine the coordinates of the vertex by letting $\theta = 0$, finding $r = \frac{5}{6}$. We determine the intercepts with the vertical axis by letting $\theta = \frac{\pi}{2}$ to get the point $\left(\frac{5}{3}, \frac{\pi}{2}\right)$ and $\theta = \frac{3\pi}{2}$ to get the point $\left(\frac{5}{3}, \frac{3\pi}{2}\right)$. Plotting these points allows us to sketch the graph as shown in Figure 9-48.

Note that the given function is not defined at $\theta = \pi$, because the denominator would be 0. Thus, no point on the parabola lies in the direction of $\theta = \pi$. This is another indication that the parabola opens to the left.

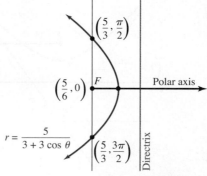

FIGURE 9-48

Self Check 2   Identify, graph, and label the conic determined by $r = \dfrac{5}{2 - 2\cos\theta}$.

Now Try Exercise 5.

EXAMPLE 3   **Graphing a Conic**

Analyze and graph the conic determined by $r = \dfrac{\sqrt{3}}{\sqrt{3} - 2\sin\theta}$.

SOLUTION   We divide the numerator and denominator of the fraction by $\sqrt{3}$ and write the equation in the form

$$r = \dfrac{1}{1 - \dfrac{2}{\sqrt{3}}\sin\theta}.$$

Comparing this result with Equation (3) in Table 1, we can determine that $e = \dfrac{2}{\sqrt{3}}$ and that $ep = 1$. From this, we determine that $p = \dfrac{\sqrt{3}}{2}$. Because $e > 1$, the conic is a hyperbola. Its directrix is parallel to the polar axis and is $\dfrac{\sqrt{3}}{2}$ units below it. The hyperbola opens upward and downward.

To graph the hyperbola, we first find its vertices by determining r for $\theta = \dfrac{\pi}{2}$ and $\theta = \dfrac{3\pi}{2}$, giving us the vertices $\left(0.46, \dfrac{3\pi}{2}\right)$ and $\left(-6.5, \dfrac{\pi}{2}\right)$. Next, we find the intercepts with the polar axis by letting $\theta = 0$ and $\theta = \pi$. This gives us the points $(1, 0)$ and $(1, \pi)$.

Note that the given equation is undefined if its denominator is 0. We determine these values excluded from the domain of the function by setting the denominator equal to 0 and solving for θ.

$$1 - \dfrac{2}{\sqrt{3}}\sin\theta = 0$$

$$\dfrac{2}{\sqrt{3}}\sin\theta = 1$$

$$\sin\theta = \dfrac{\sqrt{3}}{2}$$

$$\theta = \dfrac{\pi}{3} \quad \text{or} \quad \theta = \dfrac{2\pi}{3}$$

No point on the hyperbola can be found in the direction $\theta = \dfrac{\pi}{3}$ or $\theta = \dfrac{2\pi}{3}$; these directions are parallel to the asymptotes of the hyperbola. Note also that for $0 \le \theta < 2\pi$, the *lower* branch of the hyperbola is traced as θ increases from $\dfrac{\pi}{3}$ to $\dfrac{2\pi}{3}$. Other values of θ determine the upper branch.

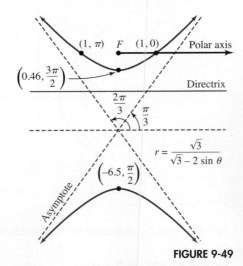

$r = \dfrac{\sqrt{3}}{\sqrt{3} - 2\sin\theta}$	
r	θ
1	0
1	π
-6.5	$\dfrac{\pi}{2}$
0.46	$\dfrac{3\pi}{2}$

FIGURE 9-49

Self Check 3   Identify, graph, and label the conic determined by $r = \dfrac{2}{1 + 2\cos\theta}$.

Now Try Exercise 13.

2. Write a Polar Equation of a Conic in Rectangular Form

EXAMPLE 4   **Writing a Polar Equation of a Conic in Rectangular Form**

Convert the polar equation

$$r = \frac{4}{3 + 3\cos\theta}$$

to an equation in rectangular form.

SOLUTION   To make the transformation from polar to rectangular, we will use the definitions $r^2 = x^2 + y^2$ and $x = r\cos\theta$. To accomplish this, our goal is to write the equation in such a way that squaring both sides will allow us to make the substitution.

$$r = \frac{4}{3 + 3\cos\theta}$$

$$3r + 3r\cos\theta = 4$$

$$3r = 4 - 3r\cos\theta \qquad \text{Write the equation with only one term containing } r \text{ on each side.}$$

$$(3r)^2 = (4 - 3r\cos\theta)^2 \qquad \text{Square both sides.}$$

$$9(x^2 + y^2) = (4 - 3x)^2 \qquad r^2 = x^2 + y^2; \ x = r\cos\theta$$

$$9x^2 + 9y^2 = 16 - 24x + 9x^2$$

$$9y^2 = 16 - 24x$$

This is an equation of a parabola.

Self Check 4   Convert the polar equation $r = \dfrac{2}{1 - \sin\theta}$ to an equation in rectangular form.

Now Try Exercise 17.

Self Check Answers   **1.** ellipse, $e = \dfrac{2}{3}, p = 2$

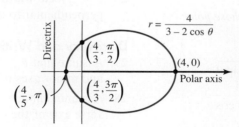

2. parabola opening to the right, $e = 1, p = \dfrac{5}{2}$

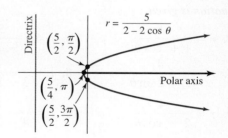

3. hyperbola, $e = 2$, $p = 1$ **4.** $4y = x^2 - 4$

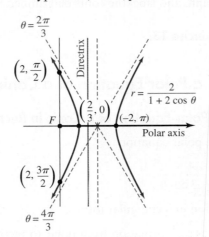

Exercises 9.6

Getting Ready

You should be able to complete these vocabulary and concept statements before you proceed to the practice exercises.

Fill in the blanks.

1. Given that the eccentricity of a conic is e, if $e = 1$, the conic is a(n) _____ ; if $e < 1$, the conic is a(n) _____ ; and if $e > 1$, the conic is a(n) _____ .

2. The conic having the polar equation $r = \dfrac{6}{1 + 2 \sin \theta}$ is a(n) _____ with eccentricity $e =$ _____, $p =$ _____, and directrix _____ units above and _____ to the polar axis.

3. For a hyperbola with equation $r = \dfrac{3}{1 - 2 \cos \theta}$, the asymptotes will be at $\theta =$ _____ and $\theta =$ _____ .

4. The graph of the polar equation $r = \dfrac{6}{3 + 3 \sin \theta}$ is a _____ .

Practice

Classify and sketch the graph of the conic whose polar equation is given.

5. $r = \dfrac{3}{1 + \cos \theta}$ **6.** $r = \dfrac{5}{1 - \sin \theta}$

7. $r = \dfrac{4}{2 - 2 \sin \theta}$ **8.** $r = \dfrac{4}{3 - 3 \cos \theta}$

9. $r = \dfrac{4}{2 - \cos \theta}$ **10.** $r = \dfrac{8}{4 - 2 \sin \theta}$

11. $r = \dfrac{3}{3 + 2 \sin \theta}$ **12.** $r = \dfrac{6}{3 + 2 \cos \theta}$

13. $r = \dfrac{4}{2 - 3 \sin \theta}$ **14.** $r = \dfrac{9}{3 + 6 \sin \theta}$

Convert each polar equation to a rectangular equation.

15. $r = \dfrac{1}{1 + \sin \theta}$ **16.** $r = \dfrac{3}{2 + \cos \theta}$

17. $r = \dfrac{8}{4 - 3 \cos \theta}$ **18.** $r = \dfrac{10}{5 - 4 \sin \theta}$

19. $r = \dfrac{9}{3 + 6 \sin \theta}$ **20.** $r = \dfrac{12}{4 + 8 \cos \theta}$

21. $r = \dfrac{8}{2 - \cos \theta}$ **22.** $r = \dfrac{8}{2 + 4 \cos \theta}$

Write a polar equation of the conic described.

23. $e = \dfrac{2}{3}$; directrix: $r \cos \theta = 5$

24. $e = 3$; directrix: $r \sin \theta = -2$

25. $e = \frac{2}{3}$; focal width $\frac{7}{3}$; directrix to the right of the pole, perpendicular to the polar axis (*Hint:* The focal width is the distance between the intercepts on the vertical axis.)

26. $e = \frac{4}{3}$; focal width 4; directrix to the left of the pole, perpendicular to the polar axis

Discovery and Writing

27. Assume that $e < 1$. Express the length of the major and minor axes of the ellipse in terms of e and p.

28. Assume that $e > 1$. Express the length of the transverse axis of the hyperbola in terms of e and p.

29. In Equation (1) of Table 1, assume that $e = 1$. Find polar coordinates of the vertex of the parabola.

30. In Equation (1) of Table 1, assume that $e > 1$. Find polar coordinates of the center of the hyperbola.

31. Assuming that $0 < e < 1$, determine the length $2a$ of the major axis and the distance c between a focus and the center of the ellipse. Then prove that $e = \frac{c}{a}$.

32. Assuming that $e > 1$, determine the length $2a$ of the transverse axis and the distance c between a focus and the center of the hyperbola. Then prove that $e = \frac{c}{a}$.

CHAPTER REVIEW

SECTION 9.1   The Circle and the Parabola

Definitions and Concepts	Examples
Standard equation of a circle with center at (0, 0) and radius r: $x^2 + y^2 = r^2$ Standard equation of a circle with center at (h, k) and radius r: $(x - h)^2 + (y - k)^2 = r^2$	See Example 1, pages 817–818
General form of a second-degree equation in x and y: $Ax^2 + Bxy + Cy^2 + Dx + Ey + F = 0$ The second-degree equation is a circle if $A = C$ and $B = 0$.	See Example 2, page 818
A **parabola** is the set of all points in a plane equidistant from a line l (called the **directrix**) and fixed point F (called the **focus**) that is not on line l. Parabolas:	See Example 3, pages 820–821

Parabola Opening	Vertex at Origin
Right	$y^2 = 4px\,(p > 0)$
Left	$y^2 = 4px\,(p < 0)$
Up	$x^2 = 4py\,(p > 0)$
Down	$x^2 = 4py\,(p < 0)$

Parabola Opening	Vertex at $V(h, k)$
Right	$(y - k)^2 = 4p(x - h)\,(p > 0)$
Left	$(y - k)^2 = 4p(x - h)\,(p < 0)$
Up	$(x - h)^2 = 4p(y - k)\,(p > 0)$
Down	$(x - h)^2 = 4p(y - k)\,(p < 0)$

See Example 4, pages 821–822

Definitions and Concepts	Examples
For a parabola that opens right or left with vertex at (h, k), the directrix is $x = -p + h$ and the focus is $(p + h, k)$. For a parabola that opens up or down with vertex at (h, k), the directrix is $y = -p + k$ and the focus is $(h, p + k)$.	See Example 5, pages 822–823

EXERCISES

Write an equation of each circle in standard form.

1. Center $(0, 0)$; radius 4
2. Center $(0, 0)$; passes through $(6, 8)$
3. Center $(3, -2)$; radius 5
4. Center $(-2, 4)$; passes through $(1, 0)$
5. Endpoints of diameter $(-2, 4)$ and $(12, 16)$
6. Endpoints of diameter $(-3, -6)$ and $(7, 10)$

Write an equation of each circle in standard form and graph the circle.

7. $x^2 + y^2 - 6x + 4y = 3$
8. $x^2 + 4x + y^2 - 10y = -13$

Write an equation of each parabola.

9. Vertex $(0, 0)$; passes through $(-8, 4)$ and $(-8, -4)$
10. Vertex $(0, 0)$; passes through $(-8, 4)$ and $(8, 4)$
11. Vertex $(-2, 3)$; passes through $(-4, -8)$; opens down

Graph each equation.

12. $x^2 - 4y - 2x + 9 = 0$
13. $y^2 - 6y = 4x - 13$

SECTION 9.2   The Ellipse

Definitions and Concepts	Examples
An **ellipse** is the set of all points P in a plane such that the sum of the distances from P to two other fixed points F and F' is a positive constant. **Ellipse with major axis on the x-axis and center at $(0, 0)$:** The standard equation of an ellipse with its center at the origin and major axis (horizontal) on the x-axis is $$\frac{x^2}{a^2} + \frac{y^2}{b^2} = 1, \text{ where } a > b > 0.$$ **Vertices (ends of the major axis):** $V(a, 0)$ and $V'(-a, 0)$ **Length of the major axis:** $2a$ **Ends of the minor axis:** $B(0, b)$ and $B'(0, -b)$ **Length of the minor axis:** $2b$ **Foci:** $F(c, 0)$ and $F'(-c, 0)$, where $c^2 = a^2 - b^2$ 	See Example 1, pages 834–835

Definitions and Concepts	Examples
Ellipse with major axis on the _y_-axis and center at (0, 0): The standard equation of an ellipse with its center at the origin and major axis (vertical) on the _y_-axis is $\dfrac{x^2}{b^2} + \dfrac{y^2}{a^2} = 1$, where $a > b > 0$. **Vertices (ends of the major axis):** $V(0, a)$ and $V'(0, -a)$ **Length of the major axis:** $2a$ **Ends of the minor axis:** $B(b, 0)$ and $B'(-b, 0)$ **Length of the minor axis:** $2b$ **Foci:** $F(0, c)$ and $F'(0, -c)$, where $c^2 = a^2 - b^2$ 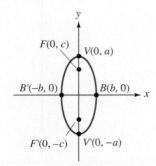	See Example 4, page 838
Ellipse with major axis horizontal and center at (_h_, _k_): The standard equation of an ellipse with center at (h, k) and major axis horizontal is $\dfrac{(x - h)^2}{a^2} + \dfrac{(y - k)^2}{b^2} = 1$, where $a > b > 0$. **Vertices (ends of the major axis):** $V(a + h, k)$ and $V'(-a + h, k)$ **Ends of the minor axis:** $B(h, b + k)$ and $B'(h, -b + k)$ **Foci:** $F(h + c, k)$ and $F'(h - c, k)$, where $c^2 = a^2 - b^2$ 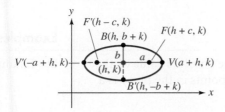	See Example 6, page 839

Definitions and Concepts	Examples
Ellipse with major axis vertical and center at (h, k): The standard equation of an ellipse with center at (h, k) and major axis vertical is $$\frac{(x-h)^2}{b^2} + \frac{(y-k)^2}{a^2} = 1, \text{ where } a > b > 0.$$ **Vertices (ends of the major axis):** $\quad V(h, a+k)$ and $V'(h, -a+k)$ **Ends of the minor axis:** $\quad B(b+h, k)$ and $B'(-b+h, k)$ **Foci:** $F(h, k+c)$ and $F'(h, k-c)$, where $c^2 = a^2 - b^2$ 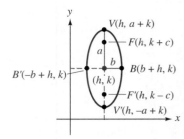	See Example 5, page 838

EXERCISES

14. Write an equation of the ellipse with its center at the origin, major axis horizontal and 12 units long, and minor axis 8 units long.

15. Write an equation of the ellipse with its center at the origin, major axis vertical and 10 units long, and minor axis 4 units long.

16. Write an equation of the ellipse centered at the point $(-2, 3)$ and passing through the points $(-2, 0)$ and $(2, 3)$.

17. Write an equation $4x^2 + y^2 - 16x + 2y = -13$ in standard form and graph it.

SECTION 9.3   The Hyperbola

Definitions and Concepts	Examples
A **hyperbola** is the set of all points P in a plane such that the absolute value of the difference of the distances from point P to two other fixed points in the plane is a positive constant. **Hyperbola with foci on the x-axis and center at $(0, 0)$:** The standard equation of a hyperbola with center at the origin and foci on the x-axis is $$\frac{x^2}{a^2} - \frac{y^2}{b^2} = 1, \text{ where } a^2 + b^2 = c^2.$$ **Vertices:** $V(a, 0)$ and $V'(-a, 0)$ **Foci:** $F(c, 0)$ and $F'(-c, 0)$ **Asymptotes:** $y = \pm\frac{b}{a}x$ 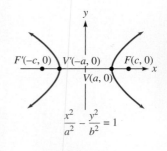	See Example 1, page 846

Definitions and Concepts	**Examples**

Hyperbola with foci on the *y*-axis and center at (0, 0):
The standard equation of a hyperbola with center at the origin and foci on the *x*-axis is

$$\frac{y^2}{a^2} - \frac{x^2}{b^2} = 1, \text{ where } a^2 + b^2 = c^2.$$

Vertices: $V(0, a)$ and $V'(0, -a)$
Foci: $F(0, c)$ and $F'(0, -c)$
Asymptotes: $y = \pm\frac{a}{b}x$

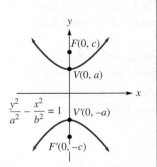

Hyperbola with transverse axis horizontal and center at (*h*, *k*):
The standard equation of a hyperbola with center at the point (h, k) and foci on a line parallel to the *x*-axis is

$$\frac{(x - h)^2}{a^2} - \frac{(y - k)^2}{b^2} = 1, \text{ where } a^2 + b^2 = c^2.$$

Vertices: $V(a + h, k)$ and $V'(-a + h, k)$
Foci: $F(c + h, k)$ and $F'(-c + h, k)$

See Example 2,
pages 847–848

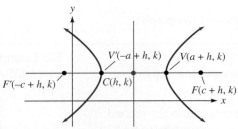

Hyperbola with transverse axis vertical and center at (*h*, *k*):
The standard equation of a hyperbola with center at the point (h, k) and foci on a line parallel to the *y*-axis is

$$\frac{(y - k)^2}{a^2} - \frac{(x - h)^2}{b^2} = 1, \text{ where } a^2 + b^2 = c^2.$$

Vertices: $V(h, a + k)$ and $V'(h, -a + k)$
Foci: $F(h, c + k)$ and $F'(h, -c + k)$

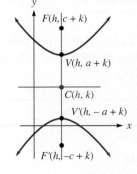

Graphing a hyperbola:
1. Write the hyperbola in standard form.
2. Identify the center, *a*, and *b*.
3. Draw the fundamental rectangle and asymptotes.
4. Sketch the hyperbola beginning with the vertices and using the asymptotes as guides.

See Example 3,
pages 849–850

EXERCISES

18. Write an equation of the hyperbola centered at the origin, passing through the points $(-2, 0)$ and $(2, 0)$, and having a focus at $(4, 0)$.

19. Write an equation of the hyperbola with its center at the origin, one focus at $(0, 5)$, and one vertex at $(0, 3)$.

20. Write an equation of the hyperbola with vertices at the points $(-3, 3)$ and $(3, 3)$ and a focus at the point $(5, 3)$.

21. Write an equation of the hyperbola with vertices at the points $(3, -3)$ and $(3, 3)$ and a focus at the point $(3, 5)$.

22. Write an equation of the asymptotes of the hyperbola $\dfrac{x^2}{25} - \dfrac{y^2}{16} = 1$.

23. Write the equation $9x^2 - 4y^2 - 16y - 18x = 43$ in standard form and graph it.

24. Graph: $4xy = 1$.

SECTION 9.4   Nonlinear Systems of Equations

Definitions and Concepts	Examples
There are three methods that can be used to solve nonlinear systems of equations: graphing, substitution, and addition. **Graphing method:** Graph each equation and identify the point(s) of intersection of the graphs.	See Example 1, page 855
Substitution method: We convert the two-variable, two-equation system into one equation and one variable using substitution. To do so, solve one equation for one variable and then substitute into the second equation. Next, solve for the one variable. Finally, use back substitution to find the second variable. Use this method when one of the equations is second degree and one is linear.	See Example 2, pages 855–856
Addition method: We eliminate one of the variables by addition and then solve the system. Use this method when a system consists of two second-degree equations of the form $ax^2 + by^2 = c$.	See Example 4, page 857

EXERCISES

25. Solve by graphing: $\begin{cases} x^2 + y^2 = 16 \\ y = x + 4 \end{cases}$.

26. Solve by graphing: $\begin{cases} 3x^2 + y^2 = 52 \\ x^2 - y^2 = 12 \end{cases}$.

27. Solve by graphing: $\begin{cases} \dfrac{x^2}{16} + \dfrac{y^2}{12} = 1 \\ x^2 - \dfrac{y^2}{3} = 1 \end{cases}$.

28. Solve by substitution or addition: $\begin{cases} 3x^2 + y^2 = 52 \\ x^2 - y^2 = 12 \end{cases}$.

29. Solve by substitution or addition:
$\begin{cases} x^2 + y^2 = 16 \\ -\sqrt{3}y + 4\sqrt{3} = 3x \end{cases}$.

30. Solve by substitution or addition: $\begin{cases} \dfrac{x^2}{16} + \dfrac{y^2}{12} = 1 \\ x^2 - \dfrac{y^2}{3} = 1 \end{cases}$.

SECTION **9.5**   Plane Curves and Parametric Equations

Definitions and Concepts	Examples
Plane curves and parametric equations: If f and g are continuous functions of t defined on an interval I, the set of ordered pairs $(f(t), g(t))$ is a **plane curve**. The equations $\quad x = f(t)$ and $y = g(t)$ are **parametric equations** for the curve, with **parameter** $t, t \in I$.	See Example 1, page 862
Eliminating the parameter: Select the equation that is simpler to solve for t and then solve for t. Substitute the expression into the other equation. Simplify the equation which is now written in terms of x and y. We also use other strategies and substitutions such as a Pythagorean Identity.	See Example 2, page 864
Finding a rectangular equation that describes a curve given by parametric equations: If y is defined as a function of x, then we can simply set $x = t$ and substitute to obtain the equation for y. If $y = f(x)$ and $x = t$, then we have the equation $y = f(t)$, and t must be in the domain of f.	See Example 5, page 866
Projectile motion: If a projectile is fired with an initial velocity of v_0 at an angle of θ above the horizontal, from a height h above the horizontal, and air resistance is assumed to be negligible, then the projectile's position after t seconds is given by the equations $\quad x = (v_0 \cos \theta)t$ and $y = -\dfrac{1}{2}gt^2 + (v_0 \sin \theta)t + h,$ where g is the acceleration due to gravity ($g \approx 32$ feet per second per second or $g \approx 9.8$ meters per second per second).	See Example 7, pages 868–869

EXERCISES

31. Sketch the curve determined by the given parametric equations.

   a. $x = 2t^2, y = 3t + 1$

   b. $x = 2 - 3t, y = 2t + 1$

For each set of parametric equations:

   a. *Sketch the curve represented by the parametric equations.*

   b. *Eliminate the parameter and write the rectangular equation for the curve. Adjust the domain of the resulting rectangular equation if necessary.*

32. $x = 3 \sin t, y = \cos t, 0 \le t \le 2\pi$

33. $x = t - 3, y = t^2$

34. Find two sets of parametric equations for the rectangular equation $y = 3x^2 - 4$. Adjust the domain of the resulting equations if necessary. Answers can vary.

Applications

35. Projectile motion Josiah passes a football to a receiver directly downfield at an initial velocity of 64 feet per second. The pass is released at a height of 7 feet from the surface of the field at an angle of 60° to the horizontal, and the receiver catches the ball at a height of 5 feet.

   a. Write a set of parametric equations for the path of the football.

   b. How long does it take for the receiver to get to the point of the reception?

   c. How far down the field does the receiver catch the ball?

SECTION **9.6**   Conic Sections in Polar Coordinates

Definitions and Concepts	Examples
Conics:	See Example 1,
For focus F and directrix d, a **conic** is the set of all points $P(r, \theta)$ that lie on the plane in any location for which the ratio of $d(FP)$ to $d(QP)$ is equal to the constant	pages 874–875
$$e = \frac{d(FP)}{d(QP)}.$$	
If $e = 1$, the conic is a **parabola**.	
If $e < 1$, the conic is an **ellipse**.	
If $e > 1$, the conic is a **hyperbola**.	
Polar equation of a conic:	
The standard form of the equation of a conic with eccentricity e, focus at the pole, and directrix perpendicular to an extension of the polar axis is	
$$r = \frac{ep}{1 - e\cos\theta}.$$	
Writing the polar equation of a conic in rectangular form:	See Example 4, page 877
To make the transformation from polar to rectangular, we will use the definitions	
$$r^2 = x^2 + y^2 \text{ and } x = r\cos\theta.$$	
To accomplish this, our goal is to write the equation in such a way that squaring both sides will allow us to make the substitution.	

EXERCISES

36. Classify and sketch the graph of the conic whose polar equation is $r = \dfrac{4}{2 + 2\cos\theta}$.

37. Convert the polar equation $r = \dfrac{5}{4 + 4\sin\theta}$ to an equivalent rectangular equation.

CHAPTER TEST

Write an equation of each circle.

1. Center $(2, 3)$; $r = 3$

2. Ends of diameter at $(-2, -2)$ and $(6, 8)$

3. Center $(2, -5)$; passes through $(7, 7)$

4. Change the equation of the circle
$x^2 + y^2 - 4x + 6y + 4 = 0$ to standard form and graph it.

Find an equation of each parabola.

5. Vertex $(3, 2)$; focus at $(3, 6)$

6. Vertex $(4, -6)$; passes through $(3, -8)$ and $(3, -4)$

7. Vertex $(2, -3)$; passes through $(0, 0)$

8. Change the equation of the parabola
$x^2 - 6x - 8y = 7$ into standard form and graph it.

Find an equation of each ellipse.

9. Vertex $(10, 0)$; center at the origin; focus at $(6, 0)$

10. Minor axis 24; center at the origin; focus at $(5, 0)$

11. Center $(2, 3)$; passes through $(2, 9)$ and $(0, 3)$

12. Change the equation of the ellipse
$9x^2 + 4y^2 - 18x - 16y - 11 = 0$ into standard form and graph it.

Find an equation of each hyperbola.

13. Center at the origin; focus at $(13, 0)$; vertex $(5, 0)$

14. Vertices $(6, 0)$ and $(-6, 0)$; $\dfrac{c}{a} = \dfrac{13}{12}$

15. Center $(2, -1)$; major axis horizontal and of length 16; distance of 20 between foci

16. Change the equation of the hyperbola $x^2 - 4y^2 + 16y = 8$ into standard form and graph it.

Solve each system using substitution or addition.

17. $\begin{cases} x^2 + y^2 = 23 \\ y = x^2 - 3 \end{cases}$
18. $\begin{cases} 2x^2 - 3y^2 = 9 \\ x^2 + y^2 = 27 \end{cases}$

Complete the square to write each equation in standard form, and identify the curve.

19. $y^2 - 4y - 6x - 14 = 0$

20. $2x^2 + 3y^2 - 4x + 12y + 8 = 0$

For each set of parametric equations:

a. Sketch the curve represented by the parametric equations.

b. Eliminate the parameter and write a rectangular equation for the curve. Adjust the domain of the resulting rectangular equation if necessary.

21. $x = 2t + 1$, $y = 4t^2$

22. $x = 2t - 4$, $y = t + 3$

23. Classify and sketch the graph of the conic whose polar equation is $r = \dfrac{10}{5 + 2\sin t}$.

 LOOKING AHEAD TO **CALCULUS**

---

1. **Using Parametric Equations to Analyze Motion of a Particle along a Curve**
   - Connect parametric equations to derivatives and vector functions— Section 9.5

---

1. Using Parametric Equations to Analyze Motion of a Particle along a Curve

Parametric equations are applied in many areas of calculus. One of these is the analysis of the motion of a particle along a curve defined by parametric equations $x(t)$ and $y(t)$. We describe the motion by a smooth vector-valued function $\mathbf{r}(t)$ in which the parameter t denotes time. This is called the **position function** or **trajectory**. We can use this function to answer the questions:

- What is the velocity of the particle at an instant in time (instantaneous velocity)?
- What is the acceleration of the particle at an instant in time (instantaneous acceleration)?
- What is the speed of the particle at an instant in time?

Using **derivatives** (Looking Ahead to Calculus, Chapter 1), we define the following:

- **Instantaneous velocity function**, $\mathbf{v}(t) = \mathbf{r}'(t)$—the derivative of the position function.
- **Instantaneous acceleration function**, $\mathbf{a}(t) = \mathbf{v}'(t)$—the derivative of the velocity function.

Since the position function is a vector function, we define the **speed** to be the magnitude of $\mathbf{v}(t)$.

- **Speed** $= |\mathbf{v}(t)| = \sqrt{(x'(t))^2 + (y'(t))^2}$

EXAMPLE 1   **Using Parametric Equations to Analyze the Motion of a Particle**

The path of a particle along a curve is given by the parametric equations

$$x = t^3 - 3t \text{ and } y = 4t^2 - 1.$$

a. Find the position, instantaneous velocity, instantaneous acceleration, and speed of the particle at time t.

b. Find the velocity, speed, and acceleration at $t = 2$.

SOLUTION   **a.** The functions are:

Position	$\mathbf{r}(t) = (t^3 - 3t)\mathbf{i} + (4t^2 - 1)\mathbf{j}$
Velocity	$\mathbf{r}'(t) = \mathbf{v}(t) = (3t^2 - 3)\mathbf{i} + 8t\mathbf{j}$
Acceleration	$\mathbf{a}(t) = \mathbf{v}'(t) = 6t\mathbf{i} + 8\mathbf{j}$
Speed	$|\mathbf{v}(t)| = \sqrt{(x'(t))^2 + (y'(t))^2} = \sqrt{(3t^2 - 3)^2 + (8t)^2}$

b. Velocity is $\mathbf{v}(2) = (3 \cdot 2^2 - 3)\mathbf{i} + (8 \cdot 2)\mathbf{j} = 9\mathbf{i} + 16\mathbf{j}$.

Speed is $|\mathbf{v}(2)| = \sqrt{(3 \cdot 2^2 - 3)^2 + (8 \cdot 2)^2} = \sqrt{337}$.

Acceleration is $\mathbf{a}(2) = (6 \cdot 2)\mathbf{i} + 8\mathbf{j} = 12\mathbf{i} + 8\mathbf{j}$.

◼

Connect to Calculus:
Using Parametric Equations to Analyze the Motion of a Particle

1. The path of a particle along a curve is given by the parametric equations

$$x = t \text{ and } y = \frac{1}{2}t^2.$$

a. Find the position, instantaneous velocity, instantaneous acceleration, and speed of the particle at time t.

b. Find the velocity, speed, and acceleration at $t = 1$.

2. The path of a particle along a curve is given by the parametric equations

$$x = 2 \cos t \text{ and } y = 2 \sin t.$$

a. Find the position, instantaneous velocity, instantaneous acceleration, and speed of the particle at time t.

b. Find the velocity, speed, and acceleration at $t = \frac{\pi}{4}$.

Sequences, Series, and Probability

10

CAREERS AND MATHEMATICS: Actuary

© Istockphoto.com/Steve Cole

Actuaries deal with the financial impact of risk and uncertainty. Actuaries are in high demand, and are well paid for their services. They work for insurance companies and estimate the probability and likely cost of events such as death, sickness, injury, disability, or loss of property. They use this information to help design insurance policies. In the finance sector, they aid companies in investing their resources to maximize the return on investment, and help ensure that pension plans are maintained on a sound financial basis.

Education and Mathematics Required

- Actuaries generally have a bachelor's degree and are required to pass a series of exams in order to become certified.
- College Algebra, Trigonometry, Calculus I and II, Linear Algebra, Probability and Mathematical Statistics, Applied Statistics, and Actuarial Mathematics are math courses required. Courses in numerical analysis, training in operations research, and substantial training in computer science are also necessary.

How Actuaries Use Math and Who Employs Them

- Actuaries estimate the probability and likely cost of an event, such as death, sickness, injury, or loss of property. To do this they must gather and correctly analyze data. Actuaries also have a broad knowledge of statistics, finance, and business.
- About 55% of all actuaries are employed by insurance carriers. Another 16% work for management, scientific, and technical consulting services. A small number are employed by government agencies.

Career Outlook and Salary

- Employment is expected to grow much faster than the average for all occupations. Employment of actuaries is expected to increase by 21% over the 2008–2018 period. Greater job growth will occur in the financial and consulting areas than in the insurance sector.
- The median annual salary is $87,210, with top salaries of more than $160,000.

For more information see http://www.bls.gov/oco.

In this chapter, we introduce a way to expand powers of binomials of the form $(x + y)^n$. This method leads to ideas needed when working with probability and statistics.

*Available online at http://www.cengagebrain.com.

10.1 The Binomial Theorem

In this section, we will learn to

1. Use Pascal's Triangle to expand a binomial.
2. Define and use factorial notation.
3. Use the Binomial Theorem to expand binomials.
4. Find a specific term of a binomial expansion.

Needlepointed triangle by William H. Mitchell

We begin this chapter by discussing a triangle, called **Pascal's Triangle**, that is named after the French mathematician Blaise Pascal. Although the Western world gives credit for the triangle to Pascal, other mathematicians in other parts of the world studied it centuries before him. Around 1068, the Indian mathematician Bhattotpala presented the first 17 rows of the triangle. Around the same time, the great Persian mathematician Omar Khayyám discussed the triangle. Today in Iran, the triangle is referred to as the Khayyám Triangle. Around 1250, the Chinese mathematician Yang Hui developed the triangle, and in China the triangle is known as Yang Hui's Triangle.

Over the years, many people have searched the triangle for the many number patterns it contains. One such person, William H. Mitchell, was so impressed with the triangle, he created a needlepoint design of it.

We introduce the triangle by discussing a way to expand binomials of the form $(x + y)^n$, where n is a natural number. Consider the following expansions:

$$(a + b)^0 = 1$$
$$(a + b)^1 = a + b$$
$$(a + b)^2 = a^2 + 2ab + b^2$$
$$(a + b)^3 = a^3 + 3a^2b + 3ab^2 + b^3$$
$$(a + b)^4 = a^4 + 4a^3b + 6a^2b^2 + 4ab^3 + b^4$$
$$(a + b)^5 = a^5 + 5a^4b + 10a^3b^2 + 10a^2b^3 + 5ab^4 + b^5$$
$$(a + b)^6 = a^6 + 6a^5b + 15a^4b^2 + 20a^3b^3 + 15a^2b^4 + 6ab^5 + b^6$$

Four patterns are apparent in the above expansions:

1. Each expansion has one more term than the power of the binomial.
2. The degree of each term in each expansion equals the exponent of the binomial.
3. The first term in each expansion is a raised to the power of the binomial.
4. The exponents of a decrease by 1 in each successive term, and the exponents of b, beginning with b^0 in the first term, increase by 1 in each successive term.

1. Use Pascal's Triangle to Expand a Binomial

To see another pattern, we write the coefficients of the previous expansions in a triangular array:

$(a + b)^0 =$						1						**Row 0**
$(a + b)^1 =$					1		1					**Row 1**
$(a + b)^2 =$				1		2		1				**Row 2**
$(a + b)^3 =$			1		3		3		1			**Row 3**
$(a + b)^4 =$		1		4		6		4		1		**Row 4**
$(a + b)^5 =$	1		5		10		10		5		1	**Row 5**
$(a + b)^6 =$	1	6		15		20		15		6	1	**Row 6**

In this array, each entry other than the 1s is the sum of the closest pair of numbers in the line above it. For example, the first 3 in the third row is the sum of the 1 and 2 above it. The 20 in the sixth row is the sum of the 10s above it.

This array, called **Pascal's Triangle** after Blaise Pascal (1623–1662), continues with the same pattern forever. The next two lines are

| $(a + b)^7 =$ | | 1 | 7 | 21 | 35 | 35 | 21 | 7 | 1 | **Row 7** |
| $(a + b)^8 =$ | 1 | 8 | 28 | 56 | 70 | 56 | 28 | 8 | 1. | | **Row 8** |

EXAMPLE 1   **Using Pascal's Triangle to Expand a Binomial**

Expand: $(x + y)^6$.

SOLUTION   The first term is x^6, and the exponents on x will decrease by 1 in each successive term. A y will appear in the second term, and the exponents on y will increase by 1 in each successive term, concluding when the term y^6 is reached. The variables in the expansion are

$$x^6, \qquad x^5y, \qquad x^4y^2, \qquad x^3y^3, \qquad x^2y^4, \qquad xy^5, \qquad \text{and} \qquad y^6.$$

We can use Pascal's Triangle to find the coefficients of the variables. Because the binomial is raised to the sixth power, we choose row 6 of Pascal's Triangle. The coefficients of the variables are the numbers in that row.

$$1 \quad 6 \quad 15 \quad 20 \quad 15 \quad 6 \quad 1$$

Putting this information together gives the expansion:

$$(x + y)^6 = x^6 + 6x^5y + 15x^4y^2 + 20x^3y^3 + 15x^2y^4 + 6xy^5 + y^6.$$

Self Check 1   Expand: $(p + q)^3$.

Now Try Exercise 21.

EXAMPLE 2   **Expanding a Binomial**

Expand: $(x - y)^6$.

SOLUTION   We write the binomial as $[x + (-y)]^6$ and substitute $-y$ for y in the result of Example 1.

Comment

In general, the signs in the expansion of $(x - y)^n$ alternate. The sign of the first term is +, the sign of the second term is −, and so on.

$$[x + (-y)]^6$$
$$= x^6 + 6x^5(-y) + 15x^4(-y)^2 + 20x^3(-y)^3 + 15x^2(-y)^4 + 6x(-y)^5 + (-y)^6$$
$$= x^6 - 6x^5y + 15x^4y^2 - 20x^3y^3 + 15x^2y^4 - 6xy^5 + y^6$$

Self Check 2   Expand: $(p - q)^3$.

Now Try Exercise 23.

2. Define and Use Factorial Notation

To use the Binomial Theorem to expand a binomial, we will need **factorial notation.**

Factorial Notation   If n is a natural number, the symbol $n!$ (read either as "***n* factorial**" or as "**factorial *n***") is defined as

$$n! = n(n - 1)(n - 2)(n - 3) \times \cdots \times (3)(2)(1).$$

EXAMPLE 3 **Evaluating Factorials**

Evaluate: **a.** 3!   **b.** 6!   **c.** 10!

SOLUTION We will apply the definition of factorial notation.

a. $3! = 3 \cdot 2 \cdot 1 = 6$

b. $6! = 6 \cdot 5 \cdot 4 \cdot 3 \cdot 2 \cdot 1 = 720$

c. $10! = 10 \cdot 9 \cdot 8 \cdot 7 \cdot 6 \cdot 5 \cdot 4 \cdot 3 \cdot 2 \cdot 1 = 3,628,800$

Self Check 3 Evaluate: **a.** 4!   **b.** 7!

Now Try Exercise 9.

ACCENT ON TECHNOLOGY **Factorials**

A graphing calculator can be used to evaluate factorials. Consider 10! The calculator operation that will be used is !.

To find 10!, first enter 10 in your calculator.

```
10█
```

Press **MATH** and move the cursor to PRB. Scroll down to 4:!.

```
MATH NUM CPX PRB
1:rand
2:nPr
3:nCr
4█!
5:randInt(
6:randNorm(
7:randBin(
```

Press **ENTER** twice and obtain a value of 3,628,800.

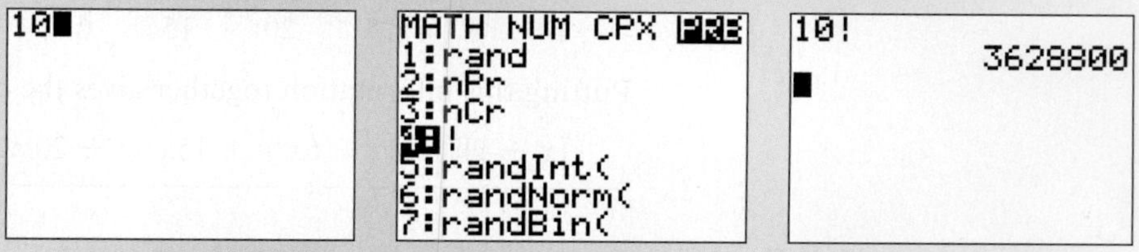

```
10!
          3628800
█
```

FIGURE 10-1

There are two fundamental properties of factorials.

Properties of Factorials
1. By definition, $0! = 1$.
2. If n is a natural number, $n(n - 1)! = n!$.

EXAMPLE 4 **Simplifying Expressions with Factorials**

Show that each expression is true. **a.** $6 \cdot 5! = 6!$ **b.** $8 \cdot 7! = 8!$

SOLUTION We will apply the definition of factorial notation.

a. $6 \cdot 5! = 6(\mathbf{5 \cdot 4 \cdot 3 \cdot 2 \cdot 1})$

$\quad\quad = 6 \cdot 5 \cdot 4 \cdot 3 \cdot 2 \cdot 1$

$\quad\quad = 6!$

b. $8 \cdot 7! = 8(\mathbf{7 \cdot 6 \cdot 5 \cdot 4 \cdot 3 \cdot 2 \cdot 1})$

$\quad\quad = 8 \cdot 7 \cdot 6 \cdot 5 \cdot 4 \cdot 3 \cdot 2 \cdot 1$

$\quad\quad = 8!$

Self Check 4 Show that $4 \cdot 3! = 4!$.

Now Try Exercise 7.

3. Use the Binomial Theorem to Expand Binomials

We can now state the Binomial Theorem.

Binomial Theorem   If n is any positive number, then

$$(a + b)^n = a^n + \frac{n!}{1!(n-1)!}a^{n-1}b + \frac{n!}{2!(n-2)!}a^{n-2}b^2$$

$$+ \frac{n!}{3!(n-3)!}a^{n-3}b^3 + \cdots + \frac{n!}{r!(n-r)!}a^{n-r}b^r + \cdots + b^n.$$

Comment

In the expansion of $(a + b)^n$, the term containing b^r is given by

$$\frac{n!}{r!(n-r)!}a^{n-r}b^r.$$

A proof of the Binomial Theorem appears in online Appendix B.

In the Binomial Theorem, the exponents on the variables in each term on the right side follow the familiar patterns:

1. The sum of the exponents on a and b in each term is n.
2. The exponents on a decrease by 1 in each successive term.
3. The exponents on b increase by 1 in each successive term.

However, the method of finding the coefficients is different. Except for the first and last terms, $n!$ is the numerator of each fractional coefficient. When the exponent on b is 2, the factors in the denominator of the fractional coefficient are $2!$ and $(n-2)!$. When the exponent on b is 3, the factors in the denominator are $3!$ and $(n-3)!$. When the exponent on b is r, the factors in the denominator are $r!$ and $(n-r)!$.

EXAMPLE 5   **Using the Binomial Theorem to Expand a Binomial**

Use the Binomial Theorem to expand $(a + b)^5$.

SOLUTION   We substitute directly into the Binomial Theorem.

$$(a + b)^5 = a^5 + \frac{5!}{1!(5-1)!}a^4b + \frac{5!}{2!(5-2)!}a^3b^2 + \frac{5!}{3!(5-3)!}a^2b^3 + \frac{5!}{4!(5-4)!}ab^4 + b^5$$

$$= a^5 + \frac{5 \cdot 4!}{1 \cdot 4!}a^4b + \frac{5 \cdot 4 \cdot 3!}{2 \cdot 1 \cdot 3!}a^3b^2 + \frac{5 \cdot 4 \cdot 3!}{3! \cdot 2 \cdot 1}a^2b^3 + \frac{5 \cdot 4!}{4! \cdot 1}ab^4 + b^5$$

$$= a^5 + 5a^4b + 10a^3b^2 + 10a^2b^3 + 5ab^4 + b^5$$

We note that the coefficients are the same numbers as in the fifth row of Pascal's Triangle. ∎

Self Check 5   Use the Binomial Theorem to expand $(p + q)^4$.

Now Try Exercise 25.

EXAMPLE 6   **Using the Binomial Theorem to Expand a Binomial**

Use the Binomial Theorem to expand $(2x - 3y)^4$.

SOLUTION   We first find the expansion of $(a + b)^4$.

$$(a + b)^4 = a^4 + \frac{4!}{1!(4-1)!}a^3b + \frac{4!}{2!(4-2)!}a^2b^2 + \frac{4!}{3!(4-3)!}ab^3 + b^4$$

$$= a^4 + \frac{4 \cdot 3!}{1 \cdot 3!}a^3b + \frac{4 \cdot 3 \cdot 2!}{2 \cdot 1 \cdot 2!}a^2b^2 + \frac{4 \cdot 3!}{3! \cdot 1}ab^3 + b^4$$

$$= a^4 + 4a^3b + 6a^2b^2 + 4ab^3 + b^4$$

We then substitute $2x$ for a and $-3y$ for b in the result.

$$(a + b)^4 = a^4 + 4a^3b + 6a^2b^2 + 4ab^3 + b^4$$

$$[2x + (-3y)]^4 = (2x)^4 + 4(2x)^3(-3y) + 6(2x)^2(-3y)^2 + 4(2x)(-3y)^3 + (-3y)^4$$

$$(2x - 3y)^4 = 16x^4 - 96x^3y + 216x^2y^2 - 216xy^3 + 81y^4$$

Self Check 6   Use the Binomial Theorem to expand $(3x - 2y)^4$.

Now Try Exercise 37.

4. Find a Specific Term of a Binomial Expansion

Suppose that we wish to find the fifth term of the expansion of $(a + b)^{11}$. It would be tedious to raise the binomial to the 11th power and then look at the fifth term. The Binomial Theorem provides an easier way.

EXAMPLE 7   **Finding a Specific Term of a Binomial Expansion**

Find the fifth term of the expansion of $(a + b)^{11}$.

SOLUTION   In the fifth term, the exponent on b is 4 (the exponent on b is always 1 less than the number of the term). Since the exponent on b added to the exponent on a equals 11, the exponent on a is 7. The variables of the fifth term are a^7b^4.

The number in the numerator of the fractional coefficient is $n!$, which in this case is 11!. The factors in the denominator are 4! and $(11 - 4)!$.

$$\frac{11!}{4!(11 - 4)!} a^7b^4 = \frac{11!}{4! \cdot 7!} a^7b^4$$

$$= \frac{11 \cdot 10 \cdot 9 \cdot 8 \cdot 7!}{4 \cdot 3 \cdot 2 \cdot 1 \cdot 7!} a^7b^4$$

$$= 330a^7b^4$$

Comment

We can also use the formula for the term containing b^r, where $r = 4$ and $n = 11$.

$$\frac{n!}{r!(n-r)!} a^{n-r}b^r = \frac{11!}{4!(11 - 4)!} a^{11-4}b^4$$

$$= \frac{11!}{4! \cdot 7!} a^7b^4$$

$$= 330a^7b^4$$

Self Check 7   Find the sixth term of the expansion in Example 7.

Now Try Exercise 43.

EXAMPLE 8   **Finding a Specific Term of a Binomial Expansion**

Find the sixth term of the expansion of $(a + b)^9$.

SOLUTION   In the sixth term, the exponent on b is 5 and the exponent on a is $9 - 5$, or 4. The numerator of the fractional coefficient is 9!, and the factors in the denominator are 5! and $(9 - 5)!$.

$$\frac{9!}{5!(9 - 5)!} a^4b^5 = \frac{9 \cdot 8 \cdot 7 \cdot 6 \cdot 5!}{5! \cdot 4!} a^4b^5$$

$$= \frac{9 \cdot 8 \cdot 7 \cdot 6}{4 \cdot 3 \cdot 2 \cdot 1} a^4b^5$$

$$= 126a^4b^5$$

Comment

We can also use the formula for the term containing b^r, where $r = 5$ and $n = 9$.

$$\frac{n!}{r!(n - r)!} a^{n-r}b^r = \frac{9!}{5!(9 - 5)!} a^{9-5}b^5$$

$$= \frac{9!}{5! \cdot 4!} a^4b^5$$

$$= 126a^4b^5$$

Self Check 8   Find the fifth term of the expansion in Example 8.

Now Try Exercise 47.

EXAMPLE 9 **Finding a Specific Term of a Binomial Expansion**

Find the third term of the expansion of $(3x - 2y)^6$.

SOLUTION We begin by finding the third term of the expansion of $(a + b)^6$.

$$\frac{6!}{2!(6 - 2)!} a^4 b^2 = \frac{6 \cdot 5 \cdot 4!}{2 \cdot 1 \cdot 4!} a^4 b^2 = 15a^4 b^2$$

We can then substitute $3x$ for a and $-2y$ for b in the expression to obtain the third term of the expansion of $(3x - 2y)^6$.

$$15a^4 b^2 = 15(3x)^4(-2y)^2$$
$$= 15(3)^4(-2)^2 x^4 y^2$$
$$= 4{,}860 x^4 y^2$$

Self Check 9 Find the fourth term of the expansion in Example 9.

Now Try Exercise 53.

Self Check Answers 1. $p^3 + 3p^2 q + 3pq^2 + q^3$ 2. $p^3 - 3p^2 q + 3pq^2 - q^3$ 3. a. 24
b. 5,040 4. $4 \cdot (3 \cdot 2 \cdot 1) = 4!$ 5. $p^4 + 4p^3 q + 6p^2 q^2 + 4pq^3 + q^4$
6. $81x^4 - 216x^3 y + 216x^2 y^2 - 96xy^3 + 16y^4$ 7. $462a^6 b^5$ 8. $126a^5 b^4$
9. $-4{,}320 x^3 y^3$

Exercises 10.1

Getting Ready

You should be able to complete these vocabulary and concept statements before you proceed to the practice exercises.

Fill in the blanks.

1. In the expansion of a binomial, there will be one more term than the _____ of the binomial.

2. The _____ of each term in a binomial expansion is the same as the exponent of the binomial.

3. The _____ term in a binomial expansion is the first term raised to the power of the binomial.

4. In the expansion of $(p + q)^n$, the _____ on p decrease by 1 in each successive term.

5. The expansion of 7! is _____.

6. $0! =$ _____.

7. $n \cdot$ _____ $= n!$.

8. In the seventh term of $(a + b)^{11}$, the exponent on a is _____.

Practice

Evaluate each expression.

9. $5!$

10. $-5!$

11. $3! \cdot 6!$

12. $0! \cdot 7!$

13. $6! + 6!$

14. $5! - 2!$

15. $\dfrac{9!}{12!}$

16. $\dfrac{8!}{5!}$

17. $\dfrac{5! \cdot 7!}{9!}$

18. $\dfrac{3! \cdot 5! \cdot 7!}{1! \cdot 8!}$

19. $\dfrac{18!}{6!(18 - 6)!}$

20. $\dfrac{15!}{9!(15 - 9)!}$

Use Pascal's Triangle to expand each binomial.

21. $(a + b)^5$

22. $(a + b)^7$

23. $(x - y)^3$

24. $(x - y)^7$

Use the Binomial Theorem to expand each binomial.

25. $(a + b)^3$

26. $(a + b)^4$

27. $(a - b)^5$

28. $(x - y)^4$

29. $(2x + y)^3$

30. $(x + 2y)^3$

31. $(x - 2y)^3$

32. $(2x - y)^3$

33. $(2x + 3y)^4$

34. $(2x - 3y)^4$

35. $(x - 2y)^4$

36. $(x + 2y)^4$

37. $(x - 3y)^5$

38. $(3x - y)^5$

39. $\left(\dfrac{x}{2} + y\right)^4$

40. $\left(x + \dfrac{y}{2}\right)^4$

Find the required term in each binomial expansion.

41. $(a + b)^4$; third term

42. $(a - b)^4$; second term

43. $(a + b)^7$; fifth term

44. $(a + b)^5$; fourth term

45. $(a - b)^5$; sixth term

46. $(a - b)^8$; seventh term

47. $(a + b)^{17}$; fifth term

48. $(a - b)^{12}$; third term

49. $\left(a - \sqrt{2}\right)^4$; second term

50. $\left(a - \sqrt{3}\right)^8$; third term

51. $\left(a + \sqrt{3}b\right)^9$; fifth term

52. $\left(\sqrt{2}a - b\right)^7$; fourth term

53. $\left(\dfrac{x}{2} + y\right)^4$; third term

54. $\left(m + \dfrac{n}{2}\right)^8$; third term

55. $\left(\dfrac{r}{2} - \dfrac{s}{2}\right)^{11}$; tenth term

56. $\left(\dfrac{p}{2} - \dfrac{q}{2}\right)^9$; sixth term

57. $(a + b)^n$; fourth term

58. $(a - b)^n$; fifth term

59. $(a + b)^n$; rth term

60. $(a + b)^n$; $(r + 1)$th term

Discovery and Writing

61. Find the sum of the numbers in each row of the first 10 rows of Pascal's Triangle. Do you see a pattern?

62. Show that the sum of the coefficients in the binomial expansion of $(x + y)^n$ is 2^n. (*Hint:* Let $x = y = 1$.)

63. Find the constant term in the expansion of
$$\left(a - \frac{1}{a}\right)^{10}.$$

64. Find the coefficient of x^5 in the expansion of
$$\left(x + \frac{1}{x}\right)^9.$$

65. If we applied the pattern of coefficients to the coefficient of the first term in the Binomial Theorem, it would be $\dfrac{n!}{0!(n - 0)!}$. Show that this expression equals 1.

66. If we applied the pattern of coefficients to the coefficient of the last term in the Binomial Theorem, it would be $\dfrac{n!}{n!(n - n)!}$. Show that this expression equals 1.

67. Define factorial notation and explain how to evaluate 10!.

68. With a calculator, evaluate 69!. Explain why you cannot find 70! with a calculator.

69. Explain how the rth term of a binomial expansion is constructed.

70. Explain the four patterns apparent in a binomial expansion.

10.2 Sequences, Series, and Summation Notation

In this section, we will learn to

1. Define and use sequences.

2. Define a sequence recursively.

3. Define series.

4. Define and use summation notation.

In this section, we will introduce a function whose domain is the set of natural numbers. This function, called a **sequence**, is a list of numbers in a specific order. Sequences are seen all around us. For example, we can unlock a digital lock by entering the correct sequence of numbers, a pilot performs the same sequence of checks before taking off, and our life is often changed by a sequence of events. We even use sequences for entertainment when we play the game of Sequence.

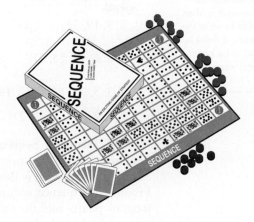

1. Define and Use Sequences

The meaning of the word *sequence* in mathematics is very similar to the meaning of the word in everyday English. It refers to a list of events in a particular order.

Sequences   An **infinite sequence** is a function whose domain is the set of natural numbers.

A **finite sequence** is a function whose domain is the set of the first n natural numbers.

Since a sequence is a function whose domain is the set of natural numbers, we can write its terms as a list of numbers. For example, if n is a natural number, the function defined by $f(n) = 2n - 1$ generates the following infinite sequence:

$$f(1) = 2(1) - 1 = 1$$
$$f(2) = 2(2) - 1 = 3$$
$$f(3) = 2(3) - 1 = 5$$
$$f(n) = 2(n) - 1 = 2n - 1$$

Writing the results horizontally, we have

$$1, 3, 5, \ldots, 2n - 1, \ldots.$$

The number 1 is the first term, 3 is the second term, 5 is the third term, and $2n - 1$ is the **general**, or **nth term**.

If n is a natural number, the function $f(n) = 3n^2 + 1$ generates the infinite sequence

$$4, 13, 28, \ldots, 3n^2 + 1, \ldots.$$

The number 4 is the first term, 13 is the second term, 28 is the third term, and $3n^2 + 1$ is the general term.

If n is one of the first four natural numbers, the function $f(n) = n^2 + 1$ generates the following finite sequence:

$$2, 5, 10, 17.$$

A constant function such as $g(n) = 1$ is a sequence, because it generates the infinite sequence

$$1, 1, 1, \ldots.$$

Comment

We seldom use function notation to denote a sequence, because it is often difficult or even impossible to write the general term. In such cases, if there is a pattern that is assumed to be continued, we simply list several terms of the sequence.

Some additional examples of infinite sequences are:

$$1^2, 2^2, 3^2, \ldots, n^2, \ldots$$

$$3, 9, 19, 33, \ldots, 2n^2 + 1, \ldots$$

$$1, 3, 6, 10, 15, 21, \ldots, \frac{n(n+1)}{2}, \ldots$$

$$2, 3, 5, 7, 11, 13, 17, 19, 23, \ldots \quad \text{(prime numbers)}$$

$$1, 1, 2, 3, 5, 8, 13, 21, \ldots \quad \text{(Fibonacci Sequence)}$$

The **Fibonacci Sequence** is named after the twelfth-century mathematician Leonardo of Pisa, also known as Fibonacci. After the two 1s in the Fibonacci Sequence, each term is the sum of the two terms that immediately precede it. The Fibonacci Sequence occurs in many applications, such as music, the growth patterns of plants, and the reproductive habits of bees. For example, in one octave of a piano keyboard, there are 2 black keys followed by 3 black keys for a total of 5 black keys. In one octave, there are 8 white keys and a total of 13 black and white keys. These are numbers in the Fibonacci Sequence.

C D E F G A B C D E F G A B C D E F G A B

The fruitlets of a pineapple are arranged in interlocking spirals, 8 spirals in one direction and 13 in the other.

2. Define a Sequence Recursively

A sequence can be defined **recursively** by giving its first term and a rule showing how to obtain the $(n + 1)$th term from the nth term. For example, the information

$$a_1 = 5 \qquad \text{The first term}$$

and

$$a_{n+1} = 3a_n - 2 \qquad \text{The rule showing how to get the } (n+1)\text{th term from the } n\text{th term}$$

defines a sequence recursively. To find the first five terms of this sequence, we proceed as follows:

$$a_1 = 5$$

$$a_2 = 3(\mathbf{a_1}) - 2 = 3(\mathbf{5}) - 2 = 13 \qquad \text{Substitute 5 for } a_1 \text{ and simplify.}$$

$$a_3 = 3(\mathbf{a_2}) - 2 = 3(\mathbf{13}) - 2 = 37 \qquad \text{Substitute 13 for } a_2 \text{ and simplify.}$$

$$a_4 = 3(\mathbf{a_3}) - 2 = 3(\mathbf{37}) - 2 = 109 \qquad \text{Substitute 37 for } a_3 \text{ and simplify.}$$

$$a_5 = 3(\mathbf{a_4}) - 2 = 3(\mathbf{109}) - 2 = 325 \qquad \text{Substitute 109 for } a_4 \text{ and simplify.}$$

3. Define Series

To add the terms of a sequence, we replace each comma between its terms with a plus sign to form a **series**. If a sequence is infinite, the number of terms in the series associated with it is infinite also. Two examples of infinite series are

$$1^2 + 2^2 + 3^2 + \cdots + n^2 + \cdots$$

and

$$1 + 2 + 3 + 5 + 8 + 13 + 21 + \cdots.$$

If a sequence is finite, the number of terms in the series associated with it is finite also. An example of a finite series is

$$3 + 7 + 11 + 15.$$

If the signs between successive terms of a series alternate, the series is called an **alternating series**. Two examples of alternating infinite series are

$$-3 + 6 - 9 + 12 - 15 + 18 - \cdots + (-1)^n 3n + \cdots$$

and

$$2 - 4 + 8 - 16 + \cdots + (-1)^{n+1} 2^n + \cdots.$$

 Series have many applications in calculus. The value of certain important functions can be approximated using something called a Maclaurin series.

4. Define and Use Summation Notation

Summation notation is a shorthand way to indicate the sum of the first n terms, or the **nth partial sum**, of a sequence. For example, the expression

$$\sum_{n=1}^{4} n \qquad \text{The symbol } \Sigma \text{ is the capital letter sigma in the Greek alphabet.}$$

indicates the sum of the four terms obtained when we successively substitute 1, 2, 3, and 4 for n.

$$\sum_{n=1}^{4} n = \mathbf{1} + \mathbf{2} + \mathbf{3} + \mathbf{4} = 10$$

The expression

$$\sum_{n=2}^{4} n^2$$

indicates the sum of the three terms obtained when we successively substitute 2, 3, and 4 into n^2.

$$\sum_{n=2}^{4} n^2 = (\mathbf{2})^2 + (\mathbf{3})^2 + (\mathbf{4})^2$$
$$= 4 + 9 + 16$$
$$= 29$$

The expression

$$\sum_{n=1}^{3} (2n^2 + 1)$$

indicates the sum of the three terms obtained if we successively substitute 1, 2, and 3 for n in the expression $2n^2 + 1$.

$$\sum_{n=1}^{3} (2n^2 + 1) = [2(\mathbf{1})^2 + 1] + [2(\mathbf{2})^2 + 1] + [2(\mathbf{3})^2 + 1]$$
$$= 3 + 9 + 19$$
$$= 31$$

EXAMPLE 1 **Evaluating a Sum**

Evaluate: $\displaystyle\sum_{n=1}^{4} (n^2 + 1)$.

SOLUTION Since n runs from 1 to 4, we substitute 1, 2, 3, and 4 for n in the expression $n^2 + 1$ and find the sum of the resulting terms:

$$\sum_{n=1}^{4} (n^2 + 1) = (1^2 + 1) + (2^2 + 1) + (3^2 + 1) + (4^2 + 1)$$

$$= 2 + 5 + 10 + 17$$

$$= 34 \qquad \blacksquare$$

Self Check 1 Evaluate: $\displaystyle\sum_{n=1}^{5} (n^2 - 1)$.

Now Try Exercise 43.

EXAMPLE 2 **Evaluating a Sum**

Evaluate: $\displaystyle\sum_{n=3}^{5} (3n + 2)$.

SOLUTION Since n runs from 3 to 5, we substitute 3, 4, and 5 for n in the expression $3n + 2$ and find the sum of the resulting terms.

$$\sum_{n=3}^{5} (3n + 2) = [3(3) + 2] + [3(4) + 2] + [3(5) + 2]$$

$$= 11 + 14 + 17$$

$$= 42 \qquad \blacksquare$$

Self Check 2 Evaluate: $\displaystyle\sum_{n=2}^{5} (3n + 2)$.

Now Try Exercise 49.

We can use summation notation to state the Binomial Theorem concisely. In Exercise 58, you will be asked to explain why the Binomial Theorem can be stated as

$$\sum_{r=0}^{n} \frac{n!}{r!(n - r)!} a^{n-r}b^r.$$

There are three basic properties of summations. The first states that *the summation of a constant as k runs from 1 to n is n times the constant.*

Summation of a Constant If c is a constant, then $\displaystyle\sum_{k=1}^{n} c = nc$.

PROOF Because c is a constant, each term is c for each value of k as k runs from 1 to n.

$$\sum_{k=1}^{n} c = \overbrace{c + c + c + c + \cdots + c}^{n \text{ number of } cs} = nc \qquad \blacksquare$$

EXAMPLE 3 **Using Summation of a Constant**

Evaluate: $\displaystyle\sum_{n=1}^{5} 13$.

SOLUTION $\displaystyle\sum_{n=1}^{5} 13 = 13 + 13 + 13 + 13 + 13$

$= 5(13)$

$= 65$

Self Check 3   Evaluate: $\displaystyle\sum_{n=1}^{6} 12.$

Now Try Exercise 47.

A second property states that *a constant factor can be brought outside a summation sign.*

Summation of a Product   If c is a constant, then $\displaystyle\sum_{k=1}^{n} cf(k) = c\sum_{k=1}^{n} f(k).$

PROOF $\displaystyle\sum_{k=1}^{n} cf(k) = cf(1) + cf(2) + cf(3) + \cdots + cf(n)$

$= c[f(1) + f(2) + f(3) + \cdots + f(n)]$ Factor out c.

$= c\displaystyle\sum_{k=1}^{n} f(k)$

EXAMPLE 4   **Using Summation of a Product**

Show that $\displaystyle\sum_{k=1}^{3} 5k^2 = 5\sum_{k=1}^{3} k^2.$

SOLUTION $\displaystyle\sum_{k=1}^{3} 5k^2 = 5(1)^2 + 5(2)^2 + 5(3)^2$

$= 5 + 20 + 45$

$= 70$

$5\displaystyle\sum_{k=1}^{3} k^2 = 5[(1)^2 + (2)^2 + (3)^2]$

$= 5[1 + 4 + 9]$

$= 5(14)$

$= 70$

The quantities are equal.

Self Check 4   Evaluate: $\displaystyle\sum_{k=1}^{4} 3k.$

Now Try Exercise 41.

The third property states that *the summation of a sum is equal to the sum of the summations.*

Summation of a Sum $\displaystyle\sum_{k=1}^{n} [f(k) + g(k)] = \sum_{k=1}^{n} f(k) + \sum_{k=1}^{n} g(k)$

PROOF

$$\sum_{k=1}^{n}\left[f(k) + g(k)\right] = \left[f(1) + g(1)\right] + \left[f(2) + g(2)\right] + \left[f(3) + g(3)\right] + \cdots + \left[f(n) + g(n)\right]$$

$$= \left[f(1) + f(2) + f(3) + \cdots + f(n)\right] + \left[g(1) + g(2) + g(3) + \cdots + g(n)\right]$$

$$= \sum_{k=1}^{n} f(k) + \sum_{k=1}^{n} g(k)$$

■

EXAMPLE 5   **Using Summation of a Sum**

Show that $\displaystyle\sum_{k=1}^{3}(k + k^2) = \sum_{k=1}^{3}k + \sum_{k=1}^{3}k^2$.

SOLUTION $\displaystyle\sum_{k=1}^{3}(k + k^2) = (1 + 1^2) + (2 + 2^2) + (3 + 3^2)$

$$= 2 + 6 + 12$$

$$= 20$$

$$\sum_{k=1}^{3}k + \sum_{k=1}^{3}k^2 = (1 + 2 + 3) + (1^2 + 2^2 + 3^2)$$

$$= 6 + 14$$

$$= 20$$

■

Self Check 5   Evaluate: $\displaystyle\sum_{k=1}^{3}(k^2 + 2k)$.

Now Try Exercise 45.

EXAMPLE 6   **Evaluating a Sum Directly and by Using Properties**

Evaluate $\displaystyle\sum_{k=1}^{5}(2k - 1)^2$ directly. Then expand the binomial, apply the properties, and evaluate the expression again.

SOLUTION   **Part 1:** $\displaystyle\sum_{k=1}^{5}(2k - 1)^2 = 1 + 9 + 25 + 49 + 81 = 165$

Part 2: $\displaystyle\sum_{k=1}^{5}(2k - 1)^2 = \sum_{k=1}^{5}(4k^2 - 4k + 1)$

$$= \sum_{k=1}^{5}4k^2 + \sum_{k=1}^{5}(-4k) + \sum_{k=1}^{5}1 \qquad \text{The summation of a sum is the sum of the summations.}$$

$$= 4\sum_{k=1}^{5}k^2 - 4\sum_{k=1}^{5}k + \sum_{k=1}^{5}1 \qquad \text{Bring the constant factors outside the summation signs.}$$

$$= 4\sum_{k=1}^{5}k^2 - 4\sum_{k=1}^{5}k + 5 \qquad \text{The summation of a constant as } k \text{ runs from 1 to 5 is 5 times that constant.}$$

$$= 4(1 + 4 + 9 + 16 + 25) - 4(1 + 2 + 3 + 4 + 5) + 5$$

$$= 4(55) - 4(15) + 5$$

$$= 220 - 60 + 5$$

$$= 165$$

Either way, the sum is 165.

■

Self Check 6   Evaluate: $\displaystyle\sum_{k=1}^{4}(2k-1)^2$.

Now Try Exercise 51.

Self Check Answers   1. 50   2. 50   3. 72   4. 30   5. 26   6. 84

Exercises **10.2**

Getting Ready
You should be able to complete these vocabulary and concept statements before you proceed to the practice exercises.

Fill in the blanks.

1. An infinite sequence is a function whose _____ is the set of natural numbers.

2. A finite sequence is a function whose _____ is the set of the first n natural numbers.

3. A(n) _____ is formed when we add the terms of a sequence.

4. A series associated with a finite sequence is a(n) _____ series.

5. A series associated with a infinite sequence is a(n) _____ series.

6. If the signs between successive terms of a series alternate, the series is called a(n) _____ series.

7. _____ is a shorthand way to indicate the sum of the first n terms of a sequence.

8. The symbol $\displaystyle\sum_{k=1}^{5}(k^2-3)$ indicates the _____ of the five terms obtained when we successively substitute 1, 2, 3, 4, and 5 for k.

9. $\displaystyle\sum_{k=1}^{5}6k^2 =$ _____ $\displaystyle\sum_{k=1}^{5}k^2$

10. $\displaystyle\sum_{k=1}^{5}(k^2+3k) = \sum_{k=1}^{5}k^2 +$ _____

11. $\displaystyle\sum_{k=1}^{5}c$, where c is a constant, equals _____.

12. The summation of a sum is equal to the _____ of the summations.

Practice
Write the first six terms of the sequence defined by each function.

13. $f(n) = 5n(n-1)$

14. $f(n) = n\left(\dfrac{n-1}{2}\right)\left(\dfrac{n-2}{3}\right)$

Find the next term of each sequence.

15. $1, 6, 11, 16, \ldots$

16. $1, 8, 27, 64, \ldots$

17. $a, a+d, a+2d, a+3d, \ldots$

18. $a, ar, ar^2, ar^3, \ldots$

19. $1, 3, 6, 10, \ldots$

20. $20, 17, 13, 8, \ldots$

Find the sum of the first five terms of the sequence with the given general term.

21. n 22. $2k$

23. 3 24. $4k^0$

25. $2\left(\dfrac{1}{3}\right)^n$ 26. $(-1)^n$

27. $3n-2$ 28. $2k+1$

Assume that each sequence is defined recursively. Find the first four terms of each sequence.

29. $a_1 = 3$ and $a_{n+1} = 2a_n + 1$

30. $a_1 = -5$ and $a_{n+1} = -a_n - 3$

31. $a_1 = -4$ and $a_{n+1} = \dfrac{a_n}{2}$

32. $a_1 = 0$ and $a_{n+1} = 2a_n^2$

33. $a_1 = k$ and $a_{n+1} = a_n^2$

34. $a_1 = 3$ and $a_{n+1} = ka_n$

35. $a_1 = 8$ and $a_{n+1} = \dfrac{2a_n}{k}$

36. $a_1 = m$ and $a_{n+1} = \dfrac{a_n^2}{m}$

Determine whether each series is an alternating infinite series.

37. $-1 + 2 - 3 + \cdots + (-1)^n n + \cdots$

38. $a + \dfrac{a}{b} + \dfrac{a}{b^2} + \cdots + a\left(\dfrac{1}{b}\right)^{n-1} + \cdots \; ; b = 4$

39. $a + a^2 + a^3 + \cdots + a^n + \cdots \; ; a = 3$

40. $a + a^2 + a^3 + \cdots + a^n + \cdots \; ; a = -2$

Evaluate each sum.

41. $\displaystyle\sum_{k=1}^{5} 2k$

42. $\displaystyle\sum_{k=3}^{6} 3k$

43. $\displaystyle\sum_{k=3}^{4} (-2k^2)$

44. $\displaystyle\sum_{k=1}^{100} 5$

45. $\displaystyle\sum_{k=1}^{5} (3k - 1)$

46. $\displaystyle\sum_{n=2}^{5} (n^2 + 3n)$

47. $\displaystyle\sum_{k=1}^{1,000} \frac{1}{2}$

48. $\displaystyle\sum_{x=4}^{5} \frac{2}{x}$

49. $\displaystyle\sum_{x=3}^{4} \frac{1}{x}$

50. $\displaystyle\sum_{x=2}^{6} (3k^2 + 2x) - 3\sum_{x=2}^{6} x^2$

51. $\displaystyle\sum_{x=1}^{4} (4x + 1)^2 - \sum_{x=1}^{4} (4x - 1)^2$

52. $\displaystyle\sum_{x=0}^{10} (2x - 1)^2 + 4\sum_{x=0}^{10} x(1 - x)$

53. $\displaystyle\sum_{x=6}^{8} (5x - 1)^2 + \sum_{x=6}^{8} (10x - 1)$

54. $\displaystyle\sum_{x=2}^{7} (3x + 1)^2 - 3\sum_{x=2}^{7} x(3x + 2)$

Discovery and Writing

55. Find a counterexample to disprove the proposition that the summation of a product is the product of the summations. In other words, prove that

$$\sum_{k=1}^{n} f(k)g(k) \neq \sum_{k=1}^{n} f(k) \sum_{k=1}^{n} g(k).$$

56. Find a counterexample to disprove the proposition that the summation of a quotient is the quotient of the summations. In other words, prove that

$$\sum_{k=1}^{n} \frac{f(k)}{g(k)} \neq \frac{\displaystyle\sum_{k=1}^{n} f(k)}{\displaystyle\sum_{k=1}^{n} g(k)}.$$

57. Explain what it means to define something recursively.

58. Explain why the Binomial Theorem can be stated as

$$\sum_{r=0}^{n} \frac{n!}{r!(n - r)!} a^{n-r} b^r.$$

10.3 Arithmetic Sequences and Series

In this section, we will learn to

1. Define and use arithmetic sequences.

2. Find arithmetic means.

3. Find the sum of the first n terms of an arithmetic series.

4. Solve problems involving sequences and series.

The German mathematician Carl Friedrich Gauss (1777–1855) was once a student in the class of a strict teacher. One day, the teacher asked the students to add together all of the natural numbers from 1 through 100. Gauss recognized that in the sum

$$1 + 2 + 3 + \cdots + 98 + 99 + 100,$$

the first number (1) added to the last number (100) is 101, the second number (2) added to the second from the last number (99) is 101, and the third number (3) added to the third from the last number (98) is 101. He reasoned that there would be 50 pairs of such numbers, and that there would be 50 sums of 101. He multiplied 101 by 50 to get the correct answer of 5,050.

This story illustrates a problem involving the sum of the terms of a sequence, called an **arithmetic sequence**, in which each term except the first is found by adding a constant to the preceding term.

1. Define and Use Arithmetic Sequences

Arithmetic Sequence   An **arithmetic sequence** is a sequence of the form

$$a, \quad a + d, \quad a + 2d, \quad a + 3d, \quad \ldots, \quad a + (n - 1)d, \ldots,$$

where a is the **first term**, $a + (n - 1)d$ is the **nth term**, and d is the **common difference**.

In this definition, the second term has an addend of d, the third term has an addend of $2d$, the fourth term has an addend of $3d$, and so on. This is why the nth term has an addend of $(n - 1)d$.

If an arithmetic sequence has infinitely many terms, it is called an **infinite arithmetic sequence**. If it has a finite number of terms, it is called a **finite arithmetic sequence**.

EXAMPLE 1   **Writing Terms of an Arithmetic Sequence**

Write the first six terms and the 21st term of an arithmetic sequence with a first term of 7 and a common difference of 5.

SOLUTION   Since the first term a is 7 and the common difference d is 5, the first six terms are

$$7, 7 + 5, 7 + 2(5), 7 + 3(5), 7 + 4(5), 7 + 5(5)$$

or

$$7, 12, 17, 22, 27, 32.$$

To find the 21st term, we substitute 21 for n in the formula for the nth term.

$$n\text{th term} = \boldsymbol{a} + (n - 1)d$$

$$21\text{st term} = \boldsymbol{7} + (\boldsymbol{21} - 1)5$$

$$= 7 + (20)5$$

$$= 107$$

The 21st term is 107.   ∎

Self Check 1   Write the first five terms and the 18th term of an arithmetic sequence with a first term of 3 and a common difference of 6.

Now Try Exercise 9.

EXAMPLE 2   **Determining a Specific Term of an Arithmetic Sequence**

Find the 98th term of an arithmetic sequence whose first three terms are 2, 6, 10.

SOLUTION   Here $a = 2$, $n = 98$, and $d = 6 - 2 = 10 - 6 = 4$. Because we want to find the 98th term, we substitute these numbers into the formula for the nth term.

$$n\text{th term} = \boldsymbol{a} + (n - 1)d$$

$$98\text{th term} = \boldsymbol{2} + (\boldsymbol{98} - 1)4$$

$$= 2 + (97)4$$

$$= 390$$ ∎

Self Check 2   Write the 50th term of the arithmetic sequence whose first three terms are 3, 8, 13.

Now Try Exercise 19.

2. Find Arithmetic Means

Numbers inserted between a first and last term to form a segment of an arithmetic sequence are called **arithmetic means**. When finding arithmetic means, we consider the last term, a_n, to be the nth term:

$$a_n = a + (n - 1)d.$$

EXAMPLE 3   **Inserting Arithmetic Means between Real Numbers**

Insert three arithmetic means between -3 and 12.

SOLUTION   Since we are inserting three arithmetic means between -3 and 12, the total number of terms is five. Thus, $a = -3$, $a_n = 12$, and $n = 5$. To find the common difference, we substitute -3 for a, 12 for a_n, and 5 for n in the formula for the last term and solve for d.

$$\mathbf{a_n} = a + (n - 1)d$$

$$\mathbf{12} = -3 + (5 - 1)d$$

$$15 = 4d \qquad \text{Add 3 to both sides and simplify.}$$

$$\frac{15}{4} = d \qquad \text{Divide both sides by 4.}$$

Once we know d, we can find the other terms of the sequence.

$$\mathbf{a + d} = \mathbf{-3} + \frac{15}{4} = \frac{3}{4}$$

$$\mathbf{a + 2d} = \mathbf{-3} + 2\left(\frac{15}{4}\right) = -3 + \frac{30}{4} = \frac{9}{2}$$

$$\mathbf{a + 3d} = \mathbf{-3} + 3\left(\frac{15}{4}\right) = -3 + \frac{45}{4} = \frac{33}{4}$$

The three arithmetic means are $\frac{3}{4}$, $\frac{9}{2}$, and $\frac{33}{4}$. ∎

Self Check 3   Find three arithmetic means between -5 and 23.

Now Try Exercise 23.

3. Find the Sum of the First n Terms of an Arithmetic Series

If we replace the commas in an infinite arithmetic sequence with plus signs, we form an **infinite arithmetic series**,

$$a + (a + d) + (a + 2d) + (a + 3d) + \cdots + [a + (n - 1)d] + \cdots.$$

If we replace the commas in a finite arithmetic sequence with plus signs, we form a **finite arithmetic series**,

$$a + (a + d) + (a + 2d) + (a + 3d) + \cdots + [a + (n - 1)d].$$

To find the sum of the first n terms of an arithmetic series, we use the following formula.

Sum of the First n Terms of an Arithmetic Series

The formula

$$S_n = \frac{n(a + a_n)}{2}$$

gives the sum of the first n terms of an arithmetic series. In this formula, a is the first term, a_n is the last (or nth) term, and n is the number of terms.

EXAMPLE 4   **Finding the Sum of the Terms of an Arithmetic Series**

Find the sum of the first 30 terms of the arithmetic series $5 + 8 + 11 + \cdots$.

SOLUTION   Here $a = 5, n = 30, d = 3$, and $a_n = a_{30} = 5 + 29(3) = 92$.
Substituting these values into the formula for the sum of the first n terms of an arithmetic series gives:

$$S_n = \frac{n(a + a_n)}{2}$$

$$S_{30} = \frac{30(5 + 92)}{2}$$

$$= 15(97)$$

$$= 1,455$$

The sum of the first 30 terms is 1,455.

Self Check 4   Find the sum of the first 50 terms of the arithmetic series $-2 + 5 + 12 + \cdots$.

Now Try Exercise 27.

4. Solve Problems Involving Sequences and Series

EXAMPLE 5   **Solving an Applied Problem**

A student deposits \$50 in a non–interest-bearing account and plans to add \$7 a week. How much will she have in the account one year after her first deposit?

SOLUTION   Her weekly balances form an arithmetic sequence,

$$50, 57, 64, 71, 78, \ldots,$$

with a first term of 50 and a common difference of 7. To find her balance in one year (52 weeks), we substitute 50 for a, 7 for d, and 52 for n in the formula for the last term.

$$a_n = a + (n - 1)d$$

$$a_{52} = 50 + (52 - 1)7$$

$$= 50 + (51)7$$

$$= 407$$

After one year, the balance will be \$407.

Self Check 5   How much will the student have in her account after 60 weeks?

Now Try Exercise 37.

EXAMPLE 6　Solving an Applied Problem

The equation $s = 16t^2$ represents the distance s (in feet) that an object will fall in t seconds.

- In 1 second, the object will fall 16 feet.
- In 2 seconds, the object will fall 64 feet.
- In 3 seconds, the object will fall 144 feet.

The object falls 16 feet during the first second, 48 feet during the next second, and 80 feet during the third second. Find the distance the object falls during the 12th second.

SOLUTION　The sequence 16, 48, 80, . . . is an arithmetic sequence with $a = 16$ and $d = 32$. To find the 12th term, we substitute these values into the formula for the last term.

$$a_n = a + (n - 1)d$$

$$a_{12} = 16 + (12 - 1)32$$

$$= 16 + 11(32)$$

$$= 368$$

During the 12th second, the object falls 368 feet. ∎

Self Check 6　How far does the object fall during the 20th second?

Now Try Exercise 41.

Self Check Answers　**1.** 3, 9, 15, 21, 27; 105　**2.** 248　**3.** 2, 9, and 16　**4.** 8,475　**5.** $463
6. 624 ft

Exercises **10.3**

Getting Ready
You should be able to complete these vocabulary and concept statements before you proceed to the practice exercises.

Fill in the blanks.

1. An arithmetic sequence is a sequence of the form

   $a, a + d, a + 2d, a + 3d, . . . , a +$ _____ d.

2. An arithmetic series is a series of the form

   $a + (a + d) + (a + 2d) + (a + 3d) + \cdots$
   $\qquad + [a + (n - 1)$ _____ $] + \cdots$.

3. If an arithmetic series has infinitely many terms, it is called a(n) _____ arithmetic series.

4. In an arithmetic sequence, a is the _____ term, d is the common _____, and n is the _____ of terms.

5. The last term of an arithmetic sequence is given by the formula _____.

6. The formula for the sum of the first n terms of an arithmetic series is _____.

7. _____ are numbers inserted between the first and last term of a sequence to form an arithmetic sequence.

8. The formula _____ gives the distance (in feet) that an object will fall in t seconds.

Practice
Write the first six terms of the arithmetic sequences with the given properties.

9. $a = 1; d = 2$

10. $a = -12; d = -5$

11. $a = 5$; third term is 2

12. $a = 4$; fifth term is 12

13. Seventh term is 24; common difference is $\dfrac{5}{2}$

14. Twentieth term is -49; common difference is -3

Find the requested term in each arithmetic sequence.

15. Find the 40th term of an arithmetic sequence with a first term of 6 and a common difference of 8.

16. Find the 35th term of an arithmetic sequence with a first term of 50 and a common difference of -6.

17. The sixth term of an arithmetic sequence is 28, and the first term is -2. Find the common difference.

18. The seventh term of an arithmetic sequence is -42, and the common difference is -6. Find the first term.

19. Find the 55th term of an arithmetic sequence whose first three terms are $-8, -1, 6$.

20. Find the 37th term of an arithmetic sequence whose second and third terms are $-4, 6$.

21. If the fifth term of an arithmetic sequence is 14 and the second term is 5, find the 15th term.

22. If the fourth term of an arithmetic sequence is 13 and the second term is 3, find the 24th term.

Find the requested means.

23. Insert three arithmetic means between 10 and 20.

24. Insert five arithmetic means between 5 and 15.

25. Insert four arithmetic means between -7 and $\frac{2}{3}$.

26. Insert three arithmetic means between -11 and -2.

Find the sum of the indicated terms of each arithmetic series.

27. $5 + 7 + 9 + \cdots$; to 15 terms

28. $-3 + (-4) + (-5) + \cdots$; to 10 terms

29. $\sum_{n=1}^{20} \left(\frac{3}{2} n + 12 \right)$ **30.** $\sum_{n=1}^{10} \left(\frac{2}{3} n + \frac{1}{3} \right)$

Solve each problem.

31. Find the sum of the first 30 terms of an arithmetic sequence with a 25th term of 10 and a common difference of $\frac{1}{2}$.

32. Find the sum of the first 100 terms of an arithmetic sequence with a 15th term of 86 and a first term of 2.

33. Find the sum of the first 200 natural numbers.

34. Find the sum of the first 1,000 natural numbers.

Applications

35. Interior angles The sums of the angles of several polygons are given in the table. Assuming that the pattern continues, complete the table.

Figure	Number of sides	Sum of angles
Triangle	3	180°
Quadrilateral	4	360°
Pentagon	5	540°
Hexagon	6	720°
Octagon	8	
Dodecagon	12	

36. Borrowing money To pay for college, a student borrows $5,000 interest-free from his father. If he pays his father back at the rate of $200 per month, how much will he still owe after 12 months?

37. Borrowing money If Juanita borrows $5,500 interest-free from her mother to buy a new car and agrees to pay her mother back at the rate of $105 per month, how much will she still owe after 4 years?

38. Jogging One day, some students jogged $\frac{1}{2}$ mile. Because it was fun, they decided to increase the jogging distance each day by a certain amount. If they jogged $6\frac{3}{4}$ miles on the 51st day, how much was the distance increased each day?

39. Sales The year it incorporated, a company had sales of $237,500. Its sales were expected to increase by $150,000 annually for the next several years. If the forecast was correct, what were its sales in 10 years?

40. Falling objects Find how many feet a brick will travel during the 10th second of its fall.

41. Falling objects If a rock is dropped from the Golden Gate Bridge, how far will it fall in the eleventh second?

42. Designing patios Each row of bricks in the triangular patio shown in the illustration is to have one more brick than the previous row, ending with the longest row of 150 bricks. How many bricks will be needed?

43. Pile of logs Several logs are stored in a pile with 20 logs on the bottom layer, 19 on the second layer, 18 on the third layer, and so on. If the top layer has one log, how many logs are in the pile?

44. Theater seating The first row in a movie theater contains 24 seats. As you move toward the back, each row has 1 additional seat. If there are 30 rows, what is the capacity of the theater?

Discovery and Writing

45. Can an arithmetic sequence have a first term of 4, a 25th term of 126, and a common difference of $4\frac{1}{4}$? Explain.

46. In an arithmetic sequence, can a and d be negative but a_n be positive?

47. Can an arithmetic sequence be an alternating sequence? Explain.

48. Between 5 and $10\frac{1}{3}$ are three arithmetic means. One of them is 9. Find the other two.

10.4 Geometric Sequences and Series

In this section, we will learn to

1. Define and use geometric sequences.
2. Find geometric means.
3. Find the sum of the first n terms of a geometric series.
4. Define and find the sum of infinite geometric series.
5. Solve problems involving geometric sequences and series.

© Istockphoto.com/Matej Michelizza

Bungee jumping is a sport invented by daredevil A. J. Hackett. His first jump was off the 43-meter high Kawarau Bridge in New Zealand. Since he was attached to a long rubber cord, he bounced up and down for a considerable time, giving him a thrilling ride. Today this sport is gaining in popularity, but it is not for the faint of heart.

Suppose a jumper is attached to a cord that stretches to a length of 100 feet. Also suppose that on each bounce, he rebounds to a height that is 60% of the distance from which he fell. If we list the distances that he falls, we will get the following sequence:

$100, 60, 36, \ldots\ldots$ Each number in the list is 60% of the number before it.

If we list the distances that he rebounds, we will get the following sequence:

$60, 36, 21.6, \ldots\ldots$ Each number in the list is 60% of the number before it.

Each of these sequences is an example of a common sequence called a *geometric sequence*.

1. Define and Use Geometric Sequences

A **geometric sequence** is a sequence in which each term except the first is found by multiplying the preceding term by a constant.

Geometric Sequence   A **geometric sequence** is a sequence of the form

$$a, ar, ar^2, ar^3, \ldots, ar^{n-1}, \ldots,$$

where a is the **first term**, ar^{n-1} is the **nth term**, and r is the **common ratio**.

Caution

> Please note that the nth term of a geometric sequence is not $a_n = ar^n$. The nth term is $a_n = ar^{n-1}$.

In this definition, the second term of the sequence has a factor of r^1, the third term has a factor of r^2, the fourth term has a factor of r^3, and so on. This explains why the nth term has a factor of r^{n-1}.

If a geometric sequence has infinitely many terms, it is called an **infinite geometric sequence**. If it has a finite number of terms, it is called a **finite geometric sequence**.

EXAMPLE 1   **Writing the Terms of a Geometric Sequence**

Write the first six terms and the 15th term of the geometric sequence whose first term is 3 and whose common ratio is 2.

SOLUTION   We first write the first six terms of the geometric sequence:

$$3, 3(2), 3(2)^2, 3(2)^3, 3(2)^4, 3(2)^5,$$

or

$$3, 6, 12, 24, 48, 96.$$

To find the 15th term, we substitute 15 for n, 3 for a, and 2 for r in the formula for the nth term.

$$n\text{th term} = ar^{n-1}$$
$$15\text{th term} = 3(2)^{15-1}$$
$$= 3(2)^{14}$$
$$= 3(16,384)$$
$$= 49,152$$

Self Check 1   Write the first five terms of the geometric sequence whose first term is 2 and whose common ratio is 3. Find the 10th term.

Now Try Exercise 11.

EXAMPLE 2   **Finding a Specific Term of a Geometric Sequence**

Find the eighth term of a geometric sequence whose first three terms are 9, 3, 1.

SOLUTION   Here $a = 9$, $r = \frac{1}{3}$, and $n = 8$. To find the eighth term, we substitute these values into the formula for the nth term.

$$n\text{th term} = ar^{n-1}$$
$$8\text{th term} = 9\left(\frac{1}{3}\right)^{8-1}$$
$$= 9\left(\frac{1}{3}\right)^7$$
$$= \frac{1}{243}$$

Self Check 2   Find the eighth term of a geometric sequence whose first three terms are $\frac{1}{3}$, 1, 3.

Now Try Exercise 17.

2. Find Geometric Means

Numbers inserted between a first and last term to form a segment of a geometric sequence are called **geometric means**. When finding geometric means, we consider the last term, a_n, to be the nth term.

EXAMPLE 3   **Inserting Geometric Means between Two Integers**

Insert two geometric means between 4 and 256.

SOLUTION   The first term is $a = 4$, and because 256 is the fourth term, $n = 4$ and $a_n = a_4 = 256$. To find the common ratio, we substitute these values into the formula for the nth term and solve for r.

$$ar^{n-1} = a_n$$

$$4r^{4-1} = 256$$

$$r^3 = 64$$

$$r = 4$$

The common ratio is 4. The two geometric means are the second and third terms of the geometric sequence:

$$ar = 4 \cdot 4 = 16$$

$$ar^2 = 4 \cdot 4^2 = 4 \cdot 16 = 64$$

The first four terms of the geometric sequence are 4, 16, 64, 256. The two geometric means between 4 and 256 are 16 and 64.

Self Check 3   Insert two geometric means between -3 and 192.

Now Try Exercise 23.

3. Find the Sum of the First n Terms of a Geometric Series

If we replace the commas in an infinite geometric sequence with plus signs, we form an **infinite geometric series**,

$$a + ar + ar^2 + ar^3 + \cdots + ar^{n-1} + \cdots.$$

If we replace the commas in a finite geometric sequence with plus signs, we form a **finite geometric series**,

$$a + ar + ar^2 + ar^3 + \cdots + ar^{n-1}.$$

To find the sum of the first n terms of a geometric series, we can use the following formula:

Sum of the First n Terms of a Geometric Series   The formula

$$S_n = \frac{a - ar^n}{1 - r} \qquad (r \neq 1)$$

gives the sum of the first n terms of a geometric series. In the formula, a is the first term, r is the common ratio, and n is the number of terms.

PROOF   We write the sum of the first n terms of the geometric series:

(1) $S_n = a + ar + ar^2 + \cdots + ar^{n-3} + ar^{n-2} + ar^{n-1}.$

Multiplying both sides of this equation by r gives

(2) $S_n r = ar + ar^2 + \cdots + ar^{n-2} + ar^{n-1} + ar^n.$

We now subtract Equation (2) from Equation (1) and solve for S_n.

$$S_n - S_n r = a - ar^n$$

$$S_n(1 - r) = a - ar^n \qquad \text{Factor out } S_n.$$

$$S_n = \frac{a - ar^n}{1 - r} \qquad \text{Divide both sides by } 1 - r.$$

EXAMPLE 4   **Finding the Sum of the Terms of a Geometric Series**

Find the sum of the first six terms of the geometric series $8 + 4 + 2 + \cdots$.

SOLUTION   Here $a = 8$, $n = 6$, and $r = \frac{1}{2}$. We substitute these values in the formula for the sum of the first n terms of a geometric series.

$$S_n = \frac{a - ar^n}{1 - r}$$

$$S_6 = \frac{8 - 8\left(\frac{1}{2}\right)^6}{1 - \frac{1}{2}}$$

$$= 2\left(\frac{63}{8}\right)$$

$$= \frac{63}{4}$$

The sum of the first six terms is $\frac{63}{4}$.

Self Check 4   Find the sum of the first eight terms of the geometric series $81 + 27 + 9 + \cdots$.

Now Try Exercise 25.

4. Define and Find the Sum of Infinite Geometric Series

Under certain conditions, we can find the sum of all of the terms in an **infinite geometric series**. To define this sum, we consider the infinite geometric series

$$a + ar + ar^2 + \cdots.$$

- The first partial sum of the series, S_1, is $S_1 = a$.
- The second partial sum of the series, S_2, is $S_2 = a + ar$.
- The nth partial sum of the series, S_n, is $S_n = a + ar + ar^2 + \cdots + ar^{n-1}$.

If the nth partial sum of an infinite geometric series, S_n, approaches some number S as n approaches ∞, then S is called the **sum of the infinite geometric series**. The following symbol denotes the sum S of an infinite geometric series, provided the sum exists:

$$S = \sum_{n=1}^{\infty} ar^{n-1}.$$

To develop a formula for finding the sum of all the terms in an infinite geometric series, we consider the formula

(3) $S_n = \dfrac{a - ar^n}{1 - r}$ $(r \neq 1)$.

If $|r| < 1$ and a is a constant, then as n approaches ∞, ar^n approaches 0 and the term ar^n in Equation (3) can be dropped. As an illustration, suppose that $a = 1$ and $r = \frac{1}{2}$. We can see from the table on the left that as n increases without bound, $ar^n = 1\left(\frac{1}{2}\right)^n$ approaches 0.

n	$1\left(\frac{1}{2}\right)^n$
1	$\frac{1}{2}$
2	$\frac{1}{4}$
3	$\frac{1}{8}$
4	$\frac{1}{16}$
10	$\frac{1}{1,024}$

This argument gives the following formula:

Sum of the Terms of an Infinite Geometric Series	If $\lvert r \rvert < 1$, the sum of the terms of an infinite geometric series is given by $$S_{\infty} = \frac{a}{1 - r},$$ where a is the first term and r is the common ratio.

Caution

> If $\lvert r \rvert \geq 1$, the terms get larger and larger in either the positive or negative direction and the sum does not approach a number. In this case, the previous theorem does not apply.

EXAMPLE 5 **Changing a Repeating Decimal to a Fraction**

Change $0.\overline{4}$ to a common fraction.

SOLUTION We write the decimal as an infinite geometric series and find its sum.

$$S_{\infty} = \frac{4}{10} + \frac{4}{100} + \frac{4}{1,000} + \frac{4}{10,000} + \cdots$$

$$= \frac{4}{10} + \frac{4}{10}\left(\frac{1}{10}\right) + \frac{4}{10}\left(\frac{1}{10}\right)^2 + \frac{4}{10}\left(\frac{1}{10}\right)^3 + \cdots$$

Since the common ratio r equals $\frac{1}{10}$ and $\left\lvert \frac{1}{10} \right\rvert < 1$, we can use the formula for the sum of an infinite geometric series.

$$S_{\infty} = \frac{a}{1 - r} = \frac{\dfrac{4}{10}}{1 - \dfrac{1}{10}} = \frac{\dfrac{4}{10}}{\dfrac{9}{10}} = \frac{4}{9}$$

Comment

Using a generalization of the method used in Example 5, we can write any repeating decimal as a fraction.

Long division will verify that $\frac{4}{9} = 0.\overline{4}$. ∎

Self Check 5 Change $0.\overline{7}$ to a common fraction.

Now Try Exercise 35.

5. Solve Problems Involving Geometric Sequences and Series

Many of the exponential-growth problems discussed in Chapter 3 can be solved using the concepts of geometric sequences.

EXAMPLE 6 **Solving an Application Problem**

A statistician knows that a town with a population of 3,500 people has a predicted growth rate of 6% per year for the next 20 years. What should she predict the population will be 20 years from now?

SOLUTION Let p_0 be the initial population of the town. After 1 year, the population p_1 will be the initial population (p_0) plus the growth (the product of p_0 and the rate of growth r).

$$p_1 = p_0 + p_0 r$$

$$= p_0(1 + r) \qquad \text{Factor out } p_0.$$

© Istockphoto.com/Denis Jr. Tangney

The population p_2 at the end of 2 years will be

$$p_2 = p_1 + p_1 r$$

$$= \mathbf{p_1}(1 + r) \qquad \text{Factor out } p_1.$$

$$= \mathbf{p_0}(1 + r)(1 + r) \qquad \text{Substitute } p_0(1 + r) \text{ for } p_1.$$

$$= p_0(1 + r)^2$$

The population at the end of the third year will be $p_3 = p_0(1 + r)^3$. Writing the terms as a sequence gives

$$p_0, \ p_0(1 + r), \ p_0(1 + r)^2, \ p_0(1 + r)^3, \ p_0(1 + r)^4, \ldots .$$

This is a geometric sequence with p_0 as the first term and $1 + r$ as the common ratio. In this example, $p_0 = 3{,}500$, $1 + r = 1.06$, and (since the population after 20 years will be the value of the 21st term of the geometric sequence) $n = 21$. We can substitute these values into the formula for the last term of a geometric sequence.

$$a_n = \mathbf{a} \mathbf{r}^{n-1}$$

$$a_{21} = \mathbf{3{,}500}(\mathbf{1.06})^{21-1}$$

$$= 3{,}500(1.06)^{20}$$

$$\approx 11{,}224.97415 \qquad \text{Use a calculator.}$$

The population after 20 years will be approximately 11,225.   ■

Self Check 6   After 30 years, what will be the approximate population?

Now Try Exercise 39.

EXAMPLE 7   **Solving an Application Problem**

A student deposits \$2,500 in a bank at 7% annual interest, compounded daily. If the investment is left untouched for 60 years, how much money will be in the account?

SOLUTION   We let the initial amount in the account be a_0 and r be the rate. At the end of the first day, the account is worth

$$a_1 = a_0 + a_0\left(\frac{r}{365}\right) = a_0\left(1 + \frac{r}{365}\right).$$

After the second day, the account is worth

$$a_2 = a_1 + a_1\left(\frac{r}{365}\right) = a_1\left(1 + \frac{r}{365}\right) = a_0\left(1 + \frac{r}{365}\right)^2.$$

The daily amounts form the geometric sequence

$$a_0, \ a_0\left(1 + \frac{r}{365}\right), \ a_0\left(1 + \frac{r}{365}\right)^2, \ a_0\left(1 + \frac{r}{365}\right)^3, \ldots ,$$

where a_0 is the initial deposit and r is the annual rate of interest.

Because interest is compounded daily for 60 years (21,900 days), the amount at the end of 60 years will be the 21,901st term of the sequence.

$$a_{21{,}901} = 2{,}500\left(1 + \frac{0.07}{365}\right)^{21{,}900}$$

We can use a calculator to find that $a_{21{,}901} \approx \$166{,}648.71$.   ■

Self Check 7   After 40 years, how much money is in the account?

Now Try Exercise 45.

EXAMPLE 8   Solving an Application Problem

A pump can remove 20% of the gas in a container with each stroke. Find the percentage of gas that remains in the container after six strokes.

SOLUTION   We let V represent the volume of the container. Because each stroke of the pump removes 20% of the gas, 80% remains after each stroke, and we have the geometric sequence

$$V, \quad 0.80V, \quad 0.80(0.80V), \quad 0.80[0.80(0.80V)], \ldots ,$$

or

$$V, \quad 0.8V, \quad (0.8)^2V, \quad (0.8)^3V, \ldots .$$

The amount of gas remaining after six strokes is the seventh term, a_n, of the sequence.

$$a_n = ar^{n-1}$$
$$a_7 = V(0.8)^{7-1}$$
$$= V(0.8)^6$$

We can use a calculator to find that approximately 26% of the gas remains after six strokes. ∎

Self Check 8   What approximate percentage would remain after seven strokes?

Now Try Exercise 47.

Self Check Answers   1. 2, 6, 18, 54, 162; 39,366   **2.** 729   **3.** 12, −48   **4.** $\dfrac{3,280}{27}$ **5.** $\dfrac{7}{9}$
6. ≈20,102   **7.** $41,100.58   **8.** 21%

Exercises 10.4

Getting Ready
You should be able to complete these vocabulary and concept statements before you proceed to the practice exercises.

Fill in the blanks.

1. A geometric sequence is a sequence of the form a, ar, ar^2, ar^3, ... The nth term is $a(\underline{\hspace{1cm}})$.

2. In a geometric sequence, a is the _____ term, r is the common _____, and n is the _____ of terms.

3. The last term of a geometric sequence is given by the formula $a_n = \underline{\hspace{1cm}}$.

4. A geometric _____ is the sum of the terms of a geometric sequence.

5. A geometric series with infinitely many terms is called a(n) _____ geometric series.

6. The formula for the sum of the first n terms of a geometric series is _____.

7. _____ are numbers inserted between a first and a last term to form a geometric sequence.

8. If $|r| < 1$, the formula _____ gives the sum of the terms of an infinite geometric series.

Practice
Write the first four terms of the geometric sequence with the given properties.

9. $a = 10; r = 2$

10. $a = -3; r = 2$

11. $a = -2; r = 3$

12. $a = 64; r = \dfrac{1}{2}$

13. $a = 3; r = \sqrt{2}$

14. $a = 2; r = \sqrt{3}$

15. $a = 2$; fourth term is 54

16. third term is 4; $r = \dfrac{1}{2}$

Find the requested term in each geometric sequence.

17. Find the sixth term of the geometric sequence whose first three terms are $\frac{1}{4}$, 1, 4.

18. Find the eighth term of the geometric sequence whose second and fourth terms are 0.2 and 5.

19. Find the fifth term of the geometric sequence whose second term is 6 and whose third term is -18.

20. Find the sixth term of the geometric sequence whose second term is 3 and whose fourth term is $\frac{1}{3}$.

Find the requested means.

21. Insert three positive geometric means between 10 and 20.

22. Insert five geometric means between -5 and 5, if possible.

23. Insert four geometric means between 2 and 2,048.

24. Insert three geometric means between 162 and 2. (There are two possibilities.)

Find the sum of the indicated terms of each geometric series.

25. $4 + 8 + 16 + \cdots$; to 5 terms

26. $9 + 27 + 81 + \cdots$; to 6 terms

27. $2 + (-6) + 18 + \cdots$; to 10 terms

28. $\frac{1}{8} + \frac{1}{4} + \frac{1}{2} + \cdots$; to 12 terms

29. $\displaystyle\sum_{n=1}^{6} 3\left(\frac{3}{2}\right)^{n-1}$ **30.** $\displaystyle\sum_{n=1}^{6} 12\left(-\frac{1}{2}\right)^{n-1}$

Find the sum of each infinite geometric series.

31. $6 + 4 + \frac{8}{3} + \cdots$ **32.** $8 + 4 + 2 + 1 + \cdots$

33. $\displaystyle\sum_{n=1}^{\infty} 12\left(-\frac{1}{2}\right)^{n-1}$ **34.** $\displaystyle\sum_{n=1}^{\infty} \left(\frac{1}{3}\right)^{n-1}$

Change each repeating decimal to a common fraction.

35. $0.\overline{5}$ **36.** $0.\overline{6}$

37. $0.\overline{25}$ **38.** $0.\overline{37}$

Applications

Use a calculator to help solve each problem.

39. Staffing a department The number of students studying algebra at State College is 623. The department chair expects enrollment to increase 10% each year. How many professors will be needed in 8 years to teach algebra if one professor can handle 60 students?

40. Bouncing balls On each bounce, the rubber ball in the illustration rebounds to a height one-half of that from which it fell. Find the total vertical distance the ball travels.

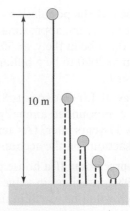

10 m

41. Bungee jumping A bungee jumper is attached to a cord that stretches to a length of 100 feet. If he rebounds to 60% of the height jumped, how far will he fall on his fifth descent? How far will he have traveled when he comes to rest?

42. Bungee jumping A bungee jumper is attached to a cord that stretches to a length of 100 feet. If he rebounds to 70% of the height jumped, how far will he travel upward on the fifth rebound? How far will he have traveled when he comes to rest?

43. Bouncing balls A Super Ball rebounds to approximately 95% of the height from which it is dropped. If the ball is dropped from a height of 10 meters, how high will it rebound after the 13th bounce?

44. Genealogy The family tree shown spans 3 generations and lists 7 people. How many names would be listed in a family tree that spans 10 generations?

45. Investing money If a married couple invests $1,000 in a 1-year certificate of deposit at $6\frac{3}{4}\%$ annual interest compounded daily, how much interest will accrue during the year?

46. Biology If a single cell divides into two cells every 30 minutes, how many cells will there be at the end of 10 hours?

47. Depreciation A lawn tractor that costs c dollars when new depreciates by 20% of the previous year's value each year. How much is the lawn tractor worth after 5 years?

48. **Financial planning** Maria can invest $1,000 at $7\frac{1}{2}\%$ compounded annually or at $7\frac{1}{4}\%$ compounded daily. If she invests the money for a year, which is the better investment?

49. **Population study** If the population of the earth were to double every 30 years, approximately how many people would there be in the year 3020? (Consider the population in 2000 to be 5 billion and use 2000 as the base year.)

50. **Investing money** If Linda deposits $1,300 in a bank at 7% interest compounded annually, how much will be in the bank 17 years later? (Assume that there are no other transactions on the account.)

51. **Real-estate appreciation** If a house purchased for $50,000 in 1998 appreciates in value by 6% each year, how much will it be worth in the year 2020?

52. **Compound interest** Find the value of $1,000 left on deposit for 10 years at an annual rate of 7% compounded annually.

53. **Compound interest** Find the value of $1,000 left on deposit for 10 years at an annual rate of 7% compounded quarterly.

54. **Compound interest** Find the value of $1,000 left on deposit for 10 years at an annual rate of 7% compounded monthly.

55. **Compound interest** Find the value of $1,000 left on deposit for 10 years at an annual rate of 7% compounded daily.

56. **Compound interest** Find the value of $1,000 left on deposit for 10 years at an annual rate of 7% compounded hourly.

57. **Saving for retirement** When John was 20 years old, he opened an individual retirement account by investing $2,000 at 11% interest compounded quarterly. How much will his investment be worth when he is 65 years old?

58. **Biology** One bacterium divides into two bacteria every 5 minutes. If two bacteria multiply enough to completely fill a petri dish in 2 hours, how long will it take one bacterium to fill the dish?

59. **Pest control** To reduce the population of a destructive moth, biologists release 1,000 sterilized male moths each day into the environment. If 80% of these moths alive one day survive until the next, then after a long time the population of sterile males is the sum of the infinite geometric series

$$1,000 + 1,000(0.8) + 1,000(0.8)^2 + 1,000(0.8)^3 + \cdots.$$

Find the long-term population.

60. **Pest control** If mild weather increases the day-to-day survival rate of the sterile male moths in Exercise 59 to 90%, find the long-term population.

61. **Mathematical myth** A legend tells of a king who offered to grant the inventor of the game of chess any request. The inventor said, "Simply place one grain of wheat on the first square of a chessboard, two grains on the second, four on the third, and so on, until the board is full. Then give me the wheat." The king agreed. How many grains did the king need to fill the chessboard?

62. **Mathematical myth** Estimate the size of the wheat pile from Exercise 57 in bushels. (*Hint:* There are about one-half million grains of wheat in a bushel.)

Discovery and Writing

63. Does 0.999999 = 1? Explain.

64. Does 0.999 . . . = 1? Explain.

10.5 Mathematical Induction

In this section, we will learn to

1. State the principle of mathematical induction.
2. Use mathematical induction to prove formulas.

Suppose you are waiting in line to get into a movie and are wondering if you will be admitted. Even if the first person in line gets in, your worries will still be justified. Perhaps there is room in the theater for only a few more people.

It would be good news to hear the manager say, "If anyone gets in, the next person in line will also get in." However, this promise does not guarantee that anyone will be admitted. Perhaps the theater is full, and no one will get in.

However, when you see the first person in line walk in, you know that everyone will be admitted because you know two things:

- The first person was admitted.
- Because of the promise, if the first person is admitted, then so is the second, and when the second person is admitted, then so is the third, and so on until everyone gets in.

This situation is similar to a game played with dominoes. Suppose that some dominoes are placed on end, as in Figure 10-2. When the first domino is knocked over, it knocks over the second. The second domino in turn knocks over the third, which knocks over the fourth, and so on until all of the dominoes fall. Two things must happen to guarantee that all of the dominoes will fall:

- The first domino must be knocked over.
- Every domino that falls must knock over the next one.

When both conditions are met, it is certain that all of the dominoes will fall.

These two examples illustrate the basic principle of mathematical induction.

FIGURE 10-2

1. State the Principle of Mathematical Induction

In Section 10.3, we saw the following formula for the sum of the first n terms of an arithmetic series:

$$S_n = \frac{n(a + a_n)}{2}.$$

If we apply this formula to the arithmetic series $1 + 2 + 3 + \cdots + n$, we have

$$S_n = 1 + 2 + 3 + \cdots + n = \frac{n(1 + n)}{2}.$$

To see that this formula is true, we can check it for some positive numbers n.

For $n = 1$: $1 = \dfrac{1(1 + 1)}{2}$ is a true statement, because $1 = 1$.

For $n = 2$: $1 + 2 = \dfrac{2(1 + 2)}{2}$ is a true statement, because $3 = 3$.

For $n = 3$: $1 + 2 + 3 = \dfrac{3(1 + 3)}{2}$ is a true statement, because $6 = 6$.

For $n = 6$: $1 + 2 + 3 + 4 + 5 + 6 = \dfrac{6(1 + 6)}{2}$ is a true statement, because $21 = 21$.

Because the set of positive numbers is infinite, it is impossible to prove the formula by verifying it for all positive numbers. To verify this sequence formula for all positive numbers n, we must have a method of proof called **mathematical induction**, first used extensively by Giuseppe Peano (1858–1932).

Mathematical Induction	If a statement involving the natural number n has the following two properties, then the statement is true for all natural numbers:

1. The statement is true for $n = 1$, and
2. If the statement is true for $n = k$, then it is true for $n = k + 1$.

Mathematical induction provides a way to prove many formulas. Any proof by induction involves two parts. First, we must show that the formula is true for the number 1. Second, we must show that, if the formula is true for any natural number k, then it also is true for the natural number $k + 1$. A proof by induction is complete only when both of these properties are established.

2. Use Mathematical Induction to Prove Formulas

EXAMPLE 1   **Using Mathematical Induction to Prove a Formula**

Use induction to prove that the following formula is true for every natural number n:

$$1 + 2 + 3 + \cdots + n = \frac{n(n + 1)}{2}.$$

SOLUTION   **Proof by induction**

Part 1:   Verify that the formula is true for $n = 1$. When $n = 1$, there is a single term, the number 1, on the left side of the equation. Substituting 1 for n on the right side, we have:

$$1 = \frac{n(n + 1)}{2}$$

$$1 = \frac{(1)(1 + 1)}{2}$$

$$1 = 1$$

The formula is true when $n = 1$. Part 1 of the proof is complete.

Part 2:   We assume that the given formula is true when $n = k$. By this assumption, called the **induction hypothesis**, we accept that

$$(1) \quad 1 + 2 + 3 + \cdots + k = \frac{k(k + 1)}{2}$$

is a true statement. We must show that the induction hypothesis forces the given formula to be true when $n = k + 1$. We can show this by verifying the statement

$$(2) \quad 1 + 2 + 3 + \cdots + k + (k + 1) = \frac{(k + 1)[(k + 1) + 1]}{2},$$

which is obtained from the given formula by replacing n with $k + 1$.

Comparing the left sides of Equations (1) and (2) shows that the left side of Equation (2) contains an extra term of $k + 1$. Thus, we add $k + 1$ to both sides of Equation (1) (which was assumed to be true) to obtain the equation

$$1 + 2 + 3 + \cdots + k + (k + 1) = \frac{k(k + 1)}{2} + (k + 1).$$

Because both terms on the right side of this equation have a common factor of $k + 1$, the right side factors, and the equation can be written as follows:

$$1 + 2 + 3 + \cdots + k + (k + 1) = (k + 1)\left(\frac{k}{2} + 1\right)$$

$$= (k + 1)\left(\frac{k + 2}{2}\right)$$

$$= \frac{(k + 1)(k + 2)}{2}$$

$$= \frac{(k + 1)[(k + 1) + 1]}{2}$$

This result is Equation (2). Because the truth of Equation (1) implies the truth of Equation (2), part 2 of the proof is complete. Parts 1 and 2 together establish that the formula is true for any natural number n. ∎

Self Check 1 Verify that the formula holds for $n = 4$.

Now Try Exercise 9.

EXAMPLE 2 **Using Mathematical Induction to Prove a Formula**

Use induction to prove that the following formula is true for all natural numbers n:

$$1 + 5 + 9 + \cdots + (4n - 3) = n(2n - 1).$$

SOLUTION **Proof by induction**

Part 1: First we verify the formula for $n = 1$. When $n = 1$, there is a single term, the number 1, on the left side of the equation. Substituting 1 for n on the right side, we have:

$$1 = \mathbf{1}[2(\mathbf{1}) - 1]$$

$$1 = 1$$

The formula is true for $n = 1$. Part 1 of the proof is complete.

Part 2: We assume that the formula is true for $n = k$. Hence,

$$(3) \quad 1 + 5 + 9 + \cdots + (4k - 3) = k(2k - 1)$$

is a true statement. To show that the induction hypothesis guarantees the truth of the formula for $k + 1$ terms, we add the $(k + 1)$th term to both sides of Equation (3). Because the terms on the left side increase by 4, the $(k + 1)$th term is $(4k - 3) + 4$, or $4k + 1$. Adding $4k + 1$ to both sides of Equation (3) gives

$$1 + 5 + 9 + \cdots + (4k - 3) + (4k + 1) = k(2k - 1) + (4k + 1).$$

We can simplify the right side and write the previous equation as follows:

$$1 + 5 + 9 + \cdots + (4k - 3) + [4(k + 1) - 3] = 2k^2 + 3k + 1$$
$$= (k + 1)(2k + 1)$$
$$= (k + 1)[2(k + 1) - 1]$$

Since this result has the same form as the given formula, except that $k + 1$ replaces n, the truth of the formula for $n = k$ implies the truth of the formula for $n = k + 1$. Part 2 of the proof is complete.

Because both of the induction requirements are true, the formula is true for all natural numbers n. ∎

Self Check 2 Verify that the formula holds for $n = 6$.

Now Try Exercise 11.

EXAMPLE 3 **Using Mathematical Induction**

Prove that $\dfrac{1}{2} + \dfrac{1}{4} + \dfrac{1}{8} + \cdots + \dfrac{1}{2^n} < 1$.

SOLUTION   **Proof by induction**

Part 1: We verify the formula for $n = 1$. When $n = 1$, there is a single term, the fraction $\frac{1}{2}$, on the left side of the equation. Substituting 1 for n on the right side, we have the following true statement:

$$\frac{1}{2} < 1.$$

The formula is true for $n = 1$. Part 1 of the proof is complete.

Part 2: We assume that the inequality is true for $n = k$. Thus,

$$\frac{1}{2} + \frac{1}{4} + \frac{1}{8} + \cdots + \frac{1}{2^k} < 1.$$

We can multiply both sides of this inequality by $\frac{1}{2}$ to get

$$\mathbf{\frac{1}{2}}\left(\frac{1}{2} + \frac{1}{4} + \frac{1}{8} + \cdots + \frac{1}{2^k}\right) < 1\left(\mathbf{\frac{1}{2}}\right),$$

or

$$\frac{1}{4} + \frac{1}{8} + \frac{1}{16} + \cdots + \frac{1}{2^{k+1}} < \frac{1}{2}.$$

We now add $\frac{1}{2}$ to both sides of this inequality to get

$$\mathbf{\frac{1}{2}} + \frac{1}{4} + \frac{1}{8} + \frac{1}{16} + \cdots + \frac{1}{2^{k+1}} < \mathbf{\frac{1}{2}} + \frac{1}{2},$$

or

$$\frac{1}{2} + \frac{1}{4} + \frac{1}{8} + \frac{1}{16} + \cdots + \frac{1}{2^{k+1}} < 1.$$

The resulting inequality is the same as the original inequality, except that $k + 1$ appears in place of n. Thus, the truth of the inequality for $n = k$ implies the truth of the inequality for $n = k + 1$. Part 2 of the proof is complete.

Because both of the induction requirements have been verified, this inequality is true for all natural numbers. ∎

Self Check 3   Verify that the formula holds for $n = 8$.

Now Try Exercise 21.

Some statements are not true when $n = 1$ but are true for all natural numbers equal to or greater than some given natural number (say, q). In these cases, we verify the given statements for $n = q$ in part 1 of the induction proof. After establishing part 2 of the induction proof, the given statement is proved for all natural numbers that are greater than q.

Self Check Answers   **1.** $10 = 10$ **2.** $66 = 66$ **3.** $\dfrac{255}{256} < 1$

Exercises **10.5**

Getting Ready

You should be able to complete these vocabulary and concept statements before you proceed to the practice exercises.

Fill in the blanks.

1. Any proof by induction requires _____ parts.
2. Part 1 is to show that the statement is true for _____.
3. Part 2 is to show that the statement is true for _____ whenever it is true for $n = k$.
4. When we assume that a formula is true for $n = k$, we call the assumption the induction _____.

Practice

Verify each formula for $n = 1, 2, 3,$ and 4.

5. $5 + 10 + 15 + \cdots + 5n = \dfrac{5n(n + 1)}{2}$

6. $1^2 + 2^2 + 3^2 + \cdots + n^2 = \dfrac{n(n + 1)(2n + 1)}{6}$

7. $7 + 10 + 13 + \cdots + (3n + 4) = \dfrac{n(3n + 11)}{2}$

8. $1(3) + 2(4) + 3(5) + \cdots + n(n + 2)$
$= \dfrac{n}{6}(n + 1)(2n + 7)$

Prove each formula by induction, if possible.

9. $2 + 4 + 6 + \cdots + 2n = n(n + 1)$

10. $1 + 3 + 5 + \cdots + (2n - 1) = n^2$

11. $3 + 7 + 11 + \cdots + (4n - 1) = n(2n + 1)$

12. $4 + 8 + 12 + \cdots + 4n = 2n(n + 1)$

13. $10 + 6 + 2 + \cdots + (14 - 4n) = 12n - 2n^2$

14. $8 + 6 + 4 + \cdots + (10 - 2n) = 9n - n^2$

15. $2 + 5 + 8 + \cdots + (3n - 1) = \dfrac{n(3n + 1)}{2}$

16. $3 + 6 + 9 + \cdots + 3n = \dfrac{3n(n + 1)}{2}$

17. $1^2 + 2^2 + 3^2 + \cdots + n^2 = \dfrac{n(n + 1)(2n + 1)}{6}$

18. $1 + 2 + 3 + \cdots + (n - 1) + n + (n - 1) + \cdots$
$+ 3 + 2 + 1 = n^2$

19. $\dfrac{1}{3} + 2 + \dfrac{11}{3} + \cdots + \left(\dfrac{5}{3}n - \dfrac{4}{3}\right) = n\left(\dfrac{5}{6}n - \dfrac{1}{2}\right)$

20. $\dfrac{1}{1 \cdot 2} + \dfrac{1}{2 \cdot 3} + \dfrac{1}{3 \cdot 4} + \cdots + \dfrac{1}{n(n + 1)} = \dfrac{n}{n + 1}$

21. $\dfrac{1}{2} + \dfrac{1}{4} + \dfrac{1}{8} + \cdots + \left(\dfrac{1}{2}\right)^n = 1 - \left(\dfrac{1}{2}\right)^n$

22. $\dfrac{1}{3} + \dfrac{2}{9} + \dfrac{4}{27} + \cdots + \dfrac{1}{3}\left(\dfrac{2}{3}\right)^{n-1} = 1 - \left(\dfrac{2}{3}\right)^n$

23. $2^0 + 2^1 + 2^2 + 2^3 + \cdots + 2^{n-1} = 2^n - 1$

24. $1^3 + 2^3 + 3^3 + \cdots + n^3 = \left[\dfrac{n(n + 1)}{2}\right]^2$

25. Prove that $x - y$ is a factor of $x^n - y^n$. (*Hint:* Consider subtracting and adding xy^k to the binomial $x^{k+1} - y^{k+1}$.)

26. Prove that $n < 2^n$.

27. There are $180°$ in the sum of the angles of any triangle. Prove that $(n - 2)180°$ is the sum of the angles of any simple polygon when n is its number of sides. (*Hint:* If a polygon has $k + 1$ sides, it has $k - 2$ sides plus three more sides.)

28. Consider the equation
$$1 + 3 + 5 + \cdots + 2n - 1 = 3n - 2.$$
   a. Is the equation true for $n = 1$?
   b. Is the equation true for $n = 2$?
   c. Is the equation true for all natural numbers n?

29. If $1 + 2 + 3 + \cdots + n = \dfrac{n}{2}(n + 1) + 1$ were true for $n = k$, show that it would be true for $n = k + 1$. Is it true for $n = 1$?

30. Prove that $n + 1 = 1 + n$ for each natural number n.

31. If n is any natural number, prove that $7^n - 1$ is divisible by 6.

32. Prove that $1 + 2n < 3^n$ for $n > 1$.

33. Prove that, if r is a real number where $r \neq 1$, then
$$1 + r + r^2 + \cdots + r^n = \dfrac{1 - r^{n+1}}{1 - r}.$$

34. Prove the formula for the sum of the first n terms of an arithmetic series:
$$a + (a + d) + (a + 2d) + [a + (n - 1)d] = \dfrac{n(a + a_n)}{2},$$
where $a_n = a + (n - 1)d$.

Discovery and Writing

35. The expression a^m, where m is a natural number, was previously defined. An alternative definition of a^m is (part 1) $a^1 = a$ and (part 2) $a^{m+1} = a^m \cdot a$. Use induction on n to prove the Product Rule for Exponents, $a^m a^n = a^{m+n}$.

36. Use induction on n to prove the Power Rule for Exponents, $(a^m)^n = a^{mn}$. (See Exercise 35.)

37. **Tower of Hanoi** A well-known problem in mathematics is the Tower of Hanoi, first attributed to Édouard Lucas in 1883. In this problem, several disks, each of a different size and with a hole in the center, are placed on a peg, with progressively

smaller disks going up the stack. The object is to transfer the stack of disks to another peg by moving only one disk at a time and never placing a disk over a smaller one.

a. Find the minimum number of moves required if there is only one disk.

b. Find the minimum number of moves required if there are two disks.

c. Find the minimum number of moves required if there are three disks.

d. Find the minimum number of moves required if there are four disks.

38. Tower of Hanoi The results in Exercise 37 suggest that the minimum number of moves required to transfer n disks from one peg to another is given by the formula $2^n - 1$. Use the following outline to prove that this result is correct using mathematical induction.

a. Verify the formula for $n = 1$.

b. Write the induction hypothesis.

c. How many moves are needed to transfer all but the largest of $k + 1$ disks to another peg?

d. How many moves are needed to transfer the largest disk to an empty peg?

e. How many moves are needed to transfer the first k disks back onto the largest one?

f. How many moves are needed to accomplish steps c, d, and e?

g. Show that part f can be written in the form $2^{(k+1)} - 1$.

h. Write the conclusion of the proof.

CHAPTER REVIEW

SECTION 10.1   The Binomial Theorem

Definitions and Concepts	Examples
Pascal's Triangle is a triangular array of numbers showing the coefficients of the expansion of $(a + b)^n$. The example shows the first several rows of Pascal's Triangle. Except for the 1s, each number in the triangle is the sum of the two numbers above it.	<table><tr><td></td><td></td><td></td><td></td><td></td><td>1</td><td></td><td></td><td></td><td></td><td></td><td>Row 0</td></tr><tr><td></td><td></td><td></td><td></td><td>1</td><td></td><td>1</td><td></td><td></td><td></td><td></td><td>Row 1</td></tr><tr><td></td><td></td><td></td><td>1</td><td></td><td>2</td><td></td><td>1</td><td></td><td></td><td></td><td>Row 2</td></tr><tr><td></td><td></td><td>1</td><td></td><td>3</td><td></td><td>3</td><td></td><td>1</td><td></td><td></td><td>Row 3</td></tr><tr><td></td><td>1</td><td></td><td>4</td><td></td><td>6</td><td></td><td>4</td><td></td><td>1</td><td></td><td>Row 4</td></tr><tr><td>1</td><td></td><td>5</td><td></td><td>10</td><td></td><td>10</td><td></td><td>5</td><td></td><td>1</td><td>Row 5</td></tr></table>
Factorial notation: $$n! = n(n - 1)(n - 2) \times \cdots \times (3)(2)(1)$$ $$0! = 1 \qquad n(n - 1)! = n!$$	See Example 3, page 892
Binomial Theorem: If n is any positive integer, then $$(a + b)^n = a^n + \frac{n!}{1!(n - 1)!}a^{n-1}b$$ $$+ \frac{n!}{2!(n - 2)!}a^{n-2}b^2 + \frac{n!}{3!(n - 3)!}a^{n-3}b^3 + \cdots$$ $$+ \frac{n!}{r!(n - r)!}a^{n-r}b^r + \cdots + b^n.$$	See Example 5, page 893

EXERCISES

Find each value.

1. $6!$

2. $7! \cdot 0! \cdot 1! \cdot 3!$

3. $\dfrac{8!}{7!}$

4. $\dfrac{5! \cdot 7! \cdot 8!}{6! \cdot 9!}$

Expand each expression.

5. $(x + y)^3$

6. $(p + q)^4$

7. $(a - b)^5$

8. $(2a - b)^3$

Find the requested term of each expansion.

9. $(a + b)^8$; fourth term

10. $(2x - y)^5$; third term

11. $(x - y)^9$; seventh term

12. $(4x + 7)^6$; fourth term

SECTION **10.2** Sequences, Series, and Summation Notation

Definitions and Concepts	Examples
An **infinite sequence** is a function whose domain is the set of natural numbers. A **finite sequence** is a function whose domain is the set of the first n natural numbers. If the commas in a sequence are replaced with plus signs, the result is called a **series**.	See pages 897 and 898
A sequence can be defined **recursively** by giving its first term and a rule showing how to obtain the $(n + 1)$th from the nth term.	See page 898
The expression $\displaystyle\sum_{k=1}^{4} n^3$ indicates the sum of the four terms obtained when 1, 2, 3, and 4 are substituted for n. If c is a constant, then $$\sum_{k=1}^{n} c = nc.$$ If c is a constant, then $$\sum_{k=1}^{n} cf(k) = c\sum_{k=1}^{n} f(k).$$ $$\sum_{k=1}^{n} [f(k) + g(k)] = \sum_{k=1}^{n} f(k) + \sum_{k=1}^{n} g(k)$$	See Example 6, page 902

EXERCISES

Write the fourth term in each sequence.

13. $0, 7, 26, \ldots, n^3 - 1, \ldots$

14. $\dfrac{3}{2}, 3, \dfrac{11}{2}, \ldots, \dfrac{n^2 + 2}{2}, \ldots$

Find the first four terms of each sequence.

15. $a_1 = 5$ and $a_{n+1} = 3a_n + 2$

16. $a_1 = -2$ and $a_{n+1} = 2a_n^2$

Evaluate each expression.

17. $\displaystyle\sum_{k=1}^{4} 3k^2$

18. $\displaystyle\sum_{k=1}^{10} 6$

19. $\displaystyle\sum_{k=5}^{8} (k^3 + 3k^2)$

20. $\displaystyle\sum_{k=1}^{30} \left(\dfrac{3}{2}k - 12\right) - \dfrac{3}{2}\sum_{k=1}^{30} k$

SECTION 10.3 Arithmetic Sequences and Series

Definitions and Concepts	Examples
An **arithmetic sequence** is of the form $a, a + d, a + 2d, \ldots, a + (n-1)d,$ where a is the first term, d is the common difference, and $a_n = a + (n-1)d$ is the nth term.	See Example 1, page 905
Numbers inserted between a first and last term to form a segment of an arithmetic sequence are called **arithmetic means**.	See Example 3, page 906
The formula $$S_n = \frac{n(a + a_n)}{2}$$ gives the sum of the first n terms of an arithmetic series, where a is the first term, a_n is the last (or nth) term, and n is the number of terms.	See Example 4, page 907

EXERCISES

Find the requested term of each arithmetic sequence.

21. $5, 9, 13, \ldots$; 29th term

22. $8, 15, 22, \ldots$; 40th term

23. $6, -1, -8, \ldots$; 15th term

24. $\dfrac{1}{2}, -\dfrac{3}{2}, -\dfrac{7}{2}, \ldots$; 35th term

Find the requested means.

25. Find three arithmetic means between 2 and 8.

26. Find five arithmetic means between 10 and 100.

Find the sum of the first 40 terms in each sequence.

27. $5, 9, 13, \ldots$

28. $8, 15, 22, \ldots$

29. $6, -1, -8, \ldots$

30. $\dfrac{1}{2}, -\dfrac{3}{2}, -\dfrac{7}{2}, \ldots$

SECTION 10.4 Geometric Sequences and Series

Definitions and Concepts	Examples
A **geometric sequence** is of the form $a, ar, ar^2, ar^3, \ldots, ar^{n-1}, \ldots,$ where a is the first term and r is the common ratio. The formula $a_n = ar^{n-1}$ gives the nth term of the sequence.	See Example 1, page 911
Numbers inserted between a first and last term to form a segment of a geometric sequence are called **geometric means**.	See Example 3, pages 911–912

Definitions and Concepts	Examples
The formula $$S_n = \frac{a - ar^n}{1 - r} \qquad (r \neq 1)$$ gives the sum of the first n terms of a geometric series, where a is the first term, r is the common ratio, and n is the number of terms.	See Example 4, page 913
If $\lvert r \rvert < 1$, the formula $$S_\infty = \frac{a}{1 - r}$$ gives the sum of the terms of an infinite geometric series, where a is the first term and r is the common ratio.	See Example 5, page 914

EXERCISES

Find the requested term of each geometric sequence.

31. $81, 27, 9, \ldots$; eleventh term

32. $2, 6, 18, \ldots$; ninth term

33. $9, \dfrac{9}{2}, \dfrac{9}{4}, \ldots$; fifteenth term

34. $8, -\dfrac{8}{5}, \dfrac{8}{25}, \ldots$; seventh term

Find the requested means.

35. Find three positive geometric means between 2 and 8.

36. Find four geometric means between –2 and 64.

37. Find the positive geometric mean between 4 and 64.

Find the sum of the first eight terms in each sequence.

38. $81, 27, 9, \ldots$ **39.** $2, 6, 18, \ldots$

40. $9, \dfrac{9}{2}, \dfrac{9}{4}, \ldots$ **41.** $8, -\dfrac{8}{5}, \dfrac{8}{25}, \ldots$

42. Find the sum of the first eight terms of the sequence $\frac{1}{3}, 1, 3, \ldots$.

43. Find the seventh term of the sequence $2\sqrt{2}, 4, 4\sqrt{2}, \ldots$.

Find the sum of each infinite sequence, if possible.

44. $\dfrac{1}{3}, \dfrac{1}{6}, \dfrac{1}{12}, \ldots$ **45.** $\dfrac{1}{5}, -\dfrac{2}{15}, \dfrac{4}{45}, \ldots$

46. $1, \dfrac{3}{2}, \dfrac{9}{4}, \ldots$ **47.** $0.5, 0.25, 0.125, \ldots$

Change each decimal into a common fraction.

48. $0.\overline{3}$ **49.** $0.\overline{9}$

50. $0.\overline{17}$ **51.** $0.\overline{45}$

Applications

52. **Investment** If Leonard invests $3,000 in a 6-year certificate of deposit at an annual rate of 7.75% compounded daily, how much money will be in the account when it matures?

53. **College enrollments** The enrollment at Hometown College is growing at the rate of 5% over each previous year's enrollment. If the enrollment is currently 4,000 students, what will it be 10 years from now? What was it 5 years ago?

54. **House-trailer depreciation** A house trailer that originally cost $10,000 depreciates in value at the rate of 10% per year. How much will the trailer be worth after 10 years?

SECTION **10.5**   Mathematical Induction

Definitions and Concepts	Examples
Mathematical induction: If a statement involving the natural number n has the following two properties, then the statement is true for all natural numbers: **1.** The statement is true for $n = 1$, and **2.** If the statement is true for $n = k$, then it is true for $n = k + 1$.	See Examples 1, 2, and 3, pages 920–922

EXERCISES

55. Verify the following formula for $n = 1$, $n = 2$, $n = 3$, and $n = 4$:

$$1^3 + 2^3 + 3^3 + \cdots + n^3 = \frac{n^2(n + 1)^2}{4}.$$

56. Prove the formula given in Exercise 55 by mathematical induction.

CHAPTER TEST

Find each value.

1. $3! \cdot 0! \cdot 4! \cdot 1!$

2. $\dfrac{2! \cdot 4! \cdot 6! \cdot 8!}{3! \cdot 5! \cdot 7!}$

Find the requested term in each expansion.

3. $(x + 2y)^5$; second term

4. $(2a - b)^8$; seventh term

Find each sum.

5. $\displaystyle\sum_{k=1}^{3} (4k + 1)$

6. $\displaystyle\sum_{k=2}^{4} (3k - 21)$

Find the sum of the first 10 terms of each sequence.

7. $2, 5, 8, \ldots$

8. $5, 1, -3, \ldots$

Find the requested means.

9. Find three arithmetic means between 4 and 24.

10. Find two geometric means between –2 and –54.

Find the sum of the first 10 terms of each sequence.

11. $\dfrac{1}{4}, \dfrac{1}{2}, 1, \ldots$

12. $6, 2, \dfrac{2}{3}, \ldots$

13. A car costing $\$c$ when new depreciates 25% of the previous year's value each year. How much is the car worth after 3 years?

14. A house costing $\$c$ when new appreciates 10% of the previous year's value each year. How much will the house be worth after 4 years?

15. Prove the following formula by induction:

$$3 + 4 + 5 + \cdots + (n + 2) = \frac{1}{2}n(n + 5).$$

16. How many six-digit license plates can be made if no plate begins with 0 or 1?

LOOKING AHEAD TO **CALCULUS**

1. **Using Maclaurin Series to Approximate Function Values**
   * Connect Maclaurin series to approximating functions—Section 10.2
2. **Using Maclaurin Series to Evaluate Limits**
   * Connect Maclaurin series to evaluating limits—Section 10.2

1. Using Maclaurin Series to Approximate Function Values

Sequences and series are used in many areas of calculus. A Taylor series is a representation of a function as a series. This series is an infinite sum of terms that are calculated from the values of its derivatives (see Looking Ahead to Calculus in Chapter 1) at a single point. If the series is centered at 0, it is called a Maclaurin series. In order to approximate a function, a finite number of items in the series are used. In Chapter 4, we looked at Taylor polynomials. A Taylor series can be regarded as the limit of a Taylor polynomial.

A Maclaurin series is centered at 0, which means it can be used to approximate the value of the function near 0. Some important Maclaurin series that have been derived are given here:

Important Maclaurin Series

$$\ln x = \sum_{n=1}^{\infty} \frac{(-1)^{n-1}(x-1)^n}{n} = (x-1) - \frac{(x-1)^2}{2} + \frac{(x-1)^3}{3} - \frac{(x-1)^4}{4} + \cdots$$

$$e^x = \sum_{n=0}^{\infty} \frac{x^n}{n!} = 1 + \frac{x}{1!} + \frac{x^2}{2!} + \frac{x^3}{3!} + \cdots$$

$$\sin x = \sum_{n=0}^{\infty} (-1)^n \frac{x^{2n+1}}{(2n+1)!} = x - \frac{x^3}{3!} + \frac{x^5}{5!} - \frac{x^7}{7!} + \cdots$$

$$\cos x = \sum_{n=0}^{\infty} (-1)^n \frac{x^{2n}}{(2n)!} = 1 - \frac{x^2}{2!} + \frac{x^4}{4!} - \frac{x^6}{6!} + \cdots$$

EXAMPLE 1

Use the given Maclaurin series to approximate $e^{0.1}$ to three decimal places. Since we are using a series to approximate the value, we can determine the accuracy by increasing the value of n. When there is no change in rounding to three decimal places, then the correct approximation has been found.

We will use $e^x = \sum_{n=0}^{\infty} \frac{x^n}{n!}$, where $x = 0.1$.

$$e^{0.1} = \sum_{n=0}^{\infty} \frac{0.1^n}{n!} \approx 1 + \frac{0.1}{1!} + \frac{0.1^2}{2!} + \frac{0.1^3}{3!} + \cdots$$

$n = 2$:

$$e^{0.1} \approx 1 + \frac{0.1}{1!} + \frac{0.1^2}{2!} = 1.105$$

$n = 3$:

$$e^{0.1} \approx 1 + \frac{0.1}{1!} + \frac{0.1^2}{2!} + \frac{0.1^3}{3!} = 1.105$$

We conclude that $e^{0.1} \approx 1.105$ correct to three decimal places. Indeed, if we check this with a calculator, we see that it is correct.

Connect to Calculus:
Using Maclaurin Series to Approximate Function Values

1. How many terms of the series are needed to evaluate $e^{0.1}$ to eight decimal places?

2. Use the given Maclaurin series to approximate cos 0.05 to five decimal places.

2. Using Maclaurin Series to Evaluate Limits

While it is useful to be able to approximate the value of a function using a Maclaurin series, their use is far more widespread. In calculus, finding integrals (discussed in the Looking Ahead to Calculus in Chapter 5) can often be difficult or even impossible using conventional methods of integration. By using a Maclaurin series, we can approximate a definite integral—which is one that has a numerical value—to any degree of accuracy desired. The actual process of integration is beyond the scope of this section, but you will learn the process when you take calculus.

In the Looking Ahead to Calculus in Chapter R, we discussed finding limits. Maclaurin series are also used to find limits.

EXAMPLE 2   Evaluate $\lim\limits_{x \to 0} \dfrac{\sin x - x + \frac{1}{6}x^3}{x^5}$ using a Maclaurin series.

Substitute the Maclaurin series for the sine function and reduce.

$$\lim_{x \to 0} \frac{\sin x - x + \frac{1}{6}x^3}{x^5} = \lim_{x \to 0} \frac{\left[x - \dfrac{x^3}{3!} + \dfrac{x^5}{5!} - \dfrac{x^7}{7!} + \cdots \right] - x + \frac{1}{6}x^3}{x^5}$$

$$= \lim_{x \to 0} \frac{\left(\dfrac{x^5}{5!} - \dfrac{x^7}{7!} + \cdots \right)}{x^5} \cdot \frac{\dfrac{1}{x^5}}{\dfrac{1}{x^5}}$$

$$= \lim_{x \to 0} \frac{\left(\dfrac{1}{5!} - \dfrac{x^2}{7!} + \cdots \right)}{1}$$

$$= \frac{1}{5!} = \frac{1}{120}$$

Connect to Calculus:
Using Maclaurin Series to Evaluate Limits

3. Use a Maclaurin series to evaluate $\lim\limits_{x \to 0} \dfrac{e^x - 1 - x}{x^2}$.

4. Use a Maclaurin series to evaluate $\lim\limits_{x \to 0} \dfrac{1 - \cos x}{1 + x - e^x}$.

Answers to Selected Exercises

Section R.1
1. root, solution   **3.** no   **5.** rational
7. no restrictions   **9.** $x \neq 0$ **11.** $x \geq 0$
13. $x \neq 3, x \neq -2$ **15.** 5; conditional equation
17. no solution; contradiction   **19.** 7; conditional equation
21. all real numbers, identity   **23.** 6; conditional equation
25. all real numbers, identity

27. 1   **29.** 9   **31.** -3 **33.** 6   **35.** -2 **37.** $\dfrac{5}{2}$

39. $-\dfrac{14}{11}$ **41.** 2   **43.** 2   **45.** no solution   **47.** -2

49. $\dfrac{1}{3}$ **51.** no solution   **53.** 2   **55.** $\dfrac{7}{4}$ **57.** $-\dfrac{2}{5}$

59. $w = \dfrac{P - 2l}{2}$ **61.** $r^2 = \dfrac{3V}{\pi h}$ **63.** $s = \dfrac{f(P_n - L)}{i}$

65. $m = \dfrac{r^2 F}{Mg}$ **67.** $y = b\left(1 - \dfrac{x}{a}\right)$ **69.** $r = \dfrac{r_1 r_2}{r_1 + r_2}$

71. $n = \dfrac{l - a + d}{d}$ **73.** $n = \dfrac{360}{180 - a}$

75. $r_1 = \dfrac{-Rr_2 r_3}{Rr_3 + Rr_2 - r_2 r_3}$ **77.** 84   **79.** 7   **81.** $2\dfrac{1}{2}$ ft

83. 20 ft by 8 ft   **85.** 20 ft

87. \$10,000 at 7%; \$12,000 at 6%   **89.** 475   **91.** \$79.95

93. 21   **95.** $2\dfrac{6}{11}$ days   **97.** $1\dfrac{1}{3}$ hr   **99.** $21\dfrac{1}{9}$ hr

101. 10 oz   **103.** 1 L   **105.** 4.5 gal   **107.** 4 L

109. 30 gal   **111.** 39 mph going; 65 mph returning

113. $2\dfrac{1}{2}$ hr   **115.** 12 mph   **117.** about 11.2 mm

Section R.2
1. imaginary   **3.** imaginary   **5.** $2 - 5i$ **7.** $12i$

9. $-4i\sqrt{6}$ **11.** $\dfrac{5}{3}i\sqrt{2}$ **13.** $-\dfrac{7}{4}i\sqrt{6}$

15. $x = 3; y = 5$ **17.** $x = \dfrac{2}{3}; y = -\dfrac{2}{9}$

19. $5 - 6i$ **21.** $-2 - 10i$ **23.** $4 + 10i$
25. $1 - i$ **27.** $-15 - 25i$ **29.** $56 + 28i$
31. $-9 + 19i$ **33.** $-5 + 12i$ **35.** $52 - 56i$

37. $-6 + 17i$ **39.** $0 + i$ **41.** $0 + \dfrac{4}{3}i$

43. $\dfrac{2}{5} - \dfrac{1}{5}i$ **45.** $\dfrac{1}{25} + \dfrac{7}{25}i$ **47.** $\dfrac{1}{2} + \dfrac{1}{2}i$

49. $-\dfrac{7}{13} - \dfrac{22}{13}i$ **51.** $-\dfrac{12}{17} + \dfrac{11}{34}i$

53. $\dfrac{6 + \sqrt{3}}{10} + \dfrac{3\sqrt{3} - 2}{10}i$ **55.** i **57.** -1 **59.** $-i$

61. 1   **63.** -1 **65.** -1 **67.** i **69.** 4
71. $21 + 12i$ volts   **73.** $6.2 - 0.7i$ volts

Section R.3
1. $ax^2 + bx + c = 0$ **3.** $\sqrt{c}, -\sqrt{c}$ **5.** $b^2 - 4ac$

7. $A = lw$ **9.** $3, -2$ **11.** $12, -12$ **13.** $2, -\dfrac{5}{2}$

15. $\dfrac{3}{5}, -\dfrac{5}{3}$ **17.** $\dfrac{3}{2}, \dfrac{1}{2}$ **19.** ± 3 **21.** $\pm 5i\sqrt{2}$

23. $\pm 3\sqrt{2}$ **25.** $\pm \dfrac{3i\sqrt{7}}{2}$ **27.** $3, -1$

29. $-1 \pm 2i\sqrt{2}$ **31.** $x^2 + 6x + 9$ **33.** $x^2 - 4x + 4$

35. $a^2 + 5a + \dfrac{25}{4}$ **37.** $r^2 - 11r + \dfrac{121}{4}$

39. $y^2 + \dfrac{3}{4}y + \dfrac{9}{64}$ **41.** 5, 3   **43.** $-4 \pm \sqrt{21}$

45. $8 \pm 2\sqrt{6}$ **47.** $-1 \pm 3i$ **49.** $-3 \pm 4i\sqrt{3}$

51. $5 \pm i\dfrac{\sqrt{10}}{2}$ **53.** $\dfrac{-2 \pm \sqrt{7}}{3}$ **55.** $\pm 2\sqrt{3}$

57. $3, -\dfrac{5}{2}$ **59.** $1, -2$ **61.** $\dfrac{-5 \pm \sqrt{13}}{6}$

63. $\dfrac{-1 \pm \sqrt{61}}{10}$ **65.** $-1 \pm i$ **67.** $-2 \pm i$

69. $\dfrac{1}{3} \pm \dfrac{1}{3}i$ **71.** $t = \pm \dfrac{\sqrt{2hg}}{g}$ **73.** $t = \dfrac{8 \pm \sqrt{64 - h}}{4}$

75. $y = \pm \dfrac{b\sqrt{a^2 - x^2}}{a}$ **77.** $a = \pm \dfrac{bx\sqrt{b^2 + y^2}}{b^2 + y^2}$

79. $x = \dfrac{-y \pm y\sqrt{5}}{2}$

81. one repeated rational root
83. two different rational roots
85. two different irrational roots

87. yes   **89.** 2, 10   **91.** $3, -4$ **93.** $\dfrac{1}{2}, -\dfrac{4}{3}$

95. $\dfrac{5}{6}, -\dfrac{2}{5}$ **97.** $\dfrac{7 \pm \sqrt{145}}{4}$ **99.** $1, -1$

101. $-\dfrac{1}{2}, 5$ **103.** -2 **105.** 4 m

107. 160 ft by 72 ft   **109.** 40.1 in by 22.6 in
111. 20 mph going and 10 mph returning
113. 7 hr   **115.** 25 sec   **117.** about 9.5 sec
119. 1.6 sec   **121.** 24.3 ft   **123.** 4 hr
125. about 9.5 hr   **127.** no
129. Maude at 7%, Matilda at 8%   **131.** 1.70 in

Section R.4
1. equal   **3.** extraneous   **5.** x
7. $x = k$ or $x = -k$ **9.** $0, -5, -4$

11. $0, \dfrac{4}{3}, -\dfrac{1}{2}$ **13.** $5, -5, 1, -1$ **15.** $6, -6, 1, -1$

17. $3\sqrt{2}, -3\sqrt{2}, \sqrt{5}, -\sqrt{5}$ **19.** 0, 1

21. $\frac{1}{8}$, -8 **23.** 1, 144 **25.** $\frac{1}{64}$ **27.** $\frac{1}{9}$ **29.** 27

31. $-\frac{1}{3}$ **33.** 2 **35.** $-\frac{3}{2}, -\frac{5}{2}$ **37.** 9 **39.** 20

41. 2 **43.** 2 **45.** $3, 4$ **47.** $\frac{1}{5}, -1$ **49.** $3, 5$

51. $-2, 1$ **53.** $2, -\frac{5}{2}$ **55.** $0, 4$ **57.** -2

59. no solution   **61.** 3 **63.** 1 **65.** 7 **67.** 0

69. 2 **71.** $\pi - 2$ **73.** $x - 5$

75. x^3 if $x \geq 0$; $-x^3$ if $x < 0$ **77.** $0, -4$

79. $2, -\frac{4}{3}$ **81.** $\frac{14}{3}, -2$ **83.** $7, -3$ **85.** no solution

87. $\frac{2}{7}, 2$ **89.** $x \geq 0$ **91.** $-\frac{3}{2}$ **93.** $0, -6$ **95.** 0

97. $\frac{3}{5}, 3$ **99.** $-\frac{3}{13}, \frac{9}{5}$ **101.** 400 ft   **103.** 8 ft

105. about $3,109

Section R.5

1. right   **3.** $a < c$ **5.** $b - c$ **7.** $>$ **9.** linear
11. $-k < x < k$

13. $(-\infty, 1)$
15. $[1, \infty)$

17. $(-\infty, 1)$
19. $[1, \infty)$

21. $(-\infty, 3]$
23. $(-10/3, \infty)$

25. $(5, \infty)$
27. $[14, \infty)$

29. $(-\infty, 15/4]$
31. $(-44/41, \infty)$

33. $(6, 9]$
35. $(8, 22]$

37. $[-11, 4]$
39. $[5, 21]$

41. $(-\infty, 0)$
43. $(0, \infty)$

45. $(3, 14/3)$
47. $(2, \infty)$

49. $(-4, 5/6)$
51. $[-2, \infty)$

53. $(-3, 9)$
55. $(-\infty, -9) \cup (3, \infty)$

57. $(-\infty, -7] \cup [3, \infty)$
59. $[-13/3, 1]$

61. $(-\infty, -3) \cup (-3, \infty)$
63. $(-1, 1/5)$

65. $(-\infty, -7/9) \cup (13/9, \infty)$
67. $(-5, 7)$

69. $(-2, -1/2) \cup (-1/2, 1)$
71. $(-7/3, -2/3) \cup (4/3, 3)$

73. $(-7, -1) \cup (11, 17)$
75. $(-18, -6) \cup (10, 22)$

77. $(-10, -7] \cup [5, 8)$

79. $\left[-\frac{1}{2}, \infty\right)$ **81.** $(-\infty, 0)$ **83.** $\left(-\infty, -\frac{1}{2}\right)$

85. $[0, \infty)$ **87.** 17 min   **89.** 12 **91.** $p \leq \$1,124.12$
93. $a > \$50,000$ **95.** anything over $1,800
97. $16\frac{2}{3}$ cm $< s < 20$ cm   **99.** $70°F \leq t \leq 86°F$
101. $|c - 0.6°| \leq 0.5°$ **103.** $|h - 55| < 17$
105. a. 26.45%, 24.76% **b.** It is less than or equal to 1%.
107. $40 + 2w < P < 60 + 2w$

Section R.6

1. quadrants   **3.** to the right   **5.** first   **7.** linear
9. x-intercept   **11.** horizontal   **13.** midpoint
15. $(2, 3)$ **17.** $(-2, -3)$ **19.** $(0, 0)$ **21.** $(-5, -5)$
23. QI   **25.** QIII   **27.** QI   **29.** x-axis

31.   **33.**

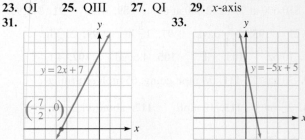

35.   **37.**

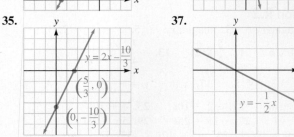

39.   **41.**

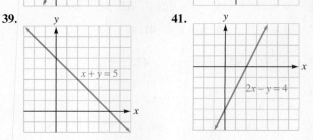

43.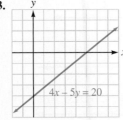

45.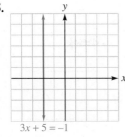

47.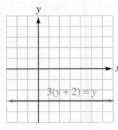

49.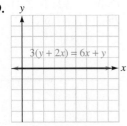

51. 1.22   **53.** 4.67   **55.** 5   **57.** $\sqrt{13}$ **59.** $\sqrt{2}$

61. 2   **63.** 5   **65.** 13   **67.** 25   **69.** $8\sqrt{2}$ **71.** 7

73. (4, 6)   **75.** (0, 1)   **77.** (0, 0)   **79.** $\left(\dfrac{\sqrt{5}}{2}, \dfrac{\sqrt{5}}{2}\right)$

81. (5, 6)   **83.** (5, 15)   **87.** $\sqrt{2}$ **91.** \$312,500

93. 200   **95.** 100 rpm   **97.** approx. 171 mi

Section R.7

1. divided   **3.** run   **5.** vertical   **7.** -1

9. m, y-intercept   **11.** 2   **13.** $-\dfrac{7}{4}$ **15.** -2

17. undefined   **19.** $\dfrac{5}{3}$ **21.** 3   **23.** $\dfrac{1}{2}$

25. $\dfrac{2}{3}$ **27.** 0   **29.** 0   **31.** undefined   **33.** positive

35. undefined   **37.** perpendicular   **39.** parallel
41. perpendicular   **43.** perpendicular   **45.** parallel
47. perpendicular   **49.** $2x - y = 0$ **51.** $4x - 2y = -7$
53. $2x - 5y = -7$ **55.** $y = -3$ **57.** $\pi x - y = \pi^2$

59. $y = x$ **61.** $y = \dfrac{7}{3}x - 3$ **63.** $y = -\dfrac{9}{5}x + \dfrac{2}{5}$

65. $y = 3x - 2$ **67.** $y = 5x - \dfrac{1}{5}$ **69.** $\dfrac{3}{2}, -4$

71. $-\dfrac{1}{3}, -\dfrac{5}{6}$ **73.** $\dfrac{7}{2}, 2$

75. 1, -1 **77.** $\dfrac{2}{3}, 2$

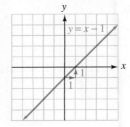

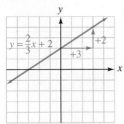

79. $-\dfrac{2}{3}, 6$

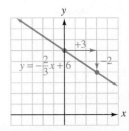

81. parallel   **83.** parallel   **85.** perpendicular

87. $y = 4x$ **89.** $y = 4x - 3$ **91.** $y = \dfrac{4}{5}x - \dfrac{26}{5}$

93. $y = -\dfrac{1}{4}x$ **95.** $y = -\dfrac{1}{4}x + \dfrac{11}{2}$ **97.** $y = -\dfrac{5}{4}x + 3$

99. $x = -2$ **101.** $x = 5$ **103.** 3.5 students per yr
105. \$642.86 per yr
107.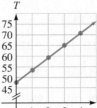

$\dfrac{\Delta T}{\Delta t}$ is the hourly rate of change in temperature.

109.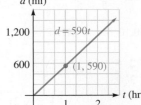

The slope is the speed of the plane.

111. $y = -3,200x + 24,300$ **113.** $y = -\dfrac{710}{3}x + 1,900$

115. \$890   **117.** \$37,200   **119.** about 838

121. $C = \dfrac{5}{9}(F - 32)$ **123.** $y = -\dfrac{9}{10}x + 47$; 11%

125. a. Chirps/min

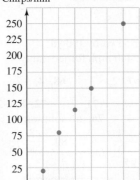

b. $y = \dfrac{23}{5}x - 210$ (answers may vary)

c. 204 (answers may vary)
127. $y = 4.44x - 196.62$

Chapter Review

1. no restrictions   **2.** $x \neq 0$ **3.** $x \geq 0$

4. $x \neq 2, x \neq 3$ **5.** $\dfrac{16}{27}$; conditional equation

6. -14; conditional equation   **7.** $\dfrac{16}{5}$; conditional equation

8. no solution, contradiction   **9.** 7; conditional equation

10. all real numbers, identity   **11.** 7; conditional equation

12. $\dfrac{1}{3}$; conditional equation   **13.** -2; conditional equation

14. 0; conditional equation   **15.** $F = \dfrac{9}{5}C + 32$

16. $f = \dfrac{si}{P_n - l}$ **17.** $f_1 = \dfrac{ff_2}{f_2 - f}$ **18.** $l = \dfrac{a - S + Sr}{r}$

19. 60%   **20.** 22.5 ft by 27.5 ft   **21.** 3 hr   **22.** 0.5 hr

23. 1.5 L   **24.** about 3.9 hr   **25.** $5\dfrac{1}{7}$ hr   **26.** $3\dfrac{1}{3}$ oz

27. \$4,500 at 11%; \$5,500 at 14%   **28.** 10   **29.** $-2 - i$

30. $-2 - 5i$ **31.** $5 - 2i$ **32.** $21 - 9i$ **33.** $0 - 3i$

34. $0 - 2i$ **35.** $\dfrac{3}{2} - \dfrac{3}{2}i$ **36.** $-\dfrac{2}{5} + \dfrac{4}{5}i$ **37.** $\dfrac{4}{5} + \dfrac{3}{5}i$

38. $\dfrac{1}{2} - \dfrac{5}{2}i$ **39.** $0 + i$ **40.** $0 - i$ **41.** $2, -\dfrac{3}{2}$

42. $\dfrac{1}{4}, -\dfrac{4}{3}$ **43.** $0, \dfrac{8}{5}$ **44.** $\dfrac{2}{3}, \dfrac{4}{9}$ **45.** $\pm\dfrac{3\sqrt{3}}{2}$

46. $\dfrac{7}{5} \pm i$ **47.** 3, 5   **48.** $-4, -2$ **49.** $\dfrac{1 \pm \sqrt{21}}{10}$

50. $0, \dfrac{1}{5}$ **51.** $2, -7$ **52.** $9, -\dfrac{2}{3}$ **53.** $\dfrac{-1 \pm \sqrt{21}}{10}$

54. $-1 \pm 2i$ **55.** $\dfrac{1}{3} \pm \dfrac{\sqrt{2}}{3}i$ **56.** $\dfrac{1}{3} \pm \dfrac{\sqrt{11}}{3}i$

57. 1   **58.** two different rational roots   **59.** $\dfrac{1}{3}$

60. 10, 2   **61.** $-\dfrac{5}{2}$ **62.** $\dfrac{8}{5}, 5$

63. either 95 yd by 110 yd or 55 yd by 190 yd
64. 320 mph for prop plane; 440 mph for jet plane

65. 1 sec   **66.** $1\dfrac{1}{2}$ ft   **67.** $2, -3$ **68.** $4, -2$

69. $1, -1$ **70.** $1, -1, 6, -6$ **71.** 9   **72.** $8, -27$
73. 5   **74.** 0   **75.** $4, -4$ **76.** no solution

77. $5, -7$ **78.** 0   **79.** $-\dfrac{4}{3}, -6$ **80.** $-\dfrac{3}{8}, \dfrac{3}{10}$

81. $(-\infty, 7)$

82. $[-1/5, \infty)$

83. $(-\infty, 5/3)$

84. $(-\infty, -12/7)$

85. $[-3, 5)$

86. $(2, \infty)$

87. $(-6, 0)$

88. $(-\infty, 2] \cup [8/3, \infty)$

89. $(-5, 1)$

90. $(-\infty, -29) \cup (35, \infty)$

91. $(2, 0)$ **92.** $(-2, 1)$ **93.** $(0, -1)$ **94.** $(3, -1)$

95.–98.

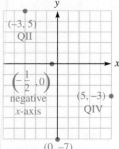

99.

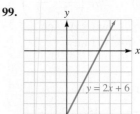

100.

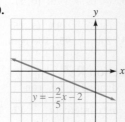

101.

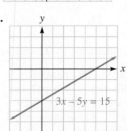

102.

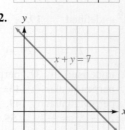

103.

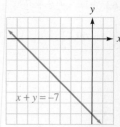

104.

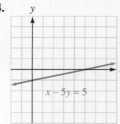

105.

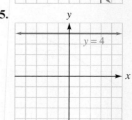

106.

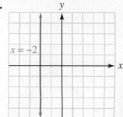

107. \$12,150   **108.** \$332,500   **109.** 10
110. $4\sqrt{2}$ **111.** 2   **112.** $2\sqrt{2}|a|$ **113.** $(0, 3)$

114. $\left(-6, \dfrac{15}{2}\right)$ **115.** $\left(\sqrt{3}, 8\right)$ **116.** $(0, 0)$ **117.** -6
118. 2   **119.** -1 **120.** 1   **121.** 3   **122.** 5   **123.** 0
124. undefined   **125.** negative   **126.** positive
127. perpendicular   **128.** neither   **129.** 7   **130.** 3
131. 200 ft per min   **132.** \$48,750 per year
133. $7x + 5y = 0$ **134.** $4x + y = -7$ **135.** $x + 5y = -3$
136. $2x + y = 9$ **137.** $y = \dfrac{2}{3}x + 3$ **138.** $y = -\dfrac{3}{2}x - 5$

139. 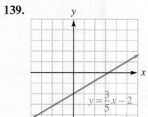   **140.**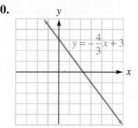

141. slope is 2; y-intercept is 0
142. slope is $-\dfrac{1}{2}$; y-intercept is 0
143. slope is $\dfrac{3}{2}$; y-intercept is -5
144. slope is $-\dfrac{1}{2}$; y-intercept is -2
145. slope is $-\dfrac{5}{2}$; y-intercept is $\dfrac{7}{2}$
146. slope is $\dfrac{3}{4}$; y-intercept is $-\dfrac{7}{2}$

147. $y = 17$ **148.** $x = -5$ **149.** $y = \dfrac{3}{4}x - \dfrac{3}{2}$

150. $y = -7x + 47$ **151.** $y = 3x + 5$ **152.** $y = \dfrac{1}{7}x - 3$
153. parallel   **154.** perpendicular

Chapter Test

1. $x \neq 0, x \neq 1$ **2.** $x \geq 0$ **3.** $\dfrac{5}{2}$ **4.** 37
5. $x = \mu + z\sigma$ **6.** $a = \dfrac{bc}{c + b}$ **7.** 87.5   **8.** \$14,000
9. $\dfrac{1}{2}, \dfrac{3}{2}$ **10.** $\dfrac{3}{2}, -4$ **11.** $x = \dfrac{-b \pm \sqrt{b^2 - 4ac}}{2a}$
12. $\dfrac{5 \pm \sqrt{133}}{6}$ **13.** $5, -3$ **14.** 8 sec   **15.** $7 - 12i$
16. $-23 - 43i$ **17.** $\dfrac{4}{5} + \dfrac{2}{5}i$ **18.** $0 + i$ **19.** i
20. 1   **21.** $2, -2, 3, -3$ **22.** $1, -\dfrac{1}{32}$ **23.** 139
24. -1
25.
$(-\infty, 2]$

2

26.
$(-\infty, 5)$

5

27. $2, -\dfrac{10}{3}$ **28.** 0
29.
$(-\infty, 3/2) \cup (7/2, \infty)$

3/2   7/2

30.
$[-9, 6]$

-9 6

31. QII   **32.** y-axis

33. 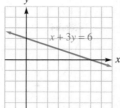   **34.**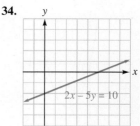

35. 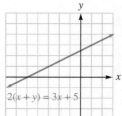   **36.**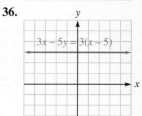

37. 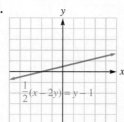   **38.**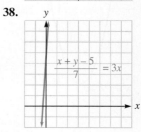

39. $\sqrt{41}$ **40.** $\pi\sqrt{2}$ **41.** $(0, 0)$
42. $\left(\sqrt{2}, 2\sqrt{2}\right)$ **43.** $-\dfrac{5}{4}$ **44.** $\dfrac{\sqrt{3}}{3}$
45. neither   **46.** perpendicular   **47.** $y = 2x - 11$
48. $y = 3x + \dfrac{1}{2}$ **49.** $y = 2x + 5$ **50.** $y = -\dfrac{1}{2}x + 5$
51. $y = 2x - \dfrac{11}{2}$ **52.** $x = 3$

Connect to Calculus

1. 4   **2.** 7   **3.** 14   **4.** 0   **5.** $-\dfrac{1}{2}$ **6.** -3
7. $\dfrac{2}{3}$

Section 1.1

1. x-intercept   **3.** axis of symmetry   **5.** x-axis
7. circle, center   **9.** $x^2 + y^2 = r^2$
11. x-intercept $= -2, 2$; y-intercept $= -4$
13. x-intercept $= 0, \dfrac{1}{2}$; y-intercept $= 0$
15. x-intercept $= -1, 5$; y-intercept $= -5$
17. x-intercept $= -2, 1$; y-intercept $= -2$
19. x-intercept $= -3, 0, 3$; y-intercept $= 0$
21. x-intercept $= -1, 1$; y-intercept $= -1$
23. 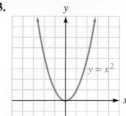   **25.**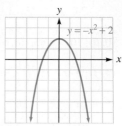

27.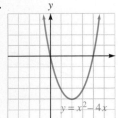

29.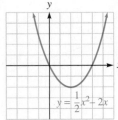

31. about the y-axis   **33.** about the x-axis
35. about the x-axis, the y-axis, and the origin
37. about the y-axis   **39.** none
41. about the x-axis   **43.** about the y-axis
45. about the x-axis

47.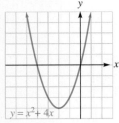

49.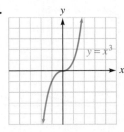

51.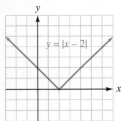

53.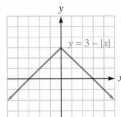

55.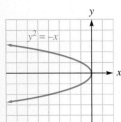

57.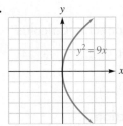

59.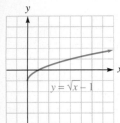

61.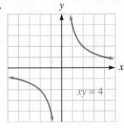

63. $(0, 0)$; 10   **65.** $(0, 5)$; 7   **67.** $(-6, 0)$; $\dfrac{1}{2}$

69. $(4, 1)$; 3   **71.** $\left(\dfrac{1}{4}, -2\right)$; $3\sqrt{5}$ **73.** $x^2 + y^2 = 25$

75. $x^2 + (y + 6)^2 = 36$ **77.** $(x - 8)^2 + y^2 = \dfrac{1}{25}$

79. $(x - 2)^2 + (y - 12)^2 = 169$

81. $x^2 + y^2 - 1 = 0$
83. $x^2 + y^2 - 12x - 16y + 84 = 0$
85. $x^2 + y^2 - 6x + 8y + 23 = 0$
87. $x^2 + y^2 - 6x - 6y - 7 = 0$
89. $x^2 + y^2 + 6x - 8y = 0$

91.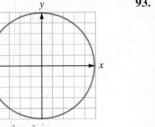

93.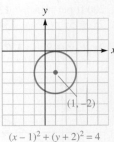

95.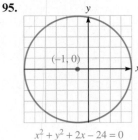

97.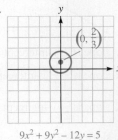

99.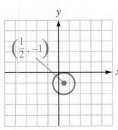

101. $(0.25, 0.88)$ **103.** $(0.50, 7.25)$ **105.** ± 2.65
107. 1.44   **109.** 4 sec
111.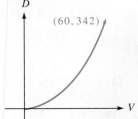

113. $x^2 + y^2 = 36$
115. $x^2 + (y - 35)^2 = 900$

117. $x^2 + y^2 - 14x - 8y + 40 = 0$

Section 1.2

1. function   **3.** domain   **5.** $y = f(x)$ **7.** dependent
9. (a) {(LSU, Tigers), (Georgia, Bulldogs), (MSU, Bulldogs), (Auburn, Tigers)}; D = {LSU, Georgia, MSU, Auburn}; R = {Tigers, Bulldogs}   **(b)** a function
11. (a) {(Jackson, Louisiana), (Jackson, Mississippi), (Jackson, Tennessee), (Alexandria, Virginia)}; D = {Jackson, Alexandria}; R = {Louisiana, Mississippi, Tennessee, Virginia}   **(b)** not a function
13. function   **15.** not a function   **17.** function
19. not a function   **21.** function   **23.** not a function
25. not a function   **27.** function   **29.** function

31. $D = (-\infty, \infty)$ **33.** $D = (-\infty, -6) \cup (-6, \infty)$

35. $D = (-\infty, -1) \cup (-1, 5) \cup (5, \infty)$

37. $D = \left[\dfrac{5}{2}, \infty\right)$ **39.** $D = (-\infty, \infty)$

41. $D = (-\infty, \infty)$ **43.** $R = [0, \infty)$ **45.** $R = [-3, \infty)$

47. $R = [0, \infty)$ **49.** $R = (-\infty, 0) \cup (0, \infty)$

51. $D = (-\infty, \infty); R = (-\infty, \infty)$

53. $D = (-\infty, \infty); R = [0, \infty)$

55. $D = (-\infty, -1) \cup (-1, \infty); R = (-\infty, 0) \cup (0, \infty)$

57. $D = [0, \infty); R = [0, \infty)$

59. $D = (-\infty, -3) \cup (-3, \infty); R = (-\infty, 1) \cup (1, \infty)$

61. $D = \left[\dfrac{1}{2}, \infty\right); R = [0, \infty)$

63. $D = (-\infty, 0) \cup (0, \infty); R = (-\infty, 1) \cup (1, \infty)$

65. $4; -11; 3k - 2; 3k^2 - 5$

67. $4; \dfrac{3}{2}; \dfrac{1}{2}k + 3; \dfrac{1}{2}k^2 + \dfrac{5}{2}$ **69.** $4; 9; k^2; k^4 - 2k^2 + 1$

71. $5; 10; k^2 + 1; k^4 - 2k^2 + 2$

73. $\dfrac{1}{3}; 2; \dfrac{2}{k + 4}; \dfrac{2}{k^2 + 3}$ **75.** $\dfrac{1}{3}; \dfrac{1}{8}; \dfrac{1}{k^2 - 1}; \dfrac{1}{k^4 - 2k^2}$

77. $\sqrt{5}; \sqrt{10}; \sqrt{k^2 + 1}; \sqrt{k^4 - 2k^2 + 2}$

79. 29 **81.** $\dfrac{17}{4}$ **83.** $3x^6 - 5x^3 + 1$ **85.** 0.28

87. 16.12 **89.** -0.12 **91.** 3 **93.** $2x + h$

95. $8x + 4h$ **97.** $2x + h + 3$ **99.** $4x + 2h - 4$

101. $3x^2 + 3xh + h^2$

103. $D = \left[\dfrac{5}{2}, \infty\right); R = [0, \infty)$

105. $D = (-\infty, 1) \cup (1, \infty); R = (-\infty, 2) \cup (2, \infty)$

107. $D = (-\infty, \infty); R = (-\infty, \infty)$

109. 117 **111.** $109{,}500$ gal   **113.** 2 ft/sec

115. $A(x) = x(x + 5) = x^2 + 5x$

117. a. $C(x) = 8x + 75$ **b.** \$755

119. a. $C(t) = 9.99 + 0.07t$ **b.** \$11.39

Section 1.3

1. vertical   **3.** range   **5.** origin   **7.** $-f(x)$

9. decreasing   **11.** function   **13.** not a function

15. function

17.

19. (a) $D = (-\infty, \infty); R = (-\infty, \infty)$ **(b)** x-intercept $= 2$;
y-intercept $= -1$ **(c)** neither   **(d)** increasing on $(-\infty, \infty)$

21. (a) $D = (-\infty, \infty); R = [-4, \infty)$

(b) x-intercepts $= -2$ and 2; y-intercept $= -4$ **(c)** even

(d) decreasing on $(-\infty, 0)$; increasing on $(0, \infty)$

23. (a) $D = (-\infty, \infty); R = (-\infty, \infty)$ **(b)** x-intercept $= 0$;
y-intercept $= 0$ **(c)** odd   **(d)** increasing on $(-\infty, \infty)$

25. (a) $D = (-\infty, \infty); R = (-\infty, 1]$

(b) x-intercepts $= -3$ and -1; y-intercept $= -1$

(c) neither   **(d)** increasing on $(-\infty, -2)$; decreasing on
$(-2, \infty)$

27. (a) $D = [-2, 1); R = (-3, 3]$ **(b)** x-intercept $= -1$;
y-intercept $= -2$ **(c)** neither   **(d)** decreasing on $(-2, 1)$

29. (a) $D = (-\infty, 0) \cup (0, \infty); R = (-\infty, 0) \cup (0, \infty)$

(b) no intercepts   **(c)** odd   **(d)** decreasing on
$(-\infty, 0) \cup (0, \infty)$

31. (a) $D = (-\infty, 0]; R = [2, \infty)$ **(b)** no x-intercept;
y-intercept $= 2$ **(c)** neither   **(d)** decreasing on $(-\infty, 0)$

33. (a) $D = (-\infty, \infty); R = \{-2, 2\}$ **(b)** no x-intercept;
y-intercept $= -2$ **(c)** neither   **(d)** constant on
$(-\infty, 0) \cup (0, \infty)$

35. even   **37.** odd   **39.** odd   **41.** even

43. neither   **45.** odd

47. (a) $D = (-\infty, \infty); R = \left[-\dfrac{1}{2}, \infty\right)$

(b) x-intercepts $= 0$ and 1; y-intercept $= 0$

(c) decreasing on $\left(-\infty, \dfrac{1}{2}\right)$; increasing on $\left(\dfrac{1}{2}, \infty\right)$

(d) local minimum $= f\left(\dfrac{1}{2}\right) = -\dfrac{1}{2}$ **(e)** $\dfrac{3}{2}$

(f) $-\dfrac{1}{2}$ and $\dfrac{3}{2}$ **(g)** 0 and 1 **(h)** neither

49. (a) $D = (-\infty, \infty); R = (-\infty, \infty)$

(b) x-intercepts $= -2, 0$ and 2; y-intercept $= 0$

(c) increasing on $\left(-\infty, -\dfrac{2\sqrt{3}}{3}\right) \cup \left(\dfrac{2\sqrt{3}}{3}, \infty\right)$;

decreasing on $\left(-\dfrac{2\sqrt{3}}{3}, \dfrac{2\sqrt{3}}{3}\right)$

(d) local minimum $= f\left(\dfrac{2\sqrt{3}}{3}\right) = -\dfrac{16\sqrt{3}}{9}$;

local maximum $= f\left(-\dfrac{2\sqrt{3}}{3}\right) = \dfrac{16\sqrt{3}}{9}$

(e) 0 **(f)** $-\dfrac{2\sqrt{3}}{3}$ **(g)** $-2, 0$, and 2 **(h)** odd

51. local maximum ≈ -1.6151; local minimum ≈ -2.3849;
$f(2) = -2; f(-\sqrt{3}) = -19.6667$

53. local maximum ≈ -0.6519; local minima $= -1$ and
$\approx -5.8481; f(-1.7) \approx -4.7179; f(1) = -1$

55. local maximum $= 1$; local minimum ≈ 0.6151;
$f\left(-\sqrt{1.3}\right) \approx -3.3479; f\left(\sqrt{3}\right) \approx 3.1523$

57. local maximum ≈ -2.3482; local minimum ≈ -4.2;
$f(2.3) \approx -2.7717; f(-3.5) \approx 6.2125$

59. (a) $D = [1, 6]; R = [700, 780]$ **(b)** increasing on
$(1, 4)$; decreasing on $(4, 6)$ **(c)** local maximum $= 780$

(d) 2004–2005 and 2006–2007   **61.** 13

63. (a) $A(x) = x(120 - 2x) = 120x - 2x^2$ **(b)** $1{,}600$ ft^2

(c) $1{,}600$ ft^2 **(d)** $x = 30; A(30) = 1{,}800$ ft^2

65. 16 purses

Section 1.4

1. linear   **3.** odd   **5.** 0

7.
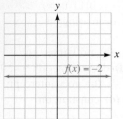

9.
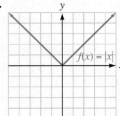

11.
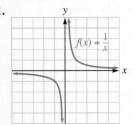

13.
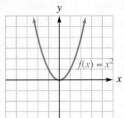

15.
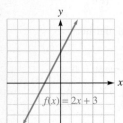

17.
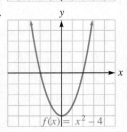

19.
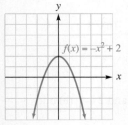

21.
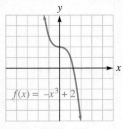

23.
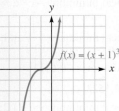

25.
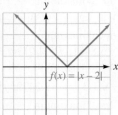

27.
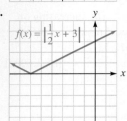

29.
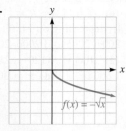

31.
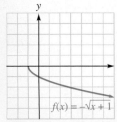

33.
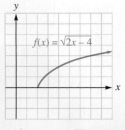

35.
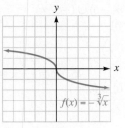

37.
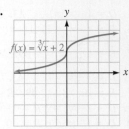

39.
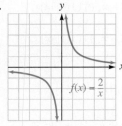

41. a. -2 **b.** 3   **43. a.** -11 **b.** -18

45. a. 2   **b.** 1   **c.** 3

47. a. -9 **b.** 16   **c.** $\dfrac{1}{6}$

49. a. 3   **b.** -4 **c.** -3

51. a. 2   **b.** 3   **c.** 4

53.
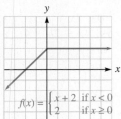

55.
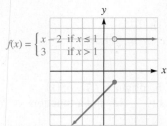

57.
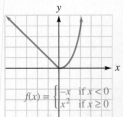

59.
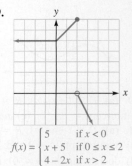

61.

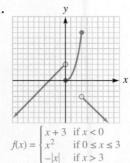

$$f(x) = \begin{cases} x+3 & \text{if } x < 0 \\ x^2 & \text{if } 0 \le x \le 3 \\ -|x| & \text{if } x > 3 \end{cases}$$

63.

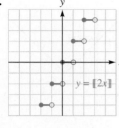

$y = [\![2x]\!]$

15. $g(x) = f(x-4) - 2$

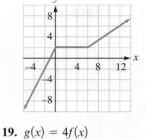

17. $g(x) = f(-x)$

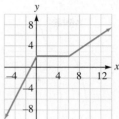

19. $g(x) = 4f(x)$

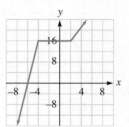

21. $g(x) = f(4x)$

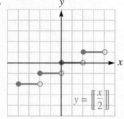

65.

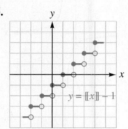

$y = [\![x]\!] - 1$

67.

$y = \left[\!\!\left[\dfrac{x}{2}\right]\!\!\right]$

69. $C(x) = 0.12x + 7$ **71.** B   **73.** $32   **75.** $1.60

77.

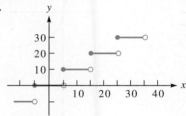

23.

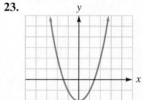

$h(x) = x^2 - 2$

25.

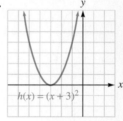

$h(x) = (x+3)^2$

79. Graphs are not the same.

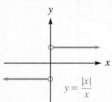

$y = \dfrac{|x|}{x}$

27.

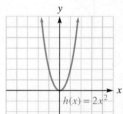

$h(x) = 2x^2$

29.

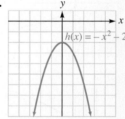

$h(x) = -x^2 - 2$

81. (a) $f(x) = \begin{cases} 0 & \text{if } 0 \le x \le 13{,}300 \\ 0.17(x - 13{,}300) & \text{if } x > 13{,}300 \end{cases}$

(b) $1,989; $5,389   **(c)** 7.956%; 11.976%

(d) $f(x) = \begin{cases} 0 & \text{if } 0 \le x \le 13{,}300 \\ \dfrac{0.17(x - 13{,}300)}{x} & \text{if } x > 13{,}300 \end{cases}$

31.

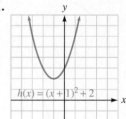

$h(x) = (x+1)^2 + 2$

33.

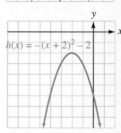

$h(x) = -(x+2)^2 - 2$

Section 1.5

1. up   **3.** to the right   **5.** two, down   **7.** y-axis
9. horizontally
Note: The answer graphs for **11–21** use a different scale from that of the original graph provided.

11. $g(x) = f(x) + 2$

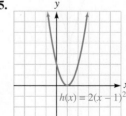

13. $g(x) = f(x + 2)$

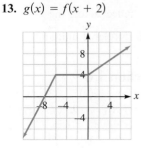

35.

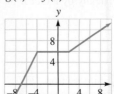

$h(x) = 2(x-1)^2$

37.

$g(x) = (x-2)^3$

39.

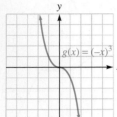

$g(x) = (-x)^3$

41.

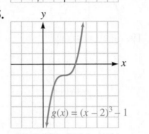

$g(x) = \frac{1}{3}x^3$

43.

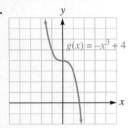

$g(x) = -x^3 + 4$

45.

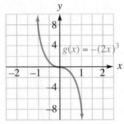

$g(x) = (x - 2)^3 - 1$

47.

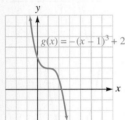

$g(x) = -(x - 1)^3 + 2$

49.

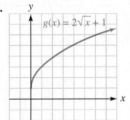

$g(x) = -(2x)^3$

51.

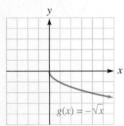

$g(x) = -\sqrt{x}$

53.

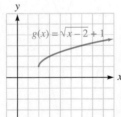

$g(x) = 2\sqrt{x} + 1$

55.

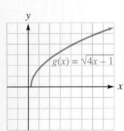

$g(x) = \sqrt{4x - 1}$

57.

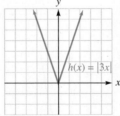

$g(x) = \sqrt{x - 2} + 1$

59.

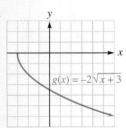

$g(x) = -2\sqrt{x} + 3$

61.

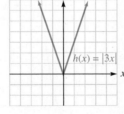

$h(x) = |3x|$

63.

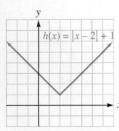

$h(x) = |x - 2| + 1$

65.

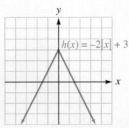

$h(x) = -2|x| + 3$

67.

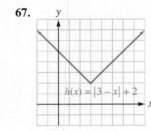

$h(x) = |3 - x| + 2$

69.

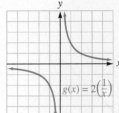

$g(x) = 2\left(\frac{1}{x}\right)$

71.

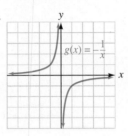

$g(x) = -\frac{1}{x}$

73.

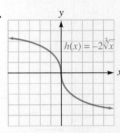

$g(x) = \frac{1}{x - 1}$

75.

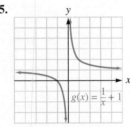

$g(x) = \frac{1}{x} + 1$

77.

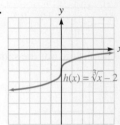

$h(x) = -2\sqrt[3]{x}$

79.

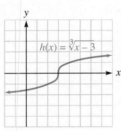

$h(x) = \sqrt[3]{x} - 3$

81.

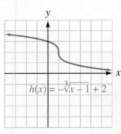

$h(x) = \sqrt[3]{x} - 2$

83.

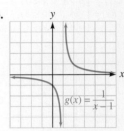

$h(x) = -\sqrt[3]{x - 1} + 2$

85.

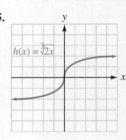

$h(x) = \sqrt[3]{2x}$

87. $f(x) = x^3; g(x) = (x - 2)^3$

89. $f(x) = \sqrt{x}; g(x) = -\sqrt{x} + 1$

91. $f(x) = x^2; g(x) = x^2 - 2$

93. $f(x) = \sqrt{x}; g(x) = \sqrt{3 - x}$

95. reflect across the x-axis and shift up 2 units

97. shift to the right 1 unit and stretch vertically by 2

99. reflect across the x-axis, shrink vertically by a factor of $\frac{1}{2}$, and shift left 1 unit and down 3 units

Section 1.6

1. $f(x) + g(x)$ **3.** $f(x)g(x)$ **5.** intersection

7. $g(f(x))$ **9.** $(f + g)(x) = 5x - 1; (-\infty, \infty)$

11. $(f \cdot g)(x) = 6x^2 - x - 2; (-\infty, \infty)$

13. $(f - g)(x) = x + 1; (-\infty, \infty)$

15. $(f/g)(x) = \dfrac{x^2 + x}{x^2 - 1} = \dfrac{x}{x - 1}; (-\infty, -1) \cup (-1, 1) \cup (1, \infty)$

17. 7   **19.** 1   **21.** 12   **23.** undefined   **25.** 0

27. -2 **29.** 0   **31.** $\dfrac{1}{4}$

33. $[-7, 7]$

35.

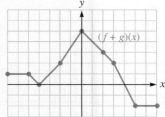

37. $f(x) = 3x^2; g(x) = 2x$ **39.** $f(x) = 3x^2; g(x) = x^2 - 1$

41. $f(x) = 3x^3; g(x) = -x$

43. $f(x) = x + 9; g(x) = x - 2$

45. 11   **47.** -17 **49.** 190   **51.** 145

53. $(-\infty, \infty); (f \circ g)(x) = 3x + 3$

55. $(-\infty, \infty); (f \circ f)(x) = 9x$

57. $(-\infty, \infty); (g \circ f)(x) = 2x^2$

59. $(-\infty, \infty); (g \circ g)(x) = 4x$

61. $[-1, \infty); (f \circ g)(x) = \sqrt{x + 1}$

63. $[0, \infty); (f \circ f)(x) = \sqrt[4]{x}$

65. $[-1, \infty); (g \circ f)(x) = x$

67. $(-\infty, \infty); (g \circ g)(x) = x^4 - 2x^2$

69. $(-\infty, 2) \cup (2, 3) \cup (3, \infty); (f \circ g)(x) = \dfrac{x - 2}{3 - x}$

71. $(-\infty, 1) \cup (1, 2) \cup (2, \infty); (f \circ f)(x) = \dfrac{x - 1}{2 - x}$

73. $f(x) = x - 2; g(x) = 3x$

75. $f(x) = x - 2; g(x) = x^2$

77. $f(x) = x^2; g(x) = x - 2$

79. $f(x) = \sqrt{x}; g(x) = x + 2$

81. $f(x) = x + 2; g(x) = \sqrt{x}$

83. $f(x) = x; g(x) = x$ **85.** 1   **87.** -2 **89.** 2

91. a. $P(x) = (R - C)(x) = 260x - 60,000$

b. $P(500) = 70,000$ **93.** $A(t) = \dfrac{9\pi t^2}{4}; 58,629,659 \text{ in}^2$

95. $P = 4\sqrt{A}$

Section 1.7

1. one-to-one   **3.** identity   **5.** one-to-one

7. not one-to-one   **9.** not one-to-one

11. not one-to-one   **13.** not one-to-one

15. one-to-one   **17.** one-to-one   **19.** not one-to-one

27. $f^{-1}(x) = \dfrac{1}{3}x$ **29.** $f^{-1}(x) = \dfrac{x - 2}{3}$

31. $f^{-1}(x) = \sqrt[3]{x} + 1$ **33.** $f^{-1}(x) = (x + 8)^3$

35. $f^{-1}(x) = \dfrac{1}{x} - 3$ **37.** $f^{-1}(x) = \dfrac{1}{2x}$

39. $f^{-1}(x) = \dfrac{1}{5}x$

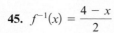

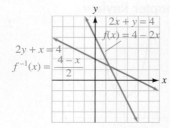

41. $f^{-1}(x) = \dfrac{x + 4}{2}$

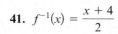

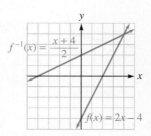

43. $f^{-1}(x) = x + 2$

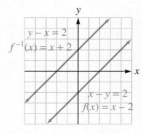

45. $f^{-1}(x) = \dfrac{4 - x}{2}$

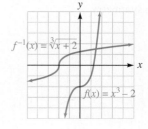

47. $f^{-1}(x) = \sqrt[3]{x + 2}$

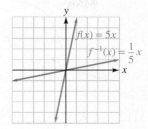

49. $f^{-1}(x) = x^3 + 4$

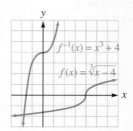

51. $f^{-1}(x) = \dfrac{1}{2x}$

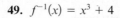
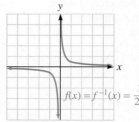

53. $f^{-1}(x) = \dfrac{x + 1}{x - 1}$

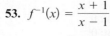
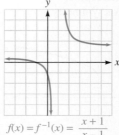

55. $f^{-1}(x) = -\sqrt{x + 3}$ $(x \geq -3)$

57. $f^{-1}(x) = \sqrt[4]{x + 8}$ $(x \geq -8)$

59. $f^{-1}(x) = \sqrt{4 - x^2}$ $(0 \leq x \leq 2)$

61. $D = (-\infty, 2) \cup (2, \infty); R = (-\infty, 1) \cup (1, \infty)$

63. $D = (-\infty, 0) \cup (0, \infty); R = (-\infty, -2) \cup (-2, \infty)$

65. $D_f = [0, \infty); R_f = (-\infty, 9]$. One possible solution is $f^{-1}(x) = \sqrt{9 - x}; D_{f^{-1}} = (-\infty, 9]; R_{f^{-1}} = [0, \infty)$.

67. $D_f = [-2, \infty); R_f = [0, \infty)$. One possible solution is $f^{-1}(x) = \sqrt{x} - 2; D_{f^{-1}} = [0, \infty); R_{f^{-1}} = [-2, \infty)$.

69. $D_f = [0, \infty); R_f = [-4, \infty)$. One possible solution is $f^{-1}(x) = |x + 4|; D_{f^{-1}} = [-4, \infty); R_{f^{-1}} = [0, \infty)$.

71.

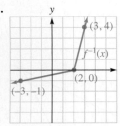

73.

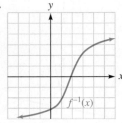

75. a. $y = 0.75x + 8.50$ **b.** $11.50   **c.** $y = \dfrac{x - 8.50}{0.75}$
d. 2

Chapter Review

1.

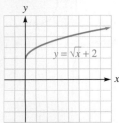

x-intercept: none;
y-intercept: 2;
y-axis symmetry

2.

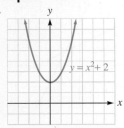

x-intercept: $\sqrt[3]{2}$;
y-intercept: -2;
no symmetry

3.

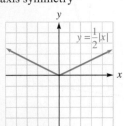

x-intercept: 0;
y-intercept: 0;
y-axis symmetry

4.

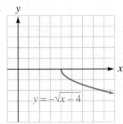

x-intercept: 4;
y-intercept: none;
no symmetry

5.

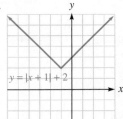

x-intercept: none;
y-intercept: 2;
no symmetry

6.

x-intercept: none;
y-intercept: 3;
no symmetry

7.

8.

9.

10.

11. $(0, 0)$; 8   **12.** $(0, 6)$; 10   **13.** $(-7, 0)$; $\dfrac{1}{2}$
14. $(5, -1)$; 3   **15.** $x^2 + y^2 = 49$
16. $(x - 3)^2 + y^2 = \dfrac{1}{25}$ **17.** $(x + 2)^2 + (y - 12)^2 = 25$

18. $\left(x - \dfrac{2}{7}\right)^2 + (y - 5)^2 = 81$
19. $(x + 3)^2 + (y - 4)^2 = 144$;
$x^2 + y^2 + 6x - 8y - 119 = 0$
20. $\left(x + \dfrac{1}{2}\right)^2 + \left(y - \dfrac{5}{2}\right)^2 = \dfrac{121}{2}$;
$x^2 + y^2 + x - 5y - 54 = 0$
21. $(x + 3)^2 + (y - 2)^2 = 9$
22. $(x - 2)^2 + (y - 4)^2 = 25$
23.

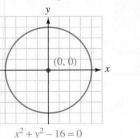

$x^2 + y^2 - 16 = 0$

24.

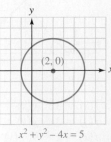

$x^2 + y^2 - 4x = 5$

25.

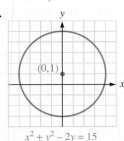

$x^2 + y^2 - 2y = 15$

26.

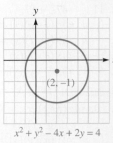

$x^2 + y^2 - 4x + 2y = 4$

27. $3.32, -3.32$ **28.** $-1, 0, 1$ **29.** $1.73, -1.73, 1, -1$
30. $4.19, -1.19$ **31.** function   **32.** function
33. not a function   **34.** function
35. $D = (-\infty, \infty); R = (-\infty, \infty)$
36. $D = (-\infty, 5] \cup [5, \infty); R = (-\infty, 3) \cup (3, \infty)$
37. $D = [1, \infty); R = [0, \infty)$ **38.** $D = (-\infty, \infty); R = [1, \infty)$
39. $8; -17; 5k - 2$ **40.** $-2; -\dfrac{3}{4}; \dfrac{6}{k - 5}$
41. $0; 5; |k - 2|$ **42.** $\dfrac{1}{7}; \dfrac{1}{2}; \dfrac{k^2 - 3}{k^2 + 3}$ **43.** 2   **44.** 5
45. $6x + 3h - 6$ **46.** $4x + 2h - 7$
47. (a) $D = (-\infty, \infty); R = (-\infty, 1]$
(b) x-intercepts: $-2, 0,$ and 2; y-intercept: 0   **(c)** even
(d) increasing on $(-\infty, -1) \cup (0, 1)$; decreasing on
$(-1, 0) \cup (1, \infty)$; constant nowhere
48. (a) $D = (-\infty, \infty); R = (-\infty, \infty)$
(b) x-intercepts: $-2, 0,$ and 2; y-intercept: 0   **(c)** odd
(d) increasing on $(-1, 1)$; decreasing on $(-\infty, -1) \cup (1, \infty)$;
constant nowhere
49. (a) $D = (-\infty, \infty); R = [2, \infty)$
(b) x-intercepts: none; y-intercept: 4   **(c)** neither
(d) increasing on $(-2, \infty)$; decreasing on $(-\infty, -2)$; constant
nowhere
50. (a) $D = (-\infty, 1) \cup (1, \infty); R = (-\infty, 1) \cup (1, \infty)$
(b) x-intercepts: 0; y-intercept: 0   **(c)** neither
(d) increasing nowhere; decreasing on $(-\infty, 1) \cup (1, \infty)$;
constant nowhere
51. even   **52.** odd   **53.** neither   **54.** even
55. (a) relative maximum: $f(2) = 3$
(b) x-intercepts: $2 \pm \sqrt{3}$; y-intercept $f(0) = -1$

(c) $D = (-\infty, \infty)$; $R = (-\infty, 3]$
(d) $f(2) = 3$; $f(-3) = -22$ **(e)** $2 \pm \sqrt{3}$ **(f)** neither

56.

$$f(x) = \begin{cases} 3x - 2 & \text{if } x < 0 \\ x^2 & \text{if } x \geq 0 \end{cases}$$

57.
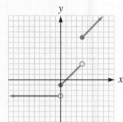
$$f(x) = \begin{cases} -3 & \text{if } x < 0 \\ x - 1 & \text{if } 0 \leq x < 4 \\ x & \text{if } x \geq 4 \end{cases}$$

58. 3 **59.** -1

60.
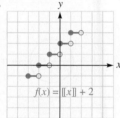
$f(x) = [[x]] + 2$

61.
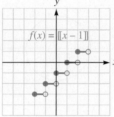
$f(x) = [[x - 1]]$

62.
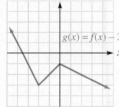
$g(x) = f(x) - 3$

63.
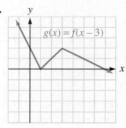
$g(x) = f(x - 3)$

64.
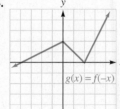
$g(x) = f(-x)$

65.
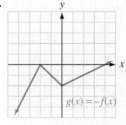
$g(x) = -f(x)$

66.
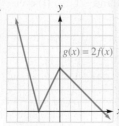
$g(x) = 2f(x)$

67.
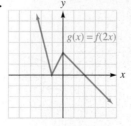
$g(x) = f(2x)$

68.
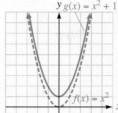
$g(x) = x^2 + 1$; $f(x) = x^2$

69.
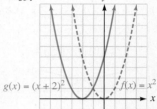
$g(x) = (x + 2)^2$; $f(x) = x^2$

70.
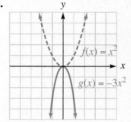
$f(x) = x^2$; $g(x) = -3x^2$

71.
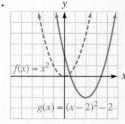
$f(x) = x^2$; $g(x) = (x - 2)^2 - 2$

72.
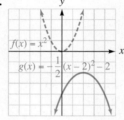
$f(x) = x^2$; $g(x) = -\frac{1}{2}(x - 2)^2 - 2$

73.
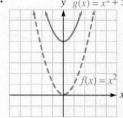
$g(x) = x^2 + 5$; $f(x) = x^2$

74.
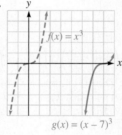
$f(x) = x^3$; $g(x) = (x - 7)^3$

75.
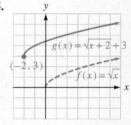
$g(x) = \sqrt{x + 2} + 3$; $(-2, 3)$; $f(x) = \sqrt{x}$

76.
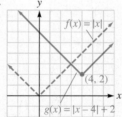
$f(x) = |x|$; $g(x) = |x - 4| + 2$; $(4, 2)$

77.
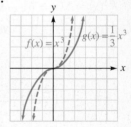
$f(x) = x^3$; $g(x) = \frac{1}{3}x^3$

78.
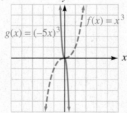
$g(x) = (-5x)^3$; $f(x) = x^3$

79.
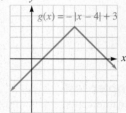
$g(x) = -|x - 4| + 3$

80.

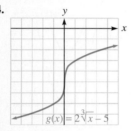

$g(x) = \frac{1}{4}|x - 4| + 1$

81.

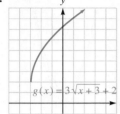

$g(x) = 3\sqrt{x + 3} + 2$

82.

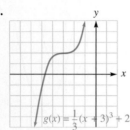

$g(x) = \frac{1}{3}(x + 3)^3 + 2$

83.

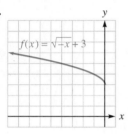

$f(x) = \sqrt{-x} + 3$

84.

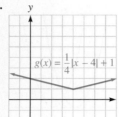

$g(x) = 2\sqrt[3]{x} - 5$

85. $(f + g)(x) = x^2 + 2x; \ D = (-\infty, \infty)$
86. $(f \cdot g)(x) = 2x^3 + x^2 - 2x - 1; \ D = (-\infty, \infty)$
87. $(f - g)(x) = x^2 - 2x - 2; \ D = (-\infty, \infty)$
88. $(f/g)(x) = \dfrac{f(x)}{g(x)} = \dfrac{x^2 - 1}{2x + 1}; \ D = \left(-\infty, -\dfrac{1}{2}\right) \cup \left(-\dfrac{1}{2}, \infty\right)$
89. $(f \circ g)(x) = f(g(x)) = 4x^2 + 4x; \ D = (-\infty, \infty)$
90. $(g \circ f)(x) = g(f(x)) = 2x^2 - 1; \ D = (-\infty, \infty)$
91. $f^{-1}(x) = \dfrac{x + 1}{7}$ \quad **92.** $f^{-1}(x) = \dfrac{2}{x}$
93. $f^{-1}(x) = \sqrt[3]{x + 6}$ \quad **94.** $f^{-1}(x) = x^3 - 4$
95. $f^{-1}(x) = \dfrac{2x - 1}{x}$ \quad **96.** $f^{-1}(x) = \dfrac{x}{1 + x}$
97. $f^{-1}(x) = \dfrac{\sqrt[3]{3x^2}}{x}$ \quad **98.** $f^{-1}(x) = \dfrac{x + 5}{2}$
99. $\left(-\infty, \dfrac{2}{5}\right) \cup \left(\dfrac{2}{5}, \infty\right)$

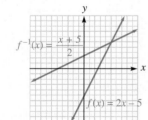

$f^{-1}(x) = \dfrac{x + 5}{2}$

$f(x) = 2x - 5$

Chapter Test
1. x-intercepts: $-4, 0,$ and 4; y-intercept: 0
2. x-intercept: 4; y-intercept: 4 \quad **3.** x-axis symmetry
4. y-axis symmetry

5.

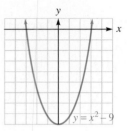

$y = x^2 - 9$

6.

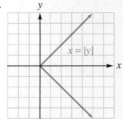

$x = |y|$

7.

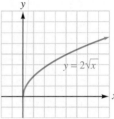

$y = 2\sqrt{x}$

8.

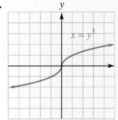

$x = y^3$

9. $(x - 5)^2 + (y - 7)^2 = 64$
10. $(x - 2)^2 + (y - 4)^2 = 32$
11.

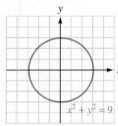

$x^2 + y^2 = 9$

12.

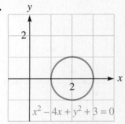

$x^2 - 4x + y^2 + 3 = 0$

13. $D = (-\infty, 5) \cup (5, \infty); \ R = (-\infty, 0) \cup (0, \infty)$
14. $D = [-3, \infty); \ R = [0, \infty)$ \quad **15.** $\dfrac{1}{2}; 2$ \quad **16.** $\sqrt{6}; 3$
17. **(a)** $D = (-\infty, \infty); \ R = (-\infty, \infty)$
(b) x-intercept: -4; y-intercept: 3 \quad **(c)** neither
(d) increasing on $(-\infty, -2) \cup (2, \infty)$; decreasing on
$(-2, -1) \cup (1, 2)$; constant on $(-1, 1)$
18. **(a)** $D = (-\infty, \infty); \ R = (-\infty, \infty)$
(b) x-intercepts: $-4, -2,$ and 3; y-intercept: -1
(c) neither
(d) increasing on $(-\infty, -3) \cup (2, \infty)$; decreasing on $(-3, -1)$;
constant on $(-1, 2)$
19. neither \quad **20.** odd

21. **(a)** relative maximum: 0; relative minimum: $-\dfrac{1}{2}$

(b) x-intercepts: $1, 0,$ and -1; y-intercept: 0

(c) $D = (-\infty, \infty); \ R = \left[-\dfrac{1}{2}, \infty\right)$

(d) $f(2) = 24; f(-3) = 144$ \quad **(e)** 1 and -1 \quad **(f)** even
22.

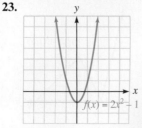

$f(x) = \begin{cases} |x| & \text{if } x \le 0 \\ \sqrt{x} & \text{if } x > 0 \end{cases}$

23.

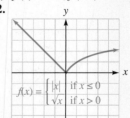

$f(x) = 2x^2 - 1$

24.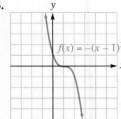

$f(x) = -(x - 1)^3$

25.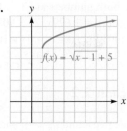

$f(x) = \sqrt{x - 1} + 5$

26. $(f + g)(x) = f(x) + g(x) = x^2 + 3x + 2$
27. $(g \circ f)(x) = g(f(x)) = 9x^2 + 2$
28. $(f/g)(x) = \dfrac{f(x)}{g(x)} = \dfrac{3x}{x^2 + 2}$
29. $(f \circ g)(x) = f(g(x)) = 3x^2 + 6$
30. $f^{-1}(x) = \dfrac{x + 1}{x - 1}$ **31.** $f^{-1}(x) = \sqrt[3]{x + 3}$
32. $(-\infty, -2) \cup (-2, \infty)$ **33.** $(-\infty, 3) \cup (3, \infty)$

Connect to Calculus
1. $y = -11x - 4; \ y = 5x - 4$ **2.** 0
3. no, 0 is not in the domain of f

4. $-1; \ y = -x - 3$ **5. a.** 10   **b.** $-\dfrac{6}{5}$ **c.** $\dfrac{15}{8}$

6. 6

Section 2.1

1. $f(x) = ax^2 + bx + c$ **3.** downward   **5.** vertex
7. $c - \dfrac{b^2}{4a}$ **9.** upward   **11.** upward; minimum
13. upward; minimum   **15.** downward; maximum
17. $(0, -1)$ **19.** $(6, 0)$ **21.** $(-1, -1)$
23. $(3, 1)$ **25.** $(2, 0)$ **27.** $(3, 1)$ **29.** $\left(\dfrac{2}{3}, \dfrac{11}{3}\right)$
31. (a) $(3, -1)$ **(b)** y-intercept: 8; x-intercepts: 2, 4
(c) $D = (-\infty, \infty); \ R = [-1, \infty)$
(d)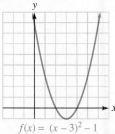

$f(x) = (x - 3)^2 - 1$

33. (a) $(-2, 1)$ **(b)** y-intercept: -3; x-intercepts: $-3, -1$
(c) $D = (-\infty, \infty); \ R = (-\infty, 1]$
(d)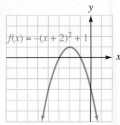

$f(x) = -(x + 2)^2 + 1$

35. (a) $(4, -2)$ **(b)** y-intercept: 30; x-intercepts: 3, 5
(c) $D = (-\infty, \infty); \ R = [-2, \infty)$
(d)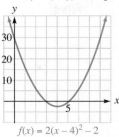

$f(x) = 2(x - 4)^2 - 2$

37. (a) $(1, 1)$ **(b)** y-intercept: $\dfrac{1}{2}$;

x-intercepts: $1 - \sqrt{2}, 1 + \sqrt{2}$
(c) $D = (-\infty, \infty); \ R = (-\infty, 1]$
(d)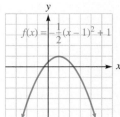

$f(x) = -\dfrac{1}{2}(x - 1)^2 + 1$

39. (a) $\left(\dfrac{1}{2}, -\dfrac{1}{4}\right)$ **(b)** y-intercept: 0; x-intercepts: 0, 1
(c) $D = (-\infty, \infty); \ R = \left[-\dfrac{1}{4}, \infty\right)$
(d)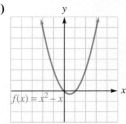

$f(x) = x^2 - x$

41. (a) $(0, 3)$ **(b)** y-intercept: 3; x-intercepts: $-\sqrt{6}, \sqrt{6}$
(c) $D = (-\infty, \infty); \ R = (-\infty, 3]$
(d)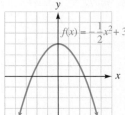

$f(x) = -\dfrac{1}{2}x^2 + 3$

43. (a) $(2, -9)$ **(b)** y-intercept: -5; x-intercepts: $-1, 5$
(c) $D = (-\infty, \infty); \ R = [-9, \infty)$
(d)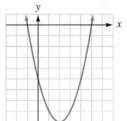

$f(x) = x^2 - 4x - 5$

45. (a) $(3, 0)$ **(b)** y-intercept: -9; x-intercept: 3
(c) $D = (-\infty, \infty)$; $R = (-\infty, 0]$
(d)

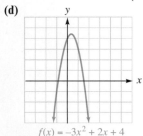

$f(x) = -x^2 + 6x - 9$

47. (a) $(3, -8)$ **(b)** y-intercept: 10; x-intercepts: $1, 5$
(c) $D = (-\infty, \infty)$; $R = [-8, \infty)$
(d)

$f(x) = 2x^2 - 12x + 10$

49. (a) $\left(\dfrac{1}{3}, \dfrac{13}{3}\right)$ **(b)** y-intercept: 4;

x-intercepts: $\dfrac{1 - \sqrt{13}}{3}, \dfrac{1 + \sqrt{13}}{3}$

(c) $D = (-\infty, \infty)$; $R = \left(-\infty, \dfrac{13}{3}\right]$
(d)

$f(x) = -3x^2 + 2x + 4$

51. (a) $(2, -3)$ **(b)** y-intercept: 1; x-intercepts:
$2 - \sqrt{3}, 2 + \sqrt{3}$ **(c)** $D = (-\infty, \infty)$; $R = [-3, \infty)$
(d)

$f(x) = x^2 - 4x + 1$

53. $f(x) = \dfrac{1}{2}x^2$ **55.** $f(x) = 5x^2 - 4$

57. $f(x) = 2(x + 1)^2 - 3$ **59.** $f(x) = -3(x - 2)^2$
61. 75 ft by 75 ft; 5,625 ft^2 **63.** 25 ft by 25 ft
65. width $= 12$ in; depth $= 6$ in
67. 12 in by 16 in   **69.** 14 in by 14 in
71. 15.4 ft   **73.** $50\sqrt{2} + 1$ **75.** 173.5 m; 8 sec
77. 160 webcams; $13,450   **79.** $1,200   **81.** $95

83. both numbers are 3
85. $f(x) = 1.679x^2 - 3.907x - 0.229$
87. 403 lb; $f(x) = 0.086616x^2 - 11.317553x + 410.484123$
89. 6 units by 4.5 units

Section 2.2

1. $f(x) = x^3$ **3.** n **5.** $(x - a)$
7. degree 4 and $a_n > 0$; end behavior like that of $f(x) = x^2$;
$f(x) \to \infty$ as $x \to -\infty$ and $f(x) \to \infty$ as $x \to \infty$; the graph rises
on the left and rises on the right
9. degree 5 and $a_n < 0$; end behavior like that of $f(x) = -x^3$;
$g(x) \to \infty$ as $x \to -\infty$ and $g(x) \to -\infty$ as $x \to \infty$; the graph
rises on the left and falls on the right
11. degree 4 and $a_n < 0$; end behavior like that of
$f(x) = -x^2$; $h(x) \to -\infty$ as $x \to -\infty$ and $h(x) \to -\infty$ as $x \to \infty$;
the graph falls on the left and falls on the right
13. degree 7 and $a_n > 0$; end behavior like that of $f(x) = x^3$;
$s(x) \to -\infty$ as $x \to -\infty$ and $s(x) \to \infty$ as $x \to \infty$; the graph falls
on the left and rises on the right
15. b   **17.** d
19. (a) 5   **(b)** negative   **(c)** $x = -3, -1, 2,$ and 4
(d) $f(x) = -2(x + 3)(x + 1)^2(x - 2)(x - 4)$
21. $x = -\dfrac{5}{2}$, multiplicity 1, crosses; $x = \dfrac{5}{2}$, multiplicity 1,
crosses
23. $x = -5$, multiplicity 1, crosses; $x = \dfrac{3}{2}$, multiplicity 1,
crosses
25. $t = -3$, multiplicity 1, crosses; $t = 0$, multiplicity 2,
touches; $t = 1$, multiplicity 1, crosses
27. $x = -\sqrt{3}$, multiplicity 1, crosses; $x = \sqrt{3}$,
multiplicity 1, crosses
29. $x = 0$, multiplicity 2, touches; $x = -4$, multiplicity 2,
touches; $x = 5$, multiplicity 1, crosses
31. $x = \dfrac{5}{2}$, multiplicity 1, crosses; $x = -3$, multiplicity 1,
crosses; $x = 1$, multiplicity 2, touches
33. $f(x) = x^3(x + 1)(x - 2)(x - 4)$
35. $f(x) = (x + 5)^2(x + 3)(x + 1)$
37. $f(x) = (x + 3)(2x + 3)(x - 4)$
39. $f(x) = 3(x - 3)(x - 1)^2(x + 2)$
41. $g(x) = -x^2(x + 2)(3x - 5)$
43. $h(x) = x^2(x + 2)^2(x - 4)$
45. $f(x) = (x + 1)(4x + 3)(x - 1)^2(x - 2)$
47. (a) falls on the left and rises on the right   **(b)** $x = -1,$
0, and 1, each with multiplicity 1   **(c)** 0
(d)

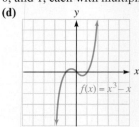

$f(x) = x^3 - x$

49. (a) rises on the left and rises on the right
(b) $x = -1$ (multiplicity 2) and 1 (multiplicity 2)   **(c)** 1
(d)

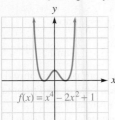

$f(x) = x^4 - 2x^2 + 1$

51. (a) rises on the left and rises on the right **(b)** $x = -3$
(multiplicity 1), 0 (multiplicity 2), and 2 (multiplicity 1)
(c) 0
(d)

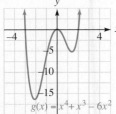

$g(x) = x^4 + x^3 - 6x^2$

53. (a) falls on the left and falls on the right   **(b)** $x = 0$
(multiplicity 3) and 1 (multiplicity 1)   **(c)** 0
(d)

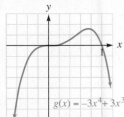

$g(x) = -3x^4 + 3x^3$

55. (a) falls on the left and rises on the right **(b)** $x = -1$
(multiplicity 1), 0 (multiplicity 1), and $\dfrac{5}{3}$ (multiplicity 1)

(c) 0

(d)

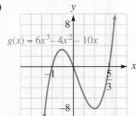

$g(x) = 6x^3 - 4x^2 - 10x$

57. (a) rises on the left and falls on the right **(b)** $x = -\dfrac{4}{3}$

(multiplicity 1), 0 (multiplicity 1), and 3 (multiplicity 1)

(c) 0

(d)

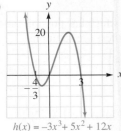

$h(x) = -3x^3 + 5x^2 + 12x$

59. (a) rises on the left and rises on the right **(b)** $x = -3$
(multiplicity 1), –1 (multiplicity 2), and 3 (multiplicity 1)
(c) -27
(d)

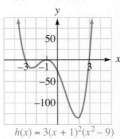

$h(x) = 3(x + 1)^2(x^2 - 9)$

61. (a) falls on the left and falls on the right **(b)** $x = -1$
(multiplicity 1), 0 (multiplicity 2), and 4 (multiplicity 3)
(c) 0
(d)

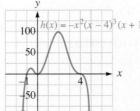

$h(x) = -x^2(x - 4)^3(x + 1)$

63. (a) falls on the left and rises on the right   **(b)** $x = -3$
(multiplicity 2), 0 (multiplicity 4), and 1 (multiplicity 1)
(c) 0
(d)

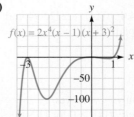

$f(x) = 2x^4(x - 1)(x + 3)^2$

65. (a) rises on the left and rises on the right   **(b)** $x = -4$

(multiplicity 1), $\dfrac{1}{2}$ (multiplicity 2), and 4 (multiplicity 1)

(c) -16

(d)

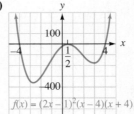

$f(x) = (2x - 1)^2(x - 4)(x + 4)$

67. $P(x) = 2x^2 + x - 3$; $P(-2) = 3$; $P(-1) = -2$; since
$-2 < 0 < 3$, there is at least one value of x such that
$-2 < x < -1$ and $P(x) = 0$
69. $P(x) = 3x^3 - 11x^2 - 14x$; $P(4) = -40$; $P(5) = 30$;
since $-40 < 0 < 30$, there is at least one value of x such that
$4 < x < 5$ and $P(x) = 0$

71. $P(x) = x^4 - 8x^2 + 15; P(1) = 8; P(2) = -1;$ since $-1 < 0 < 8$, there is at least one value of x such that $1 < x < 2$ and $P(x) = 0$
73. (a) $V(x) = x(20 - 2x)(24 - 2x) = 4x^3 - 88x^2 + 480x$
(b)

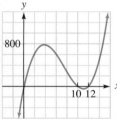

$V(x) = x(20 - 2x)(24 - 2x) = 4x^3 - 88x^2 + 480x$

(c) $D = [0, 10]$ **(d)** maximum volume ≈ 774.2 in^3, when $x \approx 3.6$ in
75. (a) $P(x) = (270 + x)(840 - 0.1x^2)$
$$= -0.1x^3 - 27x^2 + 840x + 226{,}800$$
(b) use the window $X\,[0, 100]$, $Y\,[-100, 290{,}000]$

(c) the zero is $x \approx 91.7$, so the domain is $D = [0, 91]$

(d) the maximum occurs at $x \approx 14.4$; $f(14) = 232{,}994$ and $f(15) = 232{,}988$; use $x = 14$ and plant 284 trees

77. (a) \$540,000   **(b)** March, July   **79.** yes
81. yes, such as $f(x) = x^2 + 1$ where $f(x) > 0$ for all values of x
83. a polynomial of odd degree must have end behavior with the left side rising and the right side falling, or the left side falling and the right side rising

Section 2.3
1. whole   **3.** any   **5.** factor   **7.** two   **9.** two
11. lower bound   **13.** -7
15. $(x - 1)(3x^2 + x - 5) - 9$
17. $(x - 3)(3x^2 + 7x + 15) + 41$
19. $(x + 1)(3x^2 - 5x - 1) - 3$
21. $(x + 3)(3x^2 - 11x + 27) - 85$
23. -1 **25.** 9   **27.** 3   **29.** true   **31.** true
33. false   **35.** true   **37.** $x = 1, -1,$ and 5
39. $x = 1, 2,$ and -1 **41.** $x = 1, 2,$ and -2
43. $x = 3, -3,$ and 2   **45.** $x = 1, -1,$ and $\dfrac{1}{2}$
47. $x = -1$ (multiplicity 2) and $\dfrac{1}{3}$
49. $x = 1, 2, 3,$ and 4   **51.** $x = 2$ and -5
53. $x = 1, -1, 2, -2,$ and -3
55. $x = 0, 2$ (multiplicity 3), and -2 (multiplicity 3)
57. $x = \dfrac{2}{3}$ **59.** $x = -1, 2, 3,$ and $\dfrac{2}{3}$
61. $x = -3, \dfrac{1}{3},$ and $\dfrac{1}{2}$ (multiplicity 2)

63. $x = 2$ (multiplicity 2), -3, $-\dfrac{1}{3}$, and $\dfrac{1}{2}$
65. $x = \dfrac{3}{2}, \dfrac{2}{3},$ and $-\dfrac{3}{5}$ **67.** $x = \dfrac{2}{3}, -\dfrac{3}{5},$ and 4
69. $x = -\dfrac{1}{2}, \dfrac{3}{2},$ and $\dfrac{6}{5}$ **71.** $x = -1, -\dfrac{2}{3},$ and 3
73. zero or two positive; one negative; zero or two nonreal
75. zero positive; one or three negative; zero or two nonreal
77. zero positive; zero negative; four nonreal
79. one positive; one negative; two nonreal
81. zero positive; zero negative; ten nonreal
83. zero positive; zero negative; eight nonreal
85. one positive; one negative; two nonreal
87. Lower bound is -2. Upper bound is 4.
89. Lower bound is -1. Upper bound is 1.
91. Lower bound is -4. Upper bound is 6.
93. Lower bound is -4. Upper bound is 3.
95. Lower bound is -2. Upper bound is 4.
97. 10 Ω, 20 Ω, and 60 Ω   **99.** 13 in
101. $(3, 7)$ or approximately $(1.54, 13.63)$ **103.** 0
105. $P(2) = 0$ means $(x - 2)$ is a factor, and $P(-2) = 0$ means $(x + 2)$ is a factor; this means that $(x + 2)(x - 2) = x^2 - 4$ is a factor

Section 2.4
1. $a - bi$ **3.** four
5. $\left\{-2, \dfrac{1}{2}, 3\right\}$ **7.** $\left\{0, -\sqrt{5}, \sqrt{5}\right\}$

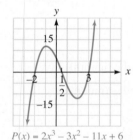

$P(x) = 2x^3 - 3x^2 - 11x + 6$

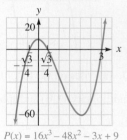

$y\ P(x) = x^3 - 5x$

9. $\left\{-1, 2, \dfrac{2 - \sqrt{3}}{2}, \dfrac{2 + \sqrt{3}}{2}\right\}$ **11.** $\left\{3, -\dfrac{\sqrt{3}}{4}, \dfrac{\sqrt{3}}{4}\right\}$

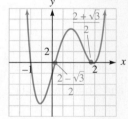

$P(x) = 4x^4 - 12x^3 + x^2 + 15x - 2$

$P(x) = 16x^3 - 48x^2 - 3x + 9$

13. $\sqrt{7}$ is a solution of $x^2 - 7 = 0$, which has possible rational zeros $x = -1, 1, -7,$ and 7; none of these are solutions
15. $\{-1, -5, 3\}$ **17.** $\left\{-\dfrac{1}{2}, 3, -3\right\}$
19. $\left\{1 \text{ (multiplicity 2)}, \sqrt{3}, -\sqrt{3}\right\}$
21. $\{-2i, 2i, 1\}$ **23.** $\{0, \sqrt{3} \text{ (multiplicity 2)}\}$

25. $\{1, 2 - 3i, 2 + 3i\}$

27. $\left\{1, \dfrac{1}{3} - i\dfrac{\sqrt{2}}{3}, \dfrac{1}{3} + i\dfrac{\sqrt{2}}{3}\right\}$

29. $\{-2, 1 \text{ (multiplicity 2)}, -2i, 2i\ \}$

31. $\{-1, 0 \text{ (multiplicity 2)}, -4i, 4i\}$

33. $P(x) = x^2 - 9x + 20$

35. $P(x) = x^3 - 3x^2 + 3x - 1$

37. $P(x) = x^3 - 11x^2 + 38x - 40$

39. $P(x) = x^4 - 3x^2 + 2$

41. $P(x) = x^3 - \sqrt{2}x^2 + x - \sqrt{2}$

43. $P(x) = x^3 - x^2 + x - 1$

45. $P(x) = x^3 - 4x^2 - 2x + 20$

47. $P(x) = x^4 - 5x^3 + 10x^2 - 10x + 4$

49. $P(x) = x^4 - 4x^3 + 6x^2 - 4x + 5$

51. $P(x) = x^3 - 2x^2 + 9x - 18$

53. $P(x) = x^3 - 2x^2 + x - 2$

55. $x = i, 1,$ and 3 **57.** $x = 0, 1, -3,$ and $1 - i$

59. $x = -1, 2,$ and $-2i$

61. 3 and $1 - i$ **63.** $-3, 3,$ and $1 - i$

65. $x^2 - 4ix - 4 = 0$ **67.** two complex

69. three real   **71.** two real, two complex

73. four real   **75.** five real

77. (a) $P(x) = (x^2 + 7)(x^2 - 7)$;

(b) $\left\{-\sqrt{7}, \sqrt{7}, -i\sqrt{7}, i\sqrt{7}\right\}$

79. (a) $P(x) = (x - 1)(x + 1)(x^2 - x + 1)(x^2 + x + 1)$;

(b) $\left\{1, -1, \dfrac{1 + i\sqrt{3}}{2}, \dfrac{1 - i\sqrt{3}}{2}, \dfrac{-1 + i\sqrt{3}}{2}, \dfrac{-1 - i\sqrt{3}}{2}\right\}$

81. $V(h) = h^3 + 9h^2 + 14h$; 3.2 ft by 5.2 ft by 10.2 ft

83. 1 mi, 5 mi, and 9 mi

85. $4 - \sqrt{2} \approx 2.6$ min, $4 + \sqrt{2} \approx 5.4$ min, and 10 min

Section 2.5

1. asymptote   **3.** vertical   **5.** x **7.** equal

9. horizontal, vertical   **11.** $D = (-\infty, 2) \cup (2, \infty)$

13. $D = (-\infty, -5) \cup (-5, 5) \cup (5, \infty)$

15. $D = (-\infty, -1) \cup (-1, 0) \cup (0, 1) \cup (1, \infty)$

17. $D = (-\infty, \infty)$ **19.** $D = (-\infty, -4) \cup (-4, -1) \cup (-1, \infty)$

21. $x = 3$ **23.** $x = 1$ and $x = -1$

25. $x = -2$ and $x = 3$ **27.** none   **29.** $y = 2$

31. $y = \dfrac{1}{2}$ **33.** $y = 0$ **35.** none

37. $D = (-\infty, -1) \cup (-1, 3) \cup (3, \infty)$

39. $x = -1$, $x = 3$, and $y = 0$ **41.** 1

43. ∞ **45.** ∞ **47.** $D = (-\infty, 0) \cup (0, 1) \cup (1, \infty)$

49. The numerator cannot equal 0.

51. $x = 0$ and $x = 1$ **53.** 0   **55.** $-\infty$ **57.** ∞

59. $y = x + 1$ **61.** $y = x - 3$ **63.** $y = 2x + 3$

65. $y = x + 2$

67.

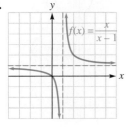

69.

71.

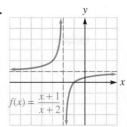

73.

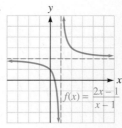

75.

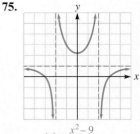

77.

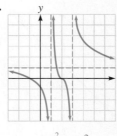

79.

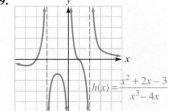

81.

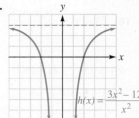

83.

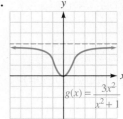

85.

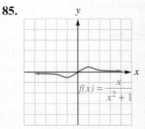

87.

89.

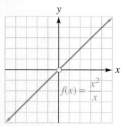

$h(x) = \dfrac{x^2 - 2x - 8}{x - 1}$

91.

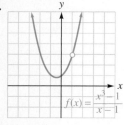

$f(x) = \dfrac{x^2}{x}$

93.

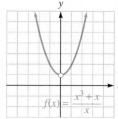

$f(x) = \dfrac{x^3 + x}{x}$

95.

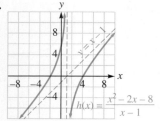

$f(x) = \dfrac{x^3 - 1}{x - 1}$

97. (a) $C(x) = 1.25x + 700$ **(b)** $\overline{C}(x) = \dfrac{1.25x + 700}{x}$

(c) \$2.65   **(d)** \$1.95   **(e)** \$1,325   **(f)** \$1.60

99. (a) $C(n) = 0.095n + 8.50$ **(b)** $\overline{C}(n) = \dfrac{0.095n + 8.50}{n}$

(c) \$0.105

101. (a) $B(n) = 300 + \dfrac{12,000}{n}$ **(b)** \$700   **(c)** \$300

103. (a) 75%   **(b)** about 85%   **(c)** The value gets closer and closer to 100%.   **(d)** Yes, as $t \to \infty$, the value of $O(t)$ approaches 100%.

Section 2.6
1. $f(x) > 0$, $f(x) \geq 0$, $f(x) < 0$, or $f(x) \leq 0$
3. graphing, using a sign chart   **5.** excluded   **7.** $[-4, 3]$
9. $(-4, -3)$ **11.** $[1, 12]$ **13.** no solution
15. $\left(-\infty, -\dfrac{4}{3}\right) \cup \left[\dfrac{1}{2}, \infty\right)$ **17.** $\{3\}$
19. $\left(-\infty, -\dfrac{1}{2}\right] \cup \left[-\dfrac{1}{3}, \infty\right)$ **21.** $(-\infty, -5] \cup [-4, \infty)$
23. $(-\infty, \infty)$ **25.** no solution
27. $\left(-\infty, -\dfrac{4}{3}\right) \cup \left(-\dfrac{4}{3}, \infty\right)$ **29.** $[-2, 1] \cup [5, \infty)$
31. $(-\infty, -3) \cup (2, \infty)$ **33.** $(-\infty, -3] \cup \{2\} \cup [5, \infty)$
35. $[-2, \infty)$ **37.** $(-\infty, -1) \cup (0, 3)$
39. $[-2, -1] \cup [1, \infty)$ **41.** $\left(-2, -\dfrac{1}{2}\right) \cup \left(\dfrac{1}{2}, \infty\right)$
43. $(-\infty, -2)$ **45.** $[-2, \infty)$ **47.** $(-2, -1)$
49. $(-\infty, -4] \cup \left[-\dfrac{1}{2}, 1\right] \cup [3, \infty)$ **51.** $(-3, 2)$
53. $(-\infty, -2] \cup (-1, 0] \cup (1, \infty)$

55. $(-3, -2) \cup (2, 3)$ **57.** $(-\infty, -3) \cup (-3, -2] \cup (2, \infty)$
59. $\left(-\dfrac{1}{3}, \dfrac{1}{2}\right)$ **61.** $\left(0, \dfrac{3}{2}\right)$ **63.** $(-\infty, 2) \cup \left[\dfrac{13}{5}, \infty\right)$
65. $\left(-\infty, -\sqrt{7}\right) \cup (-1, 1) \cup \left(\sqrt{7}, \infty\right)$
67. $(-2, -1] \cup (0, 1]$ **69.** $(-\infty, -1) \cup \left(\dfrac{1}{3}, 2\right)$
71. $(-\infty, -0.637] \cup [0, 2.5] \cup [3.137, \infty)$
73. $(-\infty, -1.414] \cup [1.414, \infty)$ **75.** $(0, 2)$ **77.** $[1, 7]$
79. $(0, 4)$ **81.** $(0, 1)$ **83.** $[200, 400]$
85. at least 61 cakes

Chapter Review
1. upward; minimum   **2.** downward; maximum
3. $(1, 6)$ **4.** $(-4, -5)$ **5.** $(-3, -13)$
6. $\left(\dfrac{1}{2}, -8\right)$
7. vertex $= (2, -3)$; $D = (-\infty, \infty)$; $R = [-3, \infty)$;
y-intercept $= 1$; x-intercepts $= 2 - \sqrt{3}$ and $2 + \sqrt{3}$;
axis of symmetry: $x = 2$

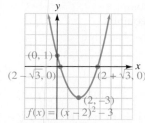

$f(x) = (x - 2)^2 - 3$

8. vertex $= (4, 4)$, $D = (-\infty, \infty)$; $R = (-\infty, 4]$;
y-intercept $= -12$; x-intercepts $= 2$ and 6;
axis of symmetry: $x = 4$

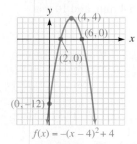

$f(x) = -(x - 4)^2 + 4$

9. vertex $= \left(\dfrac{1}{2}, -\dfrac{1}{4}\right)$; $D = (-\infty, \infty)$; $R = \left[-\dfrac{1}{4}, \infty\right)$;

y-intercept $= 0$; x-intercepts $= 0$ and 1;

axis of symmetry: $x = \dfrac{1}{2}$

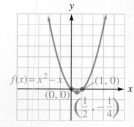

$f(x) = x^2 - x$

10. vertex $= \left(\dfrac{1}{2}, \dfrac{1}{4}\right)$; $D = (-\infty, \infty)$; $R = \left(-\infty, \dfrac{1}{4}\right]$;

y-intercept $= 0$; x-intercepts $= 0$ and 1;

axis of symmetry: $x = \dfrac{1}{2}$

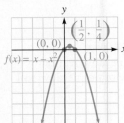

11. vertex $= \left(\dfrac{3}{2}, -\dfrac{25}{4}\right)$; $D = (-\infty, \infty)$; $R = \left[-\dfrac{25}{4}, \infty\right)$;

y-intercept $= -4$; x-intercepts $= -1$ and 4;

axis of symmetry: $x = \dfrac{3}{2}$

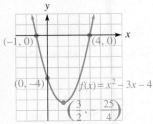

12. vertex $= \left(\dfrac{4}{3}, -\dfrac{25}{3}\right)$; $D = (-\infty, \infty)$; $R = \left[-\dfrac{25}{3}, \infty\right)$;

y-intercept $= -3$; x-intercepts $= -\dfrac{1}{3}$ and 3;

axis of symmetry: $x = \dfrac{4}{3}$

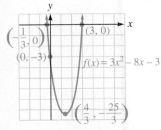

13. vertex $= (1, -2)$; $D = (-\infty, \infty)$; $R = (-\infty, -2]$;
y-intercept $= -3$; axis of symmetry: $x = 1$

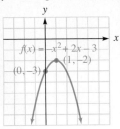

14. vertex $= (-1, 4)$; $D = (-\infty, \infty)$; $R = (-\infty, 4]$;
y-intercept $= 3$; x-intercepts $= -3$ and 1;
axis of symmetry: $x = -1$

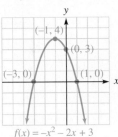

15. 300 ft   **16.** both numbers are $\dfrac{1}{2}$

17. 350 ft by 350 ft; 122,500 ft^2
18. 50 cameras; \$1,100
19. degree 4 and $a_n > 0$; end behavior like that of $f(x) = x^2$; $f(x) \to \infty$ as $x \to -\infty$ and $f(x) \to \infty$ as $x \to \infty$; the graph rises on the left and rises on the right
20. degree 5 and $a_n > 0$; end behavior like that of $f(x) = x^3$; $f(x) \to -\infty$ as $x \to -\infty$ and $f(x) \to \infty$ as $x \to \infty$; the graph falls on the left and rises on the right
21. degree 5 and $a_n < 0$; end behavior like that of $f(x) = -x^3$; $g(x) \to \infty$ as $x \to -\infty$ and $g(x) \to -\infty$ as $x \to \infty$; the graph rises on the left and falls on the right
22. degree 6 and $a_n < 0$; end behavior like that of $f(x) = -x^2$; $g(x) \to -\infty$ as $x \to -\infty$ and $g(x) \to -\infty$ as $x \to \infty$; the graph falls on the left and falls on the right
23. (a) falls on the left and rises on the right

(b) $x = 0, \dfrac{1}{2},$ and 3   **(c)** $y = 0$

(d)

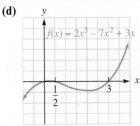

24. (a) rises on the left and rises on the right **(b)** $x = -2$ (multiplicity 2), 0, and 2   **(c)** $y = 0$
(d)

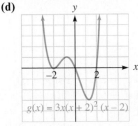

25. (a) falls on the left and rises on the right   **(b)** $x = -5$, -2, and $\frac{3}{2}$ **(c)** $y = -30$

(d)

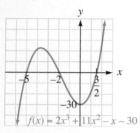

26. (a) rises on the left and falls on the right   **(b)** $x = 0$ (multiplicity 2), 3 (multiplicity 2), and -2 **(c)** $y = 0$

(d)

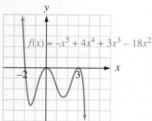

27. (a) rises on the left and rises on the right   **(b)** $x = -2$ and 2   **(c)** $y = -4$

(d)

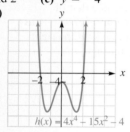

28. (a) falls on the left and falls on the right   **(b)** $x = -5, -3,$ 3, and 5   **(c)** $y = -225$

(d)

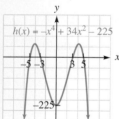

29. $3x^4 + 2x^2 + 3x + 7 = (3x^3 + 9x^2 + 29x + 90)(x - 3) + 277$
30. $2x^4 - 3x^2 + 3x - 1 = (2x^3 + 4x^2 + 5x + 13)(x - 2) + 25$
31. $5x^5 - 4x^4 + 3x^3 - 2x^2 + x - 1$
$$= (5x^4 - 14x^3 + 31x^2 - 64x + 129)(x + 2) - 259$$
32. $4x^5 + 2x^4 - x^3 + 3x^2 + 2x + 1$
$$= (4x^4 - 2x^3 + x^2 + 2x)(x + 1) + 1$$
33. 1   **34.** 66   **35.** 241   **36.** 34   **37.** false
38. false   **39.** true   **40.** true   **41.** 151
42. -113 **43.** $\frac{13}{8}$ **44.** $-6i - 1$

45. $\left\{-2, \frac{1}{2}, 3\right\}$ **46.** -2 (multiplicity 2), $-\sqrt{5}, \sqrt{5}$
47. $x = -\frac{1}{2}, \frac{1}{3}$, and 2   **48.** $x = -\frac{3}{2}, -\frac{2}{5}$, and 3
49. $x = -3, \frac{1}{2}$, and 1 (multiplicity 2)
50. $x = -1$ (multiplicity 2), $-\frac{1}{3}$, 1, and 5
51. zero   **52.** two   **53.** one   **54.** two
55. two or zero positive real zeros; one negative real zero
56. two or zero positive real zeros; three or one negative real zeros
57. $x^3 - 2x + 4 = 0$
58. $x^3 - 3x^2 + 4x - 12 = 0$
59. $x = -1, 3i$, and 2   **60.** $x = -\frac{1}{2}$ and $2 - i$
61. $\left\{0, \dfrac{3\sqrt{5} - \sqrt{61}}{2}, \dfrac{3\sqrt{5} + \sqrt{61}}{2}\right\}$
62. $\left\{-\dfrac{1}{3}, -4i, 4i, \dfrac{1}{2}\right\}$ **63.** $\left\{-1, \dfrac{3}{2}, 2\right\}$
64. $\left\{-3, \dfrac{1}{2}, 1\right\}$ **65.** $\{-5, -i, i, 2\}$ **66.** $\{-3, -i, i, 2\}$
67. 10 m by 6 m; 12 m by 5 m; 15 m by 4 m   **68.** 17.27 ft
69. $D = (-\infty, -2) \cup (-2, 0) \cup (0, \infty)$
70. $R = (-\infty, 0) \cup (0, \infty)$ **71.** 1
72. $f(0)$ is not defined   **73.** $x = -2, x = 0$ **74.** $y = 0$
75. 0   **76.** 0   **77.** ∞ **78.** $-\infty$ **79.** ∞ **80.** ∞
81. $D = (-\infty, -5) \cup (-5, 5) \cup (5, \infty)$ **82.** $D = (-\infty, \infty)$
83. $x = 1, x = -1$ **84.** $x = -7$ **85.** $x = 3, x = -2$
86. $x = 4, x = -1$ **87.** $y = \dfrac{1}{2}$ **88.** $y = -5$
89. $y = 0$ **90.** none   **91.** $y = 2x + 3$ **92.** none
93. V.A.: $x = 1$, H.A.: $y = 0$

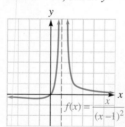

94. V.A.: $x = -1, x = 4$, H.A.: $y = 0$

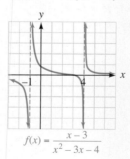

95. V.A.: $x = -4$, H.A.:$y = 3$

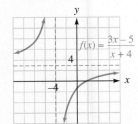

96. V.A.: $x = -3$, $x = 2$, H.A.:$y = 2$

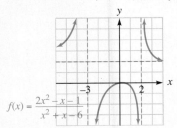

97. V.A.: $x = 3$; O.A.: $y = x + 3$

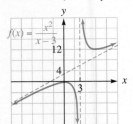

98. V.A.: $x = -2$; O.A.: $y = x - 4$

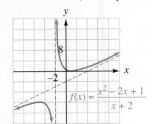

99. 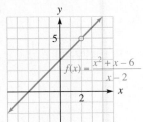   **100.**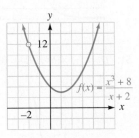

101. no solution   **102.** $\left(-\dfrac{5}{2}, \dfrac{1}{3}\right)$

103. $(-\infty, -3] \cup \left[-1, \dfrac{1}{2}\right]$ **104.** $\left(-\dfrac{5}{3}, 1\right) \cup (2, \infty)$

105. $\left(0, \dfrac{2}{3}\right)$ **106.** $(-2, -1] \cup (2, 3]$

Chapter Test

1. $(4, -10)$ **2.** $(-2, 9)$ **3.** $\dfrac{25}{8}$ sec   **4.** $\dfrac{625}{4}$ ft

5. 10 ft   **6.** 110 ft

7. (a) falls on the left and rises on the right   **(b)** $x = -2, 0$
(multiplicity 2) and 1 (multiplicity 2)   **(c)** 0
(d)

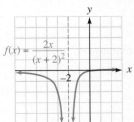

8. $x = -3, -\dfrac{1}{2}$, and 2

9. $x = -\dfrac{1}{2}$, 0 (multiplicity 2), 2, and 3

10. $x = -3, -2i$, and $\dfrac{2}{3}$

11. $x = -1, \dfrac{1}{2}$, and $1 + 2i$ **12.** $\{2, 3 - 2i, 3 + 2i\}$

13. $\{-2, 3, -2i, 2i\}$
14. V.A.: $x = -2$, H.A.: $y = 0$

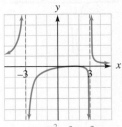

15. V.A.:$x = -3$, $x = 3$, H.A. $y = 1$

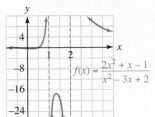

16. V.A. $x = 1$, $x = 2$, H.A. $y = 2$

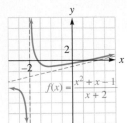

17. V.A. $x = -2$, O.A. $y = x - 1$

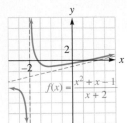

18.

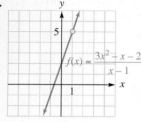

$f(x) = \dfrac{3x^2 - x - 2}{x - 1}$

19.

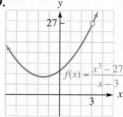

$f(x) = \dfrac{x^3 - 27}{x - 3}$

33. $D = (-\infty, \infty)$; $R = (0, \infty)$; $y = 0$

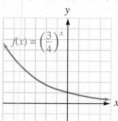

$f(x) = \left(\dfrac{1}{5}\right)^x$

20. $(-\infty, -5) \cup \left(-\dfrac{5}{2}, \infty\right)$ **21.** $[-2, 2] \cup [5, \infty)$

22. $(-\infty, 1) \cup \left(\dfrac{11}{6}, \infty\right)$

35. $D = (-\infty, \infty)$; $R = (0, \infty)$; $y = 0$

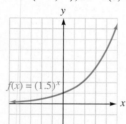

$f(x) = \left(\dfrac{3}{4}\right)^x$

Connect to Calculus

1. increasing on $(-\infty, -4) \cup (3, \infty)$; decreasing on $(-4, 3)$; local maximum $= 208$ at $x = -4$; local minimum $= -135$ at $x = 3$

2. increasing on $(-\infty, -1) \cup (1, \infty)$; decreasing on $(-1, 0) \cup (0, 1)$; local maximum $= 5$ at $x = -1$; local minimum $= 1$ at $x = 1$

3. increasing on $\left(-\sqrt{2}, 0\right) \cup \left(\sqrt{2}, \infty\right)$; decreasing on $\left(-\infty, -\sqrt{2}\right) \cup \left(0, \sqrt{2}\right)$; local minimum $= 2$ at $x = -\sqrt{2}$ and $x = \sqrt{2}$; local maximum $= 6$ at $x = 0$

37. $D = (-\infty, \infty)$; $R = (0, \infty)$; $y = 0$

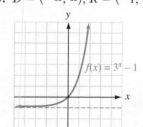

$f(x) = (1.5)^x$

4.

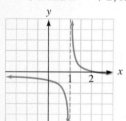

5.

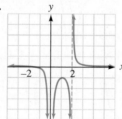

6. $x \approx 3.13040$ **7.** $x \approx 2.08150$

39. yes   **41.** no   **43.** $\dfrac{1}{2}$ **45.** 2   **47.** e

49. $D = (-\infty, \infty)$; $R = (-1, \infty)$; $y = -1$; y-intercept $= 0$

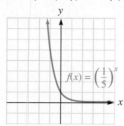

$f(x) = 3^x - 1$

Section 3.1

1. exponential   **3.** $(-\infty, \infty)$ **5.** $(0, \infty)$ **7.** asymptote
9. increasing   **11.** 2.718   **13.** increasing
15. 11.035665   **17.** 451.807873   **19.** 10.045109
21. 0.123407   **23.** 9.974182   **25.** 1,401.610208
27. 581.498631   **29.** 2.946972
31. $D = (-\infty, \infty)$; $R = (0, \infty)$; $y = 0$

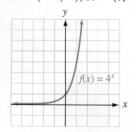

$f(x) = 4^x$

51. $D = (-\infty, \infty)$; $R = (0, \infty)$; $y = 0$; y-intercept $= \dfrac{1}{3}$

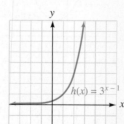

$h(x) = 3^{x-1}$

53. $D = (-\infty, \infty)$; $R = (0, \infty)$; $y = 0$; y-intercept $= 4$

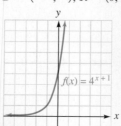

$f(x) = 4^{x+1}$

55. $D = (-\infty, \infty)$; $R = (-2, \infty)$; $y = -2$; y-intercept $= -1$

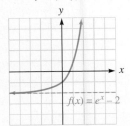

57. $D = (-\infty, \infty)$; $R = (0, \infty)$; $y = 0$;
y-intercept $= e^{-2} \approx 0.135$

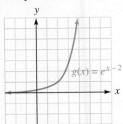

59. $D = (-\infty, \infty)$; $R = (0, \infty)$; $y = 0$; y-intercept $= 1$

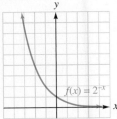

61. $D = (-\infty, \infty)$; $R = (-\infty, 1)$; $y = 1$; y-intercept $= 0$

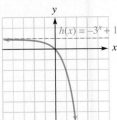

63. $D = (-\infty, \infty)$; $R = (-1, \infty)$; $y = -1$; y-intercept $= 3$

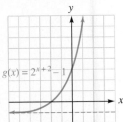

65. $D = (-\infty, \infty)$; $R = (0, \infty)$; $y = 0$; y-intercept $= 3$

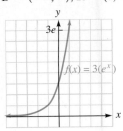

67. $D = (-\infty, \infty)$; $R = (0, \infty)$; $y = 0$; y-intercept $= 1$

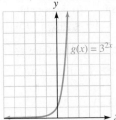

69. $D = (-\infty, \infty)$; $R = (-1, \infty)$; $y = -1$; y-intercept $= 0$

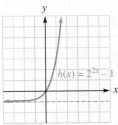

71. $D = (-\infty, \infty)$; $R = (0, \infty)$; $y = 0$; y-intercept $= 3$

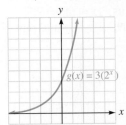

73. y-intercept $= 5$ **75.** y-intercept $= 1$
77. y-intercept $= 1$; x-intercept ≈ -0.693147
79. y-intercept $= -3$; x-intercept ≈ 1.668088
81. a. 24.498510   **b.** 0.149361   **c.** 45.462787
d. 69.422078   **83.** \$22,080.40   **85.** \$15.79
87. \$2,273,996.13   **89.** \$1,263.77   **91.** \$13,375.68
93. \$7,647.95 from continuous compounding; \$7,518.28 from annual compounding
95. \$291.27   **97.** \$12,155.61   **99.** \$560,122.19
101. Answers will vary.   **103.** 125

Section 3.2
1. birth, death   **3.** 0.1868 g   **5.** about 10 kg
7. about 35.4%   **9.** 0.1575 unit   **11.** 2 lm
13. 8 lm   **15.** about 56,570   **17.** 167,904,600
19. 9.44×10^5 **21.** 10.6 billion   **23.** 2.6
25. 23.3%   **27.** 0.076   **29.** 4,305   **31.** 61.9°C
33. about 24,060   **35.** about 492   **37.** 19.0 mm
39. 49 m/s   **41.** about 7 million   **43.** about 72.2 years
45. $y = 1.035264924^x$

47. $6.2276305089733 + 1.2566611357475^x$
49.

51. 1,200,000; a little over 14 years; a little over 21 years

Section 3.3

1. $b^y = x$ **3.** range **5.** inverse **7.** exponent

9. $(1, 0), (b, 1)$ **11.** $\log_e x$ **13.** $(-\infty, \infty)$ **15.** 10

17. $\log_8 64 = 2$ **19.** $\log_4 \dfrac{1}{16} = -2$ **21.** $\log_{1/2} 32 = -5$

23. $\log_x z = y$ **25.** $3^4 = 81$ **27.** $\left(\dfrac{1}{2}\right)^3 = \dfrac{1}{8}$

29. $4^{-3} = \dfrac{1}{64}$ **31.** $\pi^1 = \pi$ **33.** 3 **35.** -3

37. 3 **39.** $\dfrac{1}{2}$ **41.** -3 **43.** 8 **45.** -1 **47.** 5

49. $\dfrac{1}{2}$ **51.** -3 **53.** 64 **55.** 5 **57.** $\dfrac{1}{25}$ **59.** 5

61. $\dfrac{3}{2}$ **63.** $\dfrac{2}{3}$ **65.** 5 **67.** 4 **69.** 4 **71.** -3

73. 7 **75.** 4 **77.** 0.5119 **79.** -2.3307

81. -0.4055 **83.** -0.2752 **85.** no value

87. 25.2522 **89.** 0.0002 **91.** 4.0645

93. 0.0245 **95.** 120.0719 **97.** $D = (4, \infty)$

99. $D = (-\infty, 2)$ **101.** $D = (-\infty, \infty)$

103. $D = (-\infty, -5) \cup (5, \infty)$ **105.** $D = (0, \infty)$

107. $x = 0$; x-intercept $= 1$ **109.** $x = 0$; x-intercept $= 1$

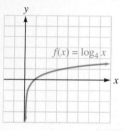
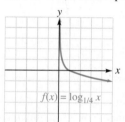

111. $x = 0$; x-intercept $= \dfrac{1}{4}$

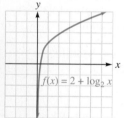

113. $x = 2$; x-intercept $= 3$

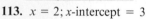
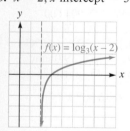

115. $x = -1$; x-intercept $= -\dfrac{3}{4}$

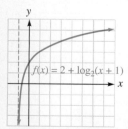

117. $x = 0$; x-intercept $= 1$

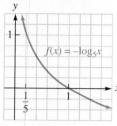

119. $x = 0$; x-intercept $= -1$

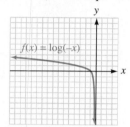

121. $x = 3$; x-intercept $= 2$

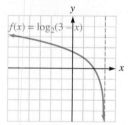

123. $x = 0$; x-intercept $= \dfrac{1}{3}$

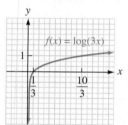

125. $x = -2$; x-intercept $= -1$

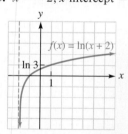

127. $x = 0$; x-intercept $= 1$

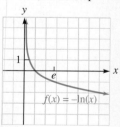

129. $x = -\pi$; y-intercept ≈ 0.49714987;
x-intercept ≈ -2.141593
131. no vertical asymptote ; y-intercept $= -2$;
no x-intercept
133. $x = -3$; y-intercept ≈ 0.47712125; x-intercept $= -2$
135. $x = -\sqrt{2}$; y-intercept ≈ -1.039721; x-intercept
≈ -0.4142136
137. b **139.** Answers may vary.
141. $y = \ln(e^x) = x$, which is a line   **143.** $g(x) = e^x$

Section 3.4

1. $20 \log \dfrac{E_O}{E_I}$ **3.** $t = -\dfrac{1}{k}\ln\left(1 - \dfrac{C}{M}\right)$

5. $E = RT\ln\left(\dfrac{V_f}{V_i}\right)$ **7.** 55 dB   **9.** 29 dB

11. 49.5 dB   **13.** 4.4   **15.** 4   **17.** no   **19.** 19.8 min
21. about 5.8 yr   **23.** about 9.2 yr   **25.** about 3,654 J
27. about 99% per year   **29.** 3 yr   **31.** about 10.8 yr
33. about 208,000 V   **35.** about 1 mile

Section 3.5

1. 0   **3.** M, N **5.** x, y **7.** x **9.** $\neq$ **11.** 0
13. 7   **15.** 10   **17.** 1   **25.** $\log_b 2 + \log_b x + \log_b y$
27. $\log_b 2 + \log_b x - \log_b y$ **29.** $2\log_b x + 3\log_b y$
31. $\dfrac{1}{3}(\log_b x + \log_b y)$ **33.** $\log_b x + \dfrac{1}{2}\log_b z$

35. $\dfrac{1}{3}\log_b x - \dfrac{1}{3}\log_b y - \dfrac{1}{3}\log_b z$ **37.** $\ln 5 - 3\ln x$

39. $\log 3 + \log x - 4$ **41.** $\ln 11 - 3$

43. $5 + \log_2 x$ **45.** $3 - \dfrac{1}{2}\log_4(x - 3)$

47. $1 + \ln 3$ **49.** 5

51. $\dfrac{1}{2}\log_2(x - 1) - \dfrac{3}{2}\log_2 x - \dfrac{1}{2}\log_2(x + 1)$

53. $4\ln x + \dfrac{1}{2}\ln(2 - x) - 4\ln(x + 3)$

55. $\log 9 + 3\log(x + 1) - 2 - \log x - \dfrac{1}{3}\log(x + 3)$

57. $\log_b \dfrac{x + 1}{x}$ **59.** $\log_b x^2\sqrt[3]{y}$ **61.** $\log_b \dfrac{\sqrt{z}}{x^3 y^2}$

63. $\log_b \dfrac{x}{y}$ **65.** $\ln 5x^3$ **67.** 5   **69.** $\ln \dfrac{x^{1/2}}{y^{1/4}}$

71. $\ln \dfrac{x}{5(x + 1)}$ **73.** true   **75.** true   **77.** true
79. false   **81.** true   **83.** 1.4472   **85.** 0.3521
87. 1.1972   **89.** 2.4014   **91.** 2.0493   **93.** 0.4682
95. 1.7712   **97.** 0.9597   **99.** 1.8928   **101.** 2.3219
103.

105.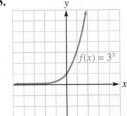

107. 4.77   **109.** 5.01×10^{-4} to 1.26×10^{-3} gram-ions/L
111. 19 dB

113. The original intensity must be raised to the fourth power.
115. V^2 L   **117.** 21   **125.** $\log 0.9 < 0$ **127.** no

Section 3.6

1. exponential   **3.** $A_0 2^{-t/h}$ **5.** -4 **7.** $\dfrac{7}{3}$ **9.** -1

11. 4   **13.** 5   **15.** 3, –1   **17.** -2 **19.** $\dfrac{5}{6}$

21. -1 **23.** 5, –5   **25.** $\dfrac{\ln 5}{\ln 4} \approx 1.1610$

27. $\dfrac{\ln 2 + \ln 13}{\ln 13} = \dfrac{\ln 26}{\ln 13} \approx 1.2702$

29. $\dfrac{\ln 2}{\ln 3 - \ln 2} = \dfrac{\ln 2}{\ln \dfrac{3}{2}} \approx 1.7095$

31. 0   **33.** 4   **35.** $\pm \dfrac{1}{\sqrt{\log 7}} \approx \pm 1.0878$

37. $\ln 10 \approx 2.3026$ **39.** $\dfrac{\ln 6}{2} \approx 0.8959$

41. 0   **43.** 0.2789   **45.** 3, 1   **47.** 0

49. 0   **51.** 0.6931   **53.** 32   **55.** $\dfrac{1}{100}$

57. e^4 **59.** 24   **61.** $\dfrac{67}{2}$ **63.** $\dfrac{e^6}{3}$ **65.** 65

67. 4   **69.** –10, 10   **71.** 5   **73.** 50   **75.** 603
77. 2   **79.** 7   **81.** 6   **83.** 5   **85.** 3, 4   **87.** –2, 3
89. 1,100   **91.** no solution   **93.** 5   **95.** 1, 7
97. ±7.456417   **99.** 5.641589   **101.** about 5.1 yr
103. about 42.7 days   **105.** about 2,900 yr
107. about 5.6 yr   **109.** about 5.4 yr
111. about 3.2 days   **113.** about 25.3 yr
115. $\dfrac{\ln 0.75}{3}$ **121.** $x = 7^{625}$

Chapter Review

1. $5^{2\sqrt{2}}$ **2.** $2^{\sqrt{10}}$
3.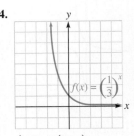
4.

5. $p = 1$, $q = 6$ **6.** $D = (-\infty, \infty)$; $R = (0, \infty)$
7.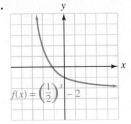
8.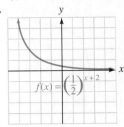

9.
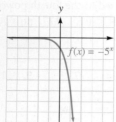
$f(x) = -5^x$

10.
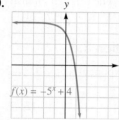
$f(x) = -5^x + 4$

11.
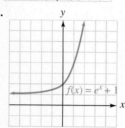
$f(x) = e^x + 1$

12.
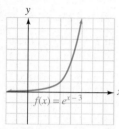
$f(x) = e^{x-3}$

13. \$2,189,703.45   **14.** \$2,324,767.37   **15.** $\dfrac{2}{3}$

16. 0.19 lm   **17.** about 635 million   **18.** about 2,708

19. $D = (0, \infty); R = (-\infty, \infty)$

20. $D = (0, \infty); R = (-\infty, \infty)$ **21.** 2   **22.** $-\dfrac{1}{2}$

23. 0   **24.** -2 **25.** $\dfrac{1}{2}$ **26.** $\dfrac{1}{3}$ **27.** 32   **28.** 9

29. 8   **30.** -1 **31.** $\dfrac{1}{8}$ **32.** 2   **33.** 4   **34.** 2

35. 10   **36.** $\dfrac{1}{25}$ **37.** 5   **38.** 3

39.
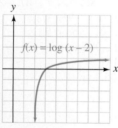
$f(x) = \log (x - 2)$

40.
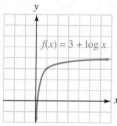
$f(x) = 3 + \log x$

41.
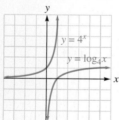
$y = 4^x$
$y = \log_4 x$

42.
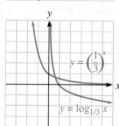
$y = \left(\dfrac{1}{3}\right)^x$
$y = \log_{1/3} x$

43. 6.1137   **44.** -0.1111 **45.** 10.3398   **46.** 2.5715

47.
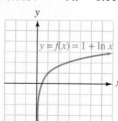
$y = f(x) = 1 + \ln x$

48.
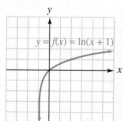
$y = f(x) = \ln(x + 1)$

49. 12   **50.** $14x$ **51.** 53 dB   **52.** 4.4   **53.** 9.5 min

54. 23 yr   **55.** 2,017 J   **56.** 0   **57.** 1   **58.** 3

59. 4   **60.** 4   **61.** 0   **62.** 7   **63.** 3   **64.** 4

65. 9   **66.** $2 \log_b x + 3 \log_b y - 4 \log_b z$

67. $\dfrac{1}{2}(\log_b x - \log_b y - 2 \log_b z)$

68. $4 \ln x - 5 \ln y - 6 \ln z$ **69.** $\dfrac{1}{3}(\ln x + \ln y + \ln z)$

70. $\log_b \dfrac{x^3 z^7}{y^5}$ **71.** $\log_b \dfrac{\sqrt{xy^3}}{z^7}$ **72.** $\ln \dfrac{x^4}{y^5 z^6}$

73. $\ln \dfrac{y^3 \sqrt{x}}{\sqrt[3]{z}}$ **74.** 3.36   **75.** 1.56   **76.** 2.64

77. -6.72 **78.** 1.7604

79. about 7.94×10^{-4} gram-ion/L   **80.** $k \ln 2$ **81.** 2

82. $-1, -3$ **83.** 2   **84.** $-3, 3$

85. $\dfrac{\log 7}{\log 3} \approx 1.7712$ **86.** $\dfrac{\log 3}{\log 3 - \log 2} \approx 2.7095$

87. $\ln 8 \approx 2.0794$ **88.** $\ln 7 \approx 1.9459$ **89.** -3

90. 8   **91.** 4, 25   **92.** 4   **93.** 2   **94.** 31

95. 3, 4   **96.** 6   **97.** $\dfrac{\ln 9}{\ln 2} \approx 3.1699$ **98.** no solution

99. $\dfrac{e}{e - 1} \approx 1.5820$ **100.** 1   **101.** about 3,300 yr

Chapter Test

1. f   **2.** c   **3.** d   **4.** a   **5.** g   **6.** b

7.
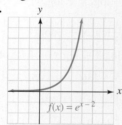
$f(x) = 2^x + 1$

8.
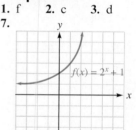
$f(x) = e^{x-2}$

9. $\dfrac{3}{64}$ g   **10.** \$1,060.90   **11.** \$4,451.08   **12.** 3

13. -3 **14.** 17   **15.** 2   **16.** -3

17.
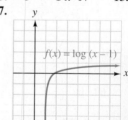
$f(x) = \log (x - 1)$

18.
$f(x) = 2 + \ln x$

19. $2 \log a + \log b + 3 \log c$

20. $\dfrac{1}{2}(\ln a - 2 \ln b - \ln c)$ **21.** $\log \dfrac{b\sqrt{a + 2}}{c^2}$

22. $\ln \dfrac{\sqrt[3]{\dfrac{a}{b^2}}}{c}$ **23.** 1.3801   **24.** 0.4259

25. $\dfrac{\ln 3}{\ln 7}$ or $\dfrac{\log 3}{\log 7}$ **26.** $\dfrac{1}{\ln \pi}$ or $\dfrac{\log e}{\log \pi}$ **27.** true

28. false   **29.** 6.4   **30.** 46 dB   **31.** $-1, 3$

32. $-5, 5$ **33.** $\dfrac{\log 3}{\log 3 - 2} \approx -0.3133$ **34.** $\ln 9 \approx 2.1972$

35. 1   **36.** 10   **37.** 9

Connect to Calculus

1. 1×10^{-10} **2.** 1×10^{10}

3. $\ln y = \dfrac{3}{4}\ln(x^3 - 2) + \dfrac{1}{2}\ln x - \dfrac{1}{3}\ln(3x - 5)$

4. $\ln y = (\ln x)^2$

Section 4.1

1. angle, vertex   **3.** acute   **5.** $\dfrac{\pi}{180°}$ **7.** $s = r\theta$

9. 42.244°   **11.** 321.006°   **13.** 12.247°

15. 14°37′30″   **17.** −217°24′46.8″   **19.** 47°4′48″

21. 2   **23.** $\dfrac{12}{19}$ **25.** 0   **27.** $\dfrac{3\pi}{2}$ **29.** $\dfrac{5\pi}{4}$

31. $\dfrac{7\pi}{6}$ **33.** $\dfrac{11\pi}{6}$ **35.** $-\dfrac{4\pi}{3}$ **37.** $-\dfrac{\pi}{3}$

39. 0.681   **41.** −3.793   **43.** 2.116   **45.** 30°

47. −450°   **49.** 150°   **51.** 405°

53. −390°   **55.** 300°   **57.** 20°   **59.** 105°

61. 200.54°   **63.** −28.65°   **65.** 151.59°

67. a.    **b.** I   **c.** acute

69. a.    **b.** IV   **c.** none

71. a.   **b.** II   **c.** obtuse

73. a.   **b.** quadrantal   **c.** right

75. a.   **b.** II   **c.** obtuse

77. a.   **b.** III   **c.** none

79. a. 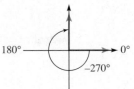   **b.** quadrantal   **c.** none

81. a. 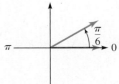   **b.** I   **c.** acute

83. a. 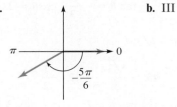   **b.** III   **c.** none

85. yes   **87.** no   **89.** yes   **91.** no   **93.** 130°

95. 240°   **97.** $\dfrac{\pi}{4}$ **99.** $\dfrac{2\pi}{3}$ **101.** $\dfrac{5\pi}{9}$ **103.** 15.708 in

105. 31.416 m   **107.** 20.054 in   **109.** 5.730 cm

111. 1.765   **113.** 0.682   **115.** 235.619 in²

117. 56.549 cm²   **119.** 1.657   **121.** 16.667 in

123. 130.9 ft   **125.** 183 mi   **127.** 220,998 mi

129. $\dfrac{100\pi}{3}$ square units   **131.** 1,065 ft²   **133.** 4π m

135. 212 in²   **137.** 1,551 yd²   **139.** 270°

Section 4.2

1. $x^2 + y^2 = 1$ **3.** cosine, secant   **5.** $\dfrac{\pi}{6}, \dfrac{\pi}{4}, \dfrac{\pi}{3}$ **7.** π

9. a. $\dfrac{\sqrt{3}}{2}$ **b.** $\dfrac{1}{2}$ **c.** $\sqrt{3}$ **d.** $\dfrac{2\sqrt{3}}{3}$ **e.** 2   **f.** $\dfrac{\sqrt{3}}{3}$

11. a. $-\dfrac{2\sqrt{2}}{3}$ **b.** $-\dfrac{1}{3}$ **c.** $2\sqrt{2}$ **d.** $-\dfrac{3\sqrt{2}}{4}$

e. −3   **f.** $\dfrac{\sqrt{2}}{4}$ **13. a.** $-\dfrac{\sqrt{6}}{3}$ **b.** $\dfrac{\sqrt{3}}{3}$ **c.** $-\sqrt{2}$

d. $-\dfrac{\sqrt{6}}{2}$ **e.** $\sqrt{3}$ **f.** $-\dfrac{\sqrt{2}}{2}$ **15. a.** $-\dfrac{4}{7}$ **b.** $\dfrac{\sqrt{33}}{7}$

c. $-\dfrac{4\sqrt{33}}{33}$ **d.** $-\dfrac{7}{4}$ **e.** $\dfrac{7\sqrt{33}}{33}$ **f.** $-\dfrac{\sqrt{33}}{4}$

17. a. −1   **b.** 0   **c.** undefined   **19. a.** 0   **b.** 1

c. 0   **21. a.** 0   **b.** −1   **c.** 0   **23. a.** 1   **b.** 0

c. undefined

25. a. $\dfrac{\sqrt{2}}{2}$ **b.** $\dfrac{\sqrt{2}}{2}$ **c.** 1

27. a. $-\dfrac{\sqrt{2}}{2}$ **b.** $-\dfrac{\sqrt{2}}{2}$ **c.** 1

29. a. $-\dfrac{\sqrt{3}}{2}$ **b.** $-\dfrac{1}{2}$ **c.** $\sqrt{3}$

31. a. $-\dfrac{1}{2}$ b. $\dfrac{\sqrt{3}}{2}$ c. $-\dfrac{\sqrt{3}}{3}$

33. a. $-\dfrac{\sqrt{3}}{2}$ b. $\dfrac{1}{2}$ c. $-\sqrt{3}$

35. $\sin\theta = \dfrac{\sqrt{2}}{2}, \cos\theta = -\dfrac{\sqrt{2}}{2}, \tan\theta = -1,$
$\csc\theta = \sqrt{2}, \sec\theta = -\sqrt{2}, \cot\theta = -1$

37. $\sin\theta = \dfrac{\sqrt{3}}{2}, \cos\theta = -\dfrac{1}{2}, \tan\theta = -\sqrt{3}, \csc\theta = \dfrac{2\sqrt{3}}{3},$
$\sec\theta = -2, \cot\theta = -\dfrac{\sqrt{3}}{3}$

39. $\sin\theta = \dfrac{1}{2}, \cos\theta = -\dfrac{\sqrt{3}}{2}, \tan\theta = -\dfrac{\sqrt{3}}{3}, \csc\theta = 2,$
$\sec\theta = -\dfrac{2\sqrt{3}}{3}, \cot\theta = -\sqrt{3}$

41. $\dfrac{\sqrt{2}}{2}$ 43. $\dfrac{\sqrt{3}}{3}$ 45. $-\dfrac{\sqrt{3}}{3}$ 47. $-\dfrac{1}{2}$

49. 1 51. $\dfrac{1}{2}$ 53. 1 55. $-\sqrt{3}$ 57. $-\dfrac{\sqrt{3}}{2}$

59. 60° 61. $\dfrac{\pi}{4}$ 63. 39° 65. $\dfrac{4\pi}{9}$ 67. $\dfrac{\pi}{6}$

69. $\dfrac{\pi}{3}$ 71. 79° 73. 70° 75. 0.258 77. 0.542

79. $-\dfrac{1}{2}$ 81. undefined 83. $-\dfrac{1}{2}$ 85. $\sqrt{3}$

87. -1 89. $\sqrt{2}$ 91. 0 93. $-\dfrac{\sqrt{3}}{2}$ 95. $\sqrt{2}$

97. $\dfrac{\pi}{6}$ and $\dfrac{5\pi}{6}$ 99. $\dfrac{\pi}{6}$ and $\dfrac{7\pi}{6}$ 101. $\dfrac{2\pi}{3}$ and $\dfrac{4\pi}{3}$

103. $\dfrac{\pi}{3}$ and $\dfrac{5\pi}{3}$ 105. $\dfrac{3\pi}{4}$ and $\dfrac{7\pi}{4}$ 107. $\dfrac{\pi}{6}, \dfrac{7\pi}{6}, \dfrac{\pi}{3},$ and $\dfrac{4\pi}{3}$

109. QII 111. QI 113. QIV 115. 30° and −330°
117. 300° and −60° 119. 210° and −150°
121. 315° and −45° 123. 135° and −225°
125. 270° and −90° 127. 90° and −270°
129. a. 63°F b. 89°F c. 76°F

131. $\dfrac{\sqrt{2}}{4}$ in 133. 0.518 units

Section 4.3

1. $D = (-\infty, \infty)$ 3. mode 5. odd 7. $-\dfrac{1}{2}$

9. $-\dfrac{\sqrt{3}}{2}$ 11. −1 13. 0 15. $-\dfrac{2\sqrt{3}}{3}$ 17. 1

19. 2 21. $\dfrac{12}{13}, -\dfrac{5}{13}, -\dfrac{12}{5}, \dfrac{13}{12}, -\dfrac{13}{5}, -\dfrac{5}{12}$

23. $\dfrac{5\sqrt{29}}{29}, \dfrac{2\sqrt{29}}{29}, \dfrac{5}{2}, \dfrac{\sqrt{29}}{5}, \dfrac{\sqrt{29}}{2}, \dfrac{2}{5}$

25. $\dfrac{\sqrt{2}}{2}, \dfrac{\sqrt{2}}{2}, 1, \sqrt{2}, \sqrt{2}, 1$

27. $-\dfrac{5\sqrt{34}}{34}, -\dfrac{3\sqrt{34}}{34}, \dfrac{5}{3}, -\dfrac{\sqrt{34}}{5}, -\dfrac{\sqrt{34}}{3}, \dfrac{3}{5}$

29. $-\dfrac{\sqrt{5}}{5}, \dfrac{2\sqrt{5}}{5}, -\dfrac{1}{2}, -\sqrt{5}, \dfrac{\sqrt{5}}{2}, -2$

31. $\cos\theta = -\dfrac{3}{5}, \tan\theta = -\dfrac{4}{3}, \csc\theta = \dfrac{5}{4}, \sec\theta = -\dfrac{5}{3},$
$\cot\theta = -\dfrac{3}{4}$

33. $\sin\theta = -\dfrac{\sqrt{7}}{4}, \tan\theta = \dfrac{\sqrt{7}}{3}, \csc\theta = -\dfrac{4\sqrt{7}}{7}, \sec\theta = -\dfrac{4}{3},$
$\cot\theta = \dfrac{3\sqrt{7}}{7}$

35. $\sin\theta = -\dfrac{\sqrt{2}}{2}, \cos\theta = \dfrac{\sqrt{2}}{2}, \csc\theta = -\sqrt{2}, \sec\theta = \sqrt{2},$
$\cot\theta = -1$

37. $\sin\theta = \dfrac{5}{7}, \cos\theta = \dfrac{2\sqrt{6}}{7}, \tan\theta = \dfrac{5\sqrt{6}}{12}, \sec\theta = \dfrac{7\sqrt{6}}{12},$
$\cot\theta = \dfrac{2\sqrt{6}}{5}$

39. $\sin\theta = \dfrac{12}{13}, \cos\theta = -\dfrac{5}{13}, \tan\theta = -\dfrac{12}{5}, \csc\theta = \dfrac{13}{12},$
$\sec\theta = -\dfrac{13}{5}$

41. $\cos\theta = \dfrac{\sqrt{33}}{7}, \tan\theta = -\dfrac{4\sqrt{33}}{33}, \csc\theta = -\dfrac{7}{4},$
$\sec\theta = \dfrac{7\sqrt{33}}{33}, \cot\theta = -\dfrac{\sqrt{33}}{4}$

43. $\sin\theta = \dfrac{\sqrt{41}}{21}, \tan\theta = \dfrac{\sqrt{41}}{20}, \csc\theta = \dfrac{21\sqrt{41}}{41},$
$\sec\theta = \dfrac{21}{20}, \cot\theta = \dfrac{20\sqrt{41}}{41}$

45. $\sin\theta = -\dfrac{24}{25}, \cos\theta = -\dfrac{7}{25}, \csc\theta = -\dfrac{25}{24},$
$\sec\theta = -\dfrac{25}{7}, \cot\theta = \dfrac{7}{24}$

47. $\sin\theta = \dfrac{12}{13}, \tan\theta = \dfrac{12}{5}, \csc\theta = \dfrac{13}{12}, \sec\theta = \dfrac{13}{5}$

49. $\sin\theta = \dfrac{4}{5}, \cos\theta = \dfrac{3}{5}, \tan\theta = \dfrac{4}{3}, \cot\theta = \dfrac{3}{4}$

51. $\sin\theta = \dfrac{40}{41}, \csc\theta = \dfrac{41}{40}, \sec\theta = \dfrac{41}{9}, \cot\theta = \dfrac{9}{40}$

53. 1 55. $\sin^2\theta$ 57. $\cos^2\theta$ 59. $\cot^2\theta$

61. 1 63. $\cos^2\theta$ 65. $\sin^2\theta$

67. $\cos\theta = \dfrac{12}{13}, \tan\theta = \dfrac{5}{12}, \csc\theta = \dfrac{13}{5}, \sec\theta = \dfrac{13}{12},$
$\cot\theta = \dfrac{12}{5}$

69. $\sin\theta = \dfrac{\sqrt{23}}{5}, \tan\theta = \dfrac{\sqrt{46}}{2}, \csc\theta = \dfrac{5\sqrt{23}}{23},$
$\sec\theta = \dfrac{5\sqrt{2}}{2}, \cot\theta = \dfrac{\sqrt{46}}{23}$

71. $\sin\theta = \dfrac{9\sqrt{97}}{97}, \cos\theta = \dfrac{4\sqrt{97}}{97}, \csc\theta = \dfrac{\sqrt{97}}{9},$
$\sec\theta = \dfrac{\sqrt{97}}{4}, \cot\theta = \dfrac{4}{9}$

73. 0.3923 75. 0.7524 77. 1.6869 79. 3.8637
81. 1.0378 83. 0.5774 85. 1 87. 0 89. 2

91. $-\dfrac{23}{4}$ **93.** $-\dfrac{1}{2}$ **95.** $\dfrac{8\sqrt{3}-\sqrt{6}}{3}$ **97.** $-\sqrt{3}$

99. 1

101. $\sin\theta=\dfrac{8}{17}, \cos\theta=\dfrac{15}{17}, \tan\theta=\dfrac{8}{15}, \csc\theta=\dfrac{17}{8},$

$\sec\theta=\dfrac{17}{15}, \cot\theta=\dfrac{15}{8}$

103. $\sin\theta=\dfrac{5}{11}, \cos\theta=\dfrac{4\sqrt{6}}{11}, \tan\theta=\dfrac{5\sqrt{6}}{24}, \csc\theta=\dfrac{11}{5},$

$\sec\theta=\dfrac{11\sqrt{6}}{24}, \cot\theta=\dfrac{4\sqrt{6}}{5}$

105. $\sin\theta=\dfrac{7}{25}, \cos\theta=\dfrac{24}{25}, \tan\theta=\dfrac{7}{24}, \csc\theta=\dfrac{25}{7},$

$\sec\theta=\dfrac{25}{24}, \cot\theta=\dfrac{24}{7}$

107. $\sin\theta=\dfrac{4\sqrt{41}}{41}, \cos\theta=\dfrac{5\sqrt{41}}{41}, \tan\theta=\dfrac{4}{5}, \csc\theta=\dfrac{\sqrt{41}}{4},$

$\sec\theta=\dfrac{\sqrt{41}}{5}, \cot\theta=\dfrac{5}{4}$

109. $\cos 41.8°$ **111.** $\cot\dfrac{3\pi}{10}$ **113.** $\sin\dfrac{\pi}{8}$

115. $\sec 38°$ **117.** true   **119.** false   **121.** 165 ft

123. 96 ft   **125.** 721 ft   **127.** 1.4 mi   **129.** 4.0°

131. distance $\approx$ 63.8 m; maximum height $\approx$ 15.9 m

Section 4.4

1. 2π **3.** key points   **5.** $\dfrac{\pi}{2},\dfrac{3\pi}{2}$ **7.** 3

9. a. amplitude $=1$; period $=\pi$

b.

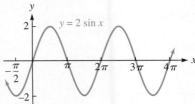

c. $x=0, \dfrac{\pi}{2}$, and π

11. a. amplitude $=2$; period $=2\pi$

b.

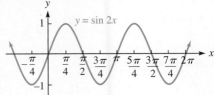

c. $x=0, \pi$, and 2π

13. a. amplitude $=4$; period $=4\pi$

b.

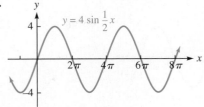

c. $x=0, 2\pi$, and 4π

15. a. amplitude $=3$; period $=2$

b.

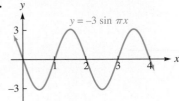

c. $x=0, 1$, and 2

17. a. amplitude $=\dfrac{1}{3}$; period $=1$

b.

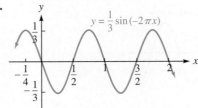

c. $x=0, \dfrac{1}{2}$, and 1

19. a. amplitude $=3$; period $=2\pi$

b.

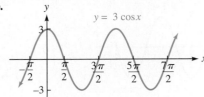

c. $x=\dfrac{\pi}{2}, \dfrac{3\pi}{2}$, and $\dfrac{5\pi}{2}$

21. a. amplitude $=1$; period $=\pi$

b.

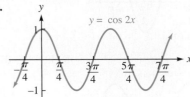

c. $x=\dfrac{\pi}{4}, \dfrac{3\pi}{4}$, and $\dfrac{5\pi}{4}$

23. a. amplitude = 2; period = 8π

b.

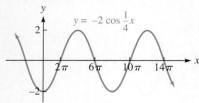

c. $x = 2\pi, 6\pi$, and 10π

25. a. amplitude = 1; period = 4

b.

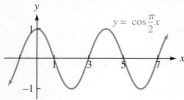

c. $x = 1, 3$, and 5

27. a. amplitude = $\dfrac{1}{2}$; period = $\dfrac{1}{2}$

b.

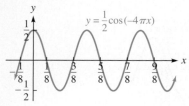

c. $x = \dfrac{1}{8}, \dfrac{3}{8}$, and $\dfrac{5}{8}$

29. a.   **b.** $\dfrac{\pi}{6}$ **c.** $\dfrac{\pi}{2}$

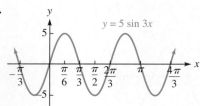

31. a.   **b.** 0   **c.** $\dfrac{\pi}{2}$

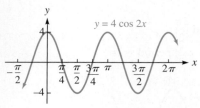

33. a.   **b.** 1   **c.** 0

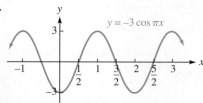

35. a.   **b.** $\dfrac{3}{2}$ **c.** $\dfrac{1}{2}$

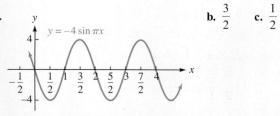

37. a.   **b.** π **c.** 0

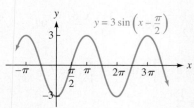

39. a.   **b.** 0   **c.** π

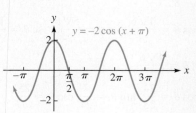

41. a. amplitude = 2; period = π

b.

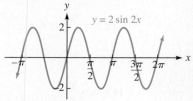

43. a. amplitude = 1; period = 8π

b.

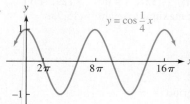

45. a. amplitude = 3; period = 4π; phase shift = $\dfrac{\pi}{2}$

b.

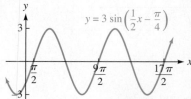

47. a. amplitude = $\dfrac{1}{2}$; period = 2; phase shift = $\dfrac{1}{3}$

b.

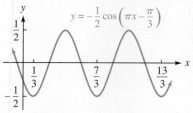

49. a. amplitude = $\dfrac{1}{2}$; period = $\dfrac{\pi}{3}$

b.

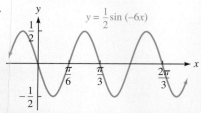

51. a. amplitude = 4; period = 6π

b.
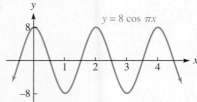

53. a. amplitude = 8; period = 2

b.
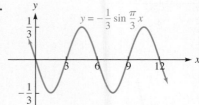

55. a. amplitude = $\dfrac{1}{3}$; period = 6

b.
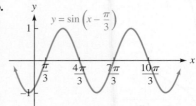

57. a. amplitude = 1; period = 2π; phase shift = $\dfrac{\pi}{3}$

b.
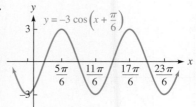

59. a. amplitude = 3; period = 2π; phase shift = $-\dfrac{\pi}{6}$

b.
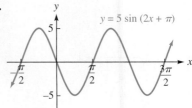

61. a. amplitude = 5; period = π; phase shift = $-\dfrac{\pi}{2}$

b.
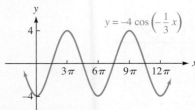

63. $y = 4 \sin 2x$ **65.** $y = \dfrac{\pi}{2} \sin \dfrac{\pi}{2} x$

67. $y = 3 \sin(2x - \pi)$ **69.** $y = 2 \sin(6\pi x + 3\pi)$

71.
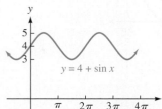

73.
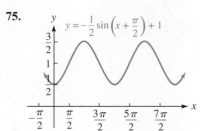

75.
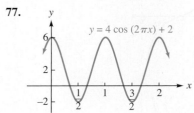

77.

79. $y = 4 \sin \dfrac{2}{3} x$ **81.** $y = -10 \cos 5\pi x$

83. $y = 2 \sin\left(\dfrac{\pi}{2} x - \dfrac{\pi}{6}\right)$ **85.** $y = -\cos\left(\dfrac{1}{3} x + \dfrac{\pi}{3}\right)$

87. $y = 5 \sin 4\pi x + 3$ **89.** 4π sec   **91.** 48 dyn/cm

93. $V = 340 \sin 100\pi t$

95. period = $\dfrac{2\pi}{\sqrt{9.8}}$ sec; frequency = $\dfrac{\sqrt{9.8}}{2\pi}$ Hz

97. 286 Hz

Section 4.5

1. the domain is all real numbers except for odd-integer multiples of $\dfrac{\pi}{2}$; the range is all real numbers

3. vertical asymptotes: $x = -\dfrac{3\pi}{2}, -\dfrac{\pi}{2}, \dfrac{\pi}{2}$, and $\dfrac{3\pi}{2}$; intercepts: $-2\pi, -\pi, 0, \pi$, and 2π

5. 0   **7.** the domain is all real numbers except for odd-integer multiples of $\dfrac{\pi}{2}$; $R = (-\infty, -1] \cup [1, \infty)$

9. $-\dfrac{3\pi}{2}, -\dfrac{\pi}{2}, \dfrac{\pi}{2}$, and $\dfrac{3\pi}{2}$

11.

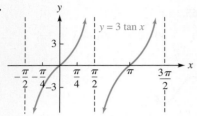

13.

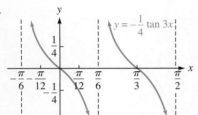

15.

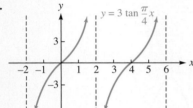

17.

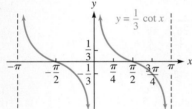

19.

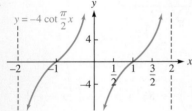

21.

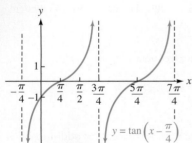

23.

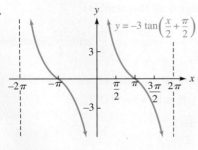

25.

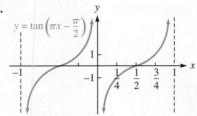

27.

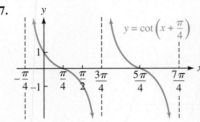

29.

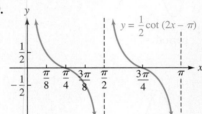

31.

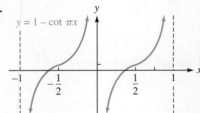

33.

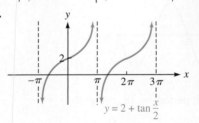

35.

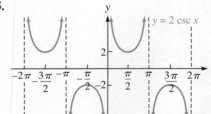

37.

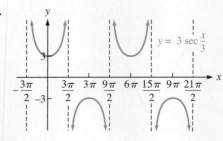

39.

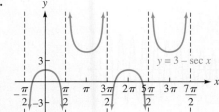

41.

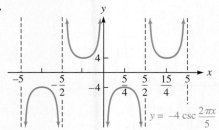

43.

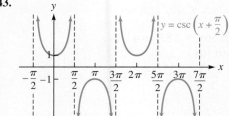

45.

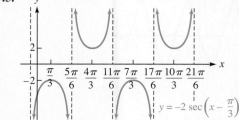

47. a. $L_1 = \dfrac{20}{\cos \theta} = 20 \sec \theta$ and $L_2 = \dfrac{25}{\sin \theta} = 25 \csc \theta$,

giving $L = L_1 + L_2 = 20 \sec \theta + 25 \csc \theta$

b. $\dfrac{\pi}{4}$, because a larger angle would prevent the log from

making the turn

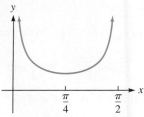

$L(\theta) = 20 \sec \theta + 25 \csc \theta$

on the interval $\left(0, \dfrac{\pi}{2}\right)$

c. 63.6 ft

Section 4.6
1. addition   **3.** damping
5.

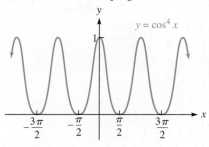

7.

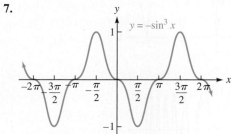

9.

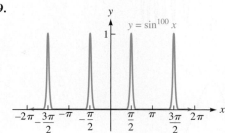

11.

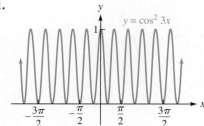

13.

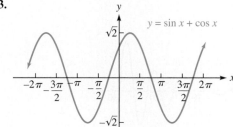

15.

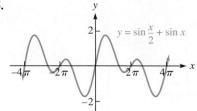

17.

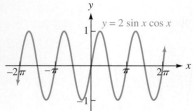

$y = 2 \sin x \cos x$

19.

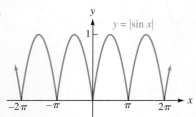

$y = |\sin x|$

21.

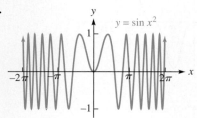

$y = \sin x^2$

23.

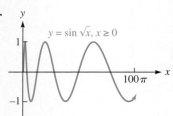

$y = \sin \sqrt{x}, x \geq 0$

25.

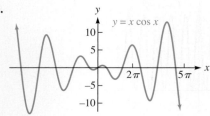

$y = x \cos x$

27.

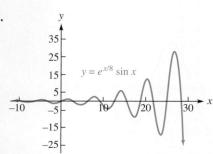

$y = e^{x/8} \sin x$

29.

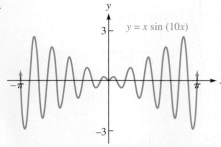

$y = x \sin(10x)$

31.

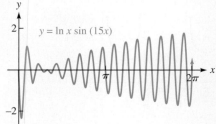

$y = \ln x \sin(15x)$

33.

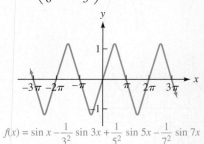

$y = 2 \sin x + 3 \sin 2x + \sin 3x$

35. $y = 0.16 \sin(0.56x - 2.83) + 1.34$
37. $y = 10.57 \sin(0.50x - 2.04) + 82.12$;

$y = 10 \sin\left(\dfrac{\pi}{6}x - \dfrac{2\pi}{3}\right) + 82$

39.

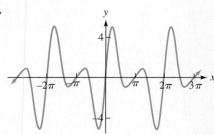

$f(x) = \sin x - \dfrac{1}{3^2} \sin 3x + \dfrac{1}{5^2} \sin 5x - \dfrac{1}{7^2} \sin 7x$

Chapter Review

1. a. (i) QI; (ii) acute   **b.** (i) QI; (ii) none of these
2. a. coterminal   **b.** not coterminal
3. a. $138°38'2.4''$ **b.** $-212°45'$

4. a. $194.405°$ **b.** $-225.751°$ **5. a.** $-\dfrac{5\pi}{4}$ **b.** $\dfrac{13\pi}{6}$

6. a. 1.560 **b.** -0.230 **7. a.** $675°$ **b.** $75°$
8. a. $265.56°$ **b.** $-140.35°$ **9. a.** 51.84 in
b. 21.20 in   **c.** 2.40 **10. a.** 125.66 in^2 **b.** 0.62
c. 41.46 in   **11.** 741 mi   **12.** 1.88 in

13. a. $\dfrac{2}{3}, \dfrac{\sqrt{5}}{3}, \dfrac{2\sqrt{5}}{5}, \dfrac{3}{2}, \dfrac{3\sqrt{5}}{5}, \dfrac{\sqrt{5}}{2}$

b. $-\dfrac{\sqrt{7}}{4}, \dfrac{3}{4}, \dfrac{\sqrt{7}}{3}, -\dfrac{4\sqrt{7}}{7}, -\dfrac{4}{3}, \dfrac{3\sqrt{7}}{7}$

14. a. $\dfrac{\pi}{4}$ **b.** $7°$ **c.** $\dfrac{4\pi}{9}$ **15. a.** $-\dfrac{1}{2}$ **b.** $-\sqrt{3}$

c. $\dfrac{\sqrt{2}}{2}$ **d.** undefined   **16. a.** $\dfrac{4\pi}{3}$ and $\dfrac{5\pi}{3}$ **b.** $\dfrac{\pi}{4}$ and $\dfrac{7\pi}{4}$

c. $\dfrac{2\pi}{3}$ and $\dfrac{5\pi}{3}$ **17. a.** $240°$ and $-120°$ **b.** $330°$ and $-30°$

18. a. $\sin \theta = -\dfrac{2\sqrt{29}}{29}, \cos \theta = \dfrac{5\sqrt{29}}{29}, \tan \theta = -\dfrac{2}{5}$,

$\csc \theta = -\dfrac{\sqrt{29}}{2}, \sec \theta = \dfrac{\sqrt{29}}{5}, \cot \theta = -\dfrac{5}{2}$

b. $\sin \theta = \dfrac{2\sqrt{13}}{13}$, $\cos \theta = -\dfrac{3\sqrt{13}}{13}$, $\tan \theta = -\dfrac{2}{3}$,

$\csc \theta = \dfrac{\sqrt{13}}{2}$, $\sec \theta = -\dfrac{\sqrt{13}}{3}$, $\cot \theta = -\dfrac{3}{2}$

19. a. $\dfrac{\sqrt{3}}{2}$ **b.** $-\dfrac{\sqrt{3}}{2}$ **c.** $\dfrac{\sqrt{3}}{3}$

20. a. $\sin \theta = \dfrac{4}{5}$, $\cos \theta = -\dfrac{3}{5}$, $\tan \theta = -\dfrac{4}{3}$, $\csc \theta = \dfrac{5}{4}$,

$\cot \theta = -\dfrac{3}{4}$

b. $\sin \theta = -\dfrac{3\sqrt{10}}{10}$, $\cos \theta = \dfrac{\sqrt{10}}{10}$, $\csc \theta = -\dfrac{\sqrt{10}}{3}$,

$\sec \theta = \sqrt{10}$, $\cot \theta = -\dfrac{1}{3}$

21. a. $\sin \theta = \dfrac{\sqrt{21}}{5}$, $\tan \theta = \dfrac{\sqrt{21}}{2}$, $\csc \theta = \dfrac{5\sqrt{21}}{21}$,

$\sec \theta = \dfrac{5}{2}$, $\cot \theta = \dfrac{2\sqrt{21}}{21}$

b. $\sin \theta = \dfrac{\sqrt{5}}{5}$, $\cos \theta = \dfrac{2\sqrt{5}}{5}$, $\csc \theta = \sqrt{5}$,

$\sec \theta = \dfrac{\sqrt{5}}{2}$, $\cot \theta = 2$

22. a. $\cos^2 \theta$ **b.** $\cot^2 \theta$ **23. a.** 0.2393   **b.** 1.2941

24. a. $\dfrac{1}{4}$ **b.** 1

25. a. $\sin \theta = \dfrac{4\sqrt{65}}{65}$, $\cos \theta = \dfrac{7\sqrt{65}}{65}$, $\tan \theta = \dfrac{4}{7}$,

$\csc \theta = \dfrac{\sqrt{65}}{4}$, $\sec \theta = \dfrac{\sqrt{65}}{7}$, $\cot \theta = \dfrac{7}{4}$

b. $\sin \theta = \dfrac{2\sqrt{14}}{9}$, $\cos \theta = \dfrac{5}{9}$, $\tan \theta = \dfrac{2\sqrt{14}}{5}$, $\csc \theta = \dfrac{9\sqrt{14}}{28}$,

$\sec \theta = \dfrac{9}{5}$, $\cot \theta = \dfrac{5\sqrt{14}}{28}$

26. 52.5 ft

27. a. amplitude = 3; period = 8

b.

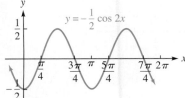

c. 0, 4, and 8

28. a. amplitude = $\dfrac{1}{2}$; period = π

b.

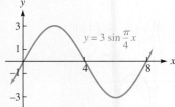

c. $\dfrac{\pi}{4}, \dfrac{3\pi}{4},$ and $\dfrac{5\pi}{4}$

29. a. amplitude = $\dfrac{3}{4}$; period = 2π; phase shift = $-\dfrac{\pi}{4}$

b.

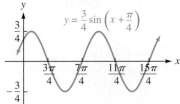

30. a. amplitude = 2; period = 2; phase shift = $\dfrac{1}{2}$

b.

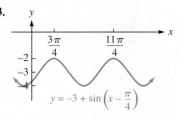

31. $y = 4 \sin \pi x$ **32.** $y = \dfrac{3}{5} \sin(4x - \pi)$

33.

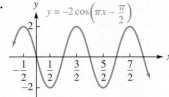

34.

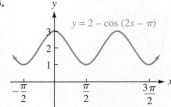

35. $y = -4 \cos 2x$ or $y = -4 \sin\left(2x + \dfrac{\pi}{2}\right)$

36.

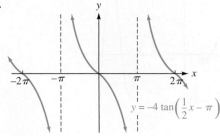

37.

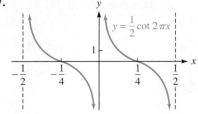

38.

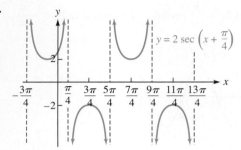

39.

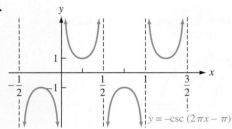

40.

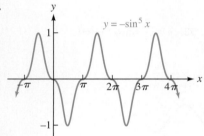

41.

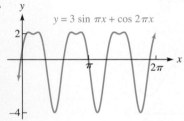

42.

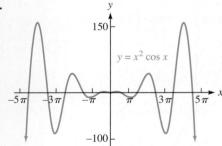

Chapter Test

1. 47.26°; 0.82 radian   **2.** 200.54° ≈ 200°32′24″

3. 10.47 in   **4.** 172.8 cm²

5. a. $\sin t = -\dfrac{\sqrt{3}}{2}$ **b.** $\cos t = \dfrac{1}{2}$ **c.** $\tan t = -\sqrt{3}$

d. $\csc t = \dfrac{-2\sqrt{3}}{3}$ **e.** $\sec t = 2$ **f.** $\cot t = -\dfrac{\sqrt{3}}{3}$

6. a. $\dfrac{\sqrt{2}}{2}$ **b.** $-\dfrac{\sqrt{3}}{2}$ **c.** $\sqrt{3}$ **7.** 315° and −45°

8. $\sin \theta = \dfrac{7\sqrt{74}}{74}$, $\cos \theta = -\dfrac{5\sqrt{74}}{74}$, $\tan \theta = -\dfrac{7}{5}$, $\csc \theta = \dfrac{\sqrt{74}}{7}$,

$\sec \theta = -\dfrac{\sqrt{74}}{5}$, $\cot \theta = -\dfrac{5}{7}$

9. $\sin \theta = -\dfrac{4\sqrt{17}}{17}$, $\cos \theta = \dfrac{\sqrt{17}}{17}$, $\csc \theta = -\dfrac{\sqrt{17}}{4}$,

$\sec \theta = \sqrt{17}$, $\cot \theta = -\dfrac{1}{4}$

10. $\sin \theta = \dfrac{\sqrt{5}}{5}$, $\cos \theta = \dfrac{2\sqrt{5}}{5}$, $\tan \theta = \dfrac{1}{2}$, $\csc \theta = \sqrt{5}$,

$\sec \theta = \dfrac{\sqrt{5}}{2}$, $\cot \theta = 2$

11.

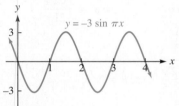

12.

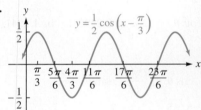

13.

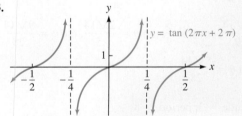

14.

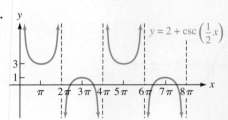

15.

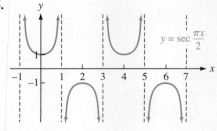

16.

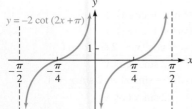

$y = -2 \cot(2x + \pi)$

17. $y = 3 \sin(4x - \pi)$

18. $y = -2 \sin 2x$ or $y = 2 \cos\left(2x + \dfrac{\pi}{2}\right)$

19.

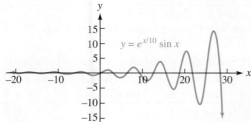

$y = e^{x/10} \sin x$

Connect to Calculus

1. a. $f'(x) = \cos x + \sec^2 x$
b. $f'(x) = -2 \csc x \cot x - 5 \sin x$
2. $v = f'(t) = -4 \sin t$
3. $\sin 55° \approx \dfrac{\sqrt{3}}{2} + \dfrac{1}{2}\left(-\dfrac{\pi}{36}\right) - \dfrac{\sqrt{3}}{4}\left(-\dfrac{\pi}{36}\right)^2 -$
$\dfrac{1}{12}\left(-\dfrac{\pi}{36}\right)^3 \approx 0.8191499732$

4. Using the series,
$\cos 50° \approx \dfrac{\sqrt{2}}{2} - \dfrac{\sqrt{2}}{2}\left(\dfrac{\pi}{36}\right) - \dfrac{\sqrt{2}}{4}\left(\dfrac{\pi}{36}\right)^2 + \dfrac{\sqrt{2}}{12}\left(\dfrac{\pi}{36}\right)^3$
≈ 0.642786. The calculator approximation is ≈ 0.6427876097.

Using the series,
$\cos 25° \approx \dfrac{\sqrt{2}}{2} - \dfrac{\sqrt{2}}{2}\left(\dfrac{\pi}{9}\right) - \dfrac{\sqrt{2}}{4}\left(\dfrac{\pi}{9}\right)^2 + \dfrac{\sqrt{2}}{12}\left(\dfrac{\pi}{9}\right)^3$
≈ 0.422213. The calculator approximation is ≈ 0.906307787, which is not even close.
5. The reason the second approximation is not close is because the polynomial given is centered at $\frac{\pi}{4}$, and 25° is too "far" from 45° for the approximation to be usable.

Section 5.1

1. $D = [-1, 1]; R = \left[-\dfrac{\pi}{2}, \dfrac{\pi}{2}\right]$ **3.** $-\dfrac{\pi}{2} \le x \le \dfrac{\pi}{2}$

5. $0 \le x \le \pi$ **7.** true   **9.** $-\dfrac{\pi}{6}$ **11.** $\dfrac{\pi}{2}$

13. $-\dfrac{\pi}{3}$ **15.** $\dfrac{\pi}{4}$ **17.** 1.253   **19.** 1.217   **21.** 1.263

23. -0.730 **25.** 1.955   **27.** -1.521 **29.** $\dfrac{\pi}{6}$

31. $-\dfrac{\pi}{4}$ **33.** 0.75   **35.** $\dfrac{5\pi}{6}$ **37.** $\dfrac{1}{2}$ **39.** 1

41. 1   **43.** $-\dfrac{\pi}{6}$ **45.** $\dfrac{2\pi}{3}$ **47.** $\dfrac{3}{4}$ **49.** 5

51. not defined, since 1.5 is not in $[-1, 1]$

53. $\dfrac{1}{2}$ **55.** $-\sqrt{3}$ **57.** $\dfrac{\sqrt{2}}{2}$ **59.** $\dfrac{5}{3}$ **61.** $\dfrac{12}{13}$

63. $-\dfrac{4}{3}$ **65.** $-\dfrac{12}{5}$ **67.** $\dfrac{6\sqrt{61}}{61}$ **69.** $\dfrac{3\sqrt{5}}{5}$

71. $\dfrac{2\sqrt{2}}{3}$ **73.** $-\dfrac{3\sqrt{7}}{7}$ **75.** $\dfrac{x\sqrt{1+x^2}}{1+x^2}$

77. $\dfrac{x\sqrt{1-x^2}}{1-x^2}$ **79.** $\sqrt{1 - 4x^2}$ **81.** $\dfrac{\sqrt{x^2-1}}{x}$

83. $\dfrac{\sqrt{x^2+9}}{3}$ **85.** 1.31   **87.** $\theta = \sin^{-1}\dfrac{y}{150}; 41.8°$

89. a. $\dfrac{\pi}{2}$ **b.** $\dfrac{2\pi}{3}$ **c.** $\dfrac{\pi}{3}$

91. $\sin^{-1}\left(-\dfrac{1}{2}\right) \ne \dfrac{11\pi}{6}$ because $\dfrac{11\pi}{6}$ is not in $\left[-\dfrac{\pi}{2}, \dfrac{\pi}{2}\right]$.

Section 5.2

1. $(-\infty, \infty), (0, \pi)$ **3.** $\dfrac{\pi}{6}$ **5.** π **7.** $-\dfrac{\pi}{2}$ **9.** $\dfrac{\pi}{3}$

11. 2.897   **13.** 1.107   **15.** 0.949   **17.** 0.128

19. 3   **21.** $\dfrac{4\sqrt{15}}{15}$ **23.** $\dfrac{40}{9}$ **25.** $\dfrac{5}{3}$ **27.** $\dfrac{1}{x}$

29. $\dfrac{\sqrt{x^2+1}}{x}$ **31.** $\sqrt{x^2-1}$ **33.** $\theta = \csc^{-1}\dfrac{160}{y}; 34.2°$

35. a. $\dfrac{\pi}{12}$ **b.** $\dfrac{7\pi}{12}$

Section 5.3

1. $\sin^2 x + \cos^2 x = 1; \tan^2 x + 1 = \sec^2 x;$
$1 + \cot^2 x = \csc^2 x$
3. $7 \sin x$ **5.** $-7 \csc^2 x$ **7.** $\sin^2 x - \sin x$
9. $\tan x - \sec x$ **11.** $\sin^2 x$ **13.** $\cot^2 x$
15. $\cos^2 x + 4 \cos x + 4$ **17.** $9 \sin^2 x + 12 \sin x + 4$
19. $\sec x \csc x$ **21.** $2 \csc x$ **23.** $\sin x(\sin x + \cos x)$
25. $\cos^2 x(\cos x + 1)$ **27.** $(\sec x - 3)(\sec x + 3)$
29. $(\sec x - \csc x)(\sec x + \csc x)$
31. $(\sin x + 6)(\sin x - 1)$ **33.** $(2 \tan x - 1)(\tan x + 1)$
35. $\cos x \tan x = \cos x \cdot \dfrac{\sin x}{\cos x} = \sin x$

39. $\sin x \sec x = \sin x \cdot \dfrac{1}{\cos x} = \tan x$

43. $\dfrac{\cot x \sin x}{\cos x} = \dfrac{\dfrac{\cos x}{\sin x} \cdot \sin x}{\cos x} = 1$

47. $\tan x \csc x = \dfrac{\sin x}{\cos x} \cdot \dfrac{1}{\sin x} = \sec x$

51. $\dfrac{\sin x}{\csc x} + \dfrac{\cos x}{\sec x} = \dfrac{\sin x}{\dfrac{1}{\sin x}} + \dfrac{\cos x}{\dfrac{1}{\sin x}} = \sin^2 x + \cos^2 x = 1$

55. $\dfrac{\sin(-x)}{\cos^2(-x) - 1} = \dfrac{-\sin x}{\cos^2 x - 1} = \dfrac{-\sin x}{-\sin^2 x} = -\csc(-x)$

59. $\dfrac{1 - \sin^2 x}{1 + \tan^2 x} = \dfrac{\cos^2 x}{\sec^2 x} = \cos^4 x$

63. $\sin^2 x \sec x - \sec x = \sec x(\sin^2 x - 1) = \sec x(-\cos^2 x)$
$$= -\cos(-x)$$

67. $\sin^2 x - \tan^2 x = \sin^2 x - \dfrac{\sin^2 x}{\cos^2 x}$

$= \dfrac{\sin^2 x \cos^2 x - \sin^2 x}{\cos^2 x} = \dfrac{\sin^2 x(\cos^2 x - 1)}{\cos^2 x} = -\sin^2 x \tan^2 x$

71. $\dfrac{1}{\sec x - \tan x} = \dfrac{1}{\sec x - \tan x} \cdot \dfrac{\sec x + \tan x}{\sec x + \tan x}$

$= \dfrac{\sec x + \tan x}{\sec^2 x - \tan^2 x} = \tan x + \sec x$

75. $(\sin x + \cos x)^2 + (\sin x - \cos x)^2 = \sin^2 x +$
$2 \sin x \cos x + \cos^2 x + \sin^2 x - 2 \sin x \cos x + \cos^2 x = 2$

79. $\dfrac{\cos x - \cot x}{\cos x \cot x} = \dfrac{\cos x - \dfrac{\cos x}{\sin x}}{\cos x \cdot \dfrac{\cos x}{\sin x}} \cdot \dfrac{\sin x}{\sin x}$

$= \dfrac{\cos x \sin x - \cos x}{\cos^2 x} = \dfrac{\cos x(\sin x - 1)}{\cos^2 x} = \dfrac{\sin x - 1}{\cos x}$

83. $\sqrt{\dfrac{\csc x - \cot x}{\csc x + \cot x}} = \sqrt{\dfrac{\left(\dfrac{1}{\sin x} - \dfrac{\cos x}{\sin x}\right)}{\left(\dfrac{1}{\sin x} + \dfrac{\cos x}{\sin x}\right)}}$

$= \sqrt{\dfrac{1 - \cos x}{1 + \cos x} \cdot \dfrac{1 + \cos x}{1 + \cos x}} = \sqrt{\dfrac{\sin^2 x}{(1 + \cos x)^2}} = \dfrac{\sin x}{1 + \cos x}$

87. $\dfrac{1}{\tan x(\csc x + 1)} = \dfrac{1}{\tan x(\csc x + 1)} \cdot \dfrac{\csc x - 1}{\csc x - 1}$

$= \dfrac{\csc x - 1}{\tan x \cdot \cot^2 x} = \tan x(\csc x - 1)$

91. $\dfrac{\csc x}{\sec x - \csc x} = \dfrac{\csc x}{\sec x - \csc x} \cdot \dfrac{\sec x + \csc x}{\sec x + \csc x}$

$= \dfrac{\csc x \sec x + \csc^2 x}{\sec^2 x - \csc^2 x} = \dfrac{\csc x \sec x + 1 + \cot^2 x}{(1 + \tan^2 x) - (1 + \cot^2 x)}$

$= \dfrac{\cot^2 x + \csc x \sec x + 1}{\tan^2 x - \cot^2 x}$

93. $x = \left[-\dfrac{\pi}{2}, \dfrac{\pi}{2}\right] \cup \left[\dfrac{3\pi}{2}, \dfrac{5\pi}{2}\right]$, etc.

95. $x = \left(0, \dfrac{\pi}{2}\right] \cup \left(\pi, \dfrac{3\pi}{2}\right]$, etc.

97. work both sides:

left $= \dfrac{\cos x + \sin x}{\cos x - \sin x} \cdot \dfrac{\cos x + \sin x}{\cos x + \sin x}$

$= \dfrac{\cos^2 x + 2 \sin x \cos x + \sin^2 x}{\cos^2 x - \sin^2 x} = \dfrac{1 + 2 \sin x \cos x}{1 - 2 \sin^2 x}$

right $= \dfrac{\csc x + 2 \cos x}{\csc x - 2 \sin x} = \dfrac{\dfrac{1}{\sin x} + 2 \cos x}{\dfrac{1}{\sin x} - 2 \sin x} = \dfrac{1 + 2 \sin x \cos x}{1 - 2 \sin^2 x}$

101.
$\dfrac{3 \cos^2 x + 11 \sin x - 11}{\cos^2 x} = \dfrac{3(1 - \sin^2 x) + 11 \sin x - 11}{1 - \sin^2 x}$

$= \dfrac{-(3 \sin x - 8)(\sin x - 1)}{(1 - \sin x)(1 + \sin x)} = \dfrac{3 \sin x - 8}{1 + \sin x}$

105.
$\dfrac{(\sin x - \cos x) + 1}{(\sin x + \cos x) - 1} \cdot \dfrac{(\sin x + \cos x) + 1}{(\sin x + \cos x) + 1}$

$= \dfrac{\sin^2 x - \cos^2 x + 2 \sin x + 1}{2 \sin x \cos x}$

$= \dfrac{\sin^2 x - (1 - \sin^2 x) + 2 \sin x + 1}{2 \sin x \cos x}$

$= \dfrac{2 \sin^2 x + 2 \sin x}{2 \sin x \cos x} = \dfrac{1 + \sin x}{\cos x}$

109.
$\dfrac{\sin x + \sin x \cos x - \sin x \cos^2 x - \sin x \cos^3 x}{\cos^4 x - 2 \cos^2 x + 1}$

$= \dfrac{\sin x(1 + \cos x) - \sin x \cos^2 x(1 + \cos x)}{(\cos^2 x - 1)^2}$

$= \dfrac{\sin x(1 + \cos x)(1 - \cos^2 x)}{(\cos^2 x - 1)^2} = \dfrac{-\sin x(1 + \cos x)}{(\cos x - 1)(\cos x + 1)}$

$= \dfrac{-\sin x}{\cos x - 1} \cdot \dfrac{\cos x + 1}{\cos x + 1} = \dfrac{-\sin x(\cos x + 1)}{-\sin^2 x} = \dfrac{\cos x + 1}{\sin x}$

$= \csc x + \cot x$

113. $\dfrac{2 \sin x \cos x + \cos x - 2 \sin x - 1}{-\sin^2 x}$

$= \dfrac{\cos x(2 \sin x + 1) - (2 \sin x + 1)}{\cos^2 x - 1}$

$= \dfrac{(2 \sin x + 1)(\cos x - 1)}{(\cos x + 1)(\cos x - 1)} = \dfrac{2 \sin x + 1}{\cos x + 1}$

117. yes **119.** no **121.** yes **123.** Answers will vary.
125. $4 \,|\sec \theta|$

Section 5.4

1. $\sin(\alpha + \beta) = \sin \alpha \cos \beta + \cos \alpha \sin \beta$
$\sin(\alpha - \beta) = \sin \alpha \cos \beta - \cos \alpha \sin \beta$
$\cos(\alpha + \beta) = \cos \alpha \cos \beta - \sin \alpha \sin \beta$
$\cos(\alpha - \beta) = \cos \alpha \cos \beta + \sin \alpha \sin \beta$

3. $135° + 30°$; $210° - 45°$

5. $\dfrac{\sqrt{6} + \sqrt{2}}{4}$ **7.** $\dfrac{\sqrt{6} - \sqrt{2}}{4}$ **9.** $-2 - \sqrt{3}$

11. $-\dfrac{\sqrt{6} + \sqrt{2}}{4}$ **13.** $-\dfrac{\sqrt{6} + \sqrt{2}}{4}$ **15.** $\dfrac{\sqrt{6} - \sqrt{2}}{4}$

17. $-\dfrac{\sqrt{6} + \sqrt{2}}{4}$ **19.** $2 + \sqrt{3}$ **21.** $\sqrt{2} + \sqrt{6}$

23. $2 + \sqrt{3}$ **25.** $\sqrt{2} - \sqrt{6}$

27. $\sin(60° + \theta) = \sin 60° \cos \theta + \cos 60° \sin \theta$

$= \dfrac{\sqrt{3}}{2} \cos \theta + \dfrac{1}{2} \sin \theta$

29. $\tan(\pi + x) = \dfrac{\tan \pi + \tan x}{1 - \tan \pi \tan x} = \dfrac{0 + \tan x}{1 - 0 \cdot \tan x} = \tan x$

31. $\cos(\pi - x) = \cos \pi \cos x + \sin \pi \sin x$

$= (-1)\cos x + 0 \cdot \sin x = -\cos x$

33. $\sin 45° = \dfrac{\sqrt{2}}{2}$ **35.** $\cos 45° = \dfrac{\sqrt{2}}{2}$

37. $\tan 135° = -1$ **39.** $\sin \dfrac{\pi}{6} = \dfrac{1}{2}$

41. $\cos \dfrac{\pi}{6} = \dfrac{\sqrt{3}}{2}$ **43. a.** $-\dfrac{56}{65}$ **b.** $-\dfrac{63}{65}$ **c.** $\dfrac{56}{33}$

45. a. $\dfrac{58\sqrt{5}}{205}$ **b.** $-\dfrac{89\sqrt{5}}{205}$ **c.** $\dfrac{58}{71}$

47. a. $-\dfrac{2\sqrt{5}}{65}$ **b.** $\dfrac{19\sqrt{5}}{65}$ **c.** $-\dfrac{2}{29}$

49. $\sin\beta = \dfrac{204}{325}$; $\cos\beta = \dfrac{253}{325}$

51.
$$\sin(30° + \theta) - \cos(60° + \theta)$$
$$= (\sin 30° \cos\theta + \cos 30° \sin\theta) - (\cos 60° \cos\theta - \sin 60° \sin\theta)$$
$$= \left(\frac{1}{2}\cos\theta + \frac{\sqrt{3}}{2}\sin\theta\right) - \left(\frac{1}{2}\cos\theta - \frac{\sqrt{3}}{2}\sin\theta\right) = \sqrt{3}\sin\theta$$

55.
$$\cos(\alpha + \beta) + \cos(\alpha - \beta)$$
$$= \cos\alpha\cos\beta - \sin\alpha\sin\beta + \cos\alpha\cos\beta + \sin\alpha\sin\beta$$
$$= 2\cos\alpha\cos\beta$$

59. $\cot(\alpha - \beta) = \dfrac{1}{\tan(\alpha - \beta)} = \dfrac{1}{\dfrac{\tan\alpha - \tan\beta}{1 + \tan\alpha\tan\beta}}$

$$= \dfrac{1 + \dfrac{1}{\cot\alpha\cot\beta}}{\dfrac{1}{\cot\alpha} - \dfrac{1}{\cot\beta}} = \dfrac{\cot\alpha\cot\beta + 1}{\cot\beta - \cot\alpha}$$

63. $\dfrac{\tan\alpha + \tan\beta}{1 + \tan\alpha\tan\beta} = \dfrac{\dfrac{\sin\alpha}{\cos\alpha} + \dfrac{\sin\beta}{\cos\beta}}{1 - \dfrac{\sin\alpha}{\cos\alpha}\cdot\dfrac{\sin\beta}{\cos\beta}}$

$$= \left(\dfrac{\dfrac{\sin\alpha}{\cos\alpha} + \dfrac{\sin\beta}{\cos\beta}}{1 - \dfrac{\sin\alpha}{\cos\alpha}\cdot\dfrac{\sin\beta}{\cos\beta}}\right)\left(\dfrac{\cos\alpha\cos\beta}{\cos\alpha\cos\beta}\right)$$

$$= \dfrac{\sin(\alpha + \beta)}{\cos(\alpha - \beta)}$$

67. 1 **69.** $\dfrac{-\sqrt{2} + \sqrt{6}}{4}$ **71.** $-\dfrac{1}{2}$ **73.** $\dfrac{3}{5}$ **75.** $\dfrac{1}{7}$

77. $\dfrac{720}{1,681}$ **79.** $x\sqrt{1 - y^2} + y\sqrt{1 - x^2}$

81. $\dfrac{x\sqrt{1 - y^2} + y}{\sqrt{1 - y^2} - xy}$ **83.** $3y\sqrt{1 - 4x^2} + 2x\sqrt{1 - 9y^2}$

93. 8.1° **95.** 74.7°

99. $\cos\alpha\cos\beta\cos\delta - \sin\alpha\sin\beta\cos\delta - \sin\alpha\cos\beta\sin\delta -$
$$\cos\alpha\sin\beta\sin\delta$$

101.
$$\sin\alpha\cos\beta\cos\delta - \cos\alpha\sin\beta\cos\delta - \cos\alpha\cos\beta\sin\delta -$$
$$\sin\alpha\sin\beta\sin\delta$$

103. $\dfrac{\tan\alpha + \tan\beta + \tan\delta - \tan\alpha\tan\beta\tan\delta}{1 - \tan\alpha\tan\beta - \tan\alpha\tan\delta - \tan\beta\tan\delta}$

Section 5.5

1. $2\sin\theta\cos\theta$

3. a. QI **b.** QI **c.** QII **d.** QII

5. $\sin 2\alpha$ **7.** $\cos 2\alpha$ **9.** $\cos 2\beta$ **11.** $\cos\beta$

13. $\sin^2 4\theta$ **15.** $\cos 18\theta$ **17.** $\tan 2\alpha$ **19.** $\cos 15°$

21. $\tan 100°$ **23.** $\tan^2 x$

25. a. $\dfrac{\sqrt{2 - \sqrt{3}}}{2}$ **b.** $\dfrac{\sqrt{2 + \sqrt{3}}}{2}$ **c.** $2 - \sqrt{3}$

27. a. $\dfrac{\sqrt{2 + \sqrt{3}}}{2}$ **b.** $\dfrac{\sqrt{2 - \sqrt{3}}}{2}$ **c.** $2 + \sqrt{3}$

29. a. $\dfrac{\sqrt{2 + \sqrt{2}}}{2}$ **b.** $-\dfrac{\sqrt{2 - \sqrt{2}}}{2}$ **c.** $-\sqrt{2} - 1$

31. a. $\dfrac{\sqrt{2 + \sqrt{2}}}{2}$ **b.** $\dfrac{\sqrt{2 - \sqrt{2}}}{2}$ **c.** $\sqrt{2} + 1$

33. a. $\dfrac{\sqrt{2 - \sqrt{2}}}{2}$ **b.** $-\dfrac{\sqrt{2 + \sqrt{2}}}{2}$ **c.** $-\sqrt{2} + 1$

35. $\dfrac{\sqrt{2}}{2}$ **37.** $\dfrac{\sqrt{2}}{2}$ **39.** $\dfrac{\sqrt{3}}{3}$ **41. a.** $\dfrac{120}{169}$

b. $-\dfrac{119}{169}$ **c.** $-\dfrac{120}{119}$ **d.** $\dfrac{2\sqrt{13}}{13}$ **e.** $\dfrac{3\sqrt{13}}{13}$ **f.** $\dfrac{2}{3}$

43. a. $\dfrac{720}{1,681}$ **b.** $-\dfrac{1,519}{1,681}$ **c.** $-\dfrac{720}{1,519}$ **d.** $\dfrac{5\sqrt{41}}{41}$

e. $-\dfrac{4\sqrt{41}}{41}$ **f.** $-\dfrac{5}{4}$

45. a. $\dfrac{4}{5}$ **b.** $-\dfrac{3}{5}$ **c.** $-\dfrac{4}{3}$ **d.** $\dfrac{\sqrt{50 + 10\sqrt{5}}}{10}$

e. $-\dfrac{\sqrt{50 - 10\sqrt{5}}}{10}$ **f.** $\dfrac{-\sqrt{5} - 1}{2}$

47. a. $-\dfrac{5\sqrt{3}}{14}$ **b.** $-\dfrac{11}{14}$ **c.** $\dfrac{5\sqrt{3}}{11}$ **d.** $\dfrac{\sqrt{98 - 7\sqrt{21}}}{14}$

e. $-\dfrac{\sqrt{98 + 7\sqrt{21}}}{14}$ **f.** $\dfrac{-2\sqrt{7} + \sqrt{3}}{5}$

49. a. $-\dfrac{24}{25}$ **b.** $-\dfrac{7}{25}$ **c.** $\dfrac{24}{7}$ **d.** $\dfrac{\sqrt{5}}{5}$ **e.** $-\dfrac{2\sqrt{5}}{5}$

f. $-\dfrac{1}{2}$

51. a. $-\dfrac{24}{25}$ **b.** $\dfrac{7}{25}$ **c.** $-\dfrac{24}{7}$ **d.** $\dfrac{3\sqrt{10}}{10}$

e. $\dfrac{\sqrt{10}}{10}$ **f.** 3

53. $\dfrac{\tan 2x}{\sin 2x} = \dfrac{\dfrac{\sin 2x}{\cos 2x}}{\sin 2x} = \dfrac{1}{\cos 2x} = \sec 2x$

57.
$$\cos 2x - \dfrac{\sin 2x}{\cos x} + 2\sin^2 x = \cos^2 x - \sin^2 x - \dfrac{2\sin x\cos x}{\cos x} +$$
$$2\sin^2 x = \cos^2 x + \sin^2 x - 2\sin x = 1 - 2\sin x$$

61. $\cot\dfrac{\theta}{2} = \dfrac{1}{\tan\dfrac{\theta}{2}} = \dfrac{1}{\dfrac{\sin\theta}{1 + \cos\theta}} = \dfrac{1 + \cos\theta}{\sin\theta}$

65. $\dfrac{\cos 2\alpha + \cos\alpha}{\cos\alpha + 1} = \dfrac{2\cos^2\alpha - 1 + \cos\alpha}{\cos\alpha + 1}$

$$= \dfrac{(2\cos\alpha - 1)(\cos\alpha + 1)}{\cos\alpha + 1} = 2\cos\alpha - 1$$

69. $\tan\left(\dfrac{\pi}{4}+\dfrac{\theta}{2}\right) = \dfrac{\tan\dfrac{\pi}{4}+\tan\dfrac{\theta}{2}}{1-\tan\dfrac{\pi}{4}\tan\dfrac{\theta}{2}} = \dfrac{1+\tan\dfrac{\theta}{2}}{1-\tan\dfrac{\theta}{2}}$

$$= \dfrac{1+\dfrac{\sin\theta}{1+\cos\theta}}{1-\dfrac{\sin\theta}{1+\cos\theta}} = \dfrac{1+\cos\theta+\sin\theta}{1+\cos\theta-\sin\theta}$$

73. $\dfrac{1-\tan^2 x}{1+\tan^2 x} = \dfrac{1-\dfrac{\sin^2 x}{\cos^2 x}}{1+\dfrac{\sin^2 x}{\cos^2 x}} = \dfrac{\cos^2 x-\sin^2 x}{\cos^2 x+\sin^2 x} = \cos 2x$

77. $\dfrac{1}{2}\sin x \tan\dfrac{x}{2}\csc^2\dfrac{x}{2} = \dfrac{1}{2}\sin x \cdot \dfrac{1-\cos x}{\sin x}\cdot\dfrac{2}{1-\cos x} = 1$

79. $\dfrac{3}{8}+\dfrac{1}{2}\cos 2x+\dfrac{1}{8}\cos 4x$

81. $\dfrac{1}{8}-\dfrac{1}{8}\cos 4x$ **83.** 1   **85.** $\dfrac{24}{25}$ **87.** $\dfrac{-119}{169}$

89. $2x\sqrt{1-x^2}$ **91.** $1-2x^2$ **93.** $\dfrac{2x}{1-x^2}$

95. $\dfrac{\sqrt{2+2x}}{2}=\sqrt{1+x}$ **97.** $\sqrt{\dfrac{1-x}{2}}=\dfrac{\sqrt{2-2x}}{2}$

99. 10.1 ft   **101.** 2   **103.** $\dfrac{1-\cos 2x}{2}$

Section 5.6

1.

$$\sin\alpha\cos\beta = \frac{1}{2}[\sin(\alpha+\beta)+\sin(\alpha-\beta)]$$
$$\cos\alpha\sin\beta = \frac{1}{2}[\sin(\alpha+\beta)-\sin(\alpha-\beta)]$$
$$\sin\alpha\sin\beta = \frac{1}{2}[\cos(\alpha-\beta)-\cos(\alpha+\beta)]$$
$$\cos\alpha\cos\beta = \frac{1}{2}[\cos(\alpha+\beta)+\cos(\alpha-\beta)]$$

3. $\dfrac{1}{4}$ **5.** $\dfrac{1}{4}$ **7.** $\dfrac{\sqrt{2}}{4}$ **9.** $\dfrac{1}{4}$ **11.** $\dfrac{\sqrt{3}-2}{4}$

13. $\dfrac{\sqrt{6}}{2}$ **15.** $-\dfrac{\sqrt{2}}{2}$ **17.** $-\dfrac{\sqrt{2}}{2}$ **19.** $\dfrac{\sqrt{6}}{2}$

21. 0   **23.** $\dfrac{1}{2}\cos 10° - \dfrac{1}{2}\cos 70°$

25. $\dfrac{1}{2}\cos\dfrac{6\pi}{7}+\dfrac{1}{2}\cos\dfrac{4\pi}{7}$ **27.** $\dfrac{1}{2}\cos 8\theta+\dfrac{1}{2}\cos 2\theta$

29. $\dfrac{1}{2}\sin\dfrac{3\alpha}{2}+\dfrac{1}{2}\sin\dfrac{\alpha}{2}$ **31.** $\dfrac{1}{2}\cos\dfrac{x}{6}-\dfrac{1}{2}\cos\dfrac{5x}{6}$

33. $2\sin 40°\cos 10°$ **35.** $-2\sin\dfrac{2\pi}{3}\sin\dfrac{\pi}{9}$

37. $2\cos 4\theta\cos\theta$ **39.** $2\cos\dfrac{3\alpha}{4}\sin\dfrac{\alpha}{4}$

41. $2\cos\dfrac{5x}{12}\cos\dfrac{x}{12}$

43. $\dfrac{\sin\alpha+\sin\beta}{\sin\alpha-\sin\beta} = \dfrac{2\sin\dfrac{\alpha+\beta}{2}\cos\dfrac{\alpha-\beta}{2}}{2\cos\dfrac{\alpha+\beta}{2}\sin\dfrac{\alpha-\beta}{2}}$

$$= \tan\frac{1}{2}(\alpha+\beta)\cot\frac{1}{2}(\alpha-\beta)$$

47. $\dfrac{\cos\alpha+\cos 5\alpha}{\cos\alpha-\cos 5\alpha} = \dfrac{2\cos\dfrac{\alpha+5\alpha}{2}\cos\dfrac{\alpha-5\alpha}{2}}{-2\sin\dfrac{\alpha+5\alpha}{2}\sin\dfrac{\alpha-5\alpha}{2}}$

$$= \dfrac{\cos 3\alpha\cos(-2\alpha)}{-\sin 3\alpha\sin(-2\alpha)} = \cot 3\alpha\cot 2\alpha$$

51. $2\cos 5\alpha\sin 2\alpha = 2\cdot\dfrac{1}{2}[\sin(5\alpha+2\alpha)-\sin(5\alpha-2\alpha)]$

$$= \sin 7\alpha - \sin 3\alpha$$

53. $10\sin(x+53.1°)$ **55.** $10\sin(x-53.1°)$

57. $\sqrt{5}\sin(x+26.6°)$ **59.** $\sqrt{2}\sin(x+45°)$

61. $\sqrt{26}\sin(x+101.3°)$ or $-\sqrt{26}\sin(x-78.7°)$

63. $2\sqrt{3}\sin(x-60°)$ **65.** $\dfrac{5}{2}$ Hz apart

67. 13 units   **69.** $\dfrac{1}{2}\sin 11\theta - \dfrac{1}{2}\sin\theta$

71.–73. Answers will vary.

Section 5.7

1. infinite   **3.** general

5. $\theta = \dfrac{\pi}{3}+2k\pi$ or $\theta = \dfrac{2\pi}{3}+2k\pi$

7. $\theta = \dfrac{3\pi}{4}+k\pi$

9. $\theta = \dfrac{5\pi}{12}+k\pi$ or $\theta = \dfrac{7\pi}{12}+k\pi$

11. $\theta = \dfrac{\pi}{4}+k\pi$ **13.** $\theta = \dfrac{\pi}{3}+2k\pi$

15. 0 and π **17.** $\dfrac{3\pi}{4}$ and $\dfrac{5\pi}{4}$ **19.** $\dfrac{\pi}{6},\dfrac{5\pi}{6},\dfrac{7\pi}{6}$, and $\dfrac{11\pi}{6}$

21. $\dfrac{\pi}{6},\dfrac{5\pi}{6},\dfrac{7\pi}{6}$, and $\dfrac{11\pi}{6}$ **23.** $\dfrac{7\pi}{12}$ and $\dfrac{23\pi}{12}$

25. $\dfrac{\pi}{4},\dfrac{11\pi}{12},\dfrac{5\pi}{4}$, and $\dfrac{23\pi}{12}$ **27.** 0

29. $\dfrac{\pi}{4},\dfrac{3\pi}{4},\dfrac{5\pi}{4}$, and $\dfrac{7\pi}{4}$ **31.** $\dfrac{\pi}{12},\dfrac{11\pi}{12},\dfrac{13\pi}{12}$, and $\dfrac{23\pi}{12}$

33. $\dfrac{\pi}{8},\dfrac{5\pi}{8},\dfrac{9\pi}{8}$, and $\dfrac{13\pi}{8}$ **35.** $\dfrac{2\pi}{3}$ and $\dfrac{4\pi}{3}$

37. $0,\dfrac{\pi}{2}$, and $\dfrac{3\pi}{2}$ **39.** $0,\dfrac{\pi}{4},\pi$, and $\dfrac{5\pi}{4}$

41. $\dfrac{\pi}{3}$ and $\dfrac{4\pi}{3}$ **43.** $0,\dfrac{\pi}{3},\pi$, and $\dfrac{5\pi}{3}$ **45.** $\dfrac{\pi}{6}$ and $\dfrac{5\pi}{6}$

47. $\dfrac{\pi}{3},\pi$, and $\dfrac{5\pi}{3}$ **49.** $0,\dfrac{\pi}{2},\pi$, and $\dfrac{3\pi}{2}$

51. $\dfrac{\pi}{2}$ and $\dfrac{3\pi}{2}$ **53.** $0,\dfrac{\pi}{4},\pi$, and $\dfrac{5\pi}{4}$

55. $\dfrac{\pi}{2}$ **57.** $\dfrac{2\pi}{3},\pi$, and $\dfrac{4\pi}{3}$ **59.** no solution

61. $0, \dfrac{\pi}{6}, \pi$, and $\dfrac{7\pi}{6}$ **63.** $\dfrac{\pi}{2}, \dfrac{7\pi}{6}, \dfrac{3\pi}{2}$, and $\dfrac{11\pi}{6}$

65. $\dfrac{\pi}{4}$ and $\dfrac{5\pi}{4}$ **67.** $\dfrac{\pi}{3}$ and $\dfrac{5\pi}{3}$ **69.** $0, \dfrac{2\pi}{3}$, and $\dfrac{4\pi}{3}$

71. 0 and π **73.** no solution   **75.** $\dfrac{\pi}{6}$ and $\dfrac{5\pi}{6}$

77. $\dfrac{3\pi}{2}$ **79.** $0, \dfrac{\pi}{6}, \dfrac{5\pi}{6}$, and π **81.** $0, \dfrac{\pi}{2}$, and π

83. $0, \dfrac{\pi}{6}, \dfrac{\pi}{2}, \dfrac{5\pi}{6}, \pi, \dfrac{7\pi}{6}, \dfrac{3\pi}{2}$, and $\dfrac{11\pi}{6}$ **85.** 0 and π

87. $\dfrac{\pi}{4}, \dfrac{3\pi}{4}, \dfrac{5\pi}{4}$, and $\dfrac{7\pi}{4}$ **89.** $\dfrac{\pi}{4}$ and $\dfrac{7\pi}{4}$

91. $\dfrac{\pi}{4}$ and $\dfrac{7\pi}{4}$ **93.** $\dfrac{\pi}{4}, \dfrac{3\pi}{4}, \dfrac{5\pi}{4}$, and $\dfrac{7\pi}{4}$

95. $\dfrac{\pi}{6}, \dfrac{\pi}{2}, \dfrac{5\pi}{6}, \dfrac{7\pi}{6}, \dfrac{3\pi}{2}$, and $\dfrac{11\pi}{6}$ **97.** $\dfrac{\pi}{2}$ and $\dfrac{3\pi}{2}$

99. $\dfrac{7\pi}{6}$ and $\dfrac{11\pi}{6}$ **101.** $0, \dfrac{\pi}{4}, \pi$, and $\dfrac{5\pi}{4}$ **103.** 0

105. $\dfrac{\pi}{4}$ and $\dfrac{5\pi}{4}$ **107.** $0, \dfrac{\pi}{2}, \pi$, and $\dfrac{3\pi}{2}$

109. $0, \dfrac{\pi}{4}, \dfrac{\pi}{2}, \dfrac{3\pi}{4}, \pi, \dfrac{5\pi}{4}, \dfrac{3\pi}{2}$, and $\dfrac{7\pi}{4}$

111. $0, \dfrac{\pi}{3}, \dfrac{\pi}{2}, \dfrac{2\pi}{3}, \pi, \dfrac{4\pi}{3}, \dfrac{3\pi}{2}$, and $\dfrac{5\pi}{3}$

113. $\dfrac{\pi}{12}, \dfrac{\pi}{6}, \dfrac{\pi}{4}, \dfrac{5\pi}{12}, \dfrac{\pi}{2}, \dfrac{7\pi}{12}, \dfrac{3\pi}{4}, \dfrac{5\pi}{6}, \dfrac{11\pi}{12}, \dfrac{13\pi}{12}, \dfrac{7\pi}{6}, \dfrac{5\pi}{4}, \dfrac{17\pi}{12},$
$\dfrac{3\pi}{2}, \dfrac{19\pi}{12}, \dfrac{7\pi}{4}, \dfrac{11\pi}{6},$ and $\dfrac{23\pi}{12}$

115. 2012   **117.** 30°   **119.** 0, π
121. Answers will vary.

Section 5.8
1. Quadratic Formula
3. No, there may be extraneous roots created by squaring the equation.   **5.** 0.412 and 2.730
7. 3.489 and 5.936   **9.** 1.318 and 4.965
11. 2.498 and 3.785   **13.** 1.004 and 4.145
15. 1.932 and 5.073   **17.** 1.107, 2.034, 4.249, and 5.176
19. 0.491, 2.651, 3.632, and 5.792   **21.** 3.380 and 6.045
23. 0.161, 1.732, 3.302, and 4.873   **25.** 1.855 and 4.428
27. 0 and $\dfrac{2\pi}{3}$ **29.** $\dfrac{\pi}{6}$ **31.** 0 and $\dfrac{4\pi}{3}$
33. 1.969 and 4.315   **35.** 1.006 and 5.277
37. 0 and 1.496   **39.** 0.769   **41.** 0.579   **43.** 0.635
45. 0.203, 1.255, 3.667, and 4.144   **47.** 0.575
49. 2.808 and 5.046   **51.** 1.297
53. 0.955, 2.187, 4.097, and 5.328   **55.** 28.8°

Chapter Review
1. $\dfrac{\pi}{4}$ **2.** $\dfrac{5\pi}{6}$ **3.** 0   **4.** 0.594   **5.** 1.918

6. −1.531   **7.** $\dfrac{4}{5}$ **8.** $-\dfrac{\pi}{4}$ **9.** $-\dfrac{\pi}{6}$

10. does not exist   **11.** $-\dfrac{12}{5}$ **12.** $\dfrac{7\sqrt{65}}{65}$

13. $\dfrac{12}{13}$ **14.** $\dfrac{3\sqrt{5}}{5}$ **15.** 1.28   **16.** $\dfrac{3\pi}{4}$ **17.** $\dfrac{\pi}{6}$

18. $\dfrac{\pi}{4}$

19. $\dfrac{1}{1 + \cos x} + \dfrac{1}{1 - \cos x} = \dfrac{1 - \cos x + 1 + \cos x}{1 - \cos^2 x}$
$$= \dfrac{2}{\sin^2 x} = 2\csc^2 x$$

20. $\dfrac{8}{\sec x + \tan x} \cdot \dfrac{\sec x - \tan x}{\sec x - \tan x} = \dfrac{8\sec x - 8\tan x}{\sec^2 x - \tan^2 x}$
$$= 8\sec x - 8\tan x$$

21. $\dfrac{1 + \tan^2 x}{1 - \sin^2 x} = \dfrac{\sec^2 x}{\cos^2 x} = \sec^4 x$

22. right $= \dfrac{\sin^2 x}{\cos^2 x} - \dfrac{\cot^2 x}{\cos^2 x} = \tan^2 x - \dfrac{\dfrac{\cos^2 x}{\sin^2 x}}{\cos^2 x}$
$$= \tan^2 x - \dfrac{1}{\sin^2 x} = \tan^2 x - \csc^2 x$$

23. $\dfrac{\sin x + \cos x}{\sin x} - \dfrac{\cos x - \sin x}{\cos x}$
$$= \dfrac{\sin x \cos x + \cos^2 x - \sin x \cos x + \sin^2 x}{\sin x \cos x}$$
$$= \dfrac{1}{\sin x \cos x} = \csc x \sec x$$

24. $\dfrac{\tan x}{1 + \sec x} + \dfrac{1 + \sec x}{\tan x} = \dfrac{\tan^2 x + 1 + 2\sec x + \sec^2 x}{\tan x(1 + \sec x)}$
$$= \dfrac{2\sec x(\sec x + 1)}{\tan x(1 + \sec x)} = \dfrac{\dfrac{2}{\cos x}}{\dfrac{\sin x}{\cos x}} = 2\csc x$$

25. $5\tan\theta$

26. $\sin x \tan^2 x = \sin x(\sec^2 x - 1) = \dfrac{\sin x}{\cos^2 x} - \sin x$
$$= \tan x \sec x - \sin x$$

27. $\dfrac{\sqrt{6} - \sqrt{2}}{4}$ **28.** $2 + \sqrt{3}$ **29.** $\dfrac{\sqrt{2} + \sqrt{6}}{4}$

30. $\dfrac{\sqrt{2} - \sqrt{6}}{4}$ **31.** −1   **32.** 1

33. a. $-\dfrac{\sqrt{10}}{10}$ b. $-\dfrac{9\sqrt{10}}{50}$ c. $-\dfrac{1}{3}$

34. a. $\dfrac{19\sqrt{5}}{65}$ b. $\dfrac{2\sqrt{5}}{65}$ c. $\dfrac{19}{22}$

35. $\dfrac{-2\sqrt{5} + \sqrt{15}}{10}$ **36.** $\dfrac{108 - 50\sqrt{5}}{19}$ **37.** 8.1°

38. 67.2°

39. a. $\dfrac{\sqrt{2 - \sqrt{3}}}{2}$ b. $\dfrac{\sqrt{2 + \sqrt{3}}}{2}$ c. $2 - \sqrt{3}$

40. a. $\dfrac{\sqrt{2 + \sqrt{3}}}{2}$ b. $-\dfrac{\sqrt{2 - \sqrt{3}}}{2}$ c. $-2 - \sqrt{3}$

41. a. $\dfrac{\sqrt{2 + \sqrt{2}}}{2}$ b. $-\dfrac{\sqrt{2 - \sqrt{2}}}{2}$ c. $-\sqrt{2} - 1$

42. a. $\dfrac{\sqrt{2 - \sqrt{2}}}{2}$ b. $\dfrac{\sqrt{2 + \sqrt{2}}}{2}$ c. $\sqrt{2} - 1$

43. a. $\dfrac{24}{25}$ b. $-\dfrac{7}{25}$ c. $-\dfrac{24}{7}$ d. $\dfrac{2\sqrt{5}}{5}$ e. $-\dfrac{\sqrt{5}}{5}$
f. −2

44. a. $-\dfrac{4\sqrt{5}}{9}$ **b.** $-\dfrac{1}{9}$ **c.** $4\sqrt{5}$ **d.** $\dfrac{\sqrt{6}}{6}$

e. $-\dfrac{\sqrt{30}}{6}$ **f.** $-\dfrac{\sqrt{5}}{5}$

45. $\dfrac{1}{4}$ **46.** -6 **47.** $-\dfrac{\sqrt{6}}{2}$ **48.** $\dfrac{\sqrt{6}}{2}$

49. $2\sqrt{29}$ units **50.** $\theta = \dfrac{\pi}{12} + \dfrac{\pi}{2}k$ **51.** $\dfrac{2\pi}{3} + k\pi$

52. 0 and π **53.** $\dfrac{\pi}{3}$ and $\dfrac{5\pi}{3}$ **54.** $0, \dfrac{\pi}{3}, \pi$, and $\dfrac{5\pi}{3}$

55. $\dfrac{\pi}{6}, \dfrac{5\pi}{6}$, and $\dfrac{3\pi}{2}$ **56.** 2016 **57.** 3.760 and 5.664

58. $0.886, 2.256, 4.028$, and 5.397

59. $0.785, 2.761, 3.927$, and 5.903

60. $0.714, 2.428, 3.855$, and 5.569

61. a. 318.2 ft **b.** 121.6 ft/sec

Chapter Test

1. $\dfrac{3\pi}{4}$ **2.** $\dfrac{\pi}{6}$ **3.** $\dfrac{5}{3}$ **4.** $-\dfrac{2\sqrt{53}}{53}$

5.
$\tan^2 x - \cot^2 x = (\sec^2 x - 1) - (\csc^2 x - 1) = \sec^2 x - \csc^2 x$

6. $\dfrac{(\sin x + \cos x)^2}{\sin x \cos x} = \dfrac{\sin^2 x + 2\sin x \cos x + \cos^2 x}{\sin x \cos x}$

$= \dfrac{1 + 2\sin x \cos x}{\sin x \cos x} = \dfrac{1}{\sin x \cos x} + \dfrac{2\sin x \cos x}{\sin x \cos x}$

$= 2 + \sec x \csc x$

7. $\dfrac{\sqrt{2 - \sqrt{2}}}{2}$ **8.** $2 - \sqrt{3}$

9. a. $\dfrac{19\sqrt{5}}{65}$ **b.** $\dfrac{2\sqrt{5}}{65}$ **c.** $-\dfrac{19}{22}$

10. a. $-\dfrac{336}{625}$ **b.** $-\dfrac{527}{625}$ **c.** $\dfrac{336}{527}$ **d.** $\dfrac{4}{5}$ **e.** $\dfrac{3}{5}$

f. $\dfrac{4}{3}$

11. $\dfrac{\pi}{6}, \dfrac{5\pi}{6}$, and $\dfrac{3\pi}{2}$ **12.** $\dfrac{\pi}{4}, \dfrac{3\pi}{4}, \dfrac{5\pi}{4}$, and $\dfrac{7\pi}{4}$

13. $\dfrac{\pi}{9}, \dfrac{5\pi}{9}, \dfrac{7\pi}{9}, \dfrac{11\pi}{9}, \dfrac{13\pi}{9}$, and $\dfrac{17\pi}{9}$ **14.** π

15. $0.464, 2.678, 3.605$, and 5.820 **16.** 1.166 and 4.307

17. 0.340 and 2.802

18. $1.107, 1.249, 1.893, 2.034, 4.249, 4.391, 5.034$, and 5.176

19. $29.9°$ **20.** $29.7°$

Connect to Calculus

1. $\dfrac{1}{2}\cos 3x - \dfrac{1}{2}\cos 7x$ **2.** $\dfrac{1}{2}\cos 12x + \dfrac{1}{2}\cos 2x$

3. $\dfrac{2}{(1 - \cos 2x)}$ **4.** $8\tan^2 \theta \sec \theta$ **5.** $\sin \theta$

6. $\dfrac{1}{125}\csc^2 \theta \sec \theta$

Section 6.1

1. $a^2 + b^2 = c^2$ **3.** right

5. $a = 12; b = 16; B = 53°$

7. $a = 5.9; c = 9.1; B = 40°$

9. $a = 206; c = 221.1; B = 21.3°$

11. $a = 9.8; A = 63°; B = 27°$

13. $a = 10.0; b = 11.1; B = 48°$

15. $b = 8.4; c = 9.5; A = 27°$

17. $b = 6.9; A = 30°; B = 60°$

19. $c = 12; A = 36.9°; B = 53.1°$

21. 685 mph **23.** 28.95 ft **25.** $2,690$ ft

27. 473.29 ft **29.** 143.27 m **31.** 60.7 ft

33. 21.5 ft **35.** 319 mi **37.** 7.72 mi

39. 5.8 mi **41.** 1.5 km **43.** $130 \, \Omega$

45. $46.4°$ **47.** $120 \, \Omega$ **49.** $31.4°$ **51.** $48.3°$

Section 6.2

1. $1,453$ ft **3.** 556 ft **5.** 631 ft **7.** 143.4 ft

9. 123 ft **11.** $45°$ **13.** $90°$ **15.** 1.31 **17.** no

19. $\dfrac{\sqrt{3}}{2}k$ **21.** 6% less **23.** $\dfrac{r}{\tan\dfrac{\theta}{2}}$

25. $b = 2k\cos\alpha$ **27.** $H = k(\tan\alpha + \tan\beta)$

29. $D(\tan\varphi - \tan\theta)$ **31.** $a = d\tan\beta$

33. $x = \dfrac{h}{\tan\theta}$

Section 6.3

1. oblique **3.** $\dfrac{a}{\sin A} = \dfrac{b}{\sin B} = \dfrac{c}{\sin C}$

5. $b \approx 66.83; c \approx 63.67; C = 71°$

7. $a \approx 7.76; c = 15; B = 75°$

9. $a \approx 3.31; b \approx 1.24; C = 65°$

11. $a \approx 5.72; c \approx 11.59; A = 28°$

13. $b \approx 3.28; c \approx 15.69; C = 96°$

15. $a \approx 96.73; c \approx 71.80; A = 60°$

17. $a \approx 99.56; b \approx 67.91; C = 48°$

19. $b \approx 32.42; c \approx 93.34; C = 30°$

21. $c \approx 176.25; B \approx 31.5°; C \approx 106.5°$

23. $b_1 \approx 78.75; B_1 \approx 152.7°; C_1 \approx 15.3°; b_2 \approx 9.88;$
$B_2 \approx 3.3°; C_2 \approx 164.7°$

25. no triangle

27. $c_1 \approx 8,862.59; A_1 \approx 14.0°; C_1 \approx 156.1°; c_2 \approx 1,564.03;$
$A_2 \approx 166.0°; C_2 \approx 4.1°$

29. $b \approx 62.52; A \approx 39.0°; B \approx 61°$

31. no triangle

33. $b_1 \approx 270.34; B_1 \approx 72.9°; C_1 \approx 62.1°; b_2 \approx 83.17;$
$B_2 \approx 17.1°; C_2 \approx 117.9°$

35. $c \approx 126.27; A \approx 50.6°; C \approx 54.4°$

37. 57.1 **39.** 40.0 **41.** 180.2 yd **43.** N $66.4°$ W

45. 508.3 ft **47.** 1.6 mi **49.** 611 ft

51. 2.5 nautical mi **53.** approximately 2:50 p.m.

55. 38.9 ft **57.** 74.7 ft **59.** $44.2°$ **61.** $44.8°$

63. $2,874$ ft or 784 ft **65.** 94 ft **67.** 618.3 ft

69. Answers will vary.

Section 6.4

1. SSS, SAS

3. The sum of any two sides must be greater than the third side. Since $2 + 3 < 10$, the sides cannot form a triangle.

5. $A \approx 25.2°; B \approx 48.2°; C \approx 106.6°$

7. $a \approx 7.29; B \approx 35.4°; C \approx 119.6°$

9. $A \approx 18.6°; B \approx 44.4°; C \approx 117°$

11. $a \approx 23.44; B \approx 39.7°; C \approx 46.3°$

13. $c \approx 5.39; A \approx 34.7°; B \approx 95.3°$

15. $b \approx 60.16$; $A \approx 43.6°$; $C \approx 37.4°$
17. $a \approx 15.15$; $B \approx 36.3°$; $C \approx 58.7°$
19. $c \approx 18.0$; $A \approx 50°$; $B \approx 50°$
21. $A \approx 21.8°$; $B \approx 38.2°$; $C \approx 120°$
23. $A \approx 60°$; $B \approx 60°$; $C \approx 60°$
25. $A \approx 36.9°$; $B \approx 53.1°$; $C \approx 90°$
27. $A \approx 26.9°$; $B \approx 110.3°$; $C \approx 42.8°$
29. 9.63 nautical miles; N 73.1° E   **31.** 302.9 mi
33. 423 mi above the earth's surface   **35.** 17.4 ft   **37.** 86.1°
39. 85.2°   **41.** 107.0°   **43.** 6 ft and 14 ft   **45.** 90°
47. $O = 83.2°$; $O' = 53.8°$; $O'' = 43°$
49. $x = R\sqrt{2 - 2\cos\theta}$
51. if $C = 90°$, then $\cos C = 0$ and $c^2 = a^2 + b^2$

Section 6.5

1. $K = \dfrac{1}{2}ab \sin C$ **3.** $K = \dfrac{c^2 \sin A \sin B}{2 \sin C}$ **5.** 13.79 units^2
7. 7.39 units^2 **9.** 61.64 units^2 **11.** 6.93 units^2
13. 192.53 units^2 **15.** 304.84 units^2 **17.** 6.50 units^2
19. 6 units^2 **21.** 42.35 units^2 **23.** 956.66 units^2
25. 14.14 units^2 **27.** 54.99 units^2 **29.** 49.74 units^2
31. 70.91 units^2 **33.** 27.10 units^2 **35.** 16.36 units^2
37. 21,340 ft^2 **39.** 154.5 m^2 **41.** 26.3 km^2
43. 126,466 ft^2 **45.** 330,196 square nautical miles
47. $K = \dfrac{b^2 \cot \dfrac{\alpha}{2}}{4}$ **49.** $K = \dfrac{1}{4}b^2\sqrt{15}$
51. Answers will vary.
53. $K = \dfrac{1}{2}r^2(\theta - \sin\theta)$ (radian measure)

Section 6.6

1. linear   **3.** $v = r\omega$ **5.** 2π radians per hr
7. $\dfrac{\pi}{1,800}$ radians per sec   **9.** $\dfrac{4\pi}{59}$ radians per day
11. $\dfrac{176}{5\pi}$ rev per sec   **13.** $\dfrac{1,716}{\pi}$ rpm
15. $\dfrac{1,125\pi}{4}$ ft per min
17. $14,400\pi$ radians per min; 90,478 in per min; 85.68 mph
19. $\dfrac{\pi}{2}$ ft per sec   **21.** $\dfrac{40}{3}$ rpm   **23.** 739 mph
25. N 30.5°   **27.** 382 mi   **29.** Answers will vary.

Chapter Review

1. $A \approx 29.1°$; $B \approx 60.9°$; $c \approx 10.30$
2. $A \approx 54.3°$; $B \approx 35.7°$; $a \approx 9.75$
3. $B = 69°$; $a \approx 3.94$; $b \approx 10.27$
4. $A = 43°$; $b \approx 15.01$; $c \approx 20.53$
5. 9.5 mph   **6.** 18.4°   **7.** 296.3
8. $C = 58.0°$; $b \approx 7.70$; $c \approx 6.60$
9. $a \approx 7.58$; $B \approx 5.24$; $C = 138°$
10. $c \approx 16.33$; $B \approx 51.5°$; $C \approx 58.5°$
11. no triangle
12. $c_1 \approx 11.65$; $B_1 \approx 66.8°$; $C_1 \approx 63.2°$; $c_2 \approx 3.77$;
$B_2 \approx 113.2°$; $C_2 \approx 16.8°$
13. 43.43 ft   **14.** 120.6 ft
15. $a \approx 86.97$; $B \approx 131.1°$; $C \approx 8.9°$
16. $A \approx 22.7°$; $B \approx 35.3°$; $C \approx 122.0°$

17. 144.9 mi; S 75.1° W   **18.** 42.76 units^2
19. 93.90 units^2 **20.** 46.31 units^2
21. 24π radians per sec; 565.49 in per sec
22. 80π radians per sec; 25π ft per sec

Chapter Test

1. $A \approx 36.3°$; $B \approx 53.7°$; $c \approx 18.6$
2. $B = 59°$; $a \approx 3.61$; $b \approx 6.00$
3. 95 ft; 460 ft
4. $C = 100.0°$; $b \approx 8.13$; $c \approx 18.26$
5. $c \approx 12.63$; $B \approx 50.6°$; $C \approx 54.4°$
6. $c_1 \approx 20.32$; $B_1 \approx 42.7°$; $C_1 \approx 113.3°$; $c_2 \approx 7.09$;
$B_2 \approx 137.3°$; $C_2 \approx 18.7°$
7. $a \approx 15.39$; $B \approx 39.1°$; $C \approx 86.9°$
8. $A \approx 39.1°$; $B \approx 47.3°$; $C \approx 93.6°$
9. 532.2 ft   **10.** 74.6 mi; S 78° W   **11.** 27.46 units^2
12. 26.91 units^2 **13.** 132.44 units^2
14. 400π radians per min; $\dfrac{20\pi}{3}$ radians per sec;
$2,400\pi$ cm per min

Connect to Calculus

1. $\dfrac{8}{3}$ ft per sec
2. $\dfrac{80\pi}{3}$ mi per min ≈ 83.8 mi per min

Section 7.1

1. polar axis, pole   **3.** False   **5.** $2\pi n$, π, $2\pi n$
7. 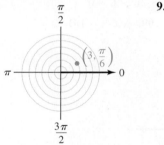   **9.**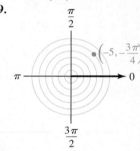

11. 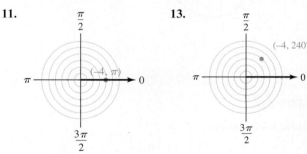   **13.**

15.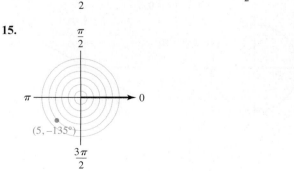

17. yes   **19.** no   **21.** yes   **23.** no

25. a. $(2, 0)$ **b.** $(2, -2\pi)$ **c.** $(-2, \pi)$

27. a. $\left(5, \dfrac{\pi}{2}\right)$ **b.** $\left(5, -\dfrac{3\pi}{2}\right)$ **c.** $\left(-5, \dfrac{3\pi}{2}\right)$

29. a. $(5, \pi)$ **b.** $(5, -\pi)$ **c.** $(-5, 0)$

31. a. $\left(4, \dfrac{3\pi}{2}\right)$ **b.** $\left(4, -\dfrac{\pi}{2}\right)$ **c.** $\left(-4, \dfrac{\pi}{2}\right)$

33. $\left(\sqrt{3}, 1\right)$ **35.** $\left(\dfrac{7}{2}, -\dfrac{7\sqrt{3}}{2}\right)$ **37.** $\left(-\dfrac{3}{2}, \dfrac{3\sqrt{3}}{2}\right)$

39. $(0, 2)$ **41.** $\left(-\sqrt{3}, -1\right)$ **43.** $\left(\dfrac{5\sqrt{2}}{2}, -\dfrac{5\sqrt{2}}{2}\right)$

45. $(0, 10)$ **47.** $\left(3\sqrt{2}, 3\sqrt{2}\right)$ **49.** $(-4, 0)$

51. $(-0.64, 3.64)$ **53.** $(4.17, 7.41)$

55. $(7, 0), (-7, -\pi)$ **57.** $\left(5, \dfrac{3\pi}{2}\right), \left(5, -\dfrac{\pi}{2}\right)$

59. $(0, 0), (0, -\pi)$ **61.** $\left(\sqrt{2}, \dfrac{\pi}{4}\right), \left(-\sqrt{2}, \dfrac{5\pi}{4}\right)$

63. $\left(2, \dfrac{\pi}{3}\right), \left(-2, \dfrac{4\pi}{3}\right)$ **65.** $\left(4, \dfrac{11\pi}{6}\right), \left(-4, \dfrac{5\pi}{6}\right)$

67. $\left(2, \dfrac{\pi}{3}\right), \left(-2, \dfrac{4\pi}{3}\right)$ **69.** $(4.05, 0.70), (-4.05, 3.84)$

71. $(5.99, 2.02), (-5.99, 5.16)$ **73.** $r \cos \theta = 3$

75. $3r \cos \theta + 2r \sin \theta = 3$ **77.** $r = 9 \cos \theta$

79. $r^2 \sin \theta \cos \theta = 12$ **81.** $r = 2 \cos \theta$

83. $r^2 = 4 \cos^2 \theta \sin^2 \theta$ **85.** $x^2 + y^2 = 9$

87. $y = 0$ **89.** $x^4 + y^4 + 2x^2 y^2 - 2xy = 0$

91. $3x + 2y = 2$ **93.** $x^2 + 2y - 1 = 0$

95. $\dfrac{3}{y} + \dfrac{2}{x} = 1$ **97.** 8   **99.** $\sqrt{5}$ **101.** 2

103.

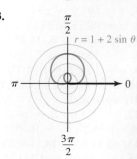

Section 7.2

1. False   **3.** polar   **5.** n

7.

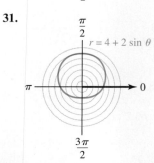

9.

11.

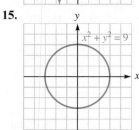

13.

15.

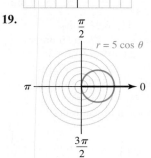

17.

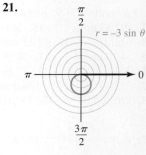

19. $r = 5 \cos \theta$

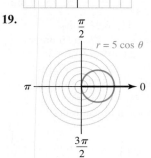

21. $r = -3 \sin \theta$

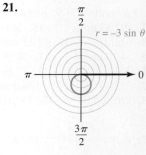

23. $r \cos \theta = 5$

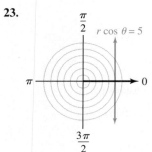

25. $r = 1 - \sin \theta$

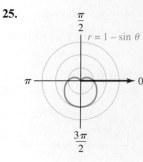

27. $r = 3 + 3 \cos \theta$

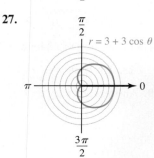

29. $r = 2 - \cos \theta$

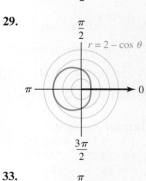

31. $r = 4 + 2 \sin \theta$

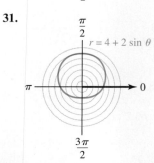

33. $r = 1 + 2 \sin \theta$

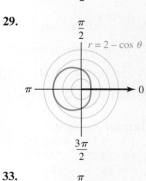

35.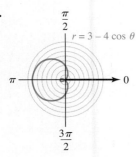

$r = 3 - 4 \cos \theta$

37.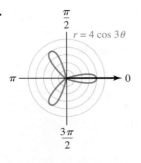

$r = 4 \cos 3\theta$

61.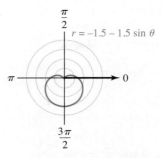

$r = -1.5 - 1.5 \sin \theta$

39.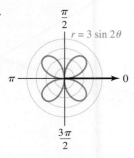

$r = 3 \sin 2\theta$

41.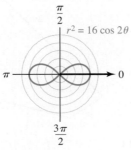

$r^2 = 16 \cos 2\theta$

63.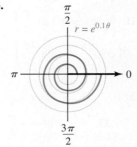

$r = e^{0.1\theta}$

65.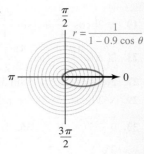

$r = \dfrac{1}{1 - 0.9 \cos \theta}$

43.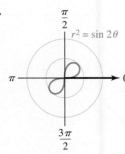

$r^2 = \sin 2\theta$

45.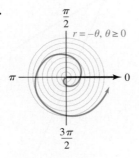

$r = -\theta,\ \theta \geq 0$

Section 7.3

1. absolute value; $\sqrt{a^2 + b^2}$; magnitude

3. $\sqrt{a^2 + b^2}$; modulus; argument

5. $r^n(\cos n\theta + i \sin n\theta)$

7. $\sqrt{13}$

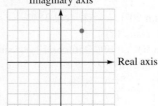

47.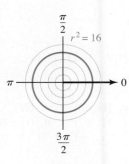

$r^2 = 16$

49.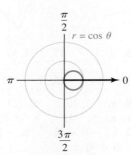

$r = \cos \theta$

9. $\dfrac{\sqrt{13}}{4}$

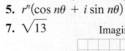

51.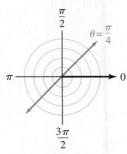

$\theta = \dfrac{\pi}{4}$

11. 7

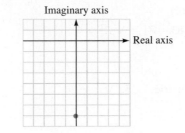

53. $r = -3 \cos \theta$ **55.** $r = 4 \sin 3\theta$

57. $r = 2 + 2 \cos \theta$ **59.** $r = 1 - 2 \sin \theta$

13. 4

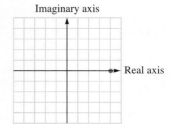

15. $6(\cos 0 + i \sin 0)$ **17.** $3\left(\cos \dfrac{3\pi}{2} + i \sin \dfrac{3\pi}{2}\right)$

19. $\sqrt{2}\left(\cos \dfrac{5\pi}{4} + i \sin \dfrac{5\pi}{4}\right)$ **21.** $6\left(\cos \dfrac{\pi}{3} + i \sin \dfrac{\pi}{3}\right)$

23. $2\left(\cos \dfrac{4\pi}{3} + i \sin \dfrac{4\pi}{3}\right)$ **25.** $2\left(\cos \dfrac{7\pi}{6} + i \sin \dfrac{7\pi}{6}\right)$

27. $5(\cos 323.13° + i \sin 323.13°)$

29. $\sqrt{29}(\cos 158.20° + i \sin 158.20°)$ **31.** $\sqrt{3} + i$

33. $0 + 7i = 7i$ **35.** $1 - i\sqrt{3}$ **37.** $-\dfrac{1}{2} + 0i = -\dfrac{1}{2}$

39. $\dfrac{3\sqrt{2}}{2} + \dfrac{3\sqrt{2}}{2}i$ **41.** $0.55 + 1.64i$

43. $8(\cos 90° + i \sin 90°)$; $2(\cos 30° - i \sin 30°)$

45. $\cos 300° + i \sin 300°$; $\cos 300° + i \sin 300°$

47. $6(\cos 2\pi + i \sin 2\pi)$; $\dfrac{2}{3}(\cos 0 + i \sin 0)$

49. $4\left(\cos \dfrac{29\pi}{12} + i \sin \dfrac{29\pi}{12}\right)$; $\cos \dfrac{11\pi}{12} + i \sin \dfrac{11\pi}{12}$

51. $27(\cos 90° + i \sin 90°)$ **53.** $(\cos 180° + i \sin 180°)$

55. $3{,}125(\cos 10° + i \sin 10°)$

57. $81(\cos \pi + i \sin \pi)$ **59.** $256(\cos 12 + i \sin 12)$

61. $64(\cos 10\pi + i \sin 10\pi)$

63. $4\sqrt{2}\left(\cos \dfrac{5\pi}{4} + i \sin \dfrac{5\pi}{4}\right)$

65. $1 + i\sqrt{3}$ **67.** $\dfrac{1}{2} + \dfrac{\sqrt{3}}{2}i$ **69.** $1 + i\sqrt{3}$

71. $1 + i\sqrt{3}, -2,$ and $1 - i\sqrt{3}$

73. $\dfrac{\sqrt{2}}{2} + \dfrac{\sqrt{2}}{2}i$ and $-\dfrac{\sqrt{2}}{2} - \dfrac{\sqrt{2}}{2}i$

75. $\sqrt{3} + i, -1 + i\sqrt{3}, -\sqrt{3} - i,$ and $1 - i\sqrt{3}$

77. $\cos 54° + i \sin 54°, \cos 126° + i \sin 126°,$
$\cos 198° + i \sin 198°, \cos 270° + i \sin 270°,$ and
$\cos 342° + i \sin 342°$

79. $2(\cos 10° + i \sin 10°), 2(\cos 130° + i \sin 130°),$ and
$2(\cos 250° + i \sin 250°)$

81. $z = 12 + 2i$ **83.** $|Z| = \sqrt{R^2 + X^2}$

Section 7.4

1. vector   **3.** geometric   **5.** $\sqrt{a^2 + b^2}$

7.

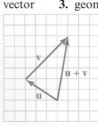

9.

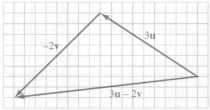

11.

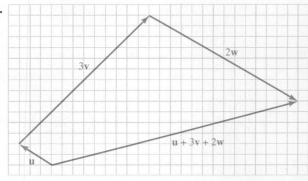

13.

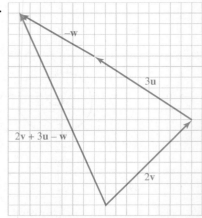

15. $\langle 0, 1 \rangle$ **17.** $\langle -4, 8 \rangle$ **19.** $\langle 9, 9 \rangle$ **21.** $\langle 8, 0 \rangle$

23. $\sqrt{10}$ **25.** 7   **27.** $\dfrac{\sqrt{61}}{15}$

29. a. $\langle 11, 10 \rangle$ **b.** $\langle 14, 11 \rangle$ **c.** $\langle 11, 13 \rangle$ **d.** $\langle 2, 3 \rangle$

31. a. $\langle 5, -4 \rangle$ **b.** $\langle 5, 5 \rangle$ **c.** $\langle -12, -7 \rangle$ **d.** $\langle 1, -7 \rangle$

33. a. $-3\mathbf{i} + 4\mathbf{j}$ **b.** $-12\mathbf{i} - 9\mathbf{j}$ **c.** $-13\mathbf{i} - 3\mathbf{j}$

d. $4\mathbf{i} + 9\mathbf{j}$

35. a. $7\mathbf{i} + 7\mathbf{j}$ **b.** $-3\mathbf{i} - 18\mathbf{j}$ **c.** $18\mathbf{i} + 15\mathbf{j}$

d. $9\mathbf{i} + 18\mathbf{j}$

37. $\sqrt{13}; \sqrt{2}; 5; \sqrt{10}; \sqrt{149}$

39. $4; \sqrt{13}; \sqrt{53}; \sqrt{37}; \sqrt{229}$

41. $\mathbf{u} = \dfrac{2\sqrt{13}}{13}\mathbf{i} - \dfrac{3\sqrt{13}}{13}\mathbf{j}$ **43.** $\mathbf{u} = \mathbf{j}$

45. $\mathbf{u} = -\dfrac{7\sqrt{85}}{85}\mathbf{i} + \dfrac{6\sqrt{85}}{85}\mathbf{j}$ **47.** $8\mathbf{i} + 8\mathbf{j}$

49. $5\sqrt{3}\mathbf{i} + 5\mathbf{j}$ **51.** $18.56\mathbf{i} + 14.09\mathbf{j}$

53. $|\mathbf{v}| = 2\sqrt{2}; \theta = 135°$ **55.** $|\mathbf{v}| = \sqrt{34}; \theta = 329.04°$

57. $|\mathbf{v}| = \sqrt{41}; \theta = 231.34°$ **59.** 0.602 **61.** 2.034

63. $40\sqrt{3}\mathbf{i} + 40\mathbf{j}$ **65.** $|\mathbf{F}| \approx 109.10$ lb; $\theta \approx 90.3°$

67. not quite; the magnitude is 109.07 lb

69. 209.65 lb   **71.** 104.50 lb; $44.6°$

73. left $= 902.31$ lb; right $= 911.39$ lb

75. 11.3 knots at N $70.5°$ W   **77.** N $10.4°$ W; 250.8 mph

79. 1,156.6 lb   **81.** 205.14 lb   **83.** 27°; 12.3 mph
85. 4.3°; 412.67 mph   **87.** 323.73 lb   **89.** 441 lb
91. $(\mathbf{F}_1)_y = 80$ lb; $(\mathbf{F}_1)_x = 95$ lb
93. N 79.9° E   **97.** $|\mathbf{r}| = 2|\mathbf{f}| \cos \dfrac{\theta}{2}$

Section 7.5

1. multiply, scalar   **3.** 0   **5.** $\text{proj}_v\mathbf{w} = \dfrac{\mathbf{w} \cdot \mathbf{v}}{\mathbf{v} \cdot \mathbf{v}}\mathbf{v}$

7. 9   **9.** 7   **11.** 0   **13.** -30 **15.** 5   **17.** -12
19. -39 **21.** $\sqrt{34}$ **23.** $\sqrt{10}$ **25.** 8   **27.** 45°
29. 150°   **31.** 36.87°   **33.** yes   **35.** no

37. yes   **39.** $\dfrac{7\sqrt{3}}{4}\mathbf{i} + \dfrac{7}{4}\mathbf{j}$ **41.** $\dfrac{7}{2}\mathbf{i} + \dfrac{7\sqrt{3}}{2}\mathbf{j}$

43. $3\sqrt{2}\mathbf{i} + 3\sqrt{2}\mathbf{j}$ **45. a.** $\dfrac{63}{13}$ **b.** $\left\langle \dfrac{315}{169}, \dfrac{756}{169} \right\rangle$

47. a. 0   **b.** 0   **49. a.** $\sqrt{2}$ **b.** $\langle 1, 1\rangle$
51. 52 ft-lb   **53.** 43 ft-lb   **55.** 450 ft-lb
57. 30 ft-lb   **59.** 720 mi   **61.** 100 lb

Chapter Review

1. a. $\left(5, \dfrac{\pi}{4}\right)$ **b.** $\left(5, -\dfrac{7\pi}{4}\right)$ **c.** $\left(-5, \dfrac{5\pi}{4}\right)$

2. a. $\left(6, \dfrac{2\pi}{3}\right)$ **b.** $\left(6, -\dfrac{4\pi}{3}\right)$ **c.** $\left(-6, \dfrac{5\pi}{3}\right)$

3. a. $\left(3, \dfrac{7\pi}{6}\right)$ **b.** $\left(3, -\dfrac{5\pi}{6}\right)$ **c.** $\left(-3, \dfrac{\pi}{6}\right)$

4. a. $\left(7, \dfrac{7\pi}{4}\right)$ **b.** $\left(7, -\dfrac{\pi}{4}\right)$ **c.** $\left(-7, \dfrac{3\pi}{4}\right)$

5. $\left(-2\sqrt{3}, -2\right)$ **6.** $\left(\dfrac{3\sqrt{2}}{2}, \dfrac{3\sqrt{2}}{2}\right)$ **7.** $(4\sqrt{3}, -4)$

8. $\left(\dfrac{5}{2}, -\dfrac{5\sqrt{3}}{2}\right)$ **9.** $\left(4\sqrt{2}, \dfrac{3\pi}{4}\right); \left(-4\sqrt{2}, \dfrac{7\pi}{4}\right)$
10. $(4, 0); (-4, \pi)$ **11.** $3r \sin \theta - 2r \cos \theta = 4$
12. $r = 7$ **13.** $7r^2 \sin \theta \cos \theta = 12$
14. $r \cos 2\theta = -3 \sin \theta$ **15.** $x^2 + y^2 + 2x + 2y = 0$
16. $x^2 - 6y = 9$ **17.** $2x + y = 2$ **18.** $x = 2$
19. 1.187   **20.** 2.050
21. $x^2 + y^2 = 9$

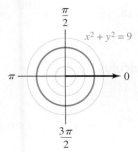

22. $y = x$

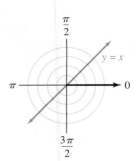

23. $x = -2$

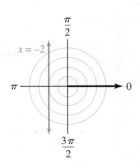

24. $x^2 + y^2 = 64$

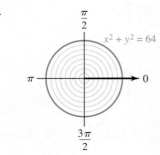

25.

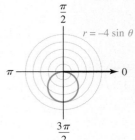

26.

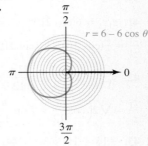

27.

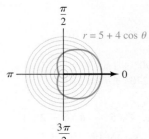

28.

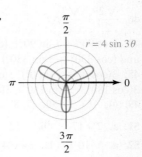

29.

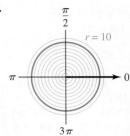

30.

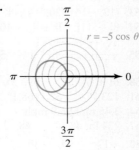

31. $r = 4 \cos \theta$ **32.** $r = 3 - 3 \cos \theta$ **33.** $r = 5 \sin 5\theta$
34. $r \sin \theta = -3$

35. $\sqrt{65}$

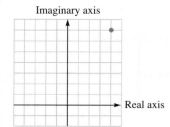

36. 9

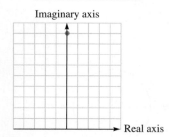

37. $5\sqrt{2}\left(\cos\dfrac{7\pi}{4} + i\sin\dfrac{7\pi}{4}\right)$ **38.** $9(\cos 0 + i\sin 0)$

39. $-2\sqrt{2} + 2i\sqrt{2}$ **40.** $4.99 + 0.29i$

41. $z_1 z_2 = 6(\cos 210° + i\sin 210°)$;

$\dfrac{z_1}{z_2} = \dfrac{2}{3}(\cos 210° + i\sin 210°)$

42. $z_1 z_2 = 4\sqrt{2}\left(\cos\dfrac{41\pi}{12} + i\sin\dfrac{41\pi}{12}\right)$;

$\dfrac{z_1}{z_2} = \sqrt{2}\left(\cos\dfrac{\pi}{12} + i\sin\dfrac{\pi}{12}\right)$

43. $-16 - 16i\sqrt{3}$ **44.** $-7 + 24i$ **45.** $2i$

46. $\dfrac{\sqrt{2}}{2} + \dfrac{\sqrt{2}}{2}i$

47. $(\cos 90° + i\sin 90°)$; $(\cos 210° + i\sin 210°)$;
and $(\cos 330° + i\sin 330°)$

48. $3(\cos 0° + i\sin 0°)$; $3(\cos 90° + i\sin 90°)$;
$3(\cos 180° + i\sin 180°)$; $3(\cos 270° + i\sin 270°)$

49. $\langle 4, -6\rangle$ **50.** $\langle 7, 8\rangle$

51. a. $2\sqrt{17}$ **b.** -12 **c.** $4\sqrt{5}$ **d.** $6 - 2\sqrt{17}$

52. a. $\sqrt{10}$ **b.** $-4\sqrt{41}$ **c.** $5\sqrt{10}$
d. $-\sqrt{10} + 2\sqrt{41}$

53. $\langle 10, -15\rangle$; $\langle 20, -4\rangle$; $\langle 4, 19\rangle$; $\langle 29, -10\rangle$

54. $8\mathbf{i} - 11\mathbf{j}$; $10\mathbf{i} + 4\mathbf{j}$; $-10\mathbf{i} + 29\mathbf{j}$; $16\mathbf{i} + 1\mathbf{j}$

55. $\dfrac{3\sqrt{13}}{13}\mathbf{i} - \dfrac{2\sqrt{13}}{13}\mathbf{j}$ **56.** $-\mathbf{j}$ **57.** $\mathbf{i} + 4\mathbf{j}$

58. $6\mathbf{i} + 6\sqrt{3}\mathbf{j}$ **59.** $5\sqrt{2}$; $315°$ **60.** $5\sqrt{2}$; $278.13°$

61. $123.7°$ **62.** $168.1°$ **63.** $18.2°$

64. 3.46 mph; 6.9 mph   **65.** $63.4°$; $54.5°$ W   **66.** no

67. 224 lb; $48.1°$ **68.** 260.6 kph; S $4.5°$ W   **69.** -8

70. -17 **71.** -30 **72.** -14 **73.** $2\sqrt{5}$

74. $\sqrt{34}$ **75.** $135°$ **76.** $171.87°$ **77.** yes   **78.** no

79. $4\sqrt{2}\mathbf{i} + 4\sqrt{2}\mathbf{j}$ **80.** $-\dfrac{3\sqrt{3}}{2}\mathbf{i} - \dfrac{3}{2}\mathbf{j}$

81. a. $-\dfrac{\sqrt{10}}{10}$ **b.** $\left\langle\dfrac{3}{10}, -\dfrac{1}{10}\right\rangle$ **82. a.** -1 **b.** $-\mathbf{j}$

83. 48 ft-lb   **84.** 475 ft-lb   **85.** 100 lb   **86.** 300 ft-lb

Chapter Test

1. $\left(-\dfrac{5\sqrt{2}}{2}, \dfrac{5\sqrt{2}}{2}\right)$ **2.** $\left(4, \dfrac{5\pi}{6}\right)$; $\left(-4, \dfrac{11\pi}{6}\right)$

3. $r = 8\tan\theta\sec\theta$ **4.** $x^2 + y^2 - 4x - 3y = 0$

5. 4.335

6. a.

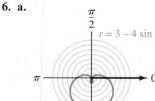

b.

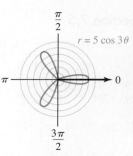

c.

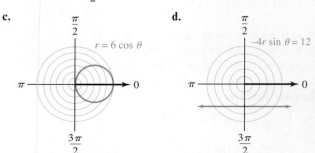

d.

7. a. $r = 4\cos 2\theta$ **b.** $r = 2 - 3\sin\theta$ **c.** $r = 5$

8. $-3\sqrt{3} - 3i$ **9.** $2(\cos 240° + i\sin 240°)$

10. $z_1 = 8(\cos 330° + i\sin 330°)$; $z_2 = 8\cos 90° + i\sin 90°$;

$z_1 z_2 = 32 + 32i\sqrt{3}$; $\dfrac{z_1}{z_2} = -\dfrac{1}{2} - \dfrac{\sqrt{3}}{2}i$

11. $-512\sqrt{3} + 512i$

12. $3\left(\cos\dfrac{3\pi}{8} + i\sin\dfrac{3\pi}{8}\right)$, $3\left(\cos\dfrac{7\pi}{8} + i\sin\dfrac{7\pi}{8}\right)$,

$3\left(\cos\dfrac{11\pi}{8} + i\sin\dfrac{11\pi}{8}\right)$, and $3\left(\cos\dfrac{15\pi}{8} + i\sin\dfrac{15\pi}{8}\right)$

13. a. 12 **b.** $-3\mathbf{i} + 13\mathbf{j}$ **c.** $22\mathbf{i} - 13\mathbf{j}$ **d.** -52

14. $|\mathbf{v}| = 2$; $300°$ **15.** $180°$

16. a. $425\mathbf{i} + 40\mathbf{j}$ **b.** 427 mph at N $84.6°$ E

17. 293 ft-lb   **18.** yes   **19. a.** $\dfrac{14}{5}$ **b.** $-\dfrac{42}{25}\mathbf{i} + \dfrac{56}{25}\mathbf{j}$

20. 302 ft-lb

Connect to Calculus

1. a. $\left\langle\dfrac{\sqrt{2}}{2}, \sqrt{2}\right\rangle$ **b.** $\langle e^4, 2\rangle$

2. a. $12t^2\mathbf{i} - \dfrac{1}{2}t^{-1/2}\mathbf{j}$ **b.** $\mathbf{r}'(t) = 12\cos 4t\mathbf{i} - 4\sin t\mathbf{j}$

Section 8.1

1. system   **3.** consistent   **5.** independent
7. consistent   **9.** dependent   **11.** is

13.

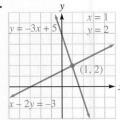

15.

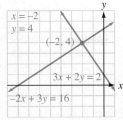

17. inconsistent system   **19.** dependent equations

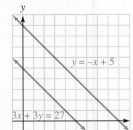

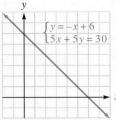

21. $(2.2, -4.7)$ **23.** $(1.7, 0.3)$ **25.** $(-1, -2)$

27. $(3, -2)$ **29.** $\left(\dfrac{1}{2}, \dfrac{1}{3}\right)$

31. no solution; inconsistent system

33. dependent equations; $(x, 3x - 6)$ **35.** $(3, 1)$

37. $(3, 2)$ **39.** $(-3, 0)$ **41.** $\left(1, -\dfrac{1}{2}\right)$

43. dependent equations; $(x, 5 - 2x)$

45. no solution; inconsistent system   **47.** $(4, -7)$

49. $(2, -3)$ **51.** $(9, -1)$ **53.** $(1, 2, 0)$

55. $\left(0, -\dfrac{1}{3}, -\dfrac{1}{3}\right)$ **57.** $(1, 2, -1)$ **59.** $(1, 0, 5)$

61. no solution; inconsistent system   **63.** $(0, 1, 0)$

65. $\left(\dfrac{2}{3}, \dfrac{1}{4}, \dfrac{1}{2}\right)$ **67.** $\left(\dfrac{1}{2}, \dfrac{1}{2}, \dfrac{1}{2}\right)$

69. dependent equations; $(x, 2 - x, 1)$

71. hamburger: \$2; fries: \$1

73. 225 acres of corn; 125 acres of soybeans

75. 210,000 cm^2 **77.** 40 g of 9% copper and 20 g of 20% copper   **79.** 10 ft

81. $E(x) = 43.53x + 742.72$; $R(x) = 89.95x$; 16 pairs per day

83. 15 hr cooking hamburgers; 10 hr pumping gas; 5 hr janitorial

85. 0–14: 1.05 million; 15–49: 1.56 million; 50+: 0.39 million

87. 30°; 50°; 100°

Section 8.2

1. matrix   **3.** coefficien   **5.** equation

7. row equivalent   **9.** interchanged

11. adding, multiple   **13.** $\left(\dfrac{13}{3}, \dfrac{8}{3}\right)$ **15.** $(7, 6)$

17. $(1, 2, 3)$ **19.** $(1, 1, 3)$ **21.** row echelon form

23. reduced row echelon form   **25.** $(2, -1)$

27. $(-2, 0)$ **29.** $(3, 1)$

31. no solution; inconsistent system

33. $(0, -7)$ **35.** $(1, 0, 2)$ **37.** $(2, -2, 1)$

39. $(1, 1, 2)$

41. dependent equations; $\left(x, -x + \dfrac{7}{2}, -\dfrac{3}{2}\right)$

43. no solution; inconsistent system   **45.** $(13, 3)$

47. $(-13, 7, -2)$ **49.** $(10, 3)$ **51.** $(0, 0)$

53. $(3, 1, -2)$ **55.** $\left(\dfrac{1}{4}, 1, \dfrac{1}{4}\right)$ **57.** $(1, 2, 1, 1)$

59. $(1, 2, 0, 1)$ **61.** $\left(\dfrac{9}{4}, -3, \dfrac{3}{4}\right)$ **63.** $\left(0, \dfrac{20}{3}, -\dfrac{1}{3}\right)$

65. dependent equations; $\left(x, \dfrac{1}{2}x - 3, 0\right)$

67. $(1, -3)$

69. dependent equations; $\left(\dfrac{8}{7} + \dfrac{1}{7}z, \dfrac{10}{7} - \dfrac{4}{7}z, z\right)$

71. dependent equations; $(1 + z, -z, -1 - z, z)$

73. no solution; inconsistent system

75. 1,300 mi

77. dictionary: 4.5 in; atlas: 3.5 in; thesaurus: 4 in

79. 2 oz of Food A; 4 oz of Food B; 6 oz of Food C

85. $(\pm 2, \pm 1, \pm 3)$

Section 8.3

1. i, j **3.** corresponding   **5.** columns, rows

7. additive identity   **9.** $x = 2$; $y = 5$

11. $x = 1$; $y = 2$ **13.** $\begin{bmatrix} -1 & 2 & 1 \\ -6 & 0 & 0 \end{bmatrix}$

15. $\begin{bmatrix} -5 & 2 & -7 \\ 5 & 0 & -3 \\ 2 & -3 & 5 \end{bmatrix}$ **17.** $\begin{bmatrix} -6 & 5 & 0 \\ 1 & -1 & 0 \end{bmatrix}$

19. $\begin{bmatrix} 15 & -15 \\ 0 & -10 \end{bmatrix}$ **21.** $\begin{bmatrix} 25 & 75 & -10 \\ -10 & -25 & 5 \end{bmatrix}$

23. $\begin{bmatrix} 18 & -1 & -4 \\ -35 & 0 & -1 \end{bmatrix}$ **25.** $\begin{bmatrix} 2 & -2 \\ 3 & 10 \end{bmatrix}$

27. $\begin{bmatrix} -22 & -22 \\ -105 & 126 \end{bmatrix}$ **29.** $\begin{bmatrix} 4 & 2 & 10 \\ 5 & -2 & 4 \\ 2 & -2 & 1 \end{bmatrix}$

31. $\begin{bmatrix} 4 & -5 & -6 \\ -8 & 10 & 12 \\ -12 & 15 & 18 \end{bmatrix}$ **33.** not possible   **35.** $\begin{bmatrix} 16 \\ 12 \\ 12 \end{bmatrix}$

37. not possible   **39.** $\begin{bmatrix} -36.29 \\ 16.2 \\ -19.26 \end{bmatrix}$

41. $\begin{bmatrix} -16.11 & 4.71 & 33.64 \\ -19.6 & 20.35 & 6.4 \\ -100.6 & 72.82 & 62.71 \end{bmatrix}$ **47.** $\begin{bmatrix} 4 & 5 \\ -7 & -1 \end{bmatrix}$

49. not possible   **51.** $\begin{bmatrix} 24 & 16 \\ 39 & 26 \end{bmatrix}$ **53.** $\begin{bmatrix} 47 \\ 81 \end{bmatrix}$

55. $QC = \begin{bmatrix} 2{,}000 \\ 1{,}700 \end{bmatrix}$ The cost to Supplier 1 is \$2,000. The cost to Supplier 2 is \$1,700.

57. $QP = \begin{bmatrix} 584.50 \\ 709.25 \\ 1{,}036.75 \end{bmatrix}$ Adult males spent \$584.50. Adult females spent \$709.25. Children spent \$1,036.75.

59. $\begin{bmatrix} 1 & 1 & 0 \\ 0 & 1 & 1 \\ 1 & 0 & 0 \end{bmatrix}$

61. $A^2 = \begin{bmatrix} 5 & 1 & 2 & 2 \\ 1 & 5 & 2 & 2 \\ 2 & 2 & 6 & 0 \\ 2 & 2 & 0 & 4 \end{bmatrix}$ A^2 indicates the number of ways two cities can be linked with exactly one intermediate city to relay messages.

63. No; if $A = \begin{bmatrix} 1 & 1 \\ 1 & 1 \end{bmatrix}$ and $B = \begin{bmatrix} 1 & 0 \\ 0 & 0 \end{bmatrix}$, then $(AB)^2 \neq A^2 B^2$.

65. Let $A = \begin{bmatrix} 1 & 2 \\ 1 & 2 \end{bmatrix}$ and $B = \begin{bmatrix} 2 & 2 \\ -1 & -1 \end{bmatrix}$; neither is the zero matrix, yet $AB = 0$.

Section 8.4
1. $AB = BA = I$ **3.** $[I \mid A^{-1}]$

5. $\begin{bmatrix} 3 & 4 \\ 2 & 3 \end{bmatrix}$ **7.** $\begin{bmatrix} 5 & -7 \\ -2 & 3 \end{bmatrix}$ **9.** $\begin{bmatrix} -2 & 3 & -3 \\ -5 & 7 & -6 \\ 1 & -1 & 1 \end{bmatrix}$

11. $\begin{bmatrix} 4 & 1 & -3 \\ -5 & -1 & 4 \\ -1 & -1 & 1 \end{bmatrix}$ **13.** not invertible

15. $\begin{bmatrix} 1 & -2 & 1 \\ 0 & 1 & -2 \\ 0 & 0 & 1 \end{bmatrix}$ **17.** not invertible

19. $\begin{bmatrix} 1 & -2 & 1 & 0 \\ 0 & 1 & -2 & 1 \\ 0 & 0 & 1 & -2 \\ 0 & 0 & 0 & 1 \end{bmatrix}$ **21.** $\begin{bmatrix} 8 & -2 & -6 \\ -5 & 2 & 4 \\ 2 & 0 & -2 \end{bmatrix}$

23. $\begin{bmatrix} -2.5 & 5 & 3 & 5.5 \\ 5.5 & -8 & -6 & -9.5 \\ -1 & 3 & 1 & 3 \\ -5.5 & 9 & 6 & 10.5 \end{bmatrix}$ **25.** $x = 23; y = 17$

27. $x = 0; y = 0$ **29.** $x = 1; y = 2; z = 2$
31. $x = 54; y = -37; z = -49$
33. $x = 2; y = 1$ **35.** $x = 1; y = 2; z = 1$
37. two of model A; three of model B   **39.** hi

41. no   **45.** $X = \begin{bmatrix} 0 \\ 0 \\ 0 \end{bmatrix}$

Section 8.5
1. $|A|$, det(A) **3.** 0   **5.** 0   **7.** 8   **9.** 1

11. $\begin{vmatrix} -2 & 3 \\ 8 & 9 \end{vmatrix} = -42$ **13.** $\begin{vmatrix} 1 & -2 \\ 4 & 5 \end{vmatrix} = 13$

15. $-\begin{vmatrix} -2 & 3 \\ 8 & 9 \end{vmatrix} = 42$ **17.** $\begin{vmatrix} 1 & -2 \\ 4 & 5 \end{vmatrix} = 13$ **19.** -54

21. -7 **23.** 86   **25.** -2 **27.** 2   **29.** 12
31. true   **33.** false   **35.** 3   **37.** 3   **39.** (1, 2)
41. (3, 0)   **43.** (1, 0, 1)   **45.** $(1, -1, 2)$

47. (6, 6, 12)   **49.** $\left(\dfrac{5}{6}, \dfrac{2}{3}, \dfrac{1}{2}, \dfrac{5}{2}\right)$ **55.** 8   **57.** -1

59. $5,000 in HiTech; $8,000 in SaveTel; $7,000 in OilCo
61. 10   **63.** 24   **65.** 8
69. $D = n \times n$ matrices; R = all real numbers   **71.** yes
73. 21.468

Section 8.6
1. first-degree, second-degree   **3.** $\dfrac{1}{x} + \dfrac{2}{x-1}$

5. $\dfrac{5}{x} - \dfrac{3}{x-3}$ **7.** $\dfrac{1}{x+1} + \dfrac{2}{x-1}$

9. $\dfrac{2}{x} - \dfrac{2}{x-2}$ **11.** $\dfrac{1}{x-3} - \dfrac{3}{x+2}$

13. $\dfrac{8}{x+3} - \dfrac{5}{x-1}$ **15.** $\dfrac{5}{2x-3} + \dfrac{2}{x-5}$

17. $\dfrac{2}{x} + \dfrac{3}{x-1} - \dfrac{1}{x+1}$ **19.** $\dfrac{1}{x} + \dfrac{1}{x^2+3}$

21. $\dfrac{3}{x+1} + \dfrac{2}{x^2+2x+3}$ **23.** $\dfrac{3}{x} + \dfrac{2}{x+1} + \dfrac{1}{(x+1)^2}$

25. $\dfrac{1}{x} + \dfrac{2}{x^2} - \dfrac{3}{x-1}$ **27.** $\dfrac{2}{x} + \dfrac{1}{x-3} + \dfrac{2}{(x-3)^2}$

29. $\dfrac{1}{x-1} - \dfrac{4}{(x-1)^3}$ **31.** $\dfrac{1}{x} + \dfrac{1}{x^2} + \dfrac{2}{x^2+x+1}$

33. $\dfrac{3}{x} + \dfrac{4}{x^2} + \dfrac{x+1}{x^2+1}$ **35.** $-\dfrac{1}{x+1} - \dfrac{3}{x^2+2}$

37. $\dfrac{x+1}{x^2+2} + \dfrac{2}{x^2+x+2}$

39. $\dfrac{1}{x} + \dfrac{x}{x^2+2x+5} + \dfrac{x+2}{(x^2+2x+5)^2}$

41. $x - 3 - \dfrac{1}{x+1} + \dfrac{8}{x+2}$ **43.** $1 + \dfrac{2}{3x+1} + \dfrac{1}{x^2+1}$

45. $1 + \dfrac{1}{x} + \dfrac{x}{x^2+x+1}$

47. $2 + \dfrac{1}{x} + \dfrac{3}{x-1} + \dfrac{2}{x^2+1}$
49. No, it's the sum of two cubes.

Section 8.7
1. half-plane, boundary   **3.** is not
5.   **7.**

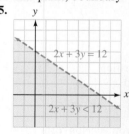

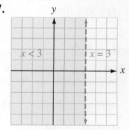

9.

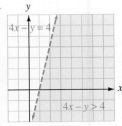

11.

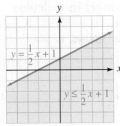

13.

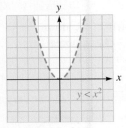

15.

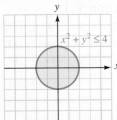

17.

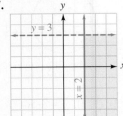

19.

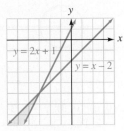

21.

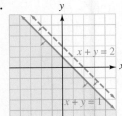

23.

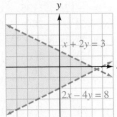

25.

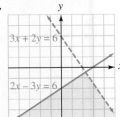

27.

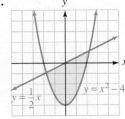

29.

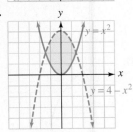

31.

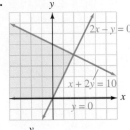

33.

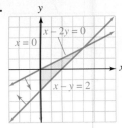

35.

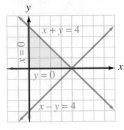

37. $\begin{cases} 6s + 4l \le 60 \\ s \ge 0 \\ l \ge 0 \end{cases}$

Chapter Review

1.

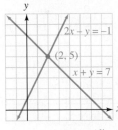

2.

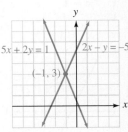

3.

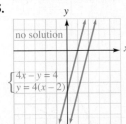

4.

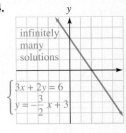

5.

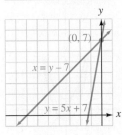

6. $(2, -1)$ **7.** $(0, -3)$ **8.** $(1, 1)$

9. infinitely many solutions; $(x, 3x - 4)$

10. no solution   **11.** $(-3, 2)$ **12.** $(-2, 5)$

13. $(2, -1)$ **14.** no solution

15. infinitely many solutions; $(x, 3x - 4)$

16. $(1, 0, 1)$ **17.** $(1, 1, -1)$ **18.** $(0, 1, 2)$

19. \$10,400

20. 900 adult tickets; 450 seniors' tickets; 450 children's tickets

21. $(1, 1)$

22. infinitely many solutions; $(x, 3x + 4)$

23. $(3, 1, -2)$ **24.** $(-10, 1, 10)$ **25.** no solution

26. $(-4, 3)$ **27.** $\begin{bmatrix} 1 & 3 & 4 \\ 4 & 0 & 2 \end{bmatrix}$ **28.** $\begin{bmatrix} 2 & 5 & 4 \\ -2 & -6 & 6 \\ -4 & 5 & -3 \end{bmatrix}$

29. $\begin{bmatrix} 4 & -1 \\ -7 & -7 \end{bmatrix}$ **30.** $\begin{bmatrix} -17 & 19 \\ 10 & -12 \end{bmatrix}$ **31.** $[5]$

32. $\begin{bmatrix} 2 & -1 & 1 & 3 \\ 4 & -2 & 2 & 6 \\ 2 & -1 & 1 & 3 \\ 10 & -5 & 5 & 15 \end{bmatrix}$ **33.** not possible

34. $[-24]$ **35.** $\begin{bmatrix} 0 \\ -6 \end{bmatrix}$ **36.** $\begin{bmatrix} 5 & -3 \\ -3 & 2 \end{bmatrix}$

37. $\begin{bmatrix} 1 & 0 & 0 \\ -\dfrac{3}{2} & \dfrac{1}{2} & \dfrac{1}{2} \\ 1 & -\dfrac{1}{2} & 0 \end{bmatrix}$ **38.** $\begin{bmatrix} 9 & 16 & -56 \\ -3 & -5 & 18 \\ -1 & -2 & 7 \end{bmatrix}$

39. no inverse   **40.** no inverse   **41.** $(1, 2, -1)$
42. $(1, 1, 0, -1)$ **43.** -7 **44.** -6 **45.** 3
46. -25 **47.** $(1, -2)$ **48.** $(1, 0, -2)$ **49.** $(1, -1, 3)$
50. $(1, 0, -1, 2)$ **51.** 21   **52.** 7
53. $\dfrac{3}{x} + \dfrac{4}{x + 1}$ **54.** $\dfrac{3}{x} + \dfrac{2}{x^2} + \dfrac{x - 1}{x^2 + 1}$
55. $\dfrac{1}{x} - \dfrac{1}{x^2 + x + 5}$ **56.** $\dfrac{1}{x + 1} - \dfrac{2}{(x + 1)^2} + \dfrac{2}{(x + 1)^3}$
57.

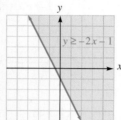

58.

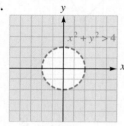

59.

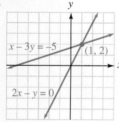

60.

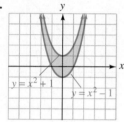

Chapter Test

1.

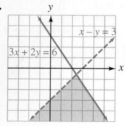

2.

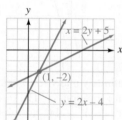

3. $(1, -3)$ **4.** $(3, 5)$
5. six L of 20% solution; four L of 45% solution
6. CD World: 100 units; Ace: 25 units; Hi-Fi: 50 units
7. $(2, 1)$ **8.** $(1, 2, 1)$ **9.** $(1, 0, -2)$
10. $\left(-\dfrac{2}{5}y + \dfrac{7}{5}, y, -\dfrac{8}{5}y - \dfrac{7}{5}\right)$ **11.** $\begin{bmatrix} 16 & -14 & 20 \\ 0 & -6 & -13 \end{bmatrix}$
12. $[-1]$ **13.** $\begin{bmatrix} -\dfrac{7}{3} & \dfrac{19}{3} \\ \dfrac{2}{3} & -\dfrac{5}{3} \end{bmatrix}$ **14.** $\begin{bmatrix} -13 & -3 & 14 \\ 4 & 1 & -4 \\ 12 & 3 & -13 \end{bmatrix}$
15. $\left(\dfrac{17}{3}, -\dfrac{4}{3}\right)$ **16.** $(-36, 11, 34)$ **17.** -12 **18.** -24
19. $-\dfrac{5}{4}$ **20.** 1   **21.** $\dfrac{3}{2x - 3} + \dfrac{1}{x + 1}$
22. $\dfrac{1}{x} + \dfrac{2x + 1}{x^2 + 2}$
23.

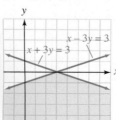

24.

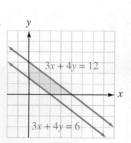

Connect to Calculus

1. a. $\langle 3, -4, 2 \rangle$ **b.** $-2\mathbf{i} + 2\mathbf{k}$
2. a. no   **b.** yes
3. $\displaystyle\int\left(\dfrac{2}{x + 5} - \dfrac{1}{x - 2}\right)dx$
4. $\displaystyle\int\left(\dfrac{2}{x} - \dfrac{1}{x^2} - \dfrac{2}{x + 1}\right)dx$

Section 9.1

1. $2, -5, 3$ **3.** $0, 0, \sqrt{5}$ **5.** left   **7.** down
9. directrix, focus   **11.** circle   **13.** parabola
15. $x^2 + y^2 = 49$ **17.** $(x - 2)^2 + (y + 2)^2 = 17$
19. $(x - 1)^2 + (y + 2)^2 = 36$
21.

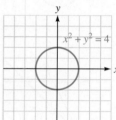

23.

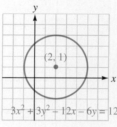

25. $(0, 0); (0, 3); y = -3$ **27.** $(0, 3); (5, 3); x = -5$
29. $(-2, 1); (-2, -5); y = 7$ **31.** $x^2 = 12y$
33. $y^2 = -12x$ **35.** $(x - 3)^2 = -12(y - 5)$
37. $(x - 3)^2 = -28(y - 5)$ **39.** $x^2 = -4(y - 2)$
41. $(y + 5)^2 = 8(x - 1)$
43. $(y - 2)^2 = -2(x - 2)$ or $(x - 2)^2 = -2(y - 2)$
45. $(x + 4)^2 = -\dfrac{16}{3}(y - 6)$ or $(y - 6)^2 = \dfrac{9}{4}(x + 4)$
47. $(y - 8)^2 = -4(x - 6)$ **49.** $(x - 3)^2 = \dfrac{1}{2}(y - 1)$
51.

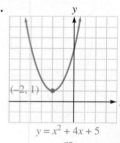

$y = x^2 + 4x + 5$
or
$y - 1 = (x + 2)^2$

53.

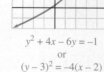

$y^2 + 4x - 6y = -1$
or
$(y - 3)^2 = -4(x - 2)$

55.

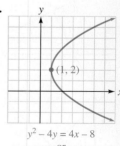

$y^2 - 4y = 4x - 8$
or
$(y - 2)^2 = 4(x - 1)$

57.

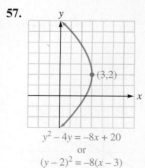

$y^2 - 4y = -8x + 20$
or
$(y - 2)^2 = -8(x - 3)$

59.

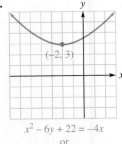

$x^2 - 6y + 22 = -4x$
or
$(x + 2)^2 = 6(y - 3)$

61.

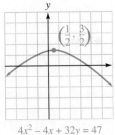

$4x^2 - 4x + 32y = 47$
or
$\left(x - \dfrac{1}{2}\right)^2 = -8\left(y - \dfrac{3}{2}\right)$

63. yes   **65.** 60 mi   **67.** $(x - 4)^2 + y^2 = 16$

69. $(x - 7)^2 + y^2 = 9$ **71.** 2 ft   **73.** $x^2 = -\dfrac{45}{2}y$

75. 1 ft   **77.** 300 ft   **79.** about 12.6 cm

81. about 520 ft   **83.** $0x^2 + 0xy + y^2 - 8x - 4y + 12 = 0$

85. $x^2 + (y - 3)^2 = 25$ **87.** $y = x^2 + 4x + 3$

Section 9.2

1. sum, constant   **3.** vertices   **5.** $a, 0, -a, 0$

7. 26 in string; thumbtacks 24 in apart   **9.** circle

11. parabola   **13.** ellipse   **15.** $\dfrac{x^2}{16} + \dfrac{y^2}{9} = 1$

17. $\dfrac{x^2}{25} + \dfrac{y^2}{16} = 1$ **19.** $\dfrac{9x^2}{16} + \dfrac{9y^2}{25} = 1$

21. $\dfrac{x^2}{7} + \dfrac{y^2}{16} = 1$ **23.** $\dfrac{(x - 3)^2}{4} + \dfrac{(y - 4)^2}{9} = 1$

25. $\dfrac{(x - 3)^2}{9} + \dfrac{(y - 4)^2}{4} = 1$

27. $\dfrac{(x - 3)^2}{41} + \dfrac{(y - 4)^2}{16} = 1$

29. $\dfrac{x^2}{36} + \dfrac{(y - 4)^2}{20} = 1$ **31.** $\dfrac{x^2}{100} + \dfrac{y^2}{64} = 1$

33.

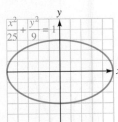

35.

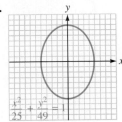

37.

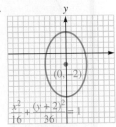

39.

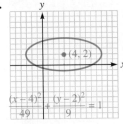

41. $\dfrac{x^2}{4} + \dfrac{(y - 1)^2}{16} = 1$ **43.** $\dfrac{(x + 1)^2}{4} + \dfrac{(y + 2)^2}{9} = 1$

45.

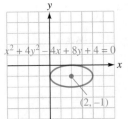

47.

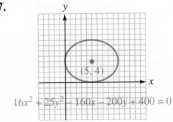

49. $\dfrac{x^2}{900} + \dfrac{y^2}{400} = 1$

51. $\dfrac{x^2}{2,500} + \dfrac{y^2}{900} = 1$; focal width: 36 m

53. about 20.8 in   **55.** 199,395 mi

61. $\dfrac{x^2}{9} + \dfrac{y^2}{8} = 1$ **65.** 4

Section 9.3

1. difference, constant   **3.** $a, 0, -a, 0$ **5.** transverse axis

7. circle   **9.** parabola   **11.** ellipse   **13.** hyperbola

15. $\dfrac{x^2}{25} - \dfrac{y^2}{24} = 1$ **17.** $\dfrac{(x - 2)^2}{4} - \dfrac{(y - 4)^2}{9} = 1$

19. $\dfrac{(y - 3)^2}{9} - \dfrac{(x - 5)^2}{9} = 1$ **21.** $\dfrac{y^2}{9} - \dfrac{x^2}{16} = 1$

23. $\dfrac{(x - 1)^2}{4} - \dfrac{(y - 4)^2}{32} = 1$ **25.** $\dfrac{x^2}{10} - \dfrac{3y^2}{20} = 1$

27. 24 sq. units   **29.** 12 sq. units

31. $\dfrac{(x + 2)^2}{4} - \dfrac{4(y + 4)^2}{81} = 1$ or $\dfrac{(y + 4)^2}{4} - \dfrac{4(x + 2)^2}{81} = 1$

33. $\dfrac{x^2}{36} - \dfrac{16y^2}{25} = 1$

35.

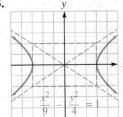

37.

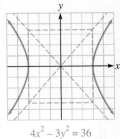

$4x^2 - 3y^2 = 36$

39.

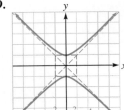

41.

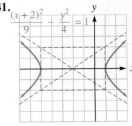

43.

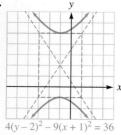

$4(y - 2)^2 - 9(x + 1)^2 = 36$

45.

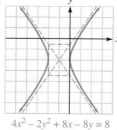

$4x^2 - 2y^2 + 8x - 8y = 8$

47.

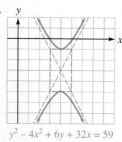

$y^2 - 4x^2 + 6y + 32x = 59$

49.

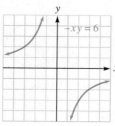

$-xy = 6$

51. $\dfrac{(x - 3)^2}{9} - \dfrac{(y - 1)^2}{16} = 1$

53. $4x^2 - 5y^2 - 60y = 0$ **55.** 24   **57.** 3 units

59. hyperbola; $\dfrac{x^2}{144} - \dfrac{y^2}{25} = 1$

61. hyperbola; $\dfrac{x^2}{36} - \dfrac{y^2}{64} = 1$

Section 9.4

1. graphs

3.

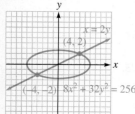

$x = 2y$ (4, 2) (-4, -2) $8x^2 + 32y^2 = 256$

5.

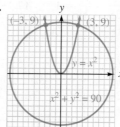

(-3, 9) (3, 9) $y = x^2$ $x^2 + y^2 = 90$

7.

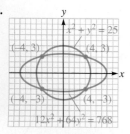

$x^2 + y^2 = 25$ (-4, 3) (4, 3) (-4, -3) (4, -3) $12x^2 + 64y^2 = 768$

9.

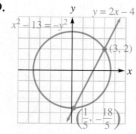

$x^2 - 13 = -y^2$ $y = 2x - 4$ (3, 2) $\left(\dfrac{1}{5}, -\dfrac{18}{5}\right)$

11.

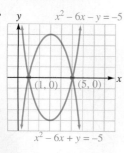

$x^2 - 6x - y = -5$ (1, 0) (5, 0) $x^2 - 6x + y = -5$

13. (1, 2) and (−1, 0)   **15.** (1, 0.67) and (−1, −0.67)
17. (3, 0) and (0, 5)   **19.** (1, 1)   **21.** (1, 2) and (2, 1)
23. (−2, 3) and (2, 3)   **25.** $\left(\sqrt{5}, 5\right)$ and $\left(-\sqrt{5}, 5\right)$
27. $\left(\sqrt{3}, 0\right)$ and $\left(-\sqrt{3}, 0\right)$
29. (2, 4), (2, −4), (−2, 4), and (−2, −4)
31. (−2, −6), (2, −6), $\left(-\sqrt{15}, 5\right)$, and $\left(\sqrt{15}, 5\right)$
33. (0, −4), (−3, 5), and (3, 5)
35. (−2, 3), (2, 3), (−2, −3), and (2, −3)   **37.** (3, 3)
39. (6, 2), (−6, −2), $\left(\sqrt{42}, 0\right)$, and $\left(-\sqrt{42}, 0\right)$
41. $\left(\dfrac{1}{2}, \dfrac{1}{3}\right)$ and $\left(\dfrac{1}{3}, \dfrac{1}{2}\right)$ **43.** 7 cm by 9 cm
45. 80 ft by 100 ft or 50 ft by 160 ft
47. \$750 at 9% or \$900 at 7.5%   **49.** (30, 3)
51. yes, at (−2, 4) and (1, 1)   **53.** about 23 mi

Section 9.5

1. plane curve, parametric, t

3. $(v_0 \cos \theta)t, \ -\dfrac{1}{2}gt^2 + (v_0 \sin \theta)t + h$

5.

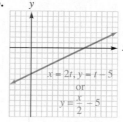

$x = 2t, y = t - 5$ or $y = \dfrac{x}{2} - 5$

7.

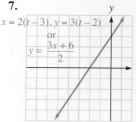

$x = 2(t - 3), y = 3(t - 2)$ or $y = \dfrac{3x + 6}{2}$

9.

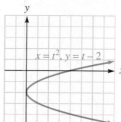

$x = t^2, y = t - 2$

11.

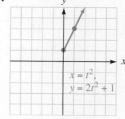

$x = t^2, y = 2t^2 + 1$

13.

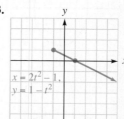

$x = 2t^2 - 1, y = 1 - t^2$

15.

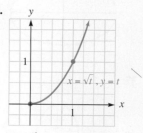

$x = \sqrt{t}, y = t$

17.

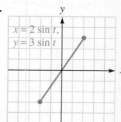

$x = 2 \sin t, y = 3 \sin t$

19. $y = \dfrac{1}{2}x + 6$

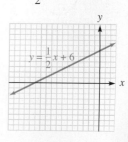

$y = \dfrac{1}{2}x + 6$

21. $x = (y - 3)^2$

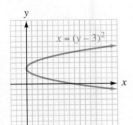

23. $y = x^2 - 3, x \geq 0$

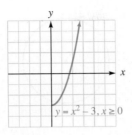

25. $y = \left|\frac{1}{2}x - 2\right|$

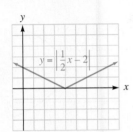

27. $y = \frac{x}{3} + 1, 3 \leq x \leq 3e^2$

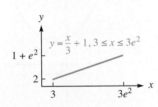

29. $y = \frac{e^{2x}}{2}, x \leq \ln 4$

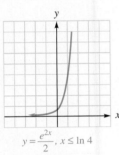

31. $\frac{x^2}{9} + \frac{y^2}{4} = 1$

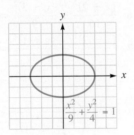

33. $\frac{x^2}{9} + \frac{y^2}{25} = 1$

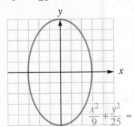

35. $xy = 1, 0 < x \leq 1$ and $y \geq 1$

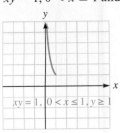

37. $x = y^3 - y, -2 \leq y \leq 2$

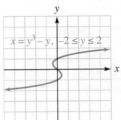

39. $x = t$ and $y = 3t + 7$; $x = t + 1$ and $y = 3t + 10$

41. $x = t$ and $y = \frac{1}{t}, t \neq 0$; $x = t + 1$ and $y = \frac{1}{t + 1}, t \neq -1$

43. $x = t$ and $y = 2t^2 - 3$; $x = t + 1$ and $y = 2t^2 + 4t - 1$

45. $x = t$ and $y = t^{2/5}, t \geq 0$; $x = t^5, y = t^2, t \geq 0$

47. a. $x = 45t$ and $y = -16t^2 + 45t\sqrt{3} + 3$ **b.** 4.91 sec

c. 97.92 ft at 2.44 sec   **d.** 220.95 ft

e. Window t:[0, 5, 0.1], X:[0, 300], Y:[0, 100]

49. a. $x = (66 \cos 65°)t$ and $y = -16t^2 + (66 \sin 65°)t + 7$

b. 3.77 sec   **c.** 105.2 ft $\approx$ 35 yd

51. a. 19.8 ft   **b.** 39.5 ft

53. a. $x_1 = 65t$ and $y_1 = 1$; $x_2 = 70(t - 4)$ and $y_2 = 2$

b. 2:22 p.m.   **c.** Shelly traveled about 544 mi;

Clara traveled about 306 mi

55. $x = a \cos t$ and $y = a \sin t$

57. $x = a \cos t$ and $y = b \sin t$

59. $y = x \tan \theta - \frac{16x^2}{v_0^2 \cos^2 \theta}$; parabola

Section 9.6

1. parabola, ellipse, hyperbola   **3.** $\frac{\pi}{3}, \frac{5\pi}{3}$

5. parabola, $e = 1, p = 3$

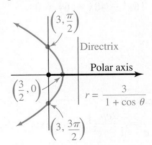

7. parabola, $e = 1, p = 2$

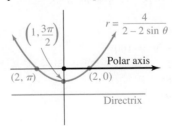

9. ellipse, $e = \dfrac{1}{2}, p = 4$

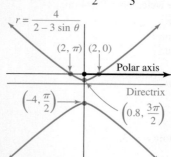

11. ellipse, $e = \dfrac{2}{3}, p = \dfrac{3}{2}$

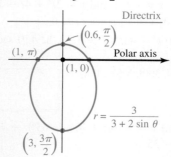

13. hyperbola, $e = \dfrac{3}{2}, p = \dfrac{4}{3}$

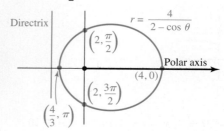

15. $x^2 = 1 - 2y$ **17.** $7x^2 + 16y^2 - 48x - 64 = 0$
19. $x^2 - 3y^2 + 12y - 9 = 0$
21. $3x^2 + 4y^2 - 16x - 64 = 0$
23. $r = \dfrac{10}{3 + 2\cos\theta}$
25. $r = \dfrac{14}{12 + 8\cos\theta} = \dfrac{7}{6 + 4\cos\theta}$
27. major axis $= \dfrac{ep}{1 + e} + \dfrac{ep}{1 - e} = \dfrac{2ep}{1 - e^2}$; minor axis $= 2ep$
29. $\left(\dfrac{p}{2}, \pi\right)$ **31.** $2a = \dfrac{2ep}{1 - e^2}; c = \dfrac{e^2p}{1 - e^2}; \dfrac{c}{a} = e$

Chapter Review
1. $x^2 + y^2 = 16$ **2.** $x^2 + y^2 = 100$
3. $(x - 3)^2 + (y + 2)^2 = 25$ **4.** $(x + 2)^2 + (y - 4)^2 = 25$
5. $(x - 5)^2 + (y - 10)^2 = 85$
6. $(x - 2)^2 + (y - 2)^2 = 89$

7. $(x - 3)^2 + (y + 2)^2 = 16$ **8.** $(x + 2)^2 + (y - 5)^2 = 16$

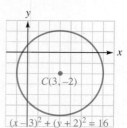

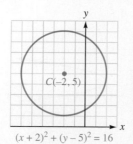

9. $y^2 = -2x$ **10.** $x^2 = 16y$
11. $(x + 2)^2 = -\dfrac{4}{11}(y - 3)$

12.   **13.**

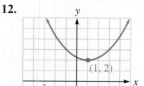

14. $\dfrac{x^2}{36} + \dfrac{y^2}{16} = 1$ **15.** $\dfrac{x^2}{4} + \dfrac{y^2}{25} = 1$
16. $\dfrac{(x + 2)^2}{16} + \dfrac{(y - 3)^2}{9} = 1$
17. $\dfrac{(x - 2)^2}{1} + \dfrac{(y + 1)^2}{4} = 1$

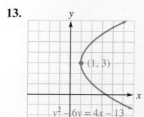

18. $\dfrac{x^2}{4} - \dfrac{y^2}{12} = 1$ **19.** $\dfrac{y^2}{9} - \dfrac{x^2}{16} = 1$
20. $\dfrac{x^2}{9} - \dfrac{(y - 3)^2}{16} = 1$ **21.** $\dfrac{y^2}{9} - \dfrac{(x - 3)^2}{16} = 1$
22. $y = \pm\dfrac{4}{5}x$
23. $\dfrac{(x - 1)^2}{4} - \dfrac{(y + 2)^2}{9} = 1$ **24.**

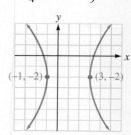

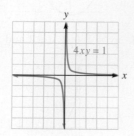

$9x^2 - 4y^2 - 16y - 18x = 43$

25.

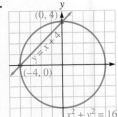

26.

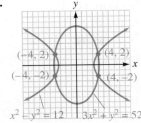

27.

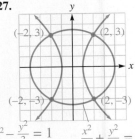

$$x^2 - \frac{y^2}{3} = 1 \qquad \frac{x^2}{16} + \frac{y^2}{12} = 1$$

28. $(-4, 2), (-4, -2), (4, 2),$ and $(4, -2)$

29. $(0, 4)$ and $\left(2\sqrt{3}, -2\right)$

30. $(-2, 3), (-2, -3), (2, 3),$ and $(2, -3)$

31. a.

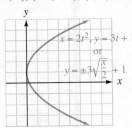

  b.

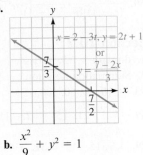

    b. $\dfrac{x^2}{9} + y^2 = 1$

32. a.

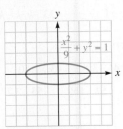

  b. $y = (x + 3)^2$

33. a.

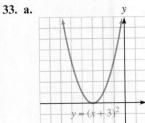

34. $x = t$ and $y = 3t^2 - 4$; $x = t + 1$ and $y = 3t^2 + 6t - 1$

35. a. $x = 32t$ and $y = -16t^2 + 32t\sqrt{3} + 7$ **b.** 3.5 sec

   c. 112 ft $\approx$ 37.3 yd

36. parabola, $e = 1, p = 2$

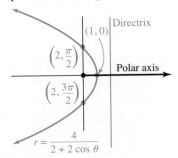

$$r = \frac{4}{2 + 2\cos\theta}$$

37. $16x^2 = 25 - 40y$

Chapter Test

1. $(x - 2)^2 + (y - 3)^2 = 9$ **2.** $(x - 2)^2 + (y - 3)^2 = 41$

3. $(x - 2)^2 + (y + 5)^2 = 169$

4.

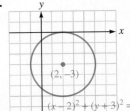

5. $(x - 3)^2 = 16(y - 2)$ **6.** $(y + 6)^2 = -4(x - 4)$

7. $(x - 2)^2 = \dfrac{4}{3}(y + 3)$ or $(y + 3)^2 = -\dfrac{9}{2}(x - 2)$

8.

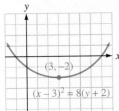

9. $\dfrac{x^2}{100} + \dfrac{y^2}{64} = 1$ **10.** $\dfrac{x^2}{169} + \dfrac{y^2}{144} = 1$

11. $\dfrac{x^2}{4} + \dfrac{y^2}{36} = 1$

12.

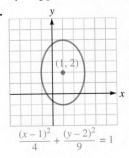

13. $\dfrac{x^2}{25} - \dfrac{y^2}{144} = 1$ **14.** $\dfrac{x^2}{36} - \dfrac{4y^2}{25} = 1$

15. $\dfrac{(x - 2)^2}{64} - \dfrac{(y + 1)^2}{36} = 1$

16.

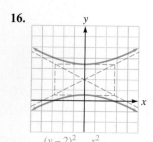

$$\frac{(y-2)^2}{2} - \frac{x^2}{8} = 1$$

17. $\left(\sqrt{7}, 4\right)$ and $\left(-\sqrt{7}, 4\right)$

18. $\left(3\sqrt{2}, 3\right), \left(-3\sqrt{2}, 3\right), \left(3\sqrt{2}, -3\right)$, and $\left(-3\sqrt{2}, -3\right)$

19. $(y - 2)^2 = 6(x + 3)$; parabola

20. $\dfrac{(x-1)^2}{3} + \dfrac{(y+2)^2}{2} = 1$; ellipse

21. a.

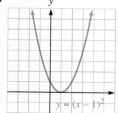

$y = (x-1)^2$

b. $y = (x - 1)^2$

22. a.

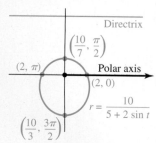

$x = 2y - 10$
or
$y = \dfrac{x + 10}{2}$

b. $x = 2y - 10$

23. Ellipse, $e = \dfrac{2}{5}, p = 5$

Directrix

$\left(\dfrac{10}{7}, \dfrac{\pi}{2}\right)$

$(2, \pi)$ — Polar axis

$(2, 0)$

$r = \dfrac{10}{5 + 2\sin t}$

$\left(\dfrac{10}{3}, \dfrac{3\pi}{2}\right)$

Connect to Calculus

1. a. $\mathbf{r} = t\mathbf{i} + \dfrac{1}{2}t^2\mathbf{j}; \mathbf{v} = \mathbf{i} + t\mathbf{j}; \mathbf{a} = \mathbf{j}; |\mathbf{v}| = \sqrt{1 + t^2}$

b. $\mathbf{v}(1) = \mathbf{i} + \mathbf{j}; |\mathbf{v}| = \sqrt{2}; \mathbf{a}(1) = \mathbf{j}$

2. a. $\mathbf{r} = 2\cos t\mathbf{i} + 2\sin t\mathbf{j}; \mathbf{v} = -2\sin t\mathbf{i} + 2\cos t\mathbf{j};$
$\mathbf{a} = -2\cos t\mathbf{i} - 2\sin t\mathbf{j}; |\mathbf{v}| = 2$

b. $\mathbf{v}\left(\dfrac{\pi}{4}\right) = -\sqrt{2}\mathbf{i} + \sqrt{2}\mathbf{j}; \mathbf{a}\left(\dfrac{\pi}{4}\right) = -\sqrt{2}\mathbf{i} - \sqrt{2}\mathbf{j}; |\mathbf{v}| = 2$

Section 10.1

1. power **3.** first **5.** $7 \cdot 6 \cdot 5 \cdot 4 \cdot 3 \cdot 2 \cdot 1$
7. $(n - 1)!$ **9.** 120 **11.** 4,320
13. 1,440 **15.** $\dfrac{1}{1,320}$ **17.** $\dfrac{5}{3}$ **19.** 18,564
21. $a^5 + 5a^4b + 10a^3b^2 + 10a^2b^3 + 5ab^4 + b^5$
23. $x^3 - 3x^2y + 3xy^2 - y^3$
25. $a^3 + 3a^2b + 3ab^2 + b^3$
27. $a^5 - 5a^4b + 10a^3b^2 - 10a^2b^3 + 5ab^4 - b^5$
29. $8x^3 + 12x^2y + 6xy^2 + y^3$
31. $x^3 - 6x^2y + 12xy^2 - 8y^3$
33. $16x^4 + 96x^3y + 216x^2y^2 + 216xy^3 + 81y^4$
35. $x^4 - 8x^3y + 24x^2y^2 - 32xy^3 + 16y^4$
37. $x^5 - 15x^4y + 90x^3y^2 - 270x^2y^3 + 405xy^4 - 243y^5$
39. $\dfrac{x^4}{16} + \dfrac{x^3y}{2} + \dfrac{3x^2y^2}{2} + 2xy^3 + y^4$ **41.** $6a^2b^2$
43. $35a^3b^4$ **45.** $-b^5$ **47.** $2,380a^{13}b^4$
49. $-4\sqrt{2}a^3$ **51.** $1,134a^5b^4$ **53.** $\dfrac{3x^2y^2}{2}$
55. $-\dfrac{55r^2s^9}{2,048}$ **57.** $\dfrac{n!}{3!(n-3)!}a^{n-3}b^3$
59. $\dfrac{n!}{(r-1)!(n-r+1)!}a^{n-r+1}b^{r-1}$ **63.** -252

Section 10.2

1. domain **3.** series **5.** infinite
7. Summation notation **9.** 6 **11.** $5c$
13. 0, 10, 30, 60, 100, 150 **15.** 21 **17.** $a + 4d$
19. 15 **21.** 15 **23.** 15 **25.** $\dfrac{242}{243}$ **27.** 35
29. 3, 7, 15, 31 **31.** $-4, -2, -1, -\dfrac{1}{2}$ **33.** k, k^2, k^4, k^8
35. $8, \dfrac{16}{k}, \dfrac{32}{k^2}, \dfrac{64}{k^3}$ **37.** an alternating series
39. not an alternating series **41.** 30 **43.** -50
45. 40 **47.** 500 **49.** $\dfrac{7}{12}$ **51.** 160 **53.** 3,725

Section 10.3

1. $(n - 1)$ **3.** infinite **5.** $a_n = a + (n - 1)d$
7. Arithmetic means **9.** 1, 3, 5, 7, 9, 11
11. $5, \dfrac{7}{2}, 2, \dfrac{1}{2}, -1, -\dfrac{5}{2}$ **13.** $9, \dfrac{23}{2}, 14, \dfrac{33}{2}, 19, \dfrac{43}{2}$
15. 318 **17.** 6 **19.** 370 **21.** 44
23. $\dfrac{25}{2}, 15,$ and $\dfrac{35}{2}$ **25.** $-\dfrac{82}{15}, -\dfrac{59}{15}, -\dfrac{12}{5},$ and $-\dfrac{13}{15}$
27. 285 **29.** 555 **31.** $157\dfrac{1}{2}$ **33.** 20,100
35. 1,080°; 1,800° **37.** \$460 **39.** \$1,587,500
41. 336 ft **43.** 210

Section 10.4

1. r^{n-1} **3.** ar^{n-1} **5.** infinite
7. Geometric means **9.** 10, 20, 40, 80
11. $-2, -6, -18, -54$ **13.** $3, 3\sqrt{2}, 6, 6\sqrt{2}$
15. 2, 6, 18, 54 **17.** 256 **19.** -162

21. $10\sqrt[4]{2}$, $10\sqrt[4]{4}$ or $10\sqrt{2}$, and $10\sqrt[4]{8}$
23. 8, 32, 128, and 512   **25.** 124   **27.** $-29,524$
29. $\dfrac{1,995}{32}$ **31.** 18   **33.** 8   **35.** $\dfrac{5}{9}$ **37.** $\dfrac{25}{99}$
39. 23   **41.** 12.96 ft; 400 ft   **43.** 5.13 m   **45.** $69.82
47. $\dfrac{1}{3}c$ **49.** 1.03×10^{20} **51.** $180,176.87
53. $2,001.60   **55.** $2,013.62   **57.** $264,094.58
59. 5,0000   **61.** 1.8447×10^{19}
63. No; answers will vary.

Section 10.5

1. two   **3.** $n = k + 1$ **29.** no
37. a. 1   **b.** 3   **c.** 7   **d.** 15

Chapter Review

1. 720   **2.** 30,240   **3.** 8   **4.** $\dfrac{280}{3}$
5. $x^3 + 3x^2y + 3xy^2 + y^3$
6. $p^4 + 4p^3q + 6p^2q^2 + 4pq^3 + q^4$
7. $a^5 - 5a^4b + 10a^3b^2 - 10a^2b^3 + 5ab^4 - b^5$
8. $8a^3 - 12a^2b + 6ab^2 - b^3$ **9.** $56a^5b^3$ **10.** $80x^3y^2$
11. $84x^3y^6$ **12.** $439,040x^3$ **13.** 63   **14.** 9
15. 5, 17, 53, 161   **16.** -2, 8, 128, 32,768   **17.** 90
18. 60   **19.** 1,718   **20.** -360 **21.** 117   **22.** 281
23. -92 **24.** $-\dfrac{135}{2}$ **25.** $\dfrac{7}{2}$, 5, and $\dfrac{13}{2}$
26. 25, 40, 55, 70, and 85   **27.** 3,320   **28.** 5,780

29. $-5,220$ **30.** $-1,540$ **31.** $\dfrac{1}{729}$ **32.** 13,122
33. $\dfrac{9}{16,384}$ **34.** $\dfrac{8}{15,625}$ **35.** $2\sqrt{2}$, 4, and $4\sqrt{2}$
36. 4, -8, 16, and -32 **37.** 16   **38.** $\dfrac{3,280}{27}$ **39.** 6,560
40. $\dfrac{2,295}{128}$ **41.** $\dfrac{520,832}{78,125}$ **42.** $\dfrac{3,280}{3}$ **43.** $16\sqrt{2}$
44. $\dfrac{2}{3}$ **45.** $\dfrac{3}{25}$ **46.** no sum   **47.** 1   **48.** $\dfrac{1}{3}$
49. 1   **50.** $\dfrac{17}{99}$ **51.** $\dfrac{5}{11}$ **52.** $4,775.81
53. 6,516; 3,134   **54.** $3,486.78

Chapter Test

1. 144   **2.** 384   **3.** $10x^4y$ **4.** $112a^2b^6$ **5.** 27
6. -36 **7.** 155   **8.** -130 **9.** 9, 14, and 19
10. -6 and -18 **11.** 255.75   **12.** about 9
13. about $0.42c$ **14.** about $1.46c$ **16.** 800,000

Connect to Calculus

1. Six terms will give the approximation correct to eight decimals; $e^{0.1} \approx 1.10517091$.
2. Two terms will give the approximation correct to five decimals; $\cos 0.05 \approx 0.99875$.
3. $\dfrac{1}{2}$ **4.** -1

Index

Ellipses

An **ellipse** is the set of all points P in a plane such that the sum of the distances from P to two other fixed points F and F' is a positive constant.

The standard equation of an ellipse with center (h, k) and horizontal major axis is

$$\frac{(x - h)^2}{a^2} + \frac{(y - k)^2}{b^2} = 1,$$

where $a > b > 0$.

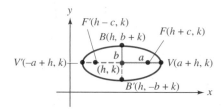

The standard equation of an ellipse with center (h, k) and vertical major axis is

$$\frac{(x - h)^2}{b^2} + \frac{(y - k)^2}{a^2} = 1,$$

where $a > b > 0$.

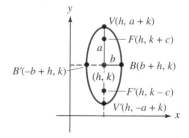

Hyberbolas

A **hyperbola** is the set of all points in a plane such that the absolute value of the difference of the distances from point P to two other points in the plane is a positive constant.

The standard equation of a hyperbola with center at (h, k) and foci on a line parallel to the x-axis is

$$\frac{(x - h)^2}{a^2} - \frac{(y - k)^2}{b^2} = 1,$$

where $a^2 + b^2 = c^2$.

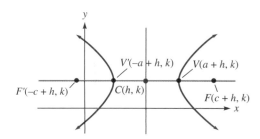

The standard equation of a hyperbola with center at (h, k) and foci on a line parallel to the y-axis is

$$\frac{(y - h)^2}{a^2} - \frac{(x - k)^2}{b^2} = 1,$$

where $a^2 + b^2 = c^2$.

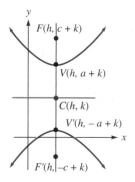

Permutations and Combinations

$$P(n, r) = \frac{n!}{(n - r)!}; \qquad P(n, n) = n!; \qquad P(n, 0) = 1$$

$$C(n, r) = \frac{n!}{r!(n - r)!}; \qquad C(n, n) = 1; \qquad C(n, 0) = 1$$

Geometry Formulas

Rectangle: $A = l \cdot w; P = 2l + 2w$

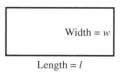

Triangle: $A = \frac{1}{2}b \cdot h$

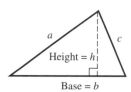

Circle: $C = 2\pi r; A = \pi r^2$

Rectangular solid: $V = l \cdot w \cdot h$

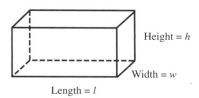

Sphere: $V = \frac{4}{3}\pi r^3$

Right circular cone: $V = \dfrac{1}{3}\pi r^2 h$

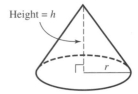

Height = h

Pythagorean Theorem
$$a^2 + b^2 = c^2$$

Arc Length and Area

Arc length:
$$s = r\theta$$

Area of a circular sector:
$$A = \dfrac{1}{2}r^2\theta$$

Area of a triangle given SAS:
$$K = \dfrac{1}{2}ab \sin C \qquad K = \dfrac{1}{2}ac \sin B \qquad K = \dfrac{1}{2}bc \sin A$$

Heron's Formula:
If a, b, and c are the three sides of a triangle and
$$s = \dfrac{a + b + c}{2},$$
then the area of the triangle is given by
$$K = \sqrt{s(s - a)(s - b)(s - c)}$$

Trigonometric Identities and Properties

Reciprocal Identities:
For any angle θ, for which the function is defined,

$$\sin\theta = \dfrac{1}{\csc\theta} \qquad \csc\theta = \dfrac{1}{\sin\theta} \qquad \sin\theta \cdot \csc\theta = 1$$

$$\cos\theta = \dfrac{1}{\sec\theta} \qquad \sec\theta = \dfrac{1}{\cos\theta} \qquad \cos\theta \cdot \sec\theta = 1$$

$$\tan\theta = \dfrac{1}{\cot\theta} \qquad \cot\theta = \dfrac{1}{\tan\theta} \qquad \tan\theta \cdot \cot\theta = 1$$

Quotient Identities:
For any angle θ, for which the function is defined,
$$\tan\theta = \dfrac{\sin\theta}{\cos\theta} \quad \text{and} \quad \cot\theta = \dfrac{\cos\theta}{\sin\theta}$$

Pythagorean Identities:
$$\sin^2\theta + \cos^2\theta = 1$$
$$\tan^2\theta + 1 = \sec^2\theta$$
$$1 + \cot^2\theta = \csc^2\theta$$

Odd–Even Properties:
$$\sin(-x) = -\sin x$$
$$\cos(-x) = \cos(x)$$
$$\tan(-x) = \tan x$$

Unit Circle

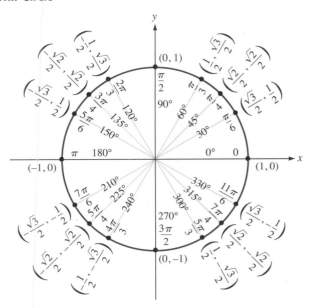

Graphs of Trigonometric Functions
$y = \sin x$

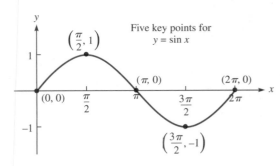

Five key points for $y = \sin x$

$\left(\dfrac{\pi}{2}, 1\right)$

$(\pi, 0)$ $(2\pi, 0)$

$(0, 0)$

$\left(\dfrac{3\pi}{2}, -1\right)$

$y = \cos x$

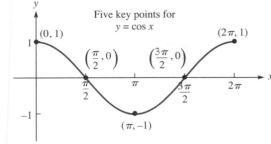

Five key points for $y = \cos x$

$(0, 1)$ $(2\pi, 1)$

$\left(\dfrac{\pi}{2}, 0\right)$ $\left(\dfrac{3\pi}{2}, 0\right)$

$(\pi, -1)$

$y = \tan x$

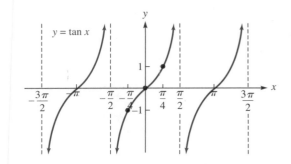

$y = \tan x$